Kuhlenbeck: The Central Nervous System of Vertebrates

Hartwig Kuhlenbeck

The Central Nervous System of Vertebrates

A General Survey of its Comparative Anatomy with an Introduction to the Pertinent Fundamental Biologic and Logical Concepts

S. Karger · Basel · München · Paris · London · New York · Sydney

Vol. 5, Part I

Derivatives of the Prosencephalon: Diencephalon and Telencephalon

262 figures comprising 765 illustrations

S. Karger · Basel · München · Paris · London · New York · Sydney

Cataloging in Publication

Kuhlenbeck, Hartwig, 1897
The central nervous system of vertebrates: a general survey of its comparative anatomy with an introduction to the pertinent fundamental biologic and logical concepts
Hartwig Kuhlenbeck. – Basel; New York: Karger, etc., 1967–1978
5 v.: ill.
1. Anatomy, Comparative 2. Vertebrates – anatomy & histology I. Title
QL 937 K96c
ISBN 3-8055-2638-5

Printed in Switzerland by Pochon-Jent AG, Bern
ISBN 3-8055-2638-5

«Die vergleichende Anatomie, Histologie, Architektonik und Embryologie des Zentralnervensystems bildet ferner einen umfangreichen Zweig und zugleich eine unentbehrliche Methode der neurobiologischen Forschung. Sie verrät die zahlreichen Wege, durch welche die Evolution der Nervensysteme der verschiedenen Tiersorten im phylogenetischen Zusammenhang ihre heutige Verschiedenartigkeit zustande gebracht hat. Vertieft man sich dabei genügend in den Zusammenhang von Form und Funktion, so gelangt man in eine wunderbare Welt der Harmonie zwischen Geist und lebendem Nervensystem...

Wer vergleichende Anatomie des Nervensystems sagt, sagt also auch vergleichende Physiologie – Psychologie und – Biologie, und das ist ein Gebiet, aus welchem die künftige Forschung mit vollen Zügen schöpfen kann...

Dass der Mensch für den Menschen sich zunächst interessiert, ist verzeihlich und naheliegend. Hat er aber einmal erkannt, dass er nur ein Glied in der Tierreihe bildet und dass sein Hirn, das Organ seiner Seele, aus dem Tiergehirn und somit aus der Tierseele stammt, so muss er doch zur Erkenntnis gelangen, dass das Studium der Neurobiologie dieser seiner Verwandten das grösste Licht auf sein eigenes Nerven- und Seelenleben werfen muss.»

August Forel
(«Die Aufgaben der Neurobiologie»)

Preface

The lectures on the central nervous system of vertebrates, given by the author during his first sojourn in Japan, 1924–1927 (Taishô 13 to Shôwa 2), intended to foster the interest in comparative neurologic studies based upon the morphologic principles established by the *Gegenbaur* or *Jena-Heidelberg School of Comparative Anatomy*. Notwithstanding their introductory and elementary nature, these lectures, published by Gustav Fischer, Jena, in 1927, included a number of advanced as well as independent concepts, and represented, as it were, the outline of a further program.

Despite various vicissitudes, and although I found the prevailing intellectual climate in the realm of biologic sciences rather unfavorable to the pursuit of investigations related to the domain of classical morphology, I have, *tant bien que mal*, carried on with my studies as originally planned, and propose to summarize my viewpoints in the present series, designed to represent a general survey, and projected to comprise five separate volumes, of which the first two are now completed. It can easily be seen that the present series follows closely the outline of my old 'Vorlesungen', meant to stress 'die grossen Hauptlinien der Hirnarchitektur und die allgemeinen Gesetzmässigkeiten, welche in Bau und Funktion des Nervensystems erkannt werden können'.

Comparative anatomy of the vertebrate central nervous system requires a very broad and comprehensive background of biological data, evaluated by means of a rational, consistent, and appropriate logical procedure. Without the relevant unifying concepts, comparative neurology becomes no more than a trivial description of apparently unrelated miscellaneous and bewildering configurational varieties, loosely held together by a string of hazy 'functional' notions. A perusal of the multitudinous literature dealing with matters involving the morphologic aspects of neurobiology reveals, to the critical observer, considerable confusion as regards many fundamental questions.

For this reason, the present attempt at an integrated overall presentation includes a somewhat detailed scrutiny of problems concerning the significance of configuration and configurational variety with respect to evolution and to correlated reasonably 'natural' taxonomic classifications. Because comparative anatomy of the central nervous system embodies the morphological clues required to infer the presumable phylogenetic evolution of the brain, a number of general questions referring to ontogenetic evolution are critically considered: it is evident that both the inferred phylogenetic sequences and the observable ontogenetic sequences represent evolutionary processes suitable for a comparison outlining the obtaining invariants.

Moreover, the comparison of organic forms involves procedures closely related to *analysis situs*. Thus, a simplified and elementary discussion of the here relevant principles of *topology* was deemed necessary.

Finally, since vertebrate comparative anatomy and vertebrate evolution, including the origin of vertebrates, cannot be properly assessed in default of an at least moderately adequate familiarity with the vast array of invertebrate organic forms, a general and elementary survey of invertebrate comparative neurology from the vertebrate neurobiologist's viewpoint, that is as seen by an 'outsider' with a modicum of first-hand acquaintance, has been included as volume two of this series. The approximately 20 pages and 12 figures dealing with this matter in my 1927 'Vorlesungen' have thus, of necessity, become rather expanded.

US N.I.H. Grant NB 4999, which is acknowledged with due appreciation, made possible the completion of Volumes 1 and 2 of this series, and, for the time being, the continuation of these studies, by supporting a 'Research Professorship' established to that effect, following my superannuation, at the Woman's Medical College of Pennsylvania.

Concluding this preamble to the present series, I may state with Cicero (*De oratore*, III, 61, 228): '*Edidi quae potui, non ut volui sed ut me temporis angustiae coegerunt; scitum est enim causam conferre in tempus, cum afferre plura si cupias non queas.*'

H.K.

Foreword to Volume 5, Part I

The large amount of material to be dealt with in volume 5 of this series made it necessary, as in the case of volume 3, to publish the present volume in two separate parts. Accordingly, the first part of volume 5, containing chapters XII and XIII, deals with the Vertebrate Diencephalon and Telencephalon. In chapter XIII, of which section 10 concerns the basic morphologic and functional aspects of the endbrain in Mammals, only the overall features of the diversified Mammalian telencephalic surface configuration and of the Mammalian cerebral cortex are here considered.

With regard to receptor structures, and supplementing chapter VII of volume 3/II as well as section 1, chapter IX of volume 4 (otic apparatus), the optic and the olfactory organs are passed under review in appropriate sections of chapters XII and XIII.

The second part of volume 5, containing chapters XIV, XV, and XVI, deals with details of Mammalian telencephalic surface morphology and of cerebral cortex, including thereto related relevant 'interdisciplinary' topics, and, in a similar manner, with the Vertebrate neuraxis as a whole.

It should again be stated that the bibliographies appended to the chapters of this series are meant to be selective, but should easily enable those interested in further particulars to find the required additional references.

As in the preceding volumes, numerous duly credited illustrations were taken from the public domain of published scientific literature also including contributions by my collaborators and myself. Illustrations without credit reference are previously unpublished originals from my own studies.

As before, I am obliged to the *Medical College of Pennsylvania* for the facilities of my '*Laboratory of Morphologic Brain Research*', and particularly grateful to the *Alumnae Association* of the whilom *Woman's Medical College of Pennsylvania*, which includes my many former students, for the continued generous contributions to the funds necessary for my work.

H.K.

Table of Contents of the Present Volume

Volume 5 Part I: Derivatives of the Prosencephalon: Diencephalon and Telencephalon

Table of Contents of the Complete Work

Volume 1 Propaedeutics to Comparative Neurology

Volume 5 Part I: Derivatives of the Prosencephalon: Diencephalon and Telencephalon

XII. The Diencephalon

XIII. The Telencephalon

Volume 5 Part II: Mammalian Telencephalon: Surface Morphology and Cerebral Cortex The Vertebrate Neuraxis as a Whole

XII. The Diencephalon

1. General Pattern and Basic Mechanisms

A. The Longitudinal Zonal System, its Principal Griseal Derivatives and their Main Communication Channels

The Vertebrate *diencephalon* or *betweenbrain* represents the caudal differentiation of the embryonic *prosencephalon*. This latter, on fairly convincing grounds, can be regarded as homologous to the *archencephalon* (v. KUPFFER) of Amphioxus, respectively to the hypothetical archencephalon of a presumptive ancestral form at a critical or nodal period of Vertebrate phylogenesis.

In contradistinction to deuterencephalon and spinal cord, the diencephalon and the telencephalon consist of neural tube components exclusively provided by *alar plate* and *roof plate*, discounting a variable but relatively slight rostral *protrusion of the basal plate* at the *tuberculum posterius* into the ventral diencephalo-mesencephalic boundary region. In certain 'lower' forms, the prosencephalic roof plate, clearly recognizable at embryonic stages, can subsequently become greatly reduced (e.g. in Myxinoids) and, particularly in the telencephalon, may not remain a relevant wall component.

The ontogenetic evolution of the *primary zonal system* (floor, basal, alar, and roof plates) as well as the subsequent evolution of the secondary diencephalic zonal system have been dealt with in chapter VI of volume 3/II. Despite some contradictory opinions, which were duly considered, the concept proposed by KINGSBURY and documented by the results of the systematic investigations undertaken by my associates and myself was evaluated as firmly established.

It will thus here be sufficient to state that the diencephalon is typically configurated by a secondary longitudinal zonal system pertaining to the alar plate. These longitudinal zones are, at least partially, oriented at an approximately right angle to the convex terminal curvature of the sulcus limitans surrounding the rostral end of the basal plate. This terminal course of the sulcus limitans can be seen at some ontogenetic stages of practically all examined Vertebrates and is even not infrequently still noticeable at the adult stage of many forms.

The morphological *grundbestandteile* of the diencephalic longitudinal zonal system are, in dorsobasal sequence, *epithalamus*, *thalamus dorsalis*, *thalamus ventralis*, and *hypothalamus*. These fundamental components provide the lateral wall and the floor of the diencephalic third ventricle. Along the ventricular wall, three fairly typical sulci, namely *sulcus diencephalicus dorsalis*, *sulcus diencephalicus medius*, and *sulcus diencephalicus ventralis* approximately indicate the boundaries between epithalamus and thalamus dorsalis, between this latter and thalamus ventralis, and between hypothalamus and thalamus ventralis, respectively (Figs. 1–5). The ontogenetic behavior of these sulci, their relationship to the transitory neuromeric segmentation, their persistence or non-persistence at the adult stage, and the complicating occurrence of accessory sulci were likewise dealt with in chapter VI of volume 3/II.

The *floor of the third ventricle* formed by the joint median portion of the antimeric hypothalamic zones includes a series of *basal midline structures*, namely *supraoptic crest*, *chiasmatic ridge*, *neurohypophysis*, and *saccus vasculosus*, of which only chiasmatic ridge and hypophysis seem to be present in *all* Vertebrates.

The *roof of the diencephalic third ventricle*, provided by the roof plate, displays various arrangements, such as an epithelial '*dorsal sac*' (second-

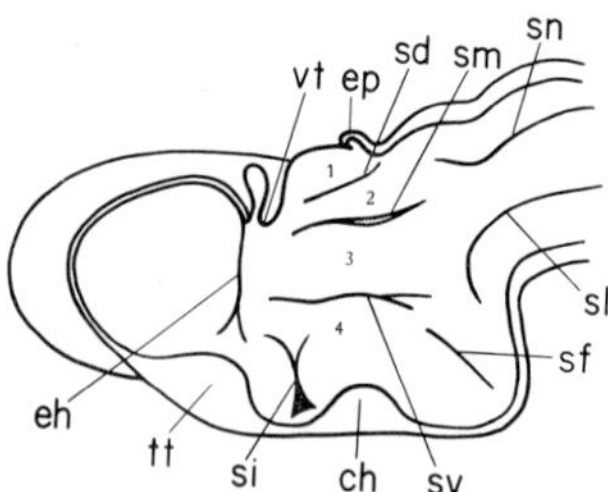

Figure 1. Semidiagrammatic sketch of Vertebrate diencephalic zones and ventricular sulci as displayed at an advanced ontogenetic stage of a tailed Amphibian (from K. and HAYMAKER, 1949). ch: chiasmatic ridge; eh: entrance into caudal evagination of telencephalon; ep: epiphysis; sd: sulcus diencephalicus dorsalis; sf: sulcus lateralis hypothalami posterioris (s. lat. infundibuli); si: sulcus intraencephalicus anterior; sl: sulcus limitans; sm: sulcus diencephalicus medius; sn: sulcus lateralis mesencephali rostrally continuous with sulcus synencephalicus; tt: torus transversus; vt: velum transversum. 1: epithalamus; 2: thalamus dorsalis; 3: thalamus ventralis; 4: hypothalamus. In the legend to Figure 184, p. 410 of vol. 3/II, sf was through a *lapsus calami* inadvertently designated as sulcus lateralis thalami instead of hypothalami.

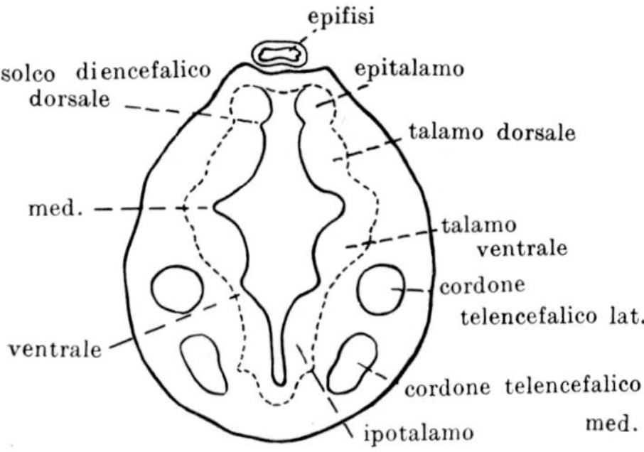

Figure 2. Diagrammatic cross-section through the diencephalon of the adult urodele Amphibian Necturus (simplified after HERRICK, from BECCARI, 1943).

ary parencephalon), a *choroid plexus*, *epiphysial structures* differentiated as pineal or 'parietal' eyes, respectively pineal gland *(epiphysis)*, and, in the diencephalo-mesencephalic boundary zone, the *subcommissural organ*. These diverse configurations and their different manifestations in the Vertebrate series were pointed out in volume 3/II *et passim*.[1]

Morphologically valid rostral and caudal *boundaries* of the Vertebrate diencephalon are definable as two curved planes cutting transversely through the neuraxis. The transverse plane of the rostral boundary passes through the embryonic *sulcus telencephalo-diencephalicus* and its correlated *torus hemisphaericus* as discussed in chapter VI, volume 3/II. At adult stages, sulcus and torus may become 'blurred'. Nevertheless, the basal midline portion of torus hemisphaericus, namely the *torus transversus* with its commissural system remains conspicuously persistent. Thus, subsequently to telencephalic evagination, at any stage and in any form, the rostral diencephalic boundary, if projected upon the midsagittal plane, includes a 'line' connecting the *velum transversum* or its remnant with the *anterior commissure* respectively the torus transversus (comprising also the commissura pallii of Amphibians). The caudal boundary includes a 'line' drawn from an undefined region of *commissura posterior* to the region of *tuberculum posterius*.[2]

[1] Cf. pp. 404–406, volume 3/II. Concerning the subcommissural organ cf. also p. 366, volume 3/I. With respect to the above-mentioned supraoptic crest of the hypothalamus, cf. pp. 358–362, volume 3/I.

[2] Both boundaries are, of course, topologically '*open neighborhood*' regions, particularly the caudal ('synencephalic') boundary, where a basal plate component ('tegmental cell cord', 'prerubral tegmentum') protrudes to a variable extent into the diencephalon (cf. also vol. 3/II, Fig. 237 A, p. 462, Fig. 146, p. 349 *et passim)*.

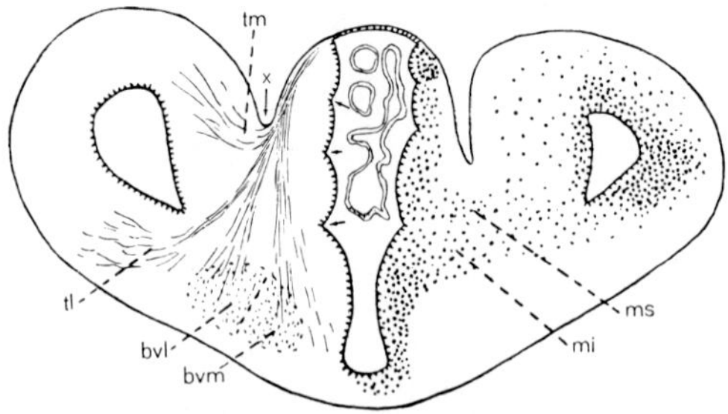

Figure 3. Sktech of cross-section through the rostral portion of the diencephalon of the urodele Amphibian Amblystoma at the level of hemispheric stalk. At left, the system of striae medullares is indicated, at right the ventral thalamic massa cellularis superior and the hypothalamic massa cellularis inferior are shown (from K., 1927). bvl: lateral forebrain bundle; bvm: medial forebrain bundle; mi, ms: massa cellularis reuniens inferior respectively superior; tl, tm: lateral and medial caudal hemispheric components of stria medullaris (so-called tractus corticohabenularis lateralis and medialis; these bundles also contribute fibers, not shown in the diagram, to the basal forebrain bundles, cf. Fig. 7). Epithalamus, dorsal thalamus, ventral thalamus, and hypothalamus (caudal part of preoptic recess) are not labelled but, together with their boundary sulci, shown by arrows, can be easily identified (cf. Fig. 4). Added arrow x: sulcus telodiencephalicus caudodorsalis or sulcus terminalis externus, the sulcus terminalis *sensu proprio (sive internus)* being a corresponding ventricular groove particularly characteristic for Mammals (cf. vol. 3/II, p. 468). It can easily be seen (cf. also Fig. 7) that the fiber systems of the caudally paired evaginated hemisphere, passing through neighborhoods of sulcus terminalis, are in part comparable to the stria terminalis system particularly developed in Mammals. Thus, to some extent, stria medullaris and stria terminalis systems overlap.

The griseum of the *epithalamus* is represented by the nucleus (or ganglion) habenulae, which is one of the most 'conservative' configurations of the Vertebrate brain, remaining essentially unchanged in the entire taxonomic series from Petromyzon to Man.[3] It is connected with the telencephalon through the so-called *stria medullaris thalami* which, in many forms, runs along the attachment of roof plate to the lateral wall *(taenia thalami)*. This bundle collects fibers from telencephalic grisea, either directly or by way of lateral and medial forebrain bundles, from undefined, mainly hypothalamic diencephalic grisea,

[3] Despite this 'conservatism' a variety of secondary differences obtain in the taxonomic series, involving e.g. asymmetries of the antimeric habenular nuclei, cytoarchitectural arrangement with minor further subdivisions, relationship to epiphysial structures, and presumably details of fiber connections respectively synaptology. Some of these features will be pointed out in the subsequent sections dealing with the diverse Vertebrate classes. The term *habenulae* (diminutivum of *habena*) denotes a small strap or bridle *(frenum)* and is derived from *post-Vesalian* human anatomy, the habenulae with their trigone (habenular ganglion) being described as 'bridles' attached to the pineal body.

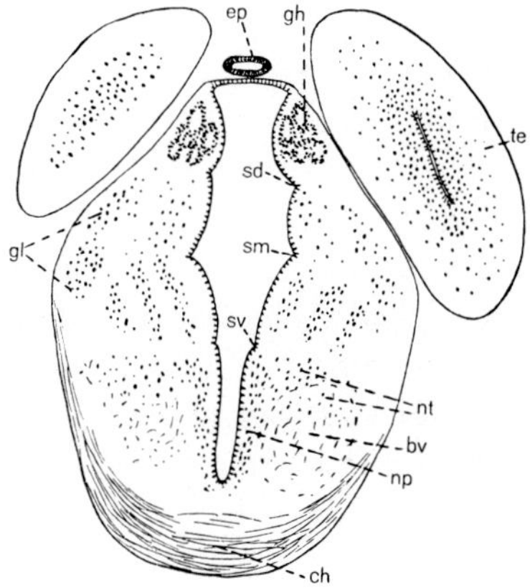

Figure 4. Sketch of cross-section through the diencephalon of the anuran Amphibian Bufo (the common toad), caudally to hemispheric stalk (from K., 1927). bv: basal forebrain bundle (lateral and medial, cf. Fig. 3); ch: optic chiasma; ep: epiphysis; gh: ganglion habenulae (epithalamus); gl: corpus geniculatum laterale (with dorsal and ventral thalamic subdivisions); np: nucleus paraventricularis (hypothalami); nt: nucleus entopeduncularis (hypothalami); sd: sulcus diencephalicus dorsalis; sm: sulcus diencephalicus medius; sv: sulcus diencephalicus ventralis; te: polus posterior telencephali.

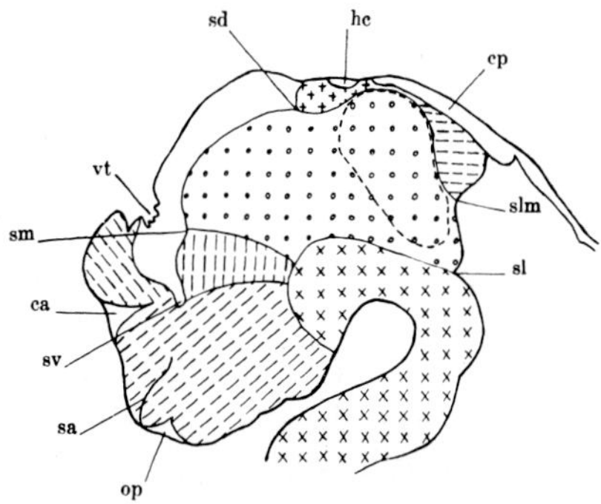

Figure 5. Diagram of diencephalic longitudinal zones as seen in a paramedian sagittal section through the brain in a Rabbit fetus of about 15 mm length (from K. and MILLER, 1942). +: epithalamus; circles: thalamus dorsalis; dotted outline within thalamus dorsalis: primordia of diencephalic pretectal nuclei; vertical hatching: thalamus ventralis; caudodorsal-ventrorostral oblique hatching: hypothalamus; rostrodorsal-caudoventral oblique hatching: telencephalic portion of lamina terminalis (commissural plate); horizontal hatching: alar plate of mesencephalon; ×: basal plate respectively tegmental cell cord; ca: commissura anterior; cp: commissura posterior; hc: habenular commissure; op: optic chiasma; sa: sulcus intraencephalicus anterior; sd: sulcus diencephalicus dorsalis; sl: sulcus limitans; slm: sulcus lateralis mesencephali; sm: sulcus diencephalicus medius; sv: sulcus diencephalicus ventralis; vt: neighborhood of velum transversum.

and, depending on the Vertebrate forms with caudally evaginated telencephalon, additional fiber systems from caudal telencephalic grisea (cf. Figs. 3, 7, 11). Although reciprocal connections can be assumed, the *stria medullaris thalami* seems to provide mainly habenular input. The *habenulopeduncular tract (fasciculus retroflexus)*, discharging into the interpeduncular nucleus, is the most conspicuous habenular output channel, through which impulses, by way of said nucleus, can be distributed to the reticular formation of the brain stem. Although the habenulo-interpeduncular tract appears as the predominant component of the fasciculus retroflexus, this latter seems to be a complex and still poorly understood system (cf. e.g. HOLMGREN and VAN DER HORST, 1925; HERRICK, 1948). Additional, but apparently less substantial separate fiber connections with tectum mesencephali obtain. The antimeric habenular grisea respectively the striae medullaris thalami are interconnected by the *habenular commissure* which is likewise a constant feature of the Vertebrate diencephalon, and, moreover, a significant morphologic landmark. In Vertebrates with pineal or parietal eyes, the *nervus parietalis s. pinealis* is connected with the habenula, but some of its fibers may proceed toward the tectum opticum, or the pretectal region.

Little can be said about the functional significance of the *epithalamus* which is relatively larger in most Anamnia than in Amniota. Taking into account its fiber connections, the habenular griseum is doubtless an intermediate station for the discharge of olfactory impulses. Thus, HERRICK (1931) considered it to be 'a center for the correlation of olfactory sensory impulses' with various somatic sensory centers. Yet, the nucleus habenulae also seems to receive various *non-olfactory impulses* and its output certainly reaches 'motor' as well as presumably 'visceral grisea'. Thus, the details concerning the functional role of the epithalamus remain obscure. Despite basically similar connections in all Vertebrates, the extent of its synaptologic relationships and its performance within the neural mechanisms of Anamnia, Sauropsida and Mammals may substantially differ.

The *thalamus dorsalis* is, in HERRICK's (1948) words, 'an important center of sensory correlation'. It receives input from all deuterencephalic afferent systems, either directly through the lemnisci, or indirectly via tectum mesencephali, pretectal grisea, and reticular formation. Direct optic input reaches a lateral portion of thalamus dorsalis, which becomes the dorsal lateral geniculate nucleus, a structure present in all Vertebrate classes from Cyclostomes to Mammals.

A caudal portion of the thalamus dorsalis becomes part of the regio praetectalis, whose grisea interrelate tectum mesencephali, thalamus, and hypothalamus. In Mammals, connections with some cortical areas obtain. The pretectal grisea are present in all Vertebrate classes,[4] their commissure being the *commissura posterior*, another important 'morphologic landmark' of the neuraxis (cf. K., 1956). A caudal relay griseum for the lateral lemniscus is the dorsal medial geniculate nucleus of Mammals. A griseum of this type, possibly homologous *qua* Reptiles, but perhaps merely analogous *qua* Birds, seems likewise recognizable in Sauropsidans.

In submammalian forms, the thalamus dorsalis is connected with the telencephalic grisea through the *lateral and medial forebrain bundles*. Thus, by way of these channels, sensory input, exclusive of that from olfactory system,[5] is transmitted to the telencephalon, and telencephalic output reaches the diencephalon. This output includes an olfactory component, which is presumably predominant and very substantial in Anamnia, but, *qua* thalamus dorsalis, becomes inconspicuous or even dubious in Reptiles, Birds and Mammals. The *lateral forebrain bundle* comprises descending as well as ascending fiber systems, or, in other words, channels for telencephalic output and input. The *medial forebrain bundle* appears to be predominantly a descending system, although the inclusion of diverse ascending components can be presumed.

At the caudal end of the diencephalon, in the pretectal and tegmental region characterizing the diencephalomesencephalic boundary zone, there are a number of fiber systems through which diencephalic grisea are connected with more caudal centers of the neuraxis. As far as the thalamus dorsalis is concerned, these channels include the lemniscus systems as well as tectopretectothalamic and thalamotegmental fiber bundles. In lower Vertebrates some of these channels overlap with

---

[4] The diencephalic *pretectal region* of Anamnia is represented by what HERRICK (1948) and others designate as '*pars intercalaris diencephali*', interpreted as a posterior division of the epithalamus. Our own studies have led to the conclusion that this 'pars intercalaris' is an intrinsic posterior (and essentially pretectal) component of thalamus dorsalis (cf.K., 1956). Mesencephalic pretectal grisea were dealt with in chapter XI of volume 4.

[5] It will be recalled that olfactory input directly reaches the telencephalon, the bulbus olfactorius being the primary afferent center. In practically all Anamnia, regardless of the non-olfactory sensory input from diencephalon into telencephalon, all grisea of this latter's lobus hemisphaericus (i.e. telencephalon exclusive of bulbus olfactorius) can be regarded as more or less 'dominated' by the olfactory input (cf. pp. 642–647, section 6, chapter VI, vol. 3/II). Some components of the nervus terminalis, associated with the olfactory system, and dealt with in chapter XIII, seem to reach directly the hypothalamus via its preoptic subdivision.

the caudal portion of the basal forebrain bundles.[6] The lemniscus systems are essentially ascending, while the other caudal channels can be considered more or less reciprocal. In addition to these fiber systems, 'vertical fibers', quite conspicuous in some Anamnia, seem to interconnect the grisea of all four longitudinal diencephalic zones (cf. Fig. 7).

The dorsal thalamus of many Anamnia including Amphibia shows a rather nondescript periventricular arrangement of the neuronal elements with a more or less distinctive lateral 'cell plate' representing the primordium of the *dorsal lateral geniculate nucleus*, apparently displaced toward the optic tract in accordance with the concept of neurobiotaxis. Although some stages of local differentiation can be recognized, HERRICK (1948) not inappropriately regards the undifferentiated dorsal thalamus 'as a single "nucleus sensitivus", which acts, in the main, as a whole, without well-defined functional localization'. However, this statement should be qualified by pointing out the differentiation of the dorsal lateral geniculate nucleus, and the rather well-defined dorsal thalamic portion of the pretectal region. Again, a fairly complex if somewhat vague nuclear pattern in the dorsal thalamus (and diencephalon in general) can be noted in certain forms, particularly in the very diversified and differentiated group of Teleosts.

Quite distinctive dorsal thalamic nuclei obtain in Sauropsida, especially in Birds, and furthermore characterize the highly developed dorsal thalamus of Mammals. In these latter, the larger part of the thalamus dorsalis becomes related to the neocortex, being, as it were, 'a cortical dependency' (HERRICK, 1948). Other dorsal thalamic grisea are here related to the corpus striatum complex, and still others are 'intrinsic thalamic centers'. The channels between dorsal thalamus and telencephalon in the Mammalian brain are the *thalamic radiations* or *stalks (peduncles)*, most of which, except for the inferior peduncle, are parts of the *capsula interna*. This latter represents a Mammalian (ne-encephalic) component of the submammalian *lateral forebrain bundle*, concomitantly developed with the expansion of the hemispheric stalk.[7] The gri-

[6] In at least various Anamnia, the extension of descending components of the basal forebrain bundles into tegmental portions of the deuterencephalon cannot be excluded. In Amniota, descending telencephalic channels reaching the brain stem assume an increased importance, culminating in the Mammalian corticopontine and cortico-bulbo-spinal tracts which can be evaluated as 'progressive' differentiation of the lateral forebrain bundle.

[7] The morphologic significance of the *hemispheric stalk* is dealt with in sections 1B, and 5, chapter VI, volume 3/II (cf. also pp. 159, 465, and Fig. 241, p. 467 loc.cit.). The significance of *hemispheric stalk* and hemispheric rotation *(Endhirnbeuge*, p. 577, vol. 3/II) for the fiber system known as *stria semicircularis s. terminalis* shall again be pointed out in section 9.

sea of the Mammalian dorsal thalamus can be roughly classified as follows (K., 1954):

(1) *cortical relay nuclei of main sensory systems* exclusive of the olfactory one, (2) *nuclei of cortical feedback systems or direct cortical modulators*, (3) *intralaminar and midline nuclei ('intrinsic nuclei', indirect cortical modulators)*, (4) *nuclei with probable subcortical connections to pallidum and striatum*, (5) *pretectal nuclei*, (6) '*intrinsic nuclei*' of dubious significance and connections.[8]

Roughly speaking, the massive expansion of dorsal thalamus in Mammals is mainly correlated with the expansion of the neocortex. In Birds, on the other hand, the very substantial expansion of thalamus dorsalis is correlated with the peculiar differentiation of the (secondary) complex of basal ganglia in the lateral wall of the cerebral hemispheres.

The *thalamus ventralis* was first recognized as a distinctive diencephalic configuration in Amphibians by Herrick (1910). Its morphologic significance within the Vertebrate diencephalic bauplan and its homologies within the taxonomic series were particularly investigated by my associates and myself, and are dealt with in section 5, chapter VI of volume 3/II.

In Anamnia, the thalamus ventralis usually extends rostralward toward the interventricular foramen, forming, at its diencephalic border, the *eminentia thalami (ventralis)*. A rostral cell group, the *massa cellularis reuniens*, *pars superior*, extends lateralward into the hemispheric stalk, blending with telencephalic cell groups (Fig. 3). More caudalward, the thalamus ventralis commonly displays a lateral component, namely the primordium of the *ventral lateral geniculate nucleus*, presumably receiving collaterals from the optic tract. The larger but nondescript medial portion of the thalamus ventralis apparently receives ascending input essentially similar to that obtaining for thalamus dorsalis. Undefined reciprocal connections with the telencephalon are mediated by the basal forebrain bundles. Descending output of thalamus ventralis reaches the deuterencephalic tegmentum, and intrinsic diencephalic connections with hypothalamus and with the dorsal grisea can be assumed.

The thalamus ventralis of Anamnia is relatively large and may approximately be as voluminous as the thalamus dorsalis. Its relative size

---

[8] Further details concerning these nuclei are dealt with in section 9. It should, moreover, be added here that this classification of 1954 merely represents a convenient first approximation which requires further qualifications insofar as some of the grisea under (3), (4), and (5) may also have cortical connections.

has decreased in Reptiles, more particularly in Birds and Mammals.[9] In this latter class, the main grisea of thalamus ventralis are *nucleus reticularis thalami*, *zona incerta*, and the ventral components of *lateral* and *medial geniculate grisea*.

The thalamus ventralis seems, in various respects, to duplicate the essential connections of thalamus dorsalis. The functional significance of the apparent duplication provided by the arrangement of these two rather similar longitudinal zones is difficult to explain, and no plausible explanation can be offered at the present time (cf. K., 1954). In phylogenetic evolution, the thalamus ventralis shows a lesser degree of progressive differentiation than the thalamus dorsalis. In Primates it becomes (relatively) 'reduced' to a quite narrow zone between dorsal thalamus and hypothalamus, laterally bordering on the fiber masses of the hemispheric stalk.

The Vertebrate *hypothalamus* is subdivided by the transverse basal midline ridge of optic chiasma and supraoptic respectively postoptic commissures into a rostral portion, the *recessus praeopticus (recessus opticus* BNA, PNA) and a caudal, *postoptic* or infundibulo-mammillary portion. The ventricular wall of the preoptic recess displays a generally dorsoventrally oriented groove or depression, the *sulcus intraencephalicus anterior* of Kupffer, formed by a remnant of the obliterated optic stalk evagination. In the ventricular wall of the caudal hypothalamus an oblique groove, commonly running in a rostrodorsal-caudobasal direction, is the *sulcus lateralis hypothalami posterioris*, also less rigorously designated as *sulcus lateralis infundibuli* (cf. Fig. 1). It is presumably related to the tendency manifested by the expansion of the lobi inferiores hypothalami particularly displayed in many Fishes.[10]

---

[9] At various stages of ontogenesis in all Amniota, however, the ventral thalamus may be as large as the dorsal thalamus, becoming then reduced in relative size by the expansion of thalamus dorsalis during the subsequent stages of ontogenetic development. In attempting to unravel the homologies of the Vertebrate thalamus ventralis in Amniota, my first approximations, based on extrapolations, were erroneous for Sauropsida as well as Mammals (K., 1927, 1930), since having missed the relevant ontogenetic [illegible], I included into the thalamus ventralis ventral components of the dorsal thalamus. Subsequently soon realizing that 'something was here wrong', I corrected this error by additional detailed studies of Reptilian and Avian diencephalic ontogeny, which particularly interested me at the time (K., 1931, 1936, 1937), while assigning the relevant investigation concerning the Mammalian diencephalon to my collaborator Miura (1933) whom I guided in his work.

[10] The sulci of the diencephalic ventricular wall, their variations by confluence, branchings, or development of accessory sulci, and 'pattern distortions', as well as their

The hypothalamus of some Anamnia such as Petromyzont Cyclostomes and a variety of Urodele Amphibians is poorly differentiated. It consists essentially of a periventricular layer and of a peripheral fiber layer occupied by the more dorsal lateral and the more ventral medial channels of the basal forebrain bundles. A few scattered lateral cell groups are usually found within the lateral forebrain bundle and represent a primordial *entopeduncular nucleus*. None of the other hypothalamic nuclei found in 'higher' Vertebrates are here clearly segregated, although some of them are suggested within a nondescript continuum. As is the case with the thalamus ventralis, a rostral, dorsolateral cell group extends into the hemispheric stalk toward the telencephalon, and represents the *massa cellularis reuniens, pars inferior* (Fig. 3). The rostral portion of entopeduncular nucleus is a derivative of this lateral hypothalamic cell population located within the hemispheric stalk.

Nevertheless, in addition to the aforementioned entopeduncular nucleus, and particularly at the levels caudally to the rostral region of preoptic recess, a dorsal and a ventral hypothalamus can be distinguished. This is especially the case in the postoptic, infundibulo-mammillary portion, where the sulcus lateralis hypothalami posterioris roughly indicates a ventricular boundary between the dorsal and ventral subdivisions. In Selachians and Osteichthyes, the caudal part of the hypothalamus provides the extensive *lobi inferiores hypothalami*, which are also recognizable, but less developed in Amphibians and Dipnoans. Particularly in many Teleosts, the posterior hypothalamus displays a highly specialized differentiation into nuclear groups. The posterior hypothalamus of Selachians and Osteichthyes, moreover, generally includes a *saccus vasculosus*, to be briefly discussed further below in dealing with the basal midline structures (subsection C).

In the hypothalamus of Reptiles, several nuclear groups can be recognized, whose homologies with the hypothalamic grisea of Birds and of Mammals seem reasonably probable, despite the divergences in diencephalic evolution manifested by the two last-named classes of Amniota.

In Mammals and Man, we found it convenient to subdivide, on the basis of ontogenetic and comparative anatomical evidence, the derivatives of the primordial hypothalamic diencephalic longitudinal zone

---

relationships to the transitory neuromeres were dealt with and depicted in sections 3 and 5, chapt. VI of volume 3/II. It seems rather evident that, apart from complications due to shrinkage artefacts, these sulci are expressions of 'folding processes' in important morphogenetic events relevant for the interpretation of morphologic homologies.

into the following three nuclear groups (K. and Haymaker, 1949; K., 1954).

(1) *Dorsal and entopeduncular group*, comprising globus pallidus and *nucleus subthalamicus of Luys* (corpus subthalamicum). The former corresponds to the submammalian nucleus entopeduncularis anterior, and the latter to nucleus entopeduncularis posterior. This group is especially related to the extrapyramidal motor system but is also closely connected with the other hypothalamic groups.

(2) *Anterior group*, including the *preoptic grisea*.

(3) *Middle group*.

(4) *Posterior group*. A particular differentiation of the dorsal posterior group is the *mammillary complex* receiving the bulk of the fornix. Although a homologous 'mammillary' griseum may be recognized in submammalian forms, a typical mammillary complex *(corpus mammillare)* is only present in Mammals.

Groups 2–4 represent the hypothalamus *sensu strictiori* in a widely used conventional terminology. Functionally, the more anterior parts of the Mammalian hypothalamus *sensu strictiori* seem to be concerned with parasympathetic outflow, and the more posterior part with sympathetic outflow.

In Fishes and some Amphibians the anterior hypothalamus is characterized by the presence of fairly large *neurosecretory elements* transmitting their secretion, by way of their neurites, to the neurohypophysis. In Amniota, there obtains further differentiation of these neurosecretory magnocellular elements, forming e.g. the preoptic paraventricular and the supraoptic nuclei.

Again, a thickened and often vascularized ependyma, representing *the ependymal organ of Kappers-Charlton*, can be found in the posterior hypothalamus, usually along the sulcus lateralis hypothalami posterioris, of various Anamnia and Amniota, including some Mammals. How general the occurrence of this '*circumventricular organ*' may be, is not exactly known, since it cannot be noted in a diversity of these Vertebrate forms. Still another hypothalamic circumventricular organ is the *supraoptic crest (organon vasculosum laminae terminalis diencephalicae)*,[11] found in Mammals, Birds, and Reptiles, and apparently in some Fishes (cf. further below in section 1C).

Summarizing the main communication channels of the 'primitive'

[11] Circumventricular, paraependymal and ependymal organs are dealt with on pp. 354–367, section 5, chapter V of volume 3/I. Comments on neurosecretion can be found on pp. 643–644 of that volume.

hypothalamus as displayed by relatively undifferentiated Anamnia, there are, rostrally, telencephalic connections through lateral and medial forebrain bundles. Caudally, there are ascending fibers of what HERRICK (1948) calls the visceral-gustatory system, moreover tecto-pretecto-hypothalamic channels, other mesencephalic connections involving torus semicircularis and reticular formation, moreover, in forms with highly developed cerebellum, lobo-cerebellar (or cerebello-lobar) channels. In addition, there are, of course, various poorly understood intradiencephalic connections.

In essential agreement with HERRICK (1948) it could be said that the hypothalamus receives input 'from almost all parts of the brain, directly or indirectly. All sensory systems are represented here', the olfactory connection being very substantial in at least a large number of Anamnia. As regards output, it is merely possible to conclude, again with HERRICK (1948), that the hypothalamus displays a wide 'range of distribution of its efferent fibers'.[12]

Generally speaking, and apart from its influence on the 'somatic motor' mechanisms, the Vertebrate hypothalamus is doubtless an important 'center' for the vegetative nervous system. Although to some extent influenced by the telencephalon, the hypothalamus may here act quite independently and can regulate vegetative activities in the absence of telencephalic control, as shown by the results of ablation experiments and by numerous observations in a variety of pathologic conditions. By and large, more details of hypothalamic functions are known with regard to Mammals and Birds than to 'lower Vertebrates' (cf. e.g. HAYMAKER *et al.*, 1969; PEARSON, 1972).

With respect to some historical aspects, it should be noted that the term '*hypothalamus*' was coined and introduced by HIS (1893) in his '*Vorschläge zur Eintheilung des Gehirns*', published in connection with the then pending formulation of the BNA. AUGUST FOREL (1877) had previously introduced the concept of a '*regio subthalamica*'. Until then, the hypothalamus was merely included in the '*centrales Höhlengrau*' of the '*Sehhügel*' or thalamus (cf. MEYNERT, 1872). FOREL's subthalamus, however, included parts of the not yet identified thalamus ventralis,

[12] Among more recent contributions to the comparative anatomy of the hypothalamus, that by CROSBY and SHOWERS (1969) and that by CHRIST (1969) in HAYMAKER's, ANDERSON's and NAUTA's treatise should be mentioned. Likewise, DIEPEN's (1962) volume on the hypothalamus in v. MÖLLENDORFF's and BARGMANN's *Handbuch der mikroskopischen Anatomie des Menschen* contains numerous data concerning hypothalamic comparative anatomy and embryology.

and, moreover, did not comprise all genuine hypothalamic grisea. In an unsystematic and rather confused way, various different concepts of a subthalamus have nevertheless persisted in the neurologic literature.[13] Except for the *nucleus subthalamicus* (or *corpus subthalamicum) of Luys*, which is morphologically a derivative of the dorsal hypothalamic zone, the term 'subthalamus' should preferably be entirely discarded (cf. K., 1948, 1954; K. and HAYMAKER, 1949).

As regards the term '*thalamus*' respectively '*thalamus opticus*', its origin is credited by SOEMMERING (1778) and other older authors to GALEN (131–201 AD). WALKER (1938) refers to the following quotation from GALEN's '*De usu partium*' (Latin translation by KUHN, 1882): '*Cavi enim intus fuerunt, quo spiritum reciperent, sursumque usque ad cerebri ventriculum ob eandem causam pertigerunt; ubi enim uterque ventriculus anterior definit ad latera, illinc nervi optici exoriuntur, ipseque ventriculorum velut thalamus propter illos nervos extitit*'. I was, so far, unable to check GALEN's original Greek text of the treatise in question, which so far remained unavailable to me.

The Greek word ὁ θάλαμος (also alternatively used in the feminine form ἡ θαλάμη) had a variety of denotations and connotations, such as cave, room, chamber, bedroom, bridal chamber, storeroom, treasury, etc. This Greek word may possibly be derived from the Egyptian *thalam*, said to denote a temple's anteroom. TRIEPEL (1921) justly remarks that GALEN apparently referred to a space *(Hohlraum)*, probably to portions of the lateral ventricles, with which he may have believed the optic nerves to be connected. Subsequently (RIOLAN, 1610), the term '*thalamus*' was used for the gray masses adjacent to lateral and third ventricle, being characterized by the adjacent optic chiasma respectively tract, and thus then called '*thalami nervorum opticorum*' (cf. also the comments by WALKER, 1938; FULTON, 1949; KEYSER, 1972; ANDERSON and HAYMAKER, 1974).

Mondino da Luzzi of Bologna (MUNDINUS, ca. 1270–1326) was apparently the first to describe the thalamus as 'buttock-shaped' masses or *anchae* on both sides of the third ventricle ('midventricle') whose

[13] Thus, despite his undeniable merit in identifying the thalamus ventralis of Amphibians and thereby of Vertebrates, HERRICK (1948) considers the pars ventralis thalami to be 'the motor zone of the thalamus, or subthalamus'. It is supposed to represent 'the primordium of the motor field of the mammalian subthalamus'. It is true that, in Mammals, the nucleus subthalamicus is related to the 'extrapyramidal motor system', but said nucleus is doubtless a derivative of the hypothalamic longitudinal zone's dorsal part, and not of the longitudinal zone represented by the thalamus ventralis in Amphibians.

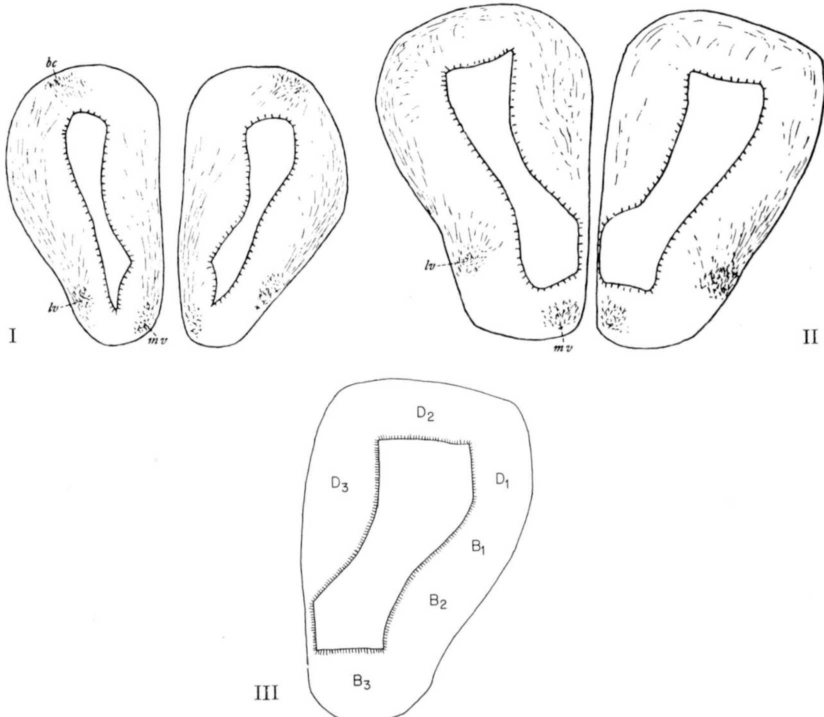

Figure 6. Two cross-sections through the lobus hemisphaericus of the urodele Amphibian Triton cristatus, showing relationships of basal forebrain bundles as displayed in *Golgi preparations.* Section I is more rostral than section II (from K., 1921). bc: location of tractus bulbo-corticalis; lv: lateral forebrain bundle; mv: medial forebrain bundle. In the added sketch III, the telencephalic longitudinal zones have been indicated by the topologic notation dealt with in chapter VI, section 6 of volume 3/II.

plexus he compared to a worm (SINGER's translation, quoted by WALKER, 1938). WALKER also states that VESALIUS (1514–1564) did neither mention the thalami, nor depict them in his illustrations. This remark, however, is not entirely correct. It is true that these configurations were not specifically described or named in the text of VESAL's classic *Fabrica* (1543) and *Epitome* (1543). Nevertheless, in Figures 7 and 8, book VII of the *Fabrica*, the thalamus, together with caudate and lentiform nucleus, as well as the capsula interna, are rather accurately depicted and marked by letters (the thalamus being D in 7, C and D in 8). Figure 10 (loc.cit.) also displays a dissection of the brain stem with parts of the thalamus, which, including the pulvinar, is designated by the letter A. However, no further relevant references to these and the other aforementioned structures are given in the text.

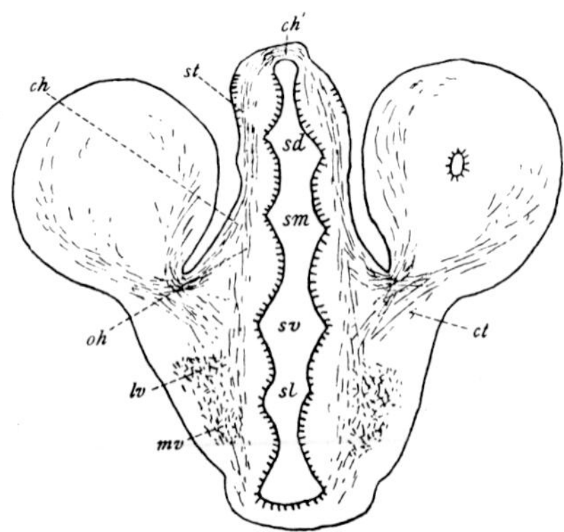

Figure 7. Cross-section through the diencephalon of Triton cristatus at the level of hemispheric stalk, polus posterior of telencephalon, and preoptic recess, showing relationships of basal forebrain bundles and of habenular channels (with exception of fasciculus retroflexus) as displayed by *Golgi preparations* (from K., 1921). ch: medial and lateral posterior telencephalic habenular channels (so-called tractus cortico-habenularis medialis and lateralis, becoming components of stria medullaris thalami in Mammals); ch′: habenular commissure; ct: posterior telecephalo-diencephalic channels (so-called tractus cortico-thalamicus medialis and lateralis); lv: lateral forebrain bundle; mv: medial forebrain bundle; oh: so-called tractus olfacto-habenularis; st: stria medullaris (taenia thalami of Mammals); sd: sulcus diencephalicus dorsalis; sm: sulcus diecephalicus medius; sl: sulcus intraencephalicus anterior (originally interpreted as sulcus limitans); sv: sulcus diencephalicus ventralis. The diencephalic longitudinal zones epithalamus (habenular nucleus or 'ganglion'), thalamus dorsalis, thalamus ventralis, and hypothalamus (at this level: caudal portion of preoptic recess) are not labelled but can easily be identified.

With regard to the *main communication channels* of the Vertebrate diencephalon, it is noteworthy that their essential features remain, for practical purposes, identical throughout the entire phylum, discounting various details of synaptology and the differentiation, within these channels, of distinctive or specific tracts.[13a]

[13a] Such tracts, insofar as they represent input channels, may be more or less restricted to neural signals pertaining to a particular 'receptor modality'. This is, e.g. the case in the deuterencephalon, where the root fibers are arranged in accordance with 'functional components'. The lateral lemniscus and its subdivisions are another example. In the diencephalon and mesencephalon, the optic tracts, and in the telencephalon the olfactory tracts represent such 'specific channels'. Some functional segregation also commonly obtains, *qua* secondary afferent fiber components, in the general spinal and bulbar lemniscus system. As regards the 'separate central representation' of 'specific sensory systems' there obtains *qua* 'central connections', a combination of segregation and correlation, that is, of separation and fusion, this latter including 'overlap'. The details of the relevant synaptic connections and circuit mechanisms remain insufficiently elucidated. The available data

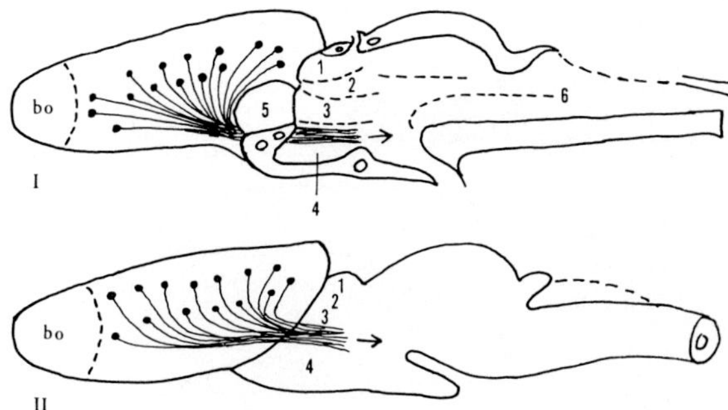

Figure 8 A. Two sketches showing the course of medial and lateral forebrain bundles in the forebrain of a 'generalized Amphibian'. bo: bulbus olfactorius; 1: epithalamus; 2: thalamus dorsalis; 3: thalamus ventralis; 4: hypothalamus; 5: ventriculus impar telencephali (aula, *cavum Monroi); 6*: sulcus limitans; I: medial aspect, showing medial forebrain bundle (the unlabelled commissures can be easily identified); II: lateral aspect, showing lateral forebrain bundle (although dorsal to the medial one in the diencephalon, and close to ventral thalamus, this bundle also runs through the hypothalamic wall). Although the descending components of the bundles are emphasized, these systems represent *reciprocal (descending* and *ascending)* communication channels.

The fundamental arrangement of the *diencephalo-telencephalic channels*, displayed in particularly simple form by some urodele Amphibians, is shown in Figures 6, 7, and 8. It will be seen that *lateral* and *medial forebrain bundles* provide connections between lateral respectively medial wall of the inverted evaginated hemispheres and the diencephalon, reaching this latter through the hemispheric stalk and allowing for decussations within the commissural plate of lamina terminalis.[14] In the

---

merely allow very generalized statements. Moreover, it also should be emphasized that, in all classes of Vertebrates from Cyclostomes to Mammals, the diencephalon, particularly the thalamus, which, to a greater or lesser degree, is reached by the two main ascending sensory channels, namely, general lemniscus and lemniscus lateralis, as well as by components of the optic tract, relays all the diverse sorts of sensory input to the telencephalon. In various submammalian forms, where the input by said channels to mesencephalic tectum and semicircular torus seems to predominate, extent and details of this diencephalic relay to telencephalon are still insufficiently known. Explicit further comments on inconclusive aspects of this topic will be omitted from some of this chapter's sections dealing with specific submammalian forms. In view of the obtaining uncertainties it should be sufficient to keep in mind this overall arrangement of the main communicating channels, also pointed out in volume 4 (pp. 149, 160, 174, 358, 792 *et passim).*

[14] The diverse configurational types of the Vertebrate lamina terminalis, and of its commissures were dealt with, from the morphogenetic aspect, in section 1 B of chapter VI, volume 3/II (cf. Figs. 62–65 loc.cit.). Further comments on the commissures were also included in section 6, pp. 619–635 of that volume.

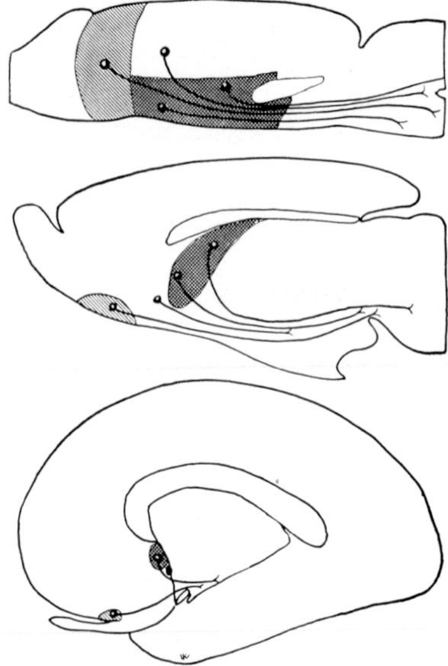

Figure 8B. Diagrams purporting to illustrate 'evolution of medial forebrain bundle' (from KRIEG, 1966). Upper sketch Amphibian, middle lower Mammal, lower sketch Human brain. Dark cross hatching 'septal nucleus' (paraterminal grisea), light cross hatching 'anterior olfactory nucleus'. It will be seen that KRIEG does not, in Mammals, include the fornix (shown in the Amphibian) as a derivative of the medial forebrain bundle (cf. Fig. 12).

diencephalic wall, the forebrain bundles, of which the lateral is located somewhat more dorsally, and the medial more ventrally, run through the hypothalamic region, but provide connections for the grisea derived from all longitudinal diencephalic zones (cf. Figs. 3, 7, 9, 10). Fibers from the paired caudally evaginated hemisphere, which do not run through the telencephalic portion of the basal forebrain bundles, connect with diencephalic[15] including the habenular grisea (epithalamus). The habenular connections form the *stria medullaris thalami* which becomes quite distinctive in Mammals. This stria medullaris, however, also receives components from the basal forebrain bundles (cf. Figs. 3, 7).

An additional rather complex but not particularly massive telencephalo-diencephalic channel, which becomes quite distinctive in Mam-

[15] In Plagiostomes, comparable channels are tractus pallii and tractus taeniae. Although Plagiostomes do not display a caudalward paired evaginated hemisphere with true polus posterior, a caudal protrusion of the unpaired hemisphere's dorsal wall obtains in various forms.

mals, is related to the amygdaloid complex located in the neighborhood of the temporal pole, and follows the curvature of sulcus terminalis caused by the hemispheric rotation. This is the *stria terminalis sive taenia semicircularis* mentioned above on p. 8 in footnote 7. Since this channel is related to grisea of the caudalward evaginated paired hemisphere its nondescript 'precursors' can be noted in Anamnia such as e.g. Amphibians and, somewhat more distinctly, in Sauropsida.

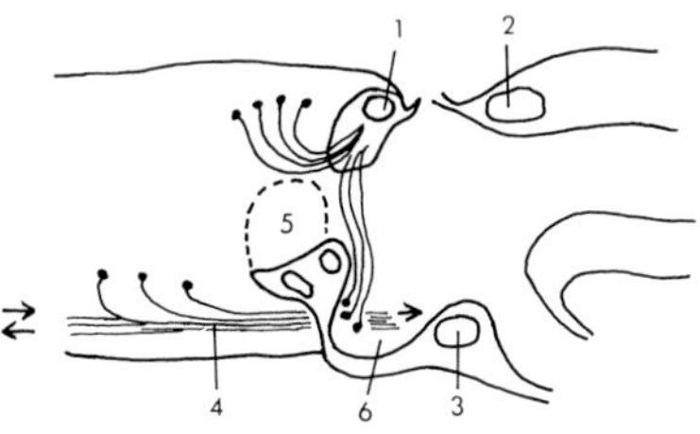

Figure 9. Sketch of the medial rostral habenular connections as seen in the medial aspect of the forebrain in a 'generalized Amphibian'. 1: habenular commissure and habenular ganglion; 2: posterior commissure; 3: optic chiasma and supra-respectively postoptic commissures; 4: medial forebrain bundle; 5: telencephalic aula; 6: preoptic recess of hypothalamus. Only the habenular *input* components are stressed. Below 5 lies the thickened commissural plate with commissura anterior and c. pallii.

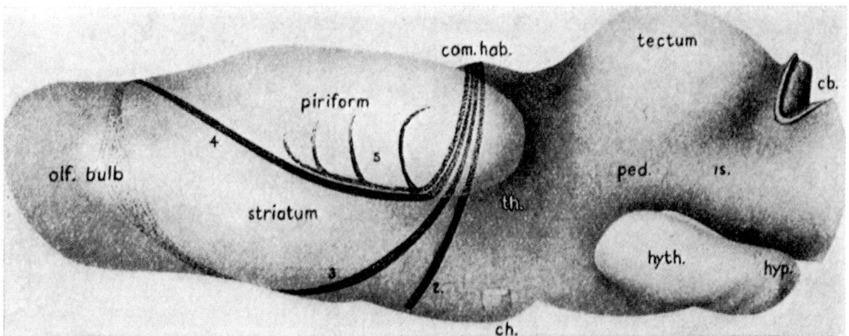

Figure 10. Sketch of lateral rostral habenular connections indicated upon a wax model of the brain of the urodele Amblystoma tigrinum (from HERRICK, 1948). 2: 'tractus olfacto-habenularis lateralis'; 3, 4: ventral and dorsal portions of 'tractus olfacto-habenularis' arising from 'nucleus olfactorius anterior'; 5: 'tractus cortico-habenularis lateralis'; cb.: cerebellum; ch.: optic nerve and location of chiasma; com. hab.: habenular commissure; hyth.: posterior portion of hypothalamus; is.: isthmus rhombencephali region; ped.: peduncle ('tegmentum', i.e. basal plate region of mesencephalon); th.: thalamus. The designation 'piriform' corresponds to telecephalic zone D_1, and 'striatum' corresponds to zones B_1 and B_2 of my terminology. The hypophysial complex is labelled hyp.

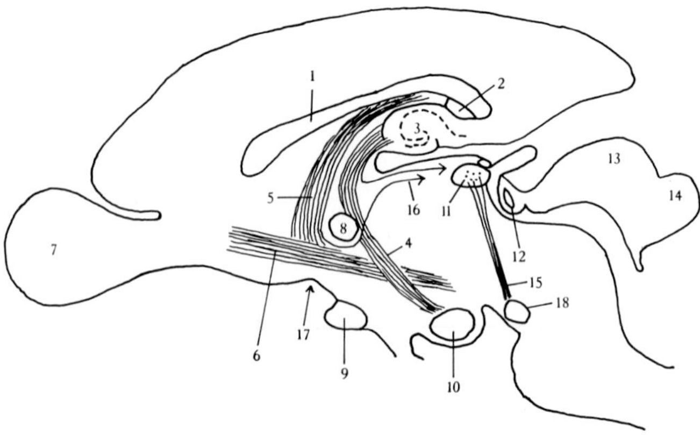

Figure 11. Sketch illustrating differentiation of some telencephalo-diencephalic communication channels indicated upon a parasagittal section through the brain of a 'lower' Eutherian Mammal (Rabbit), and based on a drawing by L. EDINGER (1912). 1: corpus callosum; 2: commissura hippocampi (psalterium); 3: hippocampal formation; 4: fornix *(sensu strictiori,* postcommissural with respect to commissura anterior); 5: fiber systems of hippocampus and paraterminal grisea running through so-called 'septum' (so-called 'precommissural fornix'); 6: conjoined basal (medial and lateral) forebrain bundle; 7: olfactory bulb; 8: anterior commissure; 9: optic chiasma and supra- respectively postoptic commissures; 10: mammillary body; 11: habenular grisea (the oval outline above 11 indicates the habenular commissure); 12: posterior commissure; 13: superior colliculus; 14: inferior colliculus; 15: fasciculus retroflexus *seu* habenulo-peduncularis; 16: stria medullaris thalami; 17: basal telencephalo-diencephalic boundary (remnant of embryonic sulcus telencephalo-diencephalicus; 18: interpeduncular nucleus. Cf. also Fig. 146 A of chapter XVI, vol. 5/II.

In Mammals, the *fornix*, whose main component is here postcommissural, becomes a specific differentiation of the primordial medial forebrain bundle, concomitantly with the differentiation of a typical 'mammillary body' (cf. Figs. 11, 12).

Again, in the Mammalian prosencephalon, the primordial lateral forebrain bundle has evolved into (a) the fiber masses of the *internal capsule* (Fig. 13), which includes most thalamic stalks, moreover the corticopontile, and the pyramidal tracts. An additional differentiation of the primitive lateral forebrain bundle is (b) the *ansa peduncularis (ansa lenticularis* and *inferior thalamic stalk)*, running more or less independently of capsula interna.

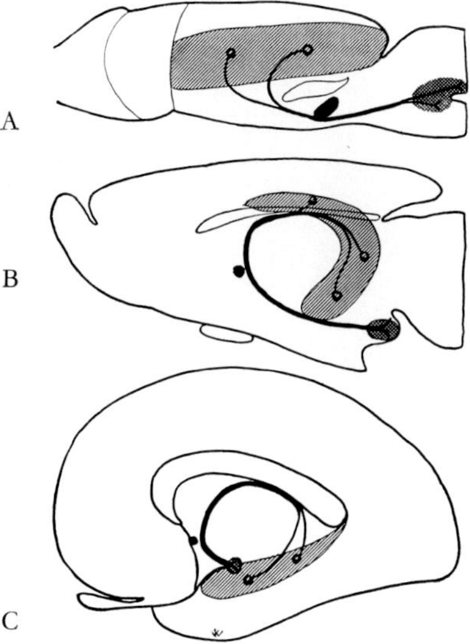

Figure 12. Simplified concept concerning the evolution of the fornix (from KRIEG, 1966). A: Amphibian; B: 'lower Mammal'; C: human brain. Black: position of anterior commissure in median plane. Hatchings indicate hippocampal formation and mammillary body in B and C. In A, the hatchings represent 'primordium hippocampi' (D_3 of my terminology) and the presumed 'forerunner' of mammillary body. The tract shown by KRIEG in A is actually a portion of the medial forebrain bundle. This component may indeed be conceived, to some extent and with various qualifications, as comparable with the Mammalian fornix.

The components of primordial forebrain bundles not included within fornix system, capsula interna, and ansa peduncularis are then crowded together into a rather inconspicuous channel usually designated as the Mammalian '*medial forebrain bundle*' but actually representing, as it were, a 'remnant' of both lateral and medial forebrain bundles of Anamnia. It essentially interconnects basal telencephalic structures such as tuberculum olfactorium, 'fundus striati', and paraterminal ('septal') grisea with the hypothalamus and probably the mesencephalic tegmentum.

The *caudal communication channels* between diencephalon and mesencephalon in 'unspecialized Anamnia' such as urodele Amphibia are provided by the *fasciculus retroflexus* and by a rather diffuse longitudinal tegmental radiation which includes some poorly understood output components as well as input components from lemniscus systems, cerebellum, reticular formation, torus semicircularis, and tectum. The

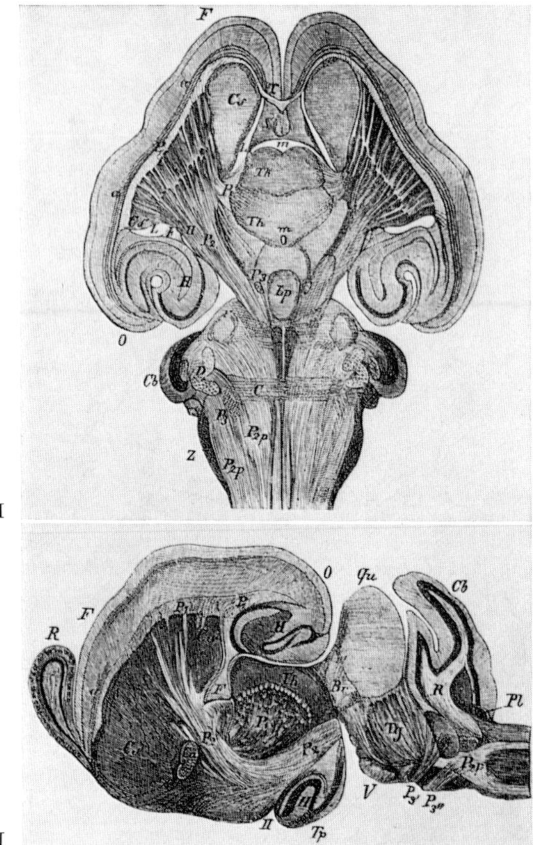

I

II

Figure 13. Two historically interesting illustrations by MEYNERT, showing the course of the lateral telencephalo-diencephalo-deuterencephalic communication channel (internal capsule, thalamic stalk, pes pedunculi) in the brain of a 'lower Mammal' (Chiropteran Vespertilio pipistrellus) in (A) horizontal and (B) sagittal sectional planes (from MEYNERT, 1872); Br: brachium conjunctivum; C: trapezoid body; Ca: anterior commissure; Cb: cerebellum; Cj, Cs: corpus striatum; D: superior olivary complex; F: frontal pole of hemisphere (in I), fornix (in II); H: hippocampus; L: lateral ventricle; Lp: interpeduncular ganglion; m: third ventricle; O: occipital pole of hemisphere; P_1, P_2: capsula interna, including its thalamic stalk (this channel and its caudal extension represent, as it were, a neencephalic differentiation of the Anamniote lateral forebrain bundle); P_{2p}, P_{2r}, P_3: unidentified tegmental fiber systems *('Bahnen der Haube im Projektionssystem der Brücke und Oblongata');* Pl: choroid plexus; Qu: mesencephalon; R: olfactory bulb (at left), cerebellar medulla (at right); S: septum (paraterminal grisea); T: corpus callosum; Tg: channels to and from tectum mesencephali (including components of lemniscus systems); Th: thalamus; Tp: temporal portion of hemisphere; V: pons; Z: medulla oblongata; II: optic tract.

rostral end of this 'tegmental radiation' overlaps with the caudal end of the basal forebrain bundles, from which it cannot be sufficiently well disentangled. In addition, there is a more dorsally located fiber field connecting the tectum opticum with the diencephalon through tracts of the pretectal region and its alar plate derivatives.

In 'higher' Vertebrate forms such as Mammals, the caudal channels are somewhat more distinctly differentiated and include, besides fasciculus retroflexus, the medial and lateral lemniscus systems, brachium conjunctivum, pretectal field, tegmental *field H of Forel* with fasciculus thalamicus (H_1) and fasciculus lenticularis (H_2), brachium quadrigeminum superius and inferius, tractus tectothalamicus dorsomedialis, components of the central tegmental tract, the mammillary peduncle, the mammillo-interpeduncular tract, and the mammillo-tegmental tract.

The diencephalon, moreover, receives *direct optic input*[16] to thalamus (dorsal and ventral lateral geniculate nuclei, pretectal grisea) and also to hypothalamus, although the mesencephalic tectum opticum, as discussed in chapter II of volume 4, may be considered the main terminal griseum of optic tract in 'lower' Vertebrates. Traditionally, the portion of the optic channel between eyebulb and its decussation in the hypothalamic floor is called the *optic nerve*. Said decussation, either essentially total, or only partial, depending on the Vertebrate forms, is the *optic chiasma*. The portion between chiasma and various diencephalic respectively mesencephalic grisea is the *optic tract* with its several subdivisions (cf. also further below on p. 61 of section 1B).

The *diencephalic commissures sensu latiori*, i.e. comprising both decussating fibers and fibers interconnecting antimeric grisea, include (1) the commissura anterior, (2) the optic chiasma, (3) the supra- or postoptic commissures within the chiasmatic ridge, (4) the so-called infundibular, and the supramammillary as well as the rostral tegmental commissures, (5) the habenular commissure, (6) the posterior commissure, and (7) in some Amniota (particularly in Crocodilians and many Mammals) a secondary fusion within the thalamic surfaces of the third ventricle's wall, the commissura media or massa intermedia, also designated as commissura mollis. This variable structure generally contains some of the so-called midline nuclei, and also affords passage for a relatively small number of commissural fibers.

---

[16] The optic input here under consideration refers to the lateral eyes. In some Vertebrates, one or two less developed dorsal (parietal, pineal) eyes are present, which display a structural arrangement differing from the lateral eyes. Comments on these eyes will be found in subsection C.

Despite various close interrelationships between telencephalon and diencephalon, which are both prosencephalic derivatives, the intrinsically telencephalic commissures provided by commissura pallii *sive* hippocampi of submammalian Vertebrates, of Prototherian and Metatherian mammals, as well as the corpus callosum of Eutheria need not be considered in this context, and shall be dealt with in chapter XIII (cf. also section 6, chapter VI, vol. 3/II).

Anterior and posterior commissures are important landmarks whose significance in drawing topologically valid anatomical telodiencephalic and diencephalomesencephalic boundaries was dealt with in volume 3/II. Both commissures, although representing highly conservative and stable transverse channels persisting in the entire Vertebrate series from Fishes to Mammals, nevertheless undoubtedly contain, in the diverse taxonomic forms, various components of dissimilar functional significance and with unlike and different connections.

The *commissura anterior*, whose median portion is a landmark of the telencephalodiencephalic boundary zone, can be evaluated as an essentially telencephalic commissure pertaining to lateral and medial forebrain bundles of Anamnia,[17] and containing a substantial olfactory component. It seems likely that through this channel decussating fibers of various functional significance interconnect the telencephalon of one side with diverse grisea of the contralateral diencephalon, in addition to the commissure's exclusively telencephalic fiber systems.

The *posterior commissure* is essentially a channel interconnecting the pretectal diencephalic and mesencephalic regions. As regards its components, many differences doubtless obtain within the Vertebrate taxonomic series, comparable, *mutatis mutandis*, to the above mentioned differences in the composition of anterior commissure. Relevant details concerning the enumerated commissures (1) to (7) shall be dealt with in the sections devoted to the respective Vertebrate groups.

B. *The Eye and its Peripheral Optic Ganglion (Retina)*

The *eye vesicle*, the therefrom resulting *eye cup (Augenbecher)* and its relevant differentiation, the *retina*, represent intrinsic components re-

---

[17] In the everted and unpaired telencephalon of Osteichthyes, the lateral forebrain bundle includes the fibers related to zone D_3, and the ventral portion of the *anterior commissure* includes here the *commissura pallii* of the inverted Anamniote brain (cf. Fig. 64, p. 175, vol. 3/II). In accordance with the modifications obtaining for the basal forebrain

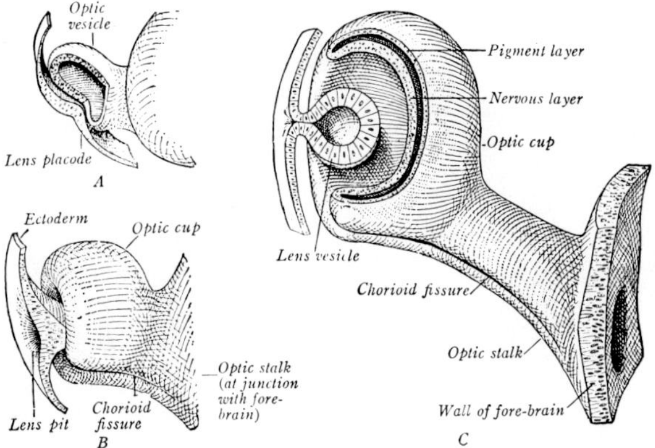

Figure 14 A. Human optic cup primordia, shown as models in partly sectioned side view (after MANN, 1928, from AREY, 1954). A: at 4.5 mm (about 4 weeks); B: at 5.5 mm (between 4 and 5 weeks); C: at 7.5 mm (nearly 5 weeks).

spectively derivatives of the primary prosencephalon. Thus, the eye is here considered to be, with respect to its neuroectodermal structures, a part of the central neuraxis, namely of the diencephalon.

In all Vertebrates, the ontogenetic morphogenesis of the eye bulb, as regards its major features, is essentially identical from Fish to Man, being characterized by the formation of an evaginated eye vesicle with its stalk. This latter becomes the optic nerve, while the vesicle, subsequently invaginated as the optic cup and commonly displaying a choroid fissure, becomes the retina (Figs. 14, 15). In addition to these neuroectodermal components, the lens arises from an external ectodermal lens placode which separates from the surface ectoderm.[18] This latter, moreover, provides the epithelium of cornea and conjunctiva. The mesoderm contributes by forming substantia propria of cornea, endothelium of anterior chamber, the vascular choroid,[19] and the compact

channels in the everted telencephalon, medial and lateral forebrain bundles tend to run at a side-by-side level through the hypothalamus rather than in a dorsoventral spatial relationship of lateral and medial bundle.

[18] Although the lens is wholly non-vascular, blood vessels usually spread over its surface during ontogenetic development. Thus, in many forms, an arteria hyaloidea represents a continuation of the arteria centralis retinae (Fig. 15 B) and reaches the posterior lens surface. A hyaloid canal, as e.g. in the adult human eye, represents a faint trace of the course of the disappeared hyaloid artery through the vitreous body.

[19] The circular *iris*, whose central opening is the *pupil*, develops from a combination of the modified anterior segments of both choroid and neuroectodermal optic cup. Opposite to the circumference of the lens, the conjoined layers commonly expand, at least in many Vertebrate groups, to form the *ciliary body* with its *ciliary processes*, to which the lens is attached through the fibers of the *zonula ciliaris*.

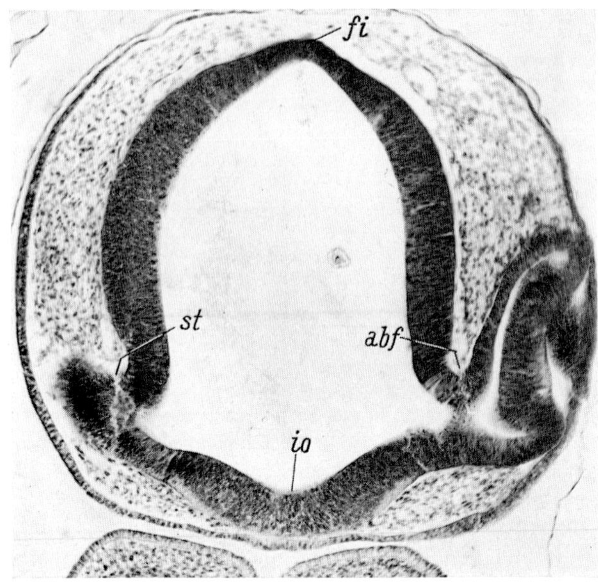

Figure 14B. Slightly oblique cross-section through the prosencephalon of an Avian embryo (Duck of 72 h incubation), showing evagination and secondary invagination of eye vesicle respectively eye cup (from K., 1936). abf: external groove of eye stalk; fi: dorsal ridge of telencephalic primordium near telencephalo-diencephalic boundary region; io: basal interoptic groove; st: beginning sulcus telodiencephalicus temporarily in close vicinity of external groove of eye stalk.

fibrous sclera. In addition, the arteria centralis retinae, with its accompanying venous channel, generally becomes included into the choroid fissure of the optic stalk and thereby into the definitive optic nerve.

Thus, from a viewpoint of morphogenesis, one might be justified to state that the Vertebrate prosencephalon gives origin, in the course of its ontogenetic differentiation, to three distinctive major derivatives, namely medial diencephalon, lateral retina (eye vesicle), and rostral telencephalon. Since, in the definitive state, the lateral eye remains configurationally connected with the diencephalon (hypothalamus) through optic 'nerve' and chiasmatic ridge, the retina shall here, somewhat arbitrarily, not be considered a major subdivision, comparable in 'rank' to endbrain and betweenbrain, but merely as the '*optic ganglion*' of the diencephalon.

Figure 16A illustrates an horizontal section through a Mammalian Primate (human) eye. The cavity of the original invaginated eye cup, posteriorly to the attachment of the lens to the ciliary body, contains

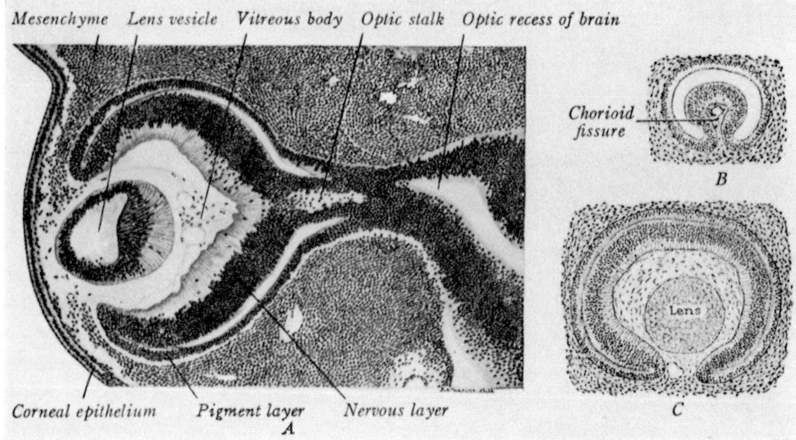

Figure 15. Optic primordia of a 10 mm human embryo (between 5 and 6 weeks) in longitudinal and transverse sections (from AREY, 1954). A 'Longitudinal' section (in plane including lens and optic stalk). B 'Transverse' section through optic stalk with choroid fissure which contains arteria and vena centrales retinae. C 'Transverse' section in plane ncluding posterior pole of lens and choroid fissure (cf. C in Fig. 14A).

the viscous or gelatinous *vitreous body (humor vitreus)* which is presumably of neuroglial derivation. The *anterior chamber*, between cornea, lens and iris, as well as the *posterior chamber*, between posterior surface of iris, lens, and lens attachment, contain the *aqueous humor*, believed to be produced by the ciliary processes of the corpus ciliare. This fluid, *qua* relatively high sodium chloride and very low protein content, is comparable to the liquor cerebrospinalis. Like this latter, it seems to be 'secreted' or 'filtered through' a neuroectodermal epithelium, which is represented by that of the ciliary body, consisting, however, of two layers. The aqueous humor, significant for the maintenance of intraocular tension, is resorbed by the '*canal of Schlemm*', a system of venous channels encircling the cornea at the sclerocorneal junction. Insufficient absorption of the aqueous humor increases intraocular pressure and causes the condition known as *glaucoma*, which may result in 'cupping' of the optic disc (exit of optic nerve), and blindness by atrophy of the retina. An actual section through a Mammalian eye (Dog) is illustrated by Figure 16 B, while Figure 16 C shows the configuration of an Amphibian eye (Anuran, Frog). The section of Figure 16 B, through the eye of a fairly aged Dog, displays, moreover, the so-called 'nucleus' of the lens. Originally a hollow vesicle formed by epithelial cells (cf. Figs. 14, 15), its posterior cells elongate into the lens fibers, while the

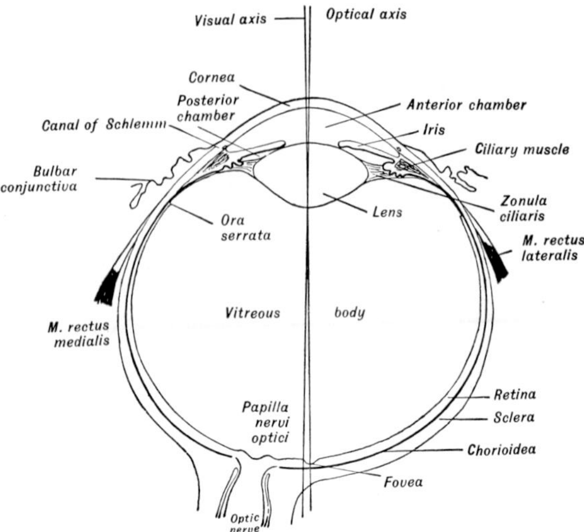

Figure 16 A. Diagram of a horizontal section, as seen from above, through the human right eye (from MAXIMOW, 1930). The ora serrata is the boundary between pars optica and pars ciliaris retinae. This latter, in turn, adjoins the pars iridica retinae at the periphery of the posterior chamber. Pars ciliaris and pars iridica together represent the pars caeca retinae. The fovea (centralis), through which the optic axis passes, is the locus of greatest visual acuity. The papilla nervi optici is devoid of neurosensory elements and represents the 'blind spot', located nasally with respect to the optic axis, and thereby in the temporal half of the visual field.

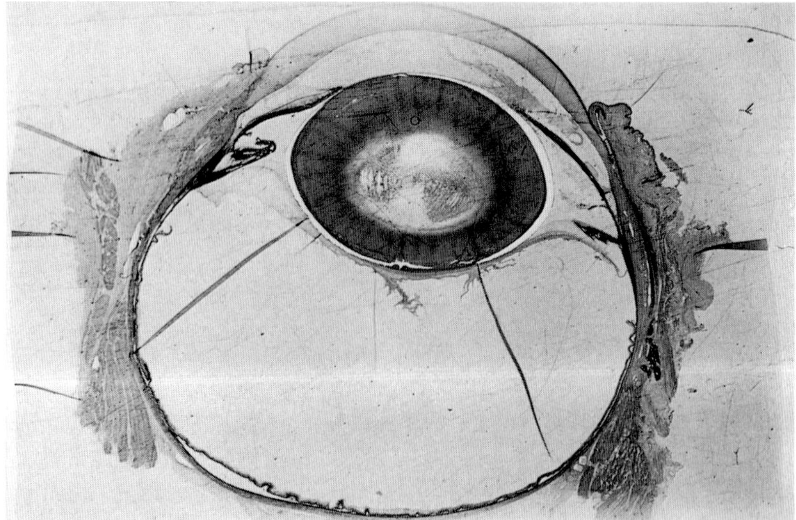

Figure 16 B. Section through the eye of a Dog, for comparison with preceding figure (routine hematoxylin-eosin preparation ×4; red. $^{2}/_{3}$). Note 'nucleus' of lens as discussed in text.

anterior cells form a layer of low cuboidal epithelium. As aging takes place, the central portion of the lens undergoes 'sclerosis', presumably because of loss of water correlated with greater density, thus becoming the 'nucleus'. This change is apparently related to the original 'compression' by the growing lens fibers which also increase until the definitive constant cell number is reached (at about 20 years of age in Man, here with a fairly accurate number of 2250 fibers). In addition, the 'nucleus', because of its distance from the periphery, may receive less 'nutrition'. Said sclerosis does not necessarily involve opacity (cataract) but doubtless predisposes to that condition. The epithelial elements of the non-vascular lens apparently produce the basement membrane (lens capsule) surrounding the lens *in toto*.

Comparative anatomy fails to provide any convincing clue to the *phylogenetic evolution of the Vertebrate eye*, since this latter, as AREY (1954) justly points out, is highly developed even in the most 'primitive' groups. Nevertheless, LANKESTER (1880, 1900), BOVERI (1904), PARKER (1908), STUDNIČKA (1918), and PLATE (1924) have elaborated some not altogether implausible speculations concerning this topic. In agreement with the gist of these concepts, NEAL and RAND (1936) consider it not improbable that the two lateral eyes of Vertebrates took their origin from paired pigmented depressions, containing neurosensory cells, in the anterior part of the neural plate respectively tube of

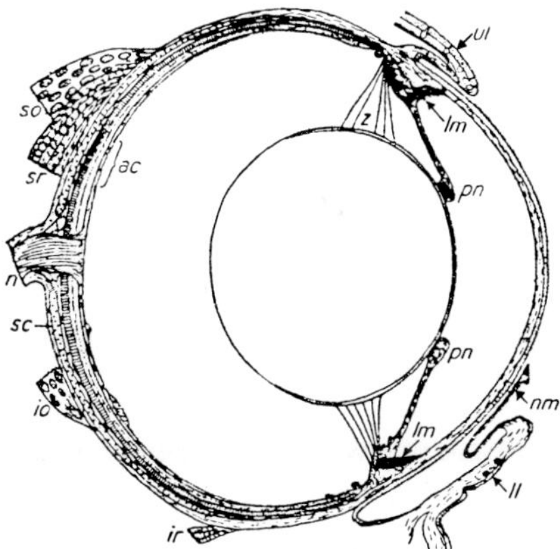

Figure 16C. Anuran Amphibian eye in vertical section (after diverse authors, from YOUNG, 1955). ac: area centralis; io: inferior oblique muscle; ir: inferior rectus; ll: lower lid; lm: lens muscles (protractors); n: optic nerve; nm: nictitating membrane; pn: pupillary rim ('nodule'); sc: scleral cartilage; so: superior oblique; sr: superior rectus; ul: upper lid; z: zonule.

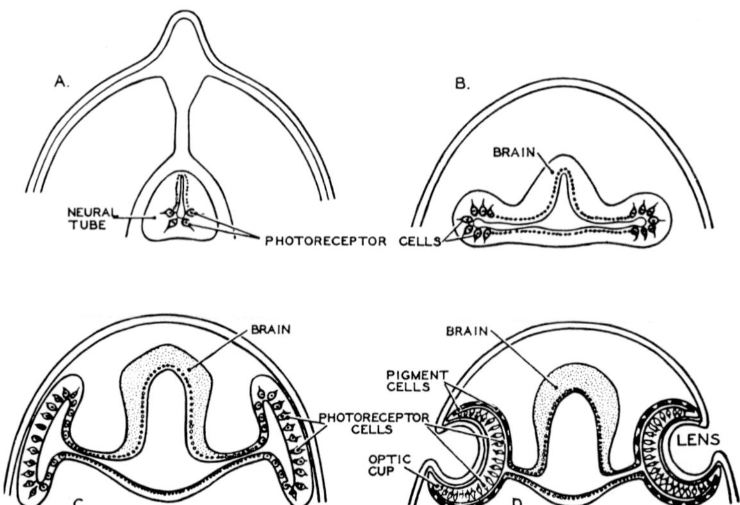

Figure 16 D. Diagrams illustrating the hypothesis that the lateral eyes of Vertebrates have evolved from lateral outgrowths of the forebrain wall containing photoreceptor elements (from NEAL and RAND, 1936).

unknown ancestral forms (cf. Fig. 16 D). The *parietal eyes* might have a similar origin from more dorsally located spots of the diencephalic prosencephalon. Thus, the parietal eyes 'look upwards', while the lateral eyes, evaginated from the primitive forebrain, became 'receptors of light from the side and below'. As regards the lateral eyes, undefined factors converted the overlying skin into a lens and led to invagination of the optic cup with an inverted position of the photoreceptors.[20] The

[20] The '*inverted*' retina of the Vertebrate lateral eyes is characterized by an arrangement such that the receptor structures of the neurosensory cells (rods or cones) are directed away from the light, which has to traverse the various retinal layers before reaching the receptor endings of said cells. In the *adverse* 'retinae' of Invertebrates such as Insects and Cephalopods, the neurosensory cells directly face the incident light, toward which the receptor endings are oriented (cf. also vol. 2, chapter IV, section 12, pp. 263–279, and Figs. 185–190). It is, moreover, of interest that the adverse (i.e. 'unreversed') pattern of retina obtains in the parietal eyes of Vertebrates. On the other hand, inverted arrangements are also displayed by the spinal cord 'eyes' of Amphioxus (cf. Fig. 34, vol. 4), and by the 'eyes' of Invertebrate Turbellarian Platyhelminthes (cf. Fig. 18, vol. 2). Still another type of inverted retina occurs in the Invertebrate Arthropod Araneae (Spiders). Both 'indirect' and 'direct' eyes may here be present in one and the same species. Of the two rows of four eyes commonly found in Spiders, the anterior median eyes are usually 'direct' and the other ones 'indirect'. These latter, moreover, are provided with a pigmented tapetum lacking in 'direct' eyes (cf. Fig. 51 C). Again, in Molluscan Pelecypoda (Pecten, Scallop), the mantle

parietal eyes, however, which are more or less rudimentary in recent Vertebrates, being, moreover, restricted to certain lower forms, differ from the lateral eyes by remaining non-invaginated vesicles, whose peripheral portions became the (neuroectodermal) lens. These eyes, to be briefly dealt with further below in subsection C, thus display a configurational pattern which, according to the concept of general homology, is analogous but not homologous to the bauplan of the lateral eyes.

Generally speaking, photoreception in Metazoa subsumes several different sorts of registration and processing by the specific transducer mechanisms. These various functional aspects can be very roughly classified as follows: (a) mere detection of the radiation, (b) registration of light direction, facilitating photo-orientation including phototaxis, (c) registration of light intensity, (d) registration of non-configurated motion (e.g. change of light direction, and motions of 'light' and 'shade'), (e) registration of shapes (configuration), (f) registration of frequency (wave length, 'color') and (g) registration of light polarization. It is evident that the enumerated functional capabilities depend on the particular morphologic, structural and physico-chemical differentiation of the organs concerned with photoreception and on the complexity of the relevant neuronal processing mechanisms. Thus, e.g. in Man, the two-dimensional picture projected upon the curved surface of the retina is transformed into a three-dimensional perception (P-event) even in monocular vision, although, in this latter case, the three-dimensional depth effect is of lesser 'vividity'.

Light is 'detected' (i.e. registered) by plants as well as by animals. Amoebae display reactions to light. Various Flagellates are provided with a photosensitive pigment spot such as the red 'stigma' of Euglena. Among Coelenterates, some Medusae have pigmented photosensitive cell groups on the margin of the umbrella; in Nausithae, a rudimentary 'lens' is associated with these spots. Photoreceptor organs of Echinoderms and Lamellibranch Mollusks are shown in Figure 16E. While organized photoreceptors thus occur in all the animal phyla,

---

eyes contain a retina with two neurosensory layers, both of which are inverted ('double retina'). The proximal layer is said to respond at the onset, and the distal one at cessation or decrease of illumination (so-called *off-on responses*, as briefly discussed for Vertebrates on pp. 594–595, vol. 3/I). In a still more simplified manner, *'image-forming eyes'* may be distinguished from eyes merely registering illumination. Among eyes of this latter type, those of Turbellarians and most ocelli of Arthropods could be mentioned. Quite evidently, as regards 'image-forming eyes', intermediate sorts of diverse eye structure, and diverse degrees of blurred or well-defined image production (resolution) obtain within the wide variety of Metazoan eyes.

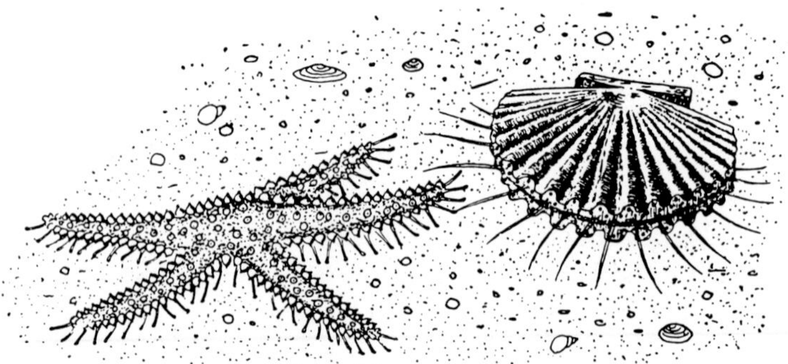

Figure 16E. Echinoderm Starfish and Molluskan Scallop equipped with optic receptors ('eyes') and 'olfactory tentacles' on the parameric arms respectively on the mantle edge (from BUDDENBROCK, 1958).

'true' 'image forming eyes' are here limited to those forms in which a 'picture', decomposed into a sufficient number of separate 'points', is projected respectively focused upon a retina (cf. above footnote 20). Signally elaborate eyes, showing striking analogies (but not homologies) with the Vertebrate lateral eyes, are found in Cephalopod Mollusks, while the complex compound eyes of Arthropods such as e.g. Insects display an altogether different type of organization (cf. vol. 2 of this series, dealing with Invertebrates).

As regards the *Vertebrate lateral eyes*, their configuration can be considered morphologically identical in all groups from Cyclostomes to Mammals despite a number of significant variations. Roughly stated, these relevant differences concern (1) eye position (extent of visual field, panoramic vision, binocular visual field), (2) relative size and overall shape of eye bulb, (3) shape of lens, (4) attachment of lens and accommodation mechanisms, (5) details of iris structure and various shapes of the pupillary opening, (6) presence of scleral cartilages or ossicles in diverse taxonomic forms, (7) internal folds or protrusions such as falciform ligament, pecten, or conus papillaris, (8) retinal vascularization, (9) retinal pattern, (10) accessory structures such as double cornea (*'brille'*), nictitating membrane, eyelids, glands, (11) variations in the basic pattern respectively development of the external eye muscles.[21]

[21] Additional differences, not directly related to those concerning pattern organization, involve 'reduction' of the lateral eyes in blind or near-blind Vertebrate forms, e.g. Myxine, various Osteichthyes, the Gymnophione Amphibians and some Urodele Amphibians, and the Insectivore Mammalian Talpa (the 'blind' Mole).

Detailed data on the comparative anatomy of the Vertebrate eye can be found in the publications by KRAUSE (1921–1923), PLATE (1924), FRANZ (1934), PRINCE (1956), YOUNG (1955, 1957), and in the comprehensive work of POLYAK (1957) on the Vertebrate visual system. In the aspect here under consideration, it will be sufficient to deal rather briefly with some of the enumerated topics. A few comments on visual fields were included in section 1, chapter XI of volume 4.

Concerning the functional aspects of the lateral eyes in Vertebrates, a purely *physical dioptric mechanism*, by means of which a more or less clearly focussed picture is projected upon the neurosensory receptor mosaic of the retina, must be distinguished from the *neurosensory transducing* and from the *neuronal processing mechanisms* transmitting the encoded signals to the optic centers of the central neuraxis. In the present context, only a few dioptric aspects related to the anatomical configurations shall be dealt with. Concerning further details such as spherical and chromatic aberration, astigmatism, improper centering, non-homogeneous refractivity within the lens, reference is made to the standard texts of physics, physiology, and ophthalmology, whose perusal is essential for an adequate understanding of the visual mechanisms.

The *refractive power* of a lens or of a comparable optic system, which depends on the curvature of the refracting media, is expressed in *diopters*. Since the focal length of a lens varies inversely with the refractive power, it becomes a convenient measure for said capacity, expressed in the focal length's reciprocal value. The standard diopter (D) is the focal length of an optic system taken as 1 m in air. Thus, the 'strength' of a lens with a focal length of 2 m is 0.5 D, and that of a lens with a focal length of 50 cm is 2 D. Roughly speaking, a biconvex (converging) lens has an anterior and a posterior focal point located on the optic axis which passes at a right angle through both surfaces and thereby through the center of the lens. Rays parallel to the optic axis and entering the anterior surface converge upon the posterior focal point and vice-versa with respect to the anterior one. The *focal length* can be conceived as the distance of the focal point from the refracting surface of a theoretical (exceedingly thin) lens, or from the reduced common nodal point of a centered optic system. The focal point corresponds to the image point of an object at 'infinity'. If the object point moves toward the lens, its image point will become displaced behind the posterior focal point such that, reaching the double anterior focal distance, the object point will lie at the double posterior focal distance. An object point at the anterior focal distance will have its image point at infinity, i.e. rays emitted from the object will become parallel after passing through the posterior lens surface. Since, in the non-accommodated eye, the posterior focal point of the refracting system lies upon the retina, it is evident that, for an accurate mapping of near objects, whose image points lie behind the posterior focal point, this latter must be moved forward by displacing the lens or by increasing the system's refractive power.[21a]

The Vertebrate lateral eye, however, does not merely depend upon the effect of its lens, but represents *a compound centered system* consisting of four refracting media with eight surfaces (cornea, aqueous humor, lens, vitreous body). Mapping the image construction by such a complex system is a very laborious mathematical procedure which has been simplified by constructing a so-called schematic eye in accordance with a theorem formulated by the famed mathematician GAUSS (1777–1855). This theorem states that every centered optical system composed of spherical surfaces has three pairs of *cardinal points*, namely two *focal points*, two *principal points*, and two *nodal points* (Fig. 17 A).

Roughly speaking, rays passing through the first medium parallel to the axis are focussed on the posterior focal point and vice versa. Oblique rays passing through the plane of the first cardinal point undergo a parallel displacement at the plane of the second. An oblique ray which, without refraction, would pass through the first nodal point, undergoes a corresponding parallel displacement and passes through the second nodal point. Rays corresponding to the optic axis do not undergo any refraction.

In a still more simplified manner, the projection of an image upon the retina of a 'schematic' human eye is illustrated by Figure 17 B. The large arrow A–B represents an object emitting, from every point, a diverging bundle of innumerable rays. Some of these rays, as shown in the diagram, are refracted to form the 'focussed' image represented by the arrow a–b. As in a photographic camera, the image is inverted and smaller than the object. The image size depends upon the angle ANB, subtended at the nodal point of the eye by an object in the visual field (*visual angle*).

[21a] Although, in this brief discussion of the refractive power of the lens, only the effect of lens curvature was considered, it should be recalled that said power (respectively the focal length of a lens) is determined not only by the curvature of the two lens surfaces, but also by the refractive index of the transparent medium (cf. chapter V, section 6, p. 371–373 of volume 3/I, and the following footnote 22 of the present chapter).

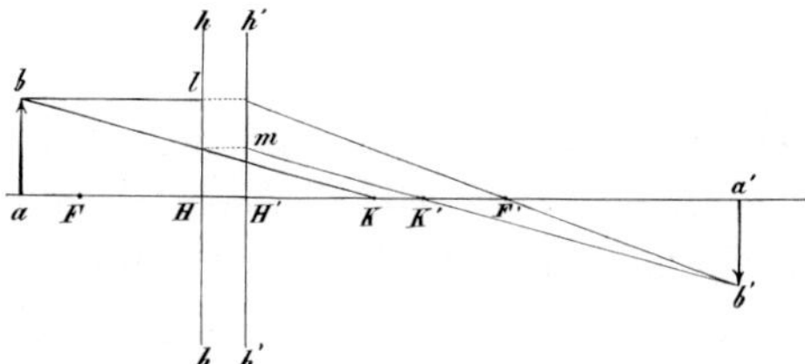

Figure 17A. Diagram illustrating the cardinal points of a centered optic system such as the simplified 'schematic human eye' in accordance with the *theorem of Gauss* (from Guttmann, 1918). F, F′: anterior and posterior focal points; H, H′: anterior and posterior principal points; K, K′: anterior and posterior nodal points; a, b: object; a′, b′: image; b–l: ray passing through posterior focal point after refraction; b–K: ray which, after refraction, becomes m–K′; h–h: anterior principal plane; h′–h′: posterior principal plane.

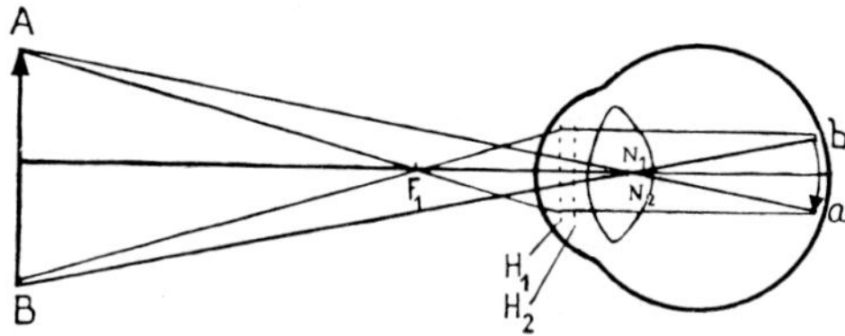

Figure 17B. Diagram illustrating the mapping of the inverted image upon the retina in a schematic human eye (from Best and Taylor, 1950). A–B: object, a–b: image; F_1: anterior focal point; H_1, H_2: anterior and posterior principal planes; N_1, N_2: anterior and posterior nodal points (for practical purposes, the two nodal points can be assumed as represented by a conjoint single one). The angle subtended at the nodal point by an object in the visual field is the visual angle. Roughly expressed, the relationship of linear object size to linear image size corresponds to the relationship of image distance to object distance as taken from the conjoint nodal point.

The *refractive power* of the resting (non-accommodated) human eye, whose posterior focal length as taken from a 'common nodal point' can be rounded off at about 17 mm, is thus approximately 58.5 D. The greatest refraction occurs at the corneal surface (42 D) the refractive power of the lens being 19 D with accommodation relaxed, and 36 D in maximal accommodation.

The farthest point from the non-accommodated eye at which an object can be clearly mapped upon the retina is the *far point*, which, for the normal (emmetropic) eye, lies at 'infinity' namely, for practical purposes, at any distance of more than 6 m. The corresponding point nearest to the accommodated eye is the *near point*, whose distance varies between 7 to 40 cm or more, depending upon age. The difference between far and near point is the *range of accommodation*. The difference between the refracting power of the eye when accommodation is completely relaxed and when at maximal accommodation is the *amplitude of accommodation*. Expressed in diopters, the reciprocal of the far point distance is termed the static refraction (R) which is zero ($\frac{100}{\infty}$ cm), and that of the near point distance expresses the *dynamic refraction* (P). The difference P–R gives, in diopters, the amplitude of accommodation. Thus, if the near point is at 6.35 cm, the dynamic refraction is $\frac{100}{6.25}$ cm=16 D. The amplitude of accommodation decreases progressively from childhood to old age, thereby resulting in *presbyopia*, as shown by the curves of figure 17 C. Discounting anomalies of configuration, such as myopia or hypermetropia, the near point of the normal human eye recedes from about 6 cm at twelve years to 40 cm or more in the fifties and can be roughly assessed at about 25 cm in middle-aged adults. The near point represents one of the various factors relevant for the '*resolving power*' of the eye, as discussed, with reference to macroscopic, microscopic, and ultrastructural space, on pp. 367–375, section 6, in volume 3/I. The tabulations of figure 17 D give various relevant numerical values pertaining to the centered optic system of the human eye.

PLATE (1924) and other authors stress the difference between eyes

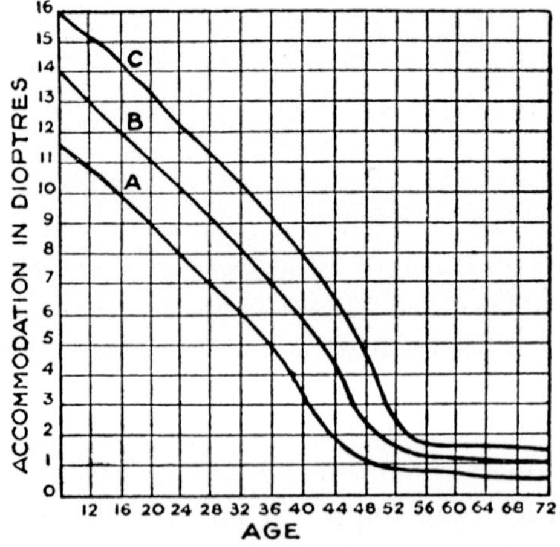

Figure 17C. Curves indicating the amplitude of human accommodation at different ages (after DUANE, from BEST and TAYLOR, 1950). A: lowest physiological values; B: average values; C: maximum values.

Anterior surface cornea	0 mm
Posterior surface cornea	0.6 mm
Anterior surface lens	3.6 mm
Posterior surface lens	7.2 mm
Retina	24.1 mm
First principal point (H)	1.7 mm
Second principal point (H′)	2.0 mm
First nodal point (K)	7.0 mm
Second nodal point (K′)	7.3 mm
Anterior focal point (F)	15.7 mm
Posterior focal point (F′)	24.1 mm

Figure 17 D. So-called constants of the human eye's optic axis, referred to anterior surface of cornea (after BEST and TAYLOR, 1950).

'adapted' to aquatic vision *(Wasseraugen)* and those 'adapted' to atmospheric vision *(Landaugen, Luftaugen)*. In atmospheric vision, the greatest refraction occurs at the corneal surface, whose curvature thus plays an important role for focussing upon the retina, the refractive indices[22] of air and cornea being 1 and about 1.37, respectively. In aquatic vision, because the refractive index of water is about 1.3, the main refractive effect is performed by the lens. This latter, in the eyes of most Fishes, has accordingly a spherical shape, which provides a maximum of refractive power. Generally speaking, the fish-eye is shortsighted (myopic). In Vertebrates with atmospheric vision, the lens tends to be flattened, especially so in Primates. It is of interest that, in Amphibians, various aquatic larvae display spherical lenses, which become relatively flattened after metamorphosis resulting in a terrestrial habitat. In Anurans the flattening is more conspicuous than in Urodeles. As a result of these and other changes, the Amphibian eye becomes more farsighted, an 'advantageous' condition less well realizable in water because of the greater opacity of this medium. The eyes of Amphibians, moreover, are of interest 'in that they show stages in the

[22] The refractive index is the ratio of the sine of the angle of incidence to the sine of the angle of refraction when light passes (obliquely) from one medium into the other, the refractive index of air being taken as 1. This index depends on the relative velocity of light in another medium compared to the velocity in air. For practical purposes, this latter velocity is taken as approximately corresponding to that *in vacuo* (i.e. to the constant c). It should be added that the refractive indices, originally obtained for monochromatic yellow light, slightly differ with respect to light of different frequencies, as demonstrated by the passage of light through a prism and by the chromatic aberration of a lens.

transition from the eyes of fish to those of higher vertebrates' (NOBLE, 1931, 1954). Arboreal and terrestrial Amphibia tend to have larger eyes than fossorial or aquatic forms.

Among Reptiles, the lens in aquatic types is essentially spherical in contradistinction to the somewhat more flattened lens in the terrestrial forms. As regards Mammals, the aquatic Cetacea and Pinnipedia (Seals and their kin) have likewise spherical lenses, but the Sirenian Dugong has retained a somewhat terrestrial lens shape.

The *attachment* of the Vertebrate lens displays a number of variations. In Cyclostomes, a specific attachment seems to be lacking; the lens is said to be held in place by the pressure from the vitreous body behind and from the anterior chamber in front. In all other taxonomic groups, the lens is attached peripherally to a conjoint choroid and (epithelial) retinal structure of various sort *('ciliary body')*. In Selachians this attachment or '*zonula*' is shaped like a circular membrane *(ligamentum suspensorium lentis)* containing a small smooth muscle within a papilla.

The *focussing of an image* upon the retina by means of the lens is called *accommodation*, requiring adjustments for near and distant vision. The lateral eyes in all Vertebrates *groups* are capable of accommodation to a greater or lesser extent. In a few *forms*, however, accommodation may be negligible or practically lacking. The mechanisms of accommodation differ with respect to the major Vertebrate groups and may be classified as follows (ROMER, 1950).

A. The position of the lens is moved.

(1) Near vision in fixed position, the lens being moved backward to focus on distant objects (Petromyzonts and Teleosts).

(2) Distant vision in fixed position, the lens being moved forward to focus on near objects (Elasmobranchs, Amphibians, and, among Amniota, Snakes).

B. The curvature of the lens is modified, fixed shape for distant vision, expanded shape for near vision (Amniota).

In the Cyclostome Petromyzonts, the lens is pushed back by a flattening of the cornea caused by an external '*musculus cornealis*' peculiar to these forms. In Elasmobranchs, a small *protractor muscle* of ectodermal (eye cup) derivation is attached to ventral rim of the lens and pulls it forward for near vision. In most Teleosts,[23] an elongated vascular

[23] In Ganoids such as the sturgeon Acipenser, no accommodation mechanism is said to be present. Much the same seems to obtain for the Dipnoans (PRINCE, 1956).

structure, the *processus falciformis*, projects through the choroid fissure into the eye cup cavity from optic nerve entrance toward the ciliary body. From the anterior edge[24] of this falciform process originates a *retractor lentis muscle* attached to the lens, which can thus be pulled backward for distant vision.

In Amphibians, the forward lens displacement is accomplished by mesodermal smooth musculature of the ciliary body, said to be present in all Amphibia as a *ventral ciliary muscle*, supplemented, in Anurans, by an additional *dorsal muscle*. In most or all Amphibia, there is an additional muscular system, the *tensor chorioideae*, developed as a series of meridionally arranged fibers in the periphery of the ciliary region. According to PLATE (1924) this muscle represents the precursor of the cillary muscle in Amniota.

In the Amniote Reptilian Snakes, the ciliary body has lost contact with the lens and has moved into the iris. When its muscle contracts, the volume of the eye bulb behind the lens is reduced, and the vitreous body then forces the lens forward and slightly medialward.

In all other Amniota, the second (B) mode of accommodation obtains. The lens is here somewhat elastic and can change its shape from a relatively flattened curvature adjusted for distant vision to a more pronounced curvature for near vision. This change is caused by contraction of the mesodermal ciliary muscle. In Reptiles and Birds, the ciliary body has a ring of processes in contact with the periphery of the lens. Upon contraction of the musculature, these processes push on the lens and thereby effect its bulging. In Mammals, on the other hand, the zonula fibers keep, as it were, the lens stretched in a flattened condition. The contraction of the ciliary muscle, by reducing the diameter of the 'ciliary ring', relax the tension exerted by the zonula fibers and the lens becomes more rounded by its own elasticity.[25] The ciliary musculature of *Sauropsidans*, like their iris musculature as dealt with below, consists of *striated muscle fibers*, in contradistinction to all other Vertebrates, including Mammals, in which these muscles are of the '*smooth*' type. The ciliary musculature of Birds, moreover, consists of three portions, the *posterior muscle of Brücke*, the *intermediate longer muscle of*

---

[24] This anterior edge is designated as the *campanula Halleri*. The vascular falciform process is interpreted to represent a 'nutritive structure for the retina'. The Teleostean musculus retractor lentis, commonly attached to the medial circumference of the lens, opposite to the bulk of suspensory ligament, pulls the lens backward and medialward.

[25] If the lens looses its elasticity with age, accommodation becomes reduced and near vision is thereby impaired *(presbyopia)*.

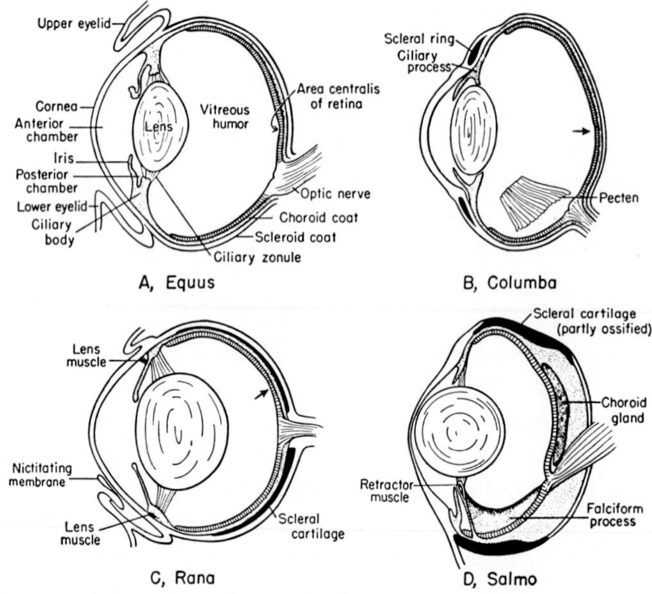

Figure 18 A. Diagrammatic vertical sections through the eye in some representative Vertebrates (Mammal, Bird, anuran Amphibian, Teleost) and displaying the organ's overall configuration (after several authors, from ROMER, 1950). Sclera and cornea unshaded; scleral ring or cartilage black; choroid, ciliary body and iris stippled; retina hatched; arrow supposedly pointing at fovea.

Müller, and the *anteromedial (distal) muscle of Crampton* which also acts on the cornea, which it tends to flatten. This represents an additional accommodation mechanism, involving the cornea.

The *iris*, present in all Vertebrate groups and formed by a combination of choroid with the two retinal layers at the rim of the optic cup, represents a pigmented circular diaphragm providing the pupillary opening for the lens. The width of the pupil seems to be fixed in some Fishes, except for changes caused by motions of lens or eyebulbs, but in numerous Fishes (e.g. Plagiostomes) and in all Tetrapod Vertebrates, muscles controlling the size of the pupil are present. These muscles are *of neuroectodermal provenance*, differentiating from the anterior outer layer of the optic cup and are commonly arranged as a circular *musculus sphincter pupillae* and as a radial *musculus dilatator pupillae.*[26]

[26] The innervation of the intrinsic eye muscles (m. ciliaris, mm. sphincter and dilator pupillae) through the vegetative nervous system is briefly dealt with in section 6, chapter VII of volume 3/II.

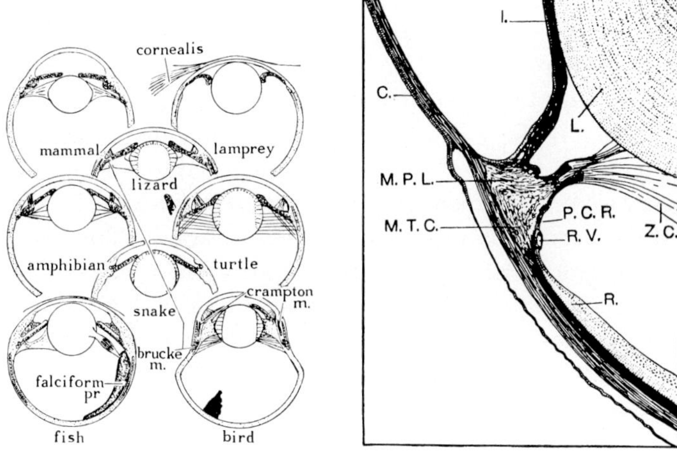

Figure 18B. Diagrams illustrating mechanisms of accommodation in diverse Vertebrates and as dealt with in the text (partly after WALLS, 1942, from PRINCE, 1956).

Figure 18C. Vertical meridian section through a Frog's eye, showing muscles of accommodation (after TRETJAKOFF, 1906, from NOBLE, 1931, 1954). C.: cornea; I.: iris; L.: lens; M.P.L.: m. protractor lentis; M.T.C.: m. tensor chorioideae; P.C.R.: pars ciliaris retinae; R.: retina; R.V.: ring vessel; Z.C.: zonula ciliaris.

Although most Vertebrates, except perhaps Cyclostomes and some Ganoids, seem to possess a *sphincter muscle*, the musculus dilatator (or dilator) is said to be lacking or to be very poorly developed in various forms whose eyes are provided with a sphincter. Be that as it may, all Vertebrate eyes with an iris musculature adjust the size of their pupillary aperture to differences of illumination by contraction in bright light and dilatation in reduced lighting or darkness. The neuronal channels and mechanisms involved in this pupillary reflex were dealt with in chapter XI of volume 4. It should here be recalled that some of the relevant pathways run through the pretectal grisea and the posterior commissure. Again, Reptiles and Birds (Sauropsida) are apparently the only Vertebrates in which both sphincter and dilator muscles consist of striated muscle fibers.

It is of interest that, with respect to the vegetative nervous system (cf. section 6, chapter VII, vol. 3/II) there obtains here, as also in the case of the Sauropsidan musculus ciliaris, a *postganglionic innervation of a striated musculature*, which latter, however, is smooth in other Vertebrates.

As regards internal folds, the falciform ligament of Teleosts was pointed out above. In Sauropsida, a vascular projection at the optic nerve 'entrance' (or rather exit) forms a vascular 'organ', called *conus papillaris*[27] in Reptiles, and *pecten* in Birds. 'Nutritive functions' have been attributed to these structures, but the peculiar shape, including parallel ridges, of the Avian pecten[28] has also suggested theories postulating an optical function of this ridge, supposed to enhance the eye's sensitivity to the movement of objects within the visual field.

With respect to some other peculiarities of various Vertebrate lateral eyes, a splitting of the cornea might be mentioned, whereby an outer layer, consisting of epithelium and transparent corium, is separated from the substantia propria corneae by a fluid-filled cleft. This *outer cornea ('spectacle'* of *'Schutzbrille')* occurs particularly in various Teleosts[29] with bottom habitat or temporarily leaving the water to crawl on the wet shore. It is also found in Dipnoans. The cornea of larval Amphibia is likewise frequently double, as in bottom-living Fishes, but both layers subsequently fuse on metamorphosis, except in some aquatic forms. Other types of '*brille*' occur in Amniota, namely in Snakes and some Lizards.

Figures 18 A–C illustrate various features of the eye bulb, as discussed above, in diverse representative Vertebrates. Concerning further details, such as scleral cartilages or ossicles, eye-lids, nictitating membrane, lacrimal duct, external eye muscles etc., of secondary interest in the aspect here under consideration, reference is made to the cited publications on the comparative anatomy of the eye. Before dealing with the *retina*, which represents the *peripheral optic ganglion*, two peculiar eye configurations might briefly be mentioned as examples of the many oddities occurring in Teleosts. Figures 19 A–C illustrate aspects of the eye in the Teleost *Anableps tetrophthalmus* (the *'four-eyed' fish)*, found along the Guyana (Guiana) coast of South America. This eye is adapted to simultaneous atmospheric and aquatic vision, two pupillary openings being formed by pigmented iris flaps joined in the midline and corresponding to an horizontal pigmented stripe of the cornea.

---

[27] Despite their mesodermal vascularization, the conus papillaris and the pecten are interpreted as being ectodermal as regards their ontogenetic origin, except for the conus of Ophidia (snakes) which is said to be mesodermal.

[28] Details concerning the pecten of the Avian eye can be found in the treatise on the Avian brain by Pearson (1972).

[29] A cover of skin spread over the eyes of Cyclostomes is also interpreted as a '*primary brille*' by Prince (1956) who distinguishes it, together with that in Amphibians, from the '*secondary brille*' of Teleosts and Dipnoans, and from the '*tertiary brille*' of some Reptiles.

The fish cruises at the surface of calm water with the upper pupil above and the lower one below the water's level. The lens is egg-shaped and focuses the light from both media. The retina displays two regions. At low tide, Anableps also crawls or jumps on the shore along the beaches. Comparable, but somewhat less conspicuous adaptation to both atmospheric and aquatic vision obtains in some other Teleosts, such e.g. as the mudskipper *Perioptbalmus* and the Perciform *Mnierpes macrocephalus*. In this latter, two flattened corneal surfaces are separated by a

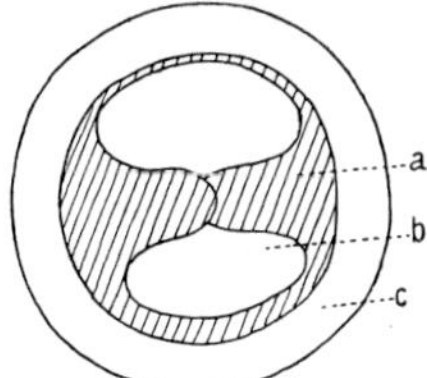

Figure 19 A. Iris and pupillary apertures in the 'four-eyed fish' Anableps tetrophthalmus (after SCHNEIDER and ORELLI, from PLATE, 1924). a: iris; b: lower pupil; c: cornea.

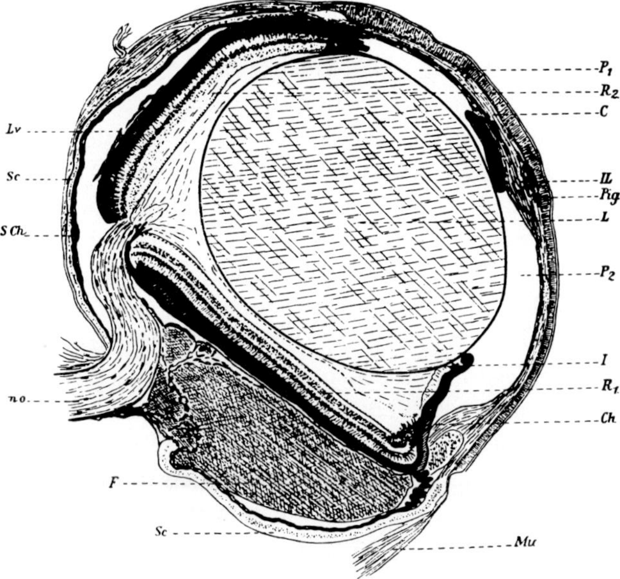

Figure 19 B. Vertical meridian section through the eye of Anableps (after KLINKOWSTRÖM, from PLATE, 1924). C: cornea; Ch: choroid; F: fatty tissue; I: iris; IL: iris flap; L: lens; Lv: lamina vasculosa; Mu: external eye muscle (rectus inferior); n.o.: optic nerve; P_1, P_2: upper and lower pupil; Pig: pigmented stripe of cornea; R_1, R_2: subdivisions of retina; Sc: sclera; SCh: suprachoroid layer.

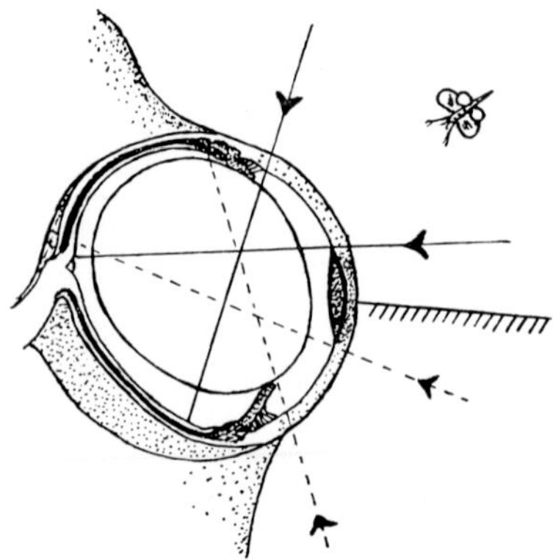

Figure 19C. Simplified sketch of vertical section through the eye of Anableps at the water surface, showing egg-shaped lens and approximate projections of visual fields upon retina. The flying insect will be registered by the lower retinal subdivision (from PRINCE, 1956).

light pigmented vertical median bar provided by a thickening of the cornea (GRAHAM and ROSENBLATT, 1970). Figure 20 shows the larva of the *deep-sea fish Stylophthalmus paradoxus*, with eye bulbs attached to a thin and excessively long stalk, which is braced by an extension of the cranial cartilage and said to be non-motile. The stalk, containing the optic nerve, includes, nevertheless, the six eye muscles. PLATE (1924) presumed that this stalk becomes reduced in the course of further ontogenetic development, about which little seems to be known.

Generally speaking, the Vertebrate *retina*, which contains the neurosensory elements and a neuronal processing apparatus including an intricate 'circuitry', is concerned with the registration of (1) *light intensity*, (2) *light frequency*, i.e. 'color', (3) *shapes*, i.e. spatial configuration, and (4) *motion of objects* in the visual field. The detailed structure of the retina has been investigated by a large number of authors. Among the older investigations, those by CAJAL (1911) are of particular significance and established many fundamental aspects displayed by the *Golgi technique*. Further relevant structural data were elucidated by the work of POLYAK whose treatise on the retina (1941) and on the vertebrate visual system (1957) also include a wealth of pertinent references. Still more recently, the relevant available data concerning the Vertebrate retina have been critically summarized in a monograph by RODIECK (1974), and a quantitative analysis of Cat retinal ganglion cells was published by HUGHES (1975).

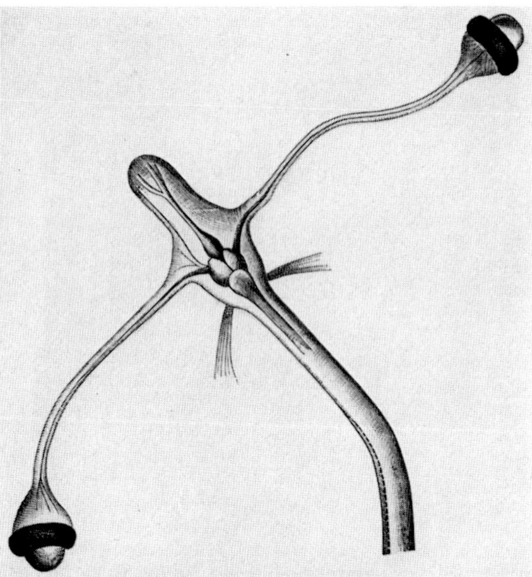

Figure 20. Stalked eyes displayed by a 42 mm long larva of the deep-sea fish Styloph-thalmus paradoxus. The thin stalks are 7 mm long (after BRAUER and CHUN, from PLATE, 1924). The well-developed subdivisions of the brain (forebrain with n. olfactorius, mesencephalon, and rhombencephalon) can easily be identified.

Figures 21 A–C illustrate routine hematoxylin-eosin preparations displaying the histologic aspects, respectively the stratification of the pars optica in the Mammalian (human) and Amphibian (frog) retina as seen in sections perpendicular to the surface of this membrane. Ten different layers are commonly distinguished: (1) the *pigment epithelium*, which is the only layer provided by the outer wall of the invaginated optic cup (cf. C in Fig. 14 A), all other layers being derivatives of the inner wall; (2) the *'bacillary layer' of rods and cones*, represented by the sensory endings of the neurosensory elements; (3) the thin *membrana limitans interna*,[30] which is a neuroglial structure comparable to the limiting membranes dealt with in section 1, 3, and 5 of chapter V of volume 3/I; (4) the *outer granular or nuclear layer*, formed by the perikarya of the neurosensory cells; (5) the *outer reticular or plexiform layer*, formed by the efferent processes of neurosensory cells, by their synapses

[30] Electron photomicrographs indicate that the outer limiting membrane of the retina, like the internal limiting membrane of the neuraxis (cf. section 3, p. 151, and section 6, p. 483, chapter V, vol. 3/I) seems to be an interface formed by ultrastructural junctional complexes between inner segments of rods and cones as well as portions of supporting *Müller cells*. It can, nevertheless be considered to represent a 'membrane-like' partition.

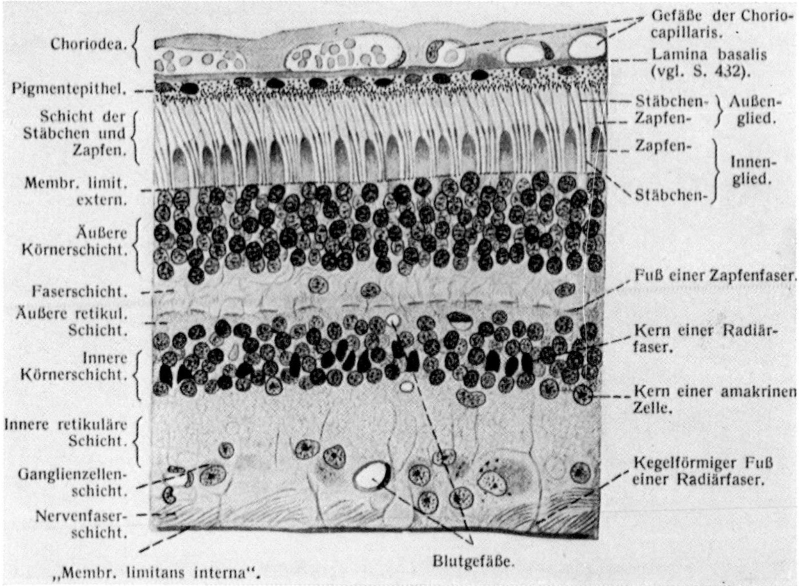

Figure 21 A. Drawing of a section through the human retina at a right angle to its plane (hematoxylin-eosin, $\times 260$; red. $^1/_1$; from Stöhr and Möllendorff, 1933).

with bipolar cells, and by afferent processes of these latter; (6) the *inner granular or nuclear layer* containing the perikarya of bipolar neurons; (7) the *inner plexiform or reticular layer*, essentially formed by synapses between efferent processes of bipolar cells and their synapses with retinal ganglion cells; (8) the *layer of ganglion cells*, from which the centripetal fibers of the optic nerve originate; (9) *the layer of optic nerve fibers*, and finally (10) the *internal limiting membrane*, of neuroglial derivation, which separates the retina from the vitreous body.

The neuroglial elements of the retina have been dealt with and depicted in section 3, chapter V of volume 3/I. The blood supply of the holangiotic retina[31] is provided by the above-mentioned *arteria centralis retinae* with its accompanying venous channels (cf. Fig. 30 B, and, as

[31] In the holangiotic retina of most Mammals, the main veins and arteries pass through the optic nerve disk. In the merangiotic retina only a small region is supplied with obvious blood vessels. The paurangiotic retina displays a few capillaries at the disk. The anangiotic retina is entirely devoid of blood vessels. The retina of Birds is anangiotic, its 'nutritive' support being presumably provided by the choroid and particularly by the pecten, whatever other functions this latter structure may have. The lack of 'obstructing' vascular components is considered as related to the high visual acuity of the Avian retina. Further details of retinal vascular pattern are dealt with by Prince (1956).

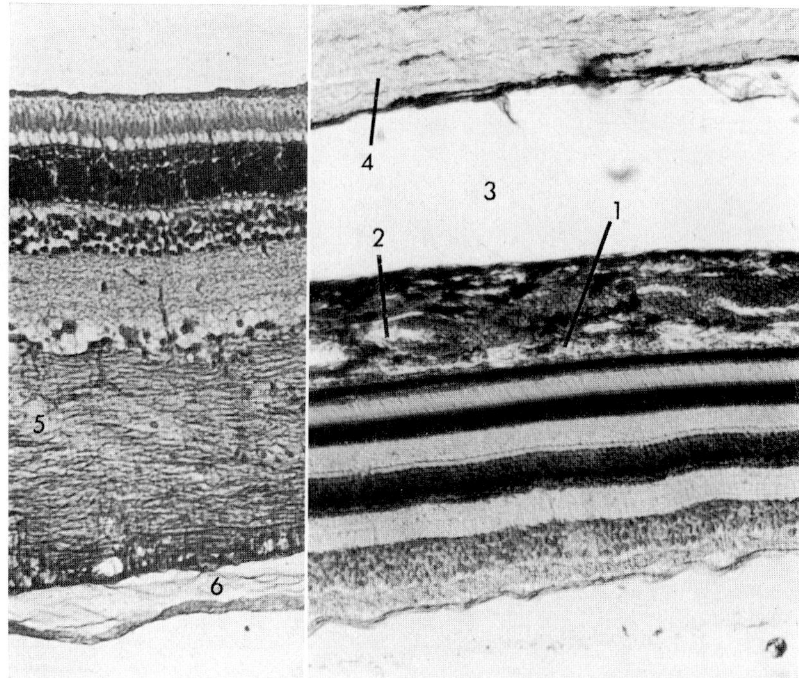

Figure 21 B. Photomicrographs of routine hematoxylin-eosin preparation of sections through human retina for comparison with Figure 21 A (I ×165, II ×66, red. $^4/_5$). 1: lamina choriocapillaris of membrana chorioidea; 2: lamina vascularis of membrana chorioidea; 3: artificial cleft between lamina vascularis and lamina suprachorioidea; 4: sclera with attached lamina fusca sclerae (pigmented lamina suprachorioidea); 5: thick layer of optic nerve fibers in vicinity of papilla nervi optici; 6: vitreous body. The retinal layers can easily be identified by comparison with Figure 21 A.

regards some retinal capillaries, Fig. 21 A). The sclera and the chorioidea, which, along the pars optica retinae, closely adjoins the retinal pigment epithelium (cf. Fig. 21 B), are supplied by a system of *ciliary arteries* reaching various parts of the eye bulb, and originating, like the arteria centralis retinae, but independently of this latter, from the ophthalmic artery.[32]

The photosensitive *neurosensory cells* of the retina are the *rods* and *cones* which can be evaluated as neuronal elements with an input portion and a 'neurite-like' output fiber connected to the terminal ramifications formed by the peripheral process ('dendrite') of a bipolar cell.

[32] '*Essentially*' independent, except for minor anastomoses between retinal and ciliary systems near their origin from a. ophthalmica. Cf. also preceding footnote.

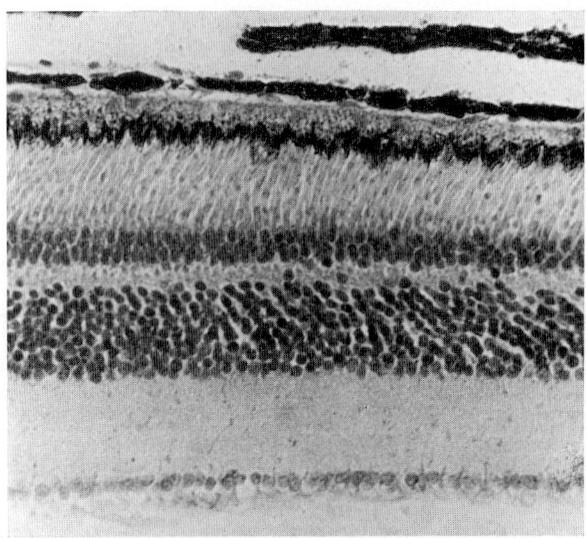

Figure 21 C. Section through an Amphibian retina (Frog) for comparison with Figures 21 A and B (hematoxylin-eosin, ×165, red. $^4/_5$). Note the larger size of the Amphibian cellular elements compared with the Human ones at identical magnification (Fig. 21 B, I). The layer of optic nerve fibers, very thin in this particular peripheral region, is slightly damaged.

Rod and cone cells are essentially similar in structure, and, for that matter, as YOUNG (1957) justly remarks, not always easy to distinguish from each other, particularly in some of the lower Vertebrates. Both neurosensory elements can be said to consist of the following four main parts: (1) an *outer segment*, which seems to be the receptor or transducing portion, comparable to a 'dendrite'; (2) an *inner segment*, whose distal portion is called *ellipsoid*, and its proximal one *myoid*, since this latter can contract in various animals; (3) a rounded *perikaryon* containing the nucleus, and (4) a short *neurite* directed toward the bipolar cells. Outer and inner segments represent the rods and cones *sensu strictiori*, which, at the boundary region between inner segment and perikaryon, are more or less firmly held in place by the outer limiting membrane. The rods of various Vertebrates may differ from the cones by displaying a smaller perikaryon and nucleus intercalated within a rod fiber originating at the base of the inner segment. Such perikaryon can be said to resemble that of a bipolar spinal or cranial nerve ganglion cell. However, some Vertebrate cones manifest a similar tendency (cf. Figs. 22 A–C). Again in contradistinction to the elongated 'bacil-

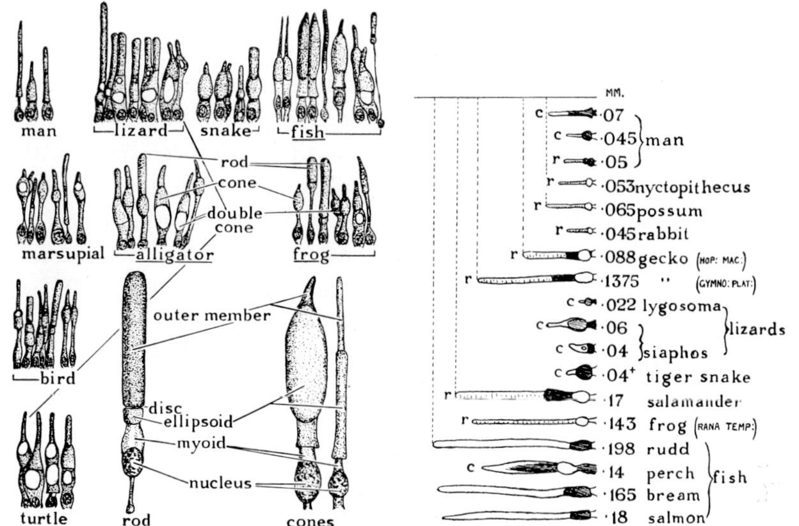

Figure 22 A. Neurosensory retinal elements (rods and cones) of some representative Vertebrates (from Prince, 1956). Man is typical of the Primates. In Lizards, Alligators, Turtles, and Frogs, the large 'paraboloids' in the cones, and the 'accessory members' of the double cones can be seen in the inner segments.

Figure 22 B. Drawing showing the disparity between the size of neurosensory retinal elements in various Vertebrates (from Prince, 1956). c: cone; r: rods; figures indicate length. Because of age differences in the diverse specimens the comparison may not be entirely accurate, since in various lower Vertebrates the continuous growth of rods and cones with increasing age is said to be not negligible.

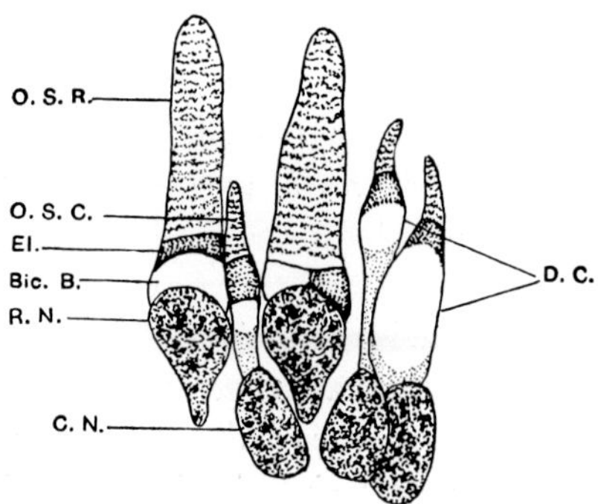

Figure 22 C. Retinal neurosensory cells of an urodele Amphibian (after Detwiler and Laurens, from Noble, 1931). Bic. B.: 'paraboloid'; C.N.: cone nucleus; D.C.: double cone; El.: 'ellipsoid'; O.S.C.: outer segment of cone; O.S.R.: outer segment of rod; R.N.: rod nucleus. The rods and cones are from the retina of a larval Amblystoma.

lary' outer segments of the rods, that of cones is usually shorter and tapering into a 'pointed' apex. Yet, elongated, rod-like outer segments of cones also occur. Aspects in differential staining and configuration of the inner segment provide here some clues to the identification *qua* cone or rod. Generally speaking, the 'neurite' of rod cells ends with a single knob-like synaptic structure ('spherule'), while the neurite of cones terminates in a likewise single, but 'flaring' expansion ('pedicle') with some thin, short protrusions. CAJAL (1911) also depicted certain cone neurites with a few collaterals. In various Vertebrates, the neurosensory cells, particularly the cones, contain oil droplets in the outer portion of their inner segments.

The numerous variations *qua* rod and cone size as well as shape displayed in the Vertebrate series present considerable difficulties for an adequate classification of these neurosensory elements. Likewise, the differences in the ratio between rods and cones appear very substantial. Thus, in diverse Fishes this may vary between one cone to several hundred rods and one cone to ten or less rods. Yet, generally speaking, the retinae of all Vertebrates appear to contain both rods and cones, although some nocturnal forms were supposed to have an all-rod retina, in contradistinction to the presumed all-cone retinae of some diurnal animals. Nevertheless, in either forms a few of the supposedly absent receptors are said to be present (cf. PRINCE, 1956). Be that as it may, rods predominate in Petromyzonts, Selachians, Ganoids, Teleosts, a variety of Reptiles, nocturnal Marsupials, and nocturnal Eutheria, while cones predominate in reptilian Chelonians, many Lizards and in diurnal Birds.

Cones and rods occur in both *single* and *double forms*, these latter being relatively more frequent in submammalian Vertebrates.[33] The double cone consists of a principal cone of normal size, commonly displaying an enlarged '*paraboloid*' *myoid*, and of a smaller accessory cone, usually containing an oil droplet in the inner segment, and attached to the larger cone by a junctional structure near the base of the myoid. A variation of the double cone is the '*twin cone*', whose two members are of approximately equal size and attached to each other for the greater part of their length. If present, the oil droplet, however, is always said to be included in the thinner 'twin'. Again, the lack of 'paraboloids' and of oil droplets is reported for the single or double cones of the retina

---

[33] In Mammals, the occurrence of double cones is apparently restricted to Prototheria (Monotremes) and Metatheria (Marsupials).

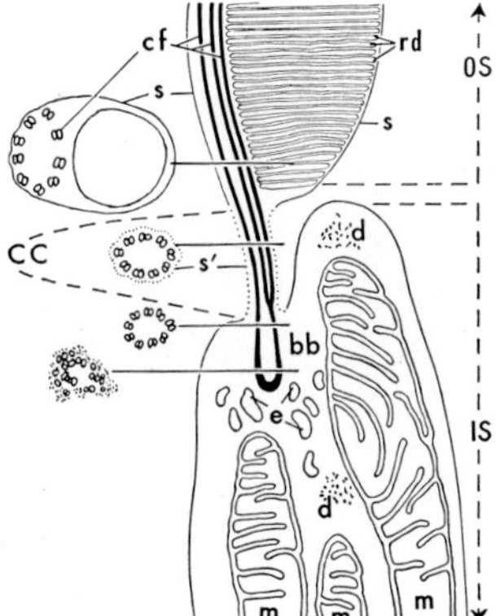

Figure 23 A. Diagram, based on electron photomicrographs, of region of junction between outer and inner segments of retinal rods (modified after De Robertis, from Bloom and Fawcett, 1962). CC: connecting cilium; IS: inner segment; OS: outer segment; cf: ciliary fibrils; d: electron-dense granules; e: vesicles of endoplasmic reticulum; m: mitochondria; rd: stacked rod disks; s: surface (plasma-)membrane; s′: dubious continuity of surface membrane covering the 'connecting cilium'. The levels of the transverse sections, shown at left, are indicated on the longitudinal section at right.

in reptilian Ophidia (snakes). Double rods are far more uncommon than double cones and occur mainly in Reptiles.

The outer segments of both rods and cones appear cross-striated in high resolution light microscopy. Electron photomicrographs of rods disclose fibrils, presumably representing a modified cilium with a basal body, joining, through a stalk, the inner to the outer segment. This latter consists of double membrane disks about 140 Å thick and 2 μ in diameter, which may be interconnected in series by short fuzzy processes. Large mitochondria aggregate in the distal portion (ellipsoid) of the inner segment. A similar structure, with minor modifications, obtains for the cones.[34] Figures 22 A–C depict Vertebrate rods and cones as seen with the light-microscope, and Figures 23 A and B illustrate some electron-microscopic aspects.

It has been observed that Vertebrate rod photoreceptor cells continually shed and renew their photoreceptive outer segments. The renewal process begins by synthesis of new membranes to form rhodopsin-containing disks at the base of the outer segment (cf. e.g. La Vail, 1976; Basinger *et al.*, 1976).

[34] Thus the membranes of some of the disk-like lamellae may display patterns of continuity, or plasma membranes may have the clefts between lamellae open to the extracellular space (cf. Fig. 23B).

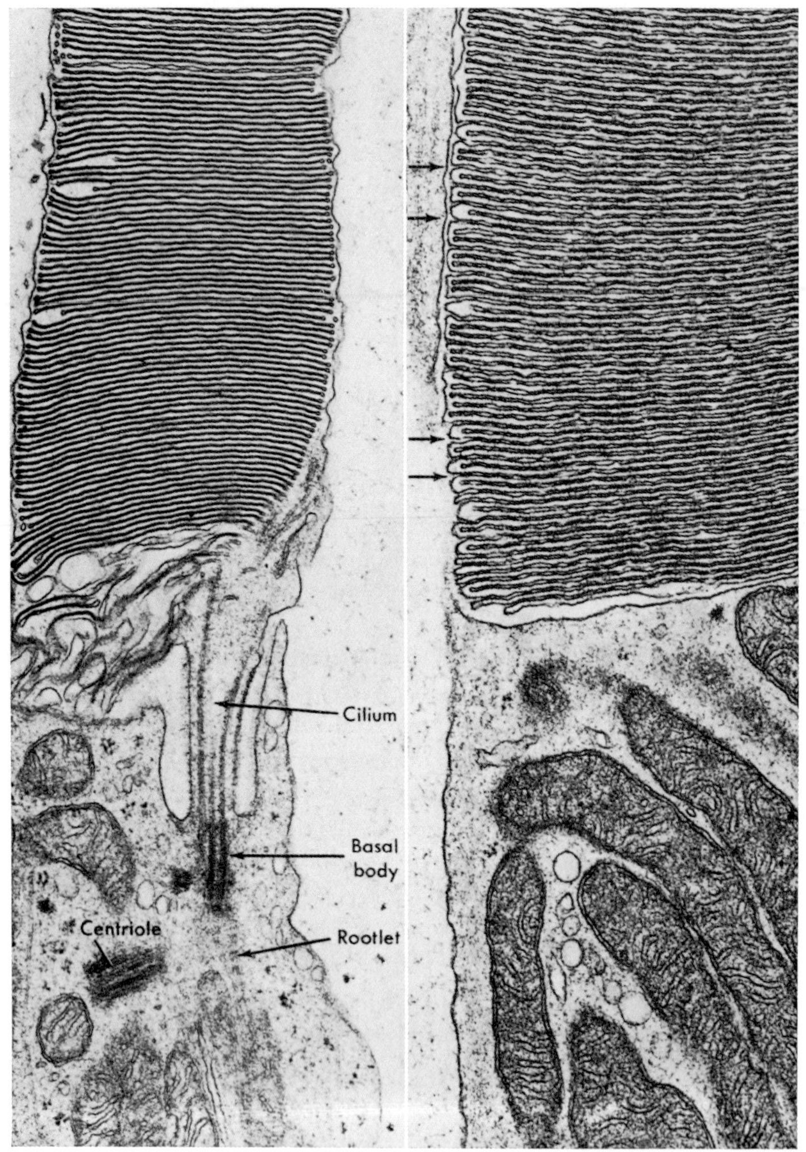

Figure 23 B. Electron photomicrograph of inner and outer segments in a rod (I) and a cone (II) of a Vertebrate retina (after KUWABARA, from BLOOM and FAWCETT, 1968). Provenance (Mammal ?) and magnification (presumably around 50 000) not given. The arrows in II show regions where some of the lamellae may directly adjoin the extracellular space.

The *bipolar cells* are short internuncial neurons whose 'dendrite' connects with the terminals of rods or cones, while their 'neurite' provides synaptic junctions with dendrites or cell body of retinal ganglion cells. The perikarya of the bipolar cells constitute most of the inner granular layer. *In toto*, the bipolar elements are oriented at a right angle to the curved plane of the retina, except in some of the foveae to be discussed further below, where their orientation becomes oblique.

KOLB and FAMIGLIETTI (1974) have recently claimed that, in the Cat retina, 'rod bipolar terminals do not synapse on ganglion cells but on two types of amacrine cell (types I and II). Cone bipolars synapse directly on ganglion cells and on type I amacrines. The type II amacrine appears to play a special internuncial role between bipolars and ganglion cells in the rod system'. *Golgi* and electron microscopic findings are presented to support this claim, which does not appear altogether convincing as a generalization even for the Cat's retina. It is, of course possible that, in the Cat's retina (as well as in the case of other Vertebrates) some such connections, perhaps related to dark adaptation, may indeed obtain as special cases.

Still more recently, DOWLING and EHINGER (1975) claim to have identified, by fluorescence and by electron microscopy, a 'new type of amine containing retinal neuron' in Goldfish and Primate (Cebus) retina. Its perikaryon is located among the amakrine cells of the inner granular or adjacent inner plexiform layer, and its processes are related to amakrine cells in the inner plexiform layer, and to horizontal cells in the outer plexiform layer. These 'interplexiform cells', as illustrated by Figure 24 D, are said to 'provide an intraretinal centrifugal pathway from inner to outer plexiform layer'.

The *horizontal cells*, located in the external portion of the inner granular layer are likewise internuncial elements whose dendrites connect with the terminals of rods[35] or cones, and whose 'horizontal' neurite makes contacts with other neurosensory cell terminals as well as presumably with bipolars. The 'horizontal' neurites criss-cross in all directions. Terminal arborizations of neurites and of dendrites pertaining to

[35] According to POLYAK (1957) the dendrites of horizontals are in contact with the terminals of a group of rods, but not with those of any cones. Thus, only the cones would activate the horizontal cells 'which, in turn, stimulate, at some distance, a group of both rods and cones'. CAJAL (1911), however, depicts dendritic connections of horizontals with both rods and cones. NELSON *et al.* (1975) claim that horizontal cells in the Cat retina receive input predominantly from cones on the cell body, while the terminal arborizations of its axon-like process predominantly receive input from rods. The significance of this arrangement cannot be said to be understood.

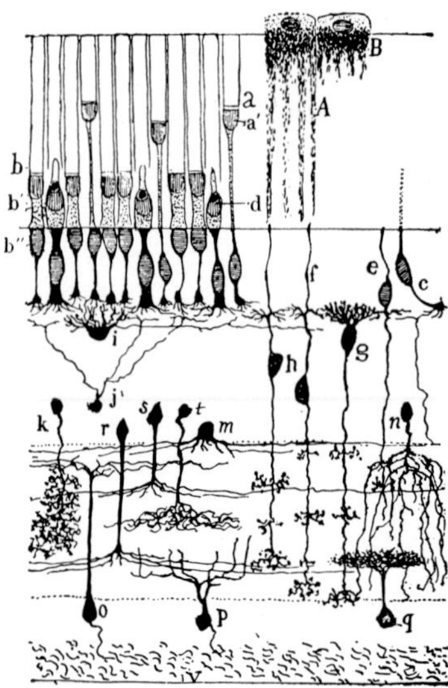

Figure 24 A. Semidiagrammatic drawing of section through the Frog's retina as displayed by the *Golgi technique* (from CAJAL, 1911). A: pigment epithelium cell with expanded processes; B: pigment cell with retracted expansions; a: 'red rod'; a′: 'ellipsoid body'; b: 'green rod'; b′: inner segment of rod; b″: perikaryon of rod; c: oblique rod; d: cones; e: displaced bipolar; f: peripheral expansion of bipolar ('club', '*massue de Landolt*'); g: large bipolar without 'club'; h: bipolar with 'club'; i: horizontal cell; j: 'stellate cell' (?); k, m, n: diverse amakrine cells; o, p: retinal ganglion cells; q: displaced amakrine; r, s, t: additional sorts of amakrines.

adjacent cells may overlap. Several 'types' of horizontal cells have been described by CAJAL (1911) and others (cf. Fig. 48, p. 72, vol. 3/I).

The *amakrine*[36] (or *amacrine*) *cells*, whose designation, introduced by CAJAL, refers to the absence of a 'typical' 'neurite', are located in the internal portion of the inner granular layer. Their branching processes are directed toward the ganglionic layer. These processes (apotiles) may effect unpolarized (asynaptic) or facultatively polarized (protosynaptic) connections of unknown significance between bipolar cells and ganglion cells. The amakrines, moreover, seem to be the main target of the efferent (centrifugal) fibers originating in the neuraxis and conveying impulses to the retina (cf. Figs. 24 A–D).

The *ganglion cells* of the retina are located in the 8th retinal layer, but a few of these elements can be displaced into the internal portion of the

[36] Cf. the comments on these cells in section 2, p. 72 (with Fig. 48), and section 8, p. 639 of chapter V in volume 3/I.

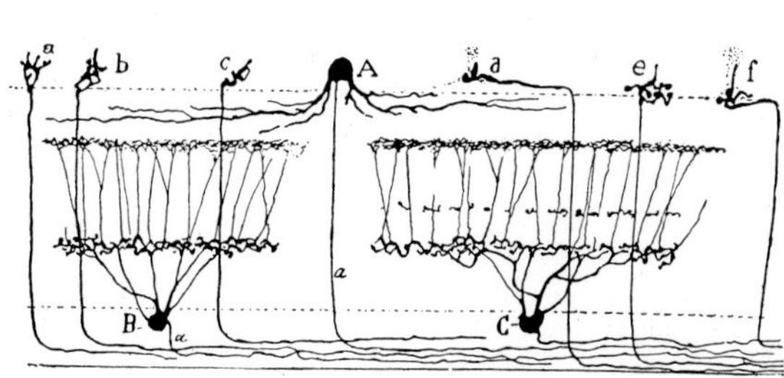

B

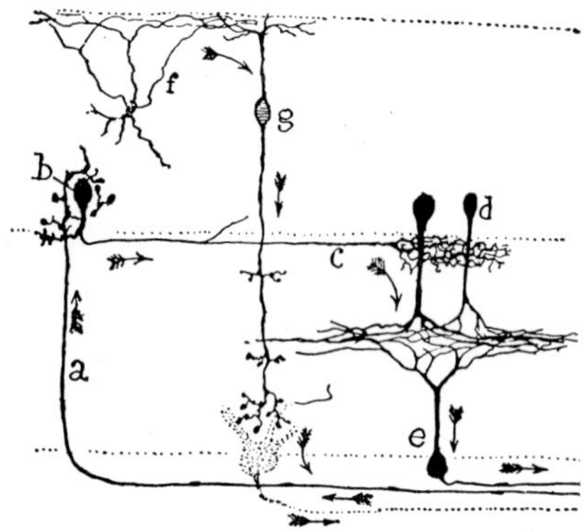

C

Figure 24B. Retinal ganglion cells and centrifugal optic nerve fibers in the Sparrow, as displayed by the *Golgi method* (from CAJAL, 1911). A: displaced ganglion cell; B, C: ganglion cells with bi- and tristratified dendritic arborizations; a–f: centrifugal optic nerve fibers with end arborizations (the two lowers letters indicate centripetal axons of ganglion cells).

Figure 24C. Diagram indicating assumed impulse conduction in centrifugal optic nerve fibers, amakrines, and retinal ganglion cells in the Avian retina (from CAJAL, 1911). a: centrifugal fibers; b: its termination on an amakrine; c: short 'neurite' *('cylindre-axe')* of amakrine; d: 'ordinary amakrines' connected with amakrine b; e: ganglion cell; f: 'stellate cell'; g: bipolar cell. The circuit a–f–g and dotted ganglion cell is supposed to represent '*la voie centrifuge supérieure*', circuit a–b–c–d–e being '*la voie centrifuge inférieure*'.

inner granular layer. Their neurites pass through the 9th layer and become the centripetal fibers (efferent *qua* retina, afferent *qua* neuraxis) of the *optic nerve*. Their dendrites are connected with the neurites of the bipolar cells. Several types of retinal ganglion cells have been described with respect to their relative size, location and dendritic branchings. These ganglion cells represent *neurons of the third order* in the neuronal chain connecting the retina with the central neuraxis, the bipolars being neurons of the second, and the neurosensory cells neurons of the first order. In addition, the dendrites and cell bodies of retinal ganglion cells display connections with the processes of amakrine cells, as mentioned above.

As regards studies dealing with ganglion cell density patterns in Mammalian retinae (e.g. ganglion cell density map of the Cat), the report by ROWE and STONE (1976) could be mentioned.

There is some recent evidence concerning different types of retinal ganglion cells, presumably related, *inter alia*, to the different sizes and conduction velocities of optic nerve fibers. Thus, X, W, and Y cells have been distinguished in the Cat (cf. e.g. ENROTH-CUGELL and ROB-

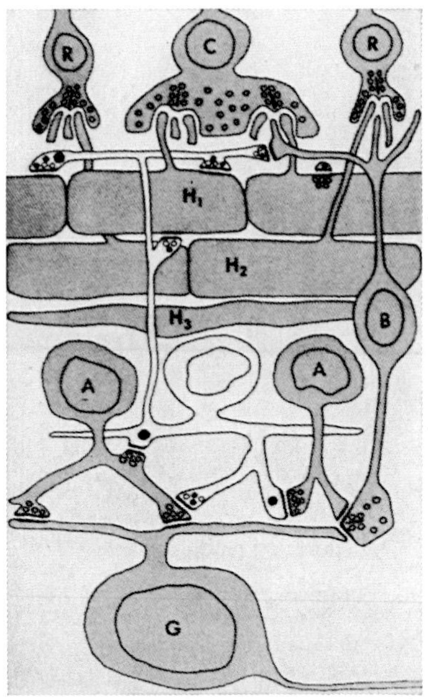

Figure 24 D. Diagram purporting to show connections of amine-containing so-called interplexiform retinal cells in the Goldfish retina (from DOWLING and EHINGER, 1975). According to these authors, the input to the amine-containing interplexiform cells 'is in the inner plexiform layer via conventional synapses of amacrine cells'. A: amacrine cell; B: bipolar cell; C: cone; G: ganglion cell of optic nerve; $H_{1, 2, 3}$: external, intermediate, and internal horizontal cell; R: rod. The amine-containing 'interplexiform cell' (light) is located between the two A cells. According to the cited authors, the input to said 'interplexiform cell' is provided 'in the inner plexiform layer via conventional synapses of amacrine cells'.

SON, 1966; STONE and FUKUDA, 1974). It is believed that X cells may subserve high resolution and stereoscopic vision, W cells being possibly related to midbrain vision, and that Y cells could play a role in peripheral vision, such as 'detection of fast-moving images' (STONE and FUKUDA, 1974).

In Primates, the smallest retinal ganglion cells, the so-called *midget ganglion cells*, are found near the fovea centralis. It is claimed that, by virtue of their position and their restricted extents and synaptic connections, they provide for the greatest visual acuity (SHEPHERD, 1974).

Moreover, with regard to the 'axonal' or 'neuronal flow' discussed in section 8, chapter V, pp. 640–643 of volume 3/I, it is of interest that

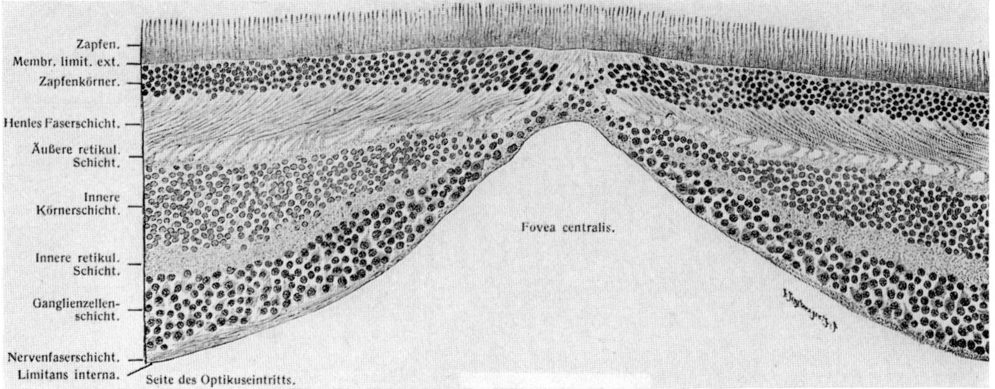

Figure 25 A. 'Horizontal' section through fovea centralis (macula lutea) of the human retina (from STÖHR and MÖLLENDORFF, 1933). The plane of the section, drawn after a preparation by Prof. HAAB, does not pass through the exact center of the fovea, in which only the central portions of the elongated cones are said to be present.

retrograde, i.e. 'cellulipetal' flow of horseradish peroxidase (HRP) from the terminals of retinal ganglion cells in the Chick's neuraxis by 'rapid transport' has been demonstrated (LAVAIL and LAVAIL, 1974). The marker enzyme was found in the neuronal cell bodies 4 h after injection, thus indicating a rare of retrograde transport of at least 84 mm/day in these neurons. Additional comments on 'neuronal flow', concerning the olfactory system, will be found in chapter XIII, section 2. Still more recent reports on findings with the aforementioned technique, and concerning the projection of the different types of retinal ganglion cells upon the optic grisea of the neuraxis can be found in a paper by KELLY and GILBERT (1975).

Except for a much smaller number of centrifugal fibers and some other components, such as retinal supporting elements and blood vessels, the neurites of ganglion cells, directed toward the optic nerve exit, form the bulk of the 9th layer. The nerve fibers in this latter are generally non-medullated but become myelinated in the optic nerve immediately after passage, at the *optic disk*, through the *lamina cribrosa* of the sclera. In various holangiotic and paurangiotic Mammals, however, some medullated nerve fibers, which appear white and opaque when seen by reflected light, run mostly along the retinal vessels toward the disk (Ungulate Camel, Rodent Squirrel, Lagomorph Rabbit, some Marsupial, also in some Ophidian Reptiles).

The terms '*fovea*', '*macula*', and '*area centralis*' refer to retinal areas

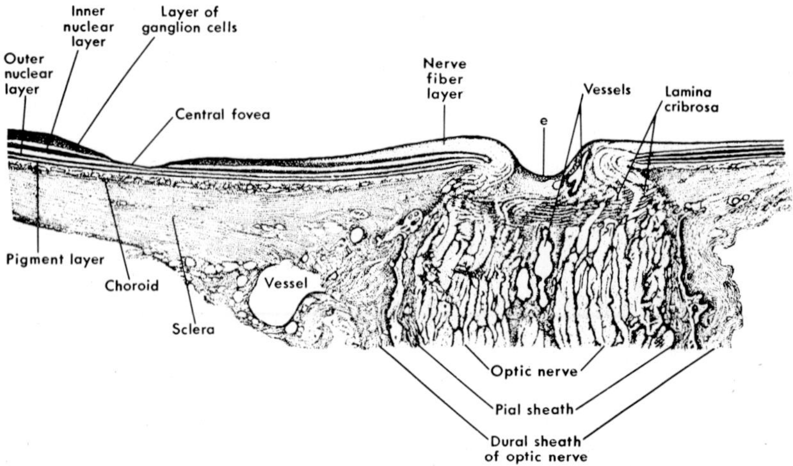

Figure 25B. 'Horizontal' section through human eye showing optic disk and fovea centralis (redrawn and slightly modified after SCHAFFER, from BLOOM and FAWCETT, 1962). e: excavation of optic disk.

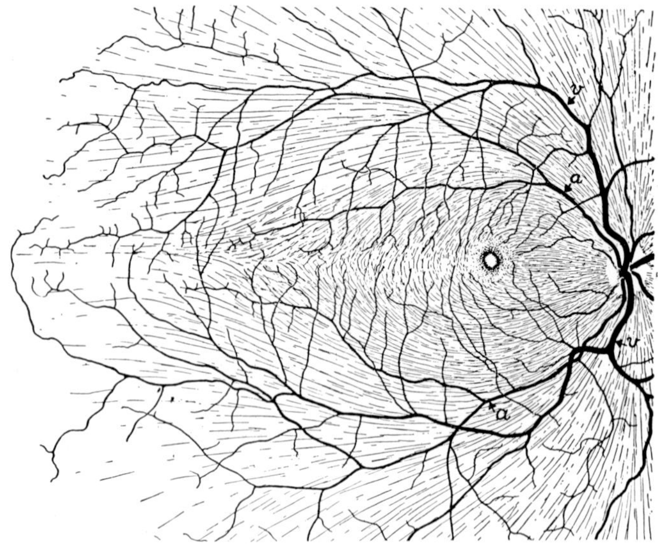

Figure 25C. Retina of right eye of adult Rhesus monkey as seen in a preparation processed with Ehrlich's intravital methylene blue technique (after POLYAK, from BLOOM and FAWCETT, 1962). a: arteries; v: veins. Optic nerve disk at right, fovea centralis at left, the dark ring surrounding it being the foveal slope (cf. also Figs. 30B, C). The original magnification, here reduced $^2/_3$, is about $\times 8$.

characterized by reduction of the retinal layers located internally to the photoreceptor elements, which latter are, moreover, here generally crowded together. By reduction of the superposed layers, greater transparency obtains and, together with the concentration of receptor endings, visual acuity becomes enhanced. Foveae or maculae (without depression) are clear of vascular obstruction. In Man and various other Vertebrates, the fovea is central with respect to the visual axis (cf. Fig. 16 A) and its receptors are exclusively elongated cones (Figs. 25 A, B).[37] Not all Vertebrates display a fovea or macula, but such areas have been identified in Mammals (mostly Primates, some Carnivores and the Horse), in Birds, various Reptiles, in Frogs, and some Teleosts, and even in a Selachian (PRINCE, 1956). Apparently all Birds and some Reptiles display two foveae in each eye, a central and a temporal one.

The 'entrance' of the optic nerve into the retina, i.e. except for the entrance of centrifugal fibers, the exit of all retinal ganglion cell neurites, is the *optic disk*, which likewise commonly affords a passage for the retinal vessels. This disk, also designated as optic papilla, is devoid of neurosensory and other neuronal elements and thus represents a '*blind spot*'. Figure 16 A indicates that, in Primates and Man, the blind spot is located in the nasal half of the retina and thereby in the temporal visual field. Except by means of a test, it is not experienced as such in the visual space of human consciousness.[38] Figures 24 A–D and 25 A–C illustrate diverse aspects of the retina, including fovea centralis and optic disk. Figures 45, 46, 48, and 52 in volume 3/I of this series depict additional histogenetic and histologic retinal features.

With regard to the neuronal flow mentioned above, the histochemical horseradish peroxidase (HRP) technique disclosed intra-axonal retrograde transport within retinal ganglion cells of the Chick from their terminals in optic tectum. The rate of retrograde transport was at least 72 mm per day. Anterograde transport from retina to tectum was likewise observed, but at the rate of about 9.2 mm per day, thus displaying slow transport in contradistinction to the retrograde rapid transport. Similar findings were reported by HANSSON (1973) in the Rat.

In the Chick, moreover, LAVAIL and LAVAIL (1974) reported retrograde transport from retina to cell bodies of the isthmo-optic nucleus

[37] In some Reptiles, however, e.g. in Sphenodon, a nearly 'all-rod' fovea is said to be present (PRINCE, 1956).

[38] Further comments on the central visual pathways and the visual fields are included in chapters XIII and XV.

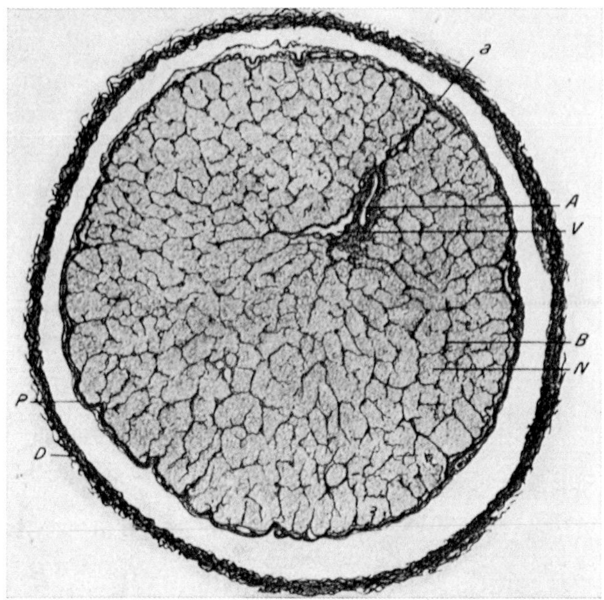

Figure 26 A. Cross-section through human nervus opticus in close vicinity to bulb (from SCHAFFER, 1933; ×30, red. $^3/_4$). A: arteria centralis retinae; B: connective tissue septa; D: sheath of dura mater; N: medullated nerve fiber bundles; P: leptomeninx; V: vena centralis retinae; a: connective tissue seam indicating original choroid fissure. The nerve fiber bundles are imbedded in neuroglia. The space between D and P is the intracranial subarachnoid space extending as far as the passage of optic nerve through sclera (cf. Fig. 25 B).

following within 4 h intravitreal injection. This would indicate a transport rate of about 84 mm per day in efferent, retinopetal neurites originating in said nucleus.

The *optic nerve* (Figs. 25 B, 26 A, B), which is not a peripheral nerve but a nerve tract of the neuraxis, does not display neurilemmal sheaths with *Schwann cells*. Its supporting elements are true neuroglia cells, and the nerve is, *in toto*, surrounded by a leptomeningeal and pachymeningeal sheath. The latter is continuous with the sclera, the former with the choroid. The nerve bundles are separated from the vascular apparatus and from the leptomeninx by typical glious limiting membranes. Generally speaking, the optic nerve fibers undergo at the *optic chiasma* an *essentially* complete decussation in all submammalian Vertebrates, but a semidecussation in Mammals (Figs. 27 A, B). In Man, perhaps 50 per cent of the fibers remain homolateral. Some qualifications, how-

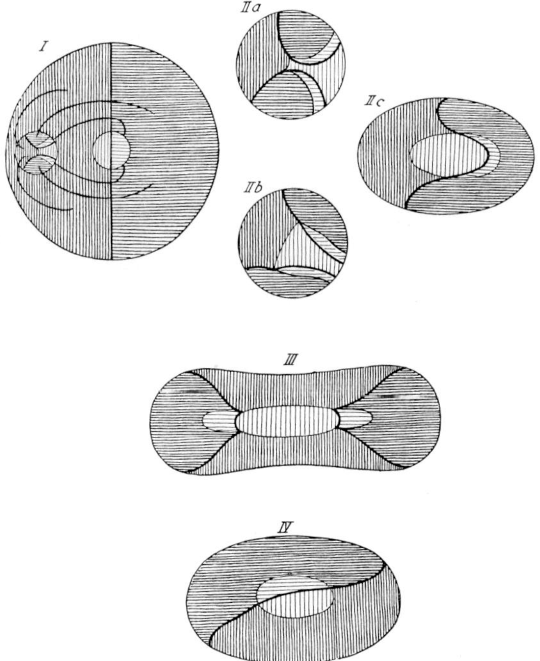

Figure 26 B. Assumed approximate location of temporal, nasal, and macular optic nerve fibers within optic nerve, chiasma, and tractus opticus of Man (from VILLIGER, 1933). I: Left retina; IIa, b, c: left optic nerve in centralward sequence; III: optic chiasma; IV: left optic tract; horizontal hatching; temporal half of retina; vertical hatching: nasal half. The fovea centralis indicated by circle and slightly spaced hatching within the chiasma, the nasal fibers decussate (not specifically indicated in diagram). Within optic tract, the nasal fibers are those from contralateral retina. Minor variations of this diagram have been elaborated by various authors.

ever, must be added to this generalization, namely (1) a variable amount of non-medullated fibers in diverse Vertebrates,[39] (2) the possibility that, at least in certain Anamnia, a few optic nerve fibers may originate from second order neurons, i.e. modified bipolars, and (3) the problem of complete versus incomplete decussation. Semidecussation has been commonly interpreted as related to binocular vision and as restricted to Mammals. There is, however, little doubt that some sort of binocular vision obtains in Birds and even certain Anamnia.

[39] In most Mammals and Birds with good vision, the number of non-medullated fibers is generally believed to be of very low percentage, but seems to reach 100 per cent in some Amphibians and Fishes with poorly developed eyes (cf. the tabulation in BLINKOV and GLEZER, 1968).

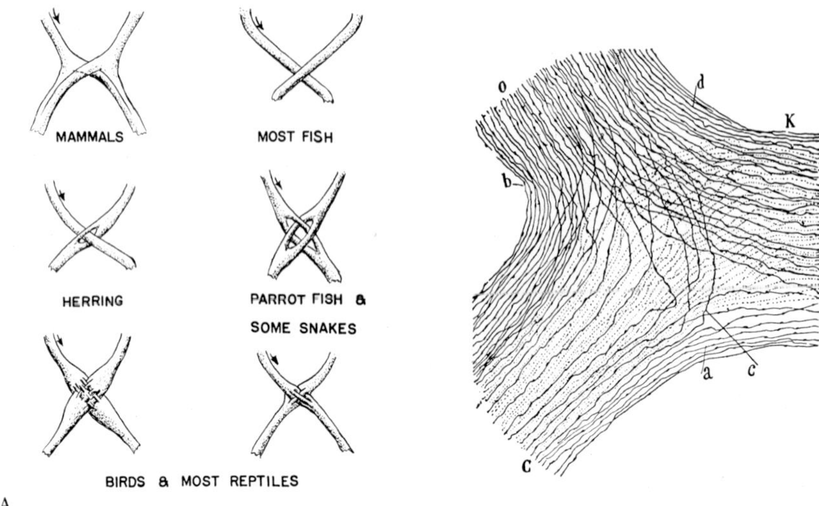

Figure 27 A. Basic patterns of optic chiasma in diverse Vertebrates (from PRINCE, 1956).

Figure 27 B. Portion of optic chiasma in a young Cat, as displayed by the *Golgi technique* (from CAJAL, 1911). C: optic tract; K: chiasma; O: optic nerve; a: *Gudden's supraoptic commissure;* b, c: non-decussating fibers; d: decussating fibers.

Whether this is related to an unknown amount of optic nerve fibers remaining homolateral, or to a bifurcation of optic nerve fibers such as occasionally noted by CAJAL (cf. Fig. 27 C) still remains an open question[39a]. Again, the optic nerve displays different calibers of its centripetal fibers, presumably and in a not entirely clarified manner, related to functional significance as well as ultimate destinations. These latter are, in the entire Vertebrate series, (1) optic tectum, (2) mesencephalic tegmentum, (3) pretectal grisea, (4) thalamic lateral geniculate grisea, and (5) some hypothalamic grisea. The endings, particularly 1–3, were discussed in chapter XI of volume 4. Further comments on the diencephalic endings are included in the section of the present chapter dealing with the diverse Vertebrate groups.

The *number of fibers* in the optic nerve, roughly corresponding to the number of retinal ganglion cells,[40] is especially great in mammalian

[39a] Findings concerning branching optic nerve fibers directed to both sides of the brain are discussed in reports by LUND *et al.* (1974), and by CUNNINGHAM (1976), who used a cobalt tracing method.

[40] Although one would expect, because of the centrifugal (retinopetal) fibers, more optic nerve fibers than retinal ganglion cells, many counts (cf. BLINKOV and GLEZER, 1968)

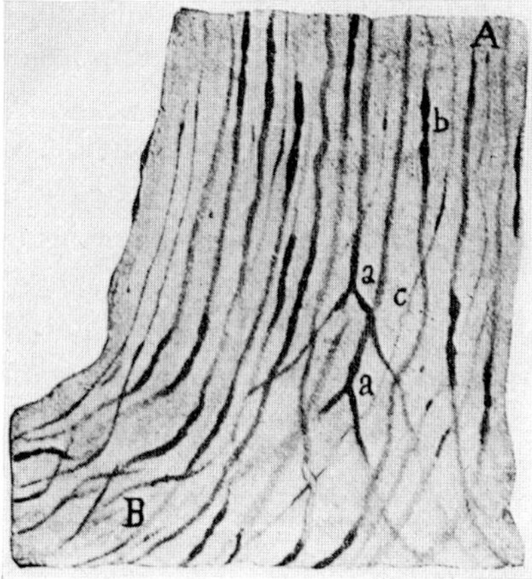

Figure 27C. Part of optic chiasma in the Rabbit, as seen by means of the *Ehrlich methylene blue technique* and displaying bifurcating fibers (from CAJAL, 1911). A: optic nerve; B: portion of chiasma; a: bifurcating fibers; b: 'strangulation' (*'Ranvier's node'*); c: thin optic nerve fiber.

Primates and in Birds, being very small in various Amphibians and Fishes, as indicated by the tabulation of Figure 28.

Concerning the Human neuraxis, it is of interest to note that the optic input channel, with about 10^6 fibers, roughly corresponds in this respect to the number of fibers in all spinal input channels. The number of fibers in the acoustic nerve is estimated at 5×10^4 (30,000 cochlear and 20,000 vestibular), the afferent fibers of all other deuterencephalic cranial nerves[40a] at about 246,000. Thus, discounting the olfactory input,[41] about 42 per cent of the total sensory input pertains here to the

have indicated some excess of ganglion cells. According to VAN BUREN, the number of ganglion cells in the human retina reaches 2×10^6, exceeding by about 66 per cent the highest results of optic nerve fiber counts. Be that as it may, several possibilities could here obtain. (1) The various hitherto extant fiber counts failed to include a substantial number of non-medullated fibers which were either not impregnated, or too thin for resolution by light microscopy. FORRESTER and PETERS (1967) estimated 117,000 fibers in the albino rat's optic nerve by means of electron microscopy, while BRUESCH and AREY (1942), by means of light microscopy, calculated 74,000 fibers. (2) The calculation estimating 10^6 ganglion cells in the human retina may be more 'accurate' than that estimating 2×10^6. (3) Some 'optic nerve ganglion cells' of the retina may not have neurites joining the optic nerve.

[40a] The afferent fibers are distributed as follows; trigeminus, 140,000; facialis 2,000; vagus and glossopharyngeus 104,000. All these figures, including those cited in the text, are of course merely tentative approximations, and are unilateral, i.e. referring to the input from one side of the body.

[41] In the macrosmatic Rabbit, the tentative number of 5×10^7 olfactory receptors with

optic system, whose importance is thereby anatomically substantiated. This system, moreover, is of comparable importance in a variety of other Vertebrates, notably Birds.

Other important *numerical relationships*, significant for the still incompletely understood details of retinal circuitry, are those between rods and cones, between these neurosensory elements and the bipolars, and the ratio between these latter and the retinal ganglion cells. With respect to rods and cones,[42] it should be kept in mind that, in addition to their overall numerical ratio, this ratio varies in different regions of one and the same retina. Thus, in Man and some other Vertebrates, the fovea centralis contains only cones, while rods predominate at the retina's periphery. In a single human eye, the cones are said to number about 4 to 7×10^6, and the rods about 11 to 7.5×10^7, the overall ratio being thus roughly 1:10. In some of the nocturnal Marsupials there are perhaps only one or two recognizable cones in the whole retina. In Birds, the cones are most numerous in diurnal species, and the rods in nocturnal ones. In various Fishes the ratio of about one cone in several hundred rods may hold, while in others one cone in ten or less rods is frequently present.[43]

As regards cones and optic nerve ganglion cells, a 1:1 relationship seems to obtain at least in some regions, such as, e.g. the human fovea. On the other hand, a number of rods (and of cones) have, by means of a bipolar, connections with a single third neuron, thereby forming a compound or subset of the retinal neuronal network.

Prince (1956) draws the conclusion that in retinae which have a

---

an identical number of *fila olfactoria* has been suggested. The *olfactory tract*, however, originating from the mitral cells of the olfactory bulb, and functionally analogous to the optic nerve, is assumed to contain not much more than 45,000 neurites. If neurites of basket cells are included, about 60,000 olfactory tract cells might be present. Again, these figures are unilateral, referring to the input from one olfactory bulb. *Mutatis mutandis*, similar numerical relationships may obtain in microsmatic Man, whose olfactory mucosa, however, is supposed to contain, in accordance with its somewhat greater area, about 4 times more receptor cells than that of the Rabbit.

[42] The rods and cones, representing receptor elements of neuronal type, are analogous to the neurosensory olfactory cells. The number of retinal and olfactory receptors, moreover, displays roughly comparable orders of magnitude. The ganglion cells of the retina, in turn, are comparable to the mitral cells of olfactory bulb, these latter, however, being neurons of the second order, while the former are of the third order. *Mutatis mutandis*, the *optic nerve* is roughly analogous to the *olfactory tract*, but, in general, contains far more fibers than its olfactory analogon (cf. section 2, chapter XIII).

[43] The rod-cone ratio for the Teleostean Eel is quoted as 1:20 by Blinkov and Glezer (1968).

Man (Primate Mammal)	1,000,000
Macaque (Primate Mammal)	1,200,000
Cat (Carnivore Mammal)	120,000
Dog (Carnivore Mammal)	155,000
Seal (Carnivore Mammal)	150,000
Seal (Ungulate Mammal)	680,000
Whale (Cetacean Mammal)	155,000
Rabbit (Lagomorph Mammal)	265,000
Rat (Rodent Mammal)	80,000
Bat* (Chiropteran Mammal)	7,000
Opossum (Metatherian Mammal)	82,000
Pigeon (Aves)	990,000
Duck (Aves)	410,000
Chicken (Aves)	415,000
Alligator (Reptilia)	105,000
Turtle* (Reptilia)	100,000
Frog (Anuran Amphibian)	29,000
Necturus (Urodele Amphibian)	360**
Amblystoma (Urodele Amphibian)	2,000**
Goldfish (Teleost)	53,000
Ameiurus (Teleost)	10,000
Amia (Ganoid)	114,000
Sturgeon (Ganoid)	12,500
Shark* (Plagiostome)	113,000
Ray* (Plagiostome	40,000
Entosphenus (Cyclostome)	5,200

Figure 28. Tabulation indicating approximate number of optic nerve fibers (from one eye) in diverse Vertebrates (based on data from various authors compiled by Blinkov and Glezer, 1968). *No relevant data concerning species (e.g. small or large) was available. **Indicates that a variable but significant percentage of non-medullated fibers obtains, as emphasized by some authors. Non-medullated fibers, although perhaps numerically negligible in Mammals (except diverse Cetaceans) and in Birds, may also be present, to an undefined extent, in some of the here not specifically indicated lower Vertebrates. As regards the Cat, Hughes and Wassle (1976) have recently reported the much higher number of 193,000 optic nerve fibers, ranging from 0.5 to 13.5 μ in diameter, on the basis of an electron microscopic investigation.

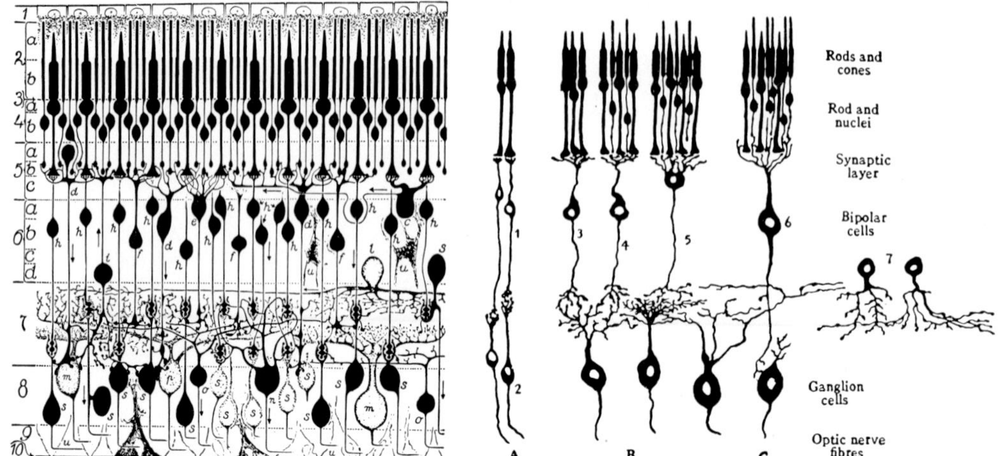

Figure 29 A. Diagram of the retinal structure in Primates, based on *Golgi preparations* as interpreted by POLYAK (from BLOOM and FAWCETT, 1962). The conventional layers 1–10, with further subdivisions (a, b, etc.) are indicated at left. Rods and cones are easily distinguishable. c: horizontal cell; d, e, f, h: centripetal bipolar cell; i: supposedly 'centrifugal' bipolar cell (presumed to discharge toward neurosensory elements); l: inner horizontal cell; m, n, o, p, s: ganglion cells of optic nerve; u: parts of *Müller's supporting cells*, whose nuclei lie in layer 6.

Figure 29 B. Various patterns of retinal circuitry, modified after POLYAK (from BEST and TAYLOR, 1950). A one-to-one cone channel; B, C: various sorts of combined rod and cone channels; 1: 'midget bipolar cell'; 2: 'midget ganglion cell'; 3, 4, 5, 6: diverse bipolar types; 7: amakrine cell. It will be noted that in this and in the preceding figure, neither the so-called peripheral clubs (*'massues'*, cf. Fig. 24 A) nor 'stellate cells' and centripetal optic nerve fibers (cf. Fig. 24 C) as depicted by CAJAL are taken into consideration. The bipolar cell indicated as discharging toward neurosensory elements in Figure 29 A could perhaps correspond to CAJAL's 'stellate cells'.

cone predominance, the outer nuclear layer (of the receptor cells) and the ganglion cell layer are about equal in size. When rods predominate, the outer nuclear layer has generally by far the greater number of cells, and the number of ganglion cells dwindles. The inner nuclear layer (of bipolars) is 'much less predictable'. Frequently its population is in inverse proportion to that of the outer nuclear layer, but it is always highly populated in a cone-dominated retina.

Based on the available data particularly elucidated by CAJAL (1911) and further elaborated in the subsequent investigations by POLYAK (1941, 1957), the overall circuitry of the retina may be interpreted as follows. Impulses from the rods and cones reach the bipolar cells which

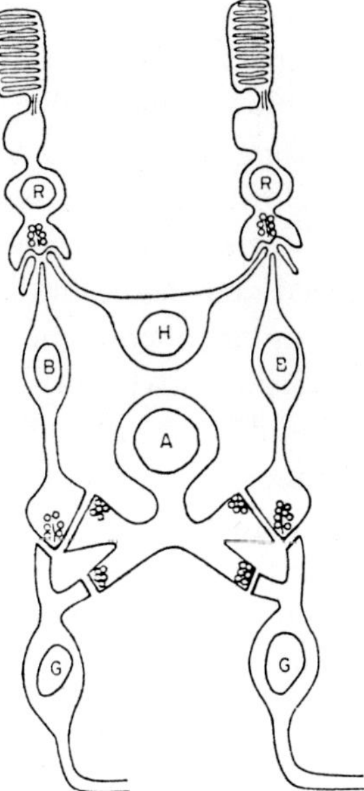

Figure 29C. Simplified concept of some reciprocal connections between Vertebrate retinal neuronal elements (after DOWLING, 1968, from PEARSON, 1972). A: amakrine cell; B: bipolar cells; G: ganglion cells; H: horizontal cell; R: receptors.

transmit the signals to the ganglion cells, whose neurites, forming the bulk of the optic nerve, channel the optic input into the neuraxis. In addition, impulses from the receptor elements are transmitted to the horizontal cells in the inner nuclear layer and are spread out to other receptors as well as to bipolars. These latter can be conceived not only as transmitters but also as relevant analyzers. The role of the amakrine cells is obscure, but these elements may interrelate bipolar as well as retinal ganglion cells, besides receiving and transmitting impulses from the centrifugal optic nerve fibers.[44] Again, small *('midget')* bipolars are believed to mediate the one-to-one connections between some cones and some retinal ganglion cells. There are, moreover, so-called giant retinal ganglion cells whose dendrites receive input from a large group of bipolars with their additional connections. Thus, various functional groupings or units or retinal elements are provided. Figures 29 A–C

[44] The functional significance of these fibers remains uncertain, but some sort of 'calibrating' or 'threshold regulating' effect on retinal transmission processes does not seem improbable. Some of the centrifugal fibers might merely be recurrent collaterals of optic nerve neurites with feedback effect, but the actual origin of centrifugal fibers *in the central neuraxis* seems most likely (cf. also GRANIT, 1955).

illustrate concepts of retinal circuitry deduced from evidence provided by the *Golgi method* (cf. also Figs. 24 A–D). The thereby established overall synaptic arrangements are compatible with the results obtained by experimental studies using procedures of electrophysiology and briefly discussed further below. Nevertheless, a detailed interpretation of these results in terms of the assumed retinal circuits still remains restricted to very generalized and vague outlines. Contemporary, and not in all respects convincing or coherent views are discussed and elaborated in a publication by SHEPHERD (1974).

Recently, FAMIGLIETTI and KOLB (1976) have reported a 'bisublaminar organization' of the inner plexiform layer in Mammals. The outer sublamina (a) is said to contain connections conveying 'off-center' signals, and the inner sublayer (b) connections conveying 'on-center' signals. The transmission of these signals is presumed to be effected by different types of bipolars and of retina ganglion cells.

The neuronal linkage receptor-bipolar-retinal ganglion cell with its collateral connections is provided by the *inner*, *invaginated layer* of the embryonic optic cup. Before turning to some relevant functional aspects of visual mechanisms, a few comments on *the outer*, *non-invaginated layer* of the optic cup seem appropriate. This layer remains unstratified, and, in the pars optica retinae, provides the low prismatic hexagonal pigment epithelium. Its cells, against which the rods and cones abut, are not uniform within the Vertebrate series and vary both *qua* size and pigment content. This latter seems most abundant in the retinae of Birds and Teleosts. The inner surface of these cells may issue processes filled with pigment granules surrounding the rod and cone extremities and shielding them from each other. Upon illumination, the pigment granules are said to move into these processes, and, in the dark, to become displaced toward the epithelial cell bodies. The pigment migration toward the bacillary layer may be accompanied by a contraction of the cones and by an extension of the rods into the pigmented region. This has been particularly noted in Frogs, whose dark-adapted eye, conversely, displays a contraction of the pigment cells and of the rods, together with a cone elongation. Although thus apparently induced by changes of illumination, the pigment migrations are also induced by various other physical effects such e.g. as temperature changes (contraction at higher temperature, cf. NOBLE, 1931, 1954).

The so-called *tapeta*, which may be absorbing *(tapetum nigrum)* or reflecting *(tapetum lucidum)* are located in the choroid beyond the juxtaretinal capillary layer *(choriocapillaris)*, being either fibrous *(tapetum*

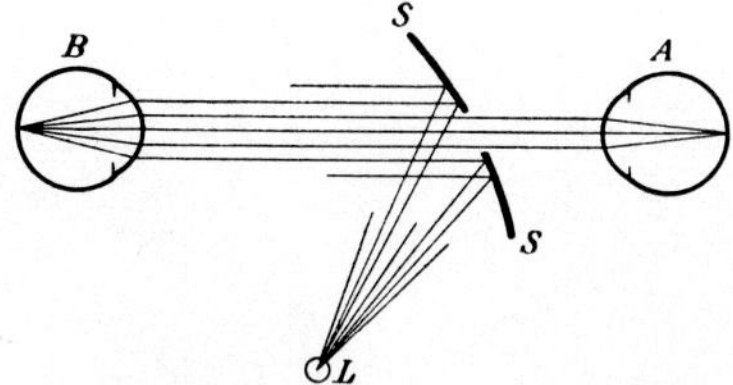

Figure 30 A. Diagram illustrating the general principle of the ophthalmoscope (from SCHENK and GÜRBER, 1918). A: eye of examiner; B: examined eye; L: source of light; S: concave mirror with central opening.

fibrosum) or cellular ('endothelial', *tapetum cellulare)*. The tapetum lucidum, commonly containing birefringent guanin crystals, reflects the light which has passed through the retina and causes the 'eye shine' or 'glow' e.g. displayed, in dim illumination, by the eyes of cats and other Carnivores, the tapetum acting here as a concave mirror.

Although some incident light is reflected back from the retina through the pupil, the retina cannot be directly examined because the examiner's head would screen out too much incident light or a light between examined and examining eye would predominate in the vision of the latter, preventing the intended observation. These difficulties are eliminated by the *ophthalmoscope*,[45] a slightly concave mirror with central perforation, reflecting light into the examined eye, while the examiner views the retina through the mirror's aperture (Fig. 30 A). The ophthalmoscopic picture (Figs. 30 B, C) displays the optic nerve disk, the fovea centralis, and the retinal vessels.

The 'adequate stimulus' to which the neurosensory elements of the Vertebrate and particularly of the human retina are adapted, is represented by one '*octave*' of electromagnetic radiation with wave lengths (λ) between roughly 0.8 and 0.4 μ (more accurately 7,700 to 3,900 Å). This range of electromagnetic radiation is commonly designated as 'visible light'.[46] The duplexity theory, proposed by MAX SCHULTZE in 1866, and reasonably well substantiated by the subsequent observations, assumes that the rods, with a low threshold particularly in dim

[45] Although HELMHOLTZ is commonly credited with the invention of the ophthalmoscope in 1851, he actually contrived an improved model of a device already originated about 2 years previously by the British mathematician CHARLES BABBAGE (1792–1871) who also invented the first computing mechanisms based on the presently used advanced mathematical and logical concepts.

[46] Further comments on the nature of light are included on pp. 375–396 of section 6, chapter V, volume 3/I.

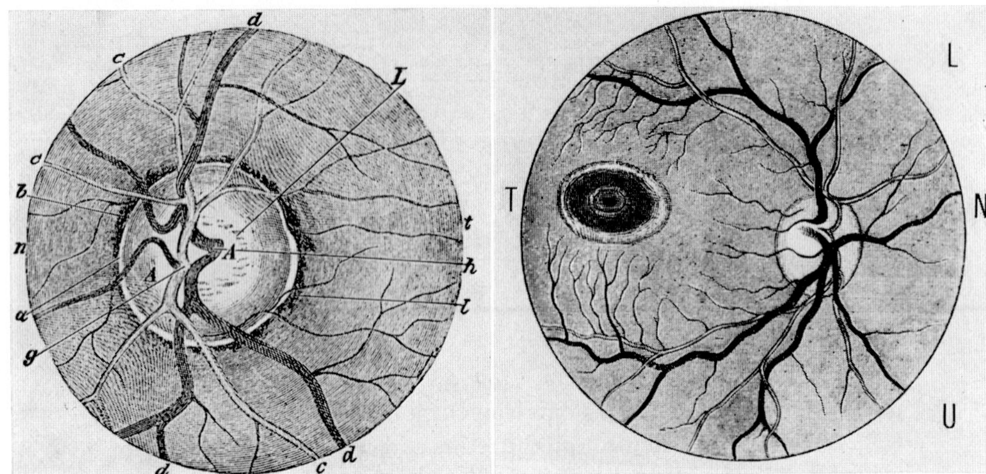

Figure 30 B. Ophthalmoscopic pictures of the human ocular fundus (after v. Jäger and Wecker, from Marle, 1927). I. By direct method (upright fundus picture). A: optic nerve disk; L: lamina cribrosa; a: scleral ring; b: choroid ring; c: arteries; d: veins; g: main bifurcation of central artery; h: main bifurcation of central vein; n: nasal side; t: temporal side. II. By indirect method (inverted fundus picture). L: lower portion of retina; N: nasal side with optic nerve disk; T: temporal side with macula; U: upper portion of retina. Both pictures are of the left eye. The 'indirect method' uses a strong magnifying glass (13 to 20D) held in front of the examined eye, which is viewed through the ophthalmoscope from a distance of about 40 cm. The thereby obtained picture is inverted and at a much lesser magnification than the upright picture viewed at very close range by means of the 'direct' ophthalmoscopic examination. In this latter, however, a much smaller portion is displayed at one time. The 'indirect' examination requires less practice with ophthalmoscopic procedure and is therefore easier to perform than the 'direct' one.

light *(scotopic vision)*, register intensity of said radiation, while the cones, with a higher threshold adapted to bright illumination *(photopic vision)*, register aspects of frequency (respectively wave lengths). Rods and cones, however, respond together over certain ranges of intensity, as shown by the curves indicated in Figure 31 A.

The mechanism whereby the energy of light, a physical R-event, becomes transduced into N-events, namely neural signals, is apparently based on the absorption of light quanta[46a] in the outer segments of the

[46a] It is well known since more than 20 years that individual Vertebrate receptors can respond to a single quantum unit of light (cf. K., 1961b, p. 182). A recent report on quantum sensitivity of rods in the Toad retina, with some bibliographic references to this topic, is the paper by Fain (1975).

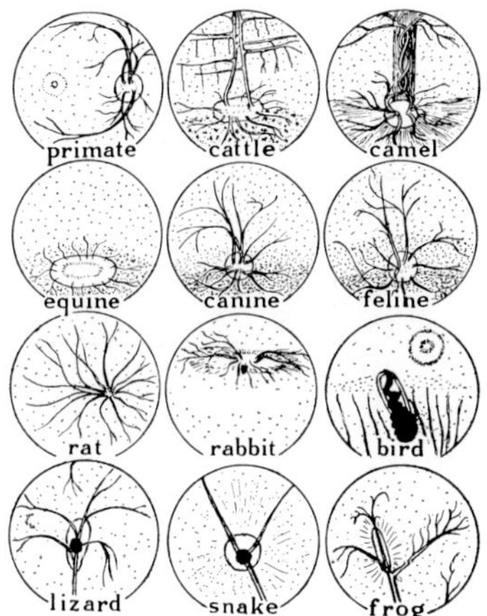

Figure 30 C. Ocular fundus diagrams in various Vertebrates, showing vascular patterns (from PRINCE, 1956). In the Bird, the large dark mass is the pecten. The vessels seen in the lower part of the fundus on both sides of that structure are choroidal, and not retinal. The fundus pictures are upright.

neurosensory elements, and seems to be a photochemical reaction mediated by 'light-sensitive' so-called visual pigments. There are perhaps four or more such photochemicals which pertain to a group of carotenoid proteins, associated with the fat-soluble vitamin A. The reddish visual pigment *rhodopsin ('visual purple')* is associated with the rods[47] and bleached by light, turning to yellow. Its chemical reaction to illumination may be the relevant stimulus triggering the receptor cell. Rhodopsin was identified approximately hundred years ago by BOLL.

A violet visual pigment, *iodopsin*, seems to be associated with the cones. It is particularly conspicuous in the *fovea centralis* and, bleached by light, produce here the yellow spot because of which said fovea was also designated as *macula lutea sive flava*. The photosensitive chemicals of cones are also called *photopsins*, and those of rods *scotopsins* (WALD, 1954). The history about diverse theories concerning visual pigments is dealt with in the publications by POLYAK (1957) and PRINCE (1956).

Although the retinal neurosensory cells are particularly adapted to the registration of 'electromagnetic waves' in the above-mentioned 'visible band' they can also be stimulated by abnormally intense radia-

[47] In the Amphibian Frogs, two types of rods are said to be present, namely those with long and with short outer segments. The first type is reported to be allied with rhodopsin, and the second with 'visual green', this latter being functionally similar to visual purple (NOBLE, 1931, 1954).

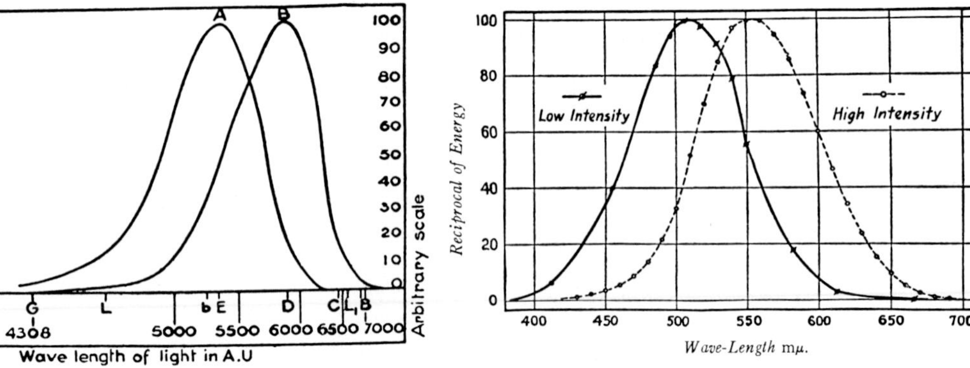

Figure 31 A. Normal human scotopic (A) and photopic (B) luminosity curves (after ABNEY and FESTING, from BEST and TAYLOR, 1950).

Figure 31 B. 'Visibility curves' for human scotopic and photopic vision based on reciprocals of energy (after HECHT, from BEST and TAYLOR, 1950).

tion of higher or lower frequencies. Apart from the threshold related to their specific adaptation, the neurosensory elements do not ordinarily respond to ultraviolet[48] or infrared radiations because these are mostly absorbed by the refractive media of the eye. Within the ultraviolet range, however, fluorescence of these media can result.[49] Moreover, electrical and mechanical stimulation can evoke crude visual responses, e.g. by a blow upon the eyeball or by pressure. Graded pressure upon the human eye bulb, besides producing luminous rings (phosphenes), can elicit the three 'primary color sensations'. 'With a pressure just sufficient to be uncomfortable violet dots appear which persist for some time after the pressure has been released. Stronger pressure causes green and still firmer pressure red figures to be seen' (BEST and TAYLOR, 1950).

If the effect of light, taken as sensory brightness *(luminosity)*, size of potential, or spike frequency at any wave length is represented by L_λ, it will be proportional to the amount of energy E_λ and the specific sensi-

48 Insects such as Bees, and possibly some other Invertebrates possess eyes responding to ultraviolet radiation. Thus, the visible spectrum of Bees is said to extend from 6,500 to 3,000 Å, thereby including 'ultraviolet' but excluding a substantial range of the 'red'. Some comments on ultraviolet vision and detection of polarized light by Invertebrates are included in section 12, chapter IV of volume 2.

49 X-rays penetrate the human eye and can stimulate the retina. These rays are not refracted. Metal letters, placed in reverse position before a closed eye in the dark can be clearly recognized as shadows (not images) against a bright ground (cf. BEST and TAYLOR 1950).

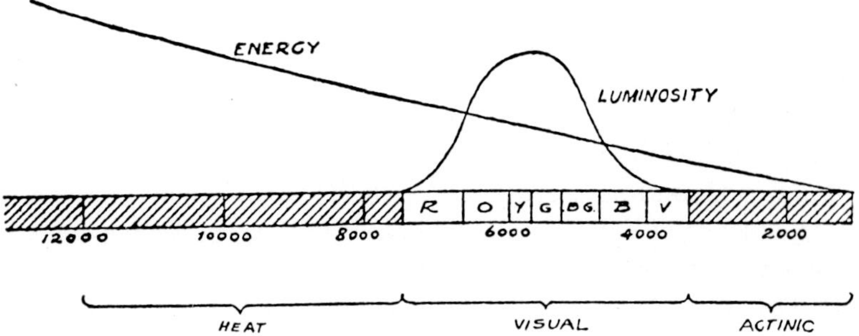

Figure 31 C. Curves showing the relative energy and luminosity of the different regions of the spectrum (after STARLING, from BEST and TAYLOR, 1950). The energy curve indicates here the relative amount of energy in the visible and invisible solar radiation at about sea-level, and should not be confused with the quantum energy at a given wave length respectively frequency.

tivity for that particular wave length S_λ, this latter being the retinal sensitivity factor. With $L_\lambda = E_\lambda \cdot S_\lambda$ the measure of S_λ becomes L_λ/E_λ. At constant L_λ the sensitivity in every wave length is indicated by $\frac{1}{E\lambda}$ (cf. GRANIT, 1955).

Figure 31 A shows normal human *scotopic* and *photopic* luminosity curves, while Figure 31 B shows corresponding '*visibility curves*', which are luminosity curves corrected to an equal-energy spectrum. Thus, the stimulating power is represented as solely a function of the wave length. For details concerning the relevant techniques and literature, reference is made to BEST and TAYLOR (1950). In scotopic vision, the fovea is almost blind, vision being essentially mediated by the more peripheral retinal regions.

It should be added that the term 'luminosity' is frequently used somewhat ambiguously with two different denotations. It may stand for the brightness of an experienced light sensation (P-event) or for the intensity of a physical illumination (R-event). In addition, the quantity of light energy necessary to stimulate the retina is not constant for all wave lengths. In the dark adapted eye (scotopic vision), the energy required to stimulate the rods is least for 'green' light of slightly more than 5,000 Å (cf. Figs. 31 A, B).

With regard to the *energy* of an electromagnetic radiation (expressed in ergs), it should be recalled that the energy of a given quantum of such radiation is proportional to its frequency and inversely proportional to its wave length. This energy is quite independent of the luminosity, as shown in figure 31 C.

Again, experienced *luminosity* or *brightness* (a P-event) is not identical with *objective light intensity* (a R-event). Human vision, moreover, cannot properly estimate diverse *degrees* of luminosity, but, on the other hand, rather accurately recognizes whether two adjacent surfaces display *identical luminosity*. The classical methods of photometry, initiated by JOHANN HEINRICH LAMBERT (1728–1777) are based upon this fact.

The amount of light illuminating a surface, i.e. the *light intensity*, is measured in *meter candles*, defined as the light orthogonally incident per second per cm^2 of a surface at one meter from a standard '*international candle*'. If the candle's distance is d, the surface illumination is $\frac{1}{d^2}$, being thus inversely proportional to the square of the distance. The illumination to be measured can be simultaneously matched with that of a standard candle at a particular distance.

The *quantity of light* reflected from a perfectly diffusing surface illuminated by 10,000 meter candles is termed a *lambert* (l, after the above-mentioned author), a *millilambert* being thus 0.001 l. The unit of retinal illumination is the *troland*, defined as that amount of light illuminating the retina from a surface with luminosity (intensity) of one standard candle per m^2 through a pupil of 1 mm^2 opening.

Following the discovery of the photoelectric effect, and the construction of photoelectric devices, it was found that the negative charge (the number of electrons) 'pulled off' by the action of light is exactly proportional to this latter's intensity. Thus, *photocells*, which can be made specifically sensitive throughout the various ranges of visible radiation, and which are directly calibrated to indicate the light intensity, are now commonly used for purposes of photometry. The retina itself is, in this sense, a photoelectric cell, but the numerous different variables affecting its response make the eye, as mentioned above, inherently unreliable for gaging degrees of light intensity.

Before discussing some retinal mechanisms providing the basis of vision, it should also again be recalled that the terms 'light', 'darkness' and 'color' refer to three entirely different domains of phenomena. Thus, *light* can (1) be conceived as a purely physical phenomenon *(R-event)* corresponding to the above-mentioned octave of electromagnetic radiation. (2) Light can be conceived in terms of physicochemical (biological) *N-events*, namely (a) as retinal and central neuronal events (signals), normally encoding the effect of retinal stimulation by physical light and (b) as comparable N-events or signals triggered by other sorts of physical respectively chemical, or neuronal stimulation. It should be stressed that these N-events are by no means exclusively retinal but also involve the central processing neuronal mechanism (e.g. the human visual cortex or area striata). (3) Light and darkness *sensu strictiori* represent private conscious phenomena or *P-events* whose occurrence can only be observed in the solipsistic (private) perceptual space-time manifold of the sentient being experiencing light respectively darkness. Such being, it can be assumed, must be endowed with an appropriate and functioning brain. *Mutatis mutandis*, the same distinction obtains for '*color*', which, *qua* physical R-events, corresponds to certain ranges of frequency respectively of wave length.

In accordance with the *duplexity theory*, the rods are concerned with scotopic vision, which is essentially colorless[50] and the cones with photopic vision, characterized by color perception. This, *prima facie*, suggests that the cones register 'color', and the rods differences in the degree of illumination (brightness, light intensity). As a first approximation, said assumption seems doubtless plausible, but must be taken with a grain of salt, since the particular parallel neural events *(Np-events)* correlated with the conscious experience of color *(P-events)* are doubtless very complex and presumably involve processed signals of rods as well as of cones.[51]

Little can be said about the 'color sense' of Invertebrates. It seems, nevertheless, certain that some sort of 'color perception' obtains in a number of these forms, particularly in diverse Insects (cf. e.g. BUDDENBROCK, 1958; v. FRISCH, 1950, and the comments in chapter IV, section 12, p. 266, vol. 2). With respect to Vertebrates, it seems permissible to assume that all forms possessing retinal cones could, to some extent, presumably detect color differences, although the cones might be conceived as photopic receptors adapted to bright illumination but not necessarily to specific frequencies.

POLYAK (1957) states that it is even now not well known to which extent color vision obtains in the Vertebrate series: 'opinions vary from those who would admit it in most species to those denying all or most color vision in by far the majority of the Vertebrates, including

---

[50] As expressed in German vernacular: *bei Nacht sind alle Katzen grau.* Or, in a more 'scientific' phraseology; if the spectrum is viewed in bright illumination by the light-adapted eye it is seen as a sequence of 'colors' with maximum luminosity in the yellow. If the illumination of the spectrum is reduced and viewed with the dark-adapted eye, the maximum luminosity shifts toward the violet end, the red and adjacent portions becoming gradually darker. This is the *Purkinje shift.* With further reduction, the spectrum becomes colorless (gray), the maximum luminosity being now displaced to the 'green' region (about *Fraunhofer's line* E).

[51] As regards the here relevant retinal N-events, cf. e.g. the curves of Figure 31 A.

most Mammals'. Good color vision, on the contrary, is said to be certain in most higher Primates, and is, of course, characteristic for Man.

According to BUDDENBROCK (1950) no animal below the level of Primates is believed to have 'a truly good color-sense' comparable to that of Man. Nevertheless, a passably well developed 'color-sense' is said to obtain in Ungulates as well as in the domestic Cat. Moreover, Fishes 'can be trained to respond to orange, or to yellowish green, or to bluish violet'. Such Fishes 'will never confuse one of these colors with its immediate neighbors in the spectrum' (BUDDENBROCK, 1950). Various degrees of color vision can likewise be assumed to obtain in diverse Amphibians, Reptiles, and particularly in Birds. Details about color perception and other aspects of vision in this latter Vertebrate class are dealt with in the treatise by PEARSON (1972).

Pioneering investigations on the physical (R-event) aspect of light and color were undertaken by NEWTON (1730). The thereby obtained data, expanded by subsequent investigators, indicate that the visible solar radiation consists of 'waves' of all length within that range, transmitted at the same velocity through space. By means of a prism it can be decomposed into a more or less continuous[52] spectral band arranged according to wave lengths respectively frequencies, and displaying intergrading colors from red to violet, conventionally described as the seven rainbow colors red, orange, yellow, green, blue, indigo, violet, or as the five colors red, orange, green, blue and violet.[53] 'Colors' corresponding to a single frequency or to a narrow frequency range are designated as homogeneous (*'homogeneal'*, NEWTON, 1730). An undefined but rather large number of different colors may be perceived by normal human vision.[54]

---

[52] Discounting here *Fraunhofer's lines* A–H, of fundamental significance for spectroscopy.

[53] Thus, in this sense, neither 'white' nor 'black' are spectral (homogeneous) colors. Neither is, e.g. 'purple', resulting from a mixture of violet and red light. Yet, as regards e.g. crayons or paints, black, white, and purple can evidently be called 'colors'.

[54] The Roman writer AULUS GELLIUS (ca. 123–169 AD, who lived at the time of the ANTONINES, elaborates in his *Noctae Atticae* (II/26) upon a philosophical discussion concerning colors. He quotes FAVORINUS (*floruit* ca. 120 AD) as stating that many more colors are detected by the eye than are expressed by words and terms (*plura sunt in sensibus oculorum quam in verbis vocibusque colorum discrimina*). The poverty in names is said to be more pronounced in Latin than in Greek. Moreover, in both languages, one and the same word is often used to denote either red, yellow, or green hues or various grayish shades. Again some American Indian tribes are said to designate different colors by the same name. The Japanese, likewise often use the word *ao* (or *aoi*) for both blue or green. Although apparently suggesting color blindness, this lack of terminologic distinction is, *per se*, very

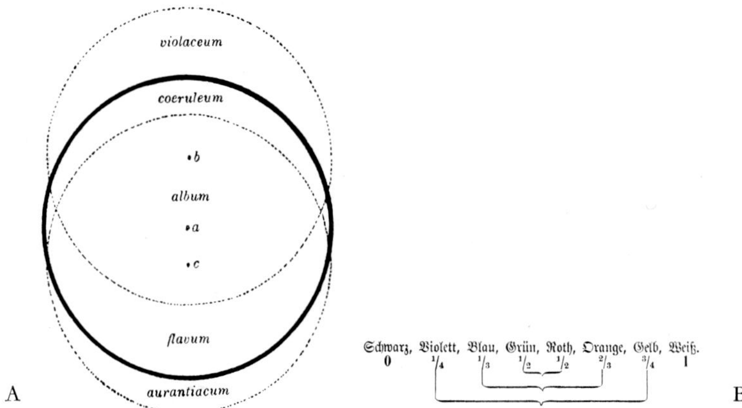

Figure 32 A. Spectrum obtained from an illuminated white disk upon a black background according to SCHOPENHAUER (from SCHOPENHAUER, 1830). a: center of middle region of refracted disk; b, c: centers of disks with greatest (b) and least (c) refraction.

Figure 32 B. SCHOPENHAUER's concept of color sensation in terms of fractionalized 'qualitative activity' of the retina (from SCHOPENHAUER, 1818, 1854). The brackets indicate complementary color pairs. White and black, not involving 'qualitative' fractionalization, are not evaluated as 'true' colors by this author.

A surface which completely scatters the rays of all 'visible' frequencies displays the 'color' white. A surface more or less absorbing all light rays displays the 'color' black. Complete lack of visible radiation ('darkness') is likewise experienced as black. White surfaces under diminishing illumination appear 'gray', fading out into black. Likewise, surfaces partially absorbing light of all frequencies are experienced as 'gray'. Transparent red media, e.g. red glass, *transmit* only red light, absorbing all other frequencies. Conversely, opaque red substances likewise absorb all frequencies, but *reflect*, or scatter red light.

If the spread spectral colors are recombined by diverse optical procedures such that appropriate recombination results (e.g. by combining two prisms with opposite orientation), white light is again obtained. This latter, however is also produced by a combination of only three more or less homogeneous 'primary colors', namely red, green, and blue (or violet), which are also designated as the *primary colors*. Combined in various appropriate proportions, not only white, but any other perceivable color can be produced. Again, if two lights pertain

---

definitely not related to said condition, nor does it even imply an insufficient appreciation of experienced color (cf. also K., 1961b, p. 444, footnote 123).

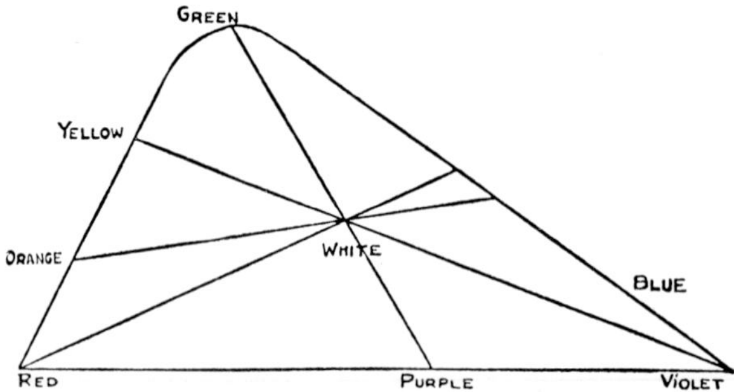

Figure 32C. Color diagram showing complementary colors in present-day interpretation (from BEST and TAYLOR, 1950). It should be recalled that purple is not a spectral 'color' but an experienced one (P-event). The other 'colors' can (but not must) be conceived in terms of physical light *and* in terms of color experience.

to appropriate ranges of the spectrum separated by a relevant distance, such as blue-violet and yellow, greenish blue and red, or blue and orange, their combination also produces the effect 'white'. Such color pairs are designated as *complementary colors* (cf. Figs. 32A, B, C).

The just mentioned effects are *additive*, i.e. requiring the actual reception of the combined 'pure' spectral colors by the retina.

Colored glass, colored liquids, and colored pigments applied on a surface produce colors by *subtractive* effect, namely by absorption of certain wave lengths. A yellow filter transmits, and a yellow pigment reflects, essentially all wave lengths except the blue range, and the remaining radiation appears yellow which, in this case, is not the 'homogeneous' spectral yellow. A combination of pigments, or of superimposed filters 'produce' color by eliminating certain wave lengths, i.e. by subtraction. Thus, if white light passes through a blue filter, which absorbs red, yellow, and violet, and then through a yellow filter, which absorbs red, blue, and violet, the subtractive effect results in essentially green light. A similar result is obtained by a mixture of blue and yellow pigment. The additive effect of light in the proper blue and yellow range, on the other hand, is white. Again, a red surface appears red in white and red light, but black in green light.

If 'light' is conceived as a perceptual 'modality', the sensations (conscious experiences) black, white, and color may be termed '*qualities*' of said modality. Further aspects of these qualities are generally designated as *hue*, *brightness* (brilliance, luminosity), and *saturation* ('purity'). Hue produced by physical light essentially depends upon wave length (frequency); factors determining brightness are (a) intensity of radiation, and (b) 'intensity' of the thereby triggered set of N-events, beginning with the transduction at the neurosensory level. Saturation seems to depend on the homogeneousness of the radiation, and may be considered inversely proportional to its admixture with 'white' light (i.e. '*noise*').

With respect to the theories of color vision, the *trialistic* aspect of 'light' (R-event, N-event, and P-event) must again be emphasized. It is, moreover, here necessary to distinguish retinal N-events from further processed central neuraxial N-events, and particularly from those which are presumably correlated with P-events[55] and can accordingly be designated as (p) N-events (or Np-events).

It is evident that a reasonably valid description ('explanation') of the relevant R-events has been provided by NEWTON, supplemented by the present-day physical theories based on the fundamental notions elaborated by JAMES CLERK MAXWELL, by MAX PLANCK *(quantum)* and by ALBERT EINSTEIN.

As regards the N-events related to *color vision*, none of the theories proposed by various authors 'is capable of accounting for all the experienced phenomena, either of normal vision or of color blindness' (BEST and TAYLOR, 1950). This is not surprising in view of the obtaining complexity involving retinal as well as central neuronal processing events. Theories considering the retinal events were elaborated by THOMAS YOUNG (1800, 1807, 1855) and by ARTHUR SCHOPENHAUER (1816, 1830) before the structure of the retina with its rods and cones was sufficiently elucidated, and, for that matter, before the cell theory became properly established.

YOUNG postulated that the 'sensitive points' of the retina are provided by a limited number of different 'particles' 'capable of vibrating in perfect unison' with a given 'undulation', restricted 'for instance, to the three principal colours, *red*, *yellow*, and *blue*'. Each sort of 'particle'

---

[55] An interesting aspect of the P-events is the continuity of the perceptual visual space across the blind spot, which is not experienced as 'empty' or as 'black', even if experienced by monocular vision, where 'compensation' by the contralateral visual field cannot obtain. The blind spot becomes subjectively demonstrable by the 'disappearance' of a small configuration in the well-known test figures and in perimetry. Again, if one or both eyes are closed, a sensation of 'black' or of 'darkness' is experienced in the corresponding visual fields. No 'darkness' or 'black', however, are experienced outside the visual fields, e.g. in the perceptual space behind one's body. Likewise, scotomas (visual field defects) caused by disruption of the input to area striata (cf. chapters XIII and XV) do not generally produce the sensation 'black', nor is cortical blindness necessarily or predominantly experienced as 'darkness'.

'is capable of being put in motion less or more forcibly, by undulations differing less or more from a perfect unison'.

SCHOPENHAUER considered *white* to result from the complete 'action', '*quoad intensionem partita*', of the illuminated retina, *black* from the lack of such illumination correlated with retinal 'inactivity', and colors from 'partial' qualitative retinal activities in given fractional quantitative relationships *('actio retinae quoad qualitatem bipartita')*, some of which combined to the effect white. He emphasized the fractional 'activity' of the retina and the therefrom resulting complementary color pairs *('Farbenpaare')*. It is of interest to compare his Figures (Figs. 32 A, B) with a contemporary diagram showing complementary colors (Fig. 32 C).

Although this 'explanation' of white and black as well as of colors does not seem *prima facie* implausible and is rather remarkable considering its time, said theory was evidently encumbered by the inherent vagueness of postulated 'intensional' and 'qualitative' 'activities' performed by a membrane whose structure and detailed *modus operandi* still remained entirely unknown. It is, moreover, strange that SCHOPENHAUER was not acquainted with the theory proposed by THOMAS YOUNG, or, if acquainted, did not refer to it. Both theories, despite various substantial differences, interpreted *color as resulting from retinal activities*. YOUNG, however, did not question the validity of NEWTON's concept of physical 'colors', which SCHOPENHAUER acrimoniously rejected.

Another weakness of SCHOPENHAUER's color theory was its exclusive emphasis on retinal activities (retinal N-events) as the basis of light and color sensations. Expressed in my terminology, SCHOPENHAUER apparently considered the retinal N-events to be Np-events, correlated with said sensations, which were then merely interpreted by the 'intellect' i.e. by the central (cerebral) Np-events respectively P-events. Thus, SCHOPENHAUER assumed that, if the auditory nerve were connected with the eye, and the optic nerve with the auditory sense organ, the latter nerve would hear, and the former see.[55a] On the basis

[55a] '*Demnach könnte auch der Gehörnerv sehen und der Augennerv hören, sobald der äussere Apparat beider seine Stelle vertauschte*' (SCHOPENHAUER, 1854). SCHOPENHAUER distinguished '*sensations*', occurring in the sense organs, from '*perception*', a function of the 'intellect', i.e. of the cerebrum. In my own terminology, I subsume experienced sensations, whether interpreted or not, under 'percepts' or 'perceptions' *sensu latiori*, respectively P-events. A percept *sensu strictiori* merely implies a higher degree of integration and interrelation than obtaining in a 'sensation', and could perhaps also be designated as *apperception* (cf. K., 1961b, fn. 1, p. 7; p. 15, p. 187, p. 193).

of the presently available data, and as far as the human brain is concerned, quite the opposite seems to be most likely. If the severed peripheral portion of the optic nerve with appended eye and retina could grow into the cut central portion of the acoustic nerve, physical light phenomena would be experienced as sounds, namely as noise comparable to static in a radio loudspeaker. Conversely, if the cut peripheral end of acoustic nerve, connected with the labyrinth, could regenerate into the central stump of the optic nerve, physical sounds would be experienced as (presumably colored) light flashes, comparable, e.g. to the registration of sound by a cathode ray oscillograph. Experienced light and experienced sounds can evidently be conceived as P-events correlated with specific localized *central Np-events*, namely *corticothalamic events* involving area striata respectively temporal auditory cortex. In other words, the optic nerve would see sounds, and the acoustic nerve would hear light through their *central* stumps.

SCHOPENHAUER investigated color vision as a disciple of GOETHE. The great poet, however, based his color theory on the aspects of subjective color experience, i.e. on the P-events, while SCHOPENHAUER, somewhat too exclusively, considered retinal events (peripheral N-events). Both SCHOPENHAUER's and GOETHE's color theories, moreover, were encumbered by a quite unnecessary polemic against NEWTON's concepts which essentially dealt with the relevant R-events. Again, while strongly supporting many of GOETHE's views, SCHOPENHAUER differed in some essential aspects from his master's interpretations.[56] GOETHE's and SCHOPENHAUER's theories, although obsolete, are nevertheless of considerable historical interest and contained relevant concepts, as justly pointed out by competent scientists such as WESSELY (1922) and W. OSTWALD (1931), but have been generally ignored. Neither SCHOPENHAUER nor GOETHE are mentioned in the historical accounts given in POLYAK's (1957) comprehensive work. Yet, SCHOPENHAUER's two treatises include highly important epistemologic considerations besides many concepts concerning upright, monocular, and binocular vision, moreover on perspective and on optical illusions. These concepts, in essentially identical formulation,

---

[56] Mildly annoyed by his disciple's differing views, GOETHE composed the following epigrams:

'*Dein Gutgedachtes, in fremden Adern,*
Wird sogleich mit dir selber hadern.'

'*Trüge gern noch länger des Lehrers Bürden,*
Wenn Schüler nur nicht gleich Lehrer würden.'

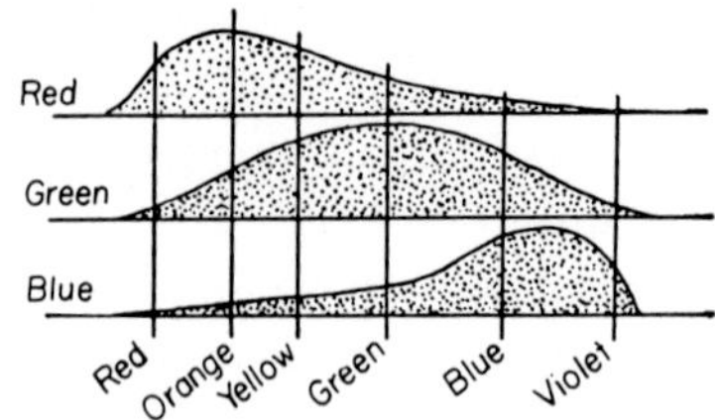

Figure 33 A. Sensitivity of the three receptor systems (cones) according to the *Young-Helmholtz theory* (from BUDDENBROCK, 1958).

but without reference to SCHOPENHAUER,[57] reappear, about 50 years later, in HELMHOLTZ' empirical theory of visual perception (cf. WESSELY, 1922; K., 1961a).

Subsequently to the early histologic studies providing data on the structure of the retina, YOUNG's original concepts were taken up by HELMHOLTZ and elaborated as the *trichromatic Young-Helmholtz theory* which postulates three types of receptors (cones), each containing different photochemical substances particularly affected by light in the frequency range of one of the primary colors (cf. Fig. 33 A).

The *theory of opposite color pairs* proposed by E. HERING about 1878 postulates altogether three different retinal receptors, namely for black-white, blue-yellow, and red-green. The first member of each color pair resulting from *anabolism* (assimilation) of the particular substance, and the second member from *breakdown* (dissimilation).[58] All frequencies of the visible spectrum have a dissimilating effect on the black-white substance, but the different frequencies to a different degree. On the other hand, only certain frequencies have a dissimilating

[57] Cf. footnote 4, p. 3, chapter I of volume 1, and footnote 73, p. 318 in K. (1957).

[58] It should be kept in mind that the concepts 'assimilation' and 'dissimilation' used by HERING, as well as the related terms 'anabolism' and 'catabolism', apparently introduced by GASKELL, are exceedingly vague, oversimplified generalizations. Assimilation *(Aufbau)* refers to the synthesis of complex 'protoplasmic' constituents by living systems, and dissimilation to the 'disintegration' *(Zerfall)* of such constituents. Anabolism is commonly conceived as requiring energy, and catabolism as 'liberating' energy (cf. also vol. 3/II, p. 23). The complexity of metabolic events, however, indicates that phenomena classified as anabolism and as catabolism are by no means mutually exclusive in protoplasmic activities, as justly pointed out by BAYLISS (1924, pp. 377, 421). As FORBES (quoted by BAYLISS) remarks, to assume that increase in 'anabolism' necessarily decreases catabolism 'is to suppose that increasing a man's salary ensures decrease of his expenditure'. Without thereby accepting HERING's hypothesis, one might nevertheless concede that, in a given receptor, the frequencies of 'opponent' radiation bands could, theoretically, trigger 'antagonistic' transduction processes of some sort.

effect on the blue-yellow or the green-red substance, while certain frequencies have an assimilating effect, and others none at all.

Mixed frequencies appear colorless if producing an equally strong dissimilation and assimilation 'moment' for the blue-yellow as well as for the red-green substances because both 'moments' then cancel each other and the effect on the black-white substance becomes predominant. Two frequency ranges producing white are therefore not 'complementary' but 'antagonistic' because they do not 'complete' each other into white but only make this latter appear because of their mutual cancellation.

Hering's model appears to be consistent with some of the actually obtaining phenomena, but does not account for various others, especially for certain types of color blindness, nor does it agree with McDougal's experiments showing that red light seen with one eye and green light simultaneously seen with the other result in the color yellow. This *binocular fusion* is evidently a *cerebral* and not a *retinal* process, which cannot here result from a 'catabolic activity' in the retina postulated by *Hering's theory*. Again, looking through a red filter on one eye, and a green filter on the other at a white surface, this latter will still appear pure white, thus likewise indicating '*cerebral fusion*'. On the whole, it seems perhaps less adequate than the *Young-Helmholtz theory* although the general concept of opponent retinal activities displayed by 'on-off-systems' has now become well substantiated.[59] *Hering's hypothesis*, moreover, agrees well with the grouping of color sensations into antagonistic pairs as emphasized in the earlier theories of Goethe and of Schopenhauer (cf. Fig. 32B).

A new approach to these problems has been initiated by the recording of the electroretinogram, which is a 'mass response', followed by studies of 'single element' responses (namely by optic nerve fibers) through microelectrodes, as undertaken by Granit (1955) and others. If the element under the electrode displays low sensitivity for light of a given frequency, much energy is needed to elicit a discharge, and vice-versa. Thus the inverse value of energy required for a threshold effect in each frequency (respectively wave length) can be plotted on the ordinate in per cent of the maximum), and the wave length on the abscissa.

Two sorts of curves were obtained by Granit, which he called *dominators* and *modulators*. The dominators display broad 'absorption

[59] Cf. pp. 594–596 of section 8, chapter V in volume 3/I.

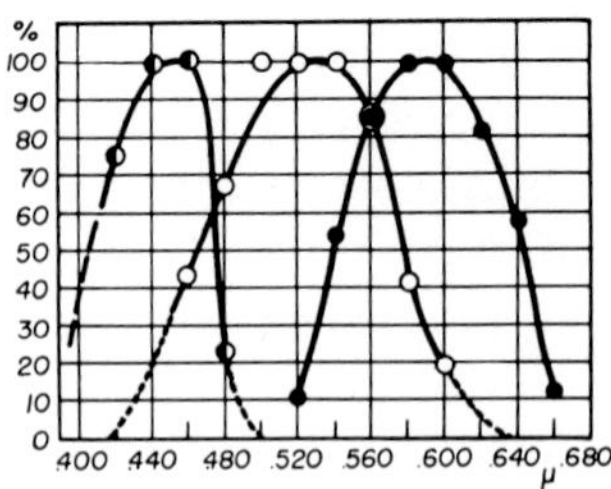

Figure 33 B. Modulator curves of the Cat's retina (after GRANIT, from BUDDENBROCK, 1958). Black circles: red; white circles: green; black and white circles: blue. Although not unlike the curves postulated by the *Young-Helmholtz theory*, the three modulator curves show a lesser degree of overlap.

bands' indicating scotopic and photopic distribution of spectral sensitivity. These dominator curves roughly correspond to those illustrated in Figures 31 A and B. The modulators (Fig. 33 B) are curves with smaller band range, centered in three spectral regions which roughly correspond to blue, green, and red regions postulated by the *Young-Helmholtz theory* (cf. Fig. 33 A). In contradistinction to this latter, however, the modulator curves show a much lesser degree of overlap. It might be assumed that the retinal ganglion cells related to the modulator curves transmit signals encoding wave length (frequency) information, and that the dominator elements, responding to a wide range of wave length, essentially transmit intensity signals, particularly insofar as the scotopic curve is concerned. It is important to realize that the concepts 'dominators' and 'modulators' refer to the information transmitted by the optic nerve to the central neuraxis, and that the preceding relevant 'analysis' or 'processing' was performed by the neural network of the retina (cf. also GRANIT, 1955).

Concerning the more peripheral retinal processes at the receptor level, the photochemical aspect, providing the 'molecular basis of visual excitation', has been studied by numerous investigators and is doubtless relevant for color vision, including 'white' and 'black'. Concerning this latter 'color', it seems likely that the 'off'-responses of retinal elements, mentioned above, play here a significant role. According to WALD (1968) normal human vision requires four different opsins, namely rhodopsin in the rods, and three others in the three different sorts of cones, each opsin being 'specified' by a particular 'gene'.

Be that as it may, the following comment by GRANIT (1955) appropriately reflects the obtaining complexities as well as difficulties: 'the color-theoretical generalizations known as trichromatic, four-component, or polychromatic theories need not concern us very much as long as it is unknown how the brain, especially the striate area, evaluates the information it receives.'[60] In addition to the *Young-Helmholtz* and *Hering theories*, as originally formulated, a number of variants and of different hypotheses have been elaborated, including *Land's* interesting suggestion, mentioned in chapter XI of volume 4, namely that color in the natural image depends on the 'random interplay' of longer and shorter wave lengths over the total visual field.[61]

In contradistinction to auditory perception, in which individual tones that compose a complex sound or a 'harmony' can be distinguished, human visual perception cannot recognize the different components of an experienced color, which may result from a '*pure*' spectral frequency band, or from diverse additive respectively subtractive *mixtures*. Much the same can be said about the 'color' white produced by the entire spectral band or by various complementary colors.

After images occur if the gaze is directed to a bright white light for a moment and the eyes are then closed or turned to a black surface. An image of the light then floats into view, becomes at first rather distinct, and then gradually fades. Similarly, if a bright colored light or surface was stared at, an image of the same color may appear. Such after images are *positive*. A *negative* after image occurs when the retina, before closing or turning the eye, was again diffusely stimulated by white light. The negative after image is then either dark following a white initial stimulus, or in the complementary color of a colored first stimulus. Negative after images are related to the phenomenon of successive contrast. Staring at a brightly illuminated colored surface and then at a gray one will make this latter appear to be tinged with the complementary color, and objects of the complementary color seem to be more brilliant or '*vivid*'. Negative after images can be 'explained' by either the *Young-Helmholtz* or the *Hering theory*. According to the former, 'fa-

[60] Thus, in the Mammalian lateral geniculate griseum and striate cortex, some elements, including 'on-off' or 'opponent color cells' seem to react to color, i.e. frequency coded retinal output.

[61] A useful publication on color vision, reviewing and considering the various theories, was prepared by Teevan and Birney (1961). The views on color vision are also discussed in the proceedings of a 1971 Symposium, published by the 'National Academy of Sciences' (1973). Data relevant to vision in general and color vision in particular, can also be found in an account of the physics of television prepared by Fink and Lutyens (1960). Although television has developed into a shoddy and (in USA) sordidly commercialized 'mass-medium', its theoretical foundations are, nevertheless, of considerable interest.

tigue' of one of the three types of cones is caused by the first stimulus. According to the latter theory, the initial catabolism or anabolism of the first stimulus is succeeded by the opposite process 'until equilibrium is restored'. Positive after images are believed to be caused by physicochemical changes in the receptors outlasting the stimulus (cf. BEST and TAYLOR, 1950). To which degree some aspects of after images represent central neuronal events remains, however, a moot question.

Intermittent stimulation of the retina by a series of light flashes or by reflecting light from a rotating disk divided in alternate black and white sectors produces, at certain frequencies, the sensation of '*flicker*', presumably related to the 'suppression of after images'. Upon further increase of frequency or rotation speed, '*fusion*' results, and the flicker is replaced by a continuous sensation, explained as 'visual persistence', such that the sensation evoked by one stimulus has not ceased before the next one is produced.[62] The frequency at which continuous sensation occurs is the *critical fusion frequency* (CFF) which depends on several variables including light intensity, frequency, size of lighted area, and states of excitability. At the fusion frequency, the experienced brightness corresponds to that which would result from a continuous stimulus of lesser intensity *(Talbot-Plateau rule)* and can be interpreted as the 'averaging', by the brain mechanisms of the intermittently occurring total illumination during a given time period.

With respect to light intensity, the critical fusion frequency rises as the intensity increases, being in the order of about 15 cps at low levels and about 60 cps at high ones. Within a certain range, it is directly proportional to the logarithm of the light intensity.[63]

Color blindness can be classified as *anomalous trichromatic*, as *dichromatic*, and as *monochromatic vision*. Persons affected with anomalous trichromatic vision seem to experience all colors but make abnormal color matches due to 'subnormal appreciation' of either red *(protanomaly* or partial *protanopia)*, or green *(deuteranomaly* or partial *deuteranopia)*. The 'defective appreciation of blue' *(tritanomaly)* is very rare.

Dichromatic vision includes *protanopia* or *red-blindness*, *deuteranopia* or *green-blindness*, and *tritanopia* or *blue-blindness*. Protanopia, also known as *Daltonism*, is the commonest form, in which red objects ap-

[62] The additive mixing of colors by rotating disks, and the principles of stroboscopy or of cinematography are well-known examples of this effect.

[63] This relationship is, *mutatis mutandis*, comparable to that expressed by the *Weber-Fechner law* (cf. K., 1957, pp. 212f.; 1971).

pear dark (gray or black). In deuteranopia both red and green are experienced as yellow. Tritanopia is very rare and seems to be a condition acquired by disease of or injury to the visual system. Still more rare is *monochromatic vision*, in which only black, gray and white, perhaps with a slight tinge of blue can be experienced. The hereditary nature of trichromatism, of protanopia, of deuteranopia, and of most forms of anomalous trichromatism seems well established. In the case of protanopia and deuteranopia, a sex-linked recessive gene is known to obtain. Altogether, about 8 per cent of males and 0.4 per cent of females show some defect of color vision. Deuteranomaly is the most common type (4.5 per cent), followed by deuteranopia (1.4 per cent).[64] Both the *Young-Helmholtz* and the *Hering theories* can be invoked for an 'explanation' of color blindness, but, on the whole, the former theory, combined with the assumption of different 'opsins', seems to provide the more satisfactory model (cf. e.g. KUFFLER, 1953; GRANIT, 1955; HUBEL and WIESEL, 1964, 1966).

Recent detailed studies on the visual system by various investigators using advanced methods of instrumentation have substantiated the significance of the retina as an intricate griseum with considerable capabilities for an initial processing of optic signals. A summary of these data and conclusions can be found in a report edited by HANDLER (1970), and in SHEPHERD's (1974) publication.

Thus, the so-called *receptive fields* provided by the layer of ganglion cells in the retina are said to be circular in shape, somewhat like a target, with an 'on' or an 'off' center and an annular antagonistic '*surround region*'. A cell with an 'on' center receptive field thus has an inhibitory 'surround region', and a cell with 'off' center receptive field is surrounded by an excitatory neighborhood. The most effective excitatory stimulus is therefore a circular spot covering the entire central 'on' region of the field. If the stimulus is enlarged to include any of the annular 'surround region', the effectiveness of the stimulus is reduced because of the just mentioned antagonism. From these properties of the receptive field it can be inferred that the thereto pertaining retinal ganglion cells may not primarily record and transmit information related to the intensity of light stimulating a group of retinal receptors, but rather the *contrast* between the intensity of illumination in the center of its receptive field as compared with the annular 'surround region'.

'On-off' discharges may play a role in brightness (white and black,

---

[64] Cf. BEST and TAYLOR (1950).

light and dark) registration, in the registration of frequencies (color), in the registration of shape, and in the registration of movements. As regards at least the human eyes, it is well known that even with the most exact fixation, small '*saccadic*' *movements* are constantly performed.[64a] These motions may counteract the refractive deficiencies of the eye as an optical instrument. This also precludes the assumption that the retinal image stimulates any set group of receptors. If, for instance, the angular distance separating two spots in the visual field is about the diameter of a cone, the two images must be shifted around so that they come to lie not on two adjacent cones but at least on two cones separated by one that is unstimulated or stimulated differently. The 'off-on' transitions related to the saccadic movements may raise a 'dynamic pattern' (GRANIT, 1955) around spatial gradients of intensity or frequency.

Further aspects of the optic system, such as 'point-to-point projection' upon lateral geniculate griseum and cortex, 'corresponding retinal points' *(horopter)*, binocular and three-dimensional vision, 'simple', 'complex' and 'hypercomplex' receptive fields, as well as optical illusions and visual field defects shall briefly be considered in the discussion of the relevant diencephalic and telencephalic grisea.

As regards the various problems dealt with in the present subsection, insofar as they concern retina and appurtenant topics related to the optic pathways, further details and discussions can be found in the treatise by WALLS (1963), and in contributions by WOLIN and MASSOPUST (1970), GIOLLI and TIGGES (1970), and NOBACK and LAEMLE (1970).

Numerous functional details of the optic pathway with respect to its relay in the lateral geniculate griseum and its termination in the visual cortex have been recorded and inferred, particularly in the Cat and in Monkeys, as results of the extensive experimental investigations by HUBEL (1960), and HUBEL and WIESEL (1961, 1966, and other publications). The optical volumes of the *Handbook of Sensory Physiology* deal with all aspects of contemporary views on photochemistry of vision (DARTNALL, 1972), on physiology of photoreceptors (FUORTES, 1972), on central processing of visual information (JUNG, 1973), and on 'visual psychophysics' (JAMESON and HURWICH, 1972).

[64a] It seems most likely that these saccadic eye movements are mediated by the fasciculus longitudinalis medialis dealt with in chapter IX and XI of volume 4, perhaps by way of the nucleus interstitialis, which, in turn, may be controlled by other pretectal grisea. These latter seem to receive direct or indirect input from the visual cortex.

C. Basal and Dorsal Midline Structures (Hypophysial Complex, Saccus vasculosus, Epiphysial Complex)

In addition to the chiasmatic ridge with its optic nerve decussation and its system of supraoptic and postoptic commissures, as pointed out further above, the *basal midline configurations* of the Vertebrate diencephalon comprise the following structures. (1) The *supraoptic crest* of Amniota and, at least in some Anamnia, e.g. in the Holocephalian Plagiostome Chimaera, a more or less comparable 'circumventricular organ', the 'organon vasculare praeopticum' of BRAAK (1963). (2) The *neurohypophysial complex*, found in all Vertebrates. (3) The *saccus vasculosus* of numerous Fishes.

In this context, further comments on the Amniote supraoptic crest, discussed in chapter V, section 5 of volume 3/I, can be omitted. As regards Anamnia, BRAAK (1963) described in the Holocephalian Plagiostome Chimaera monstrosa an ependymal structure in close apposition to the rostral portion of optic chiasma. He designated this vascularized configuration of ependymal and subependymal elements as organon vasculosum praeopticum. Said organ, represented by a slight protrusion into the preoptic portion of the third ventricle, is unpaired, but rostrally bilateral symmetric with a median longitudinal sulcus which disappears at more caudal levels (Figs. 34 A–C). Regardless of its more caudal location, the organ can be evaluated as (heterotopically) homologous[65] to the Mammalian supraoptic crest, being, as is the case for this latter, a neighborhood of the preoptic recess, which, in Chimaera, becomes greatly elongated. The organ can also be seen in Figures 29 and 30 (71 C and D of the present chapter) of our own report on the brain of Chimaera (K. and NIIMI, 1969) which, however, contains no reference to that structure, since BRAAK's communication, not directly related to the problems we were concerned with, had escaped our attention. A faint indication of a perhaps comparable ependymal swelling rostral to the chiasmatic ridge can also be noted in the preoptic recess of the Amphibian Gymnophione Schistomepum (cf. Fig. 98 B). It is not impossible that a detailed study of the preoptic recess in various Anamnia might disclose further comparable ependymal configurations which could be related to the group of ependymal, paraependymal or circumventricular organs.

---

[65] Cf. the homology qualifications enumerated on pp. 64–65, section 14, chapter VI of volume 3/II.

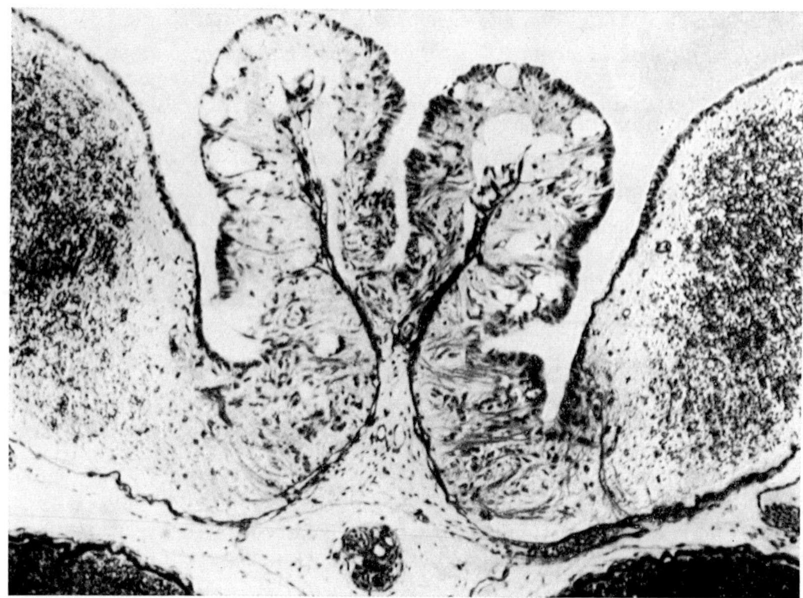
A

The Vertebrate *neurohypophysis*, although of neuraxial neuroectodermal provenance, is, except for the nerve fibers of the mainly but apparently not exclusively neurosecretory hypothalamo-hypophysial channels and their endings, an essentially non-neuronal configuration of the postchiasmatic hypothalamic floor. Generally speaking, its relevant structural elements are modified glia cells designated as *pituicytes* (Fig. 35). In addition, other sorts of glial elements are commonly present, but nerve cells are, as a rule, absent, except for rare instances of a few displaced ones, usually in the more proximal portions bordering on the hypothalamus proper.

Endings of *hypothalamo-hypophysial fibers*, swollen by the transported neurosecretory material,[66] appear frequently, at least in many Vertebrate forms, as bulbous swellings originally described as *Herring bodies*. In addition to the neurosecretory fibers, non-neurosecretory ones may likewise reach parts of the neurohypophysis. Such fibers could provide, as the case might be, either hypothalamic output or input with respect to the neurohypophysis, i.e. be either efferent or afferent.

---

[66] This material is particularly well demonstrable by the *Gomori-stain*.

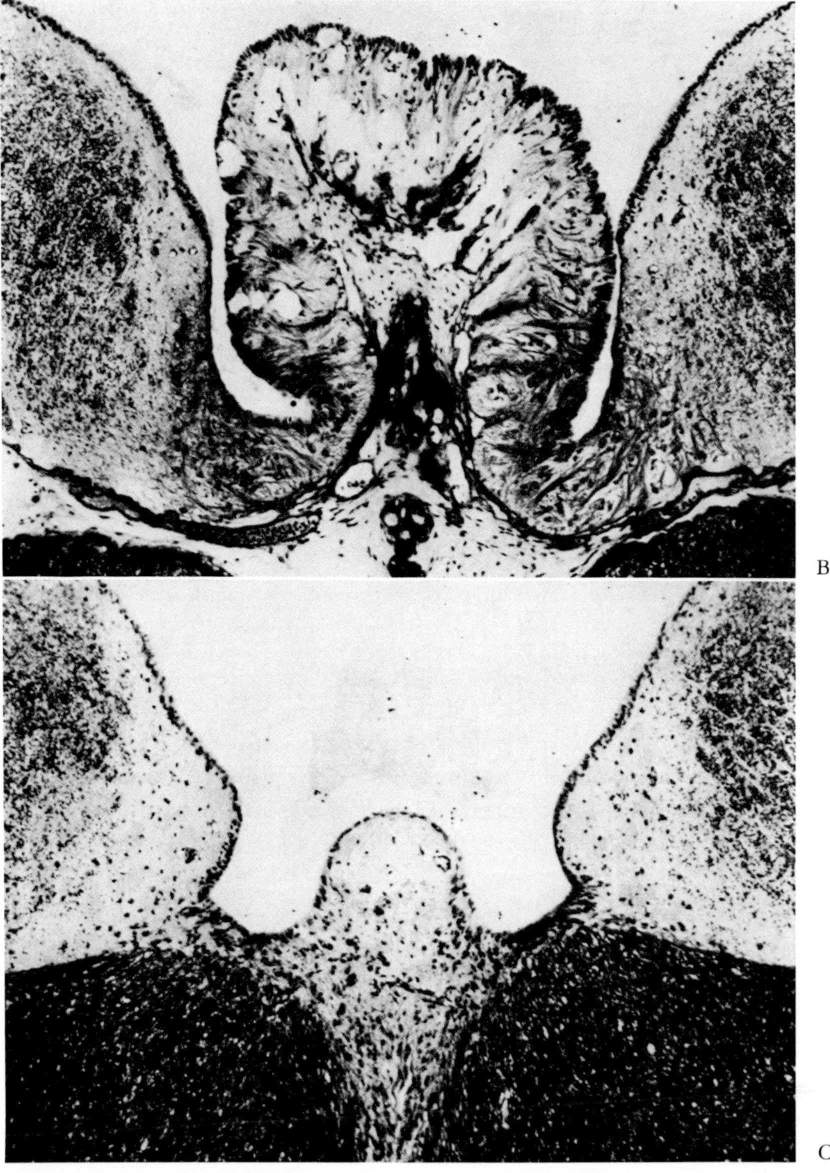

Figure 34 A–C. Three cross-sections, in rostro-caudal sequence, through the organum vasculosum praeopticum of the Holocephalian Plagiostome Chimaera monstrosa (from Braak, 1963; luxol fast blue azocarmin stain, ×100, red. $^1/_1$).

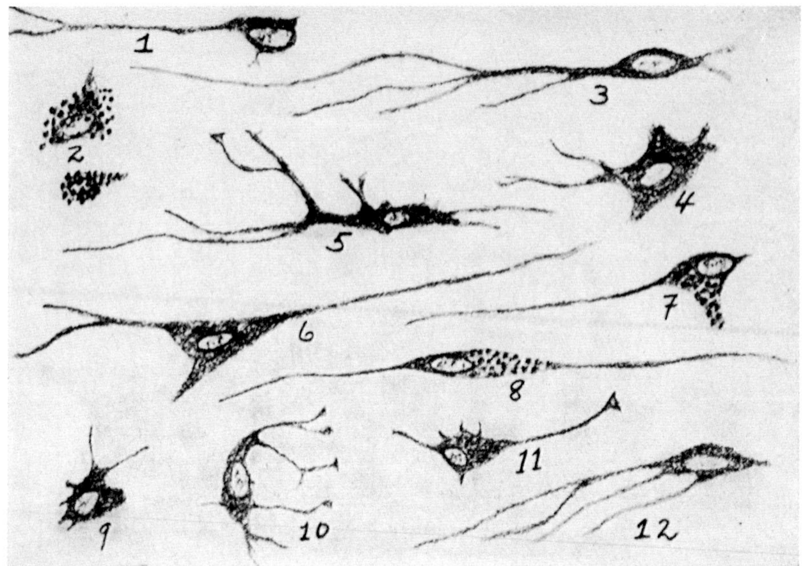

Figure 35. Diverse aspects of pituicytes from the Mammalian neurohypophysis (Ox, Bos taurus); cells 1, 3, 5, 6, 8, 10, 11, 12 display conspicuous processes, some of which with 'terminal expansions', and cells 2, 7, 8 contain pigment granules (after BUCY, from MAXIMOW, 1930; *Penfields silver carbonate impregnation*, × ca. 700, red. $^1/_1$).

Three parts of the Vertebrate neurohypophysis[67] may be distinguished, namely (a) the proximal *eminentia mediana*, (b) the intermediate *infundibular stem* (*infundibulum*, BNA, PNA of human anatomy), and (c)

[67] The neurohypophysis is also called *lobus nervosus* or *pars nervosa* of the entire hypophysial complex, which, as 'hypophysis' *sensu latiori*, is subsumed under the hypothalamus in both BNA and PNA. The latter recognize a lobus anterior with pars tuberalis and pars intermedia, and a lobus posterior, which presumably corresponds to the infundibular lobe enumerated in the present text. The lobus posterior of the BNA was frequently and very improperly understood to include the neurohypophysial infundibular lobe and the adenohypophysial pars intermedia, i.e. two anatomically and ontogenetically quite differing configurations. Both official terminologies, moreover, fail to include their 'infundibulum' into the neurohypophysis, and ignore this latter's eminentia medialis as well as to list the adenohypophysial pars distalis. As regards in particular the *eminentia medialis*, numerous detailed but in part inconclusive or unconvincing data can be found in a recent symposium edited by KNIGGE *et al.* (1972). Detailed accounts on the comparative anatomy and physiology of the Vertebrate hypophysial complex can be found in the volumes edited by HARRIS and DONOVAN (1966), and particularly, *qua* comparative anatomy, in the contribution by WINGSTRAND (1966), and in a more recent treatise by HOLMES and BALL (1974).

the *infundibular lobe* or *neural lobe sensu strictiori*. The eminentia mediana, also less rigorously termed eminentia medialis, is that part of the neurohypophysis directly continuous with the neuronal parenchyma of the postchiasmatic hypothalamus. In most Anamnia the eminentia mediana has an essentially, although somewhat modified, ependymal structure. In diverse Amniota, the eminentia mediana slightly bulges on the external brain surface, being here separated from the hypothalamus *sensu strictiori* by a shallow groove *(sulcus tubero-infundibularis* or *sulcus hypophysio-hypothalamicus)*. The infundibular stem or infundibulum, rather conspicuous in Man and other Mammals, is often inconspicuous or poorly developed in various other Vertebrates. It interconnects the eminentia mediana with the infundibular lobe. Where the eminentia mediana has an exclusively rostral location, the caudal part of infundibular stem interconnects the infundibular lobe with the neuronal parenchyma in the region of the mammillary recess.[68] Again, in Fishes possessing a saccus vasculosus, this latter develops in the topologic neighborhood corresponding to the caudal portion of infundibular stem. It should also be recalled that the neurohypophysis pertains to the neuraxial structures with a peculiar permeability of the blood-brain barrier as discussed on pp. 345 f, section 5, chapter V of volume 3/I.

The neuroectodermal *neurohypophysis* is closely connected with the *adenohypophysis*, a derivative of the body ectoderm in the region of the stomodeum. Neurohypophysis and adenohypophysis jointly form the *hypophysial complex (pituitary gland*, or hypophysis *sensu completo)*. In all gnathostome Vertebrates, the adenohypophysis is, despite various differences of details, formed in an essentially identical manner, characterized by the ingrowth, known as *Rathke's pouch*, from the ectoderm of the stomodeum. This adenohypophysial anlage commonly arises as an infolding with a lumen, i.e. as a true 'pouch', or (e.g. in Teleosts and Amphibians) as a solid epithelial bud.[68a] In either case, the ingrowth

---

[68] This region is frequently loosely designated as 'infundibular region', its ventricular groove, the sulcus lateralis hypothalami (posterioris), being also called 'sulcus lateralis infundibuli'. In the legend to Figure 184, p. 410 of volume 3/II this sulcus (sf) was erroneously termed sulcus lateralis thalami posterioris, the dropping out of hypo(thalami) remaining undetected upon proof reading.

[68a] One is here reminded of the two modes by which the Vertebrate neural tube can develop: (a) generally by the infolding of a plate into a groove which subsequently closes to become a tube with central canal; (b) in some cases (e.g. Cyclostomes, Teleosts, and in the tail bud of Birds) by the formation of a solid neural cord which subsequently acquires

loses its original connection and becomes applied to the neurohypophysis. It has been shown that, at least in Amphibians, the adenohypophysial ingrowth does not develop normally if the postchiasmatic hypothalamic floor region is experimentally destroyed, while the neurohypophysis does not properly differentiate in the absence of the adenohypophysial ingrowth. Both subdivisions of the pituitary complex, although of totally different origin, appear thus mutually dependent on one another for their development (NOBLE, 1931, 1954).

Considering the Tunicates (Hemichorda) to be Invertebrate Chordates, the *neural gland* of Ascidians[69] can doubtless be evaluated as a (heteroeidetic, heterotypic, heteropractic, and heterotopic)[70] Invertebrate kathomologon of the Vertebrate neurohypophysis. As regards the Cephalochordate Amphioxus, the so-called infundibular organ and *Hatschek's pit* may represent kathomologa of the neurohypophysis and of the adenohypophysis, respectively (cf. chapter VIII, section 3, vol. 4).

The Cyclostome hypophysial complex differs from the Gnathostome one by the very simple structure of the neurohypophysis and the peculiar ontogenetic development of adenohypophysis. In Petromyzonts, the neurohypophysis is a thickened part of the postchiasmatic floor, consisting of a modified ependymal framework, and containing neurosecretory terminals. The adenohypophysis develops from the ectodermal naso-hypophysial pit, analogous but not homologous to *Rathke's pouch*, and consists of 'pro-', 'meso-' and 'meta-adenohypophysis', this latter being in contact with the neurohypophysis (cf. Fig. 36). The neurohypophysis of Myxinoids is somewhat more differentiated, since eminentia medialis, infundibular stem, and infundibular lobe are vaguely distinguishable. The adenohypophysis, on the other hand, is less differentiated than in Petromyzonts and consists of a cellular aggregation between neurohypophysis and the naso-pharyngeal duct. This latter corresponds to the naso-hypophysial pit and sack of Petromyzonts, but differs from said sack by opening into the pharynx.

---

a lumen and becomes organized into a typical neural tube (cf. vol. 3/I, section 1, p. 12). It will also be recalled (cf. vol. 4, chapter VIII, section 6, Figs. 72, 73, pp. 126, 127) that some Teleosts possess what is called a *neurohypophysis caudalis* or *urohypophysis*. This '*spinal cord hypophysis*' differs from the '*cerebral hypophysis*' by being entirely neurohypophysial (i.e. neuroectodermal), that is to say, lacking a corresponding adenohypophysial component.

[69] Cf. vol. 2, chapter IV, section 13, p. 295.

[70] These qualifications of morphologic homology are explained in volume 3/II, chapter VI, section 1 A, pp. 64–65.

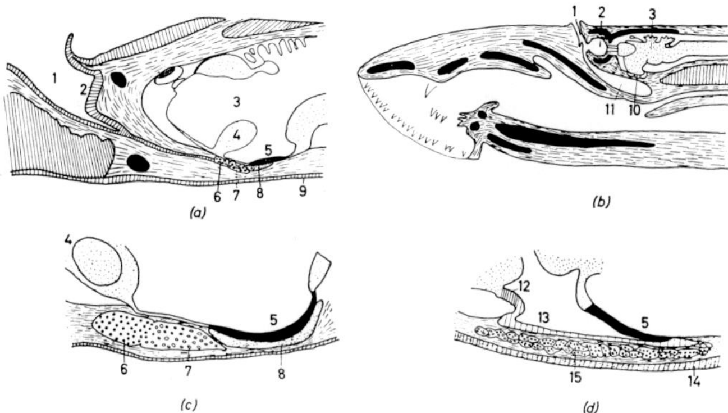

Figure 36. Median sagittal sections through head respectively pituitary region (a) of larval Petromyzon, (b, c) adult Petromyzon and (d) of Myxine (from Wingstrand, 1966). 1: naso-hypophysial pit; 2: olfactory organ; 3: brain; 4: chiasmatic ridge; 5 neurohypophysis; 6, 7, 8: pro-, meso-, and meta-adenohypophysis; 9: epithelium of oral roof; 10 (in b): 'pituitary'; 11: naso-hypophysial sac; 12: 'median eminence'; 13: 'infundibulum' of Myxine; 14: epithelium of naso-hypophysial duct in Myxine; 15: adenohypophysis of Myxine.

Figures 37–45 depict the configurational relationships of the hypophysial complex in the different groups of Gnathostome Vertebrates.[70a] It can be seen that, despite substantial diversities in topographic arrangement and degree of differentiation *qua* components in the various taxonomic forms, the fundamental morphologic pattern is preserved.[70b] As regards the *adenohypophysis*, a *pars tuberalis*, a *pars distalis*, and a *pars intermedia* are commonly distinguished.[71] This latter part,

[70a] Interesting peculiarities of the hypophysis in the Holocephalian Plagiostome Chimaera have been reported by Fujita (1963). As regards the hypophysis of Gymnophiona, cf. in addition to the author's report (K., 1970), that by Marcus and Laubmann (1925), which had escaped my attention and was thus omitted from my 1970 list of references.

[70b] A very peculiar configurational arrangement of the hypophysial complex, however, obtains in the highly aberrant Coelacant Latimeria (Millot and Anthony, 1965). The 'infundibulum' (*'pédoncule hypophysaire'*, neurohypophysis) is directed rostralward basally to optic chiasma and to telencephalon. The adenohypophysis is connected with the palate by a vascularized, partly hollow stalk containing islets of adenohypophysial tissue and presumably corresponding to *Rathke's pouch*.

[71] Despite various structural and presumably functional differences, the so-called meta-, meso-, and pro-adenohypophysis of Petromyzonts and of Gnathostome Fishes can be evaluated as morphologically corresponding (i.e. homologous) with pars intermedia, pars distalis, and pars tuberalis of the Tetrapod (Anamniote Amphibian and Amniote) adenohypophysis. Detailed elaborations on the problems here involved can be found in the

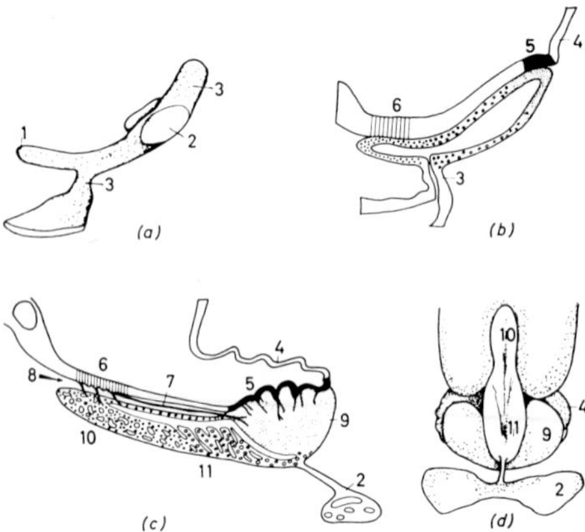

Figure 37. Development and configuration of the Plagiostome pituitary complex (from WINGSTRAND, 1966) (a): model of *Rathke's pouch* in a 28 mm Squalus embryo according to BAUMGARTNER; (b): midsagittal section through (a); (c): midsagittal section through adult Shark pituitary, combined after various authors; (d) basal aspect of Squalus pituitary according to BAUMGARTNER; 1: anterior process; 2: lateral in (a), inferior lobe in (c) and (d); 3: epithelial stalk; 4: saccus vasculosus; 5: 'neuronal lobe' (infundibular lobe); 6: median eminence; 7, 8: portal vessels; 9: pars intermedia (meta-adenohypophysis); 10, 11: pro- and meso-adenohypophysis.

however, although present in the Sauropsidan Reptiles, does not become differentiated in Birds, whose adenohypophysis is briefly discussed further below.

The original lumen of *Rathke's pouch*, as the case may be in the diverse Vertebrate taxonomic forms, can become completely obliterated or remain either as a single cleft or is represented by a few cavities, usually in the pars intermedia. Frequently, this latter subdivision is located caudally adjacent to the residual lumen.

A *vascular plexus (mantle plexus)* between neurohypophysis and adenohypophysis can be seen in Cyclostomes, Plagiostomes, and Osteichthyes. In various Gnathostome Fishes, said plexus displays a rather distinct arrangement characterized by a *hypophysio-portal circulation.* More or less clearly differentiated hypophysio-portal vessels occur in Amphibians and Dipnoans. This portal system is still more complex in Amniota. Generally speaking, the primary net of the hypophysio-por-

---

publications of WINGSTRAND (1966) and others, critically evaluated by the just cited author. Our own incidental observations on the hypophysis are restricted to the Teleost Corydora (MILLER, 1944), to Man (K., 1954), and to some Gymnophione Amphibians (K., 1970).

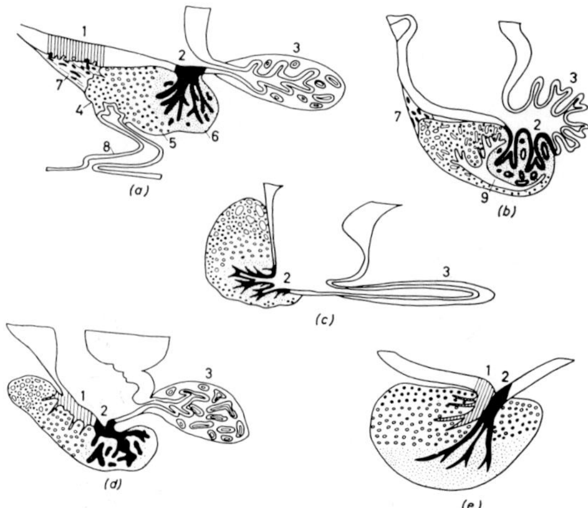

Figure 38. Midsagittal sections through Osteichthyan pituitary complex (after WINGSTRAND, 1966, combined from his own findings and those of other authors). (a): Polypterus; (b): Acipenser; (c): Amia (a–c: Ganoids); (d): Anguilla; (e): Carassius (d, e: Teleosts); 1: median eminence or, in d and e, its presumptive equivalent; 2: neural lobe (infundibular lobe); 3: saccus vasculosus; 4, 5, 6: pro-, meso-, and meta-adenohypophysis; 7: 'vascular ligament' (rudimentary portal system ?); 8: duct to the buccal cavity (remnant of *Rathke's pouch?);* 9: hypophysial cavity (likewise remnant of *Rathke's pouch* lumen ?).

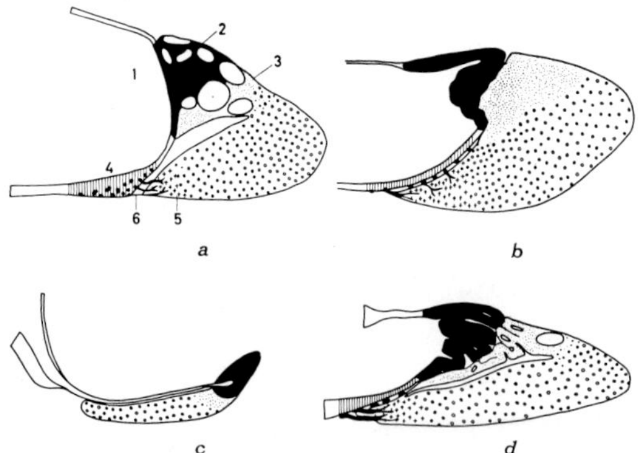

Figure 39. Midsagittal sections through the hypophysial complex in Amphibians and a Dipnoan (from WINGSTRAND, 1966). (a): Anuran (Bufo); (b): Urodele (Amblystoma); (c): Gymnophione (Hypogeophis, redrawn after STENDELL); (d): Dipnoan (Protopterus); 1: 'saccus infundibuli'; 2: 'neural lobe' (infundibular lobe); 3: pars intermedia (adenohypophysis); 4: median eminence; 5: 'zona tuberalis' (adenohypophysis); 6: portal vessels. The pars distalis lies between 3 and 5. As regards Gymnophiona, Figure 183B, p. 408, of vol. 3/II may be compared.

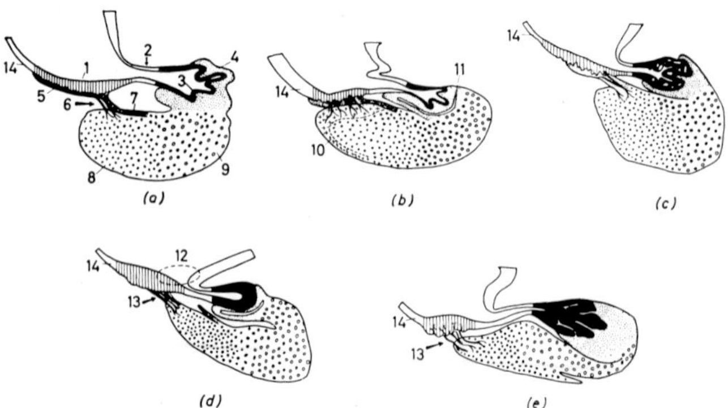

Figure 40. Midsagittal sections through the Reptilian pituitary complex (from WINGSTRAND, 1966, based on that author's original observations combined, in (a) with descriptions by de BEER and WYETH and ROW). (a): Sphenodon; (b): Chelonian (Testudo); (c): Crocodilian (Alligator); (d): Lacertilian (Lacerta); (e): Ophidian (Python); 1: median eminence; 2: infundibular stem; 3: 'neural lobe' (infundibular lobe); 4: pars intermedia; 5: juxta-neural pars tuberalis; 6: porto-tuberal tract with portal vessels; 7: 'pars tuberalis interna'; 8: cephalic lobe of pars distalis; 9: caudal lobe of pars distalis; 10 (in b): 'zona tuberalis'; 11: hypophysial cavity (remnant of *Rathke's pouch* lumen ?); 12 (in d): locus of 'tuberal plates' in Lacerta; 13: 'pars terminalis' with portal vessels; 14: 'pars oralis tuberis' (hypothalamus).

tal system seems mainly to be located in the eminentia medialis or adjacent parts of infundibular stem, and the flow in the portal vessels is assumed to be directed toward the adenohypophysis. A portal circulation of this type appears thus to be a rather constant feature of Vertebrates, establishing a link between neuraxis of which the neurohypophysis is a part, and the adenohypophysis (cf. GREEN, 1966). Additional details on the complex and only partly understood vascular system related to the hypophysial structures can be found in the publications by XUEREB *et al.* (1954), TÖRÖK (1954), K. (1954), ENGELHARDT (1956), KORITKÉ and DUVERNOY (1962), DUVERNOY and KORITKÉ (1968), DUVERNOY *et al.* (1970, 1971), and DUVERNOY (1972).

Again, some blood passing through eminentia medialis may also reach adjacent neighborhoods of the hypothalamus proper, and redescend into capillaries leading toward the infundibular stalk (cf. e.g. ANDERSON and HAYMAKER, 1974).

In addition to the vascular connections, some hypothalamic nerve fibers, both neurosecretory and non-neurosecretory, may reach parts of the adeno-hypophysis, particularly in Anamnia. Thus, in Teleosts, some authors believe that adenohypophysial cells could be affected by

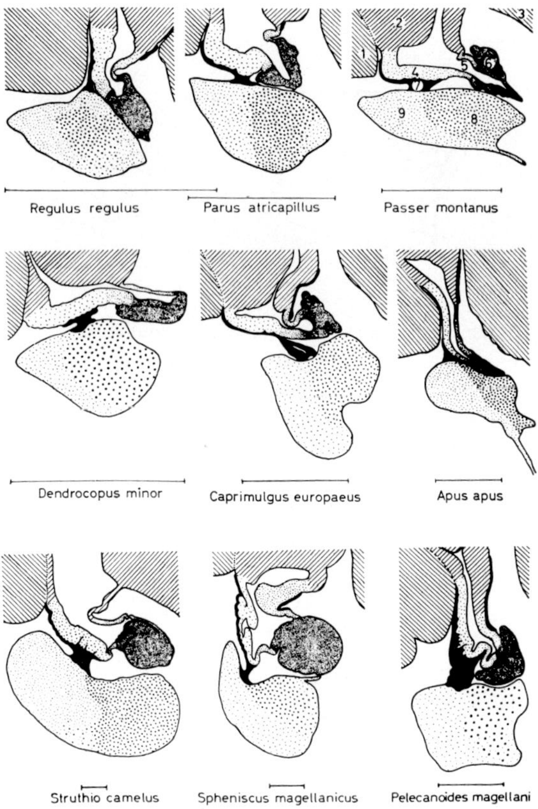

Figure 41. Midsagittal sections through the Avian pituitary complex (from WINGSTRAND, 1966). 1: chiasmatic ridge; 2: tuber (hypothalamus); 3: basis of brain stem; 4: eminentia mediana; 5: infundibular stem; 6: 'neural lobe' (infundibular lobe); 7: pars tuberalis; 8, 9: caudal and rostral ('cephalic') lobes of pars distalis; Scale under each drawing: 1 mm. The range of variations *qua* taxonomic forms is conspicuous.

direct hypothalamis neuronal control. Some of the nerve fibers ending in the adenohypophysis, however, are doubtless peripheral nerve fibers of the vegetative (essentially of the sympathetic) nervous system, which, on the other hand, is also, but less directly, subject to hypothalamic control. Many of the relevant problems are still poorly elucidated and references to their status can be found in the publications edited by HARRIS and DONOVAN (1966), HAYMAKER *et al.* (1969), BRODISH and REDGATE (1973), SWAAB and SCHADÉ (1974), and TIXIER-VIDAL and FARQUHAR (1975).

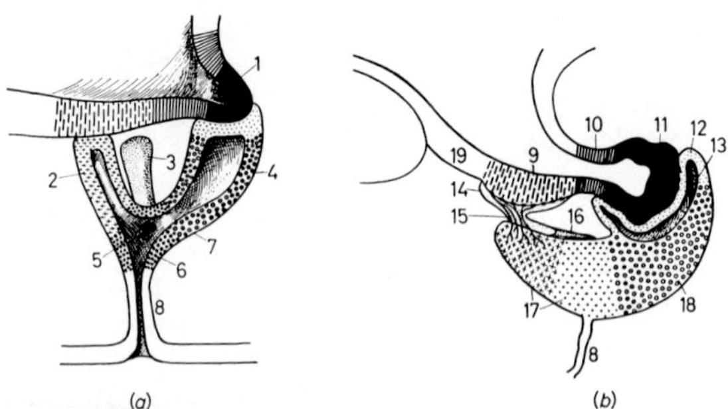

Figure 42. Diagrams indicating origin (a) and adult configuration (b) of a generalized Amniote pituitary complex, (a) being particularly based on Reptilian morphogenesis (from WINGSTRAND, 1966). 1: 'saccus infundibuli'; 2: 'anterior process'; 3: 'lateral lobe'; 4: 'aboral lobe'; 5: 'opening of the lateral lobe cavity'; 6: 'oral lobe'; 7: constriction of *Rathke's pouch;* 8: 'epithelial stalk'; 9: median eminence; 10: infundibular stem; 11: 'neural lobe' (infundibular lobe); 12: pars intermedia; 13: hypophysial cleft (remnant of *Rathke's pouch* lumen ?); 14: juxtaneural pars tuberalis; 15: porto-tuberal tract; 16: 'pars tuberalis interna'; 17, 18: rostral (cephalic) and caudal lobes of pars distalis; 19: 'pars oralis tuberis' (hypothalamus).

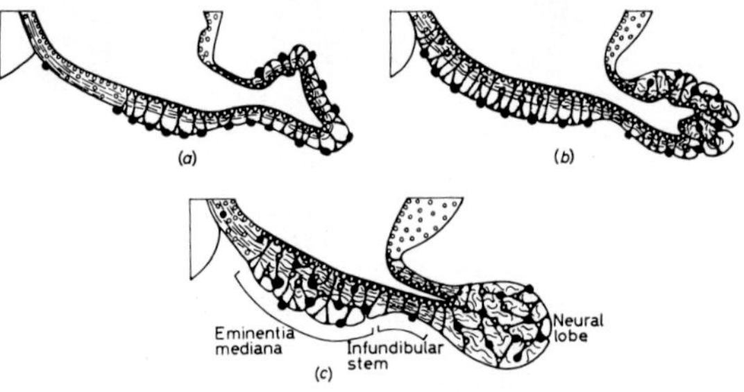

Figure 43. Diagrams showing the differentiation of the neurohypophysis in Amniota (from WINGSTRAND, 1966). (a): 'primitive type' (Sphenodon, many Lacertilia, Chelonia); (b): with 'somewhat more proliferated median eminence' as in many Reptiles and most Birds; (c) Mammalian level of differentiation. Ependymal cells and pituicytes stippled, blood vessels solid black, coarse nerve fibers indicated by lines.

Inspection of the relevant Figures (particularly Figs. 38 and 40) will reveal that finger-like protrusions of the neurohypophysial infundibular lobe into the pars intermedia of adenohypophysis occur in some Vertebrate forms. Moreover, in diverse Vertebrates, such as e.g. various Teleosts, the infundibular lobe becomes completely surrounded by the adenohypophysis (cf. e.g. MILLER, 1944).

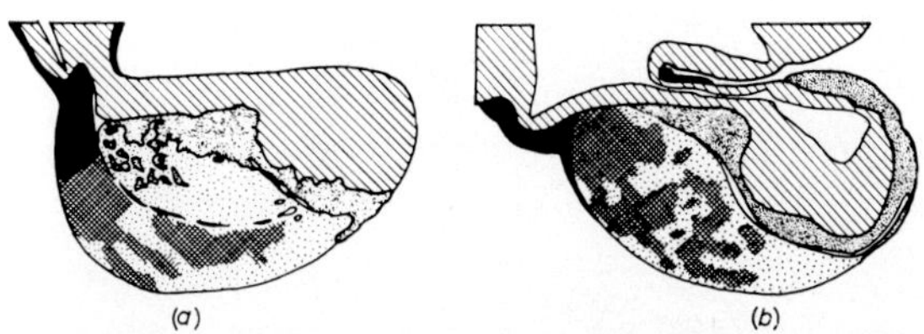

Figure 44. Midsagittal sections through the hypophysial complex of Mammals (from WINGSTRAND, 1966). (a) Rabbit, after DAWSON; (b): Cat; solid black: pars tuberalis; coarse stippled: 'pars distalis proper'; fine stippled: pars intermedia.

Again, in the adenohypophysial complex of Birds, the pars intermedia fails to develop (cf. Fig. 41). In young Avian embryos, the dorsal region of *Rathke's pouch* becomes attached to the neurohypophysis as in Reptiles and Mammals, but the juxtaneural wall of the pouch does not differentiate into a pars intermedia. Instead, said wall is separated from the neurohypophysis by connective tissue, and then differentiates like the pars distalis. In the adult, it is merged with the caudal portion of pars distalis, and cannot longer be recognized as a distinct entity.

The Avian pars distalis, which can be histologically separated into a caudal and a rostral lobe, is generally separated from infundibular stem and median eminence by a wide cleft filled with loose connective tissue and bridged by the vessels of the porto-tuberal tract. A variously configurated pars tuberalis is present.[72] As regards the configuration of the Mammalian pituitary gland, reference to Figures 43–45 will here be sufficient.

The Vertebrate hypophysial complex represents a compound endocrine organ, which, because of the multiplicity of its effects, is commonly designated as a 'master gland'. It acts on other endocrines and is in turn influenced by them. In addition, it may also act directly on the body cells of various systems. The hormones produced by the hypophysial complex are generally proteins influencing e.g. the metabolism of water, proteins, carbohydrates and fats, thereby also affecting the synthesis of protoplasm as well as growth processes. The pituitary complex provides thus 'a focal point at which much of the bodily activity is adjusted' (YOUNG, 1957).

Despite intensive studies by numerous authors, and an ever in-

---

[72] Further details on the ontogenesis and on the comparative anatomy of the Avian hypophysis, not relevant in the generalized aspect here under consideration, will be found in the publications by GRIGNON (1956) and by WINGSTRAND (1966).

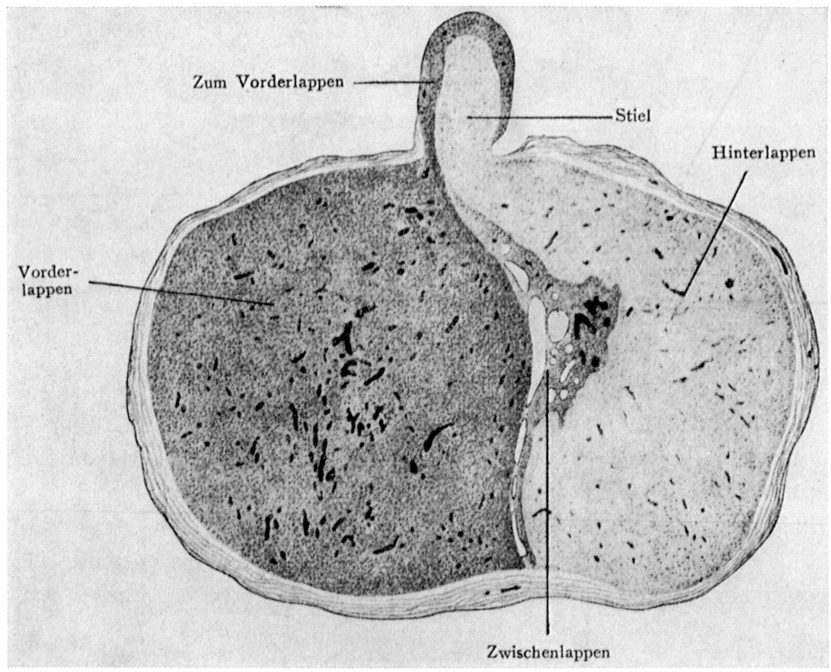

Figure 45. Midsagittal section through the human hypophysis (after STÖHR and MÖLLENDORFF, from K., 1927). Only infundibular stalk and infundibular lobe of neurohypophysis are shown. The portion labelled '*Zum Vorderlappen*' is the pars tuberalis. '*Vorderlappen*' and '*Zwischenlappen*' are pars distalis respectively intermedia (× ca. 9; red. $^1/_1$).

creasing amount of detailed data it is still at present, as stated about 15 years ago by YOUNG, 'not easy to reduce this bewildering variety of effects to order'. There is little doubt that these effects as well as the number and exact composition of the hormones differ in accordance with the various taxonomic Vertebrate groups. On the whole, the pituitary activities of Mammals are perhaps best known. In these latter Vertebrates, the neurohypophysis stores and releases hormones with an antidiuretic, vasopressor and oxytocic effect. These substances seem to be produced in the neurosecretory and paraventricular hypothalamic nuclei, being transmitted by way of the neurites whose terminals are closely related to the pituicytes which may play a role in the release of the hormones by way of the blood stream. The oxytocic principle causes contraction of uterine musculature during coitus and at the time of delivery. It likewise causes contraction of the myo-

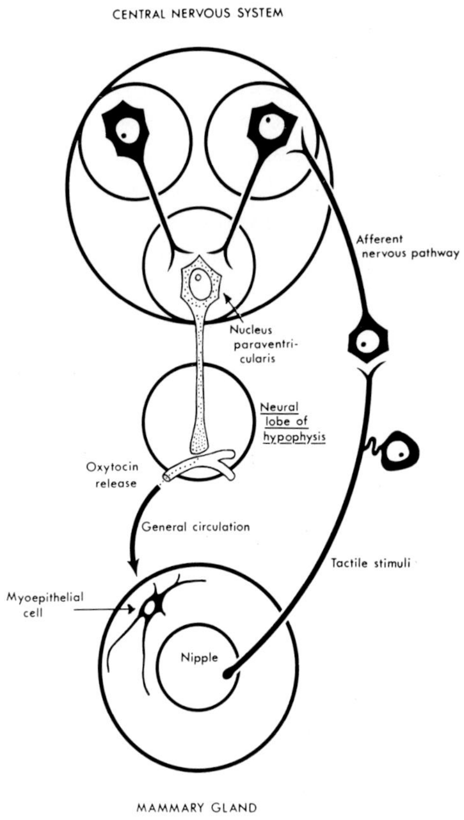

Figure 46. Diagram illustrating E. SCHARRER's concept of the 'milk-ejection reflex' (after E. SCHARRER, from BLOOM and FAWCETT, 1968).

epithelial cells in the mammary gland ('milk ejection reflex upon suckling in lactating animals', cf. Fig. 46). The vasopressor principle stimulates contraction of smooth musculature and small vessels.[73] Moreover, it conserves body water by promoting reabsorption in the distal convoluted tubules of the kidney. This effect appears functionally more important than the pressor effect. Thus, damage to hypothalamus or neurohypophysis causes the human clinical condition of *diabetes insipidus* with polydipsia which can also be produced in experimental animals and has also been recorded as a hereditary condition in a strain of Rats. The osmoregulatory function may be

[73] Thus, some authors (e.g. KROGH, 1926) suggest that the maintenance of capillary tonus is 'a normal function' of pituitary hormonal activity.

coupled with the activity of osmoreceptor cells in the supraoptic nucleus.

Vasopressin and oxytocin are cyclic polypeptides said to consist of eight different amino acids. The hormones seem to be formed in the hypothalamic neurosecretory nerve cell bodies and bound to an intra-axonal carrier protein (neurophysine, consisting of two hormone-binding proteins). The bound hormones move by neuronal (axonal) flow toward the terminals of the neurites and reach the capillaries respectively sinusoids of the neurohypophysis, being thus released into posterior lobe veins or into the hypophysio-portal circulation (cf. e.g. Zimmerman *et al.*, 1973). An effect of neurohypophysial hormones upon adenohypophysial activity is surmised. Thus, e.g. a concentration of a thyrotropin releasing hormone (TRH) has been recorded in the neurohypophysial median eminence, from which it might somehow, perhaps through the hypophysio-portal system, reach the adenohypophysis. TRH has also been demonstrated in various hypothalamic as well as in the so-called 'septal' (paraterminal) nuclei. This substance might play a double role as a 'releasing hormone' in the hypophysial complex, and as a 'transmitter substance' in various grisea. In addition to the supraopticohypophysial tract of 'magno-cellular secretory system', a tuberohypophysial tract has been established. It arises from the 'infundibular' or 'arcuate' nucleus of the hypothalamus and perhaps other hypothalamic grisea and presumably carries 'releasing hormones' as well as catecholamines into the zona externa of the infundibular stem, thus reaching the capillary system and the portal vessels. It is believed that through this '*parvicellular neurosecretory system*' a control of adenohypophysial functions is mediated.

Comparable neurohypophysial functions have been well established in Birds as well as, to some extent, in other Vertebrates, e.g. affecting the water balance of the Frog. In Birds, oxytocin is known also to exert a vasodepressor effect (cf. e.g. Pearson, 1972).

As regards the active antidiuretic vasopressor neurohypophysial principles, arginine vasopressin, lysine vasopressin, and arginine vasotocin have been identified. Their presumed distribution in the Vertebrate taxonomic series is indicated by Figure 47 A.

The limited scope of a neurobiological summary of diencephalic structure and function precludes a detailed discussion concerning histology as well as hormones of the *adenohypophysis*. It will merely be recalled that at least six different hormones with eleven or more types

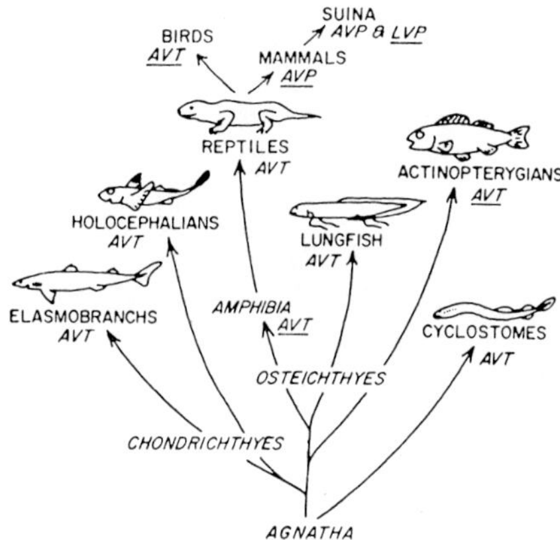

Figure 47A. Assumed distribution of antidiuretic-vasopressor neurohypophysial principles within the diverse Vertebrate classes (from SAWYER, 1967). AVP: arginine vasopressin; AVT: arginine vasotocin; LVP: lysine vasopressin. Underlining of the abbreviation indicates that the peptide has been chemically identified in one member of the group. This diagram, in addition, purports to indicate phylogenetic relationships (cf. e g. Fig. 6, p. 112, Fig. 7B, p. 148, and Fig. 8, p. 150 of vol. 1).

of physiological effects have been identified for the pars distalis. These include growth hormone (STH, somatotrophic hormone), lactogenic hormone (prolactin and luteotrophic: LTH), adrenocorticotrophic hormone (ACTH), gonadotrophic hormones (follicle stimulating: FSH; luteinizing: LH), and thyrotrophic hormone (TSH). Additional effects are diabetogenic and ketogenic (perhaps associated with growth hormone), glycotropic, parathyrotropic, and pancreatrophic. The significance of ACTH in stress and in the so-called general adaptation syndrome has been emphasized by SELYE (1950). The combined interactions of periphery, hypothalamus, neurohypophysis, adenohypophysis and adrenal cortex, in which glucocorticoids are released, is illustrated by the diagram of Figure 47B. Again, FARNER and OKSCHE (1962) assumed that, at least in Birds, the release of gonadotropin from the adenohypophysial pars distalis is controlled by a comparable effect of hypothalamic nuclei (n. supraopticus and perhaps n. paraventricularis).

The *cellular elements* of the pars distalis[73a] can be roughly distinguished

[73a] Common adenohypophysial tumors in Human pathology are adenomas. Endocrine disturbances associated with chromophobe tumors are those of 'hypopituitarism'. Acidophil-like adenomas cause 'hyperpituitarism' resulting in gigantism if occurring

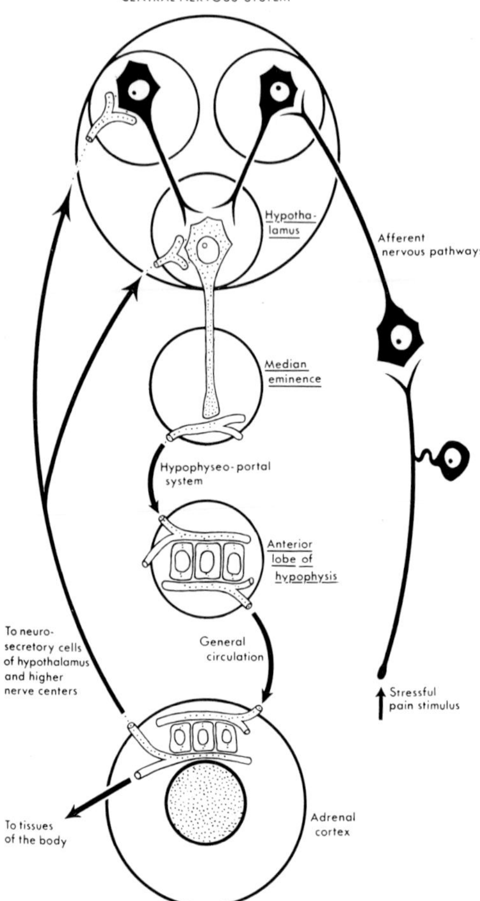

Figure 47B. Diagram illustrating E. SCHARRER's concept of some of the pathways involved in the control of glucocorticoid release by the adrenal cortex (after SCHARRER, from BLOOM and FAWCETT, 1968).

as chromophobes, acidophils, and basophils, but this classification seems no longer adequate. Nevertheless, it appears that STH, TSH, and prolactin produced by cells of the acidophil type, while basophil cells produce FSH, LH, and perhaps TSH. Nothing certain has been elucidated concerning the elements producing ACTH.

---

before puberty, and in acromegaly if developing at a later period, after the planes of growth in the long bones have closed. Basophil adenomas generally remain microscopic and do not seem to cause conspicuous symptoms.

The effects of the Amphibian pars distalis on the thyroid gland, on growth and metamorphosis as well as on gonadic and sexual activities are well established and a discussion of their details can be found in NOBLE's (1931, 1954) treatise.

The lobus intermedius of Mammals, often separated from the neurohypophysis by a cleft representing a residual lumen of *Rathke's pouch*, and frequently included in the so-called 'lobus posterior', is of variable relative size, constituting perhaps 19 per cent of the pituitary in some rodents, but only about 2 per cent in Man (BLOOM and FAWCETT, 1968). In the Whale, the Porpoise, and the Manatee, it appears to be entirely lacking. This, as mentioned above, is also the case in Birds.

The *lobus intermedius s. pars intermedia* seems to produce a melanocyte stimulating[74] hormone (MSH) with apparently several 'fractions'. It is believed that the pigmentation occurring in *Addison's disease* of man (involving the adrenal cortex) may be caused by the hypophysial release of excess ACTH and MSH, both of which stimulate the production of melanocytes.[75]

In Amphibians and at least some Reptilians and Teleosts, the effect of MSH is manifested by expansion of the melanocyte processes with their pigment, thereby causing darkening of the skin.[76] In Amphibians, moreover, the hormonal influence of the pars intermedia induces a contraction of the lipophores and possibly also of the guanophores. It appears certain that, at least in some species of Anurans, the pars intermedia plays the chief role of color-tone regulator, although additional factors, such as direct nervous control and illumination are here involved.

Experimental studies in the Frog seem to have established that nerve fibers originating in the neuraxis (hypothalamus) reach the Anuran pars intermedia. Two types of spontaneously firing neuronal units were recorded, of which one sort is inhibited by, and the other is indifferent to, increases in illumination. The receptor for the light-

---

[74] In contradistinction to those of many lower Vertebrates, the Mammalian integumental melanocytes do not 'actively' contract and expand in a manner producing 'rapid' color changes.

[75] A similar, but much less pronounced pigmentation effect presumably caused by a comparable increase of the cited hormones is occasionally observed in human pregnancy.

[76] Thus, hypophysectomized specimens of the Teleostean Minnow Fundulus are commonly lighter than normal ones. Direct nervous control by neurite terminals on integumental pigment cells is also found in isolated groups of Vertebrates such as some Teleosts and some Reptilians (cf. YOUNG, 1955).

inhibited units appears to be the pineal organ. Transection experiments indicated that the axons of the two types of units are separately grouped in the 'infundibular floor'. The long latency (30 sec) of the electrical response in the pars intermedia to external changes in illumination suggested that a humoral step, presumably outside the pituitary, is involved in the response (OSHIMA and GORBMAN, 1969).

The functional significance of the pars tuberalis, which is commonly well vascularized and traversed by components of the hypophysio-portal system remains insufficiently clarified.

In summarizing the overall relations between hypothalamus and hypophysis, SPATZ (1951) discussed four possibilities: (1) humoral-centrifugal path or neurosecretion (*Gomori-positive* fibers of hypothalamo-hypophysial channels). (2) Humoral-centripetal path (transport of substances from neurohypophysis to ventricular fluid).[77] (3) Centrifugal innervation (innervation of adenohypophysis by peripheral sympathetic fibers or by direct hypothalamic fibers). (4) Centripetal innervation (assumed chemoreception through *Gomori-negative* afferent fibers originating in hypothalamic grisea, e.g. nucleus arcuatus, and terminating with 'sensory endings' in the infundibular lobe).

At the time of this writing, it is customary to distinguish three levels of hormonal relationships pertaining to hypothalamus and pituitary, namely (1) hypothalamic factors, (2) pituitary hormones, and (3) peripheral hormones. The hypothalamic factors (1) can again be classified as (a) releasing, and (b) inhibiting. Among the releasing ones the following are said to be present: growth hormone releasing factor (GRF), prolactin releasing (PRF), melanocyte stimulating releasing (MRF), corticotropin releasing (CRF), follicle stimulating releasing (FSRF), luteinizing releasing (LRF), and thyrotropin releasing (TRF).

Among the inhibiting factors, the following are inferred: growth hormone inhibiting factor (GIF), prolactin inhibiting factor (PIF), and melanocyte inhibiting factor (MIF).

[77] A direct effect of neurohypophysial hormone upon vegetative centers in the vicinity of the (Mammalian) 3rd ventricle was demonstrated by SPIEGEL and SAITO (1924) who recorded upon intraventricular injection a vasodilator effect of small, intravenously ineffective amounts of pituitrin. This effect was subsequently obtained by other authors. It is of interest to note the opposite effects of intraventricular and of intravous injection of 'pituitrin'. Intraventricular injection causes a fall in blood pressure, intravenous injection a rise.

The assumed pituitary hormones (2) were tentatively enumerated above. As regards (3) important peripheral hormones of the system under consideration, there are adrenal steroids, gonadal steroids, and thyroxine.

The peripheral hormones, in turn, are assumed to exert a positive or negative 'hormonal feedback' either directly upon hypothalamus or upon pituitary. Since cerebral grisea are not only influenced by hormones, but also to some extent 'produce' hormones and, in particular, control endocrine secretion, endocrinologists, overemphasizing this aspect of neuraxial activity, are wont to consider the brain as 'an endocrine organ'. Some further aspects of relevant hypothalamic activities shall again be pointed out in section 9 of the present chapter.

The ultimate level of hormonal activity is, of course, the cellular one, whereby, at the 'target site', intracellular events are triggered by acting upon 'cell surface receptors', 'cytoplasmic receptor sites', and, finally, upon intranuclear events. Adenosine triphosphate (ATP) and cyclic adenosine monophosphate (AMP) are assumed to be significant substances participating in the hormonal effects at the 'target level'.

Summaries of the rapidly expanding studies of neurosection and its relationship to hormonal control are contained in the publication edited by KNOWLES and VOLLRATH (1974). It should here be added that, on the basis of recent data, certain neurosecretory neurons of the hypothalamus concerned with the control of anterior pituitary are perhaps 'bipolar', with one process secreting a hypothalamic hormone into the hypophysioportal vessels, and another process presumably secreting the same hormone in the anterior hypothalamus as a synaptic transmitter.

Another recent development which has initiated intense research activity was the detection of specific opiate receptors in neuronal membranes, particularly but not exclusively in the 'limbic system'. It was then found that these receptors interact with endogenous 'morphine-like substances' (opioid peptides, respectively enkephalins), designated as *endorphins*, and present in brain but especially in the pituitary, where they could be demonstrated, with different concentrations, in 'anterior', intermediate', and 'posterior' hypophysial subdivisions. Said peptides include at least three somewhat different 'endorphins', some of which could perhaps be neurotransmitter substances or 'neuromodulators'. The enkephalins are small peptides, related to a polypeptide called β-lipotropin, while some endorphins are of large molecular size.

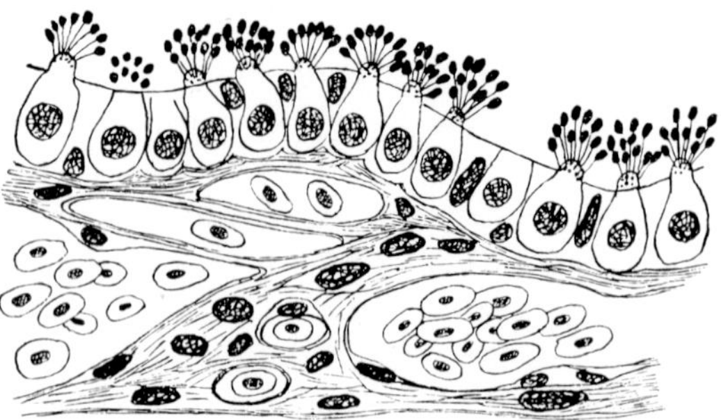

Figure 48. Part of sagittal section through the wall of the saccus vasculosus in a Teleost, showing 'Krönchenzellen' and blood vessels (simplified after DAMMERMAN, 1910, from K., 1927).

Nothing certain, up to the time of this writing, has been elucidated concerning the functional significance of these 'endorphins', which, nevertheless, according to diverse speculations, may affect the activities of the 'limbic system' (cf. e.g. GOLDSTEIN, 1976). The limbic system will be discussed in the pertinent sections of chapters XIII and XV.

The *saccus vasculosus* is a peculiar ventricular organ characteristic for most, but not all Gnathostome Fishes. It is not found in Cyclostomes, and its presence in Latimeria[78] as well as in Dipnoans is rather questionable.[79] The organ is formed by an usually highly folded modified ependymal wall closely connected with an extraventricular richly vascularized mesodermal investment containing *capillaries* and '*sinusoids*'. The irregular lumen of the saccus vasculosus seems to appear directly continuous with that of the third ventricle, although, in some forms with complex foldings, this continuity may be difficult to trace. Precipitates within the lumen might indeed result from the activities of 'secretory' saccus cells, but, on the whole do not essentially differ from the precipitates and coagulates frequently found in the ventricular system.

The main neuroectodermal elements of saccus vasculosus are the '*crown cells*' (*Krönchenzellen, cellules en couronne*) particularly investigated by DAMMERMAN (1910). These cells (Fig. 48) are each provided with approximately a score of cilia ending with small knob-like swellings

[78] Cf. p. 409, section 5, chapter VI, vol. 3/II.

[79] DORN (1955) brings a detailed critical review of the saccus vasculosus, including a list of Fishes in which this organ was or was not recorded. It seems to be present in all examined Selachians, but, although well developed in very many Osteichthyes, apparently missing in a number of these latter forms.

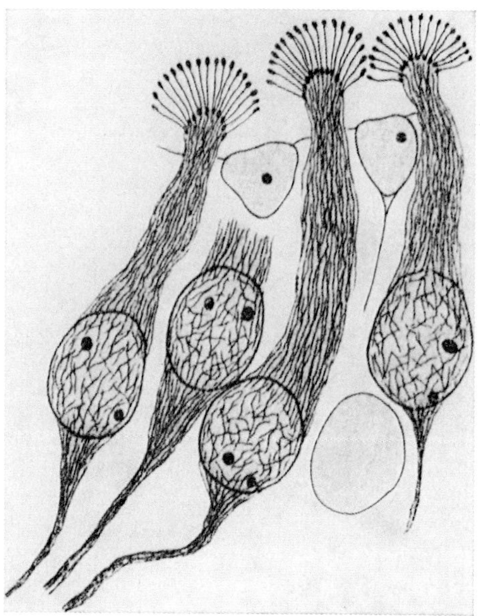

Figure 49 A. Detail of neurosensory '*Krönchenzellen*', showing basal origin of neurite, in the saccus vasculosus of an adult Trout (from DAMMERMAN, 1910).

and protruding into the ventricular fluid.[80] At the basal end of the crown cells, a neurite originates (Fig. 49 A). The structure of these elements thus strongly suggests that they represent receptory neuro-epithelial cells of neuronal type, in this respect comparable to the olfactory cells and to the retinal rods and cones.[81]

Some recent authors, however (e.g. ALTNER and ZIMMERMANN, 1972), have denied that the coronet cells are provided with axons (neurites), but findings such as depicted in Figure 49 A appear, nevertheless, rather convincing.

In addition, ALTNER and ZIMMERMANN have described elements apparently representing bipolar neurons differing from coronet cells, and designated as 'bipolar CSF-contact neurons' in at least some Osteichthyes and Chondrichthyes (Fig. 49 C). The apical processes of these cells extend into the cerebrospinal fluid (CSF). Their basal perikarya, besides possessing a neurite, seems to have presumably

[80] BARGMANN (1956) and SALAND *et al.* (1974) recorded some ultrastructural data concerning crown cells by means of electron microscopy.

[81] Cf. the classification of neuronal elements on p. 90, section 2, chapter V of volume 3/I.

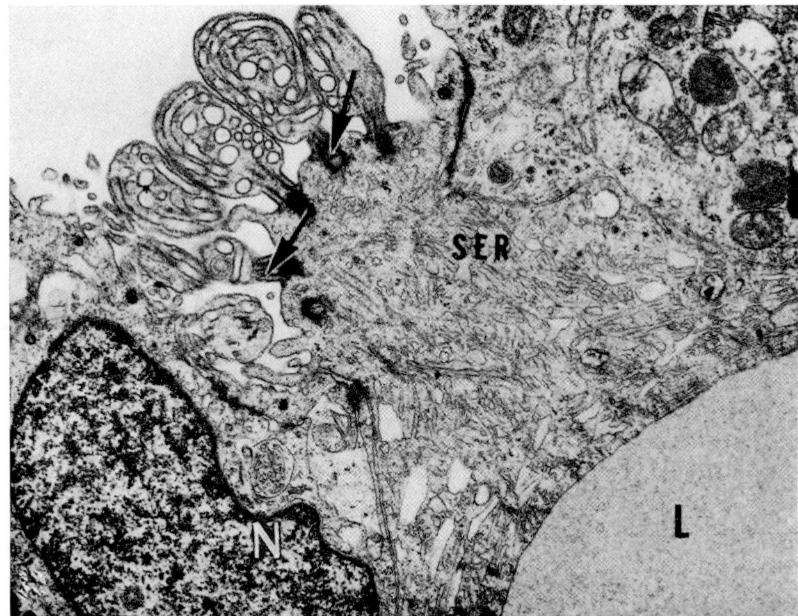

Figure 49 B. Portion of a 'crown cell' in the saccus vasculosus of an adult Reef Shark (Carcharhinus melanopterus) as seen by means of electronphotomicrography (from SALAND *et al.*, 1974). L: lipid droplet in cytoplasm of 'crown cell'; N: nucleus of 'supporting cell'; SER: so-called 'smooth endoplasmic reticulum'; arrows point to modified cilia protruding into the ventricular lumen from 'basal bodies'. ×13 600, red. $^3/_4$.

synaptic contacts with terminals of extraneous nerve fibers reaching the saccus vasculosus.

A third type of cells is supposed to subsume supporting cells, including perhaps further poorly identifiable elements, some of which may be 'secretory' (cf. Fig. 49 B).

The fibers originating from the saccus vasculosus (the so-called tractus sacco-thalamicus of DAMMERMAN)[82] are said to reach undefined thalamic and hypothalamic grisea (cf. Figs. 50 A, B). A 'ganglion sacci vasculosi' has been described in the posterior hypothalamus ventral to tuberculum posterius, i.e. in the 'mammillary' neighborhood. Fibers ending in the saccus vasculosus (so-called 'thalamo-saccular' tract) seem to originate in the postchiasmatic hypothalamus and appear essentially related to the saccular blood vessels. Some neuro-

[82] Some of these, before reaching the hypothalamus itself, may form a short 'extra-cerebral' '*nervus sacci vasculosi*'.

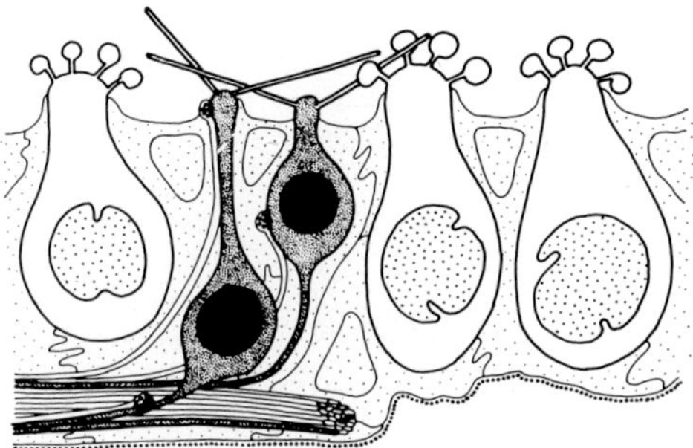

Figure 49C. Semidiagrammatic sketch of cell types and synaptic connections in the epithelium of the saccus vasculosus of the Teleost Perca, as interpreted by ALTNER and ZIMMERMANN (1972). CSF-contact neurons dark with black nuclei; coronet cells light with dotted nuclei; supporting cell dotted.

secretory fiber bundles reaching the hypophysis may have been mistaken for parts of the hypothalamo-saccular system. The topologic neighborhood in which the saccus vasculosus differentiates corresponds to a neural tube wall portion between recessus mammillaris and infundibular lobe of neurohypophysis. In some Vertebrates which do not possess a saccus, this neighborhood may contribute to the formation of the posterior wall of the infundibular stem. Little can be said concerning the functional significance of the saccus vasculosus, which DAMMERMAN interpreted as a sense organ for water pressure ('depth registration'). Be that as it may, the receptor endings are doubtless immersed in the ventricular fluid. Nevertheless, diffusion processes between capillaries and ventricular fluid do not seem impossible. Because of the lack of relevant data, neither *Dammerman's hypothesis*, nor those suggesting registration of diverse specific blood, liquor, or water conditions appear provable respectively disprovable. Yet, the organ is evidently related to some aspects of aquatic life.

ALTNER and ZIMMERMANN (1972) have reviewed, with reference to their own findings and particularly to the then more recent investigations, the problems concerning the significance of the saccus vasculosus with particular emphasis on the 'secretion hypothesis' and on that postulating a regulation of Na and K concentration in the cerebrospinal fluid. Yet, referring to DORN's (1955) report, ALTNER and ZIMMERMANN are compelled to admit that despite a great accumulation of subsequent detailed data 'we are still far from clarifying the organ's function'.

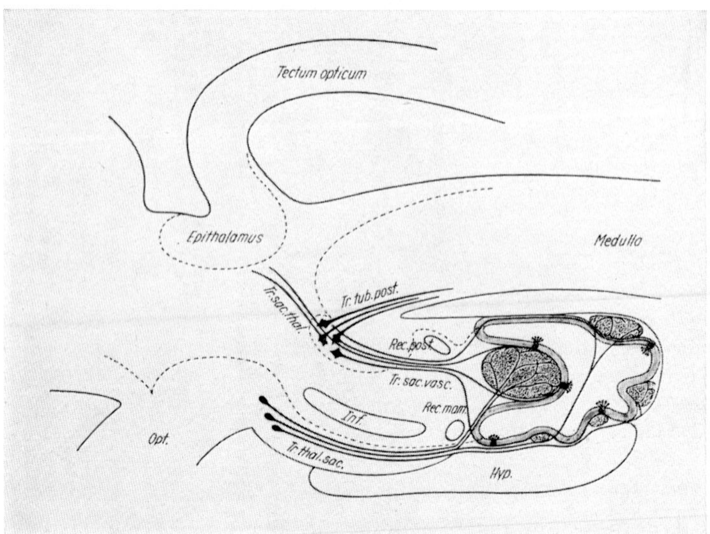

Figure 50 A. Diagram showing DAMMERMAN's original concept of saccus vasculosus innervation (from DAMMERMAN, 1910). Hyp: hypophysis; Inf.: 'infundibulum'; Opt.: optic nerve respectively chiasma; Tr. sac. vasc.: tractus sacci vasculosi; Tr. sac. thal.: tractus sacco-thalamicus; Tr. tub. post.: 'tractus tubero-posterius' *(sic)*; Tr. thal. sac.: tractus thalamo-saccularis; stippled structures are blood vessels.

Concerning the homologies of the saccus vasculosus neighborhood in other Vertebrates (Amphibia, Amniota), some comments were included on pp. 410–411, section 5, chapter VI of volume 3/II. This neighborhood is evidently included in the caudal wall of the so-called 'infundibular region' of posterior hypothalamus. Although a saccus vasculosus does not become differentiated in said Vertebrates, transitory vesicles may nevertheless occasionally be formed on the external 'infundibular surface' at some embryonic stages and in some of the higher Vertebrate forms, as recorded by CHIARUGI (1922). This could indeed be interpreted as an 'atavistic' tendency remanent in the 'genome'. CHIARUGI (1922) states: '*Sono inclinato ad ammettere che nei Mammiferi (Cavia) durante lo sviluppo siano transitoriamente roconoscibili trace del sacco vasculoso, le quale si troverebbo non soltante, come fu detto da alcuni, nel processo dell'infundibulo (e più precisamente nella parete posteriore di questo), ma anche nel contiguo stratto rostrale della parete posteriore del Diencefalo, que più tardi farà parte del tuber cinereum. Questo tratto rostrale non è identificabile, nè per posizione, nè per struttura col recesso sacculare del* MAZIÈRE; *la eminenza sacculare del* RETZIUS *non ha con esso*

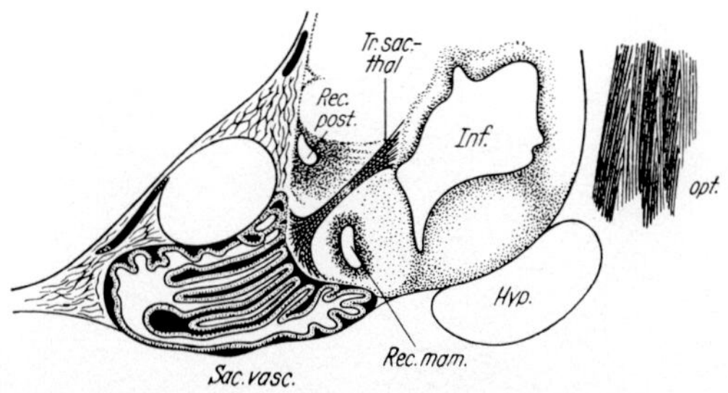

Figure 50 B. Paramedian sagittal section through the saccus vasculosus of an adult Trout (from DAMMERMAN, 1910). Rec. mam.: 'recessus mammillaris'; Rec. post.: 'recessus posterior'; Sac. vasc.: saccus vasculosus; other abbr. as in Fig. A.

alcun rapporto genetico. Ritengo accettabile la supposizione del PERNA, *che il bulbo dell'infundibulo quando esiste possa rappresentare permanentemente un residuo (peraltro molto parziale) dell'abbozzo del sacco vasculoso'.*

The dorsal wall of the Vertebrate diencephalon, a derivative of the embryonic roof plate, provides several characteristic midline structures, which display a number of variations in the diverse taxonomic groups. These structures are: (1) the caudal leaf of the *velum transversum*, whose ventral edge represents a significant dorsal landmark of the telencephalo-diencephalic boundary; (2) an *epithelial roof* derived from the parencephalic neuromere, and either developed as a thin dorsal sac in diverse Anamnia, or, more generally, with or without dorsal sac, providing a paired respectively bilaterally symmetric unpaired choroid plexus of the third ventricle; (3) *the epiphysial or pineal complex*, originating from a caudal part of the parencephalic neuromere. Moreover, the parencephalic roof rostrally to the epiphysial complex becomes bridged by the *habenular commissure*,[83] the synencephalic roof caudal to the pineal evagination being traversed by the *posterior commissure* which represents the dorsal portion of the diencephalo-mesencephalic boundary zone. This latter roof plate region includes the subcommissural organ, dealt with in section 5, chapter V of volume 3/I and also repeatedly pointed out in chapter XI of volume 4. In the present context, only the epiphysial complex of the roof plate will be considered.[83a]

[83] More rostrally, the diencephalic roof, as e.g. in some Reptiles, may be traversed by the fibers of the telencephalic '*commissura pallii posterior*' *seu* '*aberrans*'.

[83a] Ambiguous and inconclusive ontogenetic findings in some Anamnia, interpreted as suggesting a vanished dorsal nervus thalamicus or a rudimentary nervus mesencephalicus in ancestral Vertebrates, were pointed out in footnote 60, p. 845. chapter VII of volume 3/II.

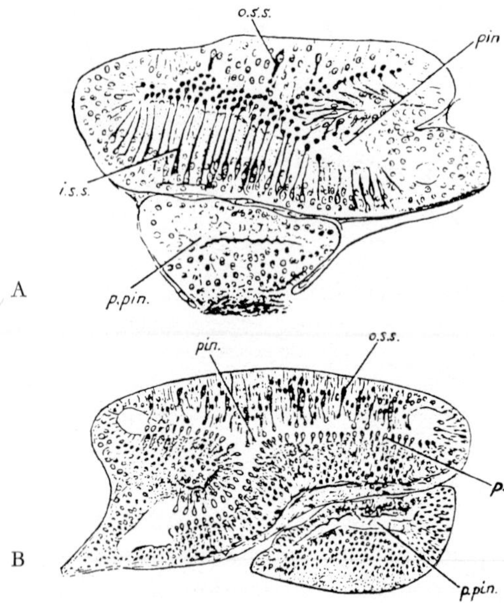

Figure 51 A, B. Pineal and parapineal organs of larval and adult Lampetra (Petromyzon) fluviatilis as seen in sagittal sections (after TRETJAKOFF, 1915, from YOUNG, 1955). A: larva; B: adult; i.s.s., o.s.s.: inner and outer neurosensory cells; pin.: pineal organ; p. pin.: parapineal organ.

The *epiphysial structures* are manifestations of the phylogenetic evolution of two parietal eyes which, in many Vertebrates, are transformed to become a preponderantly endocrine organ. Nevertheless, in some Vertebrates one or two fairly typical parietal eyes are still displayed (Figs. 51 A, B). Their image-forming capacity, however, appears entirely negligible despite the presence of a lens, and was most likely at best also quite mediocre even at their unknown but perhaps highest state of development in some ancestral forms. The still extant eyes of this type seem to be receptors merely registering presence, degree, and possibly direction of illumination (cf. above, p. 31, and footnote 20). Generally speaking, the retinae of the Vertebrate parietal eyes appear to be of the direct, i.e. adverse type. Again, in comparing said parietal eyes with the 'direct' and 'indirect' eyes of Arachnids (cf. Fig. 51 C), it may be recalled that Scorpions have only 'direct' eyes, while in Araneae (Spiders) both sorts of eye, namely with inverted and with adverse type of retina are commonly present.

It is, moreover of interest that, although generally at least faintly

I 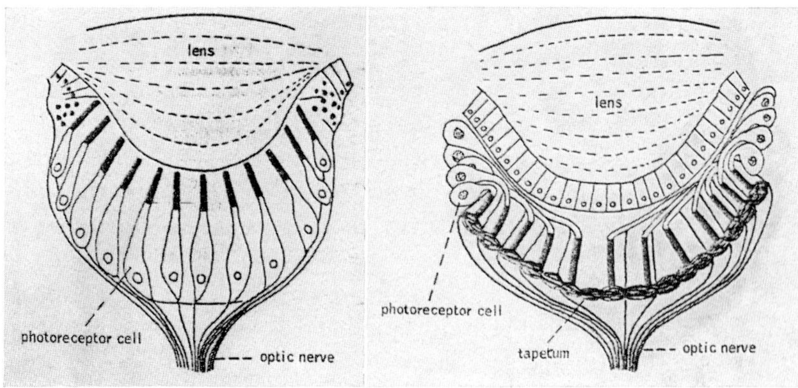 II

Figure 51 C. Direct and indirect lateral eyes of Arachnids. I. Direct eye of a Scorpion. II. Indirect eye of a Spider (after various authors, from BARNES, 1963).

suggested as an 'anlage' at ontogenetic stages, an epiphysis may be completely missing or almost entirely reduced in a number of very different Vertebrate forms, such e.g. as the Cyclostome Myxinoids, the Selachian Torpedo, the Sauropsidan Crocodilus, the Mammalians Dasypus (Edentate), Phocaena (Cetacean) and Elephant (Proboscidean).[84]

In *Cyclostome Petromyzonts*, two parietal eyes are present, a larger and somewhat better developed dorsal *pineal organ*, and a smaller ventral *parapineal organ* (Figs. 51 A, B, cf. also Fig. 183, p. 405, vol. 3/II). In contradistinction to the Vertebrate lateral eyes, the parietal eyes are not invaginated but remain vesicles, such that the distal wall is analogous to the lens, while the proximal wall, continuous with the stalk, is analogous to the retina.[85] In Petromyzonts, however, both walls contain neurosensory cells with neurite-like processes. These latter apparently end mostly within the organ, the central connections being made by ganglion cells likewise present in the organ.[86] In

[84] Cf. e.g. CREUTZFELDT (1912), and HALLER v. HALLERSTEIN (1934). Although CREUTZFELDT and also KRABBE (1921, Kgl. Danske Vidensk. Selskab Biol. Meddedelser, 3) have claimed the presence of a small epiphysis in Elephants, neither DEXLER (1907) nor JANSSEN and STEPHAN (1956) could find this structure in Indian respectively African Elephants (bibl. ref. cf. chapter XIV). A young Indian Elephant examined by myself likewise lacked an epiphysis. Cf., however, addendum at end of volume.

[85] This is particularly conspicuous in the Reptilian parietal eye (cf. Fig. 56). Topologically, the lens is homologous to the retina of the invaginated lateral eye, and the parietal eye retina to the lateral eye's layer of pigment epithelium.

[86] These cells could thus be evaluated as elements combining, in one neuron, the conductive function of both bipolar and retinal ganglion cells of the lateral eye retina.

addition, there are pigment cells and nondescript supporting cells. Some of these latter might, in fact, be secretory elements.

The pineal organ is connected with the larger right habenular ganglion, the smaller left ganglion being connected with the parapineal organ. According to KAPPERS (1947) and others, some parietal eye nerve fibers might also terminate farther caudad within the optic tectum. Parietal nerve fibers may also decussate in posterior or habenular commissures.

There is little doubt that the Petromyzont parietal eyes, although not uncommonly buried beneath a pigmented skin, are photosensitive organs. In the Ammocoetes larva there is a daily rhythm of color-change, the animals becoming dark in the day-time and pale at night. After removal of the parietal eyes said change no longer occurs (YOUNG, 1955). This effect on the color is presumably produced by the action of nerve-impulses from the parietal eyes somehow transmitted to the pituitary complex. YOUNG suggests that the control may be effected by way of the efferent habenular channel, components of which reach the hypothalamus. Again, when a bright spot of light is directed upon the parietal eye region of a stationary Ammocoetes larva, movement is usually initiated, but after a latency of 'many seconds'. Yet, these movements can also be elicited after the parietal and the lateral eyes have been removed. Thus, photosensitive structures must also be present either in the integument[87] or, as YOUNG (1955) suggests, in the wall of the diencephalon (or for that matter, in that of the rostral mesencephalon). The photosensitive function of the Petromyzont pineal complex, moreover, does not exclude concomitant secretory activities.

Although representing *prima facie* unpaired midline configurations, the upper (dorsal) pineal and the lower (ventral) parapineal eyes of Petromyzonts are evidently bilaterally paired with respect to their connections, the pineal eye being right-sided and the parapineal left-sided. It seems therefore not improbable that some ancestral Verte-

[87] YOUNG also points out that, in a Teleost (the minnow Phloxinus), which has a transparent patch on the head in the region of the pineal body, it is possible to train the fish to give appropriate responses upon changes of said spot's illumination. The pineal body is here not differentiated as an eye-like structure. Moreover, these responses can be obtained even after removal of the pineal and of the lateral eyes. While the photosensitive cells could indeed be located in the brain, cutaneous photosensitive elements cannot be excluded. The pathway for the response remains uncertain and could involve a 'humoral' effect secondarily transduced by peripheral nerve endings or by neuronal effects within the neuraxis.

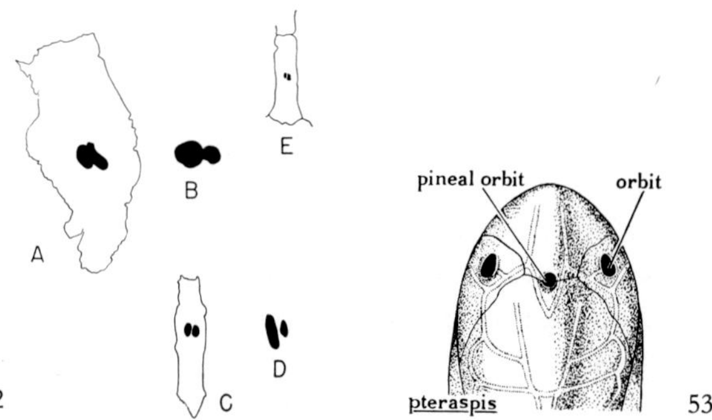

Figure 52. Paired pineal (parietal) foramina in the fossil fish orders Arthrodira (A–D) and (E) Stegoselachii (from T. EDINGER, 1956).

Figure 53. The pineal (parietal) foramen in the skull of the fossil Agnathous fish Pteraspis (after ROMER and others, from PRINCE, 1956).

brates had paired (i.e. right and left parietal eyes). This assumption is well supported by the presence of paired parietal (pineal) foramina in the fossil skulls of extinct Gnathostome fishes (cf. Fig. 52), namely of Placoderms from the Palaezoic era,[88] as particularly pointed out by T. EDINGER (1956). It is furthermore of interest that, in the Agnathous fish Pteraspis (Ordovician-Devonian) an apparently unpaired parietal orbit is present (Fig. 53). One could thus assume that the tendency toward median displacement with gradual elimination of one parietal eye, followed by still later elimination of the remaining one, manifested itself earlier in Agnatha than in Gnathostomes. It should also be recalled that the Agnathous Myxinoids have lost the entire epiphysial complex.[88a] Be that as it may, in the course of phylogenetic events, and among recent Vertebrates, some Amphibia and Reptilia have retained a distinctive single, parietal eye. This is the *pineal eye* in Amphibians, and the *parapineal eye* in Reptiles, as briefly discussed further below.

[88] The Gnathostome Arthrodira of the Silurian-Devonian (roughly 300 million years ago) were not Shark-like, and might represent a distinct class. The Stegoselachii (Devonian-Carboniferous), somewhat Skate-like in shape, might or might not be more closely related to the Elasmobranchs. Cf. also the comments on taxonomy and evolution (sections 2 and 6) in chapter II, volume 1 of this series.

[88a] A single epiphysial evagination has been rather convincingly demonstrated at ontogenetic stages by KUPFFER (1906; cf. e.g. Fig. 69D, p. 193, vol. 3/II). In adult Myxinoid specimens, all traces of an epiphysial complex generally seem to have disappeared, although EDINGER (1906) described, in one of his Myxine specimens, a clearly recognizable vesicular remnant, whose apparently rare occurrence may represent an individual variation.

At early ontogenetic stages in various gnathostome Vertebrates, including Mammals and Man, the roughly speaking single pineal diverticulum more or less distinctly displays a rostral and a caudal anlage. These closely adjacent primordia then commonly fuse into a single one. Nevertheless, and regardless of a phylogenetically remote bilaterally (side-by-side) paired arrangement, the rostral portion of the anlage may be interpreted as corresponding to the parapineal organ, and the caudal one to the pineal organ of Petromyzonts. Because of the tendency toward fusion, a conjoint epiphysial configuration cannot always be properly interpreted in terms of its presumptive bipartite, i.e. anterior parapineal, and posterior pineal anlage. Yet, in some Fishes, Amphibians, and Reptiles, it seems possible to identify relevant parapineal respectively pineal derivatives.

Roughly speaking, the epiphysial complex of Anamnia may be considered to consist of three parts, namely, (a) a proximal or 'cerebral' portion, (b) an intermediate 'stalk' generally containing the *parietal (pineal) nerve*, and (c) a terminal 'vesicle' (HALLER V. HALLERSTEIN, 1934). This subdivision, however, is frequently rather indistinct. If the stalk is open, secretion might evidently reach the ventricular fluid. If it is closed, all secretory activity might be of typical endocrine type, being presumably taken up by the blood stream. In all groups of Gnathostome Anamnia, neurosensory cells of 'retinal' type (photoreceptors) have been recorded by light as well as by electron microscopy.[89] In addition, as in Petromyzonts, nerve cells are present, from which the parietal (or pineal) nerve presumably originates. This nerve, reaching the habenular region or that of posterior commissure, and thus distributed to both sides of the neuraxis, may be a more or less distinct single or double (paired) bundle.

In *Plagiostomes* and *Ganoids*, the proximal portion displays frequently epithelial folds, the intermediate stalk may have a narrow lumen or can be a solid strand, the terminal vesicle may or may not be distinctly developed. If present, it may pierce the cartilaginous roof of the skull,

[89] According to J.A. KAPPERS (1971), the epiphysial photoreceptor elements 'are neurosensory, or primary sensory cells, not neurons'. In contradistinction to this statement, I prefer to classify all primary sensory elements, provided with a neurite-like cell process for output, as true neurons (olfactory cells, saccus vasculosus cells, retinal rods and cones, and epiphysial photoreceptors: cf. section 2, chapter 5, vol. 3/I, p. 90). This difference of opinion, however, concerns here merely an arbitrary question of semantics (cf. also pp. 72–73, vol. 3/II with respect to 'explanation' and 'description').

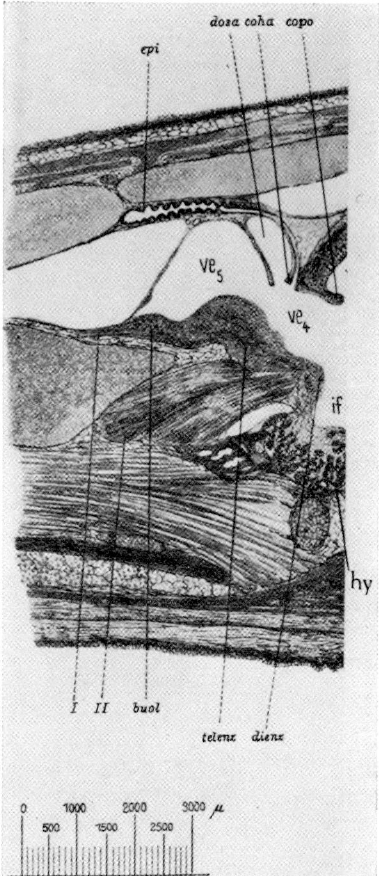

Figure 54 A. Midsagittal section through the epiphysial complex of the Teleost Esox lucius (from KRAUSE, 1932). I: olfactory fila; II: optic nerve; buol: bulbus olfactorius; coha: habenular commissure; copo: posterior commissure; dienz: diencephalon; dosa: saccus dorsalis; epi: epiphysial complex; if: 'infundibulum'; telenz: telencephalon; ve_4: third ventricle (diencephalic ventricle); ve_5: telencephalic ventricle (ventriculus communis).

or lie very closely to the inner cranial surface.[90] Neurosensory elements of 'retinal type' have been demonstrated in the vesicle (ALTNER, 1965). According to TILNEY and WARREN (1919), the Plagiostome epiphysial complex is of pineal origin.

The *Teleostean* epiphysial complex is frequently well developed as a flattened tubular structure with numerous epithelial folds (Figs. 54A, B), adjacent to the inner skull surface and terminating in the neigh-

[90] The Plagiostome habenular ganglia, which seem to be connected with the insufficiently clarified parietal nerve, are, as in Petromyzonts, *asymmetric*, but in an opposite manner, the left habenular griseum being larger than the right one. The parietal nerve in the pineal stalk can here perhaps be considered an unpaired channel, which bifurcates upon joining the central neuraxis.

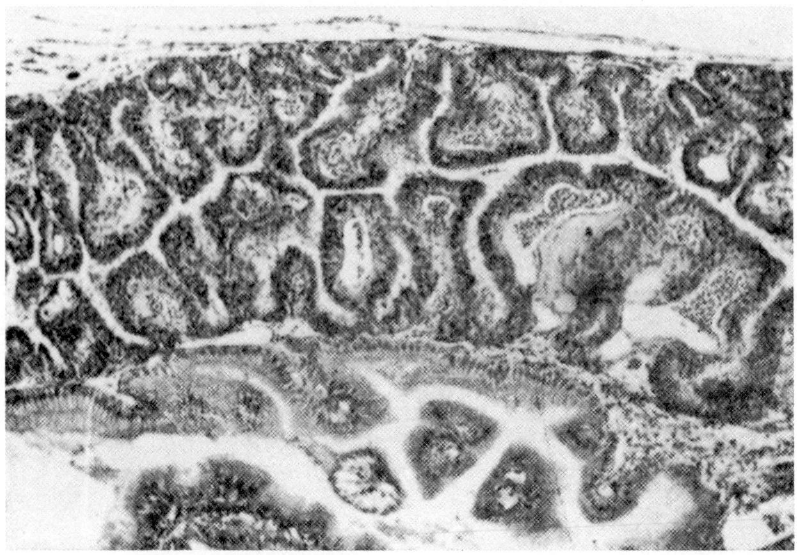

Figure 54B. Sagittal section through the epiphysis of Esox lucius (from VAN DE KAMER, 1956).

borhood corresponding to a foramen parietale. Being generally connected with the brain by a narrow stalk with a parietal nerve, it roughly corresponds to HALLER's 'terminal vesicle'. Photoreceptor and supporting cells, some of which may be secretory, are present. Since, during ontogenesis of various Teleosts, a smaller 'parapineal' vesicle makes its appearance and apparently subsequently degenerates, the adult Teleostean epiphysial complex may be of essentially pineal origin.

In *Dipnoans*, the poorly developed epiphysial complex is said to be of pineal origin. The *Amphibian* epiphysial complex is variously developed in the diverse taxonomic forms (TILNEY and WARREN, 1919; KLEINE, 1929; VIALLI, 1929; NOBLE, 1931, 1954). In some Urodeles it becomes a closed epithelial vesicle detached from the brain except for the fibers of a thin parietal nerve. In various aquatic Urodele and Anuran forms (Triturus, Pipa) the vesicle is lacking or rudimentary (NOBLE, 1931, 1954). In a number of Anurans, including the frog Rana esculenta, a parietal organ, apparently developing from the posterior subdivision of the epiphysial anlage, and thus presumably

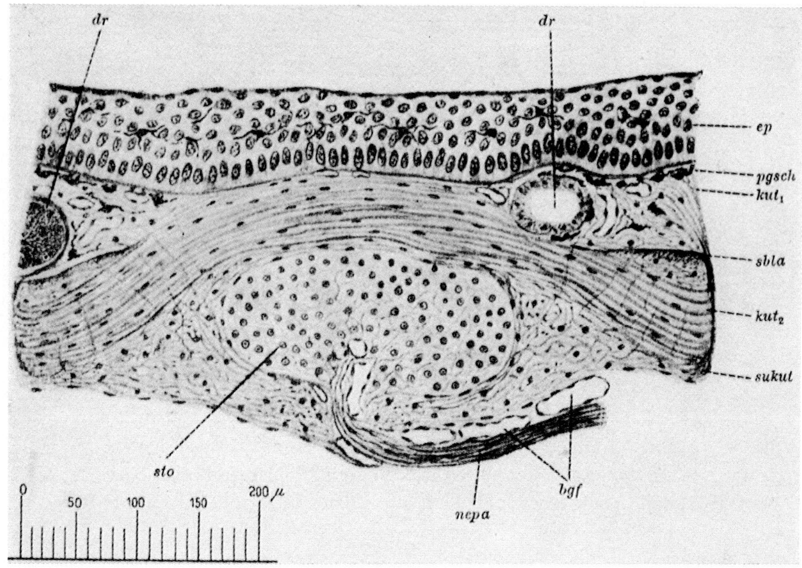

Figure 55. Midsagittal section through the pineal organ (frontal organ, Stirnorgan) of an adult Frog (from KRAUSE, 1923). bgf: blood vessels; dr: cutaneous glands; ep: epidermis; $kut_{1,2}$: loose and dense cutis; nepa: nervus parietalis; pgsch: pigment layer; sbla: 'cribriform lamella'; sto: 'frontal organ'.

representing the pineal organ, is located externally to the cranium, being covered by skin, which, in some instances, displays here a pigmentless, translucent spot. Although developing as a vesicular parietal eye vesicle, it frequently loses its lumen, but remains connected with the brain by a parietal nerve traversing the cranium (Fig. 55, cf. also Fig. 183B, p. 405, vol. 3/II). The Anuran parietal eye is also known as the so-called frontal organ *(Stirnorgan)*. Neurosensory cells of photoreceptor type have been identified. Whether the proximal portion of the epiphysial complex is a 'parapineal' (anterior) derivative, or whether the two original anlagen are completely fused in Amphibians, remains a moot question. TILNEY and WARREN assume for the entire complex a pineal origin.

The epiphysial complex of many *Reptiles* clearly displays an anterior parapineal and a posterior pineal anlage, both being closely adjacent, respectively representing subdivisions of a common neighborhood. In the Rhynchocephalian Sphenodon, and in various Lacertilia, the parapineal anlage develops into a conspicuous parietal eye, connected with the brain (habenular and posterior commissure region) through

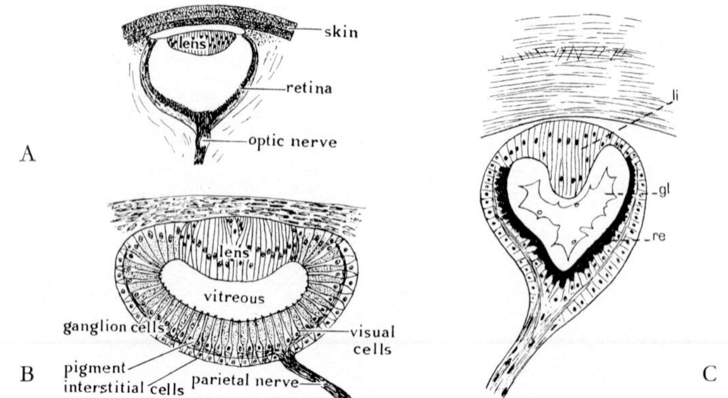

Figure 56 A, B. Semidiagrammatic drawings of the parietal eye (A) in Sphenodon, and (B) in Anguis fragilis (after Nowikoff, 1910, from Prince, 1956).

Figure 56 C. Simplified sketch of the parietal eye in Sphenodon (after Spencer, 1886, Dendy, 1911, and others, from K., 1927). gl: vitreous body; li: lens; re: retina.

a parietal nerve[91] (Figs. 56 A–C). A lens and a retina, both analogous but not homologous to those of lateral eye, are present, the retina containing neurosensory cells, ganglion cells and additional supporting or 'interstitial' cells. A parietal foramen occurs, and the parietal eye of Sphenodon is particularly well differentiated.[92] A small vesicular accessory parietal organ has occasionally been described in Lacertilians and is perhaps split off from either the parapineal or the pineal anlage, perhaps representing 'random' developmental processes related to 'phylogenetically labile' configurational events. The pineal structure is differentiated as a saccular epiphysis, whose lumen commonly communicates with the third ventricle through a hollow stalk (cf. Fig. 56 D). This pineal epiphysis, nevertheless, seems also 'still' to contain some neurosensory elements.

In Chelonia and Ophidia, on the other hand, only a saccular epiphysis is displayed, which may derive from the pineal or from the

[91] The variously developed (single, double, reduced, or apparently missing) nervus parietalis, is, of course, not a peripheral nerve, but, like the optic 'nerve', a tract of the neuraxis. The main parietal nerve is said to be connected with the right habenula, while a second one may be connected with the left epithalamic griseum. Kappers *et al.* (1936, with further bibliographic references to Dendy, Klinckowström, Novikoff and others) comment on the paired relations and point out that the epiphysial complex on either side may develop into a parietal eye.

[92] It is, however, externally visible only in young specimens (Young, 1955).

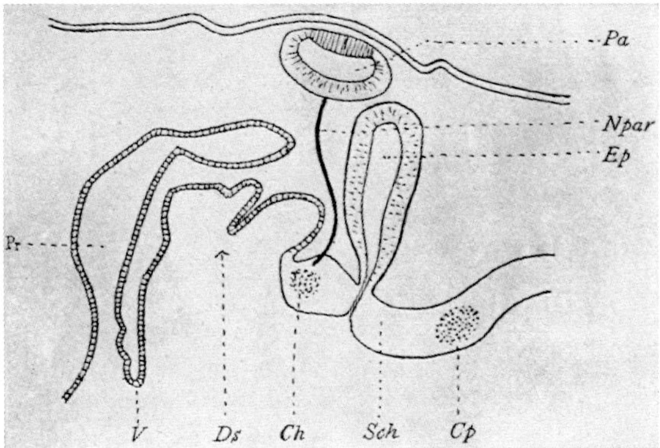

Figure 56 D. The pineal (epiphysial) complex in a 27 mm embryo of Anguis fragilis (after BÉRANECK, 1892, from TILNEY and WARREN, 1919). Ch: habenular commissure; Cp: posterior commissure; Ds: dorsal sac; Ep: pineal organ; Npar: 'nervus parapinealis' (nervus parietalis); Pa: parapineal organ; Pr: paraphysis; Sch: 'pars intercalaris posterior' (rostral synencephalic neighborhood); V: velum transversum.

conjoint (fused) epiphysial anlage. It is frequently nearly compact, particularly in Ophidia, but scattered lumina may be retained. It seems possible that some reduced neurosensory elements and nerve cells, besides 'interstitial' or 'secretory' neuroectodermal cells are present, although this is denied for Ophidia. In Crocodilia, as mentioned above, an epiphysial complex is entirely missing. Even an early embryonic anlage apparently fails to appear or is, at most, barely suggested, whose traces may then remain as KAPPERS (1947) states '*une tige épiphysaire atrophiée*'.

The *Avian* epiphysial complex, presumably arising from a conjoint (parapineal and pineal) anlage in ontogeny[93] is commonly a sacciform or club-shaped near solid or follicular[94] configuration, connected with the diencephalic roof through a thin stalk which may or may not be solid (Figs. 57 A, B). Numerous variations *qua* relative size and details of shape occur.

Most of the Avian epiphysial cellular elements are considered secretory, but there is some evidence that at least certain of these cells may represent modified or regressed neurosensory elements. Scattered nerve cells and an ill-defined parietal nerve may likewise be present.

[93] Rudiments suggesting a small parapineal vesicle have been occasionally reported.

[94] The follicles are presumably of secondary origin and not remnants of the original saccular lumen.

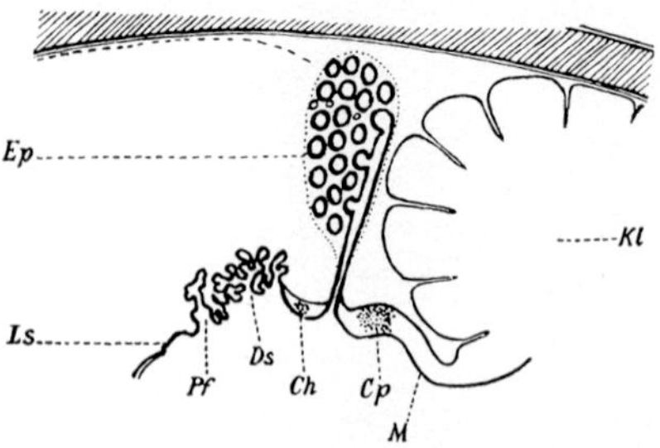

Figure 57A. Semidiagrammatic sketch of the epiphysial complex in Birds (after STUDNIČKA, 1905, from TILNEY and WARREN, 1919). Ch: habenular commissure; Cp: posterior commissure; Ds: dorsal sac; Ep: epiphysis; Kl: cerebellum; Ls: lamina terminalis; M: mesencephalon; Pf: paraphysis.

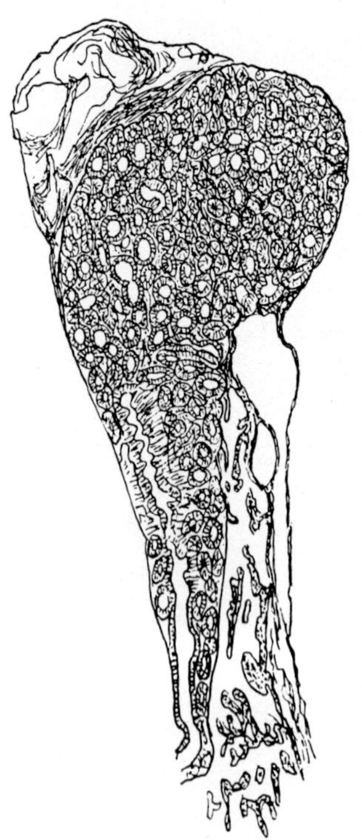

Figure 57B. Sketch of the pineal gland in a Bird, Coccothraustes vulgaris (after STUDNIČKA, from TILNEY and WARREN, 1919).

In addition, a peripheral innervation by noradrenergic sympathetic fibers, entering the pineal with the blood vessels, has been reported. Further data and comments concerning these not sufficiently clarified questions can be found in the publications by J. A. KAPPERS (1971), KAPPERS and SCHADÉ (1965), PEARSON (1972), and QUAY (1965).

The Avian epiphysial complex may exert a still poorly understood effect on the reproductive system, somehow related to illumination, but perhaps only representing a component in a complex multifactorial, redundant system of activities. Since pineals transplanted to the anterior chamber of the eye are reported as capable of restoring rhythmicity to pinealectomized birds in constant darkness (ZIMMERMAN and MENAKER, 1975); it is claimed by the cited authors that the Avian pineal does not appear to be neurally coupled to other components of the circadian system. It is suggested that the Avian pineal acts as a self-sustained oscillator, or as a coupling device within the system.

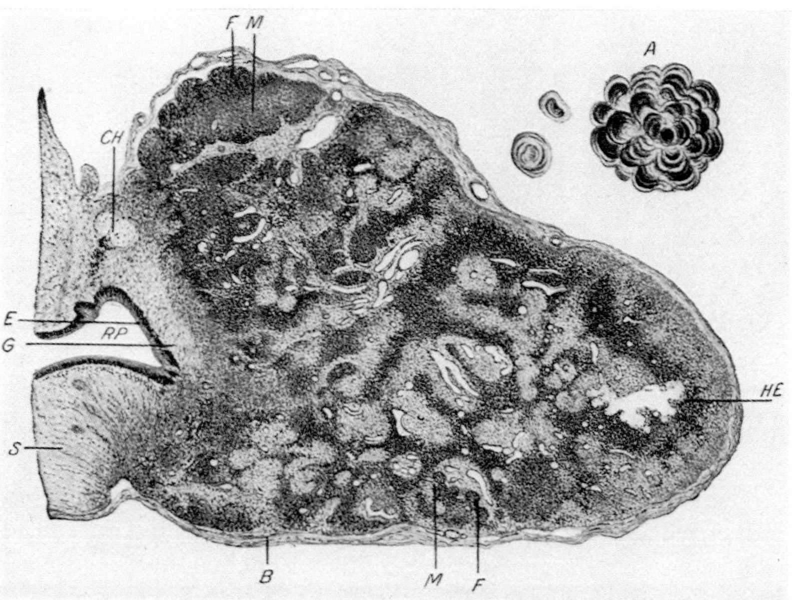

Figure 58 A. Midsagittal section through the pineal gland of a human newborn (after SCHAFFER, 1933; ×32, red. $^{9}/_{10}$). B: leptomeningeal sheath; CH: habenular commissure; E: ependyma of third ventricle roof plate; F: strands of cells with little cytoplasm; G: 'glial elements'; HE: caudal tip of pineal; M: strands of cells with substantial cytoplasm (parenchymal elements, pinealocytes); RP: recessus pinealis; S: portion of commissura posterior. The inset A depicts acervulus grains (×160; red. $^{9}/_{10}$) from the pineal body of a 69 years old woman.

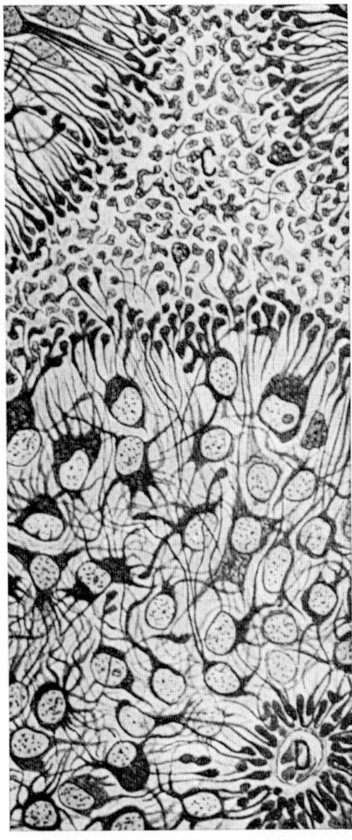

Figure 58 B. Details of pineal structure in a young boy as seen by means of *Hortega's technique* (after DEL RIO-HORTEGA, 1922, from BLOOM and FAWCETT, 1962). C: 'interlobular tissue'; D: vessel. The pinealocytes with their processes are specifically displayed by this technique.

The *Mammalian* epiphysial complex is a solid 'parenchymatous' body, frequently with a short pineal recess at its attachment (Fig. 58 A). In contradistinction to other Vertebrates, it is generally directed caudalward, being then located above the pretectal region. As previously mentioned, it may be absent or quite rudimentary in a few forms. The histologic structure of the Mammalian pineal body is generally characterized, particularly in Primates and Man, and especially at certain ontogenetic stages, by cell clusters ('lobules', islands) of neuroectodermal elements surrounded by an at least partly mesodermal (vascular) stroma forming trabeculae or septa (Fig. 58 C). The islands,

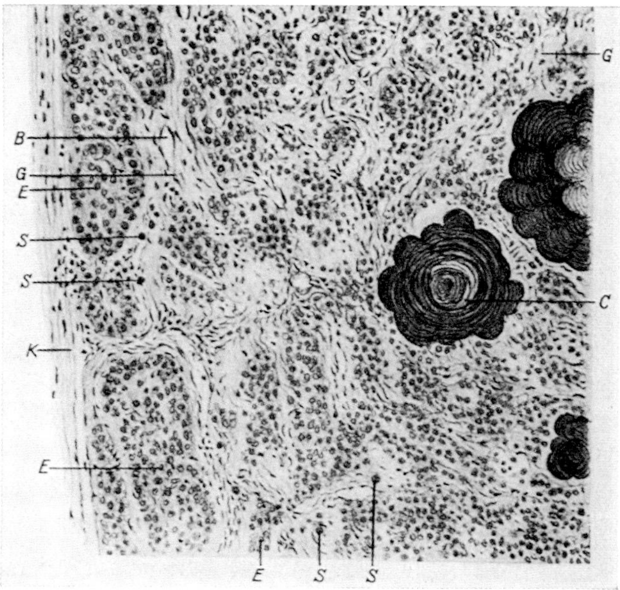

Figure 58 C. Portion of pineal gland in a 37 years old woman (from SCHAFFER, 1933; hemalum-eosin stain, ×120, red. $^1/_1$). B: blood vessel; C: concentrically layered acervulus; E: parenchymal lobules or islands; G: glial and mesodermal stroma, relatively poor in cells; K: leptomenigeal outer sheat; S: small globular acervulus grains. This Figure should be compared with the photomicrograph of a pinealoma shown by Fig. 411 A, p. 776 of volume 3/I.

in addition to other elements, contain the large *pineal cells (pinealocytes)*, well described by HORTEGA (1922) and presumably representing modified phylogenetic derivatives of the neurosensory cells in lower Vertebrates[95] (cf. e.g. J.A. KAPPERS, 1971). The fairly long processes end with 'terminal buds', frequently in the vicinity of blood vessels respectively within the vascular stroma (Fig. 58 B). In addition, glia cells, mostly of astrocytic type, but also small neuroectodermal elements of a somewhat dubious category, are present.

The stroma likewise seems to contain glia cells besides the mesodermal elements related to the vascular apparatus, whose smallest vessels ('capillaries') are generally rather wide, i.e. of 'sinusoid' type. The me-

[95] Remnants of the 'ciliary apparatus' characteristic for neurosensory elements of 'retinal' type, are, however, occasionally present (cf. J.A. KAPPERS, 1971). The Mammalian pineal cells are also designated as 'parenchymal cells' or 'chief cells'.

sodermal elements include fibrocytes respectively fibroblasts and often rather numerous cells of lymphocytic type, besides some plasma cells, mast cells, and melanocytes.[96]

The presence of nerve cells, originally interpreted as 'heterotopias' is now generally accepted for Primates and some other Mammals. These cells are said to be cholinergic, but their significance remains still unclarified. The nerve fibers within the pineal are considered to be mostly postganglionic sympathetic fibers arising in the upper cervical ganglion and entering the pineal with the vessels, in part also through paired so-called *nervi conarii* running through the tentorium cerebelli. Preganglionic cranial parasympathetic fibers, perhaps passing through the petrosal nerves, and related to the 'cholinergic' intrapineal nerve cells, may also reach the pineal gland.[97] In contradistinction to submammalian forms, a parietal (or pineal) nerve, i.e. a neuronal channel originating in the epiphysis and reaching diencephalo-mesencephalic grisea is denied by J. A. KAPPERS (1971) who, nevertheless admits that 'fibers of the habenula and posterior commissure may intermingle with the pineal parenchyma'. This problem, however, can be evaluated as still unsettled. The pineal cells are presumed to provide the hormonal secretion briefly dealt with further below.

It seems evident that the Mammalian epiphysis represents the effect of a long and complex sequence of phylogenetic events characterized by the transformation of two parietal eyes into an 'endocrine' organ, the 'retinal' neurosensory elements thereby becoming the pineal cells.[98] It is, moreover, not unlikely that these changes are associated with an ill-defined 'lability' of the nuclear genome of either pineal cells or also of other cellular elements involved in the ontogenetic development of

---

[96] A detailed account of pineal histology and cytology has been given by BARGMANN (1943).

[97] The autonomic (sympathetic and parasympathetic) innervation may mainly involve the 'pineal cells' but is perhaps also related to the pineal blood vessels.

[98] FRAUCHIGER and WILDI (1965), however, assume a '[illegible] origin' of the Mammalian pineal, comparable, *mutatis mutandis*, to that of the hypophysial complex. Mesodermal elements, providing the pineal cells, are supposed to migrate into the neuroectodermal pineal anlage. Although the Mammalian pineal does indeed display a substantial intermingling of mesodermal (meningovascular) and neuroectodermal tissue, I am not inclined to accept FFAUCHIGER's hypothesis. It should, nevertheless, be added that said hypothesis would not be inconsistent with the views of FRIEDMAN (1947) on the genesis of certain pinealomas, as quoted and briefly discussed on pp. 773–775, section 9, chapter V of volume 3/I.

the Mammalian pineal. At least two peculiarities, namely the formation of *acervuli ('brain sand')*, and the propensity toward teratoid or teratomatous growth phenomena might be evaluated as manifestations of said 'lability'.

Acervulus bodies (cf. Figs. 58 A, C) are laminated or convoluted concretions consisting mainly of phosphates and carbonates of calcium or magnesium. They are best known from observations of the human pineal, where small bodies of this type occur already in early childhood, being rather constant after the 16th year, and generally increasing in size with age. Although they may occasionally be absent in the adult, these calcifications are here commonly seen in roentgenograms and provide a suitable landmark for craniocerebral topography in connection with stereotaxic procedures.[99] In other Mammals (Hare, Sheep, Bovines, Donkey, Horse), acervuli were likewise recorded and seem also be present in young individuals, even at late fetal stages of the Calf. The question of their origin, by a 'hyaline degeneration' of pineal cells, other elements, or of 'intercellular substance', remains insufficiently clarified. Although evidently a 'regressive' or 'degenerative' process, the presence of acervuli can be regarded as 'normal' rather than 'pathological', any controversy concerning said evaluation being merely a question of arbitrary semantics.

Pineal neoplasms, discussed in section 9, chapter V, volume 3/I, are not very common, but, in addition to typical pinealomas, relatively frequently include teratoid tumors and true teratomas. In one instance of these latter, five human fetuses were recorded.

The various observations reporting conspicuous striated muscle fibers in apparently otherwise 'normal' Mammalian pineal glands (e.g. DIMITROVA, 1901; cf. also K., 1927, p. 236) can also be interpreted as recording teratoid manifestations. Because of the genome's potencies, such pineal muscle fibers do not necessarily indicate a provenance from mesodermal pineal elements but could also be produced by neuroectodermal ones.

Typical pinealomas may, but do not always, display effects on the gonads, such as either precocious or delayed puberty, gonadal enlarge-

---

[99] Pineal calcifications are not to be confused with likewise roentgenologically significant calcification in the habenular region and commissure of human adults (cf. K., 1954, pp. 25–26). Moreover, distortions in the position of pineal calcification have been shown to be an important diagnostic feature for the recognition of neoplasms, particularly of those in the cerebral hemispheres.

ment, or gonadal failure. Ectopic pinealomas in the hypothalamus have been recorded. It seems more probable that they result from disturbed histogenetic 'field effects' than from displaced embryonic 'pineal rests'.

In summarizing the presumptive phylogenetic evolution of the Vertebrate pineal complex, it can be safely stated that a pair of dorsal (parietal) eyes has been 'progressively' transformed into an 'endocrine', 'neurohumoral' organ.[100] Nothing certain can be said about these eyes in the ancestral Vertebrate forms. It seems likely, however, that the parietal eyes, perhaps originally arranged side by side (left-right, cf. Fig. 52), mediated information concerning illumination rather than spatial configuration (visual shapes, pictures). Thus, the parietal eye of some Reptiles has been interpreted as 'a dosimeter of solar energy exposure' (cf. e.g. EAKIN, 1974).

The 'lability' presumably obtaining in phylogenetic evolution and pointed out above on p. 130 is evidenced by the considerable diversity displayed *qua* differentiaton of the epiphysial structures in the Vertebrate series. Thus, the substantial differences between the tendencies toward pineal differentiation in Amphibians, and toward parapineal differentiation in Reptiles, are a conspicuous instance of said diversity.

In all *recent* Vertebrates possessing one or two conspicuous parietal eyes, their photosensory elements display signs of a secretory function apparently combining photosensory and additional secretory effects. These latter were perhaps also independently assumed by the supporting or 'auxiliary' neuroectodermal elements.

J. A. KAPPERS (1965, 1971) and KAPPERS *et al.* (1974), who have reviewed in great detail the innervation as well as the presumptive evolution of the Vertebrate pineal complex, believe that the pineal photosensory cell gradually lost its photoreceptive capacity and became a preponderantly secretory element which, however, 'possibly remains directly photosensitive or becomes indirectly photosensitive by a circuitous route, via the retinas and a complicated neural pathway'. In Sauropsida, the 'chief cells' in one pineal epithelium 'may show different gradations of the process of transformation, but, in principle, there is only one single cell type. A clear distinction between two basically

---

[100] J. A. KAPPERS (1971) objects to the term 'neurohumoral output', because the pinealocytes 'are not true neurons'. I prefer, however, to classify retinal photoreceptors as neurons (cf. fn. 89) and the pinealocytes may, accordingly, be considered 'modified neurons'.

different types of pineal chief cells cannot be made' (J.A. KAPPERS, 1971).[101]

The cited author, in addition, stresses a gradual loss of pineal neuronal output ('pinealo-fugal sensory innervation', i.e. discharge through the parietal nerves),[102] combined with an increasing development of extrinsic pineal innervation by way of the peripheral vegetative nervous system. This type of innervation, 'already distinct in some fishes and amphibians and very evident in Sauropsida', is believed to be the only one present in the Mammalian pineal.

Three stages are distinguished (a) invasion of the organ by presumably sympathetic postganglionic fibers along the perivascular spaces and remaining therein; (b) the sympathetic fibers reach the pineal elements *sensu strictiori;* (c) to these fibers are added fibers running independently of vessels through the nervi conarii; moreover, to the postganglionic sympathetic fibers may be added presumably parasympathetic preganglionic fibers connecting with intramural postganglionic neurons within the pineal. In Fishes, only stage (a) is supposed to obtain, in Amphibians and Sauropsida both stages (a) and (b) are reached, while stages (a), (b) and (c) are said to occur in the Mammalian pineal. J.A. KAPPERS' concept of the neurohumoral or 'neuroendocrine' pineal transducing system of stage (c), as compared with that of the neurohypophysis, is illustrated by Figure 59.

As regards the *pineal endocrine output* 'transduced' by the pineal elements, *melatonin* has been identified as the relevant active substance (or group of substances) in Birds and Mammals, being apparently more abundant in the former. Structurally, the melatonins are 5-methoxy N-acetyltryptamine, and the synthetizing enzyme is hydroxy-indole-O-methyl transferase (HIOMT). Again, a precursor of melatonin is *serotonin* (5-hydroxytryptamine, 5-HT), which is also known to occur in the pineal complex of at least a variety of lower Vertebrates (Anamnia and Reptilia). In Rats, a 24-hour rhythm seems to be displayed by the turnover of norepinephrine in sympathetic nerves innervating the

---

[101] Although I agree with many of J.A. KAPPERS' conclusions, I am inclined to believe that, even if only a single type of 'pineal chief cells' should obtain, these elements might have arisen from a 'convergent' evolution of both photosensory and of *ab initio* secreting 'accessory' elements.

[102] In various Vertebrates, e.g. in Frogs, fibers of the parietal nerves may reach not only the pretectal region and the tectum opticum, but also basal mesencephalic grisea (cf. OKSCHE, 1971).

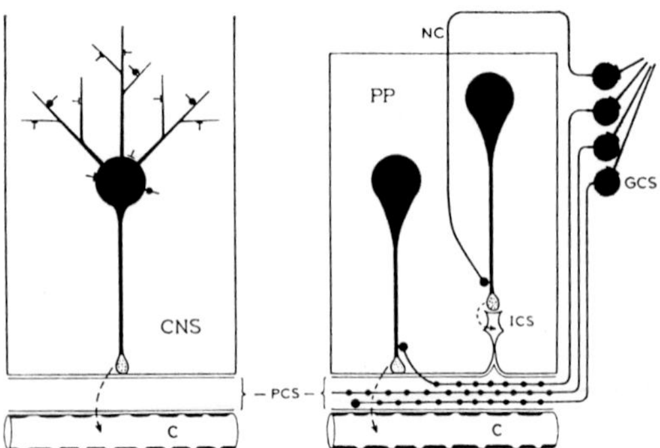

Figure 59. Diagram illustrating J. A. KAPPERS' concept of the pineal neurosecretory arrangement (right), compared with that of the neurohypophysis (left), and depicting the so-called neuro-haemal respectively parenchymohaemal contact areas (from J. A. KAPPERS, 1971). C: capillary; CNS: central nervous system; GCS: superior cervical ganglion of sympathetic trunk; ICS: intercellular space; NC: nervus conarii; PCS: pericapillary space; PP: pineal parenchyma. 'Endings of either cholinergic or aminergic nerve fibers associated with terminals of neurosecretory fibres are not drawn.'

pineal gland. This rhythm is said to persist in blinded animals but is suppressed in normal ones by light. Again, said rhythm is supposed to generate the rhythmus in pineal indolamines and N-acetyltransferase (BROWNSTEIN and AXELROD, 1974). The interrelationship of these substances have been investigated by various authors, particularly AXELROD *et al.* (1964).

Despite the multitudinous data obtained by numerous '*pinealologists*' (KAPPERS and SCHADÉ, 1965; WURTMAN *et al.*, 1968; WOLSTENHOLME and KNIGHT, 1971) the various activities of the Vertebrate pineal complex[103] remain poorly elucidated *qua* well-defined details.

It can be assumed that in Vertebrates with parietal eyes (some Anamnia and Reptilia) the pineal complex functions somewhat differently from that in Vertebrates without such eye,[104] but still with well-defined parietal nerve. Again, in Mammals, whose parietal nerve or equivalent

[103] *Pinealologists* (e.g. WURTMAN *et al.*, 1968) even claim that the pineal of Mammals 'is not part of the brain' since its innervation is supposed to be exclusively extrinsic (peripheral). Be that as it may, even the Mammalian pineal, like the neurohypophysis and the retina, are differentiations of the central neuraxis, and thus in my opinion very definitely represent subdivisions of the brain, namely of the diencephalon. As regards the pineal's biochemistry, cf. also the publications by QUAY (1965, 1970, 1972, 1974).

[104] The evidence for pineal 'control' of motility and adaptive pigmentation in 'lower Vertebrates' was discussed above in connection with the Petromyzont pineal complex.

is lacking or perhaps of negligible degree of differentiation, the pineal mechanisms may display further modifications. MOORE *et al.* (1968) have elaborated on a possible postulated pathway for the visual control of the Mammalian pineal gland by way of the basal optic tract and its connections with hypothalamus or reticular formation or both. These grisea, in turn, might control the peripheral (vegetative, i.e. sympathetic and perhaps parasympathetic) innervation of the pineal.

As regards *Birds* and *Mammals*, it seems certain that the pineal has an effect on the gonads, which, depending on poorly elucidated additional factors, may be inhibitory *qua* gonad growth and weight.

Likewise, epiphysio-hypophysial interactions are suggested, including interference with pituitary ACTH secretion, i.e., at least under some circumstances, an epiphysio-hypophysial antagonism. By way of the hypophysis, an influence on the thyroid is also assumed.

Other vaguely defined epiphysial activities may concern cold adaptation and also particularly the control of biological or circadian rhythms. Thus, the pineal is supposed to act as a 'biologic clock' or as a 'self-sustaining oscillator' driving, respectively interacting with, other 'oscillator'-systems.

AXELROD (1974), who has particularly investigated the biochemistry of the Mammalian pineal complex, summarizes his opinion as follows. 'There are circadian rhythms in serotonin, serotonin N-acetyltransferase, N-acetylserotonin and melatonin in the pineal which persist in continuous darkness and are abruptly abolished by exposure to light. These rhythms are generated by diurnal changes in the release of the neurotransmitter, noradrenaline, from sympathetic nerve terminals innervating the pineal. An increased discharge of noradrenaline at night stimulates the β-adrenergic receptor, which causes increased synthesis of serotonin N-acetyltransferase molecules inside the cell of mediation of an adenylate cyclase system. As the activity of N-acetyltransferase rises during the night, the concentration of its substrate, serotonin, falls, and that of the product, N-acetylserotonin, rises. Increased synthesis of the pineal hormone melatonin then follows as a result of O-methylation of N-acetylserotonin by hydroxyindole O-methyltransferase. The responsiveness of the pineal β-adrenergic receptor and the consequent synthesis of N-acetyltransferase change; the receptor becomes supersensitive after decreased exposure to the catecholamines noradrenaline and isoproterenol and subsensitive after increased exposure to the catecholamines. The circadian rhythm in pineal amines appears to arise from a biologic clock present in or near the suprachiasmatic nucleus in the hypothalamus. This clock in turn is modulated by inhibition by environmental light.'

Be that as it may, AXELROD discounts here the still somewhat uncertain possibility that the dubious nerve cells, recorded in the pineal by various authors, might be parasympathetic preganglionics.

Before concluding the comments on the epiphysial system, it is perhaps not without interest to quote what '*pinealologists*' (WURTMAN *et al.*, 1968) considered to be a description of the pineal 'written ten years ago' (i.e., about 1958).

'The pineal body is a part of the epithalamus; it is connected to this brain region in all species by nerve tracts. The pineal has evolved from a primitive photoreceptor, or 'third eye' which is common in extinct species. However, in the modern mammal it has lost all functional relationship to light, and persists only as a vestige which mysteriously calcifies at the time of puberty. Pineal tumors are frequently associated with precocious sexual development in young boys. It had once been believed that the human pineal was a gland, and secreted a hormone which inhibited the gonads. Under this formulation, precocious puberty developed because the tumors destroyed the pineal's ability to secrete this hormone. However, experiments designed to test the glandular function of the mammalian pineal have yielded equivocal data. Hence pineal tumors probably produce their endocrine sequelae solely as a result of the pressure that they exert on other brain areas.'

According to the cited gentlemen, 'most of the above conclusions have been shown to be erroneous'. Now, in my summary on the human diencephalon (K., 1954), not particularly concerned with the pineal gland, the following statements concerning said structure can be found on p. 39.

'The epiphysis or pineal body is frequently described as a component of the epithalamus but should be regarded as a distinct diencephalic structure since it arises as an evagination of the roof plate and not of the dorsal edge of the alar plate from which the epithalamus derives. The anlage of the epiphysis develops between the habenular commissure and the posterior commissure. In the adult, a variable but shallow pineal recess may remain between these two commissures, protruding into the pineal stalk.

Although some dissenting opinions have been expressed there can be little doubt that the pineal body represents the homologon of the rudimentary parietal eyes found in lampreys, some amphibia, and some reptiles. In Petromyzon, two such rudimentary structures, the pineal organ and the parapineal organ, are present. In some anura, there is an epiphysis and a pineal organ, in some reptiles a parapineal organ is found in addition to the epiphysis. Even in man, during the ontogenetic development of the epiphysis, there is a short transitory stage indicating that two separate primordia may have secondarily fused. It should be added that a pineal body is missing or exceedingly small in some mammals (edentates, whales, elephant), and in the alligator.

The human pineal body contains the large pineal cells described in detail by Hortega (1922). These specific neuroectodermal elements are clustered in lobules separated by connective tissue trabeculae and septa. Glia cells, small mononuclear cells interpreted as lymphocytes, and mast cells are also present. The nerve fibers connecting the pineal organ of lower vertebrates with the habenular nuclei have not been confirmed in man, but a rich network of vegetative (presumably sympathetic) fibers enters the human epiphysis with the blood vessels. Although the structure of the pineal body suggests that it might be more than a vestigial organ, its assumed endocrine function has not been clearly established. It contains the calcified concretions known as acervulus (brain sand) and may be the seat of neoplasms, including teratomas. Occasional nerve cells can be interpreted as heterotopias.'

Looking backward, there are at present only two modifications required for this summary statement, namely (1) the assumed endocrine function of the pineal, although far from being sufficiently elucidated, has been conclusively demonstrated; (2) the scattered nerve cells, if actually present,[105] can be interpreted as (perhaps parasympathetic) postganglionics.

Despite its interesting presumptive phylogenetic evolution from a pair of parietal eyes to its differentiation as a neuroendocrine paraventricular 'gland', whose complicated functions pose many intricate problems, the pineal complex can be evaluated as an 'organ' of rather minor importance.[106]

The various extirpation experiments in 'lower' Vertebrates yielded more or less nondescript results, defying a clear-cut formulation of their effect. In Birds and Mammals, more 'definite' 'modifications of gonad growth and function' were observed upon extirpation, but the

---

[105] Quay (1972) denies the occurrence of nerve cells in the pineal of the Primate Orangutan, but reports 'extensive penetration by loops of myelinated fibers from the habenular and posterior commissures'.

[106] *Mutatis mutandis*, this relative unimportance can be compared with that of the cerebellum, discussed on p. 660, section 1, chapter X of volume 4.

results of this procedure displayed rather variable sorts of changes, evidently depending on a diversity of insufficiently recognized parameters (cf. e.g. WURTMAN *et al.*, 1968). Much the same could be said about the effects of administered 'pineal extracts', of melatonin, and of other 'pure pineal compounds'.

Finally, the lack of a pineal complex or its negligible development in diverse, widely differing forms, are another case in point.[107]

One could thus evaluate the pineal complex of recent Vertebrates as providing an accessory and quite dispensable mechanism within the highly redundant 'very large system' of functions and adjustments performed by the central neuraxis. This mechanism, respectively its failure, caused by experimental elimination of the pineal complex or by tampering with its function through administration of diverse substances, might therefore have numerous different effects based on interplay with the correlated functional systems, and depending upon prevailing relevance of an undefined number of parameters (*Zustandsbedingungen*).

2. Cyclostomes

In its comparison with the diencephalon of Gnathostome Anamnia, that of the Agnathous Cyclostomes has presented several difficulties for morphologic interpretation, which were discussed in section 5, chapter VI of volume 3/II. These difficulties, related to the pattern distortion displayed by Petromyzonts, become still more accentuated by further distortion and by the exceedingly blurred configuration obtaining in Myxinoids.

The overall configuration of the *Petromyzont* diencephalon, as interpreted on the basis of our investigations (K., 1929, 1956; SAITO, 1930) is indicated by Figure 60 A. It can be seen that, because of the peculiar dorsalward rotation of the diencephalic longitudinal axis,[108] the preoptic recess becomes located dorsally to the posterior ('mammillo-infun-

---

[107] This is again paralleled (cf. above fn. 106) by the reduction or the poor, respectively indifferent, development of the cerebellum in various Vertebrates. In connection with his hypothesis concerning the pineal complex, FRAUCHIGER (FRAUCHIGER and WILDI, 1965) expressed his scepticism about the actual lack of a pineal in certain Vertebrate forms (cf. the discussion, loc.cit., on p. 664). This lack of a pineal, however, appears well documented; I can confirm it myself in Torpedo, Crocodiles, and Elephant. FRAUCHIGER's objections as well as overall theory remain rather unconvincing.

[108] Cf. volume 3/II p. 412 and the discussion to the paper by ADAM (1956).

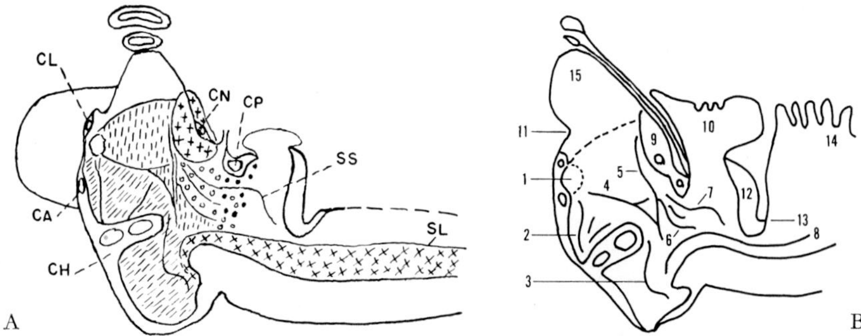

Figure 60A. Semidiagrammatic reconstruction of longitudinal zones in the diencephalon of adult Petromyzon or Entosphenus (from K., 1956). +: epithalamus; circles: thalamus dorsalis; black dots: pretectal component neighborhoods of dorsal thalamus; vertical hatching: thalamus ventralis; oblique hatching: hypothalamus; ×: basal plate of deuterencephalon (including tegmental cell cord); CA: commissura anterior; CH: chiasmatic ridge; CL: commissura pallii (s. dorsalis); CN: commissura habenulae; CP: commissura posterior; SL: sulcus limitans; SS: sulcus lateralis mesencephali (internus) continuous with sulcus synencephalicus; the vertical sulcus indicated by double line, and extending from epithalamus into thalamus ventralis, is the *'pseudosulcus' of Saito;* the other sulci ('boundary sulci', hypothalamic sulci, and accessory ones) are easily identifiable by their respective position and are unlabeled in order not to overload the diagram (cf. Figs. 60B and C).

Figure 60B. Semidiagrammatic sketch showing configuration of diecephalic ventricular sulci in the adult Petromyzont diencephalon according to the author's interpretation, combined from original observations and data recorded by Saito and *Graf* Haller, for comparison with Figures 60A and C. 1: interventricular foramen; 2: system of sulcus intraencephalicus anterior grooves; 3: sulcus lateralis hypothalami posterioris ('s. lat. infundibuli'); 4: sulcus diencephalicus ventralis; 5: *'pseudosulcus' of Saito;* 6: sulcus diencephalicus medius; 7: combined sulcus synencephalicus and sulcus lateralis mesencephali (internus); the unlabeled sulcus between 6 and 7 is an accessory sulcus; 8: sulcus limitans; 9: epithalamus (ganglion habenulae) its rostroventral boundary being sulcus diencephalicus dorsalis; 10: mesencephalic ventricle with lamina epithelialis and choroid plexus; 11: remnant of velum transversum; 12: caudal portion of mesencephalon; 13: cerebellum; 14: choroid plexus of 4th ventricle; 14: saccus dorsalis.

dibular') portion of the hypothalamus, the two major hypothalamic subdivisions being separated from each other by the nearly 'horizontally' oriented chiasmatic ridge with its commissural system. This rotation also involves a peculiar rostral expansion of the ventral thalamic zone. Thus, at rostral transverse levels of the diencephalon, only thalamus ventralis (eminentia thalami)[109] and hypothalamus are present

[109] Heier (1948) and various other authors interpret our eminentia thalami ventralis (K., 1927, 1929; Saito, 1930) as 'primordium hippocampi').

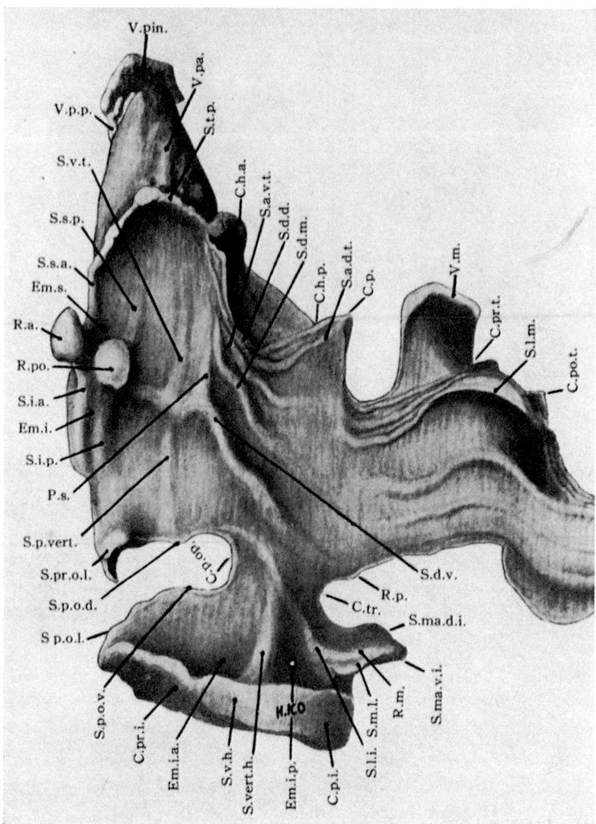

Figure 60C. Wax model of ventricular lumen in diencephalon and mesencephalon of Entosphenus (from SAITO, 1930). C.h.a., C.h.p.: anterior and posterior parts of habenular commissure; C.p.: commissura posterior; C.p.i.: comm. postinfundibularis; C.po.t.: comm. posttectalis; C.pr.t.: comm. praetectalis; C.tr.: comm. transversa; Em.i.: 'eminentia interventricularis'; Em.i.a.: 'eminentia infundibuli anterior; Em.i.p.: 'eminentia infundibuli posterior'; Em.s.: 'eminentia suprainterventricularis'; P.s.: 'pseudosulcus' (1); R.a.: recessus anterior ventriculi lateralis; R.m.: 'recessus mammillaris' (3); R.p.: 'recessus posterior'; R.po.: recessus posterior ventriculi lateralis; S.a.d.t.: sulcus accessorius dorsalis thalami dorsalis; S.a.v.t.: s. acc. ventralis thalami dorsalis; S.d.d.: s. diencephalicus dorsalis; S.d.m.: s. dienc. medius; S.d.v.: s. dienc. ventralis; S.i.a.: s. infrainterventricularis anterior (2); S.i.p.: s. infrainterventricularis posterior (2); S.l.i.: s. lateralis infundibuli (3); S.l.m.: s. lat. mesencephali (internus); S.ma.v.i.: s. mammillaris dorsalis internus; S.m.l.: s. mammillaris lateralis (3); S.p.o.d.: s. postopticus dorsalis; S.p.o.l.: s. postop. lateralis; S.p.o.v.: s. postop. ventralis; S.p.vert.: s. praeopticus verticalis (2); S.pr.o.l.: s. praeop. lateralis (2); S.s.a., S.s.p.: s. suprainterventricularis anterior et posterior (1); S.vert.h.: s. verticalis hypothalami (3); S.v.t.: s. verticalis thalami (1); V.m.: ventriculus mesencephali; V.pa.: 'ventriculus parencephalicus';

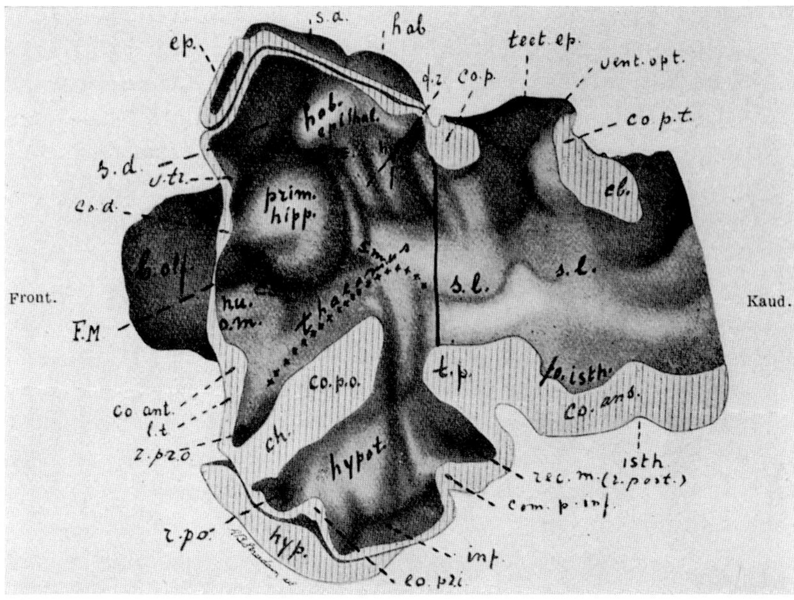

Figure 60 D. Wax model reconstruction of the diencephalic and adjacent ventricular surfaces of the Petromyzont Ichthyomyzon concolor (after HERRICK and OBENCHAIN, 1913, from KAPPERS, 1921, as labelled by this latter author). F.M.: 'foramen Monroi'; b.olf.: bulbus olfactorius; cb.: cerebellum; ch.: optic chiasma; co.ans.: comm. ansulata; co.ant.: comm. anterior; co.d.: comm. dorsalis; co.p.: comm. posterior; co.p.o.: comm. postoptica (sive supraoptica); co.pr.i.: comm. praeinfundibularis; co.p.t.: comm. posttectalis; cs: 'corpus striatum'; ep.: 'epiphysis' (pineal eye); fo. isth.: fovea isthmi; fr.: fasciculus retroflexus; hab.: ganglion habenulae; hyp.: adenohypophysis; hypot.: hypothalamus; inf.: 'infundibulum'; isth.: isthmus rhombencephali region; l.t.: lamina terminalis; nu.o.m.: 'nucleus olfactorius medialis'; prim.hipp.: 'primordium hippocampi'; rec.m.: 'recessus mammillaris' seu posterior; r.p.o.: recessus postopticus; r.pr.o.: recessus praeopticus; s.d.: saccus dorsalis; s.l.: s. limitans (whose rostral extension, indicated by crosses, is supposed to end in the preoptic recess; s.m.: s. dienc. medius; s.s.h.: subhabenular sulci; tect.ep.: ependymal roof of mesencephalic ventricle; t.p.: tuberculum posterius; vent.opt.: mesencephalic ventricle; v.tr.: velum transversum. The line drawn from posterior commissure to tuberculum posterius indicates the approximate boundary between diencephalon and mesencephalon in accordance with concepts which can be fully upheld.

V.pin.: ventriculus pinealis; V.p.p.: ventriculus parapinealis. The sulci numbered in parenthesis (1) pertain to the 'pseudosulcus' system; (2) pertain to the sulcus intraencephalicus anterior system; (3) pertain to the sulcus lateralis hypothalami posterioris system.

(Figs. 61 A–C). Their superior (dorsal) and inferior (ventral) massae cellulares reunientes extend toward the caudally evaginated portion of the telencephalon (Fig. 61 A). The epithalamus (habenular nuclei), and the thalamus dorsalis are located caudally to the eminentia thalami ventralis, together with the caudalward gradually diminishing posterior portion of the ventral thalamus (Figs. 62 A, B).

The main sulci related to the diencephalic zonal system, namely sulcus diencephalicus dorsalis, sulcus diencephalicus medius and sulcus diencephalicus ventralis, moreover the rostral end of sulcus limitans, are relatively easily identifiable at certain developmental stages of Ammocoetes (K., 1929; cf. also section 5, chapter VI, vol. 3/II). In adult Petromyzonts, particularly in large, older specimens, various accessory sulci, presumably related to subsequent growth processes, present a confusing picture which, nevertheless, can be properly interpreted on the basis of the configuration obtaining in Ammocoetes. Saito (1930) made a painstaking effort to visualize, by means of ventricular wax-models, the complex pattern of ventricular grooves obtaining in adult Entosphenus.[110] Some of the accessory sulci[111] doubtless display individual variations and might also become overemphasized by shrinkage processes related to the technical procedures, although few, if any, seem to represent outright artifacts. A comparison of Figures 60 A, B, C, and D indicates the essential differences between our interpretation and that of Herrick and Obenchain (1913), and Kappers (1921), which, in its main features, was also retained by Heier (1948) and Adam (1956). A detailed study on the brain of larval and adult Petromyzonts, well documented by numerous illustrations and references to the pertinent literature, was published by Schober (1964), who discussed the diverse problems of interpretation. This author, although

[110] The observations made in the course of our own investigations convincingly indicated that at least many of the ventricular sulci in the Vertebrate neuraxis, as previously recorded by His, Herrick, and others, represent significant aspects of morphogenetic growth processes related to the differentiation of 'specific' zonal neuronal configurations which could be evaluated as '*gestalt*' entities (subsequently recognized as definable topologic neighborhoods). Independently and from a different viewpoint, the late Prof. K. Okajima (1882–1936), who was particularly interested in the morphology of the Urodele Hynobius and of other Amphibians of Japan, recognized the relevance of ventricular sulci which were emphasized by his student Sumi (1926a). This latter author (1926b) also worked partly under my direction.

[111] It is not unlikely that the accessory sulcus in the thalamus dorsalis is related to expansive growth of the fasciculus retroflexus.

cautiously and with some reservations, tends to follow HEIER's concept of Petromyzont forebrain configuration. Figure 63 A depicts, in cross-sectional aspect, and in a simplified manner, the essential morphologic features of the Petromyzont diencephalon exclusive of the pretectal region, which is shown in Figure 63 B.

The *epithalamus* consists of the habenular ganglion[112] with its cell masses, the right ganglion being considerably larger than the left one, as mentioned above (section 1 C) in the discussion of the epiphysial complex.[113] The two habenulae are fused in the midline, thereby forming the habenular commissure (superior commissure of HEIER, 1948) which may have a rather long rostrocaudal extent.[114]

In cross-sections, the habenular neuronal elements commonly tend to assume a ring-like arrangement around centrally located neuropil and fiber bundles, including portions of the fasciculus retroflexus. A similar arrangement likewise obtains in the habenula of many Anamnia from Plagiostomes to Amphibia (cf. Figs. 4, 63 A, 70 A). HEIER (1948) distinguished several habenular nuclei (n. hab. ventralis, centralis, dorsalis, parietalis), of which the median nucleus parietalis, pertaining to both habenulae, seems related to the parietal nerves, which, however, as mentioned in the preceding section, have doubtless additional connections.

The *thalamus dorsalis* (Figs. 60A, B, 62A, B, C, 63A, B) displays an essentially periventricular arrangement of its cellular elements. A dorsal and a ventral subdivision, in part related to accessory sulci, can be variously distinguished (SAITO, 1930; HEIER, 1948; SCHOBER, 1964).[115] The caudal end of the dorsal thalamus pertains to the pretectal region and includes the fairly well developed nucleus commissurae posterioris. In addition to the periventricular cell population, a more

---

[112] The term habenula (diminutivum of *habena)* originally signifies a little strap or rein-like structure *(frenulum)* but became applied to the triangular or knot-like swelling, which contains the epithalamic griseum, at the caudal end of the human taenia thalami, being also commonly designated as 'ganglion habenulae' in view of said swelling.

[113] Neither epiphysial nor hypophysial complex, considered in section 1C, shall be dealt with in this and the following sections of chapter XII.

[114] SAITO (1930) thus distinguished an anterior and a posterior habenular commissure in adult Entosphenus.

[115] The pioneering investigations by the older authors, such as L. EDINGER, J. B. JOHNSTON, F. MAYER, K. SCHILLING, G. STERZI, D. TRETJAKOFF are listed and were duly considered in the contributions of SAITO, HEIER, and SCHOBER. Some of the relevant references were also included in volume 3/II dealing with the overall morphologic pattern.

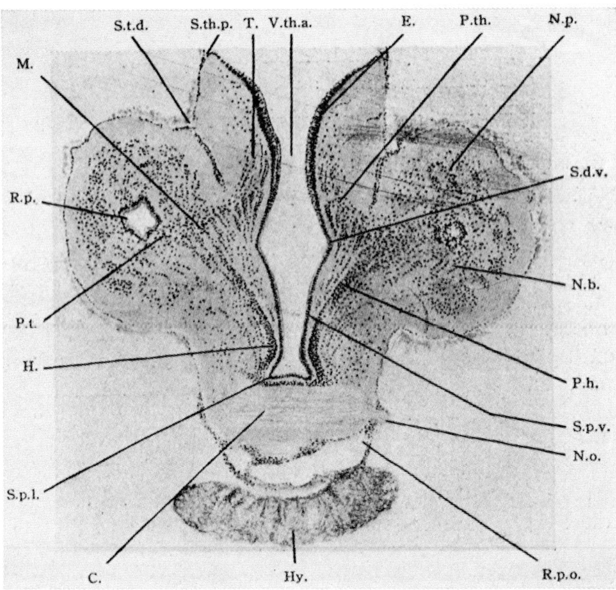

Figure 61 A. Cross-section through the caudal portion of telencephalon and rostral portion of the diencephalon of Entosphenus japonicus at the level of chiasmatic ridge and hemispheric stalk (from SAITO, 1930). C.: chiasmatic ridge; E.: eminentia thalami (ventralis); H.: hypothalamus; Hy.: adenohypophysis; M.: massa cellularis reuniens; N.b.: nucleus basalis (basis, B); N.o.: optic nerve; N.p.: nucleus pallialis (pallium, D); P.h.: 'pars hypothalami reuniens'; P.t.: 'pars telencephali reuniens'; P.th.: pars thalami ventralis reuniens; R.p.: recessus posterior telencephali; R.p.o.: rostral end of recessus postopticus; S.d.v.: sulcus diencephalicus ventralis; S.p.l.: sulcus praeopticus lateralis; S.p.v.: 'sulcus praeopticus verticalis' (part of s. intraencephalicus anterior system); S.t.d.: sulcus telencephalodiencephalicus; S.th.p.: sulcus thalamoparencephalicus; T: thalamus ventralis; V.th.a.: rostral part of third ventricle.

diffusely arranged lateral cell group of the thalamus dorsalis displays a 'neurobiotactic' displacement toward the optic tract and other fiber systems. It represents the dorsal lateral geniculate nucleus. HEIER (1948) describes several subdivisions within this nondescript 'cell plate', but such parcellation cannot be evaluated as altogether convincing.

The *thalamus ventralis*, whose cells likewise manifest an essentially periventricular arrangement, begins rostrally with the large and peculiar, but not very cell-rich eminentia thalami (Figs. 60 A, 61). This is the so-called 'primordium hippocampi *autorum*', which, in full agreement with the views of HOLMGREN (1922, 1946) and BERGQUIST (1932), and on the basis of our own studies (K., 1929; SAITO, 1930) must be in-

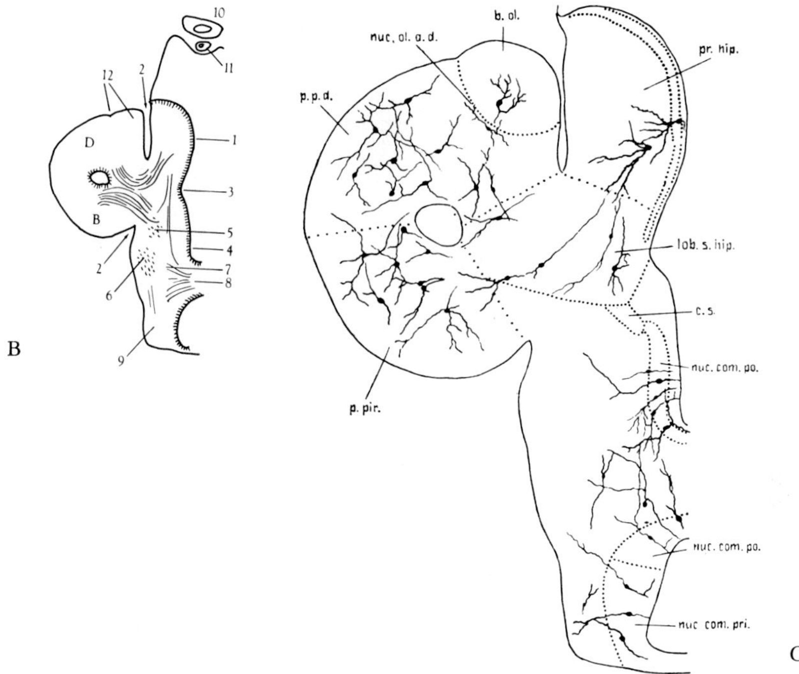

Figure 61 B. Sketch of cross-section through the forebrain of Petromyzon fluviatilis, approximately corresponding to the level of the preceding figure (simplified and reinterpreted after a figure by HEIER, 1948). 1: eminentia thalami; 2: sulcus telo-diencephalicus; 3: sulcus diencephalicus ventr.; 4: hypothalamus (preoptic recess); 5: basal forebrain bundle; 6: tractus opticus; 7: tractus opticus axialis; 8: commissura postoptica; 9: hypothalamus (postoptic recess); 10: pineal eye; 11: parapineal eye; 12: caudal end of bulbus olfactorius, separated from lobus hemisphaericus by a sulcus; D, B: telencephalic pallial and basal grundbestandteile. The roughly sketched fiber tracts (communication channels) should be compared with those of Figure 7.

Figure 61 C. Cross-section *(Golgi impregnation)* through forebrain of Petromyzon fluviatilis at level of preceding figure (from HEIER, 1948). b.ol.: caudal end of bulbus olfactorius; c.s.: 'corpus striatum' (dorsal part of preoptic hypothalamus); lob.s.hip.: 'lobus subhippocampalis'; nuc.com.po.: nucleus commissurae postopticae (the dorsal portion of this nucleus is part of the preoptic nucleus); nuc.com.pri.: nucleus commissurae praeinfundibularis; nuc.ol.a.d.: 'nucleus olfactorius anterior dorsalis'; p.p.d.: primordium pallii dorsalis (topologic zone D); p.pir.: 'primordium piriforme' (topol. zone B); pr.hip.: 'primordium hippocampi' (eminentia thalami ventralis). The divergences between HEIER's and our interpretations are evidenced by the cited author's different designations.

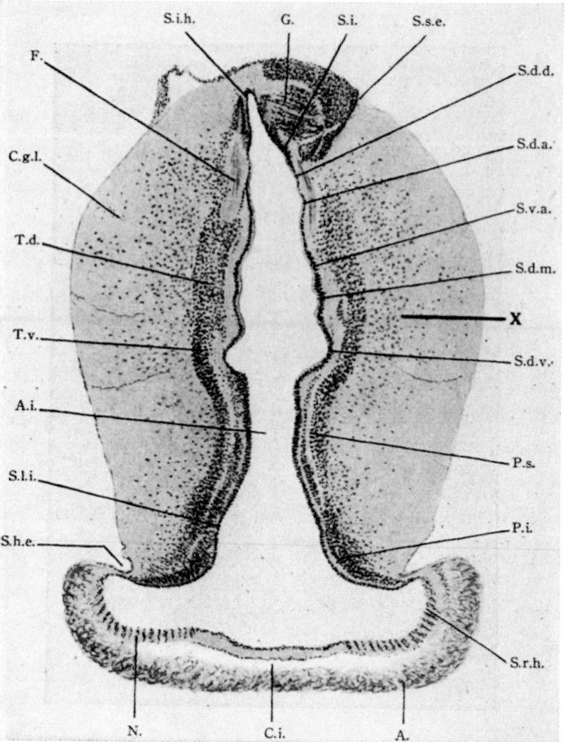

Figure 62 A. Cross-section through the diencephalon of Entosphenus japonicus at a postchiasmatic level (from SAITO, 1930). A.: adenohypophysis; A.i.: 'aditus infundibuli; C.g.l.: (dorsal) lateral geniculate nucleus; C.i.: cleft between adeno- and neurohypophysis; F.: fasciculus retroflexus; G.: ganglion habenulae (dextrum); N.: lateral neighborhood of neurohypophysis; P.i.: pars inferior hypothalami; P.s.: pars superior hypothalami (posterioris); S.d.a.: s. dorsalis accessorius; S.d.d.: s. dienc. dorsalis; S.d.m.: s. dienc. medius; S.d.v.: s. dienc. ventralis; S.h.e.: s. hypothalamicus externus; S.i.: s. intrahabenularis; S.i.h.: spatium interhabenulare; S.l.i.: s. lateralis infundibuli; S.r.h.: branch of s. lateralis infundibuli system; S.s.e.: s. subhabenularis externus; T.d.: thalamus dorsalis; T.v.: thalamus ventralis. Added x: ventral lateral geniculate nucleus.

Figure 62 B, C. Two corresponding cross-sections through the diencephalon of Petromyzon fluviatilis at the postchiasmatic level, showing (B) cytoarchitecture and fiber systems, and (C) neuronal elements displayed by *Golgi impregnation* (from HEIER, 1948). c.gen.a.: 'corpus geniculatum anterius' (this is presumably a somewhat dorsally displaced

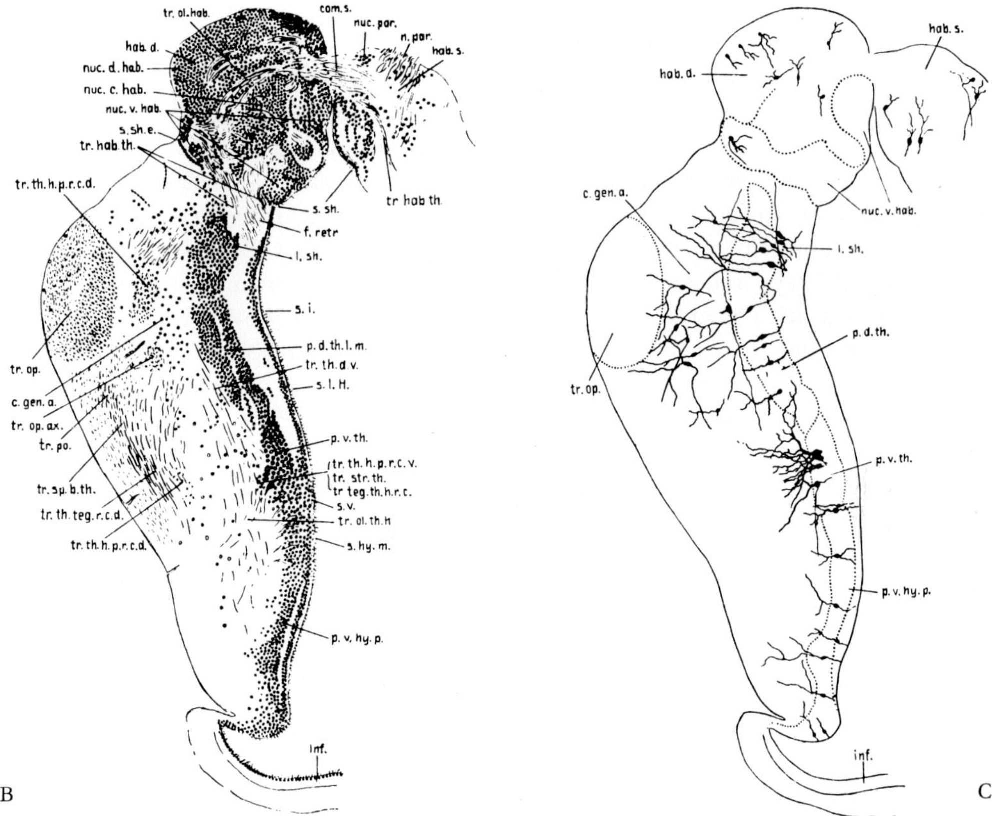

posterior component of the ventral thalamic lateral geniculate nucleus, the rostral part of the dorsal lateral geniculate nucleus being at the tip of lead tr. th.h.p.r.c.d.); com.s.: habenular commissure; f.retr.: fasciculus retroflexus; hab.d., s.: habenula dextra, sinistra; inf.: infundibulum; l.sh.: 'lobus subhabenularis'; nuc.c.hab.: nucleus centralis habenulae; nuc.d.(v.)hab.: n. dorsalis (ventralis) habenulae; nuc.par.: 'n. parietalis'; p.d.th.: pars dorsalis thalami; p.d.th.l.m.: 'pars dorsalis thalami, lobus medius'; p.v.hy.p.: pars ventralis hypothalami, lobus posterior; p.v.th.: pars ventralis thalami; s.sh.: s. subhabenularis; s.sh.e.: s. subhab. externus; s.hy.m.: 's. hypothalamicus medius'; s.i.: 's. intermedius diencephali'; s.l.H.: 's. limitans His' (s.d.m. of Saito); s.v.: s. dienc. ventralis; tr.hab.th.: tractus habenulo-thalamicus; tr.ol.hab.: tr. olfactohabenularis; tr.ol.th.h.: tr. olfacto-thalamicus et hypothalamicus; tr.op.: tr. opticus; tr.op.ax.: tr. op. axialis; tr.op.: 'tr. postopticus'; tr.sp.b.th.: tr. spino-bulbo-thalamicus; tr.str.th.: 'tr. striothalamicus et hypothalamicus; tr.teg.th.h.r.c.: tr. tegmento-thalamicus et hypothalamicus rectus et cruciatus; tr.th.d.v.: tr. thalamicus dorsoventralis; tr.th.h.p.r.c.d.(v): tr. thalamohypo-thalamicus et peduncularis rectus et cruciatus dorsalis (ventralis); tr.th.teg.r.c.d.: tr. thalamotegmentalis rectus et cruciatus dorsalis.

terpreted as corresponding to the eminentia thalami ventralis of Gnathostome Anamnia (cf. also JEENER, 1930). The caudally adjacent portions of thalamus ventralis diminish rapidly in their extension as subsequent transverse levels of habenula and thalamus dorsalis (Fig. 62 A), disappearing, at least in most adult specimens, at the pretectal levels (Fig. 63 B). Said caudal end of thalamus ventralis abuts on the tegmental cell cord (nucleus tuberculi posterioris of HEIER, 1948), seemingly merging with it in adult specimens. As in the thalamus dorsalis, a scattered lateral cell group, displayed caudally to the eminentia thalami, represents a ventral lateral geniculate nucleus (Fig. 63 A). In adult specimens, it tends to become somewhat dorsally displaced, giving thereby the erroneous impression of pertaining to the ventral portion of thalamus dorsalis (Fig. 62 B).

The *hypothalamus*, subdivided into the dorsorostral preoptic and the caudobasal 'mammillo-infundibular' portion, is also characterized by a nondescript periventricular neuronal arrangement with some scattered more peripheral elements. These latter, rostralward continuous with pars inferior massae cellularis reunientis, can be evaluated as a primordial entopeduncular nucleus. SAITO (1930), HEIER (1948), and SCHOBER (1964) moreover recognize, in both the rostral and caudal subdivisions of the hypothalamus, a further dorsal (superior) and ventral (inferior) portion of the periventricular cell population.[116] The cited authors, as well as CROSBY and SHOWERS (1969), have made an attempt to recognize several distinctive hypothalamic nuclei. In the preoptic region, the dorsal part of the paraventricular nucleus displays some relatively large, apparently neurosecretory elements, presumably pertaining to the hypothalamo-neurohypophysial tract (Figs. 63 A, C). Likewise, the basal, suprachiasmatic portion of the preoptic cell layer includes elements of this type (n. praeopticus magnocellularis of CROSBY and SHOWERS, 1969). This cell group can be evaluated as a primordial supraoptic nucleus. SCHOBER (1964) also describes in this region a nucleus of the postoptic commissure (cf. Fig. 63 C). Beyond this, more detailed parcellations appear rather unconvincing.[117]

---

[116] The subdivision into pars superior and pars inferior of the posterior hypothalamus which I had adopted (K., 1929) corresponds, as SCHOBER points out, to his own subdivision of this region into 'nucleus dorsalis' and 'nucleus ventralis'.

[117] SCHOBER (1964) justly points out the difficulties, '*die einzelnen Kerngebiete von Nachbarkernen ohne weiteres abzutrennen*'.

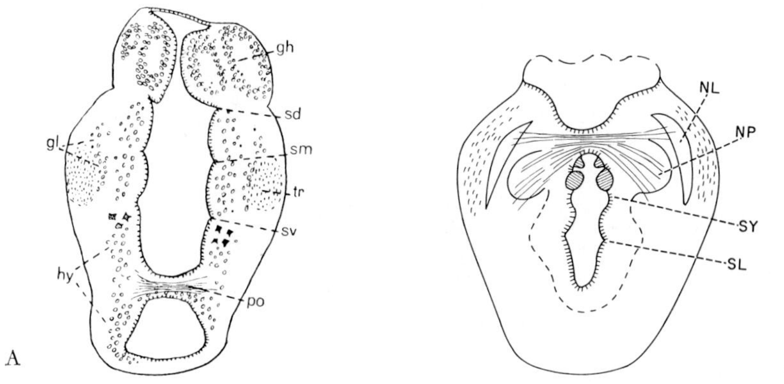

Figure 63 A. Simplified drawing of a cross-section through the diencephalon in Petromyzon, illustrating the overall morphologic pattern (from K., 1927). gh: ganglion habenulae (dextrum); gl: nucleus geniculatus lateralis (upper lead to dorsal thalamic, lower lead to ventral thalamic subdivision); hy: hypothalamus (with large neurosecretory cells in dorsal paraventricular portion); po: postoptic commissure; sd: sulcus diencephalicus dorsalis; sm: sulcus diencephalicus medius; sv: sulcus diencephalicus ventralis; tr: tractus opticus.

Figure 63 B. Semidiagrammatic cross-section through the diencephalo-mesencephalic boundary region of Petromyzon, showing posterior commissure and pretectal components (modified after K., 1956). NL: nucleus lentiformis mesencephali; NP: nucleus commissurae posterioris; SL: sulcus limitans; SY: sulcus synencephalicus. Optic tract lateral to NL, subcommissural organ (oblique hatching) basal to posterior commissure. Above that commissure, rostral portion of mesencephalic ventricle with epithelial roof plate.

The difficulties in recognizing the detailed composition of the diencephalic fiber connections are likewise considerable. The *rostral communication channels* are provided by the basal forebrain bundle and by fiber bundles from both basal and pallial telencephalon running dorsalward toward eminentia thalami as far as the habenular complex (Fig. 61 B).

The basal forebrain bundle does not display a clear-cut distinction between lateral and medial subdivisions. It presumably contains descending and ascending connections between the telencephalon and dorsal, ventral thalamic as well as hypothalamic cell groups.[118] Although the telencephalon is doubtless 'dominated' by olfactory input, ascending channels from the diencephalon, including the lateral geniculate nuclei, presumably channel all other sorts of sensory ('somatic'

[118] Tractus olfactothalamicus et hypothalamicus dorsalis et ventralis, tractus striothalamicus et hypothalamicus, tractus thalamobulbaris, tractus thalamofrontalis of HEIER (1948).

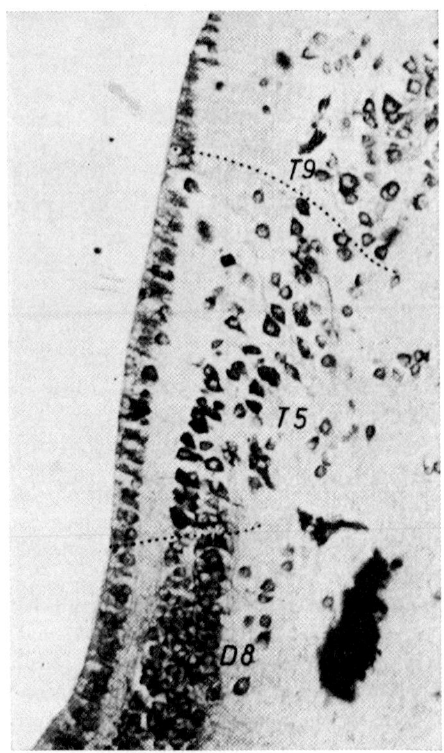

Figure 63 C. Cytoarchitecture of preoptic region in Petromyzon fluviatilis as seen in cross-section (from SCHOBER, 1964; cresyl violet; ×200, red. $^3/_5$). D8: 'nucleus commissurae postopticae'; T5: nucleus praeopticus (presumably neurosecretory cells of n. paraventricularis); T9: 'corpus striatum' of SCHOBER (thalamus ventralis of our interpretation). The dark mass lateral to D8 is apparently a precipitation artefact commonly occurring in cresyl violet stains.

and 'visceral') input into the telencephalon. The connections between telencephalon and habenular complex are mediated by the dorsalward coursing fibers mentioned above (tractus olfactohabenularis of HEIER). Although descending channels may predominate, these connections are most probably reciprocal.

Intradiencephalic connections are, *inter alia*, provided by 'vertical' (dorsobasally) oriented fibers, which include presumably reciprocal habenulothalamic, habenulohypothalamic, and thalamohypothalamic channels.

The *optic tract*,[119] running toward the tectum mesencephali, probably also gives off input to the thalamic lateral geniculate nuclei, pretectal region, and perhaps also to hypothalamus as well as to the rostral mesencephalic portion of formatio reticularis tegmenti, but a distinct 'basal optic tract' cannot be identified with certainty. HEIER (1948) has described a small tractus opticus axialis, whose decussation lies dorsally to chiasma proper and postoptic commissure (Fig. 61 B). It seems to reach the pretectal nucleus commissurae posterioris.

The *caudal diencephalic channels*, in part merging with the caudal end of the basal forebrain bundle, seem to include ascending fibers from the general spinobulbar lemniscus and from the vestibulolateral lemniscus. Descending channels reach the reticular formation and, according to HEIER (1948), may also join the fasciculus longitudinalis medialis. A connection between posterior hypothalamus and the rudimentary cerebellum is said to be present (so-called tr. lobo-cerebellaris). Although the topologic neighborhood of the mammillary body of higher Vertebrates is vaguely represented in the posterior hypothalamus, a 'nucleus mammillaris' cannot be properly delimited. HEIER (1948) nevertheless describes 'mammillo-tegmental', 'mammillo-peduncular', and related fibers, which can preferably be included in the actually nondescript fiber bundles of the caudal diencephalic channel system. There are, moreover, connections between mesencephalon, especially tectum, and pretectal thalamic region as well as habenulae and other diencephalic cell groups. The fasciculus retroflexus, whose main component is probably the habenulo-interpeduncular tract, seems to include additional connections with thalamus and hypothalamus.

It should be stressed that the detailed descriptions and namings of tracts in the Cyclostome diencephalon and neuraxis in general still remain highly conjectural, apart from the well-established but somewhat vaguely general features enumerated in the present section and in section 1 A of this chapter. Much uncertainty remains with respect to crossed and uncrossed components of these channels.

With respect to the *decussating fiber systems*, the following nine commissures are related to the Petromyzont diencephalon.

---

[119] The decussation in chiasma opticum proper and in tractus opticus axialis can be evaluated as essentially complete. In contradistinction to the rudimentary eyes of Myxinoids, those of adult Petromyzonts are well developed, but at larval (Ammocoetes) stages the eyes remain buried beneath pigmented skin. Ammocoetes larvae are said to make no movements when light is shone on to the eye region (YOUNG, 1955).

1. The *commissura pallii* ('commissura interbulbaris' of HEIER), although located entirely in the telencephalon, rostral to the rudimentary velum transversum, seems to contain, in addition to purely telencephalic systems, some fibers of the so-called tractus olfactohabenularis and of the basal forebrain bundle.

2. The *commissura anterior* ('commissura supraoptica' of HEIER), an important landmark of the telodiencephalic boundary, is the main commissure for the crossed components of the basal forebrain bundle. Like the commissura pallii, it also gives passage to intrinsic telencephalic fibers.

3. The *commissura habenulae* (commissura superior of HEIER, 1948) pertains to the habenular complex and also includes a variety of poorly elucidated crossing fiber systems.

4. The *commissura posterior*, an important landmark of the diencephalomesencephalic boundary, pertains to diencephalic and mesencephalic pretectal region. In addition to other insufficiently elucidated systems, tecto-diencephalic channels decussate in that commissure.

5. The *optic chiasma* and the accessory decussation of the axial optic tract were referred to above and require no further comments.

6. The *postoptic commissure* or commissura transversa occupies the caudodorsal extension of the chiasmatic ridge. It contains components of basal forebrain bundle, of intrinsic diencephalic fiber systems, and of caudal channels related to the mesencephalon.

7. The *commissura preinfundibularis* of HEIER is a not very conspicuous system of insufficiently identified fibers crossing in the hypothalamic floor between chiasmatic ridge and neurohypophysis.

8. The *commissura postinfundibularis* of HEIER, likewise inconspicuous and of uncertain composition, lies in the hypothalamic floor caudally to the hypophysial complex, and basally to the tuberculum posterius.[120]

9. The *commissura tuberculi posterioris* of HEIER, although extending into the caudal (synencephalic) portion of the diencephalon, seems to be merely the rostral end of the complex basal deuterencephalic system of decussating fibers known as *commissura ansularis s. ansulata*. Undefined components of the caudal diencephalic communication channels,

---

[120] The location of this commissure corresponds to the general neighborhood in which the saccus vasculosus of Plagiostomes and Osteichthyes becomes differentiated. As mentioned above in section 1C, a saccus vasculosus has not been recorded in Cyclostomes.

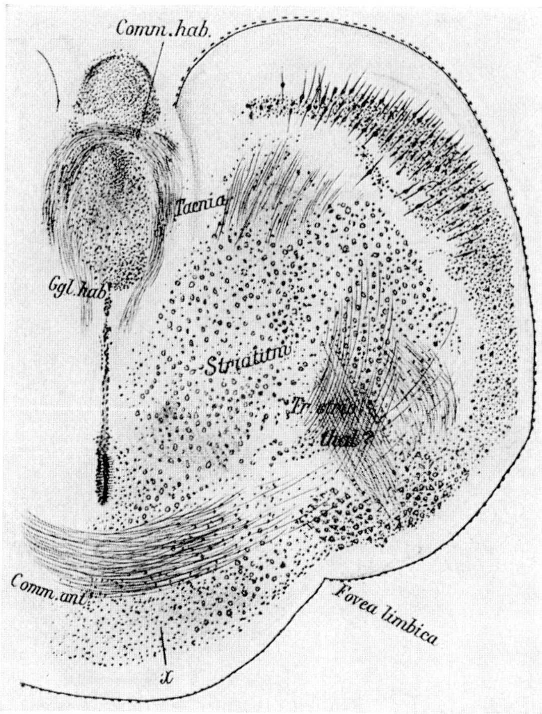

Figure 64. Cross-section through the prosencephalon of Myxine glutinosa at the level of anterior commissure and rostral extension of diencephalic topologic neighborhoods toward telencephalic ones (from EDINGER, 1906). Comm. ant.: anterior commissure; Comm. hab.: rostral portion of habenular commissure; Fovea limbica: basal part of sulcus telencephalo-diencephalicus; Striatum: rostral part of thalamus ventralis (eminentia thalami; so-called primordium hippocampi autorum); Taenia, at right: tractus telencephalo-diencephalicus dorsalis s. tractus olfactorius lateralis of JANSEN, 1930; at left: tractus habenulo-thalamicus of JANSEN, 1930; Tr. strio-thal?: basal forebrain bundle. Added x: hypothalamus (preoptic region). Cf. also Figure 244 II, p. 479 of volume 3/II which illustrates a comparable level depicted by JANSEN (1930), but passing basalward caudally to anterior commissure.

including the so-called tractus thalamotegmentalis, decussate in this commissure, which tends to merge basalward with the so-called post-infundibular commissure.

Of the here enumerated nine commissures, 1 and 2 pertain to the lamina terminalis, while 3 and 4 pass through the diencephalic respectively diencephalo-mesencephalic roof plate. The commissures 5–8 are hypothalamic midline structures of the prosencephalic alar plate. Commissure 9 lies within the rostral end of the tegmentum protruding into

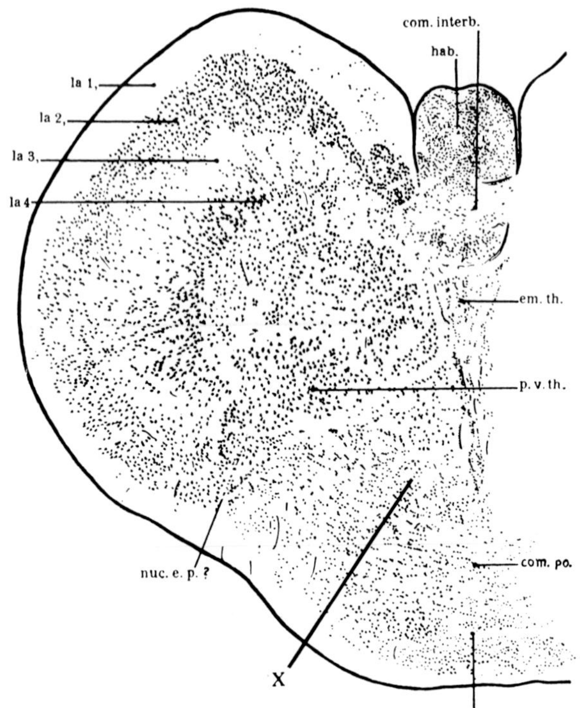
com. interb.
hab.
la 1,
la 2,
la 3,
la 4
em. th.
p. v. th.
com. po.
nuc. e. p. ?
X
nuc. post.

A

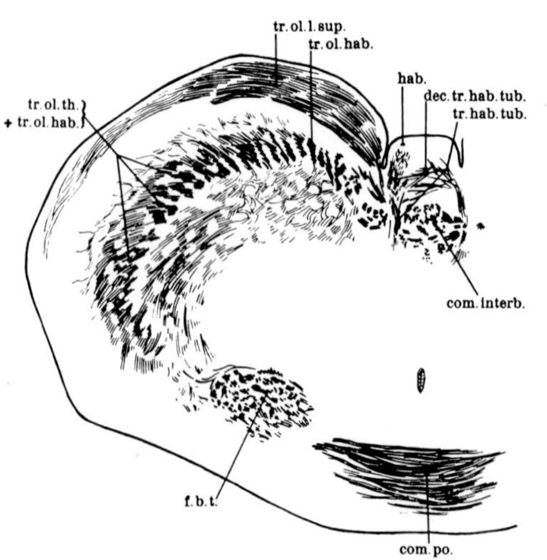
tr. ol. l. sup.
tr. ol. hab.
hab.
dec. tr. hab. tub.
tr. hab. tub.
tr. ol. th.
+ tr. ol. hab.
com. interb.
f. b. t.
com. po.

B

the synencephalic diencephalon and thus pertains to the neuraxial basal plate.

In adult *Myxinoids*, the pronounced pattern distortion involves a 'telescoping', 'sliding' or 'shoving' into each other, of telencephalic and diencephalic configurations, combined with considerable obliteration of the ventricular system. To some extent, there is also a 'telescoping' of diencephalic neighborhoods.[121] Although the pattern distortions maintain the invariants, i.e. the topologic connectedness under the transformation, the identification of the topologic neighborhoods presents some difficulties because of the nondescript outlines of the relevant cell populations, whereby a proper distinction of the basic components becomes to some degree indecisive. Nevertheless, the few embryologic data, available through the reports of Kupffer (1906), Conel (1929, 1931), and Holmgren (1946) permit some reasonably probable extra- and interpolations (Figs. 64–66).

In contradistinction to Petromyzonts, the *epithalamus* with its large nucleus habenulae extends rostrally into telencephalic levels, being wedged between the hemispheres. The antimeric habenulae are fused, but a slight groove may indicate a boundary between the two, of which, as in Petromyzonts, the left habenula is smaller than the right one.

The *thalamus dorsalis* does not extend as far rostralward as the thalamus ventralis and makes its appearance at the caudal end of the telence-

[121] Cf. volume 3/II, chapter VI, section 1B, pp. 191–195, also volume 4, chapter IX, section 3, Figs. 193A, B, 194, p. 371 loc.cit.

Figure 65. Cross-sections through the prosencephalon of Myxine glutinosa at level of commissura postoptica (diencephalon) and caudal part of telencephalon, displayed (A) by a modified *Nissl stain* and (B) by the *Bielschowsky silver impregnation* (from Jansen, 1930). com. interb.: 'commissura interbulbaris' (telencephalic component of habenular commissure); com.po.: commissura postoptica; dec.tr.hab.tub.: decussation of tractus habenulo-tubercularis (component of fasciculus retroflexus reaching tuberculum posterius); em.th.: 'eminentia thalami' (medial portion of thalamus ventralis); f.b.t.: basal forebrain bundle; hab.: habenula; la 1–4: lamination of telencephalic cell masses; nuc.e.p.?: nucleus entopeduncularis; nuc.post.: nucleus postopticus (hypothalami); p.v.th.: pars ventralis thalami (at more rostral levels, cf. p. 479, Fig. 244 II, vol. 3/II, this cell group represents the so-called 'primordium hippocampi' *autorum*); tr.hab.tub.: tractus habenulo-tubercularis (cf. above dec.tr.hab.tub.); tr.ol.hab.: tractus olfacto-habenularis; tr.ol.l.sup.: tractus olfactorius lateralis superficialis; tr.ol.th.: tractus olfacto-thalamicus. Added x: hypothalamus.

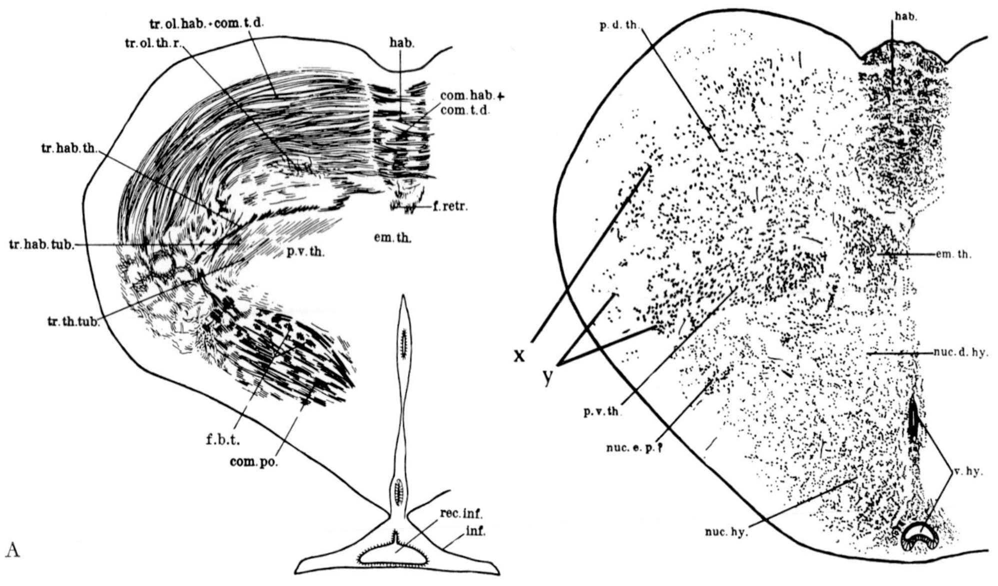

Figure 66. Cross-sections through the diencephalon of Myxine glutinosa caudally to postoptic commissure (from JANSEN, 1930). A Level of recessus infundibuli, *Bielschowsky impregnation.* B Comparable level, modified *Nissl stain.* C Level of diencephalo-mesencephalic boundary zone, modified *Nissl stain.* com.hab.: comm. habenularis; com.t.d.: comm. tecti diencephali (component of habenular commissure); corp.i.: n. interpeduncularis (ganglion interpedunculare); f.retr.: fasciculus retroflexus; hyp.p.g.: adenohypophysis; hyp.p.n.: neurohypophysis; inf.: 'infundibulum'; nuc.d.hy.: 'n. diffusus hypothalami' (dorsal portion of hypothalamus); nuc.e.p. ?: n. entopeduncularis (although JANSEN added a question mark, this cell group has indeed all morphologic characteristics of the Anamniote entopeduncular nucleus); nuc.hy.: 'n. hypothalamicus (ventral portion of posterior hypothalamus); nuc.prof.: 'n. profundus' of dorsal thalamus (as shown in C

phalic hemispheres, from which it is separated by a transverse manifestation of telencephalo-diencephalic sulcus. The cell masses of the dorsal thalamus are arranged in a peripheral nondescript cell layer whose ventrolateral portion corresponds to a dorsal lateral geniculate nucleus. A more centrally located grouping of cells is JANSEN's (1930) nucleus profundus, which, at caudal levels, seems to represent the nucleus commissurae posterioris. The caudal portion of the peripheral cell layer (Fig. 66 C) likewise can be considered a dorsal thalamic pretectal cell group.

The *thalamus ventralis* with its eminentia thalami (so-called primordium hippocampi *autorum*) extends rostrad toward the telencephalic

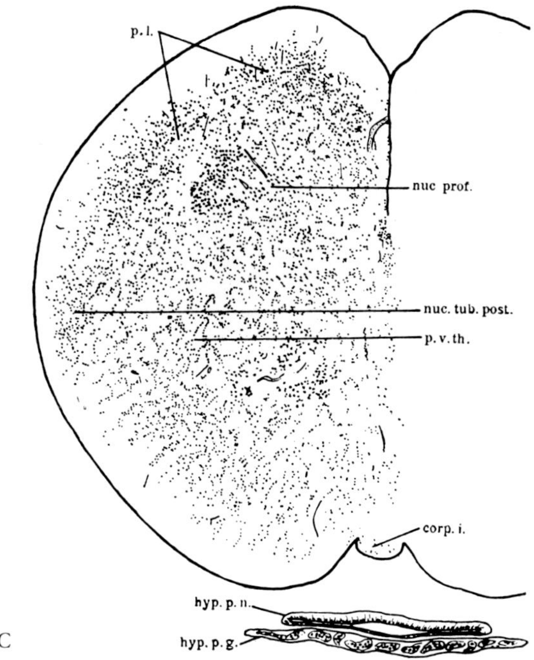

it is probably the n. commissurae posterioris); nuc.tub.post.: 'n. tuberculi posterioris' (probably caudal portion of ventral lateral geniculate nucleus); p.d.th.: pars dorsalis thalami (thalamus dorsalis); p.l.: peripheral layer of pars dorsalis thalami; p.v.th.: pars ventralis thalami (thalamus ventralis, the portion labelled em.th. being its medial portion); rec.inf.: 'recessus infundibuli'; tr.hab.th.: tr.habenulo-thalamicus; tr.ol.th.r.: tr. olfacto-thalamicus rectus; tr.th.tub.: 'tr. thalamo-tubercularis'; v.hy.: reduced hypothalamic portions of third ventricle. Other abbreviations as in Figure 65. Added designations x: dorsal lateral geniculate nucleus; y: ventral lateral geniculate nucleus.

levels concomitantly with the habenulae. Its rostral extension corresponds thus to that displayed by Petromyzonts, in which latter, however, the epithalamus has assumed a caudal position (cf. Fig. 60 A). A caudolateral portion of the cell masses within the eminentia thalami, protruding into the telencephalic neighborhoods, corresponds to the pars superior of the massa cellularis reuniens. Caudally to the hemispheric levels, the most lateral cell group of the lateralward extending part of the thalamus ventralis represents the ventral lateral geniculate nucleus. A medial cell group of the ventral thalamic diencephalic cell zone is JANSEN's eminentia thalami (Figs. 65, 66), rostralward continuous with the 'primordium hippocampi' which, in full agreement with HOLMGREN (1946) must be interpreted as the actual eminentia thalami.

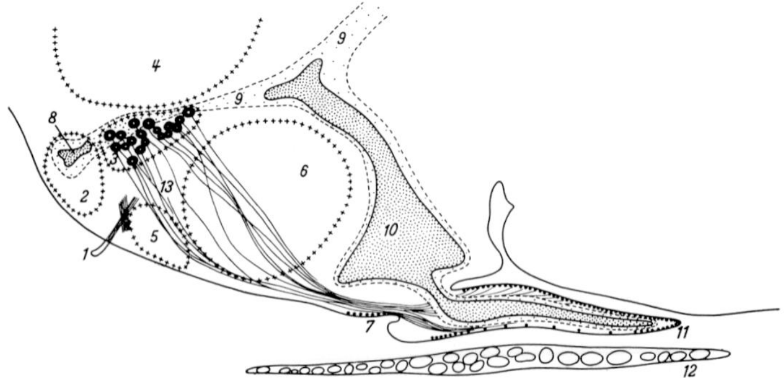

Figure 67. Diagrammatic median reconstruction of the hypothalamus and adjacent regions of Myxine (from Olsson, 1959; ×50, red. $^{1}/_{2}$). 1: optic nerve, chiasma, and tract; 2: 'nucleus praepticus parvocellularis anterior'; 3: 'nucleus praeopticus magnocellularis'; 4: 'primordium hippocampi' (eminentia thalami ventralis); 5: nucleus postopticus; 6: commissura postoptica; 7: eminentia mediana (part of neurohypophysis); 8: recessus praeopticus; 9: ventricle rudiments; 10: hypothalamic ventricle; 11: neurohypophysis; 12: adenohypophysis; 13: tractus praeoptico-hypophyseus. Neurosecretory cells black, accumulations of neurosecretory material shown by black dots.

The *hypothalamus* consists rostrally of a preoptic region with diffusely distributed cell masses, including a nondescript nucleus praeopticus and a massa cellularis reuniens, pars inferior, which is caudalward continuous with a rudimentary entopeduncular nucleus (Fig. 66 B) pertaining to the dorsolateral subdivision of the hypothalamus. Diffusely distributed cellular elements in the dorsal portion of the preoptic region appear to be neurosecretory elements (Fig. 67), from which a preopticohypophysial tract arises, reaching the neurohypophysis[122] (Olsson, 1959). The posterior, infundibulo-mammillary portion of the hypothalamus is rather large and subdivided into a dorsal and ventral region. The postoptic commissural system is very massive and contains numerous interstitial neuronal elements (Fig. 65). The rostral cell groups (nucleus postopticus of Jansen, 1930, and Olsson, 1959) protrude toward the preoptic region (Fig. 67).

The main *fiber tracts* correspond to those present in Petromyzon and, generally speaking, to the overall system of communication channels discussed above in section 1 A.

---

[122] The hypothalamo-hypophysial system of Myxine and of Bdellostoma was also investigated by Adam (1959).

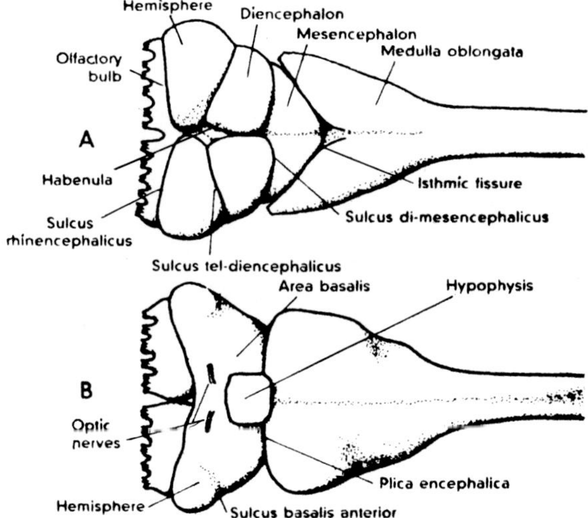

Figure 68. The brain of adult Myxine as seen (A) in dorsal, and (B) in ventral view (simplified after JANSEN, 1930, from BONE, 1963). The 'sulcus basalis anterior' is a ventral portion of the sulcus telo-diencephalicus; the 'area basalis' is the ventral surface of the diencephalon, essentially pertaining to the hypothalamus.

As regards *rostral fiber systems*, a ventral channel is represented by the common (joint medial and lateral) basal forebrain bundle (Figs. 65 B, 66 A) providing mainly the reciprocal connections between telencephalon, thalamus ventralis, and hypothalamus, with additional connections to the other diencephalic zones and to mesencephalic tegmentum. It includes the crossed and uncrossed olfactohypothalamic tracts of JANSEN and that author's 'tractus olfacto-tegmentalis'. The strongly developed 'dorsolateral system' of JANSEN (Figs. 65 B, 66 A), more or less corresponding to the stria medullaris thalami system, mainly provides the connections between telencephalon, epithalamus, and thalamus dorsalis, with additional connections to other diencephalic zones, and, via pretectal region, to tectum mesencephali. It includes tractus olfacto-habenularis and olfacto-thalamicus rectus.

The *caudal communication channels* are represented by fasciculus retroflexus with several components, tractus habenulo-thalamicus, tractus thalamo-bulbaris and tractus hypothalamo-tegmentalis (with 'tractus mammillo-tegmentalis') as described by JANSEN (1930).

The ascending components are provided, *inter alia*, by ill-defined lemniscus systems perhaps essentially similar to those of Petromyzonts.

Although the eyes of Myxinoids are rudimentary and, being generally covered by fairly thick surface layers,[123] of very dubious functional capacity, a thin optic nerve, chiasma, and tract were recorded by HOLMGREN (1919), JANSEN (1930) and others. The tract seems to reach the dorsal and ventral geniculate nuclei and possibly also tectum mesencephali as well as hypothalamus. Yet, it is most likely that, in view of the very reduced optic system, the geniculate nuclei with their further undefined connections subserve a variety of non-optic functions. The *intradiencephalic communication channels* are provided by poorly understood and nondescript fiber systems (e.g. 'tractus thalamo-hypothalamicus', 'tractus habenulo-thalamicus') essentially running in a dorsoventral respectively ventrodorsal direction, with some decussations in the diencephalic commissures.

The *commissural system* includes (1) the extensive commissura habenulae (commissura tecti diencephali) including an undefined variety of channels. (2) The posterior commissure with diencephalic pretectal and with mesencephalic components. (3) The likewise extensive commissura postoptica with telencephalic, diencephalic (particularly hypothalamic) and deuterencephalic components. (4) The much less conspicuous commissura postinfundibularis which seems essentially to interconnect the antimeric posterior hypothalamic regions. Figures 68 A, B illustrate overall gross configurational aspects of the Myxinoid diencephalon in relation to the adjacent neuraxial neighborhoods.

In conclusion it could be said that although Cyclostomes, i.e. agnathous Anamnia presumably comprised, in phylogenetic evolution, the ancestors of the Gnathostome Vertebrates which culminated in Primates and Man, the recent Cyclostomes doubtless represent highly aberrant and modified forms. Nevertheless, both Petromyzonts and Myxinoids are of considerable morphologic interest, since their neuraxis displays, despite pattern distortions, all significant features characteristic for the entire Vertebrate series (EDINGER, 1906).

3. Selachians

In comparison with the 'typical' Vertebrate diencephalic configuration of Amphibians, the Selachian diencephalon, like that of Cyclos-

[123] These often rather thick surface layers commonly may include pigmented skin in adults. Nevertheless, be it through the mediation of the rudimentary eyes, or through undefined photosensitive cutaneous structures, Myxinoids 'are sensitive to changes of illumination' (YOUNG, 1955).

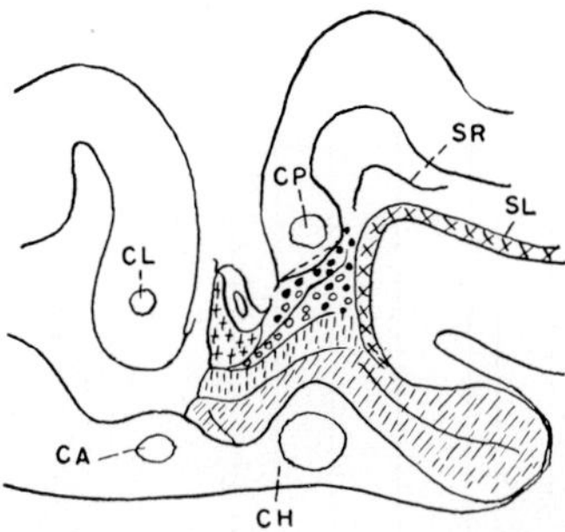

Figure 69 A. Semidiagrammatic reconstruction of longitudinal zones in the diencephalon of adult Acanthias (from K., 1956). The self-explanatory designations are identical with those of Figure 60 A. SR: sulcus lateralis mesencephali. The unlabelled sulci are: sulcus intraencephalicus anterior and sulcus lateralis hypothalami posterioris in preoptic respectively postoptic hypothalamus; sulcus diencephalicus ventralis as dorsal boundary of hypothalamus; sulcus diencephalicus medius as dorsal boundary of thalamus ventralis; sulcus synencephalicus within thalamus dorsalis; sulcus diencephalicus dorsalis as caudal boundary of epithalamus. Rostralward, the three last named sulci fuse into a common sulcus diencephalicus dorsalis (*seu* subhabenularis).

tomes, displays some degree of pattern distortion by 'telescoping' into neighboring brain subdivisions. In contradistinction to Cyclostomes, however, this is restricted to a diencephalo-mesencephalic 'overlap'. An opposite distortion, namely by 'stretching' or 'pulling away' from the telencephalon, is manifested by the pre-chiasmatic diencephalon. An extreme degree of stretching occurs in some Holocephalians (Chimaera), where the diencephalic wall is reduced to a thin membrane-like structure or peduncle containing the fiber stalks of the telencephalon, and essentially representing an elongated preoptic recess. To a lesser degree, this elongation is also quite noticeable in Squalidae such as Scyllium, and remains recognizable, although still less conspicuously, in Acanthias and Rajidae. Figures 69 A–D illustrate these pattern relationships. The peculiar stretching of the preoptic diencephalon in some Holocephalians, evident in Figure 69 D, is shown by the cross-sections of Figures 71 A–D.

The *epithalamus*, consisting of the two antimeric habenular nuclei (or 'ganglia'), extends from preoptic to postoptic levels. Its overall location is thus less caudal than in Petromyzonts, and far less rostral than in Myxinoids. A marked antimeric asymmetry of the habenulae obtains, as is the case in Cyclostomes, but in contradistinction to these latter, the left habenula is substantially larger than the right one. Both habenulae are joined in the midline by the habenular commissure, but

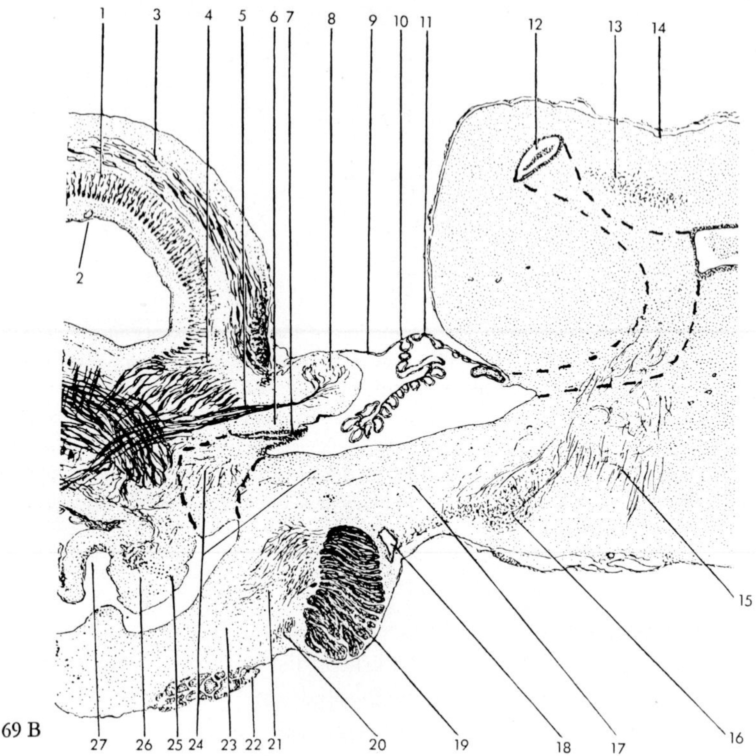
1
2
3
4
5
6
7
8
9
10
11
12
13
14
15
16
17
18
19
20
21
22
23
24
25
26
27

69 B

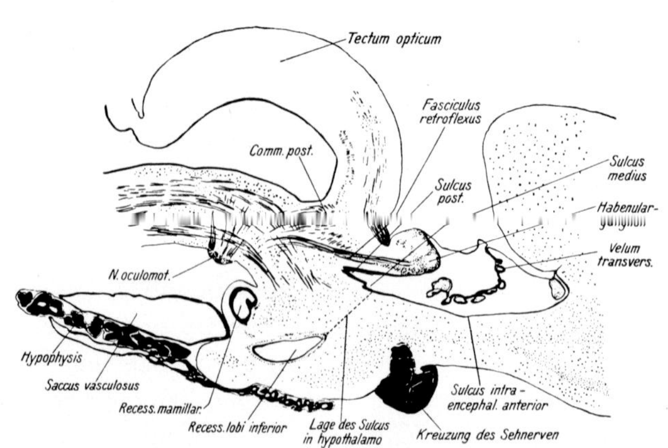
Tectum opticum
Fasciculus retroflexus
Comm. post.
Sulcus post.
Sulcus medius
Habenular-
Velum transvers.
N. oculomot.
Hypophysis
Saccus vasculosus
Recess. mamillar.
Recess. lobi inferior
Lage des Sulcus in hypothalamo
Kreuzung des Sehnerven
Sulcus intra-encephal. anterior

69 C

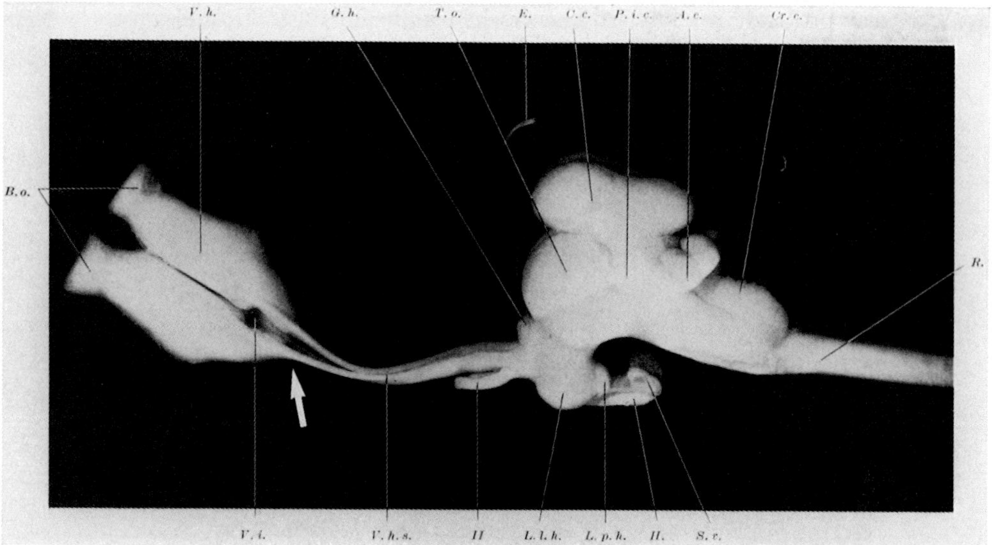

Figure 69 D. The brain of adult Chimaera monstrosa in lateral view. The telencephalon with its preoptic diencephalic stalk has been twisted in order to show telencephalon and part of the preoptic recess in dorsal view (from BRAAK, 1963). A.c.: auricula cerebelli; B.o.: bulbus olfactorius; C.c.: corpus cerebelli; Cr.c.: crista cerebelli; E.: epiphysis (with distortion of damaged stalk, the distal end is actually directed rostrad); G.h.: Ganglia habenulae; H.: hypophysial complex; L.l.h.: lobus lateralis hypothalami; L.p.h.: lobus posterior hypothalami; P.i.c.: 'pars inferior cerebelli'; R.: rhombencephalon and spinal cord; S.v.: saccus vasculosus; T.o.: tectum opticum; V.h.s.: preoptic diencephalon (telencephalic stalk); V.i.: ventriculus impar telencephali; II: optic nerve. Added arrow: diencephalo-telencephalic boundary region. V.h.: telencephalon.

Figure 69 B. Paramedian sagittal section (myelin stain) through the diencephalon and adjacent regions of Scyllium canicula (from KAPPERS, 1921). 1: stratum medullare of tectum mesencephali; 2: nucleus radicis mesencephalicae trigemini; 3: tractus opticus; 4: commissura posterior; 5: fasciculus retroflexus; 6: pretectal region; 7: sulcus diencephalicus medius; 8: ganglion habenulae; 9: parencephalic roof plate; 10: velum transversum; 11: paraphysis; 12: dorsal recess of telencephalic ventricle; 13: commissura pallii; 14: transverse dorsal sulcus of telencephalon; 15: 'tractus medianus'; 16: commissura anterior and basal forebrain bundle; 17, 18: preoptic hypothalamus and preoptic recess; 19: optic chiasma; 20: decussation of tractus pallii; 21: caudal portion of basal forebrain bundle; 22: adenohypophysis; 23: posterior hypothalamus; 24: thalamus ventralis; 25: general neighborhood of tuberculum posterius; 26: 'nucleus sacci vasculosi'; 27: recessus posterior ('mammillaris'). The hypothalamic region of saccus vasculosus is not included (cf. Fig. 183 C, p. 408, vol. 3/II).

Figure 69 C. Paramedian sagittal section through diencephalon and neighboring regions of adult Scyllium (from *Graf* HALLER, 1929). Sulcus post.: sulcus synencephalicus.

may protrude rostralward beyond its extent (cf. Figs. 70 A, B versus 71 D, E). The epiphysial stalk originates caudally to the habenulae between habenular and posterior commissure. In some forms, a medial, a lateral, and ventral habenular nucleus are clearly differentiated (TANAKA, 1959). In these instances, the 'circular' arrangement of the habenular neuronal elements is essentially restricted to the medial habenular subnucleus.

The *thalamus dorsalis*, which does not extend as far rostralward as epithalamus and thalamus ventralis, begins at fairly caudal habenular levels and expands at posthabenular levels dorsocaudalward ventrally to commissura posterior in the diencephalic region covered by the rostral expansion of tectum opticum (Figs. 70 C, D). The cell populations of thalamus dorsalis display a rather diffuse periventricular arrangement (nucleus internus of GERLACH, 1947) preclusing the unambiguous distinction of well-defined nuclei. Nevertheless, within the nondescript pretectal grisea, a *nucleus commissurae posterioris* can be recognized. In addition, laterally displaced cell groups represent a *dorsal lateral geniculate nucleus*.

The *thalamus ventralis* begins rostrad at habenular levels with an inconspicuous eminentia thalami and extends caudalward as far as the mesencephalic tegmentum. A diffuse periventricular cell group and a clearly distinguishable but poorly developed *ventral lateral geniculate nu-*

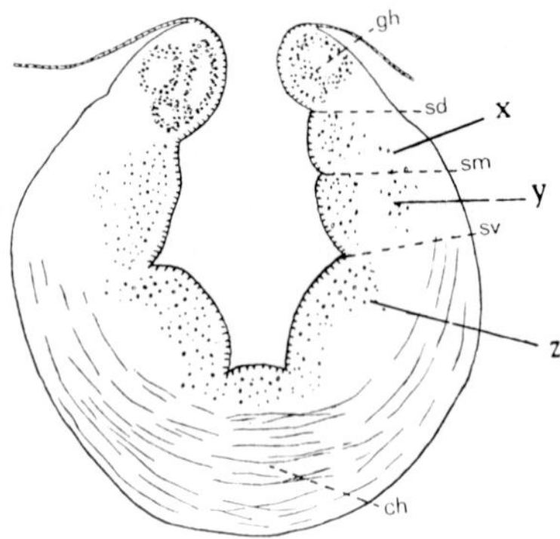

Figure 70 A. Sketch of cross-section through the diencephalon of Acanthias at caudal chiasmatic level (from K., 1927). ch: chiasma opticum; gh: ganglion habenulae; sd: sulcus diencephalicus dorsalis; sm: sulcus diencephalicus medius; sv: sulcus diencephalicus ventralis; x: dorsal lateral geniculate nucleus; y: ventral lateral geniculate nucleus; z: primordium of entopeduncular nucleus. On the left side, the section passes rostrally to thalamus dorsalis, whose rostral end is shown on right side.

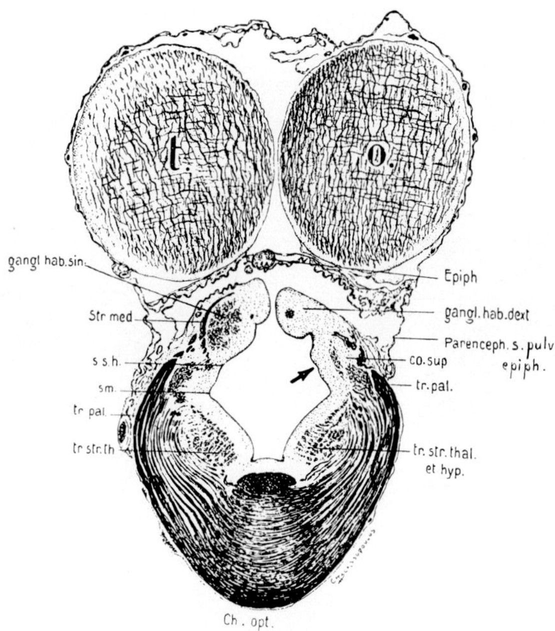

Figure 70B. Cross-section (myelin stain) through diencephalon of Acanthias vulgaris at caudal chiasmatic level (from KAPPERS, 1921). CO.sup.: commissura superior (non-medullated); s.m.: sulcus diencephalicus ventralis ('medius' of KAPPERS); Str.med.: tractus taeniae (stria medullaris); s.s.h.: sulcus diencephalicus dorsalis (subhabenularis); tr.pal.: tractus pallii; tr.str.th., tr.str.thal.et hyp.: basal forebrain bundle (tractus 'strio-thalamicus and hypothalamicus'). Added arrow: sulcus diencephalicus medius. Other abbreviations self-explanatory. Dorsally to diencephalon, the rostral tips of tectum opticum are shown.

cleus can be recognized. In contradistinction to the dorsal thalamic one, the ventral lateral geniculate is generally more distinctly developed at rostral than at caudal levels.

The *hypothalamus* consists of a rostral (pre-chiasmatic, preoptic) part, and of a postoptic caudal portion. This latter includes the *lobi inferiores*, with the lateral recesses, the neurohypophysis, and the large saccus vasculosus.[124] In the preoptic region, fairly large paraventricular cells of neurosecretory type are found, from which the hypothalamo-hypophysial tract originates. In the Holocephalian Chimaera, this

[124] The hypophysial complex, the saccus vasculosus, and, as regards Chimaera, the peculiar 'organon vasculare praeopticum' of BRAAK (1963) were dealt with in section 1C of this chapter.

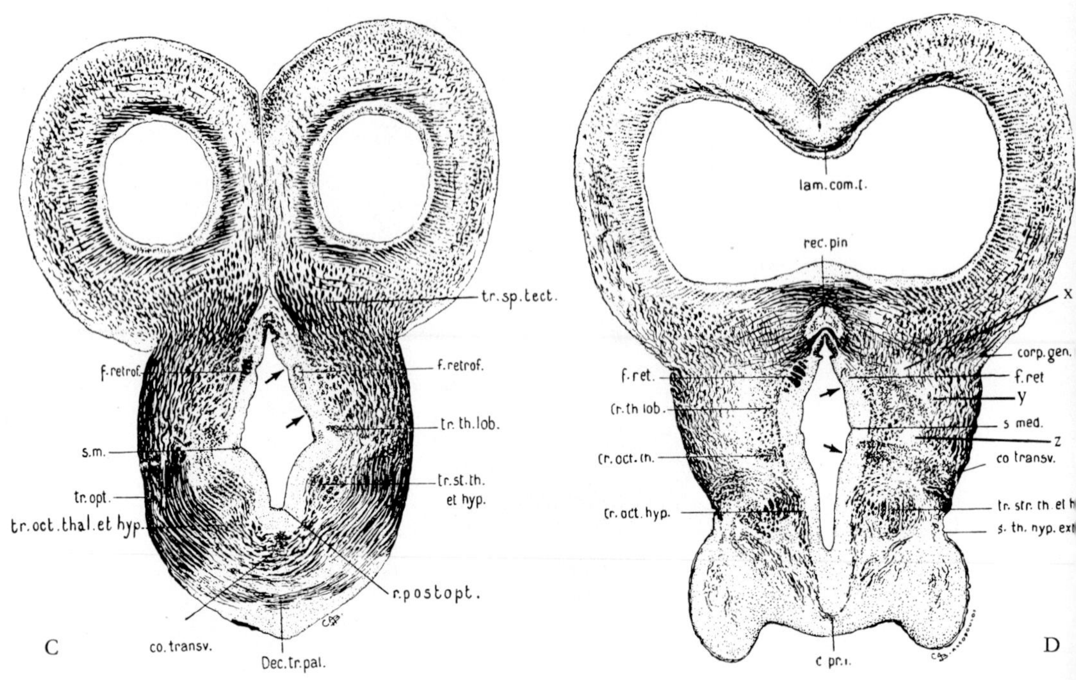

Figure 70C. Cross-section (myelin stain) through posthabenular diencephalon of Acanthias at level of commissura transversa (from KAPPERS, 1921). co.transv.: commissura transversa; Dec.tr.pal.: decussation of tractus pallii; f.retrof.: fasciculus retroflexus (note difference between left and right); r.postopt.: rostral beginning of recessus postopticus; s.m.: sulcus diencephalicus ventralis; tr.oct.thal. et hyp.: 'tractus octavo-thalamicus et hypothalamicus'; tr.sp.tect.: 'tractus spino-tectalis'; tr.th.lob.: 'tractus thalamolobaris'. Added arrows sulcus synencephalicus (upper); sulcus diencephalicus medius (lower).

Figure 70 D. Cross-section (myelin stain) through the caudal diencephalon of Acanthias, basally to common mesencephalic ventricle (from KAPPERS, 1921). c.pr.i.: commissura praeinfundibularis; co.transv.: commissura transversa; corp. gen.lat.: nucleus lentiformis mesencephali; lam.com.t.: lamina commissuralis tecti; rec.pin.: intracommissural recess of commissura posterior; s.med.: sulcus diencephalicus medius; s.th.hyp.ext.: external hypothalamic sulcus; tr.oct.hyp., th.: 'tractus octavo-hypothalamicus et thalamicus'; tr.str.th.et hyp.: caudal portion of basal forebrain bundle; tr.th.lob.: 'tractus thalamolobaris'. Added designations: x: nucleus commissurae posterioris; y, z: dorsal and ventral lateral geniculate nucleus; lower arrow: sulcus diencephalicus ventralis (at this level shallower than at preceding, where indicated by KAPPERS in Figure 70 C as s.m.); upper arrow: sulcus synencephalicus.

tract is considerably altered, since neurosecretory cells do not seem to occur in the peculiarly modified preoptic region. Nerve cells in the neurohypophysis, reported by FUJITA (1963) may, according to this author, represent an equivalent of the neurosecretory preoptic nucleus.

Because of diffuse and essentially periventricular arrangement of the hypothalamic cell masses, specific nuclei cannot be convincingly described. A dorsolateral extension of the hypothalamic cells, including some elements scattered within the basal forebrain bundle, may be evaluated as an *entopeduncular nucleus* (Fig. 70 A). In the neighborhood of the tuberculum posterius, GERLACH (1947) described a *nucleus tractus sacci vasculosi* (dorsalis), which, however, may presumably be a differentiation of the deuterencephalic tegmental cell cord (Figs. 72 C).

TANAKA (1959) who included various Rajidae (Hypotremata), such as Rhinobatus, Discobatus, Dasybatus, and Myliobatus in his study of comparative cytoarchitecture of the Selachian diencephalon, depicts an interthalamic ventricular concrescence ('*commissura interthalamica*' in Myliobatus; Fig. 71 F). This '*Verwachsung*' or '*adhaesio interthalamica*' is comparable to similar midline fusions obtaining, e.g. in the diencephalon of various Reptiles and Mammals (cf. chapter VI, section 5, vol. 3/II).

The *diencephalic ventricular sulci*, although displaying diverse individual and species variations, are commonly reasonably well identifiable. The sulcus diencephalicus dorsalis *seu* subhabenularis indicates rostrally the boundary between epithalamus and thalamus ventralis. Slightly more caudally, with the appearance of sulcus diencephalicus medius, it becomes the boundary between epithalamus and thalamus dorsalis. Still more caudally the sulcus synencephalicus joins the sulcus diencephalicus dorsalis, but this latter turns dorsad, forming a 'posthabenular' boundary sulcus, while the sulcus synencephalicus runs through the posthabenular dorsal thalamus. The sulcus diencephalicus ventralis, indicating the dorsal boundary of the hypothalamus, is rostrally well outlined but may become shallow caudalward. Within the preoptic hypothalamus, there is a variable sulcus intraencephalicus anterior, while the caudal hypothalamus displays a sulcus lateralis hypothalami posterioris (sulcus in hypothalamo of *Graf* HALLER, 1929; sulcus lateralis infundibuli) with accessory sulci in the lobi inferiores. A dorsal and ventral posterior hypothalamic region is thereby outlined. The rostral end of sulcus limitans, bending along the tuberculum posterius, is frequently well discernible. Figures 72 A–C illustrate configurational aspects of the posterior hypothalamus.

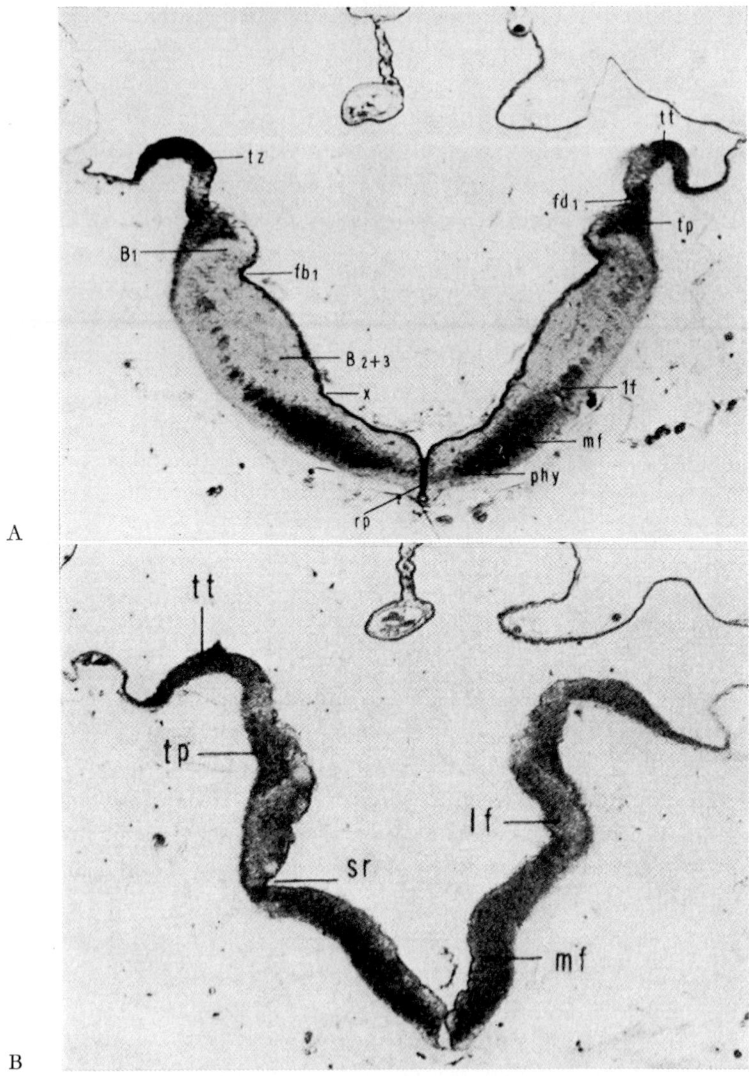

Figure 71 A. Cross-section through rostral end of recessus praeopticus in Chimaera (from K. and Niimi, 1969). B_1, B_{2+3}: telencephalic longitudinal zones; fb_1, fd_1: telencephalic sulci; lf: lateral forebrain bundle; phy: rostral portion of pars preoptica hypothalami; tp: tractus pallii; tt, tz: tractus taeniae at transition between disappearing telencephalic D-zones and taenia of diencephalic third ventricle; rp: bottom of recessus praeopticus; x: sulcus between telencephalon and preoptic hypothalamus. Although an ordinary hematoxylin-eosin stain, the medullated fiber tracts are conspicuously displayed.

Figure 71 B. Cross-section through the 'membranous' portion of the preoptic recess in Chimaera (from K. and Niimi, 1969). sr: sulcus lateralis recessus praeoptici. Other abbreviations as in Figure 79 A. mf in A and B: medial forebrain bundle.

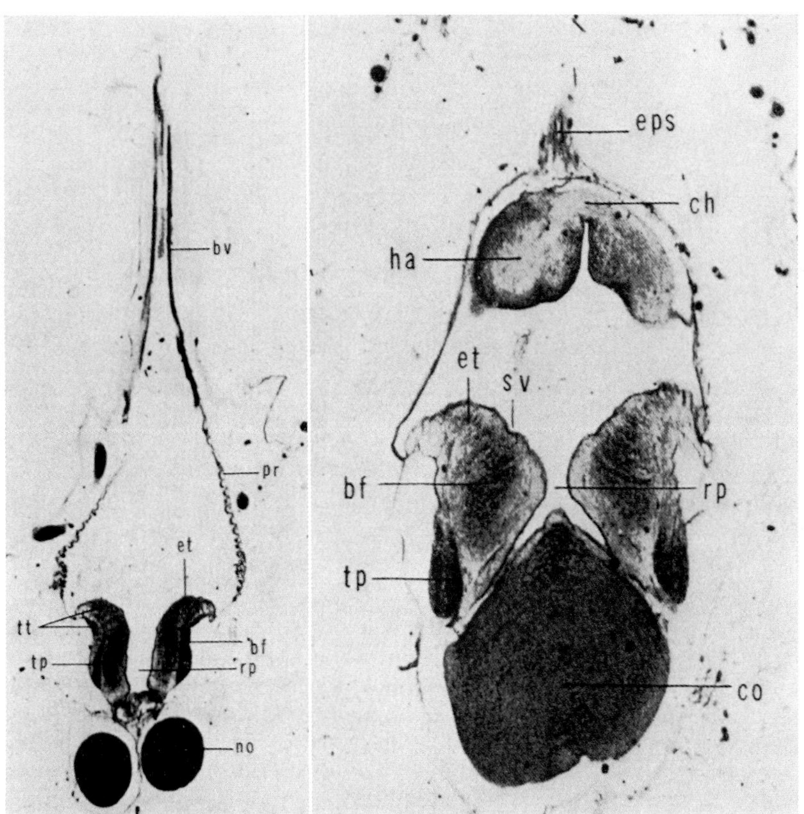

Figure 71 C. Cross-section through caudal, thickened part of recessus praeopticus in Chimaera (from K. and Niimi, 1969). bf: joint medial and lateral forebrain bundle; et: eminentia thalami ventralis; no: nervus opticus; pr: lamina epithelialis parencephalica: rp: recessus praeopticus, in whose floor the organon vasculosum praeopticum of Braak can be faintly recognized. Other abbreviations as in Figure 71 A. bv: blood vessel.

Figure 71 D. Cross-section through diencephalon of Chimaera at rostral level of optic chiasma (from K. and Niimi, 1969). ch: commissura habenulae; co: optic chiasma; eps: epiphysial stalk; ha: nucleus habenulae; sv: sulcus diencephalicus ventralis. The small midline knob above the chiasma is the caudal end of organon vasculosum. Other abbreviations as in Figures 71 A–C.

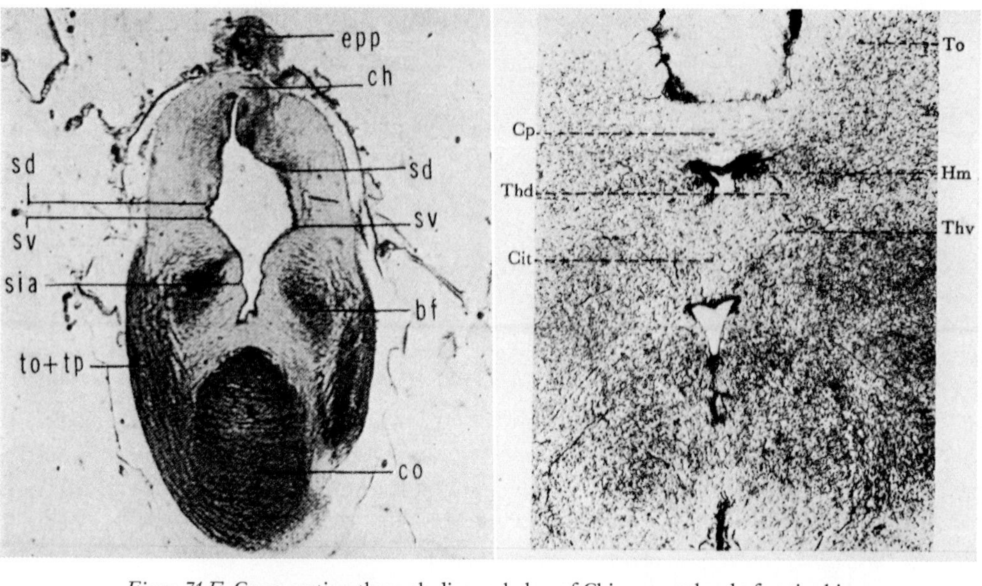

Figure 71 E. Cross-section through diencephalon of Chimaera at level of optic chiasma and rostral beginning of optic tract (from K. and NIIMI, 1969). epp: epiphysial stalk near origin; sd: sulcus diencephalicus dorsalis; sia: part of sulcus intraencephalicus anterior system; to+tp: tractus opticus and tractus pallii. Other abbreviations as in Figures 71 A–D.

Figure 71 F. Cross-section *(Nissl stain)* through the diencephalon of the Rajid Plagiostome Myliobatus (from TANAKA, 1959). Cit: 'commissura interthalamica' (median concrescence of thalamic ventricular wall); Cp: commissura posterior; Hm: nucleus habenularis medialis; Thd: thalamus dorsalis; Thv: approximate boundary between thalamus dorsalis and ventralis; To: tectum opticum.

The *diencephalic communication channels* include rostrally the basal forebrain bundle, the tractus taeniae, and the tractus pallii. The *basal forebrain bundle*, partly decussating within the anterior commissure, is a large system composed of medullated and non-medullated fibers which interconnect dorsal as well as basal telencephalic grisea with hypothalamus and thalamus.[125] A clear distinction between lateral and medial forebrain bundle, related to lateral respectively medial telencephalic wall cannot be easily made in the diencephalon, both components forming a common system running through the dorsal part of the hy-

[125] KAPPERS (1921), KAPPERS *et al.* (1936) and various other authors designate this system as tractus strio-thalamicus et hypothalamicus. It contains, however, doubtless pallial connections, including, *qua* medial forebrain bundle components, a kathomologon of the 'fornix'.

pothalamus, caudalward slanting toward tuberculum posterius and posterior hypothalamus. The *tractus medianus* of EDINGER (1893, 1908), KAPPERS *et al.* (1936), GERLACH (1947) and other authors can be interpreted as a component of the medial forebrain bundle. It interconnects the pallial area D_3 ('primordium hippocampi') with the hypothalamus in general, and in particular also with its infundibulo-mammillary region, thus presumably representing a kathomologon of the fornix. The basal forebrain bundle doubtless contains descending as well as ascending channels. Some of these latter may include, in addition to thalamo-telencephalic fibers mediating sensory input from the deuterencephalon and perhaps spinal cord, a moderate amount of optic input from dorsal and ventral geniculate nuclei. It is doubtful whether the basal forebrain bundle contains direct channels from telencephalon to deuterencephalic tegmentum, but this possibility cannot be excluded.

The *tractus taeniae* is a complex medullated and non-medullated system, running through the epithalamus into the nucleus habenulae. It represents stria medullaris and contains connections between this latter and dorsal as well as basal telencephalic grisea, receiving contributions from the basal forebrain bundle. The conventionally enumerated com-

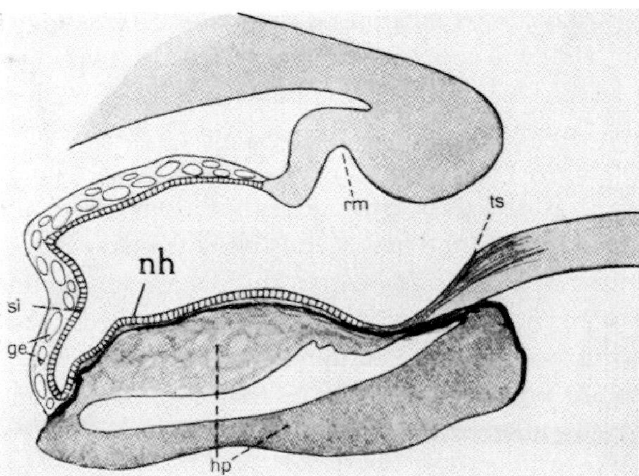

Figure 72 A. Sketch of midsagittal section showing relationships of saccus vasculosus and posterior hypothalamus in Acanthias (adapted from DAMMERMAN, 1910 and slightly modified after K., 1927). ge: blood vessels of saccus vasculosus; hp: adenohypophysis; nh: neurohypophysis; rm: 'recessus mammillaris'; si: neuroepithelium of saccus vasculosus; ts: (inferior) tractus sacci vasculosi, probably intermingled with hypothalamo-hypophysial tract. Compare with Figure 37 of section 1C.

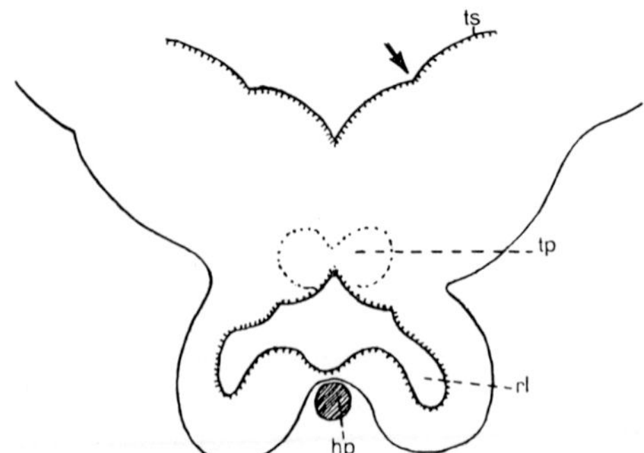

Figure 72 B. Sketch of cross-section through hypothalamic lateral recesses in Acanthias (from K., 1927). hp: adenohypophysis; rl: lateral recess; tp: tuberculum posterius; ts: torus semicircularis of mesencephalon; arrow: sulcus limitans.

ponents of tractus taeniae are tractus olfacto-habenularis medialis and lateralis, tractus and cortico-habenularis with lateral and medial subdivisions.

The medullated *tractus pallii* (Figs. 70 B, C), peculiar for Plagiostomes, connects posterodorsal portions of the telencephalon with diencephalon. It runs along the caudolateral surface of the telencephalon, passes laterally to tractus opticus, or partially intermingled with its fibers, through the diencephalon, substantially decussating caudally to chiasma and basally to commissura transversa in the hypothalamus, disappearing in the region of tuberculum posterius. It may contain ascending and descending fibers (cf. also Figs. 71 A–E).

The *optic tract*, whose fascicles mainly reach the tectum opticum mesencephali, doubtless also provides connections to lateral geniculate grisea, pretectal region and presumably also hypothalamus.

The *intrinsic diencephalic channels*, most of which are running in an essentially dorsoventral (transverse) plane, interconnect the four zonal diencephalic grisea, but are intermingled with fibers pertaining to the basal forebrain bundle and to the caudal communication channels. There are thus fibers running from preoptic recess to epithalamus (tractus olfactohabenularis medialis posterior of JOHNSTON), and, among others, a 'tractus thalamohabenularis'.

The longitudinal neurosecretory *tractus preoptico- (or supraoptico-) hypophyseus*[126] is essentially non-medullated. It may become intermin-

[126] Details on the neurosecretory hypothalamo-hypophysial system in the Plagiostome spinax niger have been recorded by BRAAK (1962).

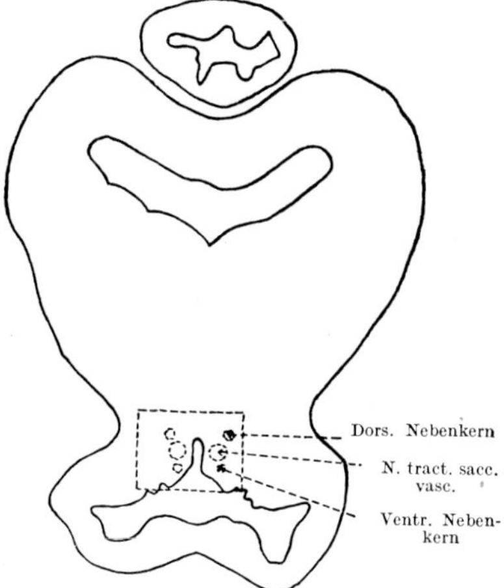

I

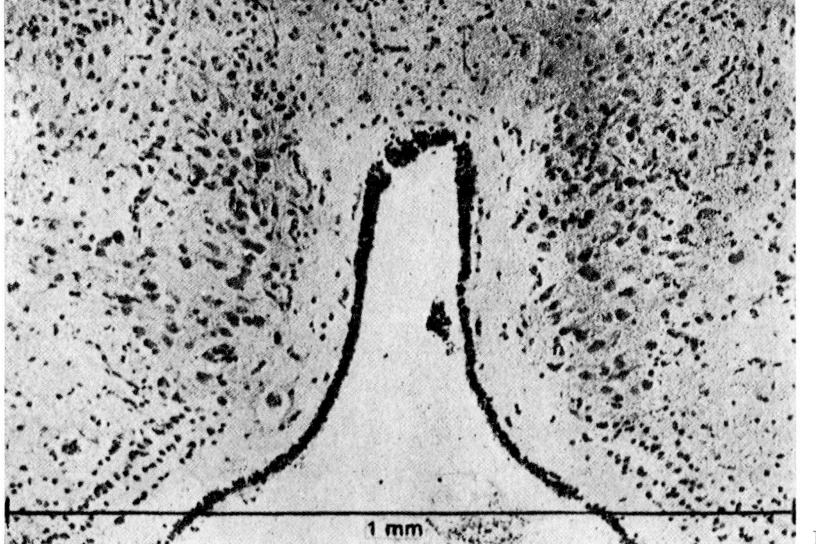

II

Figure 72C. Nucleus tractus sacci vasculosi in Galeus canis (from GERLACH, 1947). I. Sketch showing location of section. II. Cytoarchitecture of nucleus (modified *Nissl stain*).

gled with the likewise non-medullated or poorly medullated (ventral) tractus sacci vasculosi mentioned above in section 1 C. In addition, there is a dorsal tractus sacci vasculosi (cf. Fig. 50 A) which, since it seems to reach not only diencephalic cell groups but also the tuberculum posterius (cf. Figs. 50 B, 72 C) can also be subsumed under the extrinsic communication channels dealt with in the next paragraph.

As regards the *caudal communication channels*, the system of the fasciculus retroflexus, with a large proportion of medullated fibers, interconnects the habenulae with the mesencephalic nucleus interpeduncularis and with caudal hypothalamic neighborhoods. Fibers between habenulae and tectum opticum, passing through pretectal grisea, are likewise present.

Ascending extensions from the general spino-bulbo-tectal lemniscus and from the lateral lemniscus system may reach both dorsal and ventral thalamus.[127] There are, in addition, tecto-thalamic channels.

The posterior ('infundibulomammillary') portion of the hypothalamus has a conspicuous medullated connection (tractus lobo-cerebellaris) with the cerebellum[128] in addition to those with tegmental formations[129] (so-called tractus lobobulbaris and mammillopeduncularis *autorum*).

Discounting the telencephalo-diencephalic commissura anterior, the following *commissural systems* can be recognized: (1) The commissura habenulae. A non-medullated portion of this commissure seems to represent a commissure of telencephalic fiber systems (so-called commissura superior of Kappers *et al.*, 1936), the remainder of the crossing fibers being decussating or true 'antimeric' commissural channels pertaining mostly to the habenular system proper. (2) The posterior commissure, pertaining to both diencephalon and mesencephalon, as discussed in sections 1 A and 2, requires no further comments. (3) The commissura postoptica *seu* transversa is large and well medullated. It is mainly related to tectal and tegmental fiber systems,[129a] either true commissural, or decussating with diencephalic connections. (4) The commissura *seu* decussatio tractus pallii, as dealt with above. (5) Nondescript, essentially hypothalamic partly medullated and partly nonmedullated preinfundibular and postinfundibular commissures of which the latter may include a non-medullated decussatio tractus sacci vasculosi. The postinfundibular commissure merges with the rostral end of commissura ansulata (commissura tegmenti ventralis).

With respect to these diverse fiber systems it should be added that, while the enumerated main channels can be more or less distinctly rec-

[127] Kappers *et al.* (1936) also depict a tractus octavo-hypothalamicus.

[128] Cf. volume 4, chapter X.

[129] Tractus cerebello-meso-diencephalicus of Gerlach (1947) apparently mostly crossing in the commissura ansulata (mesencephalon) and related decussations. Gerlach reported that some of its fibers also reach the ventral thalamus.

[129a] Cf. volume 4, chapter XI.

ognized by means of the available techniques such as myelin stains, silver impregnations and degeneration methods, the details of their connections and of their ascending or descending course are far less clarified than most descriptions of particular tracts seem to imply. Only a very generalized or 'overall' arrangement of pathways appears reasonably well established, and the discrepancies in the interpretations by different investigators are thus easily understandable as due to the inherent difficulties and uncertainties. Further references to the relevant literature can be found in the treatises by KAPPERS *et al.* (1936), BECCARI (1943), KAPPERS (1947), and in the papers by GERLACH (1947), TANAKA (1959), K. and NIIMI (1969), and in the chapter by CROSBY and SHOWERS (1969).

4. Ganoids and Teleosts; Latimeria

The diencephalon of *Ganoids* is in numerous respects comparable to that of Plagiostomes. As in these latter, the habenular nuclei are commonly asymmetric, the left being somewhat larger than the right one. Nuclear subdivisions within the habenular ganglion are, if at all suggested, quite inconspicuous.

Both thalamus dorsalis and thalamus ventralis display an essentially periventricular arrangement. Nevertheless, some scattered nerve cells displaced toward the optic tract represent primordia of dorsal and ventral lateral geniculate nuclei. The thalamus dorsalis is caudally more developed than at rostral levels, where it does not extend as far as the thalamus ventralis, which reaches here the ventral boundary of the habenula, being separated from this latter by the sulcus diencephalicus

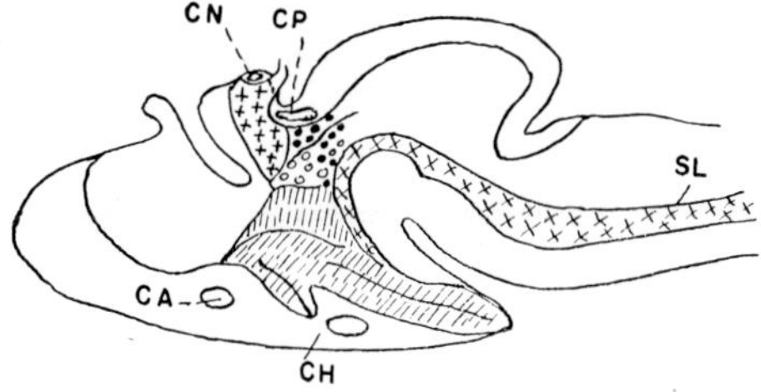

Figure 73 A. Semidiagrammatic reconstruction of longitudinal zones in the diencephalon in a young (adult) specimen of the Ganoid Amia (from K., 1956). CA: commissura anterior; CH: chiasmatic ridge; CN: commissura habenulae; CP: commissura posterior; SL: sulcus limitans. Longitudinal zones designated as in Figures 60 A and 69 A.

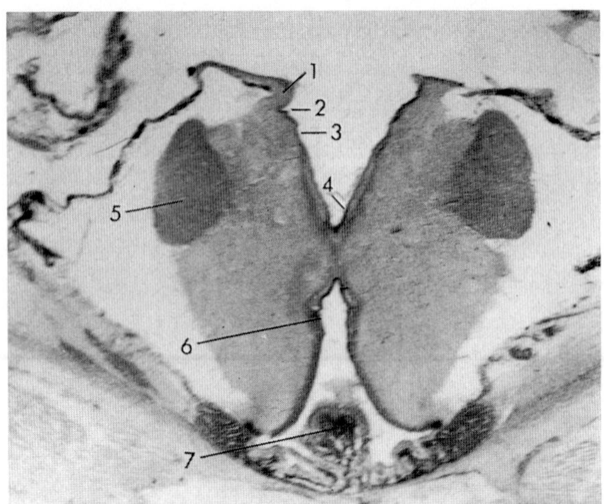

I

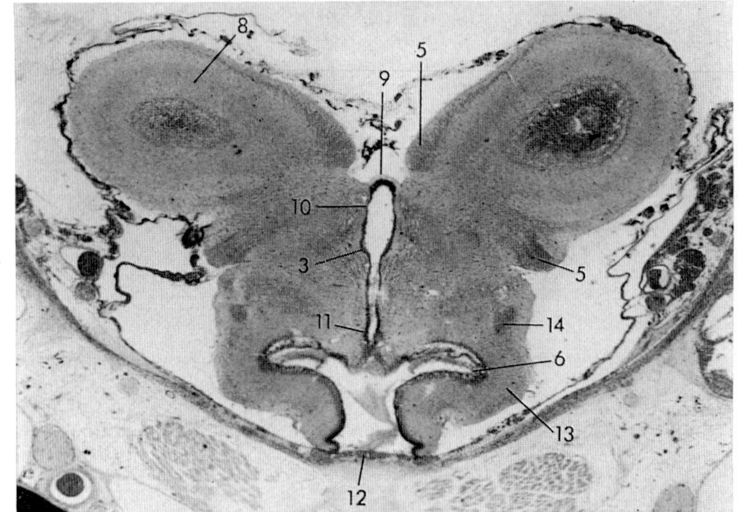

II

Figure 73 B. Cross-sections through the diencephalon of the Ganoid Amia calva, (I) at caudal level of postoptic commissures, and (II) at rostral beginning of commissura posterior (hematoxylin-eosin; I ×29, II ×23; red. $^2/_3$). 1: rostral tip of epithalamus (ganglion habenulae); 2: sulcus diencephalicus dorsalis; 3: sulcus dienc. medius; 4: sulcus dienc. ventralis; 5: tractus opticus; 6: system of sulcus lateralis hypothalami posterioris; 7: hypophysial complex (cf. Fig. 38 C); 8: tectum opticum *(sive* mesencephali): 9: rostral beginning of commissura posterior with subcommissural organ; 10: sulcus synencephalicus; 11: rostral portion of sulcus limitans above rostral end of tuberculum posterius; 12: rostral beginning of saccus vasculosus; 13: lobi inferiores hypothalami; 14: fiber bundle pertaining to commissura transversa. In comparing with Figure 73 A, based on a very young specimen, in which, moreover, the rostral recess of the posterior hypothalamus was not developed, the two cross-sectional planes BI and II would be highly oblique (dorsorostral to ventrocaudal).

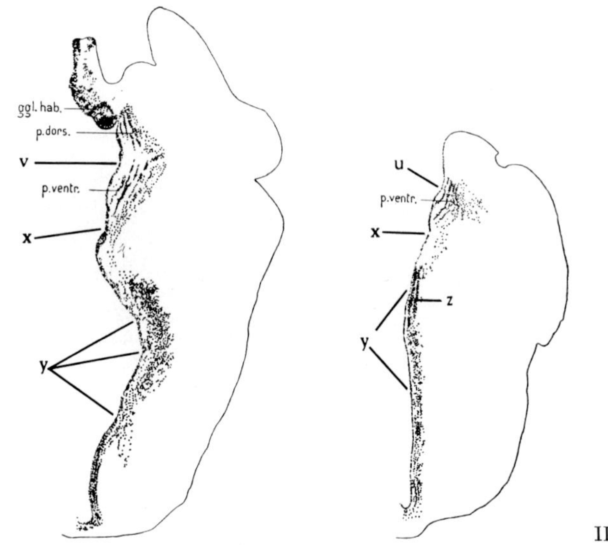

Figure 73 C. Cross-sections through the diencephalon of the Ganoid Polypterus bichir, (I) at level of ganglion habenulae, and (II) at level of preoptic nucleus (from JEENER, 1930). ggl.hab.: ganglion habenulae; p.dors.: thalamus dorsalis; p.ventr.: thalamus ventralis. The added leads are: u: sulcus diencephalicus dorsalis; v: s. dienc. medius; x: s. dienc. ventralis; y: system of s. lateralis hypothalami inferioris; z: 'magnocellular' preoptic nucleus. In I, the unlabelled groove between epithalamus (habenula) and dorsal thalamus is the sulcus dienc. dorsalis.

dorsalis. The dorsal thalamus contains caudally a nucleus of the posterior commissure and, laterally adjacent, a nondescript dorsal thalamic cell group. HOOGENBOM's (1929) corpus geniculatum laterale can be interpreted as the nucleus lentiformis mesencephali, whose mediocaudal portion is apparently the cited author's 'nucleus praetectalis'.[130]

The hypothalamus displays a 'magnocellular' preoptic nucleus, presumably related to the hypothalamo-hypophysial neurosecretory tract, and a moderately well developed posterior region with lobi inferiores respectively lateral recesses. An extensive saccus vasculosus is commonly present. An entopeduncular nucleus is suggested by nerve cells displaced toward, and partly into, the basal forebrain bundle along the dorsal part of anterior and posterior hypothalamus. A 'ganglion sacci

[130] Figures 30, 31 and 33 of HOOGENBOOM (1929). The 'nucleus praetectalis' of that author's figure 32 is presumably the rostral end of the mesencephalic torus semicircularis, while her so-called nucleus intergeniculatus is presumably a median portion of nucleus lentiformis mesencephali.

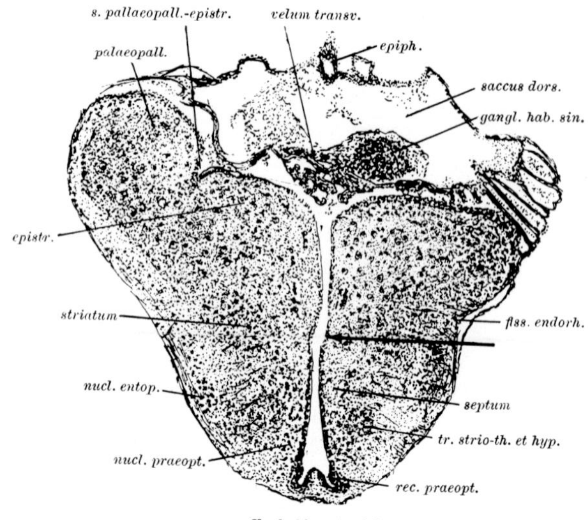

s. pallaeopall.-epistr.
velum transv.
epiph.
palaeopall.
saccus dors.
gangl. hab. sin.
epistr.
striatum
fiss. endorh.
nucl. entop.
septum
tr. strio-th. et hyp.
nucl. praeopt.
rec. praeopt.

Vorderhirn, kaudaler Teil.
Vergr. 27fach.

A

tectum opt. p. ant.
saccus epiph.
gangl. hab. dext.
gangl. hab. sin.
sulcus subhab.
sulcus med. th.
tr. strio-th. et hyp.
tr. strio-th. et. hyp.
decc. N. N. opt.

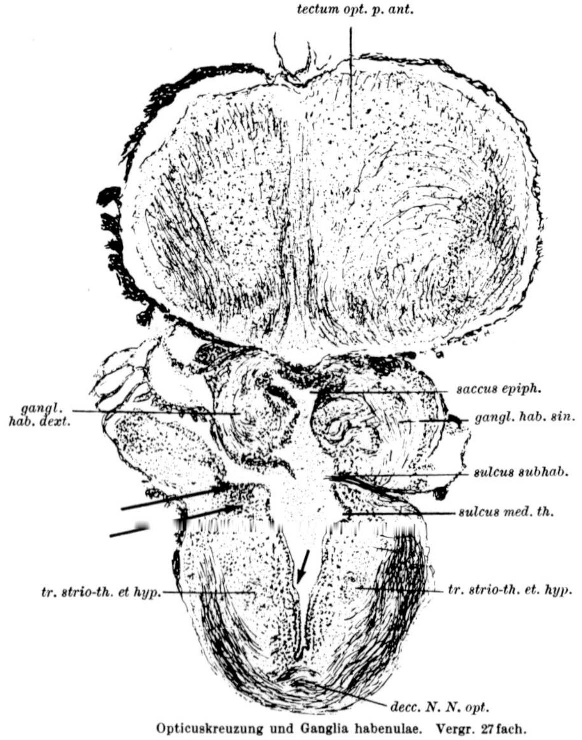

B

Opticuskreuzung und Ganglia habenulae. Vergr. 27fach.

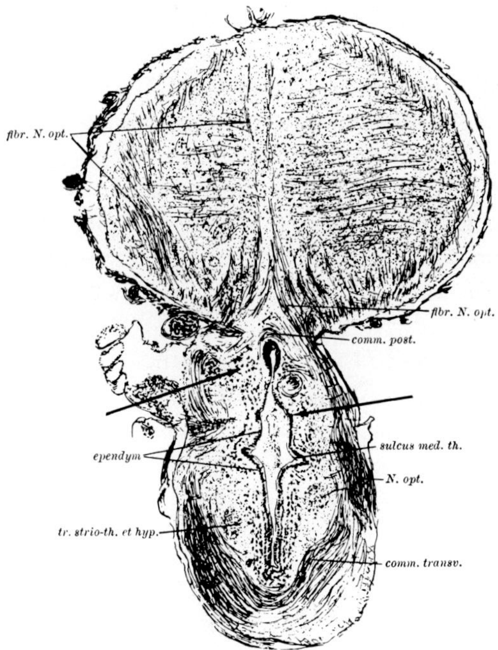

Figure 74 C. Cross-section through brain of Polyodon at level of postoptic commissures (from HOOGENBOOM, 1929). Below posterior commissure, the subcommissural organ can be identified. N.opt.: optic tract; sulcus med.th.: our s. dienc. ventralis. Added arrows: nucleus commissurae posterioris (left); sulcus diencephalicus medius (right). For posterior inferior hypothalamus and lobi inferiores cf. Figures 376 A–C of vol. 4.

Figure 74 A. Cross-section through the forebrain of the Ganoid Polyodon at the diencephalo-telencephalic boundary zone (from HOOGENBOOM, 1929). The right side of the figure is slightly more caudal than the left. epistr.: our zone B_1; fiss.endorh.: our sulcus telo-diencephalicus; palaeopall.: our D-zones; septum: dorsal portion of preoptic recess; s.pallaeopal.-epistr.: our fb_1; striatum: massa cellularis reuniens, pars inferior; added arrow: our sulcus terminalis, related to the rostral end of the diencephalic zones, and separating telencephalic from diencephalic grisea. Other self-explanatory abbreviations in agreement with our views.

Figure 74 B. Cross-section through brain of Polyodon at level of optic chiasma (from HOOGENBOOM, 1929). On the figure's left side, the caudal end of telencephalon is seen dorsal to sulcus terminalis; saccus epiph.: recess between habenulae and habenular commissure, near origin of epiphysis; sulcus med. th.: our sulcus diencephalicus ventralis; sulcus subhab.: our sulcus dienceph. dorsalis. Added arrows from above downward: sulcus terminalis; eminentia thalami ventralis; sulcus intraencephalicus anterior system.

vasculosi' was recorded by HOOGENBOOM (1929) in the boundary of tuberculum posterius and hypothalamic lobi inferiores.

The diencephalic ventricular wall displays sulcus diencephalicus dorsalis (sulcus subhabenularis), a shallow posthabenular sulcus diencephalicus medius merging with the sulcus synencephalicus system, a commonly well outlined sulcus diencephalicus ventralis (HOOGENBOOM's 'sulcus medius thalami'), a preoptic sulcus intraencephalicus anterior system, and the system of sulcus lateralis hypothalami posterioris. The sulcus limitans can be traced as far as the region of tuberculum posterius.

Figures 73 A–C and 74 A–C illustrate the overall diencephalic configuration obtaining in Amia, Polypterus, and Polyodon (cf. also Figs. 376 A–C of vol. 4). As regards the neuronal *communication channels*, HOOGENBOOM (1929) justly points out the rather diffuse arrangement of the relevant fiber systems, which, on the whole, are comparable to those described above for Plagiostomes. Because of the everted telencephalic pattern, the basal forebrain bundle is somewhat differently organized, *qua* distribution of its fiber systems, than that of Plagiostomes.[131] Thus, the fiber systems reaching the habenula (tractus taeniae system of Selachians) join, at levels of the lamina terminalis, the telencephalic basal forebrain bundles from which they then turn dorsomedialward toward the habenula at the telencephalodiencephalic boundary, thereby forming a short 'stria medullaris' (cf. Fig. 74 B). Details about the presumably obtaining optic input to thalamic and hypothalamic grisea remain insufficiently elucidated.

The *caudal communication channels* include fasciculus retroflexus, and, in a rather nondescript and poorly identifiable way, the main fiber systems described in section 3 for Plagiostomes. Besides caudal extensions of the basal forebrain bundle system, a partly crossed tractus cerebello-lobaris, a tractus thalamo-tectalis, a tractus lobo-mesencephalicus, a tractus spino-bulbo-thalamicus et hypothalamicus, and a tractus mammillopeduncularis were tentatively identified by KAPPERS (1907).

---

[131] The telencephalic portions of the basal forebrain bundles and the anterior commissure of the everted Osteichthyan brain shall be dealt with in section 5 of chapter XIII. A tractus pallii, commonly characteristic for the Plagiostome forebrain, is not present in Osteichthyes, although analogous fiber connections become included in the basal forebrain bundle.

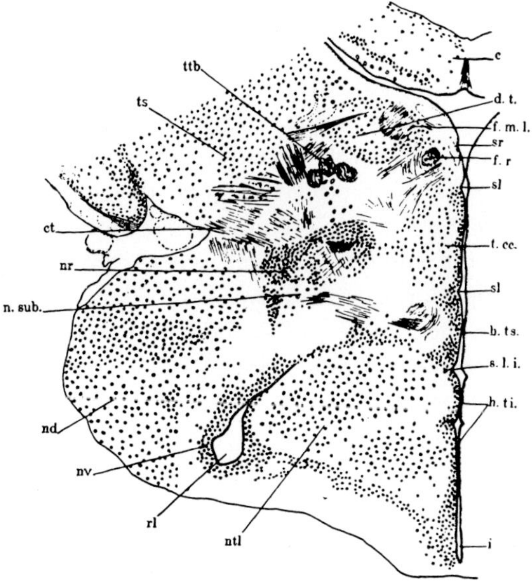

Figure 76B. Cross-section through the diencephalon of Corydora at transition to mesencephalon (from MILLER, 1940).

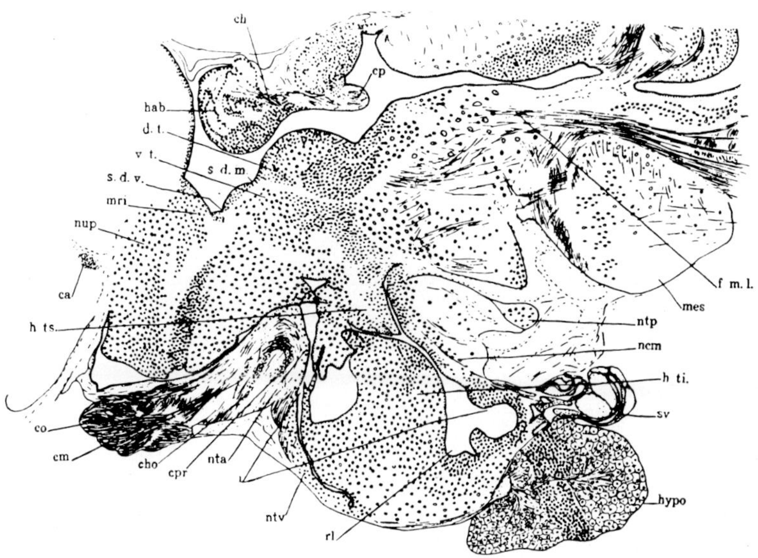

Figure 76C. Paramedian sagittal section through the diencephalon of Corydora (from MILLER, 1940).

pretectal cell groups which can be subdivided into nucleus praetectalis and area praetectalis. The nucleus lentiformis mesencephali (nucleus corticalis, Holmgren, 1920; Miller, 1940) lies laterally or dorsally to the diencephalic pretectal region.[135] The ventrolateral subdivision is particularly but not exclusively related to the optic tract and represents a variously differentiated dorsal lateral geniculate griseum.

The *ventral thalamus* begins rostrally at habenular or even pre-habenular levels with the eminentia thalami *sensu proprio*. A massa cellularis reuniens, pars superior, extends toward the telencephalon, but is less conspicuous than the more rostrally and basally located hypothalamic massa cellularis reuniens pars inferior. More caudally, a medial and a lateral portion of thalamus ventralis can be distinguished. The lateral cell groups form a ventral lateral geniculate nucleus with variable further subdivisions. A usually large medial portion of the lateral subdivision is the so-called nucleus anterior (Holmgren, 1920), 'corpus glomerulosum pars anterior' or nucleus rotundus of Jansen (1929) and Franz (1912). It forms part of the ventral lateral geniculate grisea,[136] which, in turn, may be closely adjacent or fused with the dorsal lateral geniculate ones (cf. Figs. 77 A–C). The lateral geniculate complex of many Teleosts is, as shown by Franz (1912), a highly differentiated and often laminated or folded griseum jointly formed by ventral and dorsal thalamus (cf. Fig. 77 E). Said complexity is directly related to the high development of the optic system in these Teleosts. It should be added that 'glomerular' synaptic endings, recorded by various authors in the lateral geniculate complex as well as in the hypothalamic nucleus rotundus of diverse Teleosts are not an obligate feature, but characterize only certain taxonomic forms (cf. e.g. Beccari, 1943). The available data do not yet seem sufficient to summarize, *qua* Teleostean groups, the distribution of 'glomerular' or 'aglomerular' synaptic structures upon morphologically homologous grisea.

---

[135] Wallenberg (1907) designates the medial part of this cell group as nucleus corticalis, and the lateral one as nucleus lentiformis (Fig. 38, p. 395, loc.cit.).

[136] Miller (1940) had included this griseum into the ventral portion of thalamus dorsalis, but subsequent observations convinced me that it is a derivative of the ventral thalamus. There obtains, nevertheless, as mentioned in the text, a close apposition if not fusion of dorsal and ventral thalamic derivatives in this diencephalic neighborhood of various Teleosts. It should be added that, because of these various complexities, the dorsal and ventral lateral geniculate Teleostean grisea are kathomologous but not orthohomologous with those of higher Vertebrates.

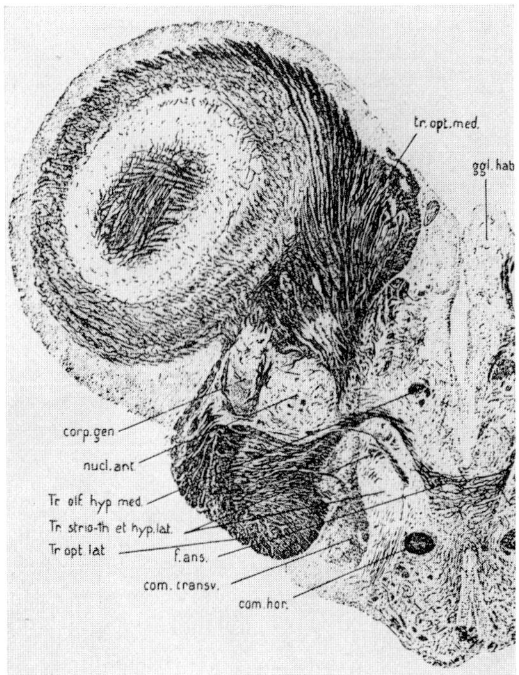

Figure 77 A. Cross-section (myelin stain) through diencephalon and tectum opticum in the Teleost Idus idus (slightly modified after JEENER, 1930, from KAPPERS, 1947). com. hor.: commissura horizontalis; f.ans.: fibrae ansulatae (components of dorsal supraoptic commissure. Other abbreviations self-explanatory. The 'nucleus anterior' is presumably a partly dorsal and partly ventral, medial component of the lateral geniculate complex. 'Corp. gen.' is its lateral component, likewise partly dorsal and partly ventral thalamic.

The *hypothalamus* of Teleosts is extensive and displays a complex as well as taxonomically rather variable nuclear differentiation. From a morphologic viewpoint, it is convenient to distinguish a rostral preoptic hypothalamus and a postoptic posterior hypothalamus, in which latter a dorsal (superior) and a ventral (inferior) subdivision are particularly well displayed.

In the *preoptic hypothalamus*, where a dorsal and a ventral portion can also be noted, although less clearly than in the posterior hypothalamus, a periventricular and a lateral subdivision are recognizable. A dorsolateral expansion of the periventricular elements forms the massa cellularis reuniens pars inferior which extends toward and into the telencephalon (cf. Fig. 196 D, chapter XIII, section 5). Caudalward and basalward, the pars inferior massae cellularis reunientis blends with the

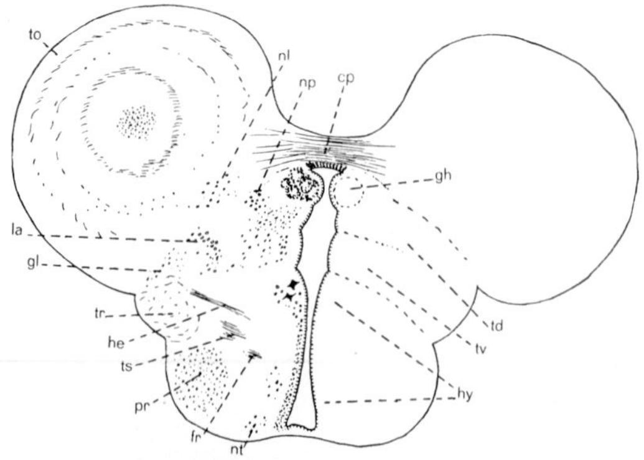

I

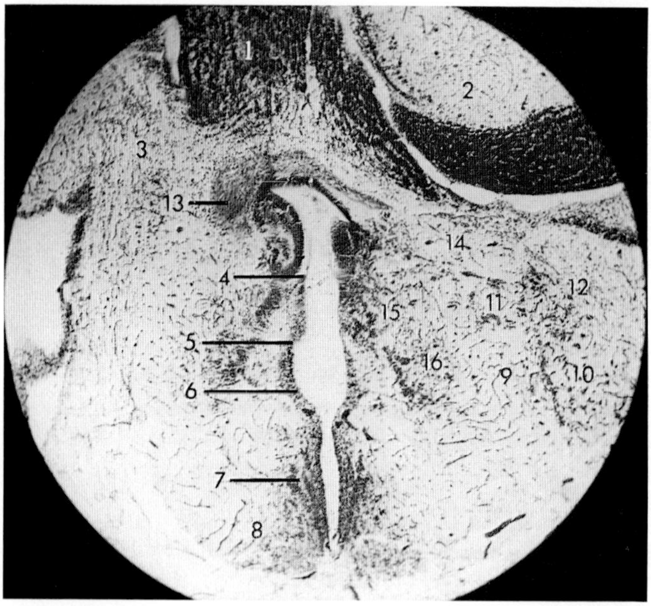

II

Figure 77B. Cross-sections through the diencephalon of the Teleosts Cyprinus auratus (I) and Carassius auratus (II) at rostral levels of posterior commissure (I from K., 1927; II modified after K., 1929). cp: comm. posterior; fr. comm. horizontalis *(Fritsch);* gh: ganglion habenulae, gl: lateral part of corpus geniculatum laterale complex, hc: comm. minor *(C.L. Herrick);* hy: hypothalamus; la: medial portion of corpus geniculatum laterale complex; nl: n. lentiformis mesencephali; np: common rostral end of n. praetectalis and n. commissurae posterioris; nt: n. tuberis; pr: n. praerotundus; td: thalamus dorsalis; to: tectum opticum; tr: tractus opticus; ts: comm. transversa; tv: thal. ventr.; 1: torus longitudinalis; 2: valvula cerebelli; 3: n. lentiformis mesencephali, pars medialis; 4: s. dienc. dorsalis; 5: s. dienc. medius; 6: s. dienc. ventralis; 7: n. hypothalamicus paraventricularis; 8: n. tuberis; 9: medial ventral thalamic component of lateral geniculate complex; 10: lateral ventral thalamic component of lateral geniculate complex; 11, 12:

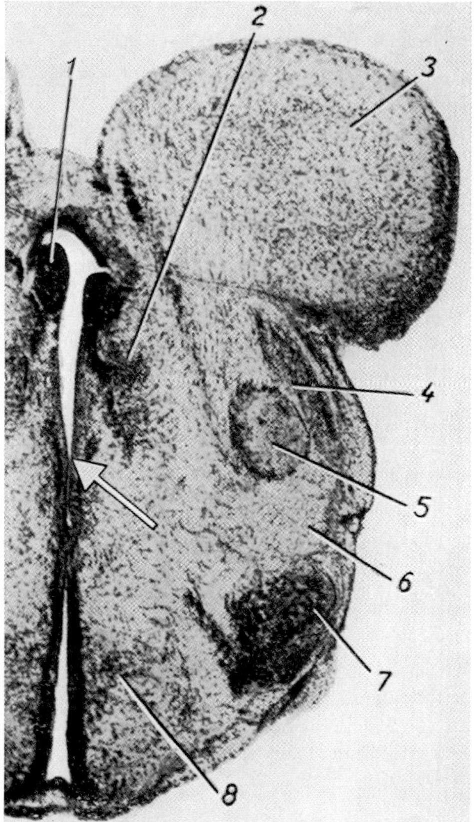

Figure 77C. Cross-section through the diencephalon of the Carp (Cyprinus carpio) at rostral level of tectum opticum (from BECCARI, 1943). 1: caudal end of n. habenulae; 2: thalamus dorsalis, medial portion; 3: tectum opticum; 4: lateral portion of corpus geniculatum laterale complex; 5: '*nucleo anteriore del talamo*' (medial portion of corpus geniculatum laterale complex); 6: optic tract; 7: nucleus praerotundus; 8: '*nucleo preottico magnocellulare*'. The added arrow points to sulcus diencephalicus ventralis. Lead 2 crosses obliquely the dark cell band of nucleus lentiformis mesencephali.

---

medial and lateral dorsal thalamic components of lateral geniculate complex; 13: n. commissurae posterioris; 14: n. praetectalis; 15: medial portion of thalamus dorsalis; 16: medial portion of thalamus ventralis. In I, the dorsal and ventral components of the lateral geniculate complex have not been separately labelled, but can be recognized. In II, the right side is further caudad than the left. Note, in I and II, caudal expansion of habenular grisea into levels of posterior commissure.

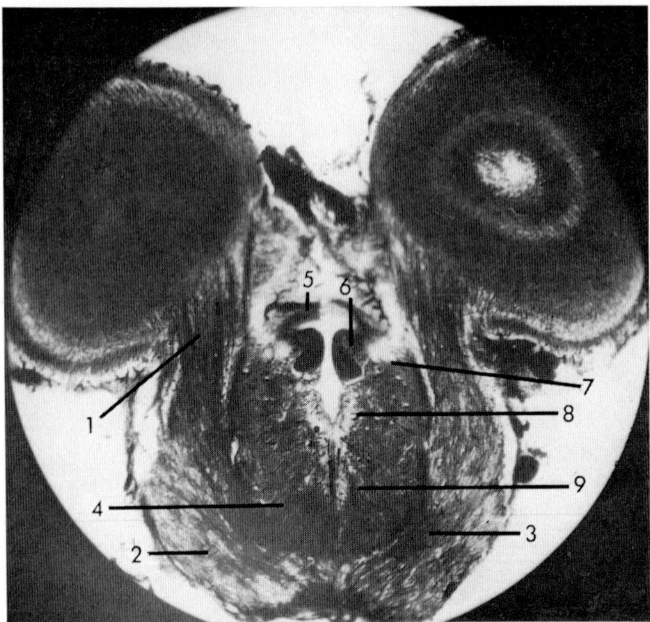

Figure 77D. Cross-section through diencephalon and rostral tectum opticum in Carassius auratus (myelin stain), showing part of the course of tractus opticus. 1: optic tract; 2: commissura supraoptica ventralis; 3: commissura supraoptica dorsalis; 4: basal forebrain bundle; 5: commissura posterior; 6: caudal end of habenular grisea; 7: thalamus dorsalis; 8: thalamus ventralis; 9: hypothalamus.

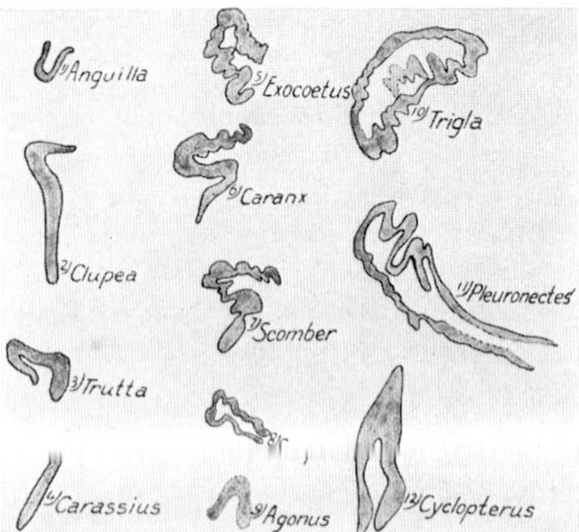

Figure 77E. Various degrees of differentiation of the corpus geniculatum laterale in some Teleosts (after FRANZ, 1912, from KAPPERS, 1921). It will be seen that neither FRANZ nor KAPPERS made a distinction between dorsal and ventral thalamic lateral geniculate body. Likewise, by comparing 4) Carassius with the present Figures 77B and C, it will be noted that the cited authors depicted here only the lateral component of the lateral geniculate body.

entopeduncular cell groups, among which, at least in some Teleosts, a suprapeduncular nucleus and interstitial nuclei of the basal forebrain bundles can be recognized as approximately delimitable entities.[136a] The rostral entopeduncular group represents the lateral subdivision of the pars preoptica hypothalami, the massa cellularis reuniens forming a transition between lateral and medial subdivision.

The medial, periventricular cell population of the pars praeoptica hypothalami usually extends rostralward into a preoptic recess located basally to the anterior commissure. A neurosecretory magnocellular preoptic 'nucleus' and a non-secretory parvocellular component can be generally distinguished, but are frequently intermingled with each other (MILLER, 1940). CROSBY and SHOWERS (1969) justly state that the magnocellular nucleus is variably and often highly developed, not uncommonly in the form of a discrete nuclear mass, which may have a more rostral, an intermediate, or a relatively caudal position (cf. e.g. Figs. 77 B, 79 A–C). No correlation could be made between the position of this 'nucleus', its differentiation, or its length, with the body and brain size of particular species of Teleosts.

In addition to the overall parvocellular preoptic nucleus, which may display dorsoventrally arranged rows of cells, i.e. appear 'laminated', various nondescript condensations, such as 'suprachiasmatic nucleus', 'diffuse supraoptic nucleus' etc. have been reported (cf. e.g. CHARLTON, 1932; NIIMI *et al.*, 1963; CROSBY and SHOWERS, 1969).

The *posterior hypothalamic region* of Teleosts is especially differentiated and complex. In both its superior (dorsal) and inferior (ventral) subdivisions, a periventricular cell population is present, often not clearly separated from the more peripheral cell groups or 'nuclei'. Its caudodorsal portion abuts on the mesencephalic tegmental cell cord (Fig. 76 B) from which it can be reasonably well distinguished. Its rostral portion, the 'nucleus paraventricularis' is continuous with the periventricular preoptic cell masses.

The most conspicuous component of the Teleostean dorsolateral posterior hypothalamus is the *nucleus rotundus complex*. This nucleus was

---

[136a] In the fairly representative Figure 257 D, p. 501 of volume 3/II (Siluroid Corydora), the lead *nen* indicates only the dorsal portion of the entopeduncular cell group, differentiated from, respectively blending with the unlabelled pars inferior massae cellularis reunientis, which can easily be identified below and laterally to lead *st*. The unlabelled main portion of the entopeduncular cell group can be identified in the lateral portion of preoptic neighborhood, basally to the concavity of *nen*.

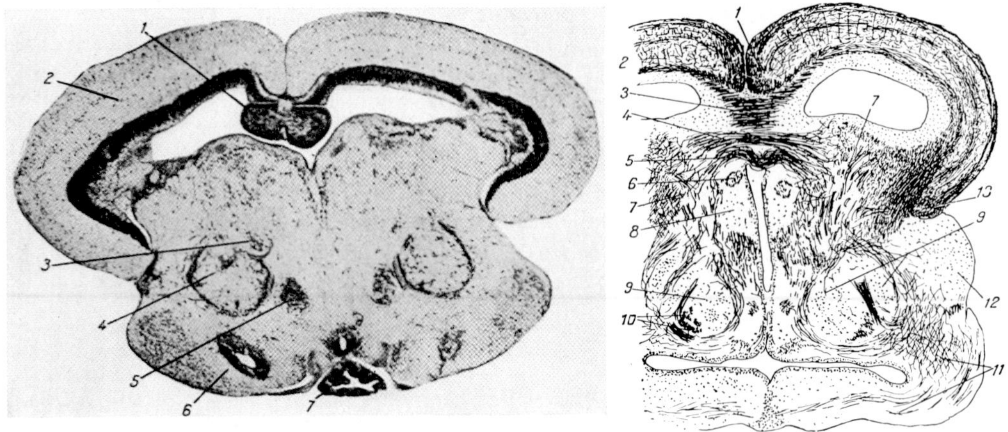

Figure 78 A. Cross-section *(Giemsa stain)* through the caudal diencephalon and rostral mesencephalon of the Teleost Gasterosteus aculeatus (from BECCARI, 1943). 1: torus longitudinalis; 2: tectum opticum; 3: nucleus suprarotundus; 4: nucleus rotundus (glomerularis); 5: nucleus praerotundus medialis; 6: lobi inferiores (hypothalami), pars lateralis, with the laterally located 'nucleus diffusus'; 7: saccus vasculosus.

Figure 78 B. Cross-sections through diencephalo-mesencephalic levels of the Teleost Anguilla anguilla. I. *Cajal silver impregnation* (from BECCARI, 1943). II. Hematoxylin-eosin stain (advanced larva, modified after K., 1929). 1: fasciculus opticus medialis; 2: tectum opticum; 3: commissura tecti; 4, 5: dorsal and ventral portions of commissura posterior; 6: fasciculus retroflexus (habenulo-peduncularis); 7: pretectal region (according to BECCARI: *estremita rostrale del tegmento del mesencefalo)*; 8: caudal portion of dorsal thalamus *('campo posthabenulare')*; 9: nucleus rotundus (aglomerularis); 10: commissura horizontalis and caudal parts of basal forebrain bundle; 11: hypothalamus (lobi inferiores); 12: 'tours lateralis' (dorsal posterior hypothalamus); 13: fasciculus opticus marginalis; 14: sulcus diencephalicus dorsalis at junction with sulcus synencephalicus; 15: sulcus diencephalicus medius; 16: sulcus diencephalicus ventralis; 17: sulcus limitans; 18: hypophysial complex; 19: tegmental cell cord in tuberculum posterius; 20: (aglomerular) nucleus rotundus complex; 21: caudal thalamus ventralis; 22: nucleus lentiformis mesencephali; 23: commissura posterior; 24: torus longitudinalis.

apparently first recognized and named by FRITSCH (1878). Because of the 'glomerular' synapses present in various species, it became also designated as corpus glomerulosum pars rotunda by FRANZ (1912) and others. It extends caudalward into the planes where mesencephalic neighborhoods overlap (cf. Figs. 76 B, 78 A–C). Characterized by a dense neuropil with scattered small and medium-sized cells, commonly forming a denser surrounding 'capsule', this nucleus blends with its adjacent cell groups nucleus praerotundus (Figs. 77 B, C), subrotundus, and n. suprarotundus. The so-called nucleus posterior hypothalami

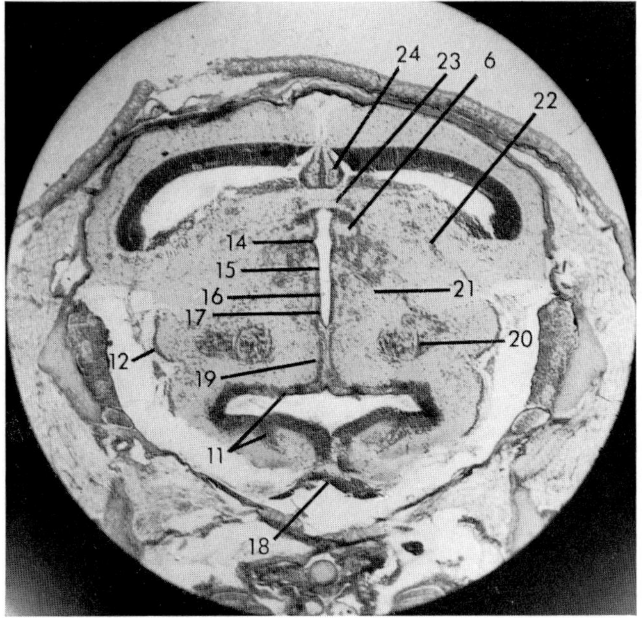

BII

(GOLDSTEIN, 1905; SHELDON, 1912; HOLMGREN, 1920) is apparently a portion of the nucleus rotundus. The entire complex has been interpreted as pertaining to the thalamus, but our observations (K., 1927; MILLER, 1940) have rather clearly established that it is a part of the dorsal hypothalamic longitudinal zone. CROSBY and SHOWERS (1969), although expressing a noncommittal opinion, have included it into their description of the hypothalamus, while BECCARI (1943) deals with this griseal complex in his discussion of the thalamus.[137] Within the caudal portion of the basal forebrain bundles scattered interstitial neuronal elements represent an extension of the entopeduncular nucleus into the posterior hypothalamus.

The *lobi inferiores* with their several rather variable recesses form the *inferior or ventral posterior hypothalamus*, which also includes the saccus vasculosus (Figs. 76 C, 78 A–C, 80 A, B). Medial (median), lateral, and posterior lobes can be distinguished. The rostral lobus medianus con-

---

[137] Further comments on the rotundus complex, with references to the bibliography and to the various terminologies, parcellations, and interpretations by the relevant authors can be found in the publications by KAPPERS *et al.* (1936), BECCARI (1943), and CROSBY and SHOWERS (1969).

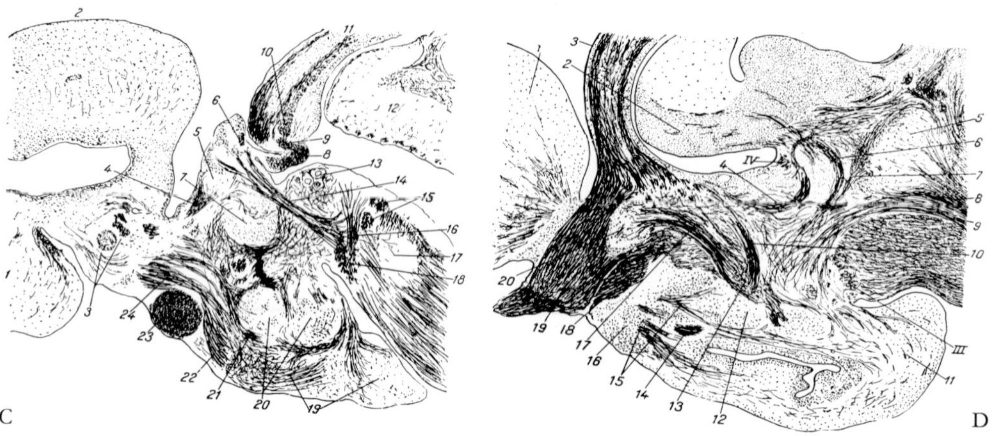

Figure 78C. Paramedian sagittal section *(Cajal silver impregnation)* through prosencephalon and mesencephalon of Anguilla (from Beccari, 1943). 1: bulbus olfactorius; 2: caudal portion of telencephalon; 3: commissura anterior; 4: stria medullaris system; 5: nucleus habenularis; 6: commissura habenulae; 7: thalamus dorsalis ('campo posthabenulare'); 8, 9: commissura posterior; 10: commissura tecti; 11: tectum opticum; 12: valvula cerebelli; 13: interstitial nucleus of fasciculus longitudinalis medialis; 14: fasciculus retroflexus; 15: 'brachium conjunctivum'; 16: radix oculomotoria; 17: interpeduncular nucleus; 18: fasciculus tectobulbaris cruciatus; 19: lobi inferiores hypothalami; 20: nucleus rotundus (aglomerularis); 21: commissura horizontalis *(Fritsch);* 22: commissura transversa; 23: tractus opticus; 24: basal forebrain bundle.

Figure 78D. Sagittal section *(Cajal silver impregnation)* through diencephalon and mesencephalon of Cyprinus carpio. The plane of the section is somewhat more lateral than that of Figure 78C (from Beccari, 1943). 1: telencephalon; 2: valvula cerebelli; 3: tectum opticum; 4: fasciculus tectocerebellaris; 5: nucleus gustatorius secundarius anterior s. superior; 6: fasciculus cerebellolobaris; 7: origin of tertiary 'gustatory tract'; 8: fascic-
ulus cerebellolobaris; 7: origin of tertiary 'gustatory tract'; 8: fasciculus gustatorius secundarius anterior; 9: fasciculus lobobulbaris; 10: dorsal fasciculus of lateral branch of commissura horizontalis; 11: lobi inferiores s. laterales hypothalami; 12: nucleus rotundus (aglomerularis); 13: 'interstitial neuropil of fasciculus thalamolobaris'; 14: ventral fasciculus of lateral branch of commissura horizontalis; 15: basal forebrain bundles; 16: nucleus praerotundus lateralis; 17: 'fasciculus thalamolobaris'; 18: commissura transversa; 19: tractus opticus; 20: dorsal and ventral thalamic grisea ('nucleo anteriore del talamo'); III, IV: intracerebral roots of oculomotor and trochlear nerves.

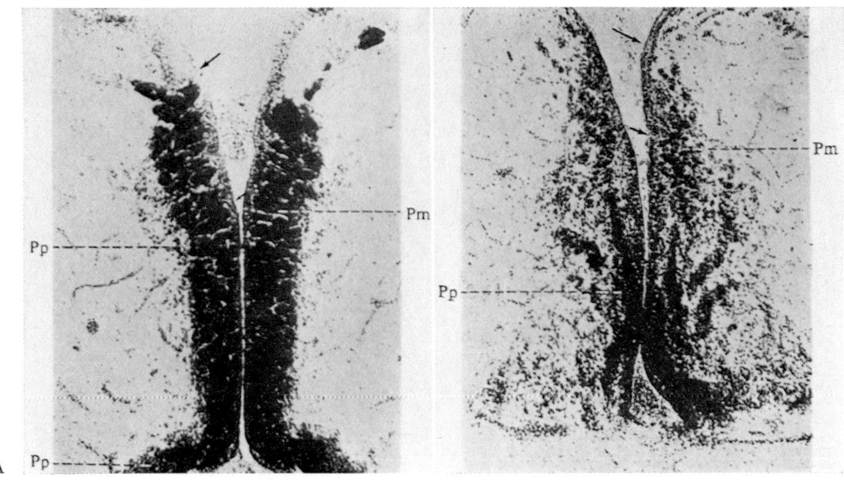

Figure 79 A. Cross-section *(Nissl stain)* through the rostral part of nucleus praeopticus magnocellularis in the Teleost Anguilla japonica (from NIIMI *et al.*, 1963b). Abbreviations Figures 79 A–C. Cho: optic chiasma; Pm: nucleus praeopticus magnocellularis; Pml, Pmp: pars lateralis respectively medialis of n. praeopt. magnocell.; Pp: nucleus praeopticus parvocellularis. The added arrows represent: (upper) sulcus diencephalicus ventralis, or, if groove flattened out, boundary of ventral thalamus and hypothalamus; (lower) branches of sulcus intraencephalicus anterior system.

Figure 79 B. Cross-section *(Nissl stain)* through caudal part of nucleus praeopticus magnocellularis in the Carp. Cyprinus carpio (from NIIMI *et al.*, 1963b).

tains a paraventricular nucleus[138] and a nucleus tuberis anterior, both of which may be neurosecretory, pertaining to the '*hypophysennahe*' neurosecretory system of DIEPEN (1953) as shown in Figure 79 D. In addition, to the periventricular cell of lobi inferiores, there is a nucleus diffusus lobi lateralis, a nucleus tuberis lateralis, a nucleus tuberis posterior, and a 'nucleus mammillaris' lateral to the rostrobasal end of the tegmental cell cord. The nucleus cerebellosus hypothalami of SHELDON (1912) may be a part of nucleus diffusus lobi lateralis. The nucleus sacci vasculosi is ill-defined. This griseum may be represented by hypothalamic cell groups in the lobus posterior (Fig. 80 A) dorsal to saccus vasculosus and ventral to the tegmental cell cord of tuberculum posterius, as well as by tegmental (i.e. deuterencephalic respectively

[138] This includes presumably basally the nucleus tuberis ventralis of SHELDON (1912) and generally the anterior hypothalamic nucleus as well as the ventromedial hypothalamic nucleus mentioned by CROSBY ans SHOWERS (1969). The so-called nucleus tuberis posterior, enumerated further below in the text, may be identical with a nucleus sacci vasculosi.

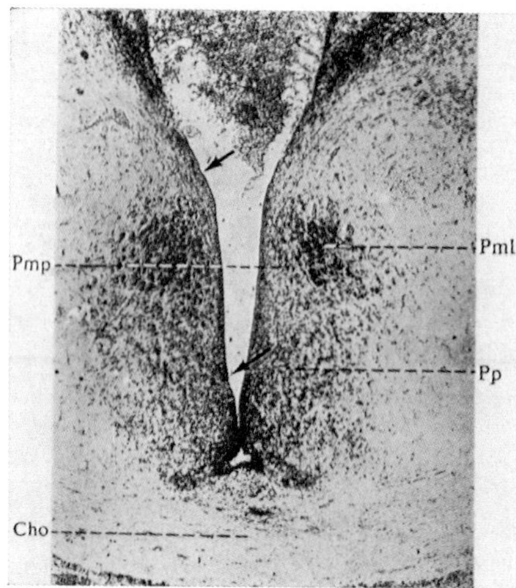

Figure 79C. Cross-section *(Nissl stain)* through the caudal part of nucleus praeopticus magnocellularis in the Catfish, Parasilurus asotus (from NIIMI *et al.*, 1963b).

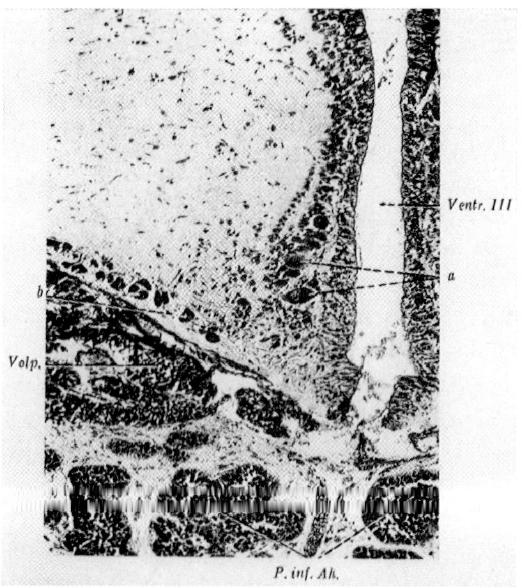

Figure 79D. Cross-section *(Nissl stain)* through the parahypophysial (hypophysennahen) portion ('tuber cinereum') of the hypothalamus in the Pike, Esox lucius (from DIEPEN, 1953). a: large, presumably neurosecretory periventricular elements; b: similar, lateral and superficially located elements of 'nucleus lateralis tuberis'; P. inf. Ah.: inferior portion of adenohypophysis; Volp.: 'oral sector' of adenohypophysis.

mesencephalic cell groups within the tuberculum posterius, not clearly delimitable from that latter's cell masses.

In the posterior hypothalamus of diverse but not all Teleosts, there is a pronounced thickening of the ventricular ependyma, which appears high columnar or even (pseudo) stratified as well as vascularized by intraepithelial vessels. This is the '*ependymal organ of Kappers-Charlton*' which is frequently located along (and on both sides) of sulcus lateralis hypothalami posterioris s. 'sulcus lateralis infundibuli'.[139] Said structure has been recorded in a number of Anamnia as well as of Amniota and can be classified as one of the ependymal or paraependymal circumventricular organs discussed in sections 3 and 5, chapter V of volume 3/I.

As regards the *diencephalic communication channels*, those connecting with the telencephalon are represented by the short 'stria medullaris' ('tractus taeniae') system reaching the epithalamus (habenula) and by the larger system of the basal forebrain bundles. Both systems are generally medullated but also contain a variable amount of non-medullated fibers. They are commonly described as mainly descending, including, e.g. the so-called olfacto-habenular, olfacto-hypothalamic, strio-thalamic and strio-hypothalamic tracts, but doubtless also contain more or less substantial ascending components.

Although some interchange respectively interweaving or redistribution of fibers may take place in the telencephalon, the medial forebrain bundle is predominantly related to the basal ($B_{2,3}$) telencephalic grisea, and the lateral forebrain bundle to the pallial grisea (D_{1-3}) as well as to the B_1 zone, which latter is also connected with the medial bundle. In the rostral portion of the diencephalon, the medial forebrain bundle tends to be located at the same level or even slightly dorsally to the lateral one (Fig. 76 A), and may, in some forms, be again subdivided into separate dorsal and ventral portions.[140] Caudalward, both medial and lateral forebrain bundles bend basalward within the

---

[139] According to KAPPERS *et al.* (1936) the ependyma may here be devoid of cilia. However, in some instances, modified ciliated processes (stereocilia ?) seem to be present. Differences in the structure of the ependyma lining the ventricular system are not without interest. It also seems possible, that, as claimed by MASAI and SATO (1963/64), paraependymal blood capillaries in the hypothalamus of Cyprinoids may become displaced into the ventricular cavity, being 'accompanied by very thin supporting tissue'.

[140] E.g. the dorsal 'tractus olfactohypothalamicus lateralis' and the adjacent ventral 'tractus olfactohypothalamicus medialis' of KAPPERS *et al.* (1936) in the Cod (Gadus morrhua).

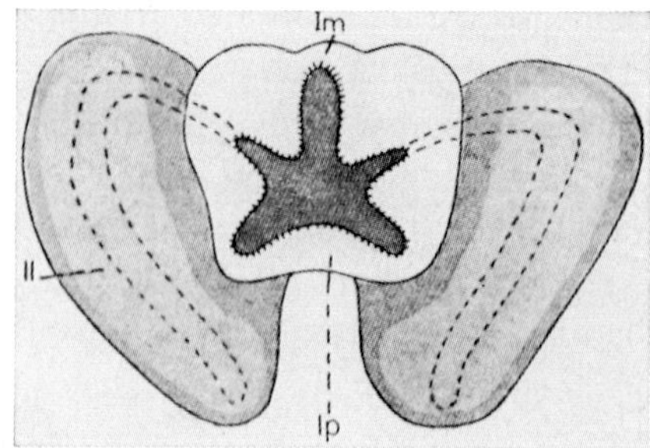

Figure 80 A. Semidiagrammatic sketch of posterior inferior hypothalamus in Cyprinus auratus, as seen in dorsal aspect (slightly modified after EDINGER, 1908, from K., 1927). ll: lobus lateralis; lm: lobus medialis; lp: lobus posterior.

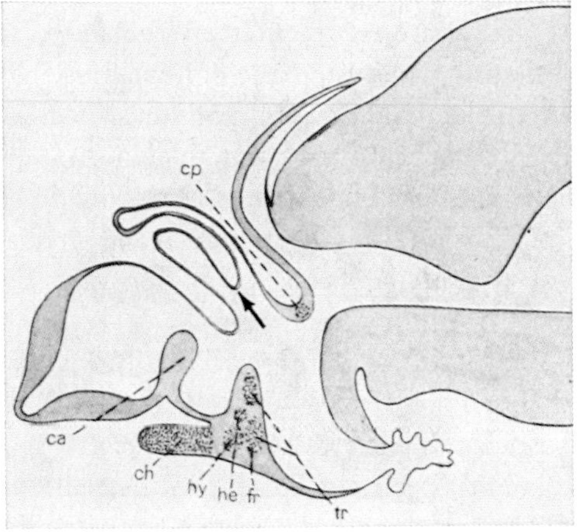

Figure 80 B. Simplified sketch illustrating the general arrangement of commissures in the diencephalic floor of the Teleost Barbus fluviatilis (based on figures by GOLDSTEIN, 1905, and slightly modified from K., 1927). ca: ridge of commissura anterior; ch: optic chiasma; cp: commissura posterior; fr: *commissura horizontalis of Fritsch;* he: *commissura minor of C. L. Herrick;* hy: commissura hypothalamica; tr: *commissura transversa of B. Haller;* arrow indicates topologic location of the here 'ignored' but well-developed commissura habenulae.

hypothalamus (Figs. 78 C, D) but also display dorsalward branchings, directed toward the mesencephalic tegmentum, and blending with the ascending caudal communication channels. The lateral forebrain bundle, commonly described as 'lateral strio-thalamic and hypothalamic tract' contains, in addition to telencephalic output, a substantial amount of ascending fibers, especially from the lateral geniculate complex as well as from thalamic and hypothalamic grisea, thus providing

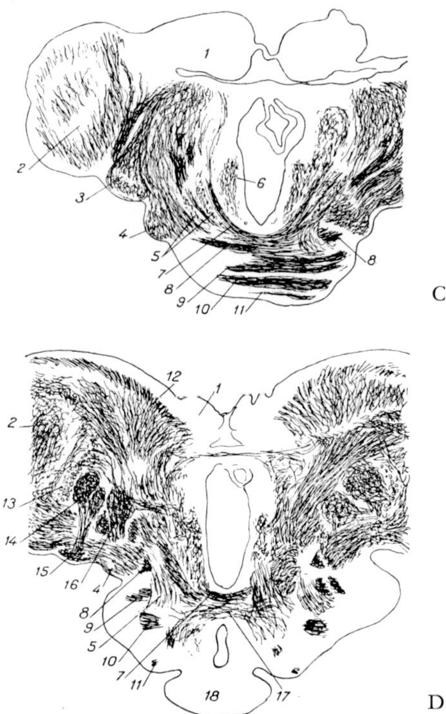

Figure 80 C, D. Cross-sections (modified *Cajal silver impregnation)* through the postoptic commissural systems of the Teleost Gambusia, C being more rostral than D (from Beccari, 1943). 1: ganglion s. nucleus habenulae; 2: rostral portions of tectum opticum; 3: part of lateral geniculate body complex; 4: tractus opticus marginalis; 5: lateral forebrain bundle; 6: medial forebrain bundle; 7: commissura postoptica dorsalis; 8: *commissura minor of C. L. Herrick;* 9: commissura transversa seu postoptica ventralis; 10: *commissura horizontalis of Fritsch;* 11: 'commissura subhorizontalis'; 12: tractus opticus medialis; 13: nucleus lentiformis mesencephali (s. nucleus corticalis); 14: nucleus praetectalis; 15: 'brachium tecti'; 16: caudal portions of thalamus dorsalis and ventralis *('nucleo anteriore del talamo'* in Beccari's interpretation respectively terminology); 17: commissura preinfundibulare; 18: hypophysial complex.

optic and other sensory input to the telencephalic grisea. Much the same can be said about the medial forebrain bundle (also designated as 'tractus olfactohypothalamicus'), which, however seems to include, if at all, only a much lesser ascending optic component from diencephalic grisea. Generally speaking, it can be stated that both forebrain bundles represent reciprocal communication channels between telencephalon and all regions of the diencephalon, with least emphasis on the epithalamus, which, through the stria medullaris system, is provided with an additional channel of its own.

Although the *optic input* through optic nerve and tract mainly reaches the optic tectum (cf. chapter XI, section 4, vol. 4), the Teleostean diencephalon receives substantial, essentially crossed but perhaps also some uncrossed optic input to the conspicuous lateral geniculate complex with dorsal and ventral thalamic components. It is, however, likely that the Teleostean lateral geniculate complex also obtains input from other sensory systems. Additional optic input into diencephalon reaches the pretectal grisea and, through an insufficiently clarified smaller basal optic root extending toward the mesencephalic tegmentum, also undefined hypothalamic grisea. The course of the optic tract through the lateral portion of the diencephalon, and some aspects of its radiation into the mesencephalic optic tectum are shown in figure 77 D (cf. also Figure 379 F, p. 856, chapter XI, vol. 4).

The *intrinsic fiber systems* of the diencephalon are, on the whole, poorly defined and rather diffuse. They can be tentatively enumerated as follows. Fibers between the preoptic hypothalamus to the habenula, joining the stria medullaris, represent a portion of the tractus olfacto-habenularis lateralis *autorum*, while another portion of these fibers seems to derive, in part perhaps as collaterals, from the basal forebrain bundles. More caudally, such 'vertical' fibers can be considered hypothalamohabenular, thalamohabenular, thalamohypothalamic, and intrathalamic connections. More or less longitudinally directed and in part fairly distinctive fiber systems are the thalamo-lobar and the rotundo-lobar or lobo-glomerular tract (Figs. 78 C, D). In addition, there obtain the essentially non-medullated neurosecretory hypothalamo-hypophysial system and the dorsal as well as the ventral tractus sacci vasculosi (Figs. 50 A, B), this latter being intermingled with, and difficult to distinguish from, the hypothalamo-hypophysial neurosecretory channel. From the lateral geniculate complex, a caudalward directed fasciculus geniculatus descendens can be recognized. In addition to intradiencephalic (particularly hypothalamic connections, it may contain fibers to respectively from tegmental, tectal and other mesencephalic grisea including e.g. input from the nucleus isthmi, and therefore in part pertaining to the system dealt with in the next paragraphs.

The complex *caudal communication channels* include the following fiber bundles. The habenulopeduncular tract or fasciculus retroflexus system, mainly directed toward the interpeduncular nucleus contains additional, poorly clarified connections (Figs. 78 B, C). Further habenulo-pretectal-tectal-subtectal (torus semicircularis) fibers are present. Although the so-called brachium tecti is mainly provided by optic tract fibers, this system of bundles in the wider sense includes pretecto-cerebellar, lobo-tectal, tecto-thalamic,[141] tecto-pretectal and lobo-cerebellar channels. The gustatory pathways, mainly reaching the hypothalamic lobar grisea, have been dealt with and were depicted in volume 4 (chapter IX, section 5, Fig. 233 A, and chapter XI, section 4). Basally, there are tegmento-lobar (or lobo-bulbar) connections as well as extensions of basal forebrain bundles into the tegmentum, particularly toward the interstitial nucleus of the fasciculus longitudinalis medialis. Fibers between dorsal posterior hypothalamus, especially the 'mammillary' region basal to tuberculum posterius provide a nondescript mammillotegmental system.

The Teleostean *diencephalic commissures* are well developed, complex, and rather variable *qua* taxonomic forms. In addition to the telencephalic commissura anterior at the telencephalo-diencephalic boundary, the commissura habenulae includes crossing telencephalic fiber systems. The commissura posterior comprises both diencephalic and mesencephalic channels.

Basally, the decussation of the optic chiasma protrudes, in various Teleosts, more or less extensively rostrad, ventrally to the preoptic recess. Its rostral tip may be located basally to telencephalic neighborhoods. At least five *postoptic commissures* can usually be distinguished within the chiasmatic ridge.[142] There is a dorsal postoptic commissure, ventrally to which the *commissura minor of C.L.* HERRICK (1891) is located. Basally to this latter passes the commissura transversa. Still more ventrally runs the *commissura horizontalis of* FRITSCH (1878), related to the nucleus rotundus.[142a] Likewise basally, but in a somewhat

---

[141] These presumably include the rostral, thalamic extensions of the general and of the vestibular lemniscus systems.

[142] MEADER (1934) distinguishes, in the Teleost Holocentrus, 8 commissures in the postchiasmatic region which he designates as commissura intertectalis ventralis, commissura intergranularis tecti, commissura intergeniculata ventralis and dorsalis, commissura transversa, commissura minor, fibrae ansulatae, and commissura horizontalis. The 'fibrae ansulatae' (cf. also Fig. 77 D) are presumably components of the dorsal supraoptic commissure and should not be confused with the more caudal mesencephalic commissura ansulata beginning in the region of tuberculum posterius.

[142a] G. FRITSCH (1878) was one of the pioneer investigators of the Fish brain. Cf. the comments on FRITSCH in footnote 87, p. 120, chapter VIII of volume 4, and in section 10, chapter XIII in the present volume.

rostral position, more adjacent to the chiasma, a hypothalamic commissure is present in at least some forms. The *commissura transversa*, as dealt with in chapter XI of volume 4, essentially interconnects mesencephalic grisea (tectum, torus semicircularis, nucleus isthmi). Commissura minor and commissura horizontalis may also in part be related to mesencephalic cell groups, while the others pertain perhaps mainly to the diencephalon. A subhorizontal commissure was recorded in some forms by Haller (1899) and Holmgren (1920). The lateral geniculate complex seems to be interconnected by an intergeniculate commissure whose fibers join those of other commissural channels, perhaps mainly the commissura transversa.

Caudally to the chiasmatic ridge proper, a lesser commissura preinfundibularis crosses the hypothalamic floor. Basally to tuberculum posterius, a postinfundibular commissure may be present, whose fibers blend with those of the mesencephalic commissura ansulata. Some general comments on the various interpretations and differing descriptions of these commissural systems, with references to the pertinent bibliography, can be found in the publications by Kappers *et al.* (1936), Beccari (1943),[143] and Crosby and Showers (1969).

The *Crossopterygian Coelacanth Latimeria*, which can be evaluated as a highly aberrant form, displays a peculiar pattern distortion of the diencephalon, which latter, nevertheless, retains the typical Vertebrate arrangement of the longitudinal diencephalic zones.[144] The most striking 'distortion' is displayed with respect to the ventral posterior (post optic) hypothalamus which, although topologically still caudobasal, has become topographically rostrobasal, being located ventrally to the chiasmatic ridge, and extending below the telencephalon. This rostrally displaced posterior inferior hypothalamus ends in the neurohypophysial '*pédoncule hypophysaire*'[145] of Millot and Anthony (1965), said *pédoncule* being, in turn, rostrally continuous with an adenohypophysial

---

[143] Beccari (1943) [illegible] the difficulties here involved, which result from the multiform variations in the differentiation of the Teleostean brain. Despite his rather detailed critical analysis, he feels constrained to remark: '*Non si può tentare una esposizione sistematica dei numerosi fasci che sono stati descritti nel talamo dei Teleostei senza entrare in lunghe discussioni sulle omologie dei nomi usati dai differenti autori, in gran parte dipendenti dalla differente interpretazione dei nn. dai quali i fasci partono e nei quali terminano. Tali discussioni esorbitano dal programma di questo libro*'.

[144] Cf. volume 3/II, chapter VI, section 4, pp. 415–417, and Fig. 199 on p. 423.

[145] Cf. footnote 70b, p. 95 in section 1C of this chapter.

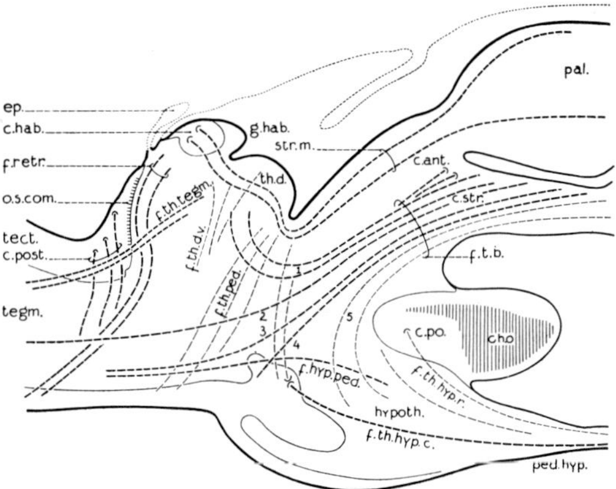

Figure 81 A. Diagram of midsagittal section through the diencephalon of the Crossopterygian Latimeria, with projection of main fiber systems upon that plane (from MILLOT *et al.*, 1964). c.ant.: anterior commissure; c.hab.: habenular commissure; ch.o.: optic chiasma; c.po.: postoptic commissures; c.post.: posterior commissure; c.str.: basal telencephalon; ep.: epiphysis; f.retr.: fasciculus retroflexus; f.t.b.: basal forebrain bundle with subdivisions 1–5; f.hyp.ped.: fasciculus hypothalamo-peduncularis; f.th.d.v.: fasciculus thalamicus dorsoventralis; f.th.hyp.c.: fasciculus thalamo-hypothalamicus caudalis; f.th.hyp.r.: fasciculus thalamo-hypothalamicus rostralis; f.th.ped.: fasciculus thalamo-peduncularis; f.th.tegm.: fasciculus thalamo-tegmentalis; g.hab.: ganglion habenulae; hypoth.: inferior hypothalamus; o.s.com.: subcommissural organ; pal.: telencephalic pallium; ped.hyp.: 'hypothalamic peduncle'; str.m.: stria medullaris system; tect.: tectum mesencephali; tegm.: tegmentum mesencephali; th.d.: thalamus dorsalis. The commissure of tuberculum posterius is indicated by a crossing mark at caudal end of the line labelled t.th.hyp.c.

peduncle (*'cordon conjonctivo-vasculaire'*). Figures 81 A–D illustrate this configurational arrangement. It can be seen that ventrally to true preoptic recess and chiasmatic ridge, a second, basal 'preoptic' but topologically postoptic recess is thereby formed. Other peculiarities, include the configuration of the tuberculum posterius, ventrally to which a posterior recess of the posterior inferior hypothalamus (lobi inferiores) is located (Fig. 81 A). The presence of a saccus vasculosus, somehow connected with the rostral basal hypothalamic floor, seems doubtful, but was tentatively assumed by MILLOT and ANTHONY (1965).[146] Again, the rather caudal position of the ganglion habenulae may be noted.

[146] The cited authors state: *'nous ne saurions l'affirmer avec certitude'*. The photomicrographs disclose that although their unique Latimeria material is adequate for highly interesting morphologic observations, it does not permit the recording of detailed fiber connections nor of well definable cytoarchitectural features.

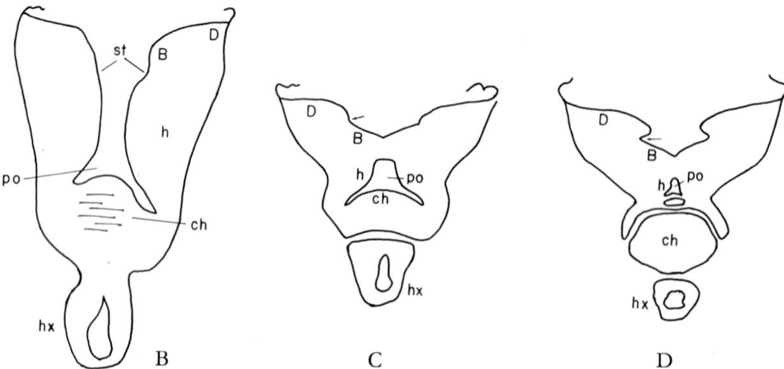

Figure 81 B–D. Sketches illustrating diencephalo-telencephalic boundary neighborhoods in Latimeria (redrawn after unlabelled photomicrographs from MILLOT and ANTHONY, 1965). B. At caudal level of optic chiasma. C. At level of commissural plate (torus transversus, anterior commissure). D. At level near rostral end of true preoptic recess. ch: optic chiasma; h: hypothalamus; hx: infra- and preoptic extension of posterior inferior hypothalamus (secondary inferior preoptic recess; po: true preoptic recess; st: sulcus terminalis telencephali; B: basal telencephalon; D: pallial telencephalon; arrow: telencephalic sulcus fd_1.

Figures 82 A and B illustrate cross-sections through the diencephalon. As far as can be inferred from the observations and illustrations published by MILLOT and ANTHONY, some parts of the ganglion habenulae show a compact cellular grouping, while thalamus dorsalis, thalamus ventralis, and hypothalamus seemingly display a rather nondescript, diffuse, and essentially periventricular cellular arrangement.

Figure 82 A. Outline of cross-section through the diencephalon of Latimeria at level of ganglion habenulae (from MILLOT and ANTHONY, 1965). g.hab.: ganglion habenulae; f.opt.: tractus opticus; s.m.H.: sulcus diencephalicus medius; str.m.: striae medullares system; thal.d.: thalamus dorsalis; thal.v.: thalamus ventralis. The lead 3^ev. into the third ventricle passes through a groove of the sulcus lateralis hypothalami posterioris system. The added arrows indicate sulcus diencephalicus dorsalis (above) and sulcus diencephalicus ventralis (below). The outlines medially to tractus opticus represent components of the basal forebrain bundle system.

Figure 82 B. Cross-section through the diencephalon of Latimeria (photomontage from unilateral section, unlabelled in the original, from MILLOT and ANTHONY, 1965). li: lobi inferiores hypothalami; tp: tuberculum posterius; arrows (in dorsoventral sequence): sulcus diencephalicus dorsalis, s.d. medius, s.d. ventralis, sulcus limitans. The ventricular indentation in thalamus dorsalis may be an artefact or an accessory 'sulcus thalami dorsalis'. The medial bulge of epithalamus is a torn portion of commissura habenulae. The plane of the section is oblique (rostrodorsal-caudoventral) to that of Figure 82 A.

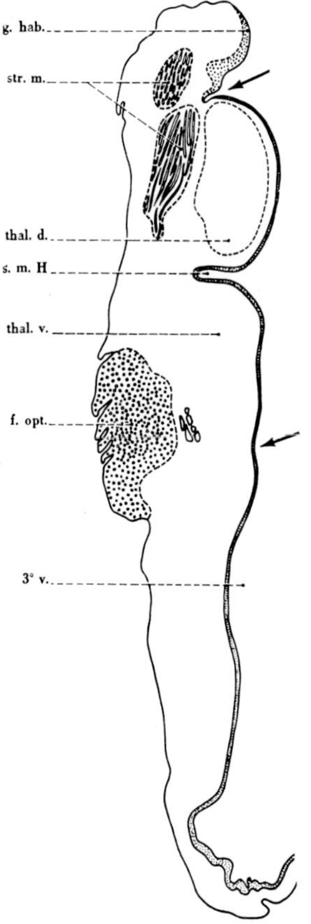

82 A

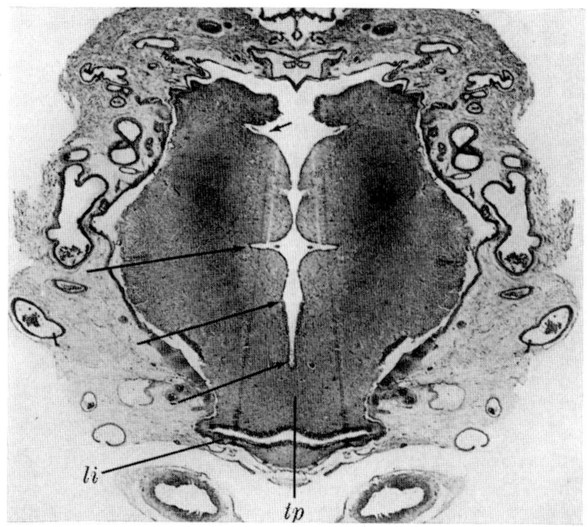

82 B

As regards *communication channels*, about which detailed observations are not available, the stria medullaris system, the basal forebrain bundle, and the fasciculus retroflexus can be easily identified. Dorsal portions of basal forebrain bundle tend to run through thalamus ventralis rather than through hypothalamus, while others run within dorsal hypothalamus through the boundary zone of hypothalamus and ventral thalamus. Figure 81 A shows, as interpreted by MILLOT *et al.* (1964) components of the diencephalic fiber system. The commissural respectively decussating systems include commissura anterior, commissura habenulae, commissura posterior, optic chiasma, and postoptic commissure. Crossing fibers seem likewise present in the neighborhood of tuberculum posterius at the basal diencephalomesencephalic boundary zone.

5. Dipnoans

Relevant data on the diencephalon of Dipnoans are provided for *Ceratodus* by the paper of HOLMGREN and VAN DER HORST (1925), and for *Protopterus* by that of GERLACH (1933). Both papers refer to the findings and publications of previous authors. The paper by SCHNITZLEIN and CROSBY (1967) on the telencephalon of Protopterus brings a few illustrations of the rostral diencephalon, since they consider the preoptic region to be telencephalic. The paper by RUDEBECK (1945)

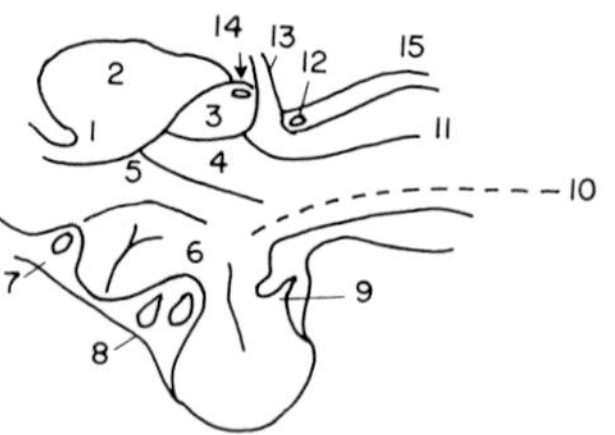

Figure 83. Diagram of midsagittal section through the diencephalon of Ceratodus, showing ventricular wall and its overall configuration in the author's interpretation modified after figures from HOLMGREN and VAN DER HORST, 1925). 1: velum transversum; 2: dorsal sac (secondary parencephalon); 3: epithalamus (ganglion habenulae); 4: thalamus dorsalis; 5: thalamus ventralis; 6: hypothalamus; 7: ridge of commissura anterior; 8: chiasmatic ridge with optic chiasma and postoptic commissures; 9: 'mammillary recess' below tuberculum posterius neighborhood; 10: sulcus limitans; 11: sulcus lateralis mesencephali and sulcus synencephalicus system; 12: commissura posterior; 13: epiphysial stalk; 14: commissura habenulae; 15: tectum mesencephali. 5 also indicates region of fusion between sulcus diencephalicus ventralis and sulcus terminalis telencephali. The three diencephalic sulci as well as sulcus intraencephalicus anterior and sulcus lateralis hypothalami posterioris, described in the text, can be easily identified.

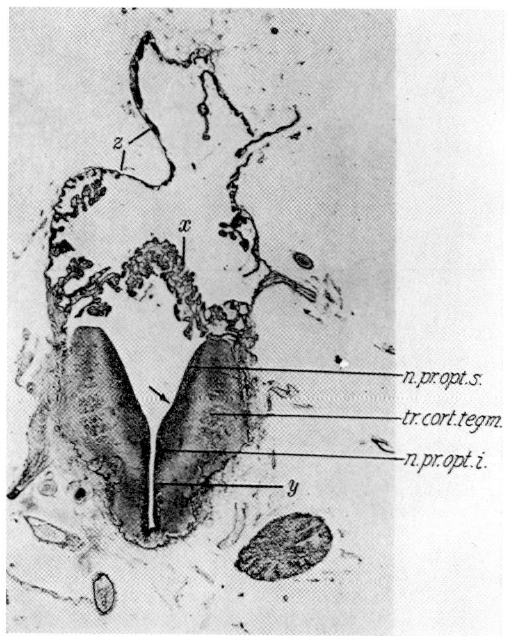

Figure 84 A. Cross-section *(Bielschowsky's silver impregnation)* through the preoptic recess of Ceratodus (from HOLMGREN and VAN DER HORST, 1925). n.pr.opt.s.: eminentia thalami ventralis; n.pr.opt.i.: nucleus praeopticus, pars superior; tr.cort.tegm.: part of basal forebrain bundle. Added designations: x: velum transversum; y: nucleus praeopticus pars inferior; z: dorsal sac; arrow: sulcus diencephalicus ventralis.

concerns mainly the telencephalon and results in a forebrain scheme substantially differing from that elaborated by our studies (cf. K. and HAYMAKER, 1949; K., 1956).

Generally speaking, the brain of *Ceratodus*, a genus pertaining to the Monopneumones, seems to be somewhat more differentiated than that of Protopterus (Dipneumones). The diencephalon in Ceratodus and that in Protopterus, moreover, display various conspicuous differences, particularly in respect to the location of the ganglion habenulae, and as regards the relationship of diencephalon to tectum mesencephali, which, in *Protopterus*, protrudes above the caudal diencephalon and contains a rostral ventricular recess (cf. Figs. 83 and 85 A).

In Ceratodus, the habenular ganglion *(epithalamus)* has a rather caudal position (Figs. 84 D, E). A difference in size between right and left ganglion is not clearly noticeable. A distinction between a dense medial and a more loosely arranged lateral habenular nucleus is faintly suggested.

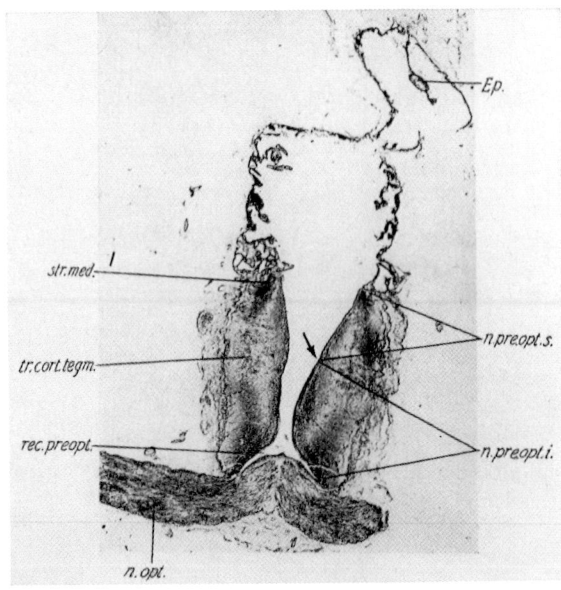

Figure 84 B. Cross-section through more caudal part of preoptic recess in Ceratodus (from HOLMGREN and VAN DER HORST, 1925). Ep.: epiphysis; n.opt.: optic nerve; n.preopt.s.: thalamus ventralis; n.preopt.i.: here pars superior and inferior of nucleus praeopticus; rec.preopt.: basal origin of sulcus intraencephalicus anterior; str.med.: stria medullaris system; tr.cort.tegm.: portion of basal forebrain bundle. Added arrow: sulcus diencephalicus ventralis.

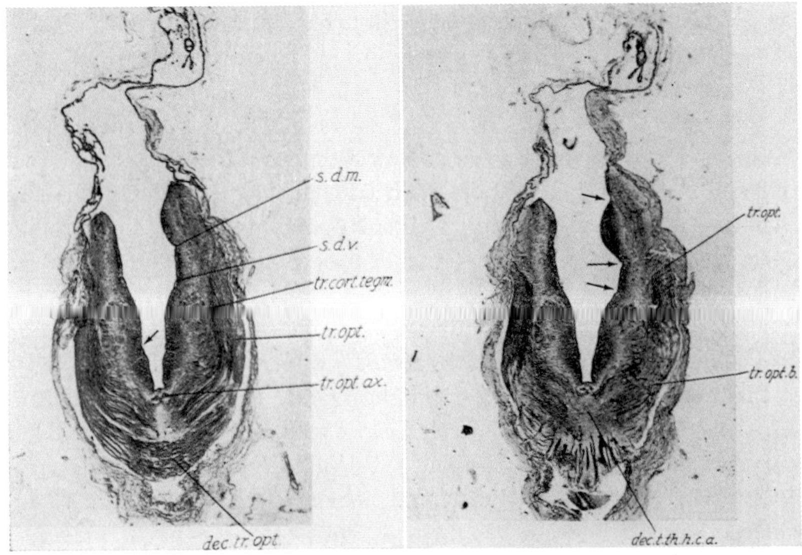

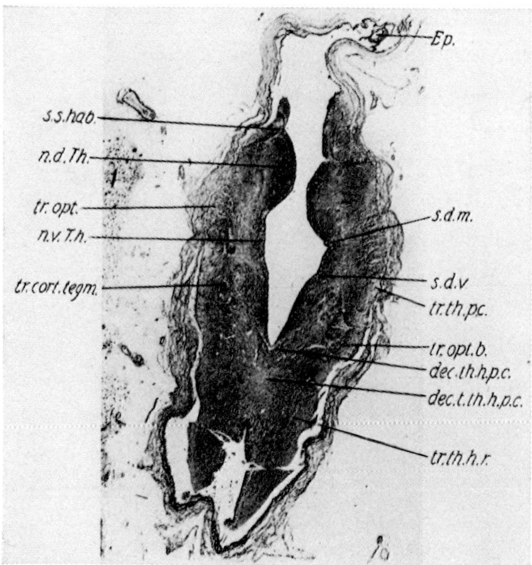

Figure 84E. Cross-section through the diencephalon of Ceratodus at the caudal level of chiasmatic ridge (from Holmgren and van der Horst, 1925). dec.t.th.h.p.c.: 'decussatio tractus tectothalamici et hypothalamici cruciati, pars posterior'; dec.th.h.p.c.: decussatio thalamohypothalamica et peduncularis cruciata (this and the preceding decussation are part of the postoptic commissural system); s.s.hab.: sulcus diencephalicus dorsalis (s. subhabenularis); tr.th.h.r.: 'tractus thalamohypothalamicus rectus; tr.th.p.c.: 'tractus thalamo-postchiasmaticus'. Other abbreviations as in Figures 84A–D.

The *dorsal thalamus* extends rostrally to the habenula and displays a conspicuous bulge into the ventricle, particularly at habenular levels (Figs. 84 C–E). The cellular arrangement is essentially periventricular, but some laterally located elements, receiving (presumably collateral) fibers from the optic tract, represent a poorly delimitable dorsal lateral geniculate nucleus. The caudal dorsal thalamus consists of the so-

---

Figure 84C. Cross-section through the diencephalon of Ceratodus at level of optic chiasma (from Holmgren and van der Horst, 1925). dec.tr.opt.: part of optic chiasma; s.d.m.: sulcus diencephalicus medius; s.d.v.: sulcus diencephalicus ventralis; tr.opt.: tractus opticus; tr.opt.ax.: tractus opticus axialis. Added arrow: caudal branch of sulcus intraencephalicus anterior.

Figure 84D. Cross-section through the diencephalon of Ceratodus slightly caudal to optic chiasma (from Holmgren and van der Horst, 1925). dec.t.th.h.c.a.: 'decussatio tractus tectothalamici et hypothalamici cruciati, pars anterior'; tr.opt.: tractus opticus; tr.opt.b.: tractus opticus basalis. Added arrows in dorsoventral sequence; sulcus diencephalicus dorsalis, s.d.medius, s.d. ventralis.

called 'pars intercalaris' *(posthabenuläres Zwischenhirngebiet*, K., 1956 *et passim)* which includes, at the mesencephalo-diencephalic boundary zone the nucleus commissurae posterioris.

The *thalamus ventralis*, whose neuronal elements are likewise periventricular, extends from the commissura anterior, where it blends with the telencephalic B-zone,[147] to the tegmental tuberculum posterius. Its intermediate and caudal portions receive apparently collateral input from the optic tract. The lateral component of the ventral thalamic cell band can here be evaluated as a poorly delimited nucleus geniculatus lateralis ventralis (Holmgren and van der Horst, 1925).

The *hypothalamus* rostral to the chiasmatic ridge contains a preoptic nucleus with a thicker superior and a thinner inferior subdivision (Figs. 84 A, B). Scattered elements within the basal forebrain bundle represent the anterior portion of a nucleus entopeduncularis. The caudal, i.e. postoptic hypothalamus displays, in its essentially periventricular arrangement, some differences between a dorsal and a ventral subdivision.

The region of the tuberculum posterius, a neighborhood transitional between the deuterencephalic tegmentum and diencephalic longitudinal zones (thalamus ventralis and hypothalamus) displays two peculiar recesses, of which Holmgren and van der Horst (1925) designate the ventral one as recessus mammillaris. Basally to this latter, the cited authors mention a saccus vasculosus, but this interpretation remains doubtful.[148] The caudal posterior inferior hypothalamus corresponds to the Osteichthyan lobi inferiores s. laterales, but is less developed, displaying far less conspicuous indications of lateral recesses (cf. Fig. 383 A, p. 867, vol. 4).

The three main diencephalic ventricular sulci of the longitudinal zonal system can be recognized (Figs. 84 A–E). The sulcus diencephalicus dorsalis (sulcus subhabenularis) delimits the epithalamus and does not extend rostralward as far as the two others. Caudalward, it joins

---

[147] This extension toward the telencephalic B-zones represents both the eminentia thalami (ventralis) and the massa cellularis reuniens, pars superior, which latter is more conspicuous in Vertebrate brains with inverted and caudalward evaginated telencephalon. Holmgren and van der Horst (1925), who include all or most of the preoptic recess into their 'telencephalon medium', have interpreted this portion of the thalamus ventralis as their 'nucleus praeopticus superior' (cf. Figs. 84 A, B).

[148] Cf. above, section 1C and footnote 79. If, as Holmgren and van der Horst believe, a saccus vasculosus should indeed be present in Ceratodus, it would, at best, be rather rudimentary or 'vestigial'.

the sulcus synencephalicus which runs through the posterior dorsal thalamus, and becomes continuous with the sulcus lateralis mesencephali (internus).[149]

The sulcus diencephalicus medius, approximately delimiting thalamus dorsalis from thalamus ventralis, does not, in turn, extend as far rostralward as the sulcus diencephalicus ventralis, which roughly indicates the dorsal boundary of hypothalamus, and rostrally ends at the telencephalo-diencephalic boundary. A poorly outlined accessory 'sulcus eminentiae thalami' may be suggested.

The preoptic hypothalamic sulcus intraencephalicus anterior has an essentially dorso-basal direction, but may display oblique rostral and caudal branches at its dorsal extremity (cf. Fig. 84 C).

The caudal sulcus lateralis hypothalami posterioris (sulcus lateralis infundibuli, sulcus hypothalamicus of HOLMGREN and VAN DER HORST) is not very prominent and runs in a more or less dorsoventral direction. Thus, as the cited authors point out, it does not clearly delimit a dorsal from a ventral part in the posterior hypothalamus. A nondescript system of ventricular grooves (sulcus mammillaris of HOLMGREN and VAN DER HORST) can be noted in the 'mammillary' region.

At levels of the tuberculum posterius, the rostral end of sulcus limitans becomes rather indistinct, but this groove is clearly identifiable at neighboring rostral mesencephalic levels (cf. Fig. 383 A, vol. 4).

The *rostral diencephalic communication channels* include the stria medullaris system (tractus taeniae) and that of the basal forebrain bundle, both with connections comparable to those of Osteichthyes. As regards optic input, a tractus opticus axialis and a tractus opticus basalis (Figs. 84 C, D) have been recorded in addition to the connections of main optic tract with the primordial lateral geniculate complex. Nothing definite is known concerning the obtaining nondescript intradiencephalic connections ('tractus dorsoventralis thalami' and comparable additional fiber bundles). The *caudal channels* include the system of fasciculus retroflexus,[150] moreover, tecto-diencephalic, and

---

[149] Cf. Figure 383 A, p. 867 of volume 4. The unlabelled sulcus below *c.p.* in that figure indicates the rostral end of sulcus lateralis mesencephali at its transitional neighborhood to sulcus synencephalicus, which could be designated either as caudal end of the latter or rostral beginning of the former.

[150] Although a difference in size between right and left habenular ganglia is not definitely noticeable, the left tractus habenulo-interpeduncularis is said to contain chiefly unmyelinated ones (HOLMGREN and VAN DER HORST, 1925).

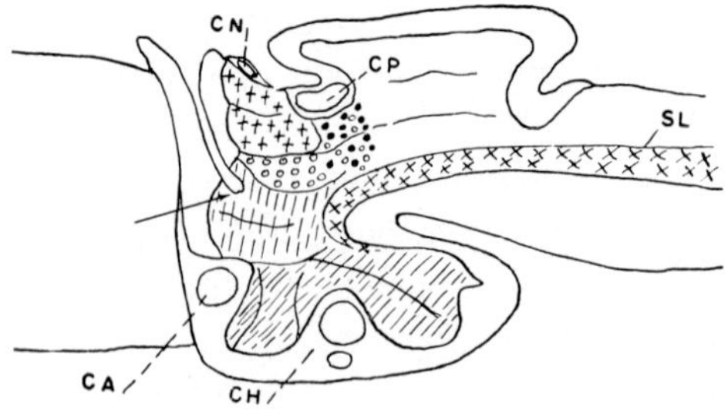

Figure 85 A. Semidiagrammatic reconstruction of ventricular aspect in the diencephalon of a young adult specimen of Protopterus (from K., 1956). CA: anterior commissure; CH: chiasmatic ridge with relatively small optic chiasma and larger postoptic respectively supraoptic commissural system; CN: commissura habenulae; CP: commissura posterior; SL: sulcus limitans. Identification of diencephalic longitudinal zones as in Figures 60A, 69A, 73A, 75A. Arrow: velum transversum. Sulci as enumerated in text.

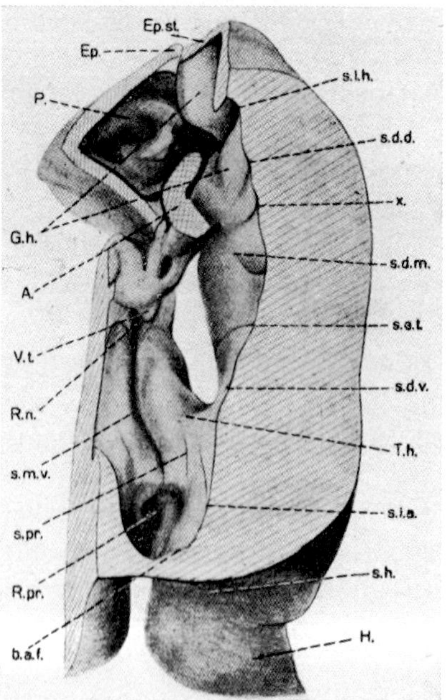

Figure 85 B. Diencephalo-telencephalic boundary region in the brain of Protopterus, as displayed by a wax model in caudal view (from GERLACH, 1933). A.: cut choroid plexus of 3rd ventricle; b.a.f.: basal optic stalk groove (system of s. intraencephalicus anterior); Ep., Ep.st.: epiphysial system (stalk); G.h.: ganglion habenulae; H.: hemisphere of telencephalon; P.: dorsal sac (secondary parencephalon); R.n.: recessus neuroporicus; R.pr.: recessus praeopticus; s.d.d.: dorsocaudal branch of s. dienc. dorsalis; s.d.m.: s.

diencephalo-peduncular fibers. It can be assumed that the overall arrangement of these ascending and descending connections, through which components of the lemniscus systems and of cerebellar output reach the diencephalon, while prosencephalic output reaches tectum and tegmentum, conforms to that obtaining in other Anamnia. HOLMGREN and VAN DER HORST (1925) made an attempt at describing some details of those medullated and non-medullated, crossed and uncrossed fiber bundles, and at distinguishing definable 'tracts'.

The *commissures* are provided by commissura anterior, habenular commissure, commissura posterior, optic chiasma, a poorly defined system of postoptic commissures, and crossing fibers in the region of tuberculum posterius ('commissura posterior tuberis', 'decussatio hypothalamica posterior').

The diencephalon of *Protopterus* differs from that of Ceratodus by the rostral position of the large epitha'amus, which protrudes, dorsally to velum transversum, into the dorsal sac (Figs. 85A, 87A; compare Fig. 85A with 83). The *habenular ganglion* is subdivided by a sulcus intrahabenularis into a dorsal and ventral portion. Caudalward these two portions become fused (Fig. 87B) and are then separated from the posthabenular dorsal thalamus (pars intercalaris, pretectal dorsal thalamus) by the shallow dorsal extension of sulcus diencephalicus dorsalis at this latter's junction with sulcus synencephalicus.[151] The cellular arrangement, at least in the young specimen at our disposal, was found essentially periventricular and rather uniformly compact. This holds, with minor qualifications to be mentioned be!ow, also for the thalamic and hypothalamic cell masses.

The *thalamus dorsalis* begins rostrally with a relatively narrow strip ventrally to ganglion habenulae (Fig. 85B) and expands caudalward in

[151] In our first investigations (K., 1929; GERLACH, 1933) we misinterpreted the sulcus synencephalicus as the caudal portion of sulcus diencephalicus dorsalis, and conceived the posthabenular region of the dorsal thalamus ('pars intercalaris', posthabenuläres Zwischenhirngebiet) as pertaining to the epithalamus. I have corrected this interpretation in a communication on the pretectal region of Anamnia (K., 1956).

dienc. medius; s.d.v.: s. dienc. ventralis (merging with s. terminalis telencephali); s.e.t.: s. eminentiae thalami; s.h.: s. hemisphaericus (*sive* diencephalo-telencephalicus externus); s.i.a.: rostral branch of s. intraencephalicus anterior; s.i.h.: s. intrahabenularis; s.m.v.: s. medianus ventralis of telencephalon impar; s.pr.: accessory s. praeopticus; T.h.: torus hemisphaericus; V.t.: velum transversum; x.: probably branch of s. synencephalicus near junction with s. dienc. dorsalis.

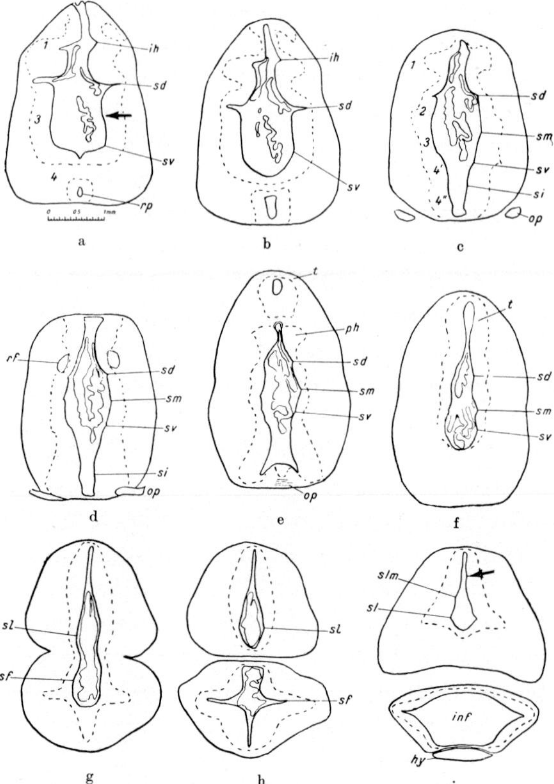

Figure 86. Outline of cross-sections, in rostro-caudal sequence, through the diencephalon of a young specimen of Protopterus (from K., 1929). hy: hypophysial complex; ih: sulcus intrahabenularis; inf: lobi inferiores hypothalami ('infundibulum'); op: optic nerve and optic chiasma, respectively; ph: posthabenular region (dorsal thalamus); rf: fasciculus retroflexus; rp: preoptic recess; sd: s. dienc. dorsalis in a and b, junction with s. synencephalicus in c, s. synencephalicus in d, e, f; sf: s. lateralis hypothalami posterioris ('s. lateralis infundibuli'); si: parts of s. intraencephalicus anterior system; sl: s. limitans; slm: s. lateralis mesencephali (internus); sm: s. dienc. medius; sv: s. dienc. ventralis; t: tectum opticum. Added arrows: s. eminentiae thalami in a; s. lateralis mesencephali accessorius in i, shown, but not labelled in Figure 83 A. In the original investigation, the posthabenular region was still erroneously interpreted as part of epithalamus, and the s. synencephalicus as a caudal portion of s. dienc. dorsalis.

Figure 87. Cross-sections through the diencephalon in Protopterus (from GERLACH, 1933). A At level of telencephalo-mesencephalic boundary neighborhood. B At level of commissura habenulae. C At level of commissura posterior. D At caudal level of chiasmatic ridge. b.a.f.: '*basale Augenstielfurche*' (system of s. intraencephalicus anterior); b.V.: basal forebrain bundle; c.h.: commissura habenulae (sive habenularum); Ch.o.: optic chiasma; C.olf.: cortex olfactoria (area ventrolateralis posterior of B_s); C.po.: commissura posterior; Ep.st.: epiphysial stalk; G.h.: ganglion habenulae; M.B.: fasciculus retroflexus *(Meynert's*

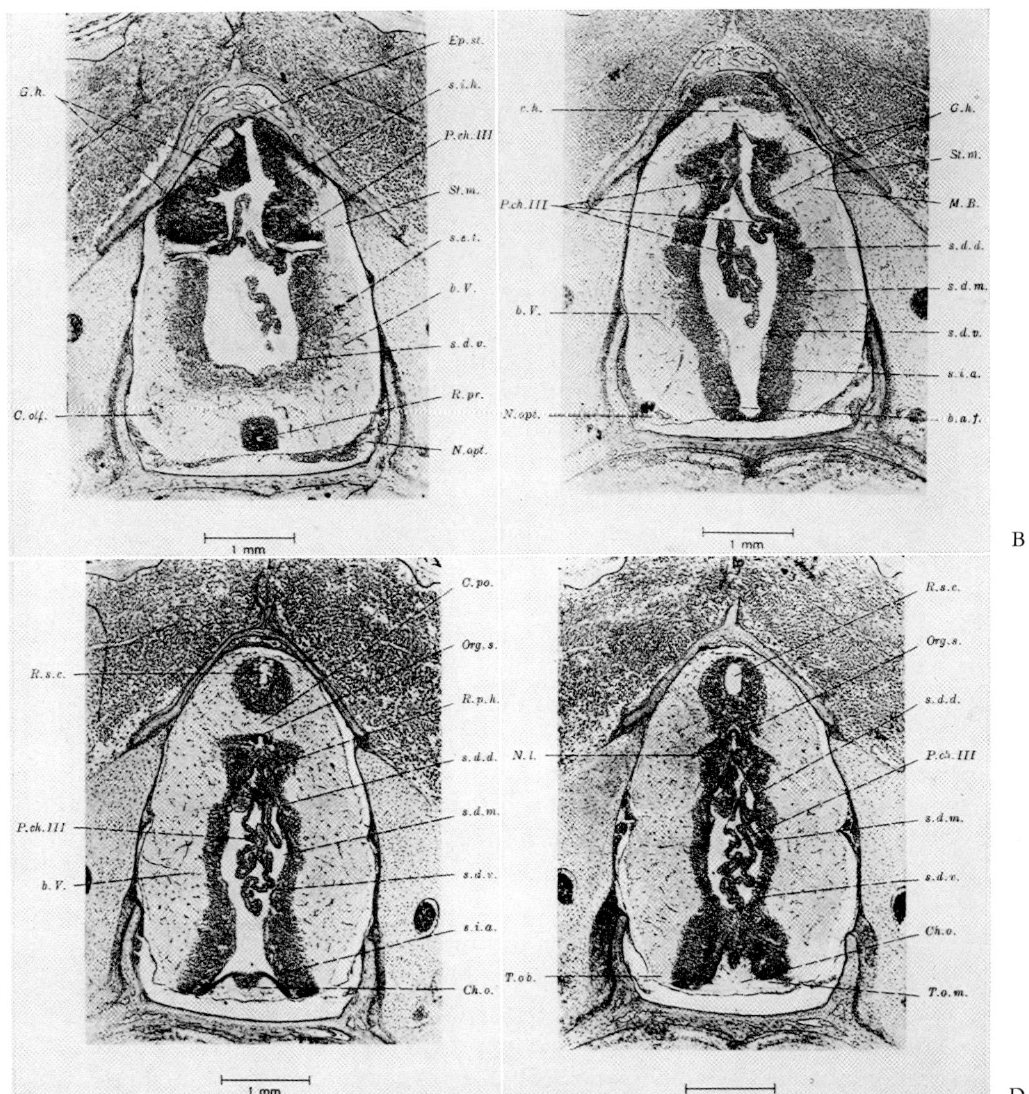

bundle); N.l.: nucleus lentiformis mesencephali; N.opt.: optic nerve; Org.s.: subcommissural organ; P.ch.III: choroid plexus of 3rd ventricle (in A probably part of velum transversum); R.p.h.: posthabenular region of dorsal thalamus; R.pr.: preoptic recess; R.s.c.: recessus supracommissuralis tecti mesencephali; s.d.d.: sulcus diencephalicus dorsalis (in C and D: s. synencephalicus); s.d.v.: sulcus diencephalicus ventralis; s.e.t.: sulcus eminentiae thalami; s.i.a.: part of sulcus intraencephalicus anterior system; s.i.h.: sulcus intrahabenularis; St.m.: stria medullaris system; T.o.b.: tractus opticus basalis; T.o.m.: tractus opticus marginalis.

the posthabenular region, dorsally and ventrally to sulcus synencephalicus. A few scattered neuronal elements lateral to the periventricular layer can be interpreted as a primordial dorsal lateral geniculate nucleus.

The *thalamus ventralis* begins rostrally with a large eminentia thalami (ventralis) abutting, at the level of torus transversus (anterior commissure) against the B-zones of the telencephalon (Figs. 85 A, B, 87 A). Its ventricular wall contains an accessory sulcus eminentiae thalami pointed out by GERLACH (1933), while the rostral end of sulcus diencephalicus ventralis reaches the commissural plate at its lateral ventricular edge. More caudally, some scattered cells, comparable to those mentioned *qua* thalamus dorsalis, provide a primordial ventral lateral geniculate nucleus.

The *hypothalamus* consists of a rostral preoptic and a caudal, posterior or postoptic portion. In the ventricular surface of the preoptic recess, the system of sulcus intraencephalicus anterior displays a rostral and a caudal branch. In the posterior hypothalamus, a sulcus lateralis hypothalami inferioris is present. It merges rostrally with sulcus diencephalicus ventralis. Caudally, it is particularly conspicuous in the moderately developed lobi laterales (*sive* inferiores). The periventricular arrangement of the neuronal elements, at least in the specimen available to us, precludes a convincing delimitation of nuclei. A few scattered elements within the course of the basal forebrain bundle seem to represent a primordial entopeduncular nucleus. The presence of a saccus vasculosus is quite doubtful.

Figure 86 depicts the overall configuration of the diencephalon and its ventricular sulci as seen in a series of cross-sections. In comparison with Ceratodus, the massive protrusion of an extensive choroid plexus into the ventricular lumen will be noted (cf. also Figs. 87 A–D).

At the basal diencephalo-mesencephalic boundary zone, the rostral end of sulcus limitans, surrounding the tuberculum posterius, could be identified by GERLACH (1933). The peculiar dorsal and ventral lips or protrusions of the said tuberculum, obtaining in Ceratodus, were not found in our specimen of Protopterus.

Concerning the fiber systems of the diencephalon, few reliable data are available for Protopterus, but our incidental observations suggest that these channels are essentially comparable to those described in greater detail for Ceratodus by HOLMGREN and VAN DER HORST (1925). The basal forebrain bundles and the stria medullaris system could be identified. As regards the optic input, GERLACH recognized tractus opticus marginalis, axialis and basalis. Among the intrinsic fiber systems, the so-called tractus habenulo-thalamicus and dorsoventralis thalami are fairly well displayed. Of the caudal communication channels, the fasciculus retroflexus is the most conspicuous,[152] but the additional dorsal and ventral fiber systems, of which the latter blend with caudal portions of the forebrain bundles are likewise present, although rather vaguely outlined in our material.

The unambiguously identifiable *commissural systems* include commissura anterior, commissura habenulae, commissura posterior, optic chiasma, postoptic commissures, and crossing fibers in the region of tuberculum posterius at the diencephalo-mesencephalic boundary region. The postoptic commissures are particularly well developed. Tectothalamic and hypothalamic, as well as tegmentothalamic and hypothalamic connections decussate within this commissural system which displays an ill-defined dorsal and ventral subdivision.

As regards the non-neuronal *roof plate grundbestandteil* of the Dipnoan diencephalon, the choroid plexus of Protopterus seems far more massively developed than that of Ceratodus. In the latter form, it extends dorsad as a component of the parencephalic roof (cf. Figs. 84A–E). In Protopterus, it fills not only the third ventricle (cf. Figs. 87A–D), but also protrudes into the mesencephalic ventricle (cf. Figs. 384C, D, pp. 872–873, vol. 4). The saccus endolymphaticus, briefly dealt with in section 1, chapter IX *et passim* of volume 4, is likewise particularly extensive in Protopterus.

6. Amphibians

The Amphibian brain, particularly that of many Urodeles, displays in a very plain manner the fundamental Vertebrate diencephalic configurational pattern. It will be recalled that C. J. Herrick (1910) was the first to identify, in Amphibians, the thalamus ventralis and its

---

[152] The exact locus of origin respectively ending of the fibers in the fasciculus retroflexus system could not be clearly ascertained, but such difficulties obtained also in the more suitable silver impregnation material of Ceratodus studied by Holmgren and van der Horst (1925). In view of the considerable development displayed by the olfactory system in both forms, as evidenced by the telencephalic 'cortical' structures, it seems not improbable that at least part of the substantial dorsal thalamic zone has here become incorporated into the olfactory habenular system and gives origin to fasciculus retroflexus components.

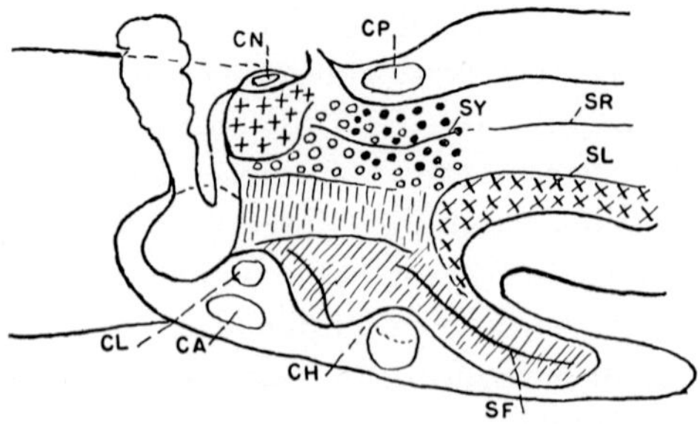

Figure 88. Semidiagrammatic reconstruction of diencephalic longitudinal zones in the diencephalon of a 'generalized' urodele Amphibian (from K., 1956). CA: anterior commissure; CH: chiasmatic ridge with its decussations; CL: commissura pallii; CN: commissura habenulae; CP: commissura posterior; SF: sulcus lateralis hypothalami posterioris ('s.l. infundibuli'); SL: sulcus limitans; SR: sulcus lateralis mesencephali (internus); SY: sulcus synencephalicus. Other details as in Figures 60A, 69A, 73A, 75A, 85A.

morphologic significance as well as that of sulcus diencephalicus ventralis and medius.[153] EDINGER (1908 *et passim)* had already stressed the import of the Urodele Amphibian brain as the simplest paradigm for a suitable morphologic understanding of the Vertebrate neuraxis. Since 1910, HERRICK devoted most of his scientific effort to an elucidation of structure and function of the Urodele brain. He presented his final conclusions and interpretations in a monograph on the brain of the Tiger Salamander (1948) which also contains many of the relevant bibliographic references. With respect to the Urodele brain, including the diencephalon, the investigations by BENEDETTI (1933; Proteus anguineus), RÖTHIG (1923, 1924; Amphibians in general, Urodeles), and by KREHT (1930, 1931; Salamandra and Proteus) also deserve special mention.

Figure 88 illustrates the general morphologic pattern of the Urodele diencephalon. As regards the arrangement of the ventricular sulci, some minor species variations and even individual ones do occur, including diverse accessory grooves which, at least in certain cases, may be artefacts. Yet, the overall configuration shown in the figure can be said to represent a reasonably constant and typical feature. On the whole, the Urodele diencephalic cell masses are characterized by a

[153] Cf. section 5, chapter VI of volume 3/II. The significance of HERRICK's findings with respect to the diencephalic pattern in the entire Vertebrate series was elaborated in our own studies (K., 1927, 1929, 1931, 1936, 1937, 1956; GERLACH, 1933, 1947; MIURA, 1933; MILLER, 1940).

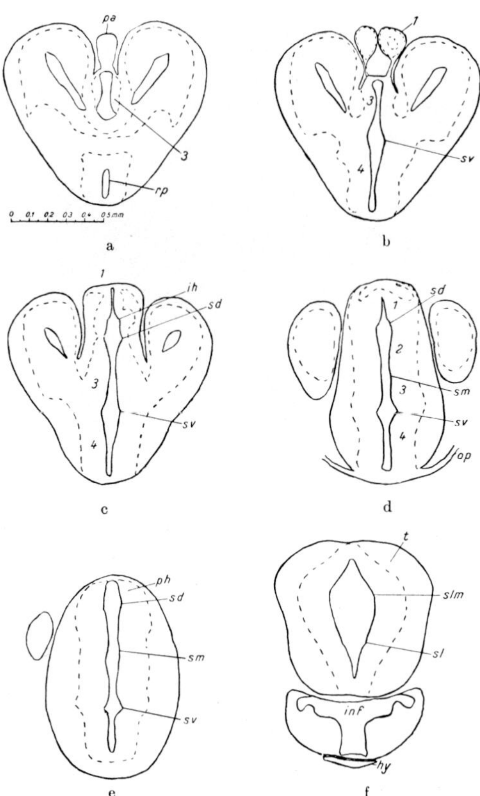

Figure 89 A. Cross-section through the diencephalon and its neighboring regions in a 27 mm larva of Triton taeniatus (from K., 1929). hy: hypophysial complex; ih: sulcus intrahabenularis; inf: lobi inferiores hypothalami ('infundibulum'); pa: paraphysis; ph: posthabenular region (pretectal portion of dorsal thalamus); rp: preoptic recess; sd: sulcus diencephalicus dorsalis (in d and e: sulcus synencephalicus); sl: sulcus limitans; slm: sulcus lateralis mesencephali (internus); sm: sulcus diencephalicus medius; sv: sulcus diencephalicus ventralis; t: tectum mesencephali; 1: epithalamus (in d: rostral beginning of dorsal thalamic pretectal region); 2: thalamus dorsalis; 3: thalamus ventralis; 4: hypothalamus.

more or less uniformly dense periventricular distribution, precluding a parcellation into circumscribed 'nuclei' but allowing for certain differences in crowding and peripheral 'scattering'.

The *epithalamus*, comprising the habenular ganglia, has a rostral position and may protrude, dorsally to velum transversum, into the space between the caudal poles of the telencephalic hemispheres, particularly in young adult or late larval specimens (cf. Figs. 89 A–C).

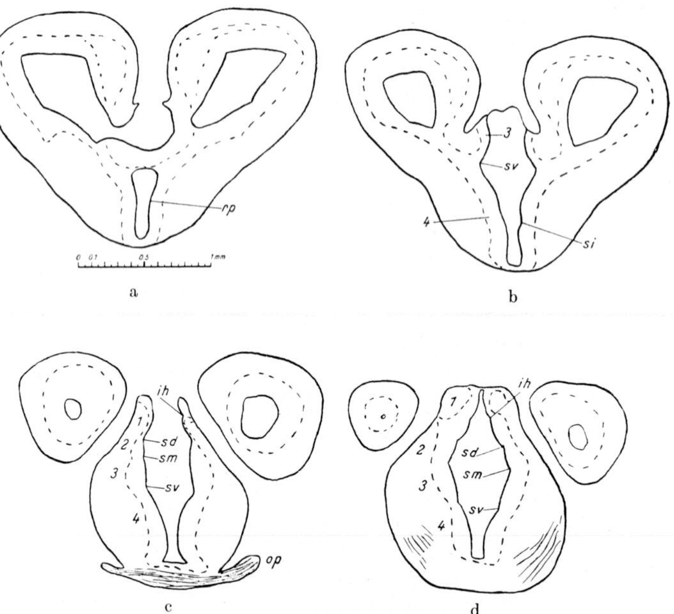

Figure 89 B. Outlines of cross-sections through the diencephalon of adult Triton taeniatus, for comparison with Figure 89 A (from K., 1929). si: sulcus intraencephalicus anterior. Other abbreviations as in Figure 89 A.

To some extent, its cells may surround fiber bundles of the stria medullaris system. A differentiation into dorsal and ventral habenular nucleus, together with an intrahabenular sulcus in the ventricular wall may or may not be conspicuous.

The *thalamus dorsalis* does not extend to the rostral limits of epithalamus and thalamus ventralis. It begins rostrally as a narrow strip between habenula and eminentia thalami (ventralis), and reaches its full expansion at posthabenular levels, on both sides of sulcus synencephalicus. Some scattered lateral neuronal elements may represent a *dorsal lateral geniculate nucleus*, but this latter is commonly included in the periventricular cell layer. This latter also generally includes, at caudal levels, the pretectal 'nucleus' commissurae posterioris, respectively the 'pretectal nucleus' of HERRICK (1948, pp. 48–49 loc.cit.).

The *thalamus ventralis* extends rostralward as far as the ventriculus impar telencephali (aula) with a substantial eminentia thalami. From the rostral thalamus ventralis, lateral cell groups protrude into basal telencephalic neighborhoods as pars superior (s. dorsalis) massae

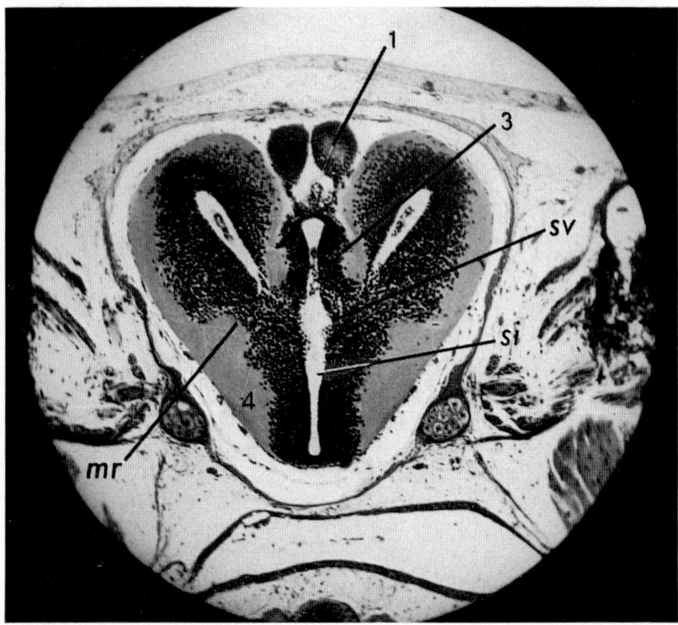

Figure 89C. Photomicrograph of cross-section through prosencephalon in the 27 mm larva of Triton. mr: massa cellularis reuniens, with merged pars superior et inferior. Other abbreviations as in Figures 89 A, B.

cellularis reunientis. Caudalward, the thalamus ventralis abuts at the rostral end of sulcus limitans, and with an indistinct transition of the merging cell masses, on the tegmental cell plate of tuberculum posterius. As in the case of thalamus dorsalis, a *ventral lateral geniculate 'nucleus'* may be suggested by some scattered elements adjoining the optic tract, or may remain included in the periventricular cell population.

The *hypothalamus*, typically subdivided by the chiasmatic ridge with its decussations into a rostral preoptic and into a caudal postoptic (posterior) region, is likewise characterized by a periventricular arrangement of its neuronal elements, thus precluding a clear parcellation into nuclei. Nevertheless, a parvocellular and a magnocellular preoptic cell group can be distinguished. This latter represents the neurosecretory component (Niimi *et al.*, 1963b; cf. Fig. 91 C). From the rostrodorsal preoptic periventricular nucleus, a massa cellularis reuniens, pars inferior, extends into the basal portion of the telencephalon. It may be more or less delimited from the ventral thalamic pars superior massae cellularis reunientis, or indistinctly merge with

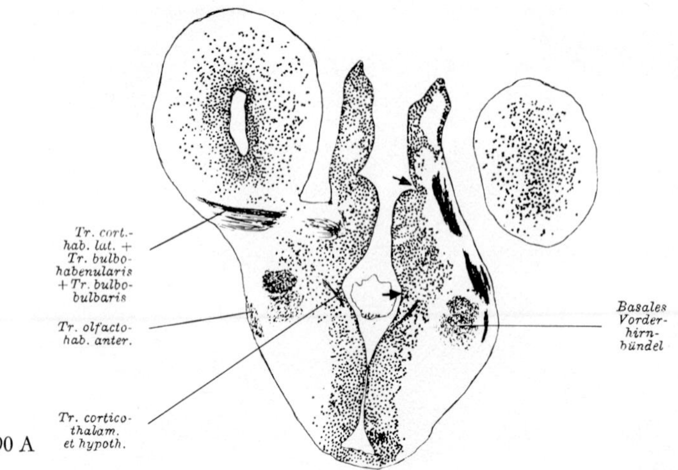
Tr. cort.-
hab. lat. +
Tr. bulbo-
habenularis
+ Tr. bulbo-
bulbaris
Tr. olfacto-
hab. anter.
Tr. cortico-
thalam.
et hypoth.
Basales
Vorder-
hirn-
bündel

90 A

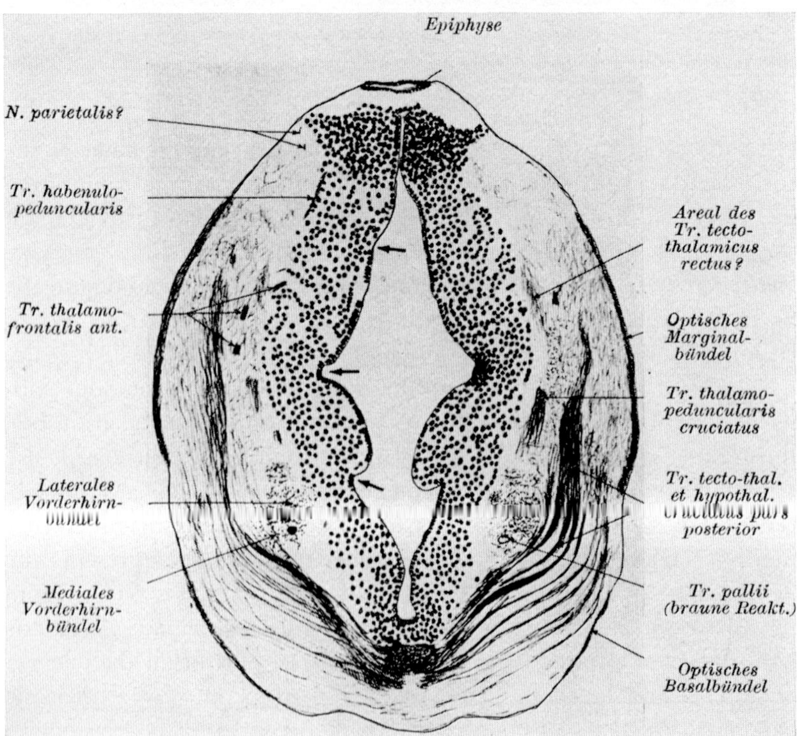
Epiphyse
N. parietalis?
Tr. habenulo-
peduncularis
Tr. thalamo-
frontalis ant.
Laterales
Vorderhirn-
Mediales
Vorderhirn-
bündel
Areal des
Tr. tecto-
thalamicus
rectus?
Optisches
Marginal-
bündel
Tr. thalamo-
peduncularis
cruciatus
Tr. tecto-thal.
et hypothal.
posterior
Tr. pallii
(braune Reakt.)
Optisches
Basalbündel

90 B

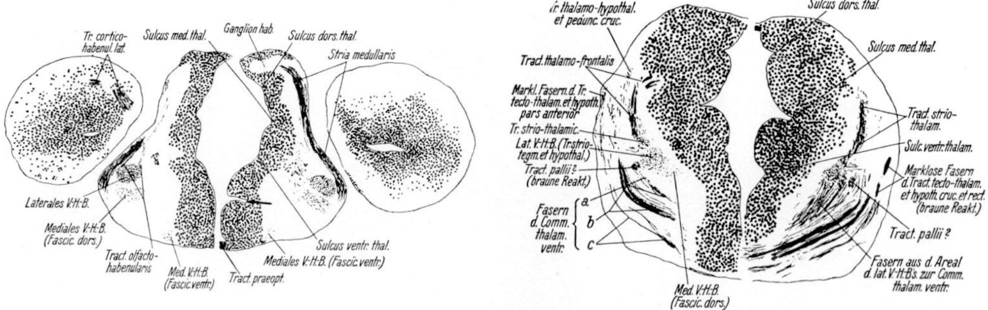

Figure 91 A. Cross-section *(Bielschowsky impregnation*, combined from two sections) through the prosencephalon of Proteus anguineus at a preoptic level (from KREHT, 1931). Added arrow: sulcus intraencephalicus anterior.

Figure 91 B. Cross-section *(Bielschowsky impregnation*, combined from two sections) through the diencephalon of Proteus. At level of chiasmatic ridge with postoptic commissures (from KREHT, 1931). The 'sulcus dorsalis thalami' (s. diencephalicus dorsalis) is probably the sulcus synencephalicus near its junction with s.d. dorsalis.

this latter (cf. e.g. Fig. 205, p. 426, vol. 3/II, and Fig. 3, section 1, present volume). In the posterior hypothalamus, a subdivision into dorsal and ventral hypothalamus can be noted, approximately delimited by the sulcus lateralis thalami posterioris. Scattered elements within, or in the vicinity of the basal forebrain bundles, mostly in the rostral two thirds of the diencephalon, can be considered a primordial nucleus entopeduncularis. Some fairly recent comments and interpretations concerning the Urodele hypothalamus can be found in the contribution by CROSBY and SHOWERS (1969).

Figures 89 to 92 depict representative cross-sections through the diencephalon of some Urodeles and show the diencephalic longitudinal

---

Figure 90 A. Cross-section *(Bielschowsky silver impregnation)* through the prosencephalon of Salamandra maculosa (from KREHT, 1930). The slightly oblique section shows caudal end of hemispheric stalk at left. Except for the basal forebrain bundle, all other tracts are components of stria medullaris system, tr. 'cortico-thalamicus et hypothalamicus' and 'tractus olfacto-habenularis anterior' being basal branches of said system. Added arrow: sulcus diencephalicus dorsalis (upper) and s.d. ventralis (lower). At left, an unlabelled intrahabenular sulcus can be identified.

Figure 90 B. Cross-section *(Bielschowsky impregnation)* through the diencephalon of Salamandra at rostral level of postoptic commissures (from KREHT, 1930). Added arrows in dorsoventral sequence: sulcus synencephalicus, s. diencephalicus medius, s.d. ventralis. The dorsal dense cell group is zone of transition between epithalamus and dorsal thalamic posthabenular region.

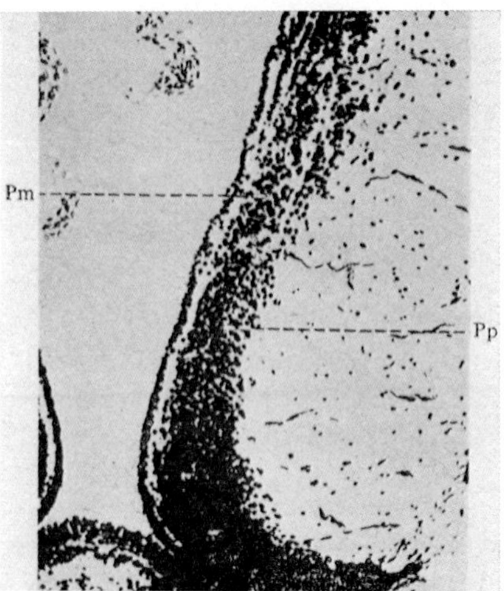

Figure 91 C. Cross-section *(Nissl stain)* through the caudal part of the preoptic nucleus in the giant Salamander Cryptobranchus japonicus (from NIIMI *et al.*, 1963b). Pm: 'nucleus' praeopticus magnocellularis; Pp: 'nucleus' praeopticus parvocellularis. The lead to Pm passes through neighborhood of sulcus intraencephalicus anterior.

zones together with the approximately delimiting sulci. These latter, together with the system of sulcus intraencephalicus anterior, with sulcus lateralis hypothalami posterioris ('sulcus lateralis infundibuli'), sulcus synencephalicus, and sulcus limitans, can also be identified, *qua* overall topologic relationships, in the diagram of Figure 88.

As in the previously discussed Anamnia, and for that matter, in all Vertebrates, the *diencephalic fiber connections* comprise comparable rostral, intrinsic, and caudal channels. In addition, there are the *commissural systems* provided by anterior, habenular, and posterior commissures, by the chiasmatic ridge with optic chiasma and post-optic commissures, and by crossing fibers in the region of tuberculum posterius as well as, to a lesser extent, in posterior hypothalamic floor adjacent to chiasmatic ridge and tuberculum posterius. These fiber systems have been analyzed in considerable detail by HERRICK (1910 and many other contributions). Because of the substantial difficulties involved[154] in studies of this sort, HERRICK's interpretations and

[154] Future investigators using contemporary techniques, such e.g. as the *Fink-Heimer technique* and those based on neuronal flow and autoradiography, will doubtless add further details, which, however, in view of the obtaining limitations, might nevertheless still remain dubious.

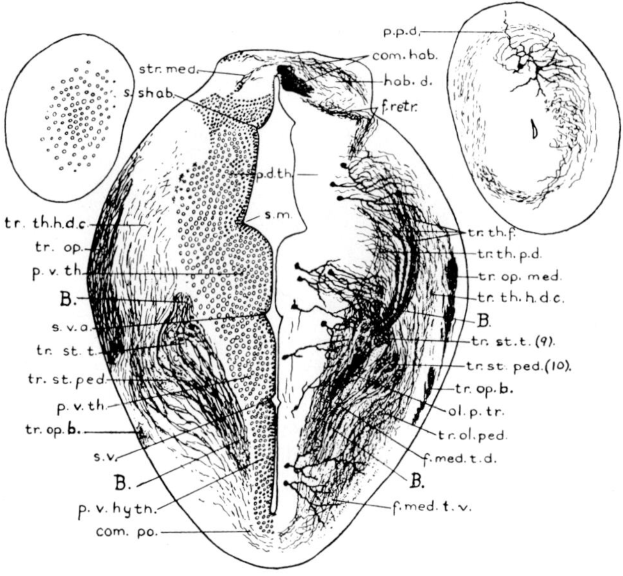

Figure 92 A. Slightly oblique cross-section (basal side inclined caudalward) through the diencephalon of Amblystoma tigrinum, combined from preparations processed by the *Golgi* and other techniques (from HERRICK, 1948). B: subdivision of 'tractus thalamotegmentalis dorsalis cruciatus'; com.hab.: habenular commissure; com.po.: postoptic commissure; f.med.t.d., f.med.t.v.: dorsal and ventral subdivisions of medial forebrain bundle; f.retr.: fasciculus retroflexus; hab.d.: dorsal 'nucleus' of habenula; ol.p.tr.: 'olfactory projection tract'; p.d.th.: thalamus dorsalis; p.p.d.: pallium of caudal tip of telencephalon; p.v.hyth.: pars ventralis hypothalami (posterioris); p.v.th.: thalamus ventralis; s.m.: sulcus diencephalicus medius; s.shab.: sulcus diencephalicus dorsalis *(sive* subhabenularis); str.med.: stria medullaris system; s.v.: sulcus diencephalicus ventralis; s.v.a.: sulcus intraencephalicus anterior ('s. ventralis accessorius thalami' HERRICK's); tr.ol.ped.: 'tractus olfactopeduncularis'; tr.op.: tractus opticus; tr.op.b.: tractus opticus basalis; tr.op.med.: tractus opticus medialis; tr.st.ped.: lateral forebrain bundle ('tr. striopeduncularis', 10 being a 'tegmental fascicle'); tr.st.t.: lateral forebrain bundle ('tr. striotegmentalis', 9 being another 'tegmental fascicle'); tr.th.f.: 'tractus thalamofrontalis'; tr.th.h.d.c.: 'tractus thalamohypothalamicus dorsalis cruciatus; tr.th.p.d.: tractus thalamopeduncularis dorsalis'.

terminology cannot be considered convincing in every respect but nevertheless represent a valuable attempt. As a useful first approximation it will again be sufficient to distinguish the *basal forebrain bundle channel*, the system of *stria medullaris*, the *'vertical' intrathalamic fiber systems*, and the various *caudal channels* as previously discussed. Concerning these latter, the lesser differentiation of lobi inferiores thalami in comparison, e.g. with Teleosts, should be kept in mind.

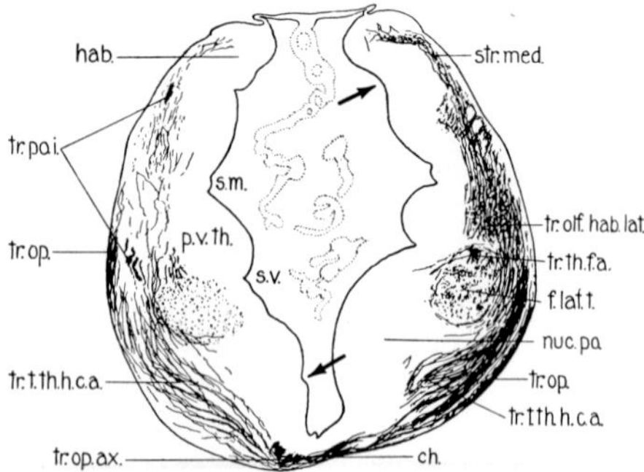

Figure 92 B. Cross-section *(Golgi impregnation)* through the diencephalon of Necturus at the level of optic chiasma (from HERRICK, 1917). ch.: optic chiasma; f.lat.t.: lateral forebrain bundle; hab.: habenula; nuc.po.: preoptic nucleus; tr.olf.hab.lat.: 'tractus olfactohabenularis lateralis'; tr.op.ax.: tractus opticus axialis; tr.po.i.: 'tractus praeoptico-intercalaris'; tr.th.fa.: 'tractus thalamofrontalis anterior'; tr.t.th.h.c.a.: 'tractus tecto-thalamicus et hypothalamicus cruciatus, pars anterior'. Added arrows: sulcus diencephalicus dorsalis (above), sulcus intraencephalicus anterior (below). Other abbreviations as in Figure 92 A.

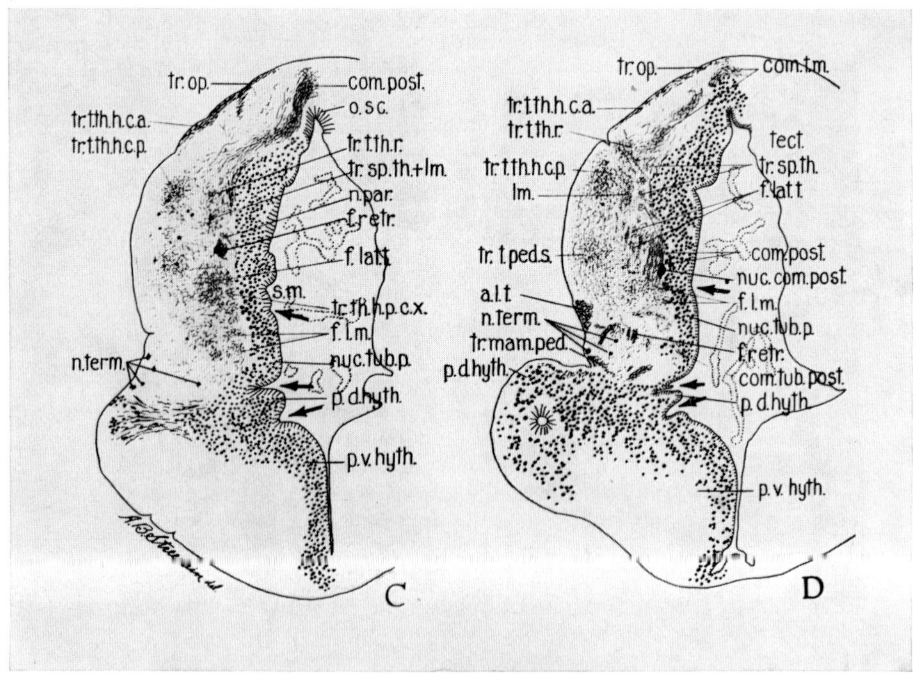

Figure 92 C, D. Two cross-sections *(Weigert stain)* through the caudal diencephalon of Necturus (from HERRICK, 1917). a.l.t.: area lateralis tegmenti; com.post.: comm. posterior; com.t.m.: comm. tecti mesencephali; com. tub.post.: comm. tuberculi posterioris (rostral tegmental decussation); f.l.m.: fasciculus longitudinalis medialis; lm.: lateral (octavo-

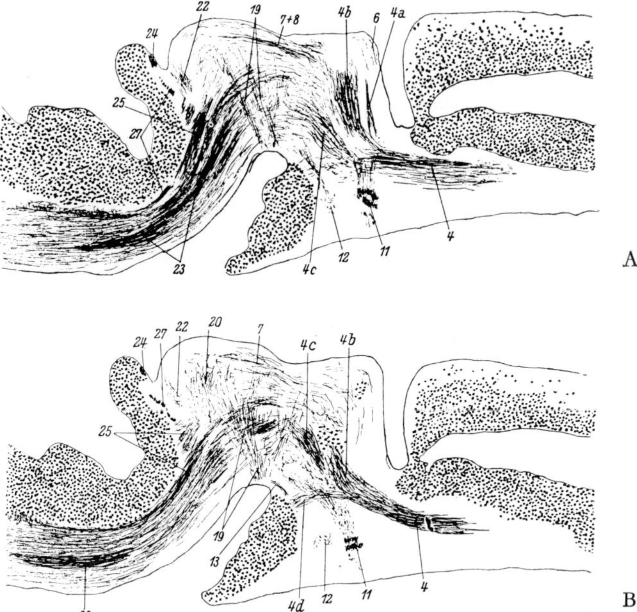

Figure 93 A, B. Two closely adjacent paramedian sagittal sections (myelin stain) through diencephalon and adjacent regions in Proteus anguineus (from KREHT, 1931). A is slightly more medial than B. 4: lateral forebrain bundle system (a: 'tractus thalamofrontalis'; b: 'tractus striothalamicus'; C: 'tr. striotegmentalis'; d: 'tr. striohypothalamicus'); 6: 'tractus thalamohypothalamicus et peduncularis cruciatus'; 7, 8: 'tractus tectothalamicus et hypothalamicus cruciatus, pars posterior', respectively 'pars anterior'; 11: 'commissura thalamica ventralis' (supra- or postoptic commissure); 12: decussatio hypothalamica ventralis (in posterior hypothalamic floor); 13: hypothalamopeduncular (mammillopeduncular ?) fibers; 19: 'tractus tectopeduncularis'; 20: 'tractus tectospinalis posterior'; 22: 'tractus bulbotectalis'; 23: lemniscus lateralis'; 24: commissura cerebellaris; 25: tractus spinotectalis et thalamicus (lemniscus medialis); 27: radix mesencephalica trigemini.

---

lateral) lemniscus; n.par: nervus parietalis; n.term.: nervus terminalis; nuc.com.post.: 'n. commissurae posterioris' (probably rostral end of torus semicircularis); nuc.tub.p.: n. tuberculi posterioris; o.sc.: subcommissural organ; p.d.hyth., p.v.hyth.: pars dorsalis respectively ventralis hypothalami posterioris (including lobi lateralis); s.m.: 's. medius thalami' (probably caudal end of s. dienc. ventr.; the true s. dienc. medius can be seen dorsal to it; still more dorsally the s. synencephalicus can be identified); tect.: tectum mesencephali; tr.mam.ped.: 'tr. mammillopeduncularis'; tr.sp.th.: 'tr. spinothalamicus' (medial lemniscus); tr.t.ped.s.: 'tr. tectopeduncularis superficialis; tr.t.th.h.c.p.: 'tr. tectothalamicus et hypothalamicus cruciatus, pars posterior'; tr.t.th.r.: tr. tectothalamicus rectus'. Added arrows: upper two: rostroconvex ending of s. limitans; lower: s. lateralis hypothalami posterioris. The sulcus above lead f.lat.t. in Figure 92D is the extension of sulcus synencephalicus into sulcus lateralis mesencephali. Other abbreviations as in Figures 92A, B.

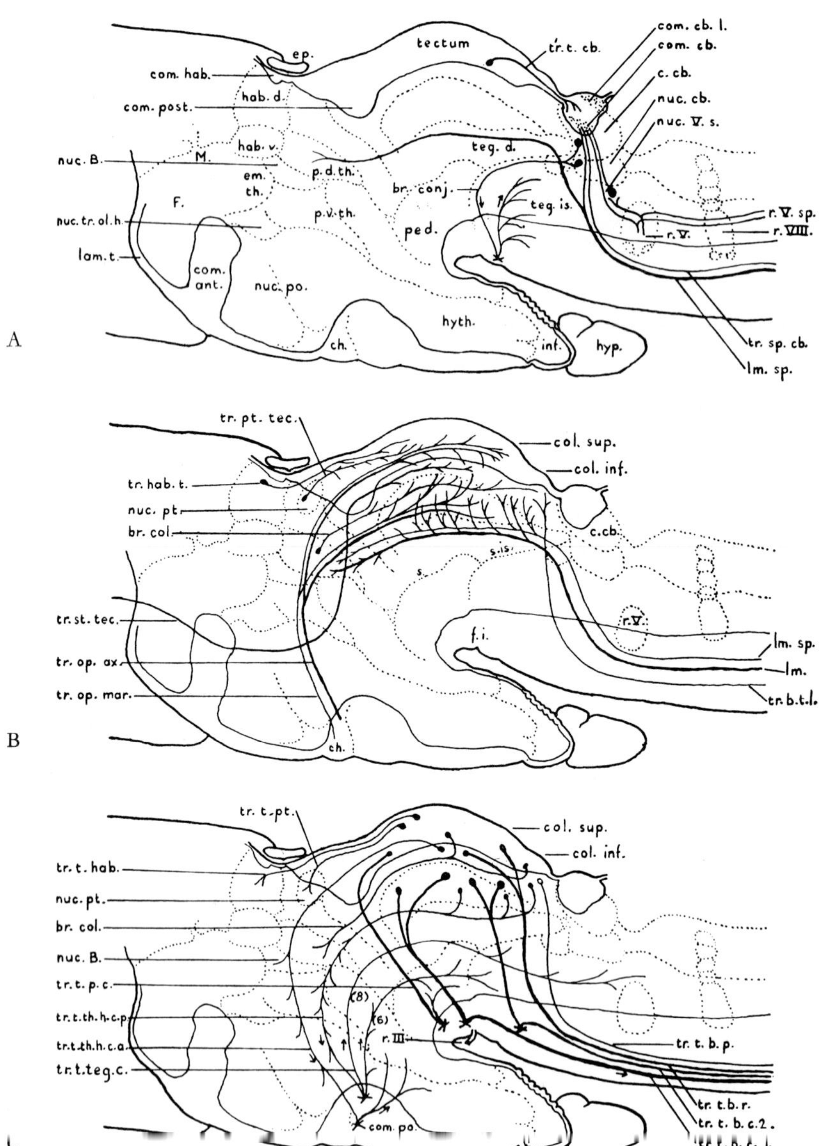

Figure 94 A–F. Diagrams of main communication channels of the diencephalon in Amblystoma tigrinum according to Herrick's interpretations (from Herrick, 1948). a.vl.p.: 'area ventrolateralis pedunculi; br.col.: brachia colliculi (mesencephali); br.conj.: 'brachium conjunctivum'; c.cb.: corpus cerebelli; ch.: chiasma opticum; col.inf.sup.: mesencephalic 'colliculi'; com.ant.: comm. anterior; com.cb.: comm. cerebelli; com.cb.l.: comm. vestibulolateralis cerebelli; com.hab.: comm. habenulae; com.po.: comm.

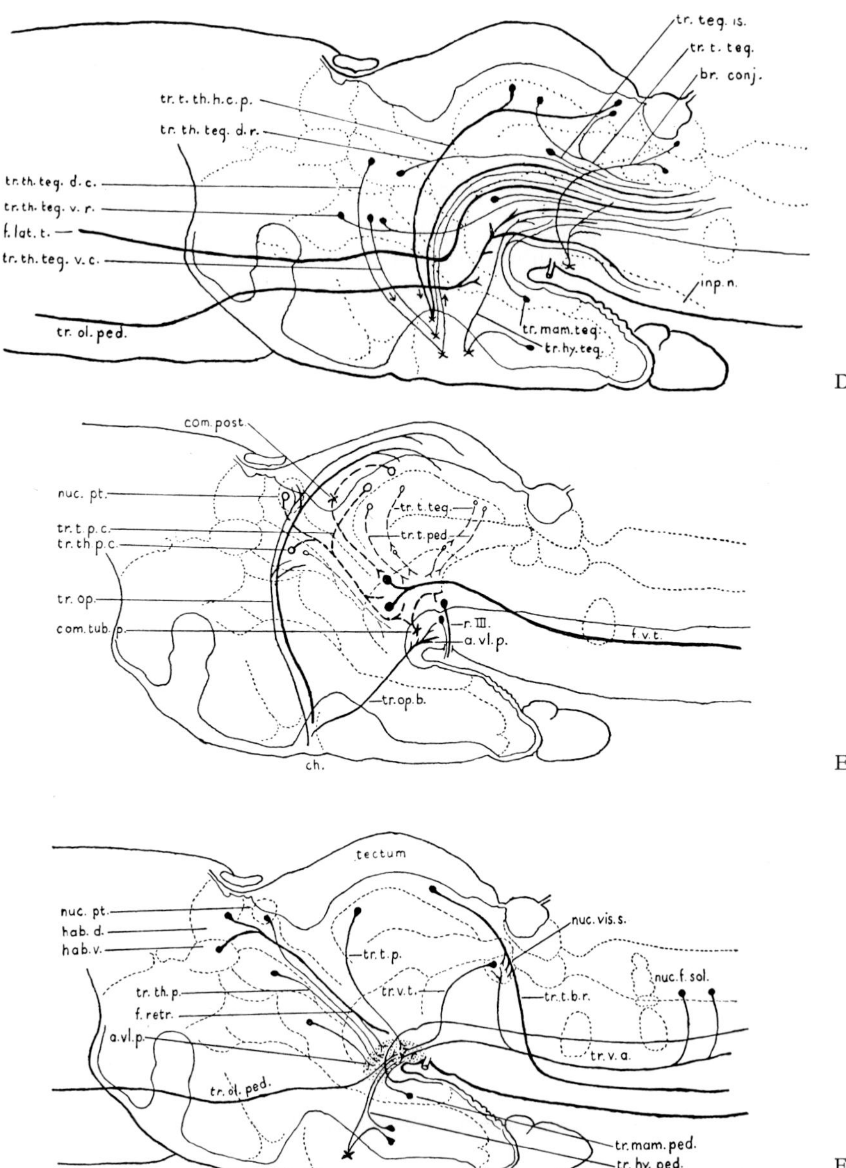

D

E

F

postoptica; com.post.: comm. posterior; com.tub.p.: comm. tuberculi posterioris; em.th.: eminentia thalami ventralis; ep.: epiphysis; F.: foramen interventriculare; f.i.: fovea isthmi; f.lat.t.: lateral forebrain bundle; f.retr.: fasciculus retroflexus; f.v.t.: 'fasciculi ventrales tegmenti'; hab.d., v.: dorsal and ventral habenular griseum; hyp.: hypophysial complex; hyth.: hypothalamus; inf.: 'infundibulum'; inp.n.: 'interpeduncular neuropil'; lam.t.: lamina terminalis; lm.: general bulbar lemniscus; lm.sp.: lemniscus

Figures 93A and B illustrate fiber systems as recorded by KREHT (1931), while Figures 94A–F show diagrams elaborated by HERRICK (1948) and indicate the self-explanatory but not always consistent[155] terminology adopted by this author in his extensive labor. Concerning optic input, HERRICK (1925, 1942) noted connections with the lateral geniculate primordia, the pretectal grisea, and hypothalamic cell groups. He distinguished a tractus opticus accessorius posterior,[156] running toward tuberculum posterius and mesencephalic tegmentum, an axial optic tract, which, at least in part, may reach pretectal grisea and tectum opticum, and a main marginal optic tract to the tectum, with connection to the thalamic grisea. Figures 92C and D indicate that HERRICK believed to have tracked the course of the telencephalic nervus terminalis as non-medullated fascicles through the diencephalon

[155] No derogatory criticism is here meant. Because of the numerous obtaining uncertainties, which nevertheless require the formulations of operationally useful first approximations, many investigators modified their interpretations and terminologies in the course of their studies. Despite my own attempts in working out valid generalized concepts, I was likewise unable to remain rigorously consistent in various secondary details.

[156] In 1948, HERRICK used the term 'basal optic tract' (cf. Fig. 94E).

Continuation of legend to Figure 94A–F

spinalis; M.: basal opening of paraphysis; nuc.B.: 'n. of Bellonci' (dorsal lateral geniculate nucleus); nuc.cb.: n. cerebelli; nuc.f.sol.: n. of fasciculus solitarius; nuc.po.: n. cerbelli; nuc.f.sol.: n. of fasciculus solitarius; nuc.po.: preoptic nucleus; nuc.pt.: 'pretectal nucleus'; nuc.tr.ol.h.: 'n. of olfacto-habenular tract'; nuc.vis.s.: superior visceral (gustatory) nucleus; nuc.V.s.: 'superior n. of trigeminus'; p.d.th.: thalamus dorsalis; ped.: 'pedunculus cerebri' (tegmental cell plate); p.v.th.: thalamus ventralis; r.III, V, VIII: radix oculomotorii, trigemini, octavi; r.V sp.: descending trigeminal root; s.: rostral end of s. limitans; s.is.: 's. isthmi'; teg.d.: 'tegmentum dorsale' (torus semicircularis); teg.is.: tegmentum isthmi; tr.b.t.l.: 'tr. bulbotectalis lateralis'; tr.hab.t.: 'tr. habenulotectalis'; tr.hy.ped.: 'tr. hypothalamopeduncularis'; tr.hy.teg.: 'tr. hypothalamotegmentalis'; tr.mam.ped.: 'tr. mammillopeduncularis'; tr. mam. teg.: 'tr. mammillotegmentalis'; tr.ol.ped.: 'tr. olfactopeduncularis'; tr.op., ax., b., mar.: tr. opticus, axialis, basalis, marginalis; tr.pt.teg.: tr. pretectotectalis; tr.sp.cb.: tr. spinocerebellaris; tr.sp.tec.: tr. spinotectalis (lemniscus spinalis); tr.st.tec.: 'tr. striotectalis'; tr.t.b.c., p., r.: tr. tectobulbaris, cruciatus, posterior, rectus; tr.t.teg.: tr. tectotegmentalis (c: cruciatus, 6, 8: further subdivisions noted by HERRICK); tr.teg.is.: tr. tegmentalis isthmi; tr.teg.b.: tr. tegmentobulbaris; tr.teg.inp.: tr. tegmento-interpeduncularis; tr.t.ped.: 'tr. tectopeduncularis; tr.t.hab.: tr. tectohabenularis; tr.t.p.c.: tr. tectopeduncularis cruciatus; tr.t.th.h.c.a., p.: tr. tectothalamicus et hypothalamicus cruciatus anterior, posterior; tr.th.p.c.: tr. thalamopeduncularis cruciatus; tr.th.teg.d.c., v.c.: tr. thalamotegmentalis dorsalis cruciatus, ventralis cruciatus; tr.v.a.: tr. visceralis ascendens; tr.v.t.: tertiary visceral tract.

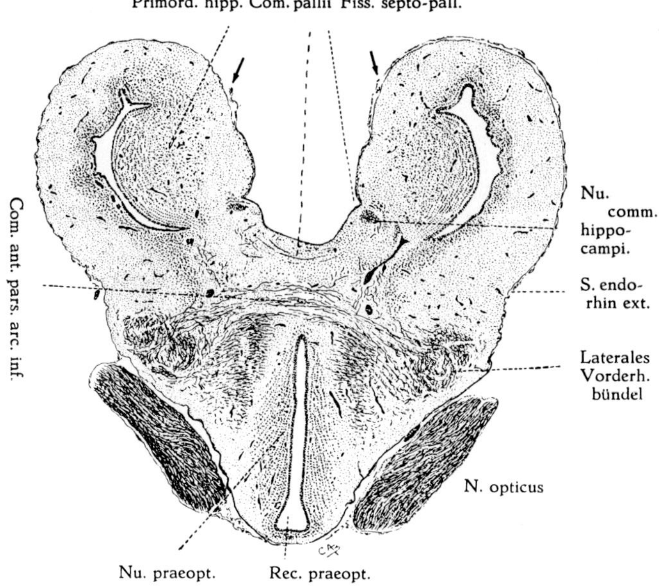

Figure 95 A. Cross-section (myelin stain) through the forebrain of Rana catesbyana at the caudal level of commissural plate (from KAPPERS and HAMMER, 1918). The 'sulcus endorhinalis externus' is part of the sulcus diencephalo-telencephalicus. The added arrows point to the attachment of lamina epithelialis (roof plate) at the transition of ventriculus impar telencephali to diencephalic (or 3rd) ventricle.

toward their apparent termination into the rostral mesencephalic tegmentum adjoining the tuberculum posterius.[157]

The configuration of the *Anuran* diencephalon, although essentially similar to that obtaining in Urodeles, differs from that latter by a more pronounced differentiation of its cell masses, which display not only a periventricular (inner) and an outer sublayer, but also a more or less distinct parcellation, i.e. the tendency toward 'grouping' of cell masses into 'nuclei'. Figures 95A to 97F illustrate various aspects of the Anuran diencephalon as recorded by diverse investigators.

In the *epithalamus*, which does not, as a rule, extend as far rostralward as the thalamus ventralis, a dorsal and a ventral habenular nucleus can usually be noted (cf. Figs. 96C, 97A). Asymmetry between right and left habenular grisea does not seem to be very pronounced

[157] Some of these fascicles are said to decussate in anterior, postoptic, posterior hypothalamic or rostral tegmental commissures. The nervus terminalis shall be dealt with in chapter XIII.

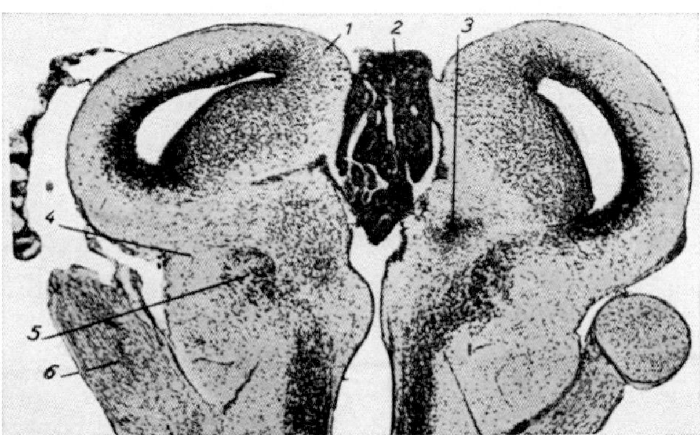

Figure 95 B. Cross-section *(Giemsa stain)* through the forebrain of Rana esculenta at a level slightly caudal to that of Figure 95 A (from Beccari, 1943). 1: caudal evagination of telencephalon; 2: paraphysis, perhaps with parts of velum transversum and saccus dorsalis *('nodo vasculoso')*; 3: 'parvocellular' anterior nucleus of thalamus ventralis; 4: hemispheric stalk; 5: massa cellularis reuniens, pars inferior s. hypothalamica; 6: optic nerve.

and can be suggested by transverse or slightly oblique cross-sections displaying local difference in cellular arrangement. Nevertheless, Morgan *et al.* (1973) noted an apparently consistent left-right asymmetry in tadpoles and adults of Rana temporaria, the left griseum being larger than the right one and containing a medial and lateral habenular nucleus. The cited authors also briefly discuss some general implications of asymmetry (and its occasional mirror-image reversions) with respect to theoretical morphology. Although, if present, the Anuran parietal 'eye' has been interpreted as being the 'pineal' organ, corresponding to that of Petromyzonts, it will be recalled that, in these latter, the right habenula is substantially larger than the left one (cf. section 1 C, p. 118).

The *thalamus dorsalis* begins rostrally at habenular levels (Figs. 96 B, C, 97 A). In the region of its greatest expansion (Figs. 97 B–D) a stratified periventricular layer, a diffuse intermediate, and an external stratum can be recognized. The rostral intermediate and external stratum jointly form the anterior dorsal *'campo diffuso'* of Beccari (1943). Slightly more caudalward the external stratum, with a conspicuous neuropil, represents the dorsal lateral geniculate nucleus, commonly designated as the *nucleus of Bellonci*[158] (cf. Fig. 97 B). Still more caudalward, a posthabenular medial portion, a nucleus commis-

[158] This nucleus, with its neuropil, was described by Bellonci (1888) and is particularly conspicuous in various Anurans. There is, in my opinion, little doubt that,

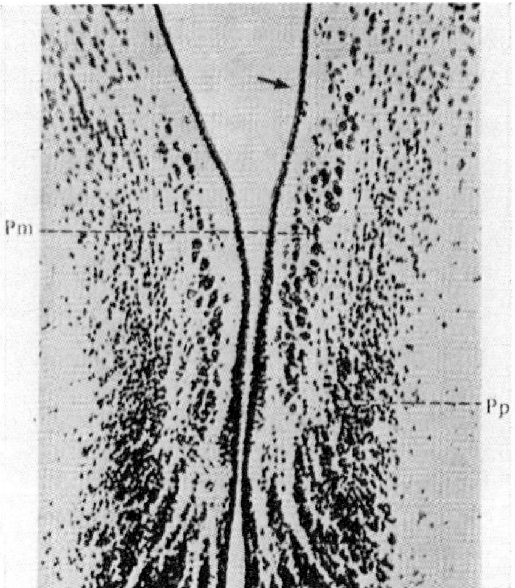

Figure 95C. Cross section *(Nissl stain)* through the preoptic hypothalamus of Rana nigromaculata (from NIIMI *et al.*, 1963b). Pm: nucleus praeopticus magnocellularis; Pp: nucleus praeopticus parvocellularis. Added arrow: approximate locus of sulcus diencephalicus ventralis. Sulc. dienc. medius not present at this level (cf. Fig. 96A).

surae posterioris, and a supracommissural area praetectalis are recognizable.

The *thalamus ventralis* begins at the level of telencephalon impar with a rather large eminentia thalami whose most rostral differentiation is the nucleus commissurae hippocampi (Fig. 95A, cf. also Fig. 272B, p. 521, vol. 3/II). Slightly more caudalward, a massa cellularis reuniens, pars dorsalis, and BECCARI's *nucleo parvocellulare anterodorsale* are present (Fig. 95B). Still more caudalward, a well-developed ventral lateral geniculate nucleus appears, separated from an intermediate diffuse cell mass, which, in turn, borders on the medial periventricular cell group designated as '*campo pluristratificato*' by

---

morphologically speaking, it represents, either *in toto* or in part, the dorsal lateral geniculate nucleus. Since the term '*nucleus of Bellonci*' refers to a peculiar differentiation of said griseum obtaining in at least some Anurans, this term should preferably not be applied in other Amphibian or Vertebrate forms (e.g. as in Urodeles by HERRICK, 1948). ADDENS (1938) designated the nucleus ovalis and the lateral geniculate complex of Reptiles (Crocodilians) as *nucleus of Bellonci*, but the nucleus ovalis is definitely a ventral thalamic derivative, caudally continuous with the ventral lateral geniculate nucleus (cf. further below section 7, p. 259). A recent publication on the central connections of the optic nerve in the Frog (SCALIA and FITE, 1974) claims to have established 'retinotopic projections' upon the lateral geniculate complex (and tectum) on the basis of the *Fink-Heimer technique*.

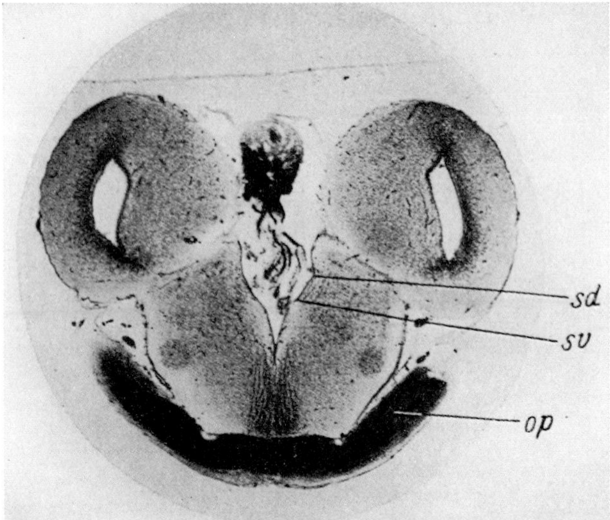

Figure 96 A. Cross-section *(Weigert stain)* through the forebrain of Rana esculenta at rostralmost level of optic chiasma (from K., 1929). op: optic nerve; sd: sulcus diencephalicus dorsalis; sv: sulcus diencephalicus ventralis.

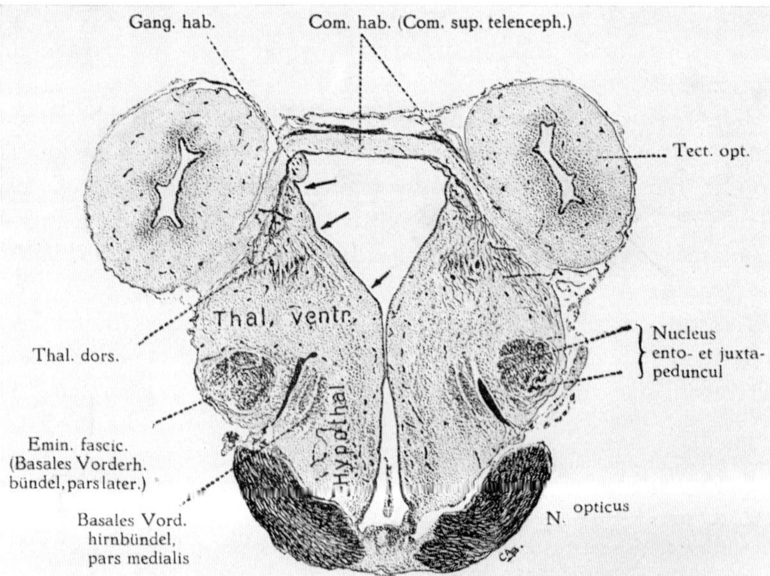

Figure 96 B. Cross-section (myelin stain) through the forebrain of Rana catesbyana at rostral level of thalamus dorsalis and habenular commissure (from KAPPERS and HAMMER, 1918). 'Tect.opt.' is evidently caudal portion of telencephalon *(quandoque dormitat bonus Homerus:* such lapsus may also occasionally be perpetrated by myself). Added arrows: sulcus diencephalicus dorsalis (top), s.d. medius (middle), s.d. ventralis (lower).

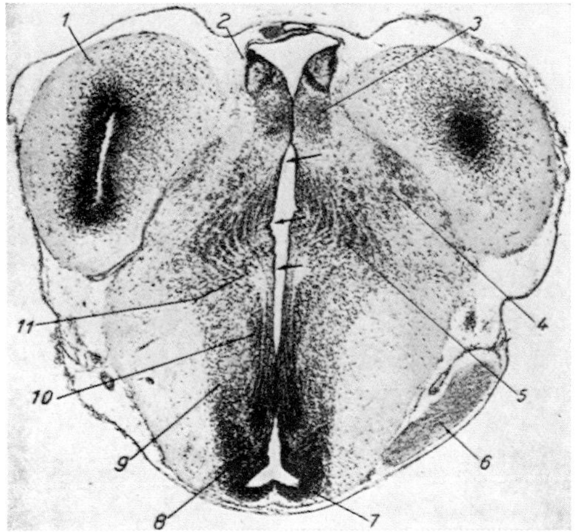

Figure 96C. Cross-section *(Giemsa stain)* through the forebrain of Bufo vulgaris at level near caudal pole of telencephalon (from BECCARI, 1943). 1: telencephalon; 2: 'dorsal habenular nucleus'; 3: 'ventral habenular nucleus'; 4: *'campo diffuso'* (lateral portions of dorsal and ventral thalamus); 5: *'campo pluristratificato'* (medial portion of ventral thalamus); 6: optic nerve; 7: recessus praeopticus; 8: inferior parvocellular preoptic nucleus; 9: lateral portion of parvocellular preoptic nucleus; 10, 11: magnocellular preoptic nucleus. Added arrows: sulcus diencephalicus dorsalis, s.d. medius, s.d. ventralis (note asymmetry between right and left, which may be either 'variation' or artefact).

BECCARI (1943). The caudalmost part of this cell group becomes narrower and ends at the tegmental cell plate (Figs. 97C, D, E, F).

As regards the caudolateral region of both thalamus dorsalis and thalamus ventralis it seems possible that griseal neighborhoods of the lateral geniculate complex receive lateral lemniscus, i.e. 'cochlear' input and thus actually correspond to the medial geniculate complex of Reptiles as mentioned further below in section 7 (p. 225). There is little doubt that at least some Salientia have a rather well-developed cochlear system, including some apparent but still rather vague griseal subdivisions in the torus semicircularis (cf. e.g. POTTER, 1965, 1969).

The *rostral hypothalamus* provides, at hemispheric stalk levels of the preoptic region, the pars inferior massae cellularis reunientis (Fig. 95B). This latter is, caudally, continuous with the entopeduncular nucleus, represented by scattered neuronal elements within the basal forebrain bundles and their neighborhood. The preoptic region likewise contains the neurosecretory magnocellular preoptic nucleus (Fig. 95C) as recorded by DIEPEN (1962), NIIMI *et al.* (1963b), CROSBY and SHOWERS (1969) and various other authors. This nucleus may, in

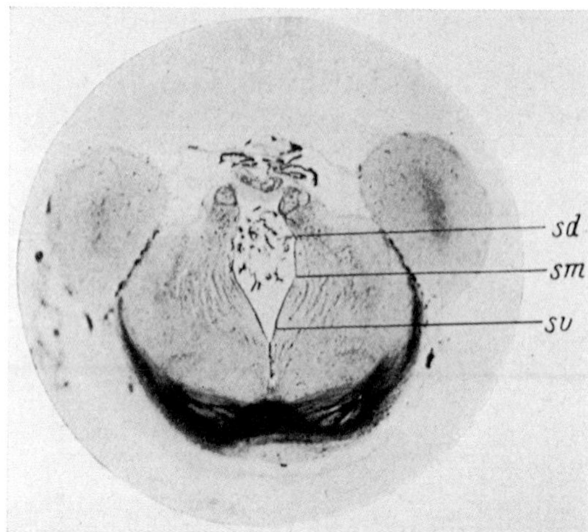

Figure 97 A. Cross-section *(Weigert stain)* through forebrain of Rana esculenta at level of polus caudalis telencephali (from K., 1929). sd: sulcus diencephalicus dorsalis; sm: s.d. medius; sv.: s.d. ventralis. Basally, the caudal portion of optic chiasma (perhaps with overlap of postoptic commissural fibers) can be identified.

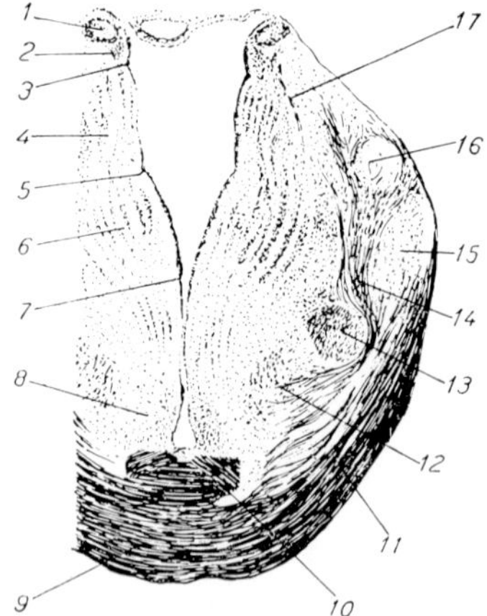

Figure 97 B. Cross-section *(Weigert stain)* through the diencephalon of Rana catesbyana at the level of '*Bellonci's nucleus*' (after ADDENS, 1938, from BECCARI, 1943). 1, 2: dorsal and ventral habenular nucleus; 3: sulcus diencephalicus dorsalis; 4: thalamus dorsalis; 5: sulcus diencephalicus medius; 6: thalamus ventralis; 7: sulcus diencephalicus ventralis; 8: preoptic hypothalamus; 9: optic chiasma; 10: 'hypothalamic optic root'; 11: tractus opticus *(fascio ottico laterale)*; 12, 13: medial and lateral forebrain bundle; 14: 'fasciculus olfactohabenularis lateralis'; 15: ventral lateral geniculate nucleus; 16: *nucleus of Bellonci* (dorsal lateral geniculate complex); 17: fasciculus retroflexus (probably here intermingled with habenulothalamic and hypothalamic fibers.

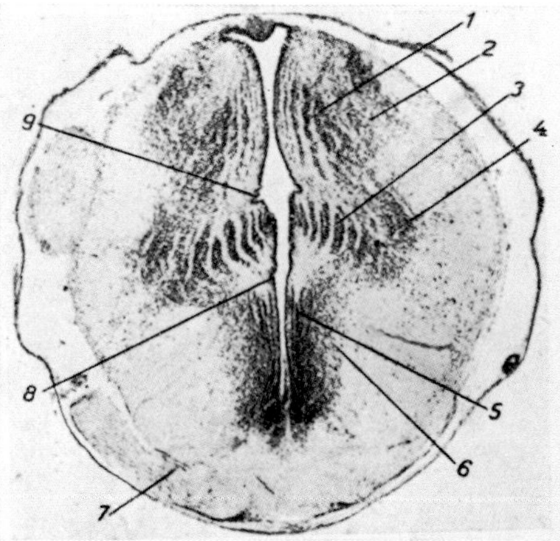

Figure 97C. Cross-section *(Giemsa stain)* through the diencephalon of Bufo vulgaris, passing dorsally through rostral posthabenular chiasma (from Beccari, 1943). 1, 2: *campo pluristificato* respectively *campo diffuso* of posthabenular dorsal thalamus; 3, 4: same subdivision of ventral thalamus; 5: nucleus praeopticus magnocellularis; 6: 'nucleus praeopticus parvocellularis lateralis'; 7: nervus opticus near chiasma; 8: sulcus diencephalicus ventralis; 9: sulcus diencephalicus medius. At left: polus posterior of telencephalon. Above roof plate: part of epiphysial stalk.

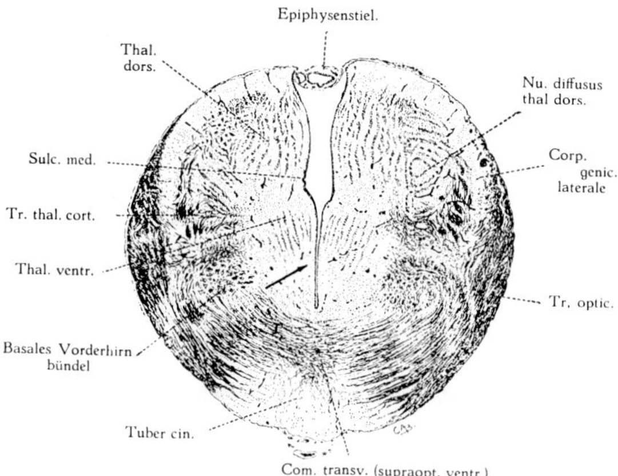

Figure 97 D. Cross-section (myelin stain) through the diencephalon of Rana catesbyana at level of supraoptic commissure (from Kappers and Hammer, 1918). Corp. genic. laterale is here joint dorsal and ventral thalamic lateral geniculate complex. Tuber cin. is rostral end of posterior hypothalamus. Added arrow: approximate locus of flattened out sulcus dienc. ventralis.

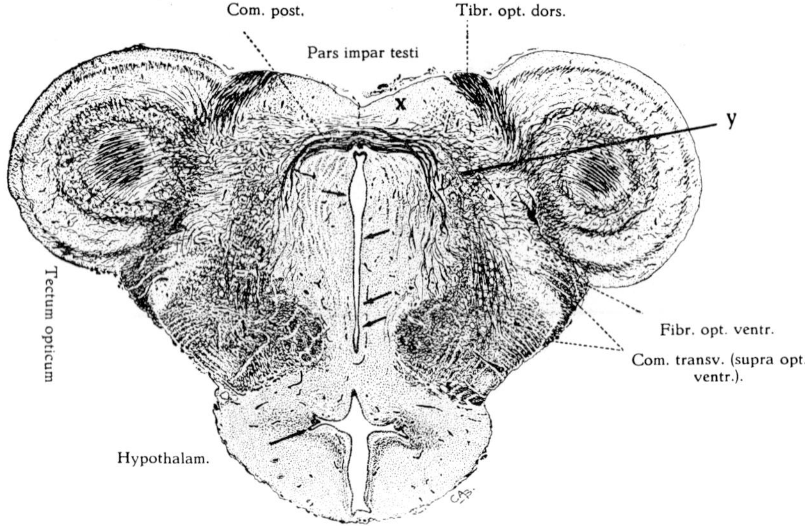

Figure 97E. Cross-section (myelin stain) through diencephalomesencephalic boundary region of Rana catesbyana (from KAPPERS and HAMMER, 1918). 'Pars impar testi' *(sic)* is still diencephalic pretectal region. Added designations: x: area praetectalis; y: nucleus commissurae posterioris; arrows in dorsoventral sequence; sulcus synencephalicus, sulcus diencephalicus medius, s.d. ventralis, s.l. limitans, s. lateralis hypothalami posterioris.

some forms, extend to or beyond the chiasmatic ridge and its commissures.

Recent authors (MCKENNA *et al.*, 1973) have claimed that catecholamine containing subependymal cells in the preoptic recess of the Toad represent a 'new', non-neuronal cell type 'specialized for the synthesis, storage and secretion of catecholamines', for which the designation 'encephalochromaffin cells' is introduced. These cells are supposed to secrete, by way of their 'apical processes' into the preoptic recess. While the presence of catecholamine within said elements seems well documented, the additional conclusions presented by the cited authors do not appear very convincing and further data on this topic must be awaited.

The *posterior hypothalamus*, with its sulcus lateralis, displays an essentially periventricular dorsal and ventral subdivision extending into the moderately developed lobi laterales *sive* inferiores (Fig. 97E). The neighborhood dorsal to sulcus lateralis hypothalami posterioris, and adjoining the tegmental cell cord of tuberculum posterius, can be evaluated as a kathomologon of the Mammalian mammillary nucleus (or complex).

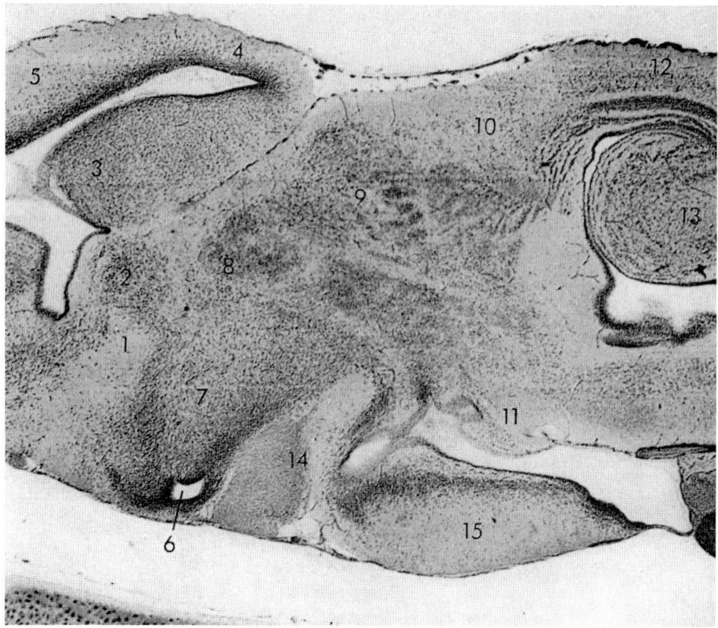

Figure 97 F. Paramedian sagittal section (cell stain) through prosencephalon and part of mesencephalon in Rana. 1: commissura anterior; 2: B_3; 3: D_3; 4: D_2; 5: D_1; 6: preoptic recess; 7: hypothalamus; 8: thalamus ventralis; 9: thalamus dorsalis; 10: nucleus lentiformis mesencephali; 11: rostral end of basal plate (tegmentum); 12: tectum opticum; 13: torus semicircularis; 14: optic chiasma and supra- respectively postoptic commissures; 15: posterior inferior hypothalamus.

Within the so-called infundibular recess of the posterior hypothalamus in the Toad, subependymal catecholamine containing cells were described by McKenna and Rosenbluth (1974), who made use of fluorescence technique and of electron microscopy. The cytological characteristics of these elements were interpreted to suggest that their primary function may be the detection of 'compounds in the cerebrospinal fluid and the transmission of 'this information to neighboring neurons via the somatodendritic synapses' (cf. above the 'encephalochromaffin cells' described in the preoptic recess).

Less details are available concerning the *diencephalic communication channels* in Anurans than in Urodeles, but, on the whole, these fiber systems can be assumed to be essentially similarly organized in both orders. For an enumeration of main tracts and commissural systems reference to their preceding description in Urodeles will therefore here be considered sufficient. With respect to thalamo-telencephalic afferent systems, various conjectures, based on still rather inconclusive evidence have been elaborated in a paper by VESSELKIN *et al.* (1971).

On the basis of *sagittal sections* (Fig. 97 F) it is, moreover, possible to distinguish, as regards both thalamus dorsalis and ventralis, a rostral, an intermediate, and a caudal subdivision, the caudal one of which may be included into the pretectal region *sensu latiori*. It must, however, be kept in mind that, because of the three-dimensional arrangement, any 'straight' sagittal plane will successively pass through more lateral components of the thalamic longitudinal zones, thereby introducing apparent differences in the longitudinal arrangement actually pertaining to the medio-lateral stratification. It is nevertheless most likely that ascending input to telencephalic grisea from diencephalon, particularly from the optic and other sensory endings in the thalamus, is better developed in Anurans than in Urodeles. This is especially the case with respect to *Bellonci's nucleus* (dorsal lateral geniculate nucleus) which seems, moreover to correlate olfactory input to preoptic hypo-

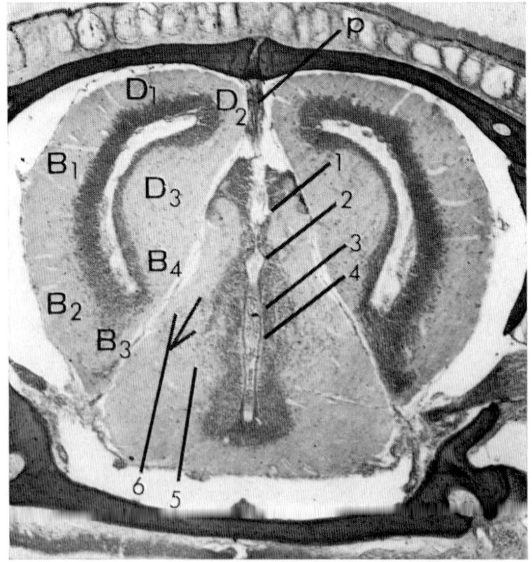

Figure 98 A. Cross-section (hematoxylin-eosin stain) through the diencephalon and adjacent configurations in the Gymnophione Amphibian Schistomepum at level of supraoptic-postoptic commissural system (slightly modified after K. *et al.*, 1966). 1: sulcus intrahabenularis; 2: sulcus diencephalicus dorsalis; 3: sulcus diencephalicus medius; 4: sulcus diencephalicus ventralis; 5: nucleus entopeduncularis; 6: dorsal and ventral lateral geniculate complex; B, D: telencephalic longitudinal zonal system; p: paraphysis.

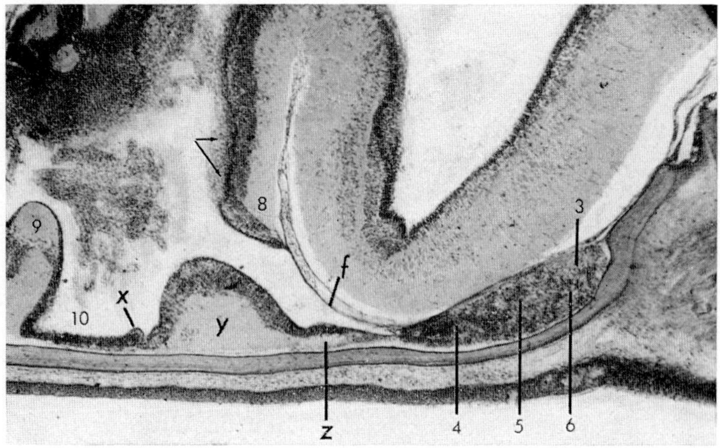

Figure 98B. Midsagittal section (hematoxylin-eosin stain) through the diencephalic floor of Schistomepum (slightly modified from K., 1970). f: part of epithelial infundibular stem; x: prechiasmatic ventricular protrusion; y: chiasmatic ridge with supra- and post-optic commissures; z: eminentia medialis (neurohypophysis); 3: infundibular process (neurohypophysis); 4–6: adenohypophysis (p. tuberalis, distalis, intermedia); 8: caudal hypothalamic wall, basally adjacent to tegmental cell plate; 9: telencephalic commissural plate (torus transversus); 10: preoptic recess. Arrows: rostral end of sulcus limitans and of basal plate.

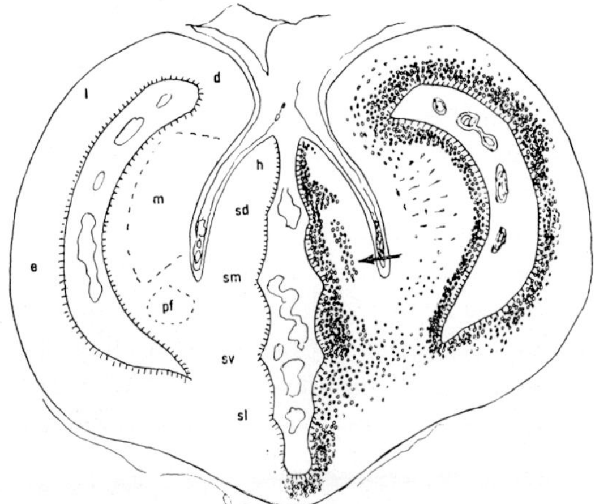

Figure 98C. Cross-section through the forebrain of Siphonops at caudal level of hemispheric stalk (from K., 1922). d: zone D_2; e: B_1; h: epithalamus (habenula); pf: B_4; l: D_1; m: D_3; sd: sulcus diencephalicus dorsalis; sl: sulcus intraencephalicus anterior (originally misinterpreted as 'sulcus limitans'); sm: sulcus diencephalicus medius; sv: s.d. ventralis. Added arrow: 'lateral geniculate complex'. The longitudinal zones can be easily identified by their relation to the sulci, likewise the contribution of thalamus ventralis and hypothalamus to massa cellularis reuniens.

thalamus and habenular nuclei with the optic input at diencephalic levels.

Few data are available on the diencephalon of *Gymnophiones*. Our own studies (K., 1922, 1929, 1970; K. *et al.*, 1966), in which the extant literature has been considered and listed, were primarily concerned with morphological problems of pattern relationships, and only incidentally recorded a few features pertaining to structural detail or main fiber systems.

The diencephalic configuration in this order conforms to the overall Amphibian bauplan[159] and, on the whole, particularly also *qua* essentially periventricular arrangement, corresponds somewhat more closely to the pattern details obtaining in Urodeles rather than in Anurans. The typical ventricular sulci remain more or less easily recognizable at the adult stage.

The *epithalamus* with its habenular ganglion displays a pronounced rostral position, extending here at least as far as the eminentia thalami ventralis toward the telencephalon impar. In some forms, an intrahabenular sulcus (Fig. 98A) is present, indicating a subdivision into dorsal and ventral habenular nucleus.

The *dorsal thalamus* likewise extends rather far rostrad toward levels of hemispheric stalk and aula. At rostral levels, there is, moreover, the tendency toward the formation of a lateral cell plate, separated from the periventricular layer, and morphological comparable with a dorsal lateral geniculate nucleus. Caudalward, this cell plate becomes confluent with a similar cell group derived from the thalamus ventralis. Still more caudalward, the cell population remains periventricular and includes the locus of posthabenular (pretectal) thalamus dorsalis together with that of the ill-defined nucleus commissurae posterioris.

The *ventral thalamus* begins at the aula as eminentia thalami, whose rostral tip provides a nucleus supracommissuralis corresponding to the 'nucleus commissurae hippocampi' of Anurans. Within the hemispheric stalk, a pars dorsalis massae cellularis reunientis is present (Fig. 98C). Caudally to hemispheric stalk, scattered lateral cells separated from the periventricular layer from an ill-defined ventral lateral geniculate nucleus with nondescript suggestions of stratification (Fig. 98A). Still more caudalward, the thalamus ventralis remains a periventricular cell layer (Fig. 98D). As regards the above-mentioned

[159] Cf. Figure 204, p. 425 of volume 3/II.

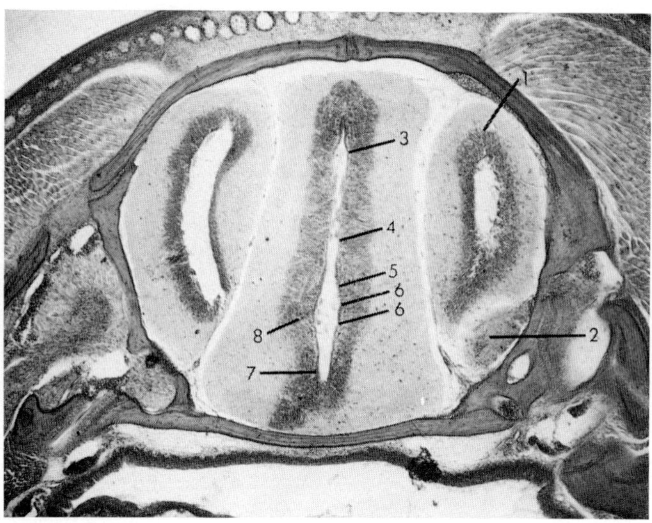

Figure 98 D. Cross-section (hematoxylin eosin stain) through caudal part of the diencephalon in Siphonops. 1: caudal hemisphere; 2: rostral tip of rhombencephalic auricula; 3: sulcus diencephalicus dorsalis; 4: s.d. medius; 5: s.d. ventralis; 6: rostral curved end of sulcus limitans (cut twice); 7: rostral end of sulcus lateralis hypothalami posterioris; 8: rostral end of tegmental cell plate.

dorsal and ventral thalamic lateral geniculate complex, morphologically homologous to that of Urodeles, Anurans, and other Anamnia, it should be pointed out that, because of the reduction of the optic system, the Caecilian lateral geniculate griseum appears dominated by the predominant olfactory system, presumably interrelating said system with other functional systems and activities of the thalamus. The Caecilian lateral geniculate complex, if compared with that of other Anamnia, can thus be considered a heteropractic homologue.[160] It should, however, a'so be here recalled that the dorsal lateral geniculate nucleus *(nucleus of Bellonci)* in Anurans, although receiving substantial input from the highly developed optic system, is also conspicuously related to the olfactory system.

Again, despite its greater similarity to the Urodele thalamus, the Caecilian one differs therefore, *qua* development of 'lateral geniculate complex' from the thalamus in both Urodeles and Anurans.

---

[160] Cf. p. 65, section 1A, chapter VI of volume 3/II.

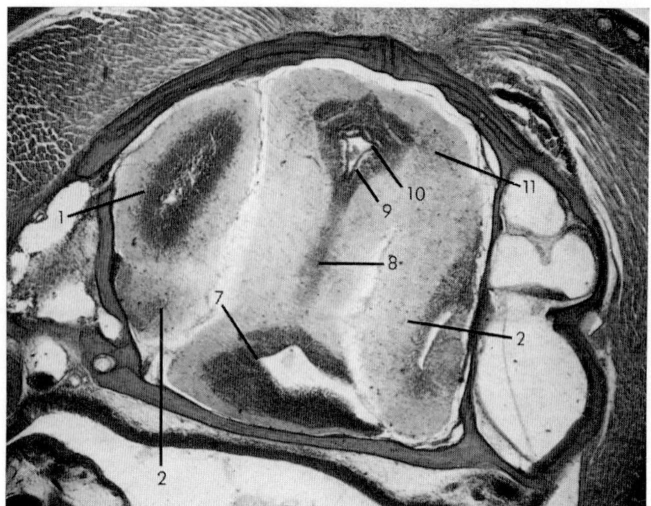

Figure 98 E. Cross-section through rostral limb of mesencephalic flexure of Siphonops, showing lobi inferiores of hypothalamus. 1, 2, 7, 8: as in Figure 98D; 9: sulcus limitans in mesencephalic ventricle; 10: sulcus lateralis mesencephali (internus); 11: nucleus lentiformis mesencephali.

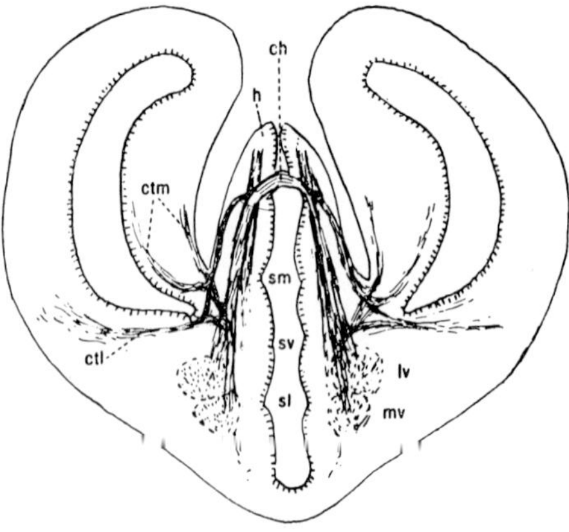

Figure 98 F. Overall arrangement of main diencephalic communication channels in Siphonops (from K., 1922). ch: habenular commissure; ctl, ctm: components of stria medullaris system (so-called cortico-habenular, -thalamic, or -hypothalamic tracts); h: habenula; lv, mv: lateral and medial forebrain bundles; sl: sulcus intraencephalicus anterior (cf. Fig. 98C, likewise for sm, sv).

The *hypothalamus* includes, at rostral preoptic levels, a pars inferior massae cellularis reunientis, extending, through the hemispheric stalk, into the basal portion of the telencephalon (Fig. 98C). Caudalward, this inferior part of the massa cellularis reuniens is continuous with a poorly defined nucleus entopeduncularis provided by scattered neuronal elements within the basal forebrain bundles. The remainder of the hypotha'amus retains a periventricular arrangement. It seems likely that, within the preoptic region, a 'magnocellular' neurosecretory cell group is included, but the presence of such cell group could not be established with certainty in our material. On the other hand, the floor of the preoptic nucleus, at least in Schistomepum, displays a characteristic ventricular protrusion (cf. Fig. 98B) which approximately corresponds to BRAAK's (1963) 'organon vasculare preopticum' in Chimaera, but remains poorly differentiated (cf. section 1C, p. 89). Its significance respectively its morphologic and functional relationships to *Braak's organ* in particular or to the circumventricular organs in general remain uncertain.

As regards the *diencephalic communication channels* which conform to the overall Amphibian pattern, no relevant detailed data are available. Figure 97F illustrates some of these tracts in Siphonops. Because of the peculiarly developed cephalic flexure, resulting in a slit-like 'interpeduncular fossa', the fasciculus retroflexus assumes a strictly rostrocaudal 'horizontal course'. In 1922, I mistook the transverse (dorsoventral) habenulothalamic and habenulohypothalamic fibers (Fig. 97F) for the habenulopeduncular tract.

In accordance with the well-developed telencephalon, dominated by the olfactory system, the anterior commissure and the commissura pallii are rather massive. The likewise substantial commissura habenulae was found in our material to pass through a ventricular midline concrescence interconnecting the antimeric habenulae (Fig. 98F) and thus not taking its course through the roof plate as is generally the case in other Amphibians respectively Vertebrates. The optic chiasma, if at all present in adult Gymnophiones, is a rudimentary or negligible structure, but the supraoptic (and postoptic) commissures in the chiasmatic ridge are of substantial size. The posterior commissure is well developed, its 'nucleus' however, remaining essentially within the periventricular layer.

With regard to the non-neuronal *roof plate* grundbestandteil of the Amphibian diencephalon in general, it can be said that, in contradistinction to Anurans, the choroid plexus of Urodeles and Gymnophiona is far more extensively developed, although some taxonomically related differences obtain here in this respect. According to HERRICK (1948), 'the enormous development of the choroid plexuses and associated endolymphatic organ of urodeles is apparently correlated with the sluggish mode of life and relatively poor provision for aeration of the blood. In the more active anurans the plexuses are smaller; but in the sluggish mudfishes, including the lungfishes, with habits similar to those of urodeles, we again find exaggerated development of these plexuses. Existing species in the border zone between aquatic and aerial respiration are all slow-moving and relatively inactive. The enlarged plexuses and sinusoids give vastly increased surfaces for passage of blood gases into the cerebrospinal fluid; and, correlated with this, the brain is thin everywhere, to facilitate transfer metabolites between brain tissue and cerebrospinal fluid. Massive thickenings of the brain wall occur in many fishes and in amniote vertebrates, but not in mudfishes and urodeles.'

It should here be added, however, that in all three orders of recent Amphibia, the saccus endolymphaticus is commonly well developed.[161] As regards Dipnoans, the difference in the arrangement of diencephalic choroid plexus were pointed out at the end of the preceding section 5. Again, in Amphibians as in other or perhaps all Vertebrates, the choroid plexus is innervated by non-medullated fibers (from peripheral nerves) of unascertained provenance. Some of these fibers might be autonomic (for the vasculature) and others could possibly be afferent (sensory).

7. Reptiles

The Reptilian diencephalon, if referred to that of Amphibians, displays a relatively much more pronounced differentiation into distinctive cell groups, respectively grisea, which can be described as 'nuclei' more or less corresponding to those obtaining in other Amniota, namely in Birds and Mammals. Yet, HUBER and CROSBY (1926), who gave a most detailed account of nuclear parcellation in a Reptilian diencephalon (Alligator), justly remark: 'even an attempt at partial plotting of the diencephalic nuclear masses in reptiles presents

[161] Thus, KRABBE (1962, edited by KÄLLEN) mistook the Gymnophione saccus lymphaticus for a part of the choroid plexus (cf. Fig. 294C, p. 461, vol. 4).

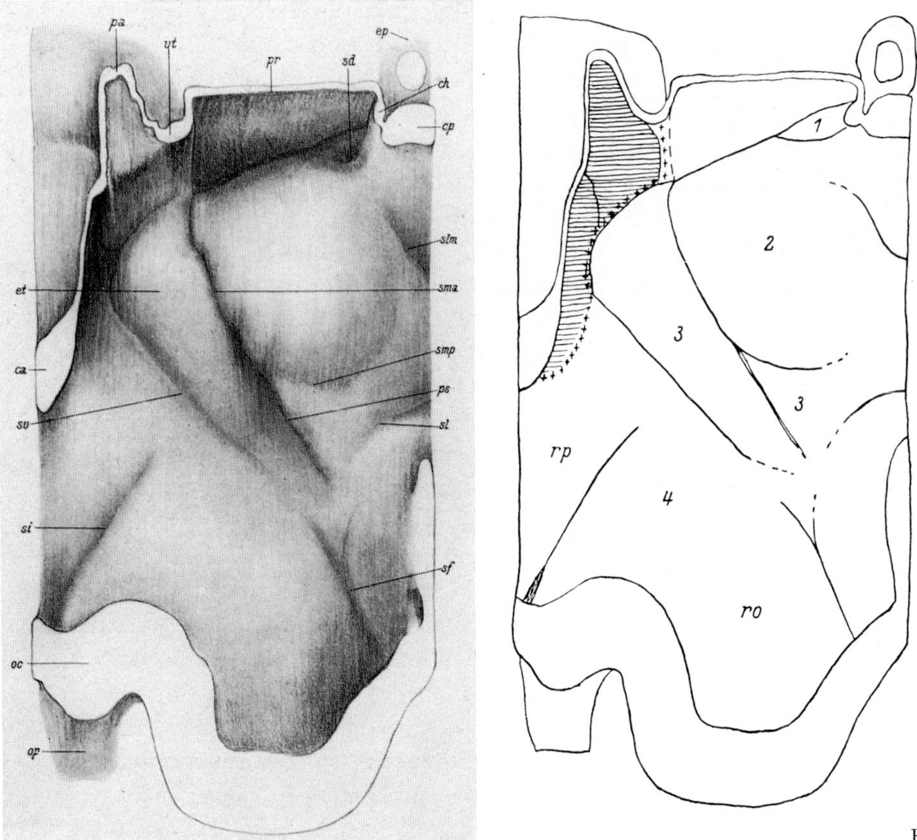

Figure 99. Ventricular aspect of the developing diencephalon in a 22 mm embryo of Lacerta agilis, which displays the fundamental longitudinal zonal arrangement retained at the 'adult' stage of Reptiles (from K., 1931). A Wax plate reconstruction. B Outline sketch. ca: commissura anterior; ch: commissura habenulae; cp: commissura posterior; ep: epiphysial stalk; et: eminentia thalami (ventralis); oc: chiasmatic ridge; op: optic nerve; pa: paraphysis; pr: 'dorsal sac' (epithelial roof of secondary parencephalon); ps: 'pseudo-sulcus'; ro: postoptic recess (posterior hypothalamus); rp: preoptic recess (anterior hypothalamus); sf: sulcus lateralis hypothalami (posterioris); si: sulcus intraencephalicus anterior; sl: sulcus limitans; slm: sulcus lateralis mesencephali (internus); sma: sulcus diencephalicus medius, pars anterior; smp: sulcus diencephalicus medius, pars posterior; sv: sulcus diencephalicus ventralis; 1: epithalamus; 2: thalamus dorsalis; 3: thalamus ventralis; 4: hypothalamus.

many difficulties and invites misinterpretations, since many of the nuclear masses in reptiles are not sharply circumscribed but grade one into the other without distinct demarcation.' The quoted authors then add 'that there is unquestionably a considerable variation (particularly in the dorsal thalamus) within the reptilian group. Undoubtedly this means the exaggeration of certain nuclear masses in one form, and their partial suppression in others.'

Despite these difficulties related to the suitable identification of distinctive open topologic neighborhoods and to the just mentioned taxonomic variations,[161a] a reasonably valid first approximation, applicable to the entire Reptilian class, can be obtained on the basis of the early griseal differentiation within the fundamental longitudinal zones as observable at certain key stages of ontogenesis (K., 1931). The peculiar transformations occurring in the course of these events, and, in particular, characterized by the behavior of the parencephalic sulcus ('pseudosulcus')[162] have been dealt with in section 5, chapter VI of volume 3/II. In the present context reference to Figure 99, which indicates the overall configuration of the fundamental longitudinal zones, will be sufficient. In the ventricular wall of adult Reptiles, the relevant sulci are on the whole poorly demarcated, especially in Lacertilians, remaining somewhat better recognizable in various Chelonians. Nevertheless, sulcus diencephalicus dorsalis, medius, ventralis, as well as sulcus synencephalicus, sulcus intraencephalicus anterior and sulcus lateralis hypothalami posterioris can frequently be sufficiently well identified. The 'pseudosulcus', if retained may, depending on the rostro-caudal level approximately correspond to either sulcus diencephalicus medius or ventralis (cf. Fig. 99). There is moreover, in certain Reptilian forms (e.g. Chelonians and Crocodilians) the tendency toward a fusion of the antimeric ventricular walls in the thalamic region (cf. Figs. 101 A–D, 102 A, B), representing a 'massa intermedia' with or without inclusion of 'nuclei reunientes'. I have never observed a 'massa intermedia' in the Squamata of my own material, although the Lacerta specimen depicted by BECCARI (1943)

[161a] BECCARI (1943) states in this respect: '*I nuclei sono stati variamente denominati e non sempre allo stesso modo identificati. Ciò forse è dipeso dalle specie differenti che sono stati preso come tipo*'.

[162] The peculiar ontogenetic 'behavior' of sulcus parencephalicus *sive* 'pseudosulcus' is also particularly conspicuous in the Avian diencephalon (K., 1936) and thus seems to represent a Sauropsidan characteristic.

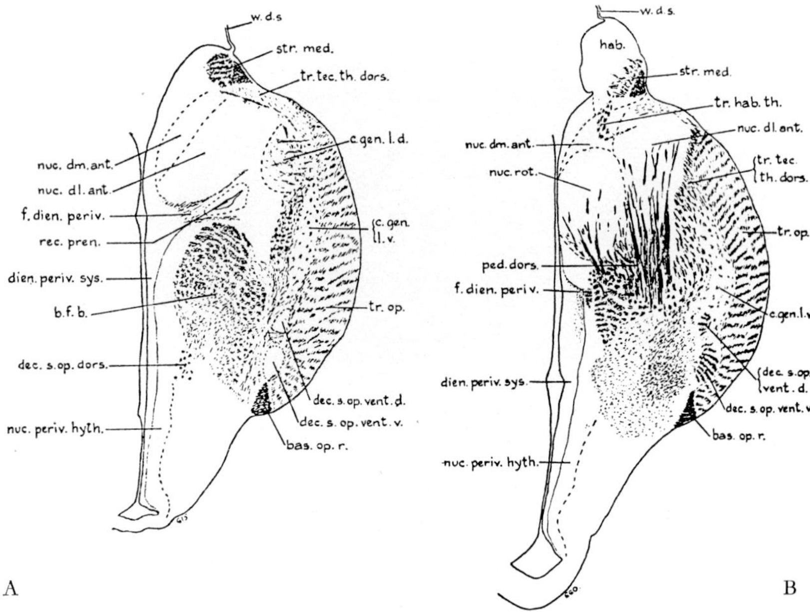

Figure 100 A. Cross-section *(Weigert stain)* through the diencephalon of adult Sphenodon punctatum rostral to habenula (from CAIRNEY, 1926). bas.op.r.: basal optic root; b.f.b.: basal forebrain bundle; c.gen.l.d., l.v.: dorsal and ventral lateral geniculate body; dec.s.op.dors.: dorsal supraoptic decussation; dec.s.op. vent.d., v.: ventral supraoptic decussation, dorsal and ventral parts; dien.periv.sys.: diencephalic periventricular system; f.dien.periv.: 'fasciculus diencephalicus periventricularis'; nuc.dl.ant.: nucleus dorsolateralis anterior; nuc.dm.ant.: nucleus dorsomedialis anterior; nuc.periv.hyth.: nucleus periventricularis hypothalami; str.med.: stria medullaris; tr.op.: tractus opticus; tr.tec.th.dors.: tractus tectothalamicus dorsalis; w.d.s.: 'wall of the dorsal sac' (attachment of roof plate).

Figure 100 B. Cross-section *(Weigert stain)* through the diencephalon of Sphenodon at level of habenula (from CAIRNEY, 1926). hab.: habenula; nuc.rot.: nucleus rotundus; ped. dors.: 'pedunculus dorsalis'. Other abbreviations as in Figure 100 A.

and here shown in Figure 100C faintly suggests a tendency toward such fusion,[163] but without a griseum reuniens.

In addition to the latest textbooks by KAPPERS *et al.* (1936), BECCARI (1943), and KAPPERS (1947), the investigations by CAIRNEY (1926), CURWEN and MILLER (1939), FREDERIKSE (1931), HUBER and CROSBY (1926), K. (1931), DELANGE (1913), PAPEZ (1935), SENN

[163] Cf. also further comments in the paper on Reptilian diencephalic *grundbestandteile* (K., 1931, pp. 310–311).

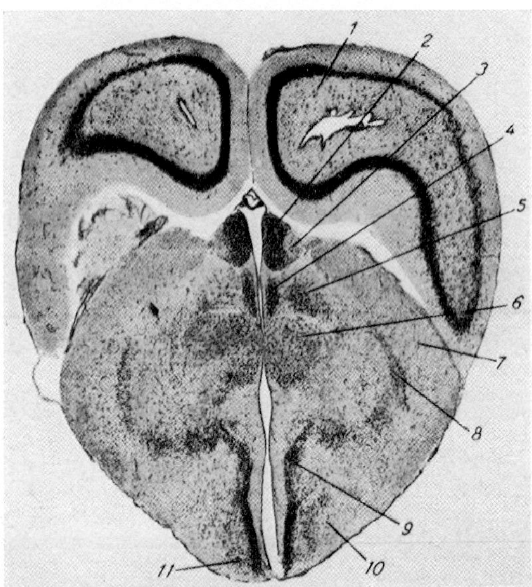

Figure 100C. Cross-section *(Nissl stain)* through the prosencephalon of Lacerta muralis (from BECCARI, 1943). 1: caudal portion of telencephalon; 2, 3: medial and lateral habenular nucleus; 4: nucleus dorsomedialis anterior; 5: nucleus dorsolateralis anterior; 6: nucleus rotundus; 7: tractus opticus marginalis; 8: nucleus geniculatus lateralis ventralis (the dorsal thalamic pars dorsalis can be recognized as a narrower cell band above lead 8); 9: 'nucleus paraventricularis parvocellularis' hypothalami; 10: '*nucleo ipotalamico laterale*' (the lead ends in the boundary region between lateral and ventral hypothalamic nucleus); 11: nucleus ventralis hypothalami. Between 9 and the lateral geniculate complex the following unlabelled grisea can be easily identified: n. dorsolateralis hypothalami, nucleus ventromedialis and nucleus ventralateralis thalami ventralis.

(1968), and WARNER (1942, 1945, 1955) may be consulted for data on various aspects of the Reptilian diencephalon and for bibliographic references to important contributions by the earlier workers concerned with this topic. Recent publications up to 1976 have not added any significant or more reliable data.[163a]

In the *epithalamus*, which assumes a rather caudal location, a lateral and a medial habenular nucleus can easily be distinguished, the medial nucleus being commonly more densely packed than the lateral one. In some forms, and at some levels, the habenular complex displays a variable degree of eversion, whereby an external subhabenular sulcus becomes accentuated (e.g. Fig. 101 D). Again, the sulcus diencephalicus

---

163a Cf. e.g. PRITZ (1974), the symposium edited by RISS *et al.* (1972), or the papers by BUTLER and NORTHCUTT (1973), and NORTHCUTT and BUTLER (1974).

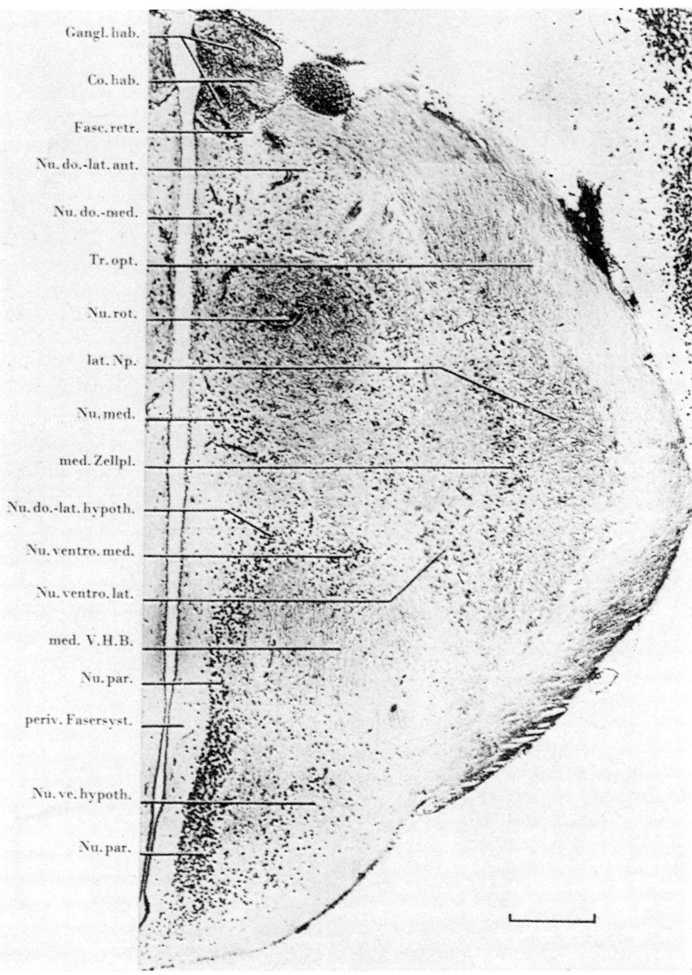

Figure 100 D. Cross-section (combined silver proteinate impregnation and cresyl violet stain) through the diencephalon of Lacerta sicula at habenular and nucleus rotundus level (from SENN, 1968). Co.hab.: commissura habenulae; Fasc.retr.: fasciculus retroflexus; Gangl.hab.: ganglion habenulae; lat.Np.: lateral neuropil of lateral geniculate complex; med.V.H.B.: medial portion of basal forebrain bundle; med. Zellpl.: medial cell plate of lateral geniculate complex; Nu.do.-lat.ant.: nucleus dorsolateralis anterior; Nu.do.-lat.hypoth.: nucleus dorsolateralis hypothalami; Nu.do.-med.: nucleus dorsomedialis; Nu.med.: nucleus medialis (thalami dorsalis); Nu.par.: nucleus paraventricularis (hypothalami); Nu.rot.: nucleus rotundus; Nu.ve.hypoth.: nucleus ventralis hypothalami; Nu.ventro.lat.: nucleus ventrolateralis (thalami ventralis); Nu.ventro.med.: nucleus ventromedialis (thalami ventralis); periv.Fasersyst.: periventricular fiber system; Tr.opt.: optic tract. The scale indicates 200 μ.

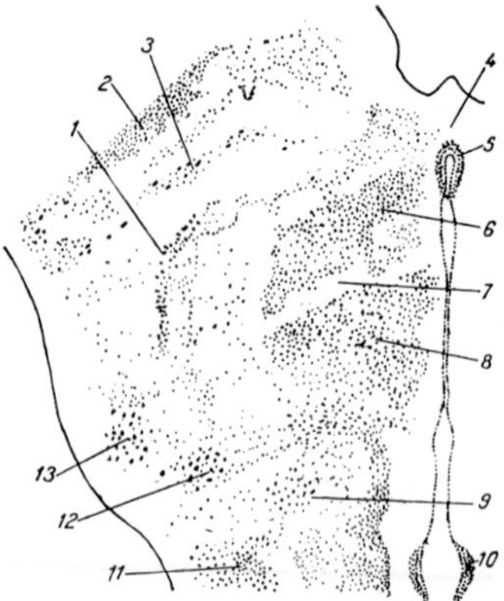

Figure 100 E. Cross-section *(Nissl stain)* through caudal diencephalic neighborhood in Lacerta muralis (from BECCARI, 1943). 1: '*genicolato pretettale*'*;* 2: periventricular layer of tectum opticum, '*sflorato rostralmente*'*;* 3: nucleus lentiformis mesencephali; 4: posterior commissure; 5: subcommissural organ; 6: nucleus posterocentralis sive posterodorsalis; 7: fasciculus retroflexus; 8: nucleus rotundus (near caudal end); 9: '*nucleo interstiziale del fascio genicolare discendente*' (presumably part of nucleus dorsolateralis hypothalami); 10: *ependymal organ of Charlton-Kappers;* 11: (posterior) entopeduncular nucleus (presumably corresponding to *n. subthalamicus of Luys);* 12: '*corpo del Luys*' (caudal portion of conjoined n. ventrolateralis and ventromedialis thalami ventralis), presumably corresponding to 'zona incerta'; 13: nucleus geniculatus medialis ventralis (nucleus delta, presumably caudal portion of n. gen. lat. ventr. complex).

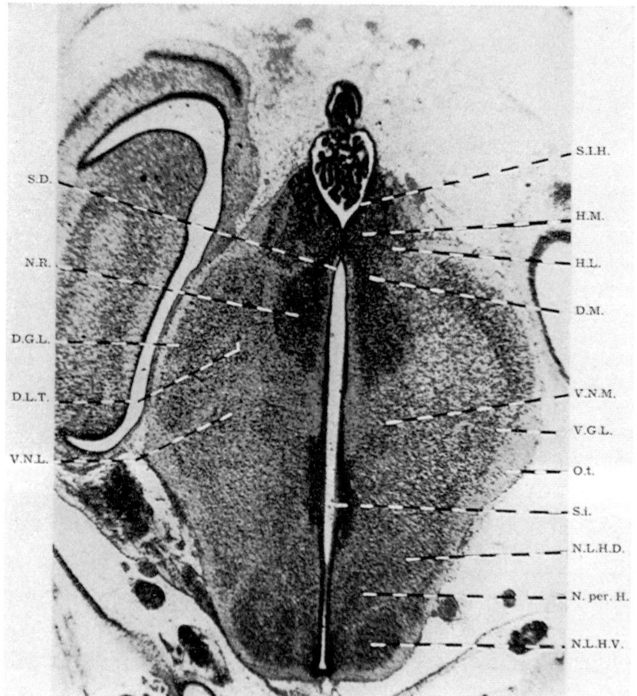

dorsalis may vanish, and an intrahabenular sulcus can be present (e.g. Fig. 102B, not labelled but easily identifiable).

The *thalamus dorsalis* does not extend rostralward as far as the interventricular foramen[164] and, at the final 'adult' stage, exceeds, as in all Amniota, the size of the ventral thalamus. It is dominated by the large *nucleus rotundus*, an oblong griseum, with rounded outlines in cross-sections, extending throughout most of dorsal thalamic zone's length, and characterized by a dense neuropil in which its cells are embedded. During ontogenesis, the early anlage of nucleus rotundus appears completely surrounded by a dense cell mass or cellular 'capsule', of which the other dorsal thalamic nuclei become derivatives respectively 'secondary differentiations'. Thus, in addition to (1) nucleus rotundus, the following nuclei can be distinguished.

The *nucleus dorsomedialis anterior* (2) and the *nucleus dorsolateralis anterior* (3) are the most rostral dorsal thalamic grisea.[165] The former extends, without very clearly distinguishable further subdivisions, caudalward dorsally to nucleus rotundus. Nevertheless, in this caudal extension, a *subhabenular nucleus* (4) and a *nucleus posterodorsalis* or *posterocentralis* (5) can roughly be delimited.[166] Its expansion within the pretectal region, represents, as it were, a rather nondescript portion of

[164] Since the pars impar telencephali with its aula is reduced to a negligible space in the Amniote prosencephalon (cf. p. 191, section 1B, chapter VI, vol. 3/II) the interventricular foramen, for practical purposes, connects the diencephalic third ventricle with the telencephalic lateral ventricles.

[165] The qualification 'anterior' was omitted in my investigation of 1931.

[166] Nucleus posterodorsalis *sive* posterocentralis corresponds to BECCARI's (1943) nucleus lentiformis thalami. It is not quite identical with the nucleus posterocentralis of HUBER and CROSBY (1926), this latter being presumably a portion of the ventrolateral cell capsule of nucleus rotundus.

Figure 100F. Cross-section *(van Gieson stain)* through the forebrain of an advanced embryo of the water snake Natrix sipedon, at level caudal to hemispheric stalk, and displaying 'adult' configuration (from WARNER, 1942). D.G.L.: dorsal lateral geniculate nucleus; D.L.T.: nucleus dorsolateralis; D.M.: nucleus dorsomedialis; H.L., H.M.: lateral and medial habenular nucleus; N.L.H.D.: nucleus lateralis hypothalami; N.L.H.V.: nucleus ventralis hypothalami; N.per.H.: nucleus paraventricularis hypothalami; N.R.: nucleus rotundus; O.t.: optic tract; S.D.: sulcus diencephalicus dorsalis; S.i.: sulcus lateralis hypothalami posterioris with *organ of Kappers-Charlton;* S.I.H.: sulcus intrahabenularis; V.G.L.: ventral lateral geniculate nucleus (the lead points to its ventral edge); V.N.L.: nucleus ventrolateralis thalami ventralis; V.N.M.: nucleus ventromedialis thalami ventralis (the lead points to its boundary against n. dorsolateralis hypothalami).

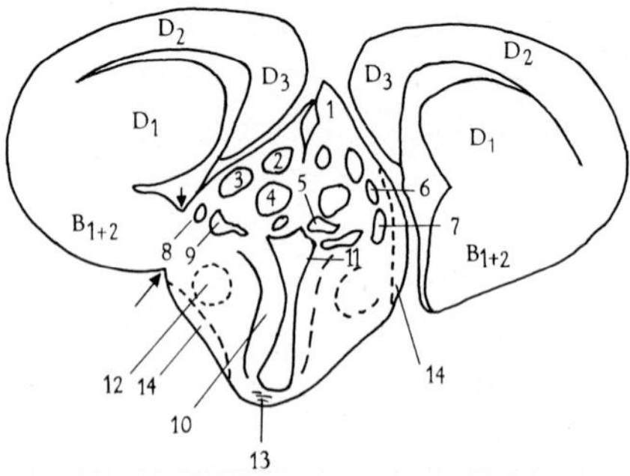

Figure 101 A. Outline sketch from photomicrograph of cross-section through the forebrain of Testudo graeca, left at caudal end of hemispheric stalk, and right slightly caudal to this latter (modified after K., 1927). 1: epithalamus (s.dienc.dorsalis, not labelled, can be identified); 2: n. dorsomedialis; 3: n. dorsolateralis; 4: n. rotundus; 5: n. medialis; 6: n. geniculatus lateralis dorsalis; 7: n. gen. lat. ventralis; 8: n. ovalis; 9: area triangularis (thalamus ventralis); 10: n. paraventricularis hypothalami; 11: n. dorsolateralis hypothalami; 12: n. entopeduncularis (anterior); 13: fibers of comm. postoptica; 14: tractus opticus. B, D, etc.: zonal system of telencephalon. Arrows: s. terminalis (left, above), s. telencephalodiencephalicus (left, below). S. dienc. ventralis combined with pseudosulcus. At this caudal level of hemispheric stalk (left), the massa cellularis reuniens, shown in Figure 101F, tends to 'fade out', and is not included in the sketch. The well-developed choroid plexus in lateral ventricles and inferior part of 3rd ventricle is likewise omitted. Nuclei 2–6 pertain to thalamus dorsalis.

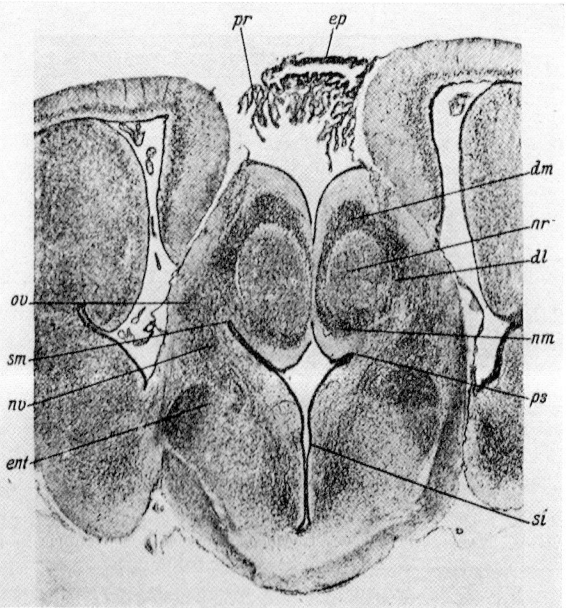

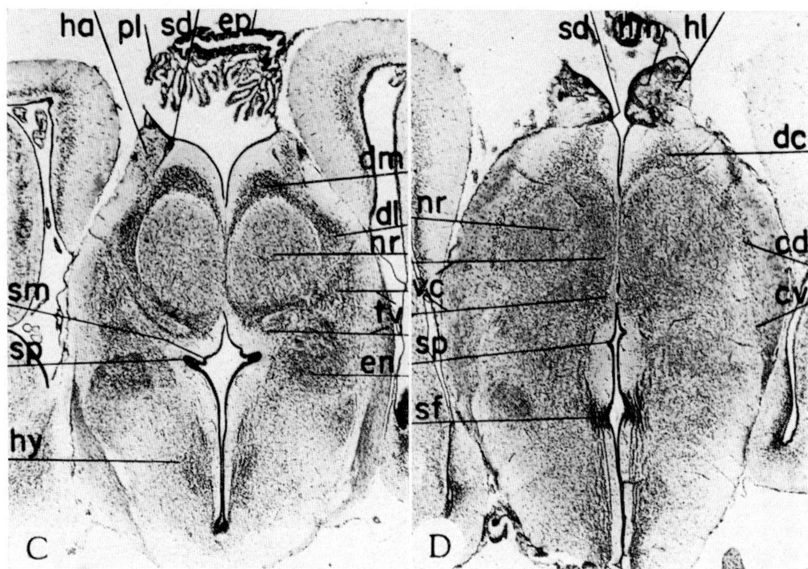

Figure 101 C, D. Cross-sections *(Nissl-stain)* through the forebrain of Chrysemys caudal to hemispheric stalk (modified after K., 1931, from K.,1951). cd: n. geniculatus lateralis dorsalis; cv: n. gen. lat. ventralis; dc: dorsal capsule of n. rotundus (n. dorsomedialis, pars posterior); dm: n. dorsomed. posterior; ha, hl, hm: habenula, n. hab. lateralis, n. hab. medialis; hy: hypothalamus; pl: choroid plexus of saccus dorsalis; sd: s. dienc. dorsalis; sf: s. lateralis hypothalami posterioris (with *organ of Kappers-Charlton);* sp: s. parencephalicus ('pseudosulcus'); tv: thalamus ventralis; vc: ventromedial cell capsule of n. rotundus (forming n. medialis). Other abbreviations as in Figure 101 B.

the dorsal and dorsolateral cell population forming a 'capsule' around nucleus rotundus.

The caudal neighborhood of ventromedial and ventrolateral cell capsule of nucleus rotundus provides a group of poorly separable nuclei which include (6) a *nucleus medialis (diagonalis of* Huber *and* Crosby, 1926) and (7) a more lateral *nucleus infrarotundus.* A portion of this latter may represent a primordium of the Mammalian dorsal medial geniculate nucleus, in essential agreement with the interpre-

---

Figure 101 B. Cross-section *(Nissl stain)* through the forebrain of the turtle Chrysemys at a level comparable to that of Figure 101 A (from K., 1931). dl: n. dorsolateralis anterior; dm: n. dorsomedialis anterior; en, ent: n. entopeduncularis (anterior); ep: epiphysial complex; nm: n. medialis; nr: n. rotundus; nv: griseum of thalamus ventralis; ov: n. ovalis; pr: dorsal sac (epithelial roof of secondary parencephalon); ps: 'pseudosulcus'; si: s. intraencephalicus anterior; sm: dienc. medialis (here joined with 'pseudosulcus').

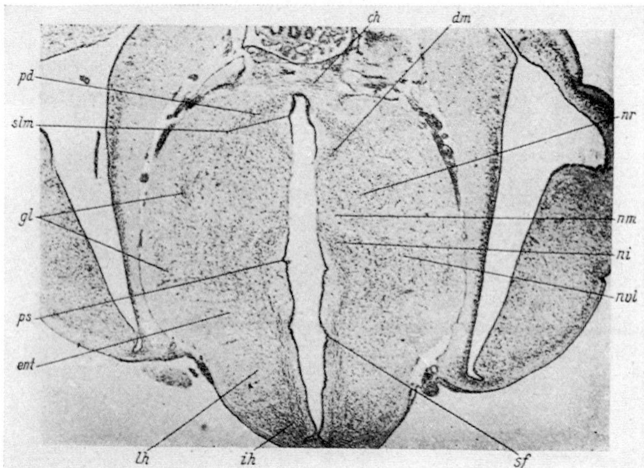

Figure 101 E. Cross-section *(Nissl stain)* through forebrain of Testudo graeca at level of habenular commissure (from K., 1931). Abbreviations for Figures 101 E–G. ch: habenular commissure (above which caudal end of habenular grisea can be seen); dm: n. dorsomedialis anterior; gl: lateral geniculate complex; ih: n. ventralis hypothalami; lh: n. lateralis hypothalami; mri, mrs: massa cellularis reuniens, pars inferior, pars superior; ni: n. ventralis, pars medialis (thalami ventralis, 'n. ventralis internus'); nm: n. medialis (thalami dorsalis); nv: griseum of thalamus ventralis ('n.' s. 'area triangularis'); nvl: n. ventralis thalami, pars lateralis; pd: n. posterodorsalis sive posterocentralis; pv: n. paraventricularis hypothalami; sf: s. lateralis hypothalami posterioris; slm: s. synencephalicus (joined to s. lateralis mesencephali); smp: s. dienc. medius, pars posterior; sv: s. dienc. ventralis. The sulcus intraencephalicus anterior, not labelled, can be identified below lead pv in Figure 101 G. Other abbreviations as in Figures 101 B–D.

tation of PAPEZ (1936). This author, however, is inclined to regard somewhat more medial neighborhoods of the caudoventral 'capsula nuclei rotundi' as the locus of said primordium. Nucleus medialis and infrarotundus correspond morphologically but not functionally to the Mammalian posterior ventral group (e.g. nucleus ventralis posteromedialis and posterolateralis, cf. K., 1954).

The caudal portion of nucleus dorsolateralis is continuous with (8) a *nucleus geniculatus lateralis dorsalis* which forms a dorsal component of the lateral geniculate complex. In the pretectal region, which includes the above-mentioned nucleus posterodorsalis complex, a conspicuous griseum, lateral to nucleus posterodorsalis, is represented by (9) the *nucleus praetectalis (geniculato pretettale of* BECCARI, 1943), which, in turn, can be subdivided into (9a) a dorsomedial and (9b) a ventrolateral portion (CURWEN and MILLER, 1939). The pretectal nucleus, as studied by BECCARI in Lacerta, is characterized by spindle-shaped elements

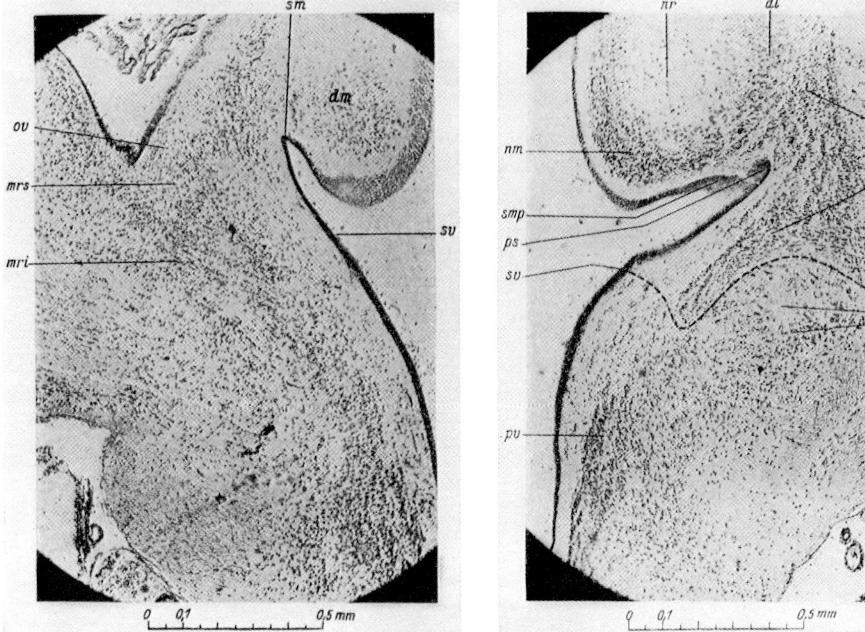

Figure 101 F. Part of cross-section *(Nissl stain)* through hemispheric stalk in Chrysemys (from K., 1931). ent: n. entopeduncularis (post., E; ant., G).

Figure 101 G. Part of cross-section *(Nissl stain)* through diencephalon of Chrysemys just caudal to hemispheric stalk (from K., 1931).

(cellule a doppio cespuglio or *doppio pennachio)* whose lateral ramifications are directed toward the optic tract, while the medial ramifications extend toward the more internal grisea. The neurites of the *geniculato pretettale* seem to provide the bulk of BECCARI's *fascio pretettale discendente.* Rostrally, the pretectal nucleus blends with the dorsal lateral geniculate nucleus.[167] Caudomedially to the *geniculato pretettale,* (10) a *nucleus of the posterior commissure* with several subgroups can be identified.[168] The caudalmost portion of the diencephalic pretectal region includes (11) the griseum of *area praetectalis,* which may extend

[167] Because of its close relationship to the dorsal portion of the lateral geniculate complex, BECCARI (1943) states that the *geniculato pretettale 'forse non ne è che una dipendenza o una suddivisione'* of the former.

[168] Dorsal nucleus of posterior commissure, superior and inferior interstitial nucleus of posterior commissure (CURWEN and MILLER, 1939). The interstitial tegmental nucleus of the posterior commissure, likewise recorded by the cited authors, can be regarded as pertaining to the mesencephalic tegmentum (cf. chapter XI, section 7, vol. 4).

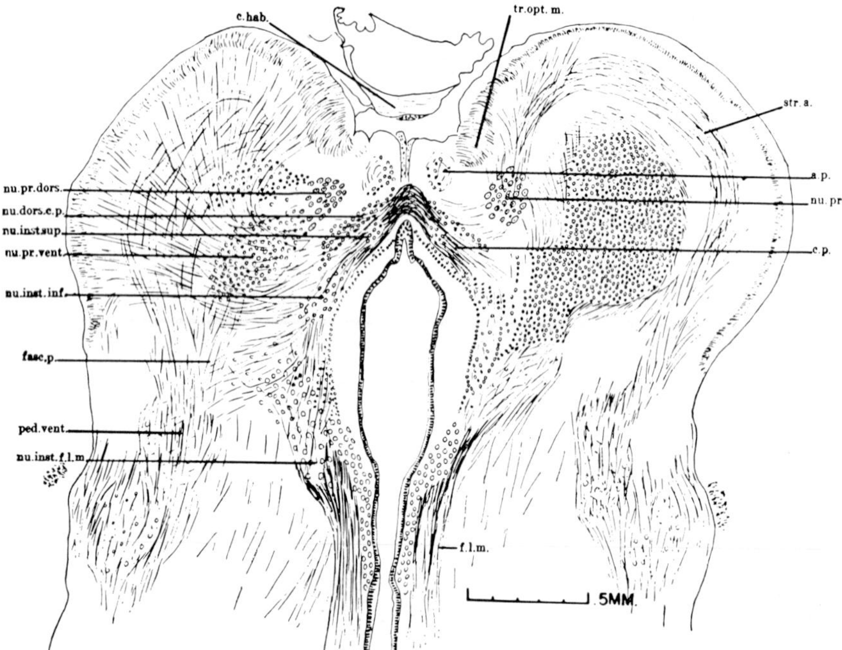

Figure 101 H. Cross-section *(Weil myelin stain)* through caudal pretectal region of Pseudemys (from Curwen and Miller, 1939). a.p.: area praetectalis; c.hab.: commissura habenulae; c.p.: commissura posterior; fasc.p.: fasciculus praetectalis descendens; f.l.m.: fasciculus longitudinalis medialis; nu.dors.c.p.: nucleus dorsalis commissurae posterioris; nu.inst.f.l.m.: nucleus interstitialis fasciculi longitudinalis medialis; nu.inst.inf., sup.: inferior and superior interstitial nuclei of posterior commissure; nu.pr., pr.dors., pr.vent.: nucleus praetectalis, n.p. dorsalis, n.p. ventralis; ped.vent.: 'ventral peduncle' (caudal extension of basal forebrain bundle system); str.a.: 'stratum album' ('deep medulla', '*tiefes Mark*') of tectum opticum; tr.opt.m.: tractus opticus marginalis, pars medialis.

dorsally to commissura posterior (Fig. 101 H). Again, in Reptiles with well-developed massa intermedia this latter may include, in some Chelonians and in Crocodilians, (12) a *nucleus reuniens* comprising several subdivisions as e.g. described by Huber and Crosby (1926) and Papez (1935). Thus, even in a cautious interpretation and parcellation about 12 if not more 'nuclei' can be distinguished in the Reptilian dorsal thalamus. With regard to the just mentioned nucleus reuniens complex of Crocodilians, its presumed relation to the ascending cochlear channel shall briefly be pointed out further below in dealing with the diencephalic fiber systems.

The *ventral thalamus* begins rostrally at the interventricular foramen with an *eminentia thalami* adjoining the basal telencephalic grisea and

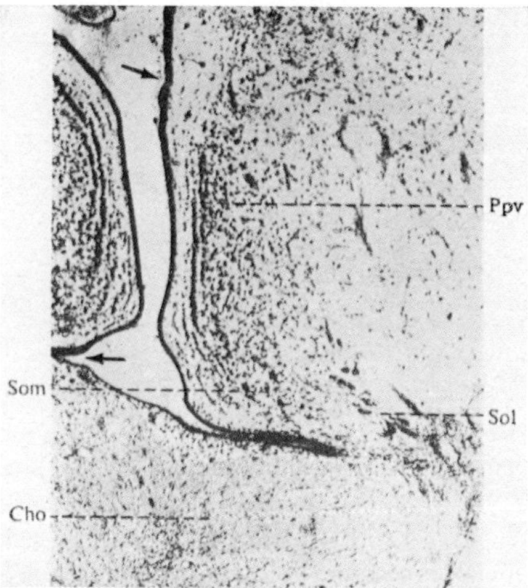

Figure 101 I. Cross-section *(Nissl stain)* through the preoptic hypothalamic region of the turtle Clemmys japonica, showing supraoptic nucleus (from NIIMI *et al.*, 1963b). Cho: optic chiasma; Ppv: nucleus praeopticus paraventricularis; Sol: nucleus supraopticus, pars lateralis; Som: nucleus supraopticus, pars medialis. Added arrows: sulcus diencephalicus ventralis (above), stem of sulcus intraencephalicus anterior system (below).

including (1) a *massa cellularis reuniens, pars superior*, extending lateralward into the hemispheric stalk. Somewhat more caudally, but still at levels of hemispheric stalk, (2) the *nucleus ovalis* can be delimited as a relatively small anterolateral cell group (cf. Fig. 101 F) which seems to represent a rostral component of the ventral lateral geniculate complex described further below. The medially adjacent cell population is (3) the *nucleus ventralis anterior (area triangularis autorum)*. Still more caudalward, (4) a *nucleus ventromedialis sive internus* and (5) a *nucleus ventrolateralis*[169] can be distinguished. Since both are located dorsally to the lateral forebrain bundle they can also be jointly designated as 'suprapeduncular nucleus'. In some forms (e.g. Lacerta, cf. BECCARI, 1943) a dorsolateralward 'detached' portion of nucleus ventrolateralis can be interpreted as (6) a primordial *ventral medial geniculate nucleus* (PAPEZ, 1936; BECCARI, 1943), *n. delta*, joined to complex (7).

[169] Also designated as nucleus ventralis thalami, pars medialis (respectively 'nucleus ventralis internus'), and pars lateralis (K., 1931). A neighborhood within pars lateralis ('nucleus ventrolateralis') corresponds to the nucleus decussationis supraopticae dorsalis of CROSBY and SHOWERS (1969).

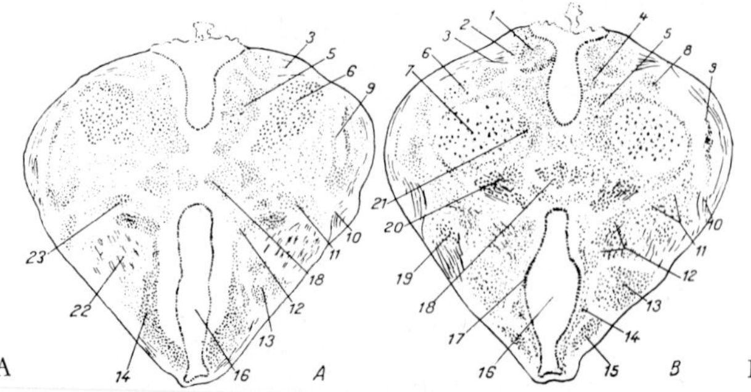

Figure 102 A, B. Cross-sections *(Nissl stain)* through the diencephalon of Alligator mississippiensis, A being more rostral than B (slightly simplified after HUBER and CROSBY, 1926, from BECCARI, 1943). 1, 2: medial and lateral habenular nuclei; 3: stria medullaris system; 4: subhabenular nucleus; 5: n. dorsomedialis anterior; 6: n. dorsolateralis anterior; 7: n. rotundus (its rostral beginning can be identified on left side of Fig. A); 8: n. posterodorsalis; 9: n. geniculatus lateralis complex; 10: fasciculus opticus marginalis; 11: 'area ventrolateralis' (thalami ventralis); 12: n. dorsolateralis hypothalami in Figure A (in Fig. B, the upper branch of lead shows n. ventromedialis thalami ventralis, and the lower branch n. dorsolateralis hypothalami); 13: n. lateralis hypothalami; 14: n. paraventricularis (posterior) hypothalami; 15: n. ventralis hypothalami; 16: 3rd ventricle; 17: *organ of Charlton-Kappers;* 18: nn. reunientes; 19: n. entopeduncularis; 20: n. medialis thalami dorsalis ('n. diagonalis'); 21: part of n. posterodorsalis group; 22: lateral forebrain bundle; 23: 'n. decussationis supraopticae dorsalis'.

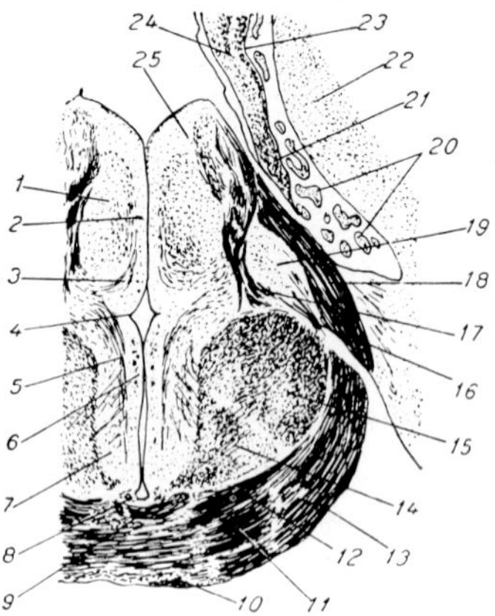

Figure 102 C. Cross-section *(Weigert-Pal-paracarmine stain)* through the diencephalon of Crocodilus porosus (after ADDENS, 1938, from BECCARI, 1943). 1: nucleus dorsolateralis anterior; 2: nucleus dorsomedialis anterior; 3: dorsal periventricular fiber system; 4: sulcus diencephalicus ventralis conjoined with 'pseudosulcus' (interpreted as s.dienc. medius by ADDENS and BECCARI); 5: 'fascio setto-talamico' (part of periventricular fiber system'); 6: magnocellular paraventricular hypothalamic nucleus; 7: nucleus paraventri-

The *ventral lateral geniculate nucleus* (7) is a conspicuous lateral and dorsolateral griseum of the thalamus ventralis, medially adjacent to the marginal optic tract. SENN (1968) has particularly investigated the differentiation of this griseum in Lacerta sicula and noted a medial cell plate, laterally to which a plate-like neuropil containing scattered cellular elements is located (cf. Fig. 100D). Together with nucleus ovalis, with the dorsal thalamic component of corpus geniculatum and the caudally adjacent, likewise dorsal thalamic *geniculato pretettale* (pertaining to the diencephalic pretectal region), the ventral lateral geniculate griseum forms a substantial lateral geniculate complex *sensu latiori*[169a] related to the well-developed optic system, from which said complex presumably receives most or much of its input. Taking the ventral lateral geniculate griseum as a morphologic unit, it can be said that about 7 'nuclei' might be distinguished in the Reptilian ventral thalamus, or, in other words, not much more than half the number of dorsal thalamic grisea.[170]

The neuronal cell aggregates of the *hypothalamus* can be roughly subdivided into a more or less continuous medial periventricular cell layer and into an array of lateral grisea. In addition, the hypothalamic wall externally to the periventricular layer is characterized by rather massive predominantly longitudinal fiber systems especially related to the basal forebrain bundles.

The *preoptic hypothalamus* contains (1) a preoptic paraventricular nucleus which includes some relatively large neuronal elements of

[169a] The stratification of this complex, and the shape of its diverse types of neuronal elements can be clearly demonstrated by suitable *Golgi impregnations* (QUIROGA, 1977).

[170] With respect to relative volume of dorsal versus ventral thalamus, no accurate figures are available. Because of the substantial expansion of ventral lateral geniculate nucleus, the relationship 13:7 *qua* 'number' of grisea presumably becomes somewhat modified in favor of thalamus ventralis *qua* actual volume.

---

cularis parvocullularis; 8: 'suprachiasmatic paraventricular griseum; 9: dorsal supraoptic decussation; 10: basal optic bundle; 11, 12: dorsal and basal portions of ventral supraoptic decussation; 13: tractus opticus; 14: medial forebrain bundle; 15: lateral forebrain bundle; 16, 17, 18: parts of olfactohabenular tracts (stria medullaris system); 19: nucleus ovalis; 20: choroid plexus of lateral ventricle; 21: fimbria hippocampi; 22: caudal epistriatum (D_1 complex); 23: hippocampal fiber system; 24: cortex medialis ('hippocampus'); 25: part of habenular griseum.

presumably neurosecretory type. Although rather scattered, these cell groups could be designated as (1a) magnocellular preoptic nucleus. Within the hemispheric stalk, (2) a massa cellularis reuniens, pars inferior, extends toward the basal grisea of the telencephalon. Caudalward this cell aggregations is continuous with (3) the nucleus entopeduncularis anterior, formed by scattered, fairly large to medium-sized cells within the basal forebrain bundles. More ventrally, (4) a diffuse nucleus praeopticus lateralis can be distinguished.[171] Dorsally to optic chiasma and extending lateralward, a magnocellular, presumably neurosecretory (5) nucleus supraopticus is present, thus displaying a medial and a lateral subdivision (cf. Fig. 101 I).

The *posterior (postchiasmatic) hypothalamus* includes (6) the posterior paraventricular nucleus. A dorsolateral portion of this griseum extends lateralward, medially and basally to thalamus ventralis, forming (7) the nucleus dorsolateralis hypothalami, a neighborhood of which presumably corresponds to BECCARI's (1943) *nucleo interstiziale del fascio genicolare discendente*. Within the caudal periventricular population basally to tuberculum posterius[172] and tegmental cell plate, an ill-defined nucleus corporis mammillaris (8) can be noted.[173] It should, however, be noted that a clear-cut boundary between preoptic, anterior (1) and postoptic, posterior (6) paraventricular nucleus cannot be recognized. Also, the nucleus dorsolateralis hypothalami (7) may extend rostralward into the preoptic region.

Among the scattered and rather diffuse cell groups lateral to the periventricular cell plate the following cell groups can be distinguished. The nucleus entopeduncularis posterior (9) has a dorsolateral position within the basal forebrain bundles. It represents a caudal, but somewhat distinctive continuation of the anterior entopeduncular nucleus (cf. Fig. 100E). Ventrally to nucleus entopeduncularis poste-

---

[171] CROSBY's and SHOWER's (1969) regio praeoptica lateralis and hypothalamica lateralis. The cell group designated by the cited authors as 'area preoptica medialis', 'nucleus suprachiasmaticus', 'area hypothalamica anterior', and 'nucleus hypothalamicus arcuatus' can be regarded as neighborhoods within the periventricular cell population.

[172] This also includes a neighborhood designated as 'nucleus infundibularis' by DIEPEN (1962, Fig. 67b, p. 102).

[173] CROSBY and SHOWERS (1969) distinguish a more compact 'nucleus mammillaris medialis' from adjacent laterally scattered cells assumed to represent a 'nucleus mammillaris lateralis'. The 'interstitial nucleus of the supramammillary decussation' mentioned by the cited authors may or may not be a component of the tegmental cell plate.

rior, scattered cells represent a nondescript external lateral hypothalami area (area hypothalamica lateralis externa) rostrally continuous with the nucleus praeopticus lateralis mentioned above. Medially and somewhat ventrally to that griseum, the nucleus lateralis hypothalami (10) is located, basally to which lies (11) the nucleus ventralis hypothalami.[173a]

In the posterior hypothalamus, along the sulcus lateralis hypothalami ('sulcus lateralis infundibuli'), i.e. extending slightly to dorsal and ventral side of that sulcus, many if not most Reptiles display the *ependymal organ of Kappers-Charlton*,[174] discussed and depicted in volume 3/I (pp. 156–157, Fig. 107, p. 161, and pp. 355–357). As regards another hypothalamic paraventricular organ, namely the supraoptic crest, which is said to be present in at least one Reptilian species (Gecko, cf. p. 362, section 5, chapter V, vol. 3/I), further observations must be awaited.[175]

The general arrangement of *communication channels* of the Reptilian diencephalon can best be understood with reference to the corresponding fiber systems in urodele Amphibians as pointed out above in section 6, and analyzed in great detail by Herrick (1948). The channels displayed by Urodeles, in turn, can be interpreted on the basis of the comparable fiber systems present in other Anamnia, and discussed in sections 2 to 5. On the other hand, many of the doubtless

---

[173a] With respect to Mammalian homologies, to be again dealt with further below in section 9, it can be assumed that nucleus entopeduncularis anterior corresponds to the globus pallidus complex, and nucleus entopeduncularis posterior to the nucleus subthalamicus *(corpus subthalamicum Luysi)* of Mammals. Nucleus lateralis hypothalami and nucleus ventralis hypothalami of Reptiles may be considered homologous to nucleus dorsomedialis and nucleus ventromedialis hypothalami in Mammals. Crosby and Showers (1969), in fact, use the Mammalian designations for the corresponding Reptilian hypothalamic nuclei. In accordance with Senn's (1968) procedure the Reptilian nomenclature of my original papers (1931, 1937) was here retained.

[174] This structure, which might also be called the *Charlton-Kappers organ*, since both investigators, working together, have about equal merits in its proper identification and recognition, pertains to the group variously called ependymal, paraependymal, or circumventricular organs. It seems to be present in all recent Reptilian groups, although perhaps with some species and even individual variations *qua* degree of differentiation.

[175] My own material, sectioned, and stained for morphologic purposes, was not particularly suited for the proper visualization of this organ. It will be recalled that its presence in the human brain was completely overlooked until properly demonstrated in 1954 (K., Anat. Rec. 118, p. 396).

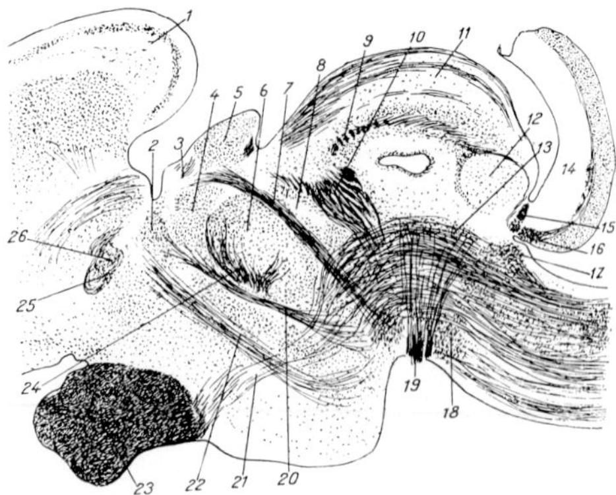

Figure 103 A. Paramedian sagittal section *(Weigert stain)* through the brain of Lacerta muralis, displaying diencephalic communication channels (from Beccari, 1943). 1: caudal pole of hemisphere; 2: nucleus dorsomedialis anterior thalami dorsalis; 3: stria medullaris; 4: nucleus dorsolateralis anterior thalami dorsalis; 5: habenular griseum; 6: nucleus rotundus; 7: fasciculus retroflexus; 8: nucleus posterocentralis; 9: commissura tecti mesencephali; 10: commissura posterior; 11: tectum; 12: torus semicircularis ('colliculus posterior'); 13: oculomotor nucleus; 14: cerebellum; 15: radix et decussatio nervi trochlearis; 16: cerebellar decussation ('decussatio veli'); 17: trochlear nucleus; 18: 'fasciculus tectobulbaris cruciatus'; 19: radix nervi oculomotorii; 20: fasciculus mammillo-thalamicus; 21: '*ansulate fibers of Bellonci*' (components of dorsal supraoptic commissure); 22: fornix system; 23: optic tract and part of chiasma; 24: dorsal peduncle of basal forebrain bundle ('fasciculus thalamostriatus'); 25: anterior commissure; 26: commissura pallii sive hippocampi.

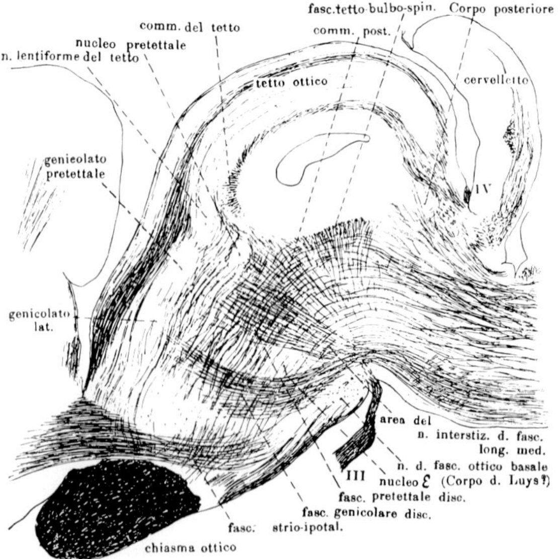

extensive 'advances' *qua* specific synaptology obtaining in the Amniote Reptilian diencephalon remain very poorly elucidated.[176]

The *rostral fiber connections* with the telencephalon provided by *lateral* and *medial forebrain bundles* may be considered reciprocal. The lateral bundle is frequently designated as the 'striothalamic' respectively 'thalamostriatal' tract. This channel, nevertheless, includes also fibers connecting with lateral and dorsal cortical regions, and, in addition, includes the 'tractus striohypothalamicus' and 'striotegmentalis', as well as 'striotectal' fibers.[177] The medial forebrain bundle contains connections of paraterminal grisea (septum) and medial cortex ('hippocampus'), representing an undefined 'fornix', including the 'septohypothalamic fibers' of KAPPERS (1947), as well as connections of dorsal cortex. A component of both forebrain bundles together with their dorsolateral hypothalamic interstitial grisea (anterior and posterior entopeduncular nucleus) may correspond to a prosencephalic portion of the Mammalian 'extrapyramidal motor system'. The ascending components of basal forebrain bundles, particularly of the lateral one, include sensory input to telencephalon mediated by dorsal

---

[176] Considerable scepticism is, in my opinion, appropriate for the evaluation of recent detailed claims based on methods of terminal fiber degeneration, evoked potential, and related physiologic procedures. The disadvantages and inherent sources of error are here no less pronounced than the obvious ones encumbering the analysis of tracts by the conventional histologic methods based on myelin stains and metallic impregnations. With regard to the recent investigations studying the distribution of monoamine-containing nerve terminals in the neuraxis, reference may be made to papers by PARENT (1973) and PARENT and POITRAS (1974) which concern the turtle Chrysemys picta, and contains relevant bibliographic references on this general topic. Numerous neurons of the catecholamine type seem to be unevenly distributed within the hypothalamic periventricular gray. Additional but rather inconclusive interpretations concerning 'organization of the turtle thalamus' can be found in papers by BELEKHOVA and KOSAREVA (1971), and HALL and EBNER (1970).

[177] Although, in the diencephalon, the lateral forebrain bundle is more or less distinctly located dorsally to the medial one, a clear-cut delimitation between the two becomes progressively difficult at levels caudal to hemispheric stalk. Thus, a dorsal and ventral 'peduncle' within this channel have been described (cf. BECCARI, 1943). There is little doubt that, in the diencephalon, fibers of the forebrain bundles effect, directly or indirectly, connections with grisea of all four fundamental longitudinal zones.

Figure 103 B. Sagittal section *(Weigert stain)* through brain of Lacerta muralis, at a more lateral level than that of Figure 103 A (from BECCARI, 1943). fasc. strio-ipotal.: medial forebrain bundle; nucleo E *(Corpo d. Luys?)*: caudal part of ventral thalamus (cf. Fig. 100 E, label 12).

and ventral thalamic grisea, the optic input being essentially transmitted by way of the lateral geniculate complex. The dorsal fibers of the basal forebrain bundle radiating into or from nucleus rotundus and nucleus dosolateralis anterior (perhaps also dorsomedialis anterior) form part of the so-called pedunculus dorsalis (cf. Fig. 100 B). Reciprocal connections between nucleus rotundus and dorsal cortex can be assumed, and seem particularly to involve the lateral subdivision of that cortex (the so-called 'primordium neopallii'), but connections with the basal and particularly the epibasal griseal complex are doubtless present.

The *stria medullaris system*, which, in the preoptic region, is joined by components of the basal forebrain bundles, is essentially but not exclusively a channel connected with the epithalamus. It also includes connections with nucleus ovalis, that is, with a griseum pertaining to the predominantly 'optic' geniculatum laterale complex, thus providing one of presumably several correlations with the olfactory system.

A further development of the *stria medullaris system*, related to the amygdaloid complex in the caudalward paired evaginated hemispheres, is the *stria terminalis system*, to be dealt with in chapter XIII. While essentially included in stria terminalis and in basal forebrain bundles of Amphibians and other Anamnia, it can be more or less clearly recognized in Reptiles, being highly modified in Birds, and characteristically developed in Mammals.

Although the *optic input channels* of the Sauropsidan brain, as is the case in Anamnia, predominantly reach the tectum mesencephali through the bulk of the optic tract, the diencephalic optic input[178] is nevertheless substantial and reaches the geniculate complex pertaining to both thalamus dorsalis and ventralis, moreover the pretectal grisea and perhaps also other thalamic cell groups, in addition to undefined hypothalamic grisea and, at least indirectly, the epithalamus. This latter and adjacent regions also receive input through the nervus parietalis (cf. section 1 C), if this latter is developed. It should be recalled that the epiphysial complex is missing, or at best barely indicated by a negligible vestige, in the Crocodilian diencephalon.

As regards *intrinsic diencephalic channels (connessioni intertalamiche* of Beccari, 1943), there are hypothalamic and thalamic periventricular and scattered more lateral fibers running in dorsoventral planes

[178] If the retina is evaluated as a peripheral component of the diencephalon (optic ganglion), the optic input to diencephalon could be subsumed under the group of 'intrinsic diencephalic channels or connections'.

(cf. Fig. 102C). BECCARI (1943) identified a thin mammillothalamic fascicle from mammillary nucleus to nucleus dorsomedialis anterior and in part also to nucleus dorsolateralis anterior.[179]

Among the *caudal diencephalic channels*, the system of fasciculus retroflexus presumably also includes intrinsic diencephalic connections. From the lateral geniculate complex, and particularly from its ventral thalamic portion, the fasciculus geniculatus descendens of BECCARI arises, running toward the mesencephalic tegmentum, and here joining the more dorsal fasciculus praetectalis descendens arising in the *genicolato pretettale* respectively pretectal nucleus (cf. Fig. 103B).

Supposedly mainly ascending, but presumably reciprocal caudal channels are the conspicuous tectothalamic bundles, of which substantial fascicles are related to nucleus rotundus ('direct dorsomedial tectothalamic bundle', also said to connect with nucleus dorsomedialis anterior). Other components of that system are the 'direct ventrolateral tectothalamic bundles' related to pretectal and lateral geniculate complex. The crossed tectothalamic fibers decussate in the postoptic or supraoptic commissural system and include particularly connections with the lateral geniculate complex.

The *ascending caudal channels* also seem to contain fibers from both lemniscus systems, namely somatic and visceral spinobulbar lemniscus and vestibulo-acoustic bulbar lemniscus, which also involves torus semicircularis respectively 'colliculi inferiores'. With respect to the cochlear subdivision of the lateral lemniscus, it has been recently claimed that, in Crocodilians, a central portion of the (dorsal thalamic) nucleus reuniens represents the relevant relay nucleus, from which an ascending pathway originates, reaching 'dorsolateral' and 'ventrolateral' telencephalic grisea (our D_1 and B_{1+2}) by way of the lateral forebrain bundle (PRITZ, 1974). Said portion ('nucleus centralis') would then be analogous to the Mammalian dorsal medial geniculate nucleus. Again, the ventral lateral geniculate nucleus (BECCARI's nucleus delta of Lacertilians, cf. Fig. 100E) seems to receive contributions from the cochlear lemniscus, whose relevant components are believed to originate from the 'central nucleus of torus semicircularis'. These various interpretations, partly based on findings by means of the *Fink-*

---

[179] KAPPERS (1947), however, states '*on ne trouve, chez les Reptiles, ni véritable noyau mamillaire, ni faisceau mamillo-thalamique (f. de Vicq d'Azyr), ni faisceau mamillo-tegmentaire*'. A rudimentary mammillary nucleus is, nevertheless, conceded by KAPPERS, who noted a *faisceau mamillo-pédonculaire*.

Heimer technique, do not appear altogether convincing, but nevertheless suggest the possibility of the inferred connections.

On the whole, the diencephalic distribution of the various ascending channels remains uncertain. The nucleus rotundus, however, seems to represent a significant griseum receiving 'sensory' input. Whether some of this input is directly provided by the medial (general) spino-bulbo-tectal lemniscus, or indirectly by way of tectorotundal channels, remains a moot question. Rostralward, the nucleus rotundus is mainly connected with the basal telencephalic grisea (D_1 and B_{1+2} complexes). Morphologically, the nucleus rotundus can be evaluated as homologous to the Mammalian nucleus medialis (dorsomedialis) thalami, but certainly not as functionally analogous. Its topologic isomorphism is thus a homomeric, homotypic, but heteropractic homology (cf. p. 65, vol. 3/II). The Mammalian cortical relay grisea provided by the posterior ventromedial and ventrolateral nuclear complex are poorly developed in Reptiles and are represented by the ventromedial (n. medialis s. diagonalis) and ventrolateral 'cell capsule' of nucleus rotundus. Although these Reptilian grisea may receive ascending 'sensory' input their synaptology most certainly differs from that obtaining for Mammals.

With respect to the *hypothalamus*, and, in particular, its posterior or caudal portion, input by way of *ascending 'visceral' channels* can be safely assumed. In addition, bundles interpreted as 'tractus lobobulbaris' and 'lobocerebellaris' are mentioned by KAPPERS (1947). Besides the channels between cerebellum and hypothalamus, cerebellar connections to dorsal and ventral thalamus by way of the 'brachium conjunctivum' system can likewise be inferred.[180]

The *descending caudal channels* comprise extensions of basal forebrain bundle components into the mesencephalic tegmentum, moreover of the above-mentioned 'lobobulbar', 'lobocerebellar', and 'mammillo-peduncular' (cf. footnote 179) systems, as well as presumed contributions to fasciculus longitudinalis medialis from the nuclei of the posterior commissure.

In addition to the dorsoventral intradiencephalic periventricular fibers mentioned in connection with the 'intrinsic channels' this periventricular system includes longitudinal fibers extending into the mesencephalon. Said system, which displays considerable variations

---

[180] Comments on these various channels, insofar as brain stem and cerebellum are concerned, were included in the hereto pertinent chapters and sections of volume 4. It is presumed that lobobulbar and lobocerebellar 'tracts' may include bulbolobar and cerebellolobar fibers.

qua distinctive differentiation in the diverse taxonomic forms, is interpreted by KAPPERS (1947) and others as a 'forerunner' of the *fasciculus longitudinalis dorsalis of Schütz*.

With respect to the *commissural systems*, the commissura anterior is related to components of the basal forebrain bundles. In addition to the essentially telencephalic commissura pallii sive hippocampi, there occurs, in Sphenodon and various Squamatae, a likewise telencephalic but 'aberrant' commissura pallii posterior which crosses through a portion of the roofplate pertaining to the diencephalic neighborhood of velum transversum (cf. Fig. 63 II, p. 174, vol. 3/II). This commissure has not been recorded in the hitherto examined Chelonians and Crocodilians.[181] Telencephalic fibers seem also to decussate in the habenular commissure. The synaptologic details of this latter as well as those of posterior commissure remain poorly elucidated.

As regards the *basal diencephalic decussations*, that of the optic chiasma can be evaluated as more or less complete. The supra- or postoptic commissures are generally well developed, and have been subdivided in various manner by different investigators whose views are carefully evaluated by BECCARI (1943). The dorsal supraoptic commissure predominantly contains crossed thalamotectal channels.[182] The ventral supraoptic commissure likewise contains tectal channels in its dorsal portion, while its ventral portion seems related to torus semicircularis (or colliculus inferior), nucleus isthmi, and ventral medial geniculate nucleus. Additional caudal fibers of the ventral postoptic commissure may be considered a 'preinfundibular commissure'. A well-defined 'postinfundibular commissure' is apparently not displayed by Reptiles, but comparable decussating fibers are presumably included in the 'supramammillary' commissural system at the level of tuberculum posterius in the basal diencephalo-mesencephalic boundary region. In Reptiles with well-developed interthalamic 'massa intermedia', this latter may include some non-medullated and medullated commissural fibers. This represents the '*commissure thalamique principale*' mentioned by KAPPERS (1947, p. 221).

---

181 '*Manca nei Cheloni e nei Loricati*' (BECCARI, 1943).

182 A griseum of thalamus ventralis, connected with the dorsal supraoptic commissure, was mentioned above in footnote 169 (p. 259). Again, BECCARI (1943) mentions '*fibre ansulate del Bellonci*' in the dorsal neighborhood of dorsal supraoptic commissure. Such fibers, supposed to be traceable from 'medulla oblongata' are not to be confused with the typical tegmental commissura ansulata of the Anamniote deuterencephalon (discussed in various sections of volume 4).

8. Birds

Although the Avian diencephalon, at its fully differentiated, i.e. adult stage, displays some conspicuous configurational features differing from that of Reptiles, a common Sauropsidan diencephalic bauplan obtains, which can easily be recognized by a comparison of ontogenetic key stages (K., 1931, 1936, 1937) as dealt with in chapter VI, section 5 of volume 3/II.

The most obvious difference in gross shape is related to the lateral displacement of the tectum mesencephali (optic lobes). In Reptiles, the diencephalon is dorsally 'covered' or 'overlapped' by the caudal telencephalon and by the mesencephalic tectum. This latter protrudes dorsally with the free surfaces of the two (anterior) tectal colliculi, as, e.g. in Amphibians, particularly Anurans. In Birds, on the other hand, the diencephalon and mesencephalon are dorsally covered by caudal telencephalon and by the cerebellum, while the optic lobes protrude laterally and even ventrally. Again, while the dorsoventral diameter of the Reptilian diencephalon generally exceeds the greatest transverse (right-left) diameter, this relationship becomes markedly reversed in Birds.[182a]

In contradistinction to various Reptiles, in which the torus semicircularis displays the tendency toward formation of the colliculi inferiores characteristic for Mammals, said torus remains entirely within the lateralward displaced portion of the mesencephalic ventricle.

The location of the diencephalic nuclei, notably that of nucleus rotundus and ventral lateral geniculate nucleus, becomes significantly modified by the aforementioned morphogenetic molding. Nevertheless, as justly pointed out by Beccari (1943), '*il talamo degli Uccelli è composto press' a poco dai medesimi nuclei principali di quello dei Rettili, e ciò sta in accordo con la vicinanza sistematica delle due classi*'.[183] Much the same could be said about the hypothalamic grisea.

[182a] This occurs in the course of early ontogeny. At 5 days of incubation in the Chick, the dorsoventral diameter (d) measures 2 mm, and the greatest transverse diameter (r) little more than 0.5 mm. The trend of this ratio ($d > r$) becomes reversed ($d < r$) between the 7th and 8th day. Cf. the growth curves depicted in Figure 42, p. 127, volume 3/II. In that diagram, greatest wall thickness (t) was given instead of greatest transverse diameter. This latter includes a variable amount of ventricular space, which may finally become almost negligible, and thus can be expressed as $r \geqslant 2 \times t$.

[183] Beccari (1943) then adds: '*Tuttavia lo sviluppo, la distribuzione topografica ed i rapporti dei nuclei alquanto differiscono, in conseguenza dello spostamente sui lati dei lobi ottici, del notevole sviluppo dei connessioni con i nuclei basale del telencefalo e della maggior complicazione del sistema ottico.*'

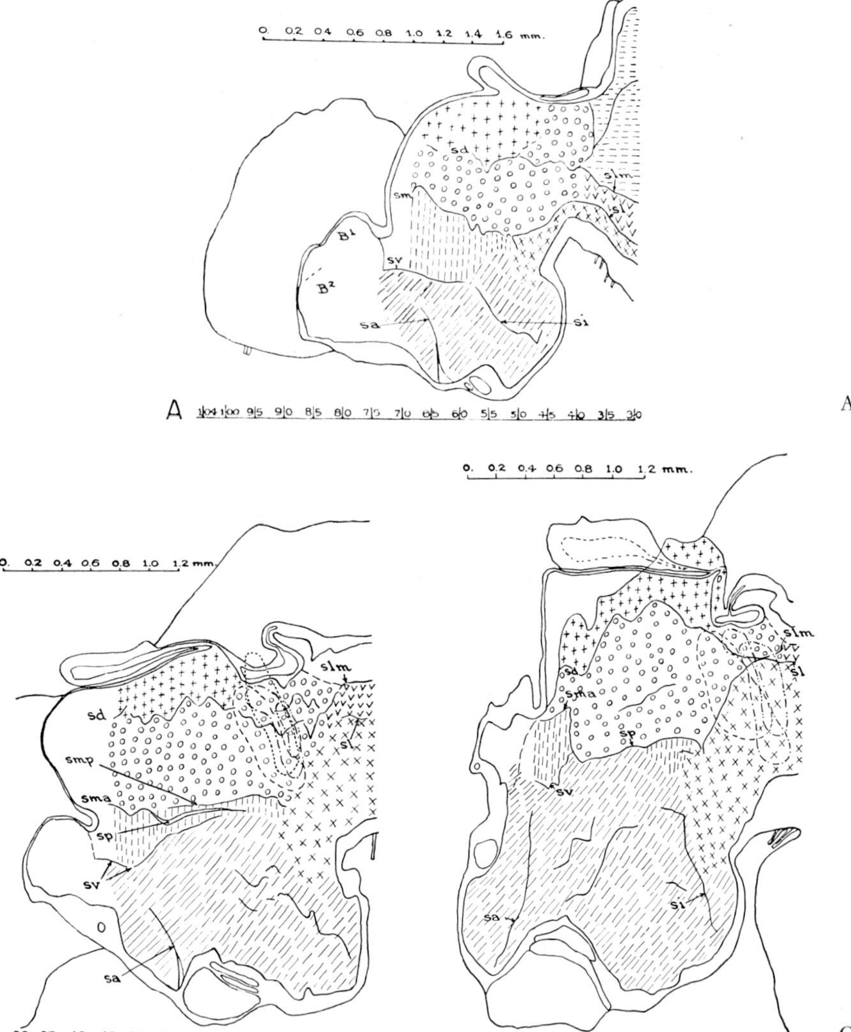

Figure 104 A–C. Mappings, based on graphic reconstructions, of the diencephalic longitudinal zones upon the ventricular wall in chick embryos at 5, 7, and 8 days of incubation (from K., 1939). +: epithalamus; circles: thalamus dorsalis; vertical hatching: thalamus ventralis; oblique hatching: hypothalamus; horizontal hatching: tectum mesencephali; V: torus semicircularis mesencephali; ×: basal plate; neighborhoods within broken lines in B and C: pretectal grisea; B_1, B_2 in A: telencephalic zones; sa: sulcus intraencephalicus anterior; sd: s. diencephalicus dorsalis; si: s. lateralis hypothalami posterioris (s.lat. infundibuli); sl: s. limitans; sm: s. diencephalicus medius; sma: s. diencephalicus medius, pars anterior; smp: s. dienc. medius, pars posterior; sp: s. parencephalicus ('pseudosulcus'); s.v.: s. diencephalicus ventralis.

At the adult stage, the diencephalic ventricle becomes relatively reduced to a narrow slit, and the typical diencephalic sulci, conspicuous during some ontogenetic stages, tend to become flattened out and almost vanish. Nevertheless, the *sulcus diencephalicus dorsalis* may remain well recognizable. A remnant of *sulcus parencephalicus* ('pseudosulcus'), which has absorbed sulcus diencephalicus medius and ventralis, is likewise commonly displayed. Another still rather well defined persistent ventricular landmark is the *sulcus lateralis hypothalami posterioris* with the *organ of Charlton-Kappers* (cf. Fig. 106D). Figures 104A–C illustrate some of the changes, occurring during ontogenesis, which affect lamina terminalis, diencephalic longitudinal zones and ventricular sulci.

Pioneering studies on diencephalic nuclei and tracts were undertaken by EDINGER and WALLENBERG (1899), also in collaboration with HOLMES (1903). Additional significant contributions are those by GROEBBELS (1924), HUBER and CROSBY (1929), and CRAIGIE (1928, 1930, 1931). Summaries of the available data, with references to additional publications can be found in the treatises by KAPPERS *et al.* (1936), KAPPERS (1947), BECCARI (1943), in a chapter by PORTMANN and STINGELIN (1961), and in the monograph on the Avian brain by PEARSON (1972). Ontogenetic data are reviewed in the treatise by ROMANOFF (1960). Our own investigations (K., 1931, 1937, 1939) which likewise contain further references, have attempted to ascertain,

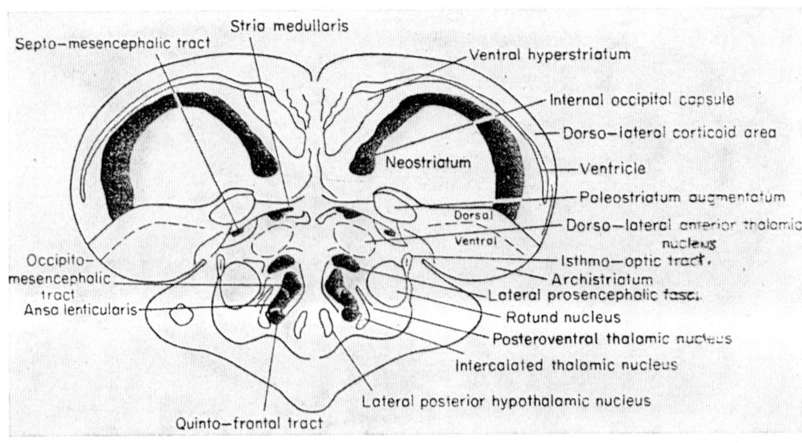

Figure 105. A generalized cross-section through the Avian diencephalon and adjacent portions of the neuraxis, indicating the overall configuration of main fiber tracts (black) and of grisea (from PEARSON, 1972).

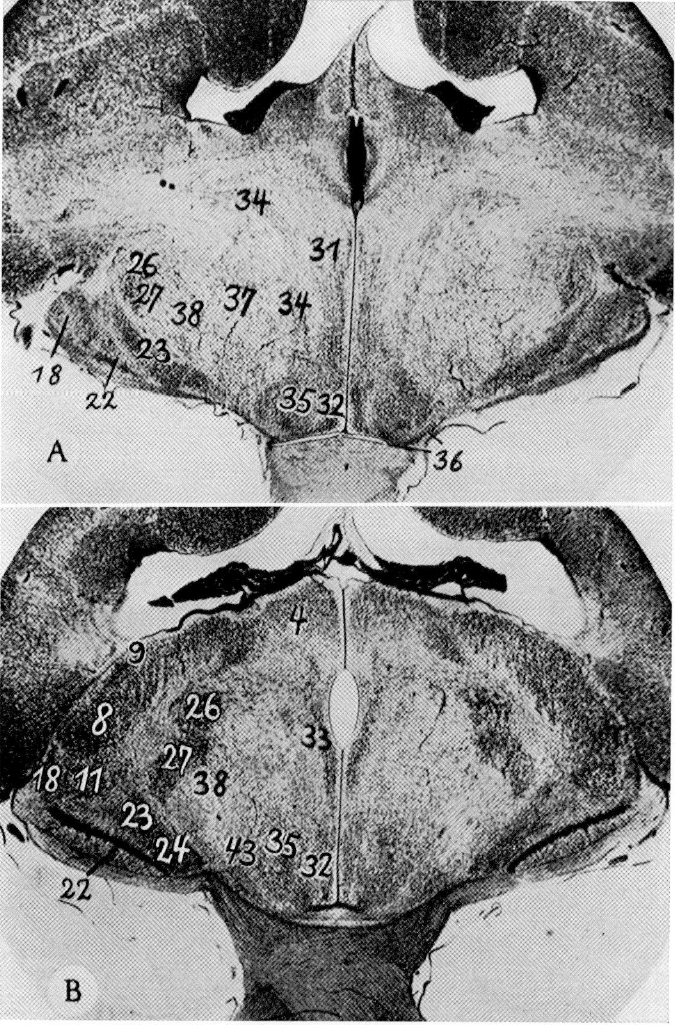

Figure 106 A–D. Cross-sections *(Nissl stain)* through the diencephalon of a newly hatched chick, displaying the definitive griseal configuration (from K., 1937). 1–3: n. habenularis lateralis, medialis (pars dorsalis), medialis (pars ventralis); 4: n. dorsomedialis; 5: n. subhabenularis medialis; 6: n. subhab. lateralis; 7: n. paraventricularis dorsalis; 8: n. dorsolateralis; 9: n. dorsolat. superficialis; 10: n. tractus septo-mesencephalici; 11: n. rotundus; 12: n. triangularis; 13: n. subrotundus; 14: n. ovoidalis; 15: n. paramedianus intermedius; 16: n. posteroventralis; 17: n. posterointermedius; 18: n. lateralis anterior; 19: n. geniculatus lateralis dorsalis ('pars principalis'); 20: n. gen. lat. dors. ('pars intercalaris'); 21: n. principalis praecommissuralis; 22: n. gen. lat. ventralis; 23: n. ventrolateralis; 24: n. decussationis supraopticae ventralis, pars anterior; 25: n. posterior decussationis supraopticae ventralis (n. gen. med. ventralis); 26: n. reticularis dorsalis;

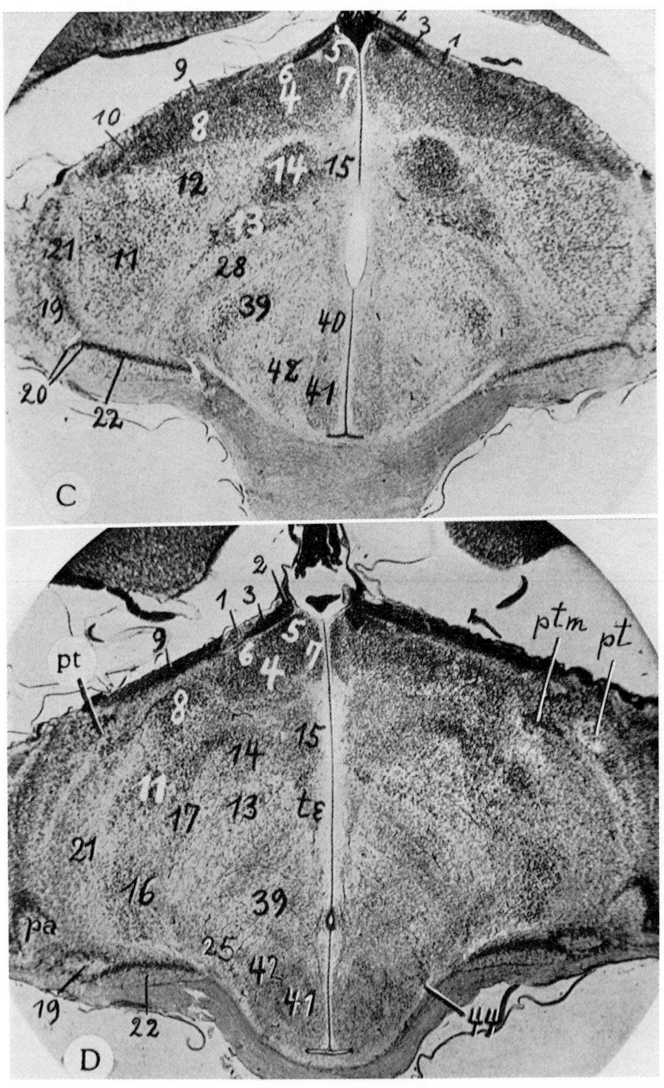

27: n. ret. ventralis; 28: n. intercalatus; 32: n. praeopticus paraventricularis inferior; 33: n. praeop. paravent. magnocellularis; 34 (in A at top): n. praeop. dorsolateralis (related to massa cellularis reuniens, pars inferior); 34 (in A more ventrally): medial subdivision of n. praeop. lateralis (37); 35: n. praeop. medialis; 36: n. supraopticus; 37: n. praeop. lateralis (lateral subdivision); 38: n. entopeduncularis interstitialis; 39: n. entoped. posterior; 40: n. paraventricularis posterior; 41: n. inferior hypothalami; 42: n. lateralis hypothalami; 43: n. lat. hypothalami anterior; 44: n. lat. externus hypothalami; pa: n. parageniculatus tecti optici; pt: n. praetectalis; te: tegmental cell groups. With respect to the numerical designations, it should be added that they do not correspond to the enumeration of the nuclei in the text, but represent a preliminary *ad hoc* notation used in my detailed investigation of 1937. ptm is n. praetectalis accessorius (medialis).

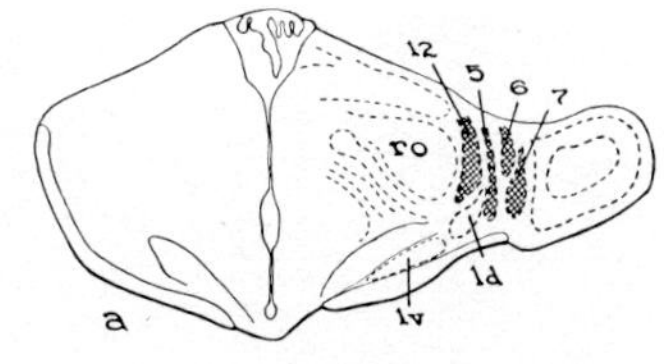

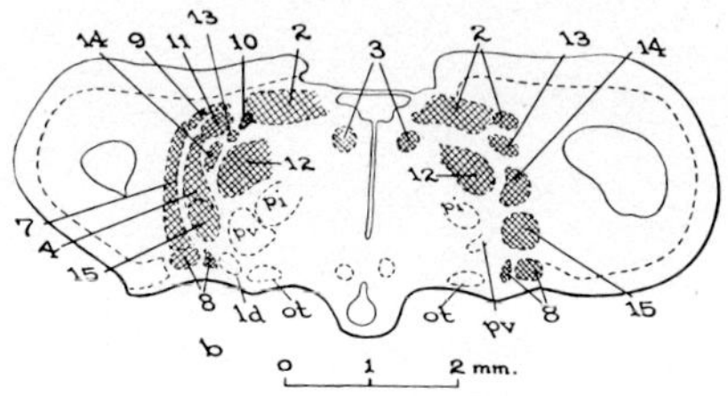

Figure 106 E. Outlines from cross-sections through the pretectal region of a Chick embryo of 12 days of incubation, displaying configurational arrangement of definitive grisea (from K., 1939). 2: n. diffusus parvocellularis commissurae posterioris; 3: n. interstitialis tegmentalis commissurae posterioris; 4: n. interstit. tractus praetecto-subpraetectalis; 5: n. laminaris praecommissuralis; 6: n. lentiformis mesencephali, pars magnocellularis; 7: n. lent. mes., pars parvocellularis; 8: n. parageniculatus tecti optici; 9: n. praetectalis lateralis; 10: n. praet. medialis; 11: n. praet. principalis; 12: n. principalis praecommissuralis; 13: n. spiriformis dorsomedialis; 14: n. spiriformis ventrolateralis; 15: n. subpraetectalis; ld: n. geniculatus lateralis dorsalis; lv: n. gen. lat. ventralis; ot: n. opticus tegmenti; pi: n. posterointermedius; pv: n. posteroventralis; ro: n. rotundus.

on the basis of the relevant ontogenetic stages, the morphologic relationship of the delimitable Avian diencephalic grisea to the fundamental longitudinal zones characterizing the Vertebrate diencephalic bauplan. The stereotaxic atlases of VAN TIENHOVEN and JUHASZ (1962; Chicken), and by KARTEN and HODOS (1967; Pigeon) include pictures of relevant diencephalic levels. A few additional appropriate references dealing with particular questions will be cited further below in the text. Figure 105 illustrates some overall features of a 'generalized' Avian diencephalon, while Figures 106 to 110 depict further details in a few forms which may be taken as 'representative'.

The relatively small size of the Avian *epithalamus* appears correlated with the reduced olfactory system in that Vertebrate class and thereby seems to confirm the preponderately olfactory function of this dien-

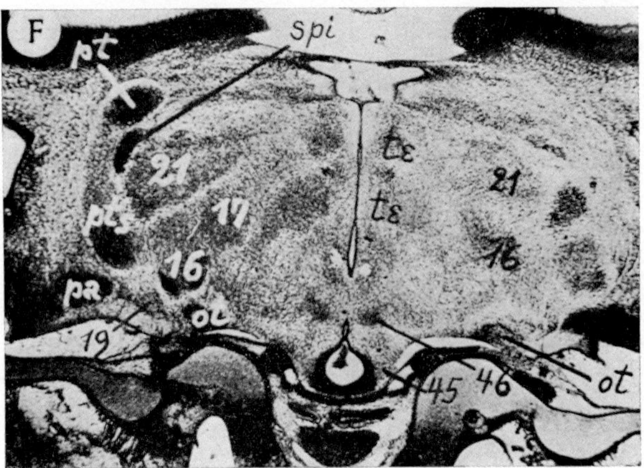

Figure 106 F. Cross-section *(Nissl stain)* through caudal portion of diencephalon at mesencephalo-diencephalic boundary zone in a Chick embryo of 12 days, showing definitive griseal configuration (from K., 1937). 45: nn. tuberis; 46: n. mammillaris; ot: n. opticus tegmenti; pts: n. subpraetectalis; spi: n. spiriformis ventralis. Other designations as in Figures 106 A–D.

cephalic *grundbestandteil.*[184] A *medial* and a *lateral habenular nucleus* can be identified. The former generally consists of smaller and more crowded cells, and could perhaps be further subdivided into pars dorsalis and pars ventralis. The less compact lateral habenular nucleus contains some larger elements and may display a further nondescript parcellation. The medial habenular nucleus extends commonly farther rostralward than the lateral one, but does not extend as far rostralward as the dorsal thalamus.

The *thalamus dorsalis* displays a considerable degree of 'nuclear differentiation' whose conceptual 'parcellation' has led to difficulties increased by the obtaining variations in the Avian taxonomic series, and resulting in differences between the terminologies propounded by the diverse investigators. As a first approximation suitable for comparison with the corresponding Reptilian nuclear configuration, the following main griseal groups may be distinguished. The anterior dorsomedial (A) and dorsolateral (B) groups, the ventrolaterally displaced nucleus rotundus group (C), the medial and ventral pararotundal group (D), the dorsal lateral geniculate complex (E), the posterocentral group (F), and the pretectal group (G).

[184] Craigie (1930) points out that the habenulae of the flightless *Kiwi* (Apteryx australis) are relatively 'of considerable size', although not as large as might be expected if all the fibers of the substantial stria medullaris system ended in them.

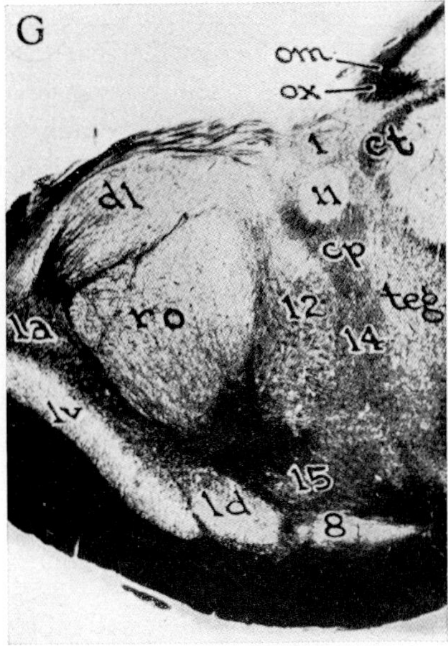

Figure 106G. Fairly lateral sagittal section *(Weigert-Pal stain)* through diencephalon and mesencephalo-diencephalic boundary region in an adult Chick (from K., 1939). 1: area praetectalis; cp: commissura posterior; ct: commissura tecti; dl: n. dorsolateralis anterior; la: n. lateralis anterior; om: tractus opticus marginalis; ox: tr. opt. axialis; teg: tegmental grisea (prerubral tegmentum, tegmental cell cord. Other designations as in Figure 106E.

The *anterior dorsomedial group* (A) may perhaps, *in toto*, represent the Reptilian anterior dorsomedial nucleus, but includes the following subdivisions (cf. K., 1937). (1) Nucleus dorsomedialis anterior *(sensu strictiori)*, nucleus subhabenularis medialis (2) and lateralis (3), and nucleus paraventricularis dorsalis (4). Again, the anterior dorsolateral group (B), similarly corresponding to Reptilian anterior dorsolateral nucleus, includes (5) nucleus dorsolateralis anterior *(sensu strictiori)*, nucleus dorsolateralis superficialis (6), and the nucleus of the septo-mesencephalic tract (7). These grisea are shown in Figures 106A–D. More caudalward, the dorsolateral and dorsomedial grisea seem to merge, displaying vague condensations designated as nn. dorsolateralis posterior, dorsointermedius posterior, and dorsomedialis posterior thalami by Karten and Hodos (1967) and extending dorsally to the nucleus paramedianus intermedius *(sive* internus) pointed out below under (D).

The *nucleus rotundus group* (C) consists of the ventrolaterally displaced nucleus rotundus (8) and of a dorsomedial extension of this latter, the nucleus triangularis (9). Despite said considerable secondary topographical displacement, there can be little doubt that, on the basis of ontogenetic evidence, the Avian nucleus rotundus and triangularis are morphologically (topologically) homologous to the Reptilian nucleus rotundus. In Birds, however, the nucleus, although quite conspicuous, does not represent the predominating dorsal thalamic configuration as displayed in Reptiles.

The *medial and ventral pararotundal group* (D) includes nucleus ovoidalis (10) and nucleus subrotundus (11). An additional medial, paraventricular griseum is the nucleus paramedianus intermedius (12). These nuclei[185] are shown in Figures 106C and D (cf. also Fig. 216, p. 439, vol. 3/II).

The *dorsal lateral geniculate complex* (E) is characterized by an elongated neuropil on the dorsomedial and medial side of the optic tract. Internally to this neuropil, and partly within it, cell groups form the nucleus

[185] Nucleus ovoidalis and subrotundus correspond morphologically to nucleus medialis and infrarotundus, respectively, of Reptiles.

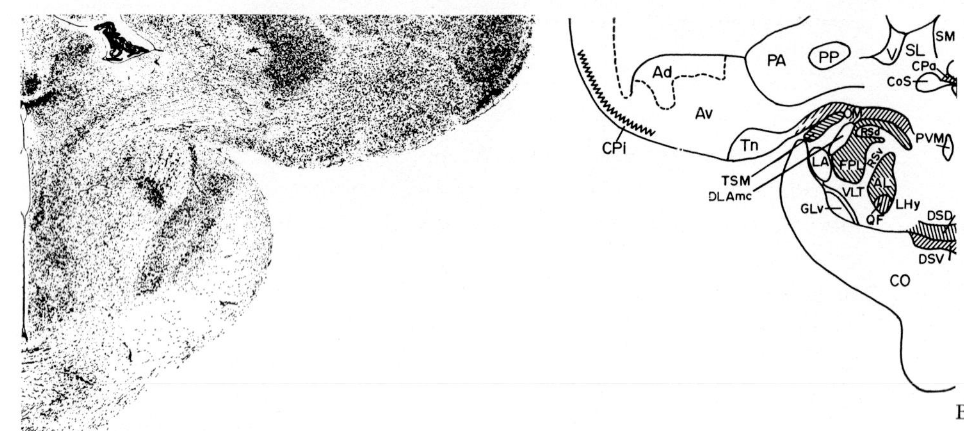

Figure 107 A–F. Cross-sections *(Nissl stain)* with explanatory outlines, through the diencephalon of the Pigeon (from Karten and Hodos, 1967). A, B at level of hemispheric stalk; C, D, E, F at more caudal levels (note rostral tip of tectum opticum). Ad: 'archistriatum, pars dorsalis'; AL: 'ansa lenticularis'; Av: 'archistriatum, pars ventralis'; CO: chiasma opticum; CoS: 'n. commissuralis septi'; CPi: 'cortex piriformis'; DLAmc: n. dorsolateralis anterior thalami, pars magnocellularis; DLL, DLM: n. dorsolat. ant. thalami (pars lateralis respectively medialis); DMA: n. dorsomedialis anterior thalami;

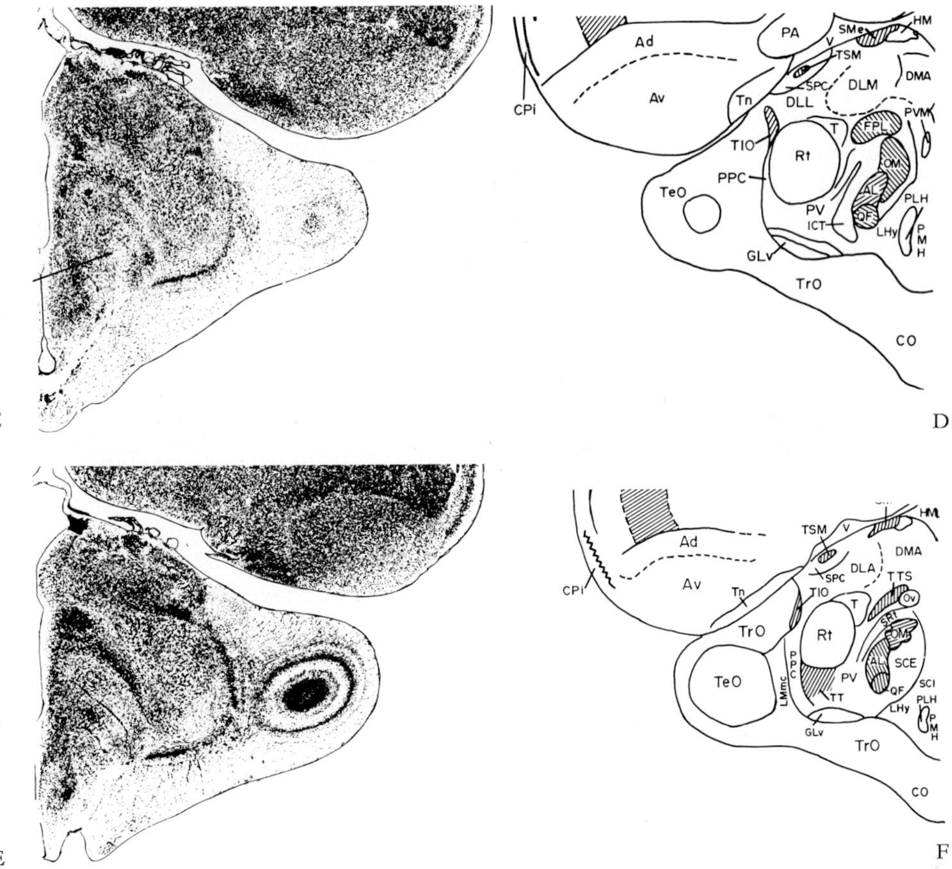

DSD, DSV: decussatio supraoptica (dorsalis respectively ventralis); FPL: lateral forebrain bundle (dorsal portion); GLv: ventral lateral geniculate nucleus (in A, part of the dorsal l.g.n. can be identified ventrally to LA); HM: n. habenularis medialis; ICT: n. hab. (thalami ventralis); LA: n. lateralis anterior (of dorsal lateral geniculate complex); LHy: n. lateralis hypothalami; LMmc: n. lentiformis mesencephali, pars magnocellularis; OM: tr. occipitomesencephalicus; Ov: n. ovoidalis; PA: 'palaeostriatum augmentatum'; PLH: n. lateralis hypothalami posterioris; PMH: n. medialis hypothalami posterioris; PP: 'palaeostriatum primitivum'; PPC: n. principalis praecommissuralis; PV: n. postero-ventralis; PVM: n. periventricularis magnocellularis (n. praeopticus magnocellularis and its caudal extension): QF: tr. quintofrontalis; RSd, RSv: n. reticularis dorsalis, ventralis; Rt: n. rotundus; SCE: n. entopeduncularis posterior; SCI: 'stratum cellulare internum' (probably griseum of prerubral tegmentum); SL, SM: 'n. septalis lateralis and medialis'; SMe: stria medullaris; SPC: n. interstitialis of septomesencephalic tract; SRt: n. sub-rotundus; T: n. triangularis; TeO: tectum opticum; TIO: tr. isthmo-opticus; Tn: n. taeniae; TrO: tr. opticus; TSM: tr. septomesencephalicus; TT: tr. tectothalamicus; TTS: 'tr. thalamostriatus'; V: ventricle; VLT: n. ventrolateralis (thalami ventralis). Arrow added to C: n. entopeduncularis posterior.

geniculatus lateralis dorsalis (13), with a rostral pars or nucleus intercalaris (14) adjacent to the ventral lateral geniculate nucleus (cf. Figs. 106C, D). The rostral portion of the dorsal lateral geniculate complex is formed by the n. lat. anterior (15), a distinctive cell aggregate caudoventrally contiguous with the dorsal geniculate nucleus. This griseum seems to be particularly well developed in the Owl (KARTEN *et al.*, 1973). It can be seen that, in the definitive adult configuration pattern, nucleus lateralis anterior may become the most rostral clearly delimitable nucleus of the thalamus dorsalis[186] (cf. Fig. 106A).

The *posterocentral group* (F) begins rostrally at caudalmost levels of nucleus rotundus and nucleus ovoidalis, and represents a griseal neighborhood, intercalated, as it were, between the dorsal thalamic groups A, B, C, D, and the pretectal group G. Said neighborhood includes nucleus posterointermedius (16), and the more ventrolaterally located nucleus posteroventralis (17). This group[187] is shown in Figures 106D and E.

The highly differentiated dorsal thalamic Avian *pretectal group* (G) can again be subdivided into (a) pretectal complex *sensu strictiori*, and (b) grisea particularly related to commissura posterior.[188] Within the pretectal complex, the nucleus principalis praecommissuralis (18) begins rostrally near the caudal end of nucleus rotundus. The nucleus laminaris praecommissuralis (19) which was noted laterally to rostral

---

[186] This is caused by the considerable changes occurring during ontogenesis, whereby a diencephalic configuration comparable to that of Amphibians and Reptiles is transformed into the Avian pattern as shown by the study of successive 'key stages' recorded in previous studies (K., 1936, 1937). At the 9th day of incubation in the chick, an eminentia thalami ventralis reaches the interventricular foramen (cf. K., 1936, Fig. 46), followed, caudally to velum transversum, by the primordia of epithalamus and those of dorsal thalamic nuclear groups (K., 1936, Figs. 47, 48). Subsequently, the ventral thalamus shifts ventrolaterad, the epithalamus caudad, and nucleus lateralis anterior, closely followed by dorsolateral and dorsomedial nuclear groups form the rostralmost dorsal thalamic cell groups. The hypothalamic pars inferior of massa cellularis reuniens together with the related interpeduncular grisea retain their location as rostralmost diencephalic cell groups within the hemispheric stalk.

[187] Ontogenetic stages of these grisea suggest their possible morphologic homology with the Reptilian nucleus posterodorsalis (posterocentralis), which pertains to the Reptilian pretectal region. In Birds, however, nn. posterointermedius and posteroventralis are rather separated from the pretectal grisea, which begin with nucleus principalis praecommissuralis. This latter, however, also seems to be a derivative of the Reptilian posterodorsal griseum.

[188] It is understood that subgroup (a) is, of course, also related to the commissura posterior.

parts of the former nucleus and medially to the nucleus lentiformis mesencephali complex (K., 1939), can be considered a subdivision of nucleus principalis praecommissuralis (cf. Fig. 106E). The nucleus praetectalis principalis (20) is a very conspicuous structure, which appears round in cross-sections, being surrounded by a ring of neuropil and medullated fibers. Medially adjacent lies the nucleus praetectalis medialis (21). Laterally adjacent, and blending with nucleus lentiformis mesencephali complex, a nucleus praetectalis lateralis (22) can be distinguished. Ventrally adjacent the nucleus interstitialis tractus praetecto-subpraetectalis (23) extends basalward together with the contiguous and still more ventral nucleus subpraetectalis (24). A dorsal and rather medial griseum, located near the external surface of diencephalon, is the area praetectalis (25) which corresponds to that of Reptilians.

The dorsal thalamic grisea of commissura posterior can be delimited as consisting of nucleus spiriformis dorsomedialis (26), nucleus spiriformis ventrolateralis (27), and nucleus diffusus parvocellularis commissurae posterioris (28). The nuclei of the pretectal region are illustrated in Figures 106E and F, as well as in Figures 398A and B of section 8, chapter XI, volume 4, where the pretectal region is also dealt with insofar as its close relations to the mesencephalon and its mesencephalic components are concerned.

It can thus be seen that, in comparison with about 12 to 13 dorsal thalamic 'nuclei' in Reptiles, 28 such 'nuclei' may be distinguished in Birds, even by an investigator averse to 'unnecessary' or 'extreme' parcellation. This number of distinctive grisea seems out of proportion with the uncertain, incomplete, and inadequate available data concerning their fiber connections, detailed synaptology, and function, as briefly discussed further below.

The *thalamus ventralis*, which, at certain stages of ontogenesis, displays a rostral eminentia thalami with a moderately developed pars superior massae cellularis reunientis, and, together with rostral end of thalamus dorsalis, reaches the level of interventricular foramen, becomes caudolateralward displaced by the subsequent considerable morphogenetic changes.[189] At the final ('adult') stage, only the hypo-

---

[189] These complex morphogenetic changes (cf. fn. 186) can be recognized by an inspection of e.g. Figures 23, 24, 43, 45, 46 (K., 1936, pp 84, 96, 97, 98), Figures 218A, 222–224 (pp. 441, 447–449) of volume 3/II, and Figures 104A–C of the present chapter. The ventral thalamic massa cellularis reuniens, pars superior, becomes reduced to a nondescript interstitial cell group within the telencephalic commissural plate (identifiable in Fig. 45, K., 1936).

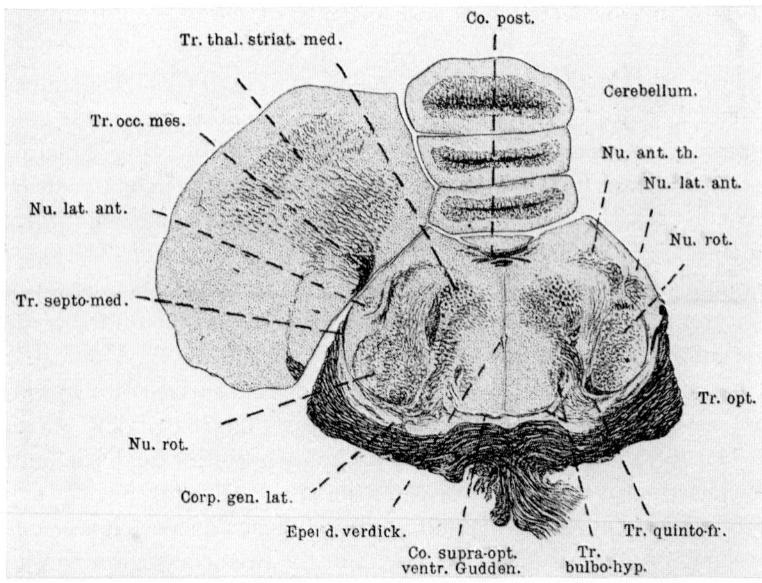

Figure 108 A. Cross-section *(Weigert stain)* through the diencephalon and adjacent regions of Pratincola rubicola (from KAPPERS, 1921). Co.post.: commissura posterior; Epend. verdick.: *organ of Kappers-Charlton;* Nu.dors.post.: leads point toward neighborhoods of nn. triangularis, ovoidalis, and subrotundus; Nu.thal.ant. (and ant.thal.): leads point to dorsomedial and dorsolateral anterior thalamic group; Nu. interc.: nucleus intercalatus (thalami ventralis); Tr.septo-med.: tractus septo-mesencephalicus. Other abbreviations self-explanatory.

thalamus and a nondescript rostralmost tip of conjoint anterior dorsomedial and dorsolateral dorsal thalamic groups extends toward the interventricular foramen.

In the adult configuration, the lateralward displaced rostral thalamus ventralis displays two closely adjacent cell groups, namely nucleus reticularis dorsalis (1) and ventralis (2), both of which presumably correspond to the Reptilian 'area triangularis'. More caudally, the nucleus intercalatus (3) is represented by a 'band' or 'plate' of cells interjacent between thalamus dorsalis and hypothalamus. The ventrolateral extension of this cell plate provides rostrally the nucleus decussationis supraopticae ventralis, pars anterior (4), and caudally the nucleus decussationis supraopticae ventralis, pars posterior (5). For these grisea, the following Reptilian homologies can be established by topologic mapping: (3) represents nucleus ventromedialis, (4) nucleus ventrolateralis, and (5) presumably the 'primordial ventral medial geniculate nucleus'.

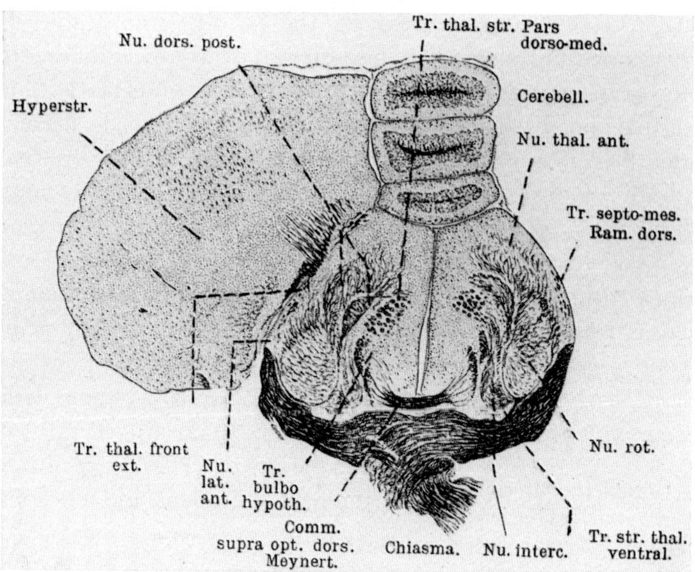

Figure 108B. Cross-section *(Weigert stain)* through the telencephalon of Pratincola rubicula caudal to preceding section (from KAPPERS, 1921). Abbreviations as in Figure 108A.

The *nucleus geniculatus lateralis ventralis* (6) is a conspicuous and relatively large formation, spindle- or lens-shaped in cross-sections. Ventrolaterally to its dense band of cells lies a neuropil adjacent to the optic tract and receiving numerous collaterals from this latter. Caudally, this nucleus becomes contiguous with the nucleus opticus tegmenti of the mesencephalon. The nucleus ventrolateralis (7) is a group of cells located medially to the preceding nucleus and extending rather far rostralward. It can be included in the ventral lateral geniculate complex and may, besides, correspond to the Reptilian nucleus ovalis.

The caudal end of the ventral geniculate complex tends to disappear rostrally to the caudal tip of dorsal lateral geniculate griseum. At these diencephalo-mesencephalic boundary levels, the dorsal lateral geniculate nucleus is intercalated between the ventral nucleus medially, and the mesencephalic nucleus parageniculatus tecti optici laterally (cf. Figs. 106D–F, and Figs. 398A, B, chapter XI, vol. 4). Ventral and dorsal thalamic lateral geniculate grisea can be considered to form a conjoint lateral geniculate complex.

The Avian *hypothalamus* is, on the whole, comparable to the Reptilian one but differs from this latter (a) by a greater differentiation of some of the grisea, and (b) insofar as the ratio of its total volume to that of thalamus has become reduced, or, in other words, because the thalamic grisea, *in toto*, have expanded more than the hypothalamic ones, both *qua* phylogenetic evolution and ontogenetic development.

Within the *preoptic region*, a nucleus praeopticus paraventricularis (1) is easily recognizable and can be further subdivided into various nondescript groupings, such as e.g. 'nucleus praeopticus anterior', 'nucleus paraventricularis superior' and 'inferior'. One of these diverse subgroups, however, the neurosecretory nucleus praeopticus magnocellularis (2), which pertains to the hypophysial system, definitely represents a particular 'entity' (cf. KUROTSU, 1935). Another distinctive group is formed by a basal condensation of cells above the lateral ventricular slit-like extension (remnant of sulcus intraencephalicus

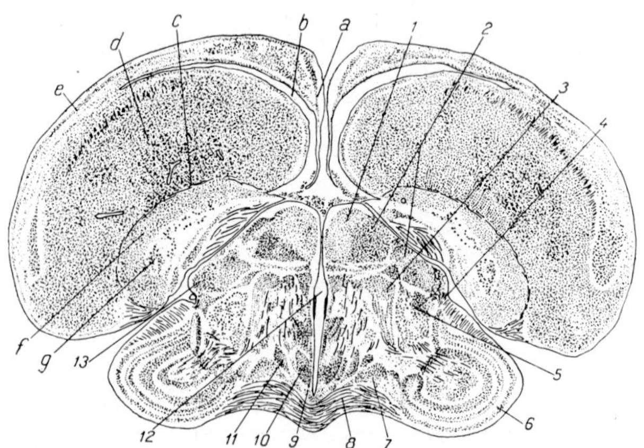

Figure 109. Cross-section through the diencephalon and adjacent brain regions in the sparrow, Passer domesticus (simplified after HUBER and CROSBY, 1929, from BECCARI, 1943). 1: nucleus dorsomedialis anterior; 2: nucleus dorsolateralis anterior; 3: nucleus ovoidalis; 4: tractus isthmo-opticus; 5: nucleus rotundus; 6: tectum opticum; 7: lateral geniculate complex; 8: caudal portion of optic chiasma; 9: nucleus inferior hypothalami; 10: nucleus lateralis hypothalami; 11: nucleus posterior decussationis supraopticae ventralis; 12: third ventricle (with *organ of Kappers-Charlton)*; 13: nucleus dorsolateralis superficialis and nucleus tractus septomesencephalici; a: 'hippocampus'; b: lateral ventricle; c: 'lamina medullaris dorsalis'; d: nucleus epibasalis centralis; e: lateral corticoid lamina; f: nucleus basalis; g: nucleus entopeduncularis (anterior).

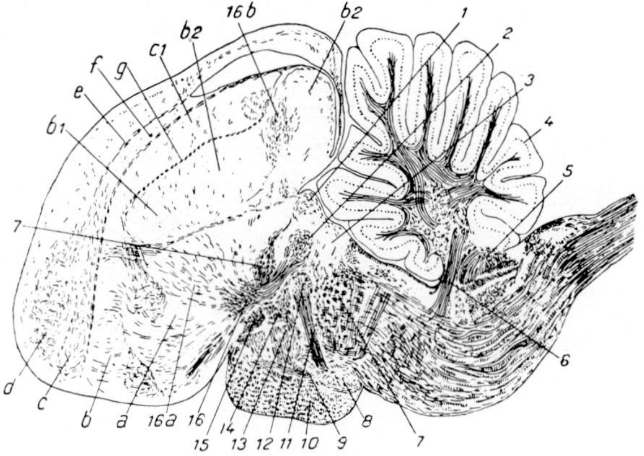

Figure 110. Paramedian sagittal section (silver impregnation) through the brain of Passer domesticus, displaying diencephalic grisea and communication channels (simplified after Huber and Crosby, 1929, from Beccari, 1943). 1: habenular griseum; 2: nucleus dorsolateralis anterior; 3: nucleus dorsolateralis, posterior portion; 4: cerebellum; 5: nucleus dorsalis magnocellularis nervi octavi; 6: brachium conjunctivum; 7: tractus occipito-mesencephalicus; 8: nucleus opticus tegmenti; 9: ventral supraoptic commissure; 10: tractus opticus; 11: tractus nuclei ovoidalis; 12: nucleus ovoidalis; 13: fasciculus thalamofrontalis intermedius; 14: tractus striotegmentalis and striocerebellaris; 15: dorsal portion of tractus septomesencephalicus; 16, 16a, 16b: fasciculus thalamofrontalis medialis with subdivisions; a: nucleus basalis; b, b_1, b_2: subdivisions of nucleus epibasalis; c, c_1: subdivisions of nucleus epibasalis dorsalis; d: rostral tip of nn. diffusi ('hyperstriatum accessorium'; e: lamina medullaris suprema; f: lamina medullaris superior; g: 'lamina hyperstriatica'.

anterior origin) just dorsal to optic chiasma. This is (3) the nucleus praeopticus paraventricularis inferior.[190]

More laterally located cell groups of the preoptic region include the following grisea. The nucleus praeopticus dorsolateralis (4), at the level of hemispheric stalk, is a lateral extension of the preoptic cell masses toward the basal grisea of the telencephalon. It represents a derivative of pars inferior massae cellularis reunientis, which also contributes to the formation of nucleus entopeduncularis (anterior), to be dealt with below. Ventrally to nucleus praeopticus dorsolateralis, scattered cell groups lateral to the peri- (or para-) ventricular gray form the nucleus praeopticus lateralis (5) with a lateral and a more medial subdivision.

[190] It presumably corresponds to the nucleus suprachiasmaticus of Crosby and Showers (1969), but the cited authors seem also to have subsumed part of nucleus praeopticus medialis (at least in the Reptilian Caiman) under the aforementioned designation.

Still more basally, and just dorsally to optic chiasma, the nucleus praeopticus medialis (6) is located. Ventrally to this nucleus, a plate-like group of larger elements forms the nucleus supraopticus (7) which, depending on the taxonomic differences, may display further subdivisions (e.g. a medial and lateral portion). Together with nucleus praeopticus magnocellularis, it pertains to the hypophysial neurosecretory system.

Still more laterad, ontogenetically dorsolateral but subsequently somewhat ventralward displaced cell groups of the dorsal hypothalamus are the entopeduncular nuclei, related to the basal forebrain bundle system, of which, at least to some extent, they represent 'interstitial nuclei'. The most rostral one, namely the *nucleus entopeduncularis anterior* (8) already mentioned above as partially derived from the embryonic massa cellularis reuniens, pars inferior, becomes secondarily completely included into hemispheric stalk and basal telencephalon, being joined to the basal ganglia and forming the so-called 'palaeostriatum primitivum' *autorum*,[191] corresponding to the Mammalian globus pallidus. Scattered multipolar cells extending caudalward within the lateral forebrain bundle system and located ventromedially to the ventral thalamic reticular group form the nucleus interpeduncularis intermedius (9) which rostrally indistinctly merges into nucleus entopeduncularis anterior.

The *nucleus entopeduncularis posterior* (10) is a still more caudal, and slightly more compact 'reticular' cell group within the lateral forebrain bundle system. It corresponds to the homonymous nucleus in Reptiles and to the Mammalian subthalamic nucleus (nucleus hypothalamicus *sive corpus subthalamicum Luysi*). Within the entopeduncular nucleus of Passerine birds, Arai (1963) and Arai *et al.* (1963) have identified a group of neurosecretory cells. Such cells were also recorded in the lateral and paraventricular hypothalamic region in addition to the generally recognized neurosecretory cell groups (2, 7) mentioned above.

The *caudal (posterior), essentially 'postoptic' hypothalamus* includes the following grisea. The nucleus paraventricularis posterior hypothalami (11) is a small-celled griseal plate representing an extension of the preoptic paraventricular nuclei, replacing the nucleus paraventricularis caudalward. It could be vaguely subdivided into additional nonde-

[191] It also includes the 'nucleus intrapeduncularis telencephali' of Karten and Dubbeldam (1973).

script cell groups, and its caudodorsal portion becomes adjacent to the rostral expansion of the tegmental cell cord's derivatives.

The nucleus lateralis hypothalami (12), laterally adjacent to the paraventricular gray, and the nucleus inferior *sive* ventralis hypothalami (13), located basally to the former, correspond to the synonymous Reptilian nuclei, respectively to nucleus hypothalamicus dorsomedialis and ventromedialis of Mammals (cf. above, section 7, footnote 173a). A still more lateral group represents the nucleus lateralis hypothalami externus (15), which can be regarded as a caudal extension of the nucleus praeopticus lateralis. This griseum is also faintly suggested in Reptiles. In the so-called 'infundibular region', a dorsal 'nucleus' mammillaris (16) is formed by a cell group ventral to the tegmental cell plate. Still more ventrally, cell groups near the external surface represent the nucleus (or nuclei) tuberis (17), as shown in Figure 106F. Thus, about 17 Avian hypothalamic 'nuclei' could perhaps be distinguished, as contrasted with 11 Reptilian ones. The neurohypophysis (cf. section 1C), the supraoptic crest,[192] and the *organ of Kappers-Charlton* (cf. Fig. 106D) might again be mentioned as additional Avian hypothalamic structures.

In accordance with the considerable expansion of the basal griseal complex within the lateral wall of the Avian cerebral hemisphere, the *lateral forebrain bundle* represents the most massive of the *rostral diencephalic communication channels*. It includes several subdivisions with a number of insufficiently understood descending and ascending components, some of which, extending into the brain stem, thereby also form part of the caudal diencephalic communication channels.

The overall arrangement of the lateral forebrain bundle system is shown in Figures 107B, D, and F. It will be seen that a ventromedial component is also designated as 'ansa lenticularis'. This portion seems to include the essentially descending 'fasciculus strio-thalamicus', 'strio-hypothalamicus', and 'strio-tegmentalis' *autorum*.[193] The most ventral portion is the so-called tractus quinto-frontalis of WALLEN-

---

[192] Cf. chapter V, section 5, p. 362 of volume 3/I.

[193] The fasciculus strio-tegmentalis is said to include a strio-cerebellar component (cf. the comments by BECCARI, 1943, p. 551). 'Ansa lenticularis' as well as 'striatum' and the therefrom derived terms are, of course, designations referring to the Mammalian brain. It is highly questionable whether an application of said terms to submammalian forms is appropriate. The configuration of basal ganglia in these forms, although displaying various degrees of homology, is entirely different from that obtaining in Mammals (cf. vol. 3/II, chapter VI, section 6).

BERG (1898) and SCHROEDER (1911), generally considered to be ascending, and directly to connect bulbar grisea, including afferent trigeminal ones, with the telencephalic basal ganglia.

Dorsal components of the lateral forebrain bundle are enumerated as lateral, intermediate, and medial 'thalamofrontal tracts' representing 'thalamo-striate tracts' to poorly defined regions of the telencephalic basal griseal complex, and the dorsally adjacent pallial grisea. Reciprocal connections of the nucleus rotundus are presumably included in that channel.

The *'occipito-mesencephalic tract'*, to a large extent apparently descending, interconnects a caudal telencephalic griseum (nucleus epibasalis caudalis)[194] with mesencephalic and rostral bulbar tegmental regions. Within the diencephalon, it assumes a medial position, dorsally to the ventral lateral forebrain channels (cf. Figs. 107D, F), and also appears to have connections with pretectal and other grisea, including hypothalamic ones. Whether some of its fibers undergo a decussation in the anterior commissure remains a moot question. A few fibers join the stria medullaris.

The *medial forebrain bundle system* is represented by the 'tractus septo-mesencephalicus' *(Bündel der Scheidewand of Edinger* and other early investigators). Its major component connects the telencephalic pallium (nuclei dorsalis diffusi, parahippocampal cortex, 'hippocampal' cortex) and paraterminal ('septal') grisea with dorsal thalamic nuclei including pretectal ones, and medial portion of optic tectum. Running through the medial wall of the hemisphere basalward toward the hemispheric stalk, it there turns laterad, describing a dorsally concave loop ventrally to the lateral forebrain bundle, and reaches the dorsolateral surface of the diencephalon. The nucleus tractus septo-mesencephalici, a derivative of the dorsolateral thalamic group, forms an 'interstitial nucleus' of said tract. A minor component detaches itself from the aforementioned loop, joining the ventromedial portion of the lateral forebrain bundle. It reaches the hypothalamus and, directly or

---

[194] Some fibers may also originate in the basomedially adjacent 'nucleus taeniae' and other neighboring grisea. Although the occipito-mesencephalic tract joins the lateral forebrain bundle system, it can also be considered a secondarily detached portion of the stria terminalis system (cf. e.g. HUBER and CROSBY, 1929). Like the 'septo-mesencephalic tract', dealt with in the next paragraph, it seems to represent a specific Avian adaptation or 'modification' of the rostral diencephalic communication channels, in some aspects also comparable to the stria medullaris system.

indirectly, also thalamic grisea. The caudalmost fibers of this component, which represents a rudimentary fornix, connect with the mammillary griseum. The entire tractus septo-mesencephalicus (*'Scheidewandbündel'*) must be interpreted as a peculiar modification, characteristic for the Avian brain, of the medial forebrain bundle[195] as displayed by Anamnia and Reptiles. The so-called 'tractus infundibuli *sive* infundibularis' seems to represent a part of the Avian medial forebrain bundle system and is perhaps related to the 'fornix'.

The *stria medullaris system sensu strictiori* is relatively reduced, concomitantly with the reduction of habenular grisea and olfactory system. It includes various lateral and medial basal telencephalic components (e.g. from the lateral nucleus taeniae and from the medial 'septal', i.e. paraterminal grisea), moreover fibers detached from the *Scheidewandbündel*, and minor contributions from the preoptic region and lateral forebrain bundle system.

The *optic tract*, which seems to consist of entirely or at least quite predominantly crossed optic nerve fibers, forms (1) a basal optic root running caudad as far as nucleus opticus tegmenti, (2) a tractus isthmo-opticus which reaches the nucleus isthmo-opticus (cf. section 8, chapter XI, vol. 4), and (3) the main or marginal optic tract to tectum mesencephali. Direct diencephalic optic connections include those to dorsal and ventral lateral geniculate complex[196] and less well distinguishable ones to undefined hypothalamic grisea. Indirect 'recurrent' optic input to nucleus rotundus, mediated by tectum opticum, appears to obtain.

Few relevant data concerning the *intrinsic diencephalic connections* are available. A neurosecretory pathway, consisting of tractus supraoptico-hypophyseus and related fibers, with several subdivisions and diverse taxonomic variations, has been identified (cf. e.g. WINGSTRAND, 1966). An at least partially intrinsic diencephalic bundle is the tractus nuclei ovoidalis, which essentially connects nucleus decussationis supraopticae ventralis, pars superior, with nucleus ovoidalis. Still another in part

[195] The fiber system designated as 'medial forebrain bundle' in the Pigeon by KARTEN and DUBBELDAM (1973, Figs. 6C, D, p. 69, Fig. 16, p. 86) represents, in my opinion, a medial component of the lateral forebrain bundle.

[196] According to KARTEN *et al.* (1973) the dorsal lateral geniculate complex projects, by way of the lateral forebrain bundle, upon the homolateral and contralateral 'hyperstriatum accessorium' and its subdivisions, and on parts of the 'hyperstriatum dorsale' (nn. diffusi and n. epibasalis dorsalis, pars superior, cf. vol. 3/II, p. 568) of the telencephalon, designated by the cited authors as the '*visual Wulst*'.

intrinsic channel seems to be the tractus praetecto-subpraetectalis between the corresponding pretectal nuclei.

The *caudal communication channels* include the caudal extensions of the lateral forebrain bundle. Their descending components are partly telencephalic,[197] and partly additional contributions from the diencephalon (e.g. from the entopeduncular nuclei). Ascending connections within this system are tractus quintofrontalis, and bulbothalamic, bulbohypothalamic, as well as presumably spinothalamic fibers, moreover rostral cerebellar extensions of the brachium conjunctivum. As in Reptiles, poorly elucidated connections of this system with dorsal and ventral thalamic as well as hypothalamic grisea can be surmised.

The *tractus tecto-thalamicus* is a fairly massive, perhaps mainly ascending complex system from optic tectum to nucleus rotundus and possibly adjacent grisea. Likewise mainly ascending connections from torus semicircularis appear to reach nucleus posterior decussationis supraopticae ventralis and nucleus ovoidalis through the abovementioned 'tract' of this nucleus.

Additional, mainly descending caudal channels are *fasciculus retroflexus* (tractus habenulo-peduncularis) from the epithalamus, and the *tractus praetectalis descendens*, which is an extension of the tractus praetecto-subpraetectalis. Ill-defined, perhaps reciprocal fiber systems connect diencephalic pretectal grisea with tectum mesencephali. The caudalward directed optic channels were already mentioned above. Descending geniculo-tegmental fibers can be inferred.

With respect to the *commissural systems*, the *commissura pallii* lies entirely within the telencephalon, but may have some sort of relationship to the medial forebrain bundle system. The *anterior commissure*, pertaining to the telencephalo-diencephalic boundary region contains various telencephalic components recently investigated by Zeier and Karten (1973). Its relationship to the basal forebrain bundle system remains uncertain.

The Avian *habenular commissure* is rather small. It seems better discernible at ontogenetic stages than in the adult condition. In those forms whose epiphysial stalk becomes a solid strand, the habenular

---

[197] E.g. tractus occipito-mesencephalicus and the 'strio-tegmental tracts'. The likewise essentially descending tractus septo-mesencephalicus can be considered a modified portion of medial forebrain bundle, as mentioned above.

commissure may merge with the *posterior commissure*. This latter, as in other Vertebrates, is predominantly related to the pretectal grisea in the diencephalo-mesencephalic boundary zone.

The *basal commissural systems* or *decussations*, in addition to the optic chiasma, include a *dorsal* and a *ventral supraoptic decussation*. Contralateral optic input to the hemispheres from dorsal lateral geniculate complex crosses in the dorsal supraoptic commissure and then joins the lateral forebrain bundle system. The ventral supraoptic decussation seems to contain connections of the torus semicircularis, directly or indirectly traceable as far as the nucleus ovoidalis. Both decussations contain tectal as well as tegmental connections and presumably also include the so-called pre- and postinfundibular decussations.

Although the Avian diencephalon, as in other Vertebrates, including Mammals, contains grisea interrelating telencephalon and caudal neuraxis (deuterencephalon, spinal cord), it should be recalled that the Mammalian telencephalon substantially differs from the Sauropsidan one. The presence of a 'true' neocortex, the peculiar differentiation of the hippocampus (cornu ammonis and fascia dentata), the configuration and structure of corpus striatum as well as of amygdaloid complex can be considered unique Mammalian features. Thus, despite the obtaining morphologic homologies, the functional significance of telencephalic and diencephalic grisea, including their details of fiber connections and synaptology, quite evidently involves considerable dissimilarities between Sauropsidan Amniota and Mammals. In attempting to interpret the Avian diencephalic neural mechanisms and their communication channels, this disparity must be kept in mind.

Thus, with regard to the spinothalamic and bulbothalamic fiber systems corresponding to the 'general sensory lemniscus', their ultimate connections with nucleus intercalatus (thalamus ventralis) and ventral 'capsule' of nucleus rotundus (i.e. with nucleus subrotundus) seem likely, but these grisea cannot possibly represent relays to a non-existent true neocortex. Likewise, the lateral (cochlear-vestibular) lemniscus system, mainly connected with torus semicircularis and apparently extended to nucleus posterior decussationis supraopticae ventralis, nucleus ovoidalis, and perhaps dorsolateral and dorsomedial grisea, cannot be relayed to an auditory neocortex comparable to that of Mammals. Although the nucleus ovoidalis might, in some respects, be analogous to the Mammalian dorsal medial geniculate nucleus, it is neither morphologically homologous nor in essential aspects functionally analogous to this latter.[198] The still very hazy generalized conclusions and inferences concerning the functional significance of the Avian diencephalon are concisely reviewed in PEARSON's treatise (1972) and, so far, subsequent investigations have not added any relevant new data for an improved interpretation of these relationships.

9. Mammals (Including Man)

The configurational aspect of the Mammalian diencephalic grisea within the overall scheme of the Vertebrate diencephalic morphologic pattern, as characterized by the four longitudinal zones epithalamus, thalamus dorsalis, thalamus ventralis, and hypothalamus was, in principle, clearly established by MIURA (1933) who investigated, at my suggestion, and under my direction, key stages of ontogenetic development in the Rabbit.[199] With regard to the zonal pattern, our

---

[198] Tactile and auditory input to the epibasal (so-called epistriate) grisea of the telencephalon has been demonstrated by various authors using the technique of 'evoked potentials', but this input does not seem to be mediated by nucleus rotundus, nor, *qua* tactile input, by any dorsal thalamic griseum. Because Avian 'striatal' tactile and auditory systems evidently differ from Mammalian neocortical primary projection areas, ERULKAR (1955) concluded that 'the existence of a dorsal thalamus in birds is in doubt'. Seen from a well-substantiated morphologic viewpoint, said statement is obviously incongruous if not outright ludicrous, and illustrates the widespread lack of basic morphologic training displayed by numerous contemporary sophisticated neurobiologic gadgeteers. On the other hand, the quoted statement would make sense and could be justified if modified as follows: experimental data appear to indicate that the grisea of the Avian dorsal thalamus are, in relevant aspects, not analogous to those of the Mammalian one.

[199] Preliminary attempts to establish, in Amniota, the homologies of the four longitudinal diencephalic zones displayed by Anamnia, particularly Amphibians, were undertaken by the present author in earlier publications (K., 1924, 1927, 1930). Following a systematic survey of conditions in Anamnia (K., 1929), a similar survey with respect to Amniota was initiated about 1930–1931. It soon became evident that my first approximation, which had failed to consider the here relevant key stages of Amniote ontogenesis, was based on inaccurate inter- and extrapolations, whereby the ventral subdivision of the thalamus dorsalis was interpreted as part of the thalamus ventralis. As soon as this error was recognized and the requisite ontogenetic key stages for accurate interpolations were determined, the author proceeded with the thus necessitated revision of the problem with respect to Sauropsida (K., 1931, 1936, 1937) which were of particular interest to him at that time, while assigning the concomitant investigation of a representative Mammalian form to MIURA (1933; cf. also K., 1935). This series of studies necessarily also included an appraisal and investigation of the neuromery problem (cf. also chapter VI, section 3, vol. 3/II).

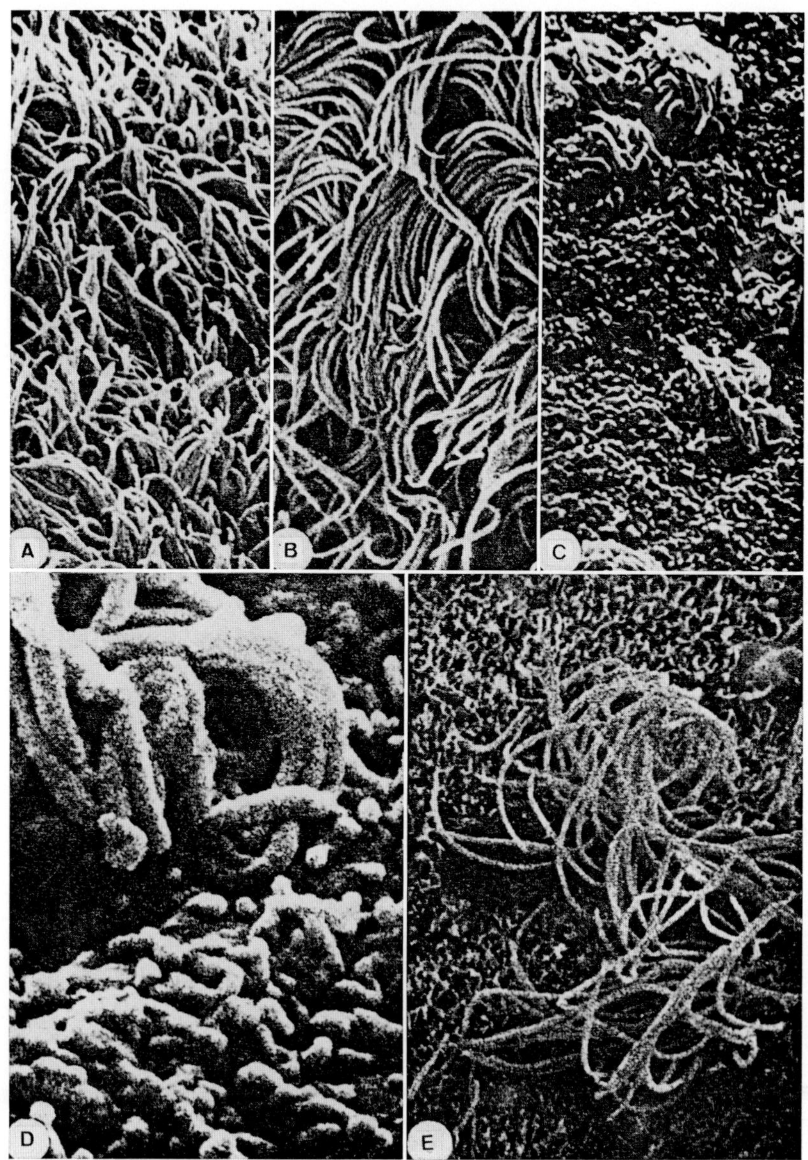

Figure 111 A–E. Scanning electron photomicrographs of the diencephalic ventricular surface in the Rabbit and Rat (from Bruni *et al.*, 1972). A Upper two thirds of the Rabbit's third ventricle (×2800). B Upper two thirds of the Rat's third ventricle (×7000). C Lower third of the Rabbit's third ventricle with non-ciliated areas (×2800). D Transitional zones in the Rabbit's diencephalic ventricular surface (×15 000). E Transitional zone in the Rat's third ventricle (×5000); (all Figures repoduced at about $^{2}/_{3}$).

joint conclusions were dealt with on pp. 444–470, section 5, chapter VI of volume 3/II. Subsequent investigations by various authors not specifically cited in said section, and, in particular, a detailed study on the development of the diencephalon of the Chinese Hamster by KEYSER (1972) which contains an extensive bibliography, can be easily interpreted in accordance with our views. Thus, these further data do not, in my opinion, require any modification of the previously elaborated concepts as summarized in the cited chapter VI of volume 3/II.

MIURA's investigation did not include two at the time still unclarified topics, namely the morphologic significance of the medial geniculate nuclei, and that of the pretectal grisea. These questions were taken up in our additional studies (K., 1935, 1954; K. and MILLER, 1942, 1949; SHINTANI-KUMAMOTO, 1959).

The Mammalian diencephalon, as a rule completely covered by the telencephalic hemispheres and to a large extent by corpus callosum and fornix, displays, in a far higher degree than that of Sauropsida, a massive connection with the telencephalon by way of the greatly enlarged *hemispheric stalk*. The free lateral surfaces of the diencephalon become thereby considerably reduced.[200]

The ventricular surface displays, in numerous Mammals, a more or less extensive midline fusion in the thalamic region, whereby a *massa intermedia* or *commissura mollis* is formed. Various so-called nuclei

[200] The ontogenetic evolution of the hemispheric stalk was dealt with and depicted (Fig. 241) on pp. 465–470, chapter VI, section 5 of volume 3/II.

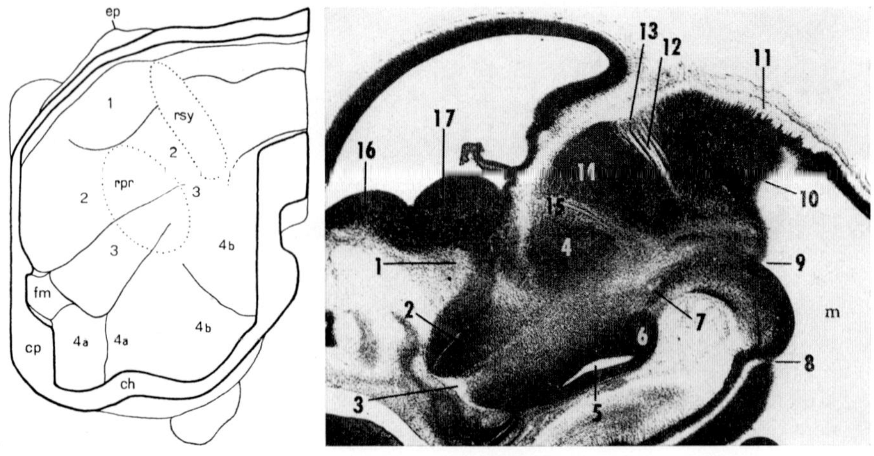

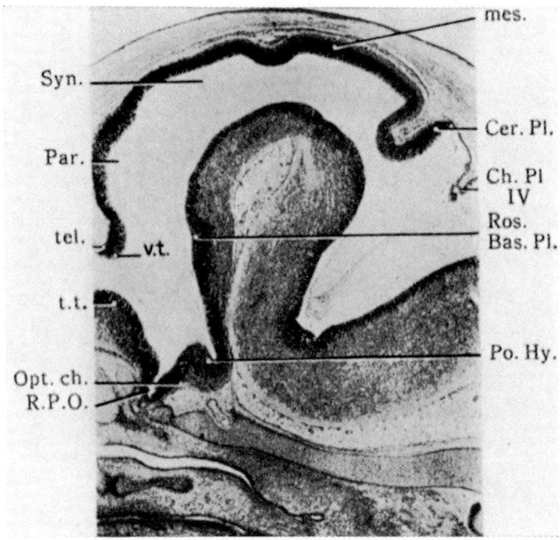

Figure 112C. Approximately midsagittal section through the diencephalon and adjoining neighborhoods of a 13.5 mm Phascolarctos phascolomys (Marsupial) embryo, showing parencephalic and synencephalic neuromeres, and rostral end of deuterencephalic basal plate (from WARNER, 1969). Cer.Pl.: cerebellar plate; Ch.Pl.IV: choroid plexus of fourth ventricle; mes.: mesencephalon; Opt.ch.: optic chiasma; Par.: parencephalon; Po.Hy.: posterior part of hypothalamus; Ros.Bas.Pl.: rostral extremity of basal plate; R.P.O.: preoptic recess; Syn.: synencephalon; tel.: dorsal telencephalo-diencephalic boundary; t.t.: torus transversus (commissural plate); v.t.: velum transversum.

Figure 112A. Outline drawing of diencephalic ventricle and its relief features in a 15 mm Rabbit embryo, displaying the configuration of the fundamental diencephalic longitudinal zones as obtaining in Mammals (from MIURA, 1933). 1: Epithalamus; 2: thalamus dorsalis; 3: thalamus ventralis; 4: hypothalamus (4a: preoptic portion on both sides of sulcus intraencephalicus anterior; 4b: postoptic or 'infundibulomammillary' portion, dorsally and ventrally to sulcus lat. hypothalami posterioris); ch: chiasmatic ridge; cp: commissural plate of lamina terminalis; ep: epiphysial anlage; fm: ventriculus impar s. communis telencephali, including *foramen interventriculare Monroi;* rpr: recessus parencephalicus; rsy: recessus synencephalicus. The fundamental ventricular sulci are not labelled but can easily be identified.

Figure 112B. Diencephalic longitudinal zonal system as seen in a slightly oblique longitudinal section through the brain of a 15 mm Rabbit embryo (slightly modified after MIURA, 1933). 1: commissural plate; 2: preoptic recess; 3: anlage of optic chiasma; 4: thalamus ventralis; 5: posterior ('infundibular') recess of hypothalamus; 6: anlage of mammillary grisea (2, 5, and 6 represent the hypothalamus); 7: rostral end of deuterencephalic basal plate; 8: fovea isthmi; 9: sulcus limitans; 10: sulcus lateralis mesencephali (internus); 11: commissura posterior; 12: fasciculus retroflexus; 13: epithalamus; 14: thalamus dorsalis; 15: zona limitans intrathalamica (precursor of lamina medullaris externa); 16: lateral ganglionic hill (D_1 of telencephalon); 17: medial ganglionic hill (B_{1+2}).

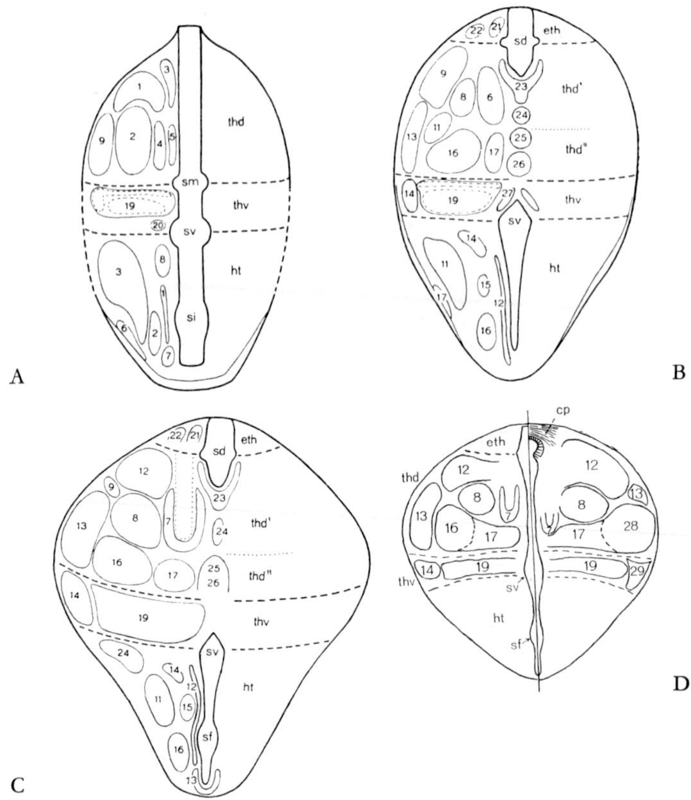

Figure 113 A–D. Diagrammatic cross-sections depicting the diencephalic longitudinal zones in the Rabbit's diencephalon and the morphologic relationships of their main griseal derivatives. This configurational pattern can be assumed to illustrate the overall Mammalian bauplan of diencephalic nuclear differentiation (A–C from MIURA, 1933, D from K., 1935). cp: comm. posterior; eth: epithalamus; ht: hypothalamus; sd: s. dienc. dorsalis; sf: s. lateralis hypothalami posterioris; si: s. intraencephalicus anterior; sm: s. dienc. medius; sv: s. dienc. ventralis. *(Numbers in epithalamus and thalamus)* 1: n. anterior dorsalis; 2: n. anterior ventralis; 3: n. parataenialis; 4: n. anterior medialis; 5: n. paramedianus; 6: n. medialis (sive dorsomedialis); 7: n. parafascicularis; 8, 9: n. lateralis dorsalis and n. lat. posterior complex (including MIURA's 'n. mediolateralis'); 11: n. ventralis anterior and ventrolateralis complex; 12: n. praetectalis and posterior complex; 13: n. gen. lat. dorsalis; 14: n. gen. lat. ventralis; 16: n. ventralis posterolateralis complex; 17: n. posteromedialis complex; 19: n. reticularis and zona incerta complex; 20: n. taeniae; 21, 22: medial and lateral habenular nuclei; 23–27: nn. reunientes group (MIURA's nn. reunientes 1–5); 28: n. gen. med. dorsalis; 29: n. gen. med. ventralis; *(Numbers in hypothalamus)* 1: paraventricular preoptic grisea; 2: n. praeopticus medialis; 3: n. praeopticus lateralis; 6: n. supraopticus; 7: n. suprachiasmaticus; 8: n. hypothalamicus anterior complex (includes here some paraventricular neighborhoods); 11: n. lateralis hypothalami; 12: intermediate and posterior paraventricular grisea; 13: n. arcuatus; 14: area hypothalamica dorsalis;

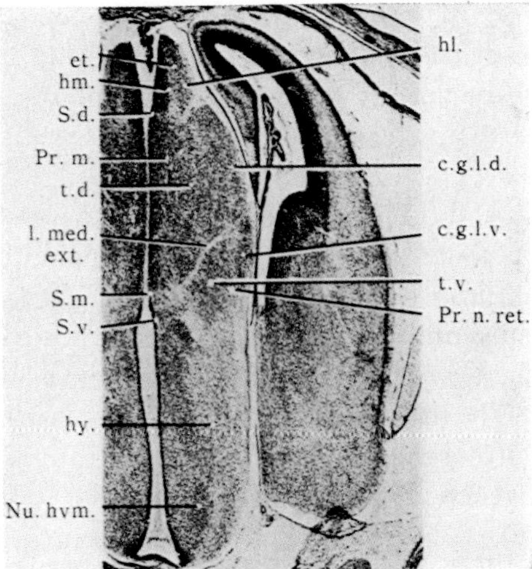

Figure 113E. Cross-sections through the diencephalon of a 30 mm Trichosurus vulpecula (Marsupial) embryo caudal to hemispheric stalk, showing the developmental arrangement of the longitudinal zones (from WARNER, 1969). c.g.l.d.: anlage of dorsal lateral geniculate nucleus; c.g.l.v.: anlage of ventral lateral geniculate nucleus; et.: epithalamus; hl., hm.: anlage of lateral respectively medial habenular nucleus; hy.: hypothalamus; l.med.ext.: anlage of lamina medullaris externa; Nu.hvm.: anlage of nucleus ventromedialis hypothalami; Pr.m.: anlage of nucleus medialis thalami; Pr.n.ret.: anlage of nucleus reticularis thalami (ventralis); S.d., S.m., S.v.: sulcus diencephalicus dorsalis, medius, ventralis; t.d.: thalamus dorsalis; tv..: thalamus ventralis.

reunientes, as pointed out further below, are closely related to, or even located within, this midline concrescence. Occasional fusions may also occur at the hypothalamic ventricular level.

With respect to the ependymal lining, some recent studies undertaken by means of the scanning and of the transmission electron microscope have disclosed a number of details (Fig. 111) elaborated in the publications by BRUNI *et al.* (1972), SCHECHTER and WEINER (1972), SCOTT *et al.* (1973), BLEIER (1975), and others as cited by the quoted authors. Changes in these aspects of the diencephalic ventricular wall of the Rat related to the estrous cycle have been claimed (BRAWER *et al.*,

---

15: n. dorsomedialis; 16: n. ventromedialis; 17: complex of nn. tuberis and tuberomammillares; 24: n. hypothalamicus *(corpus subthalamicum Luysi)*. The nomenclature in the legend corresponds to that adopted in the present treatise, and the numerical notations, based on the initial studies by MIURA and the author, differ somewhat from those of the enumeration in the text. thd: thal. dors.; thv: thal. ventralis.

1974). In addition, the scanning electron microscope has revealed so-called supraependymal cells which are presumably macrophages or histiocytes, perhaps derived from the choroid plexus, and wandering along the ventricular ependymal surface (cf. e.g. COATES, 1975, in Macaque Monkeys).

Figures 112A and B illustrate the configuration of the fundamental Vertebrate diencephalic longitudinal zones as manifested at a key stage of Mammalian ontogenetic development. The diencephalic 'neuromeres' parencephalon and synencephalon, whose gradual fading overlaps with the differentiation of the longitudinal zones, are shown in Figure 112C. The major diencephalic grisea or 'nuclei' at the 'adult' stage, as derived from the longitudinal zones and already delimitable in advanced ontogenetic stages of differentiation, are indicated in Figures 113A–D. This arrangement, as first recorded in the Rabbit, can be considered to exemplify the fundamental diencephalic griseal pattern obtaining in the entire Mammalian series, although with several qualifications necessitated by the numerous secondary variations *qua* degree and details of development related to the taxonomic diversities.[201]

The *epithalamus* with its habenular 'ganglion' is relatively well developed in macrosmatic forms, but smaller, although not excessively reduced in the microsmatic ones (e.g. Primates including Man). It is reduced, but definitely present in those considered anosmatic (e.g. Dolphins). A medial and a lateral habenular nucleus is usually present, the former commonly consisting of smaller, densely arranged nerve cells, and the latter containing larger, mostly stellate elements (CAJAL, 1911; BECCARI, 1943).

---

[201] Concerning the previous investigations, as e.g. summarized by KAPPERS *et al.* (1936) and also dealt with in a monograph on the human diencephalon (K., 1954), the following comments by BECCARI (1943) seem especially relevant. '*Tale analisi, quasi sempre monografica sopra una specie, mentre ha portato indubbiamente un contributo cospicuo alla migliore conoscenza della struttura e dei rapporti del talamo, ha d'altro canto, con il frazionamento dei nuclei e con l'uso di una nomenclatura complicata, reso sempre più difficile un orientamento d'insieme, che da qualche Autore è stato peraltro tentato, per es. dal* LE GROS CLARK *('32). Il vasto materiale anteriore al 1936 si trova ora in gran parte raccolto nel grande trattato di* ARIËNS KAPPERS, HUBER *e* CROSBY *('36); ma, come gli stessi Autori avvertono, ci dobbiamo anche oggi accontentare di una enumerazione di dati, classificati con criterio topografico e in parte fisiologico, rinunziando al desiderio di stabilire eventuali omologie con i nuclei del talamo degli altri vertebrati. Tuttavia sequendo la enumerazione si scorge un piano organico che probabilmente è fondamentale in tutti i Mammiferi, come risulta dalle considerazioni, fra gli altri, del* LE GROS CLARK, *precedentemente ricordato, e del* KUHLENBECK *('29, '30 e '37).*'

From a morphological viewpoint, and in contradistinction to the functional classification of Mammalian dorsal thalamic cell groups pointed out on p. 9 of the introductory section 1 A, these grisea can be conveniently subdivided into the following seven groups. (A) *Anterior (rostral) group*, (B) *medial group*, (C) *nuclei of the midline*, (D) *intralaminar group*, (E) *ventrolateral group*, (F) *posterior group*, and (G) *pretectal group*.

The *anterior group* (A) includes (1) nucleus anterior dorsalis (anterodorsalis) and (2) nucleus anterior ventralis (anteroventralis) as main components. Somewhat more medial grisea, more or less closely related to the ventricular wall, and variously delimited as well as named by diverse authors, are (3) nucleus parataenialis, (4) nucleus anterior medialis (anteromedialis or medialis anterior), and (5) nucleus paramedianus. A transition to the ventrolateral group is represented by nucleus lateralis dorsalis, which extends caudalward and can be interpreted as derived from the dorsocaudal portion of the primordium from which nucleus anterior ventralis has arisen.

In Monotremes, and in some Insectivores, the anterior group may give the impression of a single griseal complex, although the neighborhoods of anterodorsal, anteromedial, and anteroventral nuclei seem roughly identifiable. In most other Mammals these grisea are rather clearly recognizable as more or less distinctive nuclei. Some authors include the paratenial nucleus within the medial group.

The *medial group* (B) comprises (6) nucleus medialis (nucleus medialis dorsalis or dorsomedialis), and (7) nucleus parafascicularis with its further derivatives.[202]

The *midline griseal group* (C) variously differentiated and interpreted, respectively designated by the diverse investigators, was described by Miura in the Rabbit, who distinguished, in dorsobasal sequence, (8–11) the nuclei reunientes 1–4.

The *intralaminar group* (D), rather indistinctly differentiated in 'lower' Mammals, but more conspicuous in 'higher' ones such as Mammals and Man, is presumably represented by (12) nucleus laminaris and (13) nucleus magnocellularis of the Rabbit.

---

[202] The nucleus of the centrum medianum *(centre médian of Luys)* in 'higher' as well as in some 'lower' Mammals is presumably a derivative of the parafascicular nucleus (K., 1954; Niimi *et al.*, 1960). Miura's nucleus mediolateralis in the Rabbit may represent a subdivision of the nucleus lateralis dorsalis or of the posterolateralis complex.

The *ventrolateral group* (E) includes (14) the nucleus lateralis dorsalis,[202a] mentioned above as an extension of the anterior group, moreover (15) nucleus ventralis anterior, (16) nucleus ventralis lateralis, (17) ventralis posterolateralis, and (18) nucleus ventralis posteromedialis. Several secondary variations or subdivisions such as nucleus submedius and nucleus subparafascicularis doubtless obtain.

Again, in the Prosimian Primate Loris, medullary laminae separate the so-called ventrobasal complex into subgroups apparently corresponding to somatotopic neighborhoods for head, forelimb and hindlimb representation (KRISHNAMURTI *et al.*, 1972).

The *posterior group* (F) includes a caudal extension of the just mentioned nucleus lateralis dorsalis, which could be delimited as (19) nucleus lateralis posterior with further subdivisions, one of which may correspond to the pulvinar of 'higher' Mammals. The other important grisea of this group are the dorsal thalamic components of lateral and medial corpora geniculata, namely (20) nucleus geniculatus lateralis dorsalis, and (21) nucleus geniculatus medialis dorsalis with the adjacent nucleus suprageniculatus (cf. also PAPEZ, 1936).

The *dorsal thalamic pretectal group* (G) includes (22) area praetectalis, and the nucleus praetectalis complex, consisting of (23) nucleus praetectalis (principalis) and (24) nucleus posterior, moreover of (25) several nuclei of the posterior commissure.[203] In comparison with Sauropsida, it could perhaps be said that the Avian dorsal thalamic pretectal grisea display, in some respects, a higher or at least more distinctive differentiation into 'nuclei' than the Mammalian ones. This is possibly due to the substantial development of the Avian optic tectum with which the pretectal grisea are closely correlated.

Thus, even on the basis of a restrained or conservative parcellation, at least 25 distinct grisea can be distinguished in the dorsal thalamus of a 'lower' Mammal. With respect to ontogenetic development, comments on stratification into external and internal cell layers were included in section 5, chapter VI of volume 3/II. Concerning the ontogenetic development of medial group (B) and of ventrolateral group (E) it should be mentioned that the former derives from a

[202a] A rostral portion of nucleus lateralis dorsalis might also be distinguished as nucleus lateralis anterior.

[203] The dorsal thalamic nuclei of the posterior commissure as also briefly discussed in chapter XI, section 9 of volume 4, comprise (a) nucleus interstitialis principalis commissurae posterioris; (b) n. interst. magnocellularis c.p., (c) n. centralis subcommissuralis c.p., (d) n. medianus subcommissuralis c.p., and (e) n. interst. supracommissuralis c.p.

dorsal subdivision of the embryonic thalamus dorsalis, described by MIURA (1933) as '*dorsale Etage*'. The ventrolateral group, which corresponds to the so-called external or lateroventral thalamic segment of the older nomenclature, derives from MIURA's '*ventrale Etage*' (cf. Figs. 113B, C: thd', thd'').

The *thalamus ventralis* of Mammals,[203a] although displaying several more or less distinctive 'nuclei' is, *in toto*, relatively much reduced if compared with the extent of the large dorsal thalamus and with that of hypothalamus. It remains, nevertheless, significantly larger than the epithalamus. The following grisea can be delimited. (1) The *nucleus reticularis thalami (ventralis)* is rostrally located, its medial portion lying ventrally to the anterior dorsal thalamic group. Its lateral portion extends, in a medialward concave semicircle, through the hemispheric stalk as far as the telencephalic sulcus terminalis. This lateral extension corresponds to the pars superior massae cellularis reunientis as obtaining in Anamnia. Several ill-defined dorsomedial and ventrolateral subdivisions are noticeable during ontogeny, but do not seem to warrant additional parcellation. A more compact medial cell cluster, prominent in at least some Mammals, can, however, be delimited as (2) *nucleus taeniae (or nucleus filiformis)*. More caudalward, the cell band of thalamus ventralis becomes more compact and represents (3) *the zona incerta*. Nondescript cell groups within this zone are the '*nucleus*' *(or* '*nuclei*'*) campi Foreli*. Caudal lateral portions of the thalamus ventralis form (4) the *nucleus geniculatus lateralis ventralis* and (5) the *nucleus geniculatus medialis ventralis*, which are components of the lateral respectively of the medial corpus geniculatum complex. Finally, (6) a perhaps inconstant medial nucleus reuniens ventralis (nucleus reuniens 5 of MIURA) pertains, as ventral thalamic component, to the group of diencephalic nuclei reunientes.

The Mammalian *hypothalamus*, as already pointed out in the introductory section 1A (p. 12) can be subdivided into (A) dorsal and entopeduncular group, (B) anterior (essentially preoptic) group, (C) middle group, and (D) posterior group.

The *dorsal and entopeduncular complex*, although not generally included into the conventional concept of hypothalamus as formulated by

---

[203a] The extension of thalamus ventralis and some of its developmental stages are shown in Figures 225–240, section 5 of volume 3/II. It should here be added that, by an oversight, the label vt indicating ventral thalamic neighborhoods in Figures 231 (232) was erroneously referred to as rt in the legend to Figure 231 on p. 455.

many authors, is nevertheless a derivative of the primordial hypothalamic zone and consists of grisea predominantly associated with the so-called extrapyramidal motor system. Groups B to D, on the other hand, are especially related to the vegetative nervous system and in part to the hypophysial system as dealt with above in section 1C. The most rostral griseum of the *entopeduncular complex* is (1) the globus pallidus, which, within the hemispheric stalk, becomes closely adjacent to the corpus striatum (respectively putamen), and is commonly described as a component of the telencephalic basal ganglia. (2) The nucleus entopeduncularis, together with (3) nucleus ansae lenticularis and (4) interstitial nucleus of the inferior thalamic peduncle, can be regarded as morphologically minor subdivisions of the globus pallidus complex. The more caudally located nucleus subthalamicus (5) or *corpus subthalamicum of Luys* is ventrally adjacent to zona incerta, being separated from this latter by *Forel's fasciculus lenticularis* or field H_2.

As regards the *hypothalamus sensu strictiori* with its three major subdivisions, MIURA's (1933) investigation only considered the overall relationships of main hypothalamic grisea to the diencephalic bauplan and merely established in this respect some fundamental configurational features. More detailed studies based upon these results were included in our later publications (K. and HAYMAKER, 1949; K., 1954; CHRIST, 1969), and substantial new data, together with a critical review of the literature, were provided by DIEPEN's (1962) contribution to *v. Möllendorff's Handbuch*. The comprehensive treatise by HAYMAKER *et al.* (1969), which, *inter alia*, contains CHRIST's chapter, deals with the anatomical, functional, and clinical aspects of the hypothalamus as elaborated by 21 contributors. It should be added that, in accordance with an emphasis on mediolateral stratification, CROSBY and SHOWERS (1969, cf. also NAUTA and HAYMAKER, 1969) have stressed a subdivision into periventricular, medial (intermediate), and lateral cell populations. This general principle of subdivision was discussed in section 2, chapter VI, pp. 285–286 of volume 3/II. Among recent contributions to hypothalamic grisea in Primates (Macaca) are those by SMIALOWSKI (1972, 1973).

Within the *preoptic group*[204] of grisea, the following 'nuclei' can be recognized. (1) Area praeoptica paraventricularis (or periventricularis) with certain subdivisions, of which (2) the *neurosecretory nucleus paraven-*

---

[204] The supraoptic crest, dealt with elsewhere (cf. section 1C, p. 12 of this chapter, has been omitted from the here given enumeration.

tricularis magnocellularis is of particular importance. The embryonic pars inferior massae cellularis reunientis participates in the formation of the globus pallidus, but a preoptic remnant of this embryonic cell population may persist as (3) 'nucleus' paraventricularis superior, representing a dorsalateral extension of the periventricular preoptic area. Further differentiations of the periventricular cell aggregates are (4) nucleus praeopticus medialis and (5) nucleus suprachiasmaticus, moreover an unpaired (6) nucleus praeopticus medianus dorsally to a rostral diverticle of preoptic recess ventrally to commissura anterior. The more lateral cell groups of the preoptic region include (7) nucleus praeopticus lateralis, a more caudomedial subdivision of which is (8) the nucleus hypothalamicus anterior, and (9) the *neurosecretory* magnocellular nucleus supraopticus. This latter shows various separable subdivisions in the different Mammalian forms. CHRIST (1969) moreover describes (10) a nucleus supraopticus diffusus which is not part of the supraoptic nucleus proper, but rather a small-celled interstitial nucleus within fibers of the dorsal supraoptic commissure.

The *middle hypothalamic group*, in addition to a caudal extension of the preoptic paraventricular cell plate (1), comprises the following grisea. The nucleus of area hypothalamica dorsalis (11) is a condensation and slight lateral expansion of the just mentioned paraventricular cell aggregate. The nucleus arcuatus (12) *sive* infundibularis, also designated as nucleus paraventricularis posterior inferior by MIURA (1933), surrounds the basal portion of the third ventricle near the entrance to the infundibular recess, and extends toward the median eminence. Somewhat more laterally to the periventricular cell population, two rounded or oval griseal aggregates form nucleus hypothalamicus dorsomedialis (13) and the ventrally adjacent nucleus hypothalamicus ventromedialis (14). This latter is commonly the most conspicuous 'nucleus' of the so-called tuberal region, usually separated by a rather cell-poor zone from the dorsomedial griseum. Still more laterally, the nucleus of the area hypothalamica lateralis (15) represents the caudal continuation of the lateral preoptic nucleus (7). Near the surface of the 'tuberal region' the nuclei tuberis laterales (16) are elongated rather small cell groups particularly conspicuous in Primates and Man, but also recorded in various 'lower' Mammals. The nucleus tuberomammillaris (17) consists of somewhat more diffusely distributed large cells basolaterally to lateral hypothalamic area and usually dorsomedially or medially to the nn. tuberis laterales, and rather close to the mammillary complex of the posterior hypothalamus.

The *posterior hypothalamic group* includes the following grisea. An ill-defined nucleus of the area hypothalamica posterior (18) is located caudally to ventromedial and dorsomedial nuclei, and is medially continuous with the nucleus (or area) periventricularis posterior (19), which is a caudal extension of area hypothalamica dorsalis (11) and general hypothalamic peri- or paraventricular cell plate (1). The nucleus perifornicalis (20) consists of an aggregate of cells surrounding the caudal portion of the fornix. Nucleus praemammillaris (21) and nucleus supramammillaris (22) are rather indistinct cell aggregations close to the mammillary complex but apparently not forming part of this latter. The mammillary complex consists of nucleus mammillaris medialis (23) which is commonly the largest mammillary griseum and may further be subdivided into pars medialis and pars lateralis with respect to cell size of its subgroups. The lateral mammillary nucleus or nucleus intercalatus (24) displays variations *qua* distinct differentiation in the diverse taxonomic forms.[205] The small-celled nucleus mammillaris cinereus *autorum* (25) may be a variable part of nucleus mammillaris medialis.

Thus, discounting the five grisea of the dorsal and entopeduncular hypothalamic group, at least about 25 different nuclei may, even in a conservative parcellation, be recognized in the Mammalian hypothalamus *sensu strictiori*. Generally speaking, the hypothalamus in the entire Mammalian series displays, although with some secondary variations *qua* distinctness and relative degree of differentiation, an identical nuclear pattern including all of the here enumerated grisea.[206] SPATZ (1951, and other publications) distinguishes two essential subdivisions of the Mammalian hypothalamus: the richly medullated and the poorly medullated portions (*markreicher Hypothalamus und markarmer Hypothalamus*). The former comprises our dorsal and entopeduncular group, and, in addition, the mammillary complex. The poorly medullated hy-

---

[205] The discrepancies in the terminologies adopted by the various authors have been carefully considered by DIEPEN (1962) who clarified the subdivisions of the mammillary complex and its paramammillary grisea (cf. e.g. the tabulation on p.128 loc.cit.). As regards differences in cell size and further details of cytoarchitecture, the interested reader is referred to the descriptions by DIEPEN (1962) and by CHRIST (1969).

[206] This evaluation disagrees with the interpretation by GRÜNTHAL (1930) who concluded that the human hypothalamus is less differentiated than that of 'lower' Mammals. The cited author distinguished 32 nuclei in the hypothalamus of the Mouse, and only 9 in the Human hypothalamus (cf. the comments in K., 1954, pp.159f, which also refer to the parcellation by BROCKHAUS 1942).

pothalamus comprises our anterior and middle group as well as the posterior group exclusive of mammillary complex. SPATZ stresses his subdivision with respect to hypophysial connections and points out that only in the poorly medullated hypothalamus can be found centers directly connected with the hypophysial complex. This subdivision is doubtless to some extent useful and valid. Nevertheless, since in this concept two morphologically and presumably also functionally heterogeneous groups such as entopeduncular grisea and mammillary body are lumped together as '*markreicher Hypothalamus*', its validity does not appear altogether convincing and requires some qualifications.

On the basis of the ontogenetic investigation undertaken by my collaborators and myself (MIURA, 1933; K., 1931, 1936, 1937) as well as by those of other authors (LE GROS CLARK, 1930, 1938; GILBERT, 1935) a first systematic attempt at establishing overall morphologic homologies of the diencephalic grisea in the three Amniote classes was undertaken in 1937 as illustrated by Figure 114. This first approximation, which, so far, has not been substantially modified by subsequently available data, can still be considered essentially valid. Among these later investigations dealing with such ontogenetic aspects, those by PAPEZ (1940), DEKABAN (1954), STRÖER (1956), MIYAKE (1958), NIIMI *et al.* (1961), KUSAKA (1962/63), WARNER (1969, 1970), and KEYSER (1972) could be mentioned.

In addition, the '*Arbeiten aus der Zweiten Abteilung des Anatomischen Instituts der Universität zu Tokushima*' of which 11 volumes appeared between 1954 and 1966, contain numerous papers (in Japanese, many with German or English summaries) concerning the Vertebrate diencephalon and providing illustrations of documentary value.

It should here again be emphasized that morphologic, i.e. topologic homology of grisea respectively nuclei does by no means necessarily imply identical hodology or synaptology nor identical functions. In fact, it seems rather evident that, in Sauropsida, which lack a neocortex, the thalamic nuclei are related to a neural mechanism substantially differing from that obtaining in Mammals.

As regards the *epithalamus*, there is, of course, no doubt that the homologies of medial and lateral habenular nuclei (1, 2) in all three classes of Amniota are well established and generally recognized.

With respect to the *thalamus dorsalis*, the anterior griseal group in Reptiles, namely nucleus dorsomedialis anterior (3) dorsolateralis anterior (4), on the basis of genetic and definitive topologic relationship, can be regarded as defective kathomologous to the anterior dorso-

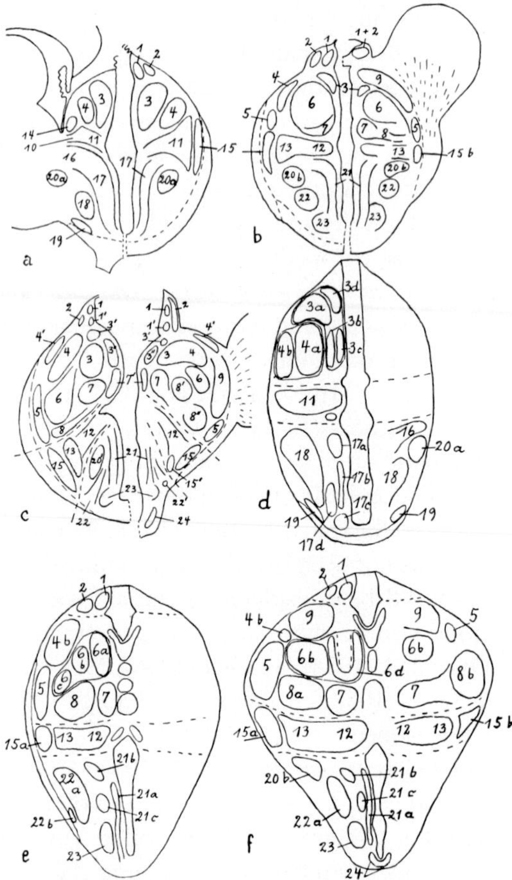

Figure 114. Diagrams illustrating a first attempt at establishing overall morphologic *(topologic)* homologies of Amniote diencephalic grisea (from K., 1937). a, b: Reptilian configuration; c: Avian configuration; d–f: Mammalian configuration. The numerical notation, which does not correspond to that of Figure 113, is explained in the text.

medial and anterior dorsolateral nuclear groups in Birds and Mammals.[207] Thus, by a one-many transformation, Reptilian griseum (3) may be mapped upon Avian nucleus dorsomedialis anterior (3), nuclei subhabenulares (3′) and nucleus paraventricularis dorsalis (3″). *Qua* Mammals, this mapping would involve nucleus anterior dorsalis (3a), nucleus anterior medialis (3b), nucleus paramedianus (3c) and nucleus parataenialis (3d).

[207] Conversely, these griseal groups in Birds and Mammals display augmentative homology with regard to the two Reptilian nuclei, upon which they may be mapped by many-one transformations (cf. vol. 1, chapter III, section 3, pp. 199–200).

Similarly, nucleus dorsolateralis anterior of Reptiles (4) may be mapped upon Avian nucleus dorsolateralis anterior (4), nucleus dorsolateralis superficialis (4′) and nucleus of tractus septomesencephalicus (likewise subsumed under 4′ in Figure 114). For comparison with Mammals, this one-many mapping would include nucleus anterior ventralis (4a) and nucleus lateralis dorsalis complex (4b). Again, the dorsal lateral geniculate griseum of Reptiles can easily be mapped upon the dorsal lateral geniculate complex of Birds and Mammals (5).

A comparison of key ontogenetic stages leaves no doubt that nucleus rotundus of Reptiles (6) and of Birds (6, including nucleus triangularis) are homologous grisea. The modified (topographic in contradistinction of topologic) ventrolateral position of the Avian nucleus rotundus can be recognized as a secondary shifting and can be followed step by step through successive ontogenetic stages. It is more difficult to recognize the Mammalian homologue. On the basis of adult topographic relationships, INGVAR (1924) assumed that the Mammalian ventrolateral complex represented the Avian nucleus rotundus, while PAPEZ (1935) assumed an homology of Reptilian nucleus rotundus and Mammalian nucleus medialis (dorsomedialis). Our own ontogenetic investigations support this interpretation by PAPEZ. The Mammalian grisea augmentatively kathomologous to Reptilian nucleus rotundus (6) appear to be nucleus medialis *sive* dorsomedialis (6a), nucleus mediolateralis (6b, which joins the nucleus lateralis posterior complex), moreover nucleus ventralis lateralis complex (6c, including n. ventralis anterior), and the nucleus parafascicularis complex (6d), of which, in turn, the nucleus of the centrum medianum seems to be a derivative.

In Reptiles, the griseal 'capsule' of nucleus rotundus displays a condensation (7) variously described as nucleus medialis (FREDERIKSE, 1931; K., 1931), nucleus reuniens anterior and posterior (PAPEZ, 1935), and nucleus diagonalis[208] (HUBER and CROSBY, 1929). A thin cell plate extending lateralward ventrally to nucleus rotundus is the nucleus infrarotundus (8). The corresponding Avian grisea are nucleus ovoidalis (7) with nucleus paramedianus internus (7′) and the griseal group of nucleus subrotundus (8), nucleus posterointermedius (8′) and postero-

---

[208] In 1931, the author had erroneously interpreted the nucleus diagonalis of the Alligator as the ventral thalamic nucleus ventralis internus. Subsequently (K., 1937) this error was corrected, and the griseum in question was identified as a subdivision of the Reptilian nucleus medialis (7, not to be confused with the Mammalian nucleus medialis sive dorsomedialis, i.e. 6a of Figure 114e).

ventralis (8″). The topologically augmentative homologous Mammalian grisea seem to be nucleus ventralis posteromedialis complex (7) and nucleus ventralis posterolateralis complex (8). More caudally, the dorsal medial geniculate nucleus (8b) can be interpreted as a further derivative,[209] separated from the remnant (8a).

Still more caudally, the Reptilian nucleus posterocentralis s. posterodorsalis (9) pertains to the dorsal thalamic pretectal complex (9) of Birds (cf. Fig. 114 c) and of Mammals (9) as already repeatedly dealt with above. It seems likely that, in particular, the Reptilian nucleus posterocentralis corresponds to the Avian nucleus principalis (and nucleus laminaris) praecommissuralis; these Sauropsidan grisea, in turn, being presumably homologous to the Mammalian nucleus posterior.

Less detailed, but somewhat different interpretations of the griseal homologies in the Mammalian dorsal thalamus with respect to the

[209] The avian nucleus posteroventralis, although homologous to Mammalian dorsal medial geniculate nucleus, is presumably not functionally analogous. To some extent, the Avian nucleus ovoidalis seems to represent such analogous griseum (cf. the preceding section 8, p. 291).

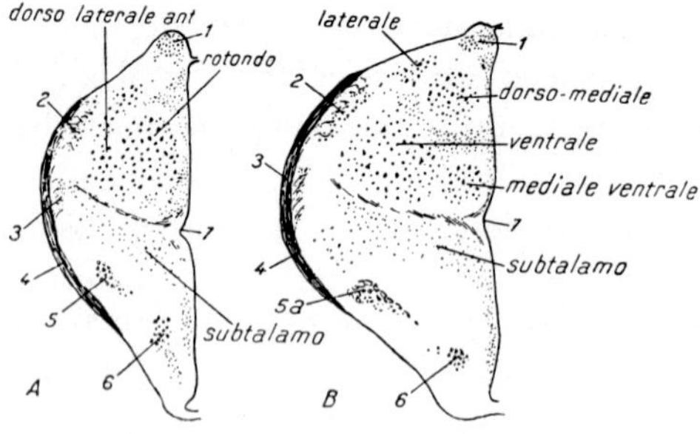

Figure 115. Transverse semidiagrammatic cross-sections through the diencephalon of a Reptile (A) and of a lower Mammal (B) illustrating Le Gros Clark's interpretation of homologies (from Beccari, 1943). 1: habenular griseum; 2, 3: dorsal and ventral lateral geniculate nucleus; 4: optic tract; 5: lateral forebrain bundle; 5a: 'cerebral peduncle'; 6: medial forebrain bundle; 7: '*sulco diencephalico medio o intermedio*' (it is rather the sulcus diencephalicus ventralis). In the diagram, as adapted by Beccari, the '*subtalamo*' lumps together thalamus ventralis and dorsal hypothalamus corresponding to Mammalian subthalamic nucleus, perhaps due to the somewhat dorsalward displaced sulcus 7 interpreted as '*medio*'.

Sauropsidan one were suggested by LE GROS CLARK (1932a) and by KAPPERS (1947). According to the former author, the ventrolateral group of Mammals would correspond to a portion of Sauropsidan nucleus dorsolateralis anterior and nucleus rotundus (cf. Fig. 115). KAPPERS, on the other hand, assumes that the Sauropsidan nucleus rotundus might either be represented by the nucleus parataenialis or, more likely by the nucleus submedius of Mammals. Both authors, like myself, thus emphasize the morphologic significance of the Sauropsidan nucleus rotundus as a key griseum for any attempt to outline details of Mammalian dorsal thalamic homologies in a comparison with the two other Amniote classes. An inspection of Figure 115 will disclose that, if the concept of a 'cell capsule' surrounding the nucleus rotundus is adopted, as pointed out above, both the interpretations of LE GROS CLARK (1932a) and of KAPPERS (1947), despite the obvious differences, are not fundamentally divergent from my own, and could be roughly adapted or combined with this latter.

ZEMAN and INNES (1963) have concisely discussed both the difficulties and the general principles obtaining in any attempt at subdividing and homologizing the Mammalian dorsal thalamic grisea. Thus, what one observer considers to represent a delimitable 'nucleus' may be interpreted by another as a mere subdivision of a larger 'nucleus' or even as a portion of quite another one.

With respect to the principles of subdivision, the neocortical fiber connections might be emphasized. This, however, leads to considerable quandaries, since a true neocortex is not present in Sauropsidans, whose complex basal ganglia doubtless have assumed functions at least in part comparable to those of Mammalian neocortex. There can be little doubt that in the course of the entirely unknown detailed phylogenetic changes from Reptilian or even Amphibian forebrain structure to the Mammalian type, substantial modifications of synaptology and hodology have occurred. Such unknown changes preclude a convincing rationale for the delimitation of the anatomically distinguishable grisea.

As regards such parcellation, the cytological characteristic of 'uniform cell population' does not represent a valid criterion, since most 'nuclei' include a diversity of cell types. At most, a roughly identical histologic pattern, 'grain', or structure provided by the various elements might be vaguely recognizable within a given 'nucleus'. This also involves aspects of cell population 'density', 'packing', or 'crowding'. In addition, there is not infrequently a more or less conspicuous delineation of such griseal neighborhoods by zones of fibers either devoid of nerve cells or with only few such elements (e.g. the various thalamic 'medullary laminae'). By a cautious application of these diverse criteria, and despite unavoidable discrepancies in interpretation by observers pertaining to different 'schools', some sort of overall agreement may nevertheless be reached.

Reverting to the *nucleus medialis* (or dorsomedialis) presumably derived and functionally modified from the Sauropsidan *nucleus rotundus*, said well delimitable Mammalian griseum, consisting of an overlapping small-celled and larger celled subdivision, generally seems to increase in relative size from 'lower' Primates to Anthropoids and Man as noted by BECCARI (1943). This can (but not must) be interpreted as indicating that the functionally modified nucleus rotundus became of at first lesser importance in 'lower' Mammals, and finally assumed a substantial significance in Man. Such view would be in agreement with its assumed function as discussed further below. On the other hand, this nucleus seems to be rather large in the Mole (cf. Fig. 121), which, however, displays some peculiarities concerning the configuration of the relevant laminae medullares.

With regard to the *thalamus ventralis*, the Reptilian massa cellularis reuniens, pars superior (10), and the area triangularis (11) correspond to the Avian reticular complex as well as to Mammalian nucleus reticularis (11) and nucleus taeniae. More caudally the Reptilian medial and lateral grisea of thalamus ventralis (12, 13) are represented by the Avian nucleus intercalatus (12), and nucleus ventrolateralis (13) with adjacent pars anterior nucleus decussationis supraopticae ventralis. In Mammals, these grisea correspond to subdivisions of zona incerta with *nuclei campi Foreli* (12, 13). The ventral grisea of lateral geniculate complex in Reptiles (15) including nucleus ovalis (14), can be identified in Birds and Mammals (15a). Likewise, a ventral medial geniculate griseum of Reptiles (15b), identified by PAPEZ (1935) can be recognized in Birds (15', nucleus posterior decussationis supraopticae ventralis) and in Mammals (15b).[210]

With respect to the *hypothalamus*, the Reptilian preoptic region com-

---

[210] The relevant concept that both the lateral and the medial geniculate complex include a ventral and a dorsal thalamic component was propounded by the author in early publications (K., 1927, pp. 219, 236, lateral geniculate body; K., 1935, medial geniculate body). However, previous authors, but without specific reference to the fundamental diencephalic longitudinal zones, had already distinguished dorsal and ventral subdivisions of geniculate grisea, particularly in Mammals.

prises the paraventricular grisea (17, including the magnocellular nucleus), the pars inferior massae cellularis reunientis (16), the lateral preoptic grisea (18), and the supraoptic nucleus (19), moreover the anterior entopeduncular nucleus (20a). Comparable, but somewhat more differentiated grisea are found in the preoptic region of Birds, as dealt with in the preceding section 8. The Mammalian homologues include diverse periventricular differentiations such as nucleus praeopticus paraventricularis superior (17a), area praeoptica periventricularis (17b), nucleus suprachiasmaticus (17c), nucleus praeopticus medialis (17c), moreover the magnocellular and the median preoptic nuclei. More laterally located corresponding grisea are nucleus praeopticus lateralis (18) and nuclcus supraopticus (19). There is also a remnant of massa cellularis reuniens, pars inferior (16) and, in addition to the globus pallidus, a smaller secondary entopeduncular nucleus (20a) remaining in the diencephalic preoptic neighborhood of the hemispheric stalk, while the globus pallidus has become displaced into the telencephalic neighborhood of said stalk.

More caudally, the Reptilian hypothalamus contains posterior paraventricular grisea (21), nucleus lateralis (22) and nucleus ventralis (23) hypothalami, a diffuse nucleus hypothalami lateralis externus, a more dorsal nucleus entopeduncularis posterior (20b), and a poorly developed mammillary griseum. Similarly, in Birds, posterior paraventricular grisea (21), nucleus lateralis (22) and nucleus ventralis hypothalami (23) represent cell groups homologous with the Reptilian ones. To these can be added nucleus lateralis hypothalami externus (22′) and posterior entopeduncular nucleus (20′). A further group, not clearly identifiable in Reptiles, is represented by the tuberal grisea (24). A rather nondescript mammillary griseum is likewise present. It should be mentioned that, in the postoptic Sauropsidan hypothalamus, a distinction between middle and posterior group is less well feasible than in Mammals. This also applies to the distinction between dorsal and interpeduncular group *versus* hypothalamus *sensu strictiori*, as introduced for Mammals by HAYMAKER and myself (K. and HAYMAKER, 1949). Reverting to the approximate mapping of Mammalian homologies with Sauropsidan grisea, there are the intermediate and posterior periventricular respectively internal cell groups (21a, b, c), nucleus dorsomedialis (22a) and ventromedialis (23) hypothalami, nucleus s. area hypothalami lateralis (22b), nucleus subthalamicus (20b), and the mammillary complex. It is doubtful whether the Mammalian nucleus arcuatus (24) is represented as a clearly distinguishable griseum in Sauropsida, but the Mammalian nuclei tuberis might possibly be considered homologous to the Avian 'tuberal grisea' mentioned above.[211]

Numerous investigators have dealt with cytoarchitecture, myeloarchitecture, fiber connections, and functional aspects of the Mammalian

[211] Figure 114, here reproduced because illustrating the first systematic attempt at establishing the morphologic homologies of Amniote diencephalic grisea on the basis of extensive ontogenetic investigations, should be properly understood as a rough first approximation, and includes some weaknesses. Thus, in e, 22a is somewhat distorted dorsolateralward, and 22b should be more extensive. Again, in f, 24 represents nucleus arcuatus, and not the 'tuberal grisea' 24 of Figure c. The Avian 'tuberal grisea' however, seem to be lateral derivatives of a matrix in part comparable to Mammalian nucleus arcuatus.

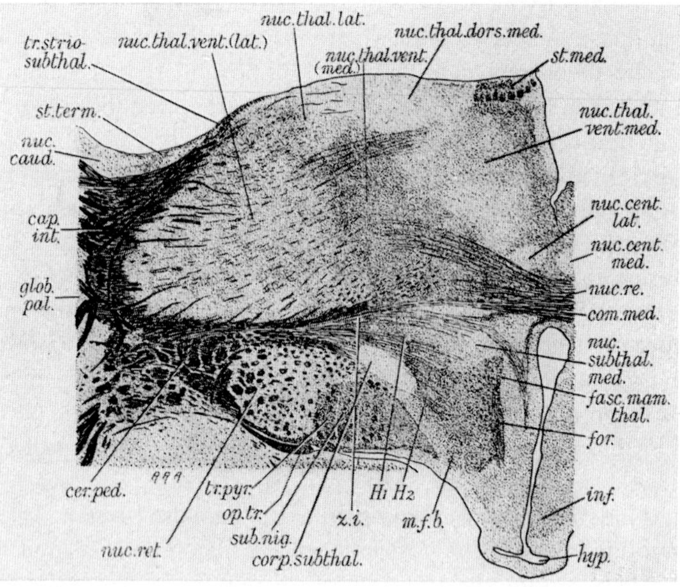

Figure 116 A. Cross-section (myelin stain) through the diencephalon of the Monotreme Echidna (from ABBIE, 1934). cap.int.: capsula interna; cer.ped.: cerebral peduncle; com.med.: 'middle commissure' (of massa intermedia); corp.subthal.: subthalamic body; fasc.mam.thal.: fasciculus mammillo-thalamicus; for.: fornix; glob.pal.: globus pallidus; hyp.: 'hypophysis'; inf.: 'infundibulum'; m.f.b.: 'medial forebrain bundle'; nuc.caud.: nucleus caudatus; nuc.cent.lat., med.: midline grisea; nuc.re.: nucleus reuniens (midline griseum); nuc.ret.: 'nucleus reticularis of cerebral peduncle' (presumably displaced n. reticularis thalami ventralis); nuc.subthal.med.: 'nucleus subthalamicus medius' presumably medial portion of zona incerta); nuc.thal.vent.med.: 'nucleus ventralis medius of thalamus' (presumably n. medialis thalami); op.tr.: optic tract; st.med.: stria medullaris; st.term.: stria terminalis; sub.nig.: substantia nigra; tr.pyr.: pyramidal tract; z.i.: zona incerta. Other abbreviations self-explanatory.

diencephalon. In addition to the data and bibliographic references included in CAJALS treatise (1911, 1955) and in the texts on comparative neurology (BECCARI, 1943; JOHNSTON, 1906; KAPPERS, 1947; KAPPERS *et al.*, 1936; KUHLENBECK, 1927; PAPEZ, 1929), the following general contributions subsequent to about 1912, and concerning a variety of Mammalian forms, as well as containing further references to the earlier studies may be mentioned:[212] ABBIE (1934), ANDO (1937), ATLAS

[212] Our own studies dealing with Human or Mammalian thalamus, hypothalamus and pretectal region have been pointed out above (K., 1948, 1951; K. and HAYMAKER, 1949; K. and MILLER, 1942, 1949; MIURA, 1933; SHINTANI-KUMAMOTO, 1959). Numerous data and references can also be found in the texts on human neuroanatomy by CLARA (1959), CROSBY *et al.* (1962), ELLIOTT (1963), KRIEG (1966), and PEELE (1961).

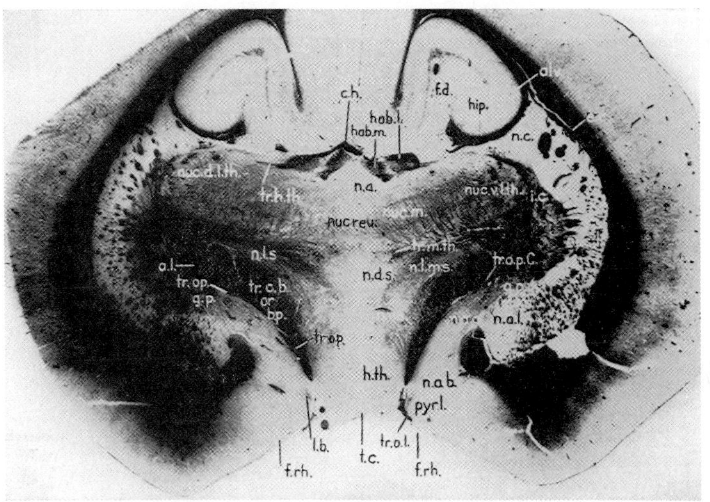

Figure 116B. Cross-section (myelin stain) through the forebrain of the Monotreme Ornithorhynchus (from HINES, 1929). a.l.: ansa lenticularis; alv.: 'alveus' (hippocampi); c.h.: commissura habenulae; e.c.: capsula externa; f.d.: fascia dentata; f.rh.: 'fissura rhinalis'; g.p.: globus pallidus; hab. 1., m.: lateral and medial habenular griseum; hip.: hippocampus; i.c.: internal capsule; l.b.: 'longitudinal bundle'; n.a.: 'nucleus anterior' thalami; n.a.b., c., l.: 'basal', 'central', and 'lateral' nuclei amygdalae; n.c.: nucleus caudatus; n.d.s.: probably nucleus reuniens of thalamus ventralis; n.l.m.s., n.l.s.: thalamus ventralis, probably zona incerta; n.v.s.: probably nucleus subthalamicus; nuc.d.l.th.: 'nucleus dorsolateralis thalami; nuc.l.: 'nucleus lentiformis'; nuc.m.: nucleus medialis thalami; nuc.reu.: nuclei reunientes; nuc.v.l.th.: 'nucleus ventrolateralis thalami; pyr.l.: piriform lobe; t.c.: 'tuber cinereum'; tr.c.b. or b.p.: tractus corticobulbaris or basis pedunculi; tr.h.th.: 'tractus habenulothalamicus'; tr.m.th.: tractus mammillo thalamicus; tr.o.l.: tractus olfactorius lateralis; tr.op: tractus opticus; tr.o.p.C.: '*tractus olfactorius projectionis Cajal*'.

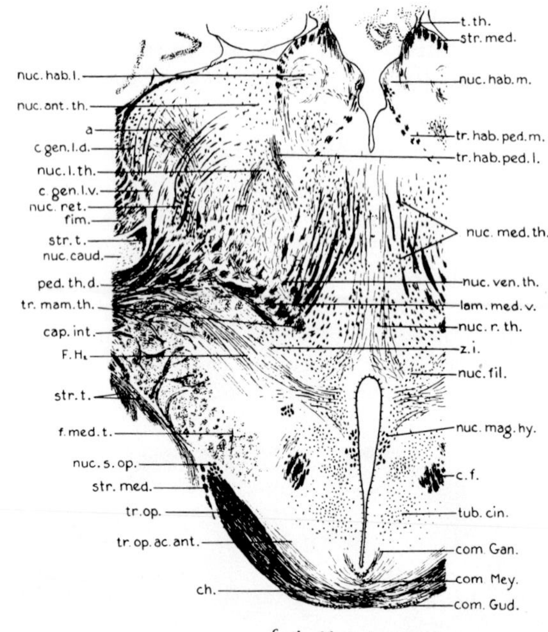

A

com.hab.
c.pineale
n.pret.
com.post.
n.lat.post.
n.p.p.infracom.
n.lat.intermed.
n.paraf.,pars p.l.
n.int.-med.dor.
n.paraf.
n.g.l.d.
tr.hab.ped.
n.g.m.marg.
n.reun.post.
n.g.l.v.
z.inc.
n.p.p.infracom.
FF
n.vent.pr.
ped.cer.
n.subthal.
n.hyp.post.
col.forn.
n.hyp.lat.
n.premam.dor.
n.premam.vent.

B

and INGRAM (1937), BODIAN (1939, 1940), CAMPBELL and RYZEN (1953); CHRIST (1969), LEGROS CLARK (1930, 1932 a, b), CRAIGIE (1925), CROUCH (1934 a, b), CROSBY *et al.* (1962), DEKABAN (1953, 1954), FEREMUTSCH (1963), GOLDBY (1943), GURDJIAN (1927), HASSLER (1959), A. HESS (1955), W.R. HESS (1954, 1968), HINES (1929), D'HOLLANDER (1913), INGRAM *et al.* (1932), ISO (1941), KEYSER (1972), KRUGER (1959), KUHLENBECK (1954), KUHLENBECK and MILLER (1942, 1949), KUREPINA (1966), LASHLEY (1941), LOO (1930), MCLARDY (1948, 1950, 1951), NIIMI *et al.* (1962), NIIMI and KUWAHARA (1973), PAPEZ (1932), PAPEZ and ARONSON (1934), PURPURA and YAHR (1966), OLSZEWSKI (1952), RIOCH (1929, 1931), J.E. ROSE (1942), SHINTANI-KUMAMOTO (1959), SOLNITZKY (1938, 1939), TSAI (1925 a, b), WALKER (1938), WARNER (1969), ZEMAN and INNES (1963).

With regard to many of these listed studies, as far as they were concerned with particular forms, such as e.g. Monotremes, Marsupials,

---

Figure 117A. Cross-section (myelin stain) through diencephalon of the Marsupial Opossum at level of posterior border of optic chiasma (from TSAI, 1925a). a: 'association fibers' between dorsal lateral geniculate griseum and dorsal thalamic grisea; cap.int.: capsula interna; c.f.: columna fornicis; c.gen.l.d., v.: dorsal and ventral lateral geniculate nucleus; ch.: optic chiasma; com. Gan., Gud., Mey.: *Ganser's*, *Gudden's*, and *Meynert's supraoptic commissure;* F.H_2: *Forel's field* H_2; fim.: fimbria s. fornix hippocampi; f.med.t.: 'medial forebrain bundle'; lam.med.v.: 'lamina medullaris ventralis' lamina medullaris externa and field H_1; nuc.ant.th.: n. anterior thalami; nuc.caud.: n. caudatus; nuc.fil.: n. filiformis; nuc.hab.l., m.: lateral and medial habenular grisey; nuc.l.th.: n. lateralis thalami; nuc.mag. hy.: n. magnocellularis hypothalami; nuc.med.th.: n. medialis thalami; nuc.ret.: n. reticularis thalami (ventralis); nuc.r.th.: griseum of nn. reunientes thalami; nuc.s.op.: n. supraopticus; nuc.ven.th.: 'n. ventralis thalami' (dorsalis); ped.th.d.: pedunculus thalami dorsalis; str.med.: stria medullaris; str.t.: stria terminalis; tr.hab.ped.l., m.: lateral and medial limb of tr. habenulopeduncularis; tr.mam.th.: tr. mammillothalamicus; tr.op.: tr. opticus; tr.op.ac.ant.: '*tr. opticus accessorius anterior of Bochenek*'*;* t.th.: taenia thalami; tub.cin.: 'tuber cinereum'; z.i.: zona incerta.

Figure 117B. Cross-section (modified *Nissl stain)* through the diencephalon of the Opossum caudally to optic chiasma (from BODIAN, 1939). c.pineale: pineal body; col.forn.: columna fornicis; com.hab.: habenular commissure; com.post.: posterior commissure; FF: *Forel's field* H; n.g.l.d., v.: dorsal and ventral lateral geniculate griseum; n.g.m.marg.: dorsal medial geniculate nucleus ('pars marginalis'); n.hyp.lat., post.: n. hypothalamicus lateralis and posterior; n.int.-med.dor.: 'n. intermedialis dorsalis'; n.lat.intermed.: n. lateralis, 'pars intermedia'; n.lat.post.: n. lat. posterior; n.paraf. (pars p.l.): n. parafascicularis (and its 'pars posterolateralis'); n.p.p. infracom.: 'n. paraventricularis posterior, pars infracommissuralis'; n.premam.dor., vent.: dorsal and ventral premammillary nuclei; n.reun.post.: n. reuniens posterior; n.subthal.: n. subthalamicus; n.vent.pr.: 'n. ventralis, pars principalis'; ped.cer.: pedunculus cerebri; tr.hab.ped.: tractus habenulopeduncularis; z.inc.: zona incerta.

Rodents, Ungulates, Carnivores or Primates, the comments by BECCARI, quoted above in footnote 201, should be recalled.

Again, with respect to delimitable grisea or so-called 'nuclei' the comments included in section 11, p. 203 f. of chapter VIII, volume 4, should be recalled as applying, *mutatis mutandis*, not only to the spinal cord, but to the entire neuraxis, and especially to the 'nuclear configuration' of the diencephalon. In this connection, the histochemical demonstration of relevant enzymes in developing as well as in adult neural structures was mentioned. The 'chemodifferentiation' of a Mammalian thalamus was carefully investigated in the Rat by EITSCHBER-

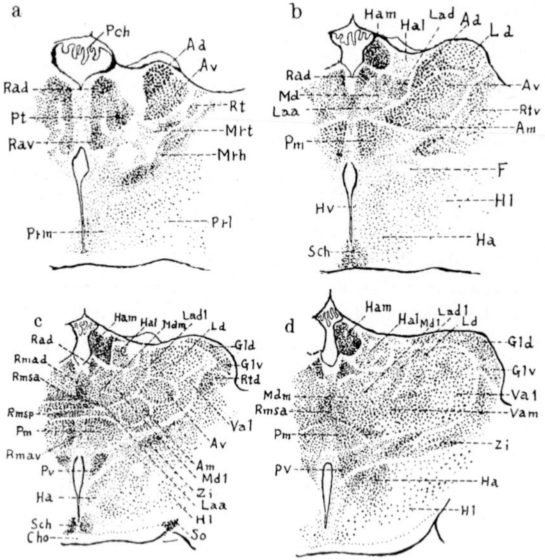

Figure 118 A. Cross-sections *(Nissl stain)* through the diencephalon of a young mouse on the 4th day after birth (from NIIMI *et al.*, 1961). Ad: n. anterior dorsalis; Am: n. ant. medialis; Av: n. ant. ventralis; Cho: optic chiasma; F: fornix; Gld, Glv: dorsal and ventral lateral geniculate nucleus; Ha: n. hypothalamicus anterior; Hal, Ham: lateral and medial habenular nuclei; Hl: n. hypothalamicus lateralis, Hv: n. hypothalamicus periventricularis; Laa, Lad, Ladl: n. laminaris, pars anterior, dorsalis, and dorsolateralis; Mdl, Mdm: n. medialis (n. med. dorsalis, n. dorsomedialis), pars medialis and pars lateralis; Mrh, Mrt: massa cellularis, p. inferior (s. hypothalami), and pars superior (s. thalami ventralis); Pch: plexus choroideus; Pm: n. paramedianus; Prl, Prm: n. preopticus lateralis and medialis; Pt: n. parataenialis; Pv: n. paraventricularis; Rad, Rav, Rmad, Rmav, Rmsa, Rmsp: nn. reunientes; Rt, Rtd, Rtv: n. reticularis (thalami ventralis); Sch: n. suprachiasmaticus; So: n. supraopticus; Val, Vam: n. ventralis, pars anterior lateralis and medialis; Zi: zona incerta.

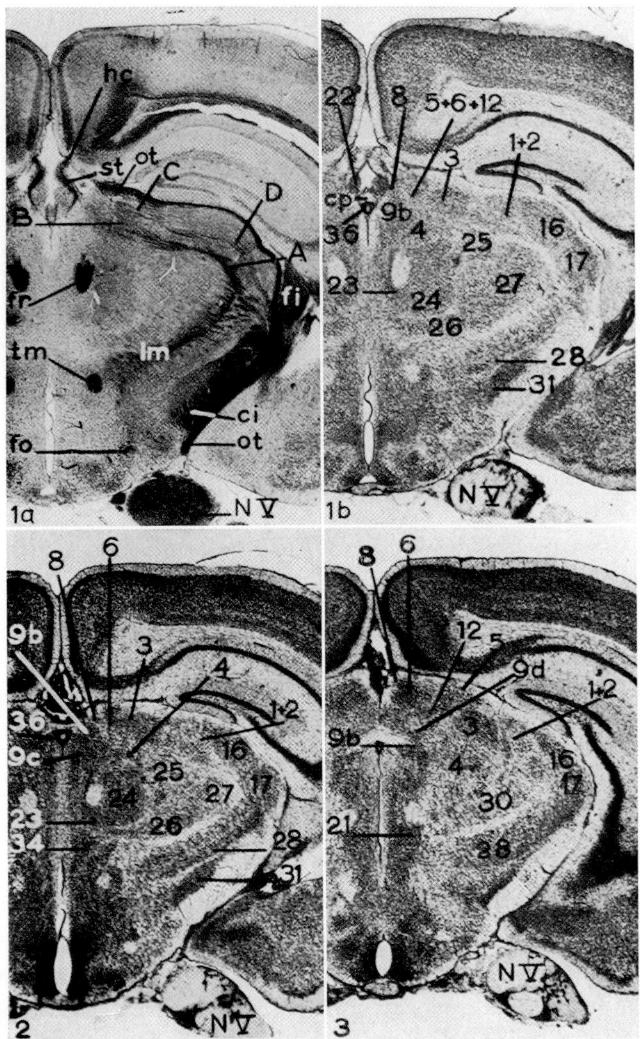

Figure 118 B. Six cross-sections (myelin and *Nissl* stains) through the diencephalon of the adult Mouse in rostrocaudal sequence, and caudal to those of Figure 118 A. 1a and 1b, 4a and 4b of adjacent sections, showing myelo- and cytoarchitecture at approximately identical levels (from Shintani-Kumamoto, 1959). A: ventromedial arciform fiber system; B: ventromedial longitudinal system; C: dorsolateral longitudinal system; D: dorsolateral arciform system; NV: trigeminal nerve and ganglion; bp: basis s. pes pedunculi; bs: brachium quadrigeminum superius; cp: commissura posterior; cs: colliculus superior; dm: dorsomedial tectothalamic tract; fi: fimbria fornicis; fo: fornix; fr: fasciculus retroflexus; hc: habenular commissure; lm: lemniscus medialis; ms: supramammillary commissure; ot: optic tract; st: stria medullaris; tm: tractus mammillothalamicus; tmt: tractus mammillo-thalamico-tegmentalis; 1+2: n. lateralis posterior complex (1: pulvinar:

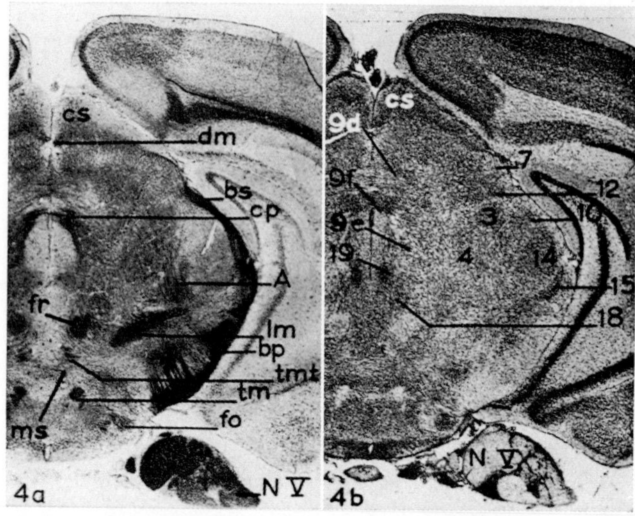

2: n. lat. dorsalis posterior thalami); 3: n. praetectalis (principalis); 4: n. posterior; 5: n. lentiformis mesencephali, pars magnocellularis; 6: n. lent. mes., pars parvocellularis; 7. n. olivaris colliculi superioris; 8: area praetectalis; 9: nn. commissurae posterioris (b: n, interstitialis principalis; c: n. centralis subcommissuralis; d: n. interstit. supracommissuralis; e: n. interstit. magnocellularis; f: n. medianus infracommissuralis)); 10: n. suprageniculatus; 12: n. sublentiformis; 14: n. gen. med. dorsalis; 15: n. gen. med. ventralis; 16: n. gen. lat. dorsalis; 17: n. gen. lat. ventralis; 18: n. interstit. fasciculi longitudinalis medialis; 19: *n. of Darkschewitsch;* 22: n. habenularis medialis; 23: n. parafascicularis; 24: n. medialis thalami; 25: 'n. mediolateralis thalami'; 26: n. ventromedialis thalami; 27: n. ventrolateralis thalami; 28: n. reticularis thalami and zona incerta (thalamus ventralis); 30: posterior end of n. ventralis thalami; 31: corpus subthalamicum (n. subthalamicus); 34: nn. reunientes; 36: subcommissural organ.

GER (1970), who recorded the 'enzyme pattern' displayed by numerous such substances. Among these latter, acid phosphatase, non-specific esterase, and cholinesterase were included. It is perhaps of especial interest that the longitudinal zones dealt with by the cited author, namely epithalamus, thalamus dorsalis, and thalamus ventralis show different reactions which corroborate, as it were, not only the morphologic but also the biochemical significance of these topologic units representing *bauplan-grundbestandteile.* Figures 120 C and D illustrate examples of 'enzyme pattern' as recorded by EITSCHBERGER.

Still more recently, a topographic atlas of catecholamine and acetylcholesterase containing grisea respectively fiber systems in the forebrain (telencephalon and diencephalon) of the Rat was prepared by JACOBOWITZ and PALKOVITS (1974). It contains bibliographic references

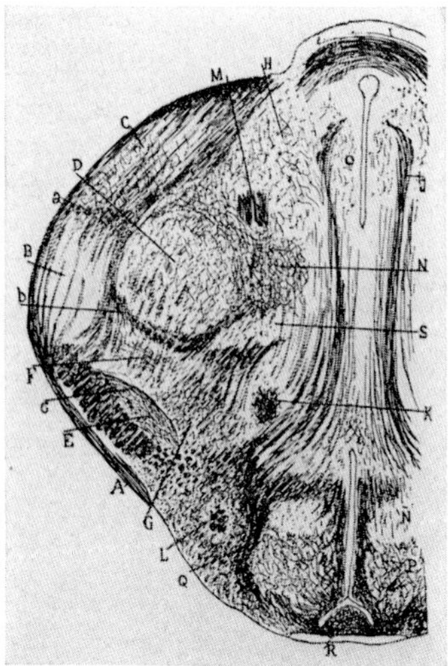

Figure 118 C. Cross-section *(Golgi impregnation)* through the diencephalon of the Mouse at the level of commissura posterior (from CAJAL, 1911, 1955). A: optic tract; B: ventral lateral geniculate body (or nucleus); C: dorsal lateral geniculate body; D: nucleus ventralis posterolateralis; E: *corpus subthalamicum of Luys;* F: zona incerta; G: medial part of zona incerta *(griseum of field of Forel);* H: nucleus of posterior commissure; I: posterior commissure; J: dorsobasal periventricular fiber system; K: mammillothalamic tract; L: fornix bundle; M: habenulopeduncular tract; N: *'noyau triangulaire'* (probably caudal end of n. medialis); N: nucleus dorsomedialis hypothalami; O: central gray; P: nucleus ventromedialis thalami; Q: basilateral extension of dorsobasal periventricular system; R: fiber system in basal raphe; S: nucleus ventralis posteromedialis; a: fiber lamina between dorsal and ventral lateral geniculate grisea; b: lamina medullaris externa (combined with extension of field H_1); c: fasciculus lenticularis (H_2). Some of CAJAL's designations are here re-interpreted in accordance with the present author's views.

to studies concerned with this topic which is only indirectly related to the morphologic approach emphasized although not considered to be exclusive, in the present treatise.

Figures 116 to 127 illustrate relevant griseal and fiber system features of the diencephalon in various Mammalian forms[213] exclusive of Primates. As regards these latter, some aspects of the Human diencephalon will be separately discussed in concluding the present chapter.

For an understanding of the presumed functional significance and the complex configurational as well as hodologic relationships dis-

---

[213] Prototheria (Echidna, Ornithorhynchus) and Metatheria (Opossum). Among Eutheria: Mouse and Rat (Rodentia,) Rabbit (Lagomorpha), Mole (Insectivora), Armadillo (Edentata), Pig (Ungulata), Cat, Dog (Carnivora), Dolphin (Cetacea).

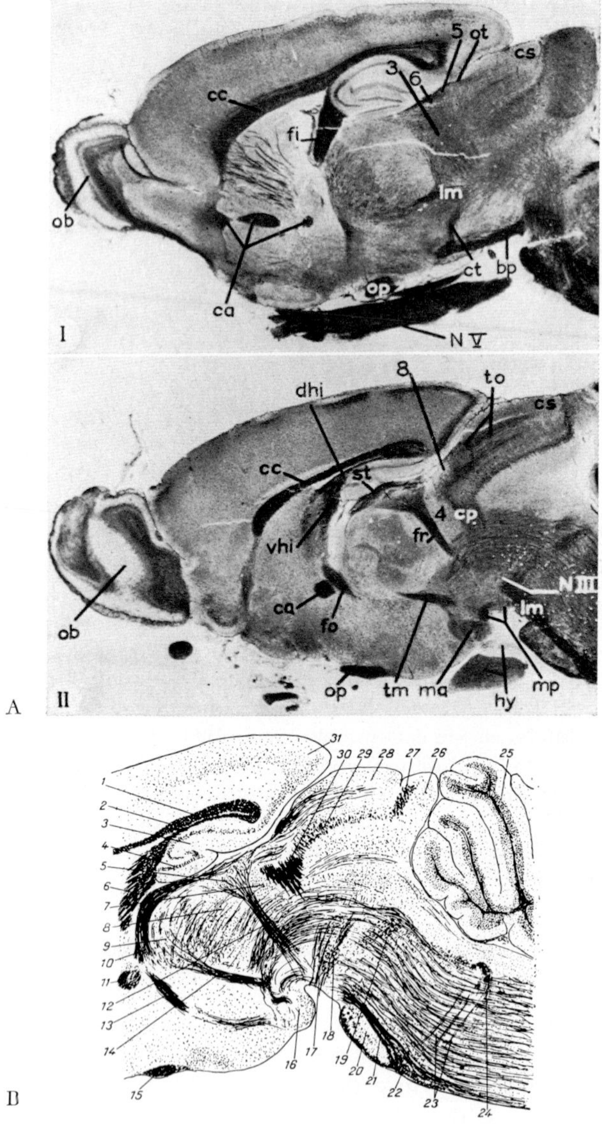

Figure 119 A. Two paramedian sagittal sections (myelin stain) through the prosencephalon of the Mouse. II is nearer to the midline than I (from SHINTANI-KUMAMOTO, 1959). bp: basis *sive* pes pedunculi; ca: anterior commissure; cc: corpus callosum; cp: commissura posterior; cs: colliculus superior; ct: *collateral corticotegmental system of Cajal;* dhi: dorsal part of hippocampal commissure; fi: fimbria fornicis *(sive* hippocampi); fo: fornix; fr: fasciculus retroflexus; hy: hypophysial complex; lm: lemniscus medialis; ma: mammillary complex; mp: mammillary peduncle; NIII, V: oculomotor and trigeminal nerve: ob: olfactory bulb; op: optic chiasma respectively optic tract; ot: optic tract

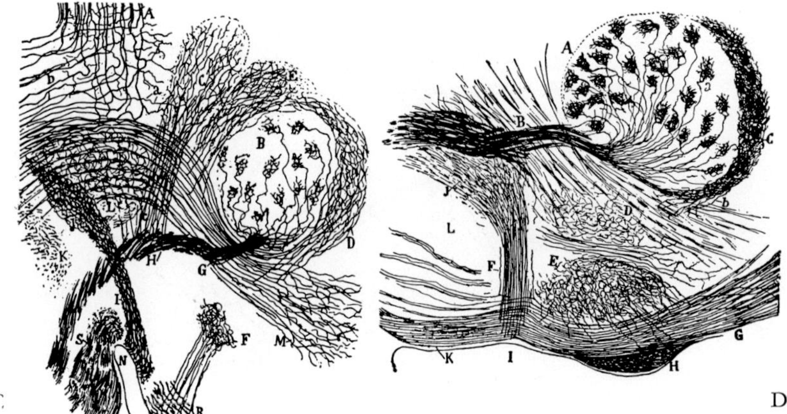

component reaching tectum mesencephali; st: stria medullaris; tm: tractus mammillothalamicus; to: tectum opticum; vhi: ventral portion of hippocampal commissure; 3, 4, 5, 6, 8: cf. legend to Figure 118B.

Figure 119B. Paramedian sagittal section *(Weigert stain)* through the brain of the Mouse, showing relevant medullated fiber system (from Beccari, 1943). 1: corpus callosum; 2: commissura hippocampi; 3: cornu ammonis; 4: fascia dentata; 5: habenular grisea; 6: fimbria fornicis; 7: pretectal grisea; 8: nucleus medialis thalami; 9: anterior thalamic grisea; 10: stria medullaris; 11: anterior commissure; 12: fasciculus retroflexus; 13: fornix; 14: mammillothalamic tract; 15: optic tract; 16: mammillary complex; 17: oculomotorius rootlets; 18: tegmental decussations; 19: brachium conjunctivum; 20: pontine grisea and fibers; 21: pyramidal tract; 22: trapezoid body in cross-section; 23: rootlets of abducens; 24: genu of nervus facialis; 25: cerebellum; 26, 27: colliculus inferior and its commissure; 28, 29: colliculus superior and its commissure; 30: posterior commissure; 31: occipital pole of telencephalon.

Figure 119C. Parasagittal section *(Golgi impregnation)* through the mesodiencephalic region of a young Mouse (from Cajal, 1911). A: posterior commissure; B: nucleus ventralis posterolateralis thalami (27 of Fig. 118B); C: pretectal grisea; D, E: dorsal thalamic grisea rostrally and caudally adjacent to B (perhaps n. ventralis lateralis, and nn. 25 and 30 of Fig. 118B); F: '*noyau spécial sous-thalamique*' (perhaps n. subthalamicus or lateral hypothalamic area); G: medial lemniscus; H: portion of medial lemniscus giving off collaterals (c) to pretectal grisea; I: tractus peduncularis transversus (basal optic root); J: nucleus opticus tegmenti; K: undefined tegmental fiber systems; L: nucleus ruber; M: rostral extensions of lemniscus medialis and of tegmental field (including H and H_1); N: interpeduncular fossa; R: caudal end of 'medial forebrain bundle'; S: pyramidal tract; a, b: rostral and caudal branches of posterior commissure.

Figure 119D. Paramedian sagittal section *(Golgi impregnation)* through diencephalic and tegmental neighborhoods of the Mouse (from Cajal, 1911). A: nucleus ventralis posterolateralis; B: lemniscus medialis; C: rostral portion of ventrolateral complex (perhaps part of n. ventralis lateralis); D: nucleus reticularis thalami (thalamus ventralis); E: subthalamic nucleus; F: collaterals of pedunculus cerebri to tegmentum; G: diencephalic portion of pedunculus cerebri; H: optic tract; I: tegmental region reached by collaterals of cerebral peduncle; K: portion of pedunculus cerebri; L: tegmental region; a: end arborizations of lemniscus medialis; b: collaterals of tegmental or tectothalamic systems.

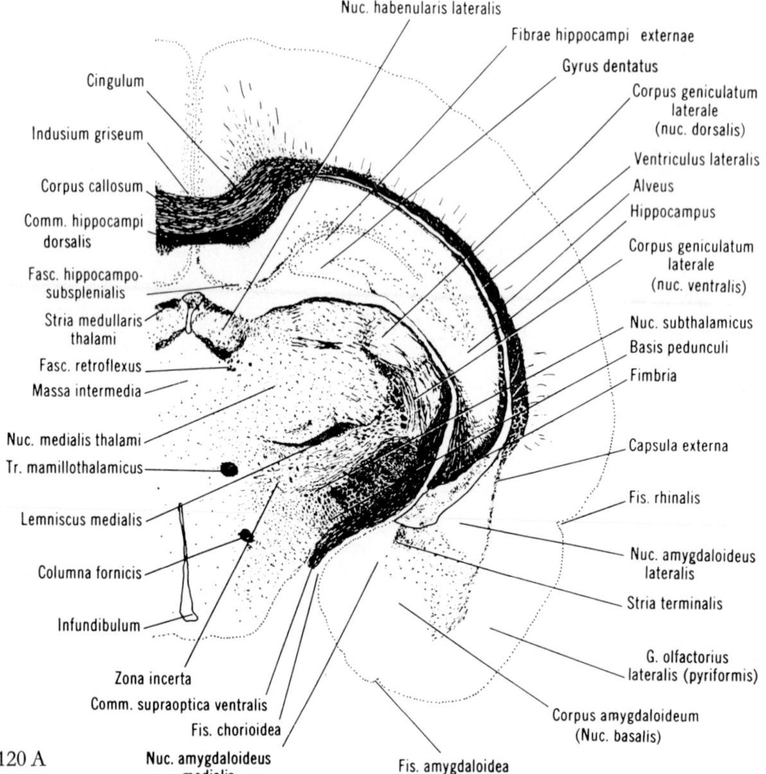

Figure 120 A. Cross-section (myelin stain) through the diencephalon of the Rat at a level caudal to optic chiasma (from ZEMAN and INNES, 1963).

Figure 120 B. Cross-section (myelin stain) through the diencephalon of the Rat at a level of optic chiasma (from ZEMAN and INNES, 1963).

Figure 120 C. Cross-section ('unspecific esterase' reaction) through the thalamus of the Rat (5th postnatal day) at level of habenular grisea (from EITSCHBERGER, 1970). cm: 'centre médian'; lh: lateral habenular nucleus; re: griseum of nuclei reunientes; tr: nucleus reticularis (thalami ventralis); tvc: nucleus ventralis posterolateralis. It can be seen that the reaction is strong in lh, tvc, and tr, being, according to EITSCHBERGER, somewhat less pronounced in cm and re. The enzyme is predominantly localized in the cell bodies.

Figure 120 D. Cross-section (cholinesterase reaction) through the rostral third of the Rat's thalamus at the 20th postnatal day (from EITSCHBERGER, 1970). CTH: dorsal group of nuclei reunientes; l: lateral portion of nucleus anterior ventralis; LAMI: 'lamina medullaris interna' (actually an accessory anterior medullary lamina); m: medial portion of nucleus anterior ventralis; tad: nucleus anterior dorsalis; tav: nucleus anterior ventralis; re: ventral group of nuclei reunientes. The reaction is intense in tad, tav and LAMI, and a histochemical difference between l and m of tav is noticeable.

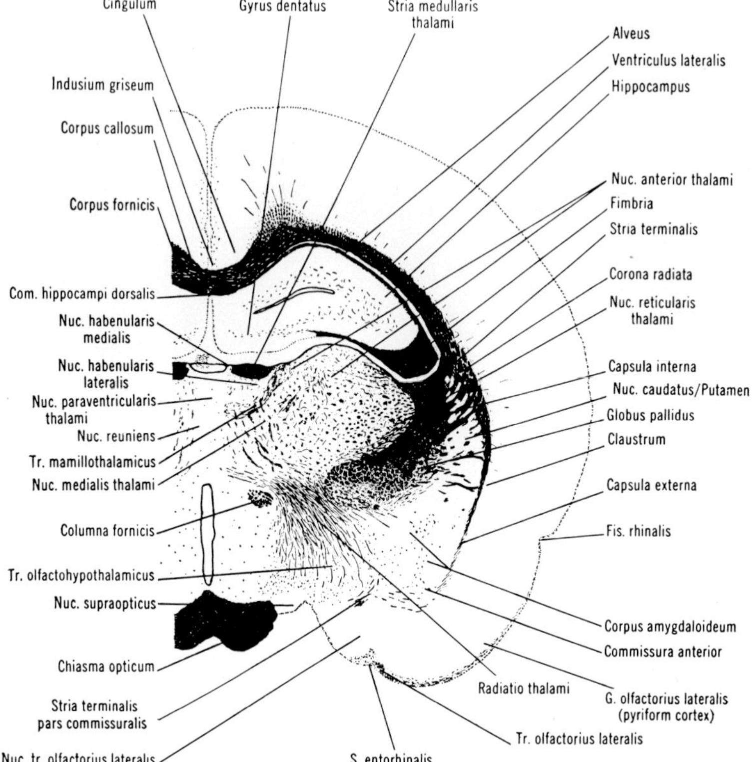

Figure 120B (legend see p. 322)

Figure 120E. Cross-section (myelin stain) through the diencephalon of the adult Rabbit at a rostral level of commissura posterior (from K. and Miller, 1942). A: ventromedial arciform fiber system; B: ventromedial longitudinal system; C: dorsolateral longitudinal system; D: dorsolateral arciform system; E: periventricular system, vertical (1) and longitudinal (2); cp: posterior commissure; fr: fasciculus retroflexus; hc: fibers of habenular commissure; tm: mammillothalamic tract; 1: 'pulvinar'; 2: n. lateralis dorsalis posterior; 3: n. praetectalis (principalis); 4: n. posterior; 5, 6: n. lentiformis mesencephali, pars magnocellularis and pars parvocellularis; 8: area praetectalis; 9: nn. commissurae posterioris (a: interstitialis, pars rostralis; b: interstitialis principalis; c: centralis subcommissuralis); 12: n. sublentiformis; 14: n. gen. med. dorsalis; 17: n. gen. lat. ventralis; 21: tegmental cell cord (or 'plate'); 23: n. parafascicularis; 24: n. medialis thalami; 25: 'n. mediolateralis'; 26: n. ventromedialis; 27: n. ventrolateralis; 28: n. reticularis (thalami ventralis); 31: corpus subthalamicum; 34: nn. reunientes.

Figure 120F. Cross-section *(Nissl stain)* through the diencephalon of the Rabbit, about 0.3 mm caudal to that of Figure 120E (from K. and Miller, 1942). For designations cf. legend to Figure 120E. 16: n. gen. lat. dorsalis.

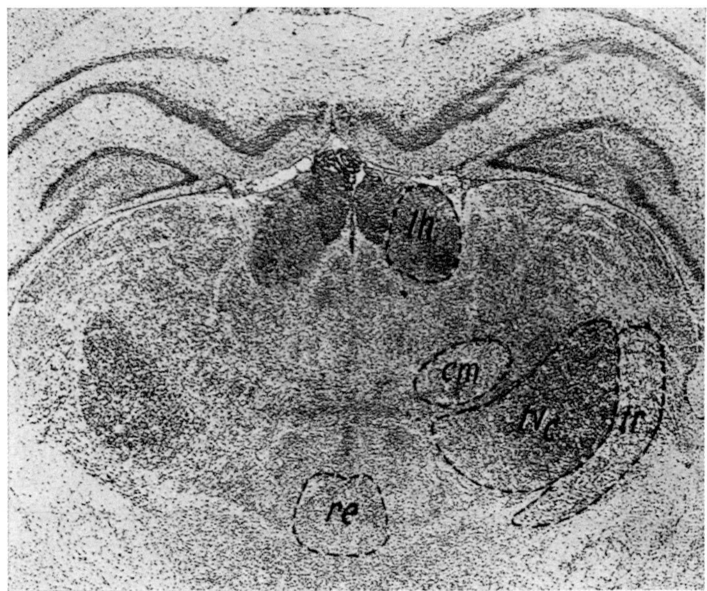

Figure 120C (legend see p. 322)

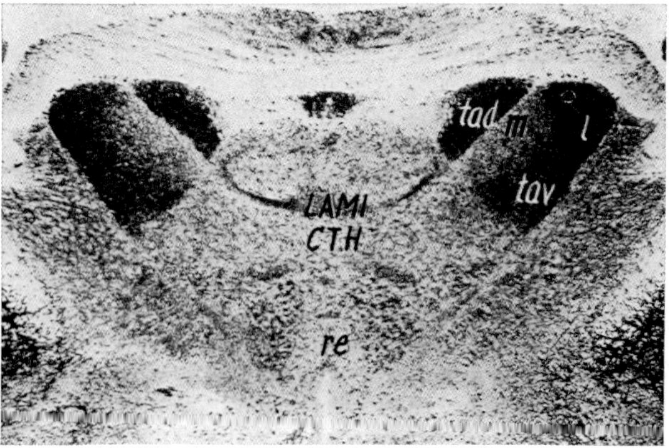

Figure 120D (legend see p. 322)

Figure 120G. Cross-section (myelin stain) through the diencephalon of the Rabbit at level of optic chiasma and habenular grisea. 1: parahippocampal cortex; 2: hippocampus; 3: medial and lateral habenular grisea; 4: midline nuclei; 6: part of supraoptic commissure above caudal end of optic chiasma; 7: optic tract and chiasma; 8: nucleus medialis

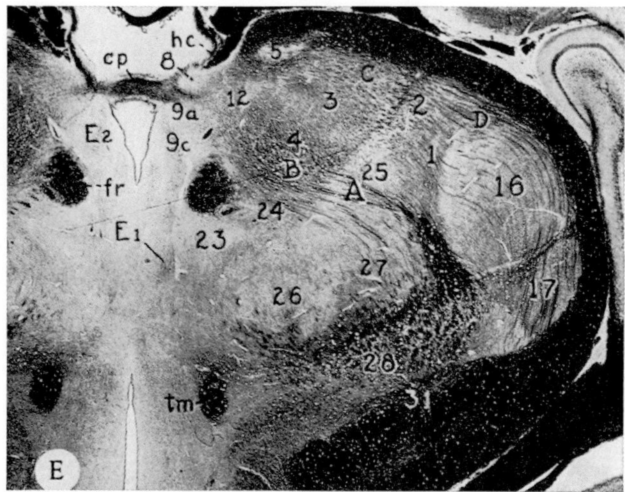

Figure 120E (legend see p. 323)

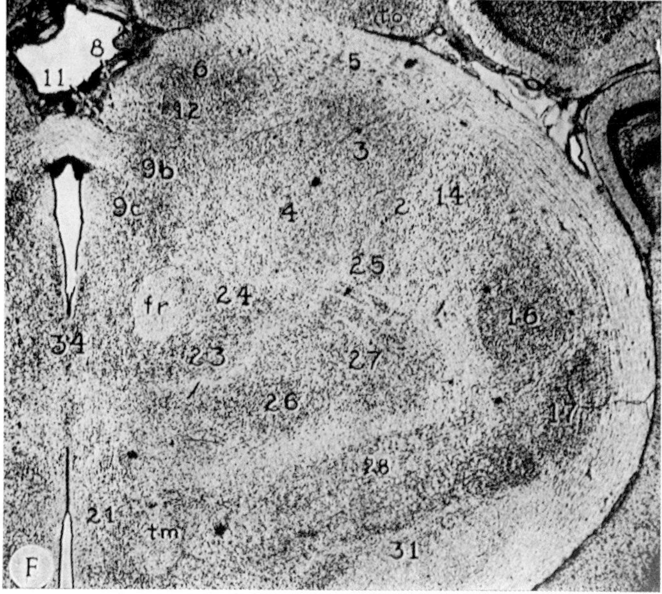

Figure 120F (legend see p. 323)

(dorsomedialis) thalami; 9: posterior lateral respectively lateral dorsal posterior thalamic grisea at transition to main pretectal group; 10: posterior ventrolateral and ventromedial dorsal thalamic griseal complex; 11: dorsal lateral geniculate nucleus; 12: ventral lateral geniculate nucleus; 13: nucleus reticularis (thalami ventralis); 14: fasciculus mammillo-

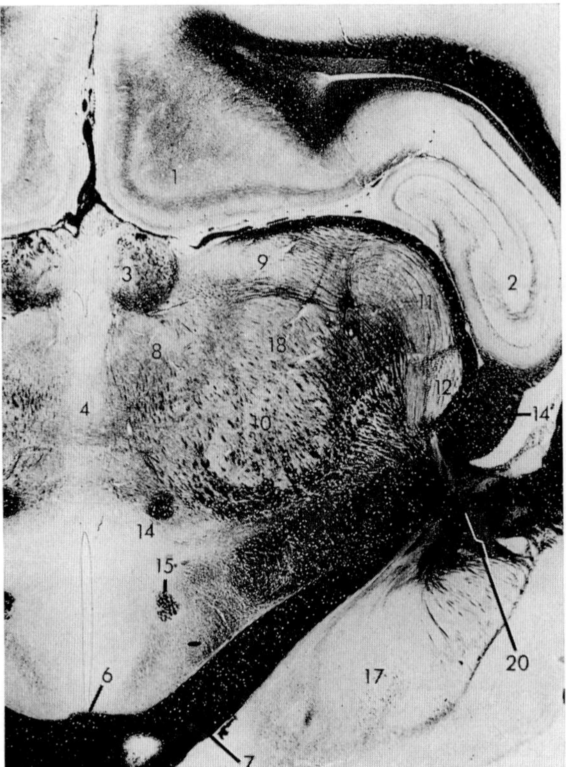

Figure 120G (legend see p. 324)

thalamicus; 14′: fimbria hippocampi; 15: fornix bundle; 16: pes pedunculi (with medially adjacent basal forebrain bundle system); 17: amygdaloid complex; 18: nucleus medialateralis; 14′: also fi. fornicis; 20: stria terminalis system. The designations apply to both Figures 120G and H. The hypothalamic grisea (cf. Figs. 113A–D) are located above 6.

Figure 120H. Cross-section *(Nissl stain)* through the Rabbit's diencephalon immediately (caudally) adjacent to section shown in Figure 120G.

Figure 121. Cross-section *(Nissl stain)* through the thalamus of the Insectivore mole Mogera at level of habenular and intralaminar grisea (from Niimi *et al.*, 1962). Am: caudal end of n. anterior medialis (the peculiar caudoventral displacement of this griseum ventrally to the intralaminar group should be noted); Cgld: n. geniculatus lateralis dorsalis; Hl, Hm: lateral and medial habenular grisea; Ladl, Lavm: dorsolateral and ventromedial portions of n. laminaris; Lda: 'n. lateralis dorsalis, pars anterior'; Mdl, Mdm: lateral and medial portions of n. (dorso) medialis; Pmd, Pmv: dorsal and ventral portions of n. paramedianus (probably grisea of nn. reunientes group); Rad: 'n. reuniens arcuatus, pars dorsalis' (perhaps portion of paratenial griseum); Rsam, Rspmv: grisea of nn. reunientes group; Rt: n. reticularis (thalami ventralis, the dorsolateral extension of this griseum is presumably a reduced ventral lateral geniculate nucleus); Val: 'n. ventralis, pars anterior lateralis'; Vamd, Vaml, Vamm, Vamv: subdivisions of n. ventralis, pars anterior.

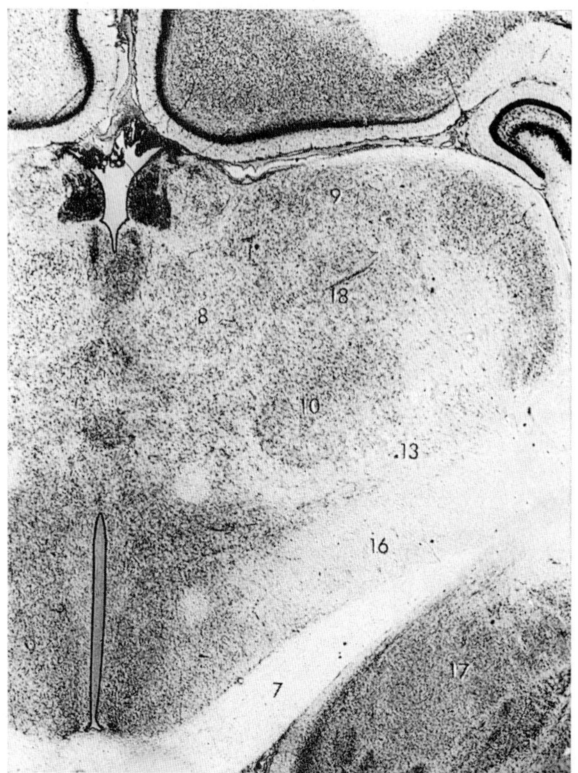

Figure 120 H (legend see p. 262)

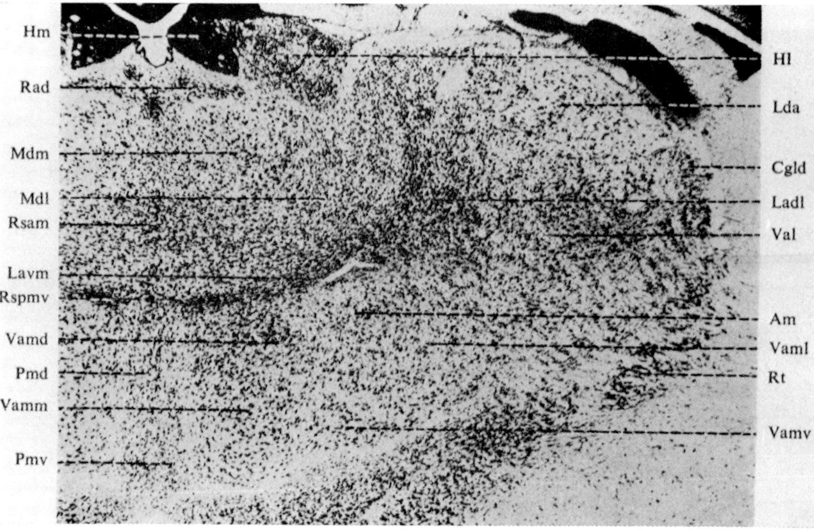

121

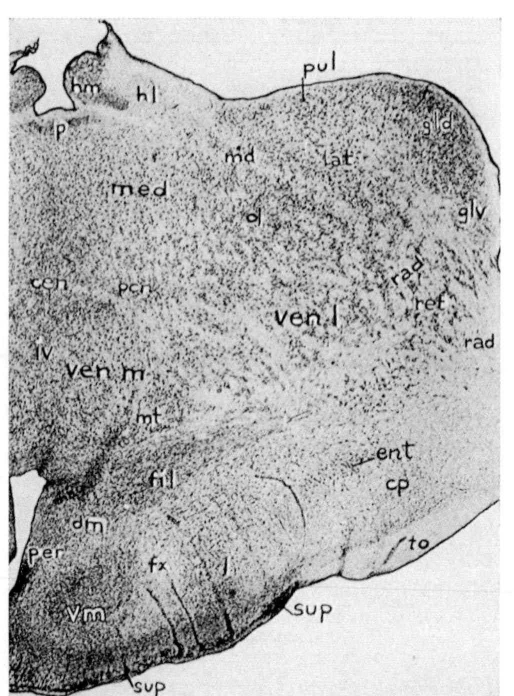

hm
hl
p
pul
gld
md
lat
med
d
glv
rad
ret
rad
ven l
ven m
mt
ent
cp
fil
dm
per
fx
to
Vm
sup
sup

122 A

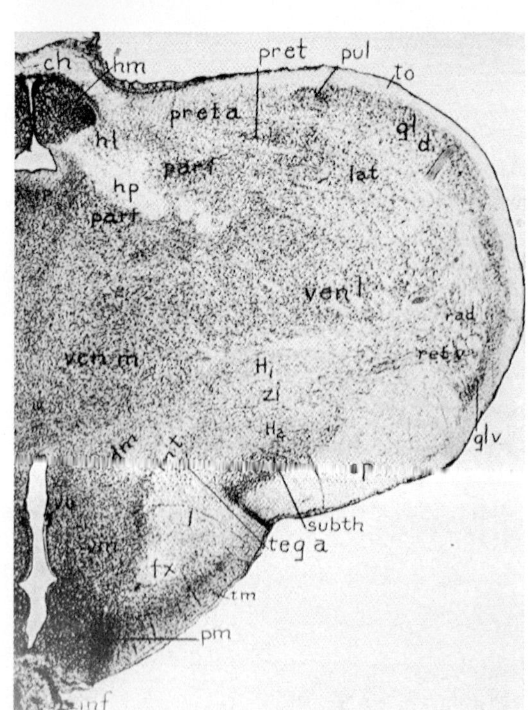

ch
hm
pret
pul
to
pret a
hl
gl d
parf
lat
hp
parf
ven l
rad
ret v
ven m
H1
zi
H2
glv
dm
mt
l
subth
fx
teg a
tm
pm
inf

122 B

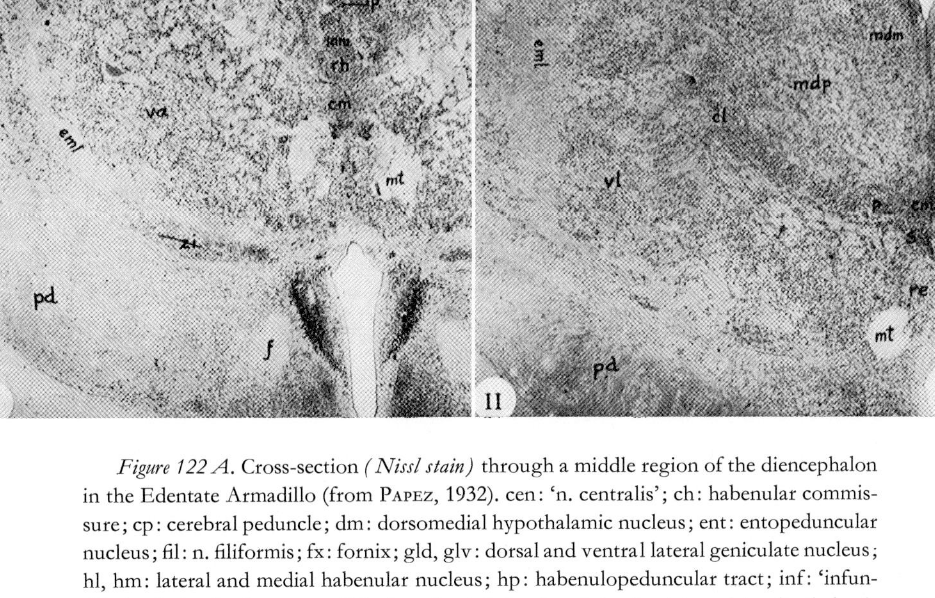

Figure 122 A. Cross-section *(Nissl stain)* through a middle region of the diencephalon in the Edentate Armadillo (from Papez, 1932). cen: 'n. centralis'; ch: habenular commissure; cp: cerebral peduncle; dm: dorsomedial hypothalamic nucleus; ent: entopeduncular nucleus; fil: n. filiformis; fx: fornix; gld, glv: dorsal and ventral lateral geniculate nucleus; hl, hm: lateral and medial habenular nucleus; hp: habenulopeduncular tract; inf: 'infundibulum'; iv: 'n. interventralis'; l: n. hypothalamicus lateralis; lat: n. lateralis thalami; mag: magnocellular hypothalamic nucleus; md, med: n. medialis thalami; mt: mammillothalamic tract; parf: n. parafascicularis; per: periventricular hypothalamic nucleus; pm: n. praemammillaris; pret a: pretectal area; pret: n. praetectalis; pul: pulvinar; rad: thalamic radiations; ret, ret v: n. reticularis thalami ventralis; subth: n. subthalamicus; sup: n. supraopticus; teg a: rostral tegmental cell group; tm: n. tuberomammillaris; to: tractus opticus; ven l: n. ventrolateralis thalami; ven m: n. ventromedialis thalami; vm: n. ventromed. hypothalami; vo: vascular organ; zi: zona incerta.

Figure 122 B. Cross-section through the diencephalon of the Armadillo at level of habenular commissure (from Papez, 1932). For designations cf. Figure 122 A.

Figure 123. Two cross-sections (cell stain) through the thalamus of the Ungulate Sus scrofa, (I) at level of anterior nuclear group, and (II) at level of maximal development of nucleus medialis (from Solnitzky, 1938). ad: n. anterior dorsalis; am: n. anterior medialis; av: n. anterior ventralis; cl: (intralaminar) 'n. centralis lateralis'; cm: n. centralis medialis'; eml: external medullary lamina; f: fornix; iad: probably part of n. parataenialis; iam, ip: grisea of reuniens group; ld: n. lat. dorsalis; lp: n. lat. posterior; mdm, mdp: portions of n. medialis thalami; mt: mammillothalamic tract; p: 'n. paracentralis'; pa: n. paraventricularis anterior; pd: cerebral peduncle; pt: n. parataenialis; r: n. reticularis (thalami ventralis); re: griseum of reuniens group; rh: griseum of reuniens group ('n. rhomboidalis'); s: griseum of reuniens group; smt: stria medullaris, va: n. ventralis anterior; vl: n. ventralis lateralis; zi: zona incerta.

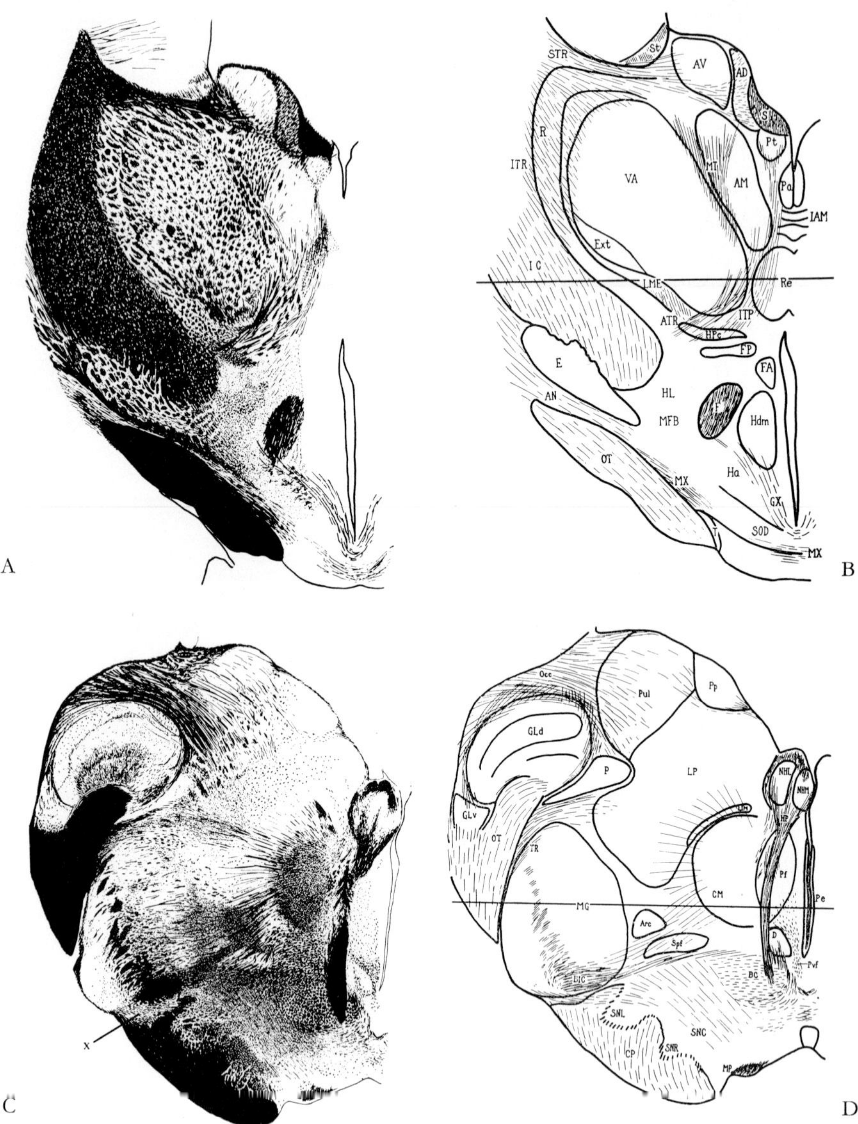

Figure 124 A–D. Cross-sections through the diencephalon of the Carnivore Cat at level of anterior thalamic grisea (A, B) and of caudal grisea (C, D). A and C myelin stain, B and D outlines of grisea as interpreted in modified *Nissl stain* preparations (from Ingram *et al.*, 1932). AD, AM: n. anterior dorsalis and anterior medialis; AN: ansa lenticularis; Arc: 'n. ventralis, pars arcuata'; ATR: anterior thalamic radiation; AV: n. anterior ventralis; BC: 'brachium conjunctivum' (probably tegmental field); BIC: brachium of inferior colliculus; CM: n. of centrum medianum; CP: cerebral peduncle;

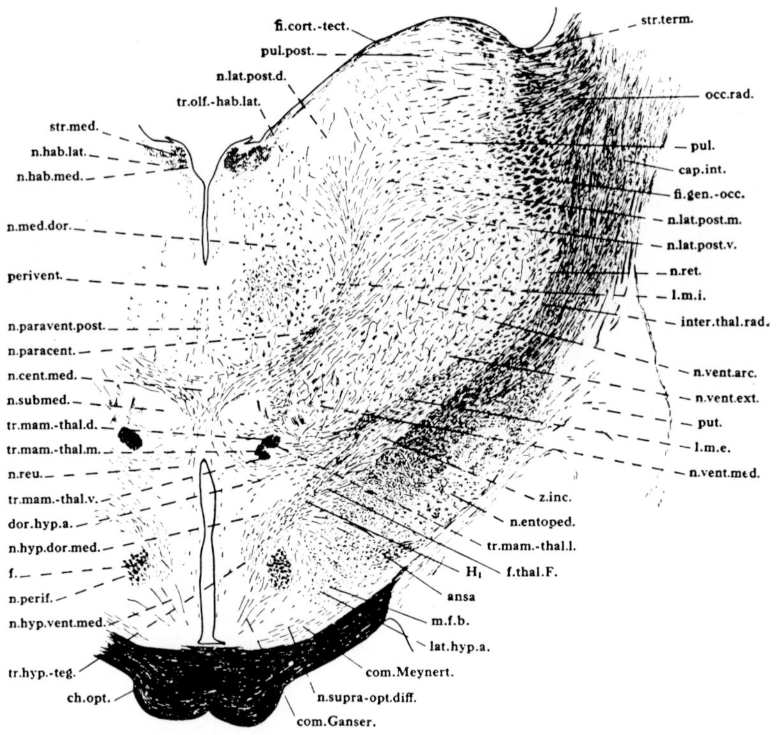

Figure 125 (legend see p. 332)

D: *n. of Darkschewitsch;* E: n. entopeduncularis; Ext: 'n. ventralis, pars externa'; F: fornix; FA: 'n. filiformis anterior'; FP: 'n. filiformis, pars principalis'; Gld, Glv: dorsal and ventral lateral geniculate nucleus; GX: '*Ganser's commissure*'; HL: n. hypothalamicus lateralis; HP: habenulopeduncular tract; HPc: 'n. hypothalamicus parvocellularis'; IAM: 'n. commissuralis interanterodorsalis'; IC: internal capsule; ITP: inferior thalamic peduncle; ITR: intermediate thalamic radiation; LIM: n. lentiformis mesencephali (magnocellularis); LME: lamina medullaris externa; LP: 'n. lateralis, pars posterior' (pretectal grisea 3 and 4 of our notation); MFB: 'medial' (basal) forebrain bundle; MG: dorsal medial geniculate griseum; MP: mammillary peduncle; MT: mammillothalamic tract; MX: '*Meynert's commissure*'; NHL, NHM: lateral and medial habenular griseum; Occ: 'occipital radiation'; OT: optic tract; P: 'n. posterior' (perhaps suprageniculate nucleus); Pa: 'n. paraventricularis anterior' (probably our n. paramedianus); Pe: 'n. paravent. posterior; Pf: n. parafascicularis; Pp: 'pulvinar pars posterior' (probably our n. lateralis dorsalis posterior); Pt: n. parataenialis; Pul: pulvinar; Pvf: periventricular fibers; R: n. reticularis (thalami ventralis); Re: n. reuniens, SNC, SNL, SNR: substantia nigra (p. compacta, lateralis, reticularis); SOD: 'n. supraopticus diffusus'; Spf: 'n. subparafascicularis'; St: stria terminalis; STR: 'superior thalamic radiation'; TR: 'temporal radiation'; VA: n. ventralis pars anterior. The line in B and D indicates the 'zero horizontal plane' passing through external auditory meatus inf. margin of orbit. Added x in C: presumably ventral medial geniculate nucleus, not identified by the cited authors.

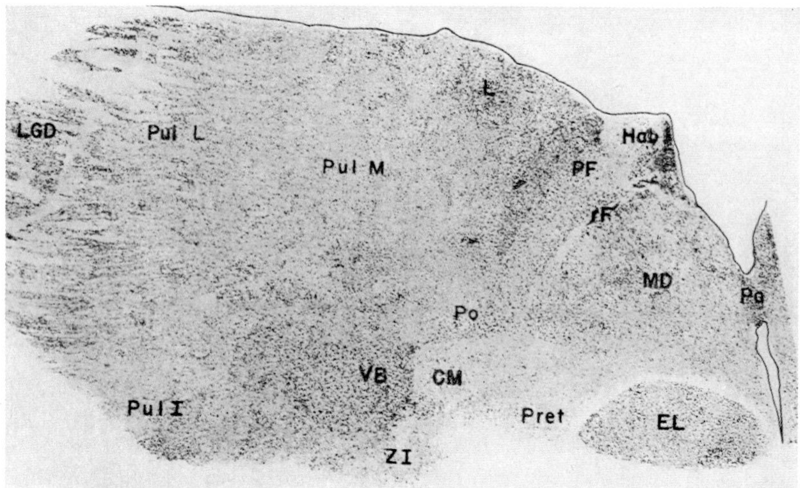

Figure 126. Cross-section *(Nissl stain)* through the thalamus of the Cetacean Dolphin at a caudal level (from KRUGER, 1959). CM: nucleus of centrum medianum; EL: nucleus ellipticus; L: 'lateral nuclear complex' (here presumably including nucleus lateralis posterior and nucleus praetectalis); LGD: dorsal lateral geniculate nucleus; MD: nucleus medialis dorsalis; Pa: 'paraventricular complex'; PF: nucleus parafascicularis; Po: 'posterior nuclear complex' (probably our nucleus posterior); Pret: 'pretectal nuclear group' (probably tegmental field and prerubral tegmentum); Pul I, L, M: pulvinar (with inferior, lateral, and medial grisea); rF: fasciculus retroflexus; VB: 'ventrobasal nuclear complex; ZI: zona incerta.

Figure 125. Cross-section (myelin stain) through the diencephalon of the Carnivore Dog at level of optic chiasma and supraoptic commissures (from RIOCH, 1931). f.thal.F.: '*fasciculus thalami Foreli*' (merely portion of H_1); fi. cort.-tect.: 'fibrae corticotectales'; fi.gen.-occ.: 'fibrae geniculo-occipitales'; inter.thal.rad.: 'intermediate thalamic radiation'; l.m.i.: lamina medullaris interna; n.perif.: nucleus perifornicalis; tr.hyp.-teg.: 'tractus hypothalamico-tegmentalis'; tr.mam.-thal.d., l., m., v.: tractus mammillotegmentalis, of which the cited author distinguishes distinctive dorsal, lateral, medial, and ventral fascicles. Other designations self-explanatory.

Figure 127 A. Cross-section *(Nissl stain)* through the hypothalamus of the Carnivore Dog (from CHRIST, 1969). ch.op.: optic chiasma; com.ant.: anterior commissure; n.interst. str.term.: nucleus interstitialis of stria terminalis; n.preop.mn.: nucleus praeopticus magnocellularis; n.supraop.: nucleus supraopticus; rec.s.o.: preoptic recess; rhinenc.: 'rhinencephalon' (probably caudal end of area ventralis anterior s. tuberculum olfactorium). Other abbreviations self-explanatory. Figures 127 A–D ×18, red. $^3/_5$.

Figure 127 B. Cross-section *(Nissl stain)* through the hypothalamus of the Dog at level of dorsomedial and ventromedial hypothalamic nuclei (from CHRIST, 1969). med.em.: median eminence (hypophysial complex); n.arc.: nucleus arcuatus. Other abbreviations self-evident.

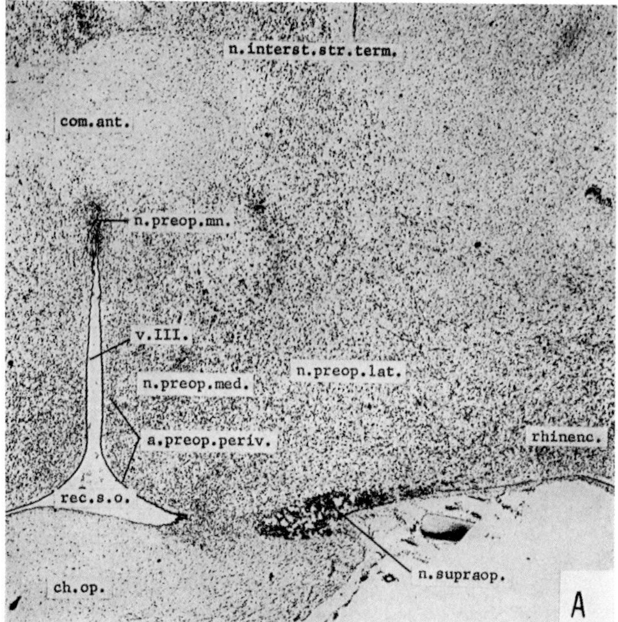
n.interst.str.term.
com.ant.
n.preop.mn.
v.III.
n.preop.med.
n.preop.lat.
a.preop.periv.
rhinenc.
rec.s.o.
n.supraop.
ch.op.
A

127 A

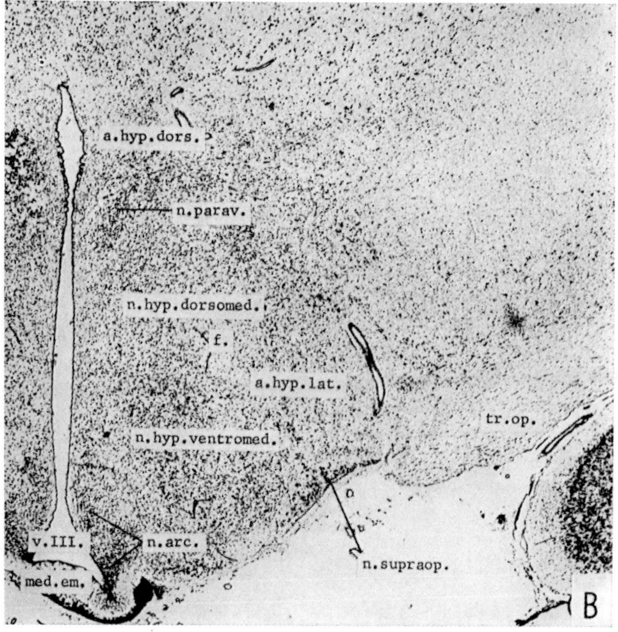
a.hyp.dors.
n.parav.
n.hyp.dorsomed.
f.
a.hyp.lat.
tr.op.
n.hyp.ventromed.
v.III.
n.arc.
med.em.
n.supraop.
B

127 B

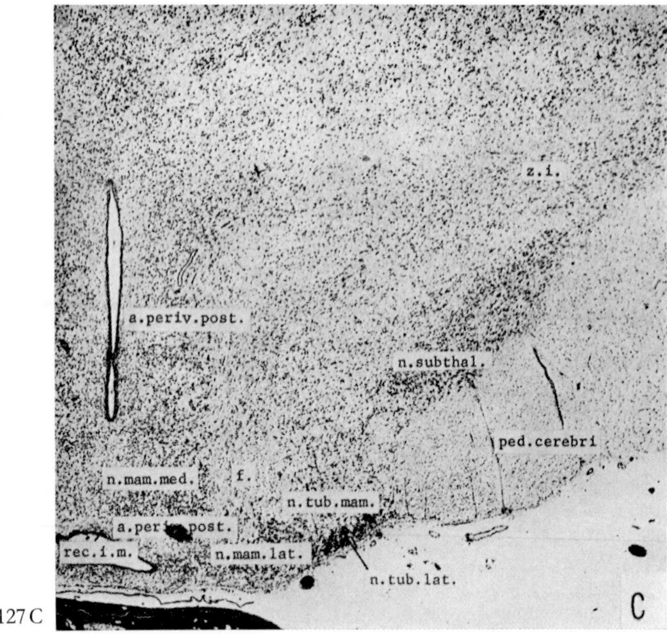
z.i.
a.periv.post.
n.subthal.
ped.cerebri
n.mam.med.
f.
n.tub.mam.
rec.i.m.
n.mam.lat.
n.tub.lat.
C

127C

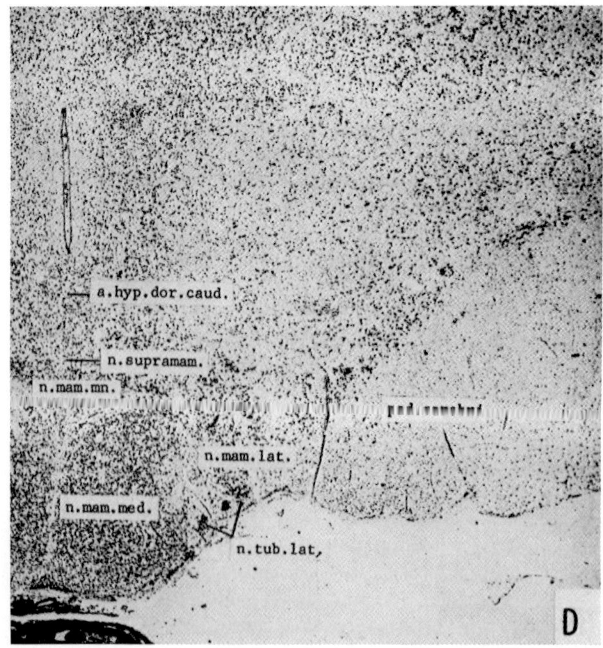
a.hyp.dor.caud.
n.supramam.
n.mam.mn.
n.mam.lat.
n.mam.med.
n.tub.lat.
D

127D

played by the Mammalian diencephalic grisea, a brief consideration of the generally recognized main communication channels may provide a suitable approach (Figs. 128 A-C).

At the caudal end of the diencephalon, in the region comprised by the diencephalomesencephalic boundary zone, there are a number of fiber systems through which epithalamic, thalamic, and hypothalamic grisea and, moreover, directly or indirectly, telencephalic grisea are connected with the more caudal centers of the neuraxis. These channels include (1) the medial lemniscus system, (2) the tegmental field, (3) the pretectal field, (4) brachium quadrigeminum superius[213a], (5) brachium quadrigeminum inferius with lateral lemniscus system, (6) tractus tectothalamicus dorsomedialis, (7) components of central tegmental tract, (8) components of fasciculus longitudinalis medialis, (9) *fasciculus longitudinalis dorsalis of Schütz* with additional more dorsal periventricular fibers, (10) the fasciculus retroflexus, (11) the mammillotegmental and mammillo-interpeduncular tracts, (12) the mammillary peduncle, and (13) the pes pedunculi which is the caudal continuation of predominantly descending channels essentially of telencephalic cortical origin, passing through the hemispheric stalk, respectively the capsula interna to be discussed further below.

Definable *intrinsic diencephalic channels* are represented by (14) the optic nerve respectively optic tract, (15) the mammillothalamic tract, (16)

[213a] Optic tract fibers reaching tectum mesencephali run within or in close vicinity to brachium colliculi superioris, while some optic input is also directed to tegmental grisea (cf. section 9, chapter XI, volume 4, and K., 1954, concerning additional details of these and other diencephalo-mesencephalic communication channels). It will be noted that the optic tract is both a caudal and an intrinsic diencephalic communication channel.

Figure 127C. Cross-section *(Nissl stain)* through the hypothalamus of the Dog at the level of tuberal nuclei (from CHRIST, 1969). a.periv.post.: posterior paraventricular area, of which a dorsal and a ventral portion are shown; n.mam.lat., med.: rostral parts of lateral and medial mammillary grisea; n.subthal.: nucleus subthalamicus *(sive* corpus subthalamicum); n.tub.lat.: lateral tuber(al) nuclei; n.tub.mam.: tuberomammillary nucleus; rec.i.m.: inframammillary recess.

Figure 127D. Cross-section *(Nissl stain)* through the hypothalamus of the Dog at the level of full expansion of mammillary grisea (from CHRIST, 1969). a.hyp.dor.caud.: dorsocaudal hypothalamic area; n.mam.mn.: median mammillary nucleus; n.supramam.: supramammillary nucleus. Other abbreviations as in Figures 127A–C.

portions of ansa lenticularis (pallidothalamic and pallidohypothalamic fibers), (17) intradiencephalic periventricular fibers (18) the hypothalamo-hypophysial pathways, as already dealt with in the introductory subsection 1 C of this chapter,[214] and (19) undefined additional intradiencephalic connections. The thalamic grisea, moreover, are generally separated from each other by myelinated fiber layers designated as medullary laminae, which contain fibers that may pertain to both in-

[214] Dubious diencephalo-epiphysial fiber connections were likewise briefly referred to in that subsection.

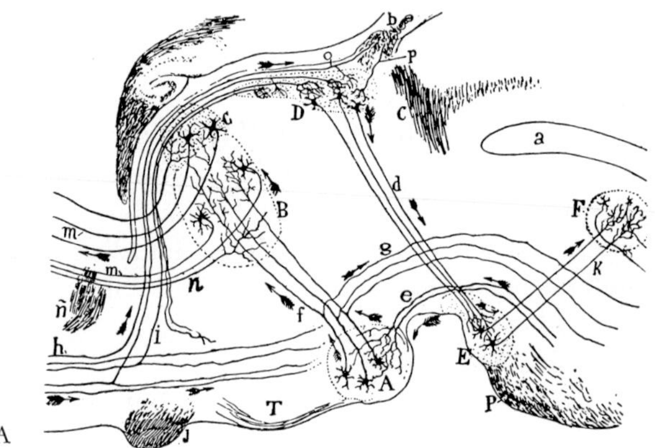

Figure 128 A. Diagram of afferent and efferent channels related to mammillary, habenular, and anterior dorsal thalamic grisea in Mammals (from CAJAL, 1911). A: medial mammillary nucleus; B: nucleus anterior ventralis (et medialis?) thalami; C: nucleus anterior dorsalis thalami (left), commissura posterior (right); D: habenular grisea; E: interpeduncular nucleus; F: nucleus dorsalis tegmenti; I: stria medullaris system; J: optic chiasma respectively tract; P: mesencephalic portion of cerebral peduncle; T: 'tuberal region of hypothalamus; a: cerebral aqueduct; d: fasciculus retroflexus; e: mammillary peduncle; f: mammillothalamic tract; g: mammillotegmental tract; h: ('medial') basal forebrain bundle; k: interpedunculotegmental tract; m: thalamocortical fibers; n: corticothalamic fibers; ñ: anterior commissure; o: fibers of striae medullaris to habenular commissure; p: fibers crossing through habenular commissure.

Figure 128 B. Diagram of the afferent sensory channels *(ruban de Reil médian et voie du trijumeau)* as related to the Mammalian prosencephalon according to CAJAL's interpretation (from CAJAL, 1911). A: nucleus ventralis posterolateralis complex *('noyau sensitif principal de la couche optique');* B, C: portions of ventrolateral complex, cf. Figs. 119 C and D *('noyaux sensitifs accessoires ou trigéminaux');* D: pretectal grisea; E: zona incerta; F: lateral mammillary griseum; G: lemniscus medialis; H: ascending secondary trigeminal pathway with additional ascending sensory channels; I: pedunculus corporis mammillaris; J: optic chiasma respectively tract; K: hippocampal formation; R: colliculus superior mesencephali; S: tectal fiber systems; T: sensorimotor neocortex; V: visual neocortex; a:

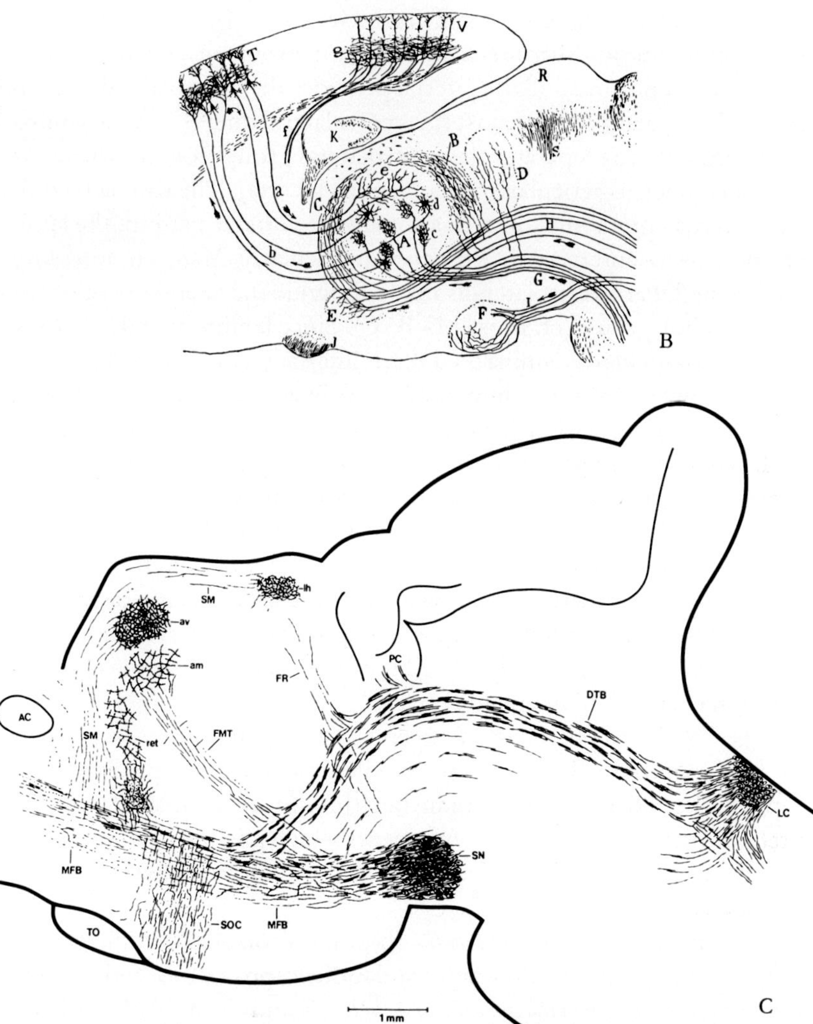

corticothalamic fibers; b: thalamocortical fibers; c: thalamic end arborizations of medial lemniscus; d: origin of thalamocortical fibers; e: endings of corticothalamic fibers; f: central (geniculocortical) optic pathway; g: fiber system of visual cortex.

Figure 128 C. Semidiagrammatic representation of the ascending 'locus coeruleus CA system' and its projection to the thalamus, based upon a combination of sagittal section in several planes (from LINDVALL *et al.*, 1974). AC: anterior commissure; DTB: dorsal periventricular bundle; FMT: fasciculus mammillothalamicus; FR: fasciculus retroflexus; LC: locus coeruleus; MFB: basal ('medial') forebrain bundle; PC: posterior commissure; SM: stria medullaris; SN: substantia nigra; SOC: supraoptic commissure; TO: optic tract; am: nucleus anterior medialis; av: nucleus anterior ventralis; lh: lateral habenular nucleus; ret: reticular nucleus (thalamus ventralis).

trinsic as well as extrinsic connections. Generally speaking, the lamina medullaris anterior surrounds the anterior group of dorsal thalamic nuclei, the lamina medullaris interna provides the ventrolateral boundary of the medial group against the ventrolateral one,[215] and the lamina medullaris externa separates the ventrolateral group from the thalamus ventralis (nucleus reticularis thalami, zona incerta). The lamina medullaris interna commonly joins, in a somewhat variable pattern, the lamina medullaris anterior; in Primates, particularly Man, an accessory lamina medullaris dorsolateralis may subdivide the lateral grisea of the ventrolateral group (cf. Figs. 143 B, C). Since lamina medullaris anterior and particularly lamina medullaris interna seem to be comparatively indistinct and variable in various subprimate Mammals, some ambiguities of interpretation have here resulted in this respect, and also with regard to the grisea of the intralaminar group, which appear located within the region of lamina medullaris interna.

From a viewpoint of *neurochemistry*, emphasizing the 'adrenergic innervation' of the diencephalon, respectively the distribution of catecholamine (CA) fibers, recent investigations with specific techniques have led to the definition of the following so-called CA systems (LINDVALL *et al.*, 1974, in the Rat) related to the caudal and intrinsic communication channels enumerated above.

The '*locus coeruleus system*' (a) ascends[216] through dorsal components of the longitudinal tegmental bundles to the diencephalo-mesencephalic boundary where it branches through tegmental (2) and (3) pretectal fields into several components which join the basal forebrain bundle and diverse diencephalic grisea extending from epithalamus to hypothalamus (cf. Fig. 128 C).

The '*dorsal periventricular bundle*' (b), said to originate in the central gray of the mesencephalon and caudal diencephalon, ascends within the *fasciculus longitudinalis dorsalis of Schütz* (9) and the more dorsally located periventricular fibers, projecting, through the diencephalic peri-

---

[215] To some extent also from the posterior group.

[216] Cf. volume 4, chapter IX, section 10, p. 544; chapter XI, section 9, p. 963. It will be recalled that the particular importance of the nucleus loci coerulei was suggested by the results of studies with techniques of histochemistry. It should also here be added that, although the traditional or 'classical' techniques of neuroanatomy did not disclose conclusive data concerning the connections of said nucleus, data obtained by LORENTE DE NÓ (1922) and others seem to indicate that it receives afferent root fibers of the trigeminus. This, however, hardly justifies the conclusion by the cited author '*que el locus coeruleus pertenece por entero al sistema del nervio trigemino*'.

ventricular system, to hypothalamic, midline thalamic, as well as to pretectal grisea and epithalamus.

Additional components of the CA system ending in paraventricular thalamic grisea (c) seem to ascend, through tegmental bundles, from cells located in the rhombencephalic reticular formation.

A system of delicate, probably dopamine-containing axons (d) is described in the 'caudal thalamus', the zona incerta, and dorsal as well as rostral hypothalamus. It seems to originate from a cell group of 'caudal thalamus' extending into the aqueduct and from caudomedial portions of zona incerta, being interpreted as 'a hitherto unknown intradiencephalic dopaminergic system', which could be subsumed under, respectively included in, channels (17) and (19) of the classification adopted for the present purpose and enumerated above.

Still not entirely elucidated axonal projections of medial preoptic and anterior hypothalamic neurons have been inferred on the basis of studies with tritiated amino acid autoradiography by CONRAD and PFAFF (1975) in the Rat.

Turning now to the *rostral communication channels* of the diencephalon, the following systems can be distinguished. Stria medullaris (20), stria terminalis[217] (21), inferior thalamic peduncle (22), telencephalic components of the above-mentioned ansa lenticularis (16), fornix (23), basal (so-called medial) forebrain bundle (24), and capsula interna system (25). This latter, in addition to channels such as the corticobulbar, corticospinal, corticopontine, corticorubral and corticonigral components, which take their course through the diencephalon, includes the frontal thalamic radiation (25a), the parietal thalamic radiation (25b), the limbic thalamic radiation (25c), the occipital thalamic radiation (25d), the insular thalamic radiation (25e), and the temporal thalamic radiation (25f). Components of (25d) and of (25f) are the optic radiation and the auditory radiation, respectively. Systems 25a to f can also be designated as the thalamic peduncles or stalks connecting with the enumerated telencephalic regions and consisting of thalamocortical and corticothalamic fibers. The above-mentioned inferior thalamic peduncle (22) is a radiation not definitely included within the capsula in-

---

[217] The stria terminalis system, particularly characteristic for Mammals, is a complex system, also designated as taenia semicircularis, arising mainly in the nucleus amygdalae complex and following the course of sulcus terminalis toward the interventricular foramen. Here, in the neighborhood of anterior commissure, various components of stria terminalis become separated.

terna but forming, together with ansa lenticularis, a loop basally to internal capsule *sensu strictiori*. The combined loop of *ansa lenticularis* and *inferior thalamic peduncle* is also known as *ansa peduncularis*, particularly in those Mammals whose inferior thalamic stalk is conspicuously developed, such as e.g. Primates, which possess a large temporal lobe with a rostral extension that includes the piriform lobe and the nucleus amygdalae. Recent, but still rather inconclusive views on ansa lenticularis and fasciculus thalamicus (H_1) can be found in a publication by KUO and CARPENTER (1973). A more or less distinctive diencephalic component of the internal capsule is provided by corticohypothalamic fibers (25 g) which, at diencephalic levels, may join the basal forebrain bundle.

The *commissural systems* of the diencephalon, discounting the essentially telencephalic commissura anterior, and the variable amount of commissural fibers passing through the likewise variable massa intermedia, comprise (1) the habenular commissure, (2) the posterior commissure, (3) the optic chiasma, (4) the supraoptic commissures, and (5) the supramammillary commissure.

The *habenular commissure*, located rostrally to the pineal recess, seems to contain decussating fiber of the stria medullaris system (including pretectal components) together with true commissural fibers interconnecting the habenular grisea.

The *posterior commissure*, likewise containing decussating and true commissural fibers, represents a boundary structure pertaining to both diencephalon and mesencephalon, being particularly related to the pretectal grisea.

The *optic chiasma*, in which a variable number of optic nerve fibers decussate, while other such fibers join the homolateral optic tract, was dealt with in section 1 B of this chapter.

The *supraoptic decussations or commissures* are formed by fibers which, although not belonging to the optic tract, cross the midline in close contiguity with optic fibers. Conventional descriptions of the supraoptic decussations, as e.g. given by KAPPERS *et al.* (1936) refer to three supraoptic decussations or commissures: the pars dorsalis of the commissura supraoptica dorsalis *(decussation of Ganser)*, the pars ventralis of the commissura supraoptica dorsalis *(decussation of Meynert)*, and the *commissura supraoptica ventralis of Gudden*. I have preferred to distinguish merely a dorsal and a ventral supraoptic commissure (K. 1954), both of which appear somewhat variable in the taxonomic Mammalian series. The dorsal supraoptic commissure may include fibers related to ansa lenticularis, zona incerta, and hypothalamic grisea. The ventral supraoptic commissure may contain tectopretectothalamic, tectopretectohypothalamic fibers, fibers of the lemniscus system,[217a] and fibers of ventral medial geniculate griseum.

Crosby *et al.* (1962) likewise refer to a dorsal and to a ventral supraoptic commissure. These authors bring a summary, with bibliographic references, of the various opinions expressed about the channels passing through said commissures, justly said to be 'among the least well-understood connections' of the brain.

The *supramammillary commissure* (posterior hypothalamic decussation), considered by Crosby *et al.* (1962) to be the most rostral of the midbrain commissures, should, although located in the diencephalo-mesencephalic boundary zone, rather be classified among the diencephalic commissures. Crossing tectopretectotegmental fibers, crossing components of tractus tectoreuniens, fibers from midline nuclei and habenulopeduncular fibers may be present in this commissure which also seems to include fibers between subthalamic nucleus and tegmentum, intermammillary fibers, decussating fornix fibers, fibers of the pallidosubthalamic system, and perhaps some fibers of the basal optic tract. The supramammillary commissure, apparently first described by Forel in 1872, and occasionally designated as *Forel's commissure*, should not be confused with *Forel's ventral tegmental decussation* of tractus rubro-bulbo-spinalis in the mesencephalon (cf. section 9, chapter XI, vol. 4).

Illustrations of representative diencephalic fiber systems, as seen in sagittal sections, and taking the Mouse as a paradigm, are shown in Figures 119 B, C, and D. Figures 128 A and B depict, in a more generalized schematic fashion, some of the main Mammalian diencephalic fiber tracts as interpreted by Cajal (1911, 1955). Figure 128 C represents a contemporary concept of catecholamine (CA) fiber tracts, including the so-called *ascending locus coeruleus CA system*.

With respect to the *communication channels* of the Mammalian prosencephalon, an extensive and very detailed dissertation was prepared by Knook (1965) at the University of Leiden.[218] On the basis of experimental work on the Rat brain with the *Nauta-Gygax method*, and using additional normal material, this author has presented a critical review

[217a] Some fibers of medial lemniscus might also be present in the dorsal supraoptic commissure (Glees, 1944).

[218] Knook (1965): The fibre-connections of the forebrain. A critical review of the hodology of the telen- and diencephalon with the adjacent mesencephalon, especially with regard to the basal ganglia, based on acute and chronic degeneration experiments in the Rat.

of the hodology in telencephalon and diencephalon with the adjacent mesencephalon. Approximately 700 pertinent publications were consulted, of which 500 were found relevant. About 1,100 connections previously described in the forebrain and adjacent mesencephalon, grouped according to their grisea of origin, were considered and checked in the experimental Rat material.

The cited author justly starts with the reasonable, although, as he judiciously remarks 'somewhat questionable' conception that the majority of the cerebral connections are present throughout the Mammalian class, and aims 'to give some idea of what can be considered as certain in the subject studied and what has still to be investigated and substantiated'. He points out that it is hardly feasible to 'invent' a connection in the forebrain that has not yet been described. According to KNOOK, 'it seems incredible that, even in the most recent textbooks of neuroanatomy, presented to our students of medicine, connections invented as long as forty years ago and frequently denied since, can be found'. He adds that, 'when the publications based on staining techniques of which the insufficiency has been proved are eliminated, together with publications of which the results cannot be controlled by lack of a complete survey of the lesions and their resulting degeneration, neuroanatomy might be liberated from the heavy burden of worthless historical material'.[219]

[219] According to KNOOK, 'objectionable' are studies based on old-fashioned methods such as the *Marchi-* and *Weigert-Pal techniques*, 'which already long have been shown inadequate and misleading, while moreover yet we have much more appropriate techniques at our service as the methods of *Nauta-Gygax* and of *Häggquist*'. While much of this criticism is doubtless justified as I have myself previously repeatedly emphasized (cf. e.g. 'appraisal of fiber tracts', p. 93, K., 1954, *et passim)*, the quoted author seems to overestimate the accuracy of the *Nauta-Gygax* and related 'terminal degeneration techniques', which include the subsequently introduced *Fink-Heimer technique*. Still more recent methods purporting to improve the tracing of connections are based on autoradiographic demonstration of anterograde axonal transport of tritium labeled proteins or on histochemical labeling involving the retrograde axonal transport of protein such as the enzyme horseradish peroxidase (cf. COWAN *et al.*, 1972; LA VAIL *et al.*, 1973). Yet, I maintain here my statement (p. 677, vol. 3/I) that an investigator postulating certain fiber connections will always be able, *per fas aut nefas*, to provide the semblance of a 'proof' by means of these techniques. *Mutatis mutandis*, comparable inadequacies, uncertainties, ambiguities, and sources of error are inherent in all other modern neurobiological research techniques, such as electron microscopy, tissue culture, histochemical methods, and the diverse procedures of electrophysiology including their microelectrode and single unit recordings etc., not to mention the 'psychologic factors' related to the personal make-up of the multitudinous investigators and the 'spirit' or 'ethos' of their 'schools' (cf. e.g. K., 1961b, §79, pp. 511f.).

Many contradictions in the literature evidently originate from misinterpretations, especially of fibers of passage, which in most instances disturb the patterns of degeneration. Moreover, the cited author is compelled to admit that, in his own material, he was unable to either confirm or deny the existence of numerous inferred connections. He considers it useless to map out a functional hypothesis from the data either confirmed or probable.[220] The framing of circuits, as is done on the basis of anatomical investigations, is adjudged 'senseless' because of the large number of evident connections whose particularly physiologic significance remains insufficiently established.[221]

Yet, despite these drawbacks, and although, figuratively speaking, everything is connected with everything else within the neuraxis, an orderly and fairly convincingly substantiated pattern of distribution as well as connection of main communication channels within the central nervous system can be inferred. Such concepts provide overall valid and operational useful diagrams of neuronal mechanisms, both on the basis of the older and of the more recent methods of approach. A sceptical attitude, stressing the flimsiness of numerous available data, is fully compatible with the formulation of such concepts in the description and 'explanation' of configuration and function displayed by the central nervous system. In this respect, I believe that KNOOK's criticism, particularly with regard to the older methods and pioneer investigators, somewhat 'overshoots the mark'.

Reverting to the *diencephalic grisea*, the *epithalamic habenular nuclei* may now be considered with respect to their *communication channels*. Figures 129 A and B illustrate medial and lateral habenular nucleus[222] with some of their fiber connections.[223] Most or many of the descending fibers in the *stria medullaris* terminate in these grisea, from which the *fasciculus retroflexus* (or tractus habenulo-interpeduncularis) arises. The

---

[220] Thus, the cited author points out that 'it is e.g. impossible to imagine which of the more than 17 anatomical possibilities a cortical impulse will use to reach the substantia nigra'.

[221] The cited author adds: 'for it has to be kept in mind that a certain fibre-system cannot be proved to terminate just on the cells forming the subsequent neuron in the hypothetical circuit'.

[222] Both habenular nuclei, at least in some Mammalian forms, can again be subdivided into cell groups, but this additional parcellation need not be considered essential in the present context.

[223] The commissura habenulae is dealt with further below in a brief discussion of the diencephalic commissural systems.

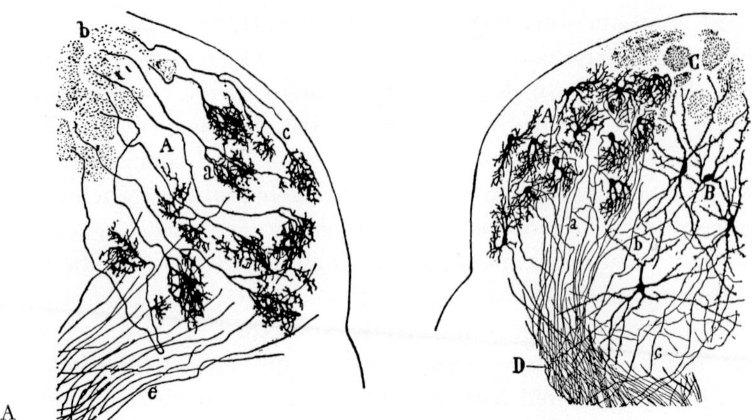

Figure 129 A. Cross-section *(Golgi impregnation)* through the habenular grisea of the Rabbit (from CAJAL, 1911). A: fibers from stria medullaris system; a: end arborizations of stria medullaris fibers in medial habenular nucleus; b: stria medullaris; c: fibers with collateral arborizations; e: fibers of fasciculus retroflexus. The medial (ventricular) surface is at right.

Figure 129 B. Cross-section *(Golgi impregnation)* through the habenular grisea of the Dog (from CAJAL, 1911). A: medial habenular nucleus; B: lateral habenular nucleus; C: stria medullaris; D: fasciculus retroflexus; a, b: origin of fasciculus retroflexus fibers from medial respectively lateral habenular nucleus; c: possibly afferent fiber of fasciculus retroflexus. The ventricular (medial) surface is here at left.

diverse components of the stria medullaris are still incompletely understood, but it seems certain that many pertain to the olfactory system, originating in the tuberculum olfactorium and its neighborhood. Others, also designated as septohabenular tract, originate in the para- and preterminal ('septal') grisea. In addition, fibers from the hippocampal formation leave the fornix and join the stria medullaris. The stria terminalis likewise contributes to the stria medullaris, which, at the rostral extremity of the diencephalon, is thus formed by branches from various channels, some of which leave the basal forebrain bundle. To this system are added fibers from hypothalamus (particularly preoptic region) and thalamus. The fasciculus retroflexus, moreover, doubtless includes ascending fibers providing input to epithalamus. Still other presumably reciprocal caudal connections obtain between epithalamus and pretectal as well as tectal grisea, mainly by way of the tractus tectothalamicus dorsomedialis. Although the epithalamus is generally considered to represent an 'olfactosomatic correlation center', this perhaps not entirely unjustified interpretation requires various qualifications. There are evidently also 'visceral', i.e. 'vegetative' interrelationships between epithalamus and hypothalamus, quite apart from relevant non-olfactory connections, including those from hippocampus respectively so-called limbic system. Thus, as already pointed out in the introductory section 1 A, and particularly with respect to the various groups of Mammals, the functional role of the epithalamus remains obscure.

With respect to the *thalamus dorsalis*, the important grisea of the ventrolateral group include those related to the medial lemniscus system and to the brachium conjunctivum component of the tegmental radiation.

The *medial lemniscus system* (1) comprises the bulbothalamic tract or lemniscus medialis *sensu strictiori* (as originating from nucleus gracilis and cuneatus), moreover, the ventral and lateral spinothalamic tracts, and the related secondary tracts from afferent cranial nerve nuclei, such as the trigeminal lemniscus and the gustatory lemniscus dealt with in volume 4. These channels, rostrally overlapping with portions of the tegmental fiber systems within the tegmental field, mainly reach nucleus ventralis posterolateralis and ventralis posteromedialis (respectively the subdivisions of these grisea) which seem to represent the cortical relay nuclei of the sensory channels included in the medial lemniscus system. This latter, however, may also provide some input to the pretectal grisea.

Some of the still unsettled questions concerning details of spinothalamic tract and medial lemniscus in general are discussed, with regard to Primates, by Glees (1952, 1957).

Recent, but by no means convincing views on cytoarchitecture and elaborate details of 'somatic sensory connectivity' in thalamic nuclei other than the ventrobasal complex (in the Cat) can be found in a paper by Jones and Burton (1974).

The *nucleus ventralis posterolateralis* seems to receive the input from extremities and trunk mediated by the lemniscus system, while the input from the head provided by that channel accordingly reaches nucleus ventralis posteromedialis. A somatotopic distribution of the type depicted in Figure 130 appears to obtain. The gustatory lemniscus may terminate in a ventral and medial subdivision of nucleus ventralis posteromedialis. The general visceral lemniscus seems also to terminate within either this or an additional neighboring medial region of nucleus ventralis posteromedialis.

Thalamocortical fibers, presumably combined with some reciprocal corticothalamic ones, connect, through the parietal thalamic radiation

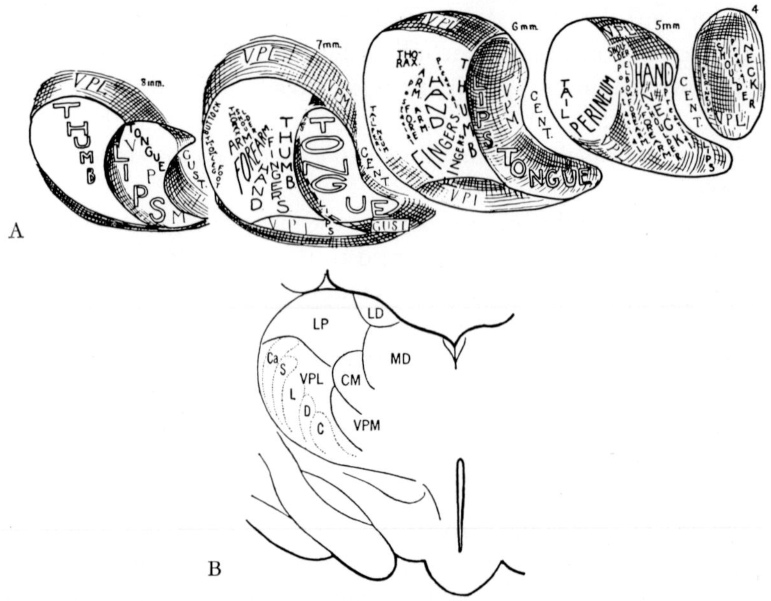

Figure 130 A. Diagram illustrating somatotopic projection of the ascending 'tactile' channel upon the nucleus ventralis posterolateralis and postermedialis complex in the thalamus of the Macaque Monkey, visualized in successive millimetric slices (adapted after MOUNTCASTLE and HENNEMAN, 1952, from KRIEG, 1966). VPL: nucleus ventralis posterolateralis; VPI: 'nucleus ventralis posterior inferior'; VPM: nucleus ventralis posteromedialis; CENT: location of centrum medianum of Luys. Caudalmost level at right, rostralmost at left.

Figure 130 B. Diagram of cross-section through diencephalon of Spider Monkey, showing somatotopic projection in nucleus ventralis posterolateralis (after CHANG and RUCH, 1947, from RUCH, 1961). CM: centrum medianum; LD: nucleus lateralis dorsalis; LP: nucleus lateralis posterior; MD: nucleus medialis (dorsalis); VPL: nucleus ventralis posterolateralis; VPM: nucleus ventralis posteromedialis. Somatotopic sequence: Ca: caudal (tail); S: sacral; L: lumbar; D: thoracic; C: cervical.

(25b), these nuclei with the 'sensory' parietal neocortex,[223a] the gustatory and visceral channel, however, seems to be directed toward the insula (cf. e.g. WOLF, 1968).

Near the rostral end of nucleus ruber tegmenti the fiber capsule of

---

[223a] The somesthetic parietal thalamic radiation also overlaps to some extent with the frontal radiation (25a) to 'motor' cortex and thus can be said to reach the 'sensorimotor cortex'. These cortical regions, as well as the presumptive significance of the insula, etc., and the topography of cortical 'projection' and 'association' areae respectively 'centers' are dealt with in chapters XIII and XV. Some basic data related to the features of the overall morphologic pattern were discussed on pp. 647–668, section 6, chapter VI of volume 3/II.

that griseum, together with fibers originating as well as passing through said nucleus, represents the tegmental field (2) which overlaps with portions of the medial lemniscus. The fibers included in the tegmental field pertain to the brachium conjunctivum system, to *Flechsig's Haubenstrahlung*,[223b] to the tegmental CA systems (a) and (c) mentioned above, to rostral components of the central tegmental tract, and to additional tegmental channels of unclarified significance. Its rostral portion, medially to zona incerta, is the tegmental field H or *Haubenfeld of Forel* (1877) which contains scattered cells of the prerubral tegmentum. It bifurcates laterad into field H_1 (fasciculus thalamicus, blending with lamina medullaris externa) and into field H_2 (fasciculus lenticularis) which extends laterad between zona incerta and nucleus subthalamicus.

The tegmental field in the wider sense of our own descriptions (K. and MILLER, 1942, 1949; K. 1954) includes the *field H of Forel*, but extends somewhat farther caudad and apparently contains other, less well distinguishable fiber systems than those passing through the field H or tegmental field *sensu strictiori (Haubenfeld)* as depicted by FOREL (1877, 1907) in his pioneering studies (cf. also Fig. 425 C, p. 985, vol. 4). Various aspects of the tegmental field and its neighborhoods are also considered and interpreted by v. MONAKOW (1918).

With respect to the many uncertain and controversial aspects of neuraxial hodology, it may be mentioned that the apparently well-documented rubrothalamic channel was recently denied in the Monkey by HOPKINS and LAWRENCE (1975). Readers interested in contemporary interpretations of nucleus ruber connections are referred to that paper which includes the thereto pertinent bibliography. Figures 130 C and D illustrate present-day concepts of overall synaptic arrangements in dorsal thalamic relay nuclei as discussed in a recent summary, by SHEPHERD (1974), of 'synaptic organization' within diverse neuraxial grisea.

As far as the grisea of the *ventrolateral group* are concerned, the fibers of the tegmental field pertaining to brachium conjunctivum system reach the nucleus ventralis lateralis through fasciculus thalamicus and lamina medullaris externa. Said nucleus, in turn, is connected by way of capsula interna through the frontal thalamic radiation (25a) with

[223b] The *Haubenstrahlung* of FLECHSIG presumably includes the 'ascending reticular pathway' and cannot be clearly separated from the other here tentatively enumerated channels. Cf. also volume 4, chapter IX, section 2, p. 342.

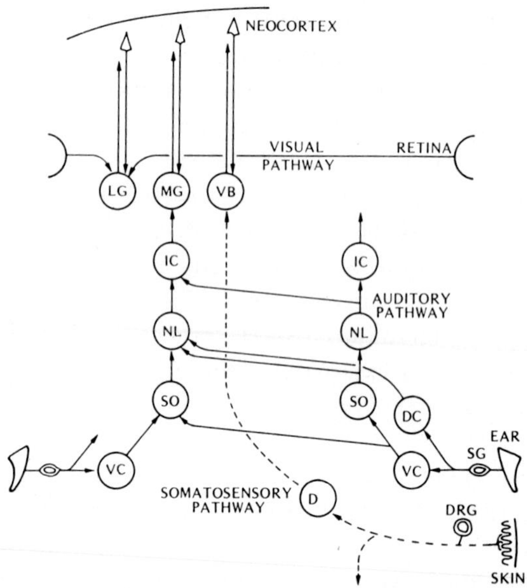

Figure 130C. Diagram purporting to illustrate the generally recognized communication channels and overall synaptic sequences related to dorsal thalamic relay nuclei and their projections to the neocortex (from SHEPHERD, 1974). D: nn. of posterior funiculi; DC: dorsal cochlear nucleus; DRG: spinal ganglion; IC: inferior colliculus; LG, MG: lateral respectively medial geniculate grisea; NL: nn. of lateral lemniscus; SG: *ganglion spirale Corti;* SO: superior olive; VB: caudal ventrolateral and medial basal griseal complex of thalamus dorsalis; VC: ventral cochlear nucleus.

premotor and motor cortex. The nucleus ventralis anterior, on the other hand, seems mainly connected with globus pallidus and striatum by way of the ansa lenticularis (16). Other fibers of this type may follow the fasciculus lenticularis (H_2) as far as field H and then bend back through fasciculus thalamicus (H_1) to reach grisea of the ventrolateral group. Nothing certain is known concerning the nucleus lateralis dorsalis, which may have reciprocal connections with regions of the parietal or of the parahippocampal cortex.

The *anterior group* of dorsal thalamic nuclei, comprising nuclei anterior dorsalis, anterior ventralis, and anterior medialis, moreover nuclei parataenialis and paramedianus, seems mainly to receive the mammillothalamic tract originating in the hypothalamic mammillary grisea (cf. Fig. 131). Direct input from the hippocampus by way of the 'postcommissural fornix' is also said to obtain (cf. e.g. ZEMAN and INNES, 1963).

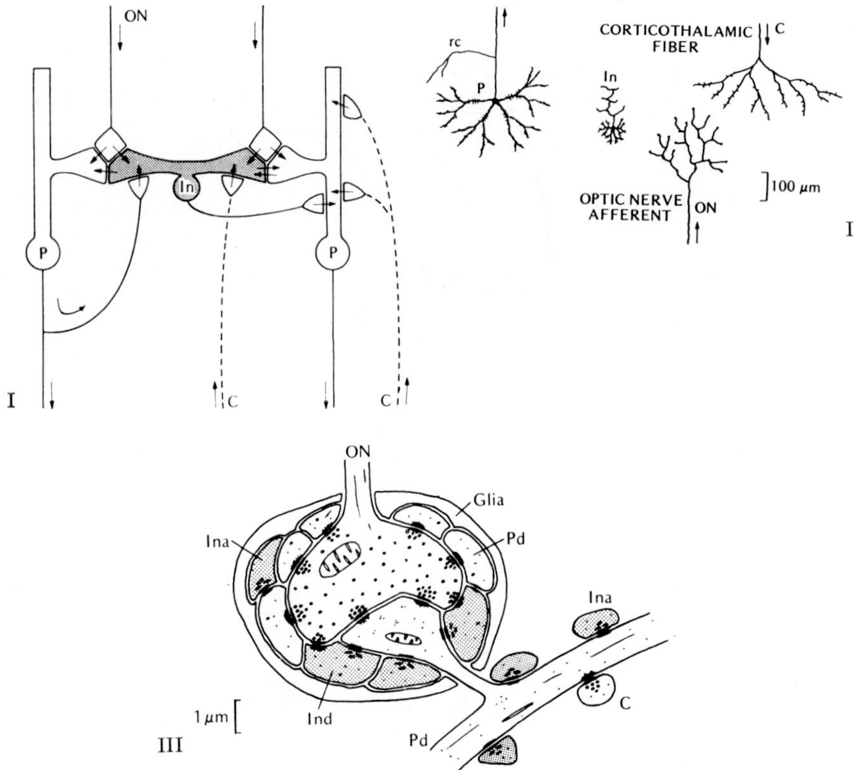

Figure 130 D. Three diagrams purporting to illustrate further details of the 'basic circuit diagram' for thalamic sensory relay nuclei with particular respect to so-called 'intrinsic cells' (combined after diverse authors, from SHEPHERD, 1974). I: Assumed basic circuitry. II: Basic types of neuronal elements and terminal arborizations. III: Synaptic connections as presumed for dorsal lateral geniculate griseum (so-called synaptic glomerulus). C: corticothalamic connections; In: 'intrinsic cell'; Ina: intrinsic cell axon; Ind: intrinsic cell dendrite; ON: optic nerve terminal; P: 'principal cell'; Pd: principal cell dendrite; rc: recurrent collateral.

Grisea of the anterior group, in turn, provide input to the parahippocampal cortex, and form part of the circuit described by PAPEZ (1937) as related to the mechanism of 'emotion', which shall be dealt with further below in connection with the Human diencephalon and with the relevant telencephalic grisea. Details concerning function and further fiber connections of the anterior nuclear group remain unclarified, but relationships of at least nn. parataenialis and paramedianus to the stria medullaris system seem probable.

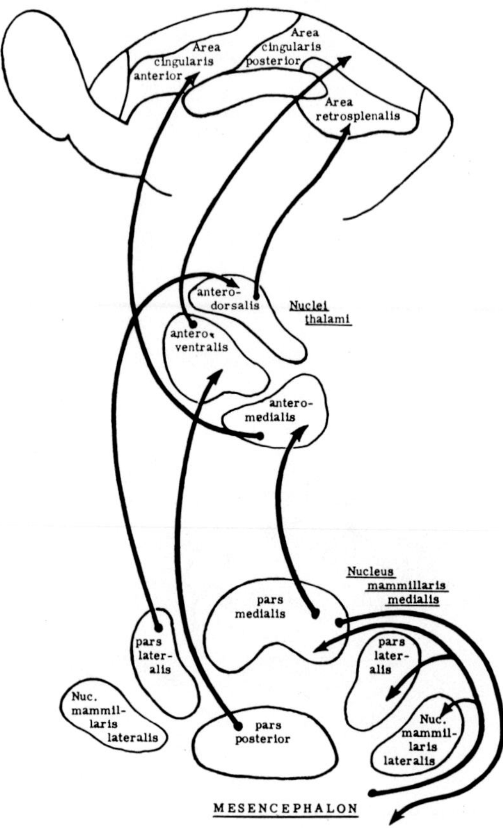

Figure 131. Diagram of the Mammalian mammillo-thalamic-limbic (cingulate) 'projection system' (after POWELL, 1958, from ZEMAN and INNES, 1963). MESENCEPHALON refers only to the two lowermost mammillary input and output fibers.

The *medial dorsal thalamic group* includes two different griseal subsets, nucleus medialis (s. dorsomedialis), and the complex of nucleus parafascicularis and centrum medianum. The nucleus medialis seems to be connected, by a reciprocal channel, with the rostral neocortex (the so-called prefrontal cortex of 'higher' Mammals).

Moreover, at least in Primates, connections with undefined regions of the temporal lobe by way of the inferior thalamic peduncle seem likely. In addition, complex intradiencephalic connections obtain, e.g. with centrum medianum, and particularly also by way of the periventricular system, with undefined hypothalamic grisea. The nucleus medialis appears related to a corticothalamic feedback circuit participating in the mechanism of '*emotion*'. Said mechanism also includes the *Papez circuit* mentioned above, as well as additional hypothalamic and telencephalic preterminal ('septal') components.

The *nucleus parafascicularis* and the *nucleus of centrum medianum*, which can be evaluated as a derivative of the parafascicular griseum are generally believed to have poorly understood connections with corpus striatum and globus pallidus, in addition to intrathalamic ones. Their caudal connections by way of tegmental and pretectal fields remain uncertain. ZORUB and RICHARDSON (1973), on the basis of electrophysiologic experiments in Cats, assume both general lemniscal and extralemniscal input into centrum medianum and pulvinar, the extralemniscal input being mediated by the 'bulbar reticular formation'.

The *midline nuclei*, variously delimited and designated by the different investigators,[224] include the diverse nn. reunientes, and seem particularly related to the periventricular fiber system as well as to other intrinsic diencephalic connections. Caudal ascending channels from tectum and tegmentum appear to reach these nuclei. Although the midline nuclei are said to influence neocortical activities, direct connections with the telencephalic cortex are somewhat dubious, and I was inclined to evaluate these grisea as 'indirect cortical modulators' (K., 1954). It may be assumed that the neocortical 'recruiting responses' are elicited by multisynaptic pathways, in part mediated by the corpus striatum (cf. e.g. POWELL, 1958; POWELL *et al.*, 1954, 1957). If the dubious nucleus submedius *autorum* is considered to be a 'midline nucleus' rather than a portion of the ventralis posteromedialis complex, connections with the 'orbital' neocortex, as e.g. claimed by KRIEG (1966) and others cannot be entirely excluded.

Various opinions concerning the delimitation of the intralaminar nuclei located within the lamina medullaris interna have been expressed (cf. e.g. D'HOLLANDER, 1913; ZEMAN and INNES, 1963). In the Rabbit, MIURA (1933) distinguished a nucleus laminaris and a nucleus

---

[224] Some authors interpret these grisea as phylogenetically 'very ancient'. Although doubtless kathomologous to the medial portion of the thalamic cell population in Anamnia, their specific Mammalian differentiation should rather be evaluated as 'new'. Typical midline nuclei are not unambiguously displayed by the Avian thalamus, while only in some Reptiles (particularly Crocodilia) ill-defined nuclei reunientes can be recognized (cf. above, section 7). Thus, the Mammalian nn. reunientes may be interpreted as manifestations of a 'progressive differentiation'.

magnocellularis.[225] It had been assumed that these grisea, presumably receiving ascending tegmental and perhaps in part tectal channels, were devoid of neocortical connections, although representing 'indirect cortical modulators' (K., 1954). Recently, however, it has been claimed that the intralaminar nuclei 'project densely to the striatum and diffusely upon the cortex' (Jones and Leavitt, 1974). The cited authors as well as some others include the above-mentioned parafascicular nucleus and the centre médian into the 'intralaminar nuclei'. The cortical projections of the intralaminar group are said to be 'non-specific', pertaining to the 'ascending activator system'.

The concept of a '*centrencephalic*' system of grisea, as expressed by Penfield (1952, 1954) and other authors, can be interpreted as somehow related to that of an 'ascending activator system' in the wider sense, i.e. including thalamic midline grisea as well as hypothalamic grisea involved in the sleep-wakefulness mechanisms pointed out further below, moreover periventricular (periaqueductal) mesencephalic grisea and reticular formation, at least with respect to some of this latter's activities.

The *posterior dorsal thalamic grisea* comprise nucleus lateralis posterior, pulvinar, dorsal lateral geniculate, dorsal medial geniculate nuclei, and nucleus suprageniculatus.

The *nucleus lateralis posterior*, like the nucleus lateralis dorsalis, mentioned above, may have reciprocal connections with the parietal or with the sensorimotor cortex and receive input from other thalamic nuclei. The *pulvinar*, poorly differentiated in lower Mammals, but substantially developed in higher ones, particularly in Primates, can be evaluated as evolved from a subdivision of the lateralis dorsalis complex. Besides its connections with neighboring thalamic nuclei, especially the pretectal group, connections with mesencephalic tectum can be assumed. Extensive relationships with the optic system[226] are thereby included. Reciprocal cortical connections with the temporoparieto-occipital region obtain. The pulvinar may be regarded as an 'association' or 'relay' griseum concerned with visual and auditory integration, moreover with thalamo-cortical feedback. A paper purporting to elaborate on the 'evolution of the pulvinar' can be found in the 'symposium' edited by Riss *et al.* (1972).

---

[225] The presumably corresponding nuclei (paracentralis and centralis lateralis) shall again be pointed out further below in dealing with the Human thalamus.

[226] Apparently including some direct input through the optic tract.

The *dorsal lateral geniculate nucleus* ('corpus geniculatum laterale' *sensu strictiori*) represents the dorsal thalamic component of the lateral geniculate complex (or 'corpus geniculatum laterale' *sensu latiori*. These thalamic grisea indeed pertain to the phylogenetically 'oldest' diencephalic nuclei, being displayed, as strictly orthohomologous configurations, in all Vertebrate groups from Petromyzon to Man.

The dorsal lateral geniculate nucleus receives input from the optic tract, presumably both by direct fibers and by collaterals, and gives origin to a channel reaching the 'visual' occipital neocortex. The intricate point to point (or rather neighborhood to neighborhood) projection of retina upon dorsal geniculate nucleus, the course of the geniculocortical (geniculostriate or geniculocalcarine radiation) and its topical projection upon the optic neocortex shall be dealt with further below in this section with particular reference to the condition in Man, and again in the relevant sections of chapters XIII and XV.

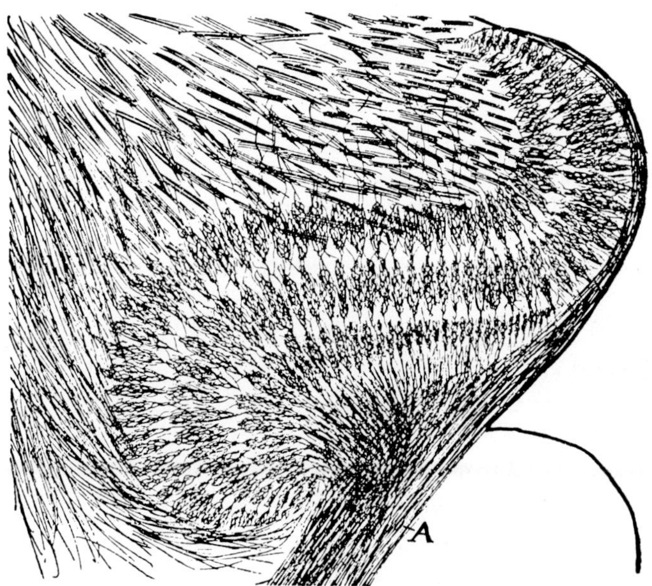

Figure 132 A. Parasagittal section *(Golgi impregnation)* through the dorsal lateral geniculate nucleus of the Cat (after TELLO, 1904, from CAJAL, 1911). A: optic tract, whose end arborizations provide three layers within the griseum. Caudal surface is at right, ventral side at bottom.

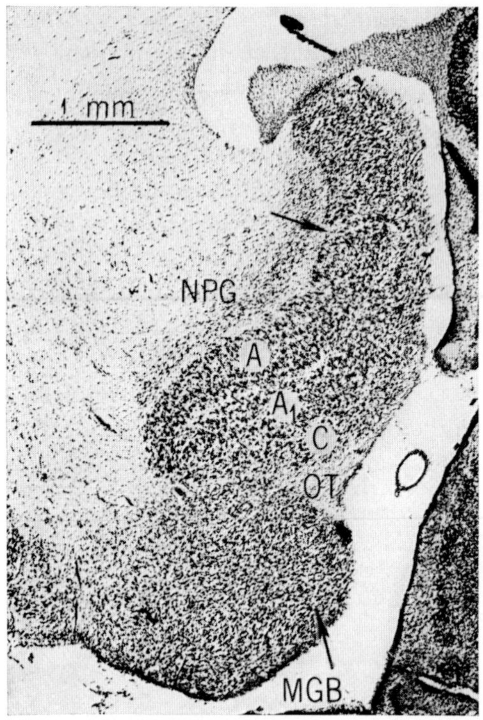

Figure 132B. Parasagittal section *(Nissl stain)* through the lateral geniculate complex of the Carnivore Ferret (from SANDERSON, 1974). A, A_1, C: laminae of dorsal lateral geniculate griseum; MGB: dorsal medial geniculate nucleus; NPG: 'perigeniculate nucleus' (portion of ventral lateral geniculate griseum); OT: optic tract. The upper arrow indicates a discontinuity in lamina A which may correspond to the representation of the optic disc. Laminae C, C_1, C_2 are not clearly distinguishable, and jointly labelled as C.

Reciprocal corticogeniculate fibers, formerly denied by some authors, seem now well established (cf. e.g. NIIMI *et al.*, 1971; GIOLLI and POPE, (1973). GILBERT and KELLY (1975), using the horseradish peroxidase technique, noted that the neurons of the Cat's visual cortex projecting to the lateral geniculate griseum are located in layer VI of said cortex. GIOLLI and CREEL (1974) have also reported on the problem of variability and inheritance of the retinogeniculate projections in albino and pigmented Rats.

Concerning the architectural features of the dorsal lateral geniculate nucleus, a 'progressive' evolution from a non-laminated pattern in lower Mammals to the conventionally recognized six laminae of Pri-

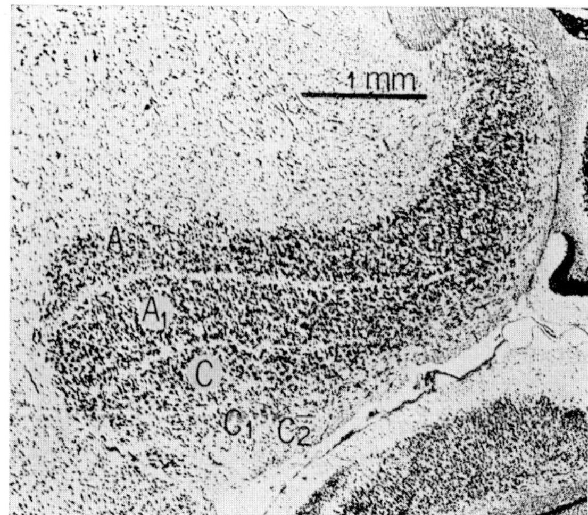

Figure 132 C. Parasagittal section *(Nissl stain)* through the dorsal lateral geniculate griseum of the Carnivore Fox (from SANDERSON, 1974). A, A_1, C, C_1, C_2: the five layers of the griseum. Rostral side is to left.

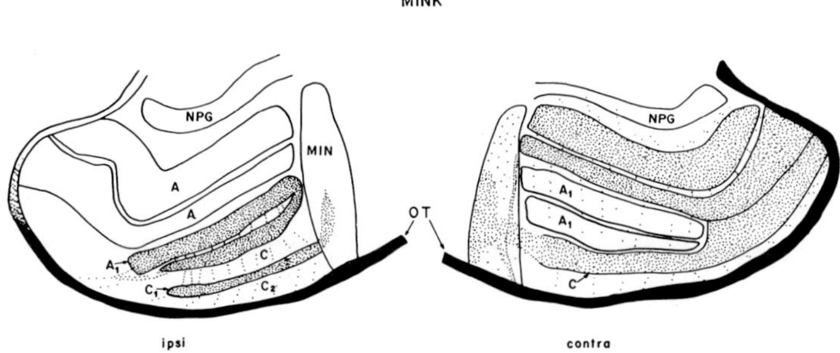

Figure 132 D. Outline drawings of horizontal sections through the lateral geniculate complex of the Carnivore Mink, showing the pattern of fiber degeneration following removal of one eye. Degenerating axons are shown as dots (from SANDERSON, 1974). A, A_1, C, C_1, C_2: laminae of dorsal lateral geniculate griseum; MIN: 'medial interlaminar nucleus' (portion of ventral lateral geniculate nucleus); OT: optic tract.

mates[226a] and Man becomes evident. In the Ungulate Sheep, and in Carnivores, an essentially trilaminal pattern appears to obtain (cf. Fig. 132 A). Recent authors (GUILLERY, 1970; SANDERSON, 1974), however, distinguish, at least in Carnivores, five layers designated as A, A_1, C, C_1, and C_2, which are variably developed (cf. Figs. 132 B–D). In some forms, A and A_1 may again be subdivided into a pair of leaflets. Experimental methods based on terminal axon degeneration seem to indicate that laminae A and C receive contralateral, and A_1 and C_1 homolateral input.[227]

According to LE GROS CLARK (1932a) there is some evidence to suggest that in the primary intrinsic differentiation of the Mammalian dorsal lateral geniculate griseum three cell layers become defined. Even the non-laminated nucleus of some 'lower' Mammals may have an irregularly striated appearance because its cells tend to be arranged in parallel bands intermingled with, or between, optic tract fibers (cf. Figs. 118 C, 120 F).[228]

Controversial questions concerning the lateral geniculate grisea are, moreover, discussed in a treatise by WALLS (1953) and in its review by HARMAN (J. comp. Neurol. *100:* 237–239, 1954).

As regards the projection of retinal quadrants upon the lateral geniculate complex, a paper by BODIAN (1937), dealing with the Opossum, also brings a general review of this topic concerning other Mammals, with relevant bibliographic references. Still more recent details and interpretations about these problems can be found in a paper by MALPELI and BAKER (1975), dealing with the geniculatum laterale of Macaca.

The individual neuronal elements within a given lamina in those Mammals with clear-cut dorsal lateral geniculate lamination generally seem to receive input from only one eye (cf. Fig. 145 D). HUBEL (1960)

[226a] A relatively simple development of the six-layered pattern is, e.g. found in the Squirrel Monkey, Saimiri (DOTY *et al.*, 1966). As regards Carnivores, a description of cyto-architecture in the Cat's lateral geniculate griseum is given in the older but detailed paper by THUMA (1928–29).

[227] The retinal projection of C_2 is said to be uncertain. The so-called perigeniculate nucleus is presumably a dorsal portion of the ventral lateral geniculate nucleus dealt with further below in connection with the thalamus ventralis. It might here also be added that, after removal of one eye in Kittens, the axons from the remaining eye grow across laminar borders into the deafferented lamina A (HICKEY, 1975).

[228] It is not improbable that investigations with sophisticated techniques might here disclose subdivisions related to topic retinal projections.

and HUBEL and WIESEL (1961, 1964, 1966) have identified, by means of experiments with microelectrodes on 'single units', such neuronal elements in Cats and Monkeys. The cited authors, moreover, traced further relevant connections of said elements with cells in the visual cortex. The implications of this arrangement, and the inferences as regards the thereto pertaining aspects of cortical function will be considered in chapter XV.

It is also claimed that lateral geniculate cells corresponding to retinal X and Y cells (cf. p. 55) can be identified (BROWN and SALINGER, 1975). The cited authors report that in adult Cats chronic immobilization of one eye by eye muscle nerve resection leads to selective but substantial loss of X cells in the binocular segment of the dorsal lateral geniculate nucleus. Again, monocular deprivation during the first 3 months in young Cats is said to show, on the basis of retrograde horseradish peroxidase axonal transport, that the arrest of cell growth in the deprived laminae involves mainly Y cells (GAREY and BLAKEMORE, 1977).

It should, moreover, be mentioned that, in addition to 'units' concerned with responses to spatial configuration and motion, lateral geniculate cells responding to wave length have been identified in the Cat, and were designated as 'opponent color cells' (PEARLMAN and DAW, 1970). Previously, similar findings in the Rhesus Monkey had been reported by HUBEL and WIESEL (1966). These latter authors reached the conclusion that a wide variety of functional cell types is present in the Monkey lateral geniculate griseum. Some of these elements seem mainly concerned with spatial variables, and others with 'color', but most appear able to handle both variables. Some have connections with rods and cones, and others with cones only.

The *dorsal medial geniculate nucleus*, which may display several subdivisions[229] (cf. Fig. 133; also Figs. 408 D and E, vol. 4), located caudally, and, depending on the taxonomic forms, also somewhat medially and either ventrally or dorsally to the preceding lateral dorsal geniculate griseum, receives mainly the cochlear input, either directly by way of the lateral lemniscus, or indirectly by way of the inferior colliculus.

This contralateral as well as homolateral caudal input tends to form a conjoint brachium quadrigeminum inferius *sive* brachium colliculi inferioris. The main output of the dorsal medial geniculate nucleus reaches, by way of the 'auditory radiation' passing through the capsula

---

[229] Cf. e.g. YOSHIDA (1924), and K. and MILLER (1942).

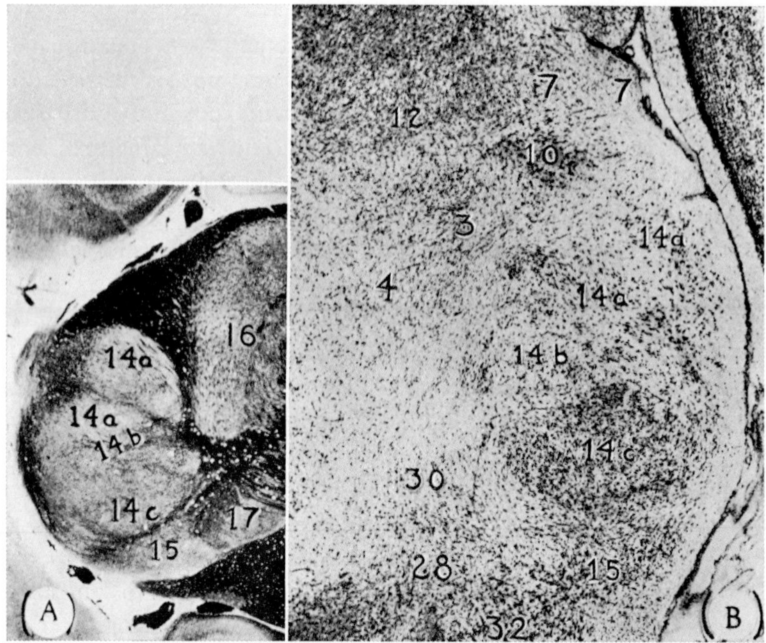

Figure 133 A. Lateral parasagittal section (myelin stain) through the geniculate griseal complexes of the Rabbit (from K. and MILLER, 1942). 14: dorsal medial geniculate nucleus and its subdivisions (a: dorsal magnocellular; b: intermediate; c: ventral parvocellular); 15: ventral medial geniculate nucleus; 16: dorsal lateral geniculate nucleus; 17: ventral lateral geniculate nucleus. The caudal surface is at left.

Figure 133B. Cross-section *(Nissl stain)* showing the cytoarchitecture of the medial geniculate complex (from K. and MILLER, 1942). 14, 15: cf. Figure 133A (it will be seen that the magnocellular subdivision displays a further parcellation); 3: nucleus praetectalis principalis; 4: nucleus posterior; 7: nucleus olivaris colliculi posterior; 10: nucleus suprageniculatus (this griseum lies medially to the plane of the sagittal section A); 12: nucleus sublentiformis; 28: thalamus ventralis (zona incerta); 30: caudal portion of ventral dorsal thalamic complex; 32: substantia nigra.

interna, the auditory temporal neocortex. The inclusion of vestibular components within these channels can be suspected. Either commissural fibers *sensu strictiori*, or merely decussating fibers, related to the medial geniculate complex, seem to pass through the variable and intricate system of postoptic respectively supraoptic commissures. Reciprocal connections with inferior and superior colliculus appear to be present, but nothing sufficiently definite is known about these and possible other, intradiencephalic connections. A more or less discrete topographic (spatial) arrangement of the auditory channels and their grisea, related to sound qualities such as pitch and intensity can be inferred on the basis of experimental results (LEWY and KOBRAK, 1936; WOOLSEY and WALZL, 1942; ROSE *et al.*, 1963; cf. also chapter IX, section 10, vol. 4). This topic shall again be pointed out in the chapters dealing with the telencephalon.

The *nucleus suprageniculatus*, described by MÜNZER and WIENER (1902), K. and MILLER (1942) in the Rabbit, and by other authors in various Mammals, including Man (cf. also K. and MILLER, 1949; K., 1954) is a relatively large griseum in close topographic relations to the dorsal medial geniculate nucleus, poorly understood as regards fiber connections and functional significance. Although KNIGHTON (1950) suspected that it might be a relay nucleus for the so-called cortical auditory area II (in the Cat), such projection has been denied by others. The suprageniculate nucleus, which appears to be a part of the medial geniculate complex, could perhaps link that griseum with pretectal region and mesencephalic colliculi but still remains a nucleus of dubious significance and connections.

The *pretectal dorsal thalamic nuclei*, within the diencephalomesencephalic boundary, likewise represent a caudal group, but, because of their close relationship to commissura posterior and to mesencephalic pretectal nuclei, are preferably separately classified as a group *sui generis*.

The *area praetectalis*, pointed out by PAPEZ (1932, 1936), and also present in Sauropsidans and some Anamnia, located caudally to the habenular grisea as well as laterally to their caudalmost portion, extends to the rostromedial extremity of superior colliculus. It can be considered a *nucleus comitans sive interstitialis tractus tectothalamici dorsomedialis.*[230]

The *nucleus praetectalis (principalis)* is located medially to the posterior dorsal thalamic group. The nucleus posterior is basally adjacent to nucleus praetectalis. Both these pretectal nuclei appear as a conjoint nuclear complex intercalated between thalamic nuclei rostrad, and mesencephalic grisea caudad. A recent paper by KANASEKI and SPRAGUE (1974) dealing with the anatomical organization of pretectal nuclei and tectal laminae in the Cat, although using the authors' own nomenclature, essentially agrees, as can be seen from their table I,

[230] An indistinct dorsal and ventral subdivision of area praetectalis is suggested in some forms.

p. 332, with the parcellation established by our own investigation (K. and MILLER, 1942, 1949; SHINTANI-KUMAMOTO, 1959).

The *dorsal thalamic nuclei commissurae posterioris*, closely related to the fibers of that commissure, form a complex subgroup of pretectal grisea, somewhat variably differentiated in the Mammalian taxonomic series. About five 'nuclear' cell groups may be distinguished within said subgroup, namely (1) nucleus interstitialis principalis commissurae posterioris, (2) n. interstitialis magnocellularis c. p., (3) n. interstitialis supracommissuralis c.p., (4) n. medianus infracommissuralis c.p., and (5) n. centralis subcommissuralis c.p. within the central gray (K. and MILLER, 1942, 1949; K., 1954; SHINTANI-KUMAMOTO, 1959). An additional mesencephalic n. interstitialis tegmentalis commissurae posterioris, close to nucleus ruber, pertains to the prerubral tegmentum.

The *mesencephalic pretectal nuclei*, such as nucleus lentiformis mesencephali pars magnocellularis[231] and pars parvocellularis, nucleus sublentiformis, and nucleus olivaris colliculi superioris, were dealt with in section 9, chapter XI of volume 4, together with some further comments on the pretectal region in general.[232]

Functional significance and exact fiber connections of the pretectal group remain poorly elucidated. Taking into consideration both the established[233] and the probable fiber connections of this griseal complex (including its mesencephalic tectal components), it could be assumed that it comprises a diversity of essentially interstitial 'centers' (i.e. centers located within fiber systems), correlating tectum mesencephali, some other thalamic grisea, epithalamus, hypothalamus, tegmental centers, and also cortical areas. Some direct reciprocal cortical connections are probable, especially with occipital, preoccipital, and temporal cortex, but these cortical connections, questioned by some authors, do not appear to be massive. Nevertheless, cortical connections arising in the visual cortex have been convincingly demonstrated by GIOLLI and POPE (1971). Using an experimental method based on axo-

---

231 This is the nucleus limitans *autorum*, also known as the large-celled nucleus of the optic tract. CROSBY *et al.* (1962, p. 299) and some other authors include said griseum into the dorsal thalamus.

232 Together with nucleus interstitialis tegmentalis commissurae posterioris, the following grisea can be included into the mesencephalic tegmental pretectal group: *nucleus interstitialis of Cajal*, *nucleus of Darkschewitsch*, nucleus medialis anterior, and tegmental cell cord.

233 Recent data on pretectal fiber connections can be found in a paper by GIOLLI *et al.* (1974).

plasmic transport of labelled protein (tritiated leucine) injected into the striate cortex of the Squirrel monkey Saimiri, Holländer (1974) believes to have identified projections of that cortex to lateral geniculate complex, posterior nucleus of thalamus, pulvinar grisea, and reticular nucleus (i.e. thalamus ventralis). Direct or indirect striatal connections cannot be excluded. Direct optic input seems conspicuous for most of the pretectal nuclei. Again, impulses from most major sensory system may reach the pretectal grisea and, in addition to the correlating activities mentioned above, complex discharge patterns can again be channelled into the various efferent pathways such as fasciculus longitudinalis medialis, central tegmental tract, the more diffuse longitudinal bundles of the reticular formation and other systems (e.g. the periventricular bundles). The pretectal mediation of the pupillary reflex, as particularly established by Spiegel, was dealt with in section 9, chapter XI of volume 4. Recent views on the relationship of the pupillary light reflex to the pretectal grisea can be found in papers by Carpenter and Pierson (1973) and Pierson and Carpenter (1974).

The functional significance of the Mammalian *ventral thalamus* is difficult to evaluate. Although very prominent during key stages of ontogenetic development (cf. Figs. 112 A, B, 113 A–E; also Figs. 226–237, vol. 3/II), it appears relatively very much reduced at the 'adult' stage (cf. e.g. Figs. 118 A, B; also Figs. 225 A–C, vol. 3/II), and becomes thus much more inconspicuous than the substantial thalamus ventralis of Anamnia.[234] With regard to the six 'nuclei' listed above in the enumeration of Mammalian ventral thalamic grisea, the very variable nucleus taeniae (s. nucleus filiformis) appears related to the stria medullaris system, and the nucleus reuniens ventralis pertains to the group of midline nuclei which may form part of the so-called 'dopaminergic system'.

The *nucleus reticularis thalami (ventralis)* surrounds in capsule- or cup-like fashion the ventrolateral dorsal thalamic griseal group. It could be related to these nuclei by 'screening' their reciprocal cortical

---

[234] It will be recalled that the thalamus ventralis had not been recognized as a distinctive diencephalic subdivision until pointed out, in 1910, by C. J. Herrick on the basis of his studies on the Amphibian brain. The homologies of the thalamus ventralis in the Vertebrate series were subsequently clarified mainly through the investigations of the present writer and his collaborators. As regards the other Amniota, the thalamus ventralis may still be relatively large in many Reptiles (cf. e.g. Fig. 101 E), but appears rather reduced in Birds (cf. e.g. Figs. 106 A–D and also Fig. 216, p. 439, vol. 3/II).

connections (cf. e.g. McLARDY, 1951). Another possibility is the participation of nucleus reticularis as an intermediate station of FLECHSIG's *Haubenstrahlung* respectively of the ascending reticular pathway, in the so-called 'unspecific' or 'non-specific' respectively 'diffuse' thalamocortical projection system postulated by various authors and discussed by McLARDY (1951) and HASSLER (1964). Still another possibility would be the interpretation of nucleus reticularis as providing, in its different regions, projections to, respectively reciprocal connections with, the various so-called 'secondary' sensory and motor cortical

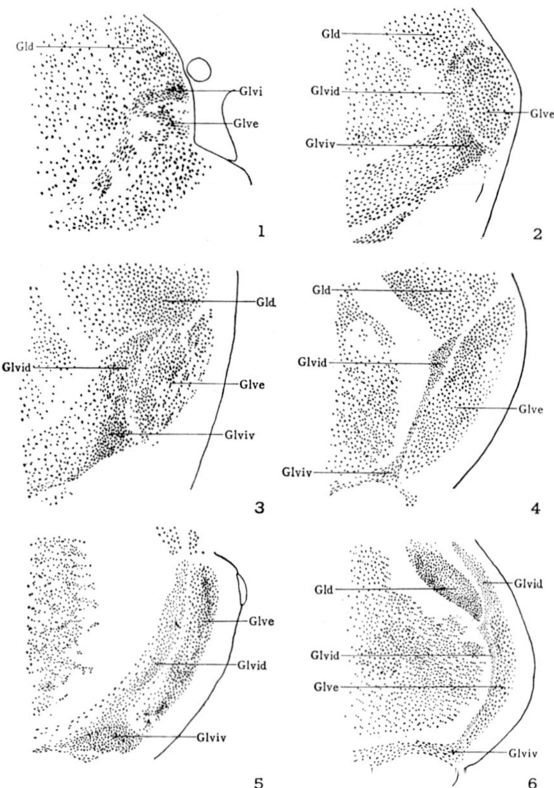

Figure 134 A. Cross-sections *(Nissl stain)* through the ventral lateral geniculate nucleus in various Mammals (from NIIMI *et al.*, 1963a). 1: Japanese Horseshoe-Bat (Chiropteran Rhinolophus); 2: Rodent Mouse; 3: Lagomomorph Rabbit; 4: Carnivore Nutria; 5: Ungulate Cow; 6: Carnivore Raccoon-Dog; Gld: dorsal lateral geniculate nucleus; Glve: external layer of ventral lateral geniculate nucleus; Glvi: internal layer of ventral l.gen.nucleus; Glvid: dorsal part of internal layer of l.gen.nucleus; Glviv: ventral part of internal layer of ventral l.gen.nucleus.

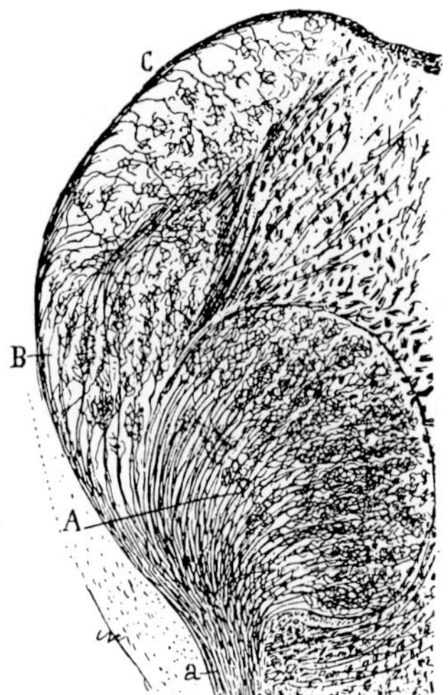

Figure 134B. Cross-section *(Golgi impregnation)* through the lateral geniculate complex of the Cat, a few days after birth (from CAJAL, 1911). A: ventral lateral geniculate nucleus, receiving direct optic input; B, C: subdivision (layers or lamellae) of dorsal lateral geniculate nucleus; a: fibers of optic tract.

areas (cf. e.g. HASSLER, 1950). None of the multitudinous recent reports dealing with thalamic and corticothalamic fiber connections on the basis of diverse experimental methods have, in my opinion, provided sufficiently conclusive or convincing evidence for a clarification of the topic here under consideration.[235]

The *zona incerta* seems to be mainly connected with and thus presumably functionally related to the entopeduncular or 'extrapyramidal motor system' of the hypothalamus. It may, however, also be connected with other, 'vegetative' parts of the hypothalamus. The so-called nucleus or nuclei of the *field of Forel ('nucleus campi Foreli')* can be included as subdivisions of the zona incerta, parts of which, particularly

[235] On the basis of recent methods depending upon axonal transport (cf. above footnote 219) JONES (1975) claims that the 'reticular complex' in Rats, Cats, and Monkeys receives collaterals from traversing thalamo-cortical and cortico-thalamic fibers, and thereby 'samples thalamo-cortical and cortico-thalamic activity in a somewhat unspecific manner'. The only efferent pathway from the 'reticular complex' is said to terminate 'in the nuclei of the dorsal thalamus'. A discussion of various interpretations concerning 'direct' and 'indirect cortical modulators' as well as 'nonspecific thalamocortical system' is included in the monograph on the human diencephalon (K., 1954). These comments, based on the lesser amount of data available at the time, more than 20 years ago, are still not entirely out of date despite many new but in part rather inconclusive data.

within its medial portion, seem also involved in the '*dopaminergic system*'. Fiber connections with globus pallidus and corpus striatum, with other components of fasciculus lenticularis and thalamicus, and with tegmentum are generally recognized. Some authors, e.g. CROSBY *et al.* (1962) refer to a striato-incerto-thalamo-olivary system, thus also assuming a relationship with the central tegmental tract.

The ventral nucleus of the lateral geniculate complex, very conspicuous in 'lower' Vertebrates, appears as a relatively lesser component of said complex in Mammals. At certain stages of ontogenetic development, its relative size and position closely correspond to the configuration obtaining in Anamnia. Subsequently, a taxonomically variable degree of secondary displacement, combined with lesser expansive growth, becomes noticeable. Nevertheless, as shown by NIIMI *et al.* (1963a), an internal and an external subdivision may be distinguished in diverse Mammalian forms (Fig. 134 A). Further parcellations can be seen in Ungulates, where the ventral lateral geniculate nucleus appears to display the highest degree of differentiation. In Rodents and Chiroptera, this nucleus, although relatively large, is poorly differentiated, being somewhat better differentiated in Carnivores. In Primates, it was found to be least developed among the Mammals investigated. One of its reduced portions being the nucleus praegeniculatus (cf. also K., 1954). Generally speaking, connections between ventral and dorsal lateral geniculate nucleus can be assumed. Although direct optic input to the Mammalian ventral nucleus was formerly doubted, there is now little doubt that such input obtains, as was already documented by the *Golgi preparations* of CAJAL (cf. Fig. 134 B). In addition, there is some evidence not only of tectal but also of telencephalic cortical connections (cf. e.g. the 'Symposium' edited by RISS *et al.*, 1972).

The *nucleus ventralis of the medial geniculate complex* seems to have connections with the dorsal grisea of the medial geniculate griseum. In addition, it appears to be connected with pretectal grisea and with the system of supraoptic or postoptic commissures; its relationships to the lateral lemniscus system and channels to cortex remain uncertain.

With regard to Mammalian *epithalamic*, *dorsal thalamic*, and *ventral thalamic grisea* in general, several criteria of classification have been proposed, which, however, can be evaluated as unsatisfactory. Concepts of this type are *neo-*, *archi-*, and *palaeothalamus*. A critique of these prefixes was included in chapter VI, pp. 663–668 of volume 3/II and in the monograph on the human diencephalon (K., 1954). No further comments are here required, except perhaps for the remark that, of the diencephalic grisea under consideration, the epithalamus, despite conspicuous telencephalic connections, doubtless displays extremely conservative features throughout the Vertebrate series. To some extent, however, this could also be claimed for the lateral geniculate complex and perhaps other grisea, despite substantial changes affecting some of their fiber connections and further phylogenetic differentiation.

There is likewise little justification for a subdivision into '*dependent*' and '*independent*' *thalamus*, supposedly based on the 'survival' of diencephalic grisea after various degrees of telencephalic ablation (cf. the comments in K., 1954, which still remain essentially valid).

Much the same could be said about a subdivision, as e.g. adopted by EITSCHBERGER (1970), and others, into '*palliothalamus*' and '*truncothalamus*'. The former is said to be connected with the telencephalic isocortex, and the 'truncothalamus' with other telencephalic regions or with brain stem grisea.

Quite apart from objections against a morphologic classification restricted to fiber connections, these latter are still not sufficiently clarified. Thus, grisea which some years ago were not believed to display cortical connections are now presumed to have such relationships. The views expressed in the recent 'symposium' edited by RISS *et al.* (1972) disclose, to the critical reader familiar with fundamental morphologic data, the unsatisfactory, inconclusive and unconvincing attempts purporting to establish sophisticated concepts of 'basic thalamic structure and function' without considering relevant morphologic criteria.

On the other hand, a functional classification of thalamic grisea in addition to their morphologic classification is evidently justified, and was proposed by the author in 1954, as pointed out in the introductory section 1A. In still more condensed fashion, these main subdivisions are represented by (a) *cortical relay nuclei of main sensory systems*, (b) *direct and indirect cortical modulators*, (c) *grisea with predominant relation to pallidum and striatum*. Additional groups are (d) the *pretectal complex*, (e) the *epithalamic habenular system*, and (f) *grisea (perhaps in part 'intrinsic') of dubious significance and connections*. The groups (d) and (e), although with telencephalic connections, doubtless deserve to be classified as griseal complexes *sui generis*. As regards contemporary views on details of thalamic synaptic organization, based on not altogether convincing interpretations of multitudinous data obtained by present-day techniques, the reader may be referred to a publication by SHEPHERD (1974) purporting to elucidate the synaptic organization of the brain.

The *dorsal and entopeduncular hypothalamus* of our terminology (K.

and HAYMAKER, 1949; K., 1954), that is to say the predominantly 'extrapyramidal' component of the hypothalamus, comprises (1) *globus pallidus*, (2) *nucleus entopeduncularis*, (3) *nucleus ansae lenticularis*, (4) *interstitial nucleus of the inferior thalamic peduncle*, and (5) *nucleus subthalamicus (subthalamic body, corpus subthalamicum of Luys)*. The first four grisea are closely associated, and can be conceived as derivatives of anterior part of entopeduncular nucleus and of massa cellularis reuniens, pars inferior, of 'lower' Vertebrates. These grisea are characterized by their relationship to the hemispheric stalk. The nucleus subthalamicus appears derived from the caudal part of the 'primordial' entopeduncular nucleus.

The *globus pallidus* is rich in medullated fibers and thus appears more whitish or 'pale' than the darker griseum of the adjacent striatum. It is characterized by loosely arranged fairly large multipolar cells, and differs from the telencephalic corpus striatum which is poor in medullated fibers and contains abundant relatively small nerve cells intermingled with scattered larger elements. In Mammals with a compact capsula interna, this latter separates the corpus striatum into a medial (adventricular) nucleus caudatus and a lateral putamen. A medullary lamina separates globus pallidus from putamen. Another medullary lamina divides the globus pallidus into a medial (inner) and a lateral (outer) segment. One or two variable additional medullary laminae may more or less distinctively segregate further subdivisions of the globus pallidus. The Mammalian *nucleus entopeduncularis*, if clearly distinguishable, seems to represent a caudomedial appendage of globus pallidus. *Nucleus ansae lenticularis*[236] and *interstitial nucleus of the inferior thalamic peduncle* can be evaluated as cell groups derived from the globus pallidus anlage, and scattered within said fiber bundles. The globus pallidus and some other grisea of the extrapyramidal motor system (substantia nigra, nucleus ruber, nucleus dentatus cerebelli, to some extent also nucleus subthalamicus) display in certain Mammals the *iron reaction of Spatz*, discussed in section 9 of chapter XI, volume 4, and section 2, chapter V, volume 3/I. In addition, peculiarities of pigmentation are here manifested in various Mammals (cf. e.g. CROSBY *et al.*, 1962, p. 260, 308, 366). Other biochemical peculiarities were dealt with in section 9, chapter XI of volume 4 and shall again be pointed out in the discussion of the human diencephalon.

[236] Not to be confused with nucleus of ansa peduncularis (nucleus basalis of MEYNERT, nuclei of substantia innominata, including nucleus subputaminalis) pertaining to the telencephalic basal ganglia and closely related to the corpus striatum complex.

The globus pallidus has abundant, presumably reciprocal connections with the corpus striatum, especially the putamen, by way of the above-mentioned lamina medullaris. Through the ansa lenticularis as well as the fields H, H_1, and H_2, reciprocal connections[237] obtain with thalamus (nucleus ventralis anterior), zona incerta, nucleus subthalamicus, substantia nigra, and undefined grisea of the 'vegetative' hypothalamus. Connections with the habenular grisea have been claimed (NAUTA, 1974). Such 'pallidohabenular pathway' might perhaps correspond to fibers proceeding, in various Anamnia, from the dorsal and entopeduncular preoptic region toward the stria medullaris system. Connections with the cerebral cortex (mainly descending fibers from premotor cortex) and with the nucleus of the centrum medianum also have been claimed. Functionally, the globus pallidus (or 'pallidum') may exert a stimulating and facilitating influence upon 'motor centers', while the striatum may have an inhibitory effect. On the other hand, cortical motor impulses might be suppressed by a complex circuit passing through the pallidum. Additional comments on the insecure concepts concerning composition and functions of these various connections can be found, with bibliographic references, in the treatises by CROSBY *et al.* (1962), KRIEG (1966), and in the author's monograph (K., 1954). Some relevant clinical data shall be considered further below in connection with the human hypothalamus.

The *nucleus subthalamicus*, conspicuously oval or lens-shaped in cross-sections through the diencephalon of Primates, but less prominent in the 'lower' Mammalian forms, represents an apparently important griseum of the extrapyramidal system dealt with in the preceding paragraph. In addition to the there mentioned overall connections, optic input is believed to reach this nucleus by way of the basal optic tract.[238] Fiber systems pertaining to nucleus subthalamicus form a substantial component of the supramammillary (posterior hypothalamic) commissure.

---

[237] The connections with nucleus ventralis anterior are perhaps predominantly ascending, while those with nucleus subthalamicus and substantia nigra may be largely descending channels of the 'extrapyramidal system'. These descending channels from globus pallidus and striatum include either direct or indirect connections with prerubral tegmentum, nucleus ruber and formatio reticularis tegmenti.

[238] Such input may also reach the substantia nigra, either directly or by way of the nucleus opticus tegmenti, which appears fairly closely related to substantia nigra, at least in some forms (cf. e.g. Fig. 407, p. 939, vol. 4).

With regard to the essentially 'vegetative' Mammalian *hypothalamus sensu strictiori*, its subdivision into anterior, middle, and posterior griseal groups, as dealt with further above, will be recalled. The detailed relationships of the recognizable fiber systems to the diverse hypothalamic nuclei are insufficiently clarified. Together with the hypothalamo-hypophysial connections discussed in section 1 C, the relatively best known data concern the mammillary complex, which, in accordance with SPATZ' concept of '*markreicher Hypothalamus*' somewhat differs from the 'vegetative hypothalamus', although closely linked with this latter, and seems to represent part of a system *sui generis*, related to the *Papez circuit* (Fig. 131) pointed out in dealing with the anterior group of dorsal thalamic nuclei. Caudal communication channels of the mammillary grisea are included in mammillary peduncle, ascending from the tegmentum and apparently also reaching some of the other hypothalamic cell groups. Caudal descending channels are provided by the mammillotegmental tract and the thinly medullated mammillo-interpeduncular bundle (cf. section 9, chapter XI, vol. 4). Commissural fibers of the mammillary complex are included in the supramammillary commissure. The main rostral input channel seems to be provided by the bulk of the fornix, and originates in telencephalic hippocampal formation, to a much lesser extent also in the para- and preterminal ('septal') grisea.[239] The fornix reaches the mammillary complex on the lateral side, but somewhat ventromedially to the mammillary peduncle. Mainly in the medial mammillary grisea, but presumably also in other subdivisions of that complex, there arises on the medial side the fasciculus mammillaris princeps as a main output channel. This heavily medullated bundle bifurcates[240] to form the tractus mammillothalamicus *(bundle of Vicq d'Azyr)* to the anterior group of thalamic nuclei, and the tractus mammillotegmentalis which reaches deuterencephalic tegmental centers. Other connections of the mammillary complex are included in the so-called medial forebrain bundle of particular impor-

---

[239] Further details shall be dealt with in chapters XIII and XV. Relevant, but inconclusive comments on composition and subdivisions of the fornix as well as on its assumed postmammillary components reaching the mesencephalic tegmentum after (partly?) crossing in the supramammillary commissure can be found in the treatise by CROSBY *et al.* (1962, p. 317).

[240] To some extent and at least in some forms, non-bifurcating, i.e. separate mammillothalamic and mammillo-tegmental fibers are said to be present (cf. K., 1954; CROSBY *et al.*, 1962).

tance for the 'vegetative' hypothalamus and may also be provided by some fibers to subthalamic nucleus as well as periventricular channels. Because of its connections with the anterior thalamic nuclei and the hippocampus, the '*markreicher Hypothalamus*' appears to be a relevant griseum of the *Papez circuit*. Since close relations with other hypothalamic centers seem to obtain, the entire hypothalamus *sensu strictiori* has been included into the so-called 'limbic system'.[241]

As regards the various grisea of the poorly medullated *('des markarmen')* hypothalamus, the ill-defined so-called medial forebrain bundle must again be pointed out as a main communication channel. Other important systems are those of stria medullaris and of stria terminalis, components of which either join the 'medial forebrain bundle' or independently reach various grisea. In addition, fibers from prefrontal cortex and components of the fornix connect with undefined hypothalamic nuclei. There is little doubt that, essentially by way of the basal optic

[241] The 'limbic system' *autorum* can be evaluated as a more sophisticated but hazy elaboration, by numerous recent investigators, of the original '*Papez-circuit*'.

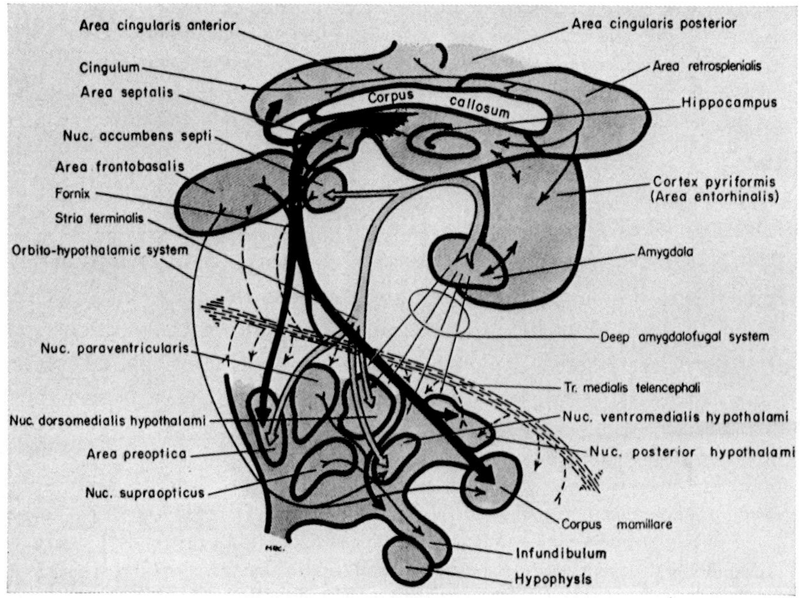

Figure 135 A. Diagram representing 'limbic-hypothalamic pathways' (modified after GLOOR, 1956, from ZEMAN and INNES, 1963).

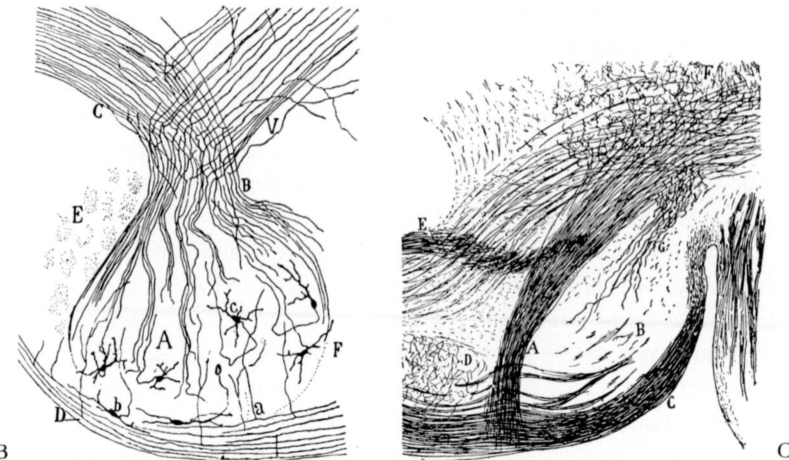

Figure 135 B. Sagittal section *(Golgi impregnation)* through medial portion of mammillary complex in the new-born Mouse (from CAJAL, 1911). A: medial mammillary nucleus; B: common stem of mammillotegmental and mammillothalamic tract; D: bundle of 'capsule' of medial mammillary nucleus; E: fornix; F: rostral portion of medial mammillary nucleus; V: mammillothalamic tract; a: collateral of 'capsule bundle'; b: 'superficial fusiform cell'; c: cell whose axon seems to join the common mammillothalamic and mammillotegmental stem.

Figure 135 C. Parasagittal section through the 'subthalamic region' of the Mouse (from CAJAL, 1911). A: collateral corticotegmental tract (at that time erroneously interpreted as '*cordon de Forel*', cf. comments in text and Fig. 119 A); B: substantia nigra; C: cerebral peduncle; D: subthalamic nucleus; E: medial lemniscus *('voie sensitive');* F: nucleus ruber tegmenti; G: efferent channel of substantia nigra *('voie issue de la substance noire').*

tract, direct optic input reaches hypothalamic grisea (cf. e.g. MOORE and LENN, 1972; PRINTZ and HALL, 1974; SWANSON and COWAN, 1975).[242] Other 'intrinsic' channels are reciprocal thalamohypothalamic connections. The caudal channels are provided by an extension of the 'medial forebrain bundle' into the tegmentum and by the periventricular system. As regards functional systems, the hypothalamus seems thus to receive more or less direct olfactory input by way of the rostral

[242] The exact distribution of this input remains uncertain. MOORE and LENN (1972) suspect the preoptic region (nucleus suprachiasmaticus), PRINTZ and HALL (1974) the ventromedial nucleus. It should be added that HAYHOW *et al.* (1960) did not find evidence of accessory optic fibers ending in hypothalamic grisea, but this may have been due to the unreliability of the terminal degeneration methods (in this case the *Nauta-Gygax technique)*. The lateral and medial terminal nuclei of the cited authors represent, in my opinion, subdivisions of the nucleus opticus tegmenti *sensu latiori* (tegmental optical grisea), while their 'dorsal terminal nucleus' seems to be an undefined portion of the pretectal grisea.

channels, direct optic input as just mentioned, and presumably more or less indirect but diversified sensory input by way of ascending caudal channels.

In a somewhat different way, and stressing that 'it is not easy to epitomize the anatomical data regarding hypothalamic fiber connections', NAUTA and HAYMAKER (1969) point out (1) few if any direct connections from generally recognized sensory pathways with the exception of the 'olfactory radiation', (2) 'massive associations' with 'limbic forebrain structures and the paramedian region of the mesencephalon', and (3) projections to 'known visceral motor nuclei', which must be established by the 'medium of the mesencephalic and bulbar reticular formation'.

Some assumed general aspects of hypothalamic communication channels are illustrated by Figure 135 A. The origin of both mammillothalamic and mammillotegmental tract is shown in Figure 135 B. Fiber systems related to nucleus subthalamicus and substantia nigra are shown in Figure 153 C. This illustration a'so displays a peculiar tract, very conspicuous in Rodents, which was first erroneously interpreted by CAJAL (1903, 1911) as the *'cordon de Forel'* or field H_2 (cf. also Fig. 119 D). But repudiating his former interpretation, CAJAL (1928) redescribed said tract as a distinct bundle consisting of collaterals from the pes pedunculi, radiating into the rostral tegmentum and representing a corticotegmental channel. Our own observations gathered in the Department of Anatomy at the former Woman's Medical College of Pennsylvania confirmed these findings and the corrected interpretation of CAJAL (cf. SHINTANI-KUMAMOTO, 1959). This latter investigator noted that the nucleus interstitialis commissurae posterioris seems to receive some of the fibers of *Cajal's corticotegmental system*.

The *vegetative hypothalamus* is commonly considered to be an important and perhaps 'dominant' 'center' for the control of vegetative nervous system in general and endocrine activities in particular. On the other hand, through an in part non-neuronal feedback mediated by circulating hormones, the activities of the hypothalamus are also influenced by way of non-neuronal channels. The relationships of hypothalamus to the hypophysial and to the epiphysial complexes have been dealt with in section 1 C of the present chapter.

The neuronal mechanisms of the hypothalamus appear to control, or at least to have a significant influence upon, a wide variety of body functions and behavioral manifestations. The dorsal and entopeduncular group, is, as already mentioned above, especially related to the 'somatic' motor apparatus generally designated as 'extrapyramidal' system,[243] but is also connected, by substantial fiber systems, with the essentially autonomic centers of the 'conventional' hypothalamus. These latter, which comprise our anterior, middle and posterior griseal groups, are believed to be involved in activities directed toward the maintenance of a constant internal environment (homeostasis), although they also receive impulses from, and in turn again discharge back into, 'somatic' motor centers, as well as into cortical grisea concerned with 'affectivity'. The relationship of the hypothalamus to the autonomic centers was first clearly established by KARPLUS and KREIDL (1909, and other publications).

The hypothalamus seems to contain a dual autonomic mechanism, the anterior group and part of the middle group of nuclei being concerned, apparently, with parasympathetic activities, and the posterior group and part of the middle group (including the lateral hypothalamic nucleus) seeming to coordinate sympathetic responses. A sharp boundary between the two functional regions cannot be drawn.

Stimulation of the anterior hypothalamus has been shown to cause bladder contraction, increased gastrointestinal peristalsis, defecation, cardiac inhibition, and constriction of pupils, while stimulation of the posterior and lateral region evokes dilatation of the pupils, rise in blood pressure, cardiac acceleration, and inhibition of gastrointestinal peristalsis.

Sweating has resulted from stimulation of the anterior region near the anterior commissure. Although the sweat glands are presumably innervated by the sympathetic division, the postganglionic fibers reaching these glands are known to be cholinergic in man and in many experimental animals.

HESS (1947) designates parasympathetic action as *endophylactic* or *trophotropic*, and sympathetic activity as *ergotropic*. He emphasizes that in the ergotropic zone functions are collectively and synergistically

---

[243] 'The extrapyramidal system' can be evaluated as a semantically helpful fictional 'entity'. It is commonly defined as comprising all 'somatic motor' centers and channels of second or higher order except the pyramidal tract and its cortex of origin. Main grisea of the extrapyramidal system are striatum, pallidum, subthalamic body, substantia nigra, nucleus ruber, and formatio reticularis tegmenti. Cerebellar nuclei such as nucleus dentatus, moreover vestibular grisea and regions of cortex cerebri, such e.g. as the premotor area, may be included. It is now generally conceded 'that the extrapyramidal and pyramidal systems are less distinct entities than they were once thought to be' (MCDOWELL and LEE, 1971).

represented in what he calls a dynamogenic zone, without distinct foci; stimuli appear to become synthesized into cooperative action. Although in the trophotropic zone, fields with diffuse boundaries exerting definite coordinative functions could be vaguely outlined, it is likewise not possible to recognize distinct centers; considerable overlap of trophotropic and ergotropic zones obtains.

Since basal telencephalic regions, especially the para- or preterminal grisea (the so-called septum), discharge into hypothalamus through the 'medial forebrain bundle', it is not surprising that stimulation of these regions rostral to the hypothalamus should be followed by autonomic responses. Likewise, stimulation of the mesencephalic tegmentum, through which efferent hypothalamic tracts take their course, presumably with additional synaptic connections, will evoke a variety of autonomic activities.

Most of the hypothalamic nuclei should not be considered as specific centers for specific functions. The diversity of function appears to depend on the different combinations of the discharging grisea, that is on the diverse impulse patterns. This viewpoint must also be kept in mind for the appraisal of fiber tracts related to the thalamus. Multiplicity of function does not necessarily depend on anatomical parcellation. Relatively few parts with complex and widespread interconnections may be sufficient. Clinical and experimental observations support this latter view which is held by numerous students of nervous activities. Again, despite a certain plasticity allowing the utilization of a variety of suitable, topographically different neuronal links, it is probable that some grisea, as nodal *centers* for the transmission of impulses, are especially significant or even essential for the performance of a given function.

The *hypothalamus*, advantageously situated to receive impulses from many sources, contains, as was well summarized in a review of ANDERSON and HAYMAKER (1948), neural mechanisms which serve to integrate impulses into action patterns – and possibly even to originate impulses – as need or circumstances arise. A '*metronomic function*', influencing and regulating periodic or fluctuating processes was emphasized by these authors and by others: regularity of breathing rhythm, constancy of pulse rate, maintenance of body temperature, balance of intake and output of fluid, the cycle of sleep, the integrity of body weight, to a certain extent also acid-base balance in the blood and body fluids, as well as the periodicity of the menstrual rhythm – all these activities of the organism seem to be affected by the functions of the hypothalamus. While the hypothalamus appears practically indispensable for some of these activities (e.g. regulation of body temperature), it is less important for others (e.g. respiration). Moreover, homeostasis implies not only neural feedback systems but also a number of peripheral regulating mechanisms such as buffer systems.

As regards cyclic processes, various so-called circadian rhythms in the Hamster are said to be abolished by bilateral lesions of the *nucleus suprachiasmaticus* (STETSON-WHITMYRE, 1976). Said nucleus is therefore suspected as representing a '*biologic clock*'. It seems nevertheless probable that various other grisea in hypothalamus as well as in other brain regions may be 'oscillators' providing comparable 'biologic clock' effects for various activities.

BOON (1938) has stressed two groups of autonomic centers in the hypothalamus, namely hypophysial centers and non-hypophysial centers. SPATZ (1951), as mentioned in the remarks on terminology and subdivision of hypothalamic nuclei, also emphasizes the relations of his 'poorly myelinated hypothalamus' to the hypophysis. While it must be granted that some structures such as supraoptic nucleus are particularly closely related to the hypophysis, I am not fully convinced that hypophysial connections alone can be used as a valid criterion for either an anatomic or a functional subdivision of the hypothalamus.

With respect to heat loss and conservation, the presence of a thermoregulatory center in the hypothalamus has been demonstrated by numerous authors since about 1880, and on the basis of animal experiments the location of this center was assumed to be in the region of the tuber cinereum (ISENSCHMIDT and SCHNITZLER, 1914; SPIEGEL, 1928). Since then it has been found to embrace also the preoptic and posterior hypothalamic regions. 'Thermodetectors' are presumed to be present in the 'preoptic' and 'supraoptic' griseal neighborhoods (cf. e.g. v. EULER, 1961).

After earlier experimental studies had shown that extensive hypothalamic lesions were followed by disturbances of the mechanism involving both heat loss and heat production, the investigations of RANSON (1940) and others led to the conclusion that the anterior part of the hypothalamus is concerned with the mechanism of heat loss. Lesions of this region are followed by hyperthermia and inability to regulate against heat. The posterior hypothalamus, on the other hand, especially the lateral area or posterior part of the nucleus hypothalamicus lateralis, is concerned with the mechanism of heat conservation. Lesions in this area cause hypothermia and incapacity to regulate against cold. Thus a homothermic animal with extensive hypothalamic lesions will become poikilothermic.[243a]

It is not possible to give an exact distribution of these activities upon the various hypothalamic nuclei beyond these general statements. Mechanisms of heat loss are sweating, peripheral vasodilation, panting, and possibly retardation of metabolic activities. Mechanisms of heat conservation are muscular rigor, shivering, peripheral vasoconstriction, acceleration of heart rate, and, to a certain degree, increased metabolic activities.

As regards *water*, *carbohydrate*, and *fat metabolism*, moreover concerning *sex functions*, the following overall remarks can be given. Water balance is partly controlled through an antidiuretic principle found in the neurohypophysis. As mentioned in the discussion of nuclei, tracts, and neurohypophysis, this principle is presumably formed by the neurosecretory functions of nerve cells in nucleus supraopticus and paraventricularis, and reaches the neurohypophysis through supraopticohypophysial and paraventriculohypophysial tracts. Destruction of 85 per cent or more of these tracts or their nuclei of origin is followed by *diabetes insipidus* (FISHER *et al.*, 1935). The anterior lobe of the hypophysis, in turn, seems to produce a diuretic principle, apparently the thyrotrophic hormone, which counterbalances the action of the neurohypophysial antidiuretic principle.

VERNEY (1947) believes that supraoptic and paraventricular nuclei contain specific receptors ('osmoreceptors') which respond to changes in osmotic pressure within the blood capillaries passing through these nuclei and thus would influence the production of antidiuretic hormone. Vesicular structures along the surface of neurosecretory cells were interpreted by that author as 'osmometers'. GAAB (1973) has recently studied these vesicles in the supraoptic nucleus of Dogs by

---

[243a] In volume 1 of this series, the distinction between poikilothermal and homoiothermal Vertebrates was briefly dealt with on pp. 79 and 144. It should here be added, however, that at least some poikilothermal Amniota (large Turtles) and Anamnia (large pelagic Teleosts) can maintain a body temperature higher than the temperature of the environment. It has been suspected that a certain amount of 'thermoregulation' is thereby implied, while, on the other hand, 'retention of heat generated from metabolism by 'thermal inertia' seems possible (cf. e.g. NEILL and STEVENS, 1974). Be that as it may, the factors involved in such temperature differences already present in certain poikilotherm Vertebrates played perhaps a role in the phylogenetic development of Avian and Mammalian homoiothermy. A recent summary and critique of theories concerning thermoregulation in Mammals and other Vertebrates has been published by BLIGH (1973).

means of light- and electron-microscopy under conditions of experimental dehydration. He showed that said vesicles whose origin and significance remains uncertain, are extracellular with respect to the aforementioned nerve cells. Yet, be that as it may, there is little doubt that osmoregulation is performed by these hypothalamic grisea.

Hyperglycemia, and *glycosuria*, transitory in nature, follows electrical stimulation of the tuberal region and the posterior part of the lateral hypothalamic area (ASCHNER, 1912; LEWY and GASSMANN, 1935). This response is generally conceded to be due largely to the discharge of sympathetic impulses, which on reaching the adrenal medulla cause adrenaline to be liberated. In the intact animal, the same mechanism is presumed to become activated in response to anger, fear and other emotional states. Sympathetic impulses, reaching the anterior lobe of the hypophysis via the superior cervical ganglion and causing overproduction of the diabetogenic hormone fraction, have also been suggested (BOON, 1938, and others).

Lesions of the anterior hypothalamus, on the other hand, especially those in the region of the paraventricular or of the periventricular nuclei, are said to be followed by *hypoglycemia* and abnormal sensitivity to insulin. The mechanism of this action is not sufficiently understood. Although it is well recognized that pontine or bulbar centers exert an important control over carbohydrate metabolism in normal animal economy, there are relatively few reliable data concerning the combined role of hypothalamus and bulbar grisea in this regard.

The 'visceral functions' and 'vegetative activities' of oblongata respectively reticular formation, and the 'visceral input' to deuterencephalic grisea were briefly dealt with in sections 1 and 2 of chapter IX of volume 4 (pp. 305, 323/24, 338, 342). The reticulo-spinal channels, in part related to the relevant 'vegetative activities', were pointed out in chapter VIII, section 11 of volume 4.

The participation of the central nervous system in the control of carbohydrate metabolism was first demonstrated by CLAUDE BERNARD (1855, 1877) who observed glycosuria resulting from small experimental lesions of the fourth ventricle's caudal floor (*piqûre diabétique*, *Zuckerstich*). This was explained as due to the stimulation of 'sympathetic centers', causing, by way of the splanchnic nerves, increased sugar production through glycogenolysis in the liver. However, since the vagus region seems to be involved by the *piqûre*, its effect might also be indirectly stimulating, by damaging antagonistic parasympathetic channels inhibiting sympathetic outflow, and thereby enhancing this latter. Again, glycogenolysis in the liver may also be mediated by sympathetic outflow to adrenal medulla, liberating epinephrine, which then reaches the liver. Liberation of epinephrine, as mentioned above, is, of course, a well-known phenomenon occurring in emotional states and involving hypothalamic activities.

The first to recognize that the *adiposity* associated with pituitary and other lesions at the base of the brain may be due to intrinsic hypothalamic damage, was probably ERDHEIM (1904). Much subsequent experimental work has served to confirm this view. The hypothalamic nucleus most concerned is probably the ventromedial, but this has been questioned by GOLD (1973) who believes that 'obesity' results from damage to neighboring 'noradrenergic' bundles or to their terminals.

Little is known of the mechanism concerned in hypothalamic control of fat metabolism and sex functions. The neural circuits of both mechanisms may either be channelled through identical hypothalamic cell groups or involve closely adjacent structures.

The destruction of the neurohypophysial eminentia medialis (DEY, 1943), as well as the destruction of the parvocellular area of tuber cinereum (nucleus periventricularis arcuatus, nucleus ventromedialis) are followed by *atrophy of the gonads* (SPATZ, 1951, and other publications). If the damage occurs early in life, puberty does not take place.

In Rabbits, ovulation has been induced by electrical stimulation of anterior hypothalamus (preoptic recess) as well as of tuber cinereum. ANDERSON and HAYMAKER (1948) in reviewing these data reach the conclusion that the hypothalamus contains a mechanism concerned in the release of gonadotrophic hormones from the anterior pituitary. SPATZ (1951) assumes that the 'parvocellular area' of the tuber cinereum, close to the neurohypophysis, and containing nucleus periventricularis arcuatus and nucleus ventromedialis, represents a hypothalamic sexual center essential for maturation and maintenance of gonads. SPATZ believes that these nuclei receive centripetal impulses through the tuberohypophysial tracts. These nuclei are thus supposed to be influenced by the hormones of the adenohypophysis through the mediation of the neurohypophysis (eminentia medialis, infundibular stem). The results of the experiments by DEY (1943) and others are interpreted as caused by the elimination of the adeno-neurohypophysial zone of contact. SPATZ also mentions observations of his associates showing that ovulation cannot be induced by electrical stimulation of tuber cinereum if the spinal cord is transected and interprets this finding in favor of his view postulating a descending neural pathway from hypothalamic sexual center to gonads.

On the basis of numerous observations and experiments it seems therefore fairly certain that a central neural mechanism involving the hypothalamus is active in sexual behavior, ovulation, luteinization, and menstruation. This mechanism may include centers of the tuber cinereum as well as median eminence and infundibular stem of neurohypophysis. As to the details of its working, only conjectures can be expressed at this time. These surmises must consider the various aspects of mediation through nerve or vascular connections with the pituitary gland, or with other endocrine and visceral organs via the autonomic system. Relevant relationships with the pituitary complex were dealt with in section 1 C of the present chapter.

With respect to an assumed hypothalamic regulation of the *sleep-wakefulness mechanism*, BEATTIE (1938) reported somnolence in experimental animals after damage to the posterior part of the hypothalamus, and similar observations were made by RANSON (1939).

HESS (1929) is the chief exponent of the view that sleep is an expression of parasympathetic activity, inasmuch as during sleep there are bradycardia, vasodilation, and other parasympathetic manifestations. Such a view would imply that hypersomnia is due to suppression of the antagonistic outflow by damage to the sympathetic areas of the hypothalamus, but the numerous available data would lead one to conclude that sleep is a passive, not an active process, due, as BEATTIE (1938) puts it, to 'a damping-down of hypothalamic activity'. This view implies that there is a 'wakefulness center' in the caudolateral part of the hypothalamus and contiguous rostral midbrain, not a sleep center. Experimental work in cats by INGRAM *et al.* (1951) tends to confirm the view that the hypothalamus normally exerts an alerting effect upon the cortex, which is involved in maintaining resting wakefulness. This is believed to be but one component, however, since strong alerting stimuli can break through the sleep of cats lacking a hypothalamus, 'indicating that accessory components may still serve to support wakefulness in the absence of the hypothalamo-cortical facilitatory mechanism'. SPIEGEL and INABA (1927) emphasized that the thalamus may also be concerned with maintenance of the sleep-wakefulness rhythm.

In a series of experiments in cats, MILLER and SPIEGEL (1940) induced sleep, occasionally followed by catatonia, the whole disturbance lasting about 7 to 9 days, by lesions of the 'subthalamus'. The region involved appears to include parts of zona incerta and subthalamic nucleus, fibers of the *H-systems of Forel* and perhaps adjacent parts of dorsal hypothalamic area. The lesions remained lateral to mammillothalamic tract and ended dorsal to the medial border of the pes pedunculi fibers. Whether the effect of the lesion was caused by damage to grisea or fiber tracts must remain in doubt, but the interpretation offered by the authors, namely that damage of pathways passing through or synapsing in that region contributed to the genesis of the symptoms, appears plausible.

In the anesthetized animal physiological sensory stimuli can elicit a generalized cortical reaction which is characterized by bilateral activation in specific and non-specific cortical areas: this response is regarded as an awakening reaction by BERNHAUT *et al.* (1953) who undertook a study in which cortical and hypothalamic potentials were recorded. Usually the specific cortical area related to the type of stimulation used shows a greater degree of excitation than the non-specific areas. Hypothalamic activation occurring simultaneously with generalized cortical responses supports the view that the hypothalamus takes an active part in the awakening reaction. The authors suggest that normally the hypothalamus is responsible for the generalized cortical or arousal reaction. It would doubtless be of interest to study the potentials of other diencephalic centers in this respect, since it must be assumed that additional, non-hypothalamic centers may play a role in this mechanism. The transitory effect of hypothalamic lesions on the state of consciousness strongly supports this view.

Concerning the '*emotional responses*', it is well known that decorticated experimental animals, on slightest provocation, exhibit phenomena of explosive and massive, predominantly sympathetic discharges, designated as *sham rage* (BARD, 1928). Such reactions cease after destruction of the posterior hypothalamus.

In recent years, stimulation experiments involving the hypothalamus (as well as other neighborhoods of diencephalon and of deuterencephalic tegmentum) have led to the concept of '*electrically controlled behavior*' which shall be dealt with in chapter XIII since the postulated thereto pertaining 'systems' substantially include telencephalic grisea presumably directly or indirectly related to the 'limbic' mechanisms.

ANDERSON and HAYMAKER (1974) bring an interesting historical review of hypothalamic and pituitary research in a volume edited by SWAAB and SCHADÉ (1974) which, in addition, contains various papers presenting contemporary views concerning the problems of 'integrative hypothalamic activity'. The comprehensive volume edited by HAYMAKER *et al.* (1969), including the editors' own contributions, takes up significant anatomical, functional, and clinical problems related to the hypothalamus in the entire Vertebrate series from Petromyzon to Man.

In accordance with the procedures followed in volume 4 concerning spinal cord and deuterencephalon, some relevant aspects of the *Human diencephalon* shall now be separately considered in concluding the present section dealing with the diencephalon of Mammals.

As regards *gross anatomical features* it should be recalled that, in the fully developed Human brain, the definitive dorsal surface of the thalamus and epithalamus, which resulted from the transformation of the lateral surface in the region of the hemispheric stalk, becomes hidden in the depth of the fissura transversa cerebri basal to corpus callosum and corpus fornicis (Fig. 136B). Upon removal of the corpus callosum, of the leptomeninges forming the so-called tela chorioidea of the third ventricle (velum interpositum), and of the choroid plexuses of lateral and third ventricles, the dorsal surface of the thalamus and the sulcus terminalis are bared as seen in Figure 136C. The torn edge of the lamina affixa is known as the taenia chorioidea ventriculi lateralis. A rostral area of the dorsal surface protrudes slightly as the tuberculum anterius thalami corresponding to the anterior nuclear group.

The *epithalamus* displays the stria medullaris thalami with its torn fringe representing the attachment of the choroid plexus of the third ventricle (taenia thalami), and the caudal enlargement designated as trigonum habenulae which contains the habenular nuclei. In the midline, attached to the habenulae laterally, to the habenular commissure rostrally, and to the posterior commissure caudally, the pineal body extends caudad, resting on the pretectal region and on the groove between the two superior colliculi. The membranous roof of the third ventricle evaginates caudad as a suprapineal recess dorsal to the pineal body.

The embryonic *lateral surface* of the thalamus caudal to the hemispheric stalk, which has become a *caudal surface* in the fully developed brain, displays the prominences of the pulvinar, of the lateral geniculate body, and of the medial geniculate body (Fig. 136D).

In addition to these free surfaces of thalamus and epithalamus, there is a moderately extensive free surface of the hypothalamus at the base of the brain. It consists of the preoptic lamina terminalis between the

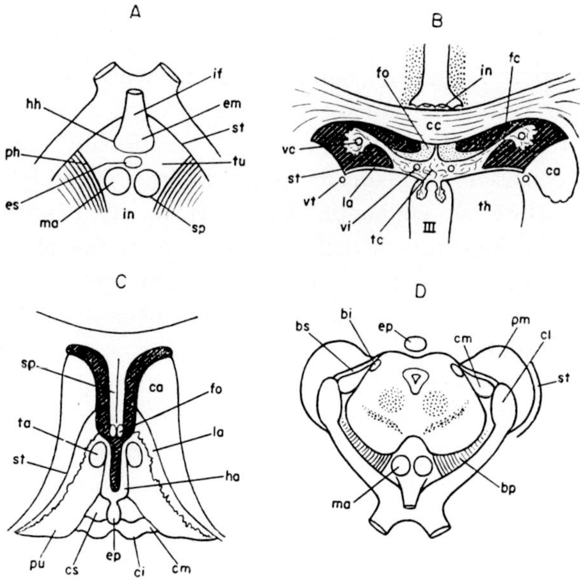

Figure 136 A–D. Diagrams illustrating gross morphologic features of the adult human diencephalon (from K., 1954). A Basal aspect, hypothalamus. B Cross-section showing the tela chorioidea ventriculi tertii (velum interpositum) in the fissura cerebri transversa and the relationships of sulcus terminalis and lamina affixa. C Dorsal aspect of the thalamus after removal of corpus callosum and velum interpositum. The pretectal region (not labeled) is a narrow strip between the habenula (ha) and the colliculus superior (cs). D Caudal surface of diencephalon. The brain stem has been cut by a section through the midbrain at the level of the colliculus superior. bi: brachium quadrigeminum inferius; bp: basis pedunculi; bs: brachium quadrigeminum superius; ca: n. caudatus; cc: corpus callosum; ci: colliculus inferior; cm: corpus geniculatum mediale; cs: colliculus superior; em: eminentia medialis; ep: epiphysis; es: eminentia saccularis; fc: fissura chorioidea; fo: fornix; ha: trigonum habenulae; hh: s. hypophysiohypothalamicus; if: infundibulum; in (Fig. A): interpeduncular fossa, (Fig. B): indusium griseum with striae longitudinales; la: lamina affixa; ma: mammillary body; ph: s. pedunculohypothalamicus; pm, pu: pulvinar; sp (Fig. A): s. postmammillaris, (Fig. C): septum pellucidum; st (Fig. A): s. posterior tractus optici, (Figs. B–D): s. terminalis; ta: tuberculum anterius thalami; tc: tela chorioidea ventriculi tertii (velum interpositum); th: thalamus dorsalis; tu: tuber cinereum (region of so-called eminentia lateralis); vc: vena chorioidea; vi: vena cerebri interna; vt: vena terminalis; III: third ventricle.

optic chiasma and the anterior commissure, of the chiasma and the optic tract, of the tuber cinereum, of the infundibulum with its eminentia medialis, and of the mammillary bodies. The telencephalo-diencephalic groove runs rostrally between medial edge of diagonal band and preoptic region, then caudalward between supraoptic nucleus and substantia perforata anterior, and still more caudally along the lateral edge of the optic tract. The mesencephalon is separated from the hypothalamus by the pedunculo-hypothalamic (tubero-peduncular) sulcus, the postmammillary sulcus, and the pedunculo-optic sulcus between pes pedunculi and medial edge of optic tract. The lateral, bulging parts of the tuber cinereum are occasionally referred to as lateral eminences and contain the nuclei tuberis. A caudal bulge has been designated as eminentia saccularis or as postinfundibular eminence. The eminentia medialis is neurohypophysial in structure; it represents the proximal bulge of the infundibulum and is separated from the rest of the hypothalamus by the sulcus hypophysio-hypothalamicus (Fig. 136 A).

As regards *craniocerebral topography*, the position of the pineal body becomes of special significance since this structure is usually filled with calcifications in elderly or even middle-aged persons and can thus be readily recognized in the roentgenogram. The attachment of the choroid plexus to the habenular commissure (taenia habenulae) and that commissure itself may furthermore contain calcifications which should not be confused with those of the pineal body. The habenular calcifications appear C-shaped or semilunar in lateral roentgenograms (Stauffer, Snow, and Adams, quoted by Spiegel and Wycis, 1952). The position of the pineal body varies in relation to the interaural plane, drawn through the center of the external auditory meatus, and perpendicular to the base line which is drawn from the inferior border of the orbit to the center of auditory meatus (Spiegel and Wycis, 1952). These variations have been emphasized by Vastine and Kinney (1927), and Davidoff and Dyke (1951); Spiegel and Wycis found that the center of the pineal body may be located between 4 and 20 mm posteriorly to the interaural plane. The last two authors, who prepared a valuable stereotaxic atlas of the human brain (1952), state that because of the variability of the human skull one must select reference points within the brain rather than rely on the skull landmarks in order to determine the coordinates of subcortical structures. The pineal body and the posterior commissure were found to be convenient reference points. Spiegel and Wycis have computed numerous data concerning the distances of diencephalic and other structures such as globus pallidus, anterior thalamic nuclear group, nucleus medialis thalami, lateral and ventral thalamic nuclei, pulvinar, geniculate bodies, nucleus ruber and others from center of pineal body as well as from posterior commissure. The range of variation was carefully noted and indicated.

If the pineal body cannot be sufficiently well demonstrated, or if the center of this structure cannot be determined, air encephalography usually shows outlines such as those of third ventricle, aqueduct or cisterns. The location of the posterior commissure and that of the anterior commissure, which likewise represents an important stereotaxic landmark, can then be identified. DAVIDOFF and DYKE (1951) have undertaken a comprehensive survey of pneumo-encephalographic anatomy. In summarizing some of my own experiences with that subject, I pointed out that cadaver encephalography with opaque contrast media might represent a useful supplementary method for the study of normal intracranial cerebral topography (K., 1940).

As regards *blood supply*, the Human diencephalon receives its blood supply from posterior cerebral, posterior communicating, middle cerebral, and anterior cerebral arteries. It must be kept in mind that both arteries and veins display many individual variations in their pattern and that occasional findings may be at considerable variance with the standard descriptions as well as with the more detailed observations of FOIX and HILLEMAND (1925) and other authors who studied the vasculature of the diencephalon.

The most proximal portion of the *posterior cerebral artery* gives off short posteromedial arteries to the tuber cinereum and the mammillary bodies. These vessels communicate with or contribute to the interpeduncular arterial plexus. Farther distally, from 4 to 6 or more posteromedial thalamic arteries arise from the posterior cerebral artery and are distributed to pulvinar and habenula. Some of these arteries run deep into the intralaminar region of the thalamus and are also known as posterior thalamoperforating arteries. Their territory comprises the region of the centrum medianum, as well as neighboring parts of nucleus medialis and nucleus ventralis lateralis. The adjacent thalamogeniculate arteries supply pulvinar, geniculate bodies, nucleus ventralis posterolateralis and possibly posteromedialis, and posterior part of internal capsule. The posterior chorioidal arteries (or branches) also arise from the posterior cerebral artery, proximal to or among the posteromedial thalamic arteries; they encircle the posterior part of the thalamus and one branch may run along the posterior part of the sulcus terminalis while one or two branches may cross the dorsal surface of the thalamus. These vessels reach the tela chorioidea of the third ventricle *(velum interpositum)*, including the chorioid plexuses of lateral and third ventricles. The posterior chorioidal arteries, jointly with branches of the anterior choroidal

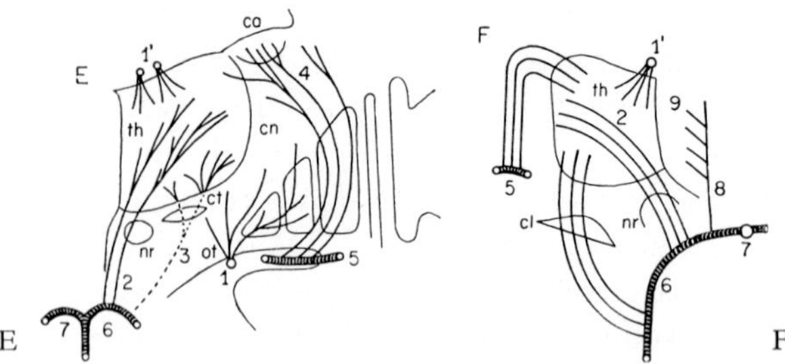

Figure 136E, F. Diagrams of human diencephalic arterial blood supply (E slightly modified after Foix and Nicolesco, 1925, and F after Foix and Hillemand, 1925, from K., 1954). 1: anterior chorioid artery; 1′: anterior and posterior chorioid arteries; 2: posterior thalamoperforating arteries; 3: thalamogeniculate arteries; 4: anterolateral striate (lenticulo-striate and lenticulo-optic) arteries; 5: A. cerebri media; 6: A. cerebri posterior; 7: A. basilaris; 8: A. communicans posterior; 9: posteromedial, including anterior thalamoperforating arteries; ca: nucleus caudatus; cl: corpus geniculatum laterale; cn: capsula interna; ct: nucleus subthalamicus; nr: nucleus ruber; ot: optic tract; th: thalamus. Two segments of globus pallidus are indicated below cn; putamen above 5, claustrum and insula indicated lateral to putamen. In Figure 136F, the vessels originating from a. cerebri media are lenticulo-optic arteries, those originating from a. cerebri posterior and passing through cl are thalamogeniculate arteries.

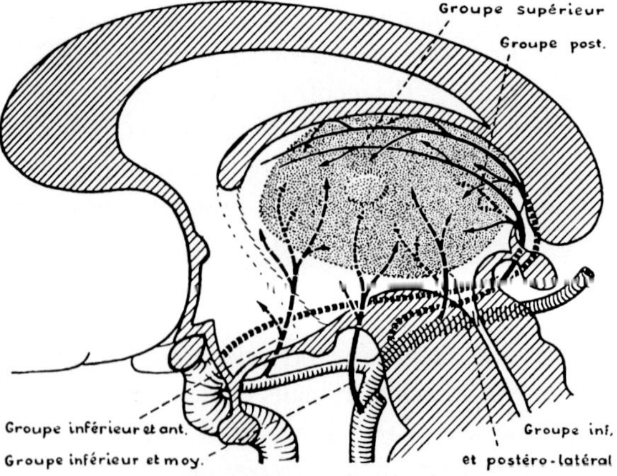

Figure 136G. Diencephalic arterial supply of the human brain as projected upon a sagittal plane (from Lazorthes *et al.*, 1962).

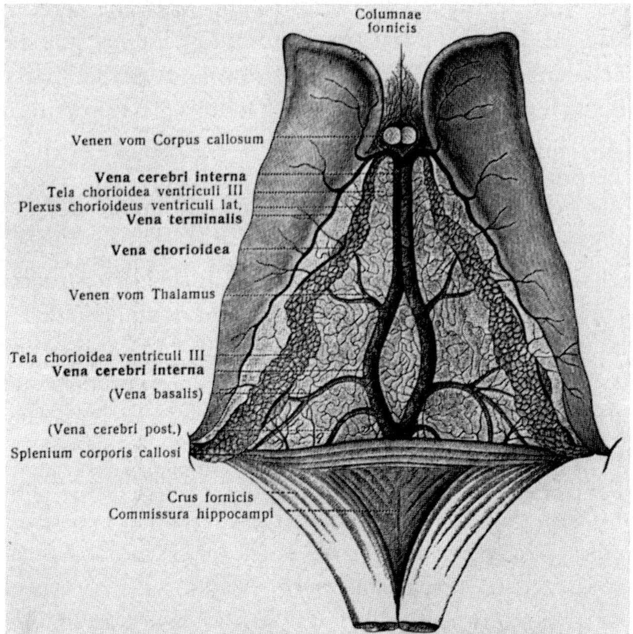

Figure 136 H. Veins in tela choroidea and related structures of the human brain. The corpus callosum has been cut, and the crus fornicis is reflected caudalward. The columns of the fornix are cut at the level of interventricular foramen (after Vicq d'Azyr, from Rauber-Kopsch, 1933). The unlabelled junction of the antimeric venae cerebri internae (at the splenium corporis callosi) is the *vena magna Galeni*, which receives the *vena basalis of Rosenthal* and then, together with inferior sagittal sinus, joins the sinus rectus (cf. Fig. 392B, p. 723 of vol. 3/II).

artery, also supply the medial and dorsal part of the thalamus (nucleus medialis, nuclei anteriores, part of nucleus lateralis dorsalis and nucleus lateralis posterior). A conspicuous but unlabelled ramification of these vessels can be seen in Figure 142C.

It should be recalled that the *posterior cerebral artery* is not infrequently (in about 8 per cent or more of studied series) a direct branch of the internal carotid. In this case a modified posterior communicating artery connects arteria cerebri posterior and arteria basilaris.

The *posterior communicating artery* provides posteromedial arteries to the hypothalamus, including the so-called internal optic arteries. It also gives off posteromedial arteries to the thalamus. These vessels, representing the anterior thalamoperforating arteries, run deep into the diencephalon and reach the anterior and medial parts of the thalamus, including nucleus ventralis lateralis. A few posteromedial arteries to the hypothalamus may arise directly from the internal carotid before its terminal bifurcation.

The *middle cerebral artery* provides anterolateral striate arteries entering through the anterior perforated substance. These vessels include the lenticulostriate arteries of which one or two larger ones, in close relation to the internal capsule, are known as *Charcot's artery or arteries of cerebral hemorrhage*. These latter vessels are shown (unlabelled) in Figures 142B and C. Among the more caudal anterolateral striate arteries variable lenticulo-optic arteries reaching the anterosuperior part of the dorsal thalamus (nucleus lateralis dorsalis, nucleus ventralis lateralis, dorsal part, anterior part of nucleus lateralis posterior) have been described but cannot be regarded as constant features.

The *anterior chorioidal artery* is a branch of the internal carotid or of the middle cerebral artery and runs caudad to enter the chorioid fissure of the inferior horn of the lateral ventricle. It also sends branches to globus pallidus, amygdaloid nucleus, optic tract, pes pedunculi, retrolenticular portion of internal capsule, pulvinar, and together with the posterior chorioid branches, to the superior surface of the thalamus.

The *anterior cerebral* and *anterior communicating artery* send anteromedial branches to the anterior perforated substance and furthermore to the anterior hypothalamus, especially the preoptic region. A variable, fairly long medial striate artery, also known as *Heubner's recurrent branch*, may supply the head of the caudate nucleus and an adjacent part of internal capsule. The thalamus proper does not, as a rule, receive any appreciable blood supply from the anterior cerebral artery.

To summarize the blood supply of the thalamus proper it can be said that its arterial vascularization derives from five sources (Figs. 136E–G): (1) thalamogeniculate arteries from arteria cerebri posterior for the caudolateral portion of the thalamus, including as a rule the greater portion of the posteroventral complex (n. ventralis posteromedialis and posterolateralis). (2) Posterior thalamoperforating arteries, likewise from arteria cerebri posterior, for the posterior thalamic region above the region supplied by thalamogeniculate vessels. (3) Posteromedial arteries from arteria communicans posterior, including anterior thalamoperforating vessels, for the anteromedial portion of the thalamus. (4) Arteria chorioidea anterior for globus pallidus, and as regards thalamus proper for pulvinar and dorsal portions of thalamus, supplemented by posterior chorioid branches. (5) Lenticulo-striate ('lenticulo-optic') arteries, variable, for a small lateral anterosuperior or laterosuperior thalamic segment.

The *venous drainage* of the diencephalon is provided on the dorsal

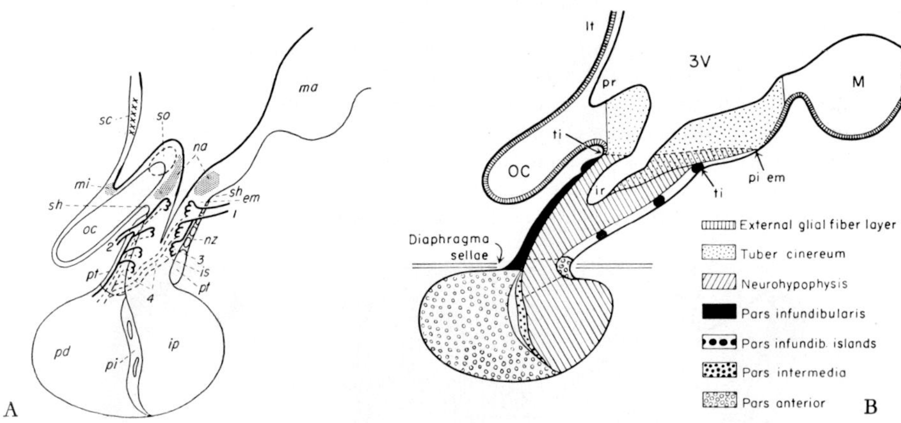

Figure 137A. Diagram of sagittal section through floor of hypothalamus illustrating hypophysio-hypothalamic relationships. The flat ependyma lining the infundibular recess and the supraoptic crest is indicated by the thinner portions of the line representing the ventricular surface. The hypophysio-portal vessels are indicated according to the concept of GREEN. (Combined after diagrams of CHRIST, GREEN, SPATZ, and original preparations, from K., 1954.) em: eminentia medialis (neurohypophysis); ip: infundibular process (neurohypophysis); is: infundibular stem (neurohypophysis); ma: mammillary body; mi: midline nucleus (inferior median preoptic nucleus); na: nucleus arcuatus; nz: neurovascular zone (posterior surface zone of infundibular stem and eminentia medialis with islands of pars tuberalis cells); oc: optic chiasma; pd: pars distalis (adenohypophysis); pi: pars intermedia (adenohypophysis); pt: pars tuberalis (adenohypophysis); sc: supraoptic crest; sh: sulcus hypophysio-hypothalamicus; so: supraoptic commissures; 1: posterior group of superior hypophysial arteries; 2: anterior group of superior hypophysial arteries; 3: tufted vessels; 4: hypophysio-portal veins (broken lines).

Figure 137B. Semidiagrammatic sagittal section illustrating additional details of hypophysial complex and hypothalamic floor (after CHRIST, from HAYMAKER, 1969). ir: infundibular recess; lt: lamina terminalis; M: mammillary grisea; OC: optic chiasma; pi em: postinfundibular eminence; pr: preoptic recess; ti: tubero-infundibular sulcus; 3V: third ventricle. Neurohypophysis and hypothalamic tuberal region are here drawn as overlapping. Pars infundibularis is pars tuberalis of adenohypophysis.

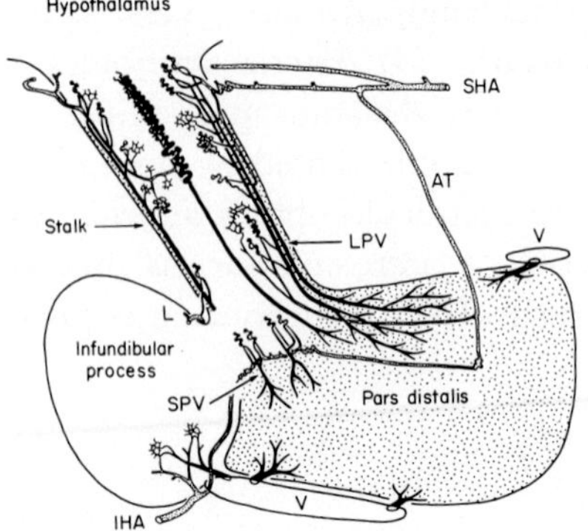

Figure 137C. Vascularization of Human hypophysial complex (after ADAMS and XUEREB *et al.*, from HAYMAKER and ANDERSON, 1971). AT: trabecular artery; IHA: inferior hypophysial artery; L: lateral branch of inferior hypophysial artery; LPV: long portal vessels; SHA: superior hypophysial artery; SPV: short portal vessels; V: venous sinuses. The arteries are said to form a primary plexus or first capillary bed. Arising from these latter are the portal vessels feeding the second capillary bed (sinusoids) of the adenohypophysis which are drained by the venous sinuses.

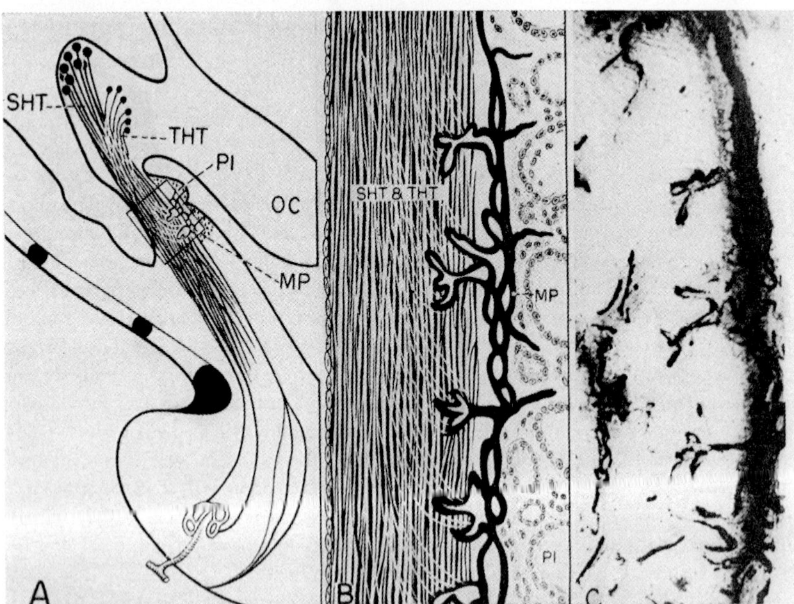

Figure 138. Hypothalamohypophysial neural connections and capillary loops of the infundibulum (from HAYMAKER and ANDERSON, 1971, C being from a preparation by CHRIST). A Overall arrangement of tracts. B, C Enlargements of the infundibular area indicated in A, MP: 'mantle capillary plexus'; OC: optic chiasma; PI: infundibular region; SHT: supraopticohypophysial tract; THT: tuberohypophysial tract.

surface by *vena terminalis*[243b] and its continuation, the *vena cerebri interna* (Fig. 136 H). Both venae cerebri internae, running in the velum interpositum, join to form the unpaired *vena cerebri magna Galeni* which begins between splenium corporis callosi and pineal body, and continues as sinus rectus within the ridge of the tentorium cerebelli along the union of this structure with the falx cerebri. Near the point where the internal cerebral veins join to form the vena cerebri magna they receive short epithalamic veins. More rostrally the internal cerebral veins receive chorioidal branches from the chorioid plexuses and adjacent diencephalic areas.

The hypothalamus and the basal portion of the thalamus are drained by veins which empty into the *basal vein of Rosenthal*, either directly or through the mediation of the venous interpeduncular plexus. The basal vein, in turn, ascends through the cisterna ambiens and empties into vena cerebri magna or into vena cerebri interna. Venous blood from hypothalamus and from basal parts of thalamus, reaching the interpeduncular plexus, may also be drained from that structure into cavernous and immediately adjacent sinuses (sphenoparietal, superior petrosal sinus). The peculiar hypophysio-portal system has been considered in the discussion of the hypophysial complex (section 1 C of this chapter). Additional details concerning its relationships in Man are shown in Figures 137 and 138.

Radio-opaque material injected into one internal carotid artery passes mostly through the vessels of the homolateral side, but nevertheless generally reaches, because of the anastomosis between the anterior cerebral arteries, also the vessels of the contralateral side, particularly the rostral ones. If the vertebral artery is injected, the distribution of the contrast material is likewise not restricted to the corresponding cerebral vessels of the homolateral side.[243c]

[243b] At the *foramen Monroi*, where the vena terminalis (commonly also called 'vena striothalamica' or 'thalamostriata' by clinicians), after taking up the vena chorioidea, becomes the vena cerebri interna and bends caudad, it is also joined by veins from the head of the caudate nucleus, and particularly by the vena septi pellucidi (cf. Fig. 136 H). With improved methods of angiography, it is possible, at the venous phase, to visualize the junction between vena septi pellucidi and vena terminalis *('Venenwinkel')*, which confluence indicates the locations of the foramen Monroi (cf. Krayenbühl and Yasargil, 1965).

[243c] The statements stressing homolateral distribution on p. 31 of my monograph on the Human diencephalon (K., 1954) were based on reported early observations with still imperfect technique and thus must be amended.

The *Monro-Kellie doctrine* postulating a constancy of intracranial volume with reciprocal relationship between brain substance, blood and cerebrospinal fluid, and the *Cushing phenomenon* characterized by a rise in systemic blood pressure following a rise in intracranial pressure, are of significance for the diencephalon as well as for the other parts of the brain. Sundry details and present-day speculations concerning 'autoregulation of cerebral blood flow' and related topics, particularly with respect to their clinical significance, can be found in the proceedings of a 'symposium' edited by FIESCHI (1972).

As regards the interpretation of *thalamic degenerations* after extensive cortical ablations, the possibility that disturbances of the vascular supply of the thalamus may introduce a source of fallacy should not be disregarded. Evaluating various experimental studies of this kind, LE GROS CLARK (1949) has stressed the probability that some of the degenerations observed in the thalamic nuclei may have been partly the direct result of an interference with their blood supply rather than the secondary effects of the cortical lesions.

A number of authors have described nerve fibers along intracerebral vessels. It is likely that the vasoconstrictor fibers are provided by

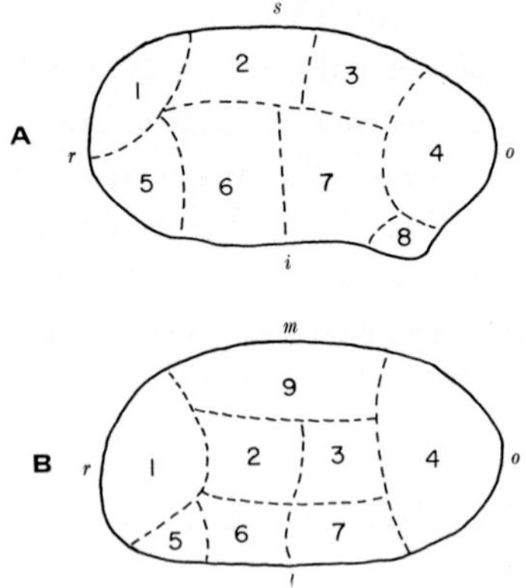

Figure 139. Diagram showing approximate overall configuration of griseal groups of human dorsal thalamus (based on and slightly modified from a plastic model prepared by the late Dr. J. PAPEZ and kindly donated to the present author). A Lateral aspect. B Dorsal aspect. 1: anterior griseal group; 2: nucleus lateralis dorsalis; 3: nucleus lateralis posterior; 4: pulvinar; 5: nucleus ventralis anterior; 6: nucleus ventralis lateralis; 7: nucleus ventralis posterolateralis; 8: nucleus geniculatus lateralis dorsalis; 9: nucleus medialis *sive* dorsomedialis; i: inferior side; l: lateral side; m: medial side; o: occipital extremity; r: rostral extremity; s: superior side.

cervical sympathetic components while vasodilator fibers are supplied by parasympathetic components of cranial nerves, especially by the facial nerve through the mediation of nervus petrosus superficialis major and pericarotid plexus. Experimental evidence seems to indicate that the extent of the vasomotor innervation of cerebral vessels may greatly differ in different species of mammals and even in different regions of the brain in a given species.

The overall configurational arrangement of *main griseal groups* within the Human thalamus, which, *in toto*, represents a rostrocaudally elongated ovoid body, is shown in the sketch of Figure 139. Figures 140 A to C illustrate the general spatial relationships of the thalamic peduncles or stalks. Figures 141 A and B show assumed projection of thalamic nuclei upon cerebral cortex. It is likely that all these projections include reciprocal connections. On the whole, observations incidental to clinical thalamotomy (cf. e.g. LARSON and SANCES, 1973) have tended to support the overall concept indicated by the depicted diagrams. Such observations, as summarized by McLARDY (1963) have also indicated that, in Man, the 'interneurones' ('microneurones', *Schaltzellen*) of thalamic grisea are easier to be identified, by their

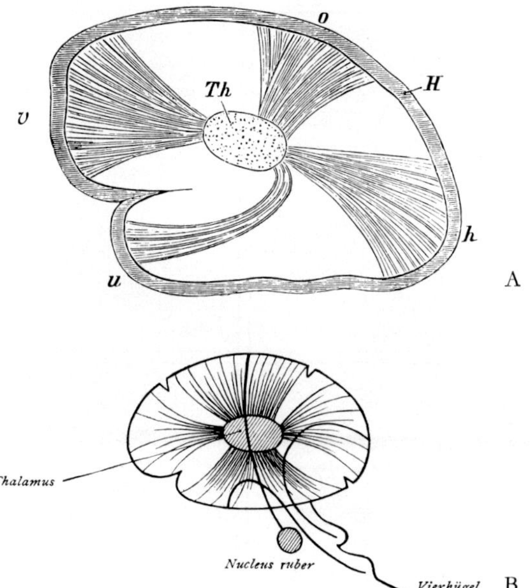

Figure 140 A, B. Oversimplified, but instructive diagrams indicating four thalamic stalks extending to frontal, parietal, occipital and temporal lobes (A, RAUBER-KOPSCH, 1914, B from VILLIGER, 1920).

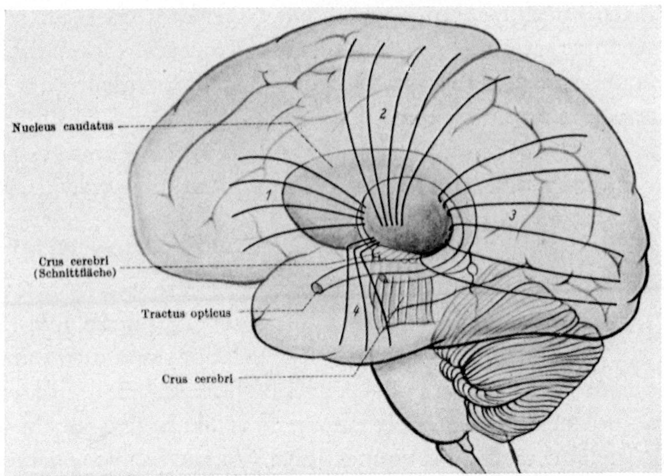

Figure 140 C. Diagram showing four thalamic stalks with somewhat more detailed relationships (from CLARA, 1959). 1: frontal thalamic stalk; 2: fronto-parietal thalamic stalk; 3: occipital thalamic stalk; 4: temporo-parietal stalk.

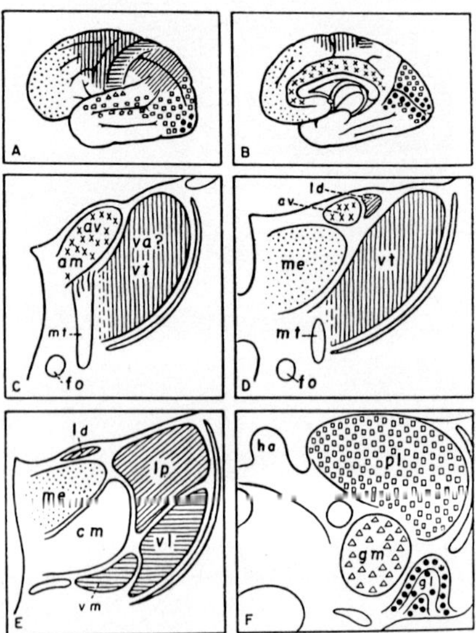

Figure 141 A. Diagram of the assumed cortical projections of Human thalamic nuclei, as based on fairly reliable data available about 20 years ago (adapted and modified after diagrams of LE GROS CLARK and of WALKER for Primates (from K., 1954). Abbreviations as for Figures 142 A to 143 G.

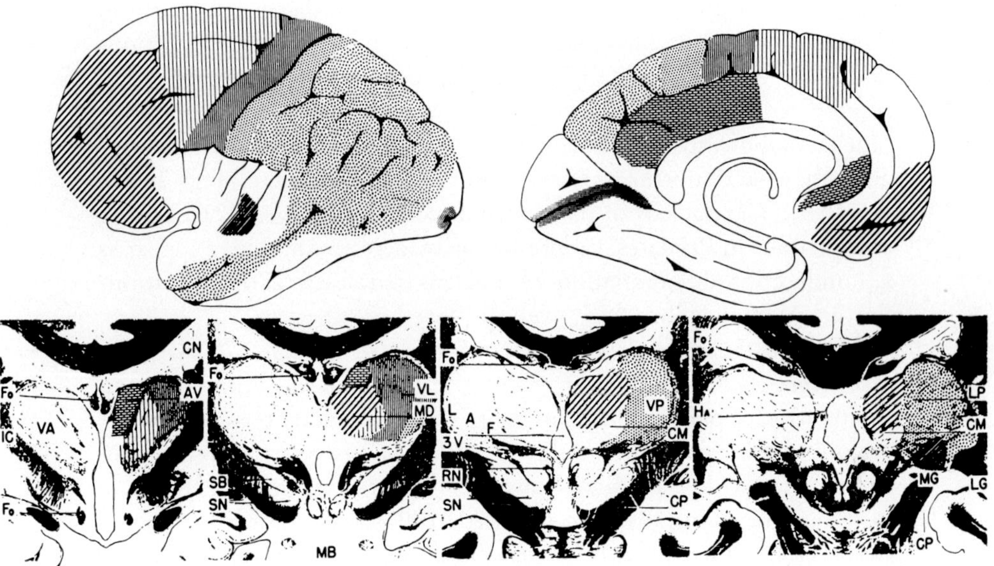

Figure 141 B. Nuclei and cortical projections of the Human thalamus as interpreted by WALKER (from WALKER, 1966). A: arm representation; AV: nucleus anterior ventralis; CM: centrum medianum nucleus; CN: nucleus caudatas; CP: pes pedunculi; F: face representation; Fo: fornix; Ha: habenular grisea; IC: internal capsule; L: leg representation; LG: lateral geniculate grisea; LP: nucleus lateralis posterior; MB: mammillary complex; MD: nucleus medialis thalami; MG: medial geniculate complex; RN: nucleus ruber; SB: subthalamic nucleus; SN: substantia nigra; VA: nucleus ventralis anterior; VL: nucleus ventralis lateralis; VP: nucleus ventralis posterolateralis complex; 3V: 3rd ventricle.

relatively larger size in comparison with glial elements, than the corresponding 'microneurons' in various other Mammals.

Figures 142A–F, and 143A–G depict cross-sections through the Human thalamus, displaying the *cytoarchitecture* and *myeloarchitecture* of the diverse nuclei as discussed above for the Mammalian thalamus in general, and should be compared with the relevant preceding Figures.[243d] It can be said that, on the whole, all thalamic grisea reasonably well delimitable in the Human diencephalon are also present in that of

[243d] Stereotaxic atlases of the human brain with detailed illustrations of thalamic grisea have been prepared by SPIEGEL and WYCIS (1952), TALAIRACH *et al.* (1957), and HASSLER (1959). The terminology and interpretation of the nuclei in the first two cited atlases substantially conforms to that adopted by myself (1954). HASSLER's terminology and interpretations, despite certain differences and in part greater details, are, nevertheless,

the other Mammals, and vice-versa, although significant differences in their relative size or differentiation, as well as minor pattern distortions doubtless obtain. Such differences are particularly noticeable with respect to midline, and intralaminar 'nuclei', as well as pulvinar. Nucleus submedius and nucleus medialis ventralis, whose delimitations are somewhat doubtful, represent another case in point (cf. Figs. 142 A, B, 143 B). The nucleus of the centrum medianum (Figs. 142 E, 143 D), conspicuous in Primates, cannot be easily identified and may appear as a nondescript differentiation of nucleus parafascicularis in various 'lower' Mammals, but fairly well displayed in others.

The *massa intermedia* (BNA), *adhaesio interthalamica* (PNA), or *commissura mollis* containing midline nuclei, and resulting from an obliteration of parts of the third ventricle by secondary concrescence or fusion of the ventricular walls, is rather extensive in a diversity of Mammalian forms. In Man, however, it is of much lesser extent and very variable, being absent in about 20 to 30 per cent of most examined case-series.[244] Occasionally, it may be double, and, in rare instances, a thin band of nerve tissue, corresponding to a strip-like massa intermedia, has been noted to cross the hypothalamic part of the third ventricle close to the preoptic region, containing an aberrant supraoptic decussation as depicted by NAUTA and HAYMAKER (1969, Fig. 4–36, p. 192).

As regards the ventrolateral nucleus, its relationship to the premotor cortex, which apparently controls the performance of skillful, patterned motor activities, probably accounts for what has been interpreted as thalamic 'speech representation' (cf. OJEMANN and WARD, 1971; BROWN, 1974).[245] 'Somatotopic' localization of body regions and of 'activation' and 'suppression' has been claimed for that griseum on the basis of experiments during stereotaxic operations (RÜMLER *et al.*, 1972).

---

essentially compatible with the here retained nomenclature of 1954. Still more recent very detailed publications on the Human thalamus are those by DEWULF (1971) and by VAN BUREN and BORKE (1972). Despite the attempts made by these authors to add some further data and to standardize some aspects of thalamic terminology, it could be said that no very significant improvement upon the concepts and interpretations of their predecessors has been achieved.

[244] A 'commissura anterior secundaria mollis' and diverse variations of the massa intermedia have been reported by FRETS (1916/17).

[245] The problem of a 'dominant' hemisphere, respectively of 'reciprocal hemispheric specialization' (cf. vol. 3/II, p. 269) seems here to play a role.

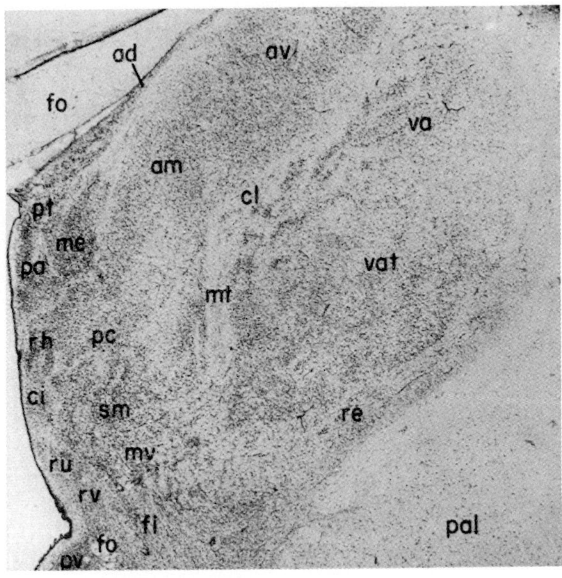

Figure 142 A. Cross-section *(Nissl stain)* through the thalamus of a newborn infant at level of rostral tip of nucleus medialis (from K., 1954; ×12, red. ²/₃). Abbreviations for Figures 142A to 143G: ad: n. anterodorsalis; al: ansa lenticularis; am: n. anteromedialis; av: n. anteroventralis; bs: brachium quadrigeminum superius; ci: n. centralis medialis; cl: n. cent. lateralis; cm: centrum medianum; dd: n. dorsales disseminati; dl: lamina medullaris dorsolateralis; dm: tractus tectothalamicus dorsomedialis; ew: *Edinger-Westphal nucleus;* fi: n. filiformis; fo: fornix; fs: fasciculus subthalamicus; gl: corpus geniculatum laterale, pars dorsalis; glv: cor. gen. lat., pars ventralis; gm: cor. gen. mediale, pars dorsalis; gmv: cor. gen. med., pars ventralis; H: *tegmental field H of Forel;* H_1: fasciculus thalamicus; H_2: fasciculus lenticularis; ha: habenula; hl: n. habenularis lateralis; hm: n. hab. medialis; ip: inferior peduncle of thalamus; it: fasciculus intrathalamicus medialis; la: lamina medullaris anterior; ld: n. lateralis dorsalis; le: lemniscus medialis and associated systems; li: lamina medullaris interna; lid: lam. med. int., pars dorsalis; liv: lam. med. int., pars ventralis; lp: n. lateralis posterior; lpi: n. lat. post., pars intermedia; lu: n. subthalamicus *(Luys);* ma: massa intermedia; me: n. medialis; mi: midline nuclei; mt: fasciculus mammillothalamicus; mv: n. medialis ventralis; nc: n. caudatus; nr: n. ruber; oc: oculomotor nucleus; pa: n. paraventricularis thalami; pal: globus pallidus; pc: n. paracentralis; pe: prerubral tegmentum in tegmental fiber field; pf: n. parafascicularis; pl: pulvinar; pli: pulvinar, pars inferior; pll: pulvinar, pars lateralis; plm: pulvinar, pars medialis; pr: n. profundus mesencephali; pt: n. parataenialis; pv: n. paraventricularis hypothalami; re: n. reticularis thalami; rh: n. rhomboidalis; ru: n. reuniens; rv: n. reuniens ventralis; sg: n. suprageniculatus; sm: n. submedius; sn: substantia nigra; sp: n. subparafascicularis; sr: stria medullaris thalami; tc: tegmental cell cord; tf: tegmental field fiber of n. ruber; va: n. ventralis anterior; vat: n. vent. ant., transition to n. vent. lateralis; vl: n. vent. posterolateralis; vm: n. vent. posteromedialis; vt: n. vent. lateralis; vtd: n. vent. lat., pars dorsalis; vtl: n. vent. lat., pars lateralis; vtm: n. vent. lat., pars medialis; vtv: n. vent. lat., pars ventralis; we: *Wernicke's field* (beginning of optic

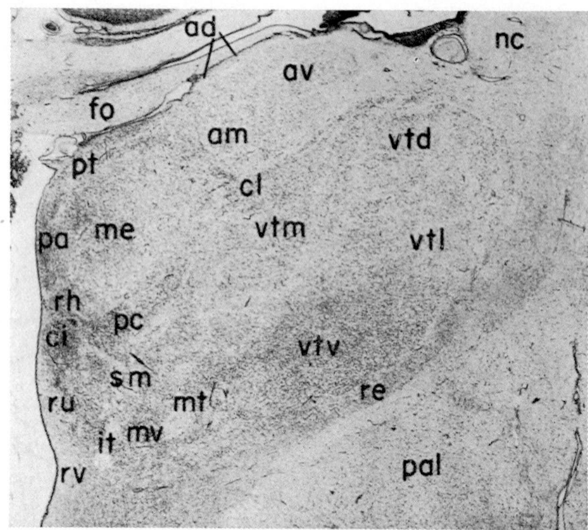

Figure 142B

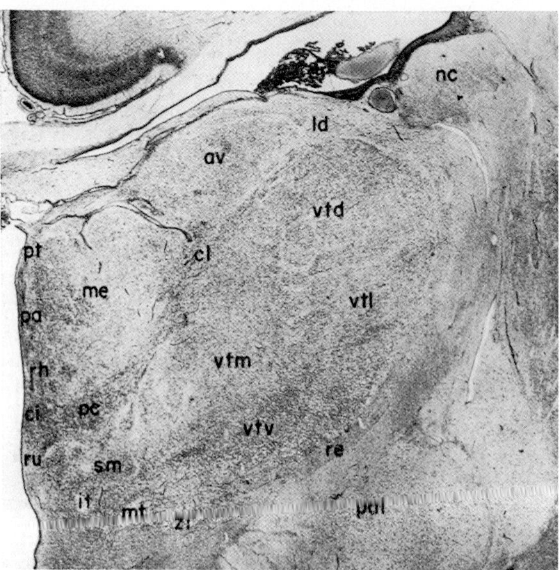

Figure 142C

radiation); za: stratum zonale; zi: zona incerta; zo: superficial intrazonal cell plate; 3: n. praetectalis (principalis); 4: n. posterior (thalami); 4′: n. posterior, pars lateralis; 5: n. lentiformis mesencephali (pars) magnocellularis; 6: n. lent. mes. (pars) parvocellularis; 7: n. olivaris colliculi superioris; 8: area praetectalis (propria); 9a: n. interstitialis commis-

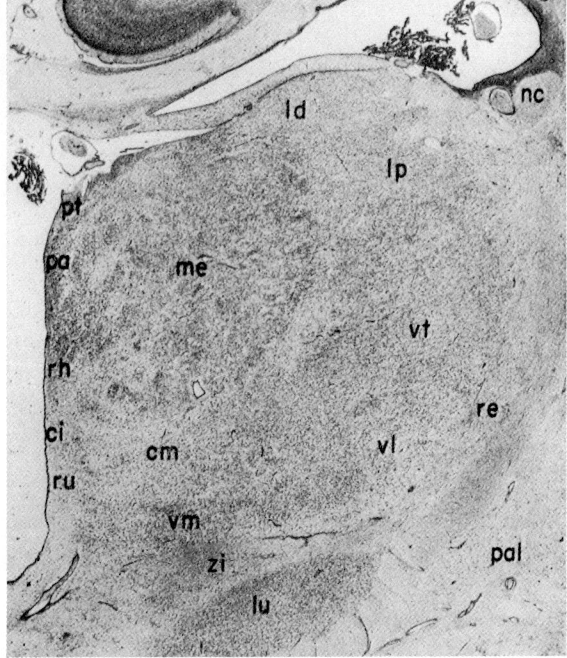

Figure 142D

surae posterioris, pars rostralis; 9b: n. interst. (principalis) comm. posterioris; 9c: n. centralis subcommissuralis commissurae posterioris; 9d: n. supracommissuralis interstitialis commissurae posterioris; 9e: n. interst. magnocellularis commissurae posterioris; 9f: n. medianus infracommissuralis commissurae posterioris; 9t: n. interst. tegmentalis commissurae posterioris; 10: n. suprageniculatus; 11: area praetectalis, pars supracommissuralis; 12: n. sublentiformis; 13: n. lateralis grisei centralis; 18: n. interst. fasciculi longitudinalis medialis; 19: *n. of Darkschewitsch;* 20: n. medialis anterior.

Figure 142B. Cross-section *(Nissl stain)* through the rostral part of the thalamus of a newborn infant, about 0.9 mm caudal to Figure 142A (from K., 1954; ×9, red. $^1/_2$).

Figure 142C. Cross-section *(Nissl stain)* through the rostral part of the thalamus of a newborn infant, about 0.6 mm caudal to Figure 142B (from K., 1954; ×10, red. $^1/_2$).

Figure 142D. Cross-section *(Nissl stain)* through the middle part of the thalamus in a newborn infant (from K., 1954; ×10, red. $^1/_2$).

Figure 142E. Cross-section *(Nissl stain)* through the posterior part of the adult Human thalamus (from K., 1954; ×9, red. $^1/_2$).

Figure 142F. Cross-section *(Nissl stain)* through the posterior part of pulvinar and the geniculate bodies in a newborn infant (from K., 1954; ×9, red. $^4/_5$). The unlabelled large artery adjacent to gyrus dentatus is the posterior cerebral artery which may here have mainly originated from arteria communicans posterior. The large vein adjacent to gm is the basal vein.

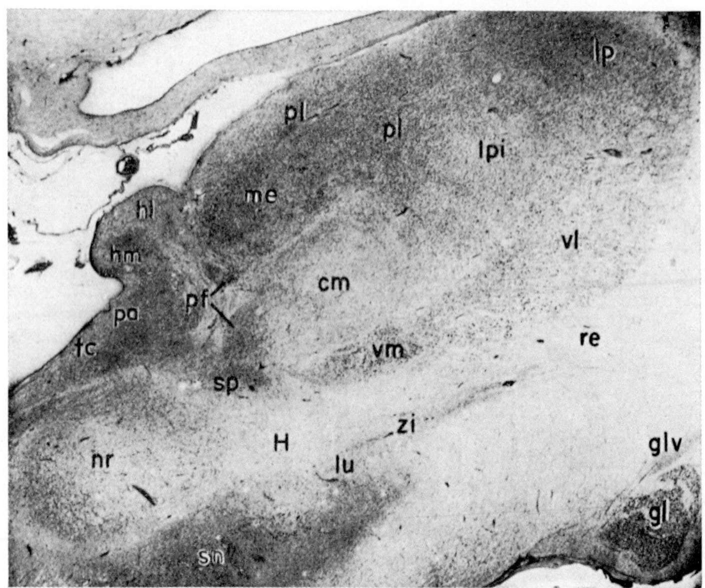

Figure 142E (legend see p. 397)

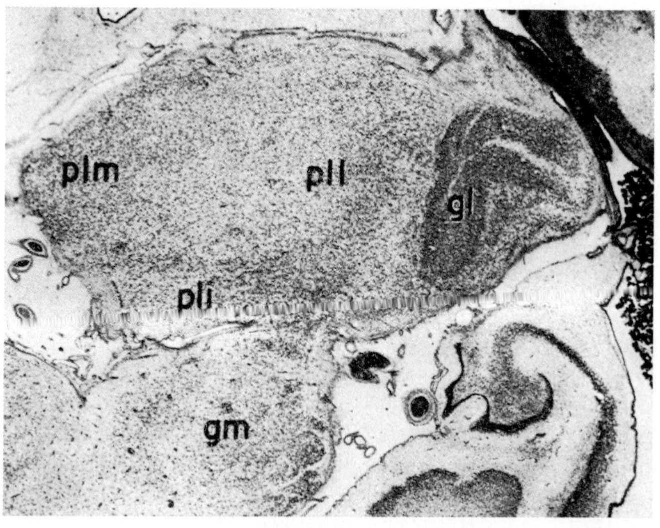

Figure 142F (legend see p. 397)

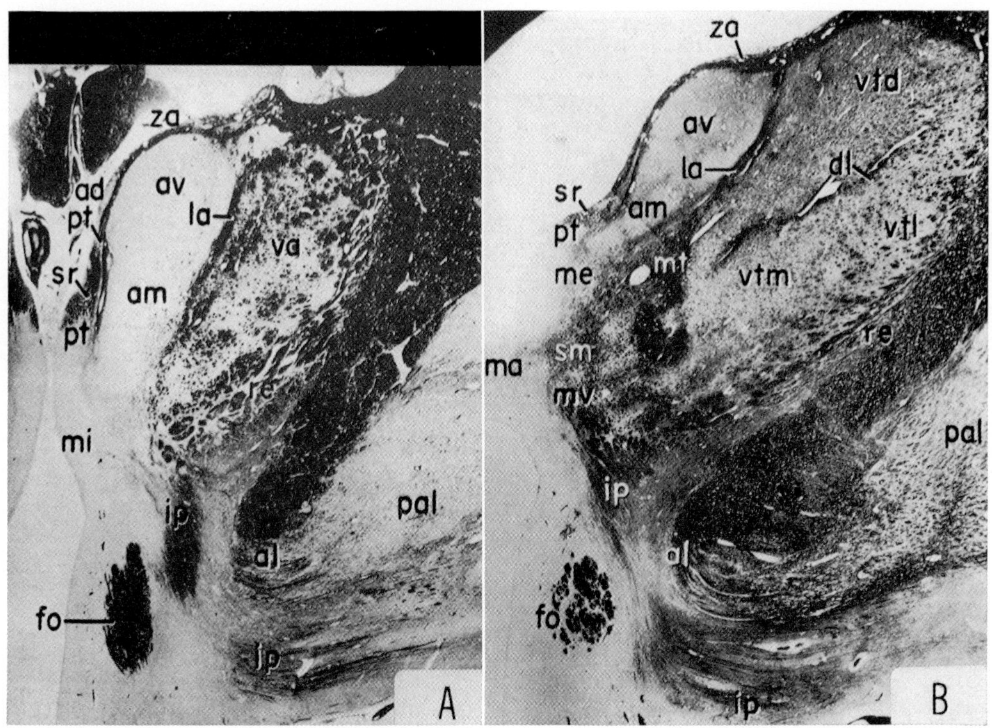

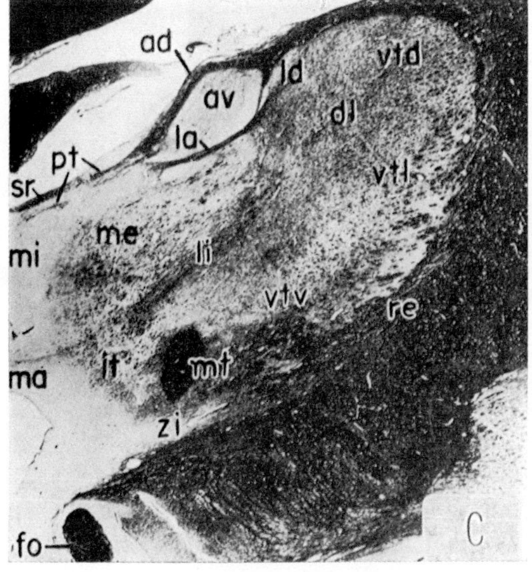

Figure 143 A. Cross-section (myelin stain) through the rostral part of the adult Human thalamus (from K., 1954; ×7, red. $^4/_5$).

Figure 143 B. Cross-section (myelin stain) somewhat caudally to preceding section and showing mammillothalamic tract entering anterior griseal group. In this and the following section the anterior nuclear group displays the tuberculum anterius indicated in Figure 136C (from K., 1954; ×7, red. $^3/_4$).

Figure 143 C. Cross-section (myelin stain) through the middle part of the adult Human thalamus (from K., 1954; ×5, red. $^3/_4$).

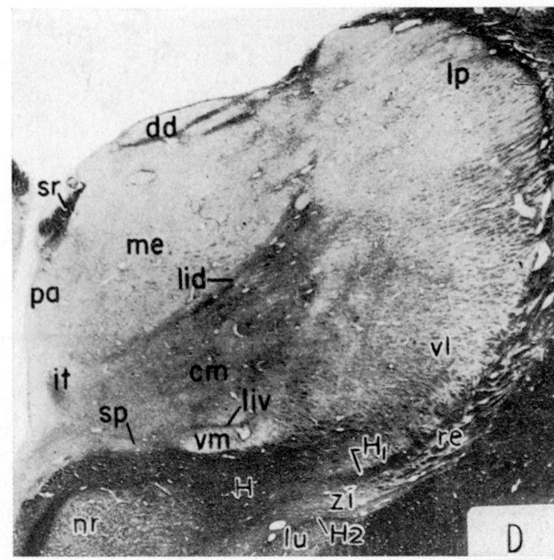

Figure 143 D. Cross-section (myelin stain) through transition of middle to caudal part of adult Human thalamus (from K., 1954; ×7, red. 3/4).

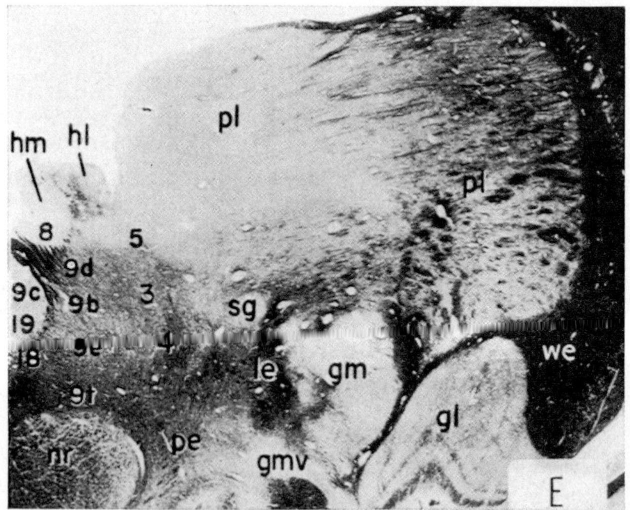

Figure 143 E. Cross-section (myelin stain) through caudal part of adult Human thalamus (from K., 1954; ×7, red. 3/4).

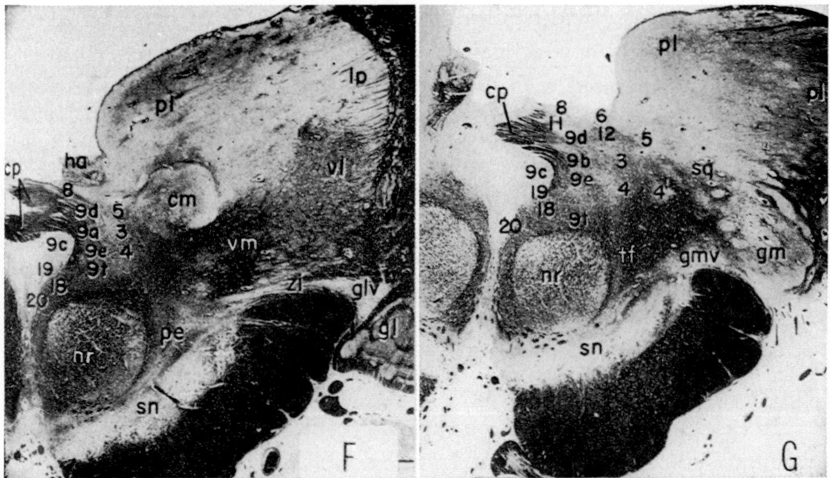

Figure 143 F, G. Two cross-sections, in rostro-caudal sequence, at diencephalo-mesencephalic transition levels, showing pattern of pretectal grisea. The plane of the section is more oblique (with basocaudal tilt) than that of Figure 143 E (from K., 1954; ×5, red. $^1/_2$).

With respect to the ventralis posteromedialis and posterolateralis complex, and supplementing data obtained in Mammals (cf. Fig. 130) observations in the course of stereotaxic surgery have indicated a roughly comparable somatotopic arrangement in Man (cf. e.g. TASKER *et al.*, 1972).[246]

The impulses via spinothalamic and secondary ascending trigeminal and related bulbothalamic tracts seem to take the circuit to contralateral thalamus (and hemisphere); the ascending impulse may also utilize homolateral channels.[246a] Proprioceptive and deep sensibility

---

[246] The cited authors conclude 'that medial lemniscus fibres may be recognized by stimulation as they course rostrally in the 8–10 mm sagittal plane to enter and relay within the ventrocaudal nucleus, where they are organized into a discrete medial-lateral contralateral homunculus'.

'Caudally and somewhat medially, the lemniscal fibres merge with spinoquintothalamic tract fibres, which, we believe, project to the caudal portion of ventrocaudal nucleus and certain immediately posterior nuclei. Here they are organized into a smaller bilateral homunculus extending caudally from the lemniscal one and somewhat at right angles to it' (TASKER *et al.*, 1972, p. 195).

[246a] In a series of experimental investigations, DUSSER DE BARENNE and SAGER (1931) reported bilateral representation for cutaneous sensibility in the thalamic relay nuclei of the Cat, but with emphasis on the contralateral side.

channels provided by the bulbothalamic fibers of the medial lemniscus appear to reach only contralateral thalamus and hemisphere. Nothing definite can be stated regarding unilateral or bilateral representation of enteroceptive sensibility or visceral pain. The possibility of individual variations in the relative amount of unilateral and bilateral representation of the different ascending pathways should also be considered. Apart from bilateral connections of ascending secondary sensory fibers, a bilateral representation might be provided by relaying commissural circuits passing through thalamic association nuclei, pretectal region, or even hypothalamus. Such commissural fibers, as described by GLEES (1944) may run through the supraoptic commissure.

In cases of *tabes dorsalis*, OBERSTEINER (1881) described the rare and peculiar symptom of *allochiria*. In this condition a stimulus applied to one extremity will be referred by the patient to a corresponding location of the contralateral extremity. It is most likely that this phenomenon, also occasionally noted in *Brown-Séquard's syndrome*, is related to the assumed bilateral sensory representation. Because of damage to the main ascending spinal pathways, the impulse may be propagated over an alternative circuit in the other half of the spinal cord and reach the homolateral sensory cortex where the sensation is referred to the opposite side. However, additional unkown factors must be involved, since a similar shift believed to occur in cortical or thalamic damage as well as following unilateral chordotomy may take place without allochiria.

Pain is a modality of consciousness, i.e. a non-registrable and 'objectively' not demonstrable nor observable *qualité pure*, which represents *sensu strictissimo* an 'introspective' private experience. The 'central origin' of pain was carefully considered by EDINGER (1891). According to HEAD (1920) it depends upon the function of the thalamus. Nevertheless, cortical participation may be assumed to be required, not only for the perception of pain, but for the occurrence of 'consciousness' in general.

It is likely that several possible pathways for impulses related to pain sensations exist: (1) the spinothalamic and bulbar system to the posterior ventral thalamic nuclear complex. (2) Spinotectal and bulbar fibers to tectum mesencephali whence the impulses may be distributed to thalamus and hypothalamus through the mediation of the pretectal grisea. (3) *Flechsig's Haubenstrahlung* which includes a relay system of internuncial neurons taking its course through the reticular formation and reaching hypothalamus as well as thalamus, possibly also through the mediation of pretectal grisea. Whether an additional pathway takes its course through the central gray or whether the central gray in the aqueductal region is merely involved as part of the pretectal discharge system remains a moot question. That hypothalamic activity may play a greater role in pain sensations than hitherto assumed appears probable on the basis of some of SPIEGEL's observations (SPIEGEL and KLETZKIN, 1952).

SPIEGEL and co-workers found that on electric stimulation of the quadrigeminal area under local anesthesia, preceding electrolysis of the spinothalamic tracts for relief of pain, patients complained of pain. In an attempt at analysis, experiments on cats were performed. Pain suggestive reactions were obtained from the superior and inferior colliculus; the periaqueductal gray proved very sensitive, particularly on mechanical stimulation. Pain suggestive reactions were also evoked from spinothalamic system, medial lemniscus, and dorsal as well as ventral parts of reticular formation (SPIEGEL *et al.*, 1953a). HASSLER's (1970) theory of pain conduction shall be discussed further below on p. 412.

Concerning the *anterior griseal group*, reference to the discussion of these nuclei in connection with the Mammalian thalamus in general will here be sufficient. Further data on the *nucleus medialis* and *centrum medianum* in the Human diencephalon can be found in the reports by MCLARDY (1948, 1950).

Little is reliably known about the Human *ventral lateral geniculate griseum*, which includes the so-called *nucleus praegeniculatus*. The *dorsal lateral geniculate nucleus* (Fig. 144) displays a conspicuous curved lamination, whose layers are separated by bands of fibers predominantly provided by the optic tract. This latter enters the griseum through the concave hilus and through the fiber bands between the laminae.

At least six laminae are easily recognized, depending on the planes of section, but some individual variations in details of development and arrangement seem to obtain, and up to eight laminae have been described. The notations used by various authors differ. Thus, CROSBY *et al.* (1962) enumerate layers 1 to 6, beginning at the hilus, while ROSE (1935) counts 7 laminae, beginning at the apex of the roughly cone-shaped structure and including a seventh layer, mentioned, but not included in their number sequence by CROSBY *et al.*, at the hilus region (cf. Fig. 144). HASSLER (1959) has termed this layer

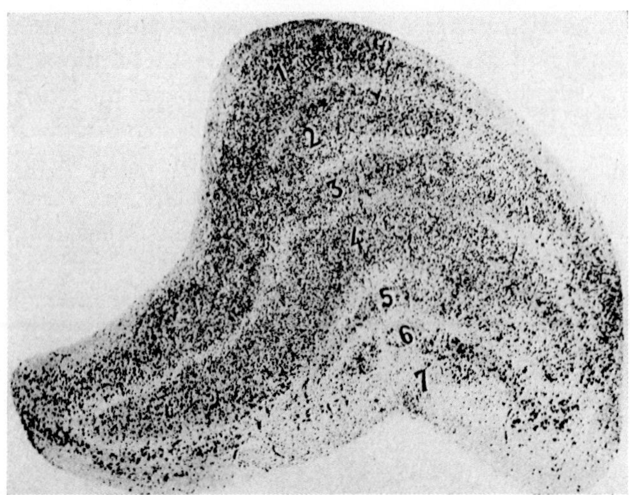

Figure 144. Somewhat oblique cross-section *(Nissl stain)* through the Human dorsal lateral geniculate nucleus showing 7 laminae (from Rose, 1935). The location of this griseum, cut in a slightly different plane, is shown in Figures 142E, F, and 143E. Notation of layers discussed in text.

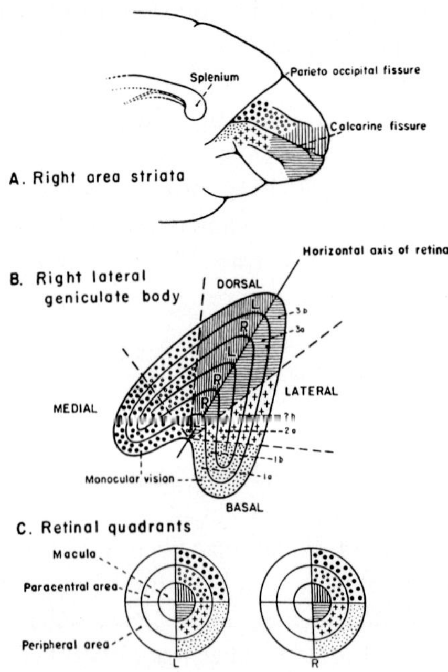

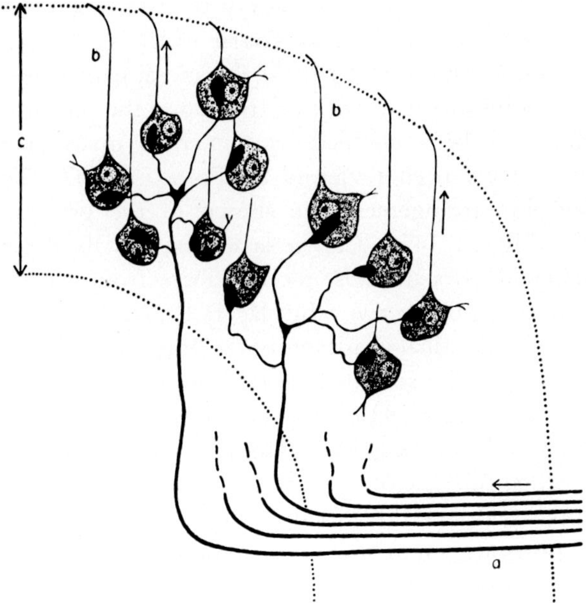

Figure 145 D. Semidiagrammatic illustration of optic nerve connections within one lamina of the Primate (including Human) dorsal lateral geniculate body, as based on the interpretation of degeneration experiments (from GLEES, 1957). a: optic fiber bundle from one retina; b: fibers of optic radiation to visual projection cortex; c: one layer of dorsal lateral geniculate griseum.

'*nullte Schicht*', and other authors, e.g. GIOLLI and TIGGES (1970), who depict said layer in various Primates, accordingly designate it as lamina 0, followed, toward the apex, by lamina 1 to 6. For practical purposes, I had adopted the notation 1a, 1b, to 3a, 3b, as shown in Figure 145 B. The output of dorsal lateral geniculate nucleus forms the optic radiation (geniculocalcarine tract) to occipital cortex (area striata), leaves at the medial convexity, at the apex, and at the lateral surface of the cone-shaped griseum. Output collaterals may reach other grisea, e.g. parts of the pulvinar, which are also believed to receive some direct optic tract input.

Figure 145 A–C. Diagram of assumed retinal projection on dorsal lateral geniculate griseum and cerebral cortex (adapted and modified after concepts of BROUWER, LE GROS CLARK, METTLER and other authors, from K., 1954). Six layers in a notation differing from that of the preceding figure, and omitting layer 7 of Rose, are here indicated.

The intricate *point to point projection* of the retina upon corpus geniculatum laterale, pars dorsalis, the course of the geniculocalcarine or optic radiation arising in that part of the lateral geniculate body, and the point to point projection of that tract upon the calcarine cortex or area striata have been clarified by numerous investigators whose contributions have been reviewed by POLYAK (1957). The general features of this arrangement are shown in the greatly simplified diagrams of Figures 145 A–C. The layers 1a, 2b, 3b of corpus geniculatum laterale, pars dorsalis, receive fibers from the nasal retinal quadrants of the opposite eye, while layers 1b, 2a, 3a receive the fibers from the temporal retinal quadrants of the homolateral eye. The lateral portion, especially the so-called lateral horn of the lateral geniculate body constitutes the projection area for the lower peripheral quadrant of the retina, while the medial and somewhat rostral part, including the so-called medial tubercle or protrusion, comprises the projection area for the upper peripheral quadrant of the retina (CHACKO, 1948). An extensive portion comprising a central segment of the laminae including the dorsal and caudal apex, opposite the ventral and rostral hilus, and representing more than half the total volume of the lateral geniculate body, receives the macular fibers. This projection area has the form of an inverted pyramid with a convex dorsal base and a flattened ventral apex pointing toward the hilus. The central projection area includes all six laminae, but in the peripheral projection areas layers 2b and 3b unite at the medial and lateral extremities of layer 3a and continue into lateral horn and medial tubercle as a fused layer (CHACKO, 1948). Layers 1a and 1b have much larger cells than those of layers 2a to 3b, and are occasionally referred to as the magnocellular laminae (cf. layers 5 and 6 of ROSE, in Fig. 144). As mentioned above, further differentiations of indications of laminar differentiation, resulting in eight layers for the macular segment in adult Man have been claimed (e.g. HARMAN, 1949), but it appears dubious whether this latter individually variable and rather indistinct additional laminar differentiation is functionally significant,[247] except perhaps for layer 0, which, at least in the Cebid Ateles, seems to have homolateral connections. Moreover, as pointed out by GIOLLI and TIGGES (1970), the retinogeniculate input among Prosimians, namely contralateral input to laminae 1, 5, and 6 (1a, 3a, 3b,) homolateral to 2, 3, 4 (1b, 2a, 2b), differs from that recorded in Anthropoids, namely

[247] Cf. the tendency toward individual variations mentioned above in the text.

contralateral 1, 4, 6 (1a, 2b, 3b) respectively homolateral 2, 3, 5 (1b, 2a, 3a). No satisfactory explanation for that apparent difference can be given.

Again, in the Prosimian Lorisiform monkey Galago, CAMPOS-HORTEGA and GLEES (1967a, b), and others, have described a binocular projection of their lamina 3. Also, in Platyrrhine Anthropoids (New World monkeys), e.g. Ateles, the lamination appears somewhat 'concealed'.

As regards the investigations of *optic pathways*, it should here be recalled that, in addition to ascending and descending degeneration displayed by the affected neuron, transneuronal degeneration, dealt with in chapter V, section 8, p. 662 of volume 3/I, has played an important role.

In each cellular layer, each optic tract fiber from the hilumward adjacent fibrous lamina seems to terminate with a circumscribed endbush, formed by a spray of 5 to 6 branches in relation to groups of 5 or 6 geniculate cells (Fig. 145D). According to GLEES and LE GROS CLARK (1941) no geniculate cell would be related to more than one bouton.[248]

The inferior quadrants of the retina are then projected upon the inferior lip of the calcarine fissure, and the superior quadrants upon the superior lip. The geniculocalcarine fibers concerned with macular vision project upon a large area in the region of the occipital pole of the hemisphere. A bilateral cortical representation of each entire macula has been assumed by some authors, but evidence in this respect is not conclusive and extensive unilateral projection of the macular quadrants may be consistent with the observed preservation of central vision in unilateral lesions. It should also be mentioned in this respect that in the optic chiasma of various mammals CAJAL (1898) and others have described bifurcating fibers which were subsequently interpreted

---

[248] GLEES (1957) added the remark: '*dies ist jedoch nur mit Vorsicht auszulegen*', since additional synaptic relationships cannot be excluded. Thus, for instance, corticogeniculate and other terminals ending on geniculate cells may be present, providing for complex interactions. Again, the question of additional internuncial cells within the geniculate griseum remains poorly elucidated (cf. e.g. the comments by CROSBY *et al.*, 1962). Moreover, the possibility of taxonomically related differences should be kept in mind. One may also find it difficult, although perhaps not impossible, to interpret the *Golgi impregnations* of Figures 132A and 134B in terms of the concept expressed by Figure 145D. The hitherto available data on numerical relations between optic nerve fibers and lateral geniculate neuronal elements pose additional problems.

as macular fibers connecting with the lateral geniculate bodies of both sides. It is a moot question whether these findings were correctly interpreted and can be assumed to apply to Primates and Man. Other questions of significance for further functional analysis of the optic pathway deserve brief mention as representing promising leads. These problems concern ratio of optic nerve fibers to cells of lateral geniculate body, total estimates of cell number, and distribution of cell size, including so-called size gradients in central and peripheral vision areas within this nucleus (CHACKO, 1949).

Data about the ratio of optic nerve fibers to lateral geniculate neuronal elements are scarce and in part contradictory (cf. BLINKOV and GLEZER, 1968). In Man, an estimate of about 10^6 unilateral optic nerve fibers seems reasonably accurate (but not precise).[249] The estimates of geniculate cells, however, vary between 6×10^5 and 1.2×10^6. BLINKOV and GLEZER tentatively assume that in Rodents there are more fibers in the optic nerve than cells in the lateral geniculate body,[250] while in the Macaque both numbers are the same, and in Dogs more cells than fibers. In Man, the number of fibers and cells are the same according to some authors, but according to others there are more fibers than cells. BLINKOV and GLEZER then add: 'Apparently the excess of the more centrally situated neurons over the more peripheral must be an indication of the higher level of organization of the system. However, this excess in the optic analyzer has so far been demonstrated only in the dog, which stands at a lower level than the primates as regards the development of visual function.' In addition, the somewhat questionable data on retinal ganglion cell numbers and on the numbers of optic nerve fibers, as pointed out in footnote 40 on p. 62 should be kept in mind. As regards the endings of the optic tract in the corpus geniculatum laterale, pars dorsalis, two different possibilities have been considered (LE GROS CLARK, 1941): (1) Each single retinal fiber may, upon arrival at the lateral geniculate body (at least in the central portion of that structure) divide into three

[249] Corresponding figures for other Vertebrates are given in the tabulation of Figure 28, p. 65, section 1B.

[250] It should be kept in mind that some of the authors concerned with this topic were presumably not familiar with the subdivision of lateral geniculate griseum into dorsal and ventral thalamic components, and did not explicitly state whether the entire griseum or only its more differential dorsal component was considered.

terminal branches, each passing to one of the corresponding cell laminae. (2) According to the alternative assumption the conducting unit in each optic nerve in respect of the geniculate body would consist of three fibers arising from three adjacent ganglion cells of the retina. Of these three fibers (in the case of the contralateral eye) one would terminate in layer 1a, a second in layer 2b, and the third in layer 3b. This latter arrangement, in the opinion of LE GROS CLARK, was supposed to represent an anatomical basis for the *Young-Helmholtz theory* of color vision, but this interpretation has been contradicted by WALLS (1953) and others.

Quite apart from the uncertainties still inherent in all color vision theories, as pointed out in section 1B, the numerous subsequent experimental studies on the optic system have, so far, not produced results unequivocally clarifying the many problems pertaining to the neural mechanisms involving the lateral geniculate complex.[251]

In view of the peculiar synaptic arrangement discussed above, it is of special interest to note that the cells in the laminae of the dorsal lateral geniculate body show the phenomenon of transneuronal or transsynaptic degeneration if the cells of origin of the optic tract located in the retina are destroyed (LE GROS CLARK and PENMAN, 1934). In addition, these cells are of course subject to retrograde degeneration (axon reaction, tigrolysis) if their axons are damaged through lesions of the visual cortex or of the optic radiation (POLIAK, 1932). It is likely that in lesions of the central nervous system transneuronal degeneration plays a greater role than hitherto assumed by most investigators. BECKER (1952) has reviewed this subject and suggests that transneuronal degeneration (which may be descending or ascending: 'antegrad-transneuronal' and 'retrograd-transneuronal') bears a significant relation to the functional import of intergriseal connections.

The *corpus geniculatum mediale*, *pars dorsalis* receives the lateral or auditory lemniscus by way of the brachium quadrigeminum inferius. Origin, course, and significance of the lateral lemniscus within the brain stem was dealt with in chapters IX and XI of volume 4. Ac-

---

[251] Cf. also the detailed monograph on the lateral geniculate body by BALADO and FRANKE (1937), and the more recent accounts by WALLS (1953), by GIOLLI and TIGGES (1970), as well as the extensive studies on the visual system by HUBEL and WIESEL (1961, etc.).

cording to Walker's (1938) finding in Monkeys, the channel carrying impulses from the apical part of the cochlea reaches the ventral and posterior portion of corpus geniculatum mediale, pars dorsalis (pars parvoccellularis), whence fibers project to the anterior and lateral portion of the auditory koniocortex. The impulses from the basal part of the cochlea are transmitted to the dorsal and anterior portion of corpus geniculatum mediale, pars dorsalis (of the parvocellular portion ?) and hence to the posterior and medial part of the auditory koniocortex. Although the details are insufficiently clarified,[252] a comparable discrete topographic (tonotopic) distribution can also be assumed in Man. Some vestibular impulses may be conducted within the ascending cochlear system. It should, moreover, be kept in mind that a secondary cortical auditory area directly adjacent to the primary one, but with reversed tonotopic projection, is present in various Mammals. Extrapolating from the available data, the secondary auditory cortex may be assumed to receive its input mainly from pars magnocellularis of the dorsal medial geniculate body. Nothing definite is known about the significance of the ventral thalamic medial geniculate griseum.

Concerning the *principal clinical symptomatology* of the Human thalamus, the following summary may be given. Although most thalamic lesions are either *vascular* or *neoplastic*, *traumatic damage* of the thalamus may be encountered in severe penetrating or perforating craniocerebral injuries. In a series of 24 selected cases of mortal brain wounds (Campbell and K., 1950; Campbell *et al.*, 1958) we found six cases with severe involvement of the thalamus. The symptoms such as contralateral hemiplegia, hemianesthesia, hemianopsia in addition to stupor or semicomatose status were difficult to evaluate in terms of purely thalamic lesions because of extensive destruction in adjacent capsula interna, basal ganglia, or cortex and hemispheric medullary center. Definite symptoms that could be interpreted as thalamic pain were not recorded in any of these cases.

With occasional exceptions (as for instance tuberculoma), the various lesions of the thalamus caused by inflammatory conditions, by encephalopathies, or by the demyelinating diseases are likewise as a

---

[252] A discussion of the controversial points at issue, with bibliographic references, can be found in the treatise by Crosby *et al.* (1962).

rule not of sufficiently restricted local character to warrant discussion in terms of thalamic pathology.

Circumscribed *vascular lesions* result in somewhat more characteristic disturbances collectively designated as the *thalamic syndrome* (DEJERINE and ROUSSY, 1906; MARIE, 1922). These disturbances may include transitory contralateral hemiplegia, complete contralateral hemianesthesia, contralateral spontaneous central pain, or excessive response to painful or even ordinary sensory stimuli, and slight contralateral hemiataxia, occasionally with choreoathetoid movements and dysarthria (SCHUSTER, 1936, 1937).

The hemiplegia is usually of a flaccid rather than a spastic type and has been attributed to involvement of the adjacent capsula interna. The hemianesthesia is presumably caused by damage to nucleus ventralis posterolateralis and posteromedialis. The frequent sparing of facial sensibility is explained by the presence of lesions not involving nucleus ventralis posteromedialis. Deep sensations and stereognosis usually remain permanently affected while superficial sensibility returns. This is explained by the bilateral representation of exteroceptive sensibility. The phenomenon of *central pain* to which EDINGER called attention as early as 1891 is still poorly understood. Release of thalamus from cortical inhibition, and irritation through scarring and gliosis have been suggested without convincing evidence. If painful sensations are assumed to be based on thalamic activity, it is rather difficult to see how thalamic destruction with complete degeneration of the nuclei concerned whould cause pain. LHERMITTE and CORNIL (1929) believe that the thalamus filters or modulates sensory impulses and that central pain may result from discharge of afferent impulses into the cortex without that modulating influence. The pain may be of exteroceptive or of visceral nature. The general concept underlying LHERMITTE and CORNIL's explanation, namely, that a profound disturbance of the thalamic discharge patterns has been created, appears plausible. In a previous text LHERMITTE (about 1924) expressed his view as follows: '*Cet appareil analyseur vient-il à être détruit dans une de ses parties essentielles? Alors s'écoulent vers la corticalité sensible du cerveau des incitations grossières, physiologiquement arrêtées par le thalamus. Ce n'est donc point la "liberation" de la couche optique qui crée le retentissement affectif douloureux des sensations mais la destruction de son appareil analyseur.*' Still, if the thalamic relay nucleus is completely destroyed, one would be compelled to assume that all sensory input into the corresponding cortical area is interrupted. However, such a distortion of thalamic activity might perhaps have a contralateral reverberation, or cause an 'imbalance' of sensory input by way of other diencephalic grisea,[253] e.g. by way of the hypothalamus.

Central pain may be extremely severe and fail to respond to analgesics. It may furthermore be combined with a raised threshold for sensory stimuli on the affected side. Effective stimuli of various types are likely to cause sensations of peculiar unpleasantness, more rarely of intense pleasure (hyperpathia, dysesthesia, overreaction). The pain may be initiated or aggravated by movements of the limbs on the affected side. It may also be spontaneous, in attacks as well as continuous, of a boring and burning type. W.R. Brain (1969) calls attention to the fact that unpleasant overreaction to painful stimuli associated with a raised threshold to such stimuli may also occur during regeneration of a peripheral sensory nerve and as a result of damage to the spinothalamic tract in spinal cord and brain stem. Brain believes that these phenomena may be the result of a reduction in the number of pain-conducting fibers or of 'defective insulation' of those that remain. It is thus possible that impaired conduction and concomitant radiation of the stimulus may evoke a diffuse and exaggeratedly painful sensation. Le Gros Clark (1949) expresses a similar view suggesting that these phenomena may be merely the result of a disturbance of the normal pattern of thalamocortical connections and parallel the abnormal sensory phenomena which we now know to be produced by a disturbance of peripheral patterns of cutaneous innervation.

According to Baudouin *et al.* (1930) central pain is likely to occur immediately after thalamic hemorrhage, while after a thrombotic lesion an interval of several weeks or months can precede the manifestations of pain.

A detailed theory concerning the thalamic circuit mechanisms pertaining to *pain conduction* in particular and to the distribution of sensory input in general has been elaborated by Hassler (1970, 1972a, b). This author emphasizes the distinction between '*Schmerz-*

---

[253] As regards such functional 'imbalance' caused by damage to grisea, which can be partly remedied by destruction of additional grisea, interesting observations have been reported by Sprague (1966a, b), and Sherman (1974) for the optic system. Again, a 'reversal' of tolerance to morphine in Rats was observed after destruction of the 'medial-thalamus' (Teitelbaum *et al.*, 1974). Present-day knowledge of the neural circuitry mechanisms and of the relevant transmitter substances is still insufficient for a significant understanding of such effects.

empfindung' and '*Schmerzgefühl*' related to fast respectively slow' pain-conducting' channels *(Zweiteilung der Schmerzleitung)*.

According to HASSLER (1972a) the cortical pain conduction inhibits or blocks the activities of subcortical pain circuits. Thus: '*Der Ausfall der kortikalen Schmerzleitung führt zum thalamischen Spontanschmerz, der immer ein Schmerzgefühl ist, schlecht lokalisiert und in seiner Stärke kaum zu differenzieren. Dieser kann durch Koagulation des Limitans beseitigt werden.*' The nucleus limitans of HASSLER's terminology corresponds to the nucleus lentiformis mesencephali complex of our terminology. Some further references to HASSLER's views will be found in section 10 of chapter XIII, in connection with a discussion of localization in the Human cerebral hemisphere.

Stereotaxic thalamotomy in the region of the ventroposterior thalamic grisea (ventralis posteromedialis and posterolateralis complex) as undertaken by SPIEGEL and WYCIS (1962) has proven relatively effective in the treatment of intractable pain originating outside the thalamus, e.g. in terminal conditions of inoperable malignant neoplasms. Recent data on results of thalamotomy can be found in the publications of SUGITA *et al.* (1972) and UEMATSU *et al.* (1974). In such thalamotomies lesions in parts of nucleus medialis, centrum medianum, and vicinity of pulvinar, as well as in ventrocaudal grisea were made. TALAIRACH (1955, as cited by SPIEGEL and WYCIS, 1962) also obtained relief in some cases of thalamic syndrome, while other neurosurgeons reported failure in attempts to alleviate that condition. Moreover, the relief of pain resulting from thalamotomy is not infrequently temporary. In addition, the danger of an occasionally developing post-operative thalamic syndrome, reported to have occurred in some instances, must be kept in mind. SPIEGEL and WYCIS (1962) therefore consider these procedures of thalamotomy only as a last resort, to be applied if all conservative methods have failed.

SPIEGEL points out that the relatively frequent association of *Dejerine's syndrome* with vegetative disturbances suggests the possibility that the abnormal reactivity to sensory and emotional stimuli as well as the appearance of vegetative symptoms in patients with thalamic lesions may be caused by changes in the state of excitation or in the excitability of the hypothalamus. Observations showed in some instances a definite overexcitability of the hypothalamus following thalamic lesions (SPIEGEL and KLETZKIN, 1952). SPIEGEL's emphasis on thalamo-hypothalamic relationships seems to provide a promising clue for the explanations of these poorly understood sensory phenomena.

As regards lesions restricted to the nucleus medialis (dorsomedialis), SPIEGEL and WYCIS (1949, 1962) have produced carefully controlled lesions in the region of that griseum. This procedure destroys the center of the frontal cortical feedback system without the drastic damage to the presumptive cortical association systems caused by other methods of 'psychosurgery'. Patients selected for this form of treatment were psychotic or were in unbearable pain; the results obtained appeared promising (SPIEGEL and WYCIS, 1949) and continue to compare favorably with the results of lobotomies (SPIEGEL *et al.*, 1953b). The location of the lesions was checked postoperatively by roentgenologic studies which demonstrated the relationship of pantopaque droplets injected at the site of the lesion to the pineal body. In some of the patients additional lesions were placed in and above the lateral hypothalamic area. Following the production of lesions in nucleus medialis thalami, SPIEGEL and his associates (1949, 1952) observed transitory somnolence, transitory disturbances of orientation, particularly for time, and transitory defects of memory affecting recent as well as remote events. There were no significant changes of pain threshold and no disturbances of the so-called psychoreflexes. Disturbances of vegetative innervation and characterized by urinary incontinence and hyperirritability of the urinary bladder were noticed. Pharmacodynamic tests revealed a shift of balance between the sympathetic and the parasympathetic system in favor of the former. Some decrease of postprandial hyperglycemia was noted. All these effects on the vegetative innervation were transitory. SPIEGEL and his collaborators suspect that they are due to the loss of impulses that are transmitted by periventricular fibers from nucleus medialis thalami to hypothalamus.

Hemiataxia as a symptom of thalamic lesions may be caused partly by loss of proprioceptive sensation, but damage to nucleus ventralis lateralis affecting the endings of fibers of brachium conjunctivum is presumably the main factor. Involvement of this region including nucleus ventralis anterior and its connections with pallidum and striatum would also explain the choreo-athetoid movements. These motor symptoms are believed to be more characteristic of the so-called syndrome of the thalamo-perforating arteries in which the anterior portion of the thalamus, the dorsal hypothalamus, and even the nucleus ruber may be affected (*'subthalamic syndrome'*). Occasional tuberculomas in that region may also cause similar symptoms. The other symptoms already enumerated are regarded as typical for the so-called syndrome of the thalamogeniculate arteries or of the posterior cerebral artery affecting the posterior part of the thalamus. All of these syndromes can be designated as thalamic syndromes in the wider sense, while the syndrome of thalamogeniculate or posterior cerebral artery corresponds to the thalamic syndrome *sensu stricto ('classical thalamic syndrome')*. A peculiar, somewhat variable condition of the hand *(main thalamique, Thalamushand)* with forced extension of some fingers or finger-joints and forced flexion of others is occasionally observed, especially if the rostral part of the thalamus is involved (SCHUSTER, 1936, 1937).

Anosognosia, a condition characterized by the fact that the patient is not conscious of the pathologic symptoms and may even deny their presence, was also observed by SCHUSTER in cases with exclusively thalamic lesions of the right side. The anosognosia concerned motor disturbances and included failure to recognize a limb as part of the patient's own body *(autotopagnosia)*. These symptoms were combined with hyperpathia. Anosognosia and autotopagnosia, however, occur also in purely cortical lesions as well as in hemiplegia and cannot be considered to be specifically thalamic symptoms. A feeling of strangeness concerning ones own fingers as well as phenomena related to *allochiria* can be experimentally elicited in the so-called '*Japanese illusion*' studied by KLEIN and SCHILDER (1929) and other authors. It is very likely that the thalamus plays an important although by no means exclusive role in the neurological mechanisms providing a postural model of the body *(Körperschema* of SCHILDER, 1923, 1935).

It is noteworthy that neoplasms of the thalamus are less likely to cause typical symptoms; BAILEY (1948) emphasizes that he has never observed a fully developed thalamic syndrome in a case of brain tumor. Nevertheless this author describes a case in which some sensory disturbances of thalamic type were present and recognizable as such. In our own series of striatothalamic tumors there was evidence of thalamic pain in one case (GLOBUS and K., 1942). This patient with a spongioneuroblastoma invading the entire left thalamus with the exception of a narrow mesial strip had, in addition to other symptoms, impairment of all sensations on the right side with hyperaffectivity on stimulation, and cramplike pains in right arm and leg. We also reported emotional overreaction in neoplastic thalamic lesions. Involvement of nucleus medialis and perhaps anterior nuclei may have been responsible for the emotional outbursts. An irritative effect of the neoplastic process must be assumed in this case, since interruption of pathways in that region should reduce emotional reactivity. As mentioned above in connection with the anterior dorsal grisea, Papez (1937) suggested that the circuit passing through the anterior nuclei plays a prominent role in the mechanism of emotive processes, but the analysis of our series of neoplastic lesions was inconclusive in that respect. Emotional overreaction has also been attributed to disturbances of hypothalamic activity and will be considered in the discussion of that region.

Damage to the *dorsal lateral geniculate body* will, of course, involve contralateral homonymous hemianopsia, to be dealt with in chapter XIII.

As regards the *Human thalamus ventralis*, and considering only the architecturally distinctive components mentioned above in dealing with the Mammalian ventral thalamus in general, three morphologically and perhaps functionally distinctive main subdivisions can be recognized: zona incerta, partes ventrales geniculorum, and nucleus reticularis thalami, whose possible functional significance was likewise pointed out above.

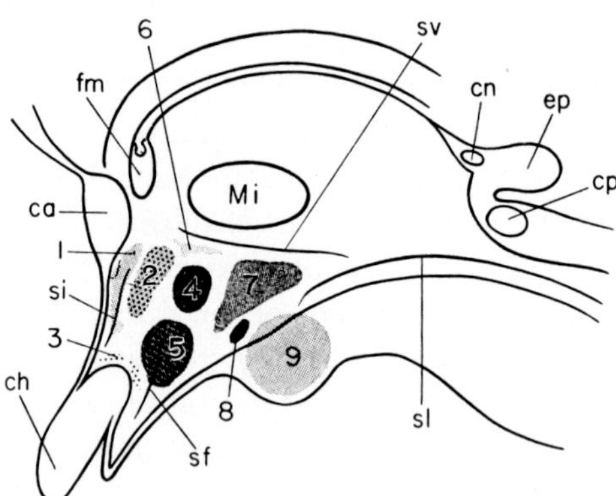

Figure 146. Simplified diagram showing the approximate location of important hypothalamic nuclei in the human brain (modified after Le Gros Clark, 1938, from K., 1954). ca: commissura anterior; ch: chiasma opticum; cn: commissura habenulae; cp: commissura posterior; ep: epiphysis; fm: foramen interventriculare *Monroi;* Mi: massa intermedia; sf: sulcus lateralis infundibuli; si: sulcus intraencephalicus anterior; sl: sulcus limitans; sv: sulcus diencephalicus ventralis; 1: nucleus praeopticus medialis and periventricularis; 2: nucleus paraventricularis; 3: nucleus supraopticus; 4: nucleus hypothalamicus dorsomedialis; 5: nucleus hypothalamicus ventromedialis; 6: dorsal hypothalamic area; 7: nucleus hypothalamicus posterior; 8: nucleus praemammillaris; 9: nuclei mammillares.

Figure 147 A, B. Cross-sections *(Nissl stain)* through the hypothalamus of a newborn infant (A) at caudal preoptic level, (B) at rostral tuberal level (from K., 1954; $\times 10$, red. $^{1}/_{2}$). Abbreviations for Figures 147 A to 148 D. ac: cortical amygdaloid nucleus;

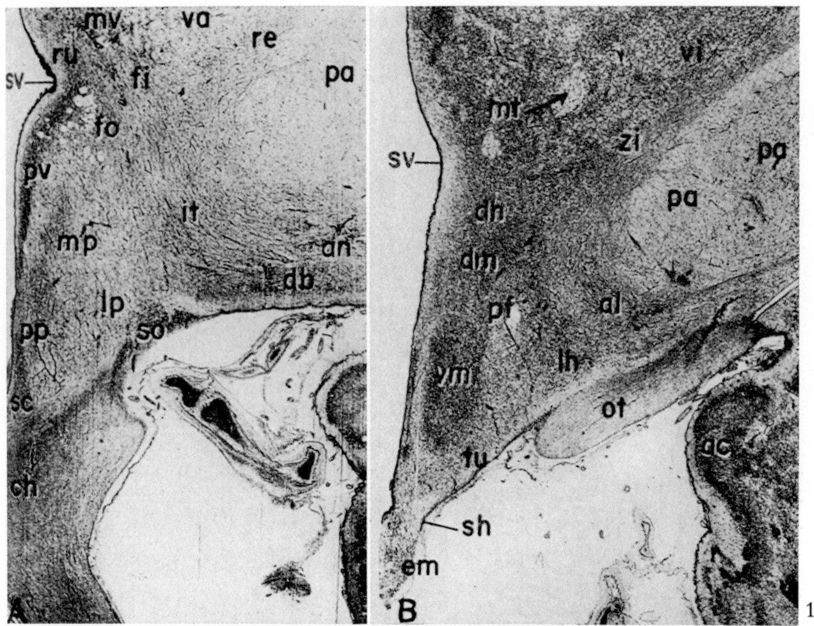

147

ah: n. hypothalamicus anterior; al: n. ansae lenticularis; an: n. of the ansa peduncularis; ap: ansa peduncularis; as: ansa lenticularis; b: subdivisions of basal amygdaloid nucleus; bl: lateral amygdaloid nucleus; ch: chiasma opticum; cm: centrum medianum; csd: decussatio supraoptica dorsalis *(Ganser's commissure)*; csm: commissura supramammillaris; csv: decussatio supraoptica ventralis *(Meynert's commissure)*; cv: fibers of decussatio supraoptica ventralis corresponding to *Gudden's commissure (?)*; d: medial amygdaloid nucleus; db: *diagonal band of Broca;* dh: dorsal hypothalamic area; di: n. supraopticus diffusus; dm: n. hypothalamicus dorsomedialis; em: eminentia medialis; en: n. entopeduncularis; fi: n. filiformis; fo: fornix; g: central amygdaloid nucleus; H: *field H of Forel;* ha: n. hypothalamicus anterior; ic: capsula interna; in: n. intercalatus; it: interstitial n. of the inferior thalamic peduncle; lh: n. hypothalamicus lateralis; lm: n. mammillaris lateralis; lp: n. praeopticus lateralis; lu: n. subthalamicus *(Luys)*; me: n. medialis thalami; mi: midline nuclei of thalamus; min: massa intermedia; ml: n. mammillaris lateralis; mm: n. mam. medialis; mmi: n. mam. med., pars medialis; mml: n. mam. med., pars lateralis; mp: n. praeopticus medialis; mt: tractus mammillothalamicus; mtt: tractus mammillothalamicus et mammillotegmentalis; mv, nv: n. medialis ventralis thalami; ot: optic tract; pa: globus pallidus; pc: n. paracentralis thalami; pf: n. perifornicalis; ph: n. hypothaamicus posterior; pm: n. praemammillaris; po: n. periventricularis posterior (n. perivent. arcuatus); pp: n. praeopticus periventricularis; pr: vessels of hypophysioportal system; pu: putamen; pv: n. paraventricularis; re: n. reticularis thalami; ru: n. reuniens ventralis; sc: n. suprachiasmaticus; sh: sulcus hypophysio-hypothalamicus; sl: n. semilunaris; sm: n. supramammillaris; sn: substantia nigra; so: n. supraopticus; su: n. submedius thalami; sv: sulcus diencephalicus ventralis; tc: tegmental cell cord; tu: n. tuberis; va: n. ventralis anterior thalami; vi: n. vent. lateralis thalami; vm: n. hypothalamicus ventromedialis; zi: zona incerta.

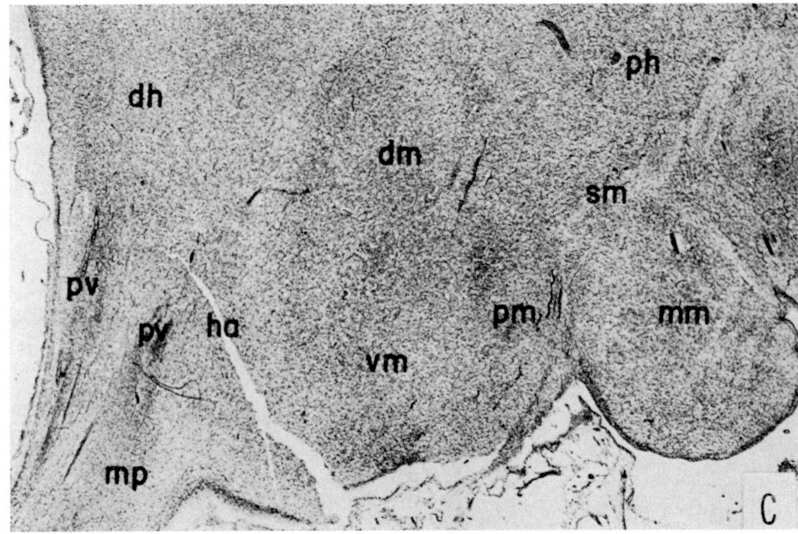

Figure 147 C. Paramedian, slightly oblique section *(Nissl stain)* through the hypothalamus of a newborn infant (from K., 1954; ×15, red. $^3/_4$).

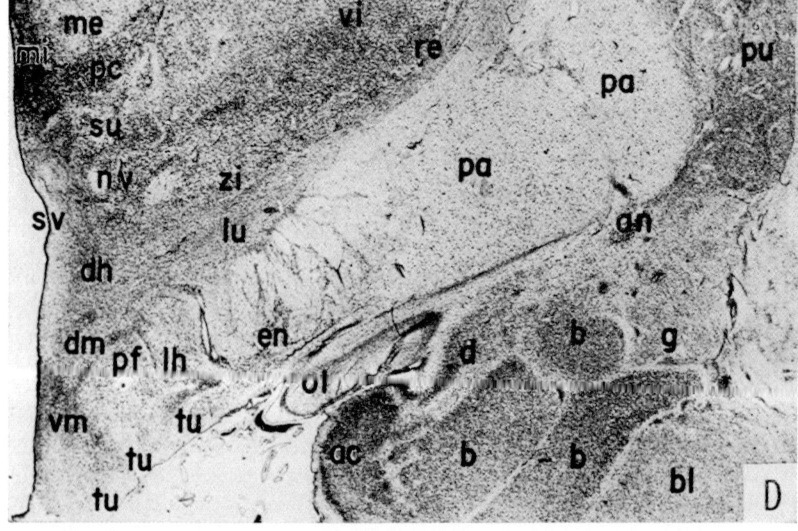

Figure 147 D. Cross-section *(Nissl stain)* at midtuberal level through the hypothalamus of a newborn infant (from K., 1954; ×6, red. $^3/_4$).

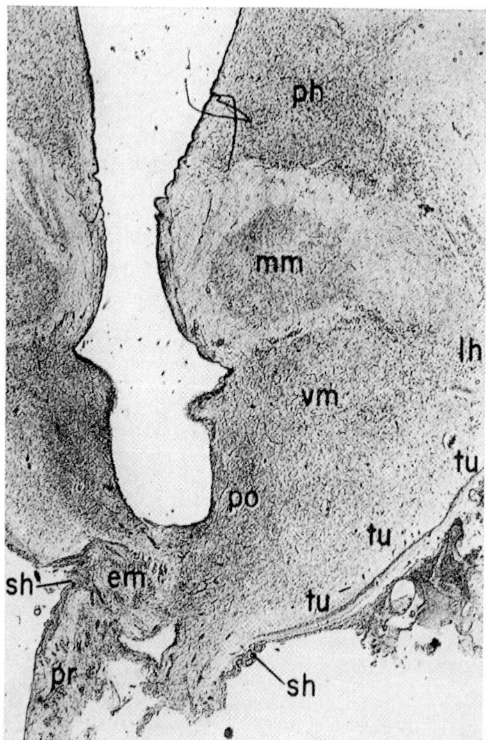

Figure 147E. Cross-section *(Nissl stain)* at posterior tuberal level through the hypothalamus of adult Man (from K., 1954; ×12, red. $^2/_3$).
The ventricular groove basal to mammillary complex represents recessus inframammillaris, the adjacent more basal groove pertains to 'recessus saccularis'. Both are caudal extensions or branches of sulcus lateralis hypothalami posterioris (s. lat. infundibuli).

Spatial relationships and overall configuration of major *Human hypothalamic 'nuclei'*, as mapped upon neighborhoods of the third ventricle, are shown in Figure 146. *Per contra*, Figures 147A–G and 148A–D illustrate some cyto- and myeloarchitectural aspects of these hypothalamic grisea, as seen in cross-sectional and sagittal planes. The delimitable nuclei correspond to those enumerated above for Mammals in general. Some of the main fiber systems can also be identified. Figure 148E, to be compared with Figure 148D, shows both similarities and slight differences between the configuration of the Human mammillary region and that of a Cercopithecid monkey (Baboon). Angioarchitectural aspects of the Primate hypothalamus, as displayed in Macacus, are illustrated by Figure 148F. The dense vascular pattern

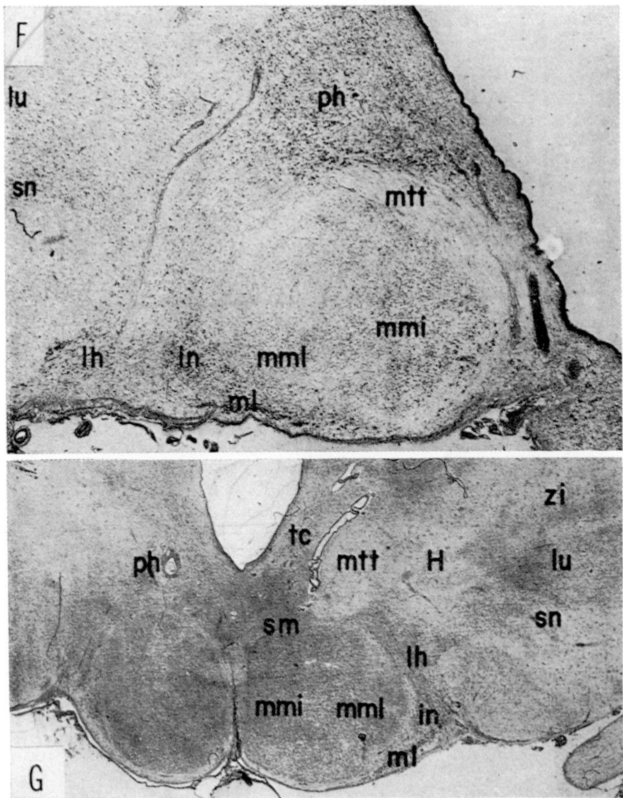

Figure 147F, G. Cross-sections *(Nissl stain)* through the mammillary grisea of adult Man (from K., 1954; A × 25, B × 12, both red. $^1/_2$). It should here be added that, according to Diepen (1962), ml (n. mammillaris lateralis of the terminology adopted by myself in 1954) actually represents the pars caudalis nuclei tuberomammillaris. The nucleus intercalatus (in) would, therefore, be the 'true' nucleus mammillaris lateralis pertaining to the mammillary complex *sensu proprio.*

of supraoptic and paraventricular grisea is particularly noticeable. An essentially similar pattern obtains for Man.

Figures 149A–F illustrate overall aspects of the entire Human diencephalon in sagittal sections. Figures 150A and B depict afferent and efferent pathways of the so-called *extrapyramidal system* as related to diverse diencephalic and additional grisea according to widely accepted contemporary interpretations consistent with clinical experience.

As far as the diencephalon is concerned, the so-called 'extrapyramidal' disturbances or diseases are particularly related to malfunction

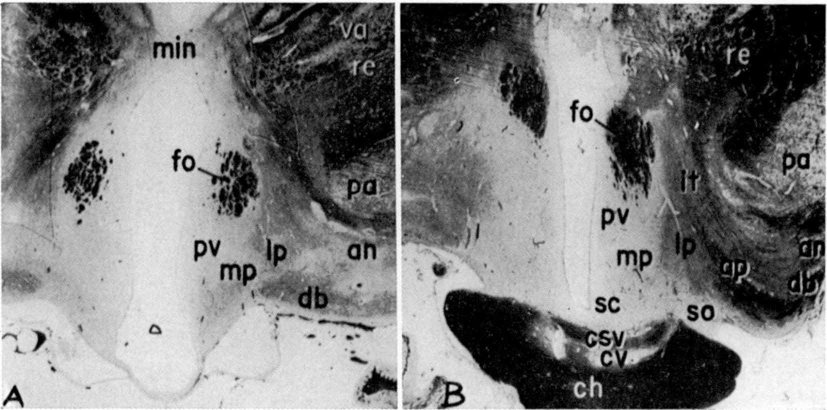

Figure 148 A, B. Cross-sections (myelin stain) through preoptic levels of the adult Human hypothalamus, (A) at prechiasmatic level, (B) at level of chiasma and supraoptic commissures (from K., 1954; ×6, red. $^{1}/_{2}$).

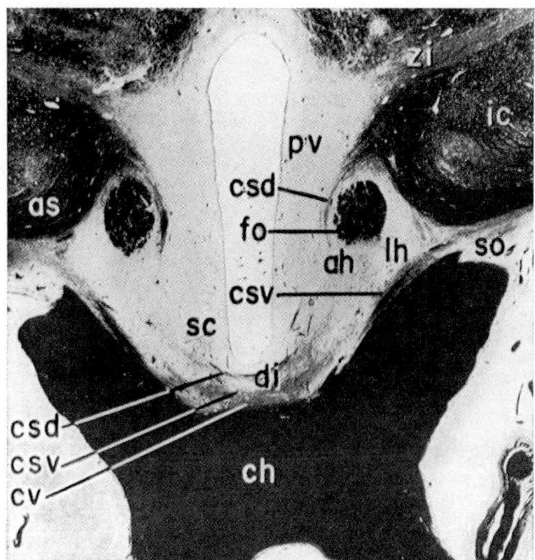

Figure 148 C. Cross-section (myelin stain) at level of optic chiasma at slightly different angle of plane, showing supraoptic commissures (from K., 1954; ×6, red. $^{1}/_{2}$). The large vessel at right is the internal carotid near its bifurcation into middle and anterior cerebral arteries.

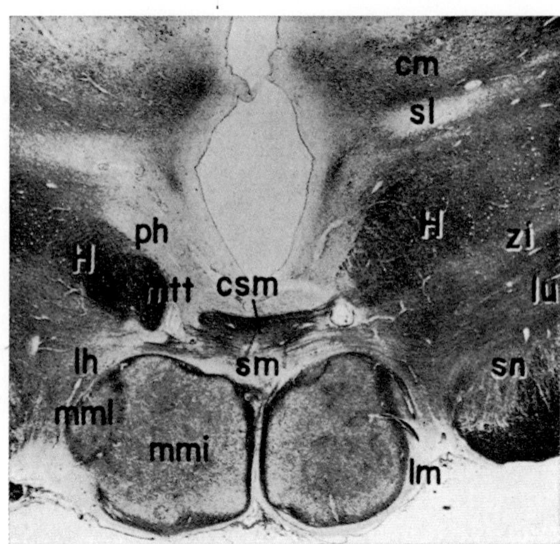

Figure 148 D. Cross-section (myelin stain) through the mammillary complex of adult Man, showing supramammillary commissure (from K., 1954; ×10, red. ²/₃).

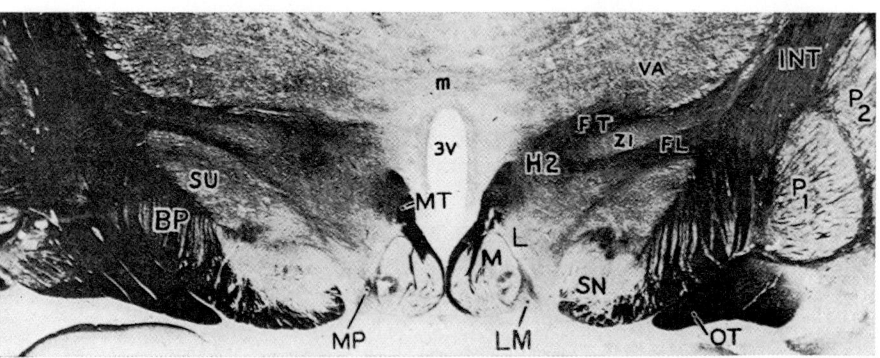

Figure 148 E. Cross-section (myelin stain) through the mammillary complex of the Cercopithecid Primate Papio (Baboon) for comparison with Figure 148D (from NAUTA and HAYMAKER, 1969). BP: pes *seu* basis pedunculi; FL: fasciculus lenticularis (H_2) FT: fasciculus thalamicus (H_1); INT: capsula interna; H_2 (it should read H): *Forel's tegmental field H;* L: lateral hypothalamic area; LM: nucleus mammillaris lateralis; M: nucleus mammillaris medialis; m: massa intermedia thalami (nn. reunientes); MP: mammillary peduncle; MT: mammillothalamic tract; OT: optic tract; $P_{1,2}$: segments of globus pallidus; SN: substantia nigra; SU: subthalamic nucleus; VA: ventral grisea of thalamus dorsalis; ZI: zona incerta; 3V: 3rd ventricle.

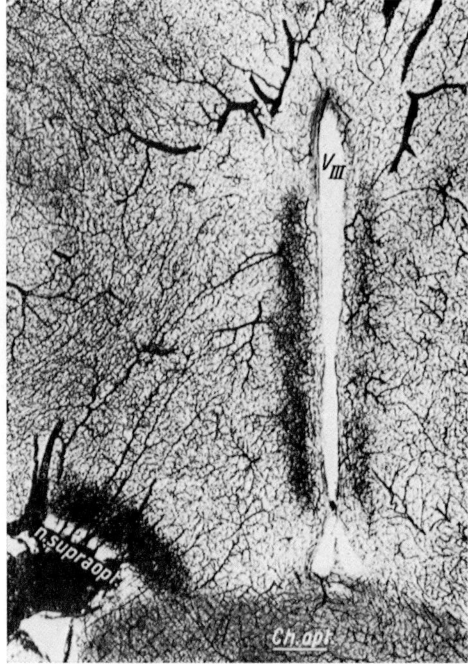

Figure 148 F. Cross-section illustrating angioarchitecture of the caudal preoptic hypothalamic region in the diencephalon of Macacus rhesus (after PFEIFER, from CLARA, 1959).

or lesions of grisea pertaining to the dorsal and entopeduncular hypothalamus of our terminology (globus pallidus, entopeduncular and related grisea, nucleus subthalamicus),[253a] although nucleus ventralis anterior and nucleus ventralis may also occasionally be implicated.[254]

The clinical manifestations, frequently but not exclusively bilateral are essentially related to the contralateral set of affected grisea. These symptoms are generally classified under the categories of (1) athetosis and chorea characterized by involuntary movements,[255] (2) tremor,

---

[253a] The diencephalic hypothalamic origin of globus pallidus in Man, and its relationship to massa cellularis reuniens, pars inferior, was particularly well documented, on the basis of well preserved embryonic material, by SCHNEIDER (1950).

[254] Telencephalic grisea involved are particularly those of corpus striatum (nucleus caudatus and putamen). As regards mesencephalic grisea, the substantia nigra may again be pointed out (cf. vol. 4, chapter IX, section 9).

[255] No semantically valid sharp distinction between athetosis and the generally somewhat more rapid fidgety motions of chorea can be made.

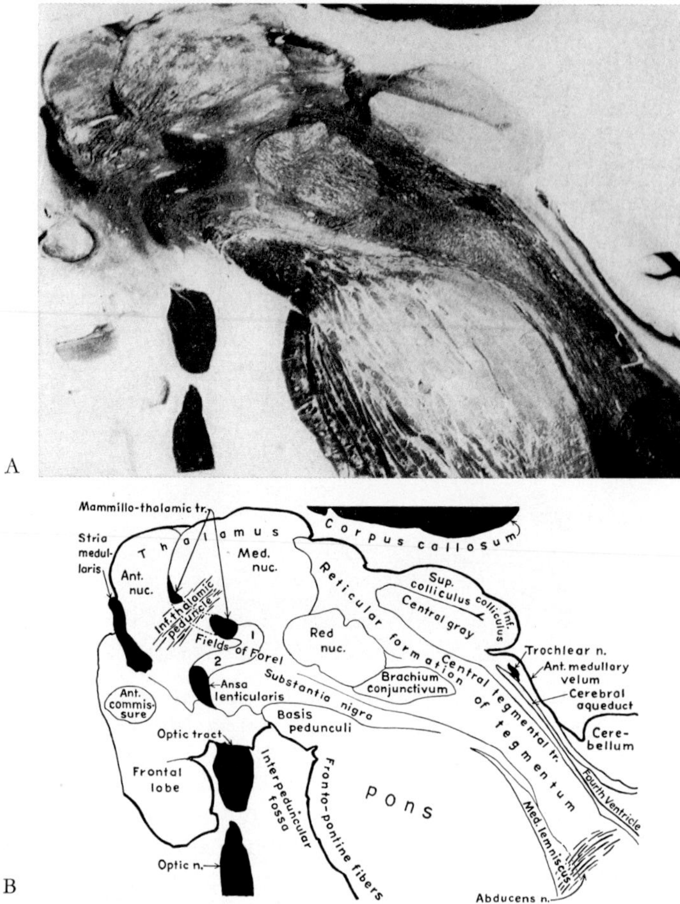

Figure 149 A, B: Parasagittal section (myelin stain) through adult Human diencephalon and mesencephalon, about 3 mm from midsagittal plane, with (B) explanatory outline (after ELLIOTT, 1963, from photograph donated by Dr. J. PAPEZ).

(3) dystonia and rigidity, respectively reduction of 'automatic associated movements' (e.g. 'mask-like face').

In contradistinction to cerebellar and other sorts of intention tremor (cf. section 9, chapter X, p. 766, vol. 4), *'extrapyramidal' tremor* occurs mainly at rest, being partially or even almost completely abolished during voluntary movements. *Dystonia* is characterized by a persistent posture with exaggerated muscle tone. *Rigidity* is manifested by resistance to passive motions which the examiner assays. As experienced by this latter, it can often be described as 'plastic rigidity' or

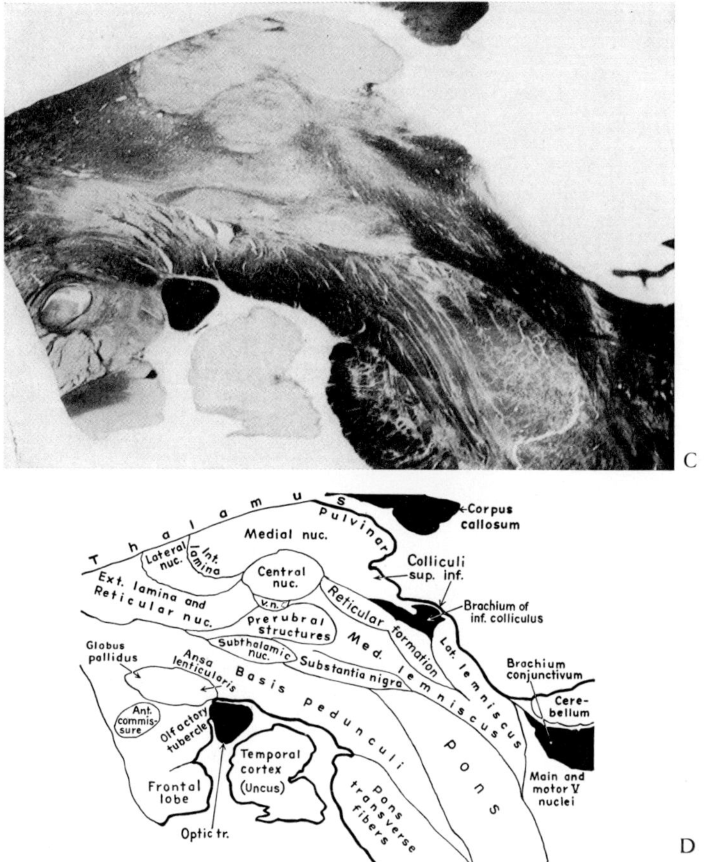

Figure 149 C, D. Parasagittal section from same series as Figure 149 A, B, about 9 mm from midline. Both section A are slightly oblique, being ventrally somewhat farther aterad from midline than dorsally (after Elliott, 1963).

'cog-wheel rigidity'. Thus, two opposite sorts of symptom groups, namely 'hyperkinesias' and 'akinesias' (hypokinesia, bradykinesia) are manifested in diversified patterns in 'extrapyramidal disturbances'.[256]

---

[256] It should also be recalled that 'extrapyramidal' pathways also include systems arising in the cerebral cortex (e.g. premotor cortex, temporoparietal lobe respectively corticopontine tracts), nucleus ruber, and reticular formation (cf. also Fig. 139, p. 244, vol. 4). The 'diaschisis' of v. Monakow, e.g. the complete flaccid paralysis immediately following transection of the spinal cord, can be assumed to result from an inability of the intact lower motor neuron to discharge if a large part of its normal synaptic input becomes abolished.

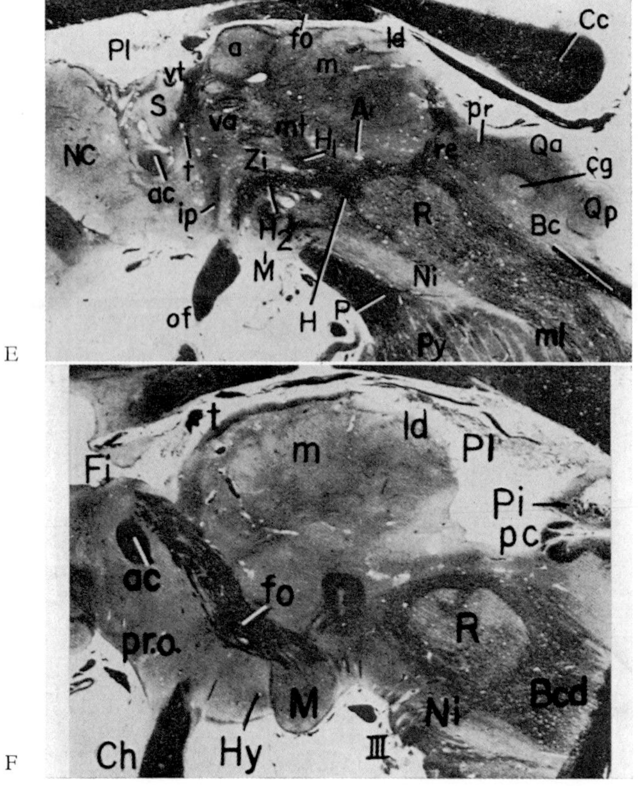

Figure 149 E, F. Parasagittal sections (myelin stain) through adult Human diencephalon and adjacent neighborhoods, E being about 2 mm, and F about 4 mm lateral from midline (after SPIEGEL and WYCIS, 1962). ac: anterior commissure; Bc, Bcd: brachium conjunctivum (and its decussation); Cc: corpus callosum, cg: central gray; Ch: optic chiasma (respectively nerve and tract); Fi: lateral ventricle in neighborhood of interventricular foramen; fo: fornix; H, H_1, H_2: tegmental field, *thalamic and lenticular fasciculi of Forel;* Hy: hypothalamus; ip: inferior thalamic peduncle; ld: nucleus lateralis dorsalis thalami; M: mammillary complex; m: nucleus medialis (sive medialis dorsalis) thalami; NC: nucleus caudatus; Ni: substantia nigra; P: pes pedunculi; pc: commissura posterior; Pi: pineal body; Pl: choroid plexus; pr: pretectal region; pr.o.: preoptic area; ot: optic nerve and chiasma; Qa, Qp: superior (anterior) and inferior (posterior) colliculus of mesencephalic tectum; R: nucleus ruber tegmenti; S: septal region of telencephalon; t: taenia *sive* stria medullaris thalami; III: rootlets of oculomotor nerve. Note the difference in the myelin staining, particularly *qua* anterior commissure, in the sections used by ELLIOTT and in those depicted by SPIEGEL and WYCIS (such differences depend on a number of variables pertaining to the specimen itself as well as to the technical procedures. Figures E and F were from the present author's material which, as mentioned in their introduction, was made freely available for the pioneering work of Drs. SPIEGEL and WYCIS).

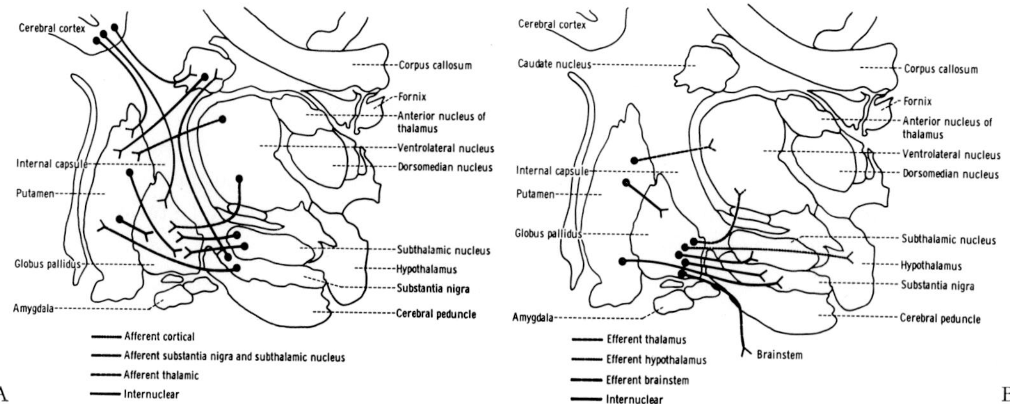

Figure 150 A, B. Diagrammatic representations of afferent (A) and efferent channels (B) pertaining to thalamic, hypothalamic (pallidum and subthalamic nucleus), and some other related grisea of so-called extrapyramidal system (from McDowell and Lee, 1971).

The generally recognized '*extrapyramidal diseases*' include *parkinsonism*, hemiballism, athetosis, *Sydenham's chorea*, *Huntington's chorea*, torsion dystonia, and *Wilson's disease* (hepatolenticular degeneration).

Parkinsonism was dealt with in section 9, chapter XI of volume 4. *Huntington's chorea* and *Wilson's disease* essentially seem to involve the corpus striatum, and are, moreover hereditary disorders related to defects of the genome, as dealt with (pp. 220–221) in subsection 1C, chapter VI of volume 3/II. It should also be added that, despite many recorded data of neuropathology and neurophysiology disclosing some functional relationships of the enumerated grisea, the actual neural mechanisms of 'extrapyramidal disorders' in general remain very poorly understood. Functionally, the globus pallidus is commonly assumed to exert a stimulating and facilitating influence upon the 'motor centers', while the corpus striatum is regarded as having an inhibitory influence.[257] It seems, however, well established that numerous quite identical symptoms of 'extrapyramidal disturbances' can result from lesions involving different combinations of the relevant grisea.

Hemiballismus, characterized by uncontrolled jerking and spastic motions of the contralateral extremities and trunk is generally caused

[257] The existence of neocortical so-called '*suppressor bands*', at one time widely accepted by diverse neurologists, can be regarded as most unlikely (cf. K., 1954).

by damage to the nucleus subthalamicus (MARTIN, 1927). However, lesions responsible for hemiballismus can occasionally be located in striatum, pallidum, or postcentral gyrus as well as within afferent and efferent connections of nucleus subthalamicus (MEYERS *et al.*, 1950).

Chorea and *athetosis* are characterized by involuntary movements which, although achieving no 'purpose', often resemble fragments of purposive movements following one another in a disorderly fashion: grimacing, rolling of the eye bulbs, twisting of the head, motions of upper and to a lesser extent of lower extremity.[258] In athetosis, the motions are slower, coarser and more writhing, but, as the term 'choreo-athetosis' indicates, cannot always be clearly distinguished from those in chorea. Lesions of the corpus striatum, notably the putamen *(état marbré, état dysmyelinique* of the VOGTS) have been implicated (cf. e.g. BRAIN, 1969). In unilateral athetosis, the contralateral grisea seem to be affected. Typical *Sydenham's chorea* is regarded as an acute toxic-infective disorder, usually occurring in childhood or adolescence, mostly following acute rheumatism, and affecting corpus striatum, subthalamic nucleus, and substantia nigra.[259] Postencephalitic chorea of diverse origin, chorea gravidarum, and senile chorea are additional forms.

Torsion dystonia (dystonia musculorum deformans) is a rare, presumably hereditary syndrome with onset in childhood or adolescence and characterized by slow, involuntary twisting movements of the whole body. Unspecific scattered changes in striatum, subthalamic nucleus, thalamus, cerebral cortex and nucleus dentatus cerebelli have been reported. *Spasmodic torticollis* (wryneck) manifesting a rotated position of the head, may be a fragmentary form of torsion dystonia, but also occurs in various other affections of the extrapyramidal mechanisms or of the nervous system (e.g. encephalitides, and also hysteria).

As regards the predominantly *autonomic hypothalamus sensu strictiori* of the Human diencephalon, the following clinical remarks, supplementing the discussion of Mammalian hypothalamic function given further above, might here be added.

---

[258] The motions are intensified by voluntary effort and excitement, but disappear during sleep.

[259] Involvement of cerebral cortex, e.g. of the frontal lobe, is also suspected (cf. e.g. McDOWELL and LEE, 1971, 1973).

With respect to *temperature regulation*, Strauss and Globus (1931), Alpers (1936), and Globus and K. (1942) reported cases of neoplastic lesions in the preoptic and tuberal regions in which there was marked hyperthermia, while Davison and Selby (1935) have recorded a case in which hypothermia was a conspicuous symptom. In this instance, the neoplasm had destroyed the posterior part of the hypothalamus and also damaged part of the paraventricular and of the supraoptic nucleus.

Concerning *basal metabolism and protein metabolism*, it seems likely that the thyroid gland may be influenced by the hypothalamus indirectly by way of the thyrotrophic hormone of the adenohypophysis or directly through the functionally somewhat dubious sympathetic innervation of the thyroid. The 'psychic' factor in the genesis of hyperthyroidism may conceivably act through the hypothalamus. Since a direct or indirect stimulating effect of thyroxin on hypothalamic centers is assumed, a vicious circle may arise.

Hypothalamic mechanisms are presumably involved in numerous other *psychosomatic conditions* such as gastric and duodenal ulcerations and similar disturbances. The rare phenomena of stigmatization, of voluntary control of heart action and similar fakir-tricks representing peculiar conditioned reflexes should also be mentioned in this connection.

Several authors, especially Lichtwitz (1936) and Hoff (1936) assume that the hypothalamus exerts an influence on the *blood picture*. Although the possibility of such an indirect or direct action cannot be denied, the available data do not appear to warrant definite conclusions in this respect. Nevertheless the observations of Farris (1938, and others) showing that emotional states of anxiety, fear, anger, and disappointment produce a marked relative lymphocytosis in healthy persons, can be interpreted as a likely instance of nervous control of leucocytosis, possibly mediated through the hypothalamus. Since a relative lymphocytosis is evident within a few minutes after emotional stimulus, a hematopoietic organ capable of producing rapid change must be involved, and Farris assumes that this role is assumed by the spleen, which is innervated by mainly sympathetic fibers taking their course through the celiac plexus.

Brain (1950) has attributed to hypothalamic lesions the *cachexia* and *emaciation* encountered in the presence of suprasellar tumors which invade the hypothalamus. The frequent occurrence of genital atrophy in association with adiposity in the presence of basal hypothalamic lesions which do not encroach on the hypophysis has been regarded as due to damage of the tuber cinereum, especially in the region of the ventromedial nucleus. SPATZ (1951) noted enlargement of the parvocellular region of the 'tuber cinereum' by hyperplastic malformation in cases of pubertas praecox.

Concerning the *sleep-wakefulness mechanisms* MAUTHNER (1890), ECONOMO (1930) and others have shown that somnolence follows lesions of the central gray of the aqueduct and the posterior part of the wall of the third ventricle. ECONOMO emphasized that not only somnolence but also inversion of the normal sleep rhythm, and even asomnia, may be caused by the periventricular lesions encountered in encephalitis lethargica. In two cases in which somnolence was a predominant symptom, GLOBUS (1940) observed vascular damage to the posterior part of the hypothalamus at the level of the mammillary bodies, including the lateral hypothalamic area. Comparable findings, characterized by atrophy and gliosis in the posterior hypothalamic region, were reported by RICHTER and TRAUT (1940) in a case of protracted somnolence due to chronic encephalitis.

These data, in conjunction with those obtained by (Mammalian) animal experiments, as mentioned further above, seem to indicate the appurtenance of at least some posterior hypothalamic mechanisms to the so-called centrencephalic system of PENFIELD (cf. p. 352). Said 'system', which should also include certain activities of the deuterencephalic reticular formation and of the periaqueductal central gray, appears to provide relevant 'activating factors' required for the occurrence of such corticothalamic events as are correlated with consciousness. In this respect, the tegmental radiation, described by FLECHSIG (1881, 1896) as *Haubenstrahlung*, and dealt with in chapter IX, section 2, p. 342 of volume 4, seems to play an important role by providing a so-called unspecific 'activating extralemniscal sensory system'.

It appears likely that a 'multifactorial' input both by the 'nonspecific' systems and by the 'specific' (lemniscal and optic) sensory channels, with variable degrees of participation by these 'unspecific' and 'specific' mechanisms, is essential for the maintenance of the waking conscious state.

A case in point is one described by STRÜMPELL (1877), also cited by GRANIT (1955) and ANDERSON and HAYMAKER (1974). The patient, a 16 years old boy, had become afflicted with complete anesthesia for all sensory input whatsoever, except for that from right eye and left ear. Whenever that eye was covered and that ear plugged, the patient would fall into a deep sleep within 2 or 3 minutes following the complete sensory deprivation. No stimuli of any kind, except illumination of the right eye, or loud sounds reaching the left ear could then awaken the patient. Left undisturbed, however, he would, in the daytime, spontaneously awake after several hours of sleep. GRANIT's (1955) comment that nevertheless the 'unspecific paths rather than the specific afferents are responsible for the task of energization' remains unconvincing. It is much more likely that both sorts of channels play a relevant, but vicariously interdependent role. As regards the sleeping conscious state, that is to say, the *dreaming state*, however, it appears most probable that this peculiar state of consciousness becomes essentially, but perhaps even here not always exclusively, activated by the 'unspecific' 'centrencephalic' systems, including formatio reticularis tegmenti (K., 1972).

Certains, perhaps rather rare types of 'seizures' may be designated as *hypothalamic (diencephalic autonomic) epilepsy*. Thus, PENFIELD (1929) described a case characterized by autonomic seizures, in which a small tumor was found within the hypothalamic part of the third ventricle, near the interventricular foramen, where it compressed rostral hypothalamic nuclei. The attacks were characterized by flushing of face, lacrimation, salivation, dilatation of the pupils, profuse sweating, slow respiration, cardiac acceleration followed by deceleration, hiccoughing, and shivering. Consciousness was lost in some of the attacks, but no convulsions were observed, although they are alleged to have occurred at an earlier time. BRAIN (1950, 1969) believes that the *vasovagal attacks of Gowers* may represent a similar type of predominantly parasympathetic paroxysmal seizure.

With respect to the neural mechanisms of *affectivity and emotion*, it is well known that patients suffering from head injuries, cerebral arteriosclerosis, or disseminated sclerosis frequently display emotional instability and exaggerated emotional reactivity in form of outbursts and of compulsive crying or laughing fits. These symptoms may be explained on the basis of an impaired control of the posterior hypothalamus by higher centers, but it must be kept in mind that the complex mechanism of emotions is assumed to include not only the hypothalamus but also thalamocortical, corticothalamic, and thalamo-hypothalamic interrelations.

SPIEGEL and WYCIS have performed hypothalamotomy in combination with dorsomedial thalamotomy or alone in cases of relapses after dorsomedial or anterior thalamotomy undertaken in psychotic or severely psychoneurotic patients (SPIEGEL *et al.*, 1953b); SPIEGEL and WYCIS, 1962). The operation, also involving the so-called 'subthalamus', was prompted by animal experiments showing that subthalamic lesions induced somnolence and diminution of reactivity, and that lesions of lateral parts of the hypothalamus transformed wild monkeys into tame ones (RANSON, 1939).

The effects of hypothalamotomy on vegetative innervation were rather slight. A transitory rise in temperature was noted for a few days. Transitory polydipsia and polyuria as well as transitory somnolence were also occasionally observed. Hypothalamotomy was originally performed in a two-stage operation. In such cases, transitory vasodilation and rise in skin temperature on the contralateral side could be detected. On the whole, the patients who were fearful, agitated, assaultive, and sometimes hallucinatory, were relieved of their anxiety, became more quiet, and more easily manageable.

The hypothalamic regions involved in the operations were the lateral hypothalamic nucleus or area, approximately in the border zone between middle and posterior group, and the dorsal hypothalamic area including the vicinity of nucleus subthalamicus and zona incerta ('subthalamus').

On the basis of observations discussed under the preceding paragraphs, various clinical syndromes of the hypothalamus have been described which can be said to represent its principal clinical symptomatology. Some of these syndromes may be associated with disturbances involving the pituitary gland and other organs besides the hypothalamus. The following hypothalamic disorders seem to be fairly well established:

(1) *Hyperthermia* due to lesions of the anterior hypothalamic region.

(2) *Hypothermia* due to lesions of the posterior hypothalamic region.

(3) *Diabetes insipidus* with polydipsia due to lesions of the supraoptic nucleus or of the supraoptico-hypophysial tract.

(4) *Adiposo-genital dystrophy (Fröhlich's syndrome)* due to lesions predominantly in the ventromedial nucleus. Whether a purely pituitary lesion (chromophobe adenoma) may cause this syndrome is still a matter of conjecture.

(5) *Hypergonadism* (pubertas praecox) associated with hyperplastic malformations of tuber cinereum or other neoplastic lesions affecting the hypothalamus. Some of these cases are associated with tumors of the pineal body, but concomitant hypothalamic involvement rather

than the pineal origin of the neoplasm is believed to be responsible for the hypergonadism. A disturbance of normal hypothalamopituitary interrelationships could be assumed as one of the possible factors.

(6) *Hypersomnia* and *narcolepsy* due to lesions in the caudal part of the hypothalamus and rostral part of the midbrain.

(7) *Hypothalamic (diencephalic autonomic) epilepsy* of predominantly sympathetic or predominantly parasympathetic type which may be caused either by destructive or irritative lesions chiefly in the anterior hypothalamus.

In addition to these established disorders, there is evidence that in some cases *Cushing's syndrome* may be due essentially to a degenerative change in rostral hypothalamic nuclei, particularly the paraventricular nucleus or its vicinity. Besides obesity which is mostly limited to head and trunk, disturbances of sex functions (feminization of male, masculinization of female), hypertension, erythremia, osteoporosis, and hyperglycemia, there may be bulimia. As is well known, the most frequently observed lesions in *Cushing's syndrome* are basophil adenoma of the pituitary and adrenal cortical hyperplasia or neoplasms. The rare hereditary condition known as *Laurence-Biedl-Moon syndrome*, characterized by obesity, hypogenitalism, mental retardation, polydactyly, and retinitis pigmentosa may at least be partly hypothalamic in origin since in such cases atrophy of nuclei in the tuberal region has been reported (Griffiths, 1938).

Brief mention should also be made of *traumatic damage* involving the hypothalamus. Hypertension, paralysis of urinary bladder and rectum, hyperhidrosis, obesity, and emotional disturbances of permanent or rhythmically-fluctuating character were recorded by Zülch (1950) in an analysis of such cases.

Lesions involving optic chiasma or optic tract, which are located on the surface of the hypothalamus, will cause various forms of hemianopsia, heteronymous in damage to chiasma, and homonymous if the optic tract is implicated.

It might also be added that Globus (1942) recorded as a newly recognized and apparently rare type of neuroectodermal neoplasm, an *infundibuloma*, presumably derived from pituicytes, and perhaps related to the spongioblastoma group (cf. chapter V, section 9, p. 764, vol. 3/I).

10. References to Chapter XII

Abbie, A. A.: The brain-stem and cerebellum of Echnidna aculeata. Philos. Trans. roy. Soc. Lond. Ser. B *224:* 1–74 (1934).

Adam, H.: Der III. Ventrikel und die mikroskopische Struktur seiner Wände bei Lampetra (Petromyzon) fluviatilis L. und Myxine glutinosa L., nebst einigen Bemerkungen über das Infundibularorgan von Branchiostoma (Amphioxus) lanceolatus Pall.; in Kappers Progress in Neurobiology, pp. 146–158 (Elsevier, Amsterdam 1956).

Adam, H.: Hypophyse und hypothalamo-neurohypophysäres Neurosekretsystem bei den Cyclostomen Myxine glutinosa und Bdellostoma stouti. Verh. dtsch. zool. Ges. 1959, pp. 157–171 (Akad. Verlagsges., Leipzig 1959).

Addens, J. L.: The presence of a nucleus of Bellonci in reptiles and mammals. Proc. kon. nederl. Akad. Wet. *41:* 1134–1145 (1938).

Alpers, B. J.: Hyperthermia due to lesions of the hypothalamus. Arch. Neurol. Psychiat. *35:* 30–42 (1936).

Altner, H.: Histologische und histochemische Untersuchung an der Epiphyse von Haien; in Kappers and Schadé Progress in Brain Research, vol. 10, pp. 154–171 (Elsevier, Amsterdam 1965).

Altner, H., and Zimmermann, H.: The saccus vasculosus; in Bourne Structure and function of nervous tissue, vol. 5, pp. 293–328 (Academic Press, New York 1972).

Anderson, E. and Haymaker, W.: Influence of the hypothalamus on sexual function. J. amer. med. Women's Ass. *3:* 402–406; 457–461 (1948).

Anderson, E. and Haymaker, W.: Breakthroughs in hypothalamic and pituitary research. Progr. Brain Res. *41:* 1–60 (1974).

Ando, S.: Zur Zytoarchitektonik des Thalamus beim Kaninchen. Folia. anat. jap. *15:* 361–410 (1937).

Arai, Y.: Diencephalic neurosecretory centers of the Passerine Bird, Zosterops palpebrosa japonica. J. Univ. Tokyo Fac. Sc., Sec. IV *10:* 249–268 (1963).

Arai, Y.; Kambara, S., and Takahashi, K.: The entopeduncular neurosecretory cell group in the diencephalon of the Passerine bird, Emberiza rustica latifascia (Japan., with English summary). Dobutsugakn Zasshi (Zool. Magaz.) *72:* 84–88 (1963).

Arey, L. B.: Developmental anatomy; 6th ed. (Saunders, Philadelphia 1954).

Aschner, B.: Über die Funktion der Hypophyse. Pflügers Arch. ges. Physiol. *146:* 1–146 (1912).

Atlas, D. and Ingram, W. R.: Topography of the brain stem of the rhesus monkey with special reference to the diencephalon. J. comp. Neurol. *66:* 263–290 (1937).

Axelrod, J.: The pineal gland: a neurochemical transducer. Science *184:* 1341–1348 (1974).

Axelrod, J.; Wurtman, R. J., and Winget, C. M.: Melatonin synthesis in the hen pineal gland and its control by light. Nature, Lond. *201:* 1134 (1964).

Bailey, P.: Intracranial tumors; 2nd ed. (Thomas, Springfield 1948).

Balado, M. und Franke, E.: Das Corpus geniculatum externum, eine anatomisch-klinische Studie (Springer, Berlin 1937).

Bard, P.: A diencephalic mechanism for the expression of rage with special reference to the sympathetic nervous system. Amer. J. Physiol. *84:* 490–515 (1928).

Bargmann, W.: Die Epiphysis cerebri; in v. Möllendorff Handb. d. mikr. Anat. d. Menschen, vol. VI/4, pp. 309–509 (Springer, Berlin 1943).

BARGMANN, W.: Der Saccus vasculosus; in KAPPERS Progress in Neurobiology. Proc. 1st. intern. Meeting of Neurobiologists, pp. 109–112 (Elsevier, Amsterdam 1956).

BARGMANN, W. and SCHADÉ, J.P. (eds.): Lectures on the diencephalon. Progress in Brain Research, vol. 5 (Elsevier, Amsterdam 1964).

BARNES, R.D.: Invertebrate zoology (Saunders, Philadelphia 1963).

BASINGER, S., HOFFMAN, R., and MATTHES, M.: Photoreceptor shedding is initiated by light in the frog retina. Science *194:* 1074–1076 (1976).

BAUDOUIN, A.; LHERMITTE, J. et LEREBOULLET, J.: Une observation anatomo-clinique d'hemorragie du thalamus. Rev. neurol. *37:* 102–109 (1930).

BAYLISS, W.M.: Principles of general physiology; 4th ed. (Longmans, London 1924).

BEATTIE, J.: Functional aspects of the hypothalamus; in LE GROS CLARK, BEATTIE, RIDDOCH and DOTT The hypothalamus, pp. 69–100 (Oliver & Boyd, Edinburgh 1938).

BECCARI, N.: Neurologia comparata anatomo-funzionale dei vertebrati compreso l'uomo (Sansoni, Firenze 1943).

BECKER, H.: Retrograde und transneuronale Degeneration der Neurone. Mainzer Akad. Wiss. Lit. Abh. math. phys. Kl. *10:* 654–811 (1952).

BELEKHOVA, M.G. and KOSAREVA, A.A.: Organization of the turtle thalamus: visceral, somatic and tectal zones. Brain Behav. Evol. *4:* 337–375 (1971).

BELLONCI, J.: Über die centrale Endigung des Nervus opticus bei den Vertebraten. Z. wiss. Zool. *47:* 1–46 (1888).

BENEDETTI, E.: Il cervello e i nervi cranici del Proteus anguineus Laur. Mem. Ist. ital. Speleol. Ser. biol. *III:* 3–79 (1933).

BERGQUIST, H.: Zur Morphologie des Zwischenhirns bei niederen Wirbeltieren. Acta zool. *13:* 57–303 (1932).

BERNARD, C.: Leçons de physiologie expérimentale appliquées à la médecine. 2 vols. (Baillière, Paris 1855/56).

BERNARD, C.: Leçons sur le diabète et la glycogenèse animale (Baillière, Paris 1877).

BERNHAUT, M.; GELLHORN, E., and RASSMUSSEN, A.T.: Experimental contributions to the problem of consciousness. J. Neurophysiol. *16:* 21–35 (1953).

BEST, C.H. and TAYLOR, N.B.: The physiological basis of medical practice. A text in applied physiology; 5th ed., 7th ed. (Williams & Wilkins, Baltimore 1950, 1961).

BLEIER, R.: Surface fine structure of supraependymal elements and ependyma of hypothalamic third ventricle. J. comp. Neurol. *161:* 555–567 (1975).

BLIGH, J.: Temperature regulation in Mammals and other Vertebrates. Frontiers in Biology, vol. 30 (Elsevier, Amsterdam 1973).

BLINKOV, S.M. and GLEZER, I.I.: The human brain in figures and tables. A quantitative handbook (Plenum Press, New York 1968).

BLOOM, W. and FAWCETT, D.W.: A textbook of histology; 8th ed., 9th ed. (Saunders, Philadelphia 1962, 1968).

BODIAN, D.: An experimental study of the optic tracts and retinal projections of the Virginia opossum. J. comp. Neurol. *66:* 113–144 (1937).

BODIAN, D.: Studies on the diencephalon of the Virginia opossum. I. The nuclear pattern in the adult. II. The fiber connections in normal and experimental material. III. The thalamo-cortical projection. J. comp. Neurol. *71:* 259–323 (1939); *72:* 200–297 (1940); *77:* 525–575 (1942).

BONE, Q.: The central nervous system; in BRODAL and FÄNGE The biology of Myxine, pp. 50–91 (Universitetes forlaget, Oslo 1963).

BOON, A. A.: Comparative anatomy and physiopathology of the autonomic hypothalamic centers (Bohn, Haarlem 1938).

BOVERI, V.: Über die phylogenetische Bedeutung der Sehorgane des Amphioxus. Zool. Jahrb. *1904:* suppl. VII (*Weismann*-Festschrift), pp. 409–428 (1904).

BOVERI, V.: Untersuchungen über das Parietalauge der Reptilien. Acta zool. *6:* 1–57 (1925).

BRAAK, H.: Über die Gestalt des neurosekretorischen Zwischenhirn-Hypophysen-Systems von Spinax niger. Z. Zellforsch. mikr. Anat. *58:* 265–276 (1962).

BRAAK, H.: Das Ependym der Hirnventrikel von Chimaera monstrosa (mit besonderer Berücksichtigung des Organon vasculare praeopticum). Z. Zellforsch. mikr. Anat. *60:* 582–608 (1963).

BRAGG, SIR WILLIAM: The universe of light (Bell, London 1931; Dover, New York 1959).

BRAIN, W.R.: Diseases of the nervous system; 4th ed., 6th ed., 7th ed. revised by the late *Lord* BRAIN and WALTON, J.N. (Oxford University Press, London 1950, 1962, 1969).

BRAWER, J.R.; PECK, S.L., and SONNENSCHEIN, C.: Morphological plasticity in the wall of the third ventricle during the estrous cycle in the Rat: a scanning electron microscopic study. Anat. Rec. *179:* 481–489 (1974).

BROCKHAUS, H.: Beitrag zur normalen Anatomie des Hypothalamus und der Zona incerta beim Menschen. Versuch einer architektonischen Gliederung. J. Psychol. Neurol. *51:* 91–196 (1942).

BRODISH, Z. and REDGATE, E.S. (eds.: Brain-pituitary-adrenal relationships. Proceedings of a symposium (Karger, Basel 1973).

BROWN, D.L. and SALINGER, W.L.: Loss of X-cells in lateral geniculate nucleus with monocular paralysis: neural plasticity in the adult cat. Science *189:* 1011–1012 (1975).

BROWN, J.W.: Language cognition and the thalamus. Confin. neurol. *36:* 33–60 (1974).

BROWNSTEIN, M. and AXELROD, J.: Pineal gland: 24-hour rhythm in norepinephrine turnover. Science *184:* 163–165 (1974).

BRUESCH, S.R. and AREY, L.B.: The number of myelinated and unmyelinated fibers in the optic nerve of vertebrates. J. comp. Neurol. *77:* 631–665 (1942).

BRUNI, J.E.; MONTEMURO, D.G.; CLATTENBURG, R.E., and SINGH, R.P.: A scanning electron microscopic study of the third ventricle of the rabbit, rat, mouse and human brain. Anat. Rec. *174:* 407–420 (1972).

BUDDENBROCK, W. v.: The senses (University of Michigan Press, Ann Arbor 1958).

BUREN, J.M. VAN: The retinal ganglion cell layer (Thomas, Springfield 1963).

BUREN, J.M. VAN and BORKE, R.C.: Variations and connections of the human thalamus (Springer, New York 1972).

BUTLER, A.B. and NORTHCUTT, R.G.: Architectonics of the diencephalon of Iguana iguana (Linnaeus). J. comp. Neurol. *149:* 439–461 (1973).

CAIRNEY, J.: A general survey of the forebrain of Sphenodon punctatum. J. comp. Neurol. *42:* 255–348 (1926/27).

CAJAL, S.R. Y: Estructura del kiasmo óptico y teoría general de los entrecruzamientos de las vias nerviosas. Rev. trim. microgr. *3:* 15–65 (1898).

CAJAL, S.R. Y: Estudios talámicos. Trab. Lab. Invest. biol. Univ. Madrid *2:* 129–143 (1903).

CAJAL, S.R. Y: Histologie du système nerveux de l'homme et des vertébrés. 2 vols. (Maloine, Paris 1909, 1911; Instituto Ramon y Cajal, Madrid, 1952, 1955).

CAJAL, S.R. Y: Sur la voie collatérale motrice du pédoncule cérébral. Trab. Lab. Invest. biol. Univ. Madrid *25:* 129–143 (1928).

CAMPBELL, B. and RYZEN, M.: The nuclear anatomy of the diencephalon of Sorex cinereus. J. comp. Neurol. *99:* 1–22 (1953).

CAMPBELL, C.B.G. and HAYHOW, W.R.: Primary optic pathways in the Echidna, Tachyglossus aculeatus: an experimental degeneration study. J. comp. Neurol. *143:* 119–136 (1971).

CAMPBELL, E. and KUHLENBECK, H.: Mortal brain wounds; a pathologic study. J. Neuropath. exp. Neurol. *9:* 139–149 (1950).

CAMPBELL, E.; KUHLENBECK, H.; CAVENAUGH, R.L., and NIELSEN, A.E.: Clinico-pathological aspects of fatal missile-caused craniocerebral injuries. Chapt. XV, pp. 335–399, Surgery in World War II, Neurosurgery, vol. 1 (Dept. of the Army, Washington 1958).

CAMPOS-HORTEGA, J.A. and GLEES, P.: The termination of ipsilateral and contralateral optic fibers in the lateral geniculate body of Galago crassicaudatus. J. comp. Neurol. *129:* 279–284 (1967a).

CAMPOS-HORTEGA, J.A. and GLEES, P.: The subcortical distribution of optic fibers in Saimiri sciureus (Squirrel monkey). J. comp. Neurol. *131:* 131–142 (1967b).

CARPENTER, M.B. and PIERSON, R.J.: Pretectal region and the pupillary light reflex. An anatomical analysis in the Monkey. J. comp. Neurol. *149:* 271–299 (1973).

CHACKO, L.W.: The laminar pattern of the lateral geniculate body in the primates. J. Neurol. Neurosurg. Psychiat. *11:* 211–224 (1948).

CHACKO, L.W.: A preliminary study of the distribution of cell size in the lateral geniculate body. J. Anat. *83:* 254–266 (1949).

CHANG, H.T. and RUCH, T.C.: Topographical distribution of spinothalamic fibers in the thalamus of the spider monkey. J. Anat. *81:* 150–164 (1947).

CHARLTON, H.H.: Comparative studies on the nucleus preopticus pars magnocellularis and the nucleus tuberi lateralis in fishes. J. comp. Neurol. *54:* 239–275 (1932).

CHIARUGI, G.: Su alcune particolarità di sviluppo del segmento posteriore del pavimento del Diencefalo e del processo del infundibulo, e sulla questione dell'esistenza di un rudimento di sacco vasculoso nei Mammiferi. Arch. ital. Anat. Embriol. *19:* 508–539 (1922).

CHRIST, J.F.: Derivation and boundaries of the hypothalamus, with atlas of hypothalamic grisea; in HAYMAKER, ANDERSON and NAUTA The hypothalamus, chapt. 2, pp. 13–60 (Thomas, Springfield 1969).

CLARA, M.: Das Nervensystem des Menschen; 3. Aufl. (Barth, Leipzig 1959).

CLARK, W.E. LE GROS: The thalamus of Tarsius. J. Anat. *64:* 371–414 (1930).

CLARK, W.E. LE GROS: The structure and connections of the thalamus. Brain *55:* 406–470 (1932a).

CLARK, W.E. LE GROS: An experimental study of thalamic connections in the rat. Philos. Trans. Ser. B *222:* 1–28 (1932b).

CLARK, W.E. LE GROS: Morphological aspects of the hypothalamus; in LE GROS CLARK, BEATTIE, RIDDOCH and DOTT The hypothalamus, pp. 1–68 (Oliver & Boyd, Edinburgh 1938).

CLARK, W.E. LE GROS: Observations on the association system of the visual cortex and the central representation of the retina. J. Anat. *75:* 225–235 (1941).

CLARK, W.E. LE GROS: Immediate problems of the anatomy of the thalamus. Rapport présenté au IV congrès neurol. internat. 1949; pp. 225–235 (1949).

CLARK, W.E. LE GROS and PENMAN, G.G.: The projection of the lateral geniculate body. Proc. roy. Soc. Lond. B *114:* 291–313 (1934).

COATES, P.W.: Scanning electron microscopy of a second type of supraependymal cell in the monkey third ventricle. Anat. Rec. *182:* 275–287 (1975).

CONEL, J.L.: The development of the brain of Bdellostoma stouti, I., II. J. comp. Neurol. *47:* 434–403 (1929); *52:* 365–499 (1931).

CONRAD, L.C.A. and PFAFF, D.W.: Axonal projections of medial preoptic and anterior hypothalamic neurons. Science *190:* 1112–1114 (1975).

COWAN, W.M.; GOTTLIEB, D.I.; HENDRICKSON, A.E.; PRICE, J.L., and WOOLSEY, T.A.: The autoradiographic demonstration of axonal connections in the central nervous system. Brain Res. *37:* 21–51 (1972).

CRAIGIE, E.H.: An introduction to the finer anatomy of the central nervous system based upon that of the albino rat (Blakiston, Philadelphia 1925).

CRAIGIE, E.H.: Observations on the brain of the humming bird (Chrysolampis mosquitus Linn., Chlorostilbon caribaeus Lawr.). J. comp. Neurol. *48:* 377–481 (1928).

CRAIGIE, E.H.: Studies on the brain of the kiwi (Apteryx australis). J. comp. Neurol. *49:* 223–357 (1930).

CRAIGIE, E.H.: The cell masses in the diencephalon of the humming bird. Proc. kon. Akad. Wetensch. Amsterdam *34:* 1038–1050 (1931).

CREUTZFELDT, H.G.: Über das Fehlen der Epiphysis cerebri bei einigen Säugern. Anat. Anz. *42:* 517–521 (1912).

CROSBY, E.C.; HUMPHREY, T., and LAUER, E.W.: Correlative anatomy of the nervous system (Macmillan, New York 1962).

CROSBY, E.C. and SHOWERS, M.J.C.: Comparative anatomy of the preoptic and hypothalamic areas; in HAYMAKER, ANDERSON and NAUTA The hypothalamus, chapt. 3, pp. 61–135 (Thomas, Springfield 1969).

CROUCH, R.L.: The nuclear configuration of the hypothalamus and subthalamus of Macacus rhesus. J. comp. Neurol. *59:* 431–449 (1934a).

CROUCH, R.L.: The nuclear configuration of the thalamus of Macacus rhesus. J. comp. Neurol. *59:* 451–485 (1934b).

CUNNINGHAM, T.J.: Early eye removal produces excessive bilateral branching in the rat: application of cobalt filling method. Science *194*: 857–859 (1976).

CURWEN, A.O. and MILLER, R.N.: The pretectal region of the turtle, Pseudemys scripta Troostii. J. comp. Neurol. *71:* 99–120 (1939).

DAMMERMAN, K.W.: Der Saccus vasculosus der Fische, ein Tiefeorgan. Z. wiss. Zool. *96:* 654–726 (1910).

DARTNALL, H.J.A. (ed.): Photochemistry of vision. Handbook of sensory physiology, vol. 7, part 1 (Springer, New York 1972).

DAVIDOFF, L.M. and DYKE, C.G.: The normal encephalogram; 3rd ed. (Lea & Febiger, Philadelphia 1951).

DAVISON, C. and SELBY, N.E.: Hypothermia in cases of hypothalamic lesions. Arch. Neurol. Psychiat. *33:* 570–591 (1935).

DEJERINE, J. et ROUSSY, G.: Le syndrome thalamique. Rev. neurol. *14:* 521–532 (1906).

DEKABAN, A.: Human thalamus. I, II. J. comp. Neurol. *99:* 639–684 (1953); *100:* 63–97 (1954).

DENDY, A.: On the structure, development, and morphological interpretation of the pineal organs and adjacent parts of the brain in the tuatara (Sphenodon punctatus). Philos. Trans. B *201:* 227–331 (1911).

DEWULF, A.: Anatomy of the normal human thalamus. Topometry and standardized nomenclature (Elsevier, New York 1971).

DEY, F. L.: Genital changes in female guinea pigs resulting from destruction of the median eminence. Anat. Rec. *87:* 85–90 (1943).

D'HOLLANDER, F.: Recherches anatomiques sur les couches optiques. La topographie des noyaux thalamiques. Mém. Acad. roy. Méd. Belg. *21:* 1–55 (1913).

DIEPEN, R.: Über das Hypophysen-Hypothalamussystem bei Knochenfischen. Verh. anat. Gs. *51:* Vers., pp. 111–122 (Fischer, Jena, 1953).

DIEPEN, R.: Der Hypothalamus; in v. MÖLLENDORFF and BARGMANN, Handb. d. mikr. Anat. d. Menschen, vol. 4, part VII (Springer, Berlin 1962).

DIMITROVA, Z.: Recherches sur la structure de la glande pinéale chez quelques mammifères. Névraxe *2:* 259–321 (1901).

DORN, E.: Der Saccus vasculosus; in v. MÖLLENDORFF and BARGMANN Handb. d. mikr. Anat. d. Menschen, vol. IV, Teil 2, pp. 140–185 (Springer, Berlin 1955).

DOTY, R.W.; GLICKSTEIN, M., and CALVIN, W.H.: Lamination in the lateral geniculate nucleus in the squirrel monkey, Saimiri sciureus. J. comp. Neurol. *127:* 335–340 (1966).

DOWLING, J.E.: Synaptic organization of the frog retina. Proc. roy. Soc. B *170:* 205–228 (1968).

DOWLING, J.E. and EHINGER, B.: Synaptic organization of the amine-containing interplexiform cells of the goldfish and Cebus monkey retina. Science *188:* 270–273 (1975).

DUSSER DE BARENNE, J.G. und SAGER, O.: Über die sensiblen Funktionen des Thalamus opticus der Katze. (Untersucht mit der Methode der örtlichen Strychninvergiftung; allgemeine Symptomatologie und funktionelle Lokalisation). Z. Neurol. Psychiat. *133:* 231–272 (1931).

DUVERNOY, H.: The vascular architecture of the median eminence; in KNIGGE, SCOTT and WEINDL Brain-endocrine interactions. Median eminence: structure and function, pp. 79–108 (Karger, Basel 1972).

DUVERNOY, H. et KORITKÉ, J.G.: Les vaisseaux sous-épendymaires du recessus hypophysaire. J. Hirnforsch. *10:* 227–245 (1968).

DUVERNOY, H.; KORITKÉ, J.G. et MONNIER, G.: Architecture du plexus primaire du système porte hypophysaire. Colloque du Centre nat. Rech. scient. No. 927 (Paris, 1970).

DUVERNOY, H.; KORITKÉ, J.G. et MONNIER, G.: Sur la vascularisation du tuber postérieur chez l'homme et sur les relations vasculaires tubéro-hypophysaires. J. neuro-visc. Relat. *32:* 112–142 (1971).

EAKIN, R. M.: The third eye (University of California Press, Berkeley 1974).

ECONOMO, C. v.: Sleep as a problem of localization. J. nerv. ment. Dis. *71:* 249–259 (1930).

EDINGER, L.: Giebt es zentral entstehende Schmerzen? (Dtsch. Z. Nervenheilk. *1:* 262–282 (1891).

EDINGER, L.: Das Zwischenhirn der Selachier und Amphibien. Abh. Senckenberg naturf. Ges. *18* (1893).

EDINGER, L.: Über das Gehirn von Myxine glutinosa. Anhang Phys. Abh. kgl. preuss. Akad. Wiss. (1906).

EDINGER, L.: Vorlesungen über den Bau der nervösen Zentralorgane des Menschen und der Tiere. I. Das Zentralnervensystem des Menschen und der Säugetiere; 8. Aufl., 1911. II. Vergleichende Anatomie des Gehirns; 7. Aufl., 1908 (Vogel, Leipzig 1908, 1911).

EDINGER, L.: Einführung in die Lehre vom Bau und den Verrichtungen des Nervensystems; 2. Aufl. (Vogel, Leipzig 1912).

Edinger, L. und Wallenberg, A.: Untersuchungen über das Gehirn der Tauben. Anat. Anz. *15:* 245–271 (1899).

Edinger, L.; Wallenberg, A. und Holmes, G.: Untersuchungen über die vergleichende Anatomie des Gehirns. 5. Das Vorderhirn der Vögel. Abh. Senckenberg. naturf. Ges. *20:* 343–426 (1903).

Edinger, T.: Paired pineal organs; in Kappers Progress in neurobiology. Proc. 1st intern. Meeting of Neurobiologists, pp. 120–129 (Elsevier, Amsterdam 1956).

Eitschberger, E.: Entwicklung und Chemodifferenzierung des Thalamus der Ratte. Ergebn. Anat. EntwGesch. *42/6:* 1–75 (1970).

Elliott, H.C.: Textbook of neuroanatomy (Lippincott, Philadelphia 1963).

Engelhardt, F.: Über die Angioarchitektonik des hypophysär-hypothalamischen Systems. Acta neuroveg. *13:* 129–170 (1956).

Enroth-Cugell, C. and Robson, J.G.: The contrast sensitivity of retinal ganglion cells of the cat. J. Physiol., Lond. *187:* 517–552 (1966).

Erdheim, J.: Über Hypophysenganggeschwülste und Hirncholeastome. Sitz. Ber. k. Akad. Wiss. Wien, math.-naturw. Cl. *113:* 537–726 (1904).

Erulkar, D.S.: Tactile and auditory areas of the brain of the pigeon. An experimental study by means of evoked potentials. J. comp. Neurol. *103:* 421–458 (1955).

Euler, C. v.: Physiology and pharmacology of temperature regulation. Pharmacol. Rev. *13:* 361–398 (1961).

Fain, G.L.: Quantum sensitivity of rods in the toad retina. Science *187:* 838–841 (1975).

Famiglietti, E.V., jr. and Kolb, H.: Structural basis for on- and off-center responses in retinal ganglion cells. Science *194:* 193–195 (1976).

Farner, D.S. and Oksche, A.: Neurosecretion in birds. Gener. comp. Endocr. *2:* 113–147 (1962).

Farris, E.J.: Increase in lymphocytes in healthy persons under certain emotional states. Amer. J. Anat. *63:* 297–323 (1938).

Feremutsch, K.: Thalamus; in Hofer *et al.* Primatologia, vol. II, part 2, Lieferung 6 (Karger, Basel 1963).

Fieschi, C. (ed.): Cerebral blood flow and intracranial pressure (Karger, Basel 1972).

Fink, R.G. and Lutyens, D.V.: The physics of television (Doubleday, Garden City 1960).

Fisher, C.; Ingram, W.R.; Hare, W.K., and Ranson, S.W.: The degeneration of the supraoptico-hypophyseal system in diabetes insipidus. Anat. Rec. *63:* 29–52 (1935).

Flechsig, P.: Zur Anatomie und Entwicklungsgeschichte der Leitungsbahnen im Grosshirn des Menschen. Arch. Anat. Physiol. anat. Abt., pp. 12–78 (1881).

Flechsig, P.: Gehirn und Seele; 2. Aufl. (Veit, Leipzig 1896).

Foix, C. et Hillemand, P.: Les artères de l'axe encéphalique jusqu'au diencéphale inclusivement. Rev. neurol. *2:* 705–739 (1925).

Foix, C. et Nicolesco, J.: Anatomie cérébrale. Les noyaux gris centraux et la région mésencéphalo-sous-optique (Masson, Paris 1925).

Forel, A.: Untersuchungen über die Haubenregion und ihre oberen Verknüpfungen im Gehirne des Menschen und der Säugethiere, mit Beiträgen zu den Methoden der Gehirnforschung. Arch. Psychiat. *7:* 393–495 (1877).

Forel, A.: Gesammelte hirnanatomische Abhandlungen mit einem Aufsatz über die Aufgaben der Neurobiologie (Reinhardt, München 1907).

Forrester, J. and Peters, A.: Nerve fibers in the optic nerve of rat. Nature, Lond. *214:* 245–247 (1967).

FRANZ, V.: Beiträge zur Kenntnis des Mittelhirns und Zwischenhirns der Knochenfische. Folia neurobiol. *6:* 402–450 (1912).

FRANZ, V.: Vergleichende Anatomie des Wirbeltierauges; in BOLK, GÖPPERT, KALLIUS and LUBOSCH Handbuch der vergleichenden Anatomie der Wirbeltiere, vol. II/2, pp. 989–1292 (Urban & Schwarzenberg, Berlin 1934).

FRAUCHIGER, E. und WILDI, E.: Zur pathologischen Anatomie tierischer Epiphysen; in KAPPERS Progress in Brain Research, vol. 10, pp. 654–664 (Elsevier, Amsterdam 1965).

FREDERIKSE, A.: The lizard's brain (Callenbach, Nijkerk 1931).

FRETS, G.P.: Zwei Fälle mit einer Commissura anterior secundaria mollis, ein Fall ohne Commissura anterior und die Variabilität der Massa intermedia. Folia. neurobiol. *10:* 19–23 (1916/17).

FRISCH, K. VON: Bees, their vision, chemical senses, and language (Cornell University Press, Ithaca 1950).

FRITSCH, G.: Untersuchungen über den feineren Bau des Fischgehirns (Gutman, Berlin 1878).

FUJITA, T.: Über das Zwischenhirn-Hypophysensystem von Chimaera monstrosa. Z. Zellforsch. *60:* 147–162 (1963).

FULTON, J.F.: Physiology of the nervous system; 3rd ed. (Oxford University Press, New York 1949).

FUORTES, M.G.F. (ed.): Physiology of photoreceptor organs. Handbook of sensory physiology, vol. 7, part 2 (Springer, New York 1972).

GAAB, M.: Zur Morphokinese der 'Verney'schen Cysten' im Nucleus supraopticus; Inaug. Diss. Würzburg (1973).

GALEN, C.: Opera omnia; in KUHN Medicorum graecorum opera (Cnobloch, Leipzig 1882).

GAREY, L.J., and BLAKEMORE, C.: Monocular deprivation: morphological effects on different classes of neurons in the lateral geniculate nucleus. Science *197*: 414–416 (1977).

GERLACH, J.: Über das Gehirn von Protopterus annectens. Anat. Anz. *75:* 310–406 (1933).

GERLACH, J.: Beiträge zur vergleichenden Morphologie des Selachierhirnes. Anat. Anz. *96:* 79–165 (1947).

GILBERT, C.D. and KELLY, J.P.: The projection of cells in different layers of the cat's visual cortex. J. comp. Neurol. *163:* 81–105 (1975).

GILBERT, M.S.: The early development of the human diencephalon. J. comp. Neurol. *62:* 81–115 (1935).

GIOLLI, R.A. and CREEL, D.J.: Inheritance and variability of the organization of the retinogeniculate projections in pigmented and albino rats. Brain Res. *78:* 335–339 (1974).

GIOLLI, R.A. and GUTHRIE, M.D.: The primary optic projections in the rabbit. An experimental degeneration study. J. comp. Neurol. *136:* 99–126 (1969).

GIOLLI, R.A. and POPE, J.E.: The anatomical organization of the visual system of the rabbit. Docum. ophthal. *30:* 9–32 (1971).

GIOLLI, R.A. and POPE, J.E.: The mode of innervation of the dorsal lateral geniculate nucleus and of the pulvinar by axons arising from the visual cortex. J. comp. Neurol. *147:* 129–144 (1973).

GIOLLI, R.A. and TIGGES, J.: The primary optic pathways and nuclei of primates; in NOBACK and MONTAGNA The primate brain. Advances in Primatology, vol. 1, pp. 29–54 (Appleton-Century-Crofts, New York 1970).

GIOLLI, R.A.; TOWNS, L.C., and HASTE, D.A.: The mode of innervation of portions of the anterior and posterior pretectal nuclei of the rabbit by axons arising from the visual cortex. J. comp. Neurol. *155:* 177–193 (1974).

GLEES, P.: The contribution of the medial fillet and strio-hypothalamic fibers to the dorsal supraoptic decussation with a note on the lateral fillets. J. Anat. *78:* 113–117 (1944).

GLEES, P.: Der Verlauf und die Endigung des Tractus spinothalamicus und der medialen Schleife, nach Beobachtungen beim Menschen und Affen. Verh. anat. Ges. 50. Vers., Erg. Bd. Anat. Anz. *99:* 48–58 (1952).

GLEES, P.: Morphologie und Physiologie des Nervensystems (Thieme, Stuttgart 1957).

GLEES, P. and CLARK, W.E. LE GROS: The termination of optic fibers in the lateral geniculate body of the monkey. J. Anat. *75:* 295–308 (1941).

GLOBUS, J.H.: Probable topographic relations of the sleep-regulating center. Arch. Neurol. Psychiat. *43:* 125–138 (1940).

GLOBUS, J.H.: Infundibuloma. A newly recognized tumor of neurohypophysial derivation with a note on the saccus vasculosus. J. Neuropath. exp. Neurol. *1:* 50–80 (1942).

GLOBUS, J.H. and KUHLENBECK, H.: Tumors of the striatothalamic and related regions. Their probable source of origin and more common forms. Arch. Path. *34:* 674–734 (1942).

GLOOR, P.: Telencephalic influences upon the hypothalamus; in FIELDS *et al.* Hypothalamic hypophysial interrelationships, pp. 74–114 (Thomas, Springfield 1956).

GOLD, R.M.: Hypothalamic obesity: the myth of the ventromedial nucleus. Science *182:* 488–490 (1973).

GOLDBY, F.: An experimental study of the thalamus in the phalanger, Trichosurus vulpecula. J. Anat. *77:* 195–224 (1943).

GOLDSTEIN, A.: Opioid peptides (endorphins) in pituitary and brain. Science *193:* 1081–1086 (1976).

GOLDSTEIN, K.: Untersuchungen über das Vorderhirn und Zwischenhirn einiger Knochenfische. Arch. mikr. Anat. *66:* 135–219 (1905).

GRAHAM, J.B. and ROSENBLATT, R.H.: Aerial vision: unique adaptation in an intertidal fish. Science *168:* 586–588 (1970).

GRANIT, R.: Receptors and sensory perception (Yale University Press, New Haven 1955).

GRIFFITHS, G.M.: The Laurence-Moon-Beadle syndrome: a pathological report. J. Neurol. Psychiat. *1:* 1–6 (1938).

GRIGNON, G.: Développement du complexe hypothalamo-hypophysaire chez l'embryon de poulet (SIT, Nancy 1956).

GREEN, J.D.: The comparative anatomy of the portal vascular system and of the innervation of the hypophysis; in HARRIS and DONOVAN The pituitary gland, vol. 1, chapt. 3, pp. 127–146 (University of California Press, Berkeley 1966).

GROEBBELS, F.: Untersuchungen über den Thalamus und das Mittelhirn der Vögel. Anat. Anz. *57:* 385–415 (1924).

GRÜNTHAL, E.: Vergleichend anatomische und entwicklungsgeschichtliche Untersuchungen über die Zentren des Hypothalamus der Säuger und des Menschen. Arch. Psychiat. Nervenkr. *90:* 216–267 (1930).

GUILLERY, R.W.: The laminar distribution of retinal fibers in the dorsal lateral geniculate nucleus of the cat. A new interpretation. J. comp. Neurol. *138:* 339–368 (1970).

GURDJIAN, E.S.: The diencephalon of the albino rat. J. comp. Neurol. *43:* 1–114 (1927).

GUTTMANN, W.: Grundriss der Physik für Mediziner; 16. Aufl. (Thieme, Leipzig 1918).

HALL, W.C. and EBNER, F.F.: Thalamotelencephalic projections in the turtle (Pseudemys scripta). J. comp. Neurol. *140:* 101–122 (1970).

HALLER, B.: Vom Bau des Wirbeltiergehirns. I. Salmo und Scyllium. Morph. Jb. *26:* 345–641 (1899).

HALLER, GRAF: Die epithelialen Gebilde am Gehirn der Wirbeltiere. Z. Anat. Entw-Gesch. *63:* 118–202 (1922).

HALLER, GRAF: Die Gliederung des Zwischen- und Mittelhirns der Wirbeltiere. Morph. Jb. *63:* 359–407 (1929).

HALLER V. HALLERSTEIN, GRAF V.: Äussere Gliederung des Zentralnervensystems; in BOLK, GÖPPERT, KALLIUS and LUBOSCH Handbuch der vergleichenden Anatomie der Wirbeltiere, vol. 2, Teil 1, pp. 1–318 (Urban & Schwarzenberg, Berlin 1934).

HANDLER, P. (ed.): Biology and the future of man (Oxford University Press, New York 1970).

HANSSON, H.A.: Uptake and intracellular bidirectional transport of horseradish peroxidase in retinal ganglion cells. Exp. Eye Res. *16:* 377–388 (1973).

HARMAN, P.J.: Cytoarchitecture of infant and adult human lateral geniculate nucleus. Abstract. Anat. Rec. *103:* 541 (1949).

HARRIS, G.W. and DONAVAN, B.T. (eds.): The pituitary gland. 3 vols. (University of California Press, Berkeley 1966).

HASSLER, R.: Die Anatomie des Thalamus. Arch. Psychiat. Z. Neurol. *184:* 249–251 (1950).

HASSLER, R.: Anatomy of the thalamus; in SCHALTENBRAND and BAILEY Introduction to stereotaxis with an atlas of the human brain, pp. 230–290 (Thieme, Stuttgart 1959).

HASSLER, R.: Spezifische und unspezifische Systeme des menschlichen Zwischenhirns; in BARGMANN and SCHADÉ Progress in Brain Research, vol. 5, pp. 1–32 (Elsevier, Amsterdam 1964).

HASSLER, R.: Dichotomy of facial pain conduction in the diencephalon; in HASSLER and WALKER Trigeminal neuralgia, chapt. 18, pp. 123–138 (Thieme, Stuttgart 1970).

HASSLER, R.: Afferente Systeme. Über die Zweiteilung der Schmerzleitung in die Systeme der Schmerzempfindung und des Schmerzgefühls; in JANZEN *et al.* Schmerz. Grundlagen, Pharmakologie, Therapie, pp. 105–120 (Thieme, Stuttgart 1972a).

HASSLER, R.: Hexapartition of inputs as a primary role of the thalamus; in FRIGYESI *et al.* Corticothalamic projections and sensorimotor activities, pp. 551–579 (Raven Press, New York 1972b).

HAYHOW, W.R.; WEBB, C., and JERVIE, A.: The accessory optic fiber system in the rat. J. comp. Neurol. *115:* 187–287 (1960).

HAYMAKER, W.: Hypothalamo-pituitary neural pathways and the circulatory system of the pituitary; in HAYMAKER *et al.* The hypothalamus, chapter 6, pp. 219–250 (Thomas, Springfield 1969).

HAYMAKER, W. and ANDERSON, E.: Disorders of the hypothalamus and pituitary gland; in BAKER Clinical neurology; 3rd ed., vol. 2, chapt. 28, pp. 1–78 (Harper & Row, New York 1971/73).

HAYMAKER, W.; ANDERSON, E., and NAUTA, W.J.H.: The hypothalamus (Thomas, Springfield 1969.)

HEAD, H.: Studies in neurology. 2 vols. (Oxford University Press, London 1920).

HEIER, P.: Fundamental principles in the structure of the brain. A study of the brain of Petromyzon fluviatilis (Håkan, Lund 1948; also Acta anat. suppl. VI).

Helmholtz, H.v.: Handbuch der physiologischen Optik. Posthum. 3rd. ed., 3 vols. (Voss, Hamburg & Leipzig 1909–1911).

Hering, E.: Grundzüge der Lehre vom Lichtsinn; in Graefe and Saemisch Handbuch der gesamten Augenheilkunde, 2. Aufl., Teil XII, pp. 1–294 (1. Aufl. 1874–80; 2. Aufl. 1899–1930; 3. Aufl. 1912–32, Engelmann, Leipzig; Springer, Berlin).

Herrick, C.J.: The morphology of the forebrain in Amphibia and Reptilia. J. comp. Neurol. *20:* 413–547 (1910).

Herrick, C.J.: The internal structure of the midbrain and thalamus of Necturus. J. comp. Neurol. *28:* 215–348 (1917).

Herrick, C.J.: The amphibian forebrain. III. The optic tracts and centers of Amblystoma and the frog. J. comp. Neurol. *39:* 433–489 (1925).

Herrick, C.J.: An introduction to neurology; 5th ed. (Saunders, Philadelphia 1931).

Herrick, C.J.: Optic and postoptic systems in the brain of Amblystoma tigrinum. J. comp. Neurol. *77:* 191–353 (1942).

Herrick, C.J.: The brain of the tiger salamander Ambystoma tigrinum (University of Chicago Press, Chicago 1948).

Herrick, C.J. and Obenchain, J.B.: Notes on the anatomy of a cyclostome brain: Ichthyomyzon concolor. J. comp. Neurol. *23:* 635–675 (1913).

Herrick, C.L.: The commissures and histology of the teleost brain. Anat. Anz. *6:* 676–681 (1891).

Hess, A.: The nuclear topography and architectonics of the thalamus of the guinea pig. J. comp. Neurol. *103:* 385–419 (1955).

Hess, W.R.: Lokalisatorische Ergebnisse der Hirnreizungsversuche mit Schlafeffekt. Arch. Psychiat. *88:* 813–816 (1929).

Hess, W.R.: Vegetative Funktionen und Zwischenhirn. Helv. physiol. pharmacol. Acta *5:* suppl. IV, pp. 5–65 (1947).

Hess, W.R.: Hypothalamus and Thalamus. Experimental-Dokumente (Thieme, Stuttgart 1954, 1968).

Hickey, T.L.: Translaminar growth of axons in the kitten dorsal lateral geniculate nucleus following removal of one eye. J. comp. Neurol. *161:* 359–381 (1975).

Hines, M.: The brain of Ornithorhynchus anatinus. Philos. Trans. roy. Soc. Lond. Ser. B *217:* 155–287 (1929).

His, W.: Vorschläge zur Eintheilung des Gehirns. Arch. Anat. EntwGesch. *1893:* 172–179 (1893).

Hoff, F.: Über die zentralnervöse Blutregulation. Fortschr. Neurol. Psychiat. *8:* 299–325 (1936).

Holländer, H.: Projections from the striate cortex to the diencephalon in the squirrel monkey (Saimiri sciureus). A light microscopic radioautographic study following intracortical injection of ³H-leucine. J. comp. Neurol. *155:* 425–440 (1974).

Holmes, R.L. and Ball, J.N.: The pituitary gland. A comparative account (Cambridge University Press, New York 1974).

Holmgren, N.: Zur Anatomie des Gehirns von Myxine. Kungl. svensk vet. Akad. Handl. 60, pt. *7:* 1–96 (1919).

Holmgren, N.: Zur Anatomie und Histologie des Vorder- und Zwischenhirns der Knochenfische. Acta zool. *1:* 137–315 (1920).

Holmgren, N.: Points of view concerning forebrain morphology in lower vertebrates. J. comp. Neurol. *34:* 391–459 (1922).

Holmgren, N.: On two embryos of Myxine glutinosa. Acta zool. *27:* 1–90 (1946).

HOLMGREN, N. and HORST, C.L. VAN DER: Contribution to the morphology of the brain in Ceratodus. Acta zool. *6:* 59–165 (1925).

HOLMGREN, U.: On the ontogeny of the pineal and parapineal organ in Teleost fishes; in KAPPERS and SCHADÉ Progress in Brain Research, vol. 10, pp. 172–182 (Elsevier, Amsterdam 1965).

HOOGENBOOM, K.J. HOCKE: Das Gehirn von Polyodon folium Lacèp. Z. mikr. anat. Forsch. *18:* 311–392 (1929).

HOPKINS, D.A. and LAWRENCE, D.G.: On the absence of a rubrothalamic projection in the monkey with observations on some ascending projections. J. comp. Neurol. *161:* 269–293 (1975).

HORTEGA, P. DEL RIO: Constitución histologica de la glandula pineal. Arch. Neurobiol. *3:* 359–389 (1922).

HUBEL, D.H.: Single unit activity in lateral geniculate body and optic tract of unrestrained cats. J. Physiol., Lond. *150:* 91–104 (1960).

HUBEL, D.H. and WIESEL, T.N.: Integrative action in the cat's lateral geniculate body. J. Neurophysiol. *155:* 385–398 (1961).

HUBEL, D.H. and WIESEL, T.N.: Responses of monkey geniculate cell to monochromatic and white spots of light. Abstract. Physiologist *7:* 162 (1964).

HUBEL, D.H. and WIESEL, T.N.: Spatial and chromatic interactions in the lateral geniculate body of the rhesus monkey. J. Physiol., Lond. *29:* 1115–1154 (1966).

HUBER, G.C. and CROSBY, E.C.: On thalamic and tectal nuclei and fiber paths in the brain of the American alligator. J. comp. Neurol. *40:* 47–227 (1926).

HUBER, G.C. and CROSBY, E.C.: The nuclei and fiber paths of the avian diencephalon, with consideration of telencephalic and certain mesencephalic centers and connections. J. comp. Neurol. *48:* 1–225 (1929).

HUGHES, A.: A quantitative analysis of the cat retinal ganglion cell topography. J. comp. Neurol. *163:* 107–128 (1975).

HUGHES, A. and WASSLE, H.: The cat optic nerve: fibre total count and diameter spectrum. J. comp. Neurol. *169:* 171–184 (1976).

INGRAM, W.R.; HANNET, F.I., and RANSON, S.W.: The topography of the nuclei of the diencephalon of the cat. J. comp. Neurol. *55:* 333–394 (1932).

INGRAM, W.R.; KNOTT, J.R.; WHEATLEY, M.D., and SUMMERS, T.D.: Physiological relationships between hypothalamus and cerebral cortex. Electroenceph. clin. Neurophysiol. *3:* 37–58 (1951).

INGVAR, S.: Zur Phylogenese des Zwischenhirns, besonders des Sehhügels. Dtsch. Z. Nervenheilk. *83:* 302–314 (1924).

ISENSCHMIDT, R. und SCHNITZLER, W.: Beiträge zur Lokalisation des der Wärmeregulation vorstehenden Zentralapparates im Zwischenhirn. Arch. exp. Path. Pharmakol. *76:* 202–233 (1914).

ISO, M.: Über die Thalamuskerne der Fledermaus. Gegenbaurs morph. Jb. *86:* 343–381 (1941).

JACOBOWITZ, D.M. and PALKOVITZ, M.: Topographic atlas of catecholamine and acetylcholinesterase-containing neurons in the rat brain. I. Forebrain (telencephalon, diencephalon). J. comp. Neurol. *157:* 13–28 (1974).

JAMESON, D. and HURWICH, L.M. (eds.): Visual psychophysics. Handbook of sensory physiology, vol. 7, part 4 (Springer, New York 1972).

JANSEN, J.: A note on the optic tract in teleosts. Proc. kon. Acad. Wetensch. Amsterdam *32:* 1104–1117 (1929).

Jansen, J.: The brain of Myxine glutinosa. J. comp. Neurol. *49:* 359–507 (1930).

Jeener, R.: Evolution des centres diencéphaliques periventriculaires des teleostomes. Proc. kon. Akad. Wetensch. Amsterdam *33:* 755–770 (1930).

Johnston, J.B.: The brain of Petromyzon. J. comp. Neurol. *12:* 1–87 (1902).

Johnston, J.B.: The nervous system of vertebrates (Blakiston, Philadelphia 1906).

Jones, E.G.: Some aspects of the organization of the thalamic reticular complex. J. comp. Neurol. *162:* 285–308 (1975).

Jones, E.G. and Burton, H.: Cytoarchitecture and somatic sensory connectivity of thalamic nuclei other than the ventrobasal complex in the cat. J. comp. Neurol. *154:* 395–432 (1974).

Jones, E.G. and Leavitt, R.Y.: Retrograde axonal transport and the demonstration of non-specific projection to the cerebral cortex and striatum from thalamic intralaminar nuclei in the rat, cat, and monkey. J. comp. Neurol. *154:* 349–377 (1974).

Jung, R. (ed.): Central processing of visual information. Handbook of sensory physiology, vol. 7, parts 3A, 3B (Springer, New York 1973).

Kamer, J.C. van de: The pineal organ in fish and amphibia; in Kappers Progress in Neurobiology. 1st intern. Meetg. of Neurobiologists, pp. 113–120 (Elsevier, Amsterdam 1956).

Kanaseki, T. and Sprague, J.M.: Anatomical organization of pretectal nuclei and tectal laminae in the cat. J. comp. Neurol. *158:* 319–337 (1974).

Kappers, C.U. Ariëns: Untersuchungen über das Gehirn der Ganoiden Amia calva und Lepidosteus osseus. Abh. Senckenberg. naturf. Ges. *30:* 449–500 (1907).

Kappers, C.U. Ariëns: Die vergleichende Anatomie des Nervensystems der Wirbeltiere und des Menschen. 2 vols. (Bohn, Haarlem 1920, 1921).

Kappers, C.U. Ariëns: Anatomie comparée du système nerveux particulièrement de celui des Mammifères et de l'Homme. Avec la collaboration de *E.H. Strasburger* (Masson, Paris 1947).

Kappers, C.U. Ariëns und Hammer, E.: Das Zentralnervensystem des Ochsenfrosches (Rana catesbyana). Psychiat. neurol. Bladen *1918* (Feestbundel *Winkler*): 368–415 (1918).

Kappers, C.U. Ariëns; Huber, G.C., and Crosby, E.C.: The comparative anatomy of the nervous system of vertebrates, including man. 2 vols. (Macmillan, New York 1936).

Kappers, J.A.: Survey of the innervation of the epiphysis cerebri and of the accessory pineal organs of vertebrates; in Kappers and Schadé Progress in Brain Research, vol. 10, pp. 87–153 (Elsevier, Amsterdam 1965).

Kappers, J.A.: The pineal organ: an introduction; in Wolstenholme and Knight The pineal gland, pp. 3–34 (Livingstone, Edinburgh 1971).

Kappers, J.A. and Schadé, J.P. (eds.): Structure and function of the epiphysis cerebri. Progress in Brain Research, vol. 10 (Elsevier, Amsterdam 1965).

Kappers, J.A.; Smith, A.R., and Vries, R.A.C. de: The mammalian pineal gland and its endocrinological implications. Progr. Brain Res. *41:* 149–174 (1974).

Karplus, J.P. and Kreidl, A.: Gehirn und Sympathicus. Pflügers Arch. ges. Physiol. 129 (1909), 135 (1910), 143 (1911) etc., as quoted in the bibliography to Kuhlenbeck (1954).

Karten, H.J.: The ascending auditory pathway in the pigeon (Columba livia). II. Telencephalic projections of the nucleus ovoidalis thalami. Brain Res. *11:* 134–153 (1968).

Karten, J.H. and Dubbeldam, J.L.: The organization and projections of the palaeostriatal complex in the pigeon (Columba livia). J. comp. Neurol. *148:* 61–89 (1973).

KARTEN, H.J. and HODOS, W.: A stereotaxic atlas of the brain of the pigeon (Columba livia) (Johns Hopkins Press, Baltimore 1967).

KARTEN, H.J. and HODOS, W.: Telencephalic projections of the nucleus rotundus in the pigeon (Columba livia). J. comp. Neurol. *140:* 35–52 (1970).

KARTEN, H.J.; HODOS, W.; NAUTA, W.J.H., and REVZIN, A.M.: Neural connections of the 'visual wulst' of the avian telencephalon. Experimental studies in the pigeon (Columba livia) and owl (Speotyto curnicularia). J. comp. Neurol. *150:* 253–277 (1973).

KELLY, J.P. and GILBERT, J.P.: The projection of a different morphological type of ganglion cells in the cat retina. J. comp. Neurol. *163:* 65–80 (1975).

KEYSER, A.: The development of the diencephalon in the Chinese hamster. Acta anat. *83:* suppl. 1 (1972).

KLEIN, E. and SCHILDER, P.: The Japanese illusion and the postural model of the body. J. nerv. ment. Dis. *70:* 241–263 (1929).

KLEINE, A.: Über die Parietalorgane bei einheimischen und ausländischen Anuren. Jena. Z. Naturw. *64:* 339–376 (1929).

KNIGGE, K.M.; SCOTT, D.E., and WEINDL, A. (eds.): Median eminence. Structure and function (Karger, Basel 1972).

KNIGHTON, R.S.: Thalamic relay nucleus for the second somatic sensory receiving area in the cerebral cortex of the cat. J. comp. Neurol. *92:* 183–191 (1950).

KNOOK, H.L.: The fibre-connections of the forebrain; Proefschrift, Leiden (van Gorkum, Assen 1965).

KNOWLES, F. and VOLLRATH, L. (eds.): Neurosecretion. The final neuroendocrine pathway. Proceedings of a symposium (Springer, New York 1974).

KORITKÉ, J.G. et DUVERNOY, H.: Les connexions vasculaires du système porte hypophysaire. Anat. Anz., Erg. Bd. 109 (1960/61): 786–806 (1962).

KOLB, H. and FAMIGLIETTI, E.V.: Rod and cone pathways in the inner plexiform layer of the cat retina. Science *186:* 47–49 (1974).

KRAUSE, R.: Mikroskopische Anatomie der Wirbeltiere in Einzeldarstellungen, vol. I–IV (De Gruyter, Berlin 1921/1923).

KRAYENBÜHL, H. and YASARGIL, M.G.: Die cerebrale Angiographie; 2nd ed. (Thieme, Stuttgart 1965).

KREHT, H.: Über die Faserzüge im Zentralnervensystem von Salamandra maculosa L. Z. mikr. anat. Forsch. *23:* 239–320 (1930).

KREHT, H.: Über die Faserzüge im Zentralnervensystem von Proteus anguineus L. Z. mikr. anat. Forsch. *25:* 376–427 (1931).

KRIEG, W.J.S.: Functional neuroanatomy; 3rd ed. (Brain Books, Evanston 1966).

KRISHNAMURTI, A.; KANAGASUNTHERAM, R., and WONG, W.C.: Functional significance of the fibrous laminae in the ventrobasal complex of the thalamus of the slow Loris. J. comp. Neurol. *145:* 515–523 (1972).

KROGH, A.: The pituitary (posterior lobe) principle in circulating blood. J. Pharmacol. exp. Ther. *29:* 177–189 (1926).

KRUGER, L.: The thalamus of the dolphin (Tursiops truncatus) and comparison with other mammals. J. comp. Neurol. *111:* 133–194 (1959).

KUFFLER, S.W.: Discharge patterns and functional organization of mammalian retina. J. Neurophysiol. *16:* 37–68 (1953).

KUHLENBECK, H.: Zur Morphologie des Urodelenvorderhirns. Jena. Z. Naturw. *57:* 463–490 (1921).

KUHLENBECK, H.: Zur Morphologie des Gymnophionengehirns. Jena. Z. Naturw. *58:* 453–484 (1922).

KUHLENBECK, H.: Über die Homologien der Zellmassen im Hemisphärenhirn der Wirbeltiere. Folia anat. jap. *2:* 325–364 (1924).

KUHLENBECK, H.: Vorlesungen über das Zentralnervensystem der Wirbeltiere. Eine Einführung in die Gehirnanatomie auf vergleichender Grundlage (Fischer, Jena 1927).

KUHLENBECK, H.: Über die Grundbestandteile des Zwischenhirnbauplans der Anamnier. Morph. Jb. *63:* 50–95 (1929).

KUHLENBECK, H.: Bemerkungen über den Zwischenhirnbauplan bei Säugetieren, insbesondere beim Menschen. Anat. Anz. *70:* 122–142 (1930).

KUHLENBECK, H.: Über die Grundbestandteile des Zwischenhirns bei Reptilien. Morph. Jb. *66:* 244–317 (1931).

KUHLENBECK, H.: Über die morphologische Stellung des Corpus geniculatum mediale. Anat. Anz. *81:* 28–37 (1935).

KUHLENBECK, H.: Über die Grundbestandteile des Zwischenhirnbauplans der Vögel. Morph. Jb. *77:* 61–109 (1936).

KUHLENBECK, H.: The ontogenetic development of the diencephalic centers in a bird's brain (chick) and comparison with the reptilian and mammalian diencephalon. J. comp. Neurol. *66:* 23–75 (1937).

KUHLENBECK, H.: The development and structure of the pretectal cell masses in the chick. J. comp. Neurol. *71:* 361–387 (1939).

KUHLENBECK, H.: Cadaver encephalography with opaque contrast medium. Anat. Rec. *77:* 145–153 (1940).

KUHLENBECK, H.: The derivatives of the thalamus ventralis in the human brain and their relation to the so-called subthalamus. Milit. Surg. *102:* 433–447 (1948).

KUHLENBECK, H.: The derivatives of thalamus dorsalis and epithalamus in the human brain: their relation to cortical and other centers. Milit. Surg. *108:* 205–256 (1951).

KUHLENBECK, H.: The human diencephalon. A summary of development, structure, function, and pathology (Karger, Basel 1954; also Confinia neurol. *14:* suppl., 1954).

KUHLENBECK, H.: Die Formbestandteile der Regio praetectalis des Anamnier-Gehirns und ihre Beziehungen zum Hirnbauplan. Folia anat. jap. *28:* 23–44 (*Nishi* Festschrift) (1956).

KUHLENBECK, H.: Brain and consciousness. Some prolegomena to an approach of the problem (Karger, Basel 1957).

KUHLENBECK, H.: Schopenhauers Bedeutung für die Neurologie (Zum 100. Todestag des Philosophen). Nervenarzt *32:* 177–182 (1961a).

KUHLENBECK, H.: Mind and matter. An appraisal of their significance for neurological theory (Karger, Basel 1961b).

KUHLENBECK, H.: A note on the hypophysis of the Gymnophione Schistometopum thomense Okaj. Folia anat. jap. *46:* 307–319 (1970).

KUHLENBECK, H.: Some comments on psychophysics. Confin. neurol. *33:* 245–257 (1971).

KUHLENBECK, H.: Schopenhauers Satz 'Die Welt ist meine Vorstellung' und das Traumerlebnis. Schopenhauer Jahrb. 53 (Festschrift *Hübscher*): 376–392 (1972).

KUHLENBECK, H. and HAYMAKER, W.: The derivatives of the hypothalamus in the human brain: their relation to the extrapyramidal and autonomic systems. Milit. Surg. *105:* 26–52 (1949).

KUHLENBECK, H.; MALEWITZ, T.D., and BEASLEY, A.B.: Further observations on the morphology of the forebrain in Gymnophiona, with reference to the topologic vertebrate forebrain pattern; in HASSLER and STEPHAN Evolution of the forebrain, pp. 9–19 (Thieme, Stuttgart 1966).

KUHLENBECK, H. and MILLER, R.N.: The pretectal region of the rabbit's brain. J. comp. Neurol. *76:* 323–365 (1942).

KUHLENBECK, H. and MILLER, R.N.: The pretectal region of the human brain. J. comp. Neurol. *91:* 369–407 (1949).

KUHLENBECK, H. and NIIMI, K.: Further observations on the morphology of the brain in the Holocephalian Elasmobranchs Chimaera and Callorhynchus. J. Hirnforsch. *11:* 267–314 (1969).

KUO, J.S. and CARPENTER, M.B.: Organization of pallidothalamic projections in the rhesus monkey. J. comp. Neurol. *151:* 201–235 (1973).

KUPFFER, K. VON: Die Morphogenie des Centralnervensystems; in HERTWIG Handbuch der vergleichenden und experimentellen Entwicklungslehre der Wirbeltiere, vol. 2, Teil 3, pp. 1–272 (Fischer, Jena 1906).

KUREPINA, M.: Cytoarchitektonik des Sehhügels (Thalamus dorsalis) der Chiroptera; in HASSLER and STEPHAN Evolution of the forebrain, pp. 356–364 (Thieme, Stuttgart 1966).

KUROTSU, T.: Über den Nucleus magnocellularis periventricularis bei Reptilien und Vögeln. Proc. kon. Akad. Wetensch. Amsterdam *38:* 784–797 (1935).

KUSAKA, Y.: The ontogenetic development of the diencephalon in the logger-headed turtle. Tokushima Arb. *9:* 57–83 (1962/63).

LANGE, S.J. DE: Das Zwischenhirn und das Mittelhirn der Reptilien. Folia neurobiol. *7:* 67–138 (1913).

LANKESTER, E.R.: Degeneration. A chapter in Darwinism (Macmillan, London 1880).

LANKESTER, E.R.: Treatise on zoology (Macmillan, London 1900).

LARSON, S.J. and SANCES, A., jr.: Cortical projection of nucleus ventralis lateralis in man. Confin. neurol. *35:* 101–109 (1973).

LASHLEY, K.S.: Thalamo-cortical connections in the rat's brain. J. comp. Neurol. *75:* 67–121 (1941).

LA VAIL, J.H. and LA VAIL, M.M.: The retrograde intraaxonal transport of horseradish peroxidase in the chick visual system: a light and electron microscopic study. J. comp. Neurol. *157:* 303–357 (1974).

LAVAIL, J.H.; WINSTON, K.R., and TISH, A.: A method based on retrograde intraaxonal transport of protein for identification of cell bodies of axons terminating within the CNS. Brain Res. *58:* 470–477 (1973).

LA VAIL, M.M.; Rod outer segment disk shedding in rat retina: relationship to cyclic lighting. Science *194:* 1071–1074 (1976).

LAZORTHES, G.; BASTIDE, G.; ROULLEAU, J. et AMARAT-GOMER, F.: Les artères du thalamus. Anat. Anz., Erg. Bd. 109 (1960/61: 828–831 (1962).

LEWY, F.H. and GASSMANN, F.K.: Experiments on the hypothalamic nuclei in the regulation of chloride and sugar metabolism. Amer. J. Physiol. *112:* 504–510 (1935).

LEWY, F.H. and KOBRAK, H.: The neural projection of the cochlear spirals on the primary acoustic centers. Arch. Neurol. Psychiat. *37:* 839–852 (1936).

LHERMITTE, J.: Les fondements biologiques de la psychologie (Gauthier-Villars, Paris n.d., ca. 1924).

LHERMITTE, J. et CORNIL, L.: Formes hémialgiques du synchrome thalamique. Gaz. Hôp., Paris *102:* 1017–1022 (1929).

LICHTWITZ, L.: Pathologie der Funktionen und Regulationen (Sijthoff, Leiden 1936).

Lindvall, O.; Bjorklund, A.; Nobin, A., and Stenevi, U.: The adrenergic innervation of the rat thalamus. J. comp. Neurol. *154:* 317–347 (1974).

Loo, Y.T.: The forebrain of the opossum, Didelphis virginiana. I. Gross Anatomy. II. Histology. J. comp. Neurol. *51:* 13–64 (1930); *52:* 1–148 (1930).

Lorente de Nó, R.: Contribución al conocimiento del nervio Trigemino. Libro en honor de *D. S. Ramón y Cajal* II, pp. 13–30 (Jiménez y Molina, Madrid 1922).

Lund, R.D., Lund, J.S., and Wise, R.P.: The organization of the retinal projection to the dorsal lateral geniculate nucleus in pigmented and albino rats. J. comp. Neurol. *158:* 383–405 (1974).

Malpeli, J.G. and Baker, F.H.: The representation of the visual field in the lateral geniculate nucleus of Macaca mulatta. J. comp. Neurol. *161:* 559–594 (1975).

Mann, I.C.: The development of the human eye (Cambridge University Press, Cambridge 1928).

Marcus, H. und Laubmann, W.: Zur Entstehung der Hypophyse bei Hypogeophis. Beitrag zur Kenntnis der Gymnophionen Nr. 8. Verh. anat. Ges., Erg. Bd. Anat. Anz. *60:* 277–278 (1925).

Marie, P.: Études sur les troubles de la sensibilité dans les syndromes thalamiques. Libro en honor de *D. S. Ramón y Cajal* II, pp. 129–135 (Jiménez y Molina, Madrid 1922).

Marle, W.: Einführung in die klinische Medizin. 3 vols. (Urban & Schwarzenberg, Berlin 1924, 1925, 1927).

Martin, J.P.: Hemichorea resulting from a local lesion of the brain (the syndrome of the body of Luys). Brain *50:* 637–651 (1927).

Masai, H. and Sato, Y.: A peculiar distribution of the blood vessels in the hypothalamus. J. Hirnforsch. *6:* 197–199 (1963/64).

Mauthner, L.: Zur Pathologie und Physiologie des Schlafes, nebst Bemerkungen über die 'Nona'. Wien. med. Wschr. *40:* 961, 1001, 1049, 1092, 1144, 1185 (1890).

Maximow, A.A.: A text-book of histology (completed by W. Bloom). (Saunders, Philadelphia 1930).

McDowell, F. and Lee, J.E.: Extrapyramidal diseases; in Baker Clinical neurology; 3rd ed., vol. 2, chapt. 26, pp. 1–53 (Harper & Row, New York 1971/73).

McKenna, O.C.; Pinner-Poole, B., and Rosenbluth, J.: Golgi impregnation of a new catecholamine-containing cell type in the toad hypothalamus. Anat. Rec. *177:* 1–14 (1973).

McKenna, O.C. and Rosenbluth, J.: Cytological evidence for catecholamine-containing cells bordering the ventricle of the toad hypothalamus. J. comp. Neurol. *154:* 133–148 (1974).

McLardy, T.: Projection of the centromedian nucleus of the human thalamus. Brain *71:* 290–303 (1948).

McLardy, T.: Thalamic projection to frontal cortex in man. J. Neurol. Neurosurg. Psychiat. *13:* 198–202 (1950).

McLardy, T.: Diffuse thalamic projection to cortex: an anatomical critique. Electroenceph. clin. Neurophysiol. *3:* 183–188 (1951).

McLardy, T.: Thalamic microneurones. Nature, Lond. *199:* 820–821 (1963).

Meader, A.G.: The optic system of the teleost, Holocentrus. J. comp. Neurol. *60:* 361–375 (1934).

Meyers, R.; Sweeney, D.B. and Schwidde, J.T.: Hemiballismus. Aetiology and surgical treatment. J. Neurol. Neurosurg. Psychiat. *13:* 115–126 (1950).

MEYNERT, T.: Vom Gehirn der Säugethiere; in STRICKERS Handbuch der Gewebelehre, Cap. XXXI, pp. 694–805 (Engelmann, Leipzig 1872).

MILLER, H.R. and SPIEGEL, E.A.: Sleep induced by subthalamic lesions with hypothalamus intact. Proc. Soc. exp. Biol. Med. *43:* 300–302 (1940).

MILLER, R.N.: The diencephalic cell masses in the teleost, Corydora paliatus. J. comp. Neurol. *73:* 345–378 (1940).

MILLER, R.N.: The hypophysis of the teleost, Corydora paliatus. J. Morph. *74:* 331–345 (1944).

MILLOT, J. et ANTHONY, J.: Anatomie de Latimeria. II. Système nerveux et organes des sens (Editions du Centre National de la Recherche Scientifique, Paris 1965).

MILLOT, J.; NIEUWENHUYS, R. et ANTHONY, J.: Le diencéphale de Latimeria chalumnae Smith (poisson Coelacanthidé). C. R. Acad. Sci. Paris *258:* 5051–5055 (1964).

MIURA, R.: Über die Differenzierung der Grundbestandteile im Zwischenhirn des Kaninchens. Anat. Anz. *77:* 1–65 (1933).

MIYAKE, O.: The development of the thalamus in the chick. Tokushima Arb. *4:* 181–236 (1958).

MONAKOW, C. v.: Der rote Kern, die Haube und die Regio hypothalamica bei einigen Säugetieren und beim Menschen. Arb. hirnanatom. Inst. Zürich *4:* 103–225 (1918).

MOORE, R.Y. and LENN, N.J.: A retinohypothalamic projection in the rat. J. comp. Neurol. *146:* 1–14 (1972).

MOORE, R.Y.; HELLER, A.; BHATNAGAR, R.K.; WURTMAN, R.J., and AXELROD, J.: Central control of the pineal gland pathways. Arch. Neurol. *18:* 208–218 (1968).

MORGAN, M.J.; O'DONNELL, J.M., and OLIVER, R.F.: Development of left-right asymmetry in the habenular nuclei of Rana temporaria. J. comp. Neurol. *149:* 203–214 (1973).

MOUNTCASTLE, V.B. and HENNEMAN, E.: The representation of tactile sensibility in the thalamus of the monkey. J. comp. Neurol. *97:* 409–439 (1952).

MÜNZER, E. und WIENER, H.: Das Zwischen- und Mittelhirn des Kaninchens und die Beziehungen dieser Teile zum übrigen Centralnervensystem, mit besonderer Berücksichtigung der Pyramidenbahn und Schleife. Mschr. Psychiat. Neurol. *12:* 241–271 (1902).

NATIONAL ACADEMY OF SCIENCES: Color vision. Proceedings of a symposium, 1971 (Natl. Acad. Sci., Washington 1973).

NAUTA, H.J.W.: Evidence of a pallidohabenular pathway in the cat. J. comp. Neurol. *156:* 19–27 (1974).

NAUTA, W.J.H. and HAYMAKER, W.: Hypothalamic nuclei and fiber connections; in HAYMAKER, ANDERSON and NAUTA The hypothalamus, chapt. 4, pp. 136–209 (Thomas, Springfield 1969).

NEAL, H.V. and RAND, H.W.: Comparative anatomy (Blakiston, Philadelphia 1936).

NEILL, W.H. and STEVENS, E.D.: Thermal inertia versus thermoregulation in 'warm' turtles and tunas. Science *184:* 1008–1010 (1974).

NELSON, R.; LUTZOW, A. v.; KOLB, H., and GOURAS, P.: Horizontal cells in cat retina with independent dendritic systems. Science *189:* 137–139 (1975).

NEWTON, SIR ISAAC: Opticks; 4th ed. (Innys, London 1730; Dover, New York 1952).

NIIMI, K.; HARADA, I.; KUSAKA, Y., and KISHI, S.: The ontogenetic development of the diencephalon in the mouse. Tokushima J. exp. Med. *8:* 203–238 (1961).

NIIMI, K.; KANASEKI, T., and TAKIMOTO, T.: The comparative anatomy of the ventral nucleus of the lateral geniculate body in mammals. J. comp. Neurol. *121:* 313–324 (1963a).

Niimi, K.; Katayama, K.; Kanaseki, T., and Morimoto, K.: Studies on the derivation of the centre médian nucleus of Luys. Tokushima J. exp. Med. *6:* 261–268 (1960).

Niimi, K.; Kawamura, S., and Ishimaru, S.: Projection of the visual cortex to the lateral geniculate and posterior thalamic nuclei in the cat. J. comp. Neurol. *143:* 279–311 (1971).

Niimi, K.; Kirimura, K.; Ikoda, T., and Kuwahara, E.: A phylogenetic study of the hypothalamic neurosecretory nuclei in vertebrates. Tokushima J. exp. Med. *10:* 63–72 (1963b).

Niimi, K. and Kuwahara, E.: The dorsal thalamus of the cat and comparison with monkey and man. J. Hirnforsch. *14:* 303–325 (1973).

Niimi, K.; Takemura, A.; Suzuki, H., and Sasaki, J.: The nuclear configuration of the dorsal thalamus in the mole. Tokushima J. exp. Med. *9:* 95–98 (1962).

Noback, C.R. and Laemle, L.K.: Structural and functional aspects of the visual pathways in primates; in Noback and Montagna The primate brain. Advances in Primatology, vol. 1, pp. 55–81 (Appleton-Century-Crofts, New York 1970).

Noble, G.K.: The biology of Amphibia (McGraw-Hill, New York 1931; Dover, New York 1954).

Northcutt, R.G. and Butler, A.B.: Evolution of reptilian visual systems: retinal projections in a nocturnal lizard, Gekko gecko (Linnaeus). J. comp. Neurol. *157:* 453–465 (1974).

Nowikoff, M.: Untersuchungen über den Bau, die Entwicklung und die Bedeutung des Parietalorgans von Sauriern. Z. wiss. Zool. *96:* 118–207 (1910).

Obersteiner, H.: On allocheiria. A peculiar sensory disorder. Brain *4:* 153–163 (1881).

Obersteiner, H.: Anleitung beim Studium der nervösen Zentralorgane im gesunden und kranken Zustande; 5th ed. (Deuticke, Wien 1912).

Ojemann, G.A. and Ward, A.A.: Speech representation in ventrolateral thalamus. Brain *94:* 669–680 (1971).

Oksche, A.: Survey of the development and comparative morphology of the pineal organ; in Kappers and Schadé Progress in Brain Research, vol. 10, pp. 3–20 (Elsevier, Amsterdam 1965).

Oksche, A.: Sensory and glandular elements of the pineal organ; in Wolstenholme and Knight The pineal, pp. 127–146 (Livingstone, Edinburgh 1971).

Olsson, R.: The neurosecretory hypothalamus system and the adenohypophysis of Myxine. Z. Zellforsch. *51:* 97–107 (1959).

Olszewski, J.: The thalamus of Macaca mulatta. An atlas for use with the stereotaxic instrument (Karger, Basel 1952).

Oshima, K. and Gorman, A.: Pars intermedia: unitary electrical activity regulated by light. Science *163:* 195–197 (1969).

Ostwald, W.: Goethe, Schopenhauer und die Farbenlehre (Unesma, Grossbothen 1931).

Papez, J.W.: Comparative neurology (Crowell, New York 1929).

Papez, J.W.: The thalamic nuclei of the nine-banded armadillo (Tatusia novemcincta). J. comp. Neurol. *56:* 49–103 (1932).

Papez, J.W.: Thalamus of turtles and thalamic evolution. J. comp. Neurol. *61:* 433–475 (1935).

Papez, J.W.: Evolution of the medial geniculate body. J. comp. Neurol. *64:* 41–61 (1936).

Papez, J.W.: A proposed mechanism of emotion. Arch. Neurol. Psychiat. *38:* 725–743 (1937).

Papez, J.W.: Connections of the pulvinar. Arch. Neurol. Psychiat. *41:* 277–289 (1939).

Papez, J.W.: The embryologic development of the hypothalamus in mammals. Proc. Ass. Res. nerv. ment. Dis. *20:* 31–51 (1940).

Papez, J.W. and Aronson, L.R.: Thalamic nuclei of Pithecus (Macacus) rhesus. Arch. Neurol. Psychiat. *72:* 1–44 (1934).

Papez, J.W. and Rundles, W.: The dorsal trigeminal tract and the centre median nucleus of Luys. J. nerv. ment. Dis. *85:* 505–519 (1937).

Parent, A.: Distribution of monoamine-containing nerve terminals in the brain of the painted turtle, Chrysemys picta. J. comp. Neurol. *148:* 153–165 (1973).

Parent, A. and Poitras, D.: Morphological organization of monoamine-containing neurons in the hypothalamus of the painted turtle (Chrysemys picta). J. comp. Neurol. *154:* 379–393 (1974).

Parker, G.H.: The origin of the lateral vertebrate eyes. Amer. Naturalist *42:* 601–609 (1908).

Pearlman, A.L. and Daw, N.W.: Opponent color cells in the cat. Science *167:* 84–86 (1970).

Pearson, R.: The avian brain (Academic Press, London 1972).

Peele, T.L.: The neuroanatomical basis for clinical neurology; 2nd ed. (McGraw-Hill, New York 1961).

Penfield, W.: Diencephalic autonomic epilepsy. Arch. Neurol. Psychiat. *22:* 358–374 (1929).

Penfield, W.: Memory mechanisms. Arch. Neurol. Psychiat. *67:* 178–198 (1952).

Penfield, W.: Studies on the cerebral cortex of man – a review and interpretation; in Delafresnaye Brain mechanisms and consciousness, pp. 284–309 (Blackwell, Oxford 1954).

Pfeifer, R.A.: Die nervösen Verbindungen der Augen mit dem Zentralorgan; in Schieck and Bruckner Kurzes Handbuch der Ophthalmologie, vol. 1, pp. 387–475 (Springer, Berlin 1930).

Pierson, R.J. and Carpenter, M.B.: Anatomical analysis of pupillary reflex patterns in the rhesus monkey. J. comp. Neurol. *158:* 121–144 (1974).

Plate, L.: Allgemeine Zoologie und Abstammungslehre. II.: Die Sinnesorgane der Tiere (Fischer, Jena 1924).

Poliak, S. (also spelled *Poljak* and *Polyak*)*:* The main afferent fiber systems of the cerebral cortex in primates (University of California Publ. in Anat. 2, Berkeley 1932).

Polyak, S.: The retina (University of Chicago Press, Chicago 1941).

Polyak, S.: The vertebrate visual system (University of Chicago Press, Chicago 1957).

Portmann, A. and Stingelin, W.: The central nervous system; in Marshall The biology and comparative physiology of birds, vol. 2, pp. 1–36 (Academic Press, London 1961).

Potter, H.D.: Mesencephalic auditory region of the bullfrog. J. Neurophysiol. *28:* 1132–1154 (1965).

Potter, H.D.: Structural characteristics of cell and fiber populations in the optic tectum of the frog (Rana catesbeiana). J. comp. Neurol. *136:* 203–232 (1969).

Powell, T.P.S.: The organization and connexions of the hippocampal and intralaminar systems. Recent Progr. Psychiat. *3:* 54–74 (1958).

POWELL, T.P.S. and COWAN, W.M.: The connexions of the midline and intralaminar nuclei of the thalamus of the rat. J. Anat. *88:* 307–319 (1954).

POWELL, T.P.S.; GUILLERY, R.W. and COWAN, W.M.: A quantitative study of the fornix-mamillo-thalamic system. J. Anat. *91:* 419–437 (1957).

PRINCE, J.H.: Comparative anatomy of the eye (Thomas, Springfield 1956).

PRINTZ, R.H. and HALL, J.H.: Evidence for a retinohypothalamic pathway in the golden hamster. Anat. Rec. *179:* 57–65 (1974).

PRITZ, M.B.: Ascending connections of a thalamic auditory area in a crocodile, Caiman crocodilus. J. comp. Neurol. *153:* 199–213 (1974).

PURPURA, D.P. and YAHR, M.D. (eds.): The thalamus (Columbia University Press, New York 1966).

QUAY, W.B.: Histological structure and cytology of the pineal organ in birds and mammals; in KAPPERS and SCHADÉ Progress in Brain Research, vol. 10, pp. 49–86 (Elsevier, Amsterdam 1965).

QUAY, W.B.: Pineal structure and composition in the orangutan (Pongo pygmaeus). Anat. Rec. *168:* 93–104 (1970).

QUAY, W.B.: Twenty-four-hour rhythmicity in carbonic anhydrase activities of choroid plexuses and pineal gland. Anat. Rec. *174:* 279–288 (1972).

QUAY, W.B.: Pineal chemistry. In cellular and physiologic mechanisms (Thomas, Springfield 1974).

QUIROGA, J. (1977). Personal communication.

RANSON, S.W.: Somnolence caused by hypothalamic lesions in the monkey. Arch. Neurol. Psychiat. *41:* 1–23 (1939).

RANSON, S.W.: Regulation of body temperature. Proc. Ass. Res. nerv. ment. Dis. *20:* 342–399 (1940).

RAUBER-KOPSCH: Lehrbuch und Atlas der Anatomie des Menschen. Abt. 3. Muskeln. Gefässe. 14. Aufl. (Thieme, Leipzig 1933).

RICHTER, R.B. and TRAUT, E.F.: Chronic encephalitis. Pathological report of a case with protracted somnolence. Arch. Neurol. Psychiat. *44:* 848–866 (1940).

RIOCH, D.M.: Studies on the diencephalon of Carnivora. I, II, III. J. comp. Neurol. *49:* 1–119, 121–154 (1929); *53:* 319–388 (1931).

RIOLAN, J., jr.: Anatome corporis humani. In Opera omnia (Plantin, Paris 1610).

RISS, W.; KOIZUMI, K., and McC. BROOKS, C. (eds.): Basic thalamic structure and function. Proceedings of a conference. Brain Behav. Evol. *6* (1972).

RODIECK, R.W.: The vertebrate retina. Principles of structure and function (Freeman, San Francisco 1974).

ROGERS, F.T.: Studies of the brain stem. XI. The effects of artificial stimulation and of traumatism of the avian thalamus. Amer. J. Physiol. *86:* 639–650 (1926).

ROMANOFF, A.L.: The avian embryo (Macmillan, New York 1960).

ROMER, A.S.: The vertebrate body (Saunders, Philadelphia 1950).

ROSE, J.E.: The thalamus of the sheep: cellular and fibrous structure and comparison with pig, rabbit and cat. J. comp. Neurol. *77:* 469–523 (1942).

ROSE, J.E.; GREENWOOD, D.D.; GOLDBERG, J.M. and HIND, J.E.: Some discharge characteristics of single neurons in the inferior colliculus of the cat. J. Neurophysiol. *26:* 294–341 (1963).

ROSE, J.E. and MOUNTCASTLE, V.B.: The thalamus tactile region in rabbit and cat. J. comp. Neurol. *97:* 441–489 (1952).

Rose, J.E. and Woolsey, C.N.: A study of thalamo-cortical relations in the rabbit. Bull. Johns Hopkins Hosp. *73:* 65–128 (1943).

Rose, J.E. and Woolsey, C.N.: Organization of mammalian thalamus and its relationship to the cerebral cortex. Electroenceph. clin. Neurophysiol. *1:* 391–404 (1949).

Rose, M.: Entwicklungsgeschichtliche Einleitung. Ontogenie des Zentralnervensystems und des Sympathikus. Phylogenie des Zentralnervensystems. Cytoarchitektonik und Myeloarchitektonik der Grosshirnrinde; in Bumke *et al.* Handbuch der Neurologie, vol. 1, pp. 1–34, 588–778 (Springer, Berlin 1935).

Röthig, P.: Beiträge zum Studium des Zentralnervensystems der Wirbeltiere. VIII. Über das Zwischenhirn der Amphibien. IX. Über die Faserzüge im Zwischenhirn der Urodelen. Arch. mikr. Anat. *98:* 616–645 (1923). Z. mikr. anat. Forsch. *1:* 5–40 (1924).

Rowe, M.H. and Stone, J.: Properties of ganglion cells in the visual streak of the cat's retina. J. comp. Neurol. *169:* 99–125 (1976).

Ruch, T.C.: Somatic sensation; in Ruch *et al.* Neurophysiology, pp. 300–349 (Saunders, Philadelphia 1961).

Rudebeck, B.: Contributions to forebrain morphology in Dipnoi. Acta zool. *26:* 9–156 (1945).

Rümler, B.; Schaltenbrand, G.; Spuler, H., and Wahren, H.: Somatotopic array of the ventro-oral nucleus of the thalamus based on electrical stimulation during stereotactic procedures. Confin. neurol. *34:* 197–199 (1972).

Saito, T.: Über das Gehirn des japanischen Flussneunauges (Entosphenus japonicus Martens). Folia anat. jap. *8:* 189–263 (1930).

Saland, L.C.; Evan, A.P., and Demski, L.S.: Ultrastructure of ependymal cells in the shark median eminence. Anat. Rec. *178:* 657–605 (1974).

Sanderson, K.J.: Lamination of the dorsal lateral geniculate nucleus in carnivores of the weasel (Mustelidae), raccoon (Procyonidae) and fox (Canidae) families. J. comp. Neurol. *153:* 239–266 (1974).

Sawyer, W.H.: Evolution of antidiuretic hormones and their functions. Amer. J. Med. *42:* 678–686 (1967).

Scalia, F. and Fite, K.: A retinotopic analysis of the central connections of the optic nerve in the frog. J. comp. Neurol. *158:* 455–477 (1974).

Schaffer, J.: Lehrbuch der Histologie und Histogenese; 3. Aufl. (Engelmann, Leipzig 1933).

Schechter, J. and Weiner, R.: Ultrastructural changes in the ependymal lining of the median eminence following the intraventricular administration of catecholamine. Anat. Rec. *172:* 643–650 (1972).

Schenck, F. und Gürber, A.: Leitfaden der Physiologie des Menschen; 15. Aufl. (Enke, Stuttgart 1918).

Schilder, P.: Das Körperschema (Springer, Berlin 1923).

Schilder, P.: The image and the appearance of the human body; studies in constructive energies of the psyche (Paul, Trench, Trubner, London 1935).

Schneider, R.: Ein Beitrag zur Ontogenese der Basalganglien des Menschen. Anat. Nachr. *1:* 115–137 (1950).

Schnitzlein, H.N. and Crosby, E.C.: The telencephalon of the lungfish, Protopterus. J. Hirnforsch. *9:* 105–149 (1967).

Schober, W.: Vergleichend-anatomische Untersuchungen am Gehirn der Larven und adulten Tiere von Lampetra fluviatilis (Linné, 1758) und Lampetra planeri (Bloch, 1784). J. Hirnforsch. *7:* 107–209 (1964).

SCHOPENHAUER, A.: Über das Sehen und die Farben (1818, 1854). Commentatio exponens theoriam colorum physiologicam (1830). Sämmtliche Werke, editio GRISEBACH, vol. VI (Reclam, Leipzig, n.d.).

SCHROEDER, K.: Der Faserverlauf im Vorderhirn des Huhnes, dargestellt auf Grund von entwicklungsgeschichtlichen (myelogenetischen) Untersuchungen, nebst Beobachtungen über die Bildungsweise und Entwicklungsrichtung der Markscheiden. J. Psychol. Neurol. *18:* 115–173 (1911).

SCHULMAN, S.: Bilateral symmetrical degeneration of the thalamus. A clinico-pathological study. J. Neuropath. expl Neurol. *16:* 446–470 (1957).

SCHULTZE, M.: Zur Anatomie und Physiologie der Retina. Arch. mikr. Anat. *2:* 165–286 (1866).

SCHULTZE, M.: Über Stäbchen und Zapfen der Retina. Arch. mikr. Anat. *3:* 215–247 (1867).

SCHUSTER, P.: Beiträge zur Pathologie des Thalamus opticus. I, II, III, IV. Arch. Psychiat. *105:* 358–432, 550–622 (1936); *106:* 13–53 (1936), 201–233 (1937).

SCOTT, D.E.; KOZLOWSKI, G.P., and DUDLEY, G.K.: A comparative ultrastructural analysis of the third cerebral ventricle in the North American mink (Mustela vison). Anat. Rec. *175:* 155–168 (1973).

SELYE, H.: Stress and general adaptation syndrome. Brit. med. J. *i:* 1383–1392 (1950).

SENN, D.G.: Bau und Ontogenese von Zwischen- und Mittelhirn bei Lacerta sicula (Rafinesque). Acta anat. *55:* suppl. 1 (1968).

SHELDON, R.E.: The olfactory tracts and centers in teleosts. J. comp. Neurol. *22:* 177–253 (1912).

SHEPHERD, G.M.: The synaptic organization of the brain. An introduction (Oxford University Press, New York 1974).

SHERMAN, S.M.: Visual fields of cats with cortical and tectal lesions. Science *185:* 355–357 (1974).

SHINTANI-KUMAMOTO, Y.: The nuclei of the pretectal region of the mouse brain. J. comp. Neurol. *113:* 43–60 (1959).

SMIALOWSKI, A.: Anterior hypothalamic nuclei in the monkey brain. Acta biol. crocoviensia, Zoolog. *15:* 15–22 (1972).

SMIALOWSKI, A.: Lateral hypothalamic nuclei in the macaque (Macaca mulatta) brain: myeloarchitectonics. Acta anat. *85:* 332–341 (1973).

SOEMMERING, S.T.: De basi encephali et originibus nervorum cranio egregentium (Vandenhoeck, Gottingen 1778).

SOLNITZKY, O.: The thalamic nuclei of Sus scrofa. J. comp. Neurol. *69:* 121–169 (1938).

SOLNITZKY, O.: The hypothalamus and subthalamus of Sus scrofa. J. comp. Neurol. *70:* 191–229 (1939).

SPATZ, H.: Neues über die Verknüpfung von Hypophyse und Hypothalamus. Acta neuroveg. *3:* 5–49 (1951).

SPENCER, W.B.: On the presence and structure of the pineal eye in Lacertilia. Quart. J. micr. Sci. *27:* 165–238 (1886).

SPIEGEL, E.A.: Die Zentren des autonomen Nervensystems (Springer, Berlin 1928).

SPIEGEL, E.A. and INABA, C.: Experimentalstudien am Nervensystem. Zur zentralen Lokalisation von Störungen des Wachzustandes. Z. ges. exp. Med. *55:* 164–182 (1927).

SPIEGEL, E.A. and KLETZKIN, M.: Thalamohypothalamic relationships. Effect of acute thalamic lesions upon the hypothalamus. Abstract. Amer. J. Physiol. *171:* 769 (1952).

SPIEGEL, E. A.; KLETZKIN, M., and SZEKELY, E. G.: Pain reactions on stimulation of the quadrigeminal region. Abstract. Fed. Proc. *12:* 136 (1953a).

SPIEGEL, E. A. und SAITO, S.: Über die hormonale Erregbarkeit vegetativer Zentren. Arb. neurol. Inst. Univ. Wien *23:* 247–260 (1924).

SPIEGEL, E. A. and WYCIS, H. T.: Physiological and psychological results of thalamotomy. Proc. roy. Soc. Med. Suppl. *42:* 84–92 (1949).

SPIEGEL, E. A. and WYCIS, H. T.: Stereoencephalotomy. I. Methods and stereotaxic atlas of the human brain. II. Clinical and physiologic applications (Grune & Stratton, New York (I) 1952, (II) 1962).

SPIEGEL, E. A.; WYCIS, H. T.; FREED, H., and ORCHINIK, C.: Thalamotomy and hypothalamotomy for the treatment of psychoses. Proc. Ass. Res. nerv. ment. Dis. *31:* 379–391 (1953b).

SPRAGUE, J. M.: Interaction of cortex and superior colliculus in mediation of visually guided behavior in the cat. Science *153:* 1544–1547 (1966a).

SPRAGUE, J. M.: Visual, acoustic, and somesthetic deficits in the cat after cortical and midbrain lesions; in PURPURA and YAHR The thalamus, pp. 391–417 (Columbia University Press, New York 1966b).

STETSON, M. H. and WATSON-WHITMYRE, M.: Nucleus suprachiasmaticus: the biologic clock in the Hamster. Science *191:* 197–199 (1976).

STÖHR, P. und v. MÖLLENDORFF, W.: Lehrbuch der Histologie; 23. Aufl. (Fischer, Jena 1933).

STONE, J. and FUKUDA, Y.: The naso-temporal division of the Cat's retina re-examined in terms of Y-, X-, and W-cells. J. comp. Neurol. *155:* 377–394 (1974).

STONE, J. and HOFFMANN, K. P.: Very slow conducting ganglion cells in the cat's retina: a major, new functional type? Brain Res. *43:* 610–616 (1972).

STRAUSS, I. and GLOBUS, J. H.: Tumor of the brain with disturbance in temperature regulation. The hypothalamus and the area about the third ventricle as a possible site for a heat-regulating center. Report of three cases. Arch. Neurol. Psychiat. *25:* 506–522 (1931).

STRÖER, W. H. F.: Studies on the diencephalon. I. The embryology of the diencephalon of the rat. J. comp. Neurol. *105:* 1–24 (1956).

STRÜMPELL, A.: Ein Beitrag zur Theorie des Schlafes. Pflügers Arch. ges. Physiol. *15:* 573–574 (1877).

STUDNIČKA, F. K.: Die Parietalorgane; in OPPEL Lehrbuch der vergleichenden mikroskopischen Anatomie, vol. 5, pp. 1–24 (Fischer, Jena 1905).

STUDNIČKA, F. K.: Das Schema der Wirbeltieraugen. Zool. Jb. (Anat.) *40:* 1–48 (1918).

SUGITA, K.; MUTSUGA, N.; TAKAOKA, Y., and DOI, T.: Results of stereotaxic thalamotomy for pain. Confin. neurol. *34:* 265–274 (1972).

SUMI, R.: Über die Morphogenese des Gehirns von Hynobius nebulosus. Folia anat. jap. *4:* 171–270 (1926a).

SUMI, R.: Über die Sulci und Eminentiae des Hirnventrikels von Diemyctylus pyrrhogaster. Folia anat. jap. *4:* 375–388 (1926b).

SWAAB, D. F. and SCHADÉ, J. P. (eds.): Integrative hypothalamic activity. Progress in Brain Research, vol. 41 (Elsevier, Amsterdam 1974).

SWANSON, L. W. and COWAN, W. M.: The efferent connections of the suprachiasmatic nucleus of the hypothalamus. J. comp. Neurol. *160:* 1–12 (1975).

TALAIRACH, J.: Chirurgie stéréotaxique du thalamus. VI. Congr. latinoamericano Neurocir. Montevideo 1955, pp. 865–925.

Talairach, J.; David, M.; Tournoux, P.; Corredor, H. et Kvasina, T.: Atlas d'anatomie stéréotaxique (Masson, Paris 1957).

Tanaka, T.: The comparative anatomy of the diencephalon in Selachians. (Japanese with English summary). Arb. 2. Abt. anat. Inst. Tokushima *6:* 115–152 (1959).

Tasker, R.R.; Richardson, P.; Rewcastle, B., and Emmers, R.: Anatomical correlation of detailed sensory mapping of the human thalamus. Confin. neurol. *34:* 184–196 (1972).

Teevan, R.C. and Birney, R.C. (eds.): Color vision (Van Nostrand, New York 1961).

Teitelbaum, H.; Catravas, G.N. and McFarland, W.L.: Reversal of morphine tolerance after medial thalamic lesions in the rat. Science *185:* 449–451 (1974).

Tello, F.: Disposición macroscopica y estructura del cuerpo geniculado externo. Trab. Lab. Invert. biol. Univers. Madrid *3:* 39–62 (1904).

Thuma, B.D.: Studies on the diencephalon of the Cat. I. The cyto-architecture of the corpus geniculatum laterale. J. comp. Neurol. *46:* 173–199 (1928–29).

Tienhoven van, A. and Juhasz, L.P.: The chicken telencephalon, diencephalon and mesencephalon in stereotaxic coordinates. J. comp. Neurol. *118:* 185–198 (1962).

Tilney, F. and Warren, L.F.: The morphology and evolutional significance of the pineal body. Amer. anat. Memoirs, vol. 9 (Wistar Institute Press, Philadelphia 1919).

Tixier-Vidal, A. and Farquhar, M.G. (eds.): The anterior pituitary. Ultrastructure in biologic systems, vol. 7 (Academic Press, New York 1975).

Török, B.: Lebendbeobachtung des Hypophysenkreislaufes an Hunden. Acta morph. Acad. Sci. hung. *4:* 83–89 (1954).

Tretjakoff, D.K.: Der Musculus protractor lentis im Urodelenauge. Anat. Anz. *28:* 25–32 (1906).

Triepel, H.: Die anatomischen Namen, ihre Ableitung und Aussprache; 10. Aufl. (Bergmann, München 1921).

Tsai, C.: The optic tracts and centers of the opossum, Didelphis virginiana. J. comp. Neurol. *39:* 173–216 (1925a).

Tsai, C.: The descending tracts of the thalamus and midbrain of the opossum. J. comp. Neurol. *39:* 217–248 (1925b).

Uematsu, S.; Konigsmark, B., and Walker, A.E.: Thalamotomy for alleviation of intractable pain. Confin. neurol. *36:* 88–96 (1974).

Vastine, J.H. and Kinney, K.: The pineal shadow as an aid in the localization of brain tumors. Amer. J. Roentgenol. *17:* 320–324 (1927).

Verney, E.B.: The antidiuretic hormone and the factors which determine its release. Proc. roy. Soc. Med. *135 B:* 25–106 (1947).

Vesalius, A.: De humani corporis fabrica libri septem (Oporinus, Basel 1543, 1555). Also Saunders, J.B. de C. and O'Malley, C.D.: The illustrations from the works of Andreas Vesalius, with annotations and translations (World Publishing, Cleveland 1950).

Vesalius, A.: De humani corporis fabrica librorum epitome (Oporinus, Basel 1543). Also Lind, L.R.: The epitome of Andreas Vesalius (transl.) (Macmillan, New York 1949).

Vesselkin, N.P.; Agayan, A.L., and Nomokonava, L.M.: A study of thalamo-telencephalic afferent systems in frogs. Brain Behav. Evol. *4:* 295–306 (1971).

Vialli, M.: L'apparato epifisario negli anfibi. Arch. zool. ital. *13:* 423–452 (1929).

Villiger, E.: Gehirn und Rückenmark; 7. Aufl. (Engelmann, Leipzig 1920).

Villiger, E.: Die periphere Innervation; 6. Aufl. (Engelmann, Leipzig 1933).

Wald, G.: The biochemistry of vision. Annu. Rev. Biochem. *22:* 497–526 (1953).

Wald, G.: The molecular basis of visual excitation. Amer. Scient. *42:* 73–95 (1954).

WALD, G.: Molecular basis of visual excitation. Science *162:* 230–239 (1968).

WALKER, A. E.: The primate thalamus (University of Chicago Press, Chicago 1938).

WALKER, A. E.: Internal structure and afferent-efferent relations of the thalamus; in PURPURA and YAHR The thalamus, pp. 1–12 (Columbia University Press, New York 1966).

WALLENBERG, A.: Eine Verbindung caudaler Hirnteile der Taube mit dem Striatum. Neurol. Cbl. *17:* 300–302 (1898).

WALLENBERG, A.: Über zentrale Endstätten des Nervus octavus der Taube. Anat. Anz. *17:* 102–108 (1900).

WALLENBERG, A.: Die basalen Äste des Scheidewandbündels der Vögel. Anat. Anz. *28:* 394–400 (1906).

WALLENBERG, A.: Beiträge zur Kenntnis des Gehirns der Teleostier und Selachier. Anat. Anz. *31:* 369–399 (1907).

WALLS, G. L.: The vertebrate eye and its adaptive radiation. Cranbrook Inst. scient. Bull. 19 (Cranbrook Press, Bloomfield Hills 1942).

WALLS, G. L.: The lateral geniculate nucleus and visual histophysiology (University of California Publ. Physiol. 9, Berkeley 1953).

WALLS, G. L.: The Vertebrate eye and its adaptive radiation (Hafner, New York 1963).

WARNER, F. J.: The hypothalamus of the opossum (Didelphys virginiana). J. nerv. ment. Dis. *70:* 485–494 (1929).

WARNER, F. J.: The development of the diencephalon of the American water snake. Trans. roy. Soc. Canada 3rd. Ser., Sect. V *36:* 53–70 (1942).

WARNER, F. J.: The fibre tracts of the forebrain of American diamond-back rattlesnake (Crotalus adamanteus). Proc. zool. Soc. *116/I:* 22–32 (1945).

WARNER, F. J.: The development of the pretectal cell masses of the American water snake (Natrix sipedon). J. comp. Neurol. *103:* 83–104 (1955).

WARNER, F. J.: The development of the diencephalon in Trichosurus vulpecula. Okaj. Folia anat. jap. *46:* 265–295 (1969).

WARNER, F. J.: The development of the pretectal nuclei in Trichosurus vulpecula. Okaj. Folia anat. jap. *47:* 73–100 (1970).

WESSELY, K.: Goethes und Schopenhauers Stellung in der Geschichte der Farbenlehre; Rektoratsrede (Springer, Berlin 1922).

WINGSTRAND, K. G.: Comparative anatomy and evolution of the hypophysis; in HARRIS and DONOVAN The pituitary gland, vol. 1, chapt. 2, pp. 58–126 (University of California Press, Berkeley 1966).

WOLF, G.: Projections of thalamic and cortical gustatory areas in the rat. J. comp. Neurol. *132:* 519–529 (1968).

WOLIN, L. R. and MASSOPUST, L. C., jur.: Morphology of the primate retina; in NOBACK and MONTAGNA The primate brain. Advances in Primatology, vol. 1, pp. 1–27 (Appleton-Century-Crofts, New York 1970).

WOLSTENHOLME, G. E. W. and KNIGHT, J. (eds.): The pineal gland (Livingstone, Edinburgh 1971).

WOOLSEY, C. N. and WALZL, E. M.: Topical projection of nerve fibers from local regions of the cochlea to the cerebral cortex of the cat. Bull. Johns Hopkins Hosp. *71:* 315–344 (1942).

WURTMAN, R. J.; AXELROD, J., and KELLY, D. E. (eds.): The pineal (Academic Press, New York 1968).

XUEREB, G. P.; PRITCHARD, M. M. L., and DANIEL, P. M.: The hypophysial portal system of vessels in man. Quart. J. exp. Physiol. *39:* 219–230 (1954).

YOSHIDA, I.: Über den Ursprung der kortikopetalen Hörbahn beim Kaninchen. Folia anat. jap. *2:* 289–296 (1924).

YOUNG, J.Z.: The life of vertebrates (Clarendon Press, Oxford 1950, 1955).

YOUNG, J.Z.: The life of mammals (Oxford University Press, New York 1957).

YOUNG, T.: On the mechanism of the eye (1801); On the theory of light and colours (1801); On physical optics (1845). A course of lectures on natural philosophy and the mechanical arts. 2 vols. (Savage, London 1807; Taylor & Welton, London 1845); Miscellaneous works of the late Thomas Young, vol. 1 (Murray, London 1855).

ZEIER, H.J. and KARTEN, H.J.: Connections of the anterior commissure in the pigeon (Columba livia). J. comp. Neurol. *150:* 201–216 (1973).

ZEMAN, W. and INNES, J.R.: Craigie's neuroanatomy of the rat revised and expanded (Academic Press, New York 1963).

ZIMMERMAN, E.A.; CARMEL, P.W.; HUSAIN, M.K.; TANNENBAUM, M.; FRANTZ, A.G., and ROBINSON, A.G.: Vasopressin and neurophysin: high concentration in monkey hypophyseal portal blood. Science *182:* 925–929 (1973).

ZIMMERMAN, N.H. and MENAKER, M.: Neural connections of sparrow pineal: role in circadian activities. Science *190:* 477–479 (1975).

ZORUB, D.S. and RICHARDSON, D.E.: An extralemniscal projection to the centrum medianum and pulvinar of the thalamus. Confin. neurol. *35:* 356–367 (1973).

ZÜLCH, K.J.: Vegetative und psychische Symptome bei umschriebenen traumatischen Zwischenhirnstörungen und ihre Beurteilung im Gutachten. Zbl. Neurochir. *10:* 73–97 (1950).

XIII. The Telencephalon

1. General Pattern and Overall Functional Significance

The Vertebrate telencephalon or endbrain represents the rostral differentiation of the embryonic prosencephalon and is exclusively a derivative of alar plate and roof plate, this latter, however, providing a lesser and rather variable endbrain component. The telencephalic bauplan comprises two first order grundbestandteile,[1] namely *bulbus olfactorius* and *lobus hemisphaericus*.

With regard to the formal aspects of the ontogenetic mechanisms resulting in the morphogenesis of the telencephalon, the following sorts of shaping events, involving growth (increase of 'mass') and configurational displacement of 'mass', can be distinguished:[2] (1) unpaired evagination (exclusively rostral), (2) paired rostral evagination, (3) paired caudal evagination (if occurring, always combined with 1 or 2), (4) inversion, (5) eversion, and (6) overall bending.

In all Vertebrates, except perhaps for a still not sufficiently clarified early unpaired component displayed by Myxinoids, the bulbus olfactorius derives its origin from paired rostral evagination. It may remain contiguous with the lobus hemisphaericus or become separated from the latter by an interconnecting olfactory stalk containing the olfactory neuronal channel of the second order *(olfactory tract)* arising in the primary olfactory center of the bulb. Both olfactory bulb and if present olfactory stalk enclose, at early ontogenetic stages, part of the original ventricular lumen (i.e. a rhinocoele), but this latter, depending on the taxonomic forms, can become partly or entirely obliterated at the definitive stage of bulb or stalk (tract) or both.

An additional, rather rudimentary, variably developed and in some forms even missing nerve, the *nervus terminalis*, likewise connects telen-

[1] The relevant morphologic and topologic definitions of *grundbestandteile* and *formbestandteile* are given in section 5, chapter III of volume 1. The definition of *bauplan* as an elementary topologic space is elaborated in section 2 of that chapter (p. 181).

[2] These formative events namely '*Wachstumsvorgänge*' and '*Gestaltungsvorgänge*', were dealt with in subsection 1B, chapter VI of volume 3/II, and definitions of the appropriate descriptive terms were given.

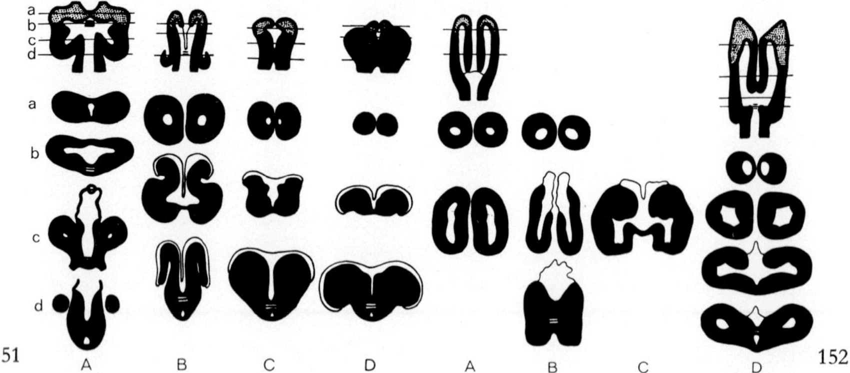

Figure 151. Telencephalic configuration in diverse Anamnia (from SCHOBER, 1966). A Petromyzon (Cyclostome). B Polypterus (Ganoid). C Salmo (Teleost). D Gasterosteus (Teleost). B–D partly after NIEUWENHUYS. Upper row: horizontal plane, with lines indicating planes of the cross-sections.

Figure 152. Telencephalic configuration in diverse Anamnia (from SCHOBER, 1966). A Protopterus (Dipnoan). B Ceratodus (Dipnoan). C Latimeria (Crossopterygian). D Salamandra (urodele Amphibian). A–C partly after NIEUWENHUYS.

cephalon or more generally speaking prosencephalon with nasal cavity respectively the olfactory organ.

The *lobus hemisphaericus*, derived, at the early ontogenetic stages, from an unpaired evagination, may remain partly or wholly unpaired, or subsequently display rostral or caudal paired evagination, or both, with or without near complete reduction of the telencephalon impar. Said reduction is related to the definitive orientation and shape of the lamina terminalis and occurs in Cyclostomes (Petromyzonts) and in all Amniota.

The lobus hemisphaericus, moreover, may be inverted or everted, or again inverted in combination with partial (intermediate) eversion. Eversion, in turn, seems restricted to unpaired evagination, while this latter is compatible with either eversion or inversion. Eversion is displayed by Osteichthyes, by the Coelacanth Latimeria, to a lesser and regionally restricted degree by some Plagiostomes (Chimaeroids) and by Dipnoans.

Since the details of these diverse configurational aspects and their relationships to the morphogenesis of lamina terminalis have been dealt with in subsection 1 B and section 6, chapter VI of volume 3/II, it will here be sufficient to illustrate the several relevant features of the diverse Vertebrate telencephalic configurations by the simplified dia-

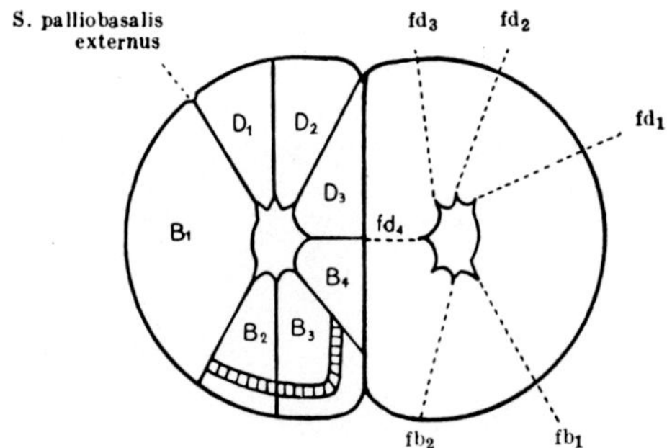

Figure 153 A. Diagrammatic cross-section through inverted, paired evaginated region of telencephalon in the Selachian Galeus, showing zonal grundbestandteile (from GERLACH, 1947).

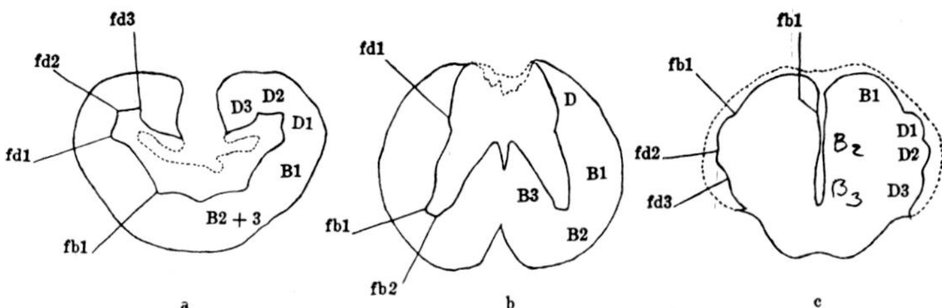

Figure 153 B. Diagrams of cross-sections through unpaired evaginated telencephalon of Anamnia, showing zonal grundbestandteile (from MILLER, 1940). a: Telencephalon impar of urodele Amphibian Salamandra; b: telencephalon impar of Dipnoan Protopterus; c: telencephalon impar of Teleost Carassius.

grams of Figures 151 to 154. Figure 155 illustrates the configurational aspect mentioned above as (6) overall bending, characterized by a peculiar, dorsally convex curvature of the lobus hemisphaericus around the hemispheric stalk. This curvature, designated as *Endhirnbeuge* or *Hemisphärenrotation*, slightly or moderately suggested in Sauropsidans and even amphibian Gymnophiona, but much more pronounced in Mammals, leads caudally to the formation of a temporal pole in addition to the posterior or occipital one. Rostrally this may, as commonly seen in Mammals and Birds, involve a pronouncedly basal position of the olfactory bulb respectively stalk.

The bauplan of the Gnathostome Vertebrate lobus hemisphaericus is characterized by a secondary longitudinal zonal system pertaining to the telencephalic alar plate. These longitudinal zones represent telence-

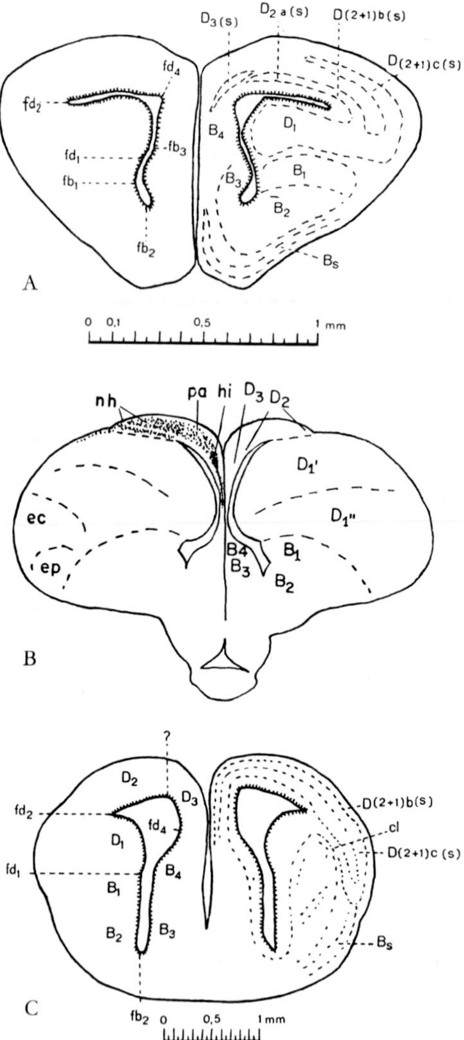

Figure 154. Semidiagrammatic cross-sections through paired evaginated telencephalon of Amniota, showing zonal grundbestandteile (from K., 1929). A Lacertilian Reptile. B Bird (Pigeon). C Mammal (advanced embryo of Insectivore Mole). cl: anlage of claustrum; ec: nucleus epibasalis centralis accessorius ('ectostriatum'); ep: nucleus epibasalis caudalis *('epistriatum' of Edinger, 'archistriatum' of Kappers);* hi: hippocampal cortex; nh: nn. diffusi (neocortical homologa); pa: parahippocampal cortex.

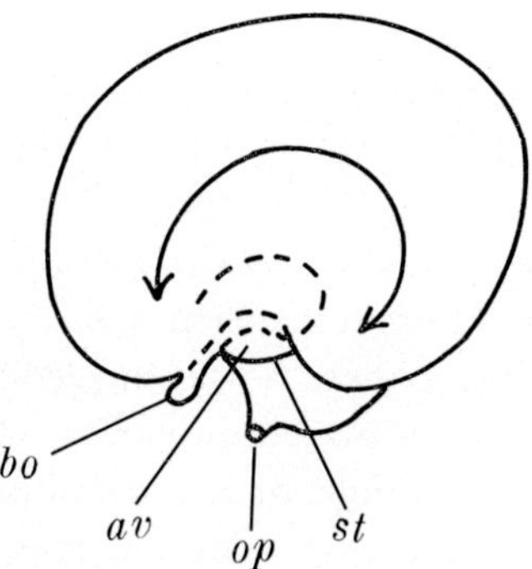

Figure 155. Sketch illustrating the bending (overall curvature, *Endhirnbeuge)* in higher Mammals. Based on a figure by HOCHSTETTER depicting the brain of a 38-mm Human embryo. av: area ventralis anterior; bo: olfactory bulb; op: optic nerve; st: part of sulcus telencephalo-diencephalicus; bent line with arrowheads represents approximate longitudinal 'axis' of bent portion of telencephalic tube, indicating directions of displacements, crudely describable as 'divergent rotations', whose joint 'transverse axis' approximately passes, at right angle to 'longitudinal axis', through a neighborhood of insula.

phalic *grundbestandteile* of the second order, for which the notations D_1, D_2, D_3, B_1, B_2, B_3, B_4 have been proposed (K., 1929). The zone B_4 is restricted to a telencephalic configuration which includes paired rostral evagination of the lobus hemisphaericus. In Cyclostomes the longitudinal zones cannot be distinguished as distinctive neighborhoods, being, either primarily or secondarily (in a phylogenetic sense) restricted to a pallial zonal neighborhood D, and to a basal zonal neighborhood B. These neighborhoods, however, by one-many transformation, can be mapped upon the neighborhoods D_1, D_2, D_3, B_1, B_2, B_3, B_4, which, conversely, by many-one transformation, can be mapped upon D and B. The Gnathostome neighborhoods, as *grundbestandteile*, may provide, by further differentiation, the substratum for additional distinctive grisea with the morphologic value of *formbestandteile*. A substantial difference obtains between the Amniote and the Anamniote lobus hemisphaericus insofar as, in the former, the *grundbestandteil* D_1, by *internation* (or *introversion*) becomes closely related to the B_1 and B_2 zones from which the basal ganglia derive. Thus, the Anamniote primary dorsal hemispheric wall (or pallium) includes D_1, D_2, D_3, while the Amniote secondary pallium is essentially restricted to D_2 and D_3 with only marginal (or 'residual') contributions from D_1. A simplified mapping of these relationships is indicated in Figures 153 and 154.

In addition to the longitudinal zones, the *paraphysis cerebri* represents a further, but apparently non-neuronal *grundbestandteil* of the Gnathostome telencephalon, being derived from a roof plate neighborhood (paraphyseal arch, J. A. Kappers, 1955) of the original pars impar telencephali, more or less adjacent to or even included in, the rostral fold (or leaf) of the velum transversum. It does not seem to be present in Cyclostomes, being there not even recognizably suggested at ontogenetic stages, but is displayed, in various degrees of differentiation, either as a permanent (adult) or as a transitory embryonic configuration, in all gnathostome Vertebrates. It is particularly well developed in various adult Amphibians and in some Reptiles.

In forms where a typical adult paraphysis is present, as an ependymal circumventricular organ (cf. vol. 3/I, chapter V, section 5, pp. 354f., and vol. 3/II, chapter VI, section 1 B, p. 157), it consists of a single layered, cuboidal or low columnar epithelium, commonly rich in glycogen, which may be secreted into the ventricular fluid. A fully differentiated paraphysis assumes the configuration of a compound racemose tubular gland whose duct opens into the ventricle of telencephalon impar.

The mesodermal (leptomeningeal) component of the paraphysis is here extremely well vascularized, whereby the basal membrane of the neuroectodermal epithelium is in immediate and extensive contact with the thin wall of a network of venous sinusoids.

In some adult Birds, a remainder of the embryonic paraphyseal anlage can be recognized as a solid nodule (Haller, 1922). In adult Mammals it is generally not a recognizable configuration, being 'transformed' into, or 'absorbed' by, the choroid plexus. Nevertheless, in Man, paraphyseal rudiments may in some rare case develop into cystic tumors (paraphyseal cysts). Very similar cysts, not of paraphyseal origin, can, however, also arise from detached and degenerated recesses of the neighboring diencephalic midline choroid folds in or adjacent to the diencephalic portion of velum transversum. Detailed studies on the paraphysis have been undertaken by J. A. Kappers (1955, 1956, also reported in Crosby *et al.*, 1962, pp. 350–351).

In accordance with views expressed by L. Edinger (1908, 1911, 1912), the Amphibian, and particularly the Urodele telencephalon may be considered a very simple and suitable paradigm of the Vertebrate telencephalon in general. In this respect, the configurational modifications related to the shape of lamina terminalis and to the extent of telencephalon impar may be discounted. Figure 156 illustrates a map-

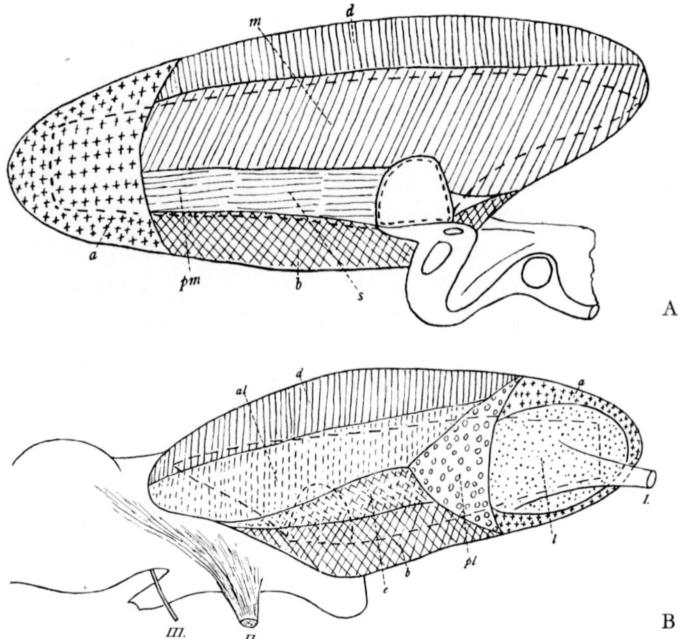

Figure 156. Cytoarchitectural maps of the telencephalon in urodele Amphibians, as recorded in Salamandra (from K., 1921a). A Medial aspect. B Lateral aspect. The dotted line indicates approximate outline of lateral ventricle. a: nucleus olfactorius anterior; al: area lateralis pallii (D_1); b: nucleus basimedialis inferior (B_3, in A), nucleus basilateralis inferior (B_2, in B); d: area dorsalis pallii (D_2) e: nucleus basilateralis superior (B_1, inappropriately designated as 'epistriatum' in 1921); l: formatio bulbaris ('lobaris' in 1921); m: area medialis pallii (D_3 *sive* 'primordium hippocampi'); pl, pm: nucleus postolfactorius lateralis et medialis (transitional neighborhoods pertaining to nucleus olfactorius anterior *sensu latiori)*; s: nucleus basimedialis superior (B_4, designated as 'septum' in 1921; its caudal extension dorsally to interventricular foramen as 'pars fimbrialis *sive* supraforaminalis septi', although recorded and discussed in the 1921 paper, was not included in this generalized map). The bulbus (or 'bulbulus') olfactorius accessorius (cf. Fig. 172E) is here included in a caudolateral neighborhood of formatio bulbaris.

ping of first order and second order *grundbestandteile*[3] in the telencephalon of Salamandra. It can be seen that, except for a relatively restricted region (aula *sive* telencephalon impar) corresponding to the paired secondary *foramen Monroi*, the typically inverted Amphibian telencephalon is rostrally as well as caudally evaginated.

The *olfactory bulb* is here directly continuous with the lobus hemi-

[3] First order *grundbestandteile:* (a) bulbus olfactorius and (b) lobus hemisphaericus. Second order *grundbestandteile:* D and B zones of lobus hemisphaericus.

sphaericus, without a connecting olfactory stalk. It represents the primary (or first-order) olfactory center, since the first order afferent fibers, being neurites of the neurosensory olfactory cells, terminate in the bulb's griseum.[4] This latter is characterized by the *mitral cells*, whose neurites represent second order olfactory fibers. In addition to the layer of mitral cells and directly adjacent cell layers, which, together, can be designated as formatio bulbaris *sensu strictiori*, there are, in Salamandra and a diversity of other Vertebrates, cell populations not included in said formation. Some such cell groups may extend toward the lobus hemisphaericus, representing transitional neighborhoods within the boundary zone of hemispheric lobe and olfactory bulb, and were designated as nucleus postolfactorius medialis et lateralis (K., 1921a, c, 1922a). All these cell groups not directly part of formatio bulbaris *sensu strictiori* may also be designated, in general agreement with views of HERRICK (1948), but discounting some details, as *nucleus olfactorius anterior*. The olfactory mechanisms and the structural details of the olfactory bulb shall be dealt with further below in section 2 of the present chapter.

Concerning the *lobus hemisphaericus* it can be seen (Fig. 156) that the longitudinal zones of the pallium, represented by area lateralis pallii (D_1), area dorsalis pallii (D_2), and area medialis pallii (D_3 *sive* primordium hippocampi) extend from the boundary zone of bulbus olfactorius as far as the caudal pole of the telencephalic hemisphere. The basal zones are represented by nucleus basilateralis superior (B_1), nucleus basilateralis inferior (B_2), nucleus basimedialis inferior (B_3) and nucleus basimedialis superior (B_4). The two basimedial zones pertain to the region which is commonly designated as *septum*, but should preferably be called *preterminal* or *paraterminal*[5] *body* (ELLIOT SMITH, 1910) because of its relationship to lamina terminalis. The nucleus basimedialis superior generally extends caudad dorsally to interventricular foramen, this variable supraterminal portion (not indicated in Fig. 156 A) being the so-called *pars supraforaminalis sive fimbrialis septi*, which fades out in the caudally evaginated part of the hemisphere. At the level of lamina ter-

---

[4] The here adopted definition of first, second, etc. order fibers and 'centers' was given on pp. 273–274 of volume 3/I (chapter V, section 4) and also discussed on p. 290 of volume 3/II (chapter VI, section 2).

[5] The term paraterminal *sensu latiori* can be used to indicate both B_3 and B_4, paraterminal *sensu strictiori* would then refer to B_4. The added qualifications preterminal, intraterminal, supraterminal and post-terminal can furthermore specify particular relationships to lamina terminals.

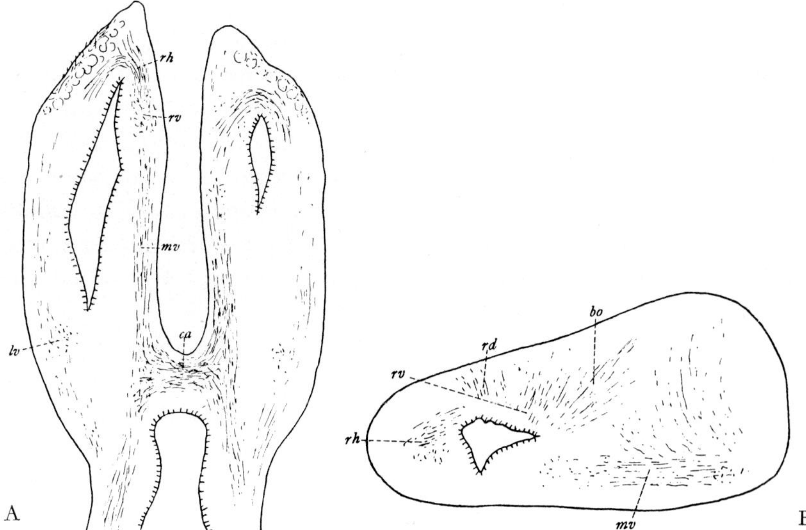

Figure 157A. Horizontal section (*Golgi impregnation* of fiber tracts) through the basal portion of the telencephalon in Salamandra maculosa (from K., 1921a). ca: commissura anterior; lv: portion of lateral forebrain bundle; mv: medial forebrain bundle; rh: radiatio olfactoria horizontalis; rv: radiatio olfactoria ventralis. These rather diffuse radiations from olfactory bulb to lobus hemisphaericus in Salamandra and most other Urodeles are condensed into the more defined olfactory tracts as displayed in a diversity of other Vertebrates.

Figure 157B. Paramedian sagittal section (*Golgi impregnation* of fiber tracts) through the medial wall of the telencephalon in Salamandra (from K., 1921a). bo: ascending portion of radiatio olfactoria ventralis (so-called 'tractus bulbo-occipitalis'); rd: radiatio olfactoria dorsalis. For other designations cf. Figure 157A.

minalis, nucleus basimedialis inferior and basilateralis inferior tend to fuse. Together with a portion of nucleus basilateralis superior, they may extend to a variable degree into the caudally evaginated hemisphere, but, in contradistinction to the three pallial zones, do not, as a fairly general rule, reach the caudal pole.[6]

The lobus hemisphaericus receives *olfactory input* from the bulbus olfactorius by way of the olfactory radiations originating in olfactory bulb and nucleus olfactorius anterior. It consists of second and third order fibers. Reciprocal connections can be assumed. All zonal subdi-

[6] This brief review and summary of the Vertebrate telencephalic zonal system as typically displayed by Urodeles omits here a discussion of the ventricular sulci approximately corresponding to zonal boundaries. The relevant pattern of sulci was dealt with in section 6, chapter VI, of volume 3/II.

visions of the Anamniote hemisphere seem to be reached by these radiations, although in various degrees of distribution, such that e.g. in caudal parts of area dorsalis pallii the olfactory input may become less pronounced. A relatively very minor input may also be provided by the *nervus terminalis*, to be dealt with further below in section 2.

In addition, the lobus hemisphaericus receives sensory input of presumably all other sorts (i.e. optic, otic, somatic and visceral of various types) by way of the diencephalon, particularly but presumably not exclusively mediated by the thalamic grisea, through the lateral and medial forebrain bundles dealt with in chapter XII.

The *telencephalic output* is likewise channeled through these bundles (Fig. 157) and, in addition, by way of the mainly but presumably not exclusively 'descending' *stria medullaris system* (Figs. 3, 7). In forms with pronounced caudal hemispheric rotation, such as Mammals, whose temporal lobe includes a substantial amygdaloid complex, a distinctive *stria terminalis system (taenia semicircularis)* is added (Figs. 237 B, CII, D). In Amphibians and other Anamnia the components of this system are distributed upon basal forebrain bundles and stria medullaris.

The further progressive development of basal forebrain bundle system as displayed in Mammals (e.g. capsula interna, cf. Fig. 13, chapt. XII) can be omitted in the present context. This topic was dealt with in sections 1A and 9 of chapter XII and shall again be considered in section 10 of the present chapter.

In accordance with views expressed by Elliot Smith (1910), Herrick (1921a), and others, the non-olfactory sensory input into the telencephalon seems to have been the relevant phylogenetic factor gradually evolving a mechanism correlating and processing all sensory input and involving storage as well as output controlling effector activities, i.e. behavior, finally taking precedence, and surpassing in bulk, the somewhat comparable mechanisms of tectum mesencephali with its relatively minor relayed olfactory input.

The term '*nosebrain*' as applied to the primitive telencephalon by Herrick (cf. e.g. Fig. 347, p. 642, vol. 3/II) appears fully justified. It should, of course, not be construed to denote an exclusively olfactory function. The cited author and others both explicitly and implicitly recognized non-olfactory input (which includes optic one) into the telencephalon. Because the details of such input were poorly understood as regards submammalian forms, these relationships were perhaps unduly 'downgraded' or not sufficiently emphasized by Herrick and others. It still remains rather uncertain how, as regards details, the non-olfactory input is distributed upon the telencephalic grisea of Anamnia and Sauropsidans.

Rensch (1956) believes that, in phylogeny, certain brain regions may have displayed a positive allometric development,[7] and that such regions subsequently became successively 'filled with functions of behavior'. In this respect, the studies of Nolte (1953) suggest that the area dorsalis pallii (D_2) of Amphibians, a region apparently not concerned with significant function, might be interpreted as the relevant neighborhood for a 'progressive' functional development of the cortex. This would well agree with the presumed derivation of the Mammalian neocortex from a neighborhood of D_2, namely D_2 al.

Rensch (1956) adds the question why such progressive development seems to occur more or less as an addition to final stages and not as an *archallaxis* (Sewertzoff, 1931), that is as an alteration beginning with the early stages. Rensch believes that after an harmoniously working structural type had evolved in phylogeny, it was easier that variants survived in which development was added to the final stages, because all earlier alterations would tend to disturb or to destroy the established harmony.

With respect to Sauropsida, who do not display a structurally differentiated neocortex, but merely a relatively small griseal area topologically homologous with the Mammalian neocortex, it seems most likely that the extensive and conspicuously differentiated lateral basal grisea (D_1, B_1, B_2) are concerned with functions at least in part comparable to those of the Mammalian neocortex. This latter, on the other hand, becomes differentiated into rostral, parietal, insular, occipital, and temporal regions, essentially concerned, in that order, with output, diverse somatic input, visceral input, optic input, and otic input (Fig. 158). One might wonder whether the dichotomy of Sauropsidan and Mammalian evolution began at an Amphibian-like rather than at a Reptilian-like stage.

Be that as it may, the remoteness from all peripheral sensory input except that related to olfaction, combined with a lack of direct output

---

[7] The term allometry refers to the observed facts that certain morphologic entities (e.g. organs, *bauplangrundbestandteile*, or *formbestandteile*) may display greater relative increase in size than the whole organism (positive allometry), lag behind in this respect (negative allometry), or correspond to the total increase (isometry). These terms apply, of course, also with regard to morphologic subsets (e.g. *grundbestandteile* of the second order versus those of first order, or, again, *formbestandteile* versus *grundbestandteile*).

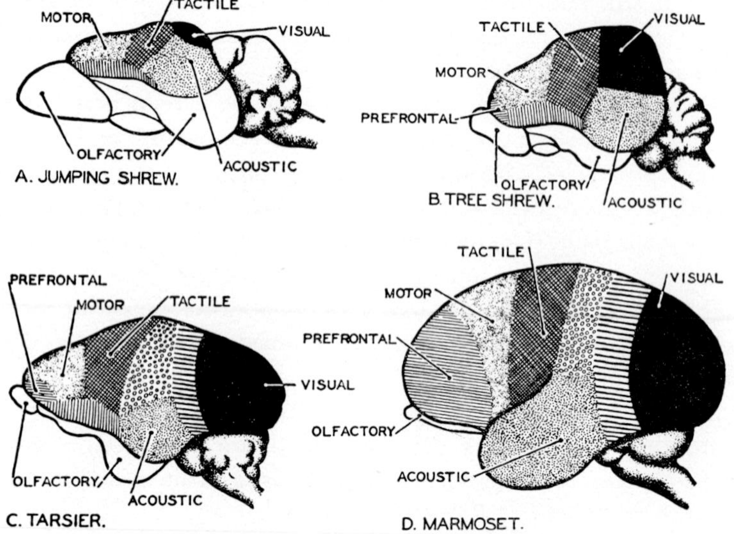

Figure 158. Brains of Mammalian Insectivore and 'lower' Primates in lateral view, showing regions of neocortex as outlined by G. ELLIOT SMITH (from NEAL and RAND, 1936). The diagrammatic sketches indicate an increasing expansion of visual cortex in proportion to the basal 'rhinencephalon' (the most caudal portion of the piriform lobe, however, is presumably 'parahippocampal'). The diagrams also show a rostral differentiation of 'prefrontal cortex', and of presumably 'tactile' (circles) and 'visual' (thick horizontal hatching) 'association areas'. The presumably 'visceral' insular area is indicated by vertical hatching. Because of the 'hemispheric rotation' *(Endhirnbeuge)* it becomes hidden in the floor of fossa Sylvii at stage D (Marmoset, Hapale, Callithricidae).

to the periphery, could perhaps be conceived as parameters *(Zustandsbedingungen)* favoring the development of the 'higher' (i.e. complex) processing mechanisms apparently culminating in the cerebral cortex of Man. Additional comments concerning the relevant significance of 'rhinencephalic' grisea for fundamental aspects of 'behavior' and 'psychology' will be found at the conclusion of the next section. Taking into due account the obtaining complexities pointed out in the foregoing paragraphs, STEINER's (1888) early conclusion, based on his experimental investigations, and antedating the concepts of ELLIOT SMITH as well as of HERRICK can evidently be fully upheld: '*Das Grosshirn der Wirbelthiere hat sich phylogenetisch aus dem Riechzentrum entwickelt*'.

In this respect, the experiments undertaken by BURR (1916a, b) on Amphibian larvae are of interest. The cited author found that regeneration of an extirpated hemispheric anlage depended upon the presence of an olfactory placode. If this latter was likewise removed, regeneration did not occur.

Experiments in Amblystoma undertaken by PIATT (1951) showed that transplanted or isolated telencephalic primordia, or those following extirpation of the nasal sac with prevention of olfactory nerve growth developed abnormally *qua* shape, symmetry, and structure. Some peripheral migration ('scattering') of the periventricular cells in the 'primordium hippocampi' was nevertheless observed, but although emphasized by PIATT, appears, in his figures, less conspicuous than in corresponding normal material. PIATT concludes that these pallial cells will migrate in the absence of olfactory or ascending non-olfactory projection fibers. This agrees with HAMBURGER's critical evaluation of neurobiotaxis as discussed on p. 549 f. (vol. 3/I) together with my own qualified acceptance of said principle. It is quite likely that related multifactorial events which have occurred in phylogeny may have become firmly stabilized in subsequent ontogenetic mechanisms not implying concepts of *Lamarckism*. On the whole, the data obtained by PIATT (1951) may be considered indecisive, as, at least to some extent, admitted by that author himself.

Comparable experiments on Amphibian larvae were also performed by KIRSCHE and KIRSCHE (1964) who also reviewed the literature on this topic, and, with respect to their own findings, reported that, after unilateral resection of the telencephalic anlage, a relatively normal size and form of the regenerating telencephalon was only achieved if the regenerating olfactory nerve grew into the regenerating brain. The cited authors noted that neuroblasts migrated by way of the regenerating nerve into the regenerating brain region and also assumed that inflowing stimuli from the olfactory nerve exerted an influence intensifying the growth processes originating in the matrix zone of the regenerating stump.

Although these various ontogenetic findings clearly indicate the significance of the olfactory organ for telencephalic development and support the above-mentioned phylogenetic interpretations, said development, particularly in 'higher' and in anosmatic Vertebrate forms, seems to depend upon insufficiently understood multiple factors (cf. also chapter VI, section 1 C, p. 256, vol. 3/II). Thus, in some forms of human arhinencephaly, quite normal hemispheres are present despite lack of olfactory fila, bulbs, and tracts. Again, the relevant multifactorial morphogenetic interactions may significantly differ in the diverse Vertebrate groups.

2. Some Remarks on the Olfactory Structures

Olfaction or smell may, *qua* private experience, be classified as a spatially non-configurated, but vaguely localized modality of consciousness. In the behavioristic semantic model, olfaction, like 'taste', is a 'sensory input' mediated by transducing chemoreceptors. NEAL and RAND (1936) justly remark that distinction between 'smell' and 'taste' is difficult to draw in those Metazoa lacking clearly differentiated sense organs. As far as aquatic animals are concerned, chemical substances, in order to reach the receptors, must be dissolved in water. Response to dissolved chemicals is a general behavioral activity of organisms. Primitive Invertebrate organisms, even without specialized chemoreceptor organs (e.g. Protozoa or Metazoan Coelenterates), respond differently to various particles or substances with different chemical properties. By and large, taste is said to be mediated by 'contact receptors', and smell by 'distance receptors'. Yet, the ectodermal taste buds in diverse Teleost might also, with some justification, be classified as 'distance receptors'.[8]

There are no convincing data suggesting that the Vertebrate olfactory organs are phylogenetically derived from any of the diverse sense organs of Invertebrates. It is nevertheless significant that the *olfactory receptors* of Vertebrates, located in the 'mucous' olfactory epithelium, are *peripheral neurons*, i.e. neurosensory cells with a 'neurite', as shall be elaborated further below. Neurosensory elements of this general type are of common occurrence among the diversified groups of Invertebrates.

The *olfactory epithelium* of most Vertebrates is a simple respectively pseudostratified epithelium in which the neurosensory elements are more or less uniformly distributed among the 'supporting', likewise ectodermal, epithelial cells. A modification is displayed by various Fishes and Amphibians, whose neurosensory elements are aggregated into 'olfactory buds' separated by areas of stratified epithelial non-sensory elements, being thereby roughly comparable to 'taste buds'[9] (cf. e.g. Fig. 166 B).

---

[8] Cf. volume 3/II, chapter VII, section 1, pp. 779ff. Although the olfactory receptors pertain to the cutaneous receptors in the wider sense (cf. vol. 3/II, chapter VI, section 1) they were omitted from the discussion in the cited chapter because of their intimate relationship to the telencephalon.

[9] It should, however, be recalled that the 'taste cells' are not neuroepithelial, that is, lack a 'neurite' or comparable central process (apotile), and thus cannot properly be classified as 'neurons'.

The '*Minor Chordata*' (Hemichorda and Urochorda) seem to lack 'olfactory' organs, but respond to chemical stimuli, and thus can be assumed to possess chemoreceptors. The Cephalochordate *Amphioxus* is provided with an unpaired *pit of Koelliker*, derived from the external anterior neuropore.[10] Said pit seems to receive terminations from the paired nervus apicis and may or may not be a chemoreceptor organ, but does not appear to include typical 'olfactory cells'.

In *Cyclostomes* as well as in *Gnathostome Vertebrates*, there are paired olfactory pits or cavities, from whose olfactory mucosa the olfactory nerves, namely the olfactory fibers or fila olfactoria arise as neuronal channels of the first order. Cyclostomes, however, are generally classified as 'monorhinal',[11] but this condition develops from a midline fusion of apparently paired anlagen, resulting in a bilaterally symmetric conjoint complex, whose paired lateral components are connected with the paired telencephalic olfactory bulbs. In Cyclostomes, the unpaired adenohypophysis develops from the ectodermal nasohypophysial pit. In Myxinoids, but not in Petromyzonts, this pit opens into the pharynx.[12]

In *Selachians*, the olfactory organs are paired pre-oral pits, with incompletely separated inlet and outlet, being unconnected with the oral cavity. In some genera, however, nasobuccal grooves extend to the corners of the mouth. Ganoids and Teleosts are likewise characterized by lateral or dorsal paired nasal pits not communicating with the oral cavity. This arrangement generally entails that, as the Fish swims, water flows into the anterior and out of the posterior opening.

In the Coelacanth Crossopterygian *Latimeria* and in *Dipnoans*, however, a connection between the olfactory pits and the oral cavity is provided by openings called internal nares or *choanae*, which are primitively in a relatively rostral position.[13] It is of interest that, in ontoge-

---

[10] Cf. volume 4, chapter VIII, section 3, pp. 59–60.

[11] Externally, the cyclostome olfactory organ appears as a single pouch, opening rather caudally on the dorsal head surface in Petromyzonts, or at the tip of the snout in Myxinoids.

[12] As far as the ambiguous paleontologic evidence is concerned, one group of Paleozoic Ostracoderms displays apparently paired and ventral 'nostrils', while two other groups show a condition comparable to that in Petromyzonts. GEGENBAUR interpreted the apparent monorhine condition in Cyclostomes as secondarily derived from amphirhine (i.e. provided with paired nasal sacs) ancestors.

[13] It will be recalled (cf. vol. 1, chapter II, section 2, p. 59), that Crossopterygians and Dipnoans have been classified as Choanichthyes. The only surviving Crossopterygian seems to be the Coelacanth Latimeria. From the viewpoint of brain morphology, however,

netic development of all Vertebrates with choanae, these latter arise by closure of the edges of nasobuccal grooves, homologous to those seen in some adult Plagiostomes (cf. e.g. KUREPINA, 1926, 1927). In Dipnoans, the connection of nasal cavity with oral cavity through choanae is correlated with the onset of the pulmonary respiration characteristic for Tetrapods to which Dipnoans, although being 'Fishes' can be regarded as representing a transition. In *Tetrapods*, the choanae tend to be displaced caudalward, and in *Mammals* they assume their pronounced caudal position through the formation of the secondary palate. A part of what was originally oral cavity becomes thereby the nasopharyngeal passages extending the nasal cavities caudalward toward the pharynx, and opening through the secondary choanae. A Mammalian remnant of the primary choana is the canalis incisivus (canalis nasopalatinus, *Stensen's foramen*).

In Fishes and in Amphibians at the strictly *aquatic stage*, the nasal sac is filled with water. In larvae of Salamanders and Frogs, the olfactory water stream passes freely from the nasal cavity to the mouth. Its return to the nasal passage is prevented in Urodele larvae by a simple flap acting as a valve, while in Anuran larvae a double fold or fringe has a more complex form but similar function.[14] During metamorphosis these choanal valves are lost and a new mechanism for closing the nasal passages develops, namely as a constrictor and two dilators of the external nares. Thus, in metamorphosed Amphibia the olfactory stream under muscular control passes freely in and out through the nasal cavity. The entering stream tends to move through the medial part of the passage, and the outgoing one through the lateral, which more or less loses its neurosensory epithelium, except for *Jacobson's organ*, dealt with further below.

In *Amniota*, as in terrestrial Amphibians, the nasal cavity becomes an air passage. Specific glands, keeping the mucosal lining moist, are generally developed. In Fishes, practically the entire nasal cavity contains sensory epithelium and displays numerous folds. Some aquatic Amphibia, notably the Perennibranchiata, agree with the Fishes in that

---

I prefer to include Latimeria (and thus Crossopterygians) into the order (or subclass) Teleostei, and to conceive the Dipnoans as representing a 'class'. It should be added that the 'choanae' in Latimeria are exceedingly small openings (cf. MILLOT and ANTHONY, 1966).

[14] In Perennibrachiata such as Siren and Amphiuma, a modification of the choanal valve obtains, permitting opening and closing, i.e. reversal of the stream. Cryptobranchus, which represents a partly metamorphosed type, has lost the choanal valves. All here mentioned forms display a well developed *Jakobson's organ*.

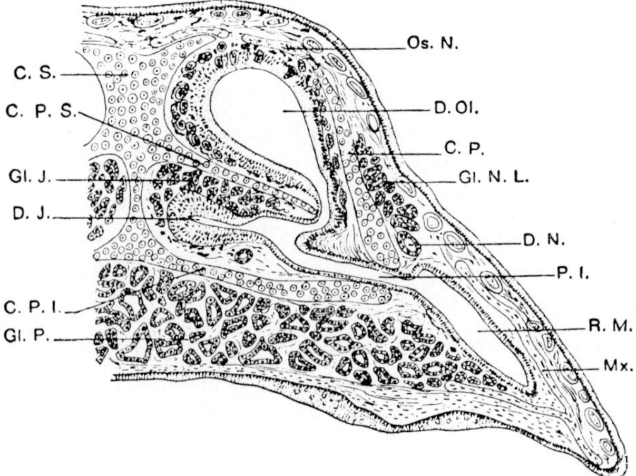

Figure 159 A. Cross-section through the nasal cavity of an anuran Amphibian (Tree Frog) showing *Jacobson's organ* and glands associated with it (after Mihalkovics, from Noble, 1931). C.P.: cartilago paranasalis; C.P.I., C.P.S.: cartilago paraseptalis (inferior, superior); C.S.: cartilago septi; D.J.: recessus medialis nasi *(ductus Jacobsoni)*; D.N.: ductus nasolacrimalis; D.Ol.: ductus olfactorius; Gl.J.: *glandula Jacobsoni;* Gl.N.L.: glandula nasalis lateralis; Gl.P.: glandula palatina; Mx.: maxilla; Os.N.: os nasale; P.I.: pars intermedia; R.M.: recessus maxillaris.

their nasal cavity is lined with folds. Generally speaking, a segregation of the nasal cavity into a dorsal olfactory and into a ventral respiratory subdivision becomes manifested in Tetrapods, the neurosensory elements being finally restricted to the olfactory portion. Concomitantly with these changes, a specialized neurosensory area develops as a separate grooved channel opening out into the roof of the mouth in a slit continuous with the choana. This is the *vomeronasal organ (Jacobson's organ)*[15] which is lined by apparently typical (neurosensory) olfactory mucosa separated from that of the main olfactory region and giving origin to a generally distinctive bundle of fila olfactoria representing the so-called *vomeronasal nerve*, which reaches a circumscribed region of the olfactory bulb frequently well identifiable as *accessory olfactory bulb (bulbus sive bulbulus olfactorius accessorius)*.

---

[15] The vomeronasal organ was apparently discovered about 1811 by Ludwig Levin Jacobson (1783–1843), a Danish anatomist, physician in Copenhagen and later military surgeon in the French and British armies as listed by Triepel (Die anatomischen Namen) and by Eycleshymer (Anatomical Names). Kolmer (1927) spells Jakobson, and quotes the paper: *Description anatomique d'un organe observé dans les Mammifères.* Ann. Mus. Hist. nat., Paris 18: 412 (1811). I have been unable to ascertain the correct spelling by checking the original publication.

A vomeronasal organ does not seem to be present in Fishes (Cyclostomes, Selachians, Ganoids, and Teleosts) although a perhaps homologous region of nasal mucosa has been described at certain Teleostean ontogenetic stages (GAWRILENKO, 1910). It may, however, be present, at a rather rudimentary stage, in Dipnoans (BROMAN, 1939; RUDEBECK, 1944). The organ's position differs somewhat in urodele, anuran and gymnophione Amphibians, being usually better developed in Anurans and particularly Gymnophiones. In Anurans it represents a distinctive blind pouch (cf. Fig. 159 A).

In the Reptilian Sphenodon this organ forms a club-shaped blind pouch opening into the choana. In Lizards and Snakes, the separation from the other nasal structures is complete. The two vomeronasal organs are here distinctive pouches opening independently into a region of the mouth rostral to the choanae. In Chelonians the organ is rather indistinct; it has disappeared in adult Crocodilians. It is likewise absent in adult Birds, although occasionally recognizable as anlage at some ontogenetic stages (ROMANOFF, 1960).

Among Mammals, the vomeronasal organ is absent or at best vestigial in Man[16] and Primates, in many Chiroptera, and in various aquatic forms. In most other adult Mammals, however, it is well developed as an oblong structure, buried in the floor of the nasal cavity toward the midline. In some Rodents, it is said to open into the main nasal cavity, but generally it connects independently with the oral cavity, as in the Reptilian Lizards and Snakes, by paired nasopalatine ducts passing through the secondary palate.

The functional significance of *Jacobson's organ* is not entirely clarified, but GEGENBAUR's suggestion that it serves to detect olfactory aspects of the oral cavity's contents, i.e. that it subserves a 'mouth-smelling', 'food-testing' function, has been widely accepted. Thus, in many Lizards and Snakes, the tongue, cleft into two prongs, and darting in and out of the mouth, does insert its tips into the vomeronasal pockets, thereby bringing adherent substances in contact with the organ's neurosensory epithelium.

As regards Mammals, it was recently claimed that, in the Hamster, sexual behavior of male animals is totally abolished by bilateral extirpation of the olfactory bulbs, while peripheral destruction of the olfactory receptors exclusive of vomeronasal organ causes anosmia but does

[16] The vestigial vomeronasal organ of Man seems to enter into the formation of the incisive canal, connecting the rostral part of the nasal passage with the oral cavity.

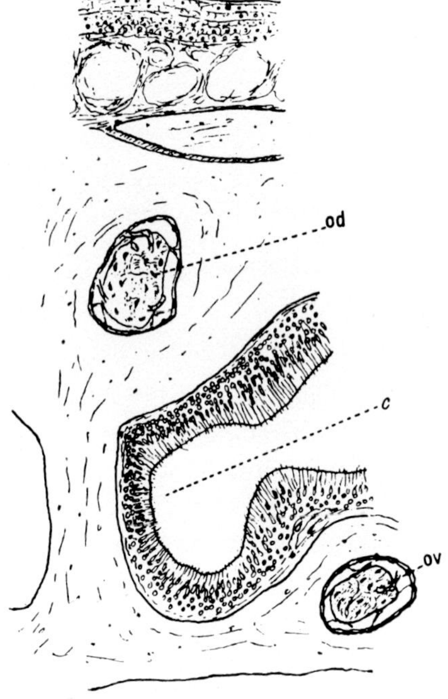

Figure 159B. Cross-section through part of the nasal cavity of the gymnophione Amphibian Ichthyophis (from K., 1922a). c: recess of cavum nasi; od: dorsal olfactory nerve; ov: ventral olfactory nerve.

not impair male Hamster mating behavior. On the other hand, peripheral destruction of the vomeronasal system, whose tubes are here only open anteriorly, near the nares, produces 'severe sexual behavior deficits' of the male animals in about one third of the cases. Combined destruction of the vomeronasal and olfactory organs ('deafferentiation') is said to eliminate copulation in 100 per cent of the affected male animals (Powers and Winans, 1975).

It should also be added that in Gymnophiona, a short protrusible and retractile tentacle, approximately located between nostril and rudimentary eye, is present in apparently all species. Its base becomes associated with the nasal passage and especially with the olfactory area of *Jacobson's organ*. By movements of the tentacles, odorous substances may be brought in contact with said sensory region which Laubmann (1927) described as a 'tactile nose' able to detect food independently of the respiratory stream. The substantial development of the olfactory region in Gymnophiones, which display two olfactory nerves on each side (Fig. 159 B), seems correlated with their burrowing habit and rudimentary eyes (cf. Noble, 1931, 1954). Concerning the topographic position of *Jacobson's organ*, its lateral location in some Amphibians (cf.

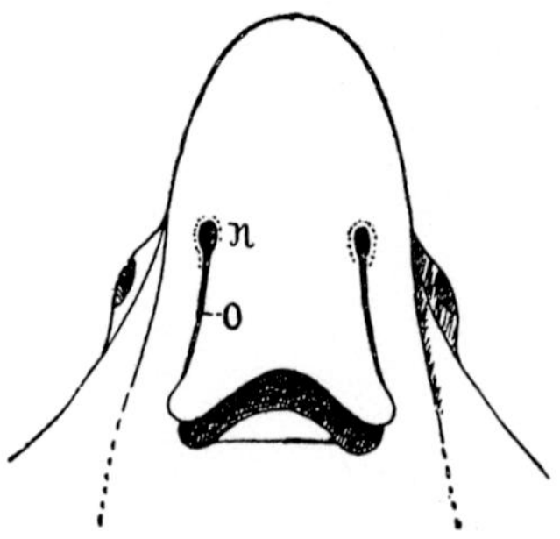

Figure 160 A. Ventral aspect of the head in a Batoid Elasmobranch Skate, showing oronasal groove (after various authors, from NEAL and RAND, 1936). n: nasal opening (naris); o: oronasal groove.

Fig. 162 A, B) and its medial location in others (cf. Fig. 159 A), as well as in Amniota, has led to some difficulties concerning homologization. Since, however, ontogenetic as well as phylogenetic rotations or torsions of the nasal passages appear to be manifested, the discrepancies of position might thus be topologically irrelevant and are e.g. discounted by NOBLE (1931). Nevertheless, LAUBMANN does not evaluate the '*Tastnase*' of Gymnophiones as homologous with *Jacobson's organ*, but rather as a differentiation *sui generis*.

As regards the *nasal passages*, a 'vestibule' becomes developed in Amniota. In Reptiles, moreover, a bony expansion, the turbinal bone or concha, projects from the lateral wall into the nasal cavity, increasing this latter's surface. The secondary palate, which leads to the formation of a distinct nasopharyngeal cavity, is here likewise more or less developed. In the course of these evolutionary changes, the vomer bone no longer remains in the roof of the oral cavity, but then lies as a septum in the nasopharyngeal passage. In Crocodilians, with a long snout and an elongated secondary palate, three pairs of conchae are present, which are also displayed by Birds. These latter, however, whose olfactory organ is rather reduced, can be classified as 'microsmatic'.

The highest development of the Vertebrate olfactory organ is reached in macrosmatic Mammals. The nasal passages and cavities are considerably enlarged, and numerous conchae (maxilloturbinals, ethmoturbinals, respectively endoturbinals and ectoturbinals) are present (cf. Fig. 163). Ventrally, these conchae seem to warm and 'filter' incoming air, dorsally, they enlarge the neurosensory surface. In Eutherian Mammals, extensions of the nasal cavity into adjacent bones commonly occur, forming the pneumatic sinuses, such e.g. as frontal, maxillary, and sphenoid sinuses, respectively ethmoid cells. A lacrimal

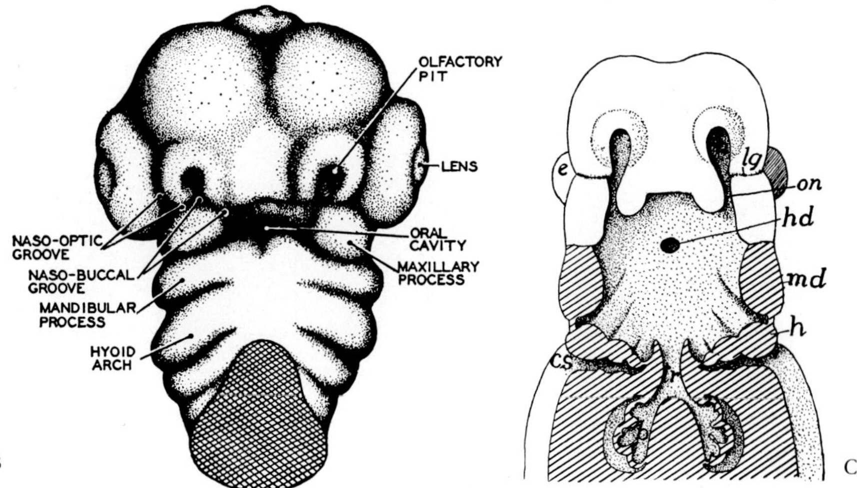

Figure 160B. Avian developmental stage of nasal organ as seen in a Chick embryo of four days (after PATTEN, from NEAL and RAND, 1936).

Figure 160C. Head of Human embryo of about 11.5 mm length, with pharyngeal floor removed, for comparison with Figure 160B (after O. HERTWIG, from NEAL and RAND, 1936). b: lung; cs: cervical sinus; e: eye; h: hyoid arch; hd: hypophysial duct *(Rathke's pouch;* lg: nasolacrimal groove; md: mandible; on: oronasal groove from naris (n in Fig. 160A); tr: trachea. The cut surfaces indicated by oblique hatching.

duct from each orbit generally opens into the nasal passage and may contribute to a moistening of the mucous membrane. It can already be noted in Amphibians (cf. Fig. 159 A).

In microsmatic Mammals, to which Man and other Primates, as well as Cetaceans pertain, the extent of the olfactory epithelium is relatively reduced, and the turbinal bones commonly display a less complex pattern (cf. e.g. Fig. 163 B). In some Cetaceans (various Odontoceti) the olfactory nerve has disappeared at the adult stage, and these forms can be classified as anosmatic.

In Man, the nasal cavity includes a vestibule adjacent to the nares, and lined by an only slightly modified extension of epidermis and corium. The olfactory epithelium is restricted to an area extending from the roof of the cavity about 8 to 10 mm downward on the septum and upon the upper nasal concha. This epithelium displays a poorly developed, ill-defined basement membrane. The lamina propria contains the branched tubulo-alveolar *glands of Bowman*, whose ducts open on the surface. The remainder of the nasal cavity is lined by columnar ciliated epithelium with goblet cells and a well-developed membrane, beneath

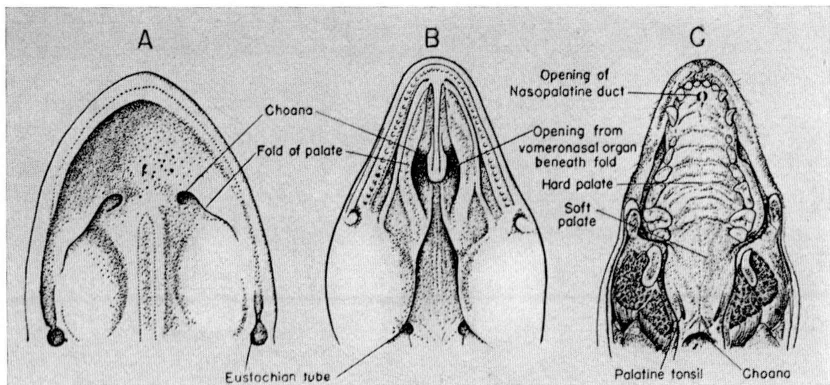

Figure 161. The roof of the mouth in an urodele Amphibian (A), in a Reptilian (B, Lizard), and in a Mammal (C, Dog), particularly showing the position of the choanae (from Romer, 1950).

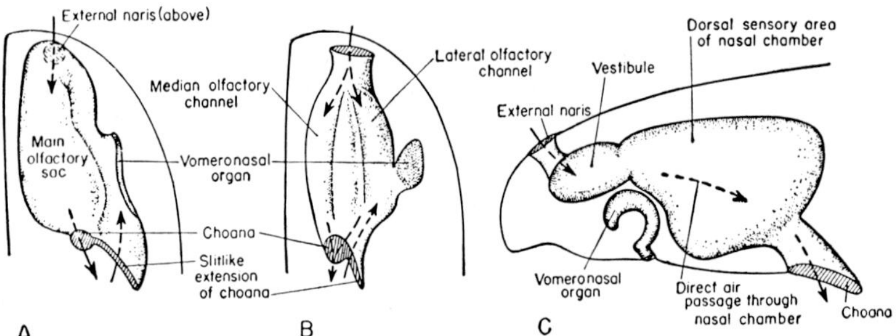

Figure 162. Aspects of the nasal cavity in urodele and anuran Amphibians, and in a lacertilian Reptile (after various authors, from Romer, 1950). A Ventral view of left palatal side in the urodele Triton. B Similar view in the anuran Toad Pipa. C Longitudinal paramedian section through the nasal region in a Lizard. The vomeronasal organ has become separated from the nasal cavity during ontogenesis, and opens independently into oral cavity through a nasopalatine duct. Arrows show main airflow inward, and outward flow toward vomeronasal organ in the two Amphibians.

which are mixed muco-serous glands within the connective tissue. This latter, particularly along the lower conchae, displays dense venous plexuses, providing a semi-cavernous, semi-erectile tissue capable of considerable engorgement, but without the specific structure of typical erectile tissue. Figures 160 to 163 illustrate various ontogenetic and comparative anatomical aspects of the Vertebrate nasal complex, about whose evolution the treatises by Noble (1931, 1954), Neal and

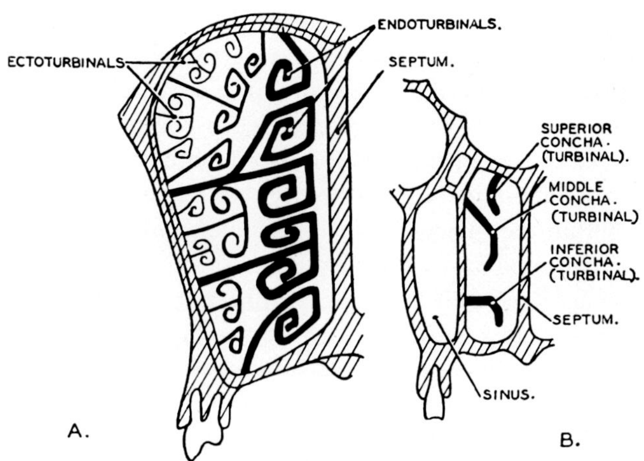

Figure 163. Diagrams of cross-sections through the nasal passages in a macrosmatic ruminant ungulate Mammal (A) and in microsmatic Man (B) showing the different patterns of turbinal bones (from NEAL and RAND, 1936).

RAND (1936) and ROMER (1950) as well as the handbook contribution by KOLMER (1927) contain relevant data and concise summaries.

In all Vertebrates with the apparent exception of Cyclostomes and Birds,[17] a separate rather thin nerve, designated as *nervus terminalis*, is associated with the olfactory mucosa. It was apparently first noticed by FRITSCH[18] (1878) as a supernumerary nerve *(überzähliger Nerv)* in a Selachian. Subsequently PINKUS (1894) described it as '*noch nicht beschriebenen Hirnnerven*' in the Dipnoan Protopterus. LOCY (1905), who introduced the term nervus terminalis, described it in Selachians, and numerous other authors identified that nerve in Osteichthyes (e.g. BROOKOVER, 1910; SHELDON, 1912), Amphibians (e.g. MCKIBBEN, 1911), Reptiles (e.g. MCCOTTER, 1917), and Mammals (e.g. HUBER and GUILD, 1913).

Ontogenetically, the nervus terminalis seems to originate from the

[17] AYERS (1919), however, believes to have identified the nervus terminalis in Petromyzonts, and as regards Birds, ROMANOFF (1960) reports that some of the fibers in the olfactory nerve 'originate from bipolar cells which are lodged within the course of the nerve itself, having previously migrated out of the olfactory placode'. Such fibers in view of the terminal nerve's ontogenetic origin, pointed out further below, might possibly represent an abortive nervus terminalis. Again, according to MILLOT and ANTHONY (1966) this nerve is not present in adult Latimeria.

[18] Cf. the reference to G. FRITSCH in footnote 87, p. 120 of chapter VIII, volume 4, in footnote 142a, p. 201 of chapter XII, and on p. 753 of the present chapter.

nasal placode. Neuroectodermal elements become detached from this latter and are transformed into bipolar nerve cells whose peripheral process remains related to the olfactory mucosa (cf. e.g. FAHRENHOLZ, 1925), while the central process enters the prosencephalon. This origin is comparable to that of the olfactory nerve with the following differences: (1) the cell bodies of the olfactory cells remain as neuroepithelial elements within the olfactory mucosa, while those of the nervus terminalis are clustered into a small ganglion or into several such ganglia within the nerve's branches; (2) the peripheral 'processes' of the olfactory cells are complex 'ciliated' structures (to be dealt with further below), while the peripheral processes of the bipolar cells of nervus terminalis appear to represent 'free endings' within the olfactory mucosa; (3) in contradistinction to the olfactory nerve (fila olfactoria), the central extremity of nervus terminalis does not, as a general rule, reach the olfactory bulb respectively the olfactory glomeruli,[19] but enters more caudally located regions of the prosencephalon.

With regard to phylogenetic interpretations respectively to a comparison with Amphioxus, which latter, despite considerable differences, seems to display some 'ancestral features', the (kat-) homology of Craniote nervus terminalis and Cephalocordate nervus apicis sive terminalis can be upheld on fairly valid although perhaps not rigorously conclusive grounds.[20]

The nervus terminalis is generally non-medullated (HERRICK, 1948) but may, exceptionally, be thinly medullated in Mammals such as the Horse (JOHNSTON, 1913a, 1914). Because of its variable differentiation and relationships to the fiber bundles of nervus olfactorius including nervus vomeronasalis (cf. Fig. 164 B), the course of the nervus terminalis is in many instances difficult to follow. Nevertheless, it seems well established that, at least in a number of different Vertebrates, the ramuli of this nerve collect into a rostral and a caudal branch, of which the former enters the telencephalic paraterminal body (so-called septum), while the caudal branch enters the hypothalamic preoptic region at the

---

[19] It is likely that apparent endings of nervus terminalis fibers in olfactory glomeruli, as e.g. described by HOLMGREN (1920) in some Teleosts, actually pertain to the olfactory nerve (so-called nervus vomeronasalis) with whose fibers the nervus terminalis frequently intermingles. Nervus vomeronasalis and nervus terminalis were occasionally mistaken for each other in the literature (cf. the comments in K., 1927, p. 248, and BECCARI, 1943, p. 519).

[20] Cf. chapter VII, section 3, p. 842 of volume 3/II, and chapter VIII, section 3, p. 60 of volume 4.

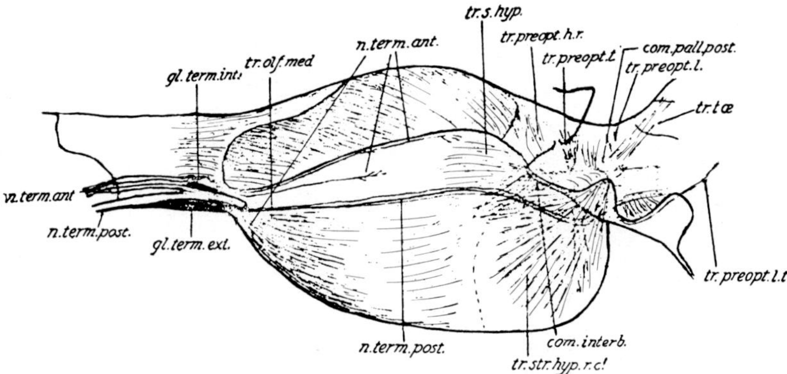

Figure 164 A. Reconstruction of the nervus terminalis and of telencephalic fiber systems in a specimen of Ceratodus, as seen from the medial side (from HOLMGREN and VAN DER HORST, 1925). com.interb.: 'commissura interbulbaris' (part of anterior commissure); com.pall.post.: 'commissura pallii posterior' (part of habenular commissure); gl.term. ext., int.: 'external' respectively 'internal' ganglion of nervus terminalis; tr.preopt.h.r., l., l.t., t.: tractus praeoptico-hypothalamicus rectus, praeoptico-lateralis, praeoptico-longitudinalis tecti, praeoptico-tubercularis; tr.s.hyp.: tractus septo-hypothalamicus; tr.str.hyp.r.c.: tractus 'strio-hypothalamicus rectus et cruciatus'; tr.tae.: tractus taeniae. Other abbreviations self-explanatory.

diencephalic portion of lamina terminalis (cf. e.g. HOLMGREN and VAN DER HORST, 1925; HERRICK, 1948).[20a] This 'posterior root' may join the diencephalic course of the basal forebrain bundle or run in close association with this latter. According to HERRICK (1948) and others, some of its fibers can be traced as far caudad as the mesencephalic nucleus interpeduncularis. As mentioned above, the bipolar cells of the nervus terminalis may be scattered along its bundles or aggregated in a variable number of small ganglia whose distribution differs either individually or *qua* taxonomic forms. Some of the nerve cells in such gan-

[20a] BURR (1924), while investigating the effect of transplantation of an additional nasal placode into the immediate neighborhood of a normal placode in Amblystoma embryos, noted the production of a slender nerve fiber bundle originating in the transplanted placode, being accompanied by ganglion cells, and establishing a connection with the diencephalon in the region of 'dorsal thalamus' ('experimentally produced aberrant cranial nerve'). BURR cautiously suggests that this might be an aberrant nervus terminalis. Since said nerve normally seems to originate from the nasal placode (cf. above p. 483), and may include ganglia, this suggestion appears quite plausible despite the fact that the normal nervus terminalis reaches the basal diencephalon (preoptic region, hypothalamus). The more dorsal diencephalic entrance of the experimentally produced might be caused by a variety of interfering factors.

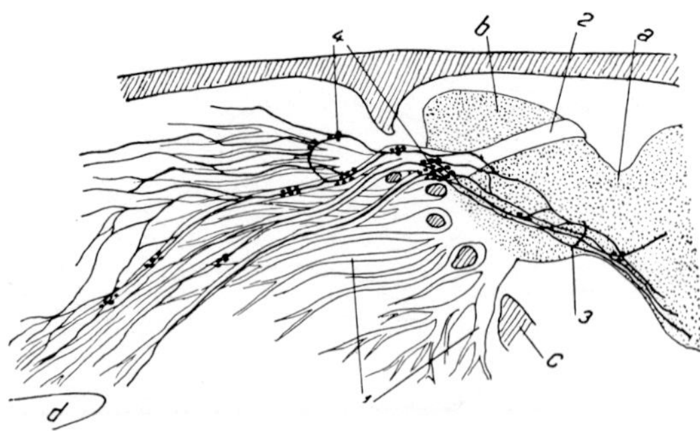

Figure 164B. Semidiagrammatic drawing of telencephalic peripheral nerves in the Rabbit (after HUBER and GUILD, 1913, from BECCARI, 1943). 1: fila olfactoria; 2: nervus vomeronasalis; 3: nervus terminalis; 4: ganglia of n. terminalis; a: olfactory peduncle; b: olfactory bulb; c: lamina cribrosa; d: *Jacobson's organ.* The right olfactory bulb is here shown from the medial side.

glia have been described as multipolar and representing postganglionic autonomic elements. It is, moreover, of interest that in the Cetacean Porpoise Phocaena, which, at the adult state, has completely lost olfactory nerves and bulbs, the nervus terminalis remains relatively well developed.

Nothing certain is known about the functional significance of the nervus terminalis, which, however, doubtless represents an afferent ('sensory') nerve derived from the nasal placode. It is related to the olfactory mucosa respectively membrane, including that of *Jacobson's organ* where this latter is present. It may or may not also be related to the non-olfactory regions (lacking olfactory cells) nasal mucosa.

It has been suggested that the nervus terminalis is an 'unspecific visceral afferent' nerve of the nasal organ, directly discharging into hypothalamus respectively paraterminal grisea (cf. BECCARI, 1943, pp. 519 and 567).[21]

With regard to the postganglionics described as intercalated in the nervus terminalis, it seems likely that they pertain to the cranial autonomic nervous system, having reached the nervus terminalis by way of various anastomoses with trigeminal branches. The trigeminus, in

[21] '*Non sembra inverosimile che possa trattarsi di un nervo arcaico connesso al labbro rostrale della primitiva placca (poi vesicola) archencefalica*'. Being related to the olfactory mucosa of Fishes '*non possiamo disconoscere che esso debba essere legato in qualche modo alla funzione di queste mucose*' (BECCARI, 1943). Yet, as the cited author implies, the presumably chemoreceptive signals mediated by said nerve must differ, *qua* significance and further processing, from those transmitted by the olfactory nerve.

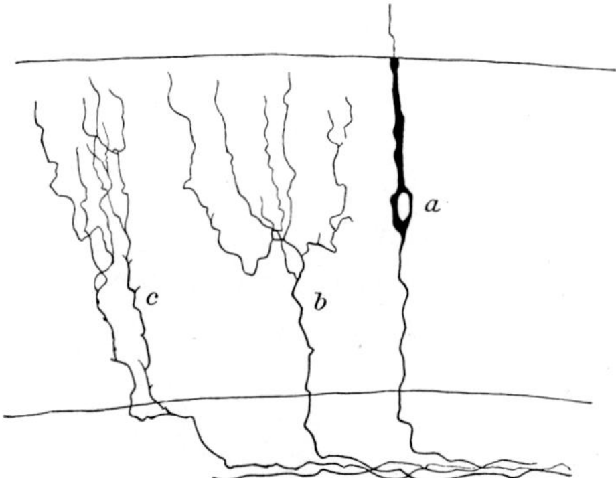

Figure 165. Golgi impregnation of olfactory mucosa in a young Mouse (after v. LENHOSSÉK, from RAUBER-KOPSCH, 1916). a: olfactory cell; b, c: free nerve terminals.

turn, receives autonomic components by way of anastomoses with the facial and other cranial nerves carrying autonomic outflow. In addition, sympathetic postganglionic fibers (of the thoracolumbar autonomic system), originating in the superior cervical ganglion of the sympathetic trunk, may join the nervus terminalis by a diversity of anastomotic ways. Thus, it could be maintained that the nervus terminalis, as such, is entirely *afferent*. Its *efferent fibers* respectively autonomic ganglion cells, found, in addition to the autochthonous bipolar sensory elements, within the small ganglia of the terminal nerve, can reasonably be interpreted as derived from the peripheral vegetative nervous system through this latter's well known numerous and multiform anastomoses (cf. e.g. SIMONETTA, 1932, and BECCARI, 1943). Figure 164 illustrates aspects of the nervus terminalis in a Dipnoan and in a Mammal.

The *nasal mucosa* of Vertebrates is thus related to the following nerves: (1) nervus olfactorius and 'nervus vomeronasalis' (if present), originating from neurosensory epithelial cells included in the olfactory mucosa, and to be dealt with *qua* relevant details further below; (2) nervus terminalis, whose bipolar pericarya, not included in said mucosa, are distributed as discussed in the preceding paragraphs; (3) afferent ('sensory') branchlets of the trigeminal nerve, which, in Tetrapods, where a separation has occurred, end in both 'respiratory' and 'olfactory' regions (cf. Fig. 165);[22] (4) efferent, parasympathetic postganglionic

[22] As regards the peripheral endings of the nervus terminalis, it seems well established that they are particularly present in *Jacobson's organ* but may also be related to the general olfactory region. It remains uncertain, however, if, or to which extent they are present in the respiratory region of Tetrapods.

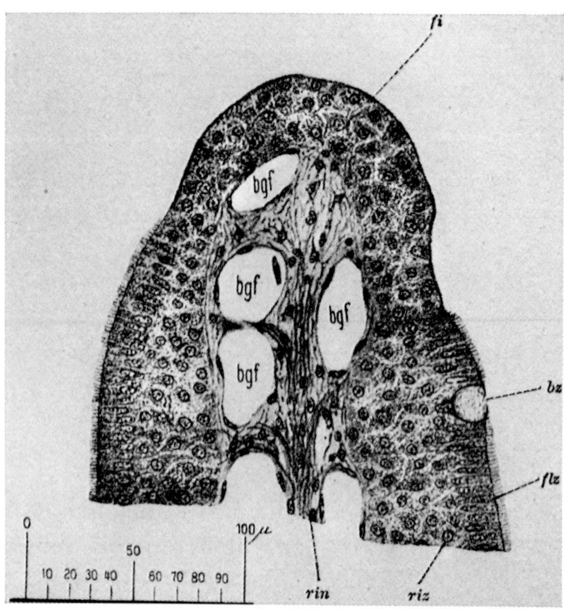

Figure 166 A. Olfactory mucosa of the Selachian Torpedo ocellata (from KRAUSE, 1923). bgf: blood vessels; bz: goblet cell; fi: border of a mucosal fold; flz: ciliated epithelial cells; rin: fila olfactoria; riz: olfactory cell.

fibers of the cranial autonomic system, which may reach the nasal mucosa either by way of nervus terminalis or trigeminus;[23] (5) efferent, postganglionic fibers of the (thoracolumbar) sympathetic, which likewise may reach the mucosa by way of trigeminal or terminalis branches, and, moreover, along the vascular adventitia. The vegetative fibers (4) and (5) are presumably secretory with respect to the nasal glands, and vasomotor (dilating respectively constricting) with respect to the blood vessels. Data on the still rather incompletely understood vegetative innervation of the nasal mucosa are concisely summarized in the treatise by KUNTZ (1947), and, to my knowledge, have not been substantially clarified by subsequent investigations.

As regards the structure of the olfactory mucosa (olfactory surface), this pseudostratified epithelial membrane contains neurosensory (neuronal) olfactory cells, supporting cells, and basal cells, from which lat-

[23] The perikarya of the parasympathetic postganglionics are presumably present not only in 'ganglions' of nervus terminalis but also in (pre)terminal ganglia of the trigeminus (e.g. ganglion sphenopalatinum of Mammals). The sympathetic postganglionic fibers can be assumed to originate in the superior cervical ganglion of the sympathetic trunk.

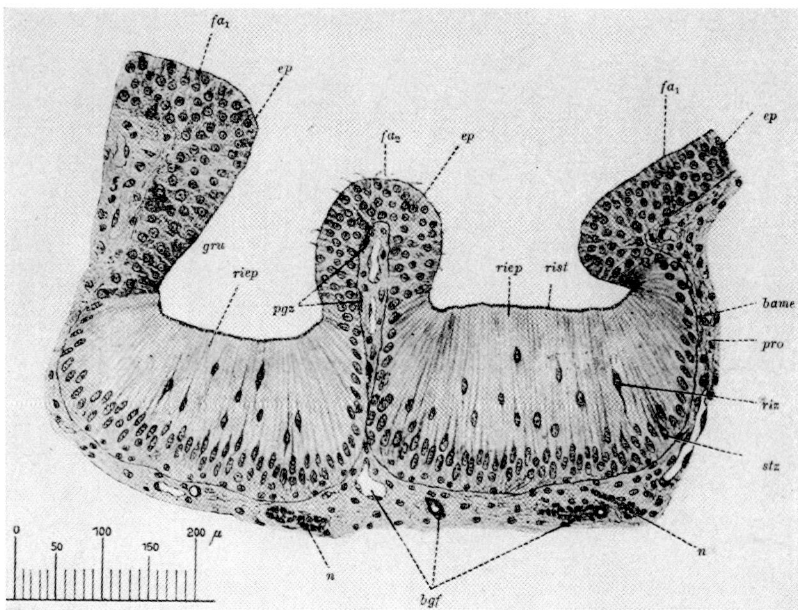

Figure 166 B. Olfactory mucosa of the Teleost Esox, showing olfactory 'buds' or 'pits' separated by non-olfactory mucosa (from KRAUSE, 1923). bame: basal membrane; bgf: blood vessels; ep: ciliated non-olfactory epithelium; fa_1, fa_2: main and accessory folds of nasal mucosa; gru: olfactory 'pit' or 'bud'; n: fila olfactoria; pro: membrana propria; riep: olfactory epithelium; rist: 'dendritic processes' *('Riechstiftchen');* riz: olfactory cell; stz: supporting cells.

ter at least the supporting cells may perhaps be replaced. Where glands, such e.g. as *Bowman's glands* mentioned above, are included in the lamina propria, their ducts pass through the epithelial membrane, which is covered by a thin layer of mucus.[24] In Tetrapods with separate respiratory and olfactory surfaces, the olfactory epithelium is generally thicker than the respiratory one. At least in Mammals, the basement membrane of the olfactory mucosa seems much thinner and less well developed than the rather conspicuous basement membrane of the 'respiratory mucosa'.

In some Mammals including Man, the olfactory surface seems characterized in the fresh state, by a yellowish-brown color contrasting

[24] In aquatic Anamnia, e.g. in aquatic forms of Amphibians, the mucus is said not to form a layer, and the olfactory 'hairs' seem to stand out in the water filling the nasal passages (NOBLE, 1931). The olfactory epithelium of some Fishes, nevertheless, may contain goblet cells (cf. Fig. 166 A).

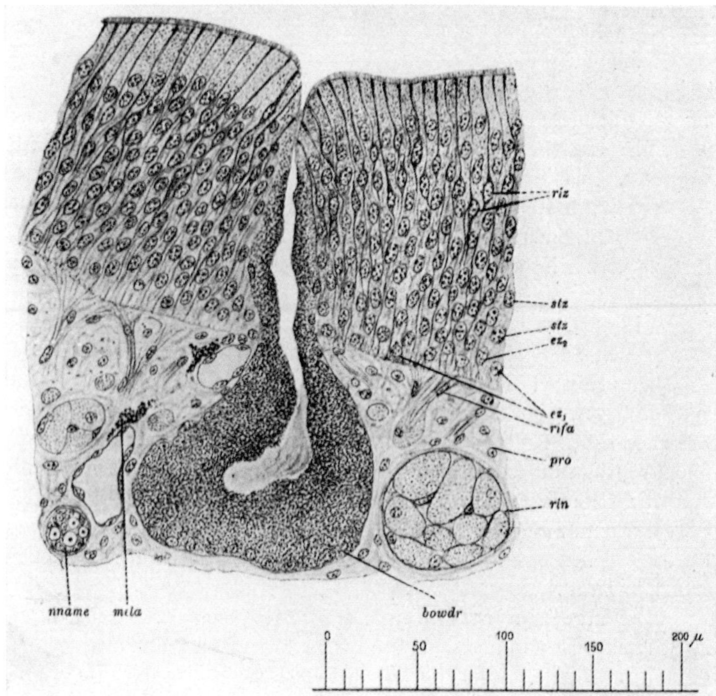

Figure 166 C. Olfactory mucosa of the Frog (from Krause, 1923). bowdr: *Bowman's gland;* ez_1, ez_2: basal cells (*'Ersatzzellen'*, 'replacement cells'); mela: melanophore; nname: branch of nervus medialis nasi (trigemini); rifa: neurite of olfactory cell; rin: filum olfactorium; stz: supporting cells.

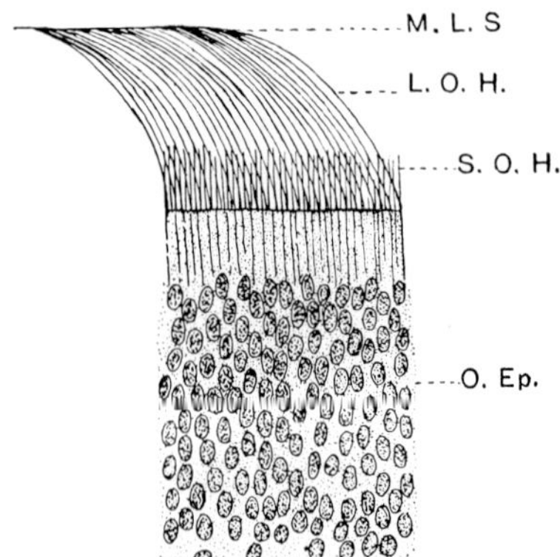

Figure 166 D. Semidiagrammatic drawing illustrating the olfactory epithelium of a Frog (after Hopkins, 1926, from Noble, 1931). L.O.H.: long, non-motile olfactory cilia, which may reach the surface of the mucus; M.L.S.: surface of mucous layer; O.Ep.: olfactory epithelium; S.O.H.: short, motile olfactory cilia.

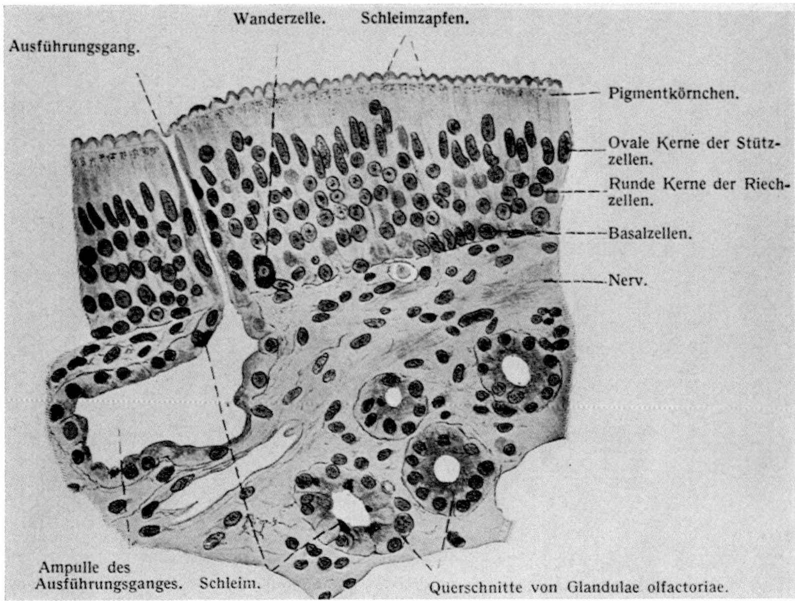

Figure 166E. Section through the human olfactory mucosa stained by *Hansen's hematoxylin*, ×400, red. $^{1}/_{1}$ (from Stöhr and Möllendorff, 1933).

with the reddish color of the respiratory surface.[25] In Man and some other Mammals, an outer zone of oval nuclei pertaining to the sustentacular cells has been distinguished from an inner zone of round nuclei pertaining to olfactory and basal cells. This distinction of sublayers, however, does not seem to be rigorously valid in all cases and for all Vertebrate forms. Figures 166 A–E illustrate overall aspects of the olfactory mucosa in diverse Vertebrates.

It had been assumed that the supporting cells were provided with motile cilia.[26] Observations by means of electron microscopy, however, appear to indicate that, in most Vertebrate forms, the surface of supporting cells is merely provided with microvilli. In Selachians, nevertheless, motile cilia of supporting cells seem to be present (Reese and Brightman, 1970).

---

[25] The 'olfactory pigment' seems to be included in the supporting cells.

[26] Thus, according to Noble (1931) in tadpoles and larval Urodeles these ciliated cells beat rapidly and help to drive a current through the nasal chamber. In view of the more recent data obtained by means of electron microscopy it seems not impossible that such motile cilia might pertain to the olfactory cells.

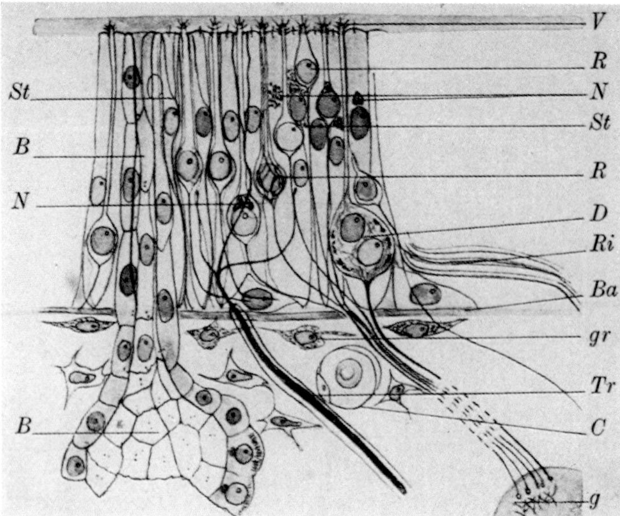

Figure 167 A. Semidiagrammatic sketch of Human olfactory mucosa structure on the basis of high resolution light microscopy (from KOLMER, 1927). B: *Bowman's gland* with duct; Ba: basal membrane; C: capillary; D: large binucleated olfactory neurosensory cell; g: ending of olfactory fila in glomerulus; gr: granulated connective tissue cell below basement membrane; N: fibrillar networks in neurosensory and in supporting cells; R: olfactory cell; Ri: neurites of olfactory cells *('Riechfasern');* St: supporting cells; Tr: fiber of trigeminal nerve, branching in mucosa; V: olfactory vesicle. The figure displays, in an olfactory cell, some of the neurofibrillar structures formerly much emphasized by investigators using certain silver impregnations.

The *olfactory cells* are the first-order neurons of the olfactory system and pertain to the category of neuroepithelial nerve cells.[27] Peripheral neurons of this type are common among Invertebrates, and also occur in Amphioxus. In craniote Vertebrates, the olfactory cells are, as far as is known, the only representatives of this subset of neurons, although four sorts of central neuroepithelial nerve cells occur, namely retinal rods and cones of lateral eyes in all classes, retinal cells in some instances of parietal eyes (cf. chapter XII, section 1 C), and neuronal saccus vasculosus cells in various Anamnia. In this respect, and discounting the fact that the retina represents an intrinsic component of the forebrain, that is, of the central neuraxis, there obtains a modicum of analogy between olfactory surface and retina.

[27] Cf. volume 3/I, chapter V, section 2, p. 90.

Figure 167B. Low power electron micrograph showing dense layer of cilia in mucous layer of Frog olfactory epithelium. Most cilia are cut in cross section (cf. Fig. 167F); small distal ciliary segments in outer, and larger proximal segments in inner layer; olfactory vesicles and microvilly at bottom (after REESE, 1965, from OTTOSON and SHEPHERD, 1967; ×3,000, red. $^9/_{10}$).

The input extremity ('dendrite')[28] of an olfactory cell displays a bulbous swelling, formerly designated as olfactory vesicle on the basis of light-microscopic observations. This 'vesicle', now also called '*Riechkopf*' (e.g. SEIFERT, 1972), contains basal corpuscles or 'centrioles' from which a variable number of cilia (about 8–20) arise.[29] These cilia, in their proximal portion, display the typical 9+2 array of subfibers, while the distal portion may contain only 2. Said cilia, which, at least in some forms, seem to be motile, are generally believed to represent the chemoreceptor transducing structures for the 'olfactory stimuli'. In Tetrapods, the cilia usually appear to bend, such that they form parallel arrays of their distal segments just below the surface of the mucus. In

[28] Now also designated as olfactory rod.

[29] Cf. volume 3/I, chapter V, section 2, pp. 121f., section 3, p. 150, section 6, p. 446, pp. 471f. The presence of basal corpuscles in the 'olfactory vesicles' was already recorded by the older investigators before the introduction of electron microscopy.

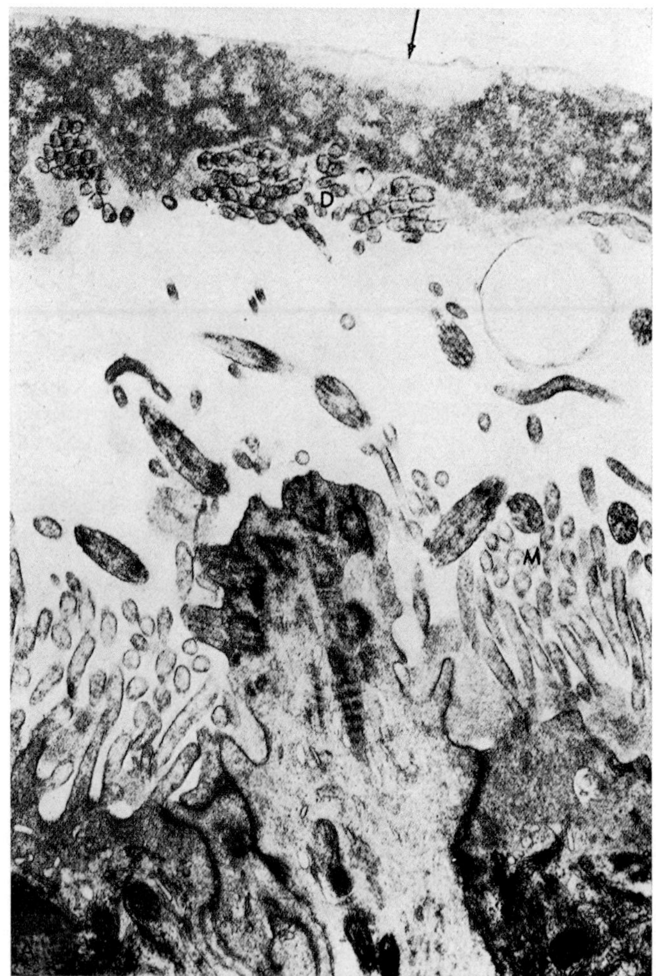

Figure 167C. Olfactory epithelium of the Macaque monkey as seen by means of electron microscopy (from REESE and BRIGHTMAN, 1970). D: distal segments of olfactory cilia clustered parallel to surface of mucus; M: microvilli of supporting cells lacking central subfibers. 'Bulbous end of apical dendrite from an olfactory bipolar cell is flanked by supporting cells'. Electron dense so-called tight junctions, not considered in the present text, but briefly dealt with in section 6, chapter V of volume 3/I, can easily be recognized. Arrow indicates surface of mucus ($\times$23,000, red. $^9/_{10}$).

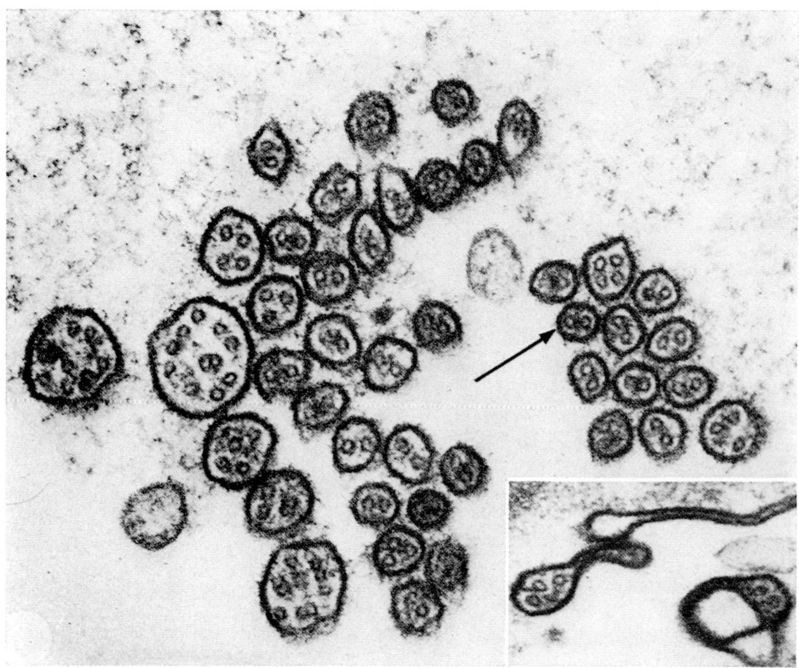

Figure 167 D. Cross-section through a cluster of olfactory cilia in the Macaque monkey as seen by means of electron microscopy (from REESE and BRIGHTMAN, 1970; ×83,000, red. $^{9}/_{10}$). In the distal ciliary segments, the tubular subfibers are reduced to a single pair (arrow). Inset shows distal segments with an included vesicle (lower right) and with redundant cell membranes (upper left).

water breathing Anamnia, however, the cilia stand out in the water. In air breathing Anurans, on the other hand, non-motile long cilia are said to penetrate the mucus, thus becoming exposed to the air, while the short cilia, exhibiting motile activity, remain below the mucus (HOPKINS, 1926; NOBLE, 1931). It should be added that the olfactory cilia of Amphibians, as stressed by NOBLE, are functional in both an air and in a water medium.

Again, a substantial difference from the overall Vertebrate arrangement is said to obtain in (at least some) Selachians, where the olfactory 'tufts' ('bulbs', 'vesicles') originate short villous protrusions extending into the water, thus lacking true olfactory cilia. *Per contra*, the supporting cells are here provided with typical motile cilia of the 9+2 pattern (REESE and BRIGHTMAN, 1970). According to the cited authors, however, in all other Vertebrate classes (Cyclostomes, Osteiichthyes, Am-

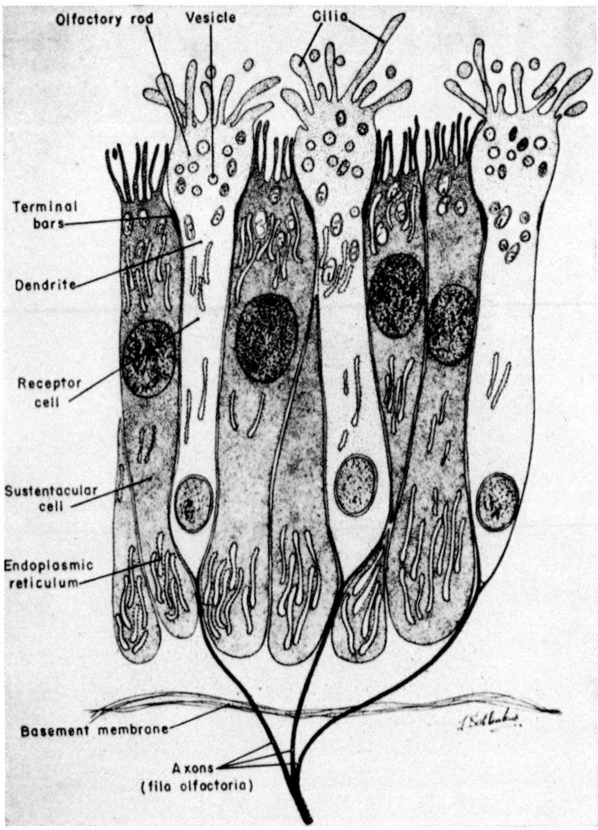

Figure 167E. Semidiagrammatic drawing, based on electron micrographs, of the Vertebrate olfactory epithelium (from De Lorenzo, 1970). The diagram indicates the terminal bars and other cytologic details omitted from the present text.

phibians, Sauropsidans, Mammals), however, true olfactory cilia seem to be displayed by the neurosensory cells. Nevertheless, some less common types of olfactory cells in Teleosts are said to lack cilia, and this lack of cilia is also claimed for 'the vertebrate vomeronasal organ'. As regards this latter in macrosmatic Mammals, however, results of an investigation undertaken by Seifert (1972) seem to disclose that the neurosensory cells of that organ differ from those of the regio olfactoria in certain minor, ultrastructural aspects. Thus, e.g. essentially the olfactory hairs are said to arise in a bundle from the 'dendritic' cell surface which lacks a typical 'vesicle' *(Riechkopf)*.

At the basal end of each olfactory cell, there arises a non-medullated

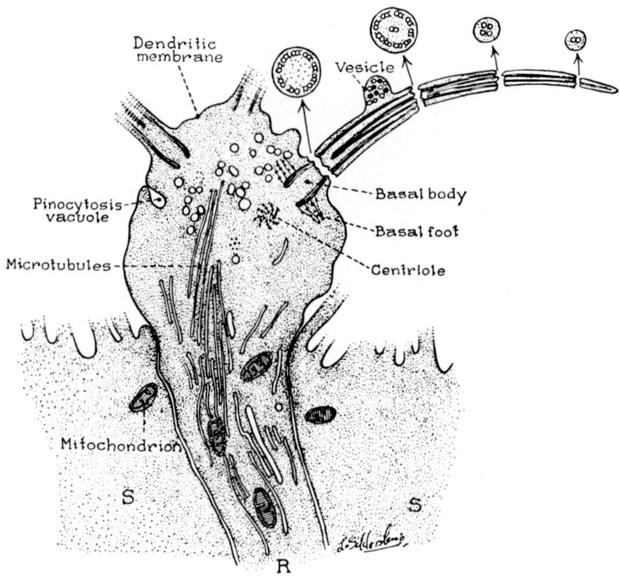

Figure 167F. Schematic drawing showing ultrastructural components of olfactory receptor (from DE LORENZO, 1970).

neurite. These neurites are gathered, below the epithelium, in bundles of roughly more or less 20 fibers, enclosed by *Schwann cells*,[30] these bundles being the fila olfactoria which reach the olfactory bulb. The olfactory cell neurites do not seem to branch along their entire course throughout which they also remain non-medullated. Figures 167 A–G depict various aspects of the olfactory surface and its neurosensory elements.

On the basis of electron microscopy, a peculiar type of presumably 'sensory' elements has been described in the olfactory mucosa of various Vertebrates (DE LORENZO, 1970; ANDRES, quoted by DE LORENZO, 1970). These cells display peripheral microvilli, and are basally connected by a synapse with a presumably 'sensory' (afferent, i.e. input) nerve fiber (cf. Fig. 168). Said cells have been compared to 'gustatory'

---

[30] This general type of enclosure, involving a 'mesaxon', whereby the neurite remains externally to the *Schwann cell* membrane, was discussed and depicted in volume 3/I, chapter V, section 6, pp. 421f., and shown in Figures 252A–C (loc.cit.). The olfactory fila, however, are particularly densely bundled within their *Schwann cell* enclosure as shown by the present Figure 167G.

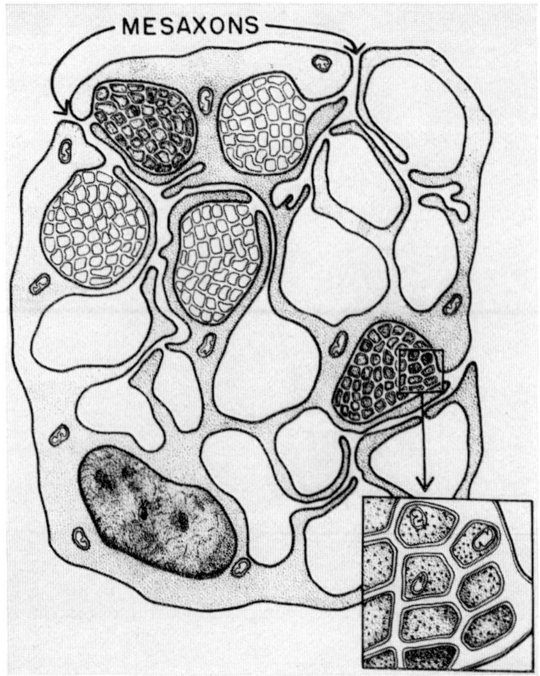

Figure 167G. Schematic drawing showing relationships of *Schwann cell* (lemnoblast) to olfactory neurites in filum olfactorium (from DE LORENZO, 1970).

or 'tactile' cells,[31] but, so far, nothing seems to be known about their functional significance. The synapses could, evidently, be provided by terminals of either nervus trigeminus or nervus terminalis.

Still another question is that concerning *regeneration* of the olfactory mucosa after destruction. Although regeneration of the receptor cells has been denied by many authors, others have reported such regeneration under favorable conditions in Frogs as well as in some Mammals (MOULTON, 1974), either from basal cells or even from elements in the *glands of Bowman*.

Reverting now to the *fila olfactoria*, that is to say, the olfactory 'nerve', their relative length which depends on the variable spatial relationships between olfactory bulb and nasal cavity may be considered first. It is well known that in some Vertebrate forms the bulb lies rather close to the olfactory mucosa, while in others some greater degree of

[31] Comparable cells have been noted in the Mammalian tracheal ('respiratory') epithelium (cf. Fig. 168C).

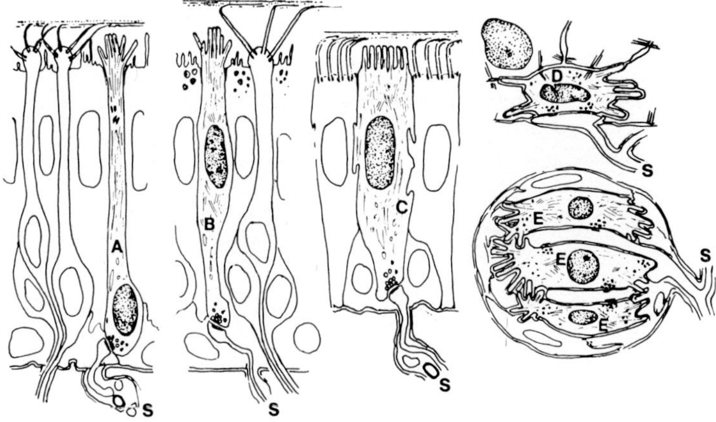

Figure 168. Drawing showing 'previously unreported sensory cells' with microvilli in the olfactory epithelium of the Cat (A), the Turtle (B), and from the mucous membrane of the Rat's trachea (C). Their ultrastructural appearance is said to resemble that of *Merkel's touch cell* (D), and the *Grandry corpuscle* (E). Cells A, B, and C have basal synaptic contacts (S) with the intraepithelial sensory nerve endings (after ANDRES, from DE LORENZO, 1970).

separation obtains. The fila olfactoria may be short and diffusely distributed, or somewhat longer and gathered in a single or (e.g. in Gymnophiona, cf. Fig. 159 B) double pair of olfactory nerves. In Mammals, such as e.g. Man, the bony lamina cribrosa separating nasal and cranial cavities, is perforated by a number of olfactory fila.

The *bulbus olfactorius*, again, may remain in close apposition to the lobus hemisphaericus, as, e.g. in Cyclostomes, some Plagiostomes (e.g. Chimaera), some Osteichthyes, apparently all Amphibians, some Reptiles, and apparently all Birds. Externally, a more or less defined approximately circular groove may delimit the bulb from the hemispheric lobe, or may be rather poorly defined as e.g. in diverse urodele Amphibians, in some of which it becomes barely suggested or is even lacking.

In some Plagiostomes, Osteichthyes, Dipnoans, and Reptiles, however, the olfactory bulbs are separated from the lobus hemisphaericus by an olfactory stalk containing the essentially secondary olfactory tracts and a very variable amount of thereto related grisea.[32] In certain

[32] Cf. the summarizing review by NIEUWENHUYS (1967). It should here be added that in some taxonomically closely related Teleostean forms the olfactory bulb may be apposite to the hemisphere or separated by a rather long stalk. Thus, diverse specimens identified

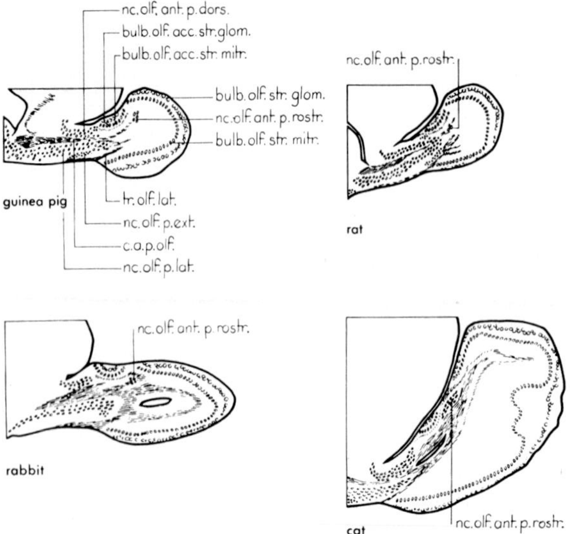

Figure 169. Drawings of sagittal sections through olfactory bulb and its continuity with lobus hemisphaericus in several Mammals, as seen by means of the *Klüver-Barrera stain* (from LOHMAN and LAMMERS, 1967). bulb.olf.: bulbus olfactorius; bulb.olf.acc.: bulbus olfactorius accessorius; c.a.p.olf.: commissura anterior, pars olfactoria; nc.olf. ant.p.dors., ext., lat., rostr.: nucleus olfactorius anterior, pars dorsalis, externa, lateralis, rostralis; str.glom., mitr.: stratum glomerulosum, mitrale; tr.olf.lat.: tractus olfactorius lateralis.

Plagiostomes, the olfactory bulb with its stalk has become displaced in the transverse plane by assuming a pronouncedly lateral position with respect to lobus hemisphaericus.

In Mammals, the olfactory bulb remains closely adjacent to the lobus hemisphaericus or (secondary) cerebral hemisphere, but the development of the neopallium results in a basal bulbar position, the bulb and its immediate neighborhoods becoming separated from the neopallium by a more or less deep groove (Fig. 169). On the basis of the relative development of the olfactory bulb and the olfactory apparatus

---

as the Silurid Corydora paliatus, and studied by MILLER (1940), differed quite noticeably by the relative length of their olfactory stalk. Footnote 4, p. 278 of our paper (K. and NIIMI, 1969) contains, however, an error (memory slip) for which I assume full responsibility. Although the length of the stalk did indeed differ, such that the bulb seemed 'closely connected' with hemispheric lobe, a typical thin stalk was always present, and in none of the specimens was there a 'directly continuous' apposite connection, as erroneously tasted in that footnote.

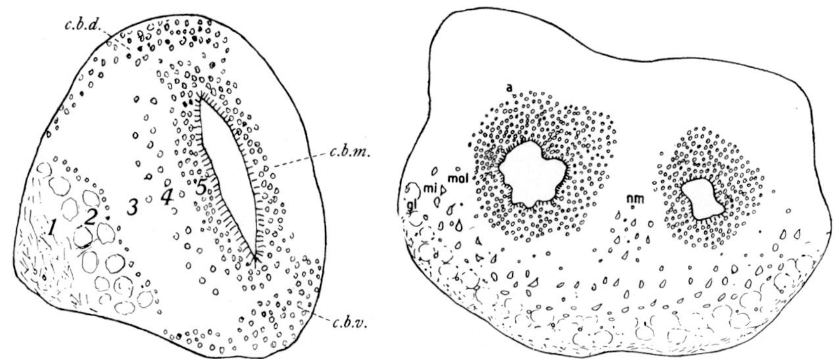

Figure 170 A. Cytoarchitecture of the olfactory bulb in the Urodele Amphibian Salamandra maculosa as seen in cross-section (from K., 1921a). c.b.d.: cellulae bulbares dorsales; c.b.m.: cellulae bulbares mediales; c.b.v.: cellulae bulbares ventrales; 1: fila olfactoria; 2: zona glomerulosa; 3: zona molecularis; 4: zona mitralis; 5: zona granularis. The lateral surface is at left. Figures 170 A–C drawn from *Nissl stain* preparations.

Figure 170 B. Cytoarchitecture of the olfactory bulb in the Anuran Amphibian Rana esculenta (from K., 1921c). a: 'nucleus olfactorius anterior'; fi: 'stratum fibrillare' (fila olfactoria); gl: stratum glomerulosum; mi: zona mitralis; mol: zona molecularis (interna); nm: 'nucleus intermedius' in secondary fusion (concrescence) of antimeric bulbs.

in general, as compared with the development of other sensory systems, Vertebrates in general, and Mammals in particular, can be roughly classified as *macrosmatic* and *microsmatic*. This evaluation involves both morphologic and functional aspects. Generally speaking, many 'lower' Mammals, such as Marsupials and Rodents, as well as 'intermediate' ones, such as Carnivores and many Ungulates, are macrosmatic, while Primates, especially Anthropoids, can be considered microsmatic. A complete reduction, i.e. disappearance of olfactory nerve, bulb, and tracts has been recorded in some Cetaceans. Such forms are thus *anosmatic*.[32a]

In macrosmatic Mammals, the adjacent rhinencephalic structures, laterally delimited from the neopallium by the sulcus rhinalis lateralis, provide a massive region of transition to the hemisphere. In micros-

---

[32a] The presence, in anosmatic Mammals, of grisea commonly pertaining to the olfactory system, such as tuberculum olfactorium, piriform lobe cortex, and hippocampal formation corroborates the comments made on pp. 642f of volume 3/II, stating that morphologically homologous grisea may be 'rhinencephalic' in some forms, but not necessarily in others. As regards the hippocampus, whose 'primordium' in Anamnia is, as a rule, doubtless closely related to olfactory input, it seems most likely, that, even in macrosmatic Mammals, the caudal (postcommissural) hippocampus is not any longer a 'rhinencephalic' griseum (cf. e.g. K. and WIENER-KIRBER, 1962).

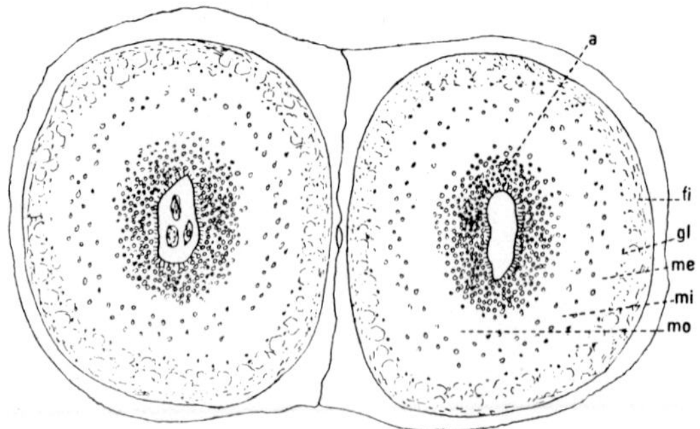

Figure 170C. Cytoarchitecture of olfactory bulb in the Gymnophione Amphibian Ichthyophis glutinosus (from K., 1922a). a: granular layer (originally described as 'nucleus olfactorius anterior'); fi: fila olfactoria; gl: stratum glomerulosum; me: stratum moleculare externum; mi: zona mitralis; mo: stratum moleculare internum.

matic forms, such as Man, the reduction of this transitional region leads to the formation of a rather narrow stalk, the tractus olfactorius (BNA, PNA) interconnecting bulb and hemisphere.

Again, in some Vertebrates, such as a few Osteiichthyes, in anuran Amphibians (cf. Fig. 170 B) and in a few Birds[33] the antimeric olfactory bulbs may be fused to a greater or lesser degree in the midline by secondary concrescence.

The Vertebrate *bulbus olfactorius* is characterized by a laminated cytoarchitectural arrangement which can be designated as formatio bulbaris[34] and displays, from the surface inward, the following fairly distinct layers: (1) stratum of fila olfactoria, (2) stratum of glomeruli olfactorii or zona glomerulosa, (3) zona molecularis sive stratum plexiforme, (4) zona mitralis containing the large mitral cells, and (5) the zona granularis provided by the granule cells.[35] A distinctive internal

[33] The Avian olfactory bulb, because of the hemispheric bend related to the expansion of the epibasal grisea, also assumes a basal position, somewhat comparable to that obtaining in Mammals.

[34] Since the Amphibian olfactory bulb was also occasionally designated as lobus olfactorius, I initially used the terms 'formatio lobaris' (somewhat inconsistently, as a beginner) for either layers 1 to 3 (K., 1921b) or layers 1–6 (K., 1922a) enumerated in the present text, reserving the term 'formatio bulbaris' for layers 1 and 2. BINDEWALD (1914) designated the entire sequence of layers as 'formatio bulbaris', and this terminology appears to be the most appropriate one (cf. also vol. 3/II, chapter VI, section 6, p. 479, and K. and NIIMI, 1969, pp. 279–280).

[35] In my first study of the Gymnophione brain (K., 1922a). I somewhat inconsistently included the granular layer into the 'nucleus olfactorius anterior'. This, of course, is not

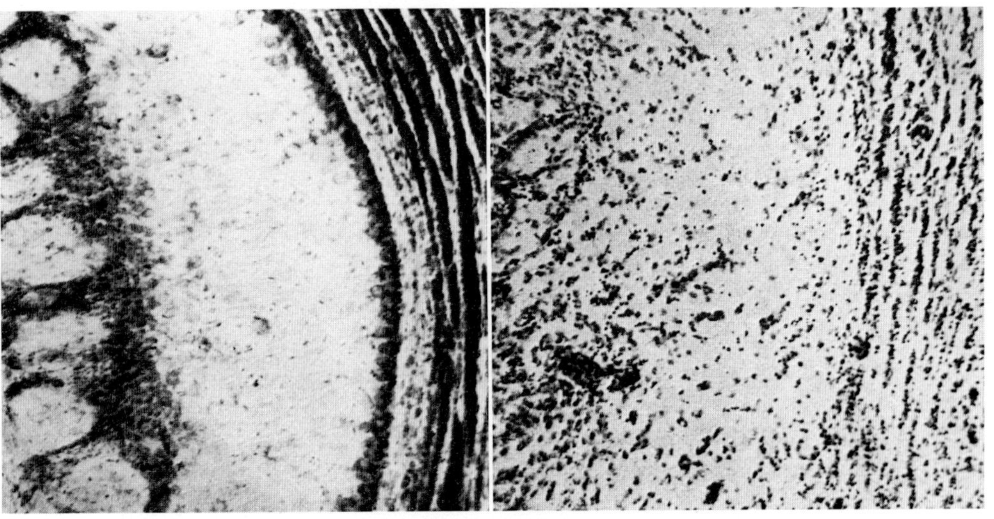

Figure 170 D. Cytoarchitecture of normal olfactory bulb in young Mouse. Glomeruli, regular layer of mitral cells, and dense 'plates' of granule cells are shown in I. In II, cytolysis of mitral cells, edema and pycnosis following intranasal instillation of VSV virus (from K. and Wiener-Kirber, 1962; *Nissl stain,* ×135, red. $^9/_{10}$).

molecular (plexiform) layer may be intercalated between zona mitralis and zona granularis, this latter then becoming layer 6, and said internal plexiform stratum the layer 5. In this respect, a comparison between urodele, anuran, and gymnophione Amphibians displays a 'progressive' differentiation, which can be interpreted in terms of neurobiotaxis (K., 1922a), as shown in Figures 170 A–C.

Again, in the Mammalian olfactory bulb (Fig. 170 D), rather numerous periglomerular neuronal elements are present, which have been designated as an 'external granular layer', although they may also be included as components of the glomerular layer. This tendency is also, rather more faintly, indicated in Sauropsida. Depending upon the taxonomic forms and the development of the olfactory bulb, a medullated layer of secondary and other olfactory as well as extrinsic fibers may be conspicuous internally to the granule cell layer (cf. Fig. 170 E), and, if a ventricular lumen (rhinocele) or its obliterated remnant are present, an ependymal layer can be recognized.

---

justified. Only if such granular cells are not part of the formatio bulbaris *sensu stricto*, they may be included into the vaguely delimited grisea of 'nucleus olfactorius anterior' as defined in section 1, p. 468 of the present chapter (cf. also vol. 3/II, chapter VI, section 6, p. 524, and K. and Niimi, 1969, pp. 279–281).

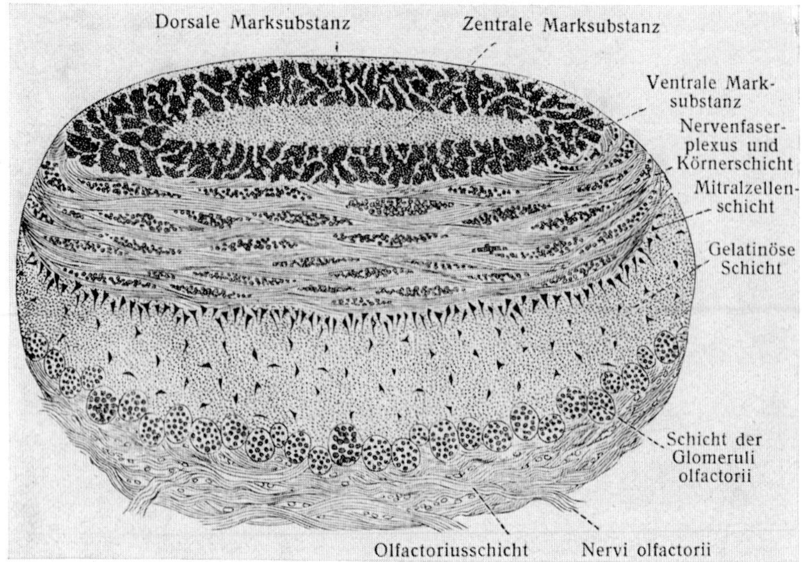

Figure 170 E. Semidiagrammatic cross-section through the human olfactory bulb (after data by HENLE, MEYNERT, and SCHWALBE, from RAUBER-KOPSCH, 1914). '*Gelatinöse Schicht*': external molecular layer; '*Zentrale Marksubstanz*': ependymal layer of obliterated rhinocele. The internal molecular layer, recognizable in Figure 170 D I, is here invaded by the 'plates' of granule cells forming '*Nervenfaserplexus und Körnerschicht*' (cf. also Fig. 172 F).

The *accessory olfactory bulb*, commonly appearing as a slight protrusion in the caudal region of the main olfactory bulb, receives the fila olfactoria of the vomeronasal nerve from *Jacobson's organ.*[36] It is perhaps present in Dipnoans (GERLACH, 1933; RUDEBECK, 1944), and can be found in Amphibians, in many Reptiles, and in many Mammals. In Amphibians the accessory bulb has a ventrolateral position.[36a] In some Mammals (e.g. Guinea pig) its location is likewise lateral, being medial in others (e.g. Rabbit). The cytoarchitecture of the accessory olfactory bulb is essentially similar to that of the main bulb, but commonly appears somewhat less differentiated respectively more 'blurred' (cf. Fig. 171).

---

[36] Since *Jacobson's organ*, respectively a true 'vomeronasal nerve' do not seem to obtain in Cyclostomes, Selachians, and Osteichthyes, a typical accessory bulb is presumably lacking in these forms. It also seems to be absent in Birds.

[36a] In Gymnophiones, the vomeronasal nerve is included within the inferior olfactory nerve.

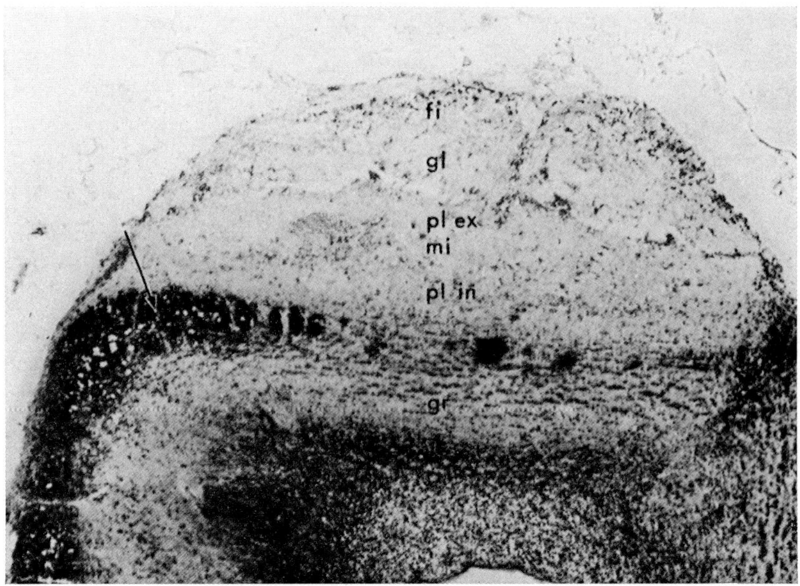

Figure 171. Accessory olfactory bulb of the Rabbit as seen by means of the *Klüver-Barrera stain* (from LOHMAN and LAMMERS, 1967; ×50, red. 9/10). fi: stratum fibrosum; gl: stratum glomerulare; gr: stratum granulare; mi: stratum mitrale; pl ex, in: stratum plexiforme externum respectively internum.

The medial respectively lateral location of the accessory olfactory bulb, like the variable location of the vomeronasal organ mentioned above on p. 478, poses an awkward question concerning the accessory olfactory bulb's homology or analogy in diverse Vertebrate forms. If, however, rotational displacement in the transverse plane is considered to be topologically irrelevant and comparable in this respect to inversion *versus* eversion, homology could be said to obtain. Otherwise, and thus depending on arbitrary evaluation, medial and lateral accessory olfactory bulbs would be analogous, but not homologous configurations.

The complex interneuronal and synaptic relationships obtaining in the olfactory bulb have been investigated by numerous authors since the fundamental studies of GOLGI initiated about 1875 (cf. GOLGI, 1894), and were subsequently particularly expanded by CAJAL (1911 and previous publications).

The *olfactory glomeruli* provide the relevant synapses between the first and second order olfactory neurons, that is to say, between the neurites of the peripheral olfactory cells, and the dendrites of the mitral cells, whose neurites form the olfactory tract respectively tracts. Moreover, at least in Mammals, the glomeruli include dendritic terminals of tufted cells, terminals of fibers pertaining to periglomerular neuronal elements, and presumably axonal respectively apotile or dendritic terminals of undetermined sources. Thus, in addition to the transmission of the olfactory impulse from the periphery, very complex and still insufficiently understood synaptic respectively interneuronal interactions seem to be mediated within the glomeruli.

Each first order olfactory fiber, which apparently does not branch in its course from mucosa to bulb, is assumed to enter only one glomerulus in Mammals and various other Vertebrates, but in some Anamnia[37] it is claimed that such fiber may branch upon reaching the bulb and terminate in two or more widely separated glomeruli.

According to LE GROS CLARK (1951) and others, the dorsal portion of the Mammalian olfactory mucosa projects upon the dorsal part of the bulb, and the ventral portion upon the bulb's ventral part, thus suggesting some sort of spatial distribution.[38]

In the Rabbit, the unilateral number of neurosensory olfactory cells is estimated at about 6 or 5×10^7, the number of glomeruli at 1,900, and that of mitral cells at 45,000 (cf. ALLISON and WARWICK, 1949; BLINKOV and GLEZER, 1968). If, as in Mammals, a primary olfactory neurite ends in only one glomerulus, then each glomerulus would, by a many-one relationship, and assuming an even distribution, receive endings from about 26,000 receptors. Again, each glomerulus (in the Rabbit) is believed to be connected with about 24 to 26 mitral cells and perhaps 68 tufted cells.

Figure 172 A illustrates CAJAL's concept of structural relationships in a Mammalian olfactory bulb. It will be seen that here the single main dendrite (CAJAL's 'primordial dendrite') of a mitral cell ends in a single glomerulus, while so-called 'accessory' or 'secondary dendrites' end in the outer plexiform layer. There is, however, no doubt that in Mammals and other Vertebrates, some mitral cells may also possess several

---

[37] In Acipenser (JOHNSTON, 1898) and Amblystoma (HERRICK, 1924b). Cf. also NIEUWENHUYS (1967).

[38] On the basis of the just quoted observations by HERRICK (cf. footnote 37) this has been questioned as far as Amphibians are concerned. It should, however, be added that LE GROS CLARK, who had at first assumed a diffuse projection of olfactory mucosa upon bulb, subsequently reached the conclusion that there is certainly some degree of regional projection upon the bulb in the Rabbit.

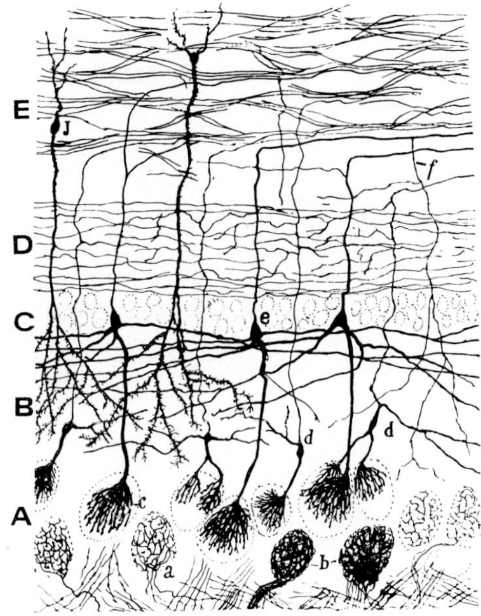

Figure 172 A. Histologic structure of olfactory bulb in a young Cat, as revealed by the *Golgi technique* (from CAJAL, 1911). A: glomerular layer; B: external molecular (plexiform) layer; C: layer of mitral cells; D: internal molecular (plexiform) layer; E: granular and internal fibrous layer; a, b, c: glomeruli; d: tufted cell; e: mitral cell; f: recurrent neurite of mitral cell; j: granule cell. Cf. also Figs. 195B, C, D, p. 288, vol. 3/I.

'primordial dendrites' ending in differing glomeruli. Moreover, since the number of these latter is generally smaller than the number of mitral cells, each glomerulus receives converging dendrites from several mitral cells.[39] Again, VAN GEHUCHTEN (1897, 1906) pointed out that, in microsmatic Amniota, the primordial dendrites of mitral cells tend to be distributed upon several glomeruli, while in macrosmatic forms these dendrites converge upon, and are predominantly restricted to, a single glomerulus (cf. Fig. 172 B). Typical mitral cells are of roughly pyramidal shape, described as comparable to a bishop's, pope's, or Tibetan lama's mitra. In contradistinction to cortical pyramidal cells, the 'primordial dendrite' arises approximately at the center of the base, the neurite originates at the opposite, centralward directed apex, and the 'secondary dendrites' extend from the 'edges' of the base.

---

[39] Thus, in the Rabbit, where about 1,900 glomeruli and approximately 45,000 mitral cells are assumed to be contained in one olfactory bulb, each glomerulus would be connected with more or less 24 mitral cells.

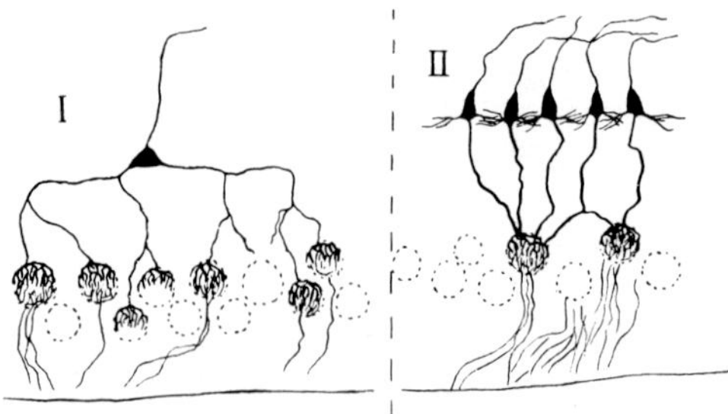

Figure 172B. Relationships of mitral cell dendrites to glomeruli in microsmatic and macrosmatic Amniota (simplified after VAN GEHUCHTEN, 1906, from K., 1927). I Bird (microsmatic). II Dog (macrosmatic Mammal).

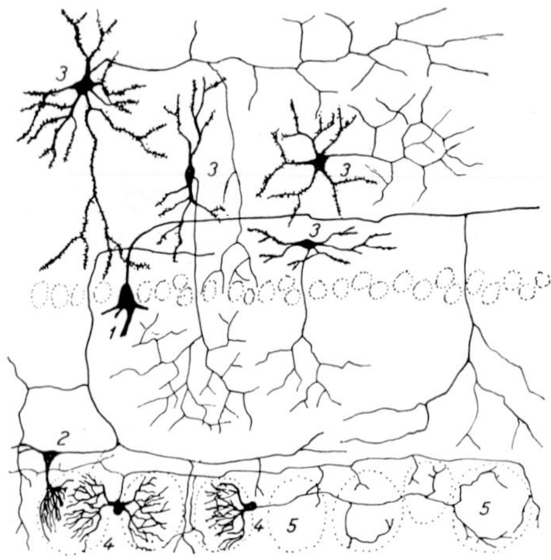

Figure 172C. Diagram showing cells with 'short neurites' in the Mammalian olfactory bulb, drawn on the basis of *Golgi preparations* (after CAJAL, 1911, from BECCARI, 1943). 1: mitral cell; 2: tufted cell *(cellule à panache)*; 3: elements of (internal) granular layer; 4: external granular elements; 5: glomeruli.

Somewhat smaller, and commonly slightly more peripherally located *tufted cells (cellules à panache, Pinselzellen, Büschelzellen)* are likewise present and said to be characteristic for Mammals. Their main peripheral dendrites join glomeruli; others, with less regular origin, are distributed in the outer plexiform layer. The centralward arising neurite joins the olfactory tract. It seems possible that some smaller, displaced mitral cells described in Reptiles, and irregular smaller, but centrally located elements in the bulb of Cyclostomes, Selachians, and Osteichthyes, also connected with the glomeruli, might be somehow comparable to the tufted cells (cf. e.g. NIEUWENHUYS, 1967). The neu-

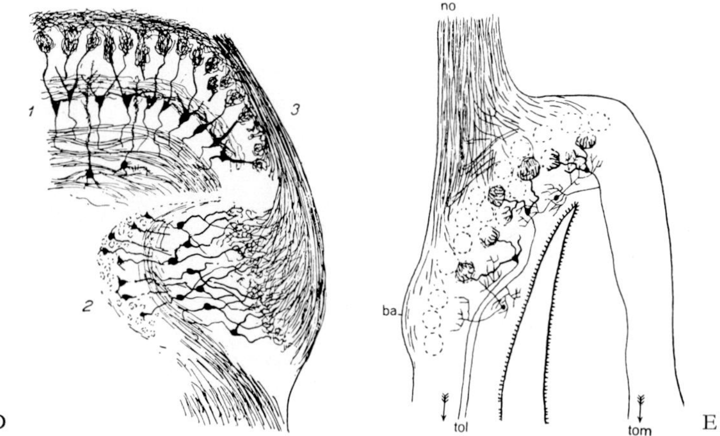

Figure 172 D. Horizontal section through olfactory bulb of a young mouse *(Golgi impregnation)*, showing accessory olfactory bulb (simplified after CAJAL, 1911, from BECCARI, 1943). 1: medial portion of main olfactory bulb; 2: accessory olfactory bulb in caudomedial location; 3: vomeronasal nerve (cf. also Fig. 171).

Figure 172 E. Bulbus olfactorius of an Urodele Amphibian *(Golgi method)*, showing the caudolaterally located accessory bulb (modified and simplified after HERRICK, 1924a, as depicted for Amblystoma, from K., 1927). ba: bulbus accessorius; no: nervus olfactorius; tol: tractus olfactorius lateralis; tom: tractus olfactorius medialis.

rites of mitral and tufted cells give off recurrent collaterals branching within the external and internal plexiform layers; other, deep collaterals, terminate within the granular layer.

The small *interglomerular cells*, also occasionally designated as *external granule cells*, are described as Golgi II cells with short axons, interconnecting several glomeruli as shown in Figure 172 C. Such cells are also present in Anamnia, as e.g. described by HERRICK (1924b) in Amblystoma.

The *(internal) granular cells* seem to include elements of diverse insufficiently distinguishable sorts, likened by CAJAL (1911) to amacrine cells, since they seem to lack a typical neurite. Such cells may have 'unpolarized' processes (neurotendrils or apotiles, cf. vol. 3/I, p. 79, 547). As a rule, at least one long process extends toward the mitral cells and the external plexiform layer (Figs. 172 A, 173 A). There are, moreover, either in the granular layer, or scattered in the internal plexiform layer, various sorts of multipolar cells[39a] with short axons spreading

[39a] These include the so-called *Blanes cells* described by that author (Rev. trimestr. micrograf. *3:* 99–127, 1897, as quoted by CAJAL, 1911).

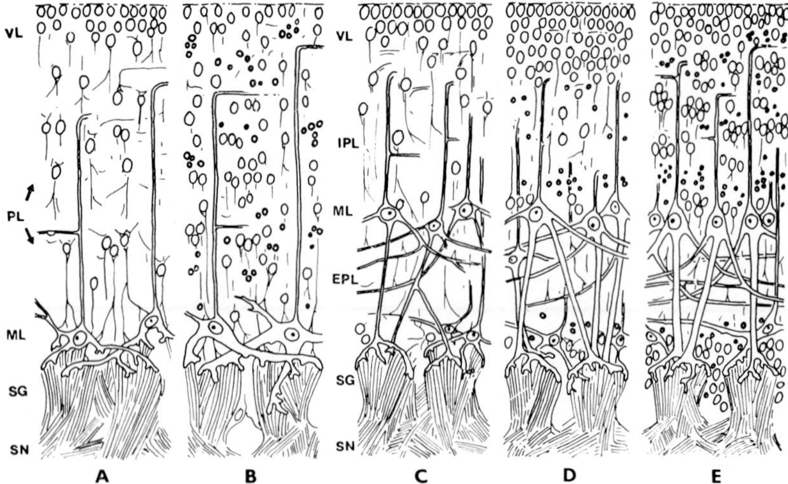

Figure 172 F. Semidiagrammatic representation of olfactory bulb structure in various Vertebrates (from ANDRES, 1970). A: Lamprey (Cyclostome); B: Selachian; C: Amphibian; D: Reptile; E: Mammal; EPL: external plexiform layer; IPL: internal plexiform layer; ML: mitral cell layer; PL: plexiform layer; SG: glomerular layer; SN: olfactory nerve layer; VL: periventricular layer. Granule cells are indicated by oval outlines of their nuclei, and their axons by fine lines. Small bold circles and dots are outlines of medullated fibers. Perikarya, dendrites and axons of mitral and tufted cells are outlined.

toward the external layers or remaining within their stratum, as shown in Figure 172 C.

As regards the (internal) granule cells, both CAJAL (1911) and HERRICK (1924b), pointing out the 'amakrine' or 'unpolarized' appearance of these elements, had assumed that they might 're-inforce' the olfactory impulses by discharging back upon the mitral cells. In my own early studies (K., 1927, 1928), I became convinced that at least some granule cells were reciprocally connected with a given mitral cell, such that a discharge from the latter would return to its cell of origin, thus establishing a true self-reexciting circuit with positive feedback, roughly comparable to the generator circuit in dynamoelectric machines (Fig. 173 A). Neither CAJAL nor HERRICK had explicitly pointed out closed loops of this type.[40] Assuming positive feedback, this would

[40] HERRICK (1924b, p. 396) explicitly states that the 'unpolarized' granule cells are capable of transmitting nervous excitations through their processes 'in either direction, thus reinforcing inactive mitral cells through collateral discharges from those that have been activated directly from the periphery'. Cf. also the comments in volume 1, chapter I, section 1, pp. 11–13.

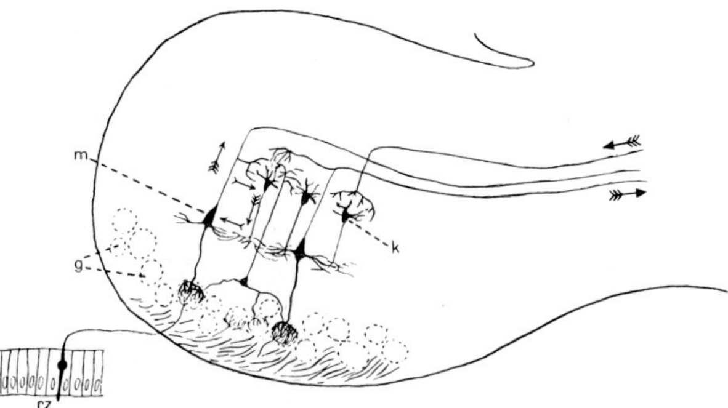

Figure 173 A. Diagram of Mammalian olfactory bulb indicating closed circuit, which might provide positive or negative feedback, between a particular mitral and a particular granule cell (from K., 1927). g: glomeruli; k: granule cell; m: mitral cell; rz: olfactory neurosensory cell (1st olfactory neuron). A tufted cell, not labelled, can be identified.

amount to both 'facilitation' and 'after-discharge' activities. Since various sorts of granule cells with a diversity of additional connections seem to obtain, I later also assumed a negative feedback, i.e. an inhibiting action of some such closed loops (K., 1957, p. 230). This, of course, would be concordant with the fairly rapid 'adaptation' to odors, which despite the continuous presence of 'smelling' substances, are no longer perceived after a certain time interval.[41] This 'adaptation' appears to be quite different from 'fatigue', since any new 'smell' is again immediately recorded.[42]

Although this concept of direct i.e. reciprocal closed neuronal circuits[43] was, with respect to the olfactory bulb, subsequently ignored, it has been re-discovered by recent authors (cf. e.g. LOHMAN and LAMMERS, 1967; OTTOSON and SHEPHERD, 1967; ANDRES, 1970; NICOLL,

[41] Cf. the comments on 'adaptation' in volume 3/I, chapter V, section 8, p. 593.

[42] It remains, of course, uncertain whether a given 'adapted' mitral cell is reactivated by the subsequent new odor. However, in view of the above-mentioned very small number of mitral cells (4.5×10^4) in comparison to olfactory cells (5×10^7), this seems rather likely. 'Adaptation', moreover, overlaps with what is called 'habituation', e.g. suppression, in 'consciousness', of the perception of a continuous noise. Such 'habituation', however, might be presumed to be cortical rather than peripheral, i.e. within the primary sensory centers.

[43] RANSON (1943) claimed that such self-reexciting circuits were an entirely new concept introduced by himself and HINSEY in 1930 (cf. also Fig. 5, p. 26, chapter VIII, vol. 4). Yet, a few years before 1943 (in 1937), I had pointed out to RANSON, in a discussion remark to one of his papers at a meeting of the American Association of Anatomists, my 1927 and 1928 formulations of said circuits.

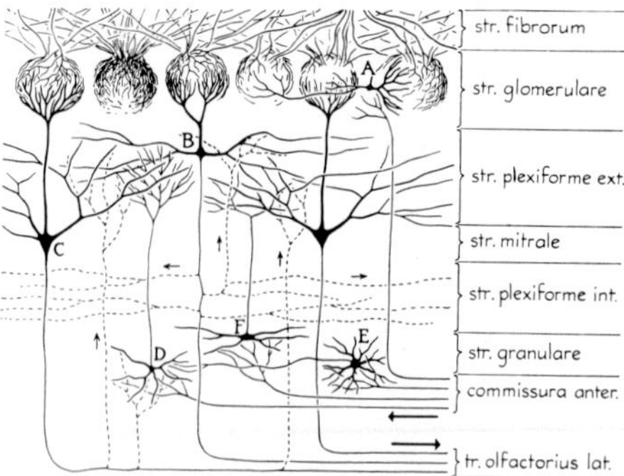

Figure 173 B. Diagram of Mammalian olfactory bulb pathways in a recent interpretation (modified after CAJAL, from LOHMAN and LAMMERS, 1967). A: external glomerular or periglomerular cell; B: tufted cells; C: mitral cells; D: internal granular cells; E: *Golgi* and *Blanes cells;* F: *Cajal cells* and horizontal fusiform cells. The collaterals of the mitral and tufted cell axons are shown by interrupted lines. It can easily be seen that the reverberating circuit C–D closely corresponds to that indicated in Figure 173 A; a collateral of C, moreover, suggest a single neuron feedback circuit.

1971), as shown in Figures 173 B and C. Without specifically implicating granule cells, NICOLL (1971) states that undefined 'secondary neurons of the olfactory bulb can be excited monosynaptically after activation of neighboring secondary neurons by antidromic and orthodromic volleys. Recurrent collaterals of secondary neurons are proposed to synapse with other secondary neurons, thus forming a direct recurrent excitatory pathway. Such a positive feedback system could strengthen the original input signal'.

Other authors (OTTOSON and SHEPHERD, 1967; REESE and BRIGHTMAN 1970; ANDRES, 1970) assume an inhibitory action of granule cells upon mitral cells. If this would not only involve a direct closed circuit, but also represent a 'lateral' inhibition of other neighboring mitral cells, in this latter case a sharpened discrimination between odors might thereby perhaps be provided. Be that as it may, a 'modulation' of mitral cells by the granule cell processes is now generally assumed (ANDRES, 1970).

Again, LANDIS *et al.* (1974) assume here reciprocal 'dendro-dendritic synapses' between granule cell 'dendrites' and mitral or tufted cell dendrites in the external plexiform layer of the olfactory bulb. These authors (cf. also vol. 4, chapter X, p. 662) claim to distinguish differences in membrane structure between excitatory and inhibitory com-

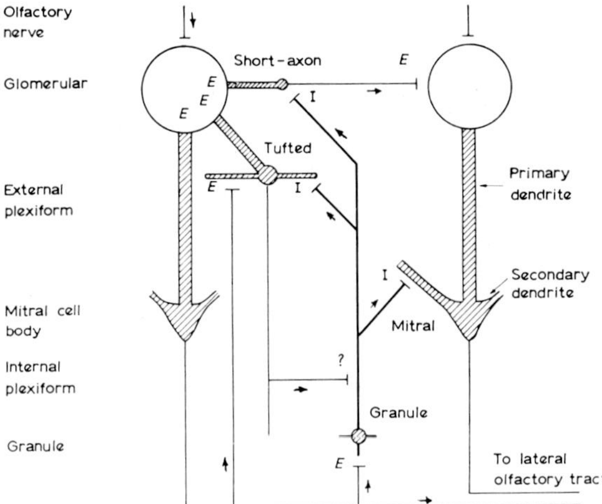

Figure 173 C. Schematic 'diagram of the main functional pathways and connections in mammalian olfactory bulb' according to recent interpretations (after SHEPHERD, 1963, from OTTOSON and SHEPHERD, 1967). E, I: presumed excitatory respectively inhibitory connections. Histological layers of the bulb indicated at left.

ponents of 'reciprocal' synapses on the basis of the so-called freeze-fracture technique in electron microscopy.

Mutatis mutandis, the complex structure of the *olfactory bulb* can be compared with that of the *retina*. The olfactory cells, which, however, are located in the mucosa and are not derived from the neuraxis, but from a placode being, moreover, not 'inverted', but 'adverse' receptors,[44] represent here the retinal rods and cones. The mitral and tufted cells, from which the olfactory tracts arise, are comparable to the retinal ganglion cells. But in contradistinction to retina, whose optic nerve and tract output is provided by neurites of third order neurons, the olfactory tract arises from neurons of the second order. The intercalated second neurons of the retina, that is to say the bipolar cells, are thus not represented by analogous elements in the olfactory bulb. This latter, nevertheless, includes a number of diverse internuncials with complex circuits[45] whose significance, on the whole, is still less well understood than those of the likewise incompletely comprehended retinal

[44] Cf. p. 30 and footnote 20 of section 1B, chapter XII.

[45] With regard to said circuitry, REESE and BRIGHTMAN (1970), ANDRES (1970), OTTOSON and SHEPHERD (1967) discuss in detail the inconclusive hypotheses based on ultrastructural (electron microscopic) data. Such findings concern 'morphological polarity'

ones. To some extent, the granule cells can evidently be compared with retinal amacrine cells.

As regards the ontogenetic differentiation of the Mammalian olfactory bulb, a recent investigation, purporting to provide further details, particularly concerning mitral cells, and based on light as well as electron microscopy, was undertaken by HINDS (1972).

By and large, the structure of the olfactory bulb does not seem to display very substantial differences in the entire Vertebrate series from Cyclostomes to Mammals. Secondary differences, e.g. less 'typical' mitral cells, nondescript appearance of the various internuncials, less well defined cytoarchitectural lamination in some Anamnia and in Birds, relationships of one mitral cell to one or to several glomeruli in diverse forms, lack of typical 'tufted cells' in submammalian forms, etc., are discussed, particularly with reference to the available data displayed in *Golgi preparations*, by NIEUWENHUYS (1967).

Comparative electron microscopic studies are said to reveal an identical ultrastructural organization of the olfactory bulb in all classes of Vertebrates, including Cyclostomes (ANDRES, 1970). Figure 172 F illustrates the cited author's concept of olfactory bulb structures in the Vertebrate series. Detailed ultrastructural observations on the Cat's olfactory bulb are described by WILLEY (1973).

Proceeding now with a discussion of the *secondary olfactory grisea* (or 'centers') receiving second order olfactory neurites and not pertaining to the olfactory bulb *sensu strictiori*, the vaguely defined '*nucleus olfactorius anterior*' should be mentioned first.[46] It may contribute third order fibers to the olfactory tracts, besides being an accessory griseum related to the circuitry of olfactory bulb.

Discounting the fibers carrying input into the olfactory bulb and to be pointed out further below, the *olfactory tracts* represent the output channel of the olfactory bulbs, essentially provided by fibers of the second order, namely by neurites of mitral and tufted cells, to which may be added the above-mentioned third order fibers from nucleus olfactorius anterior.

In some Anamnia (e.g. Petromyzonts, Dipnoans, and Amphibians) the olfactory channels connecting olfactory bulb and lobus hemisphaericus do not form well delimitable 'tracts' but several rather dif-

---

of synaptic contacts (including those between dendrites as so-called 'dendro-dendritic synapses'), neuro-neuronal maculae adherentes, behavior of the fuzz, 'dense-cored synaptic vesicles', 'membrane complexes', etc.

[46] Cf. section 1, p. 468, and footnote 35 of the present section.

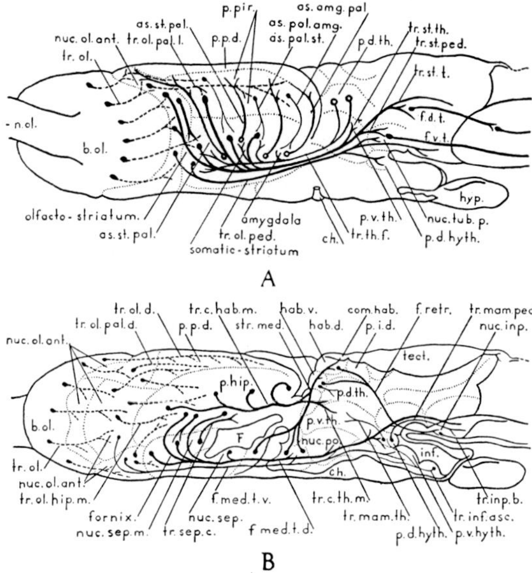

Figure 174 A, B. Lateral (A) and medial (B) aspects of the forebrain in the urodele Amphibian Necturus, indicating olfactory pathways and connections in HERRICK's interpretation (from HERRICK, 1948). as.amg.pal.: 'amygdalo-pallial association'; as. pal.st.: 'pallio-striate association'; as.st.pal.: 'strio-pallial association'; b.ol.: olfactory bulb; ch.: optic chiasma; com.hab.: habenular commissure; F: interventricular foramen; f.d.t.: 'fasciculi dorsales tegmenti'; f.med.t.d.(v.): medial forebrain bundle, dorsal (and ventral) subdivision; f.retr.: fasciculus retroflexus; f.v.t.: 'fasciculi ventrales tegmenti'; hab.d.(v.): dorsal (and ventral) habenular griseum; hyp.: hypophysial complex; inf.: 'infundibulum'; n.ol.: nervus olfactorius; nuc.inp.: interpeduncular nucleus; nuc.ol.ant.: n. olfactorius anterior; nuc.po.: preoptic grisea; nuc.sep. (med.): grisea of 'septum' (respectively 'n. septi medialis'); nuc.tub.p.: n. tuberculi posterioris; p.d.(v.) hyth.: pars dorsalis (respectively ventralis) hypothalami; p.d.th.: pars dorsalis thalami; p.hip.: 'primordium hippocampi'; p.p.d.: 'primordium pallii dorsalis'; p.pir.: 'primordium piriforme' ('n. olfactorius dorsolateralis'); p.v.th.: pars ventralis thalami; str.med.: stria medullaris thalami; tect.: tectum mesencephali; tr.c.hab.m.: 'tr. cortico-thalamicus medialis'; tr.inf.asc.: 'tr. infundibularis ascendens'; tr.inp.b.: 'tr. interpedunculo-bulbaris'; tr.mam.ped.: tr. mamillo-peduncularis; tr.mam.th.: 'tr. mamillo-thalamicus; tr.ol.: tr. olfactorius; tr.ol.d.: tr. olfactorius dorsalis; tr.ol.pal.l.: tr. olfactorius dorsolateralis; tr.ol.ped.: tr. olfacto peduncularis; tr.st.th.: 'tr. strio-thalamicus' (part of lateral forebrain bundle); tr.st.t.: tr. strio-tegmentalis (part of lateral forebrain bundle. Other designations self-explanatory. Olfactory tracts proper are drawn in broken lines.

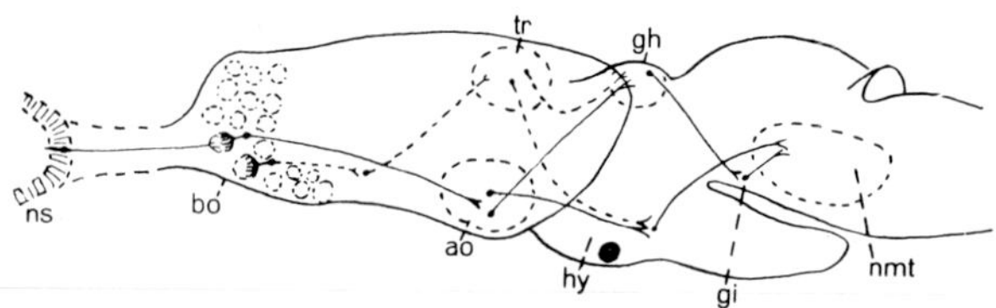

Figure 174C. Simplified sketch of diverse olfactory connections generally characteristic for Vertebrates, as displayed in an Amphibian brain (from K., 1927). ao: area olfactoria posterior (area ventrolateralis posterior); bo: bulbus olfactorius; gh: ganglion habenulae; gi: nucleus interpeduncularis; hy: hypothalamus; nmt: nucleus motorius tegmenti; ns: olfactory mucosa; tr: tertiary olfactory center ('primordium hippocampi'). Within the telencephalon, medial channels are in broken lines, lateral ones in solid lines.

fuse 'olfactory radiations' (K., 1921a). In Myxine, whose lobus hemisphaericus displays a conspicuous lamination, a superficial and a deep olfactory radiation ('tract') are displayed (cf. Fig. 184 B). In most Vertebrates, however, a lateral and a medial olfactory tract can be distinguished, although a variable basal intermediate 'olfactory radiation' may still be displayed, e.g. reaching the tuberculum olfactorium and neighboring basal grisea of Mammals.

In various Selachians and Osteiichthyes including Latimeria, as well as in diverse Reptiles, the olfactory tracts are included in often rather long and thin stalks (peduncles). Again, in certain instances, a part of said stalk's wall may be membranous, perhaps representing an extension of the telencephalic roof plate.

In all Anamnia, the entire extent of the lobus hemisphaericus seems to be reached by second, respectively third or higher order olfactory fibers, thus representing, as it were, in this sense a 'rhinencephalon'.[47] Certain pallial (apparently also some basal) regions, however, receive, in addition, more or less substantial non-olfactory sensory input by way of the basal forebrain bundle, and the olfactory input into such regions, although still present, may be substantially diminished. The pallial zone D_3 (so-called primordium hippocampi) in perhaps all Anamnia seems to be an essentially tertiary olfactory griseum ('center'), receiving third or perhaps even some higher order olfactory fibers, being, nevertheless, also connected with the regions obtaining non-olfactory sensory input (Figs. 174A–C).

[47] Cf. the comments on the definition of 'rhinencephalon' on pp. 642f., section 6, chapter VI of volume 3/II, and also the author's remarks in the 'General discussion on the terminology of the rhinencephalon' included in Bargmann and Schadé (1963, pp. 242–244).

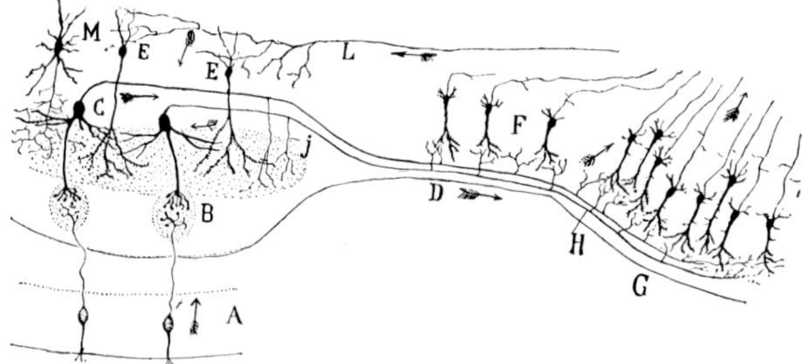

Figure 174 D. Diagram illustrating a simplified concept of some Mammalian olfactory connections as elaborated by Cajal (after CAJAL, from RAUBER-KOPSCH (1914). A: olfactory mucosa; B: olfactory glomeruli; C: mitral cells; D: tractus olfactorius lateralis; E: granule cells (it will be seen that CAJAL discounted here the direct bulbar feedback mechanisms); F, G: pyramidal elements of anterior piriform lobe cortex; H: collaterals of tractus olfactorius lateralis; j: collaterals of mitral cells (suggesting indirect bulbar feedback mechanisms); L: bulbopetal fibers; M: *Golgi-type cell* of bulb.

Tertiary, perhaps also to some extent secondary olfactory input seems to reach the hypothalamus, particularly the preoptic region, thalamic grisea (especially thalamus ventralis) and the habenular grisea, as pointed out in various sections of chapter XII. In Myxinoids, according to JANSEN (1930), a direct (secondary) olfactory projection to the mesencephalic tegmentum may be present. Generally speaking, some olfactory input of undefined order reaching hypothalamus and tegmentum seems to obtain in all Vertebrates. The tegmentum, of course, receives also olfactory input by way of fasciculus retroflexus from the epithalamus, which latter also provides such input for tectum mesencephali. Indirect olfactory impulses presumably also reach the Vertebrate cerebellum.

In Reptiles, the grisea particularly receiving olfactory input are basal cortex, parts of nucleus basilateralis inferior, paraterminal grisea[48] (n. basimedialis inferior and superior), and apparently nucleus epibasalis caudalis. In the pallium, the cortex lateralis, homologous to most of the cortex of Mammalian piriform lobe, can be considered a secondary olfactory area, the cortex medialis ('hippocampal cortex') presumably being still, at least to a substantial degree, a tertiary olfactory area.

---

[48] According to NIEUWENHUYS (1967) experimental studies may be interpreted as showing that direct bulbo-septal projections are 'absent in reptiles and mammals'. Be that as it may, at least relayed olfactory projections are, in my opinion, very definitely present.

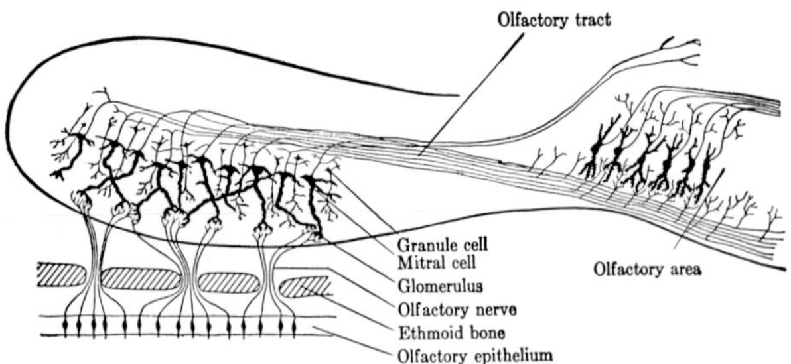

Figure 174E. HERRICK's adaptation of CAJAL's diagram (from HERRICK, 1931).

In Birds, olfactory input seems to reach the poorly developed basal cortex (area ventralis anterior and ventrolateralis posterior), parts of the nucleus basalis, and the rostral portion of the lateral corticoid field homologous to Mammalian prepiriform cortex, probably also parts of the paraterminal complex. Further olfactory connections cannot be excluded, but remain problematic, although undefined indirect connections of some functional significance may be assumed.

In Mammals, the output from olfactory bulb and nucleus olfactorius anterior is gathered (1) in the commonly fairly massive lateral olfactory tract, (2) in a diffuse and rather nondescript intermediate olfactory radiation[49] directly reaching the tuberculum olfactorium *sive* area ventralis anterior with its basal olfactory cortex, and (3) in a generally much less developed medial olfactory tract reaching paraterminal grisea and precommissural hippocampal formation (Figs. 174 D–F).

As regards the Mammalian tuberculum olfactorium, it should here be added that BECCARI (1943, and various previous communications) minimizes its olfactory input, although admitting that some secondary olfactory fibers might reach it. It is indeed likely that said tuberculum has diverse non-olfactory connections (e.g. presumably those of EDINGER's *oral sense* depicted in Figure 237 E), but despite BECCARI's reservations, which I do not entirely share, I would still include basal cortex respectively tuberculum olfactorium into the 'rhinencephalon'. Evidently, in microsmatic, and particularly in anosmatic Mammals, the tuberculum olfactorium will assume functions differing from those in macrosmatic forms.

The *lateral olfactory tract*, also designated as *stria olfactoria lateralis*,

[49] This includes both the *radiazione olfattiva profonda* and the *strie accessorie* of BECCARI (1943).

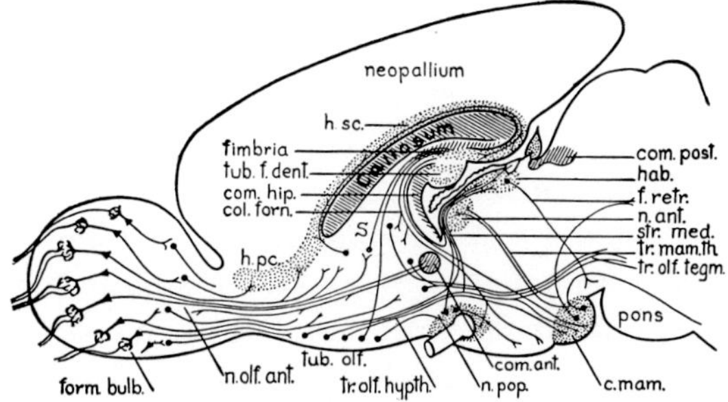

Figure 174F. HERRICK's concept of main olfactory channels in the Rat's brain, omitting the lateral, but emphasizing intermediate olfactory radiation and medial olfactory tract (from HERRICK, 1931). c.mam.: mammillary body; col.forn.: column of fornix; com. ant.: commissura anterior; com.hip.: hippocampal commissure; com.post.: posterior commissure; form.bulb.: bulbar formation; f.retr.: fasciculus retroflexus; hab.: habenulae; h.pc.: precommissural hippocampus; h.sc.: supracommissural hippocampus; n. ant.: anterior thalamic grisea; n.olf.ant.: nucleus olfactorius anterior; n.pop.: preoptic grisea; S: 'septum' (paraterminal grisea); str.med.: stria medullaris thalami; tr.mam.th.: mammillo-thalamic tract; tr.olf.hypth.: 'tractus olfacto-hypothalamicus'; tr.olf.tegm.: olfacto-tegmental tract; tub.f.dent.: postcommissural hippocampus; tub.olf.: tuberculum olfactorium (area ventralis anterior, cortex basalis). It will be seen that HERRICK still included the postcommissural hippocampus in his olfactory system ('rhinencephalon').

runs within the sulcus endorhinalis but may also in part expand over the molecular layer of the anterior portion of piriform lobe, whose dorsolateral limit is the sulcus rhinalis lateralis approximately indicating the boundary between piriform lobe cortex and insular neocortex. The lateral olfactory tract supplies thus cortical grisea of the piriform lobe, besides giving off fibers to the tuberculum olfactorium (cortex basalis, area ventralis anterior) as shown in Figures 174 E, F. A recent investigation by PRICE and SPRICH (1975) indicated, in the Rat, an average of about 42,000 axons in the lateral olfactory tract at rostral levels, and of about 32,000 axons near the tracts caudal termination.

Caudobasally, the tract ends within the amygdaloid complex, perhaps essentially within the cortical amygdaloid nucleus. The so-called nucleus tractus olfactorii lateralis represents a poorly delimitable cell population at the entrance of said tract into cortical amygdaloid nucleus and can be evaluated as merely a rostromedial neighborhood of this latter. A recent report by SCALIA and WINANS (1975) claims that in several Mammals (Opossum, Rat, Rabbit) the posteromedial portion of the cortical amygdaloid nucleus is related to vomeronasal input, while the olfactory input reaches the posterolateral cortical amygdaloid nucleus. Another report, by BROADWELL (1975), claims, on the basis of autoradiographic experimental findings, that the accessory olfactory bulb connects with portions of the amygdala which do not receive input from the main olfactory bulb. This is interpreted as 'evidence for the existence of two distinct and separate olfactory systems'.

There is little doubt that the anterior, and probably that a part of the posterior piriform lobe[50] pertain to the Mammalian rhinencephalon. Into this latter the basal cortex of tuberculum olfactorium and the cortical amygdaloid nucleus are likewise included. The cortex of tuberculum olfactorium with its islets of Calleja, and presumably basal parts of nucleus caudatus ('fundus striati') also receive olfactory input through the nondescript intermediate olfactory radiation. The *diagonal band of Broca* is a poorly understood fiber system connecting the amygdaloid complex with paraterminal and perhaps other grisea, running in a superficial location more or less transversely through caudal neighborhoods of the paraterminal region. It most probably includes fibers related to the rhinencephalic cortical amygdaloid nucleus (so-called periamygdalar cortex).

The less well defined *medial olfactory tract*, running toward the paraterminal grisea and the precommissural hippocampal formation includes the system of fibers designated, in agreement with HERRICK (1924a), as *fascio olfattivo dorso-mediale* by BECCARI (1943).[51] Again, there is little doubt that the paraterminal grisea and the rostral hippocampal formation,[52] on the basis of their olfactory input, can be considered part of the 'rhinencephalon'. Although the entire Mammalian hippocampal formation was formerly interpreted as an olfactory center and thus as pertaining to the rhinencephalon (cf. e.g. CAJAL, 1911; VILLIGER, 1920; HERRICK, 1931, etc.), this view can no longer be upheld (cf. PAPEZ, 1937; BRODAL, 1947a; K., 1954, 1957). The rostral portion of the hippocampal formation, nevertheless, seems to remain closely associated with the olfactory system, and some rather indirect olfactory input into the other parts of the hippocampal formation can-

[50] It is doubtful to which extent the parahippocampal cortex included in the caudal portion of piriform lobe (cf. e.g. Fig. 316B, p. 581, vol. 3/II) should be evaluated as part of the 'rhinencephalon'.

[51] This fascicle is believed to be formed by tertiary fibers arising in nucleus olfactorius anterior: '*insieme alla stria olfattiva mediale, raggiunge l'ippocampo rostrale, arrivando fino l'ippocampo dorsale*' (BECCARI, 1943).

[52] The morphologic significance of these grisea, together with a suitable terminology, was dealt with in chapter V, section 6 of volume 3/II.

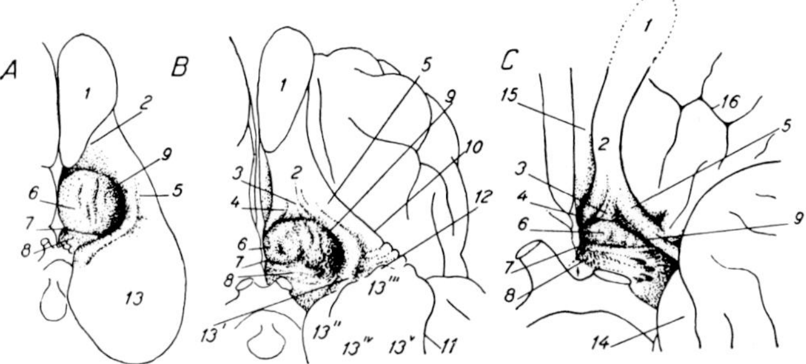

Figure 175 A–C. Basal aspect of the telencephalon in three Mammals, showing macroscopic features of the 'rhinencephalon' (from BECCARI, 1943). A Marsupial (Didelphys). B Carnivore (Tiger). C Primate (Troglodytes niger, Orang). 1: olfactory bulb; 2: olfactory stalk; 3: olfactory trigone; 4: medial olfactory gyrus respectively tract; 5: lateral olfactory tract respectively gyrus (anterior portion of piriform lobe); 6: tuberculum olfactorium; 7: '*solco diagonale*' (separating tuberculum olfactorium from less protruding basal griseal neighborhood); 8: '*spazio parolfattorio basale*' (extratubercular basal griseum); 9: sulcus endorhinalis; 10: sulcus rhinalis lateralis; 11: sulcus rhinalis posterior *(fessura rinica posteriore)*; 12: '*solco rinico trasverso*' (between portions of anterior piriform lobe); 13: piriform lobe; 13^{I}: 'gyrus intermedius' (transition of lateral olfactory tract to cortical amygdaloid nucleus); 13^{II}: gyrus semilunaris (cortical n. amygdalae); 13^{III}: gyrus ambiens (end of anterior piriform lobe); 13^{IV}: rostrobasal end of hippocampus; 13^{V}: gyrus sagittalis lateralis (posterior piriform lobe); 14: gyrus hippocampi (parahippocampal cortex); 15: gyrus rectus; 16: orbital sulci.

not be entirely excluded. Figures 174 to 175 illustrate diverse assumed aspects of olfactory connections in the Amphibian and Mammalian brain. In this latter, and depending upon taxonomical relationships, there obtains considerable variations in the morphologic expansion of the 'rhinencephalic configurations', as characterized by macrosmatic, microsmatic, and anosmatic aspects of telencephalic development. The morphologic behavior of hippocampal formation, essentially unaffected by these variations, rather clearly suggests its independence from the so-called rhinencephalon. Much the same could be said about the non-cortical components of the amygdaloid complex, and, to at least a certain degree, about the paraterminal grisea.

With respect to microsmatic Man, Figure 175 D illustrates the author's concept of what could be designated as the human 'rhinencephalon', which, concomitantly with the expansion of neocortex and non-olfactory telencephalic grisea, is crowded into a relatively small basal region of the hemisphere. It consists of (1) bulbus and tractus olfactorius with trigonum olfactorium, (2) substantia perforata anterior

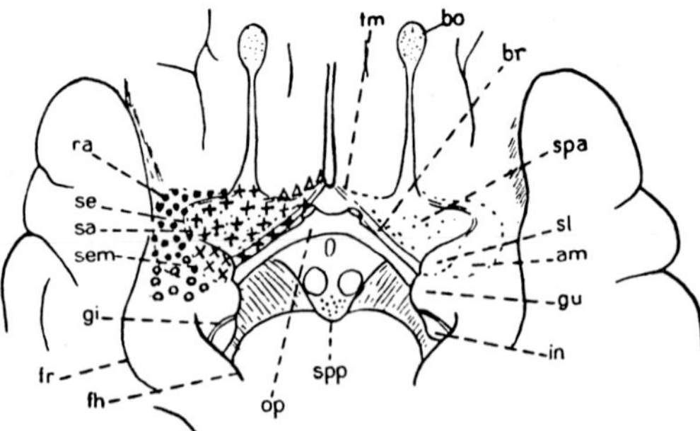

Figure 175 D. Semidiagrammatic view of human brain showing location of so-called rhinencephalon (modified after RETZIUS and after K., 1927, from K., 1957). am: gyrus ambiens; bo: bulbus olfactorius; br: Broca's diagonal band; fh: fissura hippocampi; fr: fissura rhinica; gi: *band of Giacomini* (end of gyrus dentatus); gu: gyrus uncinatus (end of cornu ammonis); in: gyrus intralimbicus (additional part of cornu ammonis, in, gi, gu representing the true 'uncus hippocampi'); op: optic chiasma; ra: sulcus rhinalis anterior (s. rhinalis lateralis); se: sulcus endorhinalis; sem: sulcus semiannularis; sl: gyrus (semi) lunaris (cortical n. amygdalae sive area ventralis posterolateralis); spa: substantia perforata anterior; spp: substantia perforata posterior; tm: tractus olfactorius medialis; coarse black dots: cortex anterior lobi piriformis; circles: cortex posterior (better: intermedius) lobi piriformis; triangles: precommissural hippocampus; + markings: tuberculum olfactorium and diagonal band (area ventralis anterior); × markings: cortical nucleus amygdalae. The parahippocampal cortex adjacent to precommissural hippocampus (rostrally) and to cortex intermedius lobi piriformis (caudally) is left unmarked. The poorly defined sulcus sa is the sulcus rhinalis arcuatus of macrosmatic Mammals. Cf. Fig. 56 A, vol. 5/II.

(tuberculum olfactorium, cortex olfactoria), (3) lateral olfactory stria with adjacent rudimentary 'prepiriform cortex' (anterior cortex of piriform lobe), (4) 'gyrus ambiens' with (intermediate) piriform lobe cortex, (5) 'gyrus semilunaris' (cortical amygdaloid nucleus), (6) components of *Broca's diagonal band* with at least in part preterminal portion of paraterminal body ('gyrus subcallosus') and to some extent septum gangliosum (intraterminal portion of paraterminal body), (7) stria olfactoria medialis with precommissural hippocampus and perhaps adjacent parts of parahippocampal cortex *(Broca's area parolfactoria)*. This, of course, does not exclude the probability that olfactory input, mediated by secondary and tertiary grisea included in the enumerated 'rhinencephalic' configurations reach undefined additional telencephalic grisea, e.g. particularly 'fundus striati' and parahippocampal cortex caudally adjacent to anterior and intermediate piriform lobe cortex. Nor is the probability excluded that some 'rhinencephalic' grisea, such e.g. as the paraterminal ones, have additional functional significance ('vegetative' or 'emotional').

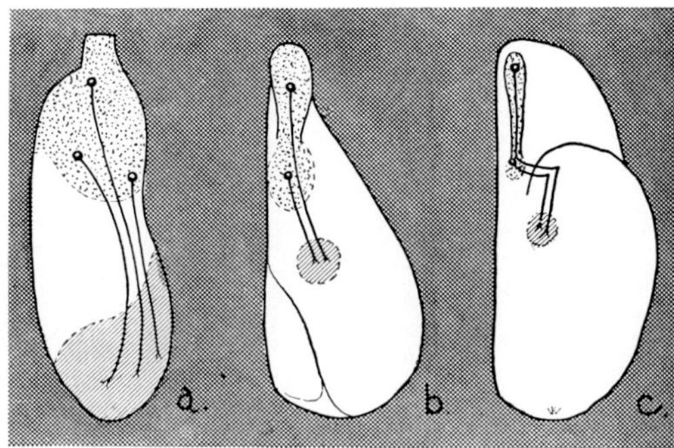

Figure 175E. Simplified diagram indicating evolution of the lateral olfactory tract ('stria') as seen in basal aspect (from KRIEG, 1966). a: Urodele Amphibium; b: Rat; c: Man. Olfactory bulb and 'anterior olfactory area' stippled, hatching indicates 'amygdala' (area ventralateralis posterior, doubtless exaggerated in a). It will be seen that KRIEG omits here most of basal cortex (partly included in his stippled anterior 'olfactory area') and the entire piriform lobe.

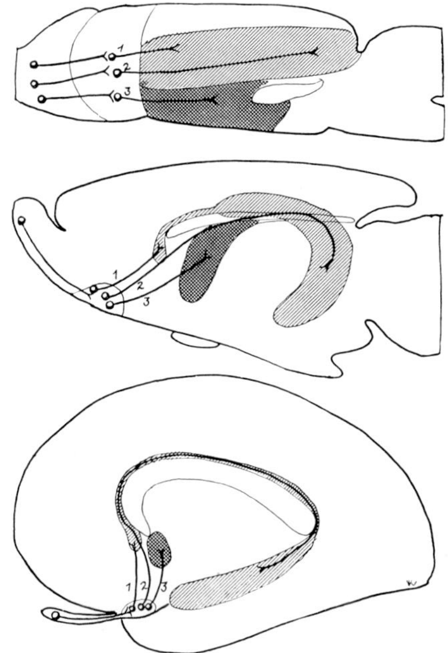

Figure 175F. Simplified diagram indicating evolution of medial olfactory tract ('stria') as seen in medial aspect (from KRIEG, 1966). Top: Urodele Amphibian, with reference to HERRICK's findings. Middle: lower Mammal. Bottom: Man. Light hatching: hippocampus; dark hatching: 'septal nuclei' (paraterminal grisea); 1: fibers to rostral hippocampus; 2: fibers to caudal hippocampus; 3: fibers to paraterminal grisea. It will be seen that KRIEG includes the Mammalian postcommissural hippocampus into the 'rhinencephalic' grisea.

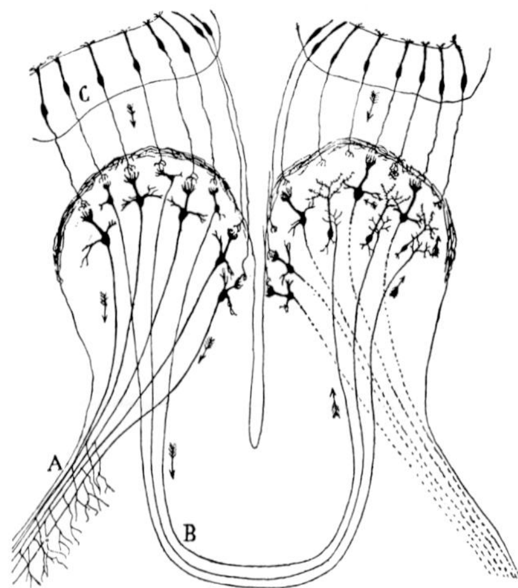

Figure 176 A. Diagram showing some olfactory channels in a lower Mammal (from CAJAL, 1911). A: lateral olfactory tract; B: bulbar olfactory portion of anterior commissure; C: olfactory mucosa. It will be seen that CAJAL assumes here an origin of the interbulbar commissural fibers from tufted cells.

The olfactory input to diencephalon (e.g. hypothalamus and epithalamus) by way of channels such as basal forebrain bundle and stria medullaris system, as dealt with in chapter XII, requires here no further comments.[53]

Again, in addition to providing output, the Vertebrate olfactory bulb receives input (bulbopetal fibers), particularly recorded for Mammals by CAJAL (1911 *et passim)* and others. Such input consists (1) of fibers from the antimeric olfactory bulb running, presumably in all Vertebrates, through the interbulbar commissure, which is a component of commissura anterior. In Petromyzonts and Selachians, interbulbar fibers may also pass through the commissura pallii. Interbulbar fibers crossing in the commissura habenulae have been reported in Myxinoids (JANSEN, 1930) and could perhaps also occur in some other Vertebrates. (2) Bulbopetal fibers arising from extrabulbar telencephalic or diencephalic (e.g. hypothalamic) grisea were recorded in various Anamnia such as Teleosts and Amphibians (SHELDON, 1912; HOLMGREN, 1920; HERRICK, 1924a) and may generally be present.

As regards Mammals, the interbulbar commissural fibers (1) were described by CAJAL (1911 *et passim*) to originate from the tufted cells

[53] The stria terminalis system will be considered further below in section 10 of the present chapter.

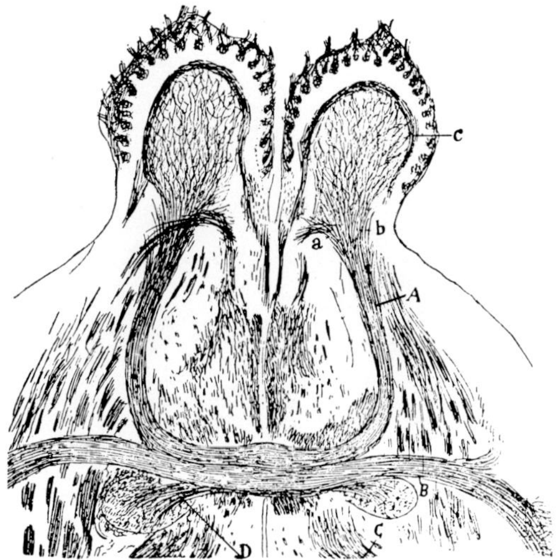

Figure 176 B. Horizontal section through the forebrain of the newborn Mouse *(Golgi impregnation*, showing two components of anterior commissure (from CAJAL, 1911). A: anterior (interbulbar) portion of anterior commissure (cf. Fig. 176 A); B: posterior portion of ant. commissure; C: column of fornix; D: stria terminalis system; a: 'superior terminal bundles of anterior commissure'; b: main bundle of interbulbar portion; c: plexus of commissural fibers in internal plexiform bulbar layer. Cf. also Fig. 230 C.

(Fig. 176 A) and to pass through the rostral limb of the anterior commissure (Fig. 176 B).[54] This commissure (Fig. 176 C) seems to include a rostral interbulbar limb, an intermediate limb forming a commissure between lateral and basal rhinencephalic grisea, and a posterior limb between posterior piriform, parahippocampal, and neocortical regions.

The bulbopetal fibers (2) from extrabulbar grisea seem mainly to originate in the olfactory tubercle and its vicinity, in the 'nucleus olfactorius anterior',[55] and perhaps in the piriform lobe cortex as well as in paraterminal grisea, reaching the bulb through lateral and medial olfactory tract and intermediate olfactory radiation. Fibers of diencephalic, particularly hypothalamic origin, entering the telencephalon

---

[54] LOHMAN and LAMMERS (1967) are of the opinion that the 'anterior olfactory nucleus' is the main origin of the fibers decussating in that limb. Be that as it may, the original observations of CAJAL appear rather convincing, although the participation of the fibers claimed by the cited recent authors seems quite probable.

[55] Evaluating here the 'nucleus olfactorius anterior' closely associated with the bulb as an 'extrabulbar' griseum.

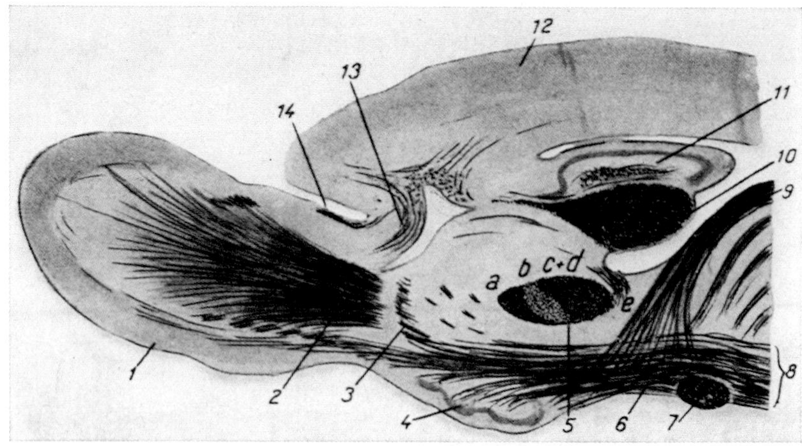

Figure 176C. Parasagittal section, fairly close to midline, through the forebrain of the Marsupial Didelphys *(Weigert stain)* showing subdivisions of anterior commissure (from BECCARI, 1943). 1: olfactory bulb; 2: anterior bundle of anterior commissure; 3: precommissural fornix; 4: tuberculum olfactorium; 5: commissura anterior (a: anterior portion, b: lateral portion; c, d: piriform and neopallial portions, e: *'parte amigdaloidea'*, i.e. stria terminalis portion); 6: *'spazio parolfattivo ventrale'* (caudal part of area ventralis anterior) and preoptic region (the lead passes just caudal to basal telencephalodiencephalic boundary sulcus); 7: tractus opticus; 8: basal forebrain bundle including *'cordone olfattivo basale'* or *'basales Riechbündel'*; 9: stria medullaris system; 10: commissura hippocampi; 11: fascia dentata; 12: neopallium; 13: *'fascio olfatto-frontale'* (presumably to precommissural hippocampus and adjacent parahippocampal cortex); 14: *'fascio olfattivo dorsomediale'* (presumably pertaining to preceding system 13).

through the basal forebrain bundle, might perhaps also reach the olfactory bulb.[56] Figures 177 A–C illustrate some aspects of bulbopetal fibers as recorded by CAJAL.

As regards the olfactory connections in a macrosmatic Mammal, we (K. and WIENER-KIRBER, 1962) recorded some data obtained by 'virus neuronography' following intranasal instillation of vesicular stomatitis virus (Fig. 170D, II). We reached the conclusion that, because of many technical difficulties and numerous uncertainties this method, at least in its present state of development, does not seem suited for generalized application. Nevertheless, descending (anterograde) and ascend-

[56] This ascending channel may also include, at least as far as tuberculum olfactorium, fibers from the rhombencephalic trigeminal region pertaining to the so-called *Oralsinn* postulated by EDINGER, and to be again pointed out in section 10 (cf. also the *'basales Riechbündel'* shown in Fig. 176C).

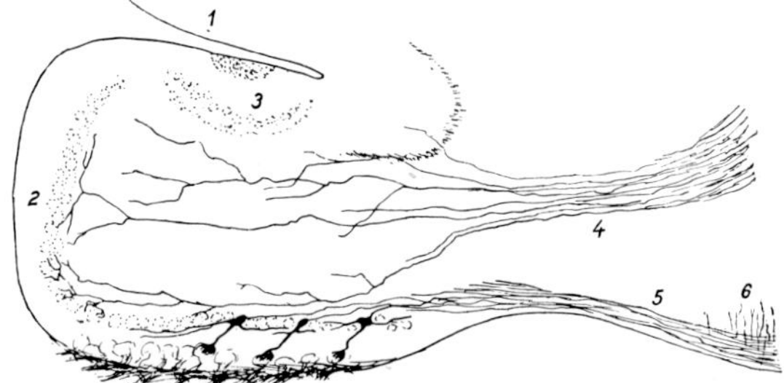

Figure 177A. Sagittal section through the olfactory bulb of a young Mouse *(Golgi impregnation)*, showing bulbopetal fibers and bulbar output fibers (after CAJAL, 1911, from BECCARI, 1943). 1: rostral pole of neopallium; 2: olfactory bulb; 3: accessory olfactory bulb; 4: anterior portion of anterior commissure; 5: lateral olfactory tract; 6: anterior piriform lobe cortex.

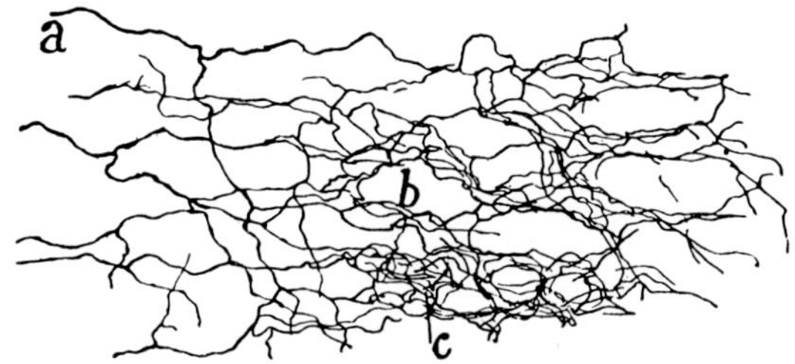

Figure 177B. Plexus of bulbopetal fibers *(fibres centrifuges)* as shown by the *Golgi impregnation* in a young Cat (from CAJAL, 1911). a: centrifugal (bulbopetal) fibers; b: locus of granule cells; c: pericellular terminal ramifications *(nids péricellulaires)*

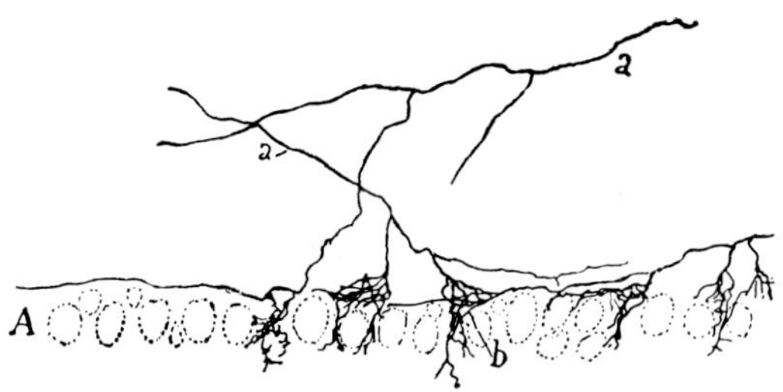

Figure 177C. Terminal arborizations *(Golgi impregnation)* of bulbopetal (centrifugal) fibers in the mitral cell layer of the olfactory bulb in a young Cat (from CAJAL, 1911). A: mitral cell layer; a: bulbopetal fibers; b: pericellular end arborizations.

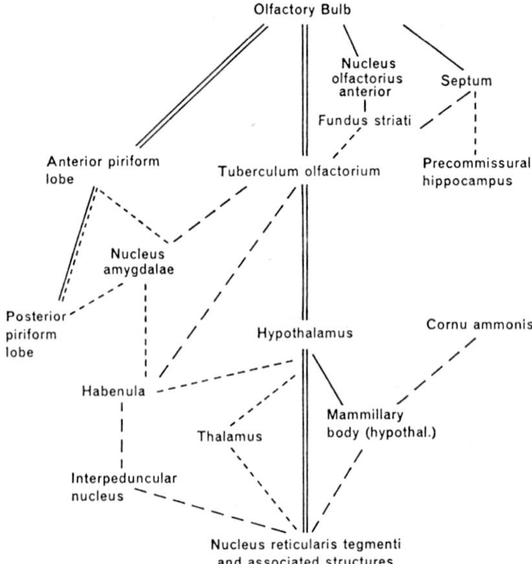

Figure 178. Diagram of neuronal transport of vesicular stomatitis virus through olfactory channels in the Mouse after intranasal inoculation. Double lines indicate main path of virus. Single or interrupted lines indicate a lesser degree of involvement. The connection of some of the indicated structures with olfactory pathways does not necessarily imply a 'rhinencephalic' status. Moreover, assuming an additional concomitant pathway for the virus via the trigeminal innervation of the nasal mucosa, the nucleus reticularis tegmenti (etc.) might be involved by 'ascending' as well as 'descending' neuronal transport of the virus (from K. and WIENER-KIRBER, 1962).

ing (retrograde), i.e. or orthodromic and antidromic transport[57] related to 'neuronal flow' and also involving transsynaptic propagation within functional systems can be shown to occur, but the erratic distribution of this 'transport' depends upon poorly definable and, so far partly uncontrollable variables.[58] Figure 178 illustrates the pathways of the neu-

57 Since it was not possible for us to check our results by means of electron microscopy, we could not obtain definitive proof whether the VSV was propagated within the axons respectively nerve fibers, or along these structures. The subsequent report by MIYOSHI *et al.* (1971), using a technique of 'immunofluorescence' likewise does not refer to electron microscopic findings with respect to propagation. Nevertheless, it has been ascertained that this virus develops within the cytoplasm and matures by budding from the plasma membrane or into cytoplasmic vesicles (DAVID-WEST and LABZOFFSKY, 1968; ZEE *et al.*, 1970). Axonal transmission within the nervous system can be regarded as highly probable.

58 Among variables partly amenable to control are differences in the virus inoculum related to tissue culture passage, egg passage, serial brain or kidney passage, and storage procedures.

ronal virus transport recorded in our experiments and, for whatever the diagram is worth, at least definitely indicates channels related to the 'olfactory system'. The findings in a subsequent investigation on VSV encephalitis in the Mouse, by MIYOSHI *et al.* (1971), although not primarily concerned with the problems considered in our own studies, conformed closely to the pertinent observations reported by ourselves.[59]

As regards the anterograde and retrograde '*neuronal flow*' which involves the virus transport, il will be recalled[60] that a slow and a rapid transport mechanism obtain. The velocity of the former remains in an approximate range of about 0.5 to 5 mm or more per day, and the rapid transport may display an order of magnitude in the general range of 20 to 100 mm per day. Both these slow and rapid 'phases' again, seem to manifest distinctive 'subclasses'. Diverse incompletely understood variables play here a role, in part related to the ultrastructural components participating in the transport, such as e.g. microtubules or mitochondria.

We estimated the transport velocity of VSV through the olfactory respectively 'rhinencephalic' neurons at ±0.2 mm per hour, thus pertaining to the category of slow transport, and compatible with estimates by other authors concerning different viruses. The virus of poliomyelitis, however (whose virions are rather smaller than those of VSV), seems to undergo a rapid transport, at least in peripheral nerves.

Another aspect of neuronal flow through the olfactory neurosensory elements has been reported by DELORENZO (1970). According to this author, colloidal gold particles placed on the nasal mucosa (of the monkey Sciurus) enter the olfactory receptor elements and are seen in the axoplasm of the fila olfactoria. They are said to reach the olfactory bulb 'by neuronal transport in mitochondria' 30 to 60 min after 'inoculation'. This, *qua* neuronal flow, would represent a conspicuous instance of 'rapid transport'. The cited author, who remarks that 'dyes placed on the olfactory mucosa are found to stain the olfactory lobes and brain in short periods of time', interprets his observations as suggesting 'a unique participation of the olfactory neuron in the blood-brain barrier system'. DELORENZO roughly defines this latter as an 'exclusion mechanism' whereby foreign substances such as proteins, viruses, particles, dyes and most drugs when injected into the blood vas-

[59] Our own investigations were prompted by the findings concerning neuroinvasiveness of VSV in the Mouse reported by SABIN and OLITZKY (1937, 1938).

[60] Cf. volume 3/I, chapter V, section 8, pp. 640–643.

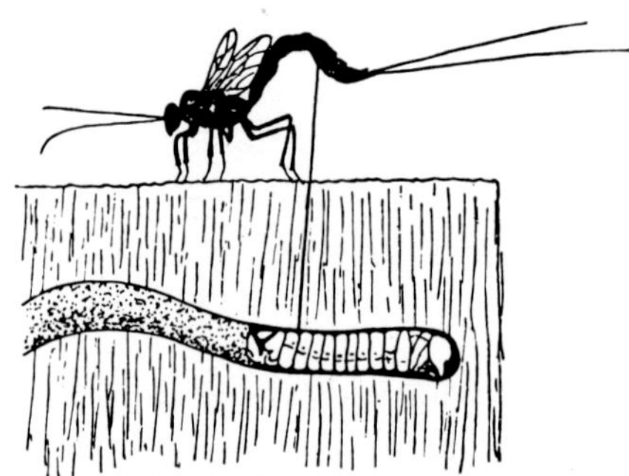

Figure 179 A. An Ephialtes wasp injecting its egg into a Wood wasp larva hidden in a piece of wood, but located by means of the aggressor's olfactory receptors (after DOFLEIN, from BUDDENBROCK, 1958).

cular system find their way into the perivascular spaces and the various body tissues, but 'have rather limited access to the central nervous system' (apparently meaning the neuroectodermal parenchyma).[61]

Although, if the different 'senses' are arbitrarily graded in terms of their 'relative importance', olfaction (and taste) can be assessed, as regards microsmatic man, to rank among the minor or to a significant extent dispensable senses. Thus, among other animals, there are completely anosmatic forms, such as, e.g., some Cetaceans. Again, complete lack of olfactory bulbs and tracts in otherwise normal human individuals has been incidentally recorded at autopsies.[62]

On the other hand, in macrosmatic Vertebrates, and in many Invertebrates, particularly Insects, the highly developed sense of smell plays a most important and perhaps even preponderant role *qua* sensory input.[63] Figure 179 A illustrates how an *Ephialtes wasp*, guided by 'olfaction', injects its egg into a *Wood wasp* larva hidden in a piece of wood.

Seen from another viewpoint, the olfactory system is evidently of considerable interest, since the morphologic evolution of the Vertebrate telencephalon, culminating, as it were, in that of Man, can be in-

---

[61] A more detailed elaboration on the so called hemato encephalic barriers is given on pp. 341-345, section 5, chapter V of volume 3/I.

[62] Cf. e.g. ISSAIEW (1932). In some such case, the individual concerned, who doubtless possessed a good sense of taste, was said to have been able to perceive 'odors'. It is not impossible that, to a certain extent, and under particular circumstances, trigeminal endings in the nasal mucosa could act as 'chemoreceptors' and mediate some aspects of 'olfaction'.

[63] Comments on the olfactory structures of Invertebrates can be found on pp. 144, 180, 187, 202, 262 et passim of volume 2. It may here be recalled that AUGUST FOREL emphasized 'the world of scent' in which Ants are presumed to live.

terpreted as taking its origin from a phylogenetic expansion of neighborhoods originally and primarily related to the olfactory bulb.[64] This evaluation seems to be supported as regards at least some developmental aspects by the observations of BURR (1916a, b) who recorded regeneration of the extirpated Amphibian telencephalon anlage if the olfactory placode was preserved, but noted failure of such regeneration if that placode was likewise extirpated.

Concomitantly with this morphologic significance, the progressive evolution of thereby modified rhinencephalic structures seems to be correlated with the development of important neuronal circuits related to various aspects of behavioral activities, as shall briefly be pointed out further below in concluding the present section.

With respect to the neural mechanism of olfaction, it can be assumed that molecules of volatile and at the same time soluble substances,[65] acting upon the peripheral endings of olfactory cells, represent the adequate stimuli (R-events). This action results a neuronal discharge of the receptor cells through the fila olfactoria.

Olfactory receptors have an exceedingly low threshold for certain adequate stimuli; thus, in some instances, a molar concentration at the order of magnitude of 10^{-13} is believed to be sufficient. In other words, only a few molecules are necessary to activate the transducer structures.

No reliable data concerning the mode of action of the chemical substances upon the receptors are available, and different, in part mutually contradictory hypotheses have been suggested. It can, however, be assumed that the permeability of the transducer surface membrane (cell membrane) is modified by the stimulating molecule. Some substances, *per contra*, may inhibit the activity of the neurosensory cells.

Again, olfactory receptors might be adequately sensitive to only one particular kind of stimulus and would thus sort out certain fundamental types of olfactory stimuli. This would require an indefinite but rather large number of functionally different olfactory cells. LE GROS CLARK (1947) suggested that apparent differences in the 'dendritic'

---

[64] Cf. pp. 642–647, section 6, chapter VI of volume 3/II.

[65] This seems to involve both water and lipid solubility. The receptor membrane doubtless includes lipid molecules. It is here of interest that the olfactory system of Fishes and larval Amphibians, living and breathing in water, shows a neural structure essentially identical with that of air-breathing Vertebrates. Again, in this respect, no significantly recognizable metamorphic change in the olfactory system seems to take place e.g. between the tadpole and the frog stage.

structures of said cells might provide a clue to such theoretically required distinction, but the available data remain inconclusive. On the other hand, biochemical differences, not detectable by ultrastructural analysis, could be the significant factor. It is thus by no means certain whether a very large or a limited, rather small number of primary receptor types must be presupposed.[65a] Nevertheless, a 'mosaic' of receptor activities may obtain within the surface of the olfactory epithelium.

Be that as it may, numerical estimates in Mammals, particularly in the Rabbit, disclose a considerable degree of many-one transformation[66] *qua* input from peripheral first order to central second order neuron, such that about 26,000 receptors may discharge upon about 24 or 26 mitral cells, whereby, discounting the tufted cells, a ratio of roughly 1000:1 would obtain (cf. above, p. 506). These signals, as pointed out above, and in a manner very roughly comparable to activities in the retina, undergo substantial processing ('nervous integration') in the olfactory bulb, where, *inter alia*, inhibiting and 'facilitating' closed circuits can be assumed.

Numerous investigations by various authors have been undertaken concerning the activities of receptors, olfactory bulb, and olfactory grisea of higher order. Olfactometers are contraptions by means of which a constant amount of the substance to be tested reaches the olfactory mucosa. Electrical potential changes from the receptor elements in that mucosa are shown by the olfactogram. Additional recordings can be obtained from fila olfactoria, olfactory bulb, and other rhinencephalic structures.

Electrophysiological studies by Adrian (1950, 1953), Ottoson and Shepherd (1967), and others, as critically discussed by Døving (1970), have provided numerous data which are still difficult to interpret.

Single unit responses of receptor cells disclosed that 40 per cent of sampled cells responded to the 26 chemicals used for stimulation. In other instances, a small number of receptors showed, in all tested units, increased spike frequency to four of the 28 used substances. Tests of four odorants on 101 different receptors indicated that the individual receptors showed a 'collective tendency' toward orderly responses (Døving, 1970).

[65a] The receptors can have widely different thresholds and 'sensitivities'. In Insects, two main types of receptors have been demonstrated, one with a relatively limited sensibility range, the other with a broad sensitivity spectrum (Ottoson and Shepard, 1967).

[66] Cf. volume 1, chapter 1, section 4, pp. 18f.

With regard to the convergence of multitudinous fila olfactoria upon a single glomerulus by many-one relationship, a study of the responses of 128 different glomeruli to 12 selected substances was undertaken (Leveteau and McLeod, 1969). It was found that the individual glomeruli responded to some but seldom to all the different 'odors', provided that relatively low stimulating concentrations were used. This is likewise interpreted as a 'tendency to order'.

According to Young (1957), the setting up of impulses in the axons of a given mitral or tufted cell presumably depends upon the stimulation of a particular set of olfactory receptors.

Electrophysiological studies by Adrian (1950, 1953) have provided numerous data which are difficult to interpret. However, a differential spatial and temporal signal pattern seems to be manifested in the olfactory bulb: different odors may stimulate different areas of that structure. As regards temporal differentiation, responses to some odorants occur with shorter latency and display a different curve in the records than the responses to others. In addition, it appears probable that selective sensibility is also manifested in the response of individual mitral cells. Odor discrimination seems to depend upon the combination of a number of factors giving information about different properties of the stimulating agent (Adrian, 1953).

Investigating the electrical activity of the rabbit's olfactory bulb, Adrian (1950) noted potential oscillations of two kinds: induced waves set up by strong olfactory stimuli, and intrinsic waves interpreted as manifesting the persistent activity of cells in the bulb. Deep anesthesia suppressed the intrinsic activity, while the olfactory signals were transmitted without interference. In moderate anesthesia the intrinsic activity became continuous, and set up a constant irregular discharge of impulses in the olfactory tract, interpreted as preventing 'the transmission of the signals from the olfactory organ'. As the anesthesia became lighter, the olfactory signals regained control to some extent. These signals are assumed to disorganize the rapid intrinsic rhythm of the bulb and to suppress the persistent discharge of impulses in favor of the olfactory discharge at each inspiration. Ultimately, however, the intrinsic activity builds up again, and the persistent discharge returns, which is believed to 'swamp' the transmission of the olfactory signals.

Since Adrian detected no sign of failure of the receptors under repeated stimulation at each breath, he suggests that the weakening and ultimate failure of sensation in man (olfactory adaptation) is due to this reappearance of the intrinsic activity after its initial disorganization. In principle, and disregarding the interpretation of the still unclarified details, this view would well agree with the postulated circular (HERRICK, 1924a) or self-reexciting (K., 1927), and perhaps also self-inhibitory discharge pattern in the olfactory bulb.

ADRIAN (1950) adds the remark that the development and the effects of the intrinsic activity of the olfactory bulb recall what happens in the cortex at different stages of anesthesia. He compares in particular the failure of olfactory sensation after repeated stimuli with our inability to pay continued attention to uninteresting sound. There is, however, in my opinion a considerable difference: we can, by shifting attention, perceive the uninteresting sounds at any time, but in the adapted olfactory condition, we cannot perceive at will the odor to which the olfactory system has become adapted. One might perhaps assume that olfactory adaptation is at least in part a peripheral process at the receptor level, but ADRIAN's observations, as quoted above, do not favor this view.

SEM-JACOBSEN *et al.* (1953) have recorded responses by depth electrography from the olfactory bulb (perhaps including olfactory fila and tract) of man, during the exploration of the frontal lobe in a psychotic patient. Definite responses to inhalation of odorants were obtained. Because of apathy and poor cooperation on the part of the patient, little subjective information, pertaining to percepts, could be elicited. The authors report that 'the measurements of the frequencies associated with the various odors showed a considerable scatter during the tests of any particular substance and a very marked overlap when the whole group is considered. It therefore seems unlikely that the frequency of response is an important factor in discrimination of smell unless the brain possesses an extremely sensitive frequency-analyzing system.'

With regard to *intensity registration*, the authors interpret their findings as indicating that there is a definite and possibly almost linear relationship between intensity of stimulus and response. It could therefore be assumed that the signals encoding intensity of smell take the form of an amplitude-modulation rather than that of the frequency-modulation prevailing in most neural communication systems.

The evidence does not, at present, seem sufficiently conclusive but nevertheless, rather relevant. It might indicate the presence of holobolic as well as heterobolic systems, that is of systems not following the all nor none principle, and would be consistent with some of the doubts concerning the universal validity of that principle in neural activities.

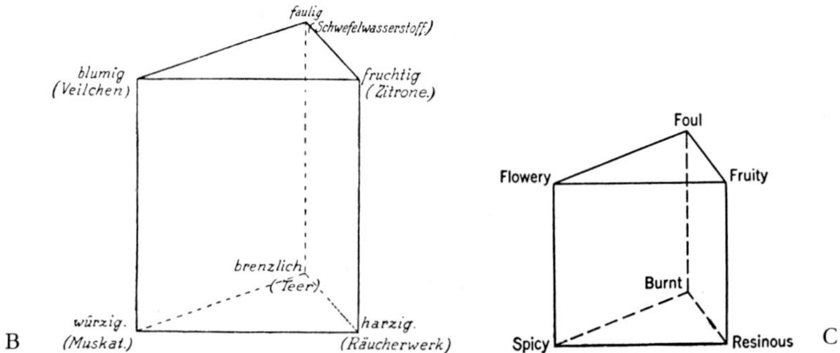

Figure 179 B, C. Two versions of so-called odor prisms as suggested by HENNING (1915/16). Mixtures of the indicated 'primaries' are supposed to be represented on the model's edges and surfaces (B from EBBINGHAUS, 1919, C from SCHIFFMAN, 1974).

Yet, logarithmic relationships, as expressed by the *Weber-Fechner law*, likewise seem to obtain in some activities of the olfactory system and, within mechanisms of such complexity, both principles do not appear mutually exclusive.

It is commonly assumed that the olfactory sense of macrosmatic Mammals, such, e.g. as the Dog, is far more developed than that of microsmatic Man. Thus, it seems well substantiated that the Dog is able to discriminate the individual odor of Humans. Experiments concerning absolute sensitivity have confirmed this difference in the Dog's olfactory threshold, but the obtained data varied as regards the numerical values for the diverse substances, indicating a sensitivity of 10 to 100 times greater than that of Man for certain substances according to some authors, while others reported a greater sensitivity of 10^5 and more times (cf. e.g. MOULTON *et al.*, 1970). Others have reached the conclusion that the absolute sensitivity between Dog and Man is not as great as might be expected, the incontestable actual difference in olfactory discrimination being related to larger area and closer packing of olfactory receptor surface and to the particular differentiation of the rhinencephalic grisea.

Many authors have commented upon the multitudinous variety of perceivable odors (i.e. olfactory P-events). The various attempts at systematization have not been particularly successful, and simplified diagrams, such as so-called *odor prisms* shown in Figures 179 B and C, conceal rather than display the extreme diversity. The different odors experienced in Human consciousness apparently outnumber the avail-

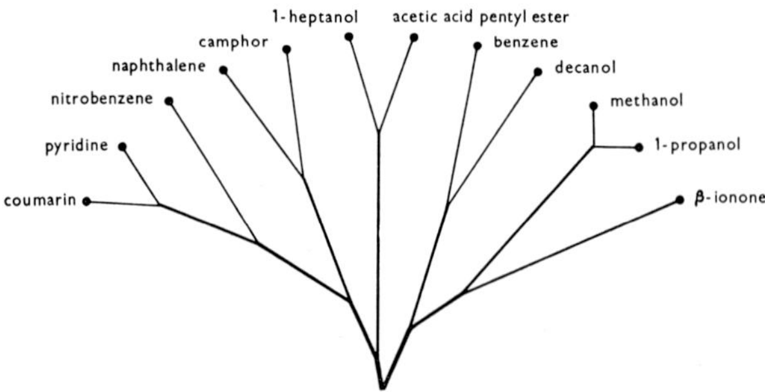

Figure 179 D. So-called 'hierarchical clustering' of some odors (after LEVETEAU and MACLEOD, 1969, from DØVING, 1970).

able vocabulary to describe them, and generalized terms such as fragrant, fruity, spicy, acid, burnt, caprylic, putrid, etc. are commonly used. Two distinct types of 'psychological scaling' can be used. The 'profile rating scales' presupposes that odor quality can be represented by verbal description, each specific odorant being again rated *qua* intensity. 'Proximity analyses', on the other hand, require a rating or ranking in terms of dissimilarities or similarities between pairs of odorants (cf. e.g. MOSKOWITZ and GERBER, 1974). Ranked similarities between odors can be expressed in form of matrices, or by a hierarchical clustering method resulting in 'tree structures' such as shown in Figure 179D.

Again, if compared with the discrimination of musical tones, or with the predictable results of color mixtures, odor mixtures produce a variety of results, namely either (1) unified sensations, or (2) 'blends' in which different odors are alternatingly perceived, or (3) 'neutralization' resulting in absence of smell, or again (4) 'masking' whereby one odor is completely suppressed by the other. In addition to adaptation, as mentioned above on p. 533, odors may change their quality with continuous stimulation. Thus, the smell of nitrobenzene may here change from a bitter almond odor to that of tar. Again, continuous stimulation by one odor can temporarily raise the threshold for certain other substances in a reciprocal manner.

There must thus be a great variety of molecular chemical action patterns (R-events) upon olfactory receptors, to which corresponds a similar variety of coded central N-events; there are again the conscious parallel events of olfactory percepts, that is of odors *sensu stricto*. It must be kept in mind that the chemical, biochemical, and physiologic

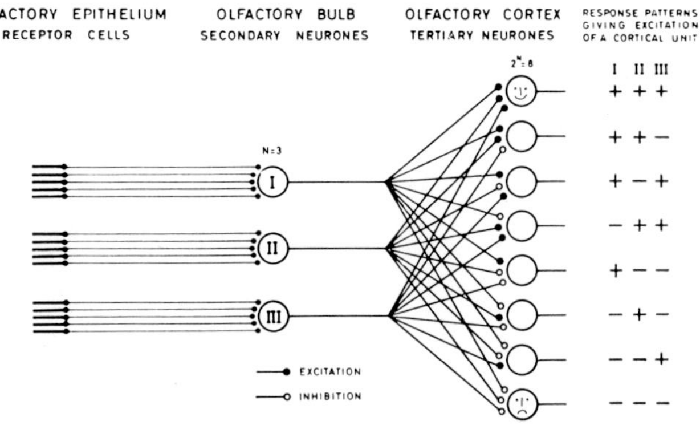

Figure 180. Tentative scheme of Vertebrate olfactory system indicating possible coding mechanisms (from DØVING, 1970). Three bulbar neurons (presumably mitral cells) are shown in synaptic contact with eight different 'cortical units'. The synapses are arranged in such a way that a certain 'odor' will evoke a combination of excitation and inhibition in the secondary neurons by means of undefined intrinsic bulbar mechanisms (not shown in the diagram) such that only one unit in the cortex will be affected in a specific response pattern. N is the number of activated glomeruli, 2^N the number of response patterns in a 'cortical unit'.

events in the nasal cavity as well as the resulting N-events are not, *per se*, odors.

In contradistinction to sight and hearing, where the relevant R-events can be rather accurately linked to certain frequencies respectively wavelengths, the physical or rather chemical (molecular) basis of olfaction and of its qualities remains poorly understood. In addition to volatility and solubility, as mentioned above, molecular configuration, including polarity of the molecule and chemical composition, and catalytic action have been considered as relevant factors.[67] Again, most or all odorant substances are said to absorb infrared radiation between λ 2 to 20 μ as energy used in 'molecular vibration'.

Be that as it may, the olfactory R-events, presumably transduced by the olfactory cilia, become transformed into N-events which become further processed in bulbus olfactorius and rhinencephalic grisea. In the course of this processing proceeding from 'lower' to 'higher' levels, fewer neuronal elements seem to be activated within the rhinencephalon. This behavior of the olfactory system and its pathways can be

[67] Detailed but inconclusive discussions of these topics can be found in the Symposia edited by ZOTTERMAN (1967) and by WOLSTENHOLME and KNIGHT (1970), as well as in the conference on odors edited by CAIN (1974).

roughly compared to the 'processing' or 'integrative' properties displayed by the visual system, where, by 'abstraction of invariants' the requirements for the activation of cell groups become more 'selective' or 'refined' as the signals 'ascend' in the central visual pathways and can, to some extent, be detected by so-called single unit electrophysiological studies of the sort undertaken by HUBEL and WIESEL (1962), and others.

Figure 180 shows a tentative scheme suggested by DØVING (1970) and illustrating the possibility of coding olfactory N-events by combinations of excitation and inhibition patterns within 'tertiary' grisea of the 'olfactory cortex'.

A more elaborate model of '*olfactory coding*', using highly schematized concepts, and extrapolating tentative numerical data obtained by ALLISON and WARWICK in the Rabbit, was elaborated by HAINER *et al.* (1954) as an 'information theory of olfaction'. Despite its purely conjectural and somewhat artificial aspect, the theory is of some value as a model or as an exercise in neuronal communication engineering.[68]

The theory is based on the all or none, that is on the digital principle, and on the validity of the *Weber-Fechner law*, whereby perception of intensity is considered to be a logarithmic function of the external chemical concentration.

In each side of the nasal cavity, 5×10^7 olfactory receptors and an identical number of fila are assumed. These fibers are distributed upon 1,900 glomeruli in each olfactory bulb; the bulb is supposed to contain approximately 45,000 mitral cells, and each glomerulus is said to be connected with 24 mitral cells. The intensity of olfaction is believed to be transmitted to the higher olfactory centers through 1,900 sets of 24 neurites, and is postulated to correspond to the degree of activation of these 1,900 identical sets. The code pattern of a distinct odor is assumed to be a given pattern of 'on-ness' on 'what might be considered a panel board of 24 lights'.

If at least 10,000 known odors exist, then 2^N must correspond to 10,000 where N is the number of neurons which must operate in paral-

[68] In this respect, much the same could be said about the adaptation of SHANNON's (1938) *circuit algebra*, based on *Boolean algebra*, to neuronal networks by McCULLOCH and PITTS (1943). Although *Shannon's circuit algebra* is fully valid for electrical networks, the complex parameters obtaining in neuronal networks require arbitrary simplifications and postulates which are incompatible with the actually prevailing conditions. Nevertheless, these simplified neural networks can be interpreted as depicting valid aspects of the logical and mathematical orderliness to which the functions of the nervous system conform.

lel for 10,000 different patterns. This number is greater than 13 and smaller than 14. But since 24 mitral cells are assumed for each glomerulus, more than 16×10^6 patterns of odor could be registered if the receptors would have equivalent chemical discrimination. Thus, on a board of 24 lights more than 16×10^6 patterns can be realized.

Because the 'board' of 24 lights must not flicker or quaver, but, to achieve the numerous distinct patterns, must 'light simultaneously' a timing or phasing device is required. By means of this device, the system must function in unison, so that simultaneous discharge of the configurational response by all cooperating neurons takes place. A short term memory or holding function of the bulb is therefore needed, retaining the pulses which, at a certain phase are then simultaneously discharged as configurational information.

The authors, who were unaware of my interpretation of feedback mechanisms or closed circuits in the olfactory bulb (K., 1927, 1928), presuppose that the granule cells receive a pulse, delay it until just the right time, and release it. The granule cell 'serves as a modified memory cell. The interesting fact is that the olfactory bulb is known to pulse or mechanically fluctuate rapidly.' While I did not, at the time, advance any specific theory of olfactory mechanisms, and although the theory set forth by HAINER and his associates may not at all be valid, either *in toto* or in some essential points, it seems nevertheless most likely that the essential pattern of granule cell circuits which I indicated in figure 181 (K., 1927, p. 249) subserves a short-term memory function and some timing, cyclic or synchronization mechanism of the sort postulated by HAINER *et al.* (Fig. 173 A).

The intensity model for olfactory response construed by these authors is by far more arbitrary than their perception model. It is assumed that 24 different kinds of end organs exist, thus giving 2×10^6 input and 1,900 output neurons for each digit of the configuration. The perception of intensity is believed to be proportional to the number of these 1,900 equivalent neurons which are discharging at a particular moment. In other words, the strength of the perception depends on the number of the 1,900 glomerular output systems (consisting of neurites of mitral cells) activated by a given odorant. The model of this intensity system does not connect the same number of primary neurons to each secondary neuron, but by contrast, connects as many as 190,000 primary neurons to but one secondary neuron, and as few as one primary to one secondary. From the number of 'primary neurons', the number of 'secondary neurons', the law of statistical fluctuation, and the *Weber-Fechner law*, three deductions are drawn: (a) there are 30 perceptual levels of intensity, (b) the effective concentration range for perception is 190,000 times the threshold for identification, and (c) a 52% increase in concentration is required for a just noticeable difference.

Various objections and doubts could be raised with respect to these views. However, any attempt at interpretation or model construction may help to clarify the problem, and lead to some useful approaches. In discussing the sense of olfaction, ADRIAN (1948) remarked: 'We know so little about the sense of smell that I must start by excusing myself for choosing such an unusual subject for a lecture. My only excuse is that it is unusual. I am sorry to say that I have not much to add to the little that is known.'[69]

Although some histologic as well as physiologic data are available, it must be admitted that these data are insufficient, and that a satisfactory theory concerning sensory and neural olfactory processes, and the coding of olfactory signals, cannot be formulated at the present time. As regards recent reports, the high concentration of carnosine (β-alanyl-L-histidine) in the olfactory epithelium, and primary olfactory pathway, including olfactory bulb of the Mouse might be mentioned (MARGOLIS, 1974). Carnosine could here be a possible neurotransmitter. Various investigations have attempted to identify particular 'biochemical markers' of the olfactory pathway, and, *inter alia*, a 'specific olfactory protein' of unknown biological activity has been recorded (cf. e.g. AGRANOFF and APRISON, 1975). Other recent studies concern the activity of olfactory bulb single units correlated with inhalation cycles and odor quality in two Rodents (Golden Hamster, Mesocricetus, and Deermouse, Peromyscus) as undertaken by MACRIDES and CHOROVER (1971), as well as the so-called 'learning-set' formation in Rats trained with odor stimuli (SLOTNICK and KATZ, 1974). These and various other investigations of that type still remain inconclusive.

In concluding the present section of rhinencephalic structures and on the sense of smell, it is perhaps appropriate to mention two additional pertinent topics.

---

[69] ADRIANS's remarks, made more than 25 years ago, still express the contemporary status of the topic. Thus, in the recent symposium (WOLSTENHOLME and KNIGHT, 1970) it was stated that there is, at present, no adequate theory of the mechanism of olfactory stimulation. In the symposium edited by ZOTTERMAN (1967), OTTOSON and SHEPHERD remark that the available knowledge of olfaction remains fragmentary, and still leaves many of the basic problems of olfaction unexplained.

1. As far as human olfactory perceptions, i.e. P-events are concerned, this sensory modality, in contradistinction to sight and various aspects of touch ('body sense', cf. section 1, chapter VI, vol. 3/II) displays indistinct localization and lack of spatial configuration ('shape perception'). Its localization is more vague than that intrinsic to sound, which likewise, in human consciousness, lacks spatial configuration. As illustrated by Figure 179 A, however, it seems evident, that at least in diverse Insects and presumably also various other organic forms, including perhaps some Vertebrates, the registration of 'odors' results in more or less accurate localization of external 'objects' or events.[70] Nothing, on the other hand, can be said with any degree of certainty about concomitant occurrence or non-occurrence of 'conscious' percepts (P-events) respectively, if occurring, their particular 'nature' or 'attributes'.

2. Presumably because of the close phylogenetic relationship between Vertebrate telencephalon and olfactory bulb respectively olfactory organ, the olfactory grisea, in addition to their specific 'sensory' function, may serve as non-specific activators for other or even most (if perhaps not all) telencephalic activities, including the 'cortical' ones of Mammals.[71] This was pointed out by HERRICK (1933), who referred to 'a generalized activity of primitive type acting on the neopallial cortex as a whole, lowering its threshold or increasing its sensitivity, or, in the case of noxious stimuli, exerting inhibitory influence'.

According to HERRICK (1933) 'this type of non-specific activity is one of the major functions of the olfactory cortex, though all parts of the neopallium also exhibit it to some degree. It comes to expression in overt behavior, learning capacity, memory, etc., as a differential influence upon other cortical and subcortical functions of those exteroceptive sensorimotor mechanisms whose specific patterns of response show well defined anatomical and physiological localization. Having no localization pattern of its own, it may act in two ways: first upon other exteroceptive systems whose localized mechanisms are adapted to execute adjustments where external orientation is demanded, and, second upon the internal apparatus of general bodily attitude, disposition and affective tone.'

[70] To some extent, of course, because of discrimination *qua* quality and intensity, the 'source' of an odor, e.g. a gas leak, may be located by human olfaction.

[71] With regard to anosmatic Mammals such as e.g. some Cetacea it could be said that the relevant grisea (hippocampus, etc.) represent here a, phylogenetically speaking, 'rhinencephalic legacy'.

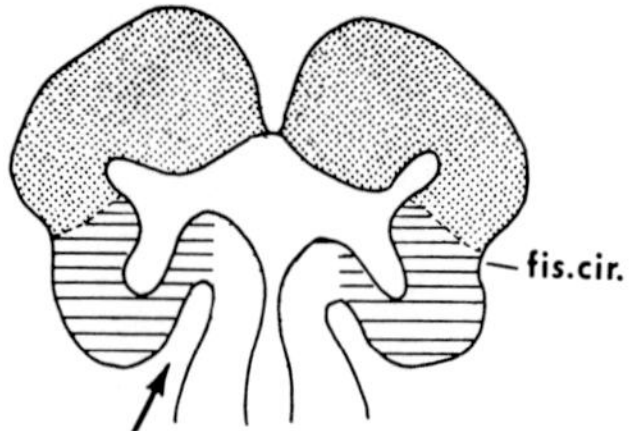

Figure 181 A. Diagrammatic horizontal section through the forebrain of a Petromyzont Cyclostome (modified after NIEUWENHUYS, 1967). Dotted portion: olfactory bulb; horizontal hatching: lobus hemisphaericus; fis.cir.: sulcus circularis. Arrow: telencephalo-diencephalic boundary.

KLEIST (1934) quite independently assumed that his 'inner brain', essentially corresponding to the 'rhinencephalon', was significant for emotional attitudes, drives, and behavior, by correlating and processing 'visceral' input from oral, anal, and genital region (as well as 'internal' organs in general) which, then, in turn, have an effect on other neural activities. Subsequently, the PAPEZ *theory of emotion* (1937) emphasized the significance of the hippocampus, phylogenetically derived from the 'rhinencephalon' as an important link (together with associated structures), in the 'mechanisms of emotion'.

KLÜVER (1952), on the basis of his own experimental work, which gave at least partial support for these unavoidably very generalized theories, presented a review of brain mechanisms and behavior with special reference to the rhinencephalon. In his opinion, the still vague aspect of said theories could be remedied by a more intensive behavioral, physiological, and biochemical analysis of the relationships between rhinencephalon, neocortex, hypothalamus and endocrines. Although, by now, about 25 years later, multitudinous new detailed but more or less disconnected and fragmentary data have been recorded, no significant progress for further significant elucidation of the topic here under consideration has been made.

A critical review of the literature concerning the Mammalian and Human olfactory bulb, the regio retrobulbaris (nucleus olfactorius anterior) and the relationship of the rhinencephalon to the cortical regions subsumed under the concept 'allocortex' is included in STEPHAN'S (1975) important handbook volume on the allocortex. This treatise, moreover, contains the results of the cited author's original investigations. Problems concerning the allocortex, however, shall be discussed in chapter XV, dealing with the Mammalian cerebral cortex.

It should finally also be recalled that, in Human experience, the 'taste' of foods and drinks, especially their 'flavor', is, to a relevant degree, not mediated by the gustatory apparatus with its relatively few qualities, (salty, sour, bitter, and sweet, cf. vol. 3/II, chapter VII, p. 790), but by the sense of olfaction, which seems to record the 'flavors'. Quite evidently, in this instance, the olfactory mucosa is affected by both the inhaled and exhaled air stream. An additional problem is, moreover, whether trigeminal endings in oral or nasal mucosa or both play a role for sensations of 'taste' or 'smell'.

3. Cyclostomes

The telencephalon of *Petromyzonts* displays inversion combined with paired rostral and caudal evagination, and reduction of the telencephalon impar, as discussed, with due consideration of different interpretations by other authors,[72] in sections 1 B and 6 of volume 3/II. The rostralward evaginated bulbus olfactorius is separated by a circular groove ('fissura circularis', NIEUWENHUYS, 1967; sulcus interolfactorius, SAITO, 1930) from the lobus hemisphaericus. This latter protrudes caudad with a polus posterior (cf. Fig. 181 A).

The *bulbus olfactorius*[73] contains the anterior recess of the lateral ventricle (Fig. 181 B). In view of the rather diffuse cytoarchitectural arrangement in the lobus hemisphaericus, it seems preferable merely to distinguish a dorsal (pallial) D-neighborhood, and a basal B-neighborhood (Fig. 181 C). There is little doubt that the entire neuronal cell population of the lobus hemisphaericus receives secondary respectively higher order olfactory input through the diffuse system of the olfactory tract through its nondescript dorsal and basal radiations. This, on the other hand, by no means excludes ascending sensory input of various sorts from the diencephalon by way of the basal forebrain bundle. The pallial and the basal cell groups are joined with those of the diencephalon by the two parts of the massa cellularis reuniens. The pars su-

---

[72] Detailed bibliographic references to investigations dealing with the Petromyzont telencephalon can be found in the papers by JOHNSTON (1902, 1912), SAITO (1930), HEIER (1948), and NIEUWENHUYS (1967). Additional Figures illustrating its aspects in Petromyzon are Figures 60A–D, 61A, and 62C of chapter XII of this volume, moreover Figure 187, p. 414, and Figure 243A, B, p. 476 of volume 3/II.

[73] Since the Vertebrate, including the Cyclostome olfactory bulb was dealt with in section 2, further details are here omitted (cf. e.g. Fig. 172F).

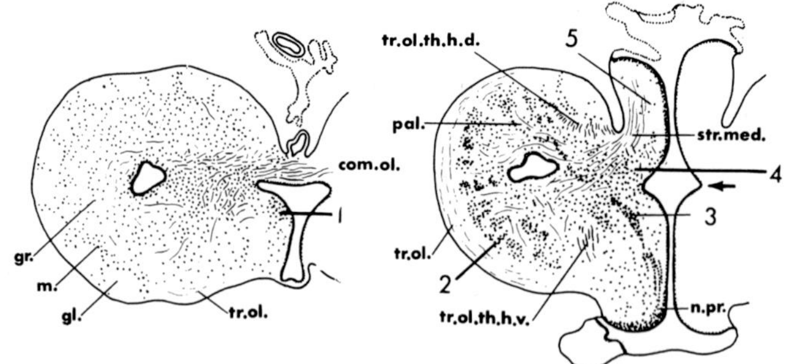

Figure 181 B. Cross-sections through olfactory bulb (a) and lobus hemisphaericus (b) of the Petromyzont Lampetra fluviatilis (from Nieuwenhuys, 1967, with modified and added designations). com.ol.: commissura olfactoria (commissura dorsalis s. pallii); gl.: glomerular layer; gr.: granular cell layer; m.: mitral cell layer; n.pr.: nucleus praeopticus; pal.: pallium (zone D); str.med.: stria medullaris; tr.ol.: tractus olfactorius; tr.ol.th.h.d. (v.): tractus olfactothalamicus and hypothalamicus dorsalis respectively ventralis; 1: rostral portion of massa cellularis reuniens, pars ventralis; 2: basal cell masses (n. basalis, zone B); 3: eminentia thalami ventralis including massa cellularis reuniens, pars ventralis; 4: massa cell. reun., pars dorsalis; 5: eminentia thalami ventralis ('primordium hippocampi' autorum). Arrow: sulcus diencephalicus ventralis.

perior joins pallium to eminentia thalami ventralis, and the pars inferior forms a transition between basis telencephali and preoptic hypothalamus.

As regards the *communication channels*, and in addition to the abovementioned *tractus olfactorius*, two main systems could be distinguished, namely that of stria medullaris, and that of basal forebrain bundle.

The *stria medullaris* includes olfactory tract fibers as well as fibers related to both pallium and basis. It passes through the eminentia thalami ventralis and runs as far as the epithalamus, becoming joined, caudally to interventricular foramen, by fibers from the preoptic hypothalamic region. The stria medullaris system corresponds roughly to the tractus olfacto-habenularis of Heier (1948), and, although apparently mainly descending, presumably includes some reciprocal (i.e. ascending) components.

The *basal forebrain bundle*, which likewise contains bulbar (i.e. olfactory tract) fibers, connects telencephalon with thalamus and hypothalamus; this overall channel extends as far as the rostral mesencephalic tegmentum. The basal forebrain bundle system includes the very vaguely outlined bundles which Heier (1948) has described as tractus olfacto-thalamicus et hypothalamicus dorsalis et ventralis, and as trac-

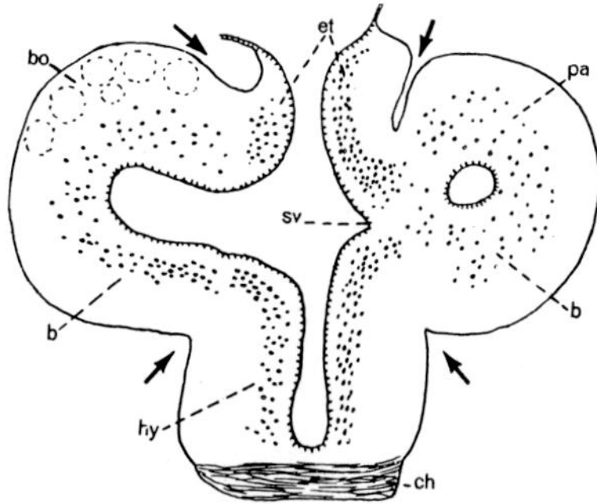

Figure 181 C. Simplified sketch showing cross-section through the forebrain of Petromyzon (from K., 1927). b: nucleus basalis; bo: bulbus olfactorius; ch: optic chiasma; et: eminentia thalami ventralis; hy: hypothalamus; pa: pallial primordium; sv: sulcus diencephalicus ventralis. Left side at level of intervertebral foramen, right side caudal to foramen. Arrows: sulcus telencephalodiencephalicus.

tus strio-thalamicus et hypothalamicus. Besides containing descending fibers, the basal forebrain bundle system provides ascending sensory input to telencephalon presumably through the mediation of diencephalic grisea. All specific details about the aforementioned channels, however, remain highly uncertain (Figs. 182A–C).

There are two *telencephalic commissures*, the *commissura dorsalis sive pallii*, and the *commissura anterior*. Because the former contains mostly fibers related to the olfactory bulb, it is also commonly designated as commissura olfactoria or commissura interbulbaris (cf. Figs. 181B, 182A, C). It includes, nevertheless, also a fair amount of fibers related to the pallial region of lobus hemisphaericus. Quite apart from this relationship, its location in a dorsal topologic neighborhood of the lamina terminalis rostroventral to the transitory velum transversum[74] clearly indicates the homology of this commissure with the commissura pallii of Gnathostome Vertebrates such as Selachians and Amniota,[75] (cf. also Fig. 183A).

The commissura anterior, designated as commissura supraoptica by Heier (1948), doubtless contains fibers related to the basal grisea of

[74] Cf. Figure 69A, p. 192, volume 3/II.

[75] In the highly everted telencephalon of Osteichthyes, as well as in that of Dipnoans and Amphibians with a fairly substantial telencephalon impar, the commissura pallii becomes topologically displaced and joined to commissura anterior, thereby manifesting analogy rather than homology with the commissura pallii of the aforementioned Vertebrate forms (cf. Figs. 62 to 65, pp. 173, vol. 3/II).

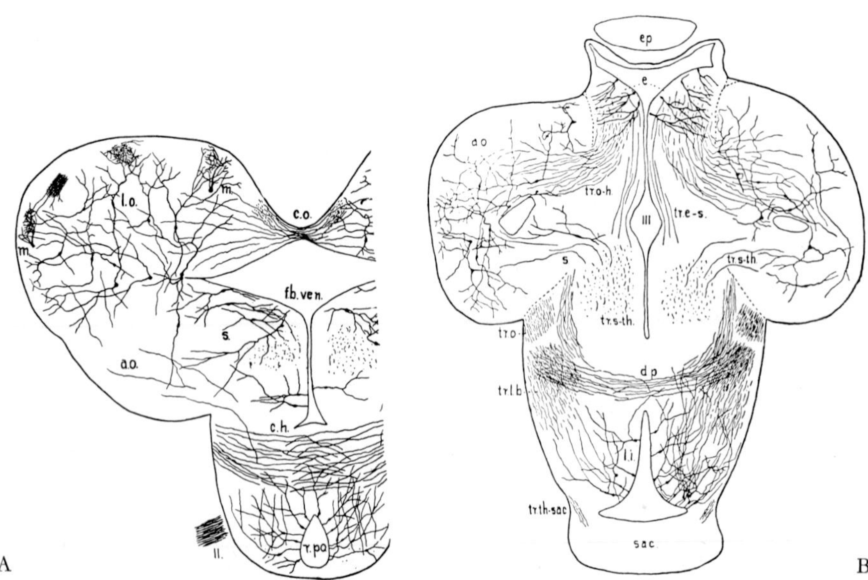

Figure 182 A. Slightly oblique cross-section *(Golgi impregnation)* through the forebrain of the Cyclostome Lampetra at level of commissura dorsalis telencephali (from Johnston, 1906). a.o.: nucleus basalis; c.h.: optic chiasma; c.o.: commissura olfactoria (commissura dorsalis); f.b.ven.: ventricle at level of interventricular foramen; l.o.: olfactory bulb; m.: mitral cell; r.po.: postoptic hypothalamic recess; s.: massa cellularis reuniens, pars inferior; II: optic nerve.

Figure 182 B. Slightly oblique cross-section *(Golgi impregnation)*, caudal to level A, through the forebrain of Lampetra (from Johnston 1906). a.o.: pallium of lobus hemisphaericus; d.p.: postoptic decussation; e.: eminentia thalami ventralis; e.p.: epiphysial structures; l.i.: posterior hypothalamus; s.: region of nucleus basalis near hemispheric stalk respectively massa cell. reun. p. inf.; sac.: 'infundibular' region of hypothalamus; tr.o.: tractus opticus; tr.e-s.: tract of eminentia thalami; tr.l.b.: 'tractus lobo-bulbaris; tr.o-p.: tractus olfactohabenularis; tr.s-th.: basal forebrain bundle; tr.th-sac.: fibers to commissura postinfundibularis; III: third ventricle at level of sulcus diencephalicus ventralis.

the lobus hemisphaericus as well as to the basal forebrain bundle. In the diencephalon, there are connections of this commissure with thalamic cell groups and particularly those of the preoptic hypothalamus (Heier's nucleus commissurae supraopticae). With respect to the channels between telencephalon and diencephalon, commissura pallii and commissura anterior were discussed on p. 152 of chapter XII.

The marked pattern distortions and the peculiarities characteristic for the *Myxinoid* brain have been repeatedly dealt with in this treatise and require no further detailed comments (cf. chapter XII, p. 155 with

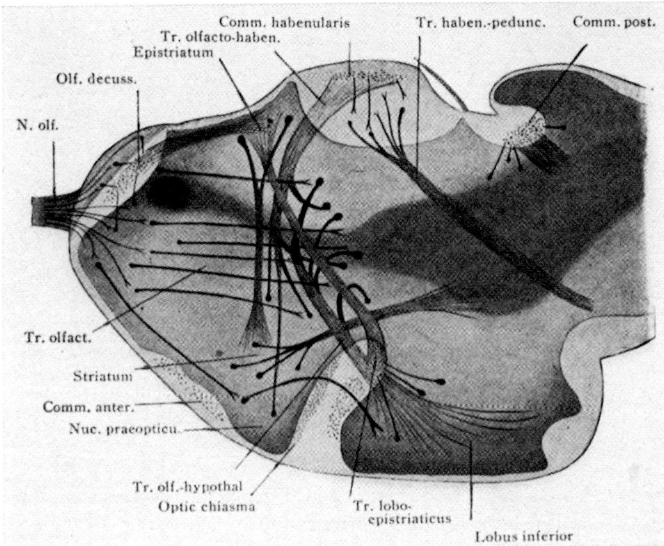

Figure 182C. Diagram indicating olfactory and additional channels in the brain of Lampetra according to Johnston's interpretation (from JOHNSTON, 1906). The eminentia thalami ventralis is here designated as 'Epistriatum', and part of the preoptic region as 'Striatum'. The ventral position of commissura anterior is exaggerated (cf. Fig. 60A). The diagram is, nevertheless, instructive and, as regards the communication channels, essentially valid.

footnote 121). It will here be sufficient to compare a relevant embryonic stage of Petromyzon and Bdellostoma as illustrated by Figure 183. Additional aspects of the Myxinoid brain are depicted in Figures 64–66, and 68 of chapter XII, Figures 193 and 194 (pp. 371–372) of volume 4, and Figures 69B–D (pp. 192–193) and 244 (pp. 478–479) of volume 3/II. In the course of ontogenesis, the telencephalic anlage, at first located at the rostral extremity of the straight brain tube axis, becomes subsequently bent in a ventral direction and still later sharply bent dorsad, as shown in Figure 183B, concomitantly with a fusion (concrescence) of lobus hemisphaericus and diencephalon. These complex transformations, combined with a scarcity of relevant embryonic material, have caused considerable difficulties for a proper interpretation of the Myxinoid's forebrain morphologic respectively topologic configuration as attempted by KUPFFER (1906), JANSEN (1930), CONEL (1931), HOLMGREN (1946), and BONE (1963).

Grossly and externally, the telencephalon is fairly well delimited from the diencephalon by a transverse sulcus telencephalo-dien-

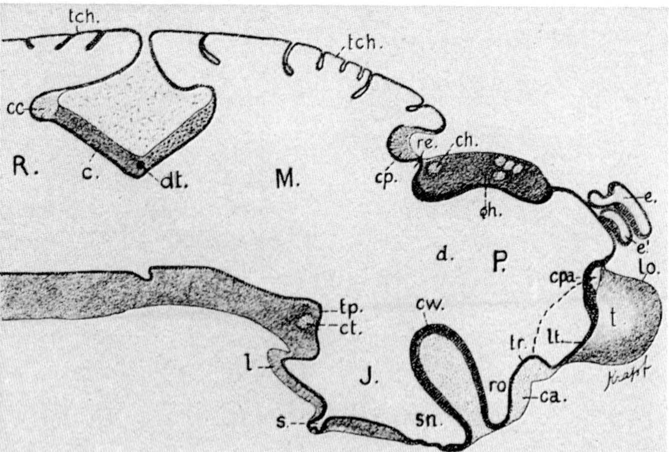

Figure 183 A. Midsagittal section, with added outline of telencephalon, through the brain of a 15 mm Petromyzont Ammocoetes larve (from KUPFFER, 1906). c: cerebellar plate; ca: commissura anterior; cc: commissura cerebellaris; ch: commissura habenulae; cp: commissura posterior; cpa: commissura pallii (commissura dorsalis); ct: commissura transversa; cw: chiasmatic ridge; d: diencephalon; dt: decussatio nervi trochlearis; e: pineal organ; e′: parapineal organ; gh: ganglion habenulae with accessory commissural fibers; J: 'infundibular' region of hypothalamus; l: lobus inferior hypothalami; lo: telencephalic hemisphere (lead points to bulbus olfactorius); lt: (telencephalic) lamina terminalis; M: mesencephalon; P: prosencephalon; R: rhombencephalon; re: recessus epiphyseos; ro: recessus preopticus; s: recessus inferior lobi inferioris hypothalami; sn: recessus postopticus; t: telencephalon with boundary in region of embryonic ventriculus impar telencephali; tch: choroid plexus epithelium; tp: tuberculum posterius; tr: torus transversus. The telencephalodiencephalic boundary has been added in accordance with the present author's interpretation.

cephalicus, which, however, flattens out at the basal surface. Bulbus olfactorius and lobus hemisphaericus are likewise delimited from each other by a transverse circular sulcus rhinencephalicus, which begins at the dorsal interhemispheric groove, and runs through dorsal and basal surfaces (Figs. 68 A, B, chapter XII, p. 159).

Internally, because of the 'telescoping' of diencephalic into telencephalic neighborhoods, the determination of the boundaries between telencephalic and diencephalic cell groups presents various difficulties, pointed out by JANSEN (1930) and other authors. A case in point is here the so-called primordium hippocampi *autorum*, which, as in Petromyzonts, HOLMGREN (1946) and myself interpret as a rostral portion of the diencephalic eminentia thalami ventralis.

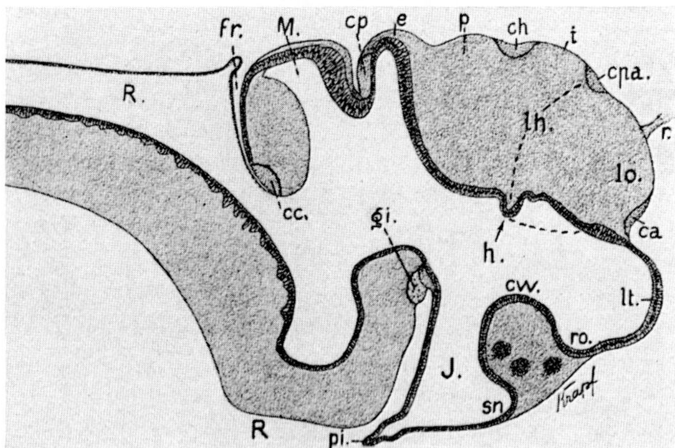

Figure 183 B. Midsagittal section through the brain of a Myxinoid Bdellostoma embryo roughly corresponding to the Petromyzont stage of Figure 183 A (from KUPFFER, 1906). e: locus of rudimentary epiphysial recess; fr: modified fissura rhombomesencephalica (the abortive corpus cerebelli, of which only the commissure is recognizable, being absorbed in the mesencephalon); gi: ganglion interpedunculare at tuberculum posterius; h: velum transversum rudiment; lh: boundary neighborhood between fused lobus hemispaericus and diencephalon; lo: bulbus olfactorius; pi: inferior recess of lobus inferior hypothalami; lt: diencephalic (preoptic) lamina terminalis; r: olfactory nerve; t: anlage of ganglion habenulae. Other abbreviations as in Figure 183 A. The dotted telodiencephalic boundary has been added in accordance with the present author's interpretation.

Figure 184 A shows, in an horizontal section, the irregular outline of the *olfactory bulb*, the nucleus olfactorius anterior, and a remnant of the obliterated lateral ventricle. JANSEN considers the Myxinoid olfactory bulb to be somewhat more highly differentiated than that of Petromyzonts and distinguishes five layers. Mitral cells and glomeruli are well developed. Less specialized stellate or fusiform cells represent the additional neuronal elements, some of which, although not of typical 'granular' aspect, are presumably the functional equivalents of olfactory granule cells.

The *lobus hemisphaericus* (Fig. 184 A, B), in which all traces of the obliterated ventricle may or may not have disappeared in adult specimens, contains, as in Petromyzonts, a dorsal (pallial) and a basal neuronal population.

The pallial region, much more differentiated than that of Petromyzonts, displays an arrangement in five layers, particularly investigated by JANSEN (1930). Two layers (*l*2 and *l*4) are cellular, the more superficial one being a 'true' cortex ('cortex lobi olfactorii' of EDINGER,

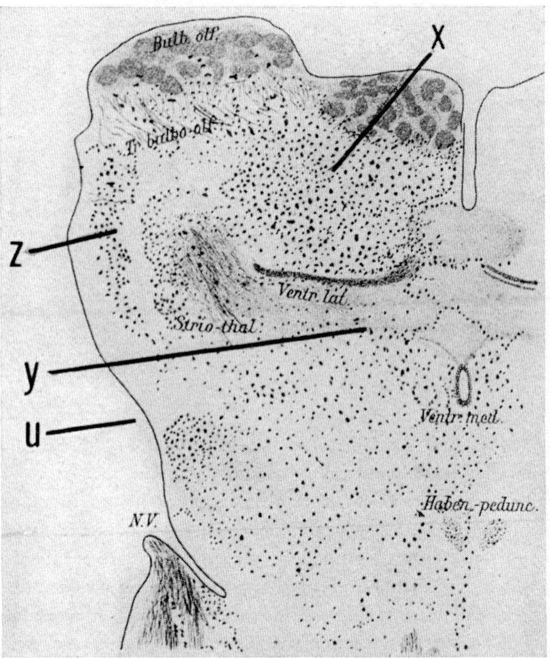

Figure 184 A. Horizontal section through the forebrain of adult Myxine glutinosa, passing through ventral portion of pallium and remnant of obliterated lateral ventricle (from EDINGER, 1906). Haben.-pedunc.: fasciculus retroflexus; N.V.: fragment of trigeminal nerve; Strio-thal.: basal forebrain bundle; Ventr.med.: remnant of 3rd ventricle. Added designations: u: sulcus telodiencephalicus; x: nucleus olfactorius anterior; y: eminentia thalami ventralis ('primordium hippocampi'); z: ventral portion of pallium (in region of basal forebrain bundle portion of nucleus basalis).

1906), while the inner one seems to correspond to the original periventricular layer. Lamina 1 is a mainly transverse fiber layer containing the tractus olfactorius lateralis superior. The fibers of layer 3 run to some extent in a sagittal direction.

The basal region is less voluminous. Rostrally, the nucleus olfactorius anterior represents a transitional cell group partly pertaining to olfactory bulb and basal lobus hemisphaericus. Caudalward, and without distinct demarcation, said nucleus is continuous with the 'primordium hippocampi'. In addition, the basal region contains two or three rather diffuse cell groups, of which the medial neighborhoods represent a massa cellularis reuniens, pars superior, blending with the rostral part of the thalami ventralis, i.e. with the primordium hippocampi *autorum*,

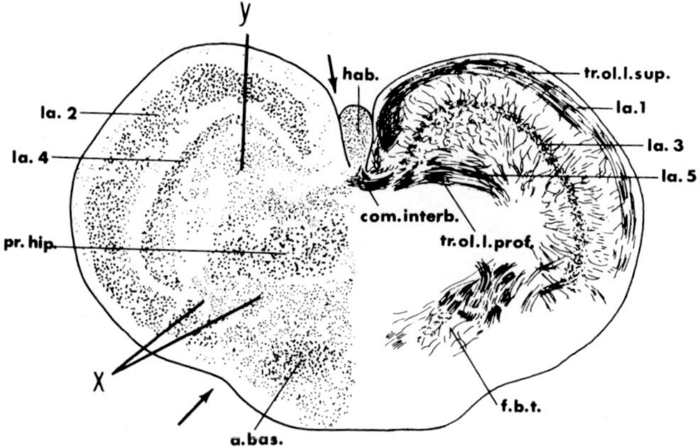

Figure 184B. Cross-section through the forebrain of Myxine, cytoarchitectural picture at left, fiber tracts at right (combined after JANSEN, 1930, from NIEUWENHUYS, 1967. a.bas.: (hypothalamic) preoptic grisea; com. interb.: commissura dorsalis sive pallii (commissura interbulbaris); f.b.t.: basal forebrain bundle; hab.: ganglion habenulae (antimeric portions fused in midline); la. 1, 2, 3, 4, 5: Jansen's layers 1–5 of pallium; pr. hip.: eminentia thalami ventralis ('primordium hippocampi'); tr.ol.l.prof. (sup.): tractus olfactorius lateralis profundus (superficialis). Added designations: x: nondescript subdivisions of nucleus basalis; y: caudal portion of nucleus olfactorius anterior. Arrows: sulcus telencephalodiencephalicus.

which can be considered homologous to the eminentia thalami ventralis of Amphibians. The area basalis of JANSEN represents, in my opinion, the rostral portion of the diencephalic, hypothalamic preoptic cell groups, the sulcus basalis anterior of JANSEN being a ventral part of sulcus telencephalo-diencephalicus. The transition of preoptic cell groups to basal telencephalic ones corresponds to the massa cellularis reuniens, pars inferior, of the nondescript extensive hemispheric stalk.

The main *communication channels* of the Myxinoid telencephalon comprise essentially the following systems. The *tractus olfactorius lateralis* consists of pars superficialis and pars profunda, corresponding, respectively, to layers 1 and 3, reaching various parts of the hemisphere. The *tractus olfactorius ventralis* largely joins the basal forebrain bundle. JANSEN (1930) distinguished three main components, namely tractus bulbo-preopticus, tractus bulbo-hypothalamicus, and tractus bulbo-tegmentalis.

In addition to these bulbar fibers, those of lobus hemisphaericus display two channels connecting with the diencephalon, and described as *dorsolateral* and *ventral fiber systems*. The former includes tractus olfactohabenularis and tractus olfactothalamicus. The ventral system participates substantially in the formation of the *basal forebrain bundle*, and comprises olfactohypothalamic, olfactotegmental, and 'olfactohippocampal' (i.e. ventral olfactothalamic components.[76] In addition, the basal forebrain bundle contains a 'striothalamic' and 'striohypothalamic' component related to the basal cell groups of lobus hemisphaericus. These cell groups are also connected with ventral thalamus ('primordium hippocampi') and habenula through tractus 'olfactocorticalis' and olfactohabenularis. It seems most probable that all the aforementioned fiber systems are, to a variable degree, reciprocal. As regards the basal forebrain bundle, there is little doubt that this channel includes ascending components, carrying non-olfactory sensory input from more caudal parts of the neuraxis into the telencephalon.

As regards the *commissural systems*, the commissura interbulbaris can be evaluated as a modified *commissura dorsalis sive pallii*. The majority of its fibers seem to be decussating components of tractus olfactorius lateralis profundus et superficialis, which include connections with the pallium. Some intrinsically pallial components cannot be excluded.

The *commissura anterior*, which represents a boundary structure between telencephalon and diencephalon includes some fibers from the olfactory tracts and particularly fibers of the basal forebrain bundle, providing crossed components of diverse channels.[77] The detailed composition of this commissure, however, is insufficiently known (Jansen, 1930).

4. Selachians

The telencephalon of Selachians (Plagiostomes, Chondrichthyes) displays paired rostral evagination together with a variable but on the whole fairly substantial telencephalon impar (unpaired rostral evagination). A true posterior pole, such as present in Cyclostomes, some Osteichthyes, and in all Tetrapod Vertebrates, is thus missing. In Chimaeroids, the telencephalon impar manifests some degree of eversion, but inversion remains here characteristic for the paired evaginated telence-

[76] The connections of 'primordium hippocampi' described as tractus corticohabenularis, tractus hippocampohypothalamicus et thalamicus by Jansen (1930) can be included into the intradiencephalic communication channels briefly mentioned on p. 160 of section 2 in chapter XII.

[77] Some tracts, however, such as the crossed component of tractus olfactohypothalamicus, may decussate in the postoptic commissural complex (tractus olfactohypothalamicus cruciatus of Jansen, 1930).

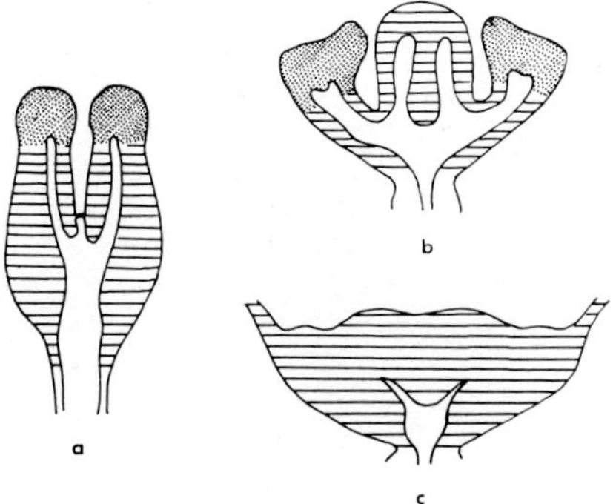

Figure 185 A. Diagrammatic horizontal sections through the telencephalon of three orders of Plagiostomes (from NIEUWENHUYS, 1967). a: Chimaera (Holocephalian); b: Scilliorhinus (Squalidan); c: Raja (Rajidan, Batoidean); dotted part: bulbus olfactorius; horizontal hatching: lobus hemisphaericus.

phalon, and, moreover is displayed by the entire telencephalon in apparently all other Plagiostome forms. Again, while in Holocephalians (Chimaeroids) the olfactory bulb retains a rostral location, it tends to become laterally displaced in Squalidae and Rajidae, with an 'olfactory stalk' of variable length, while parts of the lobus hemisphaericus protrude in a rostromedial direction (cf. Fig. 185A).

Both hemispheres are caudalward joined in an extensive supraneuroporic lamina terminalis which protrudes dorsocaudalward. In some instances, e.g. as seen in Hexanchus (Fig. 185B) the joint caudal prominence is topped by a fold of lamina terminalis enclosing a recessus dorsalis. JOHNSTON (1911a) believed that the caudal junction was caused by secondary fusion of the medial hemispheric walls, and BÄCKSTRÖM (1924) assumed that a migration of neuronal elements into the unpaired roof plate rostral to velum transversum was involved in said fusion (cf. also NIEUWENHUYS, 1967). My own observations convinced me that the connection between the two hemispheres is the result of an essentially intussusceptional dorsocaudal expansion of the lamina terminalis pertaining to the primary telencephalon impar.[78] In many instances, considerable rostral appositional growth takes place, such that the two hemispheric lobes appear fused in the midline (cf.

[78] Cf. Figures 49D (p.140), 51C–F (pp.144–146), 62 (p.173), 64 (p.175), 70 (p.194) of volume 3/II. Other relevant Figures depicting the Plagiostome telencephalon are Figures 245 to 250 (pp.481–491) of the cited volume. The epithelial roof plate becomes merely displaced rather than 'invaded' as apparently assumed by BÄCKSTRÖM (1924).

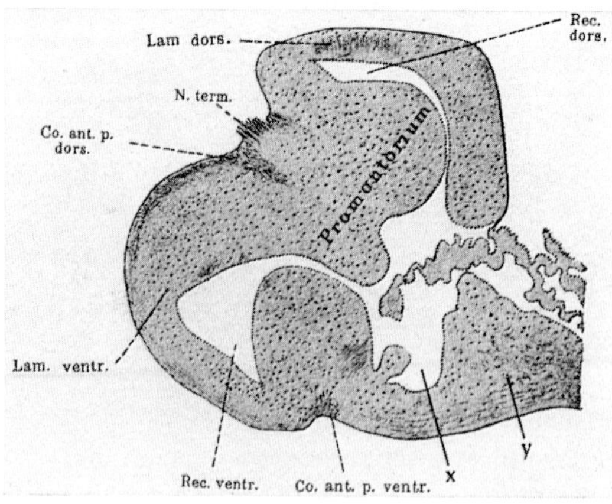

Figure 185 B. Sagittal section, slightly lateral to midline, through telencephalon and adjacent diencephalic neighborhoods in the Squalidan Hexanchus griseus (from KAPPERS, 1947). The extensive lamina terminalis beginning ventrally at commissura anterior, pars ventralis, includes lamina ventralis, promontorium, and lamina dorsalis. Added leads are x: preoptic recess; y: chiasmatic ridge.

Figs. 185, 186, 190, 191). The lateral ventricles may remain relatively wide, as in many Squalidae, or become narrow slits, as e.g. in various Holocephalians, or again, become narrow tubular channels, as in various Batoids. The ventricle of the olfactory bulb may be retained (e.g. in the Squalid Galeus) but more commonly tends to become obliterated. The choroid plexus of the hemispheric ventricle displays numerous taxonomically related variations *qua* extent and development.

Relevant data on the Plagiostome telencephalon, with numerous bibliographic references to additional authors, can be found in the publications by BÄCKSTRÖM (1924), BECCARI (1943), FAUCETTE (1969), GERLACH (1947), HAFFERL (1926), JOHNSTON (1906, 1911a), KAPPERS (1947), KAPPERS *et al.* (1936), K. and NIIMI (1969),[79] and NIEUWENHUYS (1967).

As regards the *olfactory bulb*, whose Plagiostome configurational features were mentioned above, and whose general structure was dealt with in section 2, it will here be sufficient to point out that in Chimae-

[79] It is of interest to compare our own paper on the brain of Chimaera (K. and NIIMI, 1969) with the about simultaneously published paper of FAUCETTE (1969) on olfactory bulb and general hemispheric configuration of Chimaera. The interested reader will note the fundamental differences in basic morphologic approach as dealt with in section 6 *et passim* of chapter VI, volume 3/II.

roids a rather distinct dorsal and ventral subdivision is displayed. Each of these receives its own bundles of olfactory fila, but a clear-cut separation into dorsal and ventral olfactory nerves, as is the case in Gymnophiona is not thereby unambiguously provided, although such distinction was made by FAUCETTE (1969). Near their caudal end, both bulbs show cellular groupings that superficially resemble an accessory olfactory bulb, but the presence of a vomeronasal organ with its nerve and a true accessory olfactory bulb is highly questionable in all Fishes, perhaps with the possible exception of Dipnoans.

Generally speaking, the *lobus hemisphaericus* of Plagiostomes, like that of Cyclostomes, displays a dorsal (pallial) and a ventral (basal) cell population, which, depending on taxonomic forms and ontogenetic stages, may, or may not be delimited from each other by relatively cell-free zonae limitantes in the lateral and medial paired evaginated and inverted hemispheric wall. An ill-defined, nondescript nucleus olfactorius anterior is commonly located in the open boundary neighborhood between olfactory bulb and lobus hemisphaericus respectively in parts of the olfactory stalk.

The *pallium* contains a periventricular layer vaguely demarcated from a slightly more dense 'corticoid' peripheral one. The 'corticoid' aspect is usually more pronounced at late ontogenetic stages than in adult forms. Although the pallial cell groups do not show a distinctive cytoarchitectural parcellation, three topologic neighborhoods can be roughly distinguished, particularly on the basis of their ontogenetic development. These neighborhoods are the medial zone D_3, which often displays a considerable thickening, the intermediate zone D_2, and the lateral D_1. The griseum mediale pallii or 'cortex medialis', which, particularly caudalward, may be conspicuously thickened (cf. e.g. Figs. 186B, C, 187), represents the 'primordium hippocampi'. The antimeric cortices mediales may fuse in the midline within the expanded dorsal portion of supraneuroporic lamina terminalis. The laterally adjacent zone D_2 or griseum dorsale pallii ('cortex dorsalis') represents the 'general pallium'. Zone D_1 or griseum laterale pallii ('cortex lateralis') is a pallial component in all Gnathostome Anamnia, which becomes included, by internation (or introversion) into the secondary basis of Amniota, as a part of the basal ganglia complex.[80]

[80] The details of these morphogenetic events were dealt with in section 6, chapter VI of volume 3/II. The basic concepts of internation (REMANE) or introversion (SPATZ), together with related concepts of morphogenesis, were discussed on p. 172, section 1B, chapter VI

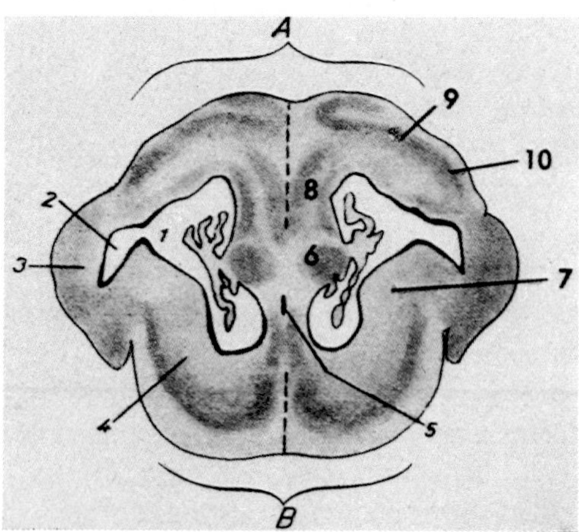

Figure 186 A. Cross-section through the telencephalon of an advanced embryo of the Squalidan Scyllium (from BECCARI, 1943). A: pallium with supraneuroporic rostral expansion of lamina terminalis; B: basis with infraneuroporic rostral expansion of lamina terminalis; 1: lateral ventricle; 2: bulbar recess (rhinocoele); 3: bulbus olfactorius; 4: griseum basilaterale inferius (n. basilat. inferior); 5: remnant of recessus neuroporicus; 6: supraneuroporic portion of griseum basimediale superius; 7: griseum basilaterale superius; 8: griseum mediale pallii ('primordium hippocampi'); 9: griseum dorsale pallii; 10: griseum laterale pallii.

The *basal (or subpallial) portion* of the lobus hemisphaericus contains laterally two only roughly separable diffuse periventricular cell aggregates, nucleus basilateralis superior (B_1) and nucleus basilateralis inferior (B_2), which are commonly referred to as 'striatal primordium'. In the medial wall, the likewise diffuse periventricular aggregates of nucleus basimedialis inferior (B_3) and basimedialis superior (B_4) can be recognized. These paraterminal or preterminal grisea are generally referred to as 'septum' or 'septal area'. In most forms, a dorsal, caudalward essentially supraneuroporic portion of nucleus basimedialis superior extends toward the ventral edge of griseum mediale pallii, with which it may or may not partly appear fused. Because of this supraneuroporic expansion of the nucleus basimedialis superior into the pallial neighborhood D_3, JOHNSTON (1913b) evaluated the superior basimedial

of volume 3/II. It will also be recalled that HOLMGREN and others, while interpreting cortex medialis as primordium hippocampi, and cortex dorsalis as general cortex, consider cortex lateralis to be the Amniote piriform cortex. This latter, in my interpretation, is a derivate of cortex dorsalis or 'general cortex' respectively 'general pallium'.

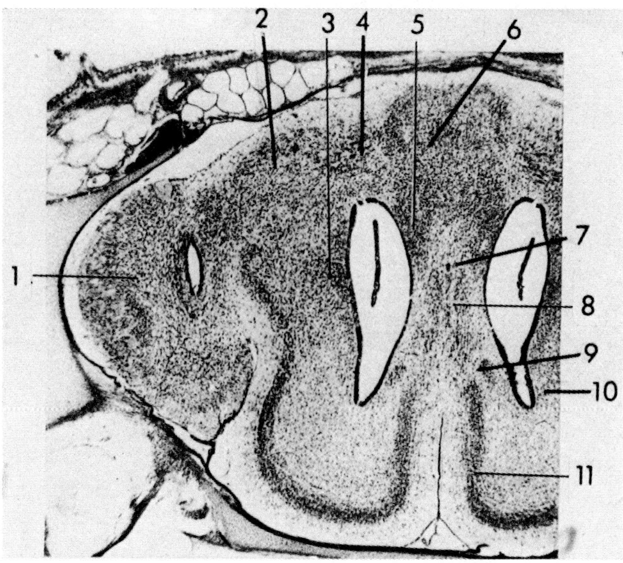

Figure 186B. Cross-section through the telencephalon of a 95 mm long embryo of Scyllium canicula (from HAFFERL, 1926, with modified and added designations). 1: bulbus olfactorius; 2: griseum laterale pallii; 3: griseum basilaterale superius; 4: griseum dorsale pallii; 5: griseum basimediale superius, pars lateralis; 6: griseum mediale pallii: 7: remnant of recessus neuroporicus; 8: griseum basimediale superius, pars (infraneuroporica) medialis; 9: griseum basimediale inferius, pars lateralis; 10: griseum basilaterale inferius; 11: cortex olfactoria (area basalis superficialis).

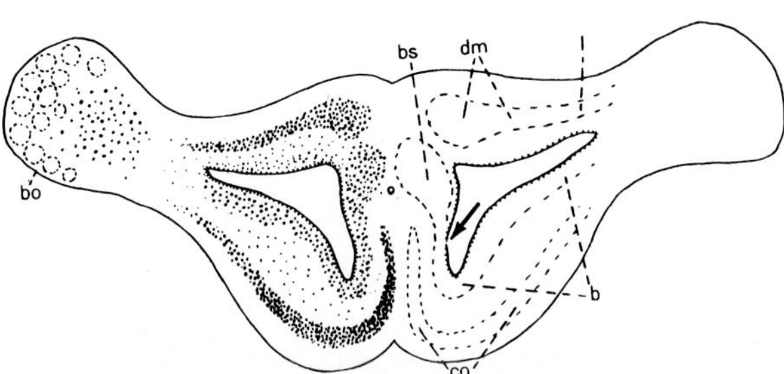

Figure 186C. Simplified drawing of cross-section through the telencephalon of adult Scyllium (from K., 1927). b: grisea basilateralia (superius et inferius); bo: bulbus olfactorius with obliterated ventricle; bs: griseum basimediale superius with supra- and infraneuroporic portion (between the antimeric grisea, the unlabelled remnant of neuroporic recess); co: cortex olfactoria; dm: griseum dorsale et mediale pallii; l: griseum laterale pallii. Added arrow indicates sulcus fb_3, the approximate boundary between superior and inferior basimedial (paraterminal) grisea.

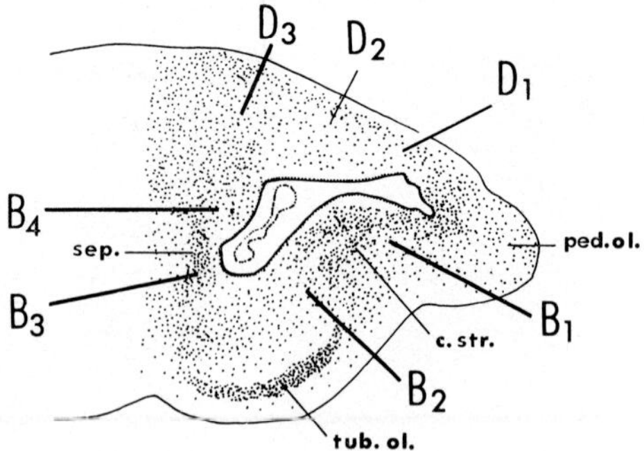

Figure 186 D. Cross-section through the telencephalon of Scylliorhinus caniculus (from NIEUWENHUYS, 1967). c.str.: 'corpus striatum'; ped.ol.: pedunculus olfactorius; sep.: 'septum'; tub.ol.: 'tuberculum olfactorium'. The designation of the pallial and basal zones in the present author's terminology have been added.

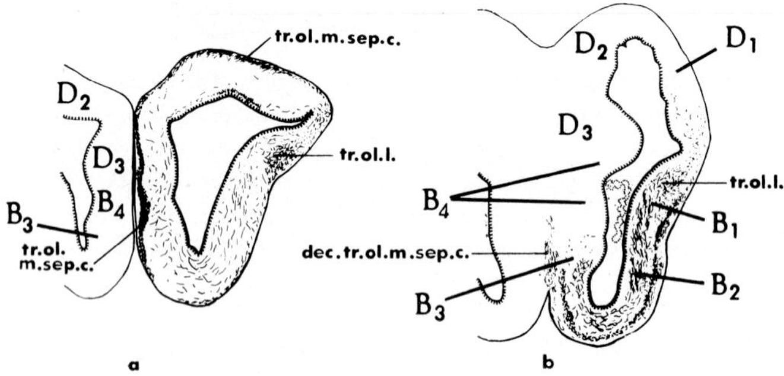

Figure 186 E. Cross-sections through lobus hemisphaericus of Squalus acanthias, drawn from *Weigert-Pal* preparations (from NIEUWENHUYS, 1967). dec.tr.ol.m.sep.c.: decussation of the tractus olfactorius medialis septi cruciatus; tr.ol.l.: tractus olfactorius lateralis; tr.ol.m.sep.c.: tractus olfactorius medialis septi cruciatus. The designations of pallial and basal zones have been added.

griseum as 'primordium hippocampi' (cf. Fig. 205 D) and, quite consistently, carried out that homologization of said griseum throughout the Vertebrate series in his highly interesting paper on the septum, hippocampus, and pallial commissures in Reptiles and Mammals. In contradistinction to this interpretation, I believe that a comparison of appropriate ontogenetic stages establishes, beyond reasonable doubt, the homology of JOHNSTON's 'primordium hippocampi' with nucleus basimedialis superior in Anamnia such as Dipnoans and Amphibians, as well as the evaluation of this griseum as a basal, and not pallial, compo-

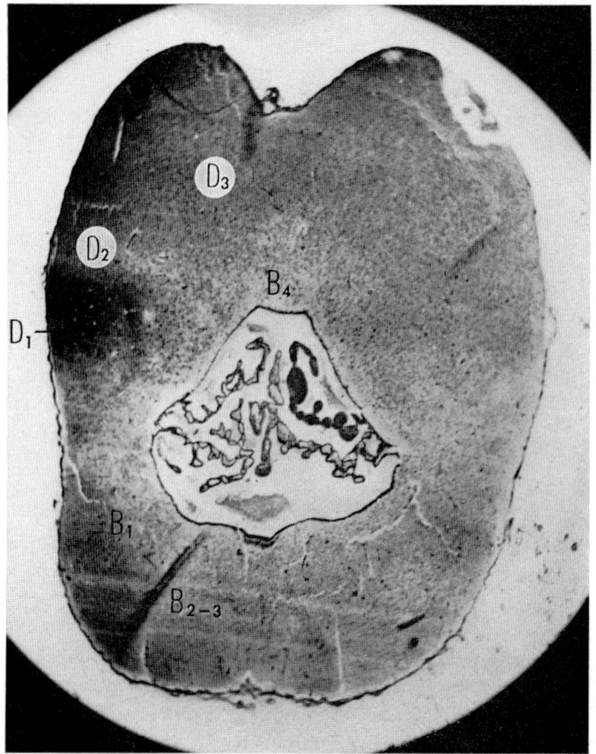

Figure 186 F. Cross-section *(Nissl stain)* through the telencephalon impar of Acanthias, showing the arrangement of D and B topologic neighborhoods. An unpaired (fused) supraforaminal part of B_4 is indicated.

nent of paraterminal body *sensu latiori* (i.e. including nucleus basimedialis inferior).

In addition to its periventricular cell population, the basis displays a commonly very conspicuous cortical or 'corticoid' layer, respectively cell plate without definite stratification, except for a peripheral 'molecular' or 'zonal' stratum. It is separated from the periventricular cell population by a zone relatively poor in cells, and is much more densely crowded and more clearly outlined than the pallial 'corticoid' layer. This basal cortex olfactoria, definitely related to olfactory input, is also referred to as 'olfactory tubercle'. Its medial edge or border tapers out dorsalward toward the nucleus basimedialis superior and is sometimes included with the paraterminal grisea as 'nucleus medialis septi'. As a topologic neighborhood, the medial portion of this cortex within the

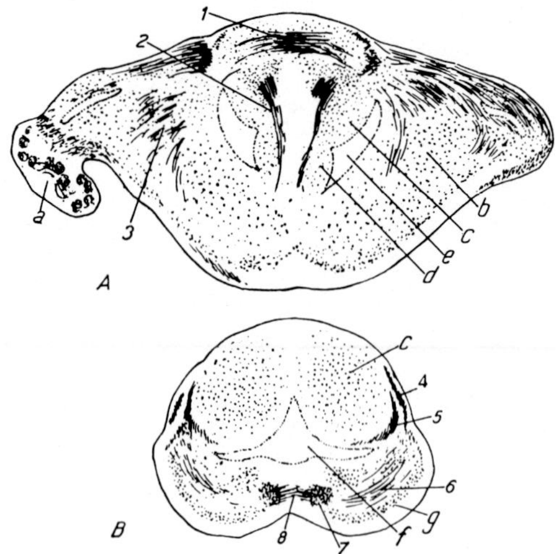

Figure 187. Cross-sections through the telencephalon of Galeus canis (redrawn after KAPPERS, 1921, from BECCARI, 1943) A Level of lateral ventricles. B Level of ventriculus impar telencephali. 1: commissura pallii; 2: 'tractus medianus' ('fornix'); 3: 'tractus olfactocorticalis lateralis'; 4: tractus pallii; 5: 'tractus taeniae dorsalis'; 6: 'tractus taeniae ventralis' and region of lateral forebrain bundle; 7: medial forebrain bundle; 8: commissura anterior; a: olfactory bulb; b: griseum basilaterale superius; c: griseum mediale pallii (in A at fusion with griseum basimediale superius); d: griseum basimediale inferius; e: lateral ventricle; f: ventriculus impar; g: cortex olfactoria.

B_3 zone corresponds to the cell populations of *Broca's diagonal band* as displayed by Mammals. Again, the cortex olfactoria represents the area ventralis *sive* ventralis anterior of the author's terminology. An area ventralis *sive* ventrolateralis posterior, corresponding to the cortical amygdaloid nucleus of Mammals, occurs only in the paired caudalward evaginated and inverted type of telencephalon as displayed by Tetrapods (Amphibia and Amniota). It is thus not present in Plagiostomes.

Figure 188 A–C. Cross-sections (hematoxylin-eosin stain) through the telencephalon of Chimaera colliei (from K. and NIIMI 1969). A at level of lateral ventricles and rostral end of lamina terminalis with recessus neuroporicus. B at opening of right lateral ventricle into ventriculus impar telencephali. C at level of everted telencephalon impar with bulging basal lobes. fo: fornix; lf: lateral forebrain bundle; lt: lamina terminalis; mf: medial forebrain bundle; rn: recessus neuroporicus; tp: tractus pallii; tt: tractus taeniae. B, D, fb, fd with subscripts indicate telencephalic longitudinal zones and ventricular grooves as dealt with in volume 3/II.

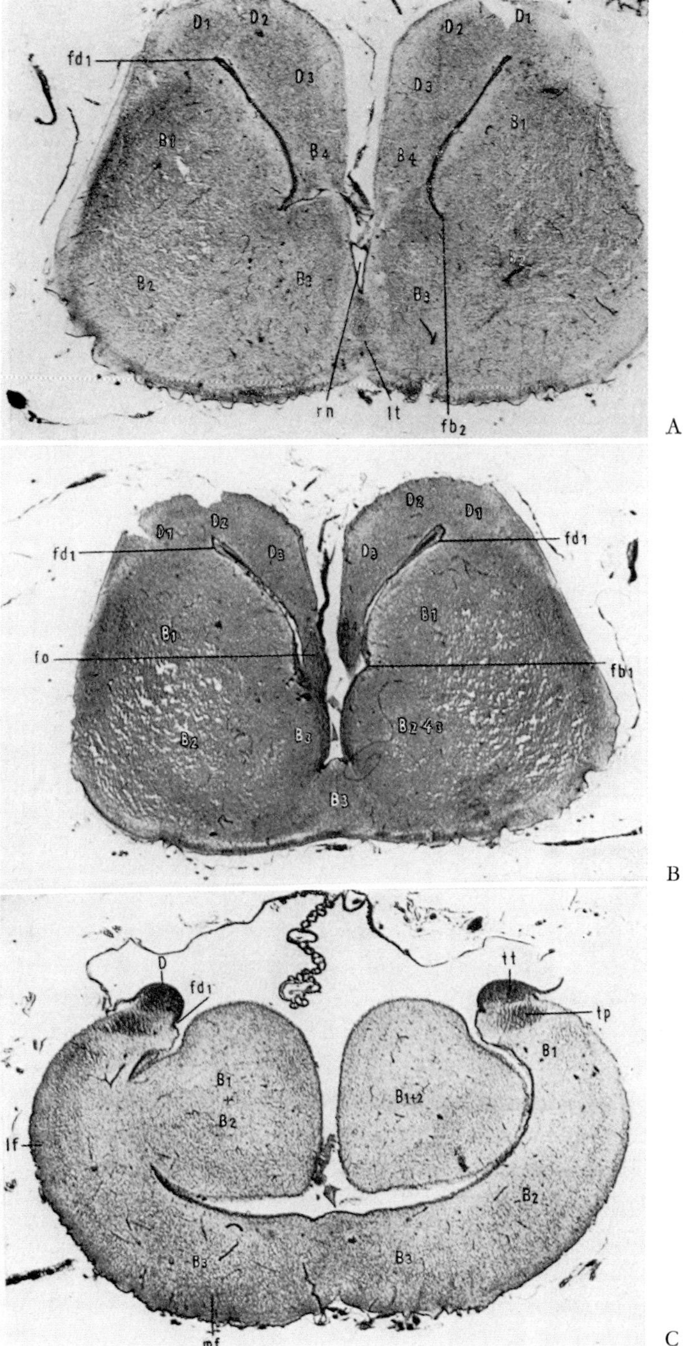
D1
D2
fd1
D3
B1
B4
B2
B3
rn
lt
fb2
A
fo
fb1
B
D
tt
tp
B1+2
lf
mf
C

Figures 186–188 display relevant features of the Chondrichthyan lobus hemisphericus as seen in cross-sections. On the whole and particularly also in Chimaeroids, it can be said that, at adult stages the grisea manifest a very diffuse cellular arrangement precluding a well-defined delimitation of cytoarchitectural 'units' but, nevertheless, permitting the identification of the typical subdivisions as topologic 'open neighborhoods' (K. and NIIMI, 1969). In addition to the partial eversion of the Chimaeroid pars impar telencephali, another peculiarity displayed by these Holocephalians is the bulging or internation of parts of nucleus basilateralis superior and inferior into the ventricular impar as so-called lobus basalis (Fig. 188 C).

As regards the Chondrichthyan telencephalic *communication channels*, the medullated and non-medullated fiber tracts are generally diffuse, displaying few well delimited tract-like bundles, as e.g. pointed out by GERLACH (1947).

The *olfactory connections* are strongly developed and doubtless reach, to a greater or lesser extent, all grisea of the lobus hemisphaericus, as e.g. also noted by NIEUWENHUYS (1967). This latter author, essentially following BÄCKSTRÖM (1924), distinguishes the following bundles shown in the diagram of Figure 189 A: tractus olfactorius lateralis, tr. olf. medialis septi cruciatus, tr. olf. medialis septi rectus, tr. olf. medialis cruciatus, moreover a pallial and a subpallial radiatio olfactoria.

The intrinsic non-medullated and medullated telencephalic fiber connections intermingle or intersect with the fibers pertaining to the long systems. GERLACH (1947) described intrinsic short systems as 'association fibers', including a tractus septo-corticalis medialis.

With regard to the *caudal communication channels*, the medial forebrain bundle represents a fairly compact and partly well medullated compound fiber system connecting the dorsomedial and medial grisea of the telencephalon with the diencephalon. It includes fibers from griseum mediale pallii reaching caudal hypothalamic grisea and can be interpreted as a primordial fornix which seems to correspond to the tractus medianus of EDINGER as described by GERLACH (1947). Some fibers from the dorsomedial and medial telencephalic grisea join the tractus taeniae described further below.

Fibers from or to the pallium (mainly griseum laterale and intermedium, i.e. D_1 and D_2, as well as parts of griseum mediale D_3), together with others related to the basilateral grisea (B_1, B_2), the cortex olfactoria, and lateral portions of griseum basimediale inferius (B_3) become included in the lateral forebrain bundle. This is a rather conspicuous,

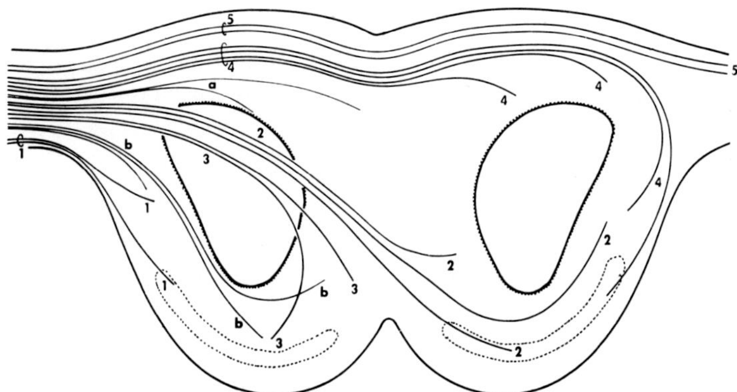

Figure 189 A. Diagram indicating the 'secondary olfactory connections' in the telencephalon of a Shark according to NIEUWENHUYS' interpretation (from NIEUWENHUYS, 1967). 1: tractus olfactorius lateralis; 2: tractus olfactorius medialis septi cruciatus; 3: tractus olfactorius medialis septi rectus; 4: tractus olfactorius medialis cruciatus; 5: commissura olfactoria; a: radiatio pallialis olfactoria; b: radiato subpallialis olfactoria. The bundles labelled 2 and 3 pass through the rostral wall of the lobus hemisphaericus.

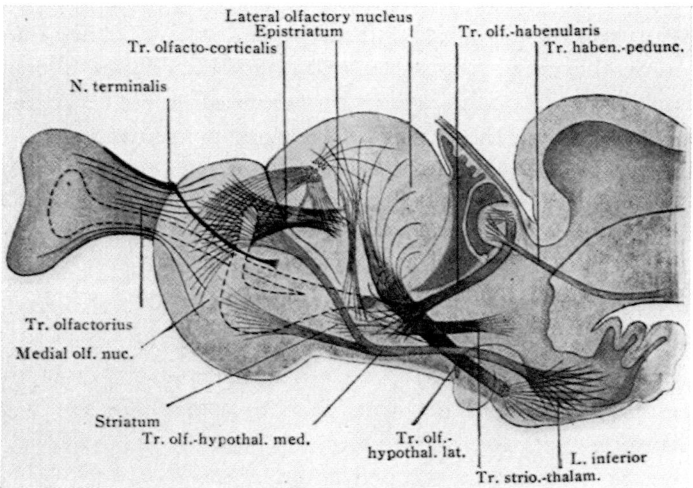

Figure 189 B. Main communication channels in the forebrain of a Squalid Plagiostome as interpreted by JOHNSTON on the basis of his own studies combined with descriptions by other authors (from JOHNSTON, 1906). The medial surface of telencephalon is drawn, and the fiber tracts are projected upon it. 'Epistriatum' is probably griseum pallii laterale, 'Lateral olfactory nucleus' includes probably dorsal and media l pallial grisea with radiations of tractus pallii, here designated as Tr.olf.-hypothal. lat.; 'Medial olf.nuc.' represents basimedial grisea, and 'Striatum' basilateral grisea. Tr.olf.-hypothal.med.: medial forebrain bundle; tr.olf.-habenularis: stria medullaris system; Tr.strio-thalam.: lateral forebrain bundle.

partly medullated fiber complex connecting with the diencephalon. Both lateral and medial forebrain bundle tend to merge into each other, thus forming a common system, the basal forebrain bundle,[81] of which some components decussate in the commissura anterior. The question whether fibers of basal forebrain bundle include direct connections between telencephalon and regions caudal to diencephalon (e.g. nucleus reticularis tegmenti) cannot be unequivocally answered on the basis of available reliable data, but some such connections cannot be excluded. Although many fibers of basal forebrain bundle connect with the hypothalamus, others are related to thalamus ventralis, thalamus dorsalis, and even epithalamus (nucleus habenulae).

The tractus taeniae includes connections of the pallium, particularly of griseum mediale (D_3 with cortex medialis), which form a conspicuous, substantially medullated, and compact tract running along the attachment (taenia) of roof plate to diencephalon.[82] Rostrally, the tractus taeniae lies a short distance medially to the tractus pallii (cf. further below). At more caudal diencephalic levels, before ending in the habenular grisea, it runs dorsally to that tract. Some exchange between the two neighboring systems seems to occur. In addition, fibers from the basal forebrain bundle and from ventral diencephalic grisea (e.g. preoptic region) join the tractus taeniae along its course.[83]

The essentially medullated tractus pallii includes connections of the pallium (D_1, D_2 and presumably also D_3) which do not join the lateral forebrain bundle but are gathered in a rather well defined tract running along the dorsolateral region of the caudal portions of telencephalon, in part fairly close to the surface. Still more caudally, it becomes located laterally to basal forebrain bundle and to tractus taeniae. Assuming a rather superficial position, it gradually shifts basad and disappears in the region of tuberculum posterius. Some of its fibers cross in the postoptic commissural system and others mingle with fibers of basal forebrain bundle. All these enumerated communication channels of the tel-

---

[81] Connections included within that channel are e.g. 'tractus septo-striohypothalamicus et hypothalamicus', 'olfacto-hypothalamicus', 'olfacto-habenularis', 'cortico-habenularis', and 'septo-habenularis'.

[82] In Chimaeroids, the tractus taeniae runs along the everted, rudimentary caudal pallial portion of the telencephalon impar.

[83] Thus, the tractus taeniae receives the fiber systems variously described as 'tractus olfactohabenularis', 'tractus septohabenularis', 'corticohabenularis', and 'habenulostriaticus'.

encephalon can be regarded as reciprocal, i.e. as including descending and ascending systems in variable and undefined proportion. The significance of ascending channels will again be pointed out further below.

The *commissural systems* consist of commissura dorsalis *sive* pallii and commissura anterior. The commissura dorsalis is variously developed with respect to the diverse taxonomic forms. Its rostral part contains decussating olfactory tract fibers (commissura olfactoria of NIEUWENHUYS, 1967, cf. Fig. 189A) while its caudal part (commissura pallii *sensu strictiori*) includes pallial interconnections[84] and apparently decussating fibers of tractus taeniae and tractus pallii (decussatio interhemisphaerica, commissura pallii anterior, commissura hippocampi of BECCARI, 1943). A caudalmost portion contains essentially non-medullated fibers (commissura pallii posterior of BECCARI). In Chimaeroids, whose pallium becomes an everted rudiment in the telencephalon impar, a commissura dorsalis *sive* pallii is entirely missing. There is, however, little doubt that, as also in Teleosts and Ganoids, fiber systems corresponding to those of commissura dorsalis *sive* pallii join the system of commissura anterior.

Discounting the just mentioned component characteristic for Holocephalians with everted telencephalon impar, the commissura anterior of Plagiostomes contains decussating fibers of medial and also lateral forebrain bundle, as well as fibers of medial and lateral olfactory tract. An interbulbar commissural component may be present. In forms with extensive infraneuroporic expansion of lamina terminalis, such as in Hexanchus, a pars ventralis and a pars dorsalis of the anterior commissure can be distinguished (cf. Fig. 185B, and KAPPERS, 1947). On the whole, the details of fiber connections passing through the telencephalic commissures and 'tracts' remain very poorly elucidated.[85] Fig-

[84] Thus, as e.g. discussed by KAPPERS *et al.* (1936) and BECCARI (1943) some authors considered certain fiber systems of commissura pallii to represent a 'corpus callosum'. Such interpretation, however, does not seem justified with respect to submammalian Vertebrates (cf. the discussion on pp. 619–623 of volume 3/II).

[85] The here adopted non-committal terminology of fiber tracts (cf. also K. and NIIMI, 1969) differs in some respects from that proposed by JOHNSTON (1911a) and KAPPERS (1921, 1947). Again, JOHNSTON's and KAPPER's terminologies are by no means in full correspondence with each other. A critical review is given by BECCARI (1943), who comments: '*Nel telencefalo dei Selaci esiste inoltre un numero considerevole di fasci che sono stati descritti e denominati in maniera assai differente dai ricercatori che li hanno studiati*'. The difficulties in disentangling the network of neuronal connections in this particular case as well as in general for the Vertebrate neuraxis are evidently extreme. Although some progress can

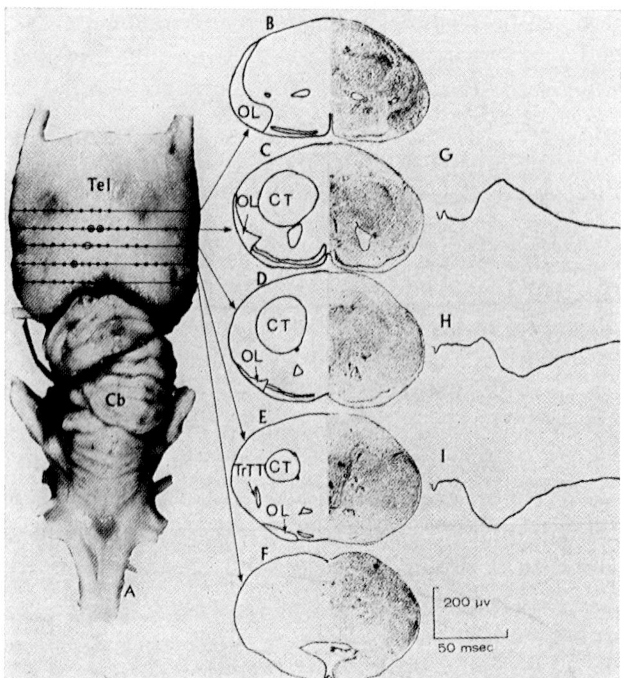

Figure 190. Dorsal view of the brain (A) in the Nurse Shark (Ginglymostoma) with schematic illustration of the sites of electrode penetrations into pallium. The transverse lines are separated by 2 mm, and the dots by 1 mm; curves are evoked field potentials, and circles the penetration from which maximal response was obtained. B to F cross-sections corresponding to transverse lines. Right halves show cellular populations in *Nissl stain.* G to I indicate ipsilateral field potentials upon optic nerve stimulation, downward deflections corresponding to negativity (from COHEN *et al.*, 1973). Cb: cerebellum; CT: 'central telencephalic nucleus'; OL: 'lateral olfactory area; Tel: telencephalon; Tr TT: 'thalamotelencephalic tract' (lateral forebrain bundle).

ures 189 A and B illustrate some acceptable interpretations concerning these channels.

Reverting to the concept, as expressed in our study of Chimaeroids (K. and NIIMI, 1969), that the basal forebrain bundle represents a complex communication channel including not only descending but also ascending fibers, i.e. input to the telencephalon, the recent electrophy-

---

doubtless be made, I have remained unconvinced by the various recent attempts to trace details, *per fas aut nefas*, by means of the terminal degeneration and other newly devised methods.

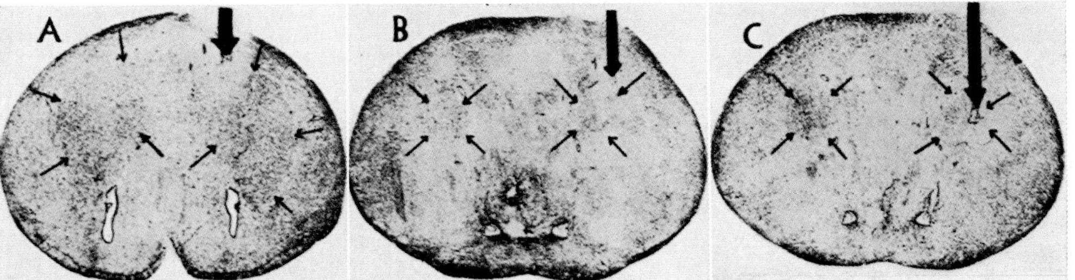

Figure 191. Cross-sections *(Nissl stain)* through lobus hemisphaericus of the Nurse Shark, indicating locus from which ipsilateral optic nerve stimulations elicited evoked potentials (from COHEN *et al.*, 1973). A Most rostral level of evoked field potential activity (positive wave), approximately corresponding to level C of Figure 190. B and C Most caudal levels (negative wave) approximately corresponding to level E of Figure 190. Small arrows indicate the so-called 'central telencephalic nucleus' (corresponding to area D_2); dorsal and ventral boundaries of active area are marked by small lesions indicated by large arrows.

siological identification of a 'visual area' in the Plagiostome telencephalon may be mentioned (COHEN *et al.*, 1973). Our own, more generalized assumption of ascending sensory input into telencephalon from diencephalon was based on the present author's early studies concerning the Amphibian brain and including the use of the *Golgi impregnation*. HERRICK (1948) likewise reached a similar conclusion on the basis of his numerous detailed investigations, and assumed that such input involved 'precursors of the ascending thalamic radiations of mammals'. He believed that they 'arise from the undifferentiated nucleus sensitivus of the thalamus', and found no evidence 'of any separation among them of projection tracts related with different functional systems'. Yet, the lateral geniculate complex of Amphibians can be regarded as clearly pertaining to the optic system. Extrapolating from Amphibians to Anamnia in general it could be stated that the thalamus represented a '*Korrelationsgebiet zwischen aufsteigenden sensiblen Bahnen (einschliesslich Opticus) und dem Telencephalon*'. '*Bei den Anamniern, bei denen die ebengenannten sensiblen Impulse nur zum geringsten Teil bis in das Endhirn geleitet werden und deren normales Verhalten hauptsächlich an die Funktion des Mittelhirnapparates geknüpft ist, bleibt der Thalamus auf einer tiefen Differenzierungsstufe stehen*' (K., 1927, pp. 221–222).[86]

---

[86] As regards this degree of differentiation, it will be sufficient to compare e.g. Figure 92A with e.g. Figure 122A. With regard to the functional significance of midbrain versus telencephalon in Anamnia for apparently 'normal' behavior, reference may be made to the

In the above-mentioned experiments by COHEN *et al.* (1973), optic nerve stimulation in the Nurse Shark (Ginglymostoma cirratum) evoked 'short latency telencephalic field potentials located in the ipsilateral posterior central nucleus' (cf. Fig. 190). The cited authors consider this griseum to be 'a well-defined visual area' and claim that their findings 'challenges classical formulations of forebrain evolution'. Although their observations can be regarded as rather convincing, one could here reply (1) that the concept of a 'well-defined visual area' remains open to various qualifications, and (2) that, for the reasons given above, their findings very definitely do not contradict nor 'challenge classical formulations of forebrain evolution'.

The grisea shown to receive optic input are designated by the cited authors as the 'central telencephalic nucleus' (Fig. 191), and seem to represent a deeper layer of the pallial region D_2, that is, of griseum dorsale pallii, namely of 'general pallium' or 'general cortex'. This is in full agreement with the view, summarized in section 6, chapter VI of volume 3/II, according to which the Mammalian neocortex derives from this topologic neighborhood. The relatively deep location of the layer receiving optic input seems, moreover, in agreement with the assumption that the griseum dorsale likewise receives substantial olfactory input which would predominantly reach the more superficial layer (cf. e.g. Fig. 189 A). If the telencephalic optic input is mediated by the lateral geniculate complex receiving fibers from the contralateral eye, the ipsilateral recording of telencephalic evoked potentials could be interpreted to indicate a decussation of fibers from lateral geniculate complex to lateral forebrain bundle of opposite side by way of supraoptic or of anterior commissure, thus restoring ipsilateral input.

5. Ganoids and Teleosts; Latimeria

The Telencephalon of Osteichthyes, including that of the Crossopterygian Latimeria, consists, as in all Vertebrates, of *bulbus olfactorius* and *lobus hemisphaericus* (Fig. 192 A). This latter, however, in contradistinction to that of all other Vertebrates, is completely everted, thereby representing a telencephalon impar with common ventricle, formed by

---

systematic work by STEINER (1885–1900) discussed on pp. 338–339, 655–659, and 808 of volume 4. Finally, it will be recalled that, although '*zum geringsten Teil*' optic and other non-olfactory sensory input was presumed (in 1927) to reach the Anamniote telencephalon.

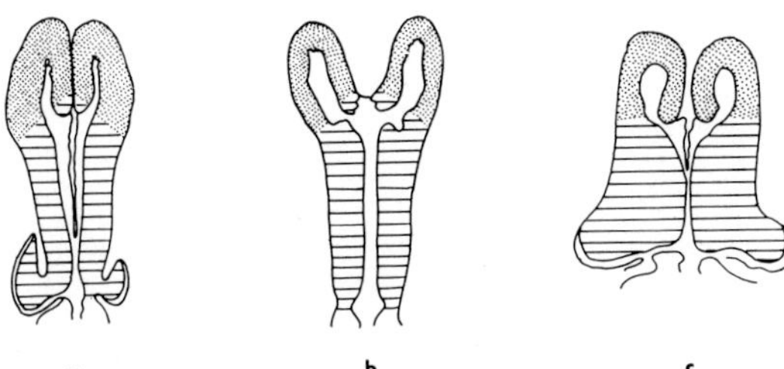

Figure 192A. Diagrammatic horizontal sections through telencephalon of three Ganoids (from NIEUWENHUYS, 1967). a: Acipenser; b: Lepidosteus; c: Polypterus. Bulbus olfactorius dotted, lobus hemisphaericus indicated by crossing lines. Polypterus is believed closely related to the extinct Crossopterygians of which Latimeria is a surviving recent form.

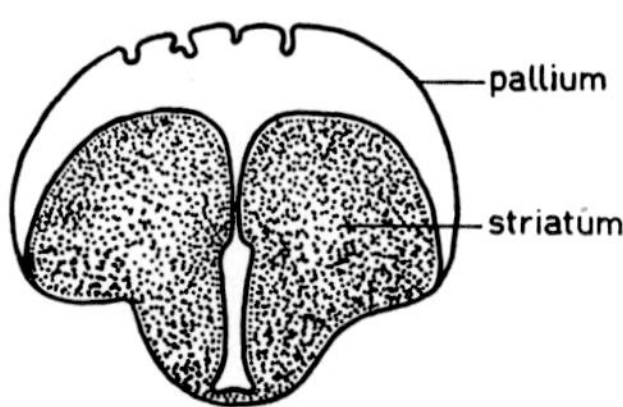

Figure 192B. Cross-section through the telencephalon of a Percid Teleost, showing RABL-RÜCKHARD's concept of 'striatum' and 'membranous pallium' (after RABL-RÜCKHARD, 1884, from NIEUWENHUYS, 1960b).

unpaired rostral evagination. The degree of eversion varies in accordance with the wide variety of taxonomic forms, and does not preclude the presence, in diverse instances, of a paired caudal evagination with true polus posterior, but nevertheless combined with eversion.[87]

Since the dorsal wall of the everted lobus hemisphaericus is provided by an expanded epithelial roof plate, topographically, but not topologically replacing the dorsal or pallial wall, RABL-RÜCKHARD (1882, 1883, 1884, 1894) and subsequently many other authors, including also EDINGER (1896b), assumed that the rather massive, thickened walls of the lobus hemisphaericus merely represented the basal ganglia ('corpus striatum') of other Vertebrates, and that the roof plate wall corres-

[87] The relevant morphologic concepts concerning inversion, eversion, etc. were dealt with in sections 1B and 6 of chapter VI, volume 3/II. It will also be recalled that a moderate partial eversion is also displayed by the unpaired portion of lobus hemisphaericus in Plagiostome Chimaeroids and in Dipnoans. 'Eversion' could here essentially be conceived as 'non-inversion'.

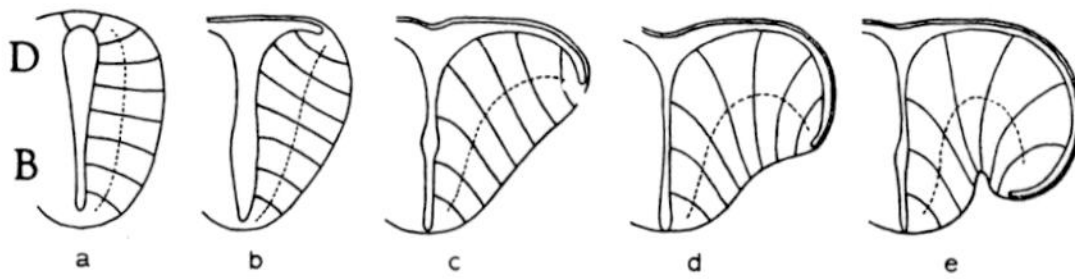

Figure 192C. Diagrams indicating the transformation of the Teleostean telencephalic wall neighborhoods of lobus hemisphaericus in the course of ontogenesis. The arbitrarily drawn lines delimiting identical wall neighborhoods correspond only in part and very roughly to the approximate limits of actual grisea (from NIEUWENHUYS, 1960b). The designations D for pallium and B for basis have been added.

ponded to a rudimentary, epithelial 'pallium' (cf. Fig. 192B). However, C. L. HERRICK (1891), GAGE (1893), and particularly STUDNIČKA (1894, 1895, 1896) reached the conclusion that the apparently 'solid' wall of the Osteichthyan lobus hemisphaericus included the pallial components of the Vertebrate telencephalon, and that the so-called membranous pallium was merely a stretched portion of the roof plate. This was confirmed by the studies of KAPPERS (1906, 1908, 1911) who stressed the concept of 'eversion' and caused EDINGER (1908) to admit, although rather cautiously, that pallial components ('Episphaerium') might be '*Seitenhälften, etwa an deren Dorsalkante*'. Further details of this eversion and relevant morphologic interpretations of the Osteichthyan telencephalic longitudinal zones were provided by the studies of HOLMGREN (1920, 1922) and by our own investigations[88] (K., 1924b, 1929; MILLER, 1940). Subsequently, the studies by NIEUWENHUYS (1960a, b, 1963, 1966), and SCHOBER (1966) recognized the well established fact that the everted telencephalon of Osteichthyans displays both a pallial and a basal neuronal cell population (Fig. 192C). The similarities as well as the divergences between the interpretations of HOLMGREN, of NIEUWENHUYS, and those resulting from our own investigations were pointed out, with further references, in section 6, chapter VI of volume 3/II. Additional details concerning the views of

[88] Because the peculiar morphogenesis of the Osteichthyan telencephalon precludes the development of a 'cerebral cortex', I omitted these forms in my early study '*Über den Ursprung der Grosshirnrinde*' (K., 1922b) but nevertheless remarked (p. 339): '*Von den Ganoiden und Teleostiern sehe ich ganz ab, da bei ihnen durch die Eversion des Vorderhirndaches die "pallialen" und die "basalen" Bestandteile sich völlig gegeneinander verschoben haben. Erst seit nicht allzu langer Zeit ist überhaupt die Auffassung* STUDNIČKAS, *dass bei diesen Typen tatsächlich ein umgestülpter "Palliumabschnitt" vorhanden ist, zur allgemeinen Anerkennung gekommen gegenüber der* RABL-RÜCKHARD'*schen Ansicht, die bei diesen Fischen nur ein Pallium membranaceum sehen wollte. Jedenfalls liegt aber hier eine von den höheren Formen gänzlich divergente Typik vor.*'

various authors who, subsequently to the analysis of the Osteichthyan telencephalon by GAGE, STUDNIČKA, KAPPERS, HOLMGREN and ourselves, failed to recognize the fundamental invariant topologic transformations resulting from eversion, are included in the papers by NIEUWENHUYS (1960a, b, and others). The argument that pallial cell groups may develop within a 'thickening *in situ*' cannot be adduced against eversion, since said thickening is clearly correlated with a complete eversion of the ependymal lining and with a corresponding lateroventral displacement of the roof plate attachment, as justly stressed by NIEUWENHUYS (1960b). Eversion is here merely combined with considerable abventricular thickening of the telencephalic alar plate wall.

In addition to the above-mentioned investigations concerning the Osteichthyan telencephalon, the following studies could be mentioned. Early attempts at an anatomical analysis were made by FRITSCH (1878) and by HALLER (1899). Among relevant subsequent investigations are those by CATOIS (1901), GOLDSTEIN (1905), HOOGENBOOM (1929), VAN DER HORST (1917), MEADER (1939), and particularly by SHELDON (1912). The considerable discrepancies in the terminologies and concepts proposed by the different authors are discussed by MILLER (1940), BECCARI (1943), KAPPERS (1947) and NIEUWENHUYS (1960b, 1967). The problems concerning configuration and topology of grisea become complicated by the considerable variations in the development of fiber systems and the uncertain interpretations as regards relevant details of their connections.

In Teleosts, for instance, HERRICK (1922, Fig. 24, p. 172) on the basis of what he considers functional factors,[89] uses the terms area olfactoria dorsalis, pars lateralis and area olfactoria dorsalis, pars dorsolateralis, for the zones D_3 and D_2, respectively. Zones D_1 and B_1 are then, together, called area olfactoria dorsalis, pars dorsomedialis. B_2 and B_3

[89] HERRICK (1922), who refers to my early investigations up to 1922, indeed recognizes that on their basis a scheme of telencephalic morphology can be carried out 'through the vertebrate series from cyclostomes to man'. According to HERRICK, however, 'this scheme is based chiefly on topographic relations with total neglect or misapprehension of the fibrous connections of the parts, and a failure to grasp essential fundamental relations in lower brains'. It can here be merely added that these 'topographic relations' subsequently were shown to be *topologic* ones, presumably encoded in the Vertebrate 'genome', and representing *configurational constraints* to which, with considerable loose 'play', respectively variations, communication channels become 'adapted'. Since these viewpoints were sufficiently elaborated in volumes 1 and 3/II of this series, no further comments are here necessary.

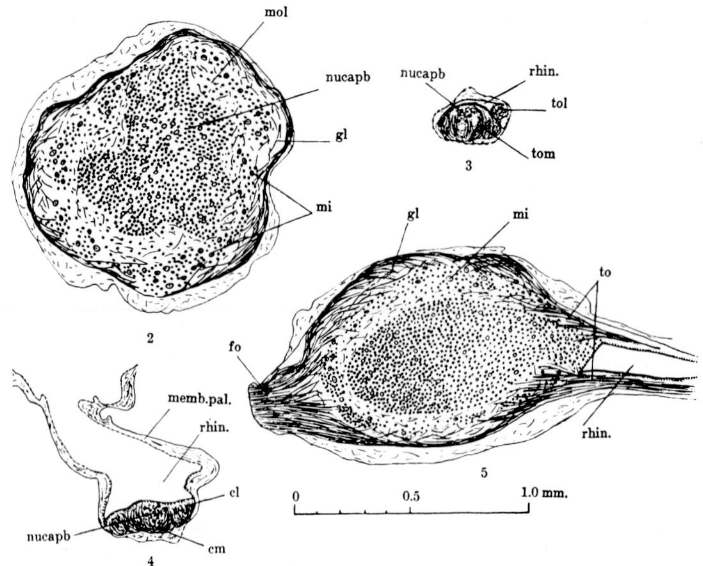

Figure 193. Three cross-sections and one sagittal section through olfactory bulb and stalk in the Siluroid Teleost Corydora paliatus (from MILLER, 1940). cl, cm: lateral and medial crus of olfactory tract; fo: fila olfactoria; gl: layer of glomeruli; memb.pal.: roof-plate component of olfactory stalk; mi: layer of mitral cells; mol: 'molecular layer'; nucapb: nucleus olfactorius anterior, pars bulbaris; rhin.: rhinocele; to: olfactory tracts; tol, tom: lateral and medial olfactory tract.

are HERRICK's area olfactoria medialis, his area olfactoria lateralis being perhaps a lateral portion of B_{2+3}. An abventricular, central region of uncertain delimitation, is his area olfacto-somatica.[89a] HERRICK's concept, particularly related to his interpretation of fiber systems, is depicted by Figure 199B. In the Ganoid Amia, however, HERRICK's area olfactoria dorsalis, pars dorsomedialis, seems restricted to B_1, area olfactoria lateralis being two portions of B_2, and area olfactoria medialis being B_3 (cf. HERRICK, 1922, Fig. 23, p. 171).

Omitting further comments on details of the bulbus olfactorius of Osteichthyes, dealt with in section 2, it will here be sufficient to state that, depending on taxonomic differences, said bulb may be directly

[89a] Since HERRICK (1922) recognizes three griseal neighborhoods where the present author distinguishes four, it is not entirely certain whether D_1 should be entirely included in HERRICK's area olfactoria dorsalis, pars dorsomedialis, or might also be partly overlapping with the cited author's area olfactoria dorsalis, pars dorsolateralis (which appears essentially to be D_2). Cf. Figure 199B.

contiguous with the lobus hemisphaericus, or be connected to this latter by an olfactory stalk (peduncle) of various and occasionally considerable length (Fig. 193). Generally speaking, the peduncles retain the original lumen characteristic for telencephalic paired evagination and inversion. Their dorsal-dorsomedial wall may be an extension of the telencephalic epithelial roof plate. The olfactory tract, commonly with a medial and lateral subdivision, is then generally located in the stalk's thickened basal wall. Some scattered neuronal elements of a 'nucleus olfactorius anterior' are commonly included in bulb or stalk or both. In some Teleosts without olfactory stalk, the caudal portion of the olfactory bulbs may appear 'fused' in the midline, but this presumably represents an extension of the bulb toward the telencephalon impar. Some olfactory tract fibers can decussate through this 'concrescence' (cf. e.g. NIEUWENHUYS, 1967).

The lobus hemisphaericus of *Ganoids* may be only moderately eversion, as e.g. in Amia and Polypterus.[90] In the former, eversion is combined with substantial thickening of the telencephalic wall, as generally characteristic for Teleosts. In Polypterus, on the other hand, the telencephalic wall remains relatively thin. A true caudal paired evagination, resulting in paired caudal ventricular spaces extending laterad to the epithalamus, and as seen in some Teleosts (cf. Fig. 61 A, p. 171, vol. 3/II), does not seem to have been recorded in the hitherto examined Ganoids.

The *pallial cell population* displays, more or less distinctly, a griseum mediale (D_1), a griseum dorsale (D_2), and a griseum laterale (D_3), which latter topologically corresponds to the griseum mediale *sive* area medialis pallii (so-called primordium hippocampi) of the inverted Gnathostome Anamniote brain. On the lateral surface, a shallow sulcus externus ('fissura endorhinalis', 'fissura externa'), which is generally much more pronounced in Teleosts, may delimit pallium and basis. Caudalward, this groove is continuous with the sulcus diencephalo-telencephalicus at the hemispheric stalk.

The pallial cell population is essentially paraventricular, but scattered cell groups, intermingled with the fiber systems, extend toward the thickened abventricular telencephalic wall and cannot be properly delimited from similar cell groups pertaining to the basis. A nonde-

[90] Figures illustrating the telencephalic configuration of Amia, Polypterus, and Lepidosteus are included in section 6, chapter VI of volume 3/II (Figs. 251, 252, 253, pp. 493, 494, 495; Fig. 61 B, p. 171, Fig. 260, p. 505; Fig. 261, p. 506).

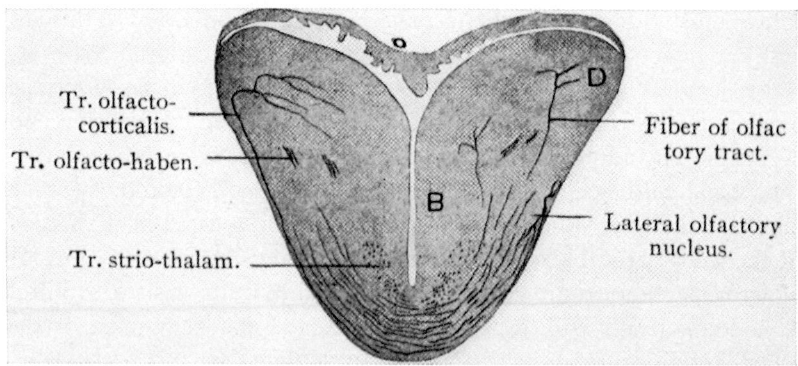

Figure 194 A. Cross-section *(Golgi impregnation)* through the telencephalon of the Ganoid Acipenser (Sturgeon) at the level of commissura anterior (from JOHNSTON, 1906). 'Lateral olfactory nucleus' is a lateral portion of nucleus basalis intermedius (B_2); 'Tr. strio-thalam.' is part of basal forebrain bundle system. Added designations: B: basis; D: pallium.

script region of these abventricular elements, frequently including relatively large nerve cells, is the above-mentioned so-called area olfacto-somatica of HERRICK (1922), which generally seems related to both forebrain bundles, although HERRICK emphasizes the connections with the lateral one.

Ventricular sulci fd_2 and fd_3, prominent in some Teleosts, are commonly missing, but fd_1, related to the so-called 'sulcus ypsiliformis' is occasionally quite well recognizable, as e.g. in Amia, or at least suggested, as e.g. in Acipenser.[90a]

The *basal cell masses* include a griseum basale superius (B_1) which is commonly strongly developed, a griseum basale intermedium (B_2), and a paraterminal griseum basale inferius (B_3). The sulci fb_1 and fb_2, indicating approximate boundaries between B_1 and B_2, respectively B_2 and B_3 are occasionally present, e.g. in Lepidosteus, but may not be recognizable in adult forms, although generally identifiable at some ontogenetic stages. The fiber systems will be dealt with following the comments on the Teleostean telencephalon.

The lobus hemisphaericus of *Teleosts* likewise contains the just enumerated pallial and basal grisea, but displays a much greater taxonomi-

[90a] The significance of the telencephalic ventricular sulci, their configuration and their relationship to the telencephalic longitudinal zonal system was dealt with in section 6, chapter VI of volume 3/II, to which reference is made for further details.

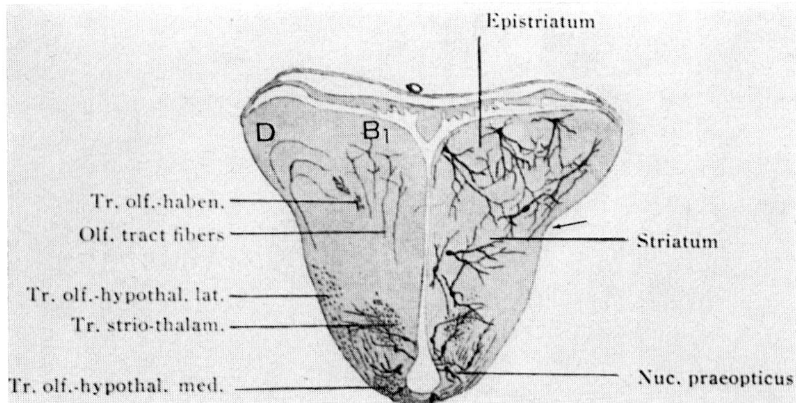

Figure 194B. Cross-section *(Golgi impregnation)* through the prosencephalon of Acipenser (from JOHNSTON, 1906). 'Epistriatum' is nucleus basalis superior (B_1); 'Striatum' is nucleus basalis intermedius and inferior (B_{2+3}); 'Tr.olf.-hypothal.med., lat., and Tr. strio-thalam.' are components of basal forebrain bundle. Added arrow: sulcus telencephalo-diencephalicus. D and B_1 are likewise added.

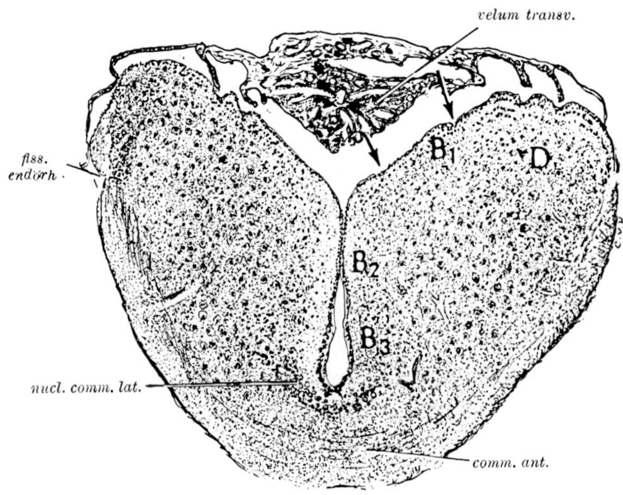

Figure 194C. Cross-section *(Weigert-Pal* paracarmine) through the telencephalon of the Ganoid Polyodon (from HOOGENBOOM, 1929). fiss.endorh.: sulcus externus; nucl. comm. lat.: interstitial nucleus of medial olfactory tract (derivative of B_3). The designations D (pallial zones D_{1-3}), B_1, B_2, B_3 have been added.

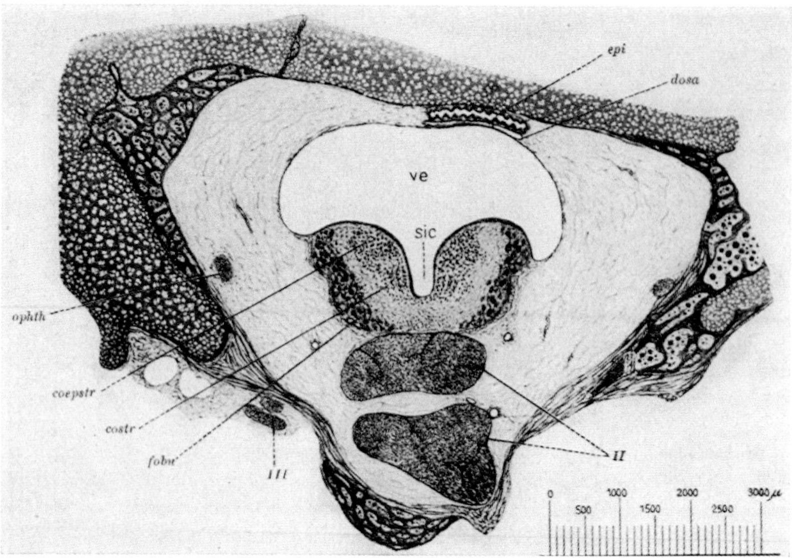

Figure 195. Cross-section through cranial cavity and telencephalon of the Teleost Esox lucius (from KRAUSE, 1923). coepstr: pallium (?); costr.: basal grisea; dosa: membranous parencephalon (dorsal sac); epi: epiphysis; fabu: caudal portion of olfactory bulb, extending into telencephalon impar; ophth: nervus ophthalmicus profundus trigemini; sic: 'sulcus intercephalicus' (midline groove of telencephalon impar); ve: ventriculus impar telencephali; II: optic nerve with part of chiasma (note the extreme rostral extension of chiasma, characteristic for many Teleosts; III: oculomotor nerve.

cally related variety with respect to details of their cytoarchitectural aspect than recorded in Ganoids, and also manifests, as a rule with some exceptions, a higher degree of differentiation within an overall pattern identical for all Osteichthyes (Figs. 194–198).

In Salmonidae such as the Trout,[91] and in Esocidae such as Esox lucius (Fig. 195) the eversion is moderate, comparable to that obtaining in Acipenser. The cytoarchitectural delimitations between the fundamental topologic components within both pallium and basis are rather indistinct. In Trutta, moreover, there is e.g. a tendency toward 'fusion', 'assimilation' or 'coalescence' of B_1 and D_{1+3} neighborhoods into a diffuse cell group surrounded by more densely crowded cells.

In Cyprinoids and Siluroids, on the other hand, the eversion, combined with wall thickening, is very pronounced. Cytoarchitectural differentiation becomes here quite noticeable, and further parcellation

[91] Cf. Figures 254, 255, pp. 496–497, volume 3/II.

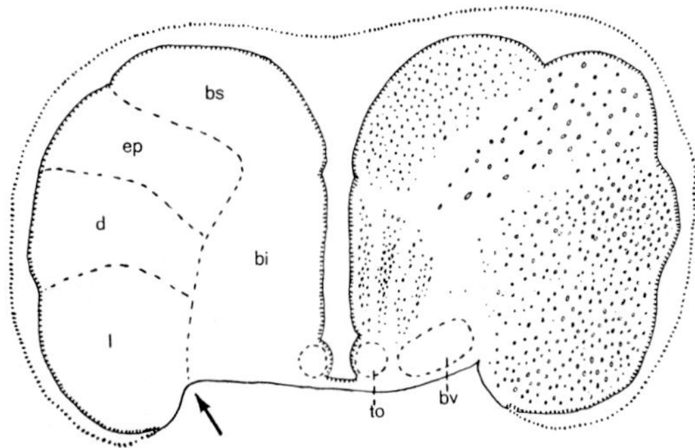

Figure 196 A. Cross-section (sketched from a *Nissl preparation)* through the telencephalon of Cyprinus auratus (from K., 1927). bs: nucleus basalis superior (B_1); bi: nuclei basales intermedius and inferior (B_{2+3}); bv: basal forebrain bundle with parts of tractus olfactorius lateralis; d: griseum dorsale pallii (D_2); ep: griseum mediale pallii (D_1); l: griseum laterale pallii (D_3, 'primordium hippocampi'); to: tractus olfactorius lateralis. Added arrow: sulcus externus.

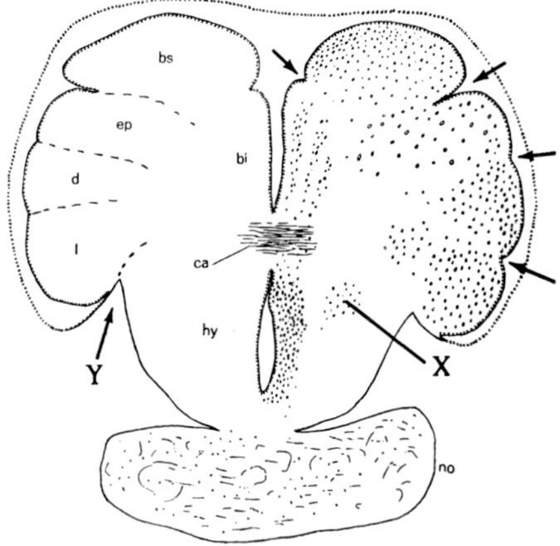

Figure 196 B. Cross-section (as in A) through the prosencephalon of Cyprinus auratus at level of anterior commissure (from K., 1927). ca: commissura anterior; hy: preoptic region (hypothalamus); no: optic chiasma and (stump of) nerve. Added designations: x: nucleus entopeduncularis; y: sulcus telencephalodiencephalicus; arrows, from left to right: sulci fb_1, fd_1, fd_2, fd_3.

within B_1, D_1, D_2, and D_3 neighborhoods obtains. As regards the paraterminal B_3 component, a caudolateral group of scattered cells, the so-called 'somatic area' of Miller (1940) and others, although not representing a 'basal cortex' absent in Osteichthyans, is nevertheless morphologically homomorphous with parts of the Selachian cortex olfac-

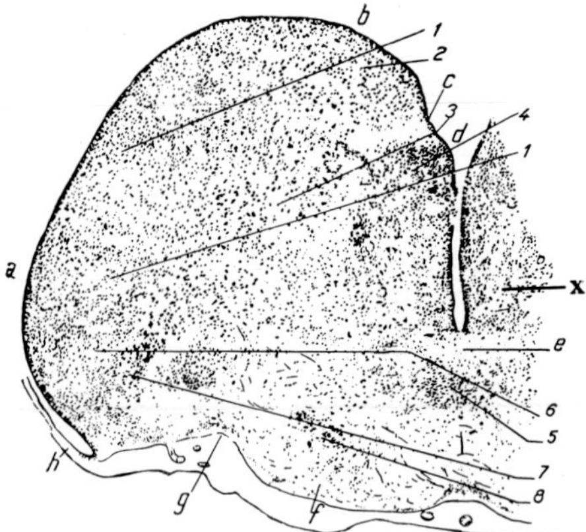

Figure 196 C. Cross-section *(Nissl stain)* through the telencephalon of Cyprinus carpio at level of anterior commissure (after SHELDON, 1912, from BECCARI, 1943). 1: nucleus olfactorius lateralis (top lead: griseum mediale pallii, D_1; lower lead: griseum dorsale pallii, D_2); 2: nucleus olfactorius dorsalis (n. bas. superior, B_1); 3: 'palaeostriatum' (perhaps internal portion of B_1); 4: supracommissural portion of corpus praecommissurale dorsale (n. bas. intermedius, B_2); 5: commissural portion of corpus praecommissurale (part of B_3); 6: 'nucleus piriformis' (griseum laterale pallii, D_3); 7: 'nucleus taeniae' (probably part of D_3); 8: nucleus of anterior commissure (probably massa cellularis reuniens, pars inferior sive hypothalamica); a: ependymal surface of pallium; b: ependymal surface of B_1; c: groove fb_1; d: zone B_2; f: commissural plate of lamina terminalis; g: sulcus externus; h: attachment of telencephalic roof plate. Designations in parenthesis and labels a–g in accordance with the present author's interpretation. Added lead x: supracommissural portion of nucleus basimedialis inferior (B_3).

toria, although presumably functionally not analogous.[92] Caudally to the commissura anterior (torus transversus), the sulcus terminalis telencephali separates the preterminal telencephalic grisea from the diencephalic preoptic cell masses (cf. Fig. 254D, p. 501, vol. 3/II).[93] Cross-sections through the telencephalon of the Gasteroid Gasterosteus aculeatus are depicted in Figure 259, p. 502 of volume 3/II.

---

[92] This 'somatic area' is telencephalic, and, although contiguous with the rostral end of the diencephalic (hypothalamic) nucleus entopeduncularis, should not be confused with this latter (cf. Figs. 198B, C).

[93] If a caudal paired evagination of the Teleostean telencephalon obtains, a branch of sulcus terminalis separates lateral surface of epithalamus from telencephalon (cf. Fig. 61A, p. 171, vol. 3/II). Additional illustrations of the telencephalon in the Siluroid Corydora are Figures 257A–D, pp. 499–501, and Figures 263A, B, p. 508 of volume 3/II.

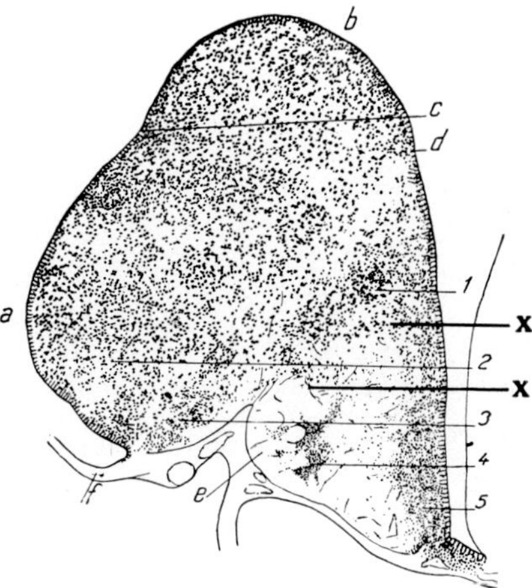

Figure 196 D. Cross-section *(Nissl stain)* through the prosencephalon of Cyprinus carpio (after SHELDON, 1912, from BECCARI, 1943). 1: intermediate part of precommissural body (B_{2+3}); 2: 'nucleus piriformis' (griseum laterale pallii, D_3); 3: nucleus taeniae (probably part of D_3); 4: nucleus entopeduncularis (hypothalami); 5: nucleus praeopticus parvocellularis; a: ependymal surface of pallium; b: ependymal layer of basis; c: sulcus ypsiliformis (fd_1); d: groove fb_1; e: preoptic hypothalamus; f: telencephalic roof plate. Designations in parenthesis and labels a–e in the present author's interpretation. Added x: pars inferior massae cellularis reunientis.

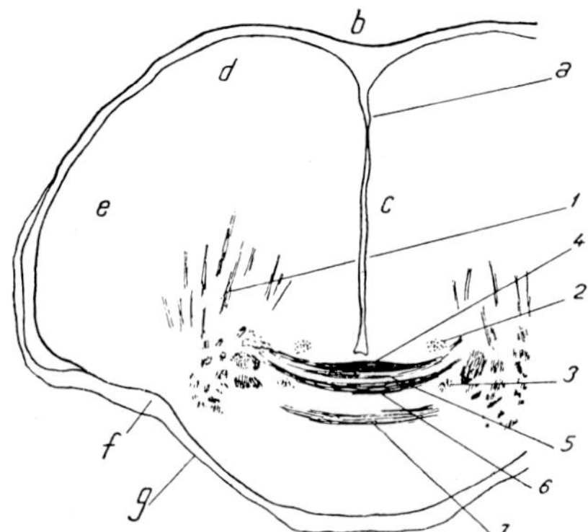

Figure 196 E. Cross-section through the telencephalon of Cyprinus carpio at level of anterior commissure (simplified after SHELDON, 1912, from BECCARI 1943). 1: 'tractus striothalamicus' (basal forebrain bundle); 2: tr. hypothalamo-olfactorius medialis (caudal portion of medial olfactory tract); 3: pars ventralis of 'tr. olfacto-thalamicus medialis' (medial portion of basal forebrain bundle); 4: commissura dorsalis (dorsal portion of comm. anterior); 5, 6: (additional components of 4); 7: (ventral portion of commissura anterior, corresponding to comm. pallii sive hippocampi of the inverted Gnathostome Anamniote telencephalon); a: groove fb_1; b: telencephalic roof plate; c: intermediate and inferior basal grisea (B_{2+3}); d: nucleus basalis superior (B_1); e: pallium (D_{1-3}); f: sulcus externus near transition to telodiencephalic sulcus; g: leptomeninx. Interpretations in parenthesis and a–f as in Figures 196 C–D.

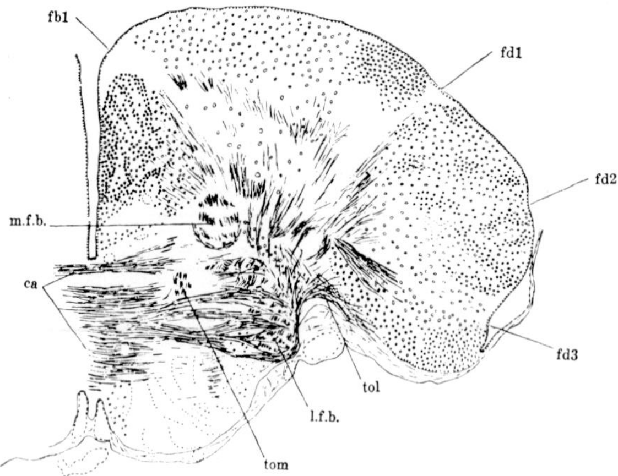

Figure 197 A. Cross-section through the telencephalon of the Siluroid Teleost Corydora at level of commissura anterior (from MILLER, 1940). ca: commissura anterior (sensu latiori, with subdivision differing from, but comparable to those in Figure 196E); l.f.b.: lateral forebrain bundle; m.f.b.: medial forebrain bundle; tol, tom: tractus olfactorius lateralis respectively medialis.

A peculiarly dense paraventricular cell plate characterizes the griseum mediale pallii (D_1) of certain Teleosts such as the Clupeiform Scleropagus (Fig. 198A), the Symbranchid Monopterus (Fig. 198B), and the Holocentrid Holocentrus (Fig. 198C), in which this formation has 'assimilated' the neighborhood D_2 (griseum dorsale pallii). Among other peculiarities of the polymorphism displayed by the Teleostean lobus hemisphaericus,[94] a secondary midline fusion or 'concrescence' of the basal grisea occurs in Monopterus, Holocentrus, and a few other forms (Figs. 198B, C). In commenting upon the findings by VAN DER HORST (1917), KAPPERS *et al.* (1936) discuss the possible interpretation of commissural bundles, presumably pertaining to the system of anterior commissure, which are taking their course through said concrescence, while some of the decussating fibers of anterior commissure retain their typical location (Fig. 198B). VAN DER HORST designated the neighborhood D_1 as 'epistriatum', medial B_{1-3} neighborhoods as 'sep-

[94] As regards the polymorphism displayed by the telencephalon, and for that matter, by the entire brain of Osteichthyes, KAPPERS *et al.* (1936), in their very detailed handbook, are compelled to remark: 'It has not been possible, without extending this account beyond reasonable limits, to discuss in the preceding pages' (1268–1289 loc.cit.) 'the many variations which are presented by the brains of ganoids and teleosts'.

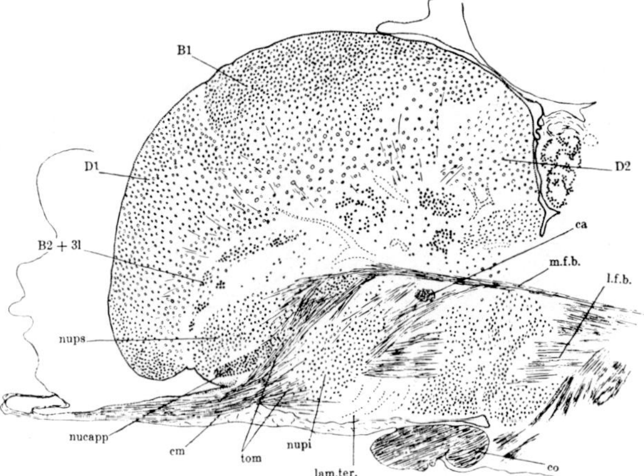

Figure 197B. Paramedian sagittal section through telencephalon and adjacent rostral diencephalon of Corydora (from MILLER, 1940). cm: medial crus of olfactory tract; co: optic chiasma; lam.ter.: lamina terminalis; nucapp: nucleus olfactorius anterior, pars supracommissuralis (probably derivative of B_3); nupi: nucleus praecommissuralis, pars inferior (probably derivative of B_3); nups: nucleus praecommissuralis, pars superior (probably derivative of B_3). Other designations as in preceding Figure.

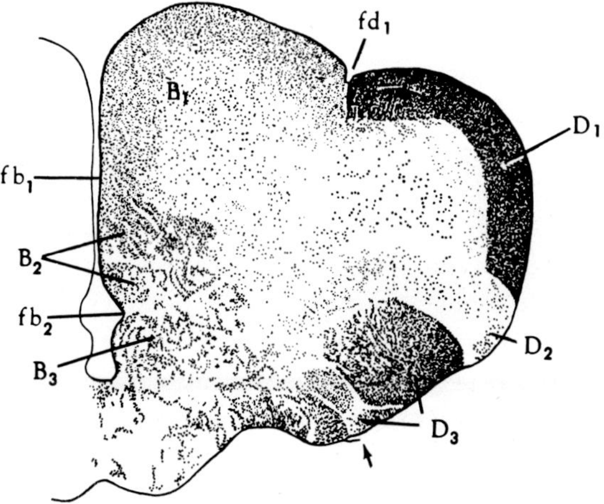

Figure 198 A. Cross-section through the telencephalon of the Teleost Scleropages formosus at level of anterior commissure (from NIEUWENHUYS, 1960b). The designations, in accordance with my interpretation, have been added to the unlabelled original. Arrow: attachment of telencephalic roof plate.

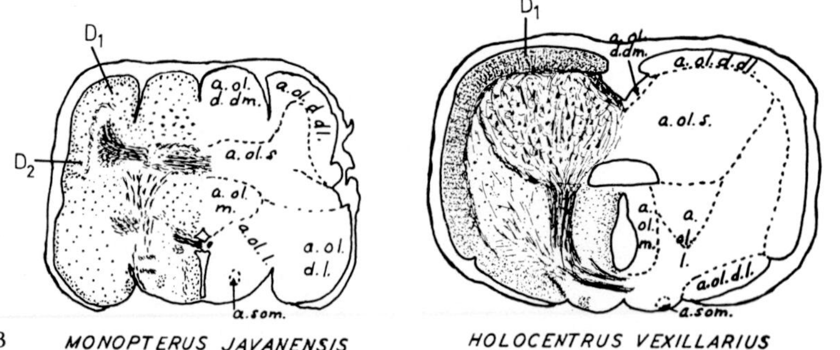

Figure 198 B, C. Cross-sections through prosencephalon respectively telencephalon of the Teleosts Monopterus javanensis and Holocentrus vexillarius (modified after MEADER, 1939, from KAPPERS, 1947). a.ol.d.dl.: 'area olfactoria dorsalis, pars dorsolateralis' (D_1); a.ol.d.dm.: 'area olf. dors., pars dorsomedialis' (B_1); a.ol.d.l.: 'area olfactoria dorsalis, pars lateralis (D_3); a.ol.l.: 'area olfactoria lateralis' (lateral subdiv. of B_{2+3}); a.ol.m.: 'area olfactoria medialis' (medial subdiv. of B_{2+3}); a.ol.s.: 'area olfacto-somatica' (mainly subdiv. of B_1); a.som.: 'area somatica' (in B ,where preoptic recess is shown, n. entopeduncularis; in C probably massa cellularis reuniens, pars inferior sive hypothalamica). Added designation in Fig. B: D_2. This neighborhood is apparently 'absorbed' by D_1 in Fig. C. Both Figures show secondary midline fusions in B-zones. In Fig. B an aberrant dorsal (pallial) component of commissura anterior.

tum', and a lateral B_{2+3} neighborhood as 'striatum'. The telencephalon of Symbranchidae, as pointed out by VAN DER HORST (1917), can be said to manifest an extreme case of the everted type, which seems to be at its maximum in these forms, combined with an excessive compactness in the development of the 'epistriatum' (D_1) and particularly of the 'septum' (B_{1-3}, especially B_1).

As regards the Osteichthyan *telencephalic communication channels* (Figs. 199 A–C), secondary and tertiary olfactory fibers (from mitral cells respectively 'nucleus olfactorius anterior' generally gather into a *medial* and into a *lateral olfactory tract*. The medial olfactory channel is mainly distributed to the B-neighborhoods, but also reaches the hypothalamus, particularly but not exclusively the preoptic region. In many Osteichthyes, the caudal part of the medial olfactory tract includes a group of interstitial neuronal elements described as a nucleus by KUDO (1928). Some rostral fibers of medial olfactory tract also reach the pallial zones. Moreover, bulbopetal fibers, as described by several authors (e.g. SHELDON, 1912), are presumed to run through medial olfactory tract. Some of these fibers may originate in the B-zones or even in the hypothalamus (perhaps preoptic region), while others seem to represent an interbulbar connection decussating in the anterior commissure.

The lateral olfactory tract distributes mainly but not exclusively to the pallial neighborhoods (D-zones).[95] There is, moreover, some intermingling of spreading medial and lateral olfactory tract fibers, such e.g. that dorsolaterally directed fibers of the medial olfactory tract join those from the lateral one.

The caudal communication channels include three main overall pathways, namely the stria medullar's system, the lateral forebrain bundle, and the medial forebrain bundle.

The *stria medullaris* (Figs. 76, 199 C) interconnects the telencephalon with the habenular grisea, and is predominantly but by no means exclusively descending. It may include bulbar olfactory fibers from tractus olfactorius medialis and lateralis, and it contains fibers originating in basal as well as in pallial grisea, particularly also as so-called tractus taeniae in the so-called nucleus taeniae, if this latter is clearly differentiated (cf. Figs. 257 C, D, pp. 500–501, vol. 3/II). Said nucleus of the Osteichthyan telencephalon appears to be a derivative of D_3 rather than of the neighboring B_3 zone. Fibers from medial and lateral forebrain bundle seem likewise to join the stria medullaris system at the level of preoptic region and may be accompanied by fibers arising in this latter.

The *medial forebrain bundle* connects mainly the basal telencephalic grisea with hypothalamus and thalamus ventralis but doubtless includes some fibers related to pallial grisea. The *lateral forebrain bundle* connects particularly the pallial grisea, but also all basal ones with thalamus ventralis and dorsalis as well as hypothalamus. Bulbar fibers from the olfactory tracts may join both forebrain bundles. HERRICK's 'tractus pallii' (Fig. 199 B) is presumably a portion of the lateral forebrain bundle, from which it cannot be properly distinguished.

[95] NIEUWENHUYS (1967), in contradistinction to some other authors, traced the lateral olfactory tract only to the caudoventral portion of the pallium and maintains that the greater part of the Teleostean pallium is 'devoid of secondary olfactory connections'. Likewise, a recent investigation of the distribution of olfactory tracts in the Bullhead Catfish (Ictalurus nebulosus) by FINGER (1975) claims that only a relatively small portion of the Teleostean telencephalon receives 'direct olfactory projections'. Be that as it may, and considering the limitations of the various experimental methods respectively the difficulties in tracing fiber systems by means of the other techniques, a wide distribution of olfactory input to the telencephalic grisea may nevertheless be upheld, regardless of considerable ascending, including optic, inputs into telencephalon, which presumably reach both pallium and basis as pointed out further below.

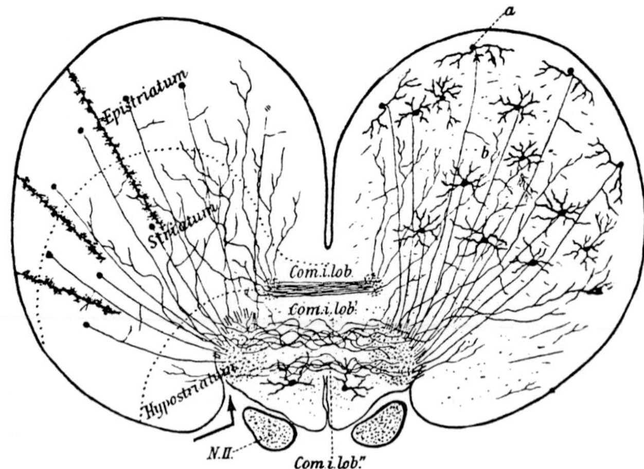

Figure 199 A. Cross-section *(Golgi impregnation)* through the prosencephalon of the Teleost Anguilla vulgaris (from Catois, 1901). This author uses a dorsobasal subdivision of the telencephalic cell masses into 'epistriatum', 'striatum', and 'hypostriatum'. a: periventricular nerve cell with collateral (b) of its neurite; Com.i.lob.: three components of commissura anterior *sensu latiori;* N. II.: optic nerve. Below anterior commissure is the hypothalamic optic recess, whose thin floor was apparently torn off. Note ependymal elements of everted telencephalon. Added arrow: sulcus telodiencephalicus.

A recent report on telencephalic output channels in two Teleosts, as interpreted from observations by means of the *Fink-Heimer technique*, was published by Vanegas and Ebbeson (1976). According to the cited authors, their striotectal bundle, roughly corresponding to a component of the stria medullaris system, reaches the tectum mesencephali. A component of lateral forebrain bundle to posterior hypothalamus is described as striolobar bundle.

The variably developed *anterior commissure* commonly displays a dorsal and a ventral subdivision (cf. Figs. 196E, 197A, 199A). Because of the eversion effect, the ventral subdivision corresponds to the commissura pallii of the inverted brain, the dorsal subdivision being the commissura anterior *sensu proprio* (cf. Fig. 64, p. 175, vol. 3/II). Both subdivisions, however, include fibers related to pallial and basal grisea as well as to the olfactory tracts. Thus, a bundle from the tractus olfactorius medialis commonly decussates in dorsal portions of the commissura anterior to reach caudolateral portions of the pallial grisea as noted by Goldstein (1905) and Sheldon (1912).

Generally speaking, the fiber systems of the Osteichthyan telencephalon are highly complex with considerable taxonomic variations,

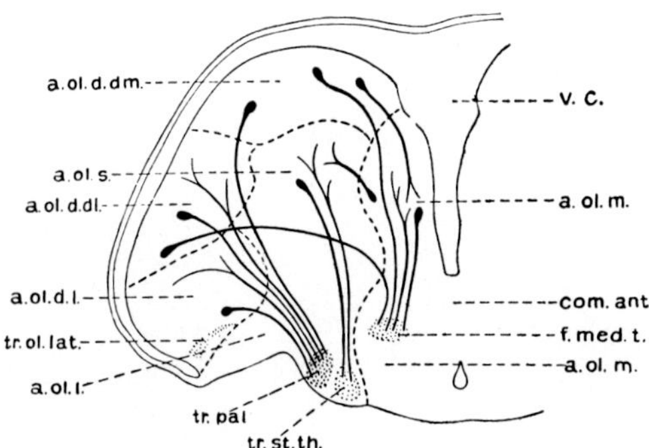

Figure 199 B. Diagrammatic section through the Carp's forebrain (caudal telencephalon medium with diencephalic preoptic recess) at level of anterior commissure, purporting to show some griseal areas and fiber connections as interpreted by HERRICK (from HERRICK, 1922, and PAPEZ, 1929). a.ol.d.dl.: area olfactoria dorsalis dorsolateralis dorsomedialis (apparently D_2); a.ol.d.dm.: area olfactoria dorsalis dorsomedialis (apparently D_1 and B_1); a.ol.d.l.: area olfactoria dorsolateralis (D_3); a.ol.i.: area olfactoria inferior (probably a lateral portion of B_3); a.ol.m.: area olfactoria medialis (apparently B_{2+3}); a.ol.s.: area olfacto-somatica (probably central part of B_1 with possible additions from D_1 and from B_2); com.ant.: commissura anterior; f.med.t.: medial forebrain bundle; tr.ol.lat.: tractus olfactorius lateralis; tr.pal.: tractus pallii (part of lateral forebrain bundle); tr.st.th.: tractus striothalamicus (part of lateral forebrain bundle); v.c.: ventriculus communis sive impar telencephali. It is not quite certain to which degree the zones D_1 of the present author may be distributed upon either the neighborhoods designated as a.ol.d.dm. and a.ol.d.l. by HERRICK.

and their description is encumbered by the multitudinous differences in the terminology adopted or coined by the diverse authors.[96] Figure 199C illustrates an oversimplified but nevertheless in various respects approximately valid concept of Teleostean prosencephalic communication channels elaborated by JOHNSTON (1906), which has not been significantly clarified by subsequent authors up to the present time.

In view of the relatively high degree of telencephalic differentiation in many Osteichthyes and particularly in Teleosts which may display elaborate fiber connections through the enumerated distinctive channels and tracts, a substantial functional significance of the telencepha-

[96] In their very detailed text, KAPPERS *et al.* (1936) are compelled to remark: 'Space does not permit here a detailed account of these connections.' Among the variations there obtain differences in the size of the recognized tracts, in the distribution of their fibers, in the degree of myelination, the relationship of medullated to non-medullated fibers, and in the relation of decussating and non-decussating components.

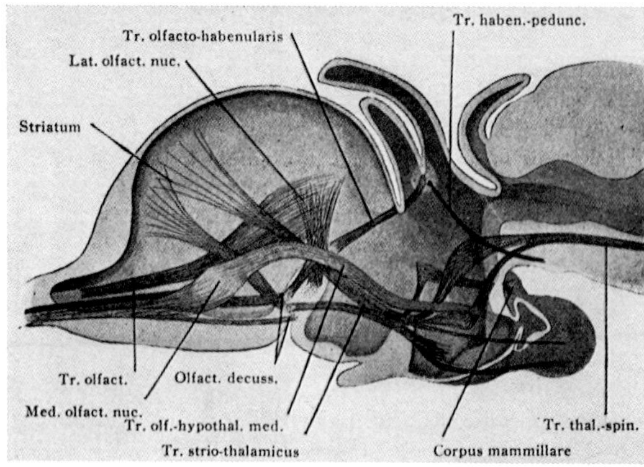

Figure 199C. Diagram of fiber tracts in the prosencephalon of a Teleost, as interpreted by JOHNSTON (from JOHNSTON, 1906). Lat.olfact.nuc.: probably undefined pallial and lateral B grisea; Med.olfact.nuc.: probably B_{2+3}; Olfact.decuss.: components of anterior commissure; Striatum: perhaps B_1; Tr.olfacto-habenularis: stria medullaris system; Tr.olf.-hypothal.med. and Tr.strio-thalamicus: components of basal forebrain bundle; Tr.thal.-spin.: probably an undefined tegmental channel, perhaps also in part fasciculus longitudinalis medialis.

lon suggests itself. On the basis of the griseal arrangements and the available inferences *qua* fiber connections the Osteichthyan telencephalon can doubtless be assumed to 'correlate' and 'integrate' its olfactory input with sensory, including optic input, transmitted, through the mediation of the diencephalon, by way of ascending channels. These latter are the above-mentioned basal forebrain bundles and perhaps, to a far lesser degree, ascending fibers in the predominantly descending stria medullaris system. Depending on the taxonomically very variable predominance of olfactory, or optic, or gustatory input into the central neuraxis, the quantitative distribution of olfactory fibers and of ascending non-olfactory ones to diverse telencephalic grisea can likewise be assumed to display considerable variations including details of synaptology. Thus, the peculiarly differentiated griseum mediale pallii (D_1) in Holocentrus and similar forms, correlated with an elaborate lateral geniculate complex, may well be a telencephalic optic 'center' (cf. e.g. ARONSON, 1963). Other pallial and basal grisea can be conceived as correlating 'taste', 'touch', visceral processes, etc., with olfactory input. Such grisea might even exert an effect depending, to some extent, more upon non-olfactory than upon olfactory input.

There is, however, hardly any justification to designate certain grisea, such as parts of the B_1 or B_2 neighborhoods, with the meaningless and vague term 'olfactosomatic area'.

Extirpation of the Osteichthyan telencephalon, as undertaken by STEINER (1888) and subsequent authors, does not affect the easily noticeable aspects of behavior (NOLTE, 1932). It is hardly possible to distinguish in an aquarium, the activities of such an operated fish from those of the normal ones, as also recently admitted by ARONSON and KAPLAN (1968). Elimination of olfaction becomes here compensated by the effects of other sensory input.

Thus, NOLTE (1932) showed that Phloxinus and Gasterosteus specimens with completely extirpated telencephalon did not loose optic associations acquired by previous training, while other specimens of these Fishes could be trained to acquire such associations subsequently to said operation.

On the other hand, more detailed observations, as undertaken by NOBLE (1936, 1937), NOBLE and BORNE (1941), SEGAAR (1960, 1961, 1965), SEGAAR and NIEUWENHUYS (1963) and others have disclosed certain changes in complex behavioral patterns, such e.g. as fighting behavior. Changes of this sort appear to differ with respect to the various Teleostean forms, and also seem to be correlated with the degree of telencephalic differentiation.[97] An effect upon learning abilities (e.g. in maze experiments with 'conditioning' stimuli) has been claimed in some Fishes (WARREN, 1961) but is, on the whole, inconspicuous and variable. 'Memory', a rather vague term for the multifactorial effects of 'storage processes', doubtless depends, as already pointed out in my old Vorlesungen (K., 1927) on neuronal events in a network whose components may be widely spaced and distributed upon different regions.

Experiments involving *telencephalic lesions* in Teleosts have particularly considered (1) aggressive behavior, (2) reproductive behavior (including mating, breeding, and so-called parental behavior), and (3) schooling behavior.

Again, electrical stimulation of telencephalic, diencephalic and mesencephalic grisea undertaken by FIEDLER (1968) resulted in nondescript behavioral reactions reported, e.g. as various motions of the fins, or in such terms as '*schwimmt*', '*wird unruhig*', '*Maulschliessen*', '*schnappt*',

[97] In addition, it has been noted, as e.g. in the experiments by Noble, that quite different lesions sometimes produce the same effect.

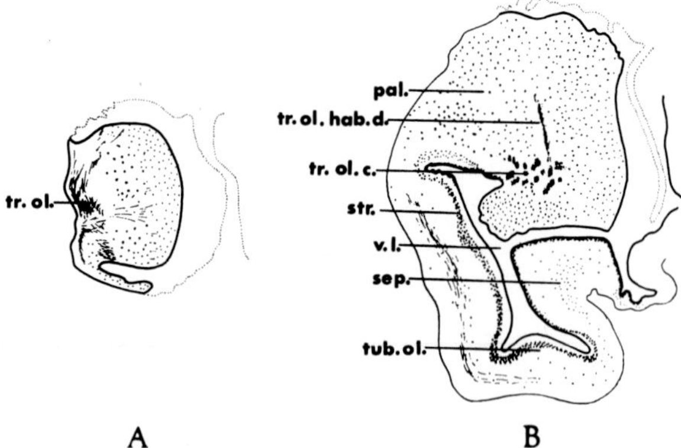

Figure 200 A, B. Cross-sections through so-called corpus rostrale (A) and through lobus hemisphaericus *sensu strictiori* of the Crossopterygian Latimeria (from NIEUWENHUYS, 1967). pal.: pallium; sep.: 'septum' (B_3); str.: 'striatum' (B_1); tr.ol.: 'tractus olfactorius habenulae dorsalis'; tub.ol.: 'tuberculum olfactorius' (B_2); v.l.: 'ventriculus lateralis' (unpaired telencephalic ventricle with recesses). Interpretations by present author in parentheses.

'*weicht zurück*', '*Ausspucken*', '*Sichschütteln*', '*Gähnen*', '*Vorwärtsstossen*', etc., etc. One is here reminded of experiments with implanted electrodes in the Chick brain undertaken by a well-known biologist, who pointed out, as a remarkable feature upon stimulation: '*das Huhn gakkert!*' Other recent experiments with electrical stimulation of telencephalon and hypothalamus in the Bluegill Lepomis were also concerned with 'evoked feeding' (DEMSKI and KNIGGE, 1971) and likewise lead to rather ill-definable results.

There is no doubt that species-typical behavior may be affected by telencephalic lesions, and it is probable that the telencephalon exerts a 'facilitating' or 'modulating' effect on diencephalic and brain stem centers mediating said behavior.[98] This control, however, if abolished by

---

[98] ARONSON (1963) and ARONSON and KAPLAN (1968) subsume this effect under the very general and rather vague concept of more or less 'unspecific arousal', as e.g. also applied to the Mammalian 'reticular formation'. The cited authors' attempt to compare some Teleostean telencephalic activities with those of the so-called limbic and associated systems of Mammals seem, however, rather unconvincing in view of the considerable morphologic structural, and functional differences, although some very generalized similarities or analogies might obtain. Recently 'stereotaxic atlases' of the forebrain in Carassius (PETER and GILL, 1975), and in Fundulus (PETER *et al.*, 1975) have been published. Their nomenclature is based on dubious morphologic concepts.

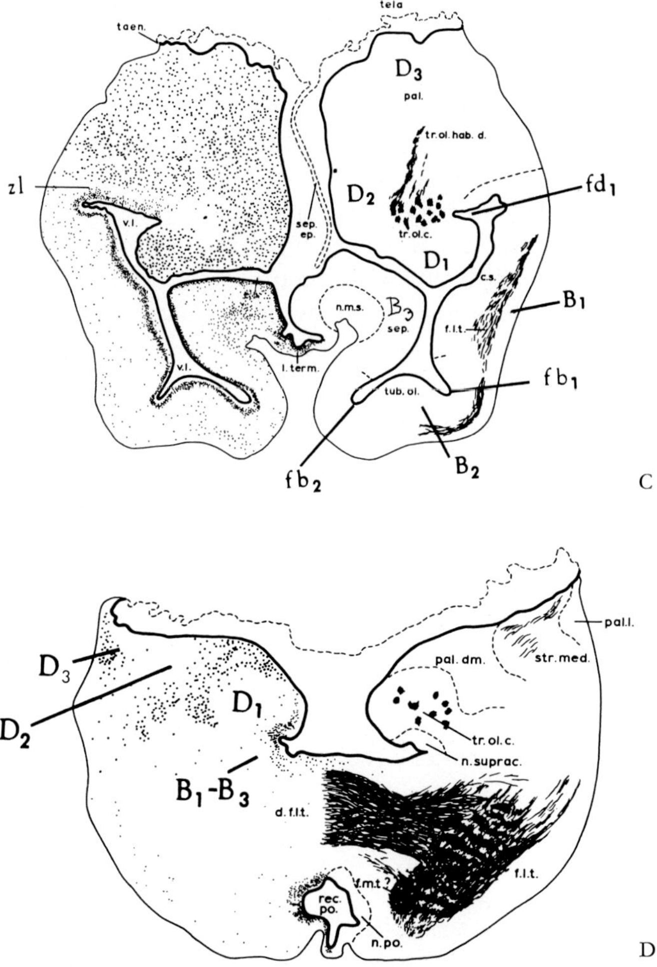

Figure 200 C, D. Cross-sections through the lobus hemisphaericus of Latimeria (A) at an intermediate level, and (B) at level of anterior commissure (from NIEUWENHUYS, 1969). c.s.: 'corpus striatum'; d.f.l.t.: commissura anterior; f.l.t.: lateral forebrain bundle; f.m.t. ?: medial forebrain bundle; l.term.: lamina terminalis; n.m.s.: 'nucleus medialis septi'; n.po.: nucleus praeopticus; n.suprac.: 'nucleus supracommissuralis; pal.: pallium; pal.dm.: pallium dorsomediale; pal.l.: pallium laterale; rec.po.: recessus praeopticus (hypothalami); sep.: 'septum'; sep.ep.: 'septum ependymale' (median sagittal fold of telencephalic roof plate); str.med.: stria medullaris; taen.: 'taenia telencephali' (attachment of roof plate); tela: (roof plate); tr.ol.c.: 'tractus olfactorius centralis'; tr.ol.hab.d.: 'tractus olfacto-habenularis dorsalis' (component of stria medullaris system); tub.ol.: 'tuberculum olfactorium'; v.l.: 'ventriculus impar telencephali'; z.l.: zona limitans. D, B, and f designations have been added, the present author's terminology given in parentheses.

extirpation or lesions, may, to a very substantial extent, become compensated by the 'redundancy effect' of the still poorly understood neuronal circuits. Extensive but nevertheless rather inconclusive discussions on this topic, with the relevant observational data and the pertinent references, can be found in the publications edited by Gilbert (1963) and by Ingle (1968).

The telencephalon of *Latimeria*, considered to be the only extant Crossopterygian, has been described by Millot and Anthony (1966). Nieuwenhuys (1967, 1969), in his interpretation of these findings, assumes that 'this ancient species stands intermediate between the lungfish and the actinopterygians' in its telencephalic configuration. As pointed out in section 6 of volume 3/II, I do not share this view and believe that Latimeria is a highly aberrant form whose peculiar and almost completely everted unpaired lobus hemisphaericus is morphologically closely related to the Teleostean one.

The *olfactory bulb* forms an ovoid '*petite masse*' at the posterior extremity of the nasal cavity. It is connected by a long and thin olfactory tract with a relatively small paired evaginated but still everted[99] so-called 'corpus rostrale', which can be interpreted as a 'pallial' nucleus olfactorius anterior (Fig. 200 A). This latter is continuous with the unpaired everted *lobus hemisphaericus sensu strictiori*.

This latter, in addition to its eversion, displays a number of peculiar foldings related to the sulci fd_1, fb_1, fb_2. In the pallium, the cell populations are distributed within the thickening. In the basis, most neuronal elements retain a rather dense periventricular arrangement, and, at the level of commissura anterior, become reduced to a supracommissural neighborhood (Figs. 200B, C, D).[100] As regards the incompletely ascertained *fiber connections*, the olfactory tract, running through the 'corpus rostrale' with which it effects connections, continues, approximately near the boundary zone of pallium and basis, as so-called tractus olfactorius centralis. In contradistinction to this fairly compact and essentially medullated tract, olfactory fibers to the ventrally located basal neighborhoods could not be traced, but can be surmised to be represented by scattered non-medullated fibers, reaching, e.g. the so-called 'tuberculum olfactorium'.

[99] This very rare instance of rostral paired evagination combined with eversion necessarily involves the telencephalic roof plate (lamina epithelialis) which thereby constitutes the dorsal, medial and ventromedial wall of the paired evagination (cf. Fig. 200 A).

[100] Cf. also Figure 264, p. 510, volume 3/II.

The caudal communication channels include a fairly compact medullated basal forebrain bundle with predominant lateral and indistinct medial subdivisions.

A stria medullaris system (so-called olfactory-habenular tract) is present, whose ill-defined components seem related in part to tractus olfactorius centralis, in part to pallium, and, more caudalward, in part to basal forebrain bundle.

The anterior commissure is rather massive. Its constituents are not ascertained, but are related to the basal forebrain bundle and presumably include a pallial as well as a basal component.

6. Dipnoans

Relevant investigations on the telencephalon in Dipnoans, with numerous references to the studies by previous authors, are those by HOLMGREN and VAN DER HORST (1925) for Ceratodus, by GERLACH (1933) for Protopterus, and by ELLIOT SMITH (1908) for Lepidosiren. Subsequent investigations, emphasizing various secondary structural details and differing in diverse aspects, as regards morphologic interpretations, from the conclusions put forward in the present treatise, are those by RUDEBECK (1945, Dipnoans in general and particularly Protopterus), by NIEUWENHUYS (1967, 1969, Ceratodus and Protopterus), by NIEUWENHUYS and HICKEY (1965, Ceratodus), and by SCHNITZLEIN and CROSBY (1967, Protopterus).

The problems concerning the homologies of the telencephalic wall neighborhoods respectively cell populations in Ceratodus, Protopterus, and Lepidosiren were dealt with in section 6, chapter VI of volume 3/II.[101]

The telencephalon of *Ceratodus*, like that of the other two extant genera of Dipnoans, displays paired rostral evagination combined with a less extensive unpaired evaginated telencephalon medium which is characterized by a moderate degree of eversion. A caudal paired evagination with true polus posterior does not occur in adult Lungfishes.[102]

---

[101] Previous contributions to this topic by the present author and referring to Lepidosiren respectively Protopterus, are included in the Vorlesungen of 1927 and in two papers (K., 1924b, 1929).

[102] Because of the peculiar expansion of the basal neighborhoods at the levels of lamina terminalis in Ceratodus, an infracommissural caudobasal bilateral recess of ventriculus

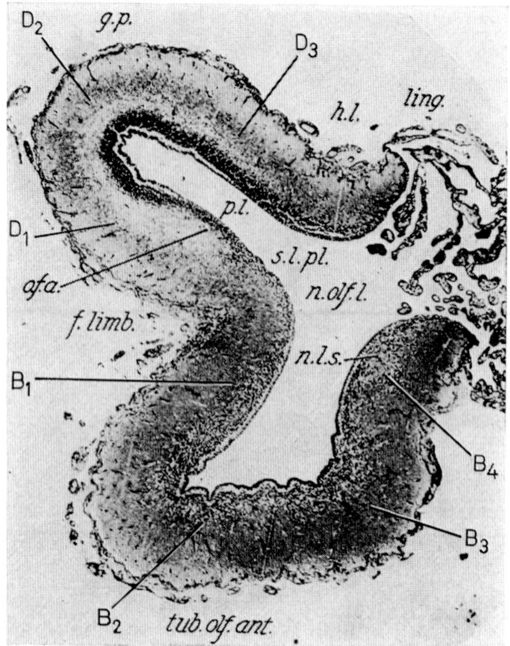

Figure 201 A. Cross-section (silver impregnation) through paired evaginated lobus hemisphaericus of Ceratodus (from HOLMGREN and VAN DER HORST, 1925). f.limb.: 'fovea limbica' (external palliobasal sulcus), g.p.: 'general pallium'; h.l.: 'hippocampal lobe'; ling.: 'lingula interolfactoria'; n.l.s.: 'nucleus lateralis septi'; n.olf.l.: 'nucleus olfactorius lateralis'; o.f.a.: 'oblong fiber amount'; p.l.: 'pyriform lobe'; s.l.pl.: 'sulcus limitans pallii lateralis' (our fd_1); tub.olf.ant.: 'tuberculum olfactorium anterius'. Our D and B notation has been added.

The *bulbus olfactorius* of Ceratodus is not directly contiguous with lobus hemisphaericus, but located rather closely to this latter's dorso-rostral extremity, the connection between bulb and lobus being effected by a short olfactory stalk. Both bulb and stalk retain their ventricular cavity (rhinocele of bulb, and lumen of stalk).

---

impar may be present (cf. Fig. 268 C, p. 515, vol. 3/II), which although corresponding to a basal caudal paired evagination, does not topologically correspond to the typical caudal paired evagination involving basal and pallial components dorsal to anterior commissure and forming a polus posterior characteristic for Tetrapods as well as for certain Teleosts. At certain ontogenetic stages, however, corresponding approximately to Figures 123 and 124, p. 330 of volume 3/II, the still unpaired telencephalon medium may display a transitory caudal recess overlapping the diencephalon, as e.g. depicted, for Ceratodus, in Figures 152 and 157 (pp. 225 and 227) in BERGQUIST's study of the Anamniote diencephalon's ontogeny (1932, cf. references to chapter XII).

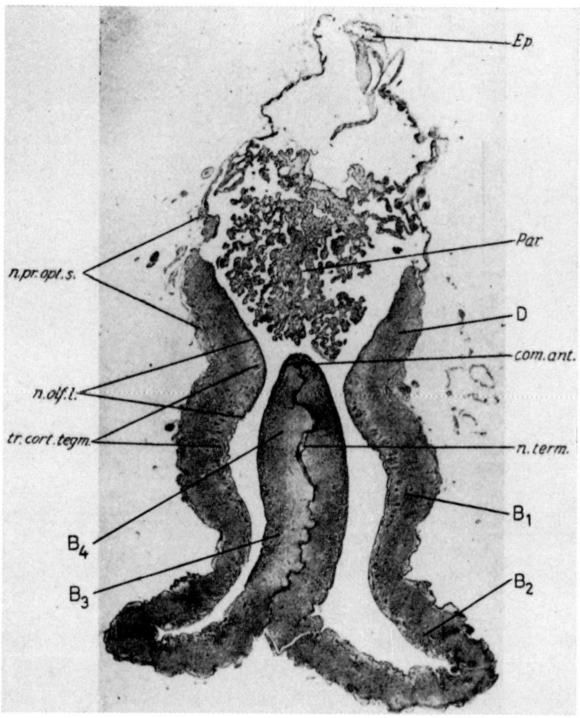

Figure 201 B. Cross-section through rostral end of unpaired evaginated lobus hemisphaericus of Ceratodus (from HOLMGREN and VAN DER HORST, 1925). com.ant.: interbulbar portion of anterior commissure; Ep: epiphysis; Par.: paraphysis; n.pr.opt.s.: 'nucleus preopticus pars superior'; n.olf.l.: 'nucleus olfactorius lateralis'; n.term.: nervus terminalis; tr.cort.tegm.: 'tractus cortico-tegmentalis complex' (component of lateral forebrain bundle). D and B notation added.

The paired portion of the *lobus hemisphaericus* of Ceratodus is not exclusively formed by an alar plate tube, but includes dorsomedially a portion of roof plate extending into the rostral paired evagination, and representing the so-called lingula interolfactoria provided with numerous choroidal folds.

The *pallium* extends from the sulcus fd_1 (sulcus limitans pallii) to the attachment of the roof plate's lamina epithelialis. Its cell masses consist of an ependymal layer and of a fairly well developed but less dense and not recognizably stratified corticoid plate, separated by a subcortical fiber layer from the periventricular cell stratum. In addition, a zonal layer of fibers lies externally to the corticoid plate. The pallial cell masses,

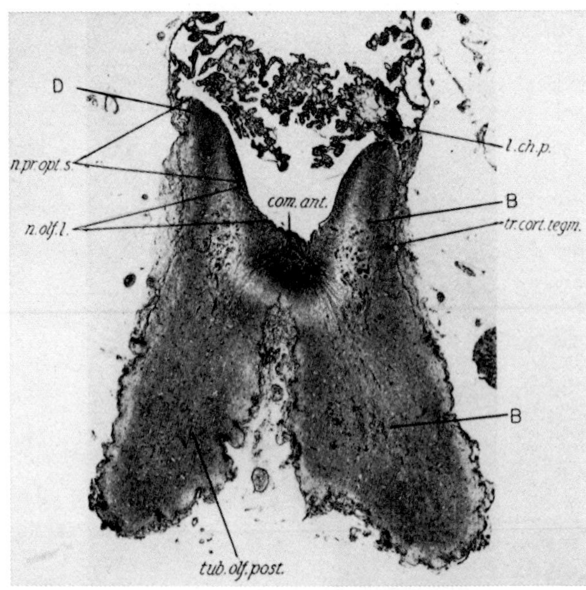

Figure 201 C. Cross section through lobus hemisphaericus of Ceratodus at a caudal level of commissura anterior (from HOLMGREN and VAN DER HORST, 1925). com.ant.: anterior commissure; l.ch.p.: 'lateral choroid plexus' (probably with part of velum transversum); tub.olf.post.: 'tuberculum olfactorium posterius' (caudalward protruding B_2 and B_3 zones). Other abbreviations as in Figure 201 B; D and B notation added.

which include the three topological neighborhoods D_1, D_2 and D_3, do not display very definite cytoarchitectural differences between each other, except for some nondescript gradients of density (cf. Fig. 201 A). The pallium decreases in extent toward the telencephalon impar, in which it becomes merely a rather narrow everted strip, comparable to the everted pallium of the Plagiostome Chimaera (cf. Figs. 201 B, C, 188 C; also Figs. 49 D, p. 140, and 250 D, p. 491, vol. 3/II).

The *basal portion* of lobus hemisphaericus includes the griseum basilaterale superius (B_1, so-called 'striatum') the ventrally adjacent griseum basilaterale inferius (B_2), which tends to expand around the floor of the ventricle, and the paraterminal ('septal') neighborhoods griseum basimediale inferius (B_3) respectively superius (B_4). This latter, in contradistinction to Protopterus, Lepidosiren, Plagiostomes, and Amphibians, does not dorsally abut on D_3, from which, in the paired evaginated lobus hemisphaericus of Ceratodus, it remains separated by the roof plate extension. As in the pallium, a periventricular layer and a

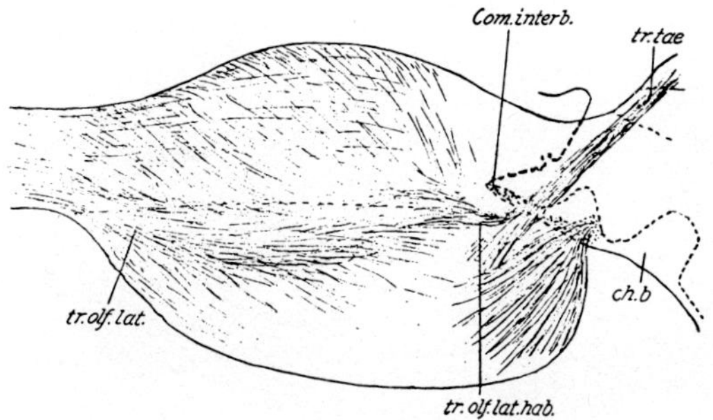

Figure 201 D. Fiber systems of Ceratodus forebrain reconstructed as seen in lateral aspect (from HOLMGREN and VAN DER HORST, 1925). ch.b: chiasmatic ridge; Com. interb.: interbulbar component of anterior commissure; tr.olf.lat.: tractus olfactorius lateralis; tr.olf.lat.hab.: tractus olfactorius lateralis habenulae; tr.tae.: 'tractus taeniae' (stria medullaris system).

corticoid plate (B_s) are present. This 'cortical' plate is particularly pronounced in some B_2 and B_3 neighborhoods, and represents the cortex of the 'tuberculum olfactorium'.[103] On the whole, the basis is somewhat more extensive than the pallium, and is correlated with a pronounced expansion of the ventral hemispheric wall and of its ventricular recess, as well as with a conspicuous dorsal extension of the paraterminal neighborhoods B_3 and B_4. The limiting groove fb_3 indicating an approximate boundary between these two neighborhoods is, at least in some regions, clearly recognizable.

The hemispheric wall dorsal to the commissural plate, consisting of a slightly everted D-rudiment[104] and of the B_{1+3} zones, was designated by HOLMGREN and VAN DER HORST (1925) as 'nucleus praeopticus superior' (cf. Fig. 201 C), but this region doubtless pertains to the telencephalic zonal system. Caudally to commissura anterior, the rostral extremity of thalamus ventralis adjoins the telencephalic grisea and was likewise included into the 'nucleus praeopticus superior' of the just cited authors (cf. chapter XII, Figs. 84 A, B, and footnote 147, p. 210). Figures 201 A–C illustrate cross-sections through lobus hemisphaericus of Ceratodus whose pattern features are also depicted by Fig-

[103] In Ceratodus, the separation of 'cortex' from periventricular layer is generally less well defined in basis than in pallium, and, for that matter, less than in the telencephalic basis of Protopterus and Lepidosiren.

[104] This D-rudiment includes fibers of the stria medullaris system and the attachment (taenia) of roof plate.

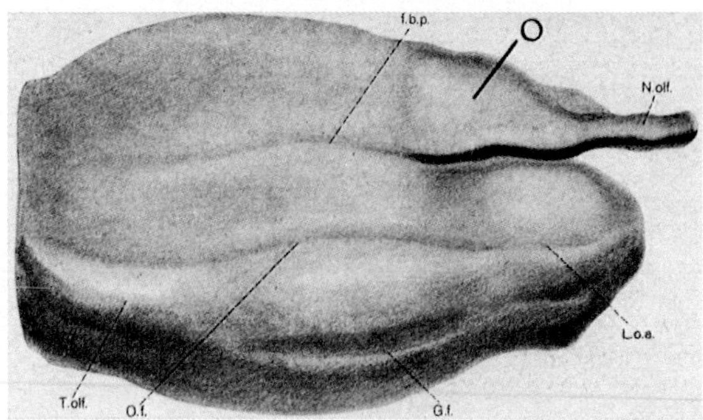

Figure 202 A. Lateral view of the telencephalon in Protopterus, based on a wax model (from GERLACH, 1933). f.b.p.: external baso-pallial sulcus; G.f.: vascular sulcus; L.o.a.: 'lobus olfactorius anterior'; N.olf.: nervus olfactorius; O: olfactory bulb; O.f.: sulcus of nervus opticus; T.olf.: 'tuberculum olfactorium'.

ure 58C, p. 167, Figure 72, p. 197, and Figure 268, p. 515 of volume 3/II.[104a]

The *communication channels* pertaining to the *olfactory bulb* spread rather diffusely over pallium and basis, and run through the external as well as the internal fiber layers separated from each other by the corticoid plates. An external, somewhat more compact component, has been described as tractus olfactorius lateralis. A vaguely outlined tractus olfactorius medialis is likewise present, and essentially nonmedullated fibers, related to 'tuberculum olfactorium' have been designated as 'tractus olfactorius ventralis'.

The *caudal communication* channels are provided by the systems of basal forebrain bundle and stria medullaris. The medial portion of basal forebrain bundle interconnects the paraterminal grisea with the diencephalon, while the lateral portion, related to pallium and basilateral grisea, effects comparable connections. Both portions tend to intermingle with the olfactory tracts and, upon reaching levels of the anterior commissure, with each other. HOLMGREN and VAN DER HORST (1925) and subsequent authors have made an attempt to distinguish several components of the basal forebrain bundle, such as tractus olfactohypothalamicus rectus et cruciatus, tractus striohypothalamicus rectus et cruciatus, tractus praeoptico hypothalamicus, tractus praeoptico-

[104a] In Figure 268, p. 515 of volume 3/II, an inadvertent error in labelling occurred, which remained undetected due to my insufficiently attentive proof reading. Beginning with B_4 on the left, the correct clockwise sequence should evidently be D_3, D_2, D_1, B_1.

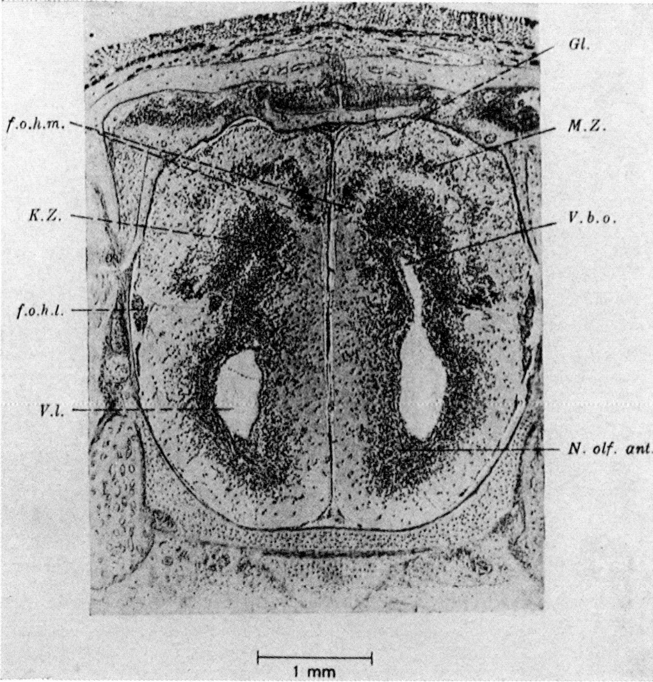

Figure 202 B. Cross-section through the telencephalon of Protopterus at the level of olfactory bulb (from GERLACH, 1933). Gl.: olfactory glomeruli; f.o.h.l.(m.): lateral (medial) external limiting sulcus of olfactory bulb; K.Z.: granule cells of bulb; M.Z.: mitral cells; N.olf.ant.: nucleus olfactorius anterior; V.b.o.: ventricle of olfactory bulb; V.l.: lateral ventricle of lobus hemisphaericus.

tubercularis, tractus striatothalamicus, tractus corticotegmentalis, tractus septohypothalamicus etc., but the distinction of these tracts remains rather uncertain. Generally speaking, as in other Anamnia, the basal forebrain bundle system can be conceived as a complex channel reciprocally interconnecting diencephalon, to some extent also mesencephalic tegmentum, and indirectly tectum mesencephali, with various regions of the telencephalon.

The stria medullaris system can likewise be conceived as a reciprocal, but predominantly descending channel interconnecting telencephalic grisea with epithalamus (ganglion s. nucleus habenulae).[105]

---

[105] HOLMGREN and VAN DER HORST (1925), who depict this system as tractus taeniae, describe also a 'tractus olfactorius lateralis habenulae and fibers detached from the

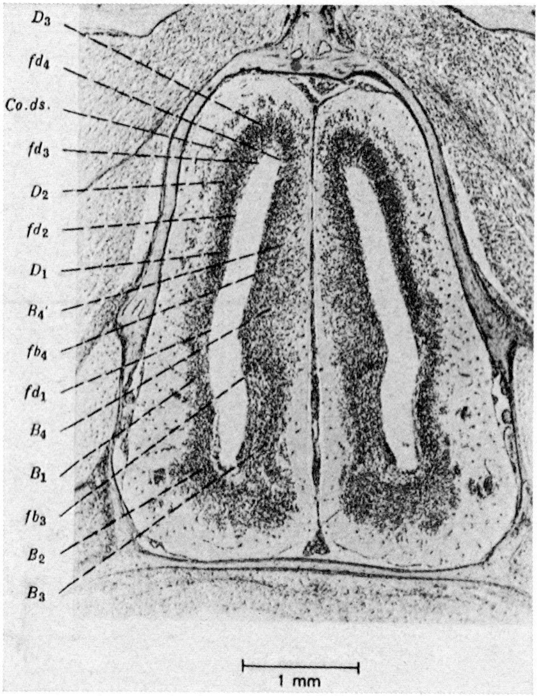

Figure 202C. Cross-section through lobus hemisphaericus of Protopterus (from GERLACH, 1933). Co.ds.: cortex pallii. Other labels conform to topologic notation adopted in text.

The anterior commissure is essentially related to the basal forebrain bundle (Fig. 201 C), but its most rostral portion (Fig. 201 B) seems to include a commissura interbulbaris. Some fibers pertaining to the stria medullaris system may also pass through the main portion of commissura anterior. Figures 164 A and 201 D illustrate relevant aspects of the telencephalic fiber systems in Ceratodus.

The telencephalon of *Protopterus*[106] differs somewhat from that of Ceratodus as regards the following features. (1) The olfactory bulb is contiguous with lobus hemisphaericus and represents the rostrodorsal

---

'corticotegmental tract' as particular components. Decussating fibers of this latter tract within the commissura habenulae are considered by these authors to represent a 'commissura pallii posterior'.

[106] Protopterus and Lepidosiren display a paired lung (Dipneumones), while Ceratodus, with one lung, is subsumed under the genus Monopneumones.

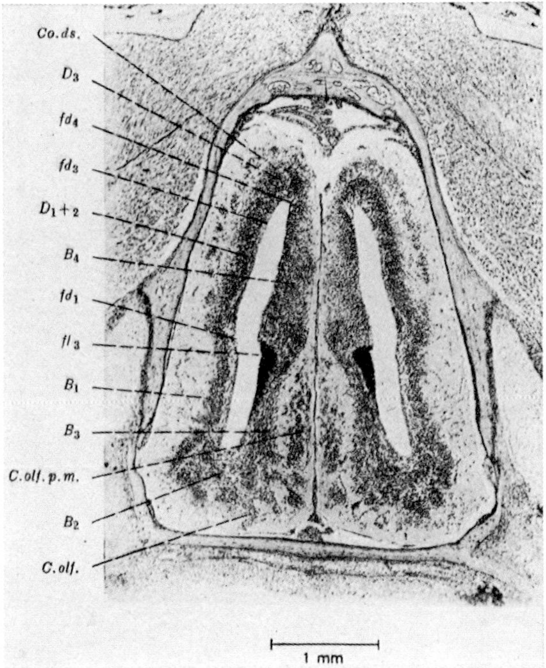

Figure 202 D. Cross-section through lobus hemisphaericus of Protopterus near caudal end of paired evaginated hemispheres (from GERLACH, 1933). C.olf.: cortex olfactoria sive basalis; C.olf.p.m.: cortex olfactoria, pars medialis. Other designations as in Figure 202C.

portion of the paired evaginated telencephalic tube (Figs. 202A, B). (2) The roof plate is restricted to the unpaired evaginated telencephalon. The paired hemispheres consist thus entirely of alar plate, whereby neighborhood B_4 becomes directly contiguous with neighborhood D_3. A limiting groove fd_4 is thus present, which runs close to the ventricular angulus dorsalis or may even be included into this latter. (3) The hemispheric wall is less folded or 'flabby' than in Ceratodus, and does not display the conspicuous caudobasal infracommissural recesses of telencephalon impar present in Ceratodus. (4) The epithalamus (habenular complex) protrudes rostralward toward the telencephalon impar from which it is separated by the velum transversum. This is still very conspicuous in young specimens (cf. Fig. 87A) but less so at older adult stages.

Omitting further details concerning the *bulbus olfactorius*, it will here

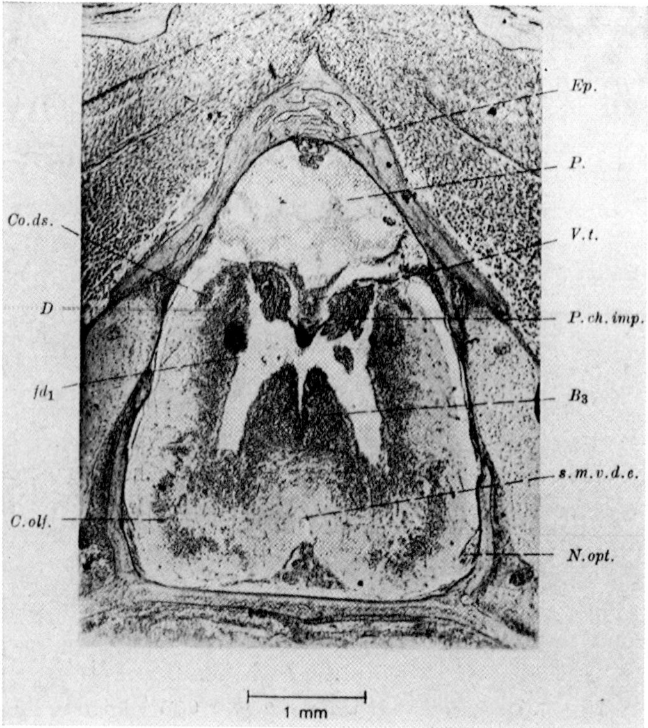

Figure 202E. Cross-section through telencephalon impar of Protopterus (from GERLACH, 1933). Ep.: epiphysis; N.opt.: optic nerve; P.: saccus dorsalis (parencephalon); P.ch.imp.: choroid plexus of telencephalon impar; s.m.v.d.e.: sulcus medianus externus telencephali imparis; V.t.: velum transversum (perhaps with part of paraphysis).

be sufficient to recall that the presence of an accessory olfactory bulb (and vomeronasal nerve) in Dipnoans is doubtful, but possible. In Protopterus, an accessory bulb was suspected by GERLACH (1933) and RUDEBECK (1944).[107] Again, the latter author described a layer of 'supracortical cells' as a special formation extending from the dorsocaudal part of the bulb toward the pallium of lobus hemisphaericus. These cells, however, are probably still pertaining to the bulb proper and merely represent the superficial elements related to the glomeruli. These cells are clearly recognizable in our Figure 202B.

[107] In the caudal ventrolateral portion of the bulb by GERLACH, and in the bulb's caudodorsal part by RUDEBECK (as well as by SCHNITZLEIN and CROSBY, 1967). Further details concerning significance of vomeronasal nerve and accessory olfactory bulb were included in section 2 of the present chapter (p. 477 and 504).

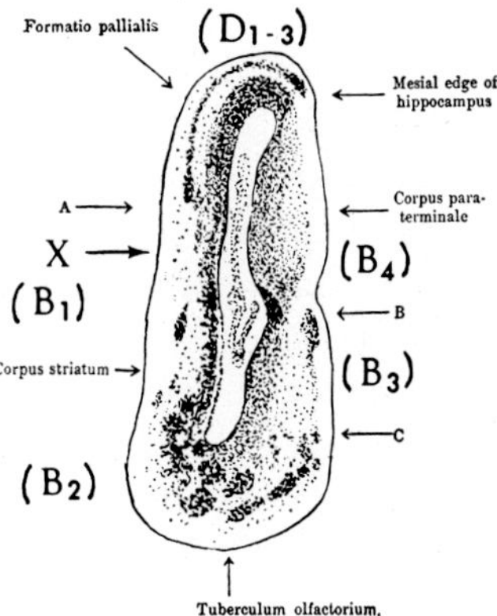

Figure 203. Cross-section through one paired evaginated lobus hemishpaericus of adult Lepidosiren (from ELLIOT SMITH, 1908) A, B, C: boundaries stressed by the cited author. Present topologic notation added in parentheses. Added X: approximately lateral boundary between pallium and basis, the basalmost part of D_1 displaying lack of pronounced cortical differentiation.

The *lobus hemisphaericus* of Protopterus includes, *mutatis mutandis*, both in the paired evaginated portion and in the telencephalon impar, the same pallial and basal grisea as enumerated above for Ceratodus (cf. Figs. 202C–E). It can be noted that, in some regions, the pallial as well as the basal 'cortical' plates are represented by disrupted clusters, reminiscent of the so-called '*islets of Calleja*' displayed by the cortex of tuberculum olfactorium in various Mammals. It should be added that the pallial and basal cortical formations in both Ceratodus and Protopterus manifest, *qua* 'cortical' arrangement, not only the above-mentioned regional differences, but also fairly substantial individual variations, in addition to differences in distinctiveness, related to, and decreasing from advanced embryonic to young adult stages and old adult stages.

At the transition to the diencephalon (Figs. 85B, 87A) the D-neighborhood disappears, and the B_{1+2} neighborhoods become replaced by the eminentia thalami ventralis which represents, as it were, an elongated massa cellularis reuniens, pars superior. The B_3 neighborhood, forming a 'bed nucleus' of commissura anterior, blends with the preoptic hypothalamus, this transition being the equivalent of a massa cellularis reuniens, pars inferior (*sive* hypothalamica).[108] Additional illustrations depicting the telencephalic configuration in Protopterus are Figure 265, p. 512, and Figure 266, p. 513 of volume 3/II.

---

[108] Since no caudal paired telencephalic evagination obtains, the hemispheric stalk remains rather straight tubular, and the massae cellulares reunientes assume a longitudinal instead of a lateralward directed (transverse) orientation.

The telencephalon of Lepidosiren, investigated by ELLIOT SMITH (1908) is, in all relevant morphologic aspects, closely similar to that of Protopterus (cf. Fig. 203).[109]

Concerning the *telencephalic communication channels* in both Protopterus and Lepidosiren the available recorded data are less detailed than for those in Ceratodus. The observations summarized by GERLACH (1933) indicate that, generally speaking, the fiber systems noted by HOLMGREN and VAN DER HORST (1925) in Ceratodus are likewise present, without substantial differences, in Protopterus. Much the same can be assumed concerning Lepidosiren.

7. Amphibians

The telencephalon of Amphibians displays rostral and caudal paired evagination, combined with a moderately large telencephalon impar (aula) whose floor is provided by a fairly extensive commissural plate pertaining to the lamina terminalis. The caudal paired evagination, with its polus posterior, protrudes, in a variable degree, laterally to diencephalon and may even, as in Gymnophiones (cf. Fig. 98 E, chapter XII, section 6, and Fig. 389 A, p. 890, vol. 4) reach as far as the mesencephalic levels. Inversion obtains throughout.

The Urodele telencephalon was particularly investigated by HERRICK, who summarized his views, with bibliographic references, in a monograph on the brain of Amblystoma tigrinum (1948). The morphologically respectively topologically relevant wall neighborhoods in Urodeles, Anurans, and Gymnophiones were mapped by the present author (K., 1921a, b, c, 1922a).[110] Instructive *Golgi-impregnation* pictures of the Anuran telencephalon were obtained by PEDRO RAMÓN (1922), and pertinent observations on morphology and fiber tracts in

---

[109] Cf. also Figure 207, p. 513 in volume 3/II.

[110] Recent histochemical observations on the telencephalon of the Bullfrog, concerning distribution of acetylcholinesterase, monoamine oxidase, and succinate dehydrogenase essentially agree with the morphologic (i.e. topologic) subdivisions (NORTHCUTT, 1974). It should, however, be emphasized that biochemical characteristics of this type may be related, as is the case of fiber connections, to factors quite independent of those relevant for the connectedness of topologic neighborhoods, i.e. for morphologic configuration. Similar histochemical characteristics may be displayed by homologous and by quite non-homologous grisea, and this applied likewise, *mutatis mutandis*, to histochemical dissimilarities. Histochemical characteristics of the type hitherto recorded cannot be used either to support or to deny morphologic homologies.

the Bullfrog are included in a study by KAPPERS and HAMMER (1918). The interpretations adopted by the *Crosby school*, with some bibliographic references, can be found in a paper by HOFFMAN (1966). Other recent papers on the Anuran telencephalon are those by CLAIRAMBAULT (1963/64) and CLAIRAMBAULT and DERER (1968).

The Amphibian telencephalon was also reviewed by PAPEZ (1929), and additional references to significant contributions by pioneering older authors, such as GAUPP, RÖTHIG, BINDEWALD, and others, are given in the treatises by KAPPERS (1921, 1947), KAPPERS *et al.* (1936), and BECCARI (1943). The telencephalic homologies, based on the topologic concept of a longitudinal zonal system, were dealt with in section 6, chapter VI, of volume 3/II.[111] Comments on the Amphibian olfactory bulb, which is directly contiguous with lobus hemisphaericus, are included in section 2 of the present chapter.

Figures 204A–D illustrate representative cross-sections through the Urodele lobus hemisphaericus. The pallium displays three longitudinal zonal areas, which are commonly distinguishable by greater or lesser differences in their cytoarchitecture, namely area lateralis pallii (D_1), area dorsalis pallii (D_2), and area medialis pallii (D_3). This latter, again, may in some instances show a less crowded dorsal, and a somewhat more densely populated ventral subdivision. A cell-free lateral and medial zona limitans frequently corresponds to the boundary between pallium and the basal neighborhoods B_1 (laterally) respectively B_4 (medially). The neuronal elements remain periventricular, with a relatively dense basal layer *(Basalschicht)* and a less compact dispersal layer *(Schwärmschicht)*, externally to which the cell-poor molecular or zonal layer is located, containing fiber tracts and neuropil. This latter, however, may extend into dispersal layer and even outer stratum of basal layer.

From a morphological viewpoint based upon topological connectedness, the area medialis pallii can indeed be interpreted to represent the 'primordium hippocampi', as is generally recognized by the diverse schools of comparative neurologists. The area lateralis pallii (nucleus olfactorius dorsolateralis *sive* primordium piriforme of HERRICK, 1948), is interpreted as homologous to the Mammalian piriform lobe, or 'palaeocortex', particularly prepiriform cortex, by KAPPERS (1947),

[111] A presentation of views concerning the homologies propounded by the *Crosby school*, and applied to the phylogeny of the Vertebrate telencephalon, is included in a paper by CROSBY *et al.* (1966).

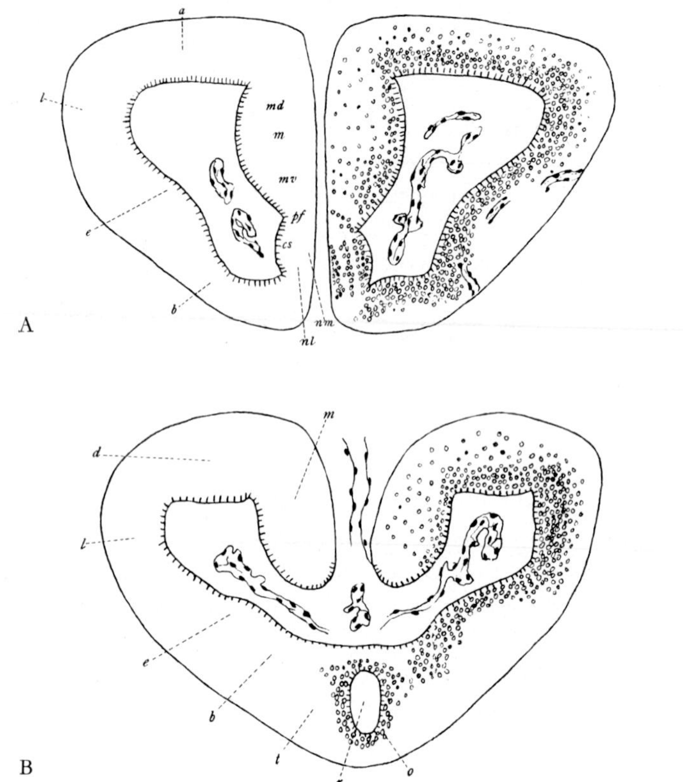

Figure 204 A, B. Cross-sections illustrating cytoarchitectural arrangement in the lobus hemisphaericus of (A) Salamandra maculosa in paired evaginated telencephalon and (B) in Triton taeniatus at level of telencephalon impar (from K., 1921a). b: nucleus basilateralis inferior; cs: 'cellulae septales' (joint nn. basimedialis inferior et superior); d: area dorsalis pallii; e: nucleus basilateralis superior; l: area lateralis pallii; m (md, mv): area medialis pallii (with dorsal and ventral subdivision); nl, nm: dispersal layers of nuclei basimediales ('nucleus medialis septi'); pf: 'pars fimbrialis septi' (dorsal extension of nucleus basimedialis superior). Cf. also Figs. 57 (p. 80) and 101 (p. 154), vol. 3/I.

HERRICK (1948), and by the *Crosby school* (CROSBY *et al.*, 1966; HOFFMAN, 1966). There is, however, little doubt that the Amphibian area lateralis pallii represents the telencephalic wall neighborhood which, in Amniota, becomes displaced by internation (introversion) to become EDINGER's epistriatum (or ELLIOT SMITH's hypopallium) of Reptiles, hyperstriatum and epistriatum of Birds, and embryonic 'lateral ganglionic hill', of Mammals, forming here part of the matrix of corpus striatum (cf. K. 1929; and section 6, chapter VI, vol. 3/II). The area lateralis pallii of Amphibians becomes thus included in the ('secondary') basis of the Amniote lobus hemisphaericus.[112]

The area dorsalis pallii must therefore be considered to represent a 'general pallium', corresponding to piriform lobe cortex, neocortex, and parahippocampal cortex of Mammals.

At the transition of bulbus olfactorius to lobus hemisphaericus, cell groups of nucleus olfactorius anterior may aggregate as nucleus postolfactorius lateralis respectively medialis in the open neighborhoods of adjoining pallial and basal zones.

The likewise periventricular basal grisea, which commonly display a much less conspicuous dispersal layer, are nucleus basilateralis superior (B_1), nucleus basilateralis inferior (B_2), nucleus basimedialis inferior (B_3), and nucleus basimedialis superior (B_4). There is not infrequently some degree of 'fusion', 'amalgamation', and 'overlap' between adjacent basal grisea, which can be considered open topologic neighborhoods.

HERRICK (1948) designates nucleus basilateralis superior[113] as 'corpus striatum, pars dorsalis', and nucleus basilateralis inferior as 'corpus striatum, pars inferior'; the caudal portion of B_{1+2}, with components of B_3, being that author's 'nucleus amygdalae'.

Nucleus basimedialis inferior (B_3) and basimedialis superior (B_4) comprise the paraterminal body *sensu latiori*, and are generally designated as 'septum'. In addition to the periventricular basal layer, representing the 'nucleus lateralis septi', scattered elements of the here usually present and conspicuous dispersal layer form the so-called 'nu-

---

[112] In my early attempts at tracing the telencephalic homologies, before determining, in sufficient detail, the essential key stages of ontogenesis in Anamnia and Amniota, which are 'blurred' by the transitory neuromery, I had erroneously homologized the Amphibian area lateralis pallii with the cortex lateralis in Reptiles, which EDINGER (e.g. 1912, Fig. 144, p. 176) interpreted as neopallium, and thus considered said area lateralis pallii to represent the primordial 'neopallium' (K., 1921a, 1922b). Subsequently, I recognized this as a double error (K., 1924b, 1927, 1929): the area lateralis pallii corresponds to the epibasal complex ('epistriatum') of Reptiles, and the cortex lateralis of these latter corresponds essentially to piriform lobe cortex, not to neopallium.

[113] In my early attempts (cf. footnote 112) I designated this griseum (B_1) as 'epistriatum', believing that it was homologous to the Reptilian 'epistriatum'. This was also done by KAPPERS and HAMMER (1918) in Anurans (cf. further below). Subsequently, I preferred to drop the term 'epistriatum' altogether, and used the term nucleus epibasalis for the D_1 derivatives in Sauropsida as well as occasionally for the corresponding D_1 zone in Anamnia.

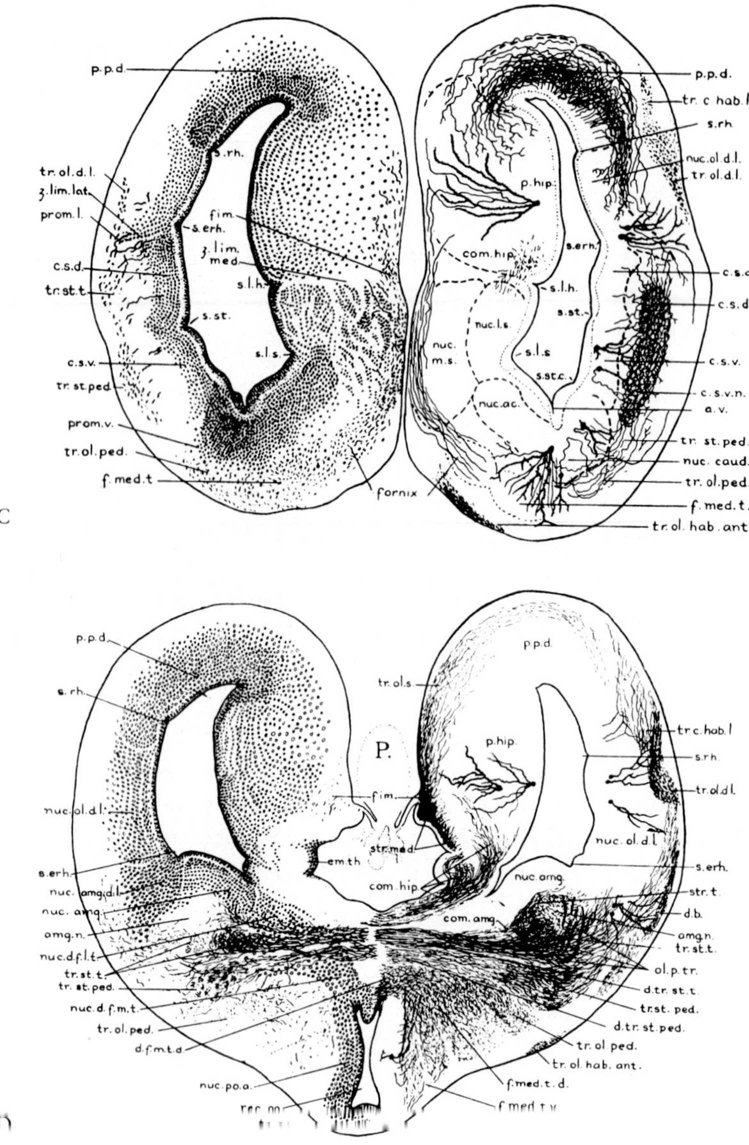

Figure 204C, D. Cross-sections, combined from cell and fiber stains, and *Golgi impregnations* through levels of paired rostrally and caudally evaginated lobus hemisphaericus in Amblystoma tigrinum (from HERRICK 1948). amg.n.: 'neuropil of amygdala'; a.v.: angulus ventralis ventriculi; com.amg.: 'commissure of amygdalae' (part of comm. anterior); com.hip.: hippocampal commissure (comm. pallii); c.s.d.(n.): 'corpus striatum, pars dorsalis' and (n.) its neuropil (n. basilat. superior); c.s.v.(n.): 'corpus striatum, pars ventralis' and (n.) its neuropil (n. basilat. inferior); d.b.: '*diagonal band of Broca*' (probably

cleus medialis septi'.[114] Dorsal portions of this latter can extend dorsally to interventricular foramen, as so-called pars fimbrialis *sive* supraforaminalis septi. This cell group should not be confused with the basally often closely adjacent rostral end of the eminentia thalami ventralis caudally to foramen interventriculare (Fig. 204D). The B_3 and B_4 components may be conjoined or rather distinctively differentiated. If the periventricular basal layer of the B_3 component of nucleus lateralis septi is particularly distinctive, it is also occasionally described as nucleus accumbens septi (cf. Fig. 204C). The basal cell masses, except for B_4, extend into the caudal paired evaginated portion of lobus hemisphaericus, but to a lesser degree than the three pallial zones, and tend to fuse into a more or less common B_{1+3} cell aggregate, which may not quite reach the polus posterior. A number of taxonomically related and even individual variations obtain within the here described overall configurational griseal pattern of the Urodele lobus hemisphaericus.

There is little doubt that, in agreement with Ludwig Edinger, the Urodele brain, and the Amphibian brain in general, particularly also with regard to the telencephalon, does indeed display the most distinct manifestation of the fundamental configurational pattern characteristic

[114] Beccari (1943) justly remarks: '*Nella regione settale si distinguono due nuclei settali, laterale e mediale: sono più campi cellulare que veri nuclei.*'

---

here fibers of lateral olfactory tract); d.f.m.t.d.: decussation of medial forebrain bundle; d.tr.st.ped.: decussation of lateral forebrain bundle; d.tr.st.t.: decussation of 'tr. strio-tegmentalis' (lateral forebrain bundle); em.th.: eminentia thalami (ventralis); fim.: 'fimbria' (fibers related to D_3); f.med.t.d.(v.): medial forebrain bundle, dorsal and (v.) ventral subdivision; nuc.ac.: 'n. accumbens septi' (n. basimed. inferior); nuc.amg. (dl.): 'n. amygdalae' and (dl.) its pars dorsolateralis (B_{2+3} and B_1); nuc. d.f.l.t.: bed nucleus of anterior commissure; nuc.caud.: 'n. caudatus' (part of n. basimed. inferior); nuc.l.s.: 'n. lateralis septi' (n. basimed. superior, paraventricular portion); nuc.m.s.: 'n. medialis septi' (n. basimed. superior, abventricular portion); nuc.ol.d.l.: 'n. olfactorius dorsolateralis sive primordium piriforme' (area lateralis pallii); nuc.po.a.: n. praeopticus, pars anterior; ol.p.tr.: 'olfactory projection tract'; p.hip.: 'primordium hippocampi' (area medialis pallii); p.p.d.: 'primordium pallii dorsalis' (area dorsalis pallii); prom.l.(v.): 'prominentia lateralis' ('ventralis') of pericellular cell plates; s.erh.: 'sulcus endorhinalis' (fd_1); s.l.h.: 's. limitans hippocampi' (fd_4); s.l.s.: 's. limitans septi' (fb_3); s.rh.: 's. rhinalis' s.st.: 's. striaticus' (fb_1); s.stc.: 'strio-caudatus' (fd_2); str.t.: 'stria terminalis'; tr.c.hab.l.: 'tractus corticohabenularis lateralis; tr.ol.d.l.: 'tr. olfactorius dorsolateralis'; tr.ol.hab.ant.: 'tr. olfacto-habenularis anterior'; tr.ol.ped.: 'tr. olfacto-peduncularis'; tr.po.: 'tr. praeopticus'; tr.st.ped.: 'tr. strio-peduncularis' (lateral forebrain bundle); tr.st.t.: 'tr. strio-tegmentalis (lateral forebrain bundle); z.lim.lat. (med.): zona limitans lateralis and (med.) medialis.

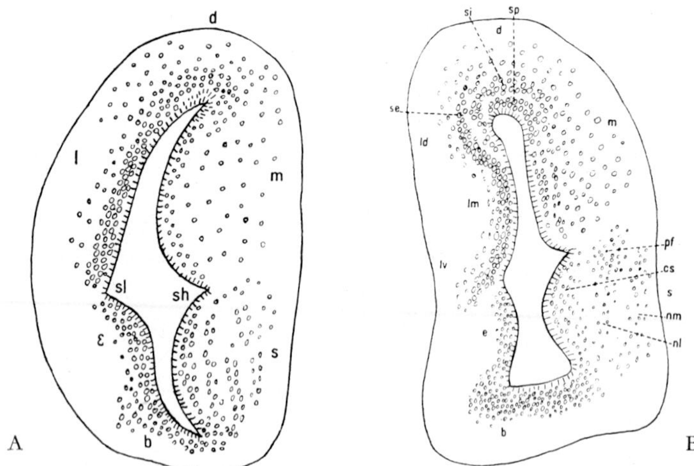

Figure 205 A, B. Drawing of cross-sections *(Nissl stain)* through lobus hemisphaericus in two different specimens of Rana esculenta (from K., 1921b, c). b: nucleus basilateralis inferior; cs: 'cellulae septales' (periventricular portion of nucleus basimedialis superior); d: area dorsalis pallii; e: nucleus basilateralis superior ('epistriatum' of original 1921 papers); l: area lateralis pallii; ld, lm, lv: dorsal, middle, and ventral subdivisions of a. lat. pall.; m: area medialis pallii; nl nm: 'nucleus lateralis' and 'nucleus medialis' septi (scattered abventricular cells of 'septum', both generally subsumed under 'n. med. septi'); pf: 'pars fimbrialis septi' (pertaining to B_4); s: 'septum' (paraterminal grisea *sensu latiori*, comprising a conjoint n. basimed. inferior and superior); se, si, sp: external, intermediate and paraependymal stratifications in regions ld and d; sh: 'sulcus limitans hippocampi' (fd_4); sl: 'sulcus limitans lateralis' (fd_1).

for the Vertebrate neuraxis. Thus, it provides a most suitable frame of reference for an understanding of both comparative anatomy and presumptive phylogenetic evolution of the forebrain (archencephalon). The question whether the periventricular arrangement of the telencephalic grisea with their basal, dispersal, and molecular layers, which arrangement is generally repeated in Vertebrate ontogeny, should be assessed as a 'primitive' or as a 'regressive' feature in recent Amphibians, remains a moot question, since the taxonomically 'lower' Selachians and Dipnoans display a far more conspicuous cerebral 'cortex'. SÖDERBERG (1922) interpreted ontogenetic stages in Amphibians, particularly Anurans, as manifesting a transitory cortex-like migration layer becoming secondarily joined to the periventricular layer in adults. This ontogenetic behavior, which the present author can confirm, is indeed suggested, but is not particularly conspicuous, and

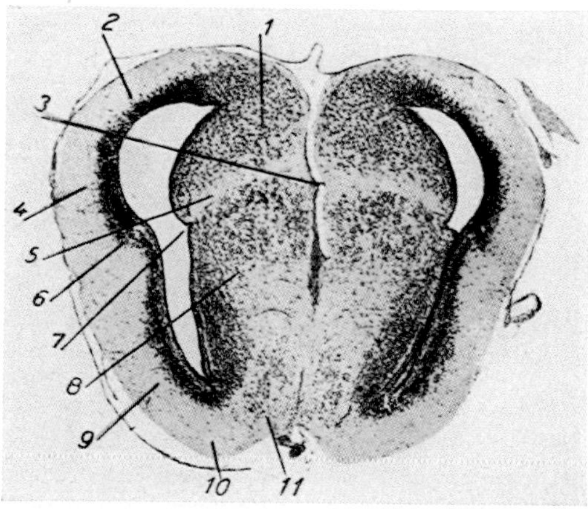

Figure 205C. Cross-section *(Giemsa stain)* through lobus hemisphaericus of Rana esculenta (from BECCARI, 1943). 1: area medialis pallii (primordium hippocampi); 2: lateral border of area dorsalis pallii; 3: external pallio-septal sulcus; 4: area lateralis pallii; 5: zona limitans medialis; 6: zona limitans lateralis (here poorly developed); 7: sulcus limitans hippocampi (fd_4); 8: paraterminal ('septal') region; 9: nucleus basilateralis inferior; 10: '*eminenza strio-settale*'; 11: abventricular layer of nucleus basimedialis inferior (so-called n. medialis septi, pars inferior).

could merely represent a transitory and secondary feature related to the histogenetic cellular displacements dealt with in section 1, chapter V of volume 3/I, or to the early stages of neuropil formation.

Be that as it may, the observed ontogenetic events do not preclude the probability that ancestral tetrapod Vertebrate forms might have possessed a telencephalon with essentially periventricular grisea of present-day Amphibian type.

Figures 205 A–206 C illustrate representative cross-sections through lobus hemisphaericus in *Anurans.* It can easily be seen that three pallial and four basal grisea are recognizable as wall neighborhoods orthohomologous with those present in Urodeles. The obtaining differences involve, *inter alia*, a greater extension of area lateralis pallii, and a lesser extension of area dorsalis. The paraterminal grisea, particularly nucleus basimedialis superior, are more developed in Anurans than in Urodeles. KAPPERS and HAMMER (1918) designated nucleus basilateralis superior as 'epistriatum', and nucleus basilateralis inferior as 'striatum'. Figure 205 D illustrates an interesting comparison between the cell populations (topologic neighborhoods) in lobus hemisphaericus of the Frog and in that of the Selachian Squalus acanthias, as made by JOHN-

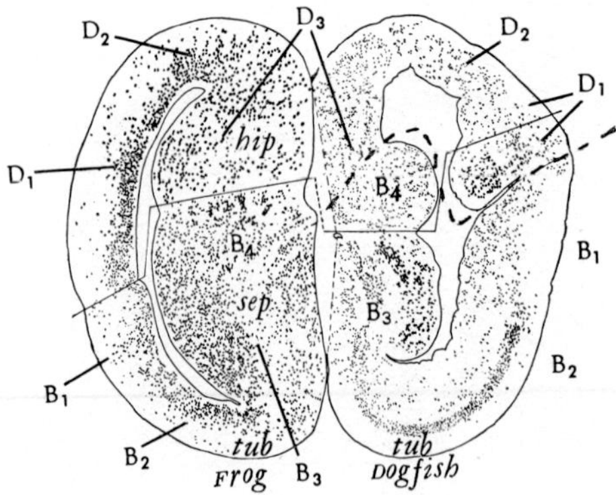

Figure 205 D. Comparison of cross-sections through lobus hemisphaericus of (left) Frog and (right) Selachian Acanthias (after JOHNSTON, from PAPEZ, 1929). hip: area medialis pallii; sep: paraterminal grisea ('septum'); tub: 'tuberculum olfactorium'. The present author's D and B notation has been added, and the broken line at right indicates my interpretation, which differs from that of JOHNSTON and PAPEZ (cf. Fig. 281, p. 536, vol. 3/II).

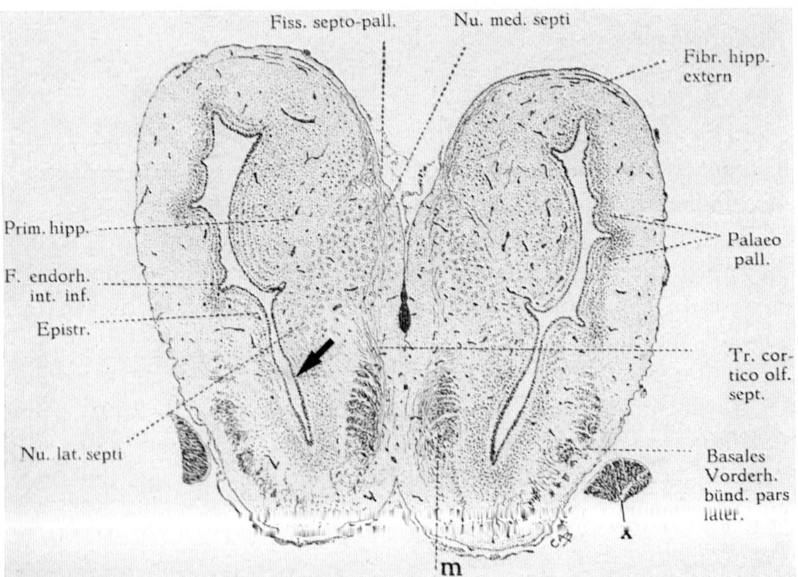

Figure 206 A. Cross-section *(Weigert-Pal stain)* through lobus hemisphaericus of Rana catesbyana (from KAPPERS and HAMMER, 1918). m: medial forebrain bundle; x: optic nerve (displaced by dissection). Added arrow shows the here faintly indicated groove (fb_3) between superior and inferior basimedial grisea. It will be seen that area lateralis pallii is interpreted by the cited authors as 'palaeopallium' and nucleus basilateralis superior as 'epistriatum'.

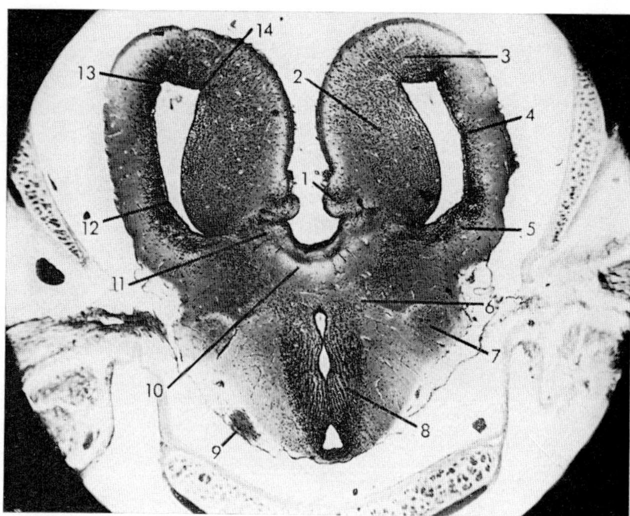

Figure 206 B. Cross-section *(Weil stain)* through the forebrain of Rana esculenta at level of commissura pallii (× ca. 14, red. $^1/_2$). 1: pars supraforaminalis nuclei basimedialis superioris; 2: area medialis pallii; 3: area dorsalis pallii; 4: area lateralis pallii; 5: nuclei basales (B_{1-3}); 6: massa cellularis reuniens, pars inferior (hypothalamica); 7: basal forebrain bundle; 8: preoptic region of hypothalamus; 9: fragment of optic nerve, 10: commissura pallii (note medullated and non-medullated portion; 11: eminentia thalami ventralis ('n. commissurae hippocampi'); 12: groove fd_1; 13: groove fd_2; 14: groove fd_3.

STON and, in a somewhat different manner, by myself on the basis of what I believe to be ontogenetic key-stages (cf. Fig. 281, p. 536, vol. 3/II). It will also be recalled, as mentioned above in section 2, that the antimeric Anuran olfactory bulbs may fuse in the midline by secondary 'concrescence' of their external surfaces. This concrescence should not be confused with a rostral appositional extension of the junction provided by the unpaired lamina terminalis as seen in Figures 205 C and 206 A.

The *Gymnophione* telencephalon displays a particularly well developed olfactory bulb contiguous with lobus hemisphaericus. In contradistinction to Urodeles and Anurans, two separate olfactory nerves, a dorsal and a ventral one, reach the rostral extremity of the bulb (cf. Figs. 159 B and 170 C, section 2). A conspicuous caudolateral bulbus accessorius is present (cf. Fig. 276, p. 523, vol. 3/II). A shallow lateral groove may run from the region of the postbulbar 'nucleus postolfactorius lateralis' along the dorsal boundary of B_2, and was designated as

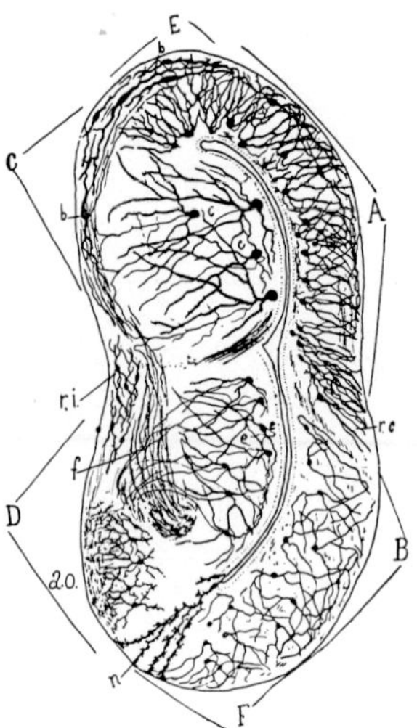

Figure 206 C. Cross-section *(Golgi impregnation)* through lobus hemisphaericus of the Frog (from PEDRO RAMÓN, 1922). A: *corteza externa* (area lateralis pallii); B: *ganglio basal o cuerpo estriado* (n. basilat. superior et inferior); D: *lámina del fórnix o ganglio superior del septum* (n. basimed. superior et inferior); E: *angulo dorsal del cerebro* (area dorsalis pallii); F: *angulo ventral* (n. basilat. inferior); a.o.: *area olfactoria* (abventricular part of n. basimed. inferior); b: *células fusiformes de la corteza* (tangential cells in zonal or dispersal layers); c: *células del ganglio superior del septum* (large neuronal elements of area medialis pallii, C); e: *células del ganglio inferior* (neuronal elements of n. basimed. superior); f: *fasciculo cortico-medialis* (medial forebrain bundle); n: *célula ependimal* (ependymal cells with pedicles at brain surface); r.c.: *region curva* (edge of area lateralis pallii at zona limitans lateralis); r.i.: *region intermedial del tabique* (medial portion of n. basilat. superior, 'n. medialis septi').

'sulcus rhinalis anterior' (K., 1922a), caudally joining the so-called 'sulcus rhinalis posterior' pointed out further below.

Figures 207 A–H illustrate cross-sectional levels in the lobus hemisphaericus of the Gymnophione Siphonops, and two corresponding levels in Ichthyophis. As in Anurans, the three pallial and four basal zones displayed by Urodeles are present.[115] Characteristic for the hitherto examined Gymnophiones is a bulge of the medial ventricular wall jointly formed by the adjacent neighborhoods D_3 and B_4. Sulcus fd_4,

[115] Additional illustrations of lobus hemisphaericus in the three orders of Amphibia are Figures 58B (p. 166), 75, 76 (pp. 200, 201), 208 (p. 427), and 269 to 276 (pp. 516–523) in section 6, chapter VI of volume 3/II). In the telencephalic *bauplan* formulae for Amphibia given on p. 524 of volume 3/II, the *grundbestandteil* B_1 was inadvertently omitted from the formula for Gymnophiona.

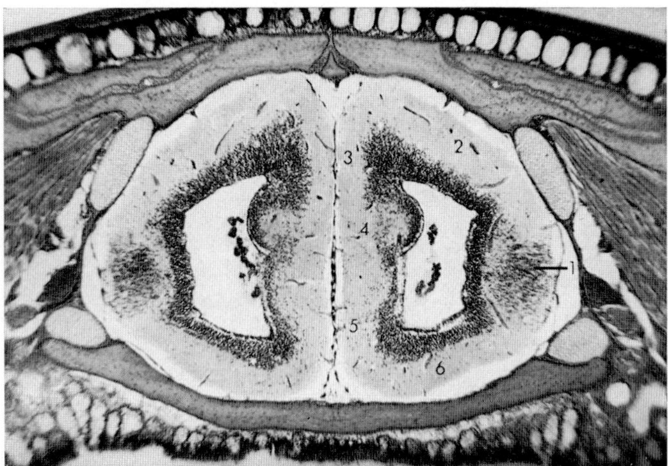

Figure 207A. Cross-section (hematoxylin-eosin stain) through lobus hemisphaericus of the Gymnophione Siphonops at level of postolfactory nuclei. 1: nucleus postolfactorius lateralis; 2: area lateralis pallii; 3: area dorsalis pallii; 4: transition of nucleus postolfactorius medialis to area medialis pallii and nucleus basimedialis superior; 5: nucleus basimedialis inferior; 6: nucleus basilateralis inferior.

clearly recognizable at some ontogenetic stages, thereby disappears. The cross-sectional plane of the lateral ventricle assumes a marked medial concavity of its outline ('inversion').

The periventricular cell population displays basal and dispersal layers. This latter is particularly developed in area medialis pallii and in nucleus basimedialis superior. In the region of nucleus basilateralis inferior and in part of nucleus basimedialis inferior, especially at levels fairly close to lamina terminalis, a distinct, although not very dense rudimentary cortical plate (cortex olfactoria, area ventralis anterior et ventrolateralis posterior) is commonly developed.[116] The caudal paired evagination usually bulges ventrolaterally and may give the impression of a 'temporal lobe', which, however, is not homologous to the essentially neocortical one of Mammals. It includes, however, a homologon of the cortical amygdaloid nucleus (area ventrolateralis posterior). On the lateral surface of this 'pseudotemporal' lobe the boundary of pallial and basal neighborhoods is generally indicated by a shallow groove

[116] Cf. e.g. the differentiation of cortex olfactoria in two different specimens of Siphonops as seen in Figures 207B and F, or in the medialward directed bend of area lateralis pallii in Figures 207D and G.

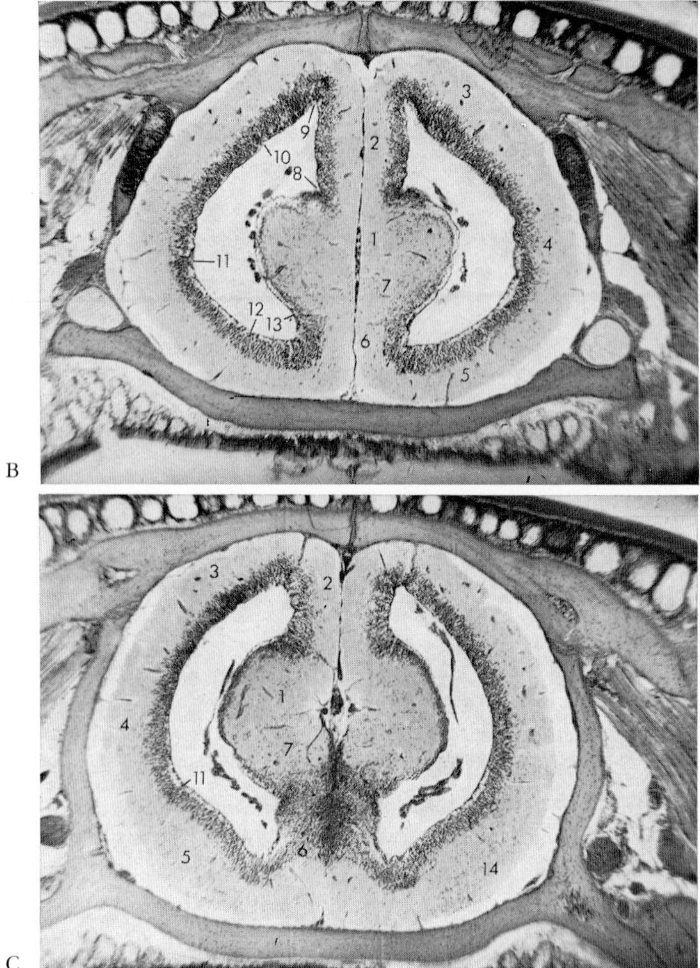

Figure 207B–E. Cross-sections, in rostrocaudal sequence, through lobus hemisphaericus of Siphonops. 1: area medialis pallii; 2: area dorsalis pallii; 3: area lateralis pallii; 4: nucleus basilateralis superior; 5: nucleus basilateralis inferior; 6: nucleus basimedialis inferior; 7: nucleus basimedialis superior; 8: groove fd_3; 9: fd_2; 10: fd_1; 11: tb_1; 12: tb_2; 13: tb_3; 14: cortex olfactoria; 15: massa cellularis reuniens, pars inferior (sive hypothalamica); 16: preoptic recess respectively griseum (rostral hypothalamus).

designated as 'sulcus rhinalis posterior' (K., 1922a), joined by the above-mentioned 'sulcus rhinalis anterior'. However, in view of the homology of area ventrolateralis posterior, said groove actually represents the sulcus endorhinalis posterior *sive* periamygdalaris; likewise,

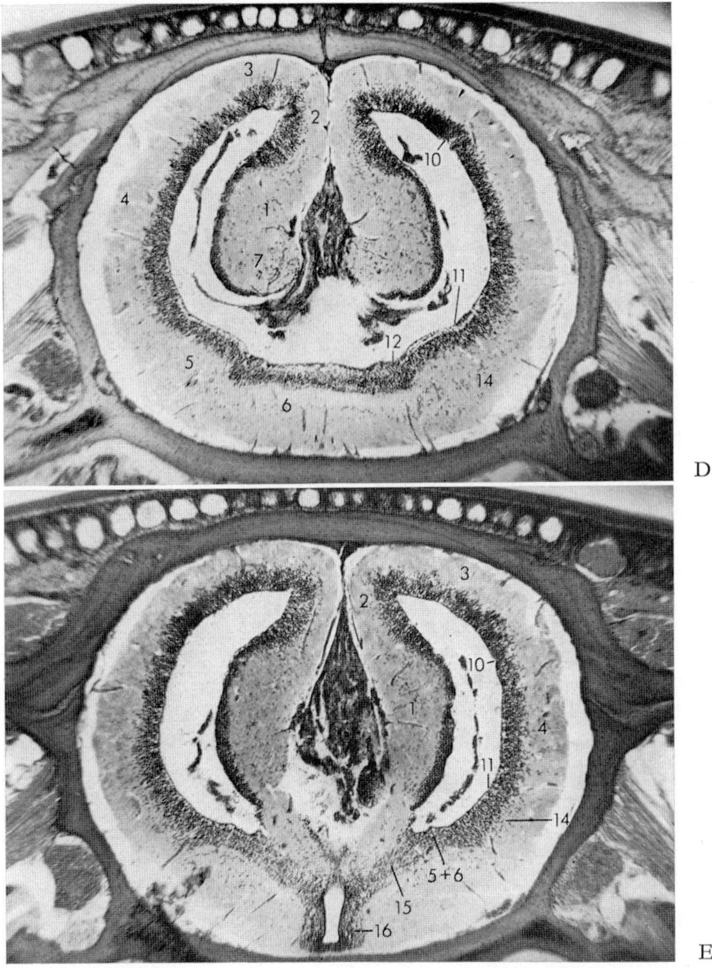

the rostral shallow groove should rather be named sulcus endorhinalis anterior. Both sulci may be barely suggested or occasionally absent.

On the whole, the basal components of the Gymnophione lobus hemisphaericus, except for the caudal paired evaginated portion, are more extensive than the pallial ones. Rostrally, a nucleus postolfactorius lateralis and a less conspicuous nucleus postolfactorius medialis are generally present. As in the other orders of Amphibians, taxonomic and individual variations within the obtaining overall configurational and cytoarchitectural pattern are commonly displayed.

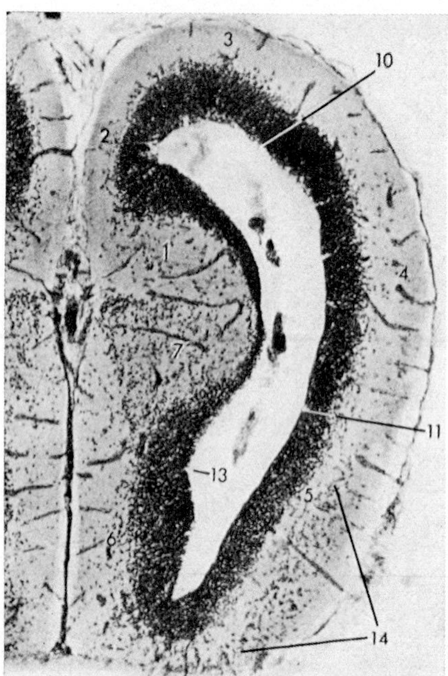

Figure 207 F. Cross-section (phosphotungstic-acid-hematoxylin, 100 μ thick) through lobus hemisphaericus of Siphonops annulatus (from CRAIGIE, 1941b, ×50, red. 1/2). This very thick section through another specimen of Siphonops displays very clearly the cytoarchitectural pattern. The designations, corresponding to those of the Figures 207 B–E, have been added to CRAIGIE's unlabelled photomicrograph.

The *fiber systems* of the Amphibian telencephalon, studied in great detail by HERRICK (1948) for Urodeles, are presumably, as regards overall distribution of the main communication channels, more or less identical in the three extant orders.[117]

The connections between olfactory bulb and lobus hemisphaericus are rather diffuse, but a tractus olfactorius *dorsolateralis* in the zonal layer of the lateral hemispheric wall, and a less massive *medial olfactory tract* have been described. Other rather nondescript bundles are dorsal,

[117] It is evident that, in correlation with the diverse modes of life and the differing behavioral activities, considerable differences in details of fiber distribution and synaptology must be assumed to obtain as regards the topologically homologous grisea in the three orders and also in the various species. But these significant details related to the actual performances of the relevant neuronal networks have still remained inaccessible to the presently available techniques of investigation (including electron-microscopy, terminal degeneration, micro-electrode, histochemical, autoradiographic, and axonal flow techniques).

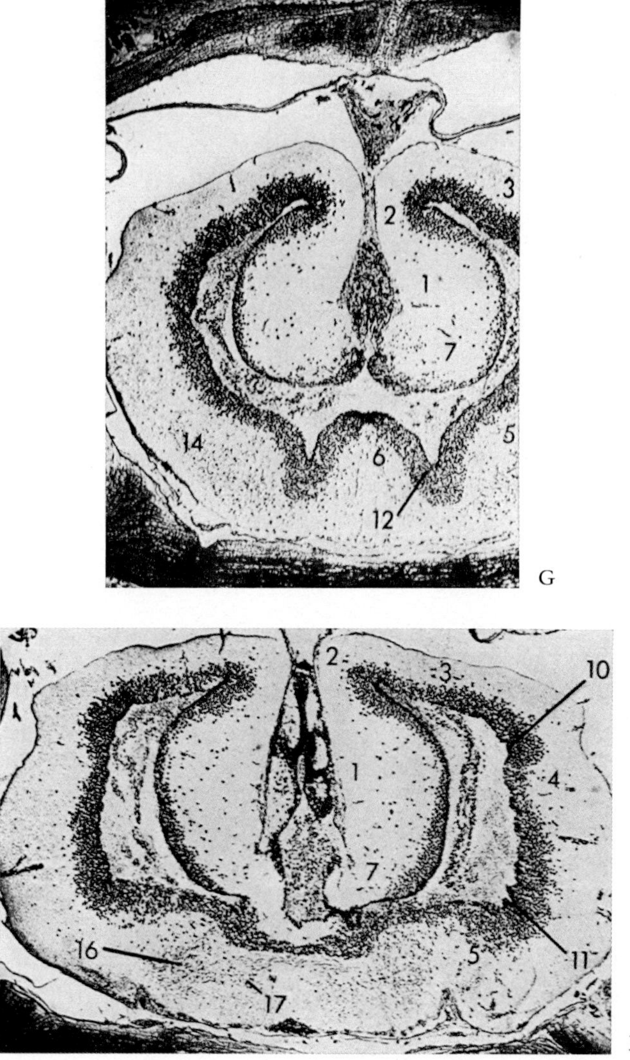

Figure 207G, H. Cross-sections *(Bielschowsky impregnation)* through the lobus hemisphaericus of Ichthyophis glutinosa at levels of (G) telencephalon impar and (H) telencephalic commissure (from RÖTHIG, 1931). The designations corresponding to those of Figures 207B–F, have been added to RÖTHIG's unlabelled photomicrographs. 16: basal forebrain bundles; 17: commissura anterior. Commissura pallii (unlabelled) can be identified above the dark basimedial cell band.

ventral and ventrolateral olfactory fasciculi.[118] It should be added that, caudalward, these 'tracts' overlap with the fibers pertaining to medial and lateral forebrain bundles. Fibers either related to these bundles or to the olfactory 'tracts' also run in the transverse plane, that is, in dorsoventral respectively ventrodorsal direction. It is probable that all grisea of lobus hemisphaericus receive, to a greater or lesser degree, input from olfactory bulb and nucleus olfactorius anterior. This, of course, does not exclude the probability that sensory input of various sorts, such as optic, vestibular, vestibulolateral, or cochlear (as the case may be), tactile, gustatory, or 'general visceral', reaches basal and dorsal telencephalic grisea through the components of both *forebrain bundles* ascending from thalamus and hypothalamus. HERRICK (1948) has described an essentially unmyelinated 'thalamofrontal tract' within the lateral forebrain bundle. Whether the ascending diencephalic-telencephalic fibers are specific and separate *qua* different functional systems, or unspecific, as HERRICK seems to assume, remains a moot question. It should also be kept in mind that, in the region of telencephalic lamina terminalis, fibers of both forebrain bundles tend to intermingle.

In addition, KICLITER and NORTHCUTT (1975) have recently claimed that 'telencephalic afferents' originate from grisea caudal to the isthmus, and, as far as 'striatal afferents' are concerned, from levels between isthmus and caudal border of thalamus. Although such connections seem possible, statements based on results by the *Fink-Heimer technique* remain open to doubt.

The *descending components of the lateral forebrain bundle*, which is essentially related to the basal and dorsal lateral hemispheric wall, and somewhat more massive than the medial bundle, run toward hypothalamus and thalamus. Some of these fibers doubtless reach the mesencephalic tegmentum, while others are believed to reach, via the pretectal region, the mesencephalic tectum.

The *descending components of the medial forebrain bundle*, which is essentially related to the basal and dorsal medial hemispheric wall, predominantly reach the hypothalamus, but presumably also thalamic and mesencephalic tegmental grisea. A portion of this channel, connecting area medialis pallii and paraterminal grisea with a caudodorsal 'mammillary'

[118] The 'ventrolateral tract' according to HERRICK (1921b) and others, is said to include a bundle, more distinct in Anurans than in Urodeles, connecting the accessory olfactory bulb with basal grisea ('striatum', 'amygdala').

hypothalamic griseum, can be evaluated as representing a primitive 'fornix'. Terms such as 'tractus strio-peduncularis', 'strio-thalamicus', 'strio-hypothalamicus', 'strio-pretectalis', 'strio-tectalis', 'septo-hypothalamicus', 'olfacto-peduncularis' etc., have been used for the description of descending fibers in the basal forebrain bundles.[119]

The *stria medullaris system* comprises predominantly descending but also some ascending fibers connecting the telencephalic medial and lateral, pallial as well as basal grisea with the epithalamus. These channels include the medial and lateral 'olfactohabenular' and 'corticohabenular' tracts described in the literature.

The *commissural systems* are provided by commissura anterior, commissura pallii (sive hippocampi) and also in part by commissura habenulae.

The *commissura anterior* is located in a portion of the telencephalic lamina terminalis which forms the floor of the ventriculus impar telencephali (aula). It includes decussating components of both forebrain bundles as well as some commissural fibers *sensu strictiori* interconnecting antimeric telencephalic grisea. Fibers of medial forebrain bundle tend to cross more rostrally than those of the lateral bundle. Fibers from the caudal paired evaginated part of lobus hemisphaericus, crossing in the anterior commissure or joining the basal forebrain bundles respectively the stria medullaris, are considered to represent a primordial stria terminalis system. In addition, fibers of nervus terminalis are said to decussate in commissura anterior.

The *commissura pallii sive commissura hippocampi* is located dorsally to commissura anterior and reaches the caudal edge (torus transversus) of commissural plate which indicates the telencephalodiencephalic boundary. In contradistinction to Cyclostomes, Selachians, and Amniota, the commissura pallii is thus not located dorsally to the interventricular foramen. These diverse morphologically important relationships were dealt with and repeatedly illustrated in sections 1 B and 6 of volume 3/II.[120] Most, but not all of the fibers of that commissure are related to the area medialis pallii (primordium hippocampi).

---

[119] These descending channels, particularly to hypothalamus and tegmentum, may include olfactory tract fibers. It should be recalled that the descending components of the 'olfactory tracts', besides consisting of second order neurites of mitral cells, are believed to contain third order neurites from the griseum of the so-called nucleus olfactorius anterior complex.

[120] Figures 62 (p. 173), 64 (p. 175), 65 (p. 179), 73 (p. 198), 75–76 (pp. 200–201), 184 (p. 410), 203 (p. 425), 336 (p. 620), 344 (p. 631) loc.cit.

Cell groups of nucleus basilateralis inferior and basimedialis inferior (B_{2+3}) form a bed nucleus of commissura anterior and to some extent also of commissura pallii. The so-called nucleus commissurae hippocampi is a rostral portion of the eminentia thalami ventralis, apparently related to fibers of the commissura pallii.

The *commissura habenulae*, although pertaining to the diencephalic epithalamus, may also be considered a telencephalic commissure, since it contains decussating stria medullaris fibers some of which may return to contralateral telencephalic grisea (e.g. as so-called commissura superior telencephali or commissura pallii posterior). Thus, fibers of olfactory tracts joining the stria medullaris are believed to form an interbulbar commissure as e.g. described by SNESSAREW (1908) and subsequent authors (tractus bulbo-bulbaris). It is not improbable that interbulbar connections also run through commissura anterior. As regards other sorts of bulbopetal fibers, namely from lobus hemisphaericus grisea to bulb, pointed out in section 2, nothing definite is known with respect to Amphibia.

The main communication channels of the Amphibian telencephalon can be outlined and depicted in reasonably valid schemes such as e.g. the diagram of Figure 208. Nevertheless, it should be kept in mind that all specific details of hodology and synaptology, despite recent 'advances', remain highly uncertain and conjectural. The relationship of medullated to non-medullated fibers in the diverse channels also displays variations *qua* taxonomic groups.

In Anurans such as the Frog, the optic system plays a substantial role, while in the wormlike, burrowing and essentially blind or near blind Gymnophiones, the olfactory, trigeminal and perhaps 'gustatory' input components predominate. The more or less 'sluggish' Urodeles can, in this respect, be evaluated as relatively 'undifferentiated', 'unspecialized', or as of a 'generalized type'.

Again, as in the case of Teleosts (cf. section 5, pp. 587 ff.) numerous ablation experiments have been performed on the Frog's telencephalon. The relevant publications by STEINER (1885), SCHRADER (1887), LOESER (1905), BURNETT (1912), and FRITZ EDINGER[121] (1912, 1913) deserve here special mention.

Deprived of its telencephalon, the Frog's behavior is not massively affected, although some inhibitory as well as excitatory (stimulating)

[121] FRITZ EDINGER, the talented son of LUDWIG EDINGER, disappeared in one of the fiendish extermination camps established by the *Nazi-regime*.

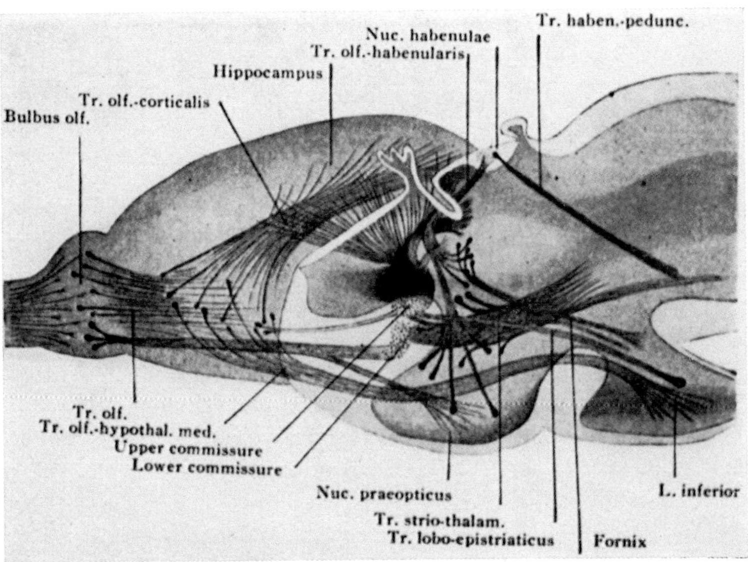

Figure 208. Diagram illustrating main prosencephalic communication channels in the brain of the Urodele Necturus in JOHNSTON's interpretation as based on his own studies combined with the results of preceding authors, and as seen in medial aspect (from JOHNSTON, 1906). Upper commissure: commissura pallii; lower commissure: commissura anterior. Cf. also Fig. 336, p. 620, vol. 3/II.

influences seem to be lost. Such 'decerebrate' Frogs with intact optic input to mesencephalon (and perhaps part of diencephalon) are said to show 'stupor' and 'lack of spontaneity' (STEINER, 1885), but the effect of the operation may not be lasting or pronounced in many instances. Such Frogs apparently do not display a 'fear reaction', but swim, feed, and breed very much as do normal animals (SCHRADER, 1887; LOESER, 1905; NOBLE, 1931, 1954). BURNETT (1912), noted that, in 'natural surroundings', normal Frogs catch more flies than 'decerebrate' ones; but if flies are introduced in a closed jar in which a normal and a 'decerebrate' Frog are confined together, most flies are said to be caught by the 'decerebrate' animal. It is believed that the normal Frog is 'disturbed' by the unfamiliar surroundings, while the operated animal has lost its inhibitions. According to HERRICK's (1926, 1931) interpretation, the distracting interference of telencephalic activity has been eliminated and the operated Frog has become 'more excitable'.

Again, BURNETT (1912) found that, in learning a maze, normal specimens of Rana pipiens and boylii would make their escape after about

20 trials with rare subsequent errors. With operated Frogs, over 100 trials were made without a successful result. This was interpreted to indicate that the required 'associations' for maze learning depend on telencephalic neuronal circuits related to significant aspects of 'learned behavior' (cf. also NOBLE, 1931, 1954).

8. Reptiles

The inverted and rostralward as well as caudalward paired evaginated Reptilian telencephalon differs from the Amphibian endbrain by the configuration of the lamina terminalis correlated with an almost complete reduction of the telencephalon medium. This, moreover, is characteristic for all three classes of Amniota. Again, as in all Amniota, the area lateralis pallii (zone D_1) of Amphibians, Dipnoans and Selachians, that is to say, of the inverted Gnathostome Anamniote telencephalon, has become medialward 'introverted' or 'internated' toward the lateral ventricle,[122] becoming thus joined to the basal topologic neighborhoods (B_1 and B_2, nucleus basilateralis superior and inferior), and thereby included into a secondary telencephalic basis. This results in a considerable thickening of the lateral basal wall in the Amniote lobus hemisphaericus. Said thickening reaches its extreme relative bulk in Birds.

While the primary Gnathostome Anamniote pallium includes neighborhoods D_1, D_2, D_3, the secondary Amniote pallium is thus essentially restricted to the neighborhoods D_2 and D_3 with very minor residual contributions by dorsolateral components of D_1. The area lateralis pallii of Amphibia is, accordingly, represented by the Reptilian epistriatum of EDINGER or hypopallium of ELLIOT SMITH. The details of this transformation were dealt with in section 6, chapter VI of volume 3/II.

Among the older basic investigations of the Reptilian telencephalon, L. EDINGER's pioneering work (1896a), summarized in his *Vergleichende Anatomie des Gehirns* (1908) is of particular importance. Another early study is that by MEYER (1892). Subsequent investigations,[123] about up to the publication of the standard treatises by KAPPERS *et al.*

[122] Cf. p. 172, chapter VI, volume 3/II.

[123] The following enumeration of contributions, also listed in the references to the present chapter, is not meant to be exhaustive, but should easily enable those interested in a more complete bibliography to find additional reports.

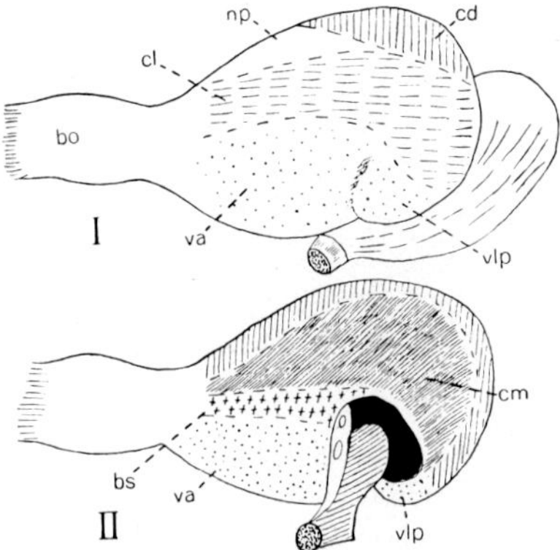

Figure 209 A. Sketches illustrating the telencephalic configuration in Chelonia, with surface projection of grisea (I) in lateral and (II) in medial aspect (from K., 1927). bo: bulbus olfactorius; bs: nucleus basimedialis superior; cd: cortex dorsalis; cl: cortex lateralis; cm: cortex medialis; np: lateral part of cortex dorsalis ('primordium neopallii'); va: area ventralis anterior (cortex basalis); vlp: area ventrolateralis posterior (caudal part of cortex basalis, 'cortical n. amygdalae'). Through an undetected lapsus, bs in Figure 316 A (II), p. 581 of volume 3/II was inadvertently designated as n. basilateralis superior instead of basimedialis superior. A comparable 'paragraphia' occurred on p. 535, line 14 from top in the cited volume, where basilateralis should correctly stand for basimedialis.

(1936), Kappers (1947) and Beccari (1943), are those by Cairney (1926), Crosby (1917), Curwen (1937–1939), Durward (1930), Elliot Smith (1919), Faul (1926), Frederikse (1931), Goldby (1934), Herman (1925), Hines (1923), Johnston (1915, 1916a, b), Kiesewalter (1922, 1925, 1928), K. and Kiesewalter (1922), de Lange (1911), Pedro Ramón (1917, 1918), Rose (1923), Shanklin (1930), Unger (1906), and Warner (1931, 1946a, b). The telencephalic zonal system in Reptiles and the correlated problems of homologies were scrutinized in contributions concerned with the Vertebrate telencephalon in general (K. 1924b, 1927, 1929).

In the following period, Schepers (1948) attempted to elaborate a not altogether convincing theory on the evolution of the forebrain based on findings in Testudo, Carey (1966) gave an account of the nuclear pattern in the telencephalon of the Blacksnake, Coluber, and Kirsche (1972), on the basis of investigations in Testudo, presented a comprehensive account on Reptilian telencephalic development, reviewing the different terminologies as well as interpretations proposed by various authors. Among other recent papers concerning structure

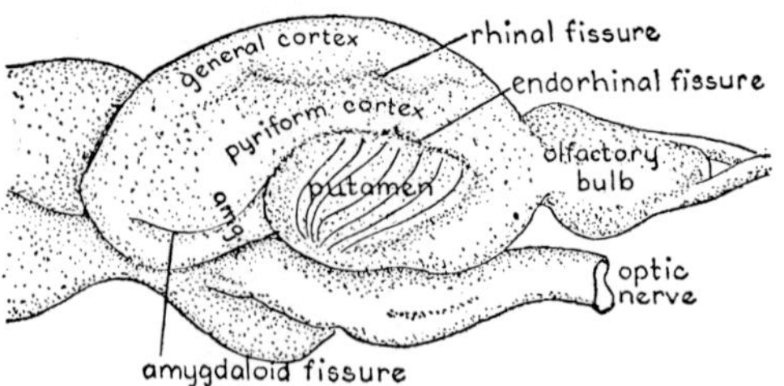

Figure 209B. Lateral aspect of telencephalon in the Box-Tortoise, Cistudo (adapted from JOHNSTON, 1915, as interpreted by HERRICK, 1926). amg. refers to amygdaloid complex partly covered by 'piriform cortex'; 'general cortex' is here cortex dorsalis; 'putamen' refers to basilateral grisea.

and configuration of the Reptilian telencephalon, without either essentially new or convincing findings, but with diverse attempts at substitutive nomenclatures, are those by FILIMONOFF (1964/65), NORTHCUTT (1967 and others), EBBESSON and VONEIDA (1969), RISS *et al.* (1969, 1972a, b), PLATEL *et al.* (1973), and ULINSKI (1974). The distribution of monoamine (catecholamine and 5-hydroxytryptamine) containing nerve terminals in the brain (including telencephalon) of the Turtle Chrysemys was investigated by PARENT (1973).

The *olfactory bulbs* in Rhynchocephalia, Squamata, and Crocodilia are located at a variable distance rostrally to lobus hemisphaericus, being connected with this latter through an olfactory stalk (cf. e.g. Fig. 213E). In Chelonia, on the other hand, the olfactory bulb is directly contiguous with lobus hemisphaericus ('sessile bulb', cf. Figs. 209A, B).

Again, in Chelonia, the olfactory bulb is a relatively large structure, being rather small in diverse Squamata such as Anolis or Chameleon. An accessory olfactory bulb, of taxonomically variable size, and located at the caudomedial portion of the main bulb, is commonly present[124] in most Reptilian forms, but seems completely absent in the order Crocodilia. The accessory bulbs are quite well developed in Snakes, being, however, most conspicuously predominant in some Lacertilia such as Varanus and Heloderma, where these configurations are larger than the main bulbs, and display eversions bulging into the bulbar ventricle (CROSBY and HUMPHREY, 1939; NIEUWENHUYS, 1967).

[124] It will be recalled that, in Amphibians, the accessory olfactory bulb has a caudolateral location with respect to the main bulb (cf. also above, section 2, p. 504).

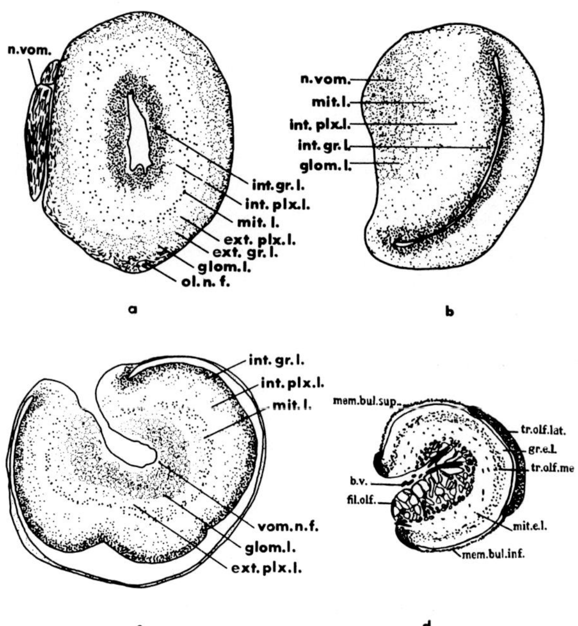

Figure 209 C. Cross-sections illustrating aspects of main and accessory olfactory bulb in some Squamate Reptiles (a–c after CROSBY and HUMPHREY, 1939, from NIEUWENHUYS, 1967; d from CURWEN, 1937). a: main olfactory bulb of the Snake Agkistrodon; b: accessory bulb of same animal; c: accessory bulb of the Lacertilian Heloderma; d: probably accessory bulb (not identified as such by the cited author) of the Lacertilian Tupinambis; b.v.: blood vessels; ext.gr.l.: external granular layer; ext.plx.l.: external plexiform layer; fil.olf.: fila olfactoria; glom.l.: glomerular layer; gr.c.l.: granule cell layer; int.gr.l.: internal granular layer; int.plx.l.: internal plexiform layer; mem.bul.inf., sup.: 'membrana bulbaris' inferior, superior; mit.c.l., mit.l.: mitral cell layer; n.vom.: nervus vomeronasalis; ol.n.f.: olfactory nerve fibers (fila olfactoria); tr.olf.lat., med.: tractus olfactorius lateralis, medialis; vom.n.f.: vomeronasal nerve fibers. Medial side at left.

Figure 209 C illustrates some aspects of Reptilian main and accessory olfactory bulbs. It will be seen that, if part of the olfactory stalk is membranous (epithelial, i.e. representing roof plate), it may or may not include fibers of the olfactory tract.

The Reptilian *lobus hemisphaericus* displays, in its secondary pallium, three distinctive, more or less abventricular laminae of cerebral cortex, which show a higher degree of differentiation than the cortex-like laminae obtaining in the pallium of various Anamnia such as Plagiostomes, Dipnoans, or, to a still lesser degree, in parts of the Gymnophione pallium.

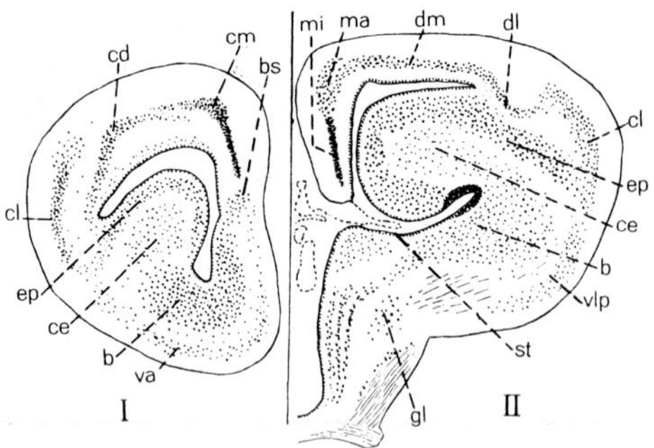

Figure 210 A. Simplified drawings of cross-sections through lobus hemisphaericus of (I) Lacerta, and (II) Testudo (from K., 1927). b: grisea basalia (B_{1-3}); bs: nucleus basimedialis superior; cd: cortex dorsalis; ce: nucleus epibasalis centralis; cl: cortex lateralis; cm: cortex medialis; dl: pars lateralis corticis dorsalis ('primordium neopallii'); dm: pars medialis corticis dorsalis (parahippocampal cortex); ep: nucleus epibasalis (paraventricularis); gl: nucleus entopeduncularis anterior (primordium of globus pallidus); ma, mi: macrocellular respectively microcellular portions of cortex medialis; st: sulcus terminalis; va: area ventralis anterior; vlp: area ventrolateralis posterior. Level I lies rostrally to lamina terminalis, level II corresponds to hemispheric stalk and region of interventricular foramen.

The *cortex medialis* (1) corresponds to the area medialis pallii *sive* 'primordium hippocampi' of Amphibians and can be regarded as heterotypic (and presumably heteropractic) homologous to the hippocampal cortex respectively 'formation' of Mammals.[125] It represents a derivative of the telencephalic longitudinal zone D_3. The dorsal or dorsolateral portion of this cortical plate generally consists of larger elements, some of which are almost typical pyramidal cells as seen in the Mammalian cerebral cortex (cf. Figs. 212 A, B, 213 B). The medial and ventromedial portion commonly displays smaller neuronal elements of similar sort, which may approach the so-called granular type. Both the zonal and the subcortical layer include fiber tracts. Scattered nerve cells, often of 'tangential' type, may be found in the zonal layer, and some nerve cells can likewise be dispersed internally to the main cortical cell plate. In comparison with Mammals, and from a topologic viewpoint, the large celled dorsolateral region of medial cortex appears homologous to cornu Ammonis, and dorsomedial respectively ventro-

[125] It might, however, be homotypic and perhaps isopractic homologous to precommissural neighborhoods of the Mammalian hippocampal formation. The qualifiers required for a formulation of the homology concept are explained on pp. 63–65 of volume 3/II.

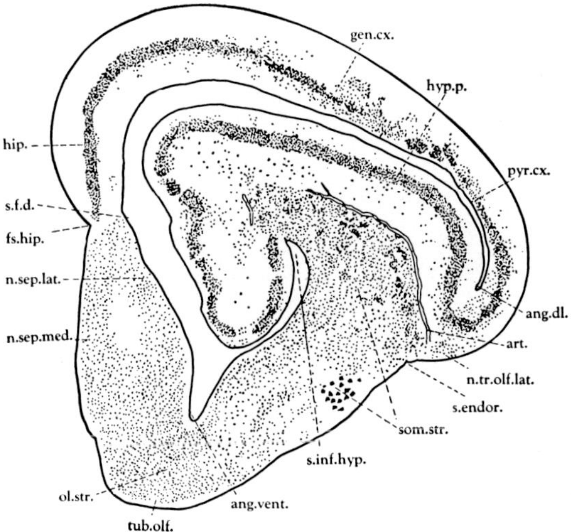

Figure 210 B. Cross-section (modified *Nissl stain)* through lobus hemisphaericus of Rhynchocephalian Sphenodon (redrawn after DURWARD, 1930, from KAPPERS *et al.*, 1936). ang.dl.: 'angulus dorsolateralis' (fd_2); ang. vent.: 'angulus ventralis' (fb_2); art.: branch of 'lateral striate artery'; fs.hip.: 'fissura hippocampi'; gen.cx.: cortex dorsalis; hip.: 'hippocampus (cortex medialis); hyp.p.: 'hypopallium posterius' (n. epibasalis); n.sep.lat.: 'nucleus septi lateralis' (periventricular portion of n. basimed. superior); n.sep.med.: 'nucleus septi medialis' (superficial cells of nn. basimedialis sup. et. inf.); n.tr.olf.lat.: nucleus of lateral olfactory tract; ol.str.: 'olfacto-striatum' (joint neighborhood of n. basimed. inf. and basilat. inf.); pyr.cx.: cortex lateralis; s.endor.: sulcus endorhinalis; s.f.d.: 'sulcus fimbrio-dentatus' (fd_4); s.inf.hyp.: 'sulcus infrahypopallialis' (fd_1); som.str.: 'somatic striatum' (n. basilat. superior, the large celled group at the lower lead, and at the open boundary of superior and inferior basilateral grisea is presumably an additional interstitial nucleus of lateral olfactory tract); tub.olf.: 'tuberculum olfactorium' (area ventralis anterior without distinguishable cortical arrangement).

medial small celled portions can be similarly evaluated as homologous to gyrus dentatus (fascia dentata).

The *cortex dorsalis* (2), a derivative of the zone D_2, is frequently demarcated from the externally overlapping cortex medialis by a cell-free zone designated as *superpositio medialis* (cf. Figs. 210 A, 212 A). Depending on rostrocaudal levels and on taxonomic forms, this superposition may be more or less pronounced or even absent. The dorsal cortical plate is commonly closer to the ventricular lining than the medial one. Its fairly large cells are likewise of approximately 'pyramidal' type. Some scattered paraependymal nerve cells may or may not be found internally to the main cell plate. In the Amphisbaenid Amphisbaena dar-

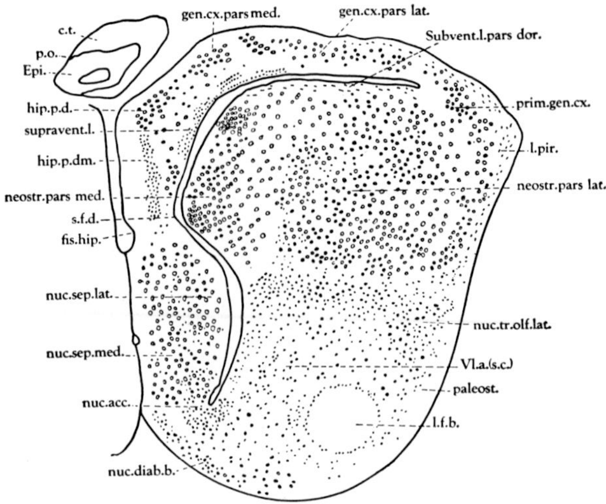

Figure 210C. Cross-section through lobus hemisphaericus of Chameleon vulgaris (redrawn after SHANKLIN, 1930, from KAPPERS *et al.*, 1936). c.t.: connective tissue; Epi.: 'epithelium' (?); fis.hip.: 'fissura hippocampi'; gen.cx.pars lat.: 'general cortex, pars lateralis' (cortex dorsalis parahippocampal cortex); gen.cx.pars med.: 'general cortex pars medialis' (perhaps part of medial cortex); hip.p.d.: 'hippocampus, pars dorsalis' (magnocellular medial cortex); hip.p.dm.: 'hippocampus, pars dorsomedialis' (parvocellular medial cortex); l.f.b.: lateral forebrain bundle; l.pir.: lobus piriformis (lateral cortex); neostr.pars med.: 'neostriatum pars medialis' (n. epibasalis medialis, pars paraventricularis); neostr.pars lat.: 'neostriatum pars lateralis' (n. epibasalis centralis); nuc.acc.: 'n. accumbens' (subdivision of n. basimed. inferior); nuc.diab.b.: 'n. of *diagonal band of Broca*' (area ventralis anterior, basal cortex); nuc.sep.lat.: 'n. lateralis septi' (periventricular part of n. basimed. superior); nuc.sep.med.: 'n. medialis septi (superficial part of n. basimed. inferior); nuc.tr.olf.lat.: n. of lateral olfactory tract; paleost.: 'paleostriatum' (part of n. basilat. inferior); p.o.: pineal complex; prim.gen.cx.: 'primordial general cortex' ('primordium pallii'); s.f.d.: 'sulcus fimbriodentatus' (fd_4); Subvent. l. pars dor.: paraventricular layer of n. epibasalis, pars dorsalis; supravent.l.: 'supraventricular layer' (periventricular region of cortex medialis); Vl.a.(s.c.): 'ventrolateral small-celled area' (portion of n. basilat. inferior).

winii heterozonata, QUIROGA (1976) has made a detailed survey of the dorsal cell plate's structure by means of the *Golgi technique* describing several cell types and layers (Fig. 212C).

In comparison with Mammals, the medial, major portion of cortex dorsalis can be evaluated as parahippocampal cortex, while the less extensive lateral portion, which displays diverse peculiarities and variations, seems to represent the neopallial primordium in accordance with

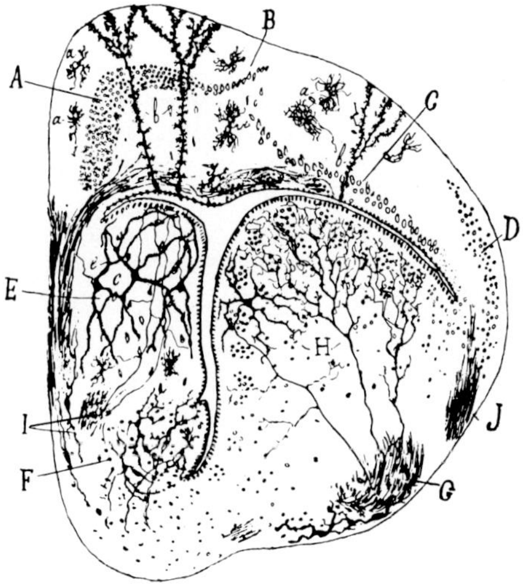

Figure 210 D. Cross-section *(Golgi impregnation)* through lobus hemisphaericus of the Lacertilian Iguana rostrally to lamina terminalis (from PEDRO RAMÓN, 1917). A: cortex medialis; B: lateral edge of cortex medialis at superpositio medialis; C: cortex dorsalis; D: cortex lateralis; E: nucleus basimedialis superior; F: nucleus basimedialis inferior; G: lateral forebrain bundle; H: nucleus epibasalis; I: system of medial forebrain bundle; J: lateral olfactory tract; a, b: glial and ependymal elements; c, e: neuronal elements in n. basimedialis superior.

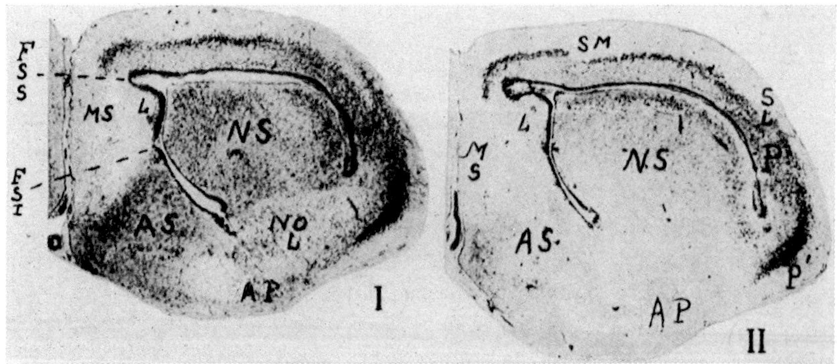

Figure 210E. Cross-sections *(Nissl stain)* through lobus hemisphaericus of the Rattlesnake Crotalus (from WARNER, 1931). AP: basal cortex respectively 'tuberculum olfactorium'; AS: nucleus basimedialis inferior; FSI: groove fb_3; FSS: groove fd_4; L: paraventricular portion of n. basimedialis superior; MS: abventricular neighborhoods of n. basimedialis superior; NOL: nn. basilaterales, superior et inferior; NS: nucleus epibasalis; P, P′: cortex lateralis; SL: superpositio lateralis; SM: superpositio medialis. Level I slightly rostral to level II.

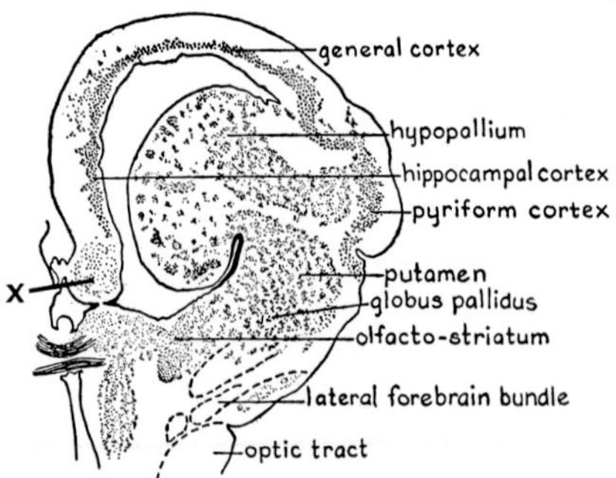

Figure 211 A. Cross-section (cell stain) through the forebrain of the Box-Tortoise at level of hemispheric stalk and anterior commissure (simplified after JOHNSTON, 1915, from HERRICK, 1926). 'putamen': nucleus basilateralis superior; 'globus pallidus': nucleus basilateralis inferior; 'olfacto-striatum': junction of massa cellularis reuniens, pars inferior with inferior basimedial and basilateral grisea; 'pyriform cortex': cortex lateralis. Added x: nucleus basimedialis superior. A portion of nucleus entopeduncularis (true homologon of globus pallidus) may be the cell mass below tip of lead 'olfacto-striatum'. The lateral forebrain bundle is actually larger than here indicated. Note 'dip' in lateral portion of cortex dorsalis, crossed by lead 'hypopallium'.

ELLIOT SMITH's (1910) and CROSBY's (1917) interpretation.[126] It is of interest that this lateral portion is restricted to levels comprising roughly between one to two rostral thirds or slightly more of lobus hemisphaericus, the caudalmost third (or fourth) being characterized by a junction (or 'fusion') of parahippocampal cortex and of cortex lateralis (cf. e.g. Fig. 209 A).

In some Reptiles, particularly various Lacertilians, the lateral edge of cortex dorsalis is overlapped by the more peripheral dorsal edge of cortex lateralis in the *superpositio lateralis* (Figs. 210 A, D, 212 B). In Lacerta, moreover, ontogenetic stages[127] indicate that the lateral edge of cortex dorsalis and most of cortex lateralis split from each other in the lateral neighborhood of D_2, while maintaining a ventral connection with the nucleus epibasalis (D_1). This ontogenetic relationship may even remain faintly indicated at adult stages (cf. Fig. 210 A).

In Crocodilians, however, the entire lateral portion of cortex dor-

[126] The somewhat ambiguous term 'general cortex' is also frequently used either for the lateral portion of cortex dorsalis or for that cortex *in toto* (cf. e.g. Figs. 210 B, C, 211 A, 211 D).

[127] Cf. Figure 284, p. 541 of volume 3/II.

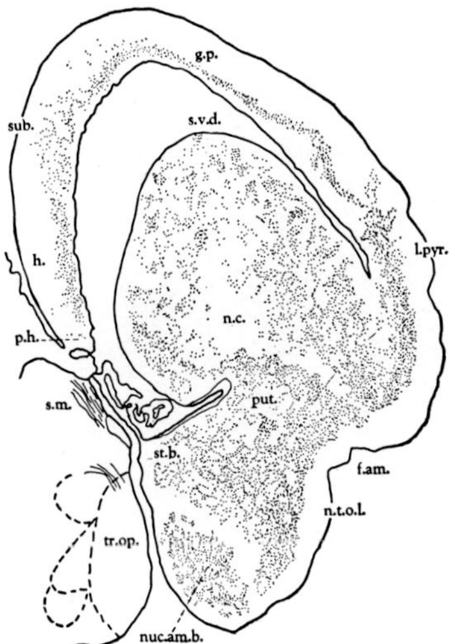

Figure 211 B. Cross-section (cell stain) through caudal paired evaginated portion of lobus hemisphaericus in the Turtle Cistudo carolina (redrawn after JOHNSTON, 1915, from KAPPERS *et al.*, 1936). f.am.: 'fissura amygdalae' (s. endorhinalis, pars posterior); g.p.: 'general pallium' (cortex dorsalis); h.: cortex medialis; l.pyr.: cortex lateralis; n.c.: nucleus epibasalis; n.t.o.l.: 'nucleus of lateral olfactory tract' (part of n. basilat. inferior); nuc.am. b.: area ventrolateralis posterior; p.h.: supraforaminal portion of nucleus basimedialis superior; put.: 'putamen' (n. basilat. superior); s.m.: stria medullaris; st.b.: bed of stria terminalis; s.v.d.: 'dorsal ventricular sulcus' (if lateral edge is meant, then fd_2); sub.: 'subiculum' (dorsal part of cortex medialis); tr.op.: tractus opticus.

salis becomes, particularly at rostral levels, 'introverted' or 'internated', corresponding to a ventricular ridge located rostrodorsally to the epibasal (D_1) ridge (cf. Figs. 211 D, I, II). More caudalward, this ridge tends to disappear (Fig. 211 D, III), but a dip in the cortical plate remains. This part of the cortical plate may display two layers of grisea.[128]

In Chelonians, the *primordium neopallii* is generally indicated by an externally concave cortical cell plate ('dip') medially connected with the remainder of cortex dorsalis, and laterally with cortex lateralis (cf. Figs. 211 A, B).[129] In some Turtles, however, a rather massive 'internation' or 'introversion' of this lateral portion of cortex dorsalis has been recorded (cf. Fig. 211 C), whereby an accessory ridge, dorsally to that of nucleus epibasalis, bulges into the lateral ventricle. At least three

[128] Cf. Figure 290, p. 550 of volume 3/II.

[129] Cf. also Figures 289 (p. 549) and 381 (p. 709) of volume 3/II).

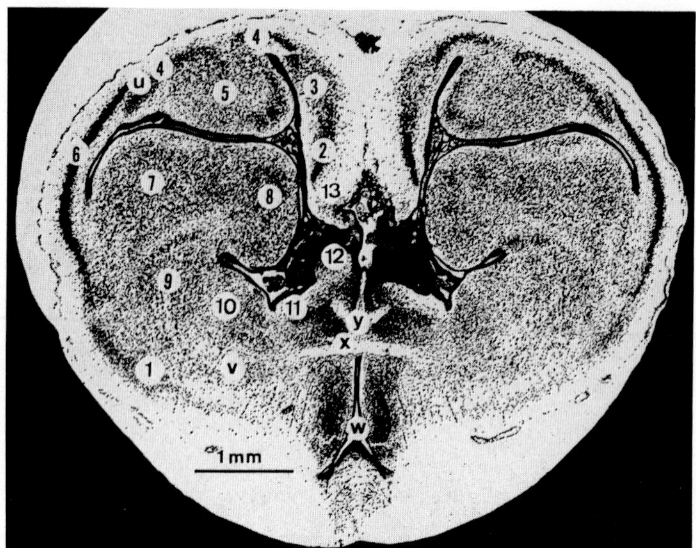

Figure 211 C. Cross-section *(Nissl stain)* through the prosencephalon of the Turtle Podocnemis unifilis at level of anterior commissure (from RISS *et al.*, 1972a). 1: basal cortex, transition between area ventralis anterior and area ventrolateralis posterior; 2: cortex medialis, pars inferior; 3: cortex medialis, pars superior; 4: cortex dorsalis, pars medialis; 5: internation of cortex dorsalis, pars lateralis; 6: cortex lateralis; 7: nucleus epibasalis, pars lateralis; 8: nucleus epibasalis, pars medialis; 9: nucleus basilateralis superior; 10: nucleus basilateralis inferior; 11: nucleus basimedialis inferior; 12: nucleus basimedialis superior; 13: supraforaminal portion of 12; u: lateral edge of cortex dorsalis with superpositio lateralis; v: massa cellularis reuniens, pars inferior, with rostral end of n. entopeduncularis anterior lower branch of lead); w: preoptic recess; x: commissura anterior; y: commissura pallii. Labels 1, 10–13, and u–y have been added to original figure, and all designations are given in accordance with the views adopted in the present treatise. The left 4 at u is pars lateralis of cort. dorsalis.

griseal groups, two of which are plate-like, and one of each is diffuse, can be noted.

These lateral differentiations of cortex dorsalis, as displayed by Crocodilia and Chelonia, are doubtless homologous to the nuclei diffusi (dorsalis and dorsolateralis) of the Avian telencephalon, dealt with in section 9 of the present chapter. Although neither extant Crocodilia nor Chelonia can seriously be considered 'ancestral' to Birds, certain Avian features of the telencephalon in said Reptilians are nevertheless remarkable.[130] Yet, such morphogenetic tendencies may have indepen-

[130] These '*Übereinstimmungen im Bauplan des Endhirns*' of Reptiles and Birds were pointed out with the following comment: '*Bei den Reptilien sind es besonders die Hydrosaurier, welche*

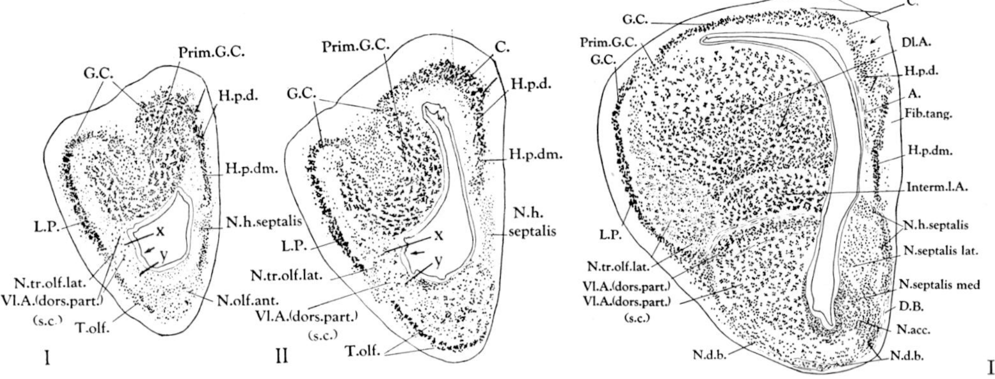

Figure 211 D. Cross-sections (modified *Nissl stain)* through the lobus hemisphaericus of Alligator mississipiensis (after Crosby, 1917, from Kappers *et al.*, 1936). A.: 'alveus' (medulla of cortex medialis); C.: 'differentiated cortex' (cortex dorsalis, parahippocampal region); D.B.: 'diagonal band of Broca' (fiber system of paraterminal body); Dl.A.: 'dorsolateral area' (n. epibasalis dorsalis, pars inferior); Fib.tang.: 'fibrae tangentiales' (zonal layer of cortex medialis); G.C.: 'general cortex' (cortex dorsalis, pars intermedia, see also below, Prim. G.C.); H.p.d., dm.: hippocampus, pars dorsalis' (cortex medialis, pars dorsalis sive magnocellularis, pars dorsomedialis *sive* parvocellularis); Interm. l.A.: 'intermediolateral area' (n. epibasalis centralis); L.P.: lobus piriformis (cortex lateralis); N.acc.: 'n. accumbens' (n. basimed. inferior); N.d.b.: 'n. of *diagonal band of Broca*' (cortex basalis *sive* area ventralis anterior); N.h.septalis: 'n. hippocampo-septalis' (n. basimed. superior); N.olf.ant.: n. olfactorius anterior; N.septalis lat., med.: 'n. septalis lateralis' respectively 'medialis' (paraventricular respectively abventricular cells of nn. basimediales); N.tr.olf.lat.: 'n. of tr. olfactorius lateralis' (n. epibasalis lateralis and perhaps part of centralis lateralis); Prim.G.C.: 'primordial general cortex' (cortex dorsalis, pars lateralis, the lower G.C. label indicates dorsal part of cortex lateralis); T.olf.: 'tuberculum olfactorium'; Vl.A. (dors.part.) (s.c.): 'ventrolateral area, dorsal part respectively small-celled portion' (subdivisions of n. basilat. superior, ventrally to which lies the unlabelled but clearly identifiable n. basilat. inferior). Level I and II are close to olfactory stalk, while III is nearer to lamina terminalis. Added designations: x: rostral tip of nucleus epibasalis (which Crosby, in I, includes in her Vl.A.); y: rostral tip of joint superior and inferior basilateral grisea. Arrow: groove fd_1. These Figures should be compared with preceding Figure 211B, and with Figure 290, p. 550 of volume 3/II.

dently originated in diverse extinct Reptilian groups. It is most likely that the internation of zone D_1, characteristic for Amniota, led to considerable morphogenetic disturbances involving the neighborhoods within the new pallio-basal junction. The variations displayed by the

---

in dieser letzteren Hinsicht zu den Vögeln überleiten, namentlich die Krokodilier, jedoch zeigen ebenfalls einige Chelonier, wie zum Beispiel Chrysemis, gewisse Anklänge an die Vogelmorphe' (K., 1929).

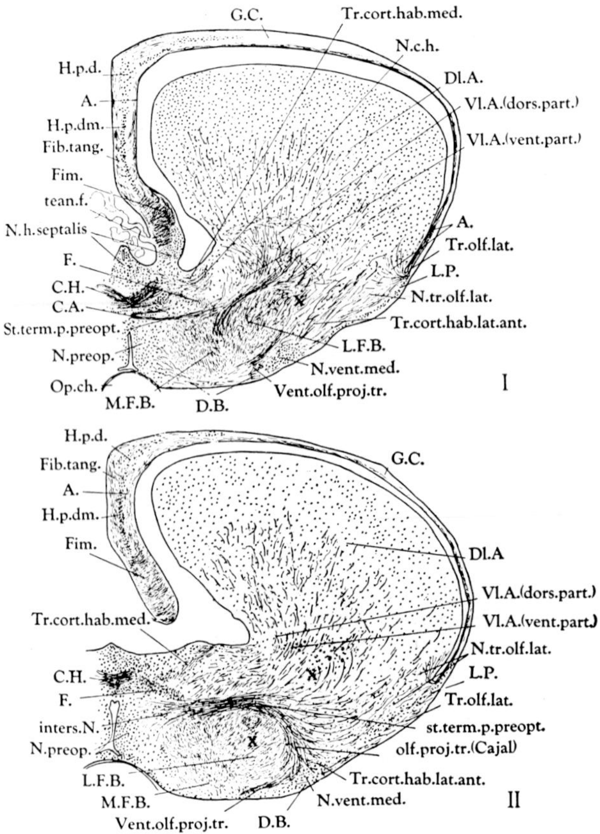

Figure 211E. Two cross-sections *(Cajal silver impregnation)* through forebrain of Alligator mississipiensis at levels of hemispheric stalk (after CROSBY, 1917, from KAPPERS *et al.*, 1936). C.A.: commissura anterior; C.H.: commissura hippocampi (commissura pallii); F.: fornix; Fim.: 'fimbria' (fornicis); inters.N.: 'interstitial nucleus' (of stria terminalis); L.F.B.: lateral forebrain bundle; M F.B.: medial forebrain bundle; N.c.h.: 'nucleus commissurae hippocampi' (probably portion of basal grisea); N.preop.: preoptic (hypothalamic) grisea; N.vent.med.: 'ventromedial nucleus' (probably part of area ventromedialis posterior); olf.proj.tr. (Cajal): '*olfactory projection tract of Cajal*'; Op.ch.: optic chiasma; st.term.p.preopt.: 'stria terminalis, pars preoptica'; tean.f.: 'taenia fornicis'; Tr.cort.hab.lat.ant.: tractus cortico-habenularis lateralis anterior; Tr.cort.hab.med.: tractus cortico-habenularis medialis; Tr.olf.lat.: lateral olfactory tract; Vent.olf.proj.tr.: 'ventral olfactory projection tract; Vl.A. (vent. part): nucleus ventrolateralis (inferior). Added x: anterior entopeduncular nucleus. This Figure should be compared with Figure 214C, II.

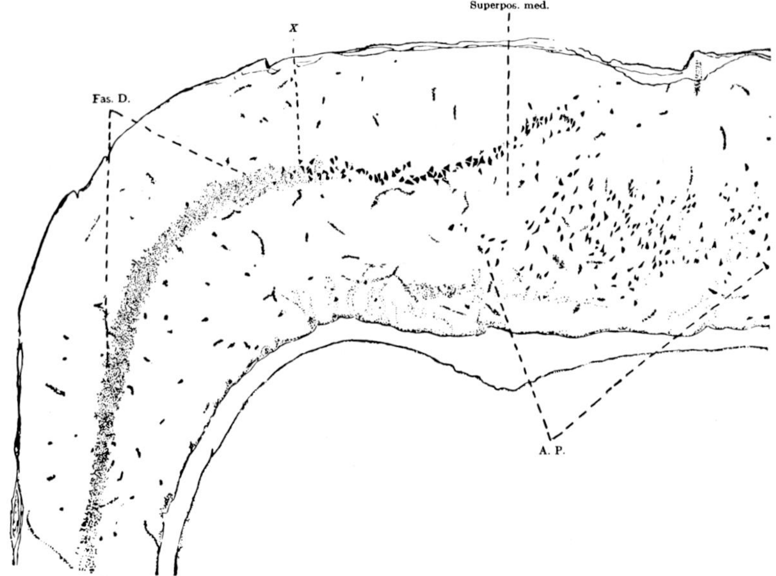

Figure 212 A. Relationships of cortex medialis and cortex dorsalis in the telencephalon of the Lacertilian Varanus salvator (after VAN'T HOOG, from KAPPERS *et al.*, 1936). A.P.: 'Ammon's pyramids' (cortex dorsalis, parahippocampal cortex); Fas.D.: 'fascia dentata' (small-celled portion of cortex medialis); Superpos. med.: superpositio medialis (boundary of hippocampal cortex medialis, and parahippocampal cortex dorsalis); X: transition of small-celled to large-celled portion of cortex medialis).

grisea of this region can be regarded as related to said phylogenetic events.[131] These latter resulted, along two divergent lines, in the expansion of the Mammalian neocortex, and in the expansion of the Avian basal grisea of the lateral telencephalic wall.

The cortex lateralis (3), whose 'fluctuating' relationship to the lateral portion of cortex dorsalis were pointed out in the preceding paragraphs, is the most lateral derivative of zone D_2, with variable but much lesser additional contributions from D_1. Its basal edge is medialward more or less directly continuous with the epibasal grisea (Figs. 210A, B, C, 211D, III). Caudalward, in the region of the hemisphere's posterior pole, the cortex lateralis blends with the parahippocampal portion of cortex dorsalis (cf. Fig. 209 A) as already mentioned above. There can be little doubt about the homology of Reptilian later-

[131] Cf. the comments, which also refer to views expressed by ELLIOT SMITH, on pp. 535–539 of volume 3/II.

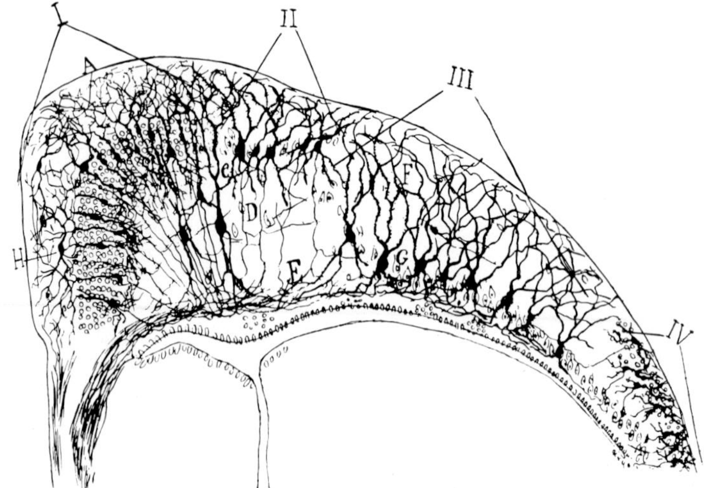

Figure 212B. Details *(Golgi impregnation)* of Lacertilian cortex pallii (from PEDRO RAMÓN, 1917). I: cortex medialis, pars parvocellularis; II: cortex medialis, pars magnocellularis; III: cortex dorsalis; IV: cortex lateralis; A: molecular (zonal) layer; C, D: deep layers of medial cortex; E: deep fibers (medullary layer) of medial and dorsal cortex; F: zonal layer of dorsal cortex; G: main layer of dorsal cortex; H: cells in zonal layer of medial cortex.

al cortex (closely related to the lateral olfactory tract),[132] and cortex of the Mammalian lobus piriformis, except for the caudal parts of this latter, which contain parahippocampal cortex (cf. Fig. 221 C). This contiguity, at caudal levels, of cortex lateralis to parahippocampal cortex obtains in all three classes of Amniota.

The *basal grisea* of the Reptilian lobus hemisphaericus include (1) a nondescript nucleus olfactorius anterior[133] in the open neighborhood between olfactory bulb, respectively stalk, and hemispheric lobe, (2) the epibasal complex, (3) nucleus basilateralis superior, (4) nucleus basilateralis inferior, (5) nucleus basimedialis inferior, (6) nucleus basimedialis superior, and (7) a variously distinctive cortex basalis.

---

[132] Under the term 'nucleus of the lateral olfactory tract', several cell groups of morphologically differing provenance have been designated by diverse authors. Such cell groups, apparently related to said tract, variously comprise basal region of cortex lateralis, adjacent neighborhoods of nucleus epibasalis complex, or adjacent ones of nucleus basilateralis superior. It is, moreover, not improbable that these differences, at least in part, reflect taxonomical variations displayed by tractus olfactorius lateralis.

[133] Cf. Figure 211 D, I.

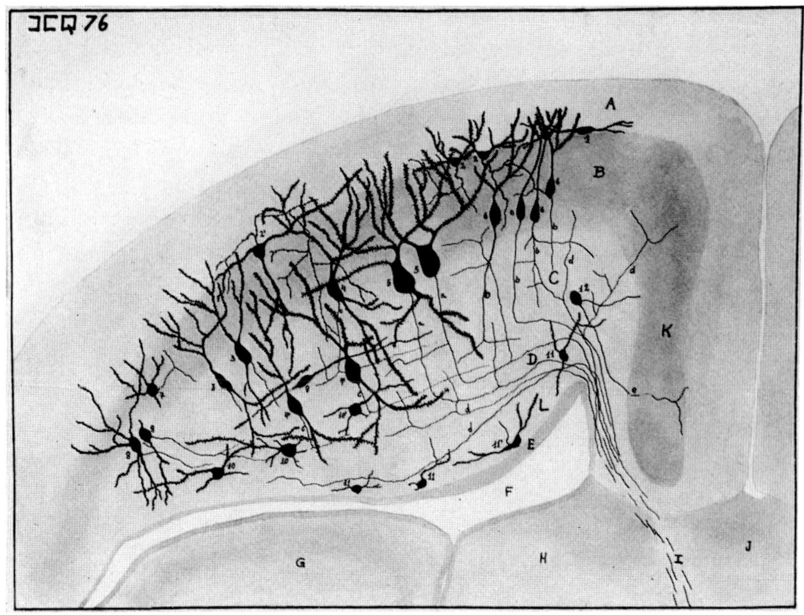

Figure 212C. Details of cortex dorsalis *(Golgi impregnation)* in the Amphisbaenid Amphisbaena darwinii heterozonata (courtesy of Dr. JUAN C. QUIROGA). A: superficial plexiform (molecular) layer; B: pyramidal cell layer; C: 'deep plexiform layer'; D: 'white matter'; E: ependymal layer; F: ventricle; G: epibasal griseum; H: lateral paraterminal griseum; I: medial forebrain bundle; K: cortex medialis; L: 'supraependymal granular layer'; 1: fusiform element; 2: 'navicular element'; 3–6: diverse types of pyramidal cells; 7, 8: cell types in region of superpositio lateralis; 9: deep horizontal cell; 10: deep cells of layer C; 11: 'supraependymal granular cells; 12: additional element of layer C; a, b: types of axon; d: afferents to cortex; e: connection of cell 12 with cortex medialis.

The conspicuous *epibasal complex* (2) corresponds to the epistriatum of EDINGER or hypopallium of ELLIOT SMITH, and forms the dorsal portion of the lateral basis telencephali of Amniota, which represents, following introversion of the originally pallial zone D_1, an 'augmented' secondary basis.

Generally speaking, a paraventricular cortex-like cell plate may be seen, which could be called (a) nucleus epibasalis dorsalis, with variable lateral, dorsal, and medial subdivisions. Basally to this lies (b) the nucleus epibasalis centralis, displaying a more diffuse arrangement. Numerous variations of this general pattern shown in Figure 210A can be noted. In Sphenodon (Fig. 210B) the nucleus epibasalis dorsalis is well developed, while nucleus centralis is inconspicuous. In the Chameleon (Fig. 210C) nucleus centralis and dorsolateral neighborhoods of nucleus dorsalis tend to fuse. In Podocnemis (Fig. 211 C), the fusion of nucleus centralis and dorsalis is more pronounced, but combined with a nondescript differentiation into lateral and medial neighborhoods. In Crocodilia (Fig. 211 D, III) the nucleus dorsalis displays an inferior, a superior, and a lateral part, segregated from nucleus centralis. This arrangement is somewhat similar to that obtaining in Birds.

The Reptilian epibasal complex, moreover, is commonly subdivided into an *anterior (rostral)* and a *caudal portion*, this subdivision being, in many cases, indicated by an oblique or transverse groove within the epibasal ridge (Fig. 213 D). The rostral portion was designated as *neostriatum* by KAPPERS, and the caudal part as *archistriatum*, said groove being therefore called fissura strio-archistriatica. In many Squamatae, the nucleus epibasalis dorsalis displays within the posterior epibasal complex an oval or elliptic arrangement including scattered elements of nucleus centralis and known as *nucleus sphaericus*.[134] The relationship of griseum epibasale posterius to the so-called amygdala shall be dealt with further below.

The ventrally adjacent *grisea basilateralia* (3, 4) may display a nucleus basilateralis superior (B_1) clearly distinguishable from a nucleus basilateralis inferior (B_2), or both cell groups may appear as a more or less homogeneous cell population.[134a]

The medial basal cell groups of lobus hemisphaericus represent the *paraterminal grisea sensu latiori* in ELLIOT SMITH's terminology.[135] Nucleus *basimedialis inferior* (5) tends lateralward to fuse with nucleus basilateralis inferior. If conspicuously distinctive by its density or cytoarchitecture, the paraventricular neighborhood of nucleus basimedialis inferior, or one of its sub-neighborhoods, represent the so-called nucleus accumbens septi.

The *nucleus basimedialis superior* (6), is closely adjacent to the basal edge of cortex medialis. At levels of lamina terminalis, its ventral portion tends to fuse with nucleus basimedialis inferior, while a dorsal portion, retaining its relationship to medial cortex, extends for a vari-

---

[134] Cf. pp. 550–553 of volume 3/II.

[134a] In Figure 286D, p. 546 of volume 3/II, the basilateral zones B_1 and B_2 of a Reptile were inadvertently labelled B_2 and B_3. This lapsus, due to inattention on my part, then remained undetected upon proof reading. Attentive readers may have recognized this error, particularly in comparing that Figure with the correctly labelled Figure 287 B.

[135] These are the so-called septal nuclei of the conventional terminology. These grisea are from the telencephalic zones B_3 (n. basimedialis inferior) and B_4 (n. basimedialis superior). This latter griseum could perhaps be designated as paraterminal body *sensu strictiori*.

able distance caudalward dorsally to foramen interventriculare, as so-called pars fimbrialis (or nucleus fimbrialis) septi, and may thus also be present at rostral levels of the caudal paired evaginated lobus hemisphaericus.

The *cortex basalis* (7) is a more or less distinctive nondescript cell plate externally covering the neighborhoods of basilateral and basimedial grisea. Its rostral portion is the *area ventralis anterior sive cortex olfactoria* ('tuberculum olfactorium'). Its caudal portion, extending into the basis of the caudal paired evaginated lobus hemisphaericus, is the *area ventrolateralis posterior*, related, as so-called nucleus amygdalae corticalis ('periamygdalar cortex'), to the amygdalar complex.

A medial extension of basal cortex, related to nucleus basimedialis inferior, is the so-called *nucleus of Broca's diagonal band.* It may even reach as far dorsad as nucleus basimedialis superior. These cell groups are, at least in part, identical with those designated 'nucleus medialis' septi, while the paraventricular basimedial cell groups represent the 'nucleus lateralis septi'. In addition to paraventricular and 'cortical' cell groups, intermediate ones may be present; moreover, the transition between superior and inferior basimedial grisea is not infrequently blurred. Thus, many discrepancies in terminology and description of 'septal nuclei' can be noted in the literature.

As regards the so-called '*amygdala*' of the Reptilian telencephalon, it should be recalled the term 'nucleus amygdalae' originally referred to a conspicuous and rather massive compound basal griseum of the Mammalian telencephalon. The corresponding, homologous Reptilian grisea are far less compact and consist (a) of the caudal portion of nucleus epibasalis (dorsalis and centralis), (b) the caudal portions of nucleus basilateralis superior and inferior, and (c) the area ventrolateralis posterior (cortex basalis). In a morphogenetic analysis of the Mammalian and Reptilian amygdaloid complex (K., 1924b, 1927, 1929), the epibasal component was designated as nucleus amygdalae β *(Mandelkernhauptkomplex)*, the basilateral component as nucleus amygdalae γ and the basal cortical component as nucleus amygdalae α (with an accessory medial subdivision δ).

A most detailed description of a Reptilian amygdaloid complex was given by Curwen (1939) for Tupinambis. On the basis of a perhaps exaggerated parcellation, seven amygdaloid nuclear masses were distinguished, namely (1) n. posterior, (2) n. internus, (3) n. centralis, (4) n. dorsalis, (5) n. marginalis, (6) n. externus, and (7) n. lateralis. Nuclei 1–5 are said to form the medial amygdaloid group, and nuclei 6 and 7

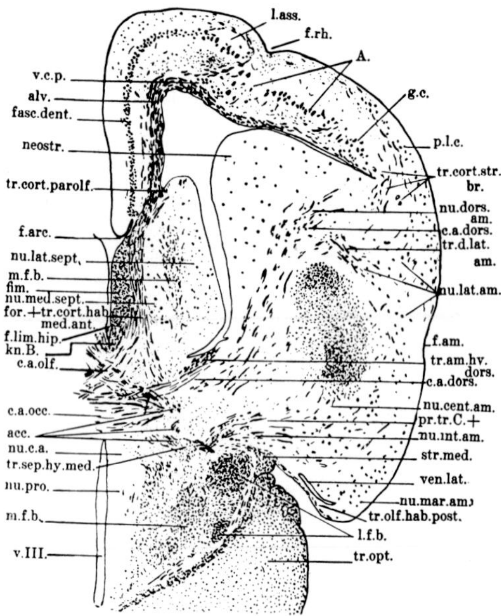

Figure 213 A. Cross-section *(Weigert stain)* through the forebrain of the Lacertilian Tupinambis at the level of anterior commissure (from Curwen, 1937). A.: 'Ammonsformation' (cortex dorsalis); acc.: 'accessory fibers of septum'; alv.: 'alveus' (medulla of cortex medialis); c.a.dors.: 'comm. archistriatica pars dorsalis' (part of comm. anterior); c.a.occ.: 'comm. archistr. pars occipitalis' (part of comm. anterior); c.a.olf.: 'comm. anterior pars olfactoria'; f.am.: 'fissura amygdalae' (s. endorhinalis, pars posterior); f.arc.: 'fissura arcuata' (external groove between cortex medialis and paraterminal grisea); f.lim.hip.: 'fissura limitans hippocampi' (external groove at basal boundary of n. basimed. superior); f.rh.: 'fissura rhinalis' (external groove between parahippocampal cortex dorsalis and hippocampal cortex medialis); fasc.dent.: 'fascia dentata'; fim.: 'fimbria'; for.: fornix; g.c.: 'general cortex' (lateral portion of cortex dorsalis); kn.B.: '*knieförmige Bündel*'; l.ass.: 'longitudinal association bundle of hippocampus'; l.f.b.: lateral forebrain bundle; m.f.b.: medial forebrain bundle; neostr.: 'neostriatum' (n. epibasalis); nu.c.a.: 'n. commissurae anterioris' (part of preoptic grisea); nu.cent.am.: 'n. centralis amygdalae' (caudal portion of basilateral grisea); nu.dors. am.:'n. dorsalis amygdalae' (caudal part of n. epibasalis centralis); nu.int.am.: 'n. internus amygdalae' (here probably part of 'n. centralis amygdalae'); nu.lat.am.: 'n. lateralis amygdalae' (lateral part of n. epibasalis grisea'); nu.mar.am.: 'n. marginalis amygdalae' (area ventrolateralis posterior); nu.med. sept.: 'n. medialis septi' (abventricular portion of n. basimed. superior); nu.pro.: preoptic grisea; p.l.c.: 'cortex piriformis, large cells' (cortex lateralis); pr.tr.C.: '*olfactory projection tract of Cajal*'; str.med.: stria medullaris; tr.am.hy.dors.: 'tr. amygdalo-hypothalamicus dorsalis'; tr.cort.hab.med.ant.: 'tr. cortico-habenularis medialis anterior'; tr.cort.parolf.: 'tr. cortico-parolfactorius'; tr.cort.str.(br.): tr. cortico-striaticus' respectively 'cort.str. brevis'; tr.d.lat.am.: 'tr. dorsolateralis amygdalae'; tr.olf.hab.post.: 'tr. olfacto-habenularis posterior'; tr.opt.: tr. opticus; tr.sep.hy.med.: 'tr. septo-hypothalamicus medialis'; v. III: third ventricle; v.c.p.: 'ventral cortical plate' (medial cells of cortex dorsalis in region of

the lateral group. With reference to the above-mentioned morphogenetic classification, CURWEN (1939) tentatively subsumes nuclei 4, 6, 7 under component β, nuclei 1, 2, 5 under component α, and nucleus 3 under component γ.[136]

Figures 210 A–211 E depict cross-sections showing the overall arrangement of pallial and basal grisea in lobus hemisphaericus of Squamate, Rhynchocephalian, Chelonian, and Crocodilian Reptiles. Figures 212 A and B, on the other hand, illustrate additional details of Reptilian pallial cortical structure.

Figure 288, p. 548 of section 6, chapter VI, volume 3/II illustrates a cross-section through the hemisphere of a Lacertilian Amphisbaenid. These legless, worm-like Reptiles with rudimentary eyes live burrowed in the ground and display a most interesting simplified Reptilian telencephalic griseal pattern, described and emphasized by JAKOB and ONELLI (1911) who mistook these animals for Gymnophiones. Through the courtesy of Dr. *Juan C. Quiroga*, Buenos Aires, I recently obtained a copy of a subsequent Argentine publication of JAKOB (1918) in which that author rectified this taxonomic mistake and described, with suitable illustrations, sections through both the Amphibian Gymnophione (Chthonerpeton) and the Reptilian Amphisbaenid telencephalon.[136a]

As regards the *communication channels* of the Reptilian telencephalon, many secondary olfactory fibers, originating from mitral, and perhaps some other bulbar neuronal elements of the second order, are said to

---

[136] The cited author's nucleus marginalis is evidently the area ventrolateralis posterior in her Figures 2 and 3 (CURWEN, 1939, pp. 619, 621) but most probably a part of her n. centralis in Figure 4 (p. 624). Her nn. posterior and internus are, in my interpretation, more likely parts of component β than parts of γ.

[136a] DR. QUIROGA informed me that he identified the Amphisbaenid in question as Amphisbaena darwinii heterozonata. This South-American species does not very much differ from Lepidosternon microcephalon, which JACOBSHAGEN and myself guessed, upon inspecting JAKOB's macroscopic illustration, as being the Amphisbaenid whose brain was depicted.

---

superpositio medialis); ven.lat.: recess of ventriculus lateralis. CURWEN's investigation on Tupinambis, carried out at Yale University, although subsequently published with my prompting and encouragement at the Department of Anatomy, Woman's Medical College of Pa., reflects the views of the *Yale group* (H. S. BURR) adherence to which was justly incumbent on the author. My differing interpretations are added in parentheses. As regards the very detailed tract specifications, I believe that they should be taken with a due measure of scepticism.

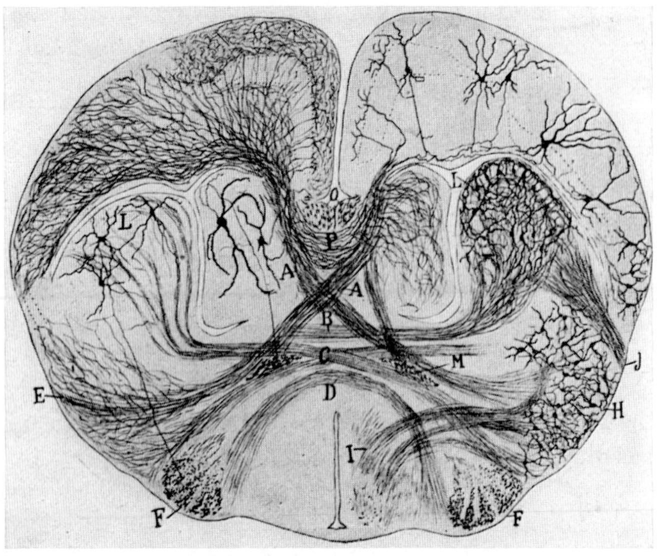

Figure 213 B. Semidiagrammatic sketch of fiber systems and cell groups in a cross-section through the Lacertilian lobus hemisphaericus at level of commissura anterior, based upon the interpretation of *Golgi impregnation* (from PEDRO RAMÓN, 1918). A: 'crossed cortico-basal fasciculi' (between comm. anterior and comm. pallii anterior); B: 'decussating commissure of nucleus sphaericus'; C: 'decussating interstriatic commissure'; D: 'non-decussating olfactory commissure' (B, C, D are components of anterior commissure); E: 'ramifications of cortico-basal fasciculus'; F: lateral forebrain bundle (the only faintly indicated, unlabelled medial forebrain bundle lies between I and F); I: 'taenia transversalis' (probably part of stria terminalis system); J: *'fasciculo bulbo-olfactivo para el nucleo esferico', 'raiz externa olfactoria' ('olfacto-epistriatic tract' of Edinger*, portion of lateral olfactory tract); H: area ventrolateralis posterior; L: nucleus epibasalis; M: 'septo-cortical fasciculi'; P: commissura pallii anterior.

terminate in nucleus olfactorius anterior. The caudalward proceeding fibers of second and third order then gather into a lateral and into a medial olfactory tract, except for some rostrally located more diffuse 'radiations'.

The larger *lateral olfactory tract*, running rather superficially along the ventral edge of cortex lateralis, distributes to this latter, and to epibasal as well as basal cell groups, including a so-called nucleus of the lateral olfactory tract (cf. above fn. 132) and the basal cortex olfactoria. A caudal portion of this channel, also described as a separate tract, since it is joined by fibers from the pallial cortical laminae, reaches the posterior epibasal grisea which include the nucleus sphaericus of Squa-

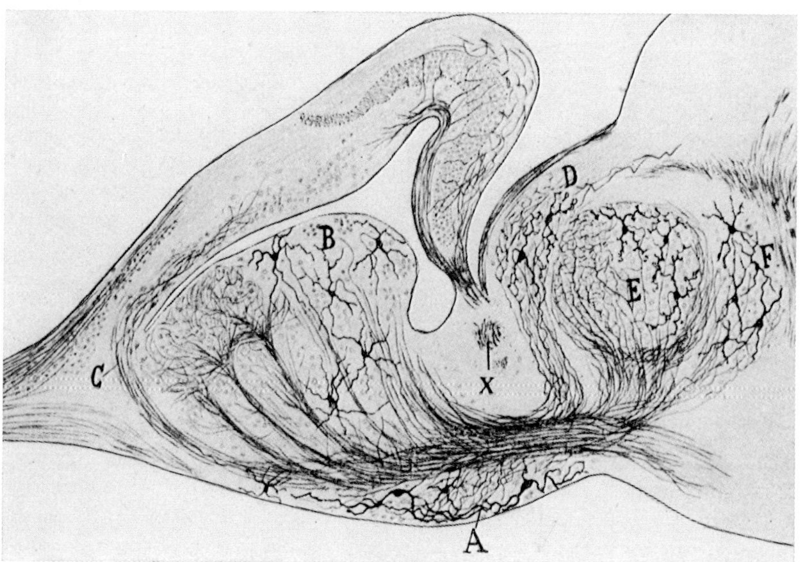

Figure 213C. Semidiagrammatic paramedian sagittal section *(Golgi impregnation)* through the forebrain of Lacerta (from PEDRO RAMÓN, 1918). A: '*foco subpeduncular*' (probably area ventrolat. posterior); B: '*cuerpo estriado*' (n. epibasalis); C: '*radiación fronto-cortical del fascículo basal*' (cortical component of basal forebrain bundle); D: anterior nuclei of thalamus (dorsalis); E: 'nucleo posterior del talamo' (n. rotundus); F: pretectal grisea. Added x: anterior commissure. The fibers above D presumably pertain to stria medullaris system. RAMÓN's interesting sketch ignores the optic tract which should have been cut somewhat caudally to region A.

matae, and represents the *tractus bulbo- et cortico-epistriaticus* respectively *archistriaticus* recorded by EDINGER (1896a, 1908), KAPPERS (1921), UNGER (1906) and others. Fibers from the accessory olfactory bulb are said to join the lateral olfactory tract and its tractus bulbo-epistriaticus. Caudal fibers of the lateral olfactory tract also join the lateral forebrain bundle.

The smaller *medial olfactory tract* runs through abventricular regions of the basimedial (or paraterminal) neighborhoods and distributes to the basimedial grisea, to the cortex medialis, and to medial parts of basal cortex olfactoria. Some fibers join the medial forebrain bundle, and others the stria medullaris system. Experimental studies on the olfactory tracts with methods of fiber degeneration were e.g. undertaken by GOLDBY (1937) and GAMBLE (1952, 1956). On the whole, if the inherent limitations of these methods are considered, their results contri-

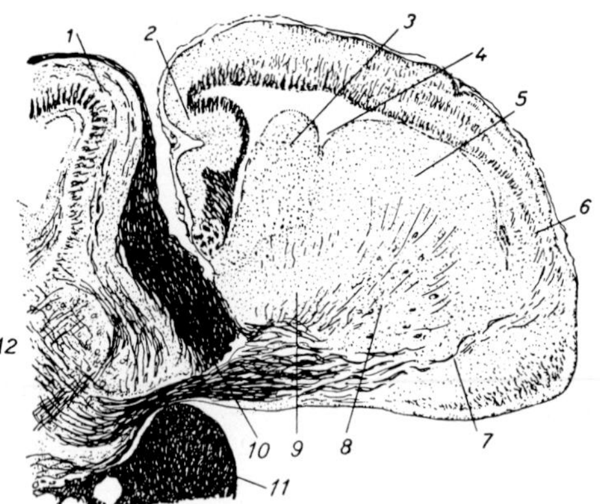

Figure 213 D. Fairly lateral sagittal section (myelin stain) through forebrain and adjacent midbrain of the Lacertilian Varanus salvator (after KAPPERS, 1921, from BECCARI, 1943). 1: tectum opticum; 2: cortex medialis; 3: nucleus epibasalis posterior; 4: '*sulco archineo-striato*' (groove between anterior and posterior portions of n. epibasalis); 5: nucleus epibasalis, pars anterior; 6: '*primordio del neopallio*' (lateral portion of cortex dorsalis); 7: 'thalamo-cortical bundle'; 8: 'thalamo-striatal bundle' (7 and 8 are components of lateral forebrain bundle); 9: '*palaeostriato*' (basilateral grisea); 10, 11: tractus and nervus opticus; 12: thalamus. The plane of the depicted section lies fairly laterally to that of Figure 213C.

bute a few details to the overall conclusions drawn from previous investigations on 'normal' material by means of myelin stains and metallic impregnations.

The *caudal communication* channels are provided by *lateral* and *medial forebrain bundles*, and by the systems of *stria medullaris* and *stria terminalis*.

The large *lateral forebrain bundle*, taking its course medially and ventromedially to the tractus olfactorius lateralis, includes reciprocal connections between diencephalon and lateral as well as dorsal neighborhoods of lobus hemisphaericus. There are, in particular, connections between nucleus rotundus and epibasal grisea[137] as well as cortex dorsalis (cf. Figs. 213C, D). PRITZ (1975), using the *Fink-Heimer technique*, describes, in the Crocodilian Caiman, an ascending component of the lateral forebrain bundle originating in nucleus rotundus thalami and reaching dorsal grisea of the epibasal complex. Since the nucleus rotundus receives input from optic tectum, the cited author interprets said vaguely circumscribed neighborhood of nucleus epibasalis (his 'rostral dorsolateral area') as a 'telencephalic visual area'.

[137] E.g. 'thalamo-striate fasciculus' etc.

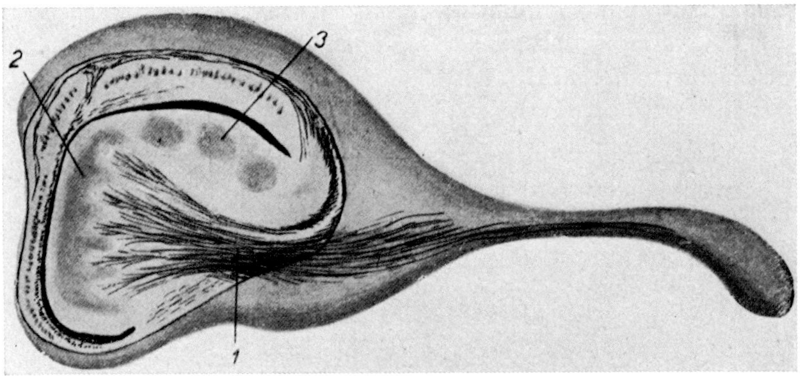

Figure 213E. Sketch of partly sectioned telencephalon of Varanus, as seen from the lateral side (after EDINGER, 1908, from BECCARI, 1943). 1: fasciculus olfacto-archistriaticus *sive* bulbo-epistriaticus of EDINGER (component of lateral olfactory tract); 2: 'archistriatum' of KAPPERS, 'epistriatum' of EDINGER (n. epibasalis, pars posterior); 3: 'neostriatum' (n. epibasalis, pars anterior). Note unlabelled olfactory bulb and stalk.

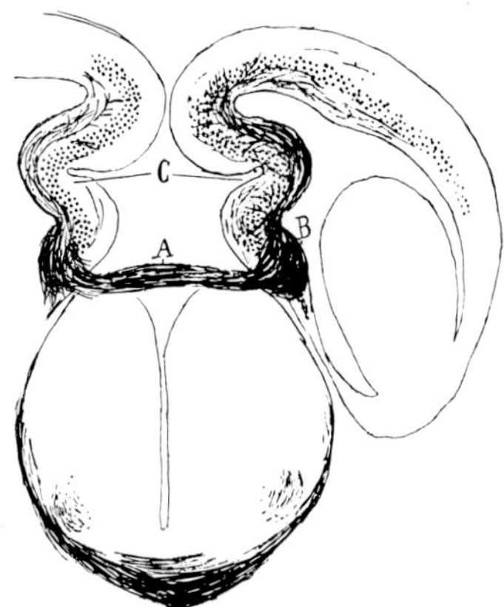

Figure 213F. Cross-section through the forebrain of the Lacertilian Iguana, showing commissura pallii posterior in diencephalic roof (from PEDRO RAMÓN, 1918). A: commissura pallii posterior; B: '*fimbria*' (fiber bundle related to cortex medialis); C: '*asta de Ammon*' (cortex medialis). For location of commissura pallii posterior, this Figure may be compared with Figure 63, II, p. 174, volume 3/II.

In addition to nucleus rotundus, other dorsal and ventral thalamic cell groups, including dorsal and ventral geniculate grisea appear to have ascending or reciprocal connections with telencephalic grisea by way of the lateral forebrain bundle system.

In the ventral portion of the lateral forebrain bundle, descending and ascending pathways seem to connect pallial, epibasal and lateral basal grisea with the diencephalic dorsal hypothalamic entopeduncular cell groups which may be considered said tract's 'interstitial nuclei', with other hypothalamic grisea, with thalamus ventralis and with deuterencephalic tegmentum.

The *medial forebrain bundle*, commonly much less conspicuous than the lateral one, runs through the paraterminal neighborhoods and includes descending and ascending connections of the dorsal cortex, moreover a perhaps predominantly descending '*fornix*' component connecting cortex medialis with caudoventral hypothalamus. Additional, partly descending and partly ascending fiber systems connect the basimedial nuclei and the medial as well as dorsal cortex with hypothalamic preoptic region and further, undefined diencephalic and perhaps tegmental grisea.

The *stria medullaris system*,[138] apparently predominantly but not exclusively descending, interconnects epithalamus (habenular nuclei) with various pallial and basal telencephalic grisea. In the diencephalon, fibers related to preoptic region join this system.

The *stria terminalis system* is essentially related to the so-called amygdaloid complex, that is to say, to epibasal and basal components of the paired caudalward evaginated portion of lobus hemisphaericus. The stria terminalis fibers, passing through the vicinity of sulcus terminalis, run mediad toward the hemispheric stalk, and dorsally cross the lateral forebrain bundle. One component joins the anterior commissure, while another bends ventralward on the homolateral side to reach preoptic and other hypothalamic grisea.[139] Some fibers, again, may join the stria medullaris system.

The *intrinsic telencephalic communication channels* include 'cortical association fibers' within the zonal layer, medial and lateral 'cortico-striate fibers' between lateral basis and pallium, and similar fibers in the paraterminal region. Among these latter, the '*diagonal band of Broca*' comprises fibers apparently joining area ventralis and ventrolateralis posterior with paraterminal grisea and perhaps cortex medialis.

The *telencephalic commissures* are represented by commissura pallii anterior, commissura pallii posterior, commissura anterior, and in

[138] E.g. 'tractus cortico-habenularis medialis', 'lateralis', and 'olfacto-habenular tracts' with further subdivisions. These latter may be related to paraterminal and basilateral grisea, including the so-called nucleus of the lateral olfactory tract.

[139] This includes the '*olfactory projection tract of Cajal*'.

part by fibers crossing in the diencephalic, epithalamic commissura habenulae.

The *commissura pallii anterior* or *hippocampal commissure* interconnects the dorsomedial cortices[140] and, in addition to commissural fibers *sensu strictiori*, may include decussating fibers with undefined connections, such, e.g. as the '*fasciculos cortico-basilares*' depicted by PEDRO RAMÓN (cf. Fig. 213B).

The *commissura pallii posterior* (Fig. 213F) found in Sphenodon and in numerous Squamates, but apparently missing in Chelonia and Crocodilia, crosses within the diencephalic roof plate, caudally to velum transversum but rostrally to, and separated from the commissura habenulae (cf. also Fig. 63, II, p. 174, vol. 3/II). It seems to interconnect the antimeric medial and dorsal cortices.

The *commissura anterior*, located ventrally to commissura pallii anterior, (cf. Figs. 211 A, C, E, 213 A, B) displays taxonomically related variations and contains a diversity of poorly elucidated components. Commissural respectively decussating fibers of 'amygdala' respectively stria terminalis appear well documented ('commissura archistriatica'). These are, in addition, decussating fibers of lateral and medial forebrain bundles. A so-called pars olfactoria interconnects basilateral grisea. Whether it does or does not contain interbulbar fibers remains a moot question.

The *commissura habenulae*, in which decussating fibers of stria medullaris and commissural fibers of habenular grisea seem to cross, doubtless also contains commissural fibers of lobus hemisphaericus[141] and may also include interbulbar connections.

As regards the enumerated communication channels and their various commissures or decussation (Figs. 213 A–F), components of olfactory tracts, 'olfacto-archistriatic tract', lateral and medial forebrain bundles, fornix, anterior and posterior pallial commissure, anterior commissure, stria medullaris and stria terminalis are commonly well medullated, but the ratio of medullated and non-medullated fibers in the diverse communication channels displays numerous taxonomically related and perhaps also individual variations.

---

140 Whether the pallial commissures include fibers of the 'primordium neopallii' remains a moot question. If so, and depending on arbitrary definition, such fibers might or might not be considered 'forerunners' of corpus callosum (cf. the comments on p. 623 and footnote 259a of volume 3/II.

141 The '*commissura telencefalica superiore*' of BECCARI (1943), not to be confused with the commissura pallii posterior.

Concerning the *functional significance* of the Reptilian telencephalon, the pioneering experimental investigations by STEINER (1900) have conclusively shown that Lizards deprived of their telencephalon, although displaying loss of 'spontaneity'[142] and particularly of any 'fear reaction', react with 'normal motility' to cutaneous 'prodding' and avoid obstacles, running, e.g. through the openings of a grill, without colliding against its bars. If exposed to the sun, such Lizards move into the shade, even skillfully climbing from the sunny bottom of a grill to its shady top. Quite evidently, the operated animals register and process optic input, presumably through mesencephalic and diencephalic grisea. If only one hemisphere is extirpated, e.g. the right one, the animal reacts to threatening motion from the right side, but not from the left, thus showing that input from the right eye reaches contralateral telencephalic grisea, and vice-versa. Extirpation of the rostral third of both hemispheres, except for the presumed loss of olfaction, had no apparent effect on the Lizard's behavior, neither had extirpation of the dorsal pallium *('Decke des Grosshirns')*. This would seem to indicate that integrity of epibasal, basal, and parts of medial cortex would be sufficient for fairly 'normal' telencephalic activities.

The *Golgi pictures* obtained by PEDRO RAMÓN (1918), as e.g. here exemplified by Figure 213C, seem to show that efferent as well as afferent fibers connect the cortex dorsalis (especially its 'primordium neopallii') as well as the epibasal complex, by way of the lateral forebrain bundle, with diencephalic and tegmental grisea. Thus, optic input doubtless reaches the dorsal cortex as well as the rostral epibasal complex. The reports by KRUGER and BERKOWITZ (1960), HALL and EBNER (1970), PRITZ (1975), and various others,[143] based on electrophysiological pro-

---

[142] This includes 'spontaneous feeding' *(spontane Nahrungsaufnahme)*, which can also be performed by Fishes and Amphibians deprived of telencephalon. In his original experiments with telencephalic extirpation in Frogs, STEINER failed to record spontaneous feeding (1885), but following experiments by SCHRADER (1887) which disclosed such feeding activities, admitted (1900, pp. 56–57) that his Frog experiments had not included observations on this question *('Ich selbst habe mich mit dieser Frage gar nicht besonders beschäftigt')*.

[143] Cf. the discussion, with additional relevant bibliographic references, by KIRSCHE (1972). Some inconclusive data are also discussed by KRUGER (1969). Previously, POWELL and KRUGER (1960) had claimed, on the basis of rather unconvincing cell degeneration experiments, that only nucleus rotundus and their 'nucleus dorsomedialis anterior' (apparently HUBER and CROSBY's nucleus diagonalis, respectively our nucleus medialis) projected to 'lateral palaeostriatum' (our basilateral grisea).

cedures or on methods of degeneration, indicate that dorsal cortex and apparently also epibasal grisea receive optic and other 'somatic' input. It remains, however, quite doubtful to which extent such input is 'discretely' respectively 'somatotopically' organized and functionally analogous to that reaching the Mammalian neocortex.[144] This qualification, in particular, also applies to the report by EBNER and COLONNIER (1975) who investigated ultrastructural aspects of what they call the Turtle's 'visual cortex'. The cited authors describe dendritic spines, electron-dense '*fuzz*', axo-somatic contacts, etc., in that part of the cortex dorsalis receiving optic input from the dorsal lateral geniculate nucleus. It is claimed that some synaptic patterns are 'remarkably similar' to those found on pyramidal cells of Mammalian neocortex.

As regards efferent fibers from the lateral portion of cortex dorsalis, reaching the mesencephalic tegmentum, it seems likely that this output component mediates the motor responses to electric stimulation observed by JOHNSTON (1916a) in the Turtle, and BAGLEY and RICHTER (1924) in the Alligator. The resulting movements of body and limbs were of a general type ('mass-movements'), very indirectly comparable to the more 'individualized' movements which can be elicited from the Mammalian motor neocortex. In addition, the report by KOPPÁNYI and PEARCY (1925) seems to indicate that such movements may also result from stimulation of the underlying rostral epibasal complex.

The experimental data reported by KRUGER and BERKOWITZ (1960) and others[145] seem to confirm the anatomically well established fact based on the arrangement of fiber tracts, namely that lateral ('piriform lobe') and medial ('hippocampal') cortices receive olfactory input. The recorded types of potential obtainable from lateral and from medial

---

[144] Although there can be little doubt that the lateral portion ('primordium neopallii') of the Reptilian cortex dorsalis is morphologically homologous to the Mammalian neopallium *in toto*, this orthohomology must be qualified as heterotypic (cf. vol. 3/II, p. 65) and does, moreover, not imply identical details of functional mechanisms. Again, it should be emphasized that findings based on histochemical behavior and fiber connections have no bearing whatsoever on problems of morphologic homology and the thereto related configurational aspects of phylogenetic evolution. It seems rather evident that, in the course of phylogeny, topologically comparable (homeomorphic or homomorphic) neighborhoods (grisea) can acquire different biochemical characteristics, as well as differences in fiber connections and in synaptologic details, becoming thus heterotypic and heteropractic ortho- or kathomologous.

[145] Some rather inconclusive elaborations on 'experimental analyses of the Reptilian nervous system', with various references to telencephalic grisea can also be found in a paper by KRUGER (1969).

cortex appear to differ. This would agree with presumptive differences in the functional significance of these cortices, the medial one being, as it were, a forerunner of the peculiar hippocampal formation in Mammals.

9. Birds

The Avian telencephalon, whose external aspect is shown in Figure 214A, displays, with respect to the configuration of grisea within the lobus hemisphaericus, a further development of features characteristic for the Reptilian one, especially as regards the expansion and differentiation of the epibasal components. This leads to a pronounced thickening of the hemispheric lobe's lateral basal wall. A conspicuous similarity between the configuration of Avian and Crocodilian telencephalon is here quite evident (cf. Figs. 211E and 214C).

The *hemispheric bend*[146] or 'rotation' *('Endhirnbeuge, Hemisphärenrotation)*, which is barely suggested in some Amphibians (Gymnophiona) but somewhat more clearly indicated in various Reptiles, becomes more pronounced in Birds, although less so than in many Mammals, where said bend results in the formation of a 'true' temporal lobe characterizing the caudalward paired evaginated hemisphere.

In the rostralward paired evaginated portion of the Avian hemisphere, however, the bend is quite conspicuous, and results in a rostroventral position of the olfactory bulbs (Figs. 214A–C, 215A–F). In contradistinction to Crocodilia, the commonly small olfactory bulbs are sessile (i.e. not stalked), as is the case in Chelonian Reptiles.[147] Fusion of the bulbs in the midline occurs to various degrees in some Birds, but not in others, even as regards taxonomically closely related forms (Fig. 214B). In the Paleognathe (Ratite) Kiwi (Apteryx) the olfactory bulb is relatively large, more so than in other Birds. Still fairly well developed bulbs are found in the Albatros, in the Duck, the Dove, and the Chick, while those in Passeriformes are rather small.[148] As regards its structural details, the Avian olfactory bulb does not, in the aspect here under consideration, significantly differ from the ar-

---

[146] The hemispheric bend is dealt with on p. 577 of volume 3/II.

[147] In the Ratite Dromiceius, the bulbs are constricted at their base and thereby display what could be called 'a short stalk'.

[148] Cobb (1960a, b) and Pearson (1972) bring tabulations of the ratio of olfactory bulb size to that of the lobus hemisphaericus in various species and orders of Birds.

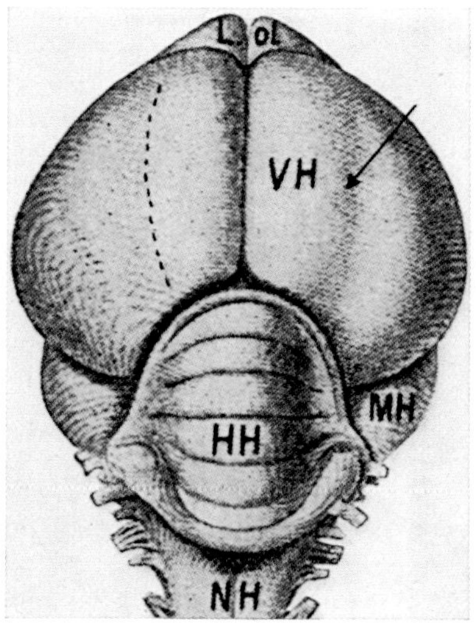

Figure 214 A. Dorsal aspect of the Pigeon's brain (after WIEDERSHEIM, from R. HERTWIG, 1912). Lol: bulbus olfactorius; HH: cerebellum; MH: laterally displaced roof of mesencephalon (lobus opticus); NH: oblongata; VH: lobus hemisphaericus. Added arrow points to vallecula, laterally limiting the sagittalwulst, whose boundary is also indicated by dotted line on left side, and corresponds to sulc. rhinalis lat. of Mammals.

rangement found in Mammals, and was dealt with in section 2 of the present chapter. An accessory olfactory bulb has not been recorded in Birds.

Among the early studies on morphology, cytoarchitecture and fiber systems of the Avian telencephalon are those by BUMM (1883), TURNER (1891), EDINGER and WALLENBERG (1899), EDINGER *et al.* (1903), WALLENBERG (1906), SCHROEDER (1911), ROSE (1914), KUENZI (1918), DENNLER (1922), and HUNTER (1924). These were followed by the detailed investigations undertaken by HUBER and CROSBY (1929) and by CRAIGIE (1928, 1929a, b, 1930, 1932, 1934, 1935a, b, c, 1936, 1939, 1940a, b, 1941a, b). Many of this latter author's papers were concerned with cytoarchitectural aspects of the Avian cerebral cortex. DURWARD (1932, 1934), moreover, studied the telencephalon of the New Zealand Kiwi, Apteryx australis, and the development of basal ganglia in the Sparrow. KAPPERS (1922) compared the ontogenetic development of Avian and Mammalian basal ganglia. The views of this author, are embodied in his standard treatises (KAPPERS *et al.*, 1936; KAPPERS, 1947).

My own investigations (K., 1924b, 1929) were mainly concerned

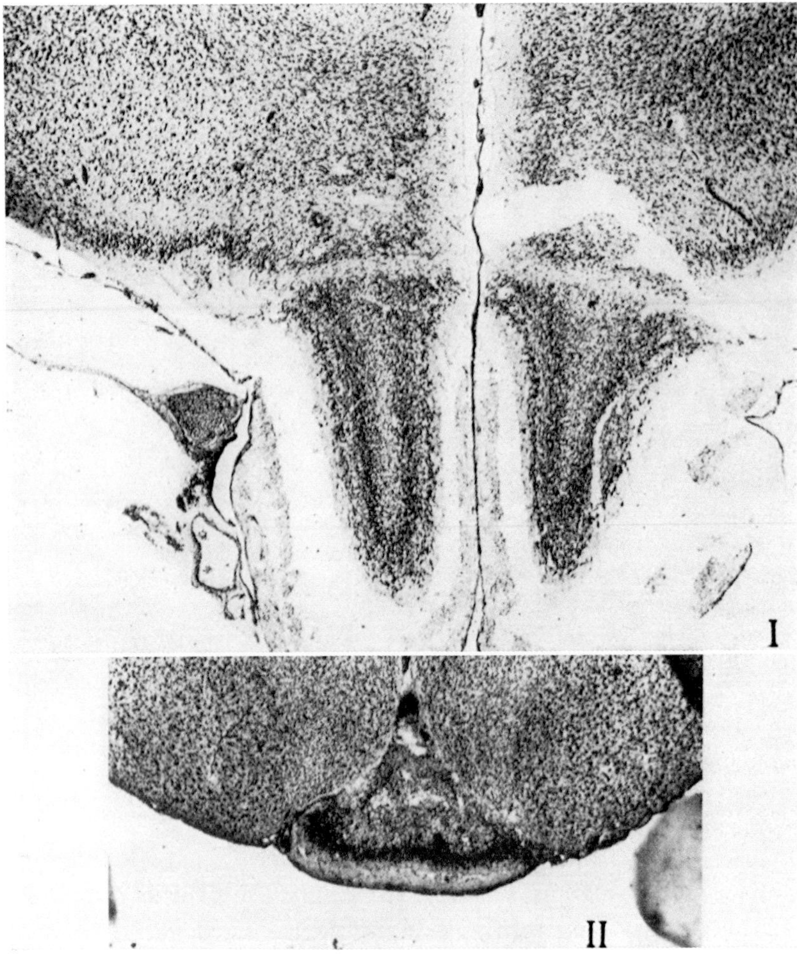

Figure 214B. Transverse sections *(Nissl stain)* through the olfactory bulbs of the Tinamids Nothura maculosa (I) and Rhynchotus rufescens (II), showing well developed separate respectively reduced and fused olfactory bulbs. The anterior lateral corticoid layer (prepiriform cortex) is well developed and clearly recognizable in I (from CRAIGIE, 1940a, ×25, red. $^1/_1$).

with the homologies of the Vertebrate telencephalic grisea (including those in Birds), and their ontogenetic derivation from a longitudinal zonal system. The attempt was made to establish a suitable unified notation applicable to these grisea in their taxonomic diversity respectively phylogenetic evolution from Cyclostomes to Mammals. An in-

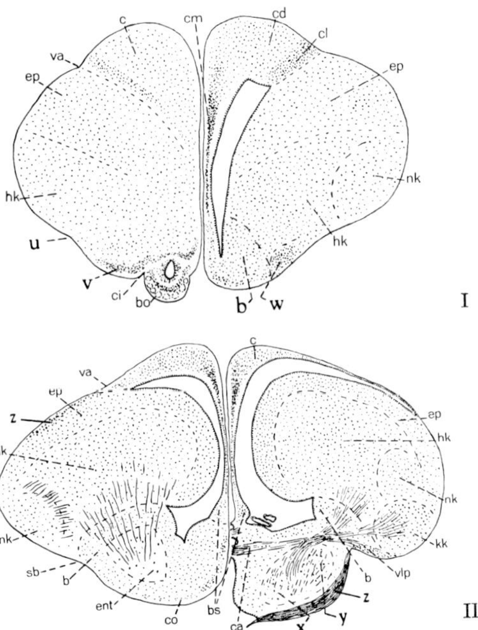

Figure 214C. Simplified sketches of cross sections through an Avian brain (pigeon) at various levels in rostro-caudal sequence (slightly modified from K., 1927). b: nucleus basalis; bo: bulbus olfactorius; bs: nucleus basimedialis superior; c, cd: region of parahippocampal cortex; ca: commissura anterior; ci: sulcus sive fissura circularis; cl: nuclei diffusi (griseum diffusum); cm: medial (hippocampal) cortex; co: area ventralis anterior (basal cortex, tuberculum olfactorium); ent: nucleus entopeduncularis; ep: griseum epibasale superius; hk: nucleus epibasalis centralis; kk: nucleus epibasalis caudalis; nk: nucleus epibasalis centralis accessorius ('ectostriatum'); sb: sulcus basalis lateralis (externus); u: rostral accessory branch of sb; v: rostral ('prepiriform') corticoid layer; va: vallecula; vlp: area ventrolateralis posterior (n. taeniae); w: nucleus epibasalis ventrolateralis; x: nucleus entopeduncularis at diencephalic (hypothalamic) level; y: nucleus reticularis thalami ventralis; z: lateral corticoid layer (intermediate region); z (at lower right): optic tract. Except for u, sb corresponds to Mammalian sulc. endorhinalis.

vestigation on the ontogenetic development of the telencephalon and particularly of its cerebral cortex in the Chick (K., 1938) was then undertaken in order to provide further documentation of the previous results.

Subsequently, Jones and Levi-Montalcini (1958), adopting my nomenclature, elucidated ontogenetic patterns of differentiation of grisea and particularly of fiber tracts in the Chick's hemisphere. Additional studies on the Avian telencephalon were undertaken by Källen

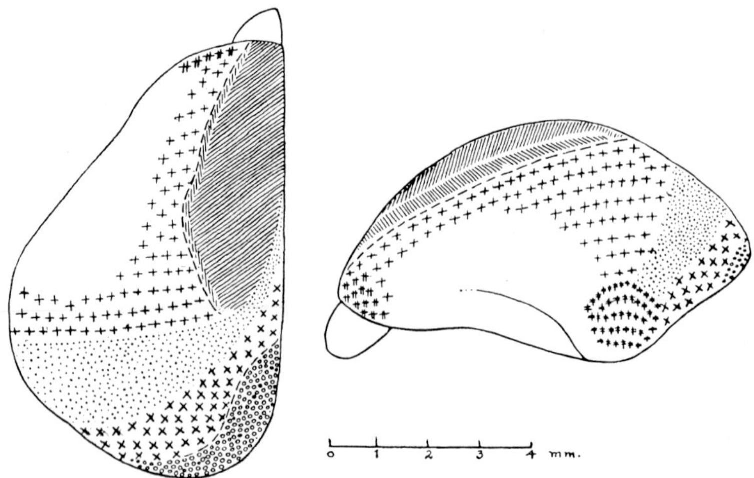

Figure 215 A. Mapping of pallial grisea upon surface of lobus hemisphaericus in Chick embryo of 19 days of incubation, already displaying definitive (adult) overall configuration (from K., 1938). Crosses with two vertical bars; rostral subdivision of lateral corticoid plate; simple crosses: intermediate subdivision; crosses with two horizontal bars: caudal subdivision; dots and x: region of parahippocampal cortex; circles: hippocampal cortex; oblique and vertical hatching: nuclei diffusi (essentially corresponding to sagittalwulst).

(1953, 1962, and others),[149] by HAEFELFINGER (1958),[149a] and by STINGELIN (1958). In his treatise on the Avian embryo, ROMANOFF (1960) has included an account of telencephalic differentiation. Other reports on ontogenetic development are those by KRABBE (1952), ROGERS (1960) and others, summarized by PEARSON (1972). COBB (1960a, b) dealt with the size of the olfactory bulb, and with some morphologic features of the Avian brain in general, as well as of telencephalon in

---

[149] Results of the studies by JONES and LEVI-MONTALCINI, and by KÄLLEN were discussed in chapter VI, section 6, pp. 555–571 of volume 3/II, together with an elaboration of my own views on that topic.

[149a] HAEFELFINGER uses my topologic D and B notations for early embryonic stages, since, as he justly remarks '*sich leider die nomenklatorischen Bezeichnungen der Adultform nicht in befriedigender Weise auf das Embryonalstadium übertragen lassen*'. As regards the differentiated adult grisea, however, he reverts to the commonly adopted so-called '*current terminology*'. He also noticed the conspicuous similarities between KÄLLEN's (1953) and my previous results, but, since KÄLLEN did not refer to my studies of 1929 and 1938 in the cited paper, HAEFELFINGER (1958, p. 21) was misled to assume that KÄLLEN had no acquaintance with said studies (*'Obgleich er von der oben zitierten Arbeit* KUHLENBECKS *keine Kenntnis hat'*). Cf., however, volume 3/II, footnotes 221 (p. 544), 226 (p. 559), 240a (p. 595).

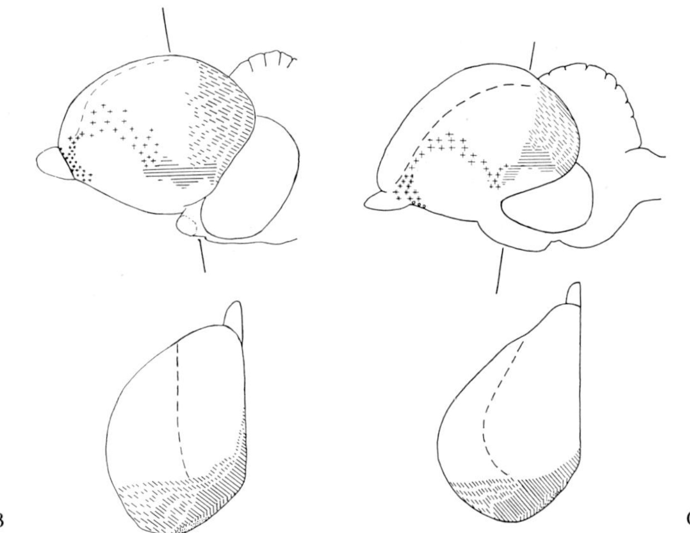

Figure 215 B. Mapping of pallial grisea upon surface of lobus hemisphaericus in the Loon, Gavia (from CRAIGIE, 1940b). Oblique and vertical hatching: parahippocampal fields; stipples: hippocampal cortex; horizontal: caudal subdivision of lateral corticoid plate; crosses as in Figure 215 A. The broken line indicates lateral border of sagittal wulst.

Figure 215 C. Mapping of pallial grisea upon lobus hemisphaericus of the Penguin, Eudyptes chrysocome (from CRAIGIE, 1941a). Markings as in Figure 215 B.

particular. Stereotaxic atlases which illustrate telencephalic grisea were published by VAN TIENHOVEN and JUHASZ (1962, Chick), by KARTEN and HODOS (1967, Dove), and by STOKES *et al.* (1974). A comprehensive treatise on the Avian brain, with extensive bibliographies, and emphasizing functional aspects as well as related experimental results, was prepared by PEARSON (1972). Some additional papers concerning specific topics will be referred to, where appropriate, further below in the text.

The *secondary pallium* of the Avian lobus hemisphaericus corresponds, as in the other two classes of Amniota (Reptiles, Mammals), to the topologic zonal neighborhoods D_3 and D_2, with minor residual or boundary contributions from D_1. This pallium includes (1) a medial or 'hippocampal' cortex, (2) a dorsomedial parahippocampal cortex, (3) dorsal non-cortical grisea, and (4) a lateral corticoid layer or plate. Region (1) corresponds to the cortex medialis of Reptiles, regions (2) and moreover also (3) correspond to Reptilian cortex dorsalis, and region

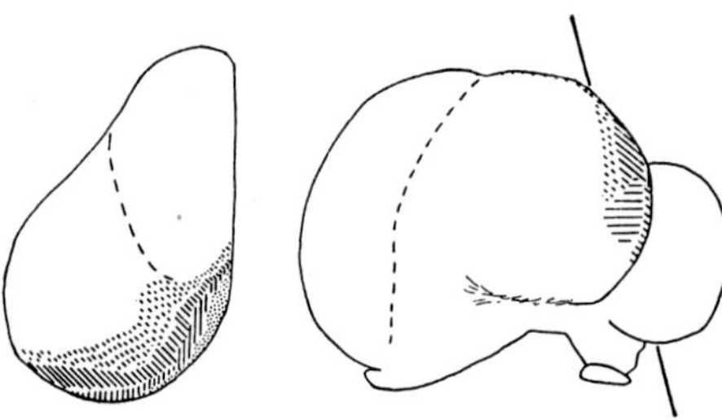

Figure 215 D. Mapping of parahippocampal and hippocampal cortex upon the lobus hemisphaericus of the Ostrich, Struthio (from CRAIGIE, 1936). Markings as in Figures 215 B and C.

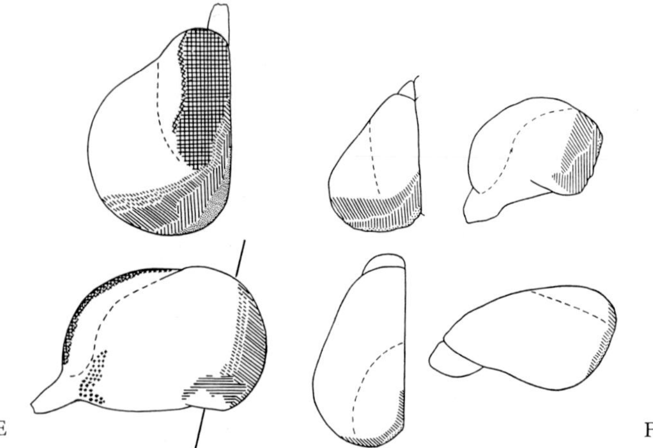

Figure 215 E. Mapping of pallial areas upon lobus hemisphaericus of the Ratite Rhea americana (from CRAIGIE, 1939). Cross hatching and x: nuclei diffusi. Other markings as in Figure 215 B–D.

Figure 215 F. Mappings of parahippocampal and hippocampal cortex upon lobus hemisphaericus of the Kiwi, Apteryx (above) and (below) of the Emu, Dromiceius (from CRAIGIE, 1935a, b, c). Markings as in Figures 215 B–E.

(4) with its rather blurred transition to (3) is homologous to Reptilian cortex lateralis.

The *cortex medialis*, derived from zone D_3, is homologous to the hippocampal cortex of Mammals. It is the region H of CRAIGIE's notation, and can be arbitrarily subdivided in subregions H_1 to H_4, H_1 being the rather small-celled ventral strip, which is comparable to a poorly differentiated Mammalian fascia dentata, as e.g. seen in the precommissural hippocampus of various lower Mammals. The Avian cor-

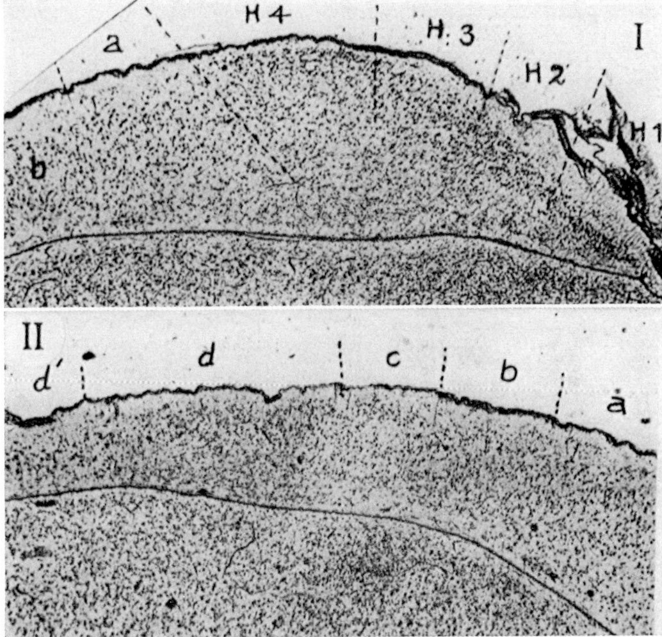

Figure 216 A. Cross-sections *(Nissl stain)* through medial (I) and lateral (II) pallial cortex of the adult Chick (from K., 1938; ×17, red. $^1/_1$). H_1–H_4: subdivisions of medial (hippocampal) cortex; a–d′: subdivisions of dorsal (parahippocampal) cortex. The rather arbitrary cortical parcellation is dealt with in the text.

tex medialis can be described as an essentially single-layered cortical plate with quite narrow subcortical layer and likewise rather narrow zonal layer. In the wider subregions H_y and H_z, however, a nondescript stratification may be manifested, the innermost cell layer becoming rather dense (cf. Fig. 217 E).

The parahippocampal cortex dorsalis corresponds to the major, medial extent of the Reptilian cortex dorsalis, and commonly displays a more or less distinct stratification as first pointed out by CRAIGIE (1929a, b). It can be arbitrarily subdivided, from the neighborhood of its contiguity with medial cortex lateralward, into regions a, b, c, d (e.g. CRAIGIE, 1935a, b; K., 1938). In addition to a narrow zonal layer (I) and to the likewise very narrow subcortical layer, CRAIGIE (1929a, b) distinguished five laminae (II–VI) which he compared to those of the Mammalian neocortex (Figs. 216 B, C). Although, as can be seen in these Figures, such stratification is indeed suggested, particularly in Apte-

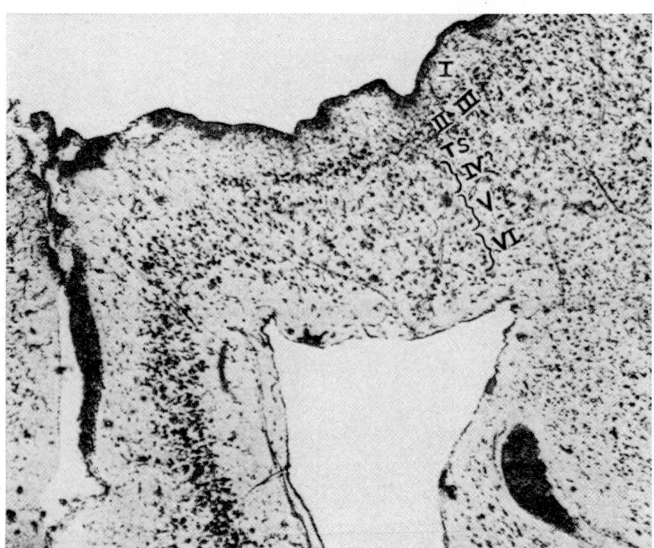

Figure 216 B. Cross-section *(Nissl stain)* through dorsomedial portion of the pallium in the Emu, at level of anterior commissure (from CRAIGIE, 1934; ×34, red. $^1/_1$). TS: fibers of septomesencephalic tract; I–VI: layers of parahippocampal cortex. The medial (hippocampal) cortex is at left.

ryx, it is less clear-cut than in the Mammalian neocortex, and, moreover, concerns parahippocampal cortex, but not the morphological homologue of Mammalian neocortex. In other Avian forms, a distinct stratification may be missing (cf. Fig. 216 A). Again, in some forms and at some cross-sectional levels, a superpositio medialis, comparable to that obtaining in Reptiles, is faintly recognizable at the zone of transition between parahippocampal and hippocampal cortex (cf. Fig. 217 E). Caudally to the sagittalwulst with its griseum diffusum, dealt with further below, the parahippocampal and the hippocampal cortex extend far lateralward, as shown in Figures 215 A–F. The hippocampal cortex thereby usually reaches the hemisphere's polus posterior.

Laterally to the parahippocampal cortex, and directly continuous with it, there is a non-cortical griseum, the *nucleus diffusus*, which corresponds to the region of CROSBY's primordium neopallii in Reptiles. This pallial thickening or pallial torus represents the so-called *sagittalwulst* (or *wulst*) of the Avian hemisphere, laterally bounded by the vallecula (cf. Fig. 214 A), and medially not distinctly delimited from the

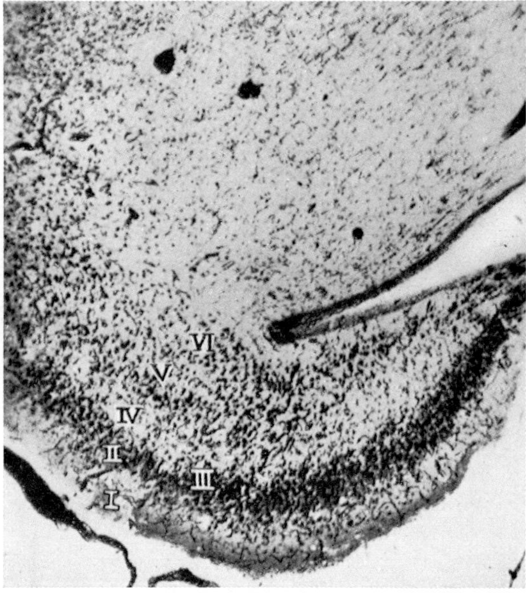

Figure 216 C. Cross-section *(Nissl stain)* through caudal and ventrolateral extension of parahippocampal cortex in the Kiwi (Apteryx), showing the cortical layers (from CRAIGIE, 1929a, b; ×: about 38, red. $^{1}/_{1}$).

parahippocampal pallium. The *sagittalwulst,* insofar as it contains the griseum diffusum, does not extend into the caudal regions of the hemisphere (cf. Fig. 215 A).[150] An external (dorsal, superficial) nucleus diffusus dorsalis, and an internal (ventral, deeper) nucleus diffusus dorsolateralis can usually be distinguished, but in some instances (cf. Fig. 219 A) an intermediate cell plate (nucleus diffusus intercalatus *sive* intermedius) can be noted.

A thin *lamina medullaris suprema accessoria* may separate nucleus diffusus dorsalis and dorsolateralis. If a n. diffusus intercalatus *sive* intermedius is present ('intercalated cells of lamina frontalis suprema' in the terminology of KAPPERS *et al.*, 1936), said lamina can appear subdivided by the intercalated nondescript cell plate. These nuclei diffusi (respectively the nucleus diffusus complex) doubtless are, on the basis of their topologic connectedness, homologous to the 'cortical dip' of some Reptilians (cf. Fig. 211 C), and to the Mammalian neocortex, ir-

[150] The taxonomically related variations in topography and extent of the sagittalwulst can be noted in the Figures 215 A–F.

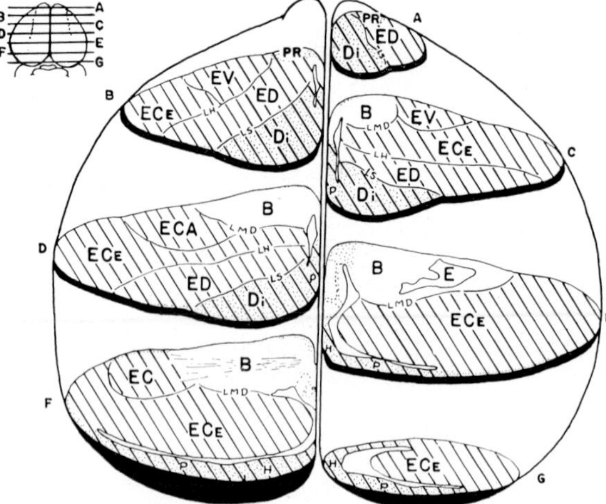

Figure 217 A. Transverse section through the telencephalon of a newly hatched Chick. The levels of the sections are indicated in the inset (after JONES and LEVI-MONTALCINI, 1958). B: nucleus basalis; Di: nuclei (palliales) diffusi: E: nucleus entopeduncularis; EC: nucleus epibasalis caudalis; ECA: nucleus epibasalis centralis accessorius; ECE: nucleus epibasalis centralis; ED: nucleus epibasalis dorsalis; EV: nucleus epibasalis ventrolateralis; H: hippocampal area; LH: 'lamina hyperstriatica'; LMD: 'lamina medullaris dorsalis'; LS: 'lamina medullaris suprema'; P: parahippocampal area; PR: 'prepiriform' area.

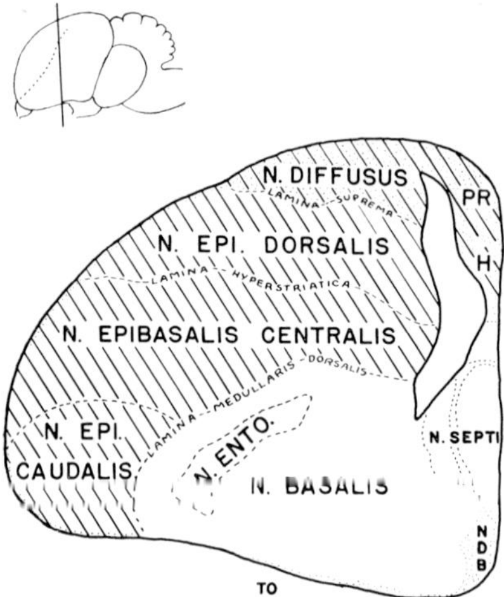

Figure 217 B. Diagrammatic representation of telencephalic grisea of a newly hatched Chick as seen in a cross-section whose plane and level is indicated by inset (from JONES and LEVI-MONTALCINI, 1958). NDB: 'nucleus of *Broca's diagonal band;* N.ENTO.: nucleus entopeduncularis; PR: parahippocampal cortex; TO: area ventralis anterior. Other designations self-explanatory and corresponding to Figure 217 A.

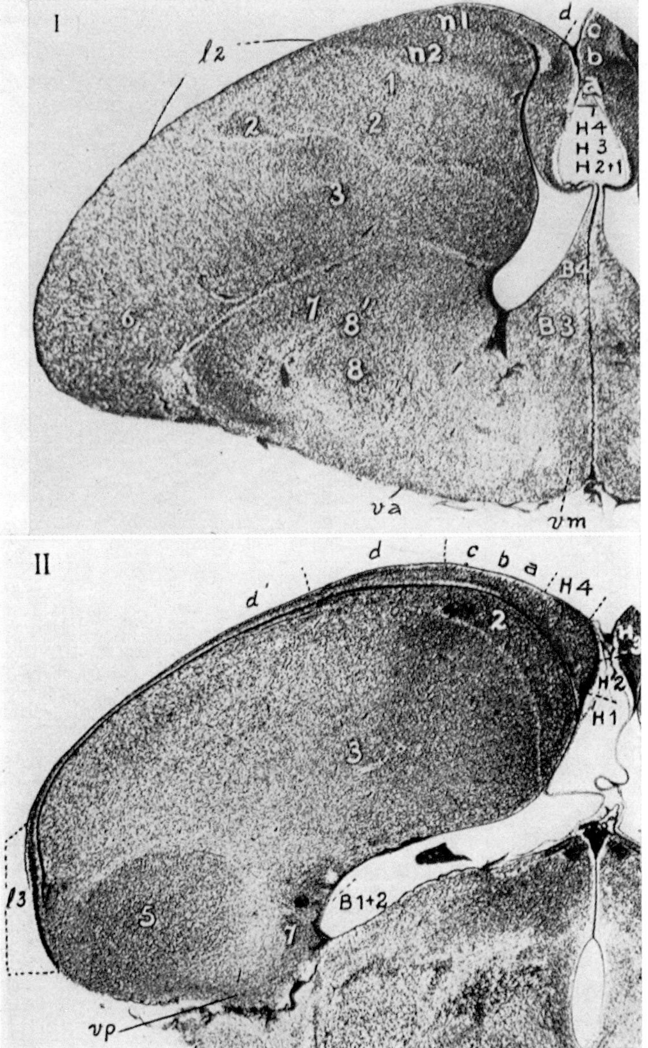

Figure 217C. Cross-section *(Nissl stain)* through telencephalon of a newly hatched Chick (I) rostrally and (II) caudally to hemispheric stalk (from K., 1938). a–d: parahippocampal fields; B_{1+2}: basilateral grisea; B_3: griseum basimediale inferius; B_4: griseum basimediale superius; l_2, l_3: intermediate and caudal portions of lateral corticoid field; n_1, n_2: griseum diffusum pallii dorsale and dorsolaterale ('nuclei diffusi'); va: area ventralis anterior; vm: '*nucleus of the diagonal band of Broca*'; vp: area ventrolateralis posterior ('cortical amygdaloid nucleus' including 'n. taeniae'); 1: nucleus epibasalis dorsalis, pars superior; 2: nucleus epibasalis dorsalis, pars inferior; 3: nucleus epibasalis centralis; 5: nucleus epibasalis caudalis; 6: nucleus epibasalis centralis accessorius; 7: nucleus basalis (B_{1+2}); 8, 8′: nucleus entopeduncularis anterior.

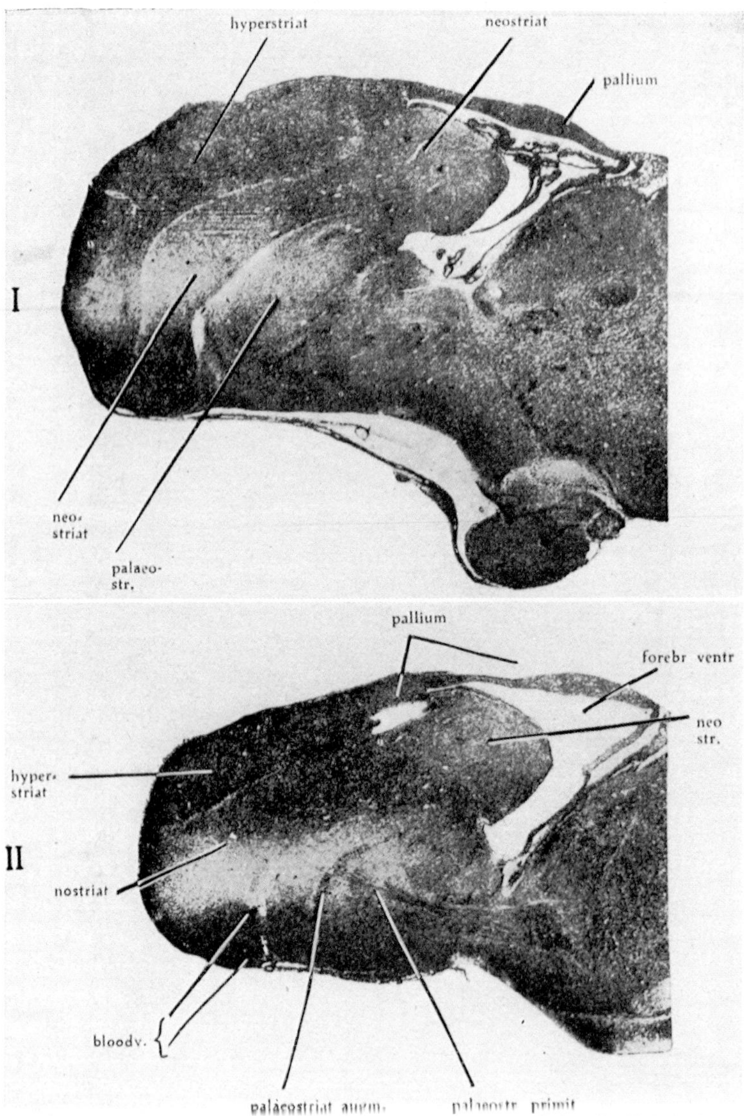

Figure 217 D. Sagittal sections through the forebrain of a Chick embryo at the 11th day of incubation (from KAPPERS, 1928). The caudal portion of 'neostriatum' corresponds to neighborhoods of our caudal epibasal complex. The 'hyperstriatum' corresponds to our dorsal epibasal grisea, in part also to our pallial grisea diffusa. The 'palaeostriatum primitivum' is our nucleus entopeduncularis anterior. The blood vessel in II corresponds to the Reptilian 'lateral striate artery' depicted in Figure 381, p. 709 of volume 3/II, and separating epibasal complex from basilateral grisea.

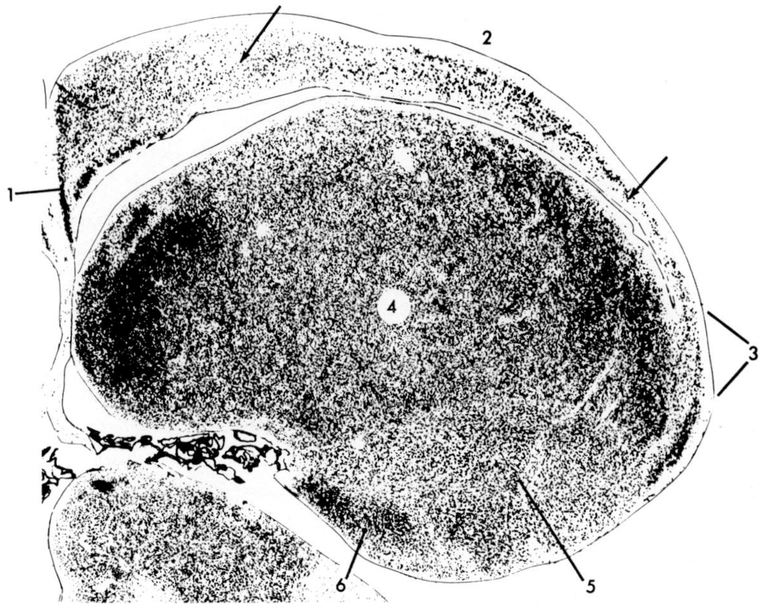

Figure 217E. Cross-section *(Nissl stain)* through the caudally evaginated portion of the Pigeons lobus hemisphaericus, i.e. at a level caudally to interventricular foramen (from KARTEN and HODOS, 1967). 1: cortex medialis (hippocampal cortex); 2: cortex dorsalis (parahippocampal cortex); 3: cortex lateralis (lateral corticoid plate); 4: caudal part of central and dorsal epibasal complex; 5: nucleus epibasalis caudalis; 6: 'nucleus taeniae'. The arrows indicate faintly suggested superpositio medialis and lateralis. All designations have been added and are in accordance with terminology and interpretation adopted in the present treatise.

respective of structural and presumably functional differences. Their designation as hyperstriatum accessorium and nucleus intercalatus hyperstriati supremi (cf. the Tabulation of Fig. 217F) does not seem appropriate since said grisea pertain to the secondary pallium, respectively to zone D_2, while the so-called hyperstriatum and its subdivisions pertain to the secondary basis, respectively to zone D_1, which provides the epibasal complex.

The lateral corticoid layer, which ontogenetically derives from lateral neighborhoods of D_2, laterally to the nucleus diffusus complex, with some variable but relatively minor contributions from D_1, displays a rostral, an intermediate, and a caudal subdivision (Fig. 215A). The rostral portion, because of the pronounced anterior curvature of the telencephalic bend, assumes a ventral position just dorsally to the olfactory bulb, and may, in various forms, extend from the lateral surface toward the basimedial surface of the hemisphere. It corresponds to the anterior (so-called prepiriform) region of the Mammalian cortex lobi piriformis. The intermediate region extends as a thin layer to a variable degree lateralward from the sagittalwulst upon the hemisphere's surface. It is likewise homologous to the anterior portion of Mammalian piriform lobe cortex. Caudally, it sweeps lateralward (and basalward). Its basalmost portion represents the subdivision, also equivocally designated as 'periamygdalar cortex', adjoining the caudolateral parahippocampal cortex.[151]

The *cell plate of the lateral corticoid region* does not display any distinct stratification. At its transition to nuclei diffusi, a slight overlap comparable with the superpositio lateralis of Reptiles may occasionally be indicated. In some forms, the caudal portion of the lateral corticoid plate, together with a lateral portion of parahippocampal cortex, is located within a rather thin, almost membranous pallium covering the epibasal complex from which it is separated by the lateral extension of a very narrow ventricular space (cf. Figs. 217 C, E).

In concluding the discussion of the Avian pallial grisea, it should be added that, at the rostral extremity of lobus hemisphaericus, namely in the region characterized by the anterior limb of hemispheric bend, the pallial grisea, extending ventralward, tend to merge into each other with rather indistinct boundaries, and include the nondescript cell aggregates designated as cortex frontalis and located within the *Frontalmark* of EDINGER.

Again, at suitable ontogenetic stages (cf. e.g. Fig. 296, p. 558, vol. 3/II) it can easily be seen that the matrix of nuclei diffusi (D_2l) forms a noticeable bulge dorsally to that of epibasal matrix (D_1) and separated from this latter by groove fd_2. Subsequently, the expansion of the nucleus epibasalis dorsalis flattens out both bulge and groove. It then may become difficult to distinguish nucleus epibasalis dorsalis, pars superior, from the pallial grisea diffusa. Thus, although HUBER

[151] It should here be added that the terms 'piriform lobe cortex' and 'cortex periamygdalaris' in Mammals are not standardized and shall be dealt with below in section 10. The piriform lobe of Mammals includes in its rostral subdivision piriform lobe cortex homologous to Reptilian cortex lateralis, and in its caudalmost portion parahippocampal cortex. The term 'periamygdalar cortex' should preferably be restricted to the cortical amygdaloid nucleus, which is not pallial, but basal, and corresponds to the area ventrolateralis posterior of Amphibia. In Birds, this is the nucleus taeniae. Cf. also Figures 217E and 221C (Mammals).

Current terminology	*Kappers et al.* (1936) *Kappers* (1947)	*Kuhlenbeck* (1938)	*Rose* (1914)	*Edinger et al.* (1903)
Accessory hyperstriatum	Hyperstriatum accessorium	Nucleus diffusus dorsalis	B	Cortex frontalis
Intercalated nucleus	Nucleus intercalatus hyperstriati supremi	Nucleus diffusus dorsolateralis	A	Frontal mark
Dorsal hyperstriatum	Hyperstriatum dorsale	Nucleus epibasalis dorsalis, pars superior	C	
Superior frontal lamina (LFS)	Nucleus intercalatus hyperstriati superioris			
Ventral hyperstriatum (dorso-ventral) (ventro-ventral)	Hyperstriatum ventrale (dorso-ventrale) (ventro-ventrale)	Nucleus epibasalis dorsalis pars inferior	D, D_1	Hyperstriatum
Frontal neostriatum	Neostriatum frontale	Nucleus epibasalis centralis, pars medialis	G_1	
Intermediate neostriatum	Neostriatum intermediale	Nucleus epibasalis centralis, pars posterior	G, G_2	
Caudal neostriatum	Neostriatum caudale		L, G_3	
Parolfactory lobe				Parolfactory lobe
Basal nucleus	Nucleus basalis	Nucleus epibasalis ventrolateralis	R	Mesostriatum laterale
Ectostriatum	Ectostriatum	Nucleus epibasalis centralis accessorius	S	Ectostriatum
Archistriatum	Archistriatum	Nucleus epibasalis caudalis	K	Epistriatum
Paleostriatum augmentatum	Paleostriatum augmentatum	Nucleus basalis	H	Mesostriatum
Paleostriatum primitivum	Paleostriatum primitivum	Nucleus entopeduncularis	J	Nucleus entopeduncularis

Figure 217F. Tabulation of nomenclature referring to Avian telencephalic grisea as adopted by different investigators (from PEARSON, 1972).

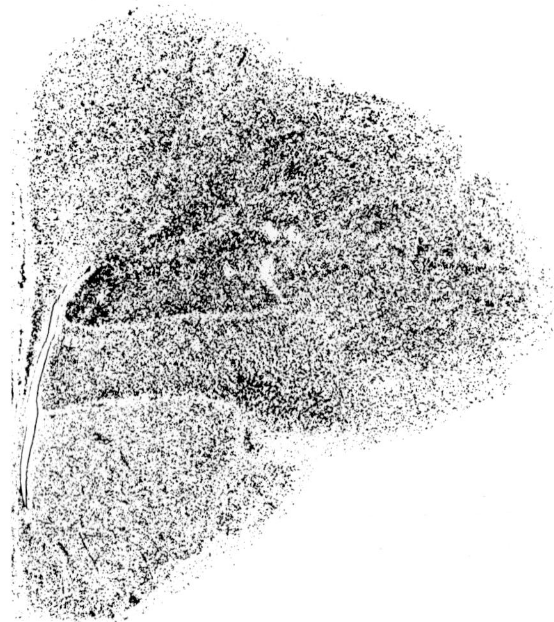

Figure 218 A. Cross-section *(Nissl stain)* through a fairly rostral portion of the Pigeon's lobus hemisphaericus, at level of nucleus epibasalis ventrolateralis (from KARTEN and HODOS, 1967)

and CROSBY's hyperstriatum dorsale generally seems to correspond to the present author's pars superior of nucleus epibasalis, it appears, in some instances (e.g. Fig. 220B, C) to represent the nucleus diffusus dorsolateralis of the here adopted terminology.

Turning now to the *basal grisea*[152] of the Avian lobus hemisphaericus, the following main components can be distinguished: (1) the epibasal complex (D_1) and its subdivisions, (2) the basilateral complex, (3) the basimedial complex, and (4) the rudimentary cortex basalis.

The massive lateral wall characterizing the Avian hemispheric lobe is essentially the result of the expansive development displayed by the *internated epibasal complex* (1). The telencephalic ventricle becomes thereby conspicuously narrowed, but exhibits, in this respect, diverse taxonomic variations, being perhaps widest in Apteryx, and particular-

[152] It will be recalled that, in all Amniota, the lateral basis telencephali has absorbed, by internation (introversion) a formelement (grundbestandteil D_1) which is pallial in Anamnia. Both pallium and (lateral) basis telencephali of Amniota are thus secondary, the former being diminished (D_2, D_3), the latter increased (D_1, B_1, B_2).

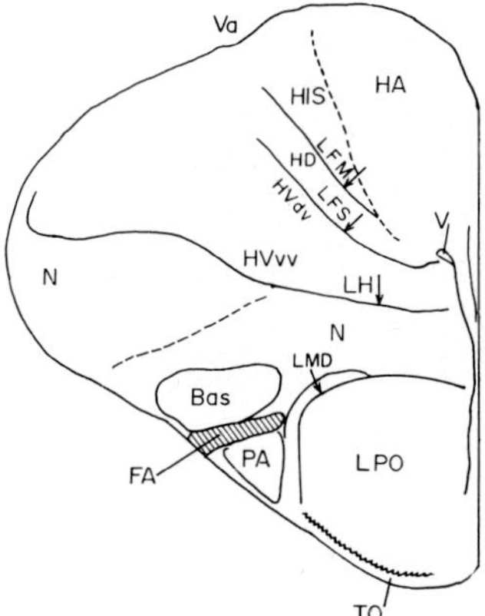

Figure 218 B. Mirror-image outline drawing of Figure 218 A, as interpreted by the cited authors (from KARTEN and HODOS, 1967). The designations also apply to Figure 218 D. Ac: nucleus accumbens; Bas: nucleus basalis (nucleus epibasalis ventrolateralis of the present author's terminology); E: ectostriatum; FA: tractus fronto-archistriatalis; HA: hyperstriatum accessorium; HD: hyperstriatum dorsale; HIS: hyperstriatum intercalatum superius; HV: hyperstriatum ventrale; HVdv: hyperstriatum ventrale dorsoventrale; HVvv: hyperstriatum ventrale ventro-ventrale; LFM: lamina frontalis suprema; LFS: lamina frontalis superior; LH: lamina hyperstriatica; LMD: lamina medullaris dorsalis; LPO: lobus parolfactorius; N: neostriatum; NI: neostriatum intermedium; PA: palaeostriatum augmentatum; PP: palaeostriatum primitivum; QF: tractus quintofrontalis; SL: nucleus septalis lateralis; TO: tuberculum olfactorium; TSM: tractus septomesencephalicus; V: ventricle; Va: vallecula. The (hippocampal) cortex medialis, not designated, can be easily identified in Figures 218 A and C, likewise the parahippocampal cortex dorsalis, which however, is less distinct in A than in C.

ly reduced in Psittaciformes. As a rule, a slit-like ventricular extension passes dorsolaterally and in part caudally to the more caudal levels of the epibasal complex.

Generally speaking, the following *epibasal derivatives* can be recognized. (1a) *Nucleus epibasalis dorsalis*, *pars superior*, commonly separated from the nucleus diffusus complex by the lamina medullaris suprema,[152a] and from (1b) *nucleus epibasalis inferior* by a variable lamina medullaris hyperstriatica accessoria, which may be split, if a nonde-

---

[152a] The terminology of these laminae medullares is not standardized. The here adopted designations agree with the overall concept illustrated by JONES and LEVI MONTALCINI as shown in Figure 217 B. PEARSON (1972) uses the term 'superior frontal lamina' for the present hyperstriatica accessoria (cf. the tabulation of Fig. 217 F), his lamina frontalis suprema, however, being apparently our lamina medullaris suprema.

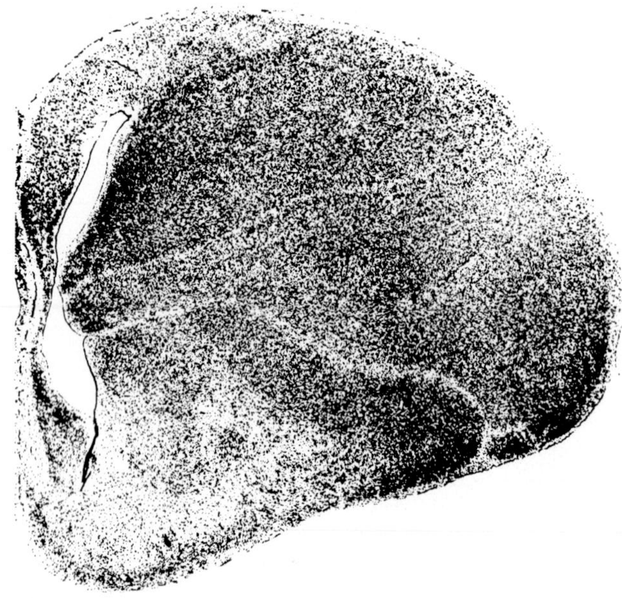

Figure 218C. Cross-section *(Nissl stain)* through a more caudal level of the Pigeon's lobus hemisphaericus, somewhat rostral to lamina terminalis (from KARTEN and HODOS, 1967).

script *nucleus epibasalis dorsalis intercalatus* is present between (1a) and (1b) as indicated in PEARSON's tabulation (Fig. 217F). It should be emphasized that, in my interpretation, the nucleus epibasalis dorsalis, pars superior (so-called dorsal hyperstriatum *autorum*) does not belong to the *sagittalwulst*. The usually conspicuous main lamina hyperstriatica separates the dorsal epibasal complex from the extensive *nucleus epibasalis centralis* (1c). A pars medialis and a pars lateralis of this griseum may frequently be distinguished as well as an anterior and a posterior neighborhood. The *nucleus epibasalis ventrolateralis* (1d) is a basally and rather superficially located griseal rostral differentiation within the anterior limb of the hemispheric bend. The *nucleus epibasalis centralis accessorius* (1e) is a lateral griseal condensation of the main central epibasal nucleus, usually separated from this latter by fiber bundles suggesting a nondescript medullary lamina.[153]

[153] Although the nucleus epibasalis centralis accessorius quite evidently corresponds to the ectostriatum of current terminology, I am inclined to trace it laterad as far as the hemispheric surface, while other authors depict it as laterally surrounded by the main

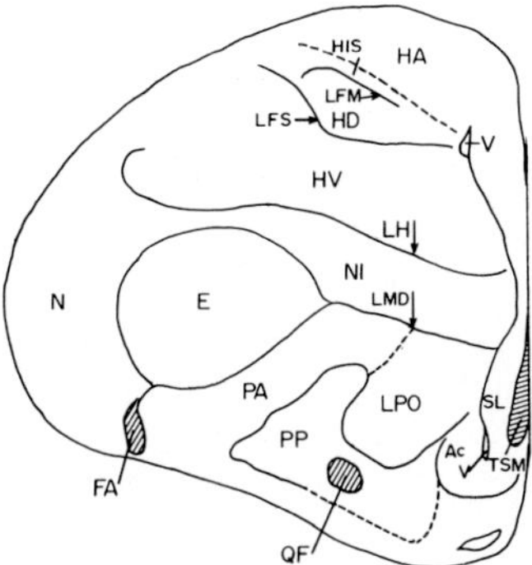

Figure 218 D. Mirror-image outline of Figure 218C, as interpreted by the cited authors. Abbreviations as in Figure 218B (from KARTEN and HODOS, 1967).

The *nucleus epibasalis caudalis* (1f), located within the posterior third of the hemisphere, is a caudal and fairly lateral epibasal derivative, assuming a rather ventral position related to the caudal limb of the hemispheric bend. A lamina medullaris occipitalis frequently separates said griseum from the central epibasal nucleus, and is, at least in part, related to the occipitomesencephalic tract. Central and caudal epibasal grisea are separated from the basilateral ones by the lamina medullaris dorsalis. ZEIER and KARTEN (1971) subdivide the overall neighborhood of nucleus epibasalis caudalis (their 'archistriatum', EDINGER's epistriatum) into several grisea, namely archistriatum anterius, intermedium, intermedium pars dorsale, archistriatum mediale, archistriatum posterius, and nucleus taeniae. Only the posteromedial 'archistriatal' grisea are supposed to be the Avian 'amygdalar homologue' (cf. also COHEN, 1975). The cited subdivisions do not seem entirely convincing.

Recently, NOTTEBOHM and ARNOLD (1976) have described sex-differences in various telencephalic grisea of Canaries and Zebra Finches, *inter alia* particularly concerning a cell group within the so-called archistriatum. This dimorphism is supposed to be correlated with sex-differences in singing behavior.

The *basilateral grisea* of the Avian hemispheric lobe are represented by an essentially uniform *nucleus basalis*, provided by a fusion of B_1 and

central nucleus or 'neostriatum', that is to say as a more or less 'internal' nucleus. It should be added that, as regards the lateral extension of 'ectostriatum', a clear-cut boundary between said nucleus and the 'neostriatum' is indeed usually difficult to trace.

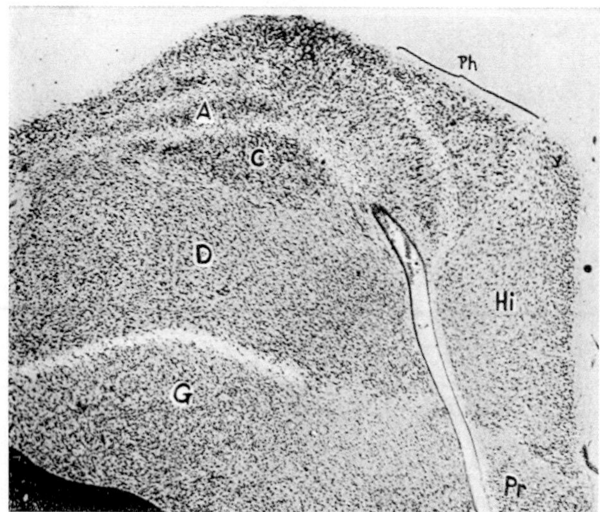

Figure 219A. Cross-section *(Nissl stain)* through dorsomedial part of lobus hemisphaericus in the Humming Bird Lampornis mango, rostrally to lamina terminalis (from Craigie, 1932; ×40, red. $^1/_1$). A: nucleus diffusus dorsolateralis, with unlabelled accessory differentiation (nucleus diffusus intercalatus); B: nucleus diffusus dorsalis; C: nucleus epibasalis dorsalis, pars superior; D: nucleus epibasalis dorsalis, pars inferior; G: nucleus epibasalis centralis; Hi: cortex medialis (hippocampus); Ph: cortex dorsalis (parahippocampal cortex); Pr: nucleus basimedialis superior. The present author's nomenclature, where differing from that of Craigie, has been substituted.

B_2 derivatives. Nevertheless, an ill-defined rostral subdivision of that complex, caudalward gradually 'fading out' medially to the main nucleus basalis, could perhaps be vaguely delimited as the so-called '*lobus parolfactorius*' (e.g. Karten and Hodos, 1967).

The *nucleus entopeduncularis (anterior)*, a diencephalic hypothalamic derivative, characterized by fairly large nerve cells, and essentially constituting an interstitial griseum of the lateral forebrain bundle, protrudes far rostrad through the hemispheric stalk into the basilateral telencephalon.[153a] It is designated as '*palaeostriatum primitivum*' in the widely used conventional terminology, being separated from the nucleus basalis by a variably distinctive ventral medullary lamina. A nondescript group of smaller cells within the entopeduncular nucleus is depicted as '*nucleus intrapeduncularis*' by Karten and Hodos (1967).

As regards the much less extensive *basimedial (paraterminal) grisea*, a

[153a] It represents the homologue of the Mammalian globus pallidus (K., 1924b).

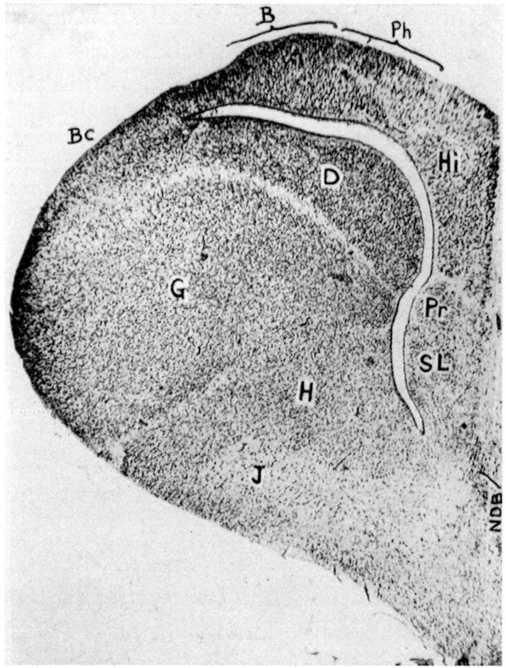

Figure 219B. Cross-section *(Nissl stain)* through the lobus hemisphaericus of Lampornis at a rostral level of lamina terminalis (from CRAIGIE, 1932; $\times 25$, red. $^1/_1$). B: joint nuclei diffusi (griseum diffusum); BC: lateral corticoid plate (continuous with n. diffusus dorsolateralis); D: joint nucleus epibasalis dorsalis (the lateral extension of pars superior can still be vaguely recognized internally to corticoid plate; H: nucleus basalis; J: nucleus entopeduncularis (anterior); NDB: 'nucleus of diagonal band'; Pr: nucleus basimedialis superior; SL: griseum basimediale inferius. Other designations as in Figure 219A.

nucleus basimedialis inferior (B_3) and a *nucleus basimedialis superior* (B_4), similar to those obtaining in Reptiles, can be recognized.[154] Medial (more superficial) cell groups of both grisea are designated as *nucleus septi medialis*, and the more adventricular ones as *nucleus septi lateralis*. The term '*nucleus accumbens*', which should preferably be used for a dense cellular adventricular condensation of *griseum basimediale inferius*, is occasionally also applied to an adjacent region of the inferior basilateral neighborhood (e.g. KARTEN and HODOS, 1967). There indeed not

[154] CRAIGIE (1932), in accordance with an interpretation of JOHNSTON concerning Reptiles, designated the Avian nucleus basimedialis superior as 'primordium hippocampi' (cf. Figs. 219A, B).

infrequently obtains a fusion of inferior basilateral and basimedial cell aggregates, as, e.g. also in the so-called 'fundus striati' of Mammals.

As in most Vertebrates, the telencephalon of Birds also includes a *basal cortex* B_s, derived from the lateral and medial B zones, and apparently related to the olfactory system. In Birds, this essentially non-stratified cortex shows a rather rudimentary condition which corresponds to the relatively minor development of olfactory components. The rostral part of the basal cortex is a thin, vaguely outlined corticoid layer covering the surface of nucleus basalis and extending toward the basimedial region. It originates from the grundbestandteile B_{1+2+3} with perhaps contributions from B_4 and represents the *area ventralis anterior*, also designated as tuberculum olfactorium.[155] The caudal portion of the basal cortex, essentially derived from B_{1+2} is the *area ventrolateralis posterior*, homologous to the cortical nucleus amygdalae that is, to the true periamygdalar cortex of Mammals, and represented by the *nucleus taeniae* of the Avian hemisphere.

Summarizing the internal configuration of the basis lobi hemisphaerici in Birds, it could be stated that the lateral basis consists of more or less laminated and rather homogeneous[156] grisea, in part separated by more or less distinct thinner medullated laminae. Although, as regards the epibasal derivatives, the six 'nuclei' (1a to 1f) enumerated above can generally be recognized, there obtain numerous taxonomic variations concerning details of extension and, in some cases, secondary additional subdivisions. The detailed investigations of HAEFELFINGER (1958) and of STINGELIN (1958) deal with this topic. The latter author also particularly emphasized cytoarchitectural aspects and some cytologic details. A summary of these data, which transcend the limits imposed on the present general survey, can also be found in the treatise by PEARSON (1972).

With respect to Reptilian homologies, there is little doubt that the complex of Avian nuclei epibasales dorsales and presumably nucleus epibasalis ventrolateralis corresponds to the Reptilian dorsal epibasal nuclear complex, and the Avian central epibasal complex to the nucleus epibasalis centralis of Reptiles. The nucleus sphaericus in these latter, respectively the caudal epibasal complex, is doubtless at least kathomologous to the Avian nucleus epibasalis caudalis.

---

155 Caudomedial components of area ventralis anterior are included in the griseum laterale of the so-called septum, respectively in the '*nucleus of Broca's diagonal band*'.

156 That is to say, with rather uniform distribution of the cellular elements.

Despite a greater degree of fusion in Birds, to some extent also noticeable in Reptiles, the homology of basilateral grisea in both Sauropsidan classes seems evident. Still more obvious are the homologies of the basimedial grisea.

In comparison with Mammals, it should be pointed out that the corpus striatum of these latter is a rather homogeneous griseal complex, derived from a fusion of D_1 and B_{1+2} components, being thus kathomologous to the greater rostral extent of the Avian epibasal and basilateral complex.[157] In this respect, the separation into nucleus caudatus and putamen as displayed by many, but not all Mammals, is merely an incidental feature. An undefined lateral neighborhood of the Avian dorsal epibasal griseum can be evaluated as kathomologous to the Mammalian claustrum. Caudal neighborhoods of the Avian epibasal complex and especially the nucleus epibasalis caudalis are, *in toto*, kathomologous to the Mammalian amygdaloid complex, except for the nucleus amygdalae corticalis, whose orthohomology with Avian nucleus taeniae was mentioned above. From a morphological viewpoint, the Avian '*amygdala*' would include, in addition to nucleus taeniae and the entire nucleus epibasalis caudalis, also undefined caudal neighborhoods of central epibasal and dorsal epibasal as well as of basilateral grisea, in contradistinction to the interpretation of COHEN (1975) mentioned further above. Figures 216 to 219 illustrate the configuration of telencephalic grisea in some representative Avian forms. The tabulation of Figure 217F indicates the differences in the nomenclature adopted by diverse groups of investigators.

The numerous taxonomically related variations, particularly difficult to interpret if their ontogenetic derivation from the fundamental telencephalic longitudinal zones are not properly taken into account, moreover the rather nondescript cytoarchitectural subdivisions with

---

[157] It should again be emphasized that this morphologic (topologic, configurational homology does not imply identical respectively similar function nor histologic structure. It seems, in fact, not unlikely that the Avian epibasal complex performs, by means of a quite different 'circuitry', at least some of the functions performed by Mammalian neocortex. The term 'neostriatum' for the Avian nucleus epibasalis centralis (and for the anterior epibasal complex of Reptiles) does not seem appropriate from either a morphologic or a phylogenetic viewpoint. There is, strictly speaking, no corpus striatum in any Vertebrate class except Mammals, where this griseum, derived from a fusion of various, phylogenetically 'old' components (D_1, B_{1+2}), becomes a 'striatum' by its relationship to bundles ('stripes') at the capsula interna. This latter, in turn, is a particular expansion, displayed only by Mammals, of the phylogenetically 'old' lateral forebrain bundle (cf. also pp. 663–668, vol. 3/II).

often very indistinct 'boundaries' (open neighborhood transitions), and the diversity of nomenclature used by the investigators may easily engender confusion. It is occasionally uncertain whether some authors using identical terms actually refer to the same griseum. Thus, in the tabulation of JONES and LEVI-MONTALCINI (1958) reproduced as Figure 304 on p. 568 of volume 3/II, the 'parolfactory lobe' of EDINGER *et al.* is interpreted as corresponding to ROSE's field R or 'nucleus basalis' of HUBER and CROSBY, that is to say to my nucleus epibasalis ventrolateralis. However, the 'parolfactory lobe' of current terminology, and probably also that of EDINGER *et al.* quite definitely represents a rostral portion of my nucleus basalis, i.e. of the 'mesostriatum' ('palaeostriatum augmentatum').

As regards the *communication channels* of the Avian lobus hemisphaericus (Figs. 220 A–G), rostral, intrinsic, and caudal fiber systems may be distinguished. The *rostral channels* are provided by the generally poorly developed olfactory connections, whose fibers vaguely gather in a slightly more distinct lateral olfactory tract and in a rather nondescript medial one.[158] The *lateral olfactory tract* is related to at least the rostral region of the lateral corticoid plate ('prepiriform cortex') and to rostral basilateral grisea ('anterior olfactory nucleus' of HUBER and CROSBY (1929) and parts of 'nucleus parolfactorius'), moreover to basal cortex ('tuberculum olfactorium'), perhaps also, as components of tractus bulbo-ponto-epistriaticus (EDINGER *et al.*, 1903) with a few fibers to nucleus taeniae and epibasalis caudalis.

The generally still less distinct *medial olfactory tract* is related to paraterminal grisea (B_{3+4}), perhaps also to rostral portions of cortex medialis, moreover to medial neighborhoods of 'tuberculum olfactorium'. It should be added that the nondescript 'nucleus olfactorius anterior' *autorum* includes not only undefined neighborhoods of griseum basilaterale inferius, but also adjacent ones of griseum basimediale inferius. Although interbulbar commissural fibers and other bulbopetal connections can be inferred, nothing certain is known with respect to their course in either medial or lateral olfactory tract, respectively decussation of interbulbar fibers in anterior or habenular commissure. It will also be recalled that a nervus terminalis and a vomeronasal component

[158] The question is here left open to which extent, respectively in which proportion, the olfactory tracts include olfactory fibers of the second order (such as mitral cell neurites), of the third order (from bulbar or 'nucleus olfactorius anterior' elements), or even of higher order from lobus hemisphaericus grisea related to said 'olfactory tracts'.

of the olfactory nerve are presumably lacking in Birds.[159] The numerous taxonomic variations in the relative development of the Avian olfactory system, as e.g. dealt with by Cobb (1960a, b) and others are concisely summarized by Pearson (1972). Again, the taxonomic differences also involve the variable ratio of medullated and non-medullated fibers within the 'olfactory channels'.

As regards the poorly understood *intrinsic channels*, the above-mentioned laminae medullares contain, in addition to fibers pertaining to the caudal channels, intrinsic connections between the grisea of the lateral hemispheric wall. Thus, a fronto-occipital channel ('tractus interstriaticus', 'tractus bulbo-fronto-epistriaticus', tractus cortico-epistriaticus'), is generally rccognized as running within the lamina hyperstriatica, particularly in its lateral portion. The so-called *frontalmark*, in the lamina frontalis suprema (also occasionally called lamina frontalis superior) and its variable accessory laminae, is likewise regarded as an 'association system' related to pallial and epibasal grisea. The '*diagonal band of Broca*' or '*fasciculus diagonalis Brocae*' (a term originally referring to an analogous Human and Mammalian fiber system) is essentially a fiber bundle interrelating caudal portions of area ventralis anterior and apparently also nucleus taeniae with paraterminal grisea ('septum') and perhaps also ventral portions of cortex medialis. The diagonal fasciculus runs, in general, superficially (medially) to the septomesencephalic tract dealt with further below as one of the caudal communication channels. Said septomesencephalic tract, however, contains a substantial intratelencephalic component, interconnecting, through dorsal fibers, medial cortex, parahippocampal cortex, and grisea diffusa of the *sagittalwulst*. Ventral components, remaining within lobus hemisphaericus, interconnect parahippocampal and medial cortex with paraterminal grisea, and, bending lateralward around the lateral ventricle, with basilateral and epibasal grisea.

The *caudal communication channels* of the Avian lobus hemisphaericus can roughly be subdivided into the following main fiber systems: (1) lateral forebrain bundle, (2) medial forebrain bundle, (3) stria terminalis, (4) stria medullaris.[160]

The extensive *lateral forebrain bundle system*, related to the expanded lateral wall of the Avian lobus hemisphaericus, is generally assumed to

---

[159] Cf. section 2, p. 483, 504 of the present chapter.

[160] Cf. also section 8 of chapter XII, in which the caudal telencephalic communication channels are dealt with as rostral diencephalic ones.

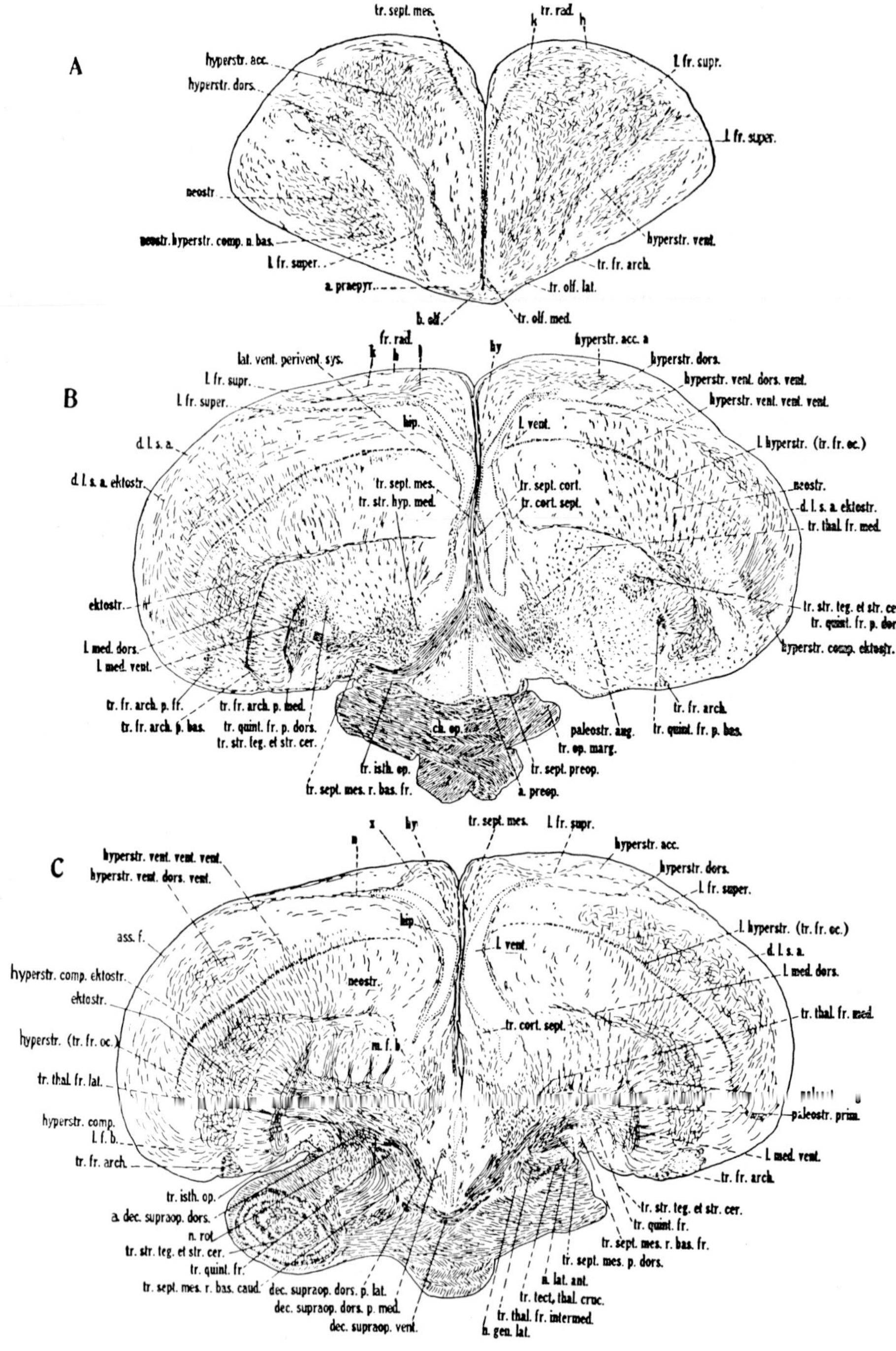
A
tr. sept. mes.
tr. rad.
k
h
hyperstr. acc.
hyperstr. dors.
l. fr. supr.
l. fr. super.
neostr.
neostr. hyperstr. comp. n. bas.
l. fr. super.
a. praepyr.
b. olf.
hyperstr. vent.
tr. fr. arch.
tr. olf. lat.
tr. olf. med.
B
fr. rad.
lat. vent. perivent. sys.
l. fr. supr.
l. fr. super.
d. l. s. a.
d. l. s. a. ektostr.
hy
hyperstr. acc. a
hyperstr. dors.
hyperstr. vent. dors. vent.
hyperstr. vent. vent. vent.
hip.
l. vent.
l. hyperstr. (tr. fr. oc.)
tr. sept. mes.
tr. str. hyp. med.
tr. sept. cort.
tr. cort. sept.
neostr.
d. l. s. a. ektostr.
tr. thal. fr. med.
ektostr.
l. med. dors.
l. med. vent.
tr. str. teg. et str. cer.
tr. quint. fr. p. dors.
hyperstr. comp. ektostr.
tr. fr. arch. p. fr.
tr. fr. arch. p. bas.
tr. fr. arch. p. med.
tr. quint. fr. p. dors.
tr. str. teg. et str. cer.
ch. op.
paleostr. aug.
tr. op. marg.
tr. fr. arch.
tr. quint. fr. p. bas.
tr. isth. op.
tr. sept. mes. r. bas. fr.
tr. sept. preop.
a. preop.
C
x
hy
tr. sept. mes.
l. fr. supr.
hyperstr. acc.
hyperstr. dors.
l. fr. super.
hyperstr. vent. vent. vent.
hyperstr. vent. dors. vent.
ass. f.
hyperstr. comp. ektostr.
ektostr.
hip.
l. vent.
neostr.
l. hyperstr. (tr. fr. oc.)
d. l. s. a.
l. med. dors.
tr. thal. fr. med.
tr. cort. sept.
hyperstr. (tr. fr. oc.)
m. f. b.
tr. thal. fr. lat.
hyperstr. comp.
l. f. b.
tr. fr. arch.
paleostr. prim.
l. med. vent.
tr. fr. arch.
tr. isth. op.
a. dec. supraop. dors.
n. rot.
tr. str. teg. et str. cer.
tr. quint. fr.
tr. sept. mes. r. bas. caud.
dec. supraop. dors. p. lat.
dec. supraop. dors. p. med.
dec. supraop. vent.
tr. str. teg. et str. cer.
tr. quint. fr.
tr. sept. mes. r. bas. fr.
tr. sept. mes. p. dors.
n. lat. ant.
tr. tect. thal. cruc.
tr. thal. fr. intermed.
n. gen. lat.

include, as *descending components*, striothalamic, striohypothalamic, striotegmental (striobulbar), striomesencephalic (striotectal), and striopontine[161] (telencephalopontine) 'tracts'. The vague prefix '*strio*' refers here to undefined epibasal grisea, perhaps even including those of the pallial wulst and also portions of basilateral grisea. The qualification 'bulbar' in striobulbar tract refers, of course, to the brain stem and should not be confused with its connotation referring to olfactory bulb (e.g. in 'tractus bulbo-fronto-epistriaticus). A poorly defined component of the lateral forebrain bundle, which becomes more distinctive at

[161] This tract presumably includes the fibers interpreted as 'striocerebellar tract' by KAPPERS and some other authors, but the existence of direct telencephalocerebellar connections, not mediated by pontine or reticular grisea, seems rather dubious.

Figure 220 A–C. Cross-sections (pyridine-silver impregnation), in rostrocaudal sequence, through telencephalon and adjacent neighborhoods in the Sparrow, Passer domesticus (modified after HUBER and CROSBY, 1929, from KAPPERS *et al.*, 1936). a. praepyr.: anterior (prepiriform) lateral cortex; a.preop.: preoptic 'hypothalamic' region; ass.f.: 'association fibers'; b.olf.: olfactory bulb; ch.op.: optic chiasma; d.l.s.a., d.l.s.a. ektostr.: lateral corticoid plate; dec.supraop.dors.p.lat. (med.), vent.: dorsal respectively ventral supraoptic decussations; ektostr.: 'ectostriatum'; fr.rad.h, k, l: so-called frontal radiations (1: from 'lamina frontalis suprema'); hip: cortex medialis (hippocampus); hy: transition of cortex medialis to cortex dorsalis; hyperstr.acc. (a.): n. diffusus dorsalis; hyperstr.comp.ektostr.: probably part of n. epibasalis dorsalis; hyperstr.comp.l.f.b.: component of lateral forebrain bundle; hyperstr.dors.: probably n. diffusus dorsolateralis; hyperstr.vent. (vent.dors.vent., vent.vent.vent.): subdivisions of n. epibasalis dorsalis; l.fr.supr., super., hyperstr. (tr.fr.oc.): 'lamina frontalis superior' 'suprema', 'hyperstriatica' (sive tr. fronto-occipitalis); l.med.dors., vent.: lamina medullaris dorsalis, ventralis; l.vent. (perivent. sys.) lateral ventricle (and its periventricular fiber system); n: fusion of superior and supreme frontal laminae; n.dec.supraop.dors.: 'n. of dorsal supraoptic commissure'; n.gen.lat.: lateral geniculate complex; n.lat.ant.: 'n. lateralis anterior'; n.rot.: n. rotundus; neostr.: n. epibasalis centralis; neostr.hyperstr.comp.n.bas.: neighborhood of n. epibasalis ventrolateralis; paleostr.aug., prim.: n. basalis respectively n. entopeduncularis (anterior); tr.cort.sept.: 'tr. corticoseptalis'; tr.fr.arch. (p.bas., p.fr., p.med.): 'gr. fronto-archistriaticus et neostriaticus' (with subdivisions); tr.isth.op.: tr. isthmo-opticus; tr.olf.lat.(med.): lateral (and medial) olfactory tract; tr.op.marg.: tr. opticus marginalis; tr.quint.fr.(p.bas., p.dors.): tr. quintofrontalis (with basal and dorsal subdivisions); tr.sept.cort.: tr. septo-corticalis; tr.sept.mes.(p.dors., r.bas.caud., r.bas.fr.): tr. septomesencephalicus (with subdivisions); tr.sept.preop.: tr. septo-praeopticus; tr.str. hyp.med.: tr. striohypothalamicus medialis; tr.str.teg. et str.cer.: tr. striotegmentalis et striocerebellaris; tr.thal.fr.(p.fr., intermed., lat., med.): tr. thalamofrontalis with subdivisions; tr.tect.thal.cruc.: tr. tectothalamicus cruciatus; x: part of parahippocampal cortex. The present author's nomenclature has been substituded where appropriate. As regards the so-called 'hyperstriatum dorsale', cf. the comment on p. 666.

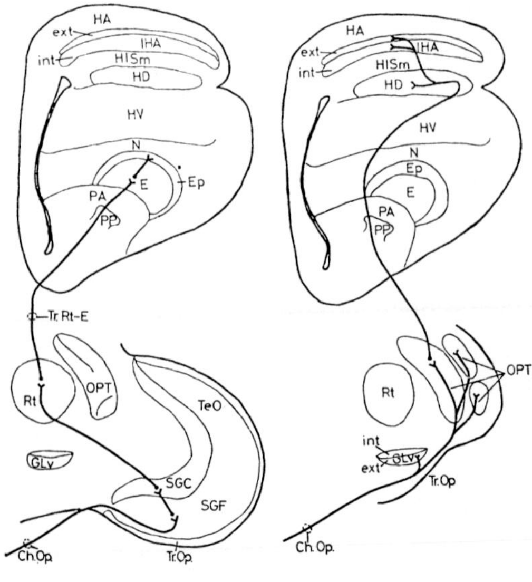

D **Tectofugal** **Thalamofugal** E

Figure 220 D, E. Diagram of assumed 'tectofugal' and 'thalamofugal' optic projections upon Avian telencephalic grisea (from KARTEN *et al.*, 1973). Ch.Op.: optic chiasma; E: n. epibasalis centralis accessorius ('ectostriatum'); Ep: 'zona periectostriatalis'; GLv: ventral lateral geniculate nucleus (with 'internal' and 'external' subdivisions); HA: n. diffusus dorsalis ('hyperstriatum accessorium'); HD: n. epibasalis dorsalis, pars superior ('hyperstriatum dorsale'); HISm: perhaps n. diffusus dorsalis ('hyperstriatum intercalatus suprema' *(sic)*); HV: n. epibasalis dorsalis, pars inferior ('hyperstriatum ventrale'); IHA: perhaps n. diffusus intercalatus sive accessorius ('n. intercalatus hyperstriatum accessorium', with 'lamina internus' and 'lamina externus' *(sic)*); OPT: dorsal lateral geniculate complex ('nn. optici principalis thalami dorsalis' *(sic)*); PA: n. basalis ('palaeostriatum augmentatum'); PP: n. entopeduncularis, pars anterior ('palaeostriatum primitivum'); Rt: n. rotundus thalami; SGC: 'stratum griseum centrale'; SGF: 'stratum griseum et fibrosum superficiale'; TeO: tectum opticum; Tr.Op.: tr. opticus; Tr.Rt.-E: 'tr. rotundo-ectostriatalis'.

diencephalic levels and presumably includes here the '[illegible] tract', is also designated as 'ansa lenticularis'.

Ascending components of the lateral forebrain bundle system include 'tracts' designated as thalamofrontal (lateral, intermediate, medial), and thalamostriate. KARTEN *et al.* (1973) have attempted to trace, essentially by means of the *Fink-Heimer technique*, ascending optic pathways into the telencephalon of the Pigeon and the Owl. Their not improbable interpretation is depicted in Figures 220 D and E. Also, their

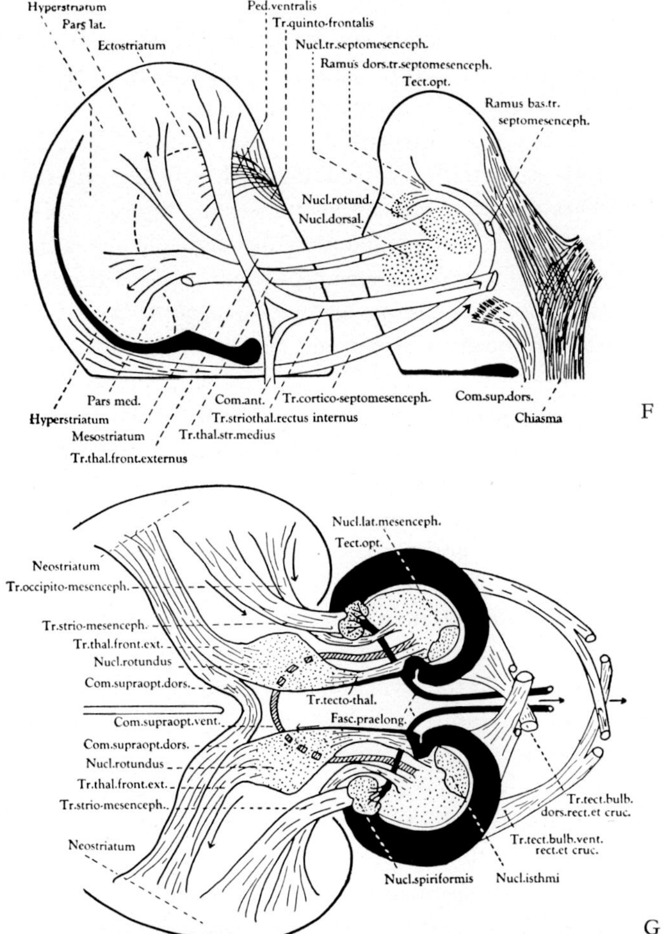

Figure 220 F, G. Two diagrams illustrating KAPPERS' concept of main Avian prosencephalo-mesencephalic communication channels (from KAPPERS *et al.*, 1936).

tractus rotundo-ectostriatalis would be included in the just mentioned thalamostriate tract of previous authors.[162] The so-called '*visual wulst*', however, is also activated by peripheral auditory stimuli (ADAMS and KING, 1967).

[162] The results of the cited authors would essentially agree with some of our own observations with evoked optic potentials (K. and SZEKELY, 1963; cf. also vol. 4, chapter XI, section 8, p. 921). Responses were obtained from dorsolateral telencephalon, including caudal portions of wulst, but involved also the more caudal neighborhoods of the hemisphere. As regards the more rostral region, KARTEN *et al.* (1973) use the term '*visual wulst*'.

A particularly relevant ventral component of the lateral forebrain bundle is the *ascending quintofrontal tract*[163] of WALLENBERG (1903, 1904) recently again investigated by ZEIGLER and KARTEN (1973). It originates essentially in the main sensory trigeminal nucleus[164] and reaches the nucleus epibasalis ventrolateralis ('basal nucleus' of current terminology). It is a partially crossed channel, whose contralateralward directed fibers mainly decussate in the isthmic region, and represents a direct connection, not mediated by diencephalic grisea, between an afferent nucleus of the oblongata and a telencephalic griseum.

The essentially descending *occipitomesencephalic tract* is a channel characteristic for the Avian brain and may be evaluated as a specific development ('axonal grouping') of fiber systems pertaining to lateral forebrain bundle, stria terminalis, and stria medullaris systems of other Vertebrates. It connects caudal epibasal and basal grisea (nucleus epibasalis caudalis, other caudal epibasal cell groups, nucleus taeniae, probably also neighborhoods of nucleus basalis, i.e. basilateral grisea) with pretectal region and perhaps tectum, as well as deuterencephalic tegmentum. At the hemispheric stalk, the bundles of this tract swing medialward, dorsally to lateral forebrain bundle, and then run medially, later dorsally to said channel, toward pretectal region and mesencephalic tegmentum. At the level of lamina terminalis, fibers are given off to the so-called interepistriatic or interarchistriatic component of anterior commissure. The occipitomesencephalic tract originates thus in a topologic neighborhood corresponding to the Reptilian and Mammalian amygdaloid complex, whose Avian grisea, however, seem to be substantially related to the optic system.

The main differentiation of the medial forebrain bundle is the *tractus septomesencephalicus*[165] or '*Bündel der Scheidewand*', dealt with above as regards its intrinsic telencephalic components. As an extrinsic, caudal communication channel, it seems predominantly descending, and is related to pallial grisea, connecting with thalamic grisea and tectum opticum as pointed out in section 8 of chapter XII, in which the loop of that tract within the hemispheric stalk and ventrally to lateral forebrain bundle was described. It contains also a minor fornix component and likewise gives off some fibers to stria medullaris ('tractus corticohabe-

---

[163] Originally interpreted as 'tractus isthmostriaticus' by WALLENBERG, who subsequently identified its origin and renamed it quintofrontal tract.

[164] Cf. chapter IX, section 9, pp. 493–494 of volume 4.

[165] Cf. chapter XII, section 8, p. 288, and volume 4, chapter XI, section 8, p. 932.

nularis'). A nondescript mostly non-medullated component of medial forebrain bundle, not included in the massive septomesencephalic tract, interconnects preterminal grisea and hypothalamus (particularly preoptic region).

The *stria medullaris system*, related to the epithalamus with its habenular griseum, is relatively small in Birds, in accordance with the minor, but taxonomically variable development of the Avian olfactory mechanisms. It includes fibers related to nucleus taeniae and caudal epibasal and epibasal grisea ('archistriatum'), to basimedial grisea ('tractus septohabenularis'), perhaps also to medial olfactory tract ('tractus olfactohabenularis'), and to *scheidewandbündel sensu latiori* ('tractus corticohabenularis').

As regards the *commissural systems*, the Avian *commissura pallii*, located in the lamina terminalis dorsally and commonly somewhat caudally to commissura anterior, is rather small and presumably related not only to cortex medialis, but also to parahippocampal cortex with neighboring more lateral pallial grisea, as well as to basimedial grisea adjacent to the commissure.

The fairly large *anterior commissure* displays a rather diffuse rostral, and a usually more compact and larger caudal branch (cf. e.g. ZEIER and KARTEN, 1973). The rostral portion is related to basilateral and epibasal grisea as well as to 'tuberculum olfactorium'. The caudal branch or 'commissura epistriatica' of KAPPERS *et al.* (1936) is mainly related to the caudal epibasal complex.

The rather small *habenular commissure*, which may merge with the posterior commissure, but is comparatively better developed in the Kiwi as described by HUNTER (1923) and CRAIGIE (1930), is related to the stria medullaris system, and represents, in part, thereby also a telencephalic commissure. Whether interbulbar fiber are contained in this commissure or in commissura anterior remains an open question. The inclusion of undefined telencephalic fiber systems in the dorsal supraoptic commissure was already mentioned in section 8 of chapter XII.

Figures 220F and G illustrate some of the main communication channels of the Avian telencephalon connecting with diencephalic and mesencephalic neighborhoods. Numerous taxonomic variations in the relative development of these channels as well as *qua* ratio of medullated to non-medullated fibers should be kept in mind. Much uncertainty obtains also with respect to ratio of decussating and of true commissural fibers in the above-mentioned commissural systems, respectively qua ratio of crossed and uncrossed fibers in diverse channels.

As regards the *functional significance* of the Avian lobus hemisphaericus, numerous experimental studies have been undertaken by diverse older[166] and recent authors. Generally speaking, Birds whose telencephalon is 'completely' removed remain in a state of 'stupor' with occasional periods of restlessness which are said to arise from 'visceral activities' (ROGERS, 1916). The Birds must be fed since they seem, in most instances or at least for a considerable length of time, unable to take up food and drink for themselves.[167] If stimulated, they are capable of flying, avoiding obstacles, and do not appear to have any gross visual defects, although normal behavioral response to visual stimuli is lost. Damage involving diencephalic, presumably hypothalamic 'centers' abolishes temperature regulation. Removal of most of the pallium, including some dorsal epibasal components, seems to result in little apparent modification of the caged Bird's behavior. With regard to the optic input into telencephalon from lateral geniculate complex and (via tectum mesencephali) from nucleus rotundus, experiments by HODOS *et al.* (1973) seem to indicate that extensive destruction of either griseum diffusum (in the *wulst*) or lateral geniculate complex produced little or no impairment of optic pattern discrimination performance, but that, in those cases in which both lateral geniculate complex and nucleus rotundus were involved, substantial loss of performance resulted.

As regards the Owl, the rostral position of the eyes, and the prey-catching skill of this bird suggest stereoscopic vision, despite the fact that few if any optic tract fibers, not decussating in the chiasma, would effect bilateral connections. PETTIGREW and KONISHI (1976), who re-

---

[166] Early systematic and detailed studies, in which the entire Avian telencephalon was extirpated, are those by FLOURENS (1824) on the functions of the Chick brain. It seems, however, evident that many of the observations reported by this author were either mistaken or highly biased. Thus, *inter alia*, FLOURENS claimed that, following extirpation of the telencephalon ('cerebral lobes'), a hen would completely lose 'sight'. This was not confirmed by the results of subsequent investigators. Discounting the evidently unobservable 'mental' (i.e. conscious) aspect of sight, its behavioristic aspect, i.e. the observable registration of optic input by Birds deprived of telencephalon is now known to be affected without being thereby abolished. Such Birds, e.g. are able to avoid obstacles. An English translation of parts of FLOUREN's treatise is included in a collection of papers on the cerebral cortex selected by v. BONIN (1960). In addition to FLOURENS, another pioneer worker on experimental studies on the nervous system was E. F. A. VULPIAN (1826–1887), who published a treatise on comparative neurophysiology (1866).

[167] Diencephalic lesions alone will also prevent spontaneous feeding, such that the Birds, unless artificially fed, will starve to death in the presence of food (ROGERS, 1919).

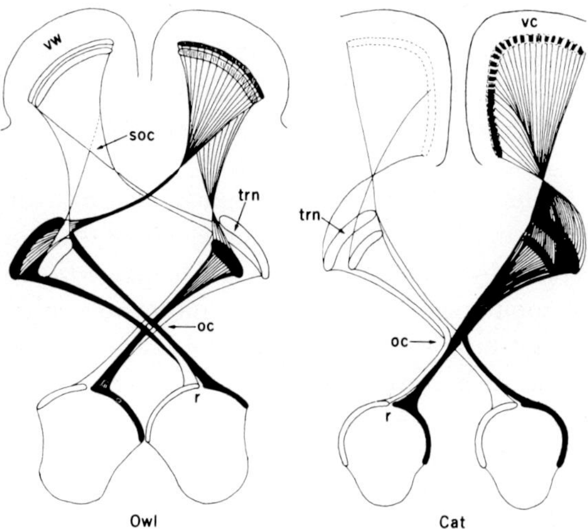

Figure 220 H. Diagram purporting to show channels for bilateral optic input into telencephalic grisea of Owl and Cat (from PETTIGREW and KONISHI, 1976). oc: optic chiasma; r: retina; soc: supraoptic commissure; trn: 'thalamic relay nucleus' (lateral geniculate grisea); vc: visual cortex; vw: '*visual wulst*'.

corded bilateral optic input to the Owl's '*visual wulst*', believe that this may be related to stereoscopic vision and assume that the homolateral optic input, in contradistinction to the contralateral one via optic chiasma and lateral geniculate grisea, may be mediated via the supraoptic decussation (Fig. 220H). This decussation would thus provide the required additional homolateral input. It is, of course, also possible that a similar bilateral projection, provided by re-crossing optic channels, might obtain for tectum opticum. It will be recalled that SZEKELY and myself (K. and SZEKELY, 1963) recorded bilateral evoked potentials both in telencephalic grisea and tectum mesencephali of the Chick upon unilateral optic stimuli (cf. also vol. 4, chapter XI, section 8, pp. 921–923).

Progressive *removal of epibasal components* results in loss of successive phases of the more complex feeding, mating and nesting behavioral activities. The nucleus epibasalis centralis accessorius ('ectostriatum') and particularly the nucleus epibasalis ventrolateralis[168] ('nucleus basalis') seem related to feeding behavior respectively its reflexes. The rhombencephalic principal sensory trigeminal nucleus, the quintofron-

[168] WITKOVSKY *et al.* (1973) have recorded 'single unit responses' from that nucleus, activated by light mechanical stimulation of beak or buccal cavity.

tal tract, and the nucleus epibasalis ventrolateralis are said to be the structures mediating eating but not drinking activities. Bilateral destruction of this system results in a reduction of the Bird's responsiveness to food or in a disruption in the neurosensory control of food intake, or both (ZEIGLER and KARTEN, 1973).

As regards the so-called Avian '*amygdala*', recent experiments by COHEN (1975) seem to indicate that the integrity of that posteromedial region as defined by the cited author, and its projection to the hypothalamus via components of tractus occipitomesencephalicus are essential for 'establishing visually conditioned heart rate changes' in defensive conditioned performances.

Integrity of *basilateral complex* (nucleus basalis, 'mesostriatum', 'palaeostriatum augmentatum') seems to be necessary for conditioned reflex formation. KALISCHER (1905) and ROGERS (1922a) reported that extirpation of the rostral (frontal) region of the telencephalon had no apparent effect on the Bird's behavior.

As regards *biochemical data*, there are numerous studies on the distribution of serotonin, dopamine, and noradrenaline in various regions of the Avian brain including telencephalon, and not all of these are in obvious agreement with one another. A review of this topic is given by PEARSON (1972). With respect to cholinesterase, it is of interest that WÄCHTLER (1973) recorded a high concentration in nucleus basalis ('palaeostriatum augmentatum'), most of nucleus entopeduncularis, and parts of basimedial grisea of the Sparrow. A much lesser concentration obtained in epibasal and pallial grisea. Essentially similar observations were made by KARTEN and DUBBELDAM (1973). In Mammals, the entire corpus striatum together with globus pallidus displays high cholinesterase concentration. Thus, KARTEN and DUBBELDAM conclude that the Avian 'palaeostriatum augmentatum' resembles the Mammalian caudate-putamen, and the Avian 'palaeostriatum primitivum' complex the Mammalian globus pallidus (cf. above footnote 153a).

This, in the biochemical aspect under consideration, is certainly true. Yet, from the morphological viewpoint, the Mammalian caudate-putamen (corpus striatum) doubtless results from a fusion of epibasal and basilateral (D_1 and B_{1+2}) components. It is thus, irrespective of the obtaining histochemical differences, homologous to 'palaeostriatum augmentatum', together with most of so-called 'neostriatum', and most of 'hyperstriatum' of Birds.[169] The obtaining histochemical as

[169] For further details cf. section 6, chapter VI of volume 3/II.

well as structural differences, on the other hand, would substantiate the view, expressed above (footnote 157), that the Avian epibasal complex is presumably concerned with various functions corresponding, in part, to those of Mammalian neocortex. In other words, the Mammalian corpus striatum is homologous, but not analogous to a topologic neighborhood formed by the fusion of B_{1+2} with D_1 derivatives.[170]

Stimulation experiments by means of electrodes, as undertaken by ROGERS (1922a) and others, elicited various effector reactions of generalized type, mostly elicited from subpallial (epibasal) grisea, such as beak opening and closing, bilateral 'winking', nondescript movements of head and neck, ruffling or depression of feathers (this latter from the frontal region) or squeaks, defaecation and urination. Stimulation of medial cortex and of caudal parahippocampal cortex resulted in vegetative responses, such as (mostly contralateral) pupillary constriction. According to MACDONALD and COHEN (1973), short latency tachycardia, hypertension and hyperpnea were elicited 'from the archistriatum, occipitomesencephalic tract, and hypothalamus'. Blood pressure decrease, followed by long latency tachycardia were elicited from the 'septal complex' although occasionally slight bradycardia resulted.

It can be regarded as rather certain that, within the hemispheres, there are no distinctive cortical (or subcortical) areas controlling skeletal muscles or muscle groups, comparable to the excitable neocortical areas of Mammals. This agrees with the reasonably well substantiated general conclusion that a channel analogous to the Mammalian pyramidal tract ist not present in Birds nor, for that matter, in Reptiles and in Anamnia.

VON HOLST and SAINT PAUL (1960, 1962, 1963) extended their experiments with probing electrodes by using electrodes remaining *in loco*, and either connected with fine wires to the stimulating apparatus, or provided with a small radio receiver and thus activated by remote control. Various, and very variable complex behavior sequences[171] could be elicited by stimulation, mostly from diencephalon and rostral brain stem. Yet, nothing of significance *qua* 'localization' of activities resulted from these experiments,[172] since the obtained data disclosed the following unpredictable or uncertain stimulation effects: (1) a stimulation

---

[170] In other words, there obtains heterotypic and in part, heteropractic orthohomology, since identical *grundbestandteile* are involved (cf. pp. 64–65, section 1A, chapter VI of volume 3/II).

[171] Cf. above p. 588: *das Huhn gackert!*

[172] Cf. HINDE (1966) and PEARSON (1972).

field may give or not give a response at different times; (2) changing behavior patterns may be evoked over a period of hours as the result of activating a particular stimulation field; (3) the same pattern of action can often be released by stimulating different fields.

Von Holst and St. Paul (1962) conclude that 'the organism comprises a bundle of drives, which support one another to greater or lesser extent. "Spontaneous" activity is the result of a continual and shifting interplay of forces in the central nervous system'.[173]

A concise discussion of pertinent functional aspects of Avian telencephalon as recorded by Rogers and preceding investigators, is included in Herrick's book 'Brains of Rats and Men' (1926). Detailed subsequent data on this topic, with numerous bibliographic references, can be found in the treatise by Pearson (1972). Another recent publication, edited by Goodman and Schein (1974), summarizes contemporary attempts, by various experimental methods such as electrical brain stimulation (ECS), implanted electrodes, electrical recordings, ablations, etc., to elucidate diverse so-called 'brain behavior relationships'. Despite the multitudinous but on the whole rather nondescript and ambiguous findings reported in said publication and accompanied by prolix comments, the obtained results can be evaluated as inconclusive and, moreover, based on highly debatable morphologic concepts. From a critical and sceptical viewpoint, one may thus be justified to maintain that, notwithstanding the profuse subsequently recorded data and the added verbiage, no very substantial progress on this topic has been accomplished since the above-mentioned summary by Herrick, compiled about 50 years ago.

10. Mammals (Including Man)

Despite substantial dissimilarities, the Mammalian telencephalon, if compared with that of submammalian Amniota, displays, as regards its overall configurational arrangement, a fairly close resemblance to the telencephalon of various Lacertilian Reptiles.[174] The most conspicuous

[173] The critical and sceptical observer, familiar with the 'uncertain nervous system' (as aptly so characterized by Burns, 1968), will hardly consider these conclusions as anything new, but merely as '*une vérité de La Palice*'.

[174] Thus, referring to the brain in general, Kappers (1947) justly states: '*La structure du système nerveux central des Mammifères dérive plutôt du type des Lézards que de celui des Oiseaux*'. This, of course also particularly applies to the telencephalon. Nevertheless, as regards

morphologic differences between the Mammalian and the Sauropsidan telencephalon concern the following features.

(1) The predominating Mammalian development of the secondary pallium (D_2 and D_3) in general, and in particular the development of a neopallium containing the neocortex. Nevertheless, the hippocampal cortex (cortex medialis of Reptiles) also manifests a considerable development, combined with a differentiation of specifically Mammalian type. In addition, both the (anterior) piriform lobe cortex (cortex lateralis of Sauropsidans) and the parahippocampal cortex display, in Mammals, a high degree of further progressive evolution. Parahippocampal cortex and neocortex correspond to subdivisions of the Reptilian cortex dorsalis.

(2) The subcortical grisea of the lateral, secondary telencephalic basis, provided by the internation of D_1 and its secondary fusion with B_{1+2} are (a) *relatively* much less voluminous than in Birds or even in Reptiles, and (b) display a *different type of organization*. This latter is characterized by the amalgamation of D_1 and B_{1+2} components into a rather homogeneous *corpus striatum*, and by the formation of two additional grisea, namly the much smaller *claustrum* (a lateral D_1 derivative), and the rather extensive caudal *amygdaloid complex*.[174a]

In contradistinction to these conspicuous differences, the basimedial Mammalian grisea (B_3 and B_4) display, *qua* configurational features, no significant changes from the arrangement obtaining in Sauropsidans.

Older as well as contemporary investigations dealing with the Mammalian, including the Human telencephalon, are exceedingly numerous and concern a multitudinous diversity of aspects. The extensive litera-

---

phylogenetic speculations, and despite numerous hypotheses propounded by diverse authors, the relationships of Reptiles to Mammals remain obscure, both with respect to the available paleontologic data, and to the inferences based on the brain morphology of recent forms. It seems still a moot question whether Reptiles and Mammals derived independently, i.e. separately from Amphibian-like ancestors, or whether Mammals derived from 'true' Reptiles. Mammals can hardly be conceived as derived from Birds, yet the Mammalian cerebellum displays far more conspicuous homologies with that of Birds than with that of Reptiles (cf. chapter X, vol. 4). The fallacy of the still widely accepted phylogenetic homology concept based on a *petitio principii* (i.e. on circular reasoning) is thereby, as in other instances, rather well evidenced.

[174a] In the Mammalian telencephalic *bauplan* formula given on p. 593 of volume 3/II, the notation for corpus striatum should correctly read $D_1+B_{(1+2)}$ m, l instead of $D_2+B_{(1+2)}$ d, l; moreover, the notation (D_1+B_{1+2}) k for the amygdaloid complex should have been included following the notation for corpus striatum.

ture on these topics defies a relevant enumeration in this context, and reference must be made to the bibliographies included in the texts by Kappers *et al.* (1936), Kappers (1947), Beccari (1943), Clara (1959), Crosby *et al.* (1962), and Krieg (1966). Some additional pertinent references useful for further orientation will be given further below in context with specific problems, as well as in the subsequent chapters XIV and XV.

The topics concerning the Mammalian telencephalon will here be taken up, in accordance with the concepts presented in the preceding volumes, under three headings related to different aspects, namely (a) overall configuration and function in the present chapter XIII, (b) surface morphology in chapter XIV (vol. 5/II), and (c) relevant details of cerebral cortex in chapter XV (vol. 5/II). Quite evidently, a certain amount of overlap and repetition, which shall be reduced to a minimum, becomes unavoidable. It is obviously impossible to deal with overal configuration without some references to surface morphology and cerebral cortex. The chapter on surface morphology, moreover, shall also deal with apposite aspects of *domestication* and '*anthropology*'.

As regards overall interpretation, Meynert's (1872) chapter on the brain of Mammals in Stricker's *Handbuch* provided an important step, followed by the further elucidations contained in Cajal's (1911) and Edinger's (1911) treatises. Contributions including relevant general morphologic data on the telencephalon of some particular Mammalian forms are, among many others, the following. For Prototheria and Metatheria, namely, Monotremes and Marsupials, the publications by Elliot Smith (1896) Ziehen (1897, 1908), Livini (1907), Obenchain (1925) Hines (1929), and Loo (1930, 1931). With respect to Eutheria, Ganser (1882) investigated the Insectivore Talpa, Humphrey (1936) and Schneider (1957) dealt with Chiroptera, and Craigie (1925) as well as Zeman and Innes (1963) with the Rodent Rat. Assuming that the tendency of neurons to be stained by luxol-fast-blue represents an index of their 'functional capacity', the telencephalic grisea of the white Rat were recently surveyed by Garcia Santos (1975) with respect to the chronologic sequence in the appearance of 'luxol-positive' elements.

Again, Elliot Smith (1899, 1903) investigated Edentates and the Prosimian Primate Lemuroids. Data on Telencephalon of Ungulates are included in the texts on veterinary anatomy by Ellenberger and Baum (1926) and by Sisson and Grossman (1959).

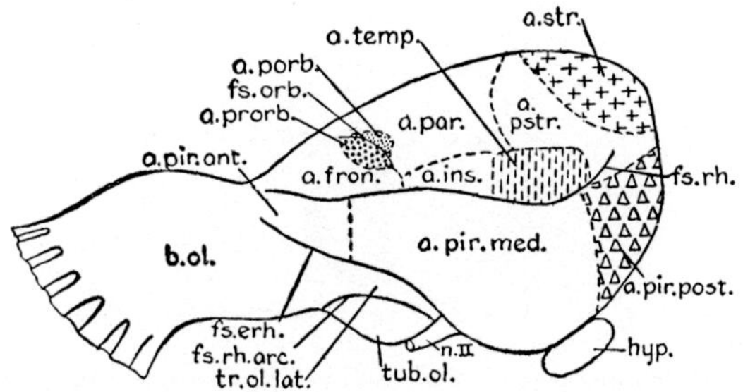

Figure 221 A. Lateral aspect of the Marsupial Opossum's telencephalon, showing cortical regions in GRAY's interpretation (after GRAY, 1924, from HERRICK, 1926). a.ins.: area insularis; a.fron.: area frontalis; a.par.: area parietalis; a.pir.ant., med.: area piriformis anterior and media ('prepiriform cortex'); a.pir.post.: area piriformis posterior (parahippocampal cortex); a.porb.: area postorbitalis; a.prorb.: area praeorbitalis; a.pstr.: area peristriata; a.str.: area striata; a.temp.: area temporalis; b.ol.: bulbus olfactorius; fs.erh.: sulcus endorhinalis; fs.orb.: sulcus orbitalis; fs.rh.: sulcus rhinalis lateralis; fs.rh. arc.: sulcus rhinalis arcuatus (accessory groove delimiting prominence of tuberculum olfactorium); hyp: hypophysis; n.II.: optic nerve; tr.ol.lat.: tractus olfactorius lateralis; tub.ol.: tuberculum olfactorium.

Non-cortical telencephalic grisea of the Lagomorph Rabbit were investigated by YOUNG (1936) and basic data on the Rabbit's as well as on the Carnivore Cat's telencephalon can be found in the atlases by WINKLER and POTTER (1911, 1914).

As regards Proboscidia, DEXLER (1907) dealt with the Indian Elephant, and JANSSEN and STEPHAN (1956) with the African Elephant. RIESE (1925), LANGWORTHY (1932), PILLERI (1962, 1964, 1966), JANSEN and JANSEN (1968) and a few others investigated Cetacea.

Among diverse fairly recent stereotaxic atlases of Mammalian brains depicting and describing telencephalic grisea, those by LIM *et al.* (1960; Carnivore Dog), and by SNIDER and LEE (1962; Primate Macaca) may be mentioned.

As regards Mammals in general, the telencephalic grisea were considered by SPIEGEL (1919), and a recent catalogue and atlas of Mammalian brains, containing data on the telencephalon has been prepared by BRAUER and SCHOBER (1970).

The overall morphologically valid configurational aspects of the Mammalian telencephalon can be dealt with by taking as suitable paradigms the arrangement displayed in some 'lower' forms such as various Marsupials, Insectivores, Rodents, and also in the Lagomorph Rabbit (Figs. 221, 222, 223 A, B).

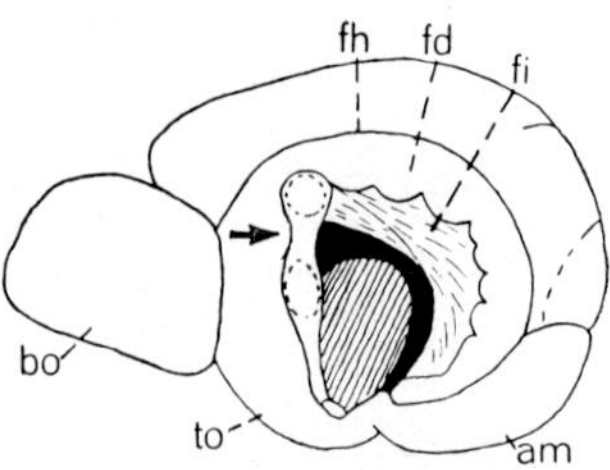

Figure 221 B. Medial aspect of telencephalon in the Marsupial Orolestes inca (slightly modified after OBENCHAIN, 1925, from K., 1927). am: area ventrolateralis posterior (amygdaloid complex); bo: olfactory bulb; fd: fascia dentata (gyrus dentatus); fh: fissura hippocampi; fi: fimbria (fornicis); to: tuberculum olfactorium (area ventralis anterior). Added arrow indicates midline recess of *Mantelspalte* which, if pronounced, results in the so-called 'open cavum septi'; in lamina terminalis above arrow: commissura hippocampi *sive* dorsalis; below arrow: commissura anterior.

Discounting the olfactory bulb, discussed in section 2 of the present chapter, the *lobus hemisphaericus* of the above-mentioned and of all other Mammalian forms can be said to display a (secondary, i.e. D_{2+3}) pallium and a (secondary, i.e. D_1+B_{1-4}) basis. The *entire* free surface of the hemispheric lobe is underlaid by pallial and by basal[175] *cerebral cortex*.

On the lateral surface, the pallium is approximately delimited from the basis by the *sulcus endorhinalis*, which roughly corresponds to the superficially located tractus olfactorius lateralis.[176] The neopallium, in turn, is approximately delimited from the pallium of piriform lobe by the *sulcus rhinalis lateralis*.

On the medial surface, the 'fissura' *sulcus hippocampi* approximately delimits the parahippocampal from the hippocampal pallium. This sulcus is particularly developed in the postcommissural (caudal paired evaginated) portion of lobus hemisphaericus, but may, depending on

[175] With regard to basal cortex, this statement should be qualified as valid if, with substantial morphologic justification, the interstitial neuronal elements within *Broca's diagonal band* and the [illegible] evaluated as medial extensions of the basal cortex. It should also be added that in most Reptiles as well as in some Anamnia (e.g. some Selachians, cf. Fig. 186 C of the present chapter and Fig. 247, p. 483, vol. 3/II, and to some degree in Dipnoans, cf. Fig. 267, p. 513, vol. 3/II and Fig. 203 of the present chapter) the surface of all or most of lobus hemisphaericus is likewise underlaid by cerebral cortex.

[176] If the tractus olfactorius lateralis is well developed and relatively wide, an additional, accessory sulcus may be present between its basal edge and the surface of tuberculum olfactorium ('fissura rhinalis arcuata' of Fig. 221 A).

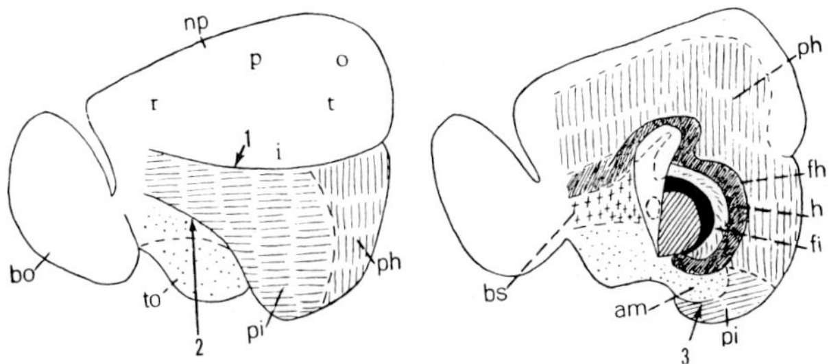

Figure 221 C. Lateral and medial aspects of the telencephalon in the Insectivore Erinaceus europaeus (with some additions, from K., 1927). am: area ventrolateralis posterior (amygdaloid complex); bo: bulbus olfactorius; bs: nucleus basimedialis superior; fh: fissura hippocampi; fi: fimbria (fornicis); h: hippocampal formation (retrocommissural: gyrus dentatus); i: insular region; np: neopallium; o: occipital region; p: parietal region; ph: parahippocampal cortex; pi: cortex anterior (et medius) lobi piriformis ('prepiriform cortex'); r: rostral (or frontal) region; t: temporal region; to: tuberculum olfactorium; 1: sulcus rhinalis lateralis; 2: sulcus endorhinalis; 3: locus of sulcus periamygdalaris if present (caudal extension of 2). Note: in the legend to Figure 316, p. 581 of volume 3/II, bs was, through a lapsus calami, erroneously designated as nucleus basilateralis superior (instead correctly as *basimedialis).*

the particular taxonomic forms, extend rostrad as a less pronounced groove. In the postcommissural region, located within the posterior limb of the hemispheric bend, the hippocampal cortex is delimited from the crus *sive* fimbria fornicis by the sulcus fimbrio-dentatus. In the precommissural region, and again depending on taxonomic variations, a rather shallow groove may delimit hippocampal pallium from paraterminal region *('sulcus septo-pallialis').* The sulci within the neopallium, characterizing the gyrencephalic Mammalian brains, shall be dealt with in chapter XIV of volume 5/II. The grooves enumerated above, on the other hand, are present, with various degrees of development, in both gyrencephalic and lissencephalic Mammals. Moreover, many submammalian forms display more or less noticeable grooves comparable to sulcus endorhinalis and 'septo-pallialis'.

Turning now to the pallial grisea, which are exclusively cortical,[177]

[177] Discounting the very variable matrix remnants which we have designated as *subependymal cell plate* (cf. vol. 3/I, chapter V, section 1). Cortices originating by matrix exhaustion are *hologenic* (or *totoparietine),* while those originating from only part of the matrix are *merogenic* (or *semiparietine).* A transitional type is *hekaterogenic* (or *bigenetic),* as pointed out in section 6, chapter VI of volume 3/II. Further comments on this topic will be found in chapter XV of volume 5/II.

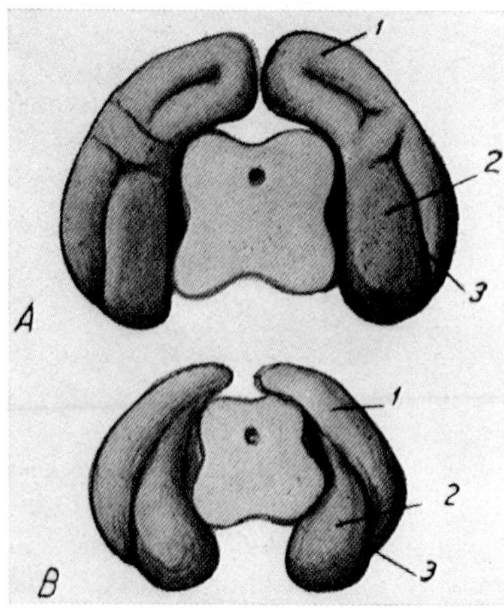

Figure 221 D. Caudal pole of lobus hemisphaericus in (A) the cat and (B) in the Guinea Pig (slightly modified after CAJAL, 1911, from BECCARI, 1943). 1: neopallium (occipital region); 2: parahippocampal region of posterior piriform lobe (so-called *noyau angulaire* or *spheno-occipitales Ganglion*, and *écorce temporal olfactive* or *sphenoidales Riechzentrum of Cajal);* 3: caudal extension of sulcus rhinalis lateralis.

four main subdivisions can be distinguished[178] in accordance with the topologic viewpoints elaborated in chapter VI, section 6, pp. 471–472 and 571–668 of volume 3/II. In addition to the *pallial cortical regions*, which are essentially hologenic, except for some variable or minor merogenic contribution to the hekaterogenic piriform lobe cortex and to adjacent parts of the insular one, there is an entirely merogenic *basal cortex*.

Thus, *in toto*, the following five main cortical regions obtain in Mammals: (1) *hippocampal formation*, (2) *parahippocampal cortex*, (3) *neocortex*, (4) *anterior piriform lobe cortex*, and (5) *basal cortex*. The general morphology of the four pallial cortices will here be discussed separately from the basal cortex, which shall be considered in dealing with the basal grisea. Details of cortical architectonics and structure, however, are given in chapter XV of volume 5/II. It will here be recalled that the cortices (1), (2), (4), and (5) are also *in toto* designated as *allocortex*, in contradistinction to the *isocortex* or *neocortex* (3). The parahippocampal cortex (2), however, at least in some of its regions, can be conceived to

[178] Subdivisions as conceived by other authors, and additional comments on cytoarchitecture and parcellation shall be dealt with in chapter XV of volume 5/II.

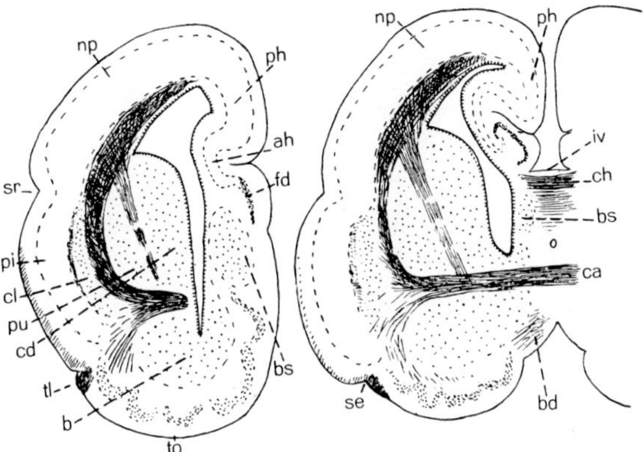

Figure 222 A. Sketches of two cross-sections through the telencephalon of the Metatherian Didelphys marsupialis, illustrating the main configurational features (from K., 1927). ah: cornu Ammonis; b: fundus striati ('n. basalis' respectively conjoint parts of nn. basimed. et basilat. inferiores); bs: nucleus basimedialis superior; ca: commissura anterior; cd: nucleus caudatus; ch: commissura dorsalis (comm. hippocampi, 'comm. pallii'); cl: claustrum; fd: fascia dentata (gyrus dentatus); iv: indusium verum; np: neopallium; ph: parahippocampal pallium; pi: cortex lobi piriformis (anterioris et medii, 'prepiriform cortex'); pu: putamen; se: sulcus endorhinalis; sr: sulcus rhinalis lateralis; tl. tractus olfactorius lateralis; to: tuberculum olfactorium; bd: *diagonal band of Broca.*

assume an intermediate status between allocortex and typical isocortex. A detailed survey of the allocortex based on numerous original studies, has been undertaken by STEPHAN (1975).

(1) The Mammalian *hippocampal formation* displays three regions, namely (a) *precommissural*, (b) *supracommissural*, and (c) *postcommissural hippocampus.*

The *precommissural hippocampus* (Figs. 221 B, C, 222 A, B, 229 B) consists of a dorsal cortical plate of larger cellular elements ventrally continuous with a small celled plate of diminishing width, whose tapering edge adjoins the paraterminal nucleus basimedialis superior. This arrangement is quite similar to that obtaining for the cortex medialis (hippocampal cortex) of Sauropsidans. A slight bend, corresponding to the rostral end of fissura hippocampi, may indicate the transition between dorsal portion (cornu Ammonis) and ventral one (fascia dentata *sive* Gyrus dentatus).

The *supracommissural hippocampus* (Fig. 223 C), as commonly well developed in Aplacentalia (Prototheria and Metatheria), is characterized by a deeper sulcus hippocampi and a separation of the lateralward con-

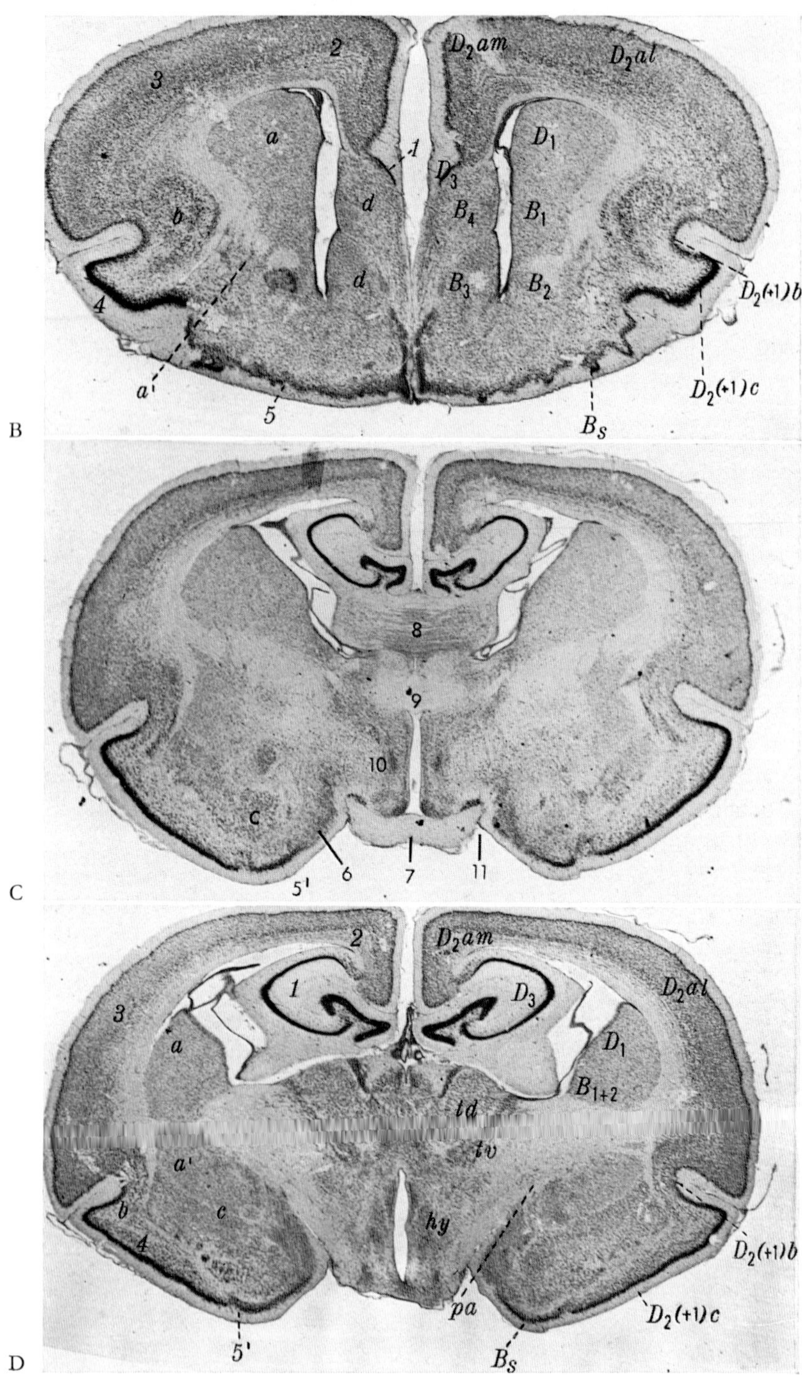
2
3
a
b
4
a'
5
d
d
1
D_2am
D_2al
D_1
D_3
B_4
B_1
B_3
B_2
$D_2(+1)b$
$D_2(+1)c$
B_s
B
8
9
10
c
5'
6
7
11
C
2
D_2am
3
1
a
D_3
D_2al
D_1
B_{1+2}
td
tv
a'
b
c
4
hy
pa
$D_2(+1)b$
$D_2(+1)c$
5'
B_s
D

vex *cornu Ammonis* from the dorsomedialward concave *gyrus dentatus*, whose cortical plate forms a hilus, into which the basal edge of cornu Ammonis protrudes. To a various, but generally much lesser degree, this arrangement is found in some 'lower' Eutheria such, e.g. as Chiroptera and Edentata. In the vast majority of Eutherian Mammals, however, and concomitantly with the 'progressive' development of corpus callosum, the supracommissural hippocampus becomes greatly reduced to an abortive continuation of the precommissural one. It is characterized by a narrow strip whose ventromedial edge extends to variable degree toward the surface of corpus callosum,[179] and includes both rudimentary cornu Ammonis (dorsolaterally) and gyrus dentatus (ventromedially). ELLIOT SMITH (1897) pointed out that, like all vestigial structures, this reduced supracommissural hippocampus varies greatly, not only in diverse forms, even within the same order, but also in individuals of the same species.

The *postcommissural hippocampal formation*, which follows the caudal limb of the hemispheric bend, and thus displays a caudal convexity, is generally the largest and most highly developed portion of the hippocampal cortex. The *cornu Ammonis* is a magnocellular layer of pyramidal cells extending, as a cortical plate with caudomedial concavity, from the medial border of parahippocampal cortex, into the hilus of the more small-celled gyrus dentatus. Depending on the level of the hemispheric bend, this hilus has a lateral, rostral, and even dorsal con-

[179] The variable and thin film of hippocampal gray matter on the surface of the corpus callosum is part of the indusium griseum, to be discussed further below.

Figure 222B–D. Cross-sections through the telencephalon of the Metatherian Marsupial Dromiciops australis *(Nissl stain,* ×17, red. $^3/_5$). 1: hippocampal formation; 2: parahippocampal cortex; 3: neocortex; 4: cortex lobi piriformis (anterioris in B, medii in C); 5: basal cortex (area ventralis anterior, tuberculum olfactorium); 5′: area ventrolateralis posterior of basal cortex (cortical amygdaloid nucleus); 6: *diagonal band of Broca* (extension of area ventralateralis posterior, 5′); 7: optic chiasma; 8: commissura dorsalis sive hippocampi aut pallii; 9: commissura anterior (which also contains pallial fibers); 10: preoptic region of hypothalamus with paraventricular, suprachiasmatic and supraoptic grisea; 11: telodiencephalic groove; a: corpus striatum (n. caudatus); a′: corpus striatum (putamen); b: claustrum; c: amygdaloid complex; d: paraterminal complex; hy: hypothalamus; pa: globus pallidus; td: thalamus dorsalis (above it the stria medullaris and above this latter, rostral portions of habenular grisea); tv: thalamus ventralis. D and B symbols on right side represent the topologic notations. B just rostral to lamina terminalis, C at level of commissura pallii and anterior, hemispheric stalk, and preoptic recess, D at more caudal level of hemispheric stalk.

cavity. The rostrobasal end of hippocampal formation becomes adjacent to the amygdaloid complex (Figs. 221 B, C). The cornu Ammonis has been subdivided into several fields to be dealt with in chapter XV of volume 5/II.

The *fascia dentata* (or *gyrus dentatus*) is delimited from the extraventricular fiber components of the fornix system by the sulcus *fimbrio-dentatus*, the edge of the crus fornicis being the *fimbria fornicis*, to which the choroid plexus of the lateral ventricle is attached. The subcortical white matter of the hippocampal formation, especially of cornu Ammonis, is designated as the *alveus*. It is prominent in the postcommissural hippocampus and also in the fully developed supracommissural hippocampus of Aplacentalia. It should also be added that in Cetaceans, the hippocampal formation, although displaying the typical Eutherian features, is generally rather small, as pointed out by BREATHNACH and GOLDBY (1954), JANSEN and JANSEN (1968), and others.

(2) The *parahippocampal cortex*, located between hippocampal formation and neocortex, corresponds to the medial and intermediate portion of Reptilian cortex dorsalis and to the Avian parahippocampal cortex. In contradistinction to the hippocampal formation, its cell layer is not only much wider, but displays a definite stratification, somewhat similar, although with various differences, from that of neocortex.[180] Again, inner (medial) and outer (lateral) subregions, as well as rostral and caudal portions can be distinguished. The outer caudal portion of parahippocampal cortex extends into the piriform lobe (cf. Figs. 221 A, C, D) and can also be designated as 'area' or '*cortex piriformis posterior*'.

The region of parahippocampal cortex adjacent to the hippocampus within the temporal lobe was named *subiculum cornu Ammonis* by most of the older authors (cf. OBERSTEINER, 1912; VILLIGER, 1920), being synonymous with the gyrus hippocampi (cf. chapter XIV). CAJAL (1911), however, used the term 'subiculum' (or *portion ammonique*) for a relatively wide portion of cornu Ammonis, i.e., of true hippocampal formation adjacent, respectively transitional, to the parahippocampal cortex. The neighboring subiculum of the older authors was then called '*presubiculum*'. ROSE (1929 *et passim*) and other recent authors have adopted CAJAL's nomenclature. It should thus be emphasized that the 'prosubiculum' and 'subiculum' of current terminology (as e.g. depict-

[180] Thus, EDINGER (1912) included the entire Mammalian parahippocampal cortex into his neopallium, and more recent authors have likewise evaluated the parahippocampal cortex, at least in part, as neo- or isocortex (cf. the comments in chapter XV, vol. 5/II).

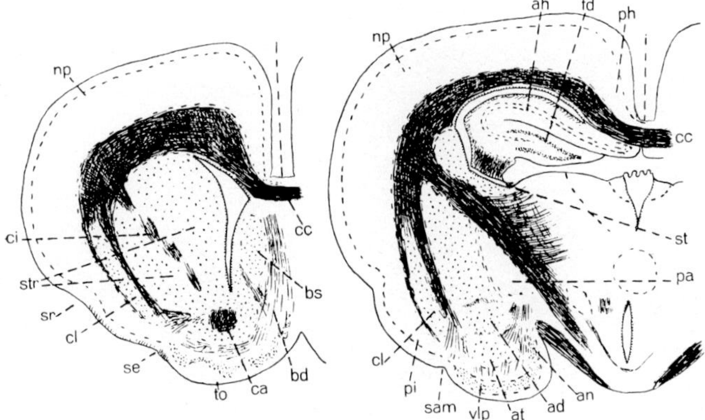

Figure 223 A. Sketch of two cross-sections through the telencephalon of the Eutherian Lagomorph Rabbit (from K., 1927). ah: cornu Ammonis; an: medial accessory griseum of amygdaloid complex; at: principal amygdaloid griseum; bd: *Broca's diagonal band;* bs: nucleus basimedialis superior; ca: commissura anterior (mainly 'pars olfactoria' with so-called tractus olfactorius intermedius); cc: corpus callosum; ci: capsula interna; cl: claustrum (between capsula externa and extrema); fd: fascia dentata; i: indusium griseum; np: neopallium; pa: globus pallidus; ph: parahippocampal cortex; pi: cortex lobi piriformis; sam: 'sulcus amygdalae' (caudal part of se); se: sulcus endorhinalis; sr: sulcus rhinalis lateralis; st: sulcus terminalis; str: corpus striatum (n. caudatus and putamen); to: tuberculum olfactorium (area ventralis of basal cortex); vlp: cortical amygdaloid nucleus (area ventrolateralis posterior of basal cortex). The parahippocampal cortex is not indicated, although present, in the more rostral section. As regards the supracommissural hippocampus, respectively *Cajal's écorce interhémisphérique,* cf. Figure 223C.

ed in Fig. 295, p. 421 by Crosby *et al.*, 1962) is, in the present treatise, considered to be an intrinsic part of the cornu Ammonis, that is, of the hippocampal formation. As regards the so-called presubiculum, I am not certain whether the various authors consistently use this term for adjacent neighborhoods of parahippocampal cortex and cornu Ammonis clearly pertaining to either the latter or to the former.

(3) The essentially *hexalaminar neocortex* (or *isocortex)* corresponding to a lateral neighborhood of the Reptilian cortex dorsalis, extends from the parahippocampal cortex medially to the piriform lobe cortex laterally and into both rostral (frontal) and caudal (occipital) pole of lobus hemisphaericus. In addition, it covers the temporal pole of those Mammals whose pronounced caudal hemispheric bend results in an expansion of that configuration.[181] Discounting details of cytoarchitec-

[181] In 'lower' Mammals (cf. Figs. 221 A, C) the temporal pole contains piriform lobe and amygdaloid (basal) cortex.

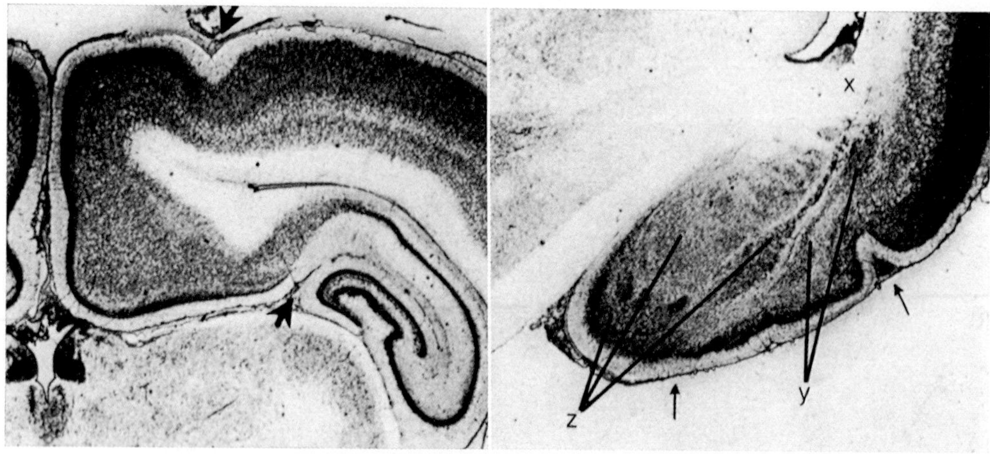

Figure 223 B. Gradients between hippocampal cortex, parahippocampal cortex, cortex of piriform lobe, and basal cortex of amygdaloid grisea in the Rabbit as seen in cross-section *(Nissl stain)* at low magnification (from K., 1957). x: internal capsule (on top of x: tail of caudate nucleus); y: claustrum; z: amygdaloid complex. Arrows: left upper: sulcus lateralis, approximately indicating boundary between parahippocampal cortex and neocortex; left lower: sulcus *sive* fissura hippocampi; right upper: sulcus rhinalis lateralis, approximate boundary between insular neocortex and piriform lobe cortex; right lower: transition of piriform lobe cortex to basal cortex of amygdaloid complex (the corresponding variable caudal extension of s. endorhinalis is here not present).

tural parcellation, five main regions of neocortex, already displayed in 'lower' mammals, may be roughly distinguished, namely, *regio rostralis* and *parietalis*, as well as *occipital*, *temporal* and *insular regions* (Figs. 221 A, C).

(4) The *anterior piriform lobe cortex*, rather indistinctly laminated, and ventrolaterally adjacent to neocortex, extends from the olfactory tract into the piriform lobe and covers much of that configuration, except for a variable portion of said lobe's caudal region, into which the parahippocampal cortex protrudes.[182]

Turning now to the *basis* of the Mammalian lobus hemisphaericus, the following grisea can be distinguished: (1) *corpus striatum complex*,

---

[182] Roughly speaking, this neighborhood of parahippocampal cortex corresponds to BRODMANN's area 28, while the anterior piriform lobe cortex corresponds to area 51, also occasionally called 'prepiriform cortex'. Yet, one could distinguish an anterior and an intermediate area within the anterior piriform cortex (area 51), the area piriformis posterior (area 28) being parahippocampal cortex (cf. Fig. 221 A and further comments in chapter XV, vol. 5/II).

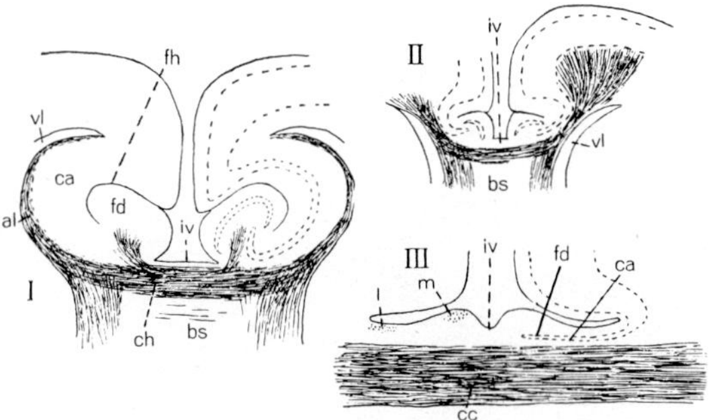

Figure 223C. Diagrams of supracommissural hippocampus in 'lower' and 'higher' Mammals (slightly modified after ELLIOT SMITH, 1897, and K., 1927). I: Aplacentalia (Prototheria, Metatheria); II: some 'lower' Eutheria (e.g. Edentata, Chiroptera); III: some 'lower', 'intermediate' and 'higher' Eutheria, including Primates. al: alveus; bs: nucleus basimedialis superior; ca: cornu Ammonis; cc: corpus callosum; ch: commissura dorsalis *sive* pallii (comm. hippocampi); fd. fascia dentata; fh: fissura hippocampi; iv: indusium verum; l, m: striae longitudinales laterales respectively mediales *(Lancisii)*; vl: lateral ventricle.

(2) *claustrum*, (3) *amygdaloid complex*, (4) *paraterminal grisea*, and (5) *cortex basalis*.

The Mammalian *corpus striatum complex* (1) derives ontogenetically and presumably also phylogenetically from a fusion of D_1 and B_{1+2} components. Its characteristic striped appearance, emphasized by the designation striatum, is due to the numerous medullated fiber bundles, pertaining to capsula interna, which traverse that griseum. In many 'lower' and apparently in all 'higher' Mammals, moreover, the bulk of capsula interna forms a compact channel which incompletely separates a medial, adventricular portion, the *nucleus caudatus*, from a lateral portion, the *putamen* (cf. e.g. Figs. 222A, 223A).

In some Mammalian forms, however, e.g. in Monotremes, in the Rat, and in the Mouse (cf. Figs. 226B, C, 229C), the fiber system of capsula interna runs through the striatum mainly in scattered bundles, such that a well definable separation into caudate nucleus and putamen does not obtain, or is, at most, only faintly indicated in the region of hemispheric stalk. Within the *hemispheric stalk*, the *globus pallidus*, a derivative of the hypothalamic anterior entopeduncular griseal complex, is closely adjacent to the striatum, respectively putamen, separated from this latter merely by a thin medullary lamina.

The structure of the striatum is, on the whole, fairly homogeneous. It contains a dense population of small to medium sized cells, essentially

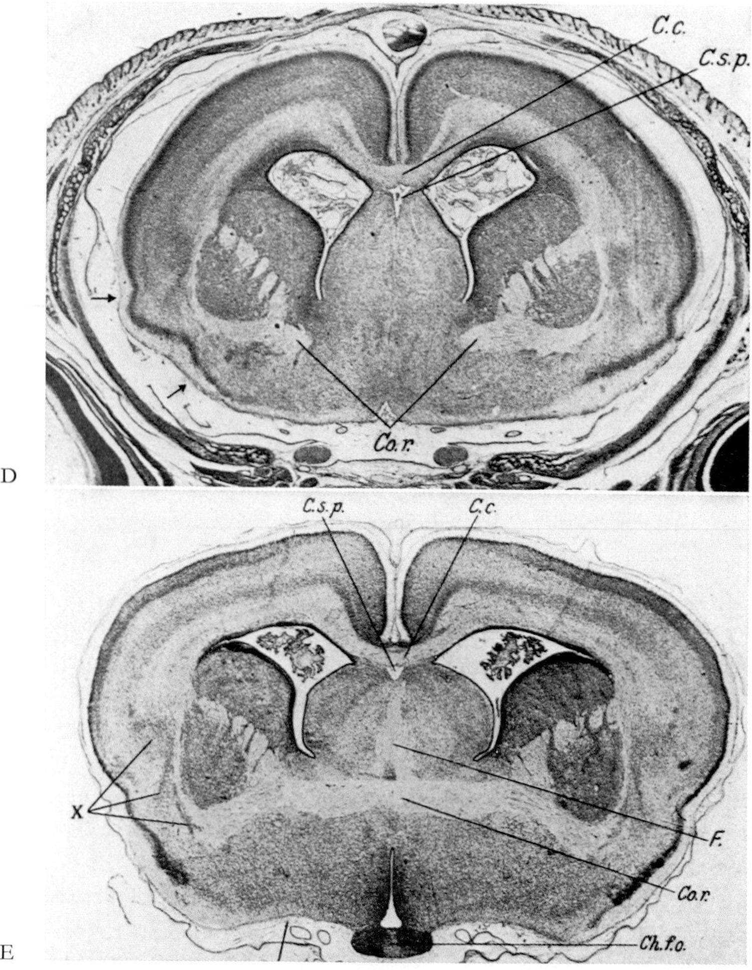

Figure 223 D, E. Ontogenetic stages of Rabbit telencephalon respectively forebrain as seen in cross-sections, and showing transitory cavum septi (from HOCHSTETTER, 1941). D Rabbit fetus of 26 mm head length, at rostral level of commissural plate. E Newborn Rabbit at level of commissura anterior and rostral hemispheric stalk. C.c.: corpus callosum; Ch.f.o.: chiasma opticum; Co.r.: commissura anterior; C.s.p.: cavum septi ('pellucidi'); F: columna fornicis. Added designations: x: claustrum; arrows: in D, upper: sulcus rhinalis lateralis; lower: sulcus endorhinalis; in E: sulcus telencephalo-diencephalicus. The cortical regions, the basal ganglia, and the internal capsule can easily be identified by comparison with preceding and subsequent Figures. The basal cortex in D is at transition between area ventralis anterior and ventrolateralis posterior, while representing the rostralmost portion of area ventrolateralis posterior in E. This latter Fig. shows a noticeable difference between neocortical and parahippocampal region. On the other hand, the developing cell aggregates in the hemispheric stalk's neighborhoods, including telen-

of *Golgi II type*, with some scattered larger elements of *Deiters-type*, whose neurites mostly enter the pallidum (Figs. 225 A, B). This latter griseum is rich in medullated fibers, between which rather large multipolar cells with long dendrites are loosely scattered. Their neurites mostly join the ansa lenticularis. Putamen and pallidum, following the BNA and PNA, are frequently designated as '*nucleus lentiformis*', conceived as part of the corpus striatum, but this terminology seems hardly appropriate.

The basalmost portion of the striatum rostrally to lamina terminalis neighborhoods and adjacent to basal cortex may display some nondescript cytoarchitectural peculiarities, such e.g. as greater crowding, and is also known as the *fundus striati* (cf. e.g. BROCKHAUS, 1942). The fundus striati may also tend to blend with the adjacent griseum basimediale inferius (nucleus accumbens).

There is also, at least in diverse forms, a rather small group of fairly large multipolar cells within the lateral region of lamina terminalis, and intercalated into the fibers of ansa peduncularis. Said griseum is located between the caudal end of basal cortical area ventralis basally, and the adjacent border of putamen and pallidum dorsally. This intercalated cell group is the *nucleus ansae peduncularis (nucleus of substantia innominata, nucleus subputaminalis, nucleus basalis of Meynert)*. It presumably represents a minor basal differentiation of the telencephalic striatal complex. JONES *et al.* (1976) claim, on the basis of their interpretations of data obtained with horseradish peroxidase and autoradiographic techniques, that, in New World and Old World Monkeys, the *nucleus basalis of Meynert* (nucleus substantiae innominatae) represents a griseum receiving inputs from a variety of brain stem sources, and projecting widely as well as diffusely upon all structures of the telencephalon.

As regards some *biochemical features of the striatum*, which derives from a fusion of D_1 and B_{1+2} neighborhoods, its strong *acetylcholinesterase reaction* is of interest.[183] If compared with the morphologically

[183] This reaction also obtains in the whole or at least in parts of the paraterminal grisea. WÄCHTLER (1973) distinguishes various degrees of said reaction's intensity. It is also here of interest that gradients of intensity can be displayed in various regions of one and the same griseal complex, including e.g. in some instances gradients in the rostro-caudal sequence of cross-sectional levels.

cephalic amygdaloid complex, massa cellularis reuniens, pars inferior (with primordium of hypothalamic globus pallidus), and with the hypothalamic preoptic grisea, are here still at the stage of a nondescript unorganized secondary matrix.

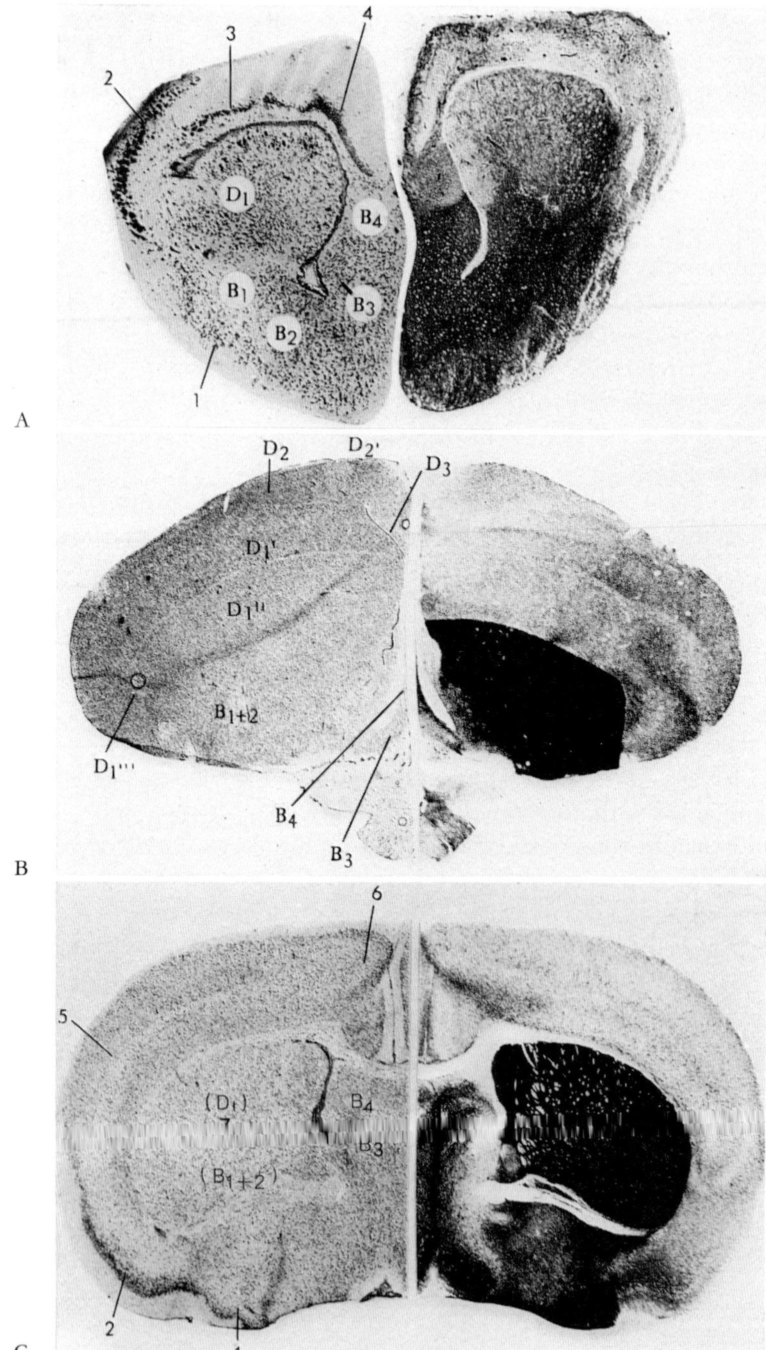
A
1
2
3
4
D1
B4
B1
B3
B2
B
D2
D2'
D3
D1'
D1''
D1'''
B1+2
B4
B3
C
6
5
(D1)
B4
B3
(B1+2)
2
1

homologous neighborhoods of Sauropsida, it can be seen that, in these latter, the strongest reaction is displayed only by the B_{1+2} components, as e.g. recorded by WÄCHTLER (1973) and here shown in Figures 224A–C. It seems evident that, as in the case of fiber connections, and also *qua* details of functional activities, morphologically homologous grisea may display different biochemical features. Conversely, these latter, despite various degrees of correlation, cannot be used as decisive criteria for the establishment of morphologic homologies.

The *claustrum* (2), whose ascertainable ontogenetic development and inferred phylogenetic derivation was discussed on pp. 587–589 of section 6, chapter VI in volume 3/II, is doubtless, as a D_1 derivative, morphologically homologous to the nucleus epibasalis lateralis of Reptiles and to lateral neighborhoods of the Avian nucleus epibasalis dorsalis complex. The caudal portions of the claustrum, moreover, are closely related to the amygdaloid complex, from which they can only be delimited in an arbitrary fashion.

The basal edge of the claustrum, particularly in its rostral portion, is continuous with the basal edge of anterior piriform cortex of which the claustrum appears to be a dorsomedially directed infolding. This is especially conspicuous at some ontogenetic stages and still remains vaguely suggested at the adult stage of various lower Mammals. In these latter, the claustrum is, in general, essentially coextensive with the 'prepiriform cortex', and, to some extent, with the insular cortex. In 'higher' Mammals, in which the expansion of the insula is concomitant with a basal 'displacement' of the rostral portions of piriform lobe cortex, the claustrum becomes predominantly coextensive with the insular cortex.

Figure 224 A–C. Distribution of cholinesterase in the basal grisea of (A) the Reptilian (Lacerta), of (B) the Avian (Passer domesticus) and of (C) the Mammalian (Mouse) telencephalon, as shown by histochemical reaction (courtesy of Priv. Doz. Dr. K. WÄCHTLER, who kindly supplied unlabelled photomicrographs of his 1973 *Habilitationsschrift*). 1: cortex basalis respectively tuberculum olfactorium; 2: cortex lateralis (Lacerta) respectively piriform lobe cortex (Mouse); 3: cortex dorsalis (Lacerta); 4: cortex medialis (Lacerta); 5: neocortex; 6: parahippocampal cortex; 7: corpus striatum (Mouse) formed by fusion of epibasal and basilateral grisea; B_1, B_2: superior and inferior basilateral grisea; B_3, B_4: inferior and superior basimedial grisea; D_1: epibasal grisea; D_1': nuclei epibasales dorsales; D_1'': nucleus epibasalis centralis; D_1''': nucleus epibasalis centralis accessorius; D_2: nuclei diffusi dorsales; D_2': parahippocampal cortex; D_3: hippocampal cortex (cortex medialis). The labels have been added in accordance with my interpretation. Anterior commissure in C easily identified on right side.

Figure 225 A. Sagittal section *(Golgi impregnation)* through corpus striatum of the newborn Mouse (from CAJAL, 1911). A: large cells with long neurite; B: smaller *(Golgi-type)* cells with short neurite; C: endings of ascending fibers; a: neurites. Rostral side of section at top.

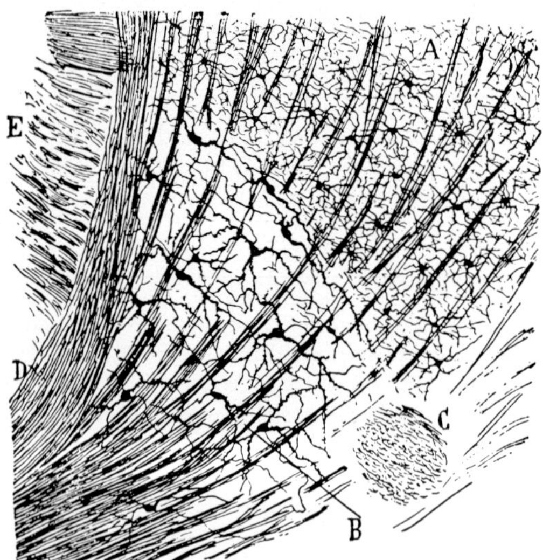

Figure 225 B. Portion of sagittal section *(Golgi impregnation)* through striatum and pallidum of a young Mouse (from CAJAL, 1911). A: striatum (only the smaller neuronal elements are impregnated); B: globus pallidus; C: anterior commissure; D: capsula interna; E: stria terminalis *(taenia semi-circulaire)*.

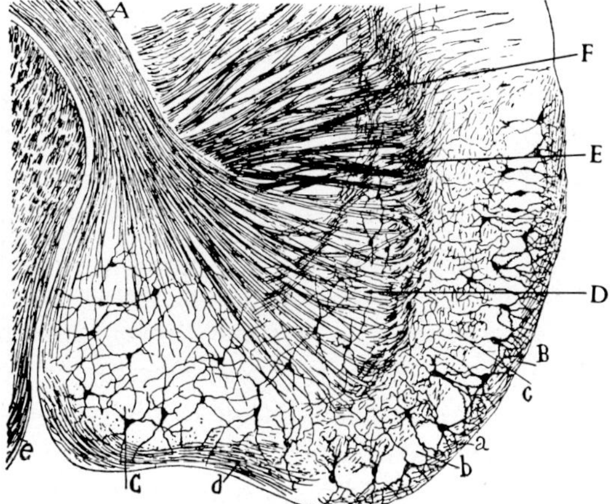

Figure 225 C. Cross-section *(Golgi impregnation)* through caudobasal portion of the lobus hemisphaericus in a young Mouse (from CAJAL, 1911). A: stria terminalis; B: cortex of (intermediate) piriform lobe; C: amygdaloid complex; D: transition of amygdaloid complex to claustrum; E: claustrum; F: striatum; a: tractus olfactorius lateralis; b: cells of piriform lobe cortex; d: zonal fibers of amygdaloid complex; e: optic tract. D, E, F have been added to CAJAL's Figure.

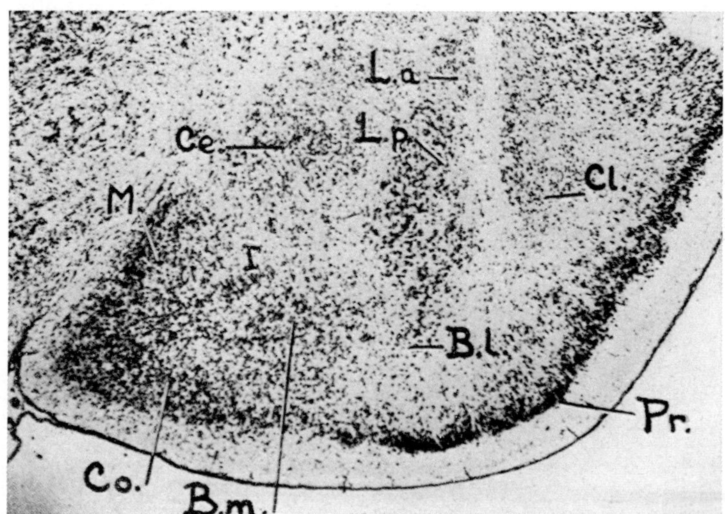

Figure 225 D. Cross-section *(Nissl stain)* through the amygdaloid complex of the Rat (from BRODAL, 1947b; ×6, red. $^9/_{10}$). B.l., B.m.: subdivisions of 'basal amygdaloid nucleus'; Ce : central amygdaloid nucleus; Cl.: claustrum; Co.: cortical amygdaloid nucleus; I.: 'intercalated masses'; La., Lp.: 'lateral amygdaloid nucleus'; M.: 'medial amygdaloid nucleus'; Pr.: cortex of piriform lobe. B.l., B.m., I., La., Lp. pertain to nucleus amygdalae beta, Ce. to gamma, Co. to alpha, and M. to delta of the terminology adopted in the present text.

The claustrum can be described as a generally thin plate of gray matter closely adjacent to the overlying cortex, from which it is separated by a variably distinctive medullary lamina, the *capsula extrema.*[184] An additional lamina, the *capsula externa*, delimits claustrum from striatum, respectively putamen.

Although the claustrum is apparently present in all examined Mammals, the width of its griseal plate, and its degree of development and 'distinctiveness' from the overlying cortex, display numerous taxonomically related variations[185]. In some 'lower' Mammals (cf. e.g. Figs. 222B, 223B) it may even give the impression of a double plate. Its neuronal elements, of variable, but generally medium size may be nondescript multipolar (stellate, polygonal, but hardly typically pyramidal) or, in some instances and regions, pronouncedly fusiform, with dorsoventral long axis parallel to brain surface. An investigation of the Cat's claustrum by Berlucchi (1927) reviews the relevant preceding literature and brings, *inter alia*, an illustration of said griseum's structure as seen in a successful *Golgi impregnation.*

The Mammalian *amygdaloid complex* (3) whose ontogenetic and presumably phylogenetic derivations were likewise discussed in volume 3/II (pp. 590–591) originates from caudal D_1+B_{1+2} neighborhoods. Its configuration and structure have been investigated by numerous authors, e.g. Völsch (1906, 1910), De Vries (1910), Landau (1919), K. (1924a, b, 1929), Hilpert (1928), Mittelstrass (1937), Brockhaus (1940a, b), Crosby and Humphrey (1941, 1944), Brodal (1947b), Sanides (1957), Smialowsky (1965), Humphrey (1968), and Kahle (1969). Since this complex includes a diversity of cell groups with nondescript boundaries predisposing to arbitrary parcellations, moreover since relevant taxonomic as well as individual variations doubtless obtain, several confusing nonstandardized nomenclatures have resulted, which are not easily brought in correspondence with each other nor with the actually observable configurations.[186] Thus,

---

[184] Thus, Brodmann (1909), perhaps particularly impressed by the coextension of claustrum and insular cortex in many Mammals, [illegible] the claustrum as a layer of the insular cortex: '*Strukturell ist die Inselrinde bekanntermassen ausgezeichnet durch die Ausbildung des Claustrums als einer besonderen aus der innersten Rindenschicht oder der Lamina multiformis (VII) abgesprengten zellulären Unterschicht.*'

[185] Crosby *et al.* (1962) enumerate, with bibliographic references, various inconclusive attempts at subdivisions, e.g. into a dorsal and a ventral portion, this latter extending as 'claustrum diffusa' *(sic)* or 'parvum' toward the amygdala.

[186] Beccari (1943) justly comments that if the diverse authors '*che sistematicamente studiarono la struttura del corpo amigdaloide in vari mammiferi (appartenenti a quasi tutti gli ordini)*

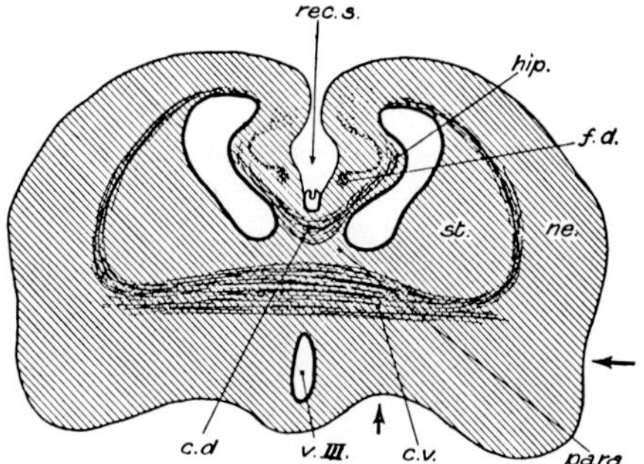

Figure 226 A. Cross-section through the forebrain in an advanced fetus of the Prototherian Monotreme Echidna at level of anterior commissure (from ELLIOT SMITH, 1910). c.d.: commissura dorsalis *sive* pallii (com. hippocampi); c.v.: commissura ventralis *(sive* anterior); f.d.: gyrus dentatus; hip.: hippocampal formation; ne.: neocortex; para.: paraterminal grisea; rec.s.: recessus superior ventriculi tertii; st.: striatum; v.III: recessus praeopticus of third ventricle. Added arrows: sulcus rhinalis lateralis (upper); sulcus telodiencephalicus (lower).

depending on the adopted criteria, between 7 to 10 or even more amygdaloid 'nuclei' can be described, most of which pertain to the nucleus amygdalae beta of the here adopted simplified subdivision.

On the basis of its ontogeny and of the topologic homologies of its principal components, the following main griseal regions may be appropriately distinguished within the amygdaloid complex (Figs. 225 C, D).

(a) *Nucleus amygdalae gamma (n. centralis* of KAHLE and CROSBY, *supraamygdaleum centrale* of BROCKHAUS), contiguous to or continuous with the striatum (putamen). It is a transitional griseum between striatum and amygdaloid complex. It was also designated as *dorsaler Nebenkern* (K., 1924b).

(b) *Nucleus amygdalae beta* or main amygdaloid complex *(Mandelkernhauptkomplex)*. It essentially corresponds to the caudal epibasal (or D_1) neighborhood in Reptiles, and can be parcellated into a number of nuclei, partly separated by medullated fiber bundles, and partly differing in cell size or type of cell populations (nucleus amygdalae basalis and lateralis of BRODAL, of CROSBY and of KAHLE, amygdaleum profundum of BROCKHAUS). A cellular band, or nondescript island-like cellu-

---

si fossero messi d'accordo sul numero dei nuclei e sulla loro nomenclatura, oggi forse non riuscirebbe difficile di esporre in forma definitiva la composizione del corpo amigdaloide; ma disgraziatamente ciò non è avvenuto'.

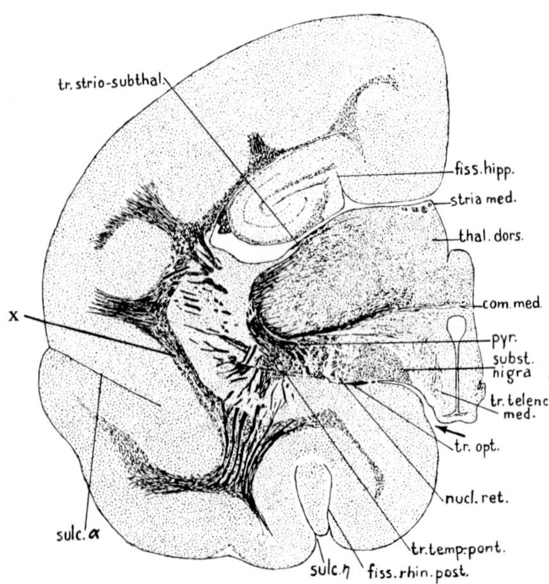

Figure 226 B. Cross-section through the forebrain of adult Echidna (from ADDENS and KUROTSU, 1936). com.med.: commissural fibers in massa intermedia interthalamica; nucl. ret.: probably portion of nucleus entopeduncularis related to globus pallidus; pyr.: pyramidal tract; fiss.rhin.post.: caudal portion of sulcus rhinalis lateralis; tr.strio-subthal.: perhaps portion of stria medullaris; tr.telenc.med.: basal forebrain bundle. Other abbreviations self-explanatory. The unlabelled striatum and amygdaloid complex are easily identifiable. Greek letter designations of neopallial sulci following ZIEHEN's (1897) notation. Added x: claustrum; arrow: appr. boundary between piriform lobe cortex and cortical amygdaloid nucleus (caudal end of sulcus endorhinalis).

lar groups connect lateral components of this subcomplex with caudal neighborhoods of the claustrum, and their inclusion in either claustrum or amygdaloid complex becomes arbitrary (claustrum preamygdaleum of BROCKHAUS).

(c) *Nucleus amygdalae alpha*, corresponding to the submammalian area ventrolateralis posterior, that is to say to a portion of *basal cortex*. It is the *nucleus amygdalae corticalis* of CROSBY and of KAHLE, amygdaleum superficiale (and periamygdaleum) of BROCKHAUS. The *nucleus amygdalae delta (medialer akzessorischer Kern;* K., 1924b) or nucleus amygdalae medialis of CROSBY and of KAHLE can be evaluated as a differentiation of the cortical amygdaloid griseum,[187] particularly related

[187] The cortical amygdaloid griseum (area ventrolateralis posterior, nn. amygdalae alpha and delta) corresponds to the area periamygdalaris intermedia and medialis (Pam 2, 3)

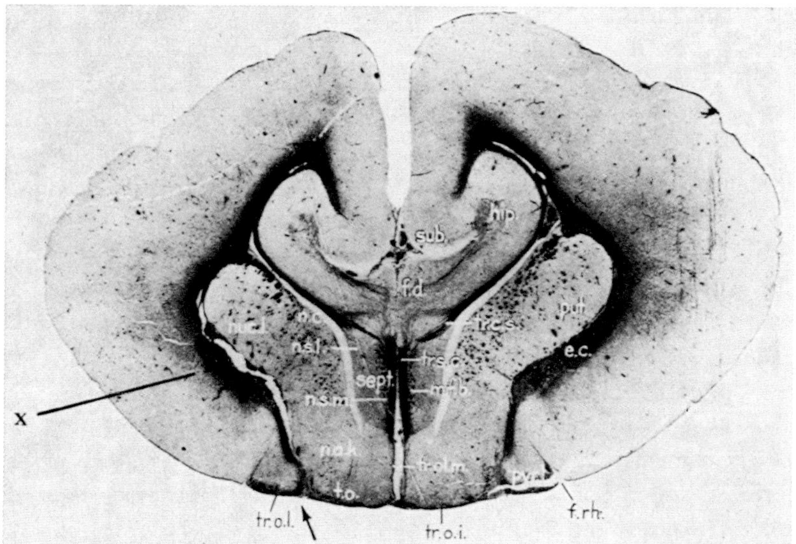

Figure 226C. Cross-section (myelin stain) through a precommissural level of the telencephalon in the Prototherian Monotreme Ornithorhynchus anatinus (from HINES, 1929). e.c.: external capsule; f.d.: fascia dentata; f.rh.: sulcus rhinalis lateralis; hip.: hippocampal formation; m.f.b.: basal forebrain bundle; n.a.k.: nucleus accumbens (fundus striati); n.c.: 'nucleus caudatus' (striatum); n.s.l., n.s.m.: nucleus lateralis respectively medialis septi; nuc.l.: 'nucleus lentiformis' (striatum); put.: 'putamen' (striatum); pyr l : piriform lobe; sept.: septum (paraterminal grisea); t.o.: tuberculum olfactorium (area ventralis anterior, basal cortex); tr.c.s.: 'tractus cortico-septalis'; tr.o.i., tr.o.l., tr.o.m.: intermediate, lateral and medial olfactory tracts; tr.s.c.: 'tractus septo-corticalis'. Added x: probable location of claustrum; arrow: sulcus endorhinalis. Note the relative smallness of both piriform lobe and tuberculum olfactorium (cf. also Fig. 340, p. 625, vol. 3/II).

to endings of the lateral olfactory tract. The basal cortical components of the amygdaloid complex may or may not be separated from the pallial cortex anterior, intermedius and posterior lobi piriformis by a caudal extension of sulcus endorhinalis (sulcus periamygdalaris, sulcus semiannularis in Man). Although the pallial piriform lobe with its anterior piriform and its posterior parahippocampal component is morphologically quite distinct from the basal amygdaloid cortex, this latter

---

of ROSE (1929) while his area periamygdalaris lateralis (Pam 1) is doubtless either intermediate or posterior piriform lobe cortex. The term periamygdalar cortex should preferably be restricted to the cortical amygdaloid nucleus.

is frequently included in the piriform lobe as depicted by many authors.

The *paraterminal grisea* (4) include (a) *nucleus basimedialis inferior* and (b) *nucleus basimedialis superior*. The former is orthohomologous to that of all Gnathostome Vertebrates, the latter only to that of Gnathostomes with inverted telencephalon, since this griseum representing a neighborhood connecting B_3 with D_3, cannot develop in the everted lobus hemisphaericus of Ganoids and Teleosts. In contradistinction to the grisea of the lateral hemispheric base, the medially located paraterminal grisea are much less voluminous and of rather uniform configurational aspect throughout the Vertebrate series.

A region of Mammalian griseum basimediale inferius, designated as *nucleus accumbens (septi)*, not infrequently tends to merge with the laterally adjacent striatum into a common so-called *fundus striati*. Both inferior and superior basimedial grisea display a preterminal (precommissural) and an intraterminal portion. In addition, a variable but rather minor dorsal intraterminal component of nucleus basimedialis superior becomes supracommissural, being the *indusium verum* of Elliot Smith (1897), the *indusium falsum sive spurium* being the rudimentary supracommissural hippocampus of 'higher' Eutheria. While both superior and inferior basimedial grisea represent the paraterminal grisea *sensu latiori*, the nucleus basimedialis superior may also be designated as paraterminal griseum *sensu strictiori*. With reference to the peculiarities of this latter neighborhood, providing the septum pellucidum of Primates and some other 'higher' Mammals, the paraterminal or basimedial grisea in their entirety have been designated as '*septum*' respectively as septal grisea by most authors.

The cytoarchitectural parcellation of the paraterminal grisea is beset with considerable difficulties because of nondescript groupings, indistinct boundaries, and numerous variations. Many fibers of fornix, of medial forebrain bundle *sensu strictiori*, and of *Broca's diagonal band* run in dorso-basal respectively baso-dorsal direction through this region. In addition, the basal cortex of area ventralis anterior extends medio-dorsally into the paraterminal neighborhoods and is generally included into the 'septum' as part of or even entirely into the so-called nuclei mediales septi. Moreover, the intraterminal portion of nucleus basimedialis superior of Mammals may be complicated by the presence of a cavum septi pellucidi.

Among investigations concerning the Mammalian paraterminal grisea those by the following authors may be mentioned as relevant in

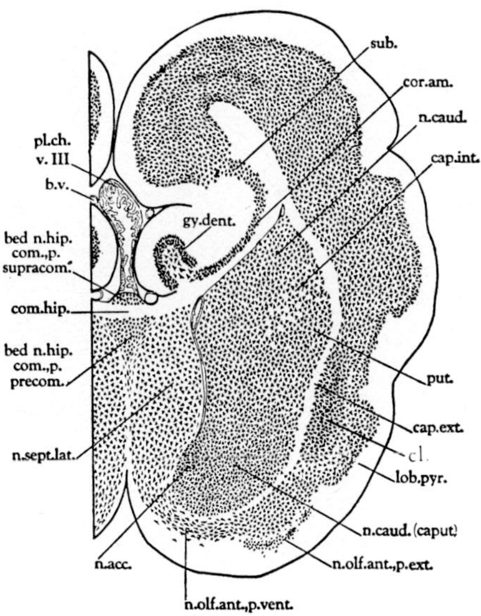

Figure 227A (legend see p. 712)

the aspect here under consideration: ELLIOT SMITH (1895, 1896, 1897), HOCHSTETTER (1919, 1935, 1941), YOUNG (1926), THOMPSON (1932a, b), STEPHAN and ANDY (1962), ANDY and STEPHAN (1964, 1968), and K. (1969).

The most detailed cytoarchitectural parcellation of the paraterminal region is that attempted by ANDY and STEPHAN (1964, 1968). These authors distinguish a dorsal, ventral, medial, and caudal group, each again with several 'nuclei' and 'nuclear subdivisions', thereby obtaining about 17 cytoarchitectural subdivisions some of which pertain to griseum basimediale superius, and others to griseum basimediale inferius.

In contradistinction, and taking into account taxonomic as well as individual variations, in addition to the somewhat random or nondescript distribution of the various cell types, I prefer to distinguish, in the Mammalian preterminal paraterminal body, a medial and a lateral subdivision. The former and superficial component pertains to the basal cortex, and includes the '*nucleus of the diagonal band*'. The lateral (adventricular) cell groups display an ill-defined dorsal and ventral subdivision of nucleus basimedialis superior, and a variable dorsal component

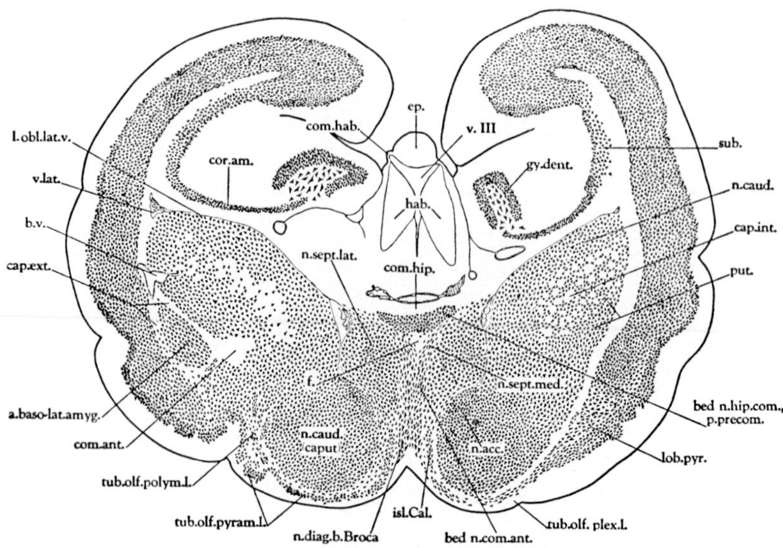

Figure 227 B

Figure 227 A–D. Drawings of cross-sections (cell stain) through telencephalon respectively forebrain of the Chiropteran Bat Tadarina mexicana (from Humphrey, 1936). a.baso-lat.amyg.: 'area basolateralis amygdalae' (in B and C probably part of claustrum); b.v.: blood vessel; bed n. (etc.): 'bed nuclei' of hippocampal and anterior commissures, and of stria terminalis (bed n. comp. p. supracom. is the indusium verum); cap. ext.: capsula externa; cap. extr.: capsula extrema; cap.int.: capsula interna; cl.: claustrum (its dorsal extension in D at left seems a dubious interpretation; at right it is doubtless at end of lead 'lob. pyr.'); com.ant.: anterior commissure; f.: fornix; fis.endorh., rhin.: s. endorhinalis respectively s. rhinalis lateralis; glob.pal.: globus pallidus; gy.dent.: gyrus dentatus; hab.: habenula; hip.: hippocampal formation; isl.Cal.: *island of Calleja;* l.obl. lat.v.: line of obliterated lateral ventricle; lob. pyr.: lobus piriformis; mas.interc.: 'massa intercalata' (part of n. amygdalae gamma); n.acc.: n. accumbens; n.amyg.cent.: n. amygdalae gamma; n.amyg.med.: part of cortical n. amygdalae; n.caud.: n. caudatus; n.diag.b.Broca: n. of the *diagonal band of Broca;* n.olf.ant. (etc.): probably part of area ventralis anterior; n.preop.mag.: probably part of n. entopeduncularis and globus pallidus complex; n.preop.ant., med., int., perivent., prin.: preoptic grisea; n.sept.lat., med.: n. septi lateralis et medialis; pl.ch.v.lat.: choroid plexus of lateral ventricle; put.: putamen; sub.: 'subiculum' (here part of cornu Ammonis); tect.opt.: tectum opticum; thal.: thalamus; tub.olf. (etc.): grisea of tuberculum olfactorium (area ventralis anterior); v. lat.: lateral ventricle; v.III: third ventricle. Because of the peculiar 'telescoping' of telencephalic, diencephalic, and mesencephalic neighborhoods (cf. also Fig. 13B of chapter XII), the interpretation of many Chiropteran prosencephalic grisea is difficult and uncertain. The differences between some of Humphrey's original interpretations and my own as here given, are based on personal studies of the Chiropteran brain.

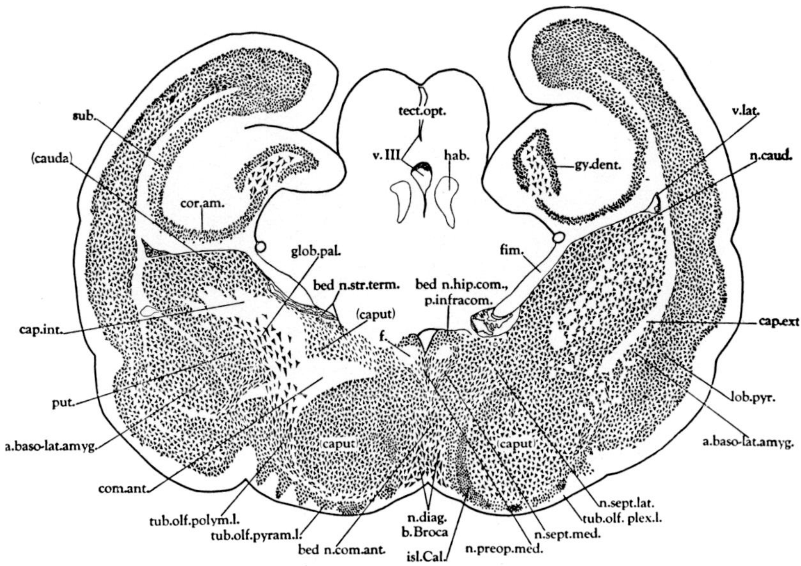

Figure 227 C

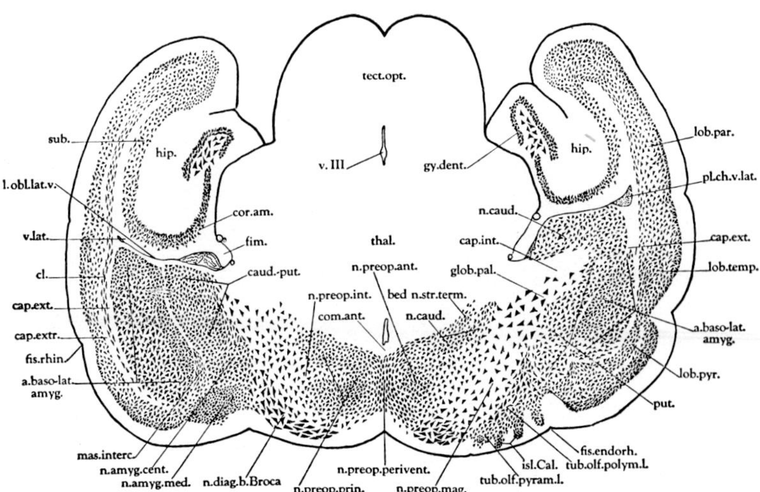

Figure 227 D

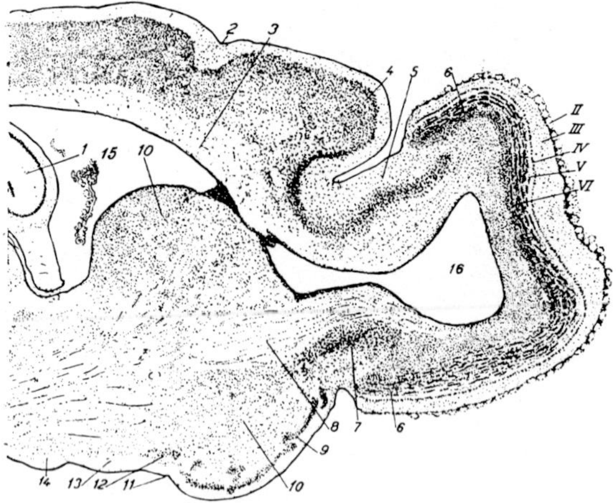

Figure 228 A. Sagittal section *(Giemsa stain)* through the telencephalon of the Insectivore Erinaceus europaeus (from BECCARI, 1943). 1: postcommissural hippocampal formation; 2: frontal sulcus; 3: corpus callosum; 4: rostral pole of neopallium; 5: olfactory stalk with dorsal part of nucleus olfactorius anterior; 6: bulbus olfactorius with glomerular (II), outer plexiform (III), mitral (IV), internal plexiform (V), and granular (VI) layers; 7: ventral part of n. olfactorius anterior; 8: rostral limb of anterior commissure; 9: tuberculum olfactorium; 10: corpus striatum; 11: groove delimiting olfactory tubercle; 12: griseum of diagonal band; 13: 'basal parolfactory space'; 14: optic tract (between 13 and 14: sulcus telencephalo-diencephalicus); 15: lateral ventricle; 16: rhinocele.

of nucleus basimedialis inferior not included in a typical nucleus accumbens. The intraterminal portion (essentially ANDY's and STEPHAN's caudal group) may vaguely display dorsomedial and dorsolateral as well as ventromedial and ventrolateral cell groups of nucleus basimedialis superior, and nondescript grisea of nucleus basimedialis inferior not included in nucleus accumbens (e.g. the so-called '*bed nuclei*' of anterior commissure and of stria terminalis).[188]

In Man, some Primates, and also in some other large Mammals with [illegible] brains, the intraterminal region of the paraterminal griseum *sensu strictiori*, subjected to extensive stretching between corpus callosum, fornix, and anterior commissure, becomes thinned, forming a membrane-like partition between the two lateral ventricles rostrally to interventricular foramen, and known as the *septum pellucidum* of human

[188] Cf. also the comments in my paper on human corpus callosum and septum pellucidum (K., 1969).

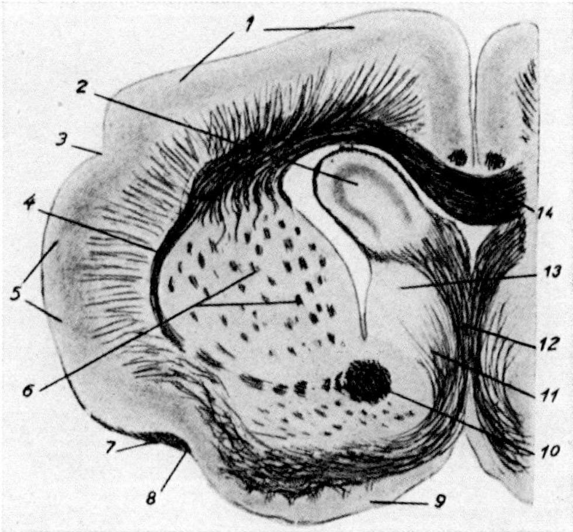

Figure 228 B. Cross-section *(Weigert stain)* through the telencephalon of Erinaceus at level of lamina terminalis (from BECCARI, 1943). 1: neopallium (and parahippocampal cortex); 2: postcommissural subsplenial) hippocampus; 3: sulcus rhinalis lateralis; 4: capsula externa; 5: cortex of piriform lobe; 6: fascicles of capsula interna; 7: tractus olfactorius lateralis; 8: sulcus endorhinalis; 9: tuberculum olfactorium; 10: bundle of commissura anterior; 11, 12: paraterminal and hippocampal fiber connections with basal grisea; 13: paraterminal grisea; 14: corpus callosum.

anatomy (BNA, PNA). Its dorsal portion, commonly entirely or partly devoid of neuronal elements *(septum gliosum)*[189] may or may not contain a closed cavity of variable size at the adult stage *(cavum septi pellucidi* BNA, PNA). Said cavity originates within the commissural plate of lamina terminalis by tissue resorption during embryonic development and, although not communicating with the ventricular system, except in cases of secondary rupture or defects, may become partly or completely lined by rather typical ependyma developed *in situ* from the spongioblastic tissue elements (OLIVEROS, 1965).

This *true cavum septi pellucidi* should not be confused with the so-called *open cavum septi*, which merely represents a caudal recess of the interhemispheric fissure which may or may not protrude a short distance into the intraterminal paraterminal body.

Concerning the cavum septi pellucidi, and with respect to the adult stage, THOMPSON (1932a, b) attempted to classify all Mammals into the

[189] Cf. K. (1969). In contradistinction to the essentially non-neuronal dorsal portion of the septum pellucidum, the basally adjacent portion, characterized by neuronal elements, was designated as septum gangliosum. Both septum gliosum and gangliosum pertain to the topologic neighborhood B_4, that is to say nucleus basimedialis superior.

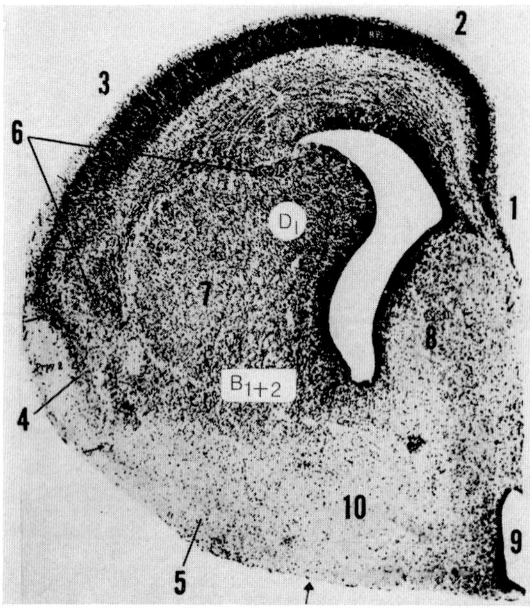

Figure 229 A. Cross-section through forebrain of a 15 mm Mouse fetus at level of hemispheric stalk (from Rose, 1935, labelled in accordance with my interpretation). 1: precommissural hippocampal cortex primordium; 2: parahippocampal cortex primordium; 3: neocortex primordium; 4: piriform lobe cortex primordium; 5: undifferentiated basal cortex primordium at transition between area ventralis anterior and ventrolateralis posterior; 6: cell plate of claustrum primordium; 7: anlage of corpus striatum; 8: paraterminal grisea primordium; 9: preoptic recess with adjacent preoptic (hypothalamic) grisea; 10: hemispheric stalk with still unorganized cell groups (cf. Fig. 223 E).

Figure 229 B–D. Cross-sections (*Nissl stain* B, D; myelin stain C) through telencephalon respectively forebrain of the adult (Eutherian, Rodent) Mouse. B Rostrally to lamina terminalis. C At level of anterior commissure and corpus callosum. D Caudally to hemispheric stalk. 1: area ventralis anterior (tuberculum olfactorium); 2: anterior piriform lobe cortex and lateral olfactory tract; 3: neocortex; 4: parahippocampal cortex; 5: precommissural hippocampus (cornu Ammonis of parasubicular type); 6: precommissural hippocampus (of fascia dentata type); 7: nucleus basimedialis superior; 8: nucleus basimedialis inferior (with rostral neighborhoods of *Broca's diagonal band* in B, and fornix bundles in C); 9: striatum (with scattered bundles of capsula interna); 10: claustrum; 11: fundus striati; 12: anterior commissure; 13: corpus callosum, 14: postcommissural hippocampus, cornu Ammonis; 15: postcomm. hip., fascia dentata; 16: amygdaloid complex; 17: nucleus amygdalae alpha sive corticalis; 18: nucleus amygdalae delta; 19: basis pedunculi (diencephalic portion of capsula interna, mediali to which nucleus subthalamicus sive corpus Luysi can be identified); 20: fimbria fornicis; 21: fasciculus retroflexus; 22: fasciculus mammillo-thalamicus; 23: pars tecta fornicis; x: optic nerve; y: trigeminal and eye muscle nerve branches; Z: diencephalon (for details consult Figures of chapter XII). Unlabeled lateral ventricles easily recognizable.

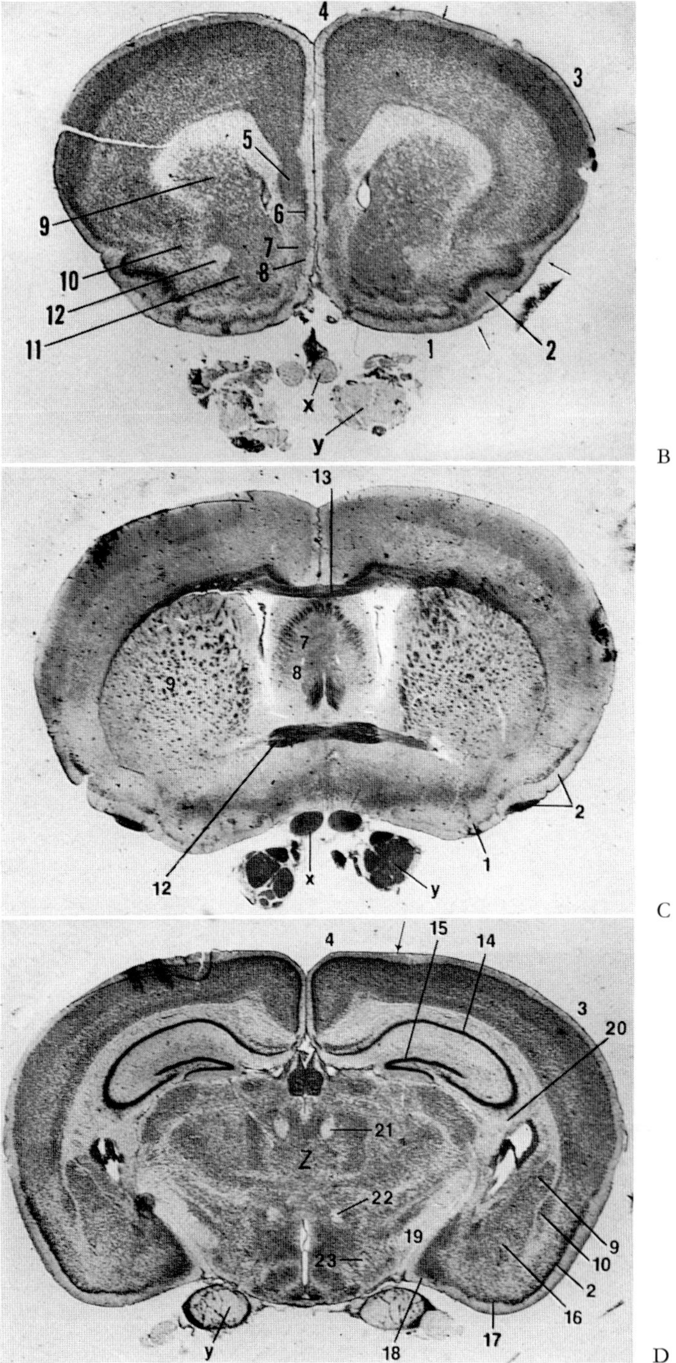
4
3
5
9
6
7
10
8
12
11
1
2
x
y
B
13
7
8
9
2
1
12
x
y
C
4
15
14
3
20
21
Z
22
19
9
23
10
2
16
y
18
17
D

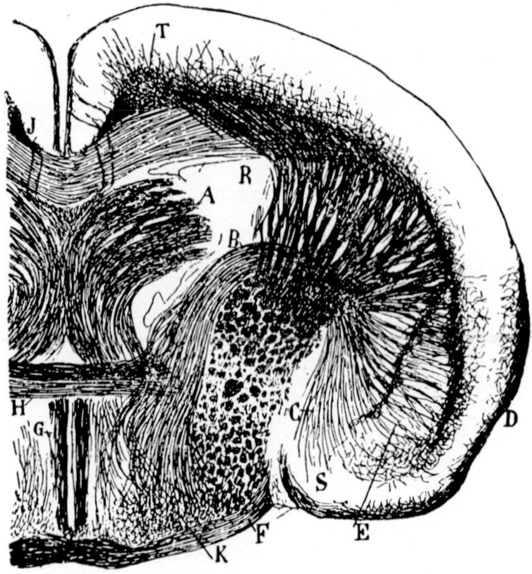

Figure 229 E. Cross-section *(Golgi impregnation)* through forebrain of a young Mouse at level of anterior commissure and hemispheric stalk (from Cajal, 1911). A: column of fornix; B, C: stria terminalis system; D: cortex of intermediate piriform lobe; E: lateral portion of main amygdaloid complex, dorsalward merging with claustrum; F: optic tract; G: periventricular fibers of preoptic recess *('faisceau médian de la cloison');* H: anterior commissure; J: perforating fibers of supracommissural hippocampus rudiment joining fornix; K: basal forebrain bundle joined by stria terminalis components; R: striatum traversed by capsula interna bundles; S: region of cortical nucleus amygdalae; T: longitudinal (sagittal) bundles of subcortical fibers. Compare this Figure with Figure 225 C.

following groups. (A) With *solid 'septum'*, respectively *absent cavum* (e.g. Marsupials and the Insectivore Mole). (B) With *cavum present*, either (a) open (e.g. Cat, Tarsius, Macacus), or (b) closed (e.g. Chimpanzee, Man).

Hochstetter (1919, 1935, 1941), whose investigations on the cavum septi pellucidi are of particular importance, and who clearly substantiated its origin from tissue resorption, noted that a true cavum septi occurs, as a transitory space at some stages of ontogenetic development, in a wide variety of Mammals lacking such cavum at the adult stage. In some Mammals with large brain, the cavum may or may not persist, but even if present, becomes generally reduced by obliteration, which results in a *glial midline raphe*. Hochstetter (1935) justly pointed out that the so-called open cavum septi is entirely unrelated to the true cavum septi *('nichts anderes ist als ein Teil der Mantelspalte und nichts mit dem bei älteren Katzenembryonen und auch bei neugeborenen Katzen noch vorhandenen Cavum septi pellucidi zu tun hat')*.[190]

HOCHSTETTER (1935) reaches the conclusion that one can, at present, merely distinguish between Mammals which, at the adult stage, display more or less extensive remnants of a cavum septi pellucidi, and those in which no trace of a (true) cavum is found. As regards these latter forms, it would require additional extensive and tedious investigations in order to distinguish those in which the paraterminal portion of the commissural plate (lamina terminalis) remains entirely solid during ontogenesis, from those whose lamina terminalis displays a transitory cavum that subsequently disappears without trace. An example of transitory true cavum in the Rabbit, as recorded by HOCHSTETTER (1941), ist shown in Figures 223D, E.

There is little doubt that both the true cavum septi in those adult Mammals displaying said cavity, and the spurious so-called 'open cavum' in adult Mammals with 'solid septum' manifest numerous individual and taxonomic variations. The location of said 'open cavum' is indicated in Figure 221B. Figures 338A–E, p. 622 of volume 3/II might also be compared. Although these diagrams by ELLIOT SMITH are topologically entirely correct, the deep recess of the Mantelspalte depicted in Figure 338E (loc.cit.) is doubtless highly exaggerated and potentially misleading, since it neglects the solid commissural plate in which a true closed cavum septi may originate and persist. My own incidental observations seem to suggest that, *qua* adult stage, a solid septum, with or without the trace of a *Mantelspaltenrezess* is generally found in Monotremes, Marsupials, Edentates, Rodents, Lagomorphs and Carnivores.[191]

A true *cavum septi* of very variable size and with diverse degrees of frequency, may be found in adult specimens of Ungulates (e.g. Horse, Bovines, and occasionally Sheep), of Cetacea, and of Primates. If a cavum septi is present, its thin walls, whose lateral surfaces pertain to the

[190] '*Jedenfalls kann aber kaum ein Zweifel darüber bestehen, dass der Igel sowie die Katze, die Maus und vielleicht auch der Maulwurf zu den Säugern gehören, die während ihrer Entwicklungszeit ein Cavum septi p. besitzen und bei denen sich dieses schliesslich wieder vollständig zurückbildet.*' '*Wann das Cavum septi pellucidi beim menschlichen Embryo den Höhepunkt seiner Entwicklung erreicht hat, vermag ich leider mit voller Sicherheit nicht zu sagen, weil ich immer noch nicht über eine genügend grosse Zahl einwandfrei konservierter älterer Föten verfüge*' (HOCHSTETTER, 1935).

[191] It is, however, uncertain whether an occasionally seen small space in the solid Carnivore septum represents a *Mantelspaltenrezess* or the remnant of a minor cavum as noted in the newborn Cat by HOCHSTETTER (1935).

anterior horn of the lateral ventricle, are the *laminae septi pellucidi* of human anatomy (BNA, PNA).[192]

The caudal portion of the 'septal' telencephalic lamina terminalis includes, moreover, a rather small, more or less nodular interventricular midline structure within the rostral border of *foramen Monroi* (telencephalon medium). This is the *subfornical organ*, at whose dorsal edge the choroid plexus is attached. It pertains to the group of circumventricular and paraependymal organs dealt with in section 5, chapter V of volume 3/I. Although not explicitly mentioned by diverse authors describing the Mammalian telencephalon, it seems nevertheless probable that the subfornical organ ist present in most if not all Mammals (cf. also the paper by COHRS, 1936). It should also be mentioned that said organ was depicted in the Bat Myotis by JOHNSTON (1913b, Fig. 45) and designated as *nodulus marginalis*. Recently, observations suggesting that 'selective accumulations of hormones' (estrogen, androgen, and glucosteroids) can be found in the vicinity of the cerebral ventricles have led to the concepts of 'hormone architectonics' and 'periventricular cerebral glands', to which the 'circumventricular organs' may pertain. 'Hormone-neurotransmitter interrelationships' and 'unspecific truncothalamic systems' are postulated, and rather unconvincing phylogenetic speculations have been propounded. Further details concerning these inconclusive hypotheses can be found in the text on 'Anatomical Neuroendocrinology' edited by STUMPF and GRANT (1975).

---

[192] As regards the various anomalies of the septum pellucidum, particularly in Man (complete absence, perforations, etc.) cf. the papers by THOMPSON (1932a, b) and HAHN and K. (1930) with further references.

Figure 230 A–C. Horizontal sections (myelin stain A, C, *Nissl stain* B) through the brain of an adult Mouse, illustrating telencephalic relationships in a dorsobasal sequence of three planes. 1: corpus callosum; 2: paraterminal grisea; 3: commissura hippocampi *sive* fornicis; 4: striatum with scattered bundles of internal capsule; 5: claustrum; 6: optic tract; 7: postcommissural hippocampus; 8: caudal parahippocampal cortex; 9: stria medullaris; 10: fasciculus retroflexus; 11: posterior commissure; 12: mesencephalon; 13: cerebellum with cerebellar nuclei; 14: bulbus olfactorius; 15: precommissural (rostral) parahippocampal cortex; 16: precommissural hippocampus; 17: fornix; 18: capsula interna; 19: globus pallidus; 20: brachium conjunctivum; 21: anterior commissure with rostral and caudal limb; 22: mammillo-thalamic tract; 23: fimbria of fornix; 24: stria terminalis; 25: pes pedunculi; 26: substantia nigra; 27: nucleus interpeduncularis; 28: trigeminal root; 29: tegmental radiation including general lemniscus.

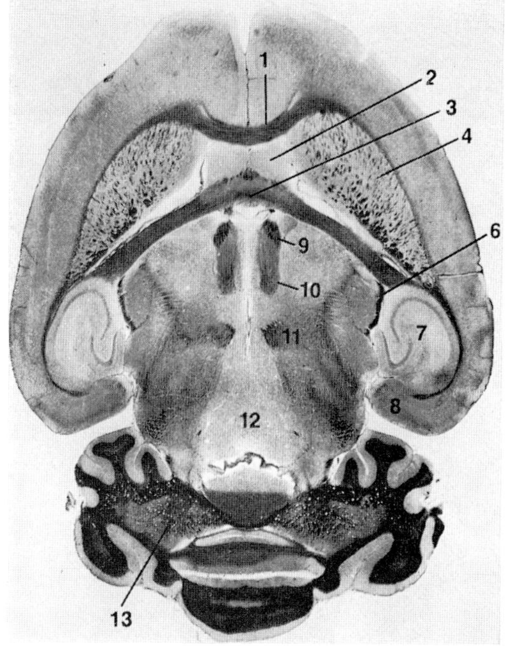

Figure 230 A

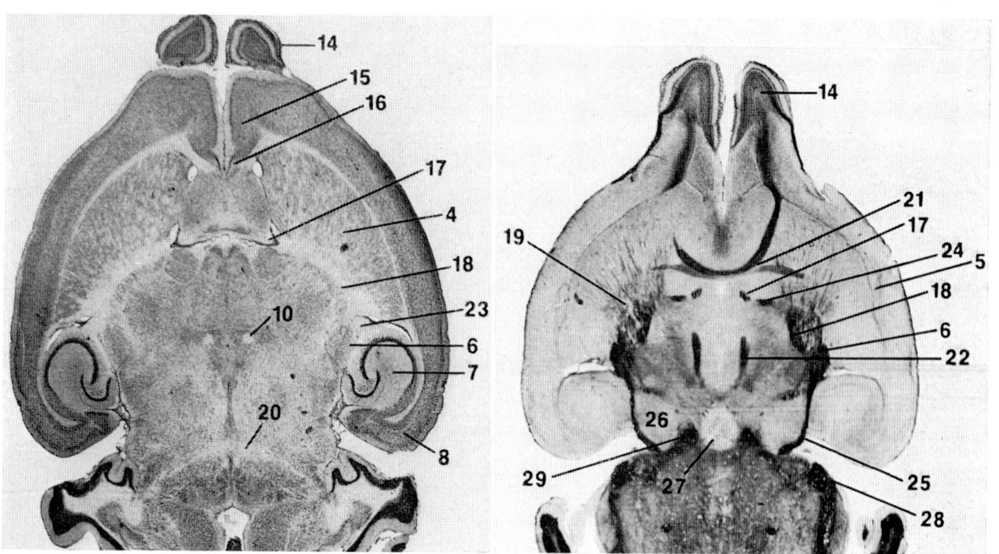

Figure 230 B, C

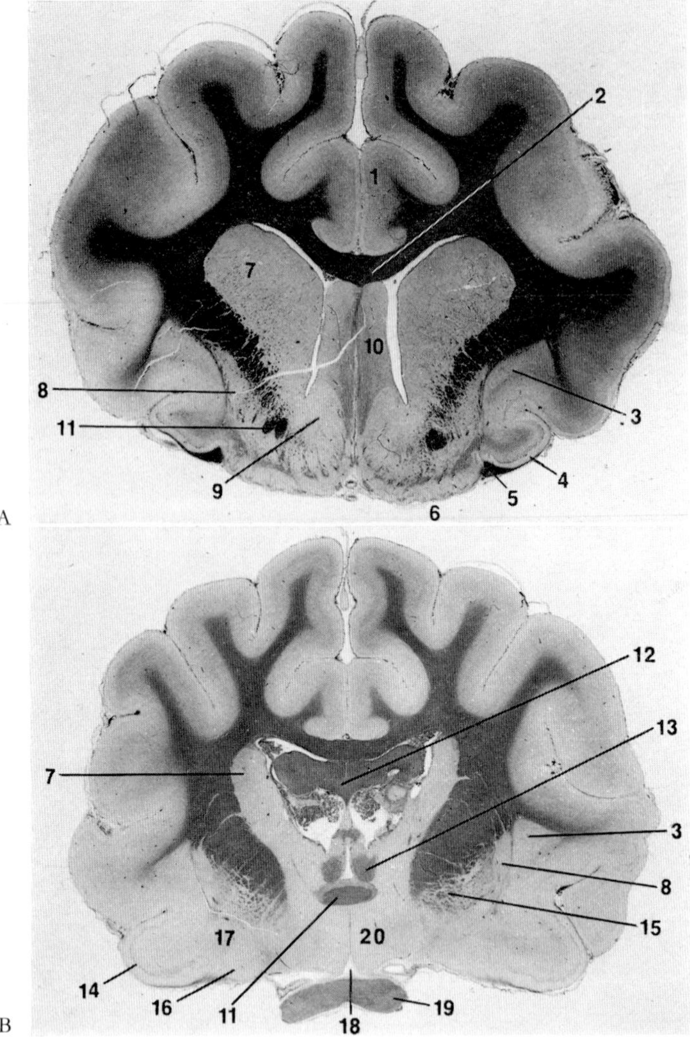

Figure 231 A–D. Cross-sections (myelin stain) through telencephalon and adjoining regions of the Cat, at four levels in rostrocaudal sequence. 1: gyrus cinguli; 2: corpus callosum; 3: claustrum; 4: anterior piriform lobe; 5: tractus olfactorius lateralis; 6: basal cortex (tuberculum olfactorium); 7: nucleus caudatus; 8: putamen (between 7 and 8: capsula interna); 9: nucleus accumbens septi; 10: septal griseum (nucleus basimedialis superior); 11: anterior commissure; 12: commissura hippocampi; 13: fornix; 14: transition between intermediate and posterior piriform lobe; 15: globus pallidus; 16: cortical

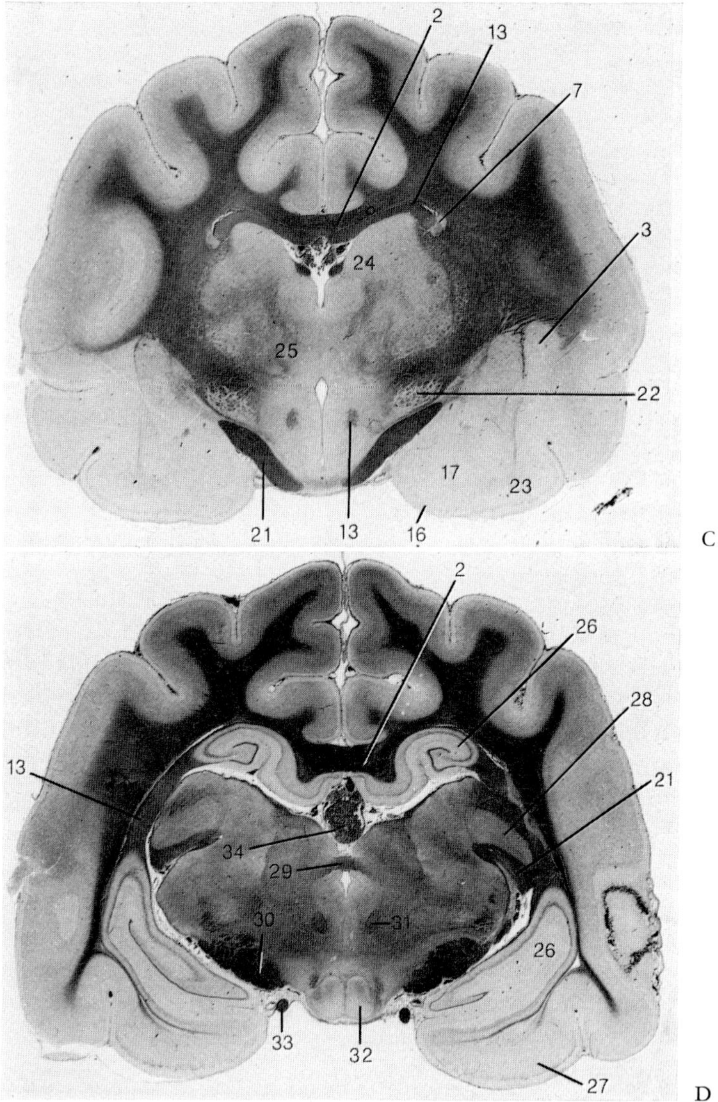

amygdaloid nucleus; 17: amygdaloid complex; 18: preoptic recess; 19: optic chiasma; 20: preoptic grisea; 21: optic tract; 22: nucleus entopeduncularis; 23: lobus piriformis, begin of posterior part; 24: stria medullaris thalami; 25: fasciculus mammillothalamicus; 26: postcommissural hippocampus; 27: posterior (parahippocampal) piriform lobe; 28: lateral geniculate complex; 29: posterior commissure; 30: pes pedunculi; 31: fasciculus retroflexus; 32: mammillary complex; 33: nervus oculomotorius; 34: pineal body. For details of diencephalic configurations, cf. the illustrations in section 9 of chapter XII.

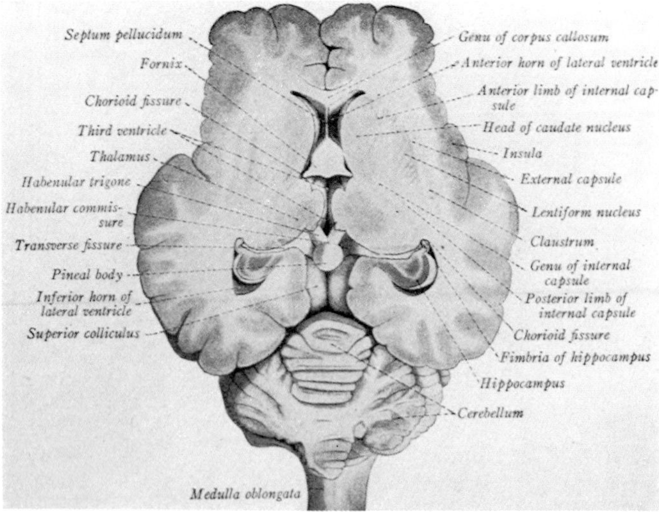

Figure 232. Horizontal section (unstained) through the Ungulate Sheep's brain, illustrating overall configurational relationships of telencephalon (from RANSON, 1943).

(5) The merogenic *cortex basalis* which, in the here adopted classification and terminology, is both the fifth major cortical subdivision and the fifth major subdivision of basal grisea, consists of two main subregions, namely, (a) *area ventralis anterior*, extending from olfactory bulb respectively tract to lamina terminalis respectively hemispheric stalk, and (b) *area ventrolateralis posterior* within the caudal paired evaginated lobus hemisphaericus, essentially caudo-ventrolaterally to hemispheric stalk and laterally respectively caudolaterally to lamina terminalis.

The *area ventralis anterior* (a), depending on the development of olfactory bulb and olfactory tract in the Mammalian taxonomic series (macrosmatic, microsmatic, or anosmatic forms), commonly merges rostrally through an ill-defined transitional neighborhood with the nondescript grisea of nucleus olfactorius anterior. Caudalward it includes the *tuberculum olfactorium* generally characterized by an essentially nonstratified, scalloped cortical cell plate,[193] more or less distinctively separated from the fundus striati by a subcortical zone relatively poor in cellular elements. In various macrosmatic Mammals, this cortical plate is disrupted into cell clusters, some of which consist of larger,

[193] CAJAL (1911) and others distinguish, however, a plexiform (molecular, zonal), an intermediate granular, and a deeper pyramidal-polymorph layer.

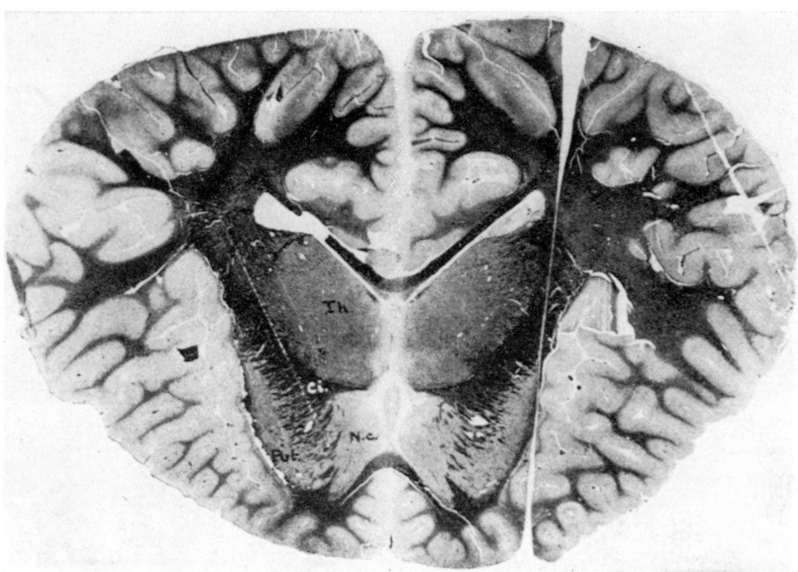

Figure 233 A. Cross-section (myelin stain) through the forebrain of the Cetacean Delphinus delphis (from RIESE, 1925). Ci.: capsula interna; N.c.: nucleus caudatus; Put.: putamen; Th.: thalamus.

and others of smaller, more granular, multipolar or irregularly pyramidal elements. These clusters are known as the *islets of Calleja.*[194] Lateralward, the cortex of area ventralis anterior, through a variable region of transition, adjoints the pallial cortex of the piriform lobe, the sulcus endorhinalis approximately indicating the boundary between the two cortical regions.[195] Medialward, the basal cortex of area ventralis anterior extends, as a more diffusely arranged cell plate, into the precommissural paraterminal region and includes the grisea of *Broca's diagonal band* as well as the so-called nuclei mediales septi, that is, structures not generally evaluated as parts of the basal cortex by most authors. A

---

[194] CALLEJA (1893; quoted after CAJAL, 1911).

[195] Although KAPPERS originally applied his terms '*palaeocortex*' respectively *palaeopallium* to the cortex of the piriform lobe (essentially to the 'prepiriform' cortex), that is, to a pallial cortex, he also included much or most of the basal cortex into his palaeocortex or palaeopallium (cf. e.g. Fig. 325, p. 598, vol. 3/II). Quite apart from well founded objections against the concept 'palaeocortex', it does not seem appropriate to subsume pallial piriform lobe cortices and the morphologically entirely different basal cortex under a common term implying their identity as a 'pallial' component.

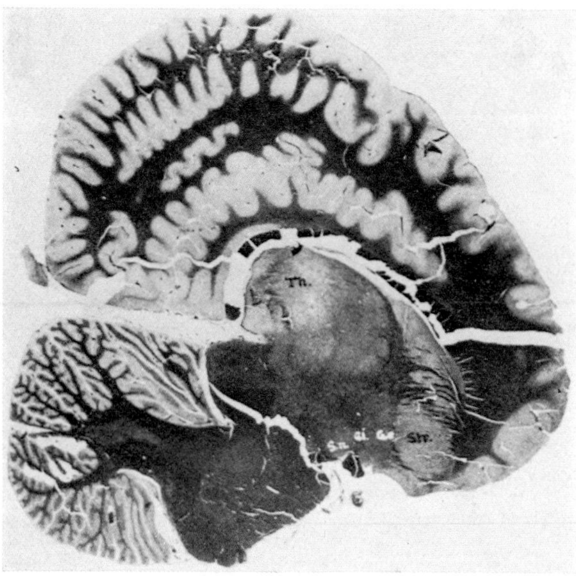

Figure 233 B. Sagittal section (myelin stain) through the brain of Delphinus delphis (from RIESE, 1925). Ge., Gi.: external and internal segment of globus pallidus; Th.: thalamus; Sn.: substantia nigra; Str.: striatum. Comparison with Figure 233 A will explain the dorsal location of thalamus, related to pronounced hemispheric bend, as seen in that cross-section.

more or less protruding tuberculum olfactorium is characteristic for macrosmatic Mammals, but the area ventralis anterior with its basal cortex is present in microsmatic as well as in anosmatic forms. In Man, it roughly corresponds to the so-called *substantia perforata anterior*.

The area *ventrolateralis posterior* (b), separated from the cortex of lobus piriformis by the caudal end of sulcus endorhinalis, is the *cortical nucleus amygdalae* dealt with above in discussing the amygdaloid complex. Due to the hemispheric bend, it becomes adjacent to intermediate anterior piriform lobe cortex, to parahippocampal cortex of posterior piriform lobe, and to the rostrobasal end of the hippocampal formation (cf. Figs. 221 B, C), being thereby included in the temporal lobe.

In addition to Figures 221 to 225 which illustrate the fundamental Mammalian telencephalic configurational and some structural relationships discussed in the preceding paragraphs, Figures 226 to 234 depict further aspects of these arrangements in diverse representative Mammals ranging from Prototherian Monotremes to the Eutherian Primate Macacus.

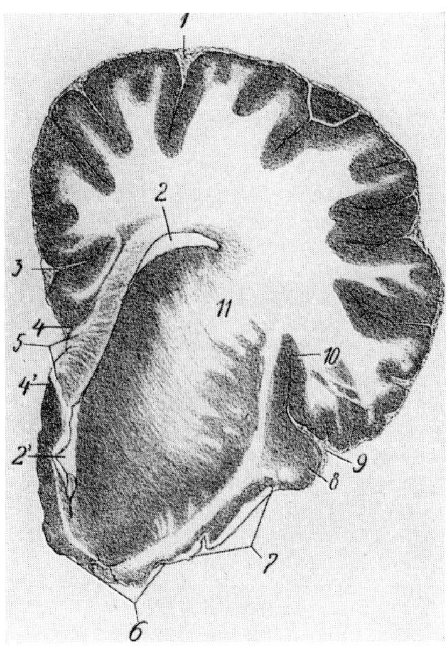

Figure 233 C. Cross-section (unstained) through right hemisphere of the Proboscidean Elephas indicus at rostralmost tip of corpus callosum (from DEXLER, 1907). 1: 'fissura longitudinalis paramediana dorsalis'; 2, 2': lateral ventricle; 3: sulcus calloso-marginalis; 4, 4': sulcus corporis callosi; 5: genu corporis callosi; 6: 'trigonum olfactorium' (basal cortex of area ventralis anterior); 7: stria (tractus) olfactoria lateralis and part of piriform lobe cortex; 8: 'gyrus olfactorius lateralis' (dorsal part of piriform lobe); 9: lateral trace of sulcus rhinalis lateralis; 10: 'bottom of fissura rhinalis' (rostral part of *fossa Sylvii);* 11: capsula interna. In the region between left lead of 6 and lead 4' there is rostral parahippocampal cortex and, near 4', precommissural hippocampal rudiment (not pointed out in the drawing).

As regards the *telencephalic communication channels*, it seems appropriate to distinguish (A) *intrinsic*, and (B) *extrinsic systems*. The former include (a) the *olfactory tracts*,[196] dealt with in section 2, (b) *fiber connections between cortical grisea*, to be considered in chapter XV, and (c) *fiber connections between cortex and basal grisea as well as between the basal grisea*.

The *extrinsic connections* (B), consisting of ascending and descending channels, were dealt with in section 9 of chapter XII with respect to the diencephalic grisea. It must, moreover, be taken into consideration that some of the main extrinsic channels, particularly those related to the basal grisea, also include intrinsic telencephalic fiber connections.

[196] These, of course, are not present in anosmatic Mammals.

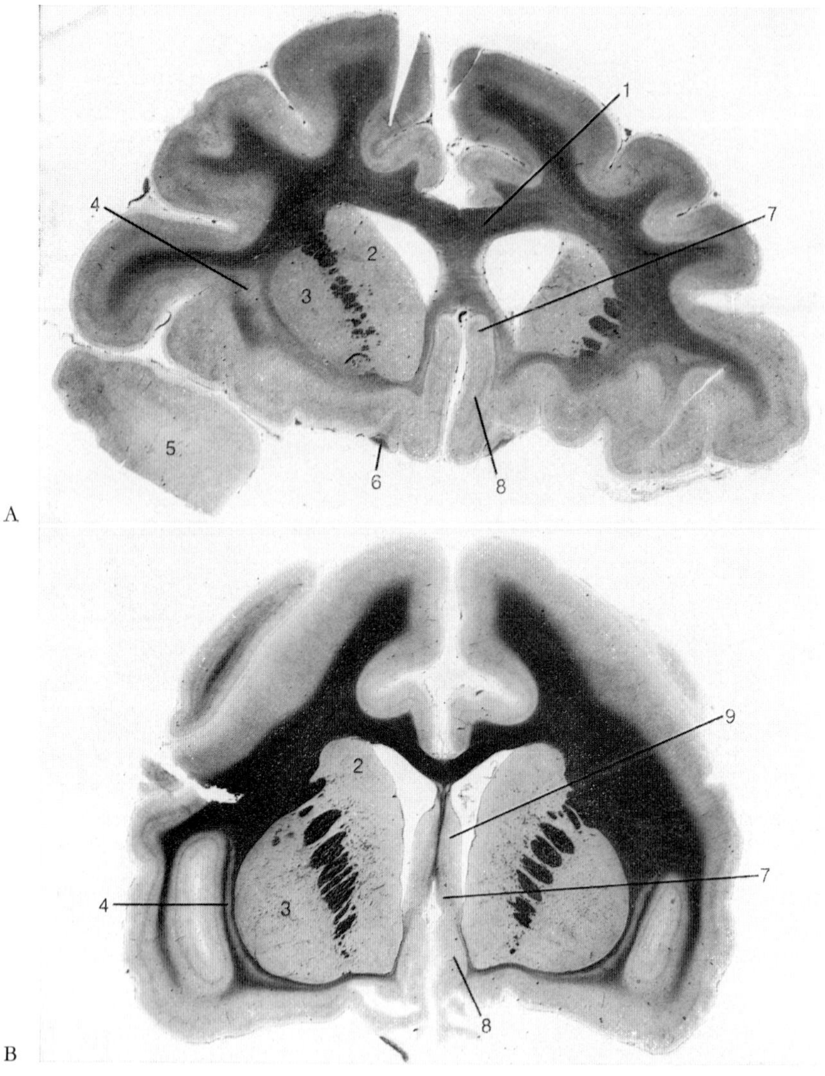

Figure 234 A–E. Cross-sections (myelin stain) through the telencephalon and adjacent regions in the Primate Macaque Monkey at five levels in rostrocaudal sequence. 1: corpus callosum; 2: nucleus caudatus; 3: putamen; 4: claustrum; 5: tip of temporal lobe; 6: olfactory tract (stalk); 7: precommissural hippocampus; 8: rostral parahippocampal cortex; 9: septal grisea; 10: fornix; 11: globus pallidus; 12: end of nucleus accumbens septi; 13: anterior commissure; 14: basal cortex; 15: tractus olfactorius lateralis; 16: anterior piriform lobe cortex; 17: optic chiasma and tract; 18: amygdaloid complex; 19: cortical amygdaloid nucleus; 20: posterior piriform lobe; 21: stria medullaris thalami; 22: postcommissural hippocampus; 23: part of lateral geniculate complex; 24: posterior

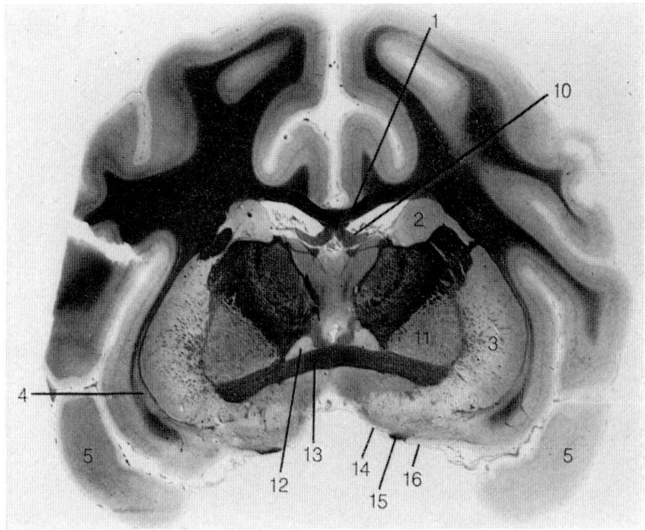

Figure 234 C

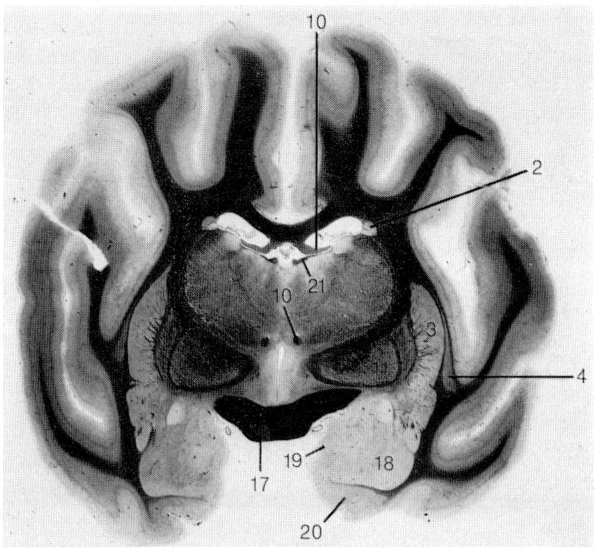

Figure 234 D

commissure; 25: substantia nigra (with pes pedunculi below and n. ruber tegmenti dorsomedially above); 26: medial geniculate complex; 27: pulvinar. Section A is from another specimen than sections B–E. For details of diencephalic configurations cf. the illustrations in section 9 of chapter XII.

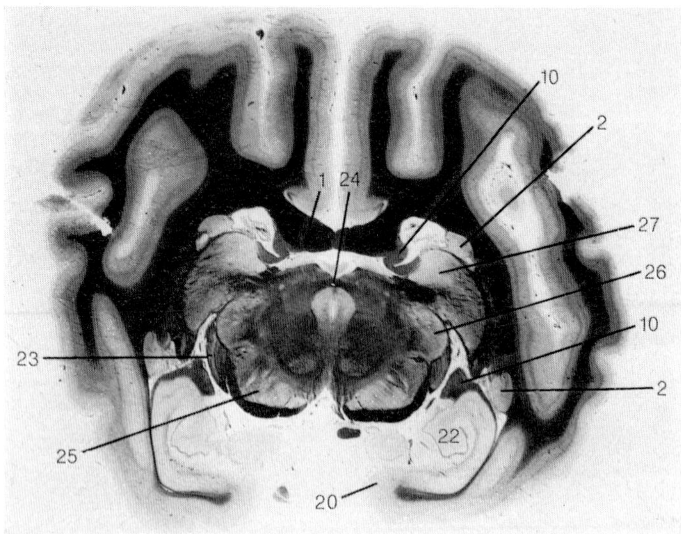

Figure 234 E (legend see p. 728)

With regard to the bilateral symmetric configuration of the telencephalon, there are, in addition, (C) the *commissural systems*, comprising (a) *commissura anterior* and (b) *commissura dorsalis sive pallii*, in which latter, at least in Eutheria, (1) *corpus callosum* and (2) *commissura hippocampi* can be clearly distinguished.

In the aspect here under consideration, discounting olfactory tracts as well as intrinsic cortical systems, and from a purely anatomical viewpoint, the main communication channels can be said to comprise (a) *capsula interna*, (b) *ansa lenticularis*, (c) *fornix system*, (d) *basal forebrain bundle*, (e) *stria medullaris system*, (f) *stria terminalis system*, and (g) *Broca's diagonal band*. There is, however, a certain degree of overlap between some of these systems. None of the various classifications as here adopted or used by other authors can be regarded as entirely satisfactory. Capsula externa and capsula extrema, mentioned above as related to the claustrum, may here be omitted. Intrinsic fibers systems passing through capsula externa shall be considered in chapter XV of volume 5/II.

(a) The *capsula interna*, rostrally either partly separating nucleus caudatus from putamen, or passing through the striatum in scattered bundles, is a complex system containing the mainly descending connections of the pallial cortex which gather in the *pes pedunculi*, and the reciprocal,

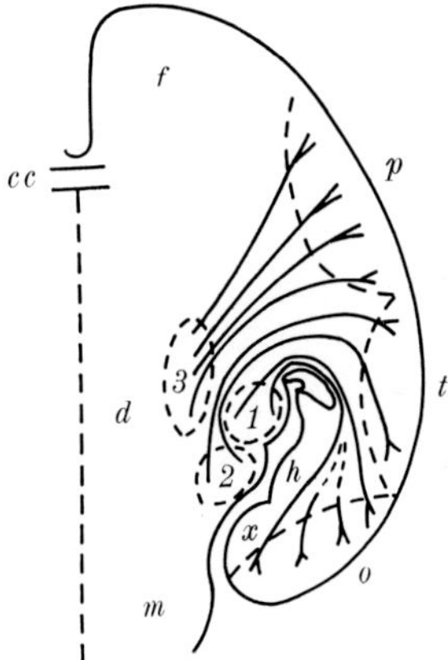

Figure 235. Diagram illustrating the overall spatial relationships of some main neopallial input channels (adapted and modified after HERRICK, 1926; K., 1927, and FORTUYN and STEFENS, 1951). cc: corpus callosum; d: diencephalon (thalamus); f: frontal (rostral) region of neopallium; h: hippocampal formation; m: mesencephalon; o: occipital neopallium; p: parietal neopallium; t: temporal neopallium; x: parahippocampal pallium; 1: lateral geniculate complex; 2: medial geniculate complex; 3: main sensory grisea of thalamus. The ascending channel to frontal region and all corticofugal channels have been omitted.

but particularly ascending systems providing neocortical and parahippocampal input from thalamic grisea. All these connections pass through the hemispheric stalk, where the capsula interna, except for retro- and sublenticular components in Mammals with extensive temporal and occipital lobes, lies between '*lenticular nucleus*' (putamen and pallidum) on the lateral side, and *caudate nucleus* as well as *thalamus* on the medial side. The dorsal subcortical white matter containing components of the capsula interna is the *corona radiata* which also includes radiating intrinsic cortical (e.g. also callosal) connections. The general arrangement of thalamic stalks was discussed in section 9 of chapter XII, and on pp. 647–650, section 6, chapter VI of volume 3/II. Figure 235 ilustrates overall spatial relationships of some main neopallial input channels.

(b) The *ansa lenticularis system*, like that of the capsula interna, can be conceived as a derivative or progressive differentiation of the lateral forebrain bundle of Reptiles and Anamnia. Its various components are poorly understood and in part controversial. Corticostriate and thalamostriate connections are believed to be present and also to reach the

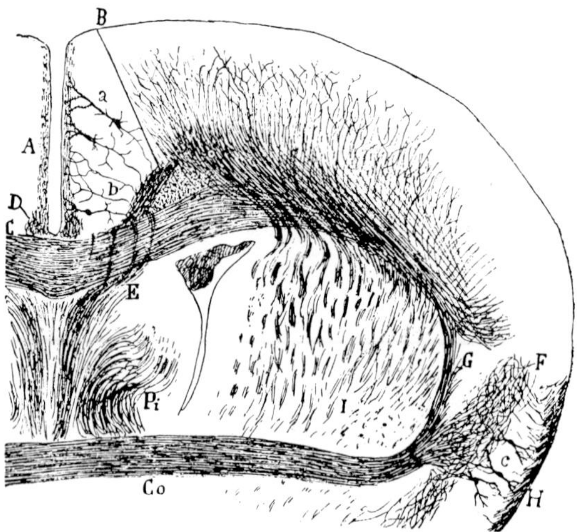

Figure 236 A. Cross-section *(Golgi impregnation)* through the telencephalon of a young Mouse at level of anterior commissure (from CAJAL, 1911). A: parahippocampal cortex; B: longitudinal subcortical bundle (cingulum); C: corpus callosum; Co: commissura anterior; D: striae longitudinales (medial and lateral components appear here blended); E: so-called fornix longus with fibrae perforantes; F: subcortical fiber plexus of piriform lobe; G: neopallial bundle of commissura anterior; H: tractus olfactorius lateralis; I: striatum with scattered bundles of capsula interna; Pi: columna fornicis; a: pyramidal cells of parahippocampal cortex; b: afferent fibers of parahippocampal cortex; c: cortical neurons of (intermediate) piriform lobe.

globus pallidus. From striatum and pallidum, channels extend to nucleus subthalamicus, griseum of the *field of Forel*,[197] grisea of hypothalamus *sensu strictiori*, as well as particularly to mesencephalic substantia nigra, moreover to nucleus ruber and prerubral tegmentum (cf. e.g. CROSBY *et al.*, 1962). Some of these various connections, essentially pertaining to the so-called *extrapyramidal* motor system, are indicated in Figures 150 A and B of chapter XII.

Reverting to the relevant relationships between the mesencephalic substantia nigra and the striato-pallidal complex, dealt with in section 9, chapter XI of volume 4, recent detailed studies by HASSLER and his associates (e.g. HAJDU *et al.*, 1973) are of some interest. In the substantia nigra, these investigators recorded, by electron-microscopy, 6 different types of synapses ('boutons') characterized by their particular

[197] Further relationships of these fiber connections to FOREL's fields H_1 *(fasciculus lenticularis)* and H_2 *(fasciculus thalamicus)* are discussed on p. 347 in section 9 of chapter XII.

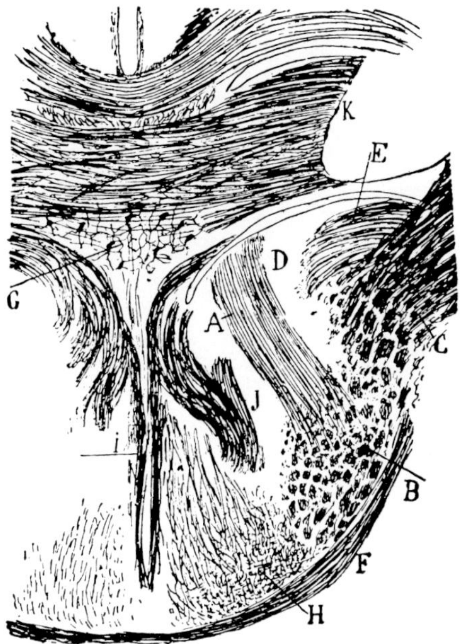

Figure 236 B. Cross-section *(Golgi impregnation)* through region of hemispheric stalk in a young Mouse (from CAJAL, 1911). A: stria medullaris system; B: intermingling of stria medullaris and capsula interna systems *('fibres de projection du cerveau')*; C: capsula interna; D: bed nucleus of stria terminalis and medullaris; E: stria terminalis; F: tractus opticus; G: caudal end of paraterminal grisea *('noyau triangulaire')*; H: basal forebrain bundle *('grande voie olfactive de projection)*; I: fornix component to rostral and intermediate hypothalamus *('faisceau ammonique destiné au tuber cinereum')*; J: main fornix bundle; K: commissura hippocampi *('psalterium ventral')* below unlabelled corpus callosum.

synaptic vesicles, one type being rather convincingly identified as *serotoninergic*. Generally speaking, it seem not improbable, on the basis of said studies, that GABA[198] represents the transmitter for striatonigral neurites, while the nigrostriatal ones appear to be *dopaminergic* (cf. KIM *et al.*, 1971; PARIZEK *et al.*, 1971; KATAOKA *et al.*, 1974).

(c) The *fornix system*, a derivative of the Anamniote medial forebrain bundle, is, to a substantial extent, an efferent channel of the hippocampal formation, apparently originating from the pyramidal cells of cornu Ammonis, including its so-called subicular portion, and gathering in the *fimbria fornicis sive hippocampi*, to which the choroid plexus

[198] Cf. volume 3/I, chapter V, section 8, p. 631.

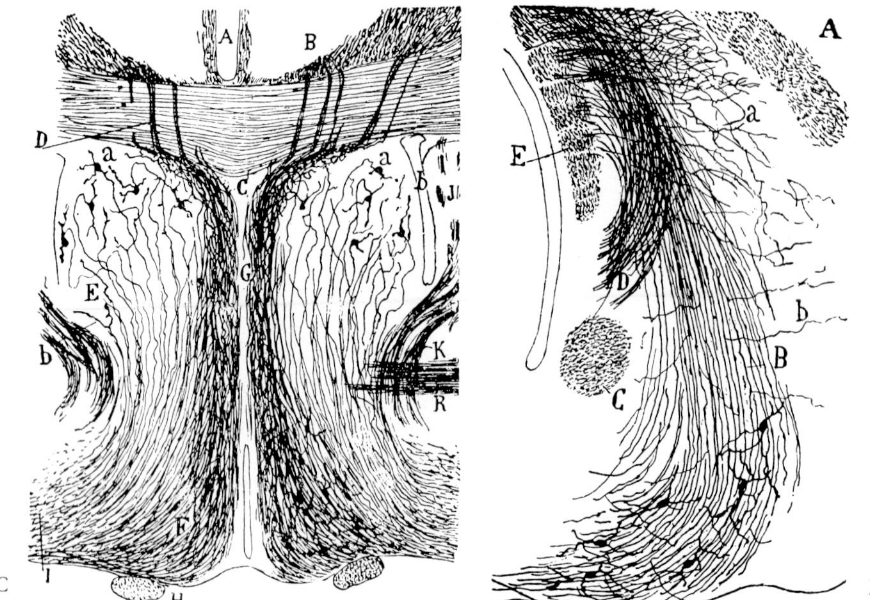

Figure 236 C. Cross-section *(Golgi impregnation)* through paraterminal grisea and preoptic recess of newborn Mouse (from CAJAL, 1911). A: interhemispheric fissure; B: cingulum with fibrae perforantes; C: fornix longus joined to *Zuckerkandl's bundle;* D: fibrae perforantes of fornix longus; E: paraterminal grisea (region of n. basimed. inferior); F: caudal portion of *Broca's diagonal band;* G: diagonal band combined with *Zuckerkandl's bundle;* H: optic nerve; I: more lateral portion of diagonal band; J: striatum; K: part of capsula interna; R: anterior commissure; a: cells of nucleus basimedialis superior; b (left): internal capsule; b (right): lateral ventricle.

Figure 236 D. Sagittal section *(Golgi impregnation)* through paraterminal grisea of young Mouse (from CAJAL, 1911). A: corpus callosum; B: *Zuckerkandl's bundle;* C: anterior commissure; D: columna fornicis; E: commissura hippocampi *('psalterium ventral');* a: psalterium collaterals to paraterminal grisea; b: collaterals of *Zuckerkandl's bundle.*

of the lateral ventricle is attached. Part of this channel forms the *crus fornicis*, which passes as *columna fornicis (pilier antérieur du trigone)* through the paraterminal region, crosses the caudal side of anterior commissure, and, taking its course through the hypothalamus, finally reaches the mammillary complex on its lateral side. Some of its fiber cross in the still poorly understood supramammillary commissure, first recorded by FOREL in 1872 as commissure Y in the Mole (1907), and dealt with in section 9 of chapter XII. Such crossed and perhaps also a

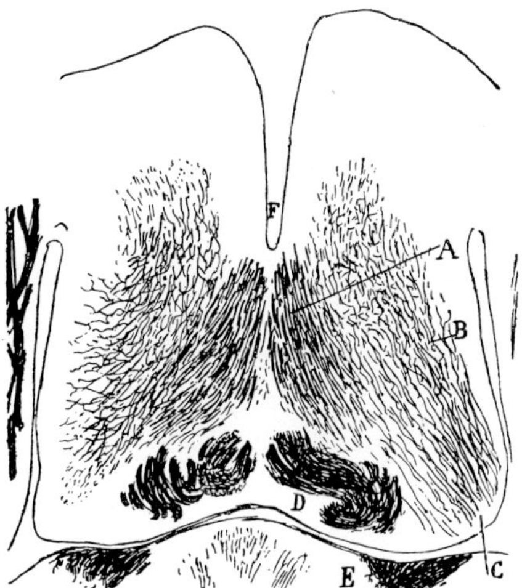

Figure 236E. Horizontal section *(Golgi impregnation)* through paraterminal grisea of young Mouse, dorsal to anterior commissure (from CAJAL, 1911). A: *Zuckerkandl's bundle;* B, C: lateral fiber systems of paraterminal grisea; D: corpus *sive* columna fornicis; E: stria medullaris on surface of diencephalon; F: interhemispheric fissure.

few uncrossed fornix fibers seem to reach tegmental (reticular) grisea of the midbrain, and have been described as *postmammillary fornix* (cf. CROSBY *et al.*, 1962).

This overall aspect of the fornix system is, however, complicated by a number of additional features, of which the following are the most conspicuous ones. (1) Fibers of the fornix join the commissura hippocampi which particularly displays differences between conditions in Eutheria and in Aplacentalia, to be discussed further below. (2) Fibers of the fornix system are distributed in variable ways to the paraterminal ('septal') grisea and may thereby have additional connections, overlapping with *diagonal band of Broca*, thus involving relations with e.g. tuberculum olfactorium and amygdaloid complex. (3) Fibers, pertaining to or originating in septal grisea, join the descending part of the fornix system, respectively the *diagonal band of Broca* into a rather inextricable bundle, also known as ZUCKERKANDL's *Riechbündel* (cf. ZUKKERKANDL, 1888) (CAJAL's *faisceau de Zuckerkandl)*, shown in Figures 236C, D, and E. In other words, essentially hippocampal components (2), and essentially paraterminal components (3) cannot be clearly distinguished from each other. (4) Some fibers of the fornix system pass into the hypothalamus as so-called *precommissural fornix.*[199] (5) Fibers of both pre- and postcommissural fornix join the preoptic and other non-mammillary hypothalamic grisea,[200] being thus not included in the channel reaching the mammillary complex. (6) Fibers of either pre- or postcommissural fornix or of both also seem to reach the anterior thalamic grisea. (7) Other, mostly postcommissural fornix fibers join the stria medullaris system (as so-called *cortico-habenular tract*). (8) Depending on Mammalian forms respectively on development of corpus callosum, fibers from supracommissural and precommissural hippocampus join in various ways the fornix system, and may be joined by parahippocampal fiber systems. (9) Longitudinal supracommissural fibers in various Eutheria represent the *striae longitudinales laterales et mediales* on the dorsal surface of corpus callosum, discussed further below. *Fibrae perforantes* from these striae commonly pass through the corpus callosum to join the subcallosal channels (cf. Figs. 236 A and C). (10) Undefined rostral portions of the fornix system, in the dorsal portion of paraterminal body, were first recorded and designated as *fornix longus* by FOREL in 1872 in his doctor's dissertation.[201] CAJAL (1911) considered the fibrae perforantes (corporis callosi) to be part of said fornix longus, which he also designated as *faisceau arqué du septum* (cf. Fig. 236 A). (11) In addition to the mainly descending fibers of the fornix system, a sizeable variety of ascending fibers, respectively of reciprocal connections may be present, and originate in the diverse grisea connected with the fornix.

The difficulties in disentangling these various systems will also be evident to the reader comparing their descriptions by CAJAL (1911), KAPPERS *et al.* (1936), BECCARI (1943), CROSBY *et al.* (1962), and ZEMAN and INNES (1963). Yet, despite numerous discrepancies and differences of interpretation, these accounts agree reasonably well with the generalized channels as given above.

A recent report by SWANSON and COWAN (1975) claims that, in the Rat, the efferent fibers of the 'hippocampus proper' are confined to the precommissural fornix, ending primarily in the 'septum'. The fibers

---

[199] The terms 'precommissural' and 'postcommissural', as applying to the fornix, refer to anterior commissure and should not be confused with precommissural, supracommissural and postcommissural hippocampus, referring to commissura pallii respectively corpus callosum.

[200] CAJAL's *faisceau médian de la cloison* in Figure 229 E, *faisceau ammonique destiné au tuber cinereum* in Figure 236 B.

[201] Included in '*Gesammelte hirnanatomische Abhandlungen*' (FOREL, 1907).

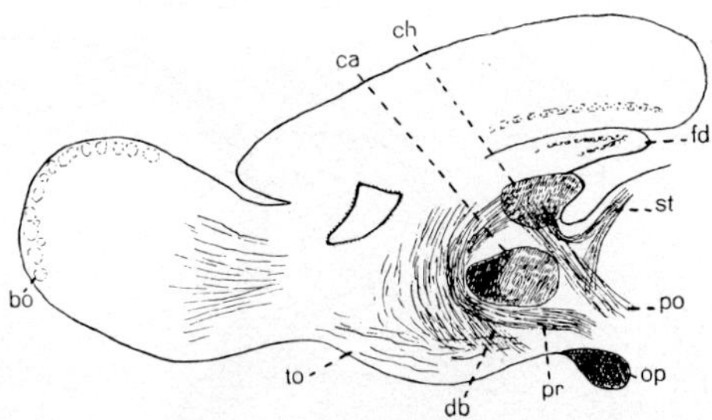

Figure 237 A. Sketch of sagittal section through forebrain of the Marsupial Didelphys illustrating various fiber systems (modified after EDINGER, 1908, from K., 1927). bo: olfactory bulb; ca: anterior commissure; ch: commissura dorsalis (comm. hippocampi); db: *Broca's diagonal band;* fd: gyrus dentatus *sive* fascia dentata; op: optic tract; po: fasciculus postcommissuralis fornicis (fornix *sensu strictiori);* pr: fasciculus praecommissuralis fornicis; st: striae medullares thalami; to: tuberculum olfactorium.

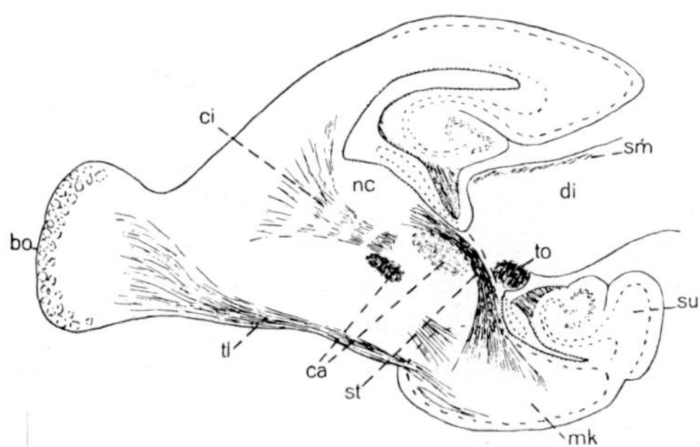

Figure 237 B. Sketch of sagittal section through forebrain of the Marsupial Didelphys, at a level somewhat lateral to that of Figure 237 A (modified after JOHNSTON, from K., 1927). bo: olfactory bulb; ca: rostral and caudal limb of anterior commissure; ci: capsula interna; di: diencephalon; mk: amygdaloid complex; nc: nucleus caudatus; sm: stria medullaris system (striae medullares thalami); st: striae terminales; su: subiculum cornu Ammonis; tl: tractus olfactorius lateralis; to: tractus opticus.

distributed to the postcommissural fornix are said to arise from the so-called 'subicular region'. The cited authors obtained their findings by means of an autoradiographic technique using injections of tritiated amino acids. This seems to be taken up by neurons being followed by subsequent transport of labeled proteins along their axons. One might, however, wonder whether the method is as accurate as the authors seem to believe.

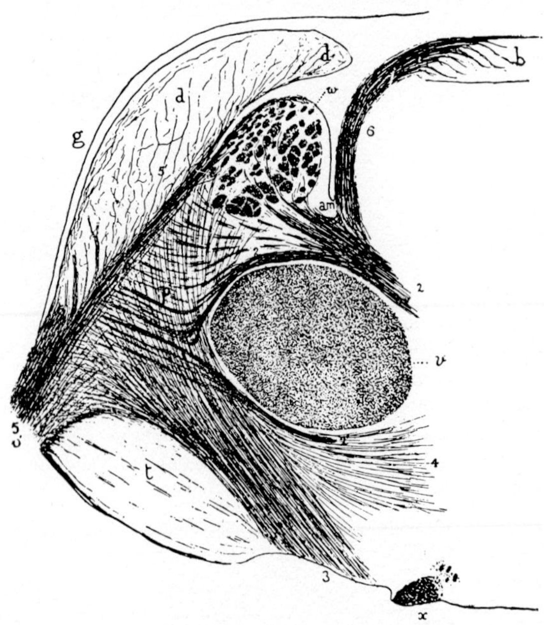

Figure 237C, I. Sagittal section through commissural and precommissural neighborhoods of the right cerebrum in the Prototherian Ornithorhynchus, showing main fiber systems (after ELLIOT SMITH, from JOHNSTON, 1906). am: tractus cortico-habenularis; b: nucleus habenulae; d: fascia dentata; g: parahippocampal cortex; o: 'olfactory peduncle'; p: precommissural (paracommissural) body; t: tuberculum olfactorium; v: anterior commissure (including neopallial fibers); w: hippocampal commissure; x: optic tract; 2: supracommissural fornix; 2′: infracommissural fornix; 3: *diagonal band of Broca;* 4: basal forebrain bundle; 5: part of medial olfactory tract reaching hippocampal formation; 6: stria medullaris thalami.

(d) The somewhat variably developed *basal forebrain bundle* of Mammals, although frequently designated as 'medial forebrain bundle', may be considered the phylogenetic remnant of those components of both lateral and medial Anamniote forebrain bundles not differentiated into internal capsule and ansa lenticularis, respectively into fornix system. It includes fibers related to area ventralis anterior (tuberculum olfactorium), fundus striati, and paraterminal grisea, connecting with the hypothalamus as pointed out in sections 1A and 9 of chapter XII (cf. Figs. 229E, 236B). This basal forebrain bundle is doubtless in part related to the olfactory system (*'voie olfactive de projection'*, *'grande voie olfactive de projection'* of CAJAL, 1911) but also includes presumably non-olfactory channels of the vegetative nervous system and is

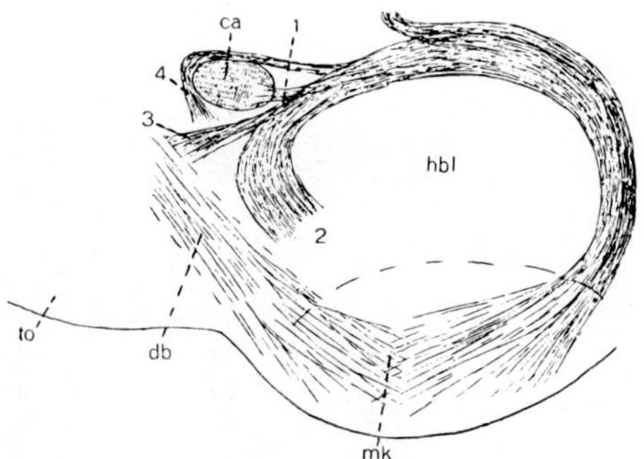

Figure 237C, II. Sketch illustrating the overall arrangement of the stria terminalis system in the Mammalian hemispheric stalk, as projected upon a sagittal plane (modified after JOHNSTON, from K., 1927). ca: commissura anterior; db: *Broca's diagonal band;* hbl: hemispheric stalk *(Hemisphärenblasenstiel);* mk: amygdaloid complex; to: tuberculum olfactorium; 1–4: components of striae terminalis enumerated in text.

present in macrosmatic, microsmatic, and anosmatic Mammals. The probable extension of the basal forebrain bundle into the mesencephalic tegmentum, as indicated on p. 21 of section 1A, chapter XII, *et passim*, should again be emphasized. It is, moreover, quite possible that, by means of this channel, ascending secondary afferent trigeminal nuclei reach tuberculum olfactorium and fundus striati *('lobus parolfactorius')* at least in some Mammals, in accordance with EDINGER's (1912) hypothesis of an *'oral sense' (Oralsinn)*. This sensory mechanism, as illustrated by Figure 237E, is assumed to correlate olfactory impulses and touch impulses of the snout, e.g. while probing the ground in search for food *(Spürsinn der Schnauze unter Zusammenwirkung von Olfactorius und Trigeminus)*. An at least partially comparable mechanism is well documented by the quintofrontal tract of Birds, pointed out in the preceding section 9. The Avian olfactory sense, however, is much less well developed than that of most Mammals, and may thus here play a lesser role than in EDINGER's Mammalian 'oral sense'.

(e) The *stria medullaris system* (Figs. 236B, E, 237A, B) gathers the channels connecting the epithalamus (habenular nuclei) with diverse telencephalic grisea (bulbar olfactory grisea, area ventralis anterior, amygdaloid grisea, piriform lobe, paraterminal grisea, hippocampal and perhaps parahippocampal grisea), as also dealt with in section 9 of chapter XII.

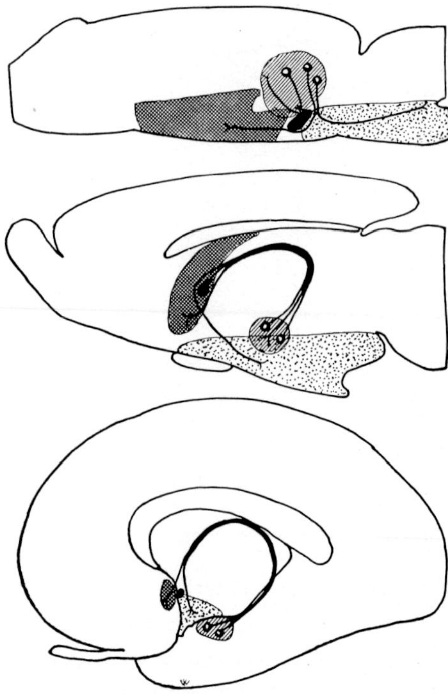

Figure 237 D. Diagrammatic sketches purporting to illustrate evolution of stria terminalis in KRIEG's interpretation (from KRIEG, 1966). Top: Amphibian stage; middle: lower Mammalian stage; bottom: Human stage; oblique hatching: 'amygdala'; cross-hatching: 'septal' grisea; stipple: hypothalamus.

(f) The *stria terminalis system* is a perhaps predominantly but not exclusively efferent channel of the amygdaloid complex (Figs. 225 C, 229 E, 236 B, 237 B, C, D). In conformity with the caudal limb of the hemispheric bend, this fiber system follows the sulcus terminalis, describing a curve around the hemispheric stalk, as shown in Figure 237 C, and is thus also referred to as *taenia semicircularis (bandelette semi-circulaire, stria cornea)*.

Generally speaking, the following 5 components can roughly be distinguished. (1) *Fasciculus commissuralis*, joining the anterior commissure. (2) *Fasciculus hypothalamicus*, reaching undefined hypothalamic grisea. (3) *Fasciculus infracommissuralis*, related to paraterminal grisea, fundus striati and perhaps tuberculum olfactorium. (4) *Fasciculus supracommissuralis* perhaps related to fundus striati and possibly also to

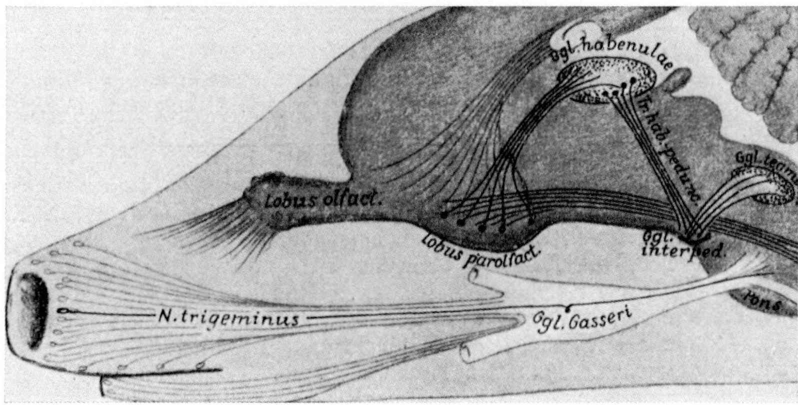

Figure 237E. Sketch illustrating the circuits involved in the oral sense *(Oralsinn)* postulated by EDINGER (from EDINGER, 1912). Lobus olfact.: bulbus olfactorius; lobus parolfact.: tuberculum olfactorium *sive* area ventralis anterior; Glg.tegmenti: nucleus dorsalis tegmenti *sive* ganglion tegmenti profundum of GUDDEN.

preoptic hypothalamus. (5) *Fasciculus habenularis*, joining the stria medullaris system.

(g) *Broca's diagonal band* (Figs. 223A, 236C, 237A, C) is a system of fibers running close to the basal brain surface. It essentially connects the paraterminal grisea with area ventralis anterior (tuberculum olfactorium), area ventrolateralis posterior (cortical amygdaloid nucleus) as well as perhaps other amygdaloid grisea and some adjacent hypothalamic grisea. The term 'diagonal' refers to the obliquely curved course of some of its components, running from cortical amygdaloid nucleus, in a rostrodorsal direction, medially to substantia perforata anterior or olfactory tubercle, toward the lamina terminalis (cf. Fig. 175D). In some Mammals, including Man, the preterminal portion of this bundle, together with its accompanying griseum, forms the so-called *gyrus subcallosus*. There are, in addition, nondescript components of diagonal band fibers related to the basal forebrain bundle system.

The Mammalian telencephalic *commissural systems* consist, as in all other Vertebrates with inverted telencephalon, of a *dorsal commissure* carrying fibers related to the pallium, and of a *ventral commissure* whose fibers are related to both basis and pallium.

As regards the *dorsal commissure*, some differences obtain between that of Prototherians and Metatherians on one hand, and that of Eutherians on the other. In the former groups, the dorsal commissure is mostly related to the hippocampal formation with its well developed supracommissural portion, and is, therefore, commonly designated as *commissura hippocampi*, although it apparently also includes a variable amount of fibers related to the parahippocampal cortex. The commissural fibers related to the neocortex take here their course through the ventral commissure (commissura anterior). KAPPERS (1921) and KAPPERS *et al.* (1936) pointed out that in some Marsupials, e.g. in Macropus, a corpus callosum is not lacking, since some 'neopallial' fibers pass through the dorsal commissure. It seems, however, likely that said fibers pertain mostly to the parahippocampal cortex which KAPPERS and many other authors consider to be part of the neocortex.

In Eutheria, concomitantly with the expansion and predominance of the postcommissural hippocampus, the supracommissural portion becomes reduced, and an increasingly substantial amount of neopallial fibers takes its course through the rostral portion of the dorsal commissure, whereby its hippocampal components are displaced caudalward and ventralward to the neopallial commissural channel. This latter represents the *corpus callosum*[202] in contradistinction to the displaced *commissura hippocampi*. The supracommissural hippocampus becomes concomitantly reduced (cf. Figs. 223 C, 238 A, B). In most Eutherian forms, this portion of the hippocampus is represented by a wedge-like cortical fringe accompanying the parahippocampal cortex.[203] This fringe may or may not extend to a variable degree upon the surface of corpus callosum as *indusium spurium*. The *indusium verum*, on the other

---

[202] Further comments on morphologic and phylogenetic aspects of the Vertebrate respectively Mammalian telencephalic commissures can be found on pp. 619–627 in section 6, chapter VI, of volume 3/II, where the ontogenetic origin and development of the corpus callosum is dealt with. Because of the ventralward displacement displayed by commissura hippocampi, ABBIE (1939) elaborated an interpretation postulating an hypothetical intermediate phylogenetic stage in which a secondary growth of callosal fibers perforated the supracommissural hippocampus (his 'subiculum'), leaving one portion above, and displacing another below this 'new' commissure. There is no ontogenetic nor comparative anatomical ('phylogenetic') evidence whatsoever for this hypothesis. The relationships of hippocampus, corpus callosum, and fornix in Eutheria qua derivation from those in Aplacentalia are clearly suggested as caused by simple caudalward displacement combined with supracallosal hippocampal reduction and a bending around the splenium related to a folding due to growth processes of the pallium at the dorsal neighborhood of caudal limb of hemispheric bend. This results in a rostral convexity of the hippocampal formation beneath the ventral splenial surface (cf. Fig. 221 C).

[203] ROSE (1929 *et passim*) as well as some other authors use the term *taenia tecta* for this fringe of reduced supracommissural hippocampal cortex, but said term originally referred to the *stria longitudinalis lateralis* (cf. e.g. VILLIGER, 1920).

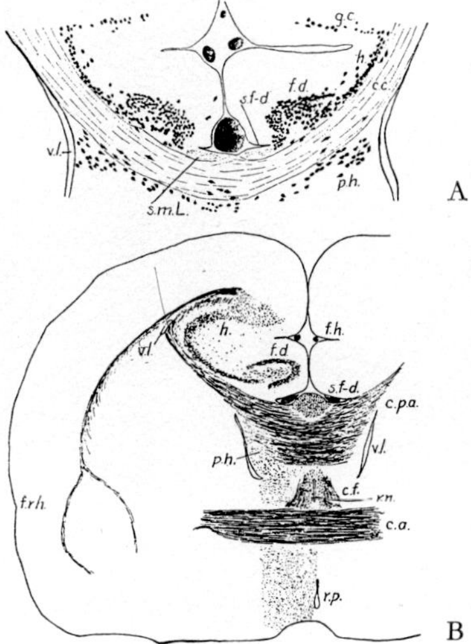

Figure 238 A. Cross-section through middle level of corpus callosum in the Chiropteran Bat Myotis showing supracommissural hippocampal formation (from JOHNSTON, 1913b). c.c.: corpus callosum; f.d.: fascia dentata; g.c.: parahippocampal cortex; h.: cornu Ammonis; p.h.: nucleus basimedialis superior (JOHNSTON's 'primordium hippocampi'); s.f-d.: 'sulcus fimbrio-dentatus'; s.m.L.: '*stria medialis Lancisii*' (here actually the conjoint striae longitudinales mediales et laterales); v.l.: lateral ventricle.

Figure 238 B. Cross-section, at level of pars transversa (horizontal limb) of anterior commissure and neuroporic recess, through forebrain of the Bat Myotis (from JOHNSTON, 1913b). c.a.: commissura anterior; c.f.: columna fornicis; c.p.a.: commissura pallii; f.h.: fissura hippocampi; f.rh.: sulcus rhinalis lateralis; r.n.: recessus neuroporicus; r.p.: recessus praeopticus (hypothalamus). Other abbreviations as in Figure 238A. The commissura hippocampi, not labelled, at level of p.h. below corpus callosum.

hand, is a variable amount of gray matter along the median dorsal surface of corpus callosum and represents the supracommissural rudiment of nucleus basimedialis superior. Two pairs of rather thin longitudinal fiber bundles likewise run along the dorsal surface of corpus callosum, the *stria longitudinalis lateralis sive taenia tecta* close to the supracommissural hippocampus, and the *stria longitudinalis medialis sive stria Lancisii* related to the indusium verum. From both paired bundles, perforating fibers passing through the corpus callosum commonly join the fornix system (cf. Figs. 229E, 236A, C).

The subdivisions of *corpus callosum* into *rostrum*, *genu*, *corpus*, and *splenium* were dealt with and depicted in section 6, chapter VI of volume 3/II. It will here be sufficient to add that a typical genu, located

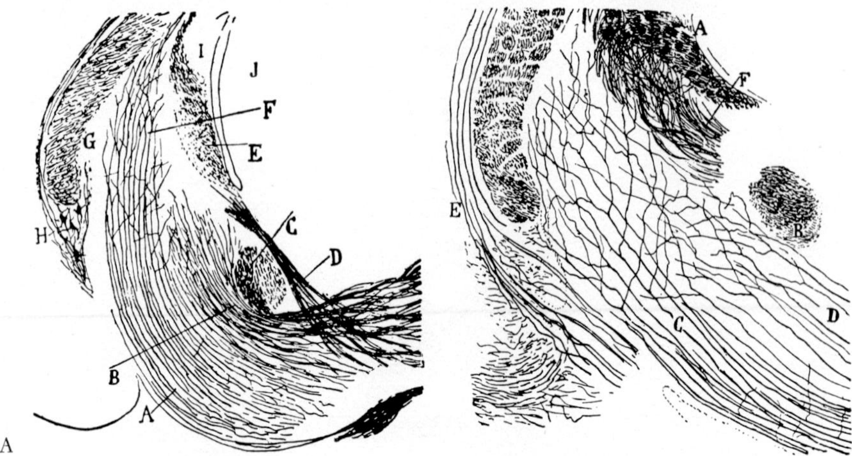

Figure 239 A. Almost midsagittal section *(Golgi impregnation)* through corpus callosum and paraterminal grisea of a young Mouse (from CAJAL, 1911). A: ascending fibers of *Zuckerkandl's bundle;* B: precommissural bundle (presumably of fornix system); C: anterior commissure; D: general hypothalamic component of fornix system *('faisceau ammonique allant au tuber cinereum');* E: ventral psalterium (commissura hippocampi *sive* fornicis); F: dorsal portion of *Zuckerkandl's bundle;* G: corpus callosum; H: portion of nucleus basimedialis superior; I: fascia dentata; J: diencephalon.

Figure 239 B. Midsagittal section *(Golgi impregnation)* through paraterminal region (lamina terminalis) of a young Mouse (from CAJAL, 1911). A: ventral psalterium; B: anterior commissure; C, D: ascending and branching fibers of *Zuckerkandl's bundle;* E: striae longitudinalis corporis callosi; F: columna fornicis.

between rostrum and corpus, is only well developed in large Mammals with prominent rostral limb of hemispheric bend, whereby the rostrum displays a corresponding curvature.

The *commissura hippocampi sive fornicis*, also called *psalterium*,[204] is related to the postcommissural hippocampal formation. In at least some Mammals, CAJAL (1911) could distinguish a *psalterium dorsale*, through which fibers from the contralateral parahippocampal cortex reach the hippocampus *(voie temporo-ammonique croisée)*, and a *psalterium ventrale*, representing a commissure of crura fornicis and which is, in CAJAL's opinion, a *commissure inter-ammonique*.[205] Figures 239 A–E illustrate di-

[204] The term *psalterium sive lyra Davidis* originally referred, in human gross anatomy, to the triangle formed by commissura fornicis and converging crura fornicis ventrally to splenium corporis callosi.

[205] This, however, does not exclude decussating mainly descending but perhaps also ascending fibers related, by way of the contralateral columna fornicis, to the diencephalic course of this fiber system.

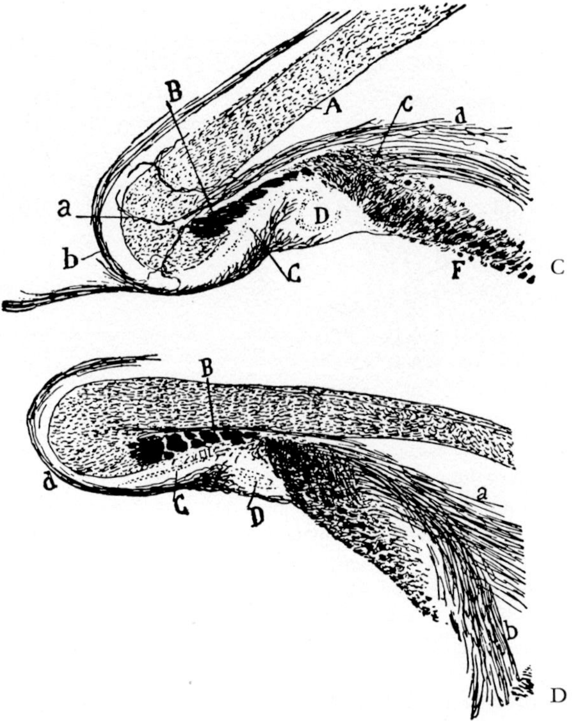

Figure 239C. Sagittal section close to midline *(Golgi impregnation)* through caudal portion of corpus callosum in a young Mouse (from CAJAL, 1911). A: corpus callosum; B: dorsal psalterium; C: cornu Ammonis; D: fascia dentata; F: fornix; a: perforating fiber in splenium; b: striae longitudinalis curving around splenium; c: psalterium ventrale; d: unidentified fornix system fibers in paraterminal grisea.

Figure 239D. Midsagittal section *(Golgi impregnation)* through caudal region of corpus callosum in a young Mouse (from CAJAL, 1911). B: dorsal psalterium *(voie temporo-ammonique croisée)*; C, D: begin of postcommissural hippocampus formation; a: collaterals of ventral psalterium; b: *Zuckerkandl's bundle;* d: striae longitudinales corpori callosi.

verse aspects of Mammalian corpus callosum and psalterium as recorded by CAJAL in exceptionally successful *Golgi preparations* of the Mouse brain.

The morphologic significance of the *anterior commissure* as midline landmark of the Vertebrate telencephalo-diencephalic boundary was pointed out on pp. 179–180 and 195–199, in section 1B, chapter VI of volume 3/II. The mammalian anterior commissure is a rather conspicuous fiber bundle whose pars transversalis crosses, ventrally to the paraterminal grisea, at a right angle to the midsagittal plane. Lateral-

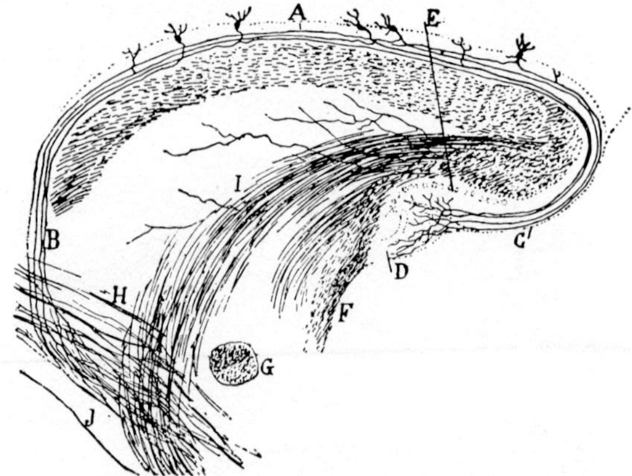

Figure 239E. Semidiagrammatic drawing of sagittal section *(Golgi impregnation)* through corpus callosum of a young Mouse (from CAJAL, 1911). A: stria longitudinalis medialis; B: 'projection fibers' of stria longitudinalis medialis; C: fibers of stria longitudinalis to fascia dentata; D: fascia dentata; E: cornu Ammonis; G: anterior commissure; H: 'projection fibers of frontal cortex'; I: *Zuckerkandl's bundle;* J: transition of area ventralis anterior to area ventrolateralis posterior.

ward, the anterior commissure displays a rostral and a caudal limb (cf. Figs. 176A, B, 229C, E, 230C). Its detailed particular composition varies in accordance with the taxonomic groups.

Generally speaking, the following components can be distinguished. (1) *interbulbar fibers;* (2) *fibers related to tuberculum olfactorium, paraterminal grisea, and fundus striati;* (3) *fibers related to piriform lobe cortex;* (4) *fibers related to amygdaloid complex and stria terminalis system;* (5) *fibers related to neopallium respectively neocortex.*

The interbulbar components (1), consisting of second order olfactory fibers from formatio bulbaris as well as third order fibers from nucleus olfactorius anterior, were pointed out in section 2 of the present chapter. Components (1) and (2) essentially represent the anterior limb, while components (3) and (4) pertain to the posterior limb. Components (5), although not necessarily restricted to the posterior limb in all Mammals, are, on the whole, related to this latter in 'higher forms', comprising here fibers of the temporal neocortex. Again, in addition to the interbulbar fibers (1), the components (2), (3), and (4) pertain largely to the olfactory system, except, of course, in anosmatic forms. Components (3), however, can also include parahippocampal fiber systems from the posterior part of piriform lobe which may not necessarily pertain to the 'olfactory system'. Much the same can be said about components (4) related to amygdaloid complex and stria terminalis system. It is, moreover, uncertain to which extent the diverse fibers passing through the anterior commissure are commissural *sensu strictiori* (i.e. interconnecting antimeric grisea), or merely decussating fibers of descending respectively ascending 'tracts'.

It should also be added that the *commissura habenulae*, although being a diencephalic (epithalamic) commissure, in which fibers of the stria terminalis system cross, may, in this respect, also represent a telencephalic commissure. To an undetermined extent, and *mutatis mutandis*, this comment also applies to the system of supra- and postoptic basal diencephalic commissures.

In concluding the preceding discussion of main telencephalic communication channels, the critical monograph by Knook (1965), referred to and discussed on pp. 341 f. of chapter XII, should again be mentioned with respect to numerous additional and still incompletely ascertained details. Other studies of importance are those by Poliak (1932) which concern the Macaque monkey, and those of Krieg (1954, 1963, 1973) dealing with Macaque, albino Rat, and Man. The fiber systems of the Human telencephalon will be dealt with separately further below in this chapter as well as in chapter XV of volume 5/II.

Turning now to the *functional significance* of the Mammalian lobus hemisphaericus, it is perhaps appropriate first to consider those animal experiments in which a *total* or almost total extirpation of the telencephalon was performed. The most relevant studies are perhaps those by Goltz (1892), by M. Rothmann (1909) and by Dresel (1924) on Dogs,[206] by Dusser de Barenne (1919) on Cats, by Rogers (1924) on Opossums, and by Karplus and Kreidl (1914) on Macacus.

Complete or nearly complete elimination of the Mammalian telencephalon is followed by very striking changes in behavior, generally describable as *stupor* or profound depression, with loss of the complex habits, and occasionally alternating with periods of restlessness.[207] The Dog, for instance, appears as not any more recognizing friends or enemies, nor the given situations as either familiar or unfamiliar. Again, after some time, certain, even slight touch stimuli may provoke outbursts of motor and vegetative reactions interpreted as '*sham rage*'.

---

[206] A detailed review and interpretation of the findings in M. Rothmann's decerebrate Dog was subsequently published by that author's son, H. Rothmann (1923).

[207] Cf. also the concise summary by Herrick (1926).

However, with the passage of time, a number of fairly complex functions reappear. The Dog seeks food and avoids obstacles. H. ROTHMANN (1923) summarizes these events as follows.

'*Bei allen diesen Leistungen geht aus der Beschreibung aber deutlich der Einfluss der ständigen Übungen hervor, und es ist erstaunlich, wie weitgehende Funktionen der grosshirnlose Hund auf allen Gebieten, sei es der Motilität und Sensibilität, sei es der Nahrungsaufnahme oder der Stimmfunktion, gezeigt hat.*

Es ist interessant zu beobachten, wie der Hund ohne Grosshirn sich aus einem seelenlosen Bewegungsautomaten, ja einem Idioten, zu einem sich aus eigener Spontaneität frei im Raume bewegenden Lebewesen wieder entwickelt hat und dies bei beinahe völligem Ausfall der höheren Sinnesorgane.'

The author evidently meant here '*höhere Sinneszentren*' (i.e. cortical sensory grisea), since none of the 'higher organs' of sense such e.g. as eye or ear were in any way affected.

'*Es ist auch deutlich einzusehen, wie nach dem Fortfall des Grosshirns nach und nach sich die niederen Zentren in ihrer Leistungsfähigkeit gesteigert haben und den grosshirnlosen Hund zu komplizierten Reflexen und Leistungen im Bereich der höheren Sinnesorgane wie der niedrigsten psychischen Regungen befähigt haben.*'

However, concerning '*psychische Regungen*' (i.e. consciousness) one may remain in doubt, and the question may be asked, as reformulated by LHERMITTE (1924) in referring to a controversy between GOLTZ, MUNK, ZIEHEN, and others: '*Doit-on considérer le chien décérébré comme un pur automate ou comme un être chez lequel persistent encore des résidus de sensations et une ébauche de conscience?*' I believe, however, that this question cannot be answered with any degree of certitude.

As regards more or less extensive ablations of cerebral cortex, and particularly of neopallial areas, relevant studies were undertaken by MUNK (1881) in the Dog and Monkey, as well as by GOLTZ (1888) on the Dog, LEYTON and SHERRINGTON (1917) in anthropoid Apes, and by some other authors.

A very systematic series of investigations has been carried out by LASHLEY (1929 and numerous other publications) in the Rat. Figuratively speaking, nature has occasionally performed such experiments on Man, as e.g. recorded by EDINGER and FISCHER (1913), GAMPER (1926), and ourselves (K. *et al.*, 1959, 1964). Reference to some aspects of these observations shall be included further below in dealing with the Human telencephalon.

LASHLEY's findings in the Rat provide a considerable amount of factual evidence which can be interpreted in several ways. The cerebral cortex of the Rat displays a number of ill-defined projection areas with characteristic subcortical (essentially thalamo-cortical) connections and apparently with limited specific functional significance. Certain visual functions are definitely related with an area of occipital cortex. Nevertheless, the entire cortex, at least to a large extent, seems to be *equipotential in learning*. LASHLEY demonstrated that no single part of the cerebrum could be considered necessary for the learning of various training habits. If a Rat with intact cortex is taught a visual discrimination habit and then the occipital cortex is removed, that habit becomes totally lost. Removal of no other part of the cortex has this effect. However, after removal of occipital cortex, the habit can be reacquired by renewed training. It appears nevertheless possible that the occipital cortex of the Rat is essential for certain aspects of pattern vision, yet a Rat without occipital cortex shows normal discrimination of light intensity, can avoid obstacles, and seems to recognize food by sight.

When blind Rats are taught to acquire the maze habit, destruction of occipital cortex impairs this habit and is followed by considerable difficulties in relearning, although the Rats could not have used visual stimuli in the initial learning. The occipital (visual) cortex of the Rat appears thus to be of significant import for the integration of spatial relationships, even if the animal is blind.

With regard to the equipotentiality of the Rat's cerebral cortex and especially of its neocortex, experiments performed by MAIER (1934) seem, moreover, to indicate that the pattern of a lesion as well as its extent may determine the degree of the resulting functional deterioration. Thus, if symmetrically paired single, double, and triple cortical lesions on both hemispheres were inflicted, the lesions being spatially separated, but their total extent (i.e. the 'mass' of destroyed tissue being equal), the subsequent behavioral deterioration was found to be inversely proportional to the ratio of the length to the width of the lesions.

Despite some undeniable localizatory principles, considerable equipotentiality doubtless obtains in the cerebral cortex of the Rat. This equipotentiality is not altogether identical with, but is closely related to the principles of *stability* and *ultrastability* as formulated by ASHBY. Equipotentiality in this connotation may be defined as the capacity of a given griseum or cortical region to perfom a variety of different functions, similar or identical to functions performed by other regions or grisea. Equipotentiality in this sense implies unspecificity or limited specificity combined with multiplicity of functions. In addition, the topographic distributions of various performances will depend on functional demands, such that whatever region is momentarily idle will become available, either *in toto* or with a set of its elements, for the execution of a given activity. Likewise, if a certain area is destroyed, at least some other areas will be able to function vicariously. Comparable manifestations of equipotentiality can easily be provided by the circuitry of electrical computing or control mechanisms. The so-called '*bus-lines*' of such mechanisms represent unspecific transmission lines. Computers may likewise automatically switch from failing circuits to properly functioning circuits. Position frames connect a problem in a given problem position to any computer that has become idle and is therefore available. Such machines also switch automatically from one problem position to others. Banks of registers, so-called panels, may perform double or triple duty, depending on conditions, as for instance storing either problem or program.

Equipotentiality is certainly also, to some extent, manifested by Human cerebral cortex and other Human brain structures, but I do not believe that the Human brain can be interpreted as narrowly in terms of Rat-brain functions as many of the views elaborated by LASHLEY seem to imply. I fully agree in this respect with most of the evaluations and critical comments expressed by C. J. HERRICK (1926).

Taking the occipital (visual) cortex as an example, vision will be much more severely impaired in a Dog than in a Rat, if that cortex is removed, although the Dog can still discriminate differences of light intensity. In Monkeys, serviceable vision seems to be completely abolished by removal of both occipital lobes (KLÜVER, 1942). It appears thus most likely, as this author suggests, that Primates and non-Primates differ considerably as to the way in which visual functions are affected by destruction of the visual cortex.

Quite apart from presumably substantial differences in detailed synaptology and biochemistry between the various Mammalian (and Vertebrate) forms, factors related to 'equipotentiality' and factors related to 'localization' seem to interplay. Thus, an apparently contradictory emphasis on their respective aspects may not actually imply their incompatibility, and merely indicate the unpredictable, probabilistic but nevertheless presumably *causally* and not *merely* statistically *determinate* behavior of the '*uncertain brain*'.

Closely related to *ablation* experiments are those performed by *transection*, as e.g. by BREMER (1935). In the Cat, transection at the mesencephalic level, just caudally to the oculomotor nucleus, produces a

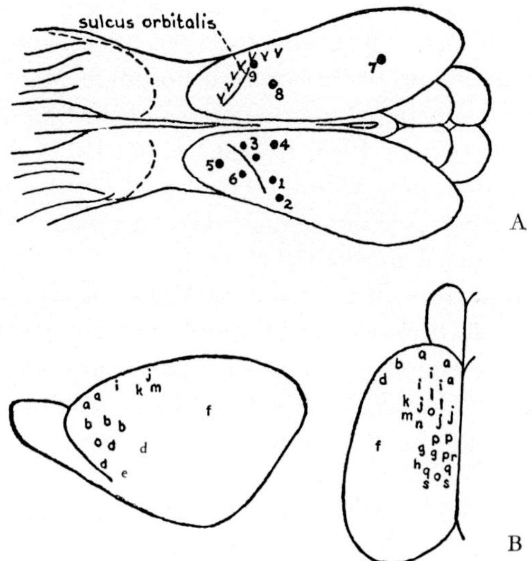

Figure 240 A. Electrically excitable cortical spots of the Opossum (after Gray and Turner, 1924, from Herrick, 1926). 1: extension of fingers; 2: flexion of fingers; 3: flexion of elbow; 4: twitching on back of forearm; 5: contraction of orbicularis oculi; 6: snout movements; 7: erection of ear; 8: retraction of vibrissae; 9: movements of tongue; v: erection of vibrissae.

Figure 240 B. Electrically excitable cortical spots of the Rat (after Lashley, from Herrick, 1926). a: head turned to opposite side; b: nose retracted; c: vibrissae moved; d: chewing movements; e: tongue protruded; f: eye closed; g: ear adducted; h: ear erected; i: shoulder drawn forward; j: forearm retracted; k: elbow flexed; l: elbow extended; m: wrist flexed; n: forearm rotated; o: back flexed to opposite side; p: hind leg drawn forward; q: contralateral leg flexed, homolateral extended; s: tail drawn to opposite side. All movements contralateral to stimulated hemisphere except where otherwise stated.

state of coma or sleep, apparently caused by substantial deafferentiation of the resulting '*cerveau isolé*'. Optic and olfactory input is generally insufficient for arousal, although very strong olfactory stimuli or direct cortical stimulation may evoke transitory EEG arousal patterns. *Per contra*, transection of brain from spinal cord, resulting in an '*encéphale isolé*', does not eliminate recordable awake behavior. This latter is now generally assumed to depend on the extra-lemniscal so-called ascending reticular activating system dealt with in section 2, chapter IX of volume 4. It remains, however, a moot question to which extent the ascending lemniscal sensory channels likewise play a substantial role for the maintenance of the waking condition (cf. p. 430, chapter XII).

Experimental *transections of corpus callosum, anterior commissure* and *optic chiasma* in Primates (BLACK and MYERS, 1964), and operations necessitating transections of the corpus callosum in Man have provided diverse data for functional interpretations of the '*split-brain*' (ETTLINGER, 1965; GAZZANIGA, 1965, 1970; CUÉNOD, 1972), to be considered further below in dealing with the Human telencephalon, as well as in section 5 of chapter XV in volume 5/II, concerned with the fiber systems of the Mammalian cerebral cortex.

Turning now from ablation and transection to *stimulation effects*, the presence of an electrically excitable motor cortex in the Mammalian telencephalon was demonstrated by the highly important experiments of FRITSCH and HITZIG (1870) in the Dog. These, and subsequent similar investigations, by FERRIER (1873), GRÜNBAUM and SHERRINGTON (1901), LEYTON and SHERRINGTON (1917), LASHLEY (1921), GRAY and TURNER (1924), and others in various Mammals, including Primates and Marsupials, showed that the caudal portion of the Mammalian neopallial *regio rostralis sive frontalis* is characterized by a somatotopic distribution of spots from which essentially contralateral activity of muscles respectively muscular groups can be elicited, and from which the pyramidal tract seems to take its origin. The obtained results varied in accordance with details of experimental procedures, including strength of the applied current. Mappings as given by GRAY and TURNER in the Opossum and by LASHLEY in the Rat are shown in Figure 240. The distinctive so-called '*frontal eye field*' of various Mammals can be regarded as a specialized part of the *premotor cortex*. The unpredictable variations in response, and further aspects of this general topic will be considered in chapter XV of volume 5/II dealing with the cerebral cortex.

The results obtained by electrical stimulation of the Human motor cortex in the course of brain operations, as recorded by FEDOR KRAUSE (1908, 1911) and others, particularly also by PENFIELD and RASMUSSEN (1950), together with the relevant interpretations, shall be discussed further below in dealing with the Human telencephalon. It should also here be mentioned that, as interpolated by the editors into KUNTZ's (1970) biographic sketch of HITZIG, it was apparently ROBERTS BARTHOLOW, of Cincinnati, Ohio, who first applied electrodes to the exposed Human cerebral cortex, using as subject a servant of his household afflicted with an eroding cancer of the scalp. Weak faradization of the cortex produced muscular contractions in the contralateral limbs and turning of the head toward that side (Amer. J. med. Sci. *67:*

305–313, 1874). Upon publication of this observation, BARTHOLOW was forced to leave Cincinnati. As an aside, and considering the relevant historical aspects, it seems perhaps appropriate to rectify and censure the totally unfounded legends concerning the experiments undertaken by FRITSCH and HITZIG in 1870. Thus, HERRICK (1931, p. 336) states: 'During the Franco-Prussian war, an army surgeon, FRITSCH, while operating on a wounded soldier, applied the galvanic electric current to the exposed surface of the brain and observed a twitching of some muscles. This was followed immediately by experimental researches upon the electrical excitability of the cerebral cortex of dogs, the first results of which were published by FRITSCH and HITZIG in 1870.' According to GREY WALTER (1953, p. 48): 'Two medical officers of the Prussian army, wandering through the stricken field of Sedan, had the brilliant if ghoulish notion to test the effect of the Galvanic current on the exposed brain of some casualties. These pioneers of 1870, FRITSCH and HITZIG, found that when certain areas of the brain were stimulated by the current, movement took place in the opposite side of the body.' KUNTZ (1953, 1970) states that 'according to PERCIVAL BAILEY, the legend goes that FRITSCH discovered in dressing a wound of the brain during the Prussian-Danish war in 1864 that irritation of the brain causes twitching of the opposite side of the body'.

Actually, FRITSCH was not a military surgeon but participated as a non-medical Guards officer of the reserve, in the Austro-Prussian war of 1866, and the Franco-Prussian war of 1870–1871, but there is no evidence that he made any medical observations on the wounded. In 1864 (Prussian-Danish war), FRITSCH was in Africa on a scientific expedition. The paper by FRITSCH and HITZIG (1870) was published before the outbreak of the Franco-Prussian war of 1870–1871. This paper indeed contains a footnote by HITZIG on p. 323, referring not to an experiment, but merely to a casual observation which he made while temporarily working at the Berlin garrison hospital in 1866, on a wounded soldier with exposed brain. Said observation, however, was not related to any motor activity but merely indicated that a brain lesion might not result in clinically detectable disturbances.[208] On p. 308 of

[208] The paper states verbatim: '*Man könnte die Erklärung der Beobachtungen von uns verlangen, die in hinreichender Anzahl über chirurgische Verletzungen des Gehirns ohne Störung irgend welcher Funktion vorliegen*[1]'. The footnote then adds: '*Auch der eine von uns* (HITZIG) *hat einen solchen Fall während seiner Tätigkeit als dirigierender Arzt am allgemeinen Garnisonlazareth zu Berlin im Jahre 1866 beobachtet. Einem Soldaten* (ANGELMEIER) *war ein Granatsplitter genau in die Glabella gedrungen und hatte dort ein dreieckiges Loch gemacht. Aus diesem Loche entleerte*

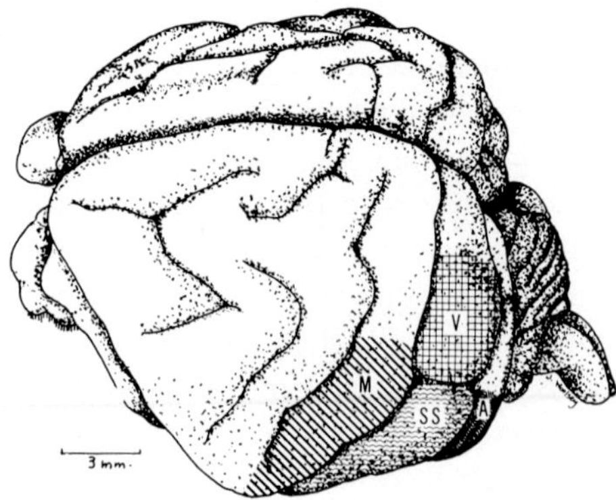

Figure 241 A. Neopallial areas in the Monotreme Echidna (from LENDE, 1969). A: auditory area; M: somatic motor area; SS: somatic sensory area; V: visual area. Note the caudal location of the projection areas and the lack of associational areas between the primary regions, also the large expanse of an apparently 'silent' rostral area.

the cited paper, on the other hand, it is stated that HITZIG had applied weak galvanic currents to the occipital region of the head in normal Human subjects and thereby obtained eye movements.[209] These latter experiments, undertaken in 1870, were mentioned as having suggested the subsequent study by FRITSCH and HITZIG on Dogs. For those recognizing the general import of the history of medicine respectively neurology, it is not without interest to note how relevant facts can become grossly distorted in careless and uncritical, if not outright malicious statements not only by slipshod (GREY WALTER) but also by otherwise competent and respectable (HERRICK) authors.[210]

---

sich während wenigstens 14 Tagen immerwährend Gehirnsubstanz. Schliesslich heilte die Wunde von selbst zu. Sehr geistreich war dieser Kranke nicht, im Gegenteil schien er trägen Verstandes. Da man ihn indessen vorher nicht gekannt hatte, so war nicht zu entscheiden, ob er nicht von Natur geistig arm war. Grobe motorische oder sensible Störungen bot er jedenfalls nicht dar.'

[209] Since the here unspecified 'eye motions' were reported to be accompanied by vertigo, it is most likely that they represented vestibular galvanic nystagmus and not movements elicited from the occipital cortex.

[210] Cf. footnote 87, p. 120 of volume 4. Again, the derogatory comments, quoted by KUNTZ (1953, 1970), and imputing to HITZIG 'incorrigible conceit and vanity complicated by Prussianism' can be considered highly gratuitous. As an ethnically perhaps incompatible foreigner, HITZIG was unable to cope with the extreme corruption and depravity then rampant at the Zürich Burghölzli Insane Asylum, vividly described by his successor FOREL (1935). This latter, as a more congenial Swiss with better understanding of the local conditions, was finally able, with considerable difficulties, to remedy that situation.

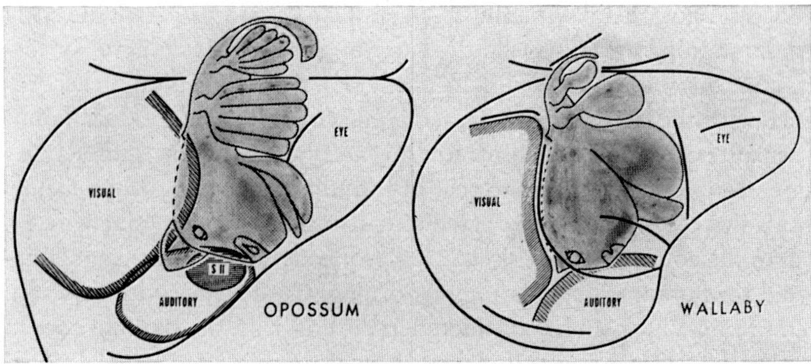

Figure 241 B. Lateral aspect of telencephalon in two Marsupials (Opossum and Wallaby) showing neopallial areas (from LENDE, 1969). SII: second sensorimotor area. According to LENDE's interpretation, somatic motor and sensory areas are here joined into a common area, whose somatotopic representation is indicated by the animalculus figures. *Eye* denotes the rostral (frontal) motor eye field.

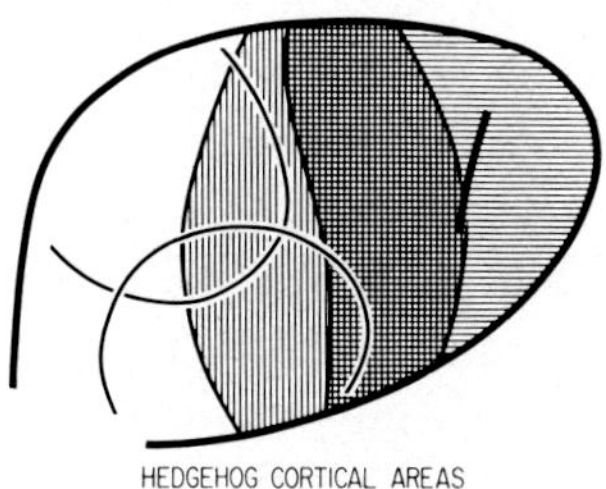

Figure 241 C. Rough outline of neopallial areas in the Insectivore Erinaceus (from LENDE, 1969). Somatic sensory; vertical hatching; somatic motor: horizontal hatching; sensory-motor overlap: cross-hatching; auditory area: left lower outline; visual area: left upper outline. Note the considerable overlap. The thick line at right indicates the orbital sulcus.

As regards the *parietal neopallial region*, especially its neighborhood adjacent to the rostral one, the recording of evoked potentials and related experimental methods have disclosed its functional significance as a *somatic sensory cortex* with somatotopic distribution of the input. This distribution seems to preserve, with distortions related to greater

---

'*Prussianism*', although occasionally displaying some aspects of rigidity, actually may be said to imply a sense of honor and duty, resulting in frugality, simplicity, reliability, incorruptibility, responsibility, honesty, justice and hard work.

and lesser peripheral 'sensitivity', overall *topologic aspects*[210a] of the body configuration, which can be depicted by so-called *'animalculi'* or *'homunculi'* outlines (cf. e.g. Figs. 241 B, D, E). Said projection, in accordance with the decussating general lemniscus system, is essentially but perhaps not exclusively contralateral, and generally corresponds, *qua* somatotopic distribution, to that of the adjacent rostral motor cortex. Both sorts of 'projection', related to cortical output respectively input, led to the concept of 'representation' respectively 'foci' or 'centers' related to muscle groups or regions of peripheral sensory innervation. This topic will again be considered in dealing with the cerebral cortex. ADRIAN (1947) demonstrated the existence, in the Cat, of an additional *secondary sensory area*, located basally to the main one. A *supplementary motor area* was detected by PENFIELD and RASMUSSEN (1950) in Man. It is located on the medial surface of the hemisphere, somewhat rostrally to the dorsal end of the main motor area. A comparable supplemental area was also noted in some other Mammals, e.g. by WOOLSEY (1958).

Because of an apparently close relationship between somatomotor and somatosensory regions, recent authors (cf. e.g. LENDE, 1969) designate the motor areas as 'motor-sensory', and the sensory ones as 'sensory-motor'. The larger areas with lower thresholds are MsI (the main motor area) and SmI, and the smaller ones as MsII and SmII, whose spatial relations are indicated in Figure 241 E. There are doubtless significant but still poorly understood differences in details of 'representation' and 'functional aspects' between various Mammalian

---

[210a] Thus, WERNER and WAITSEL (1968) as well as other authors write about the 'topology of the body representation in the Mammalian somatosensory area'. Yet, although these and additional authors mention *topology* with respect to various aspects of neurological and morphological topics, no relevant *elucidation* of the actual principles to topology applicable to *morphologic problems* seems to have been given previously to my elaboration in section 2, chapter III of volume 1 (1967) and section 1a, chapter VI of volume 3/II (1973). Intuitively, everybody senses the '*sameness*' displayed by homologous morphologic configurations such, as in diverse Vertebrates, humerus or femur, which represent very simple cases. The topologic transformations appear here trivial and obvious to the superficial observer, who does not attempt to formulate said transformations into precise terms. The questions to be answered are, in particular: (1) what is the topologic meaning of morphological sameness, and (2) how can such sameness be established. The obviousness or '*Selbstverständlichkeit*' are deceiving, and all definitions of homology based on either phylogenetic derivation or on 'identical' developmental mechanisms (*Entstehung unter vergleichbaren organisierenden Beziehungen*, whose causal aspects, based on physiochemical interactions, are entirely unknown) involve several gross errors in logic. These latter are *petitio principii* and *ignoratio elenchi*, combined with the fallacy of ὕστερον πρότερον.

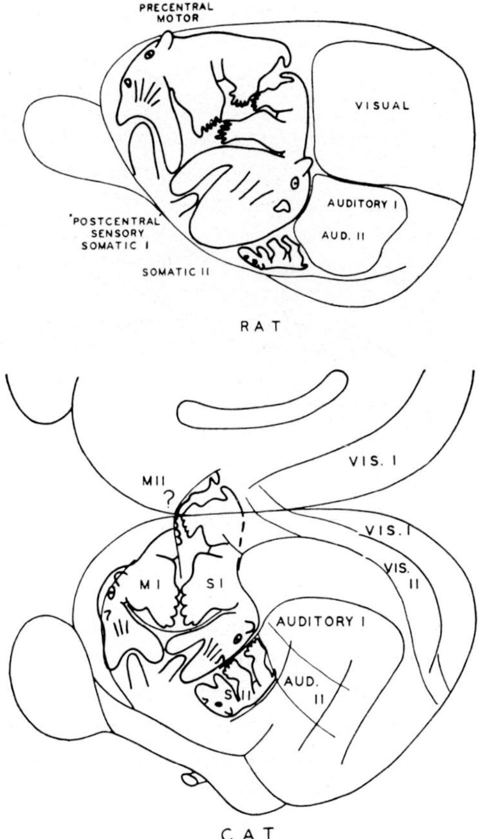

Figure 241 D. Diagram indicating functional neopallial areas in the Rat and in the Cat according to WOOLSEY's interpretation (from WOOLSEY, 1960). MI: 'precentral motor'; SI: 'postcentral sensory'.

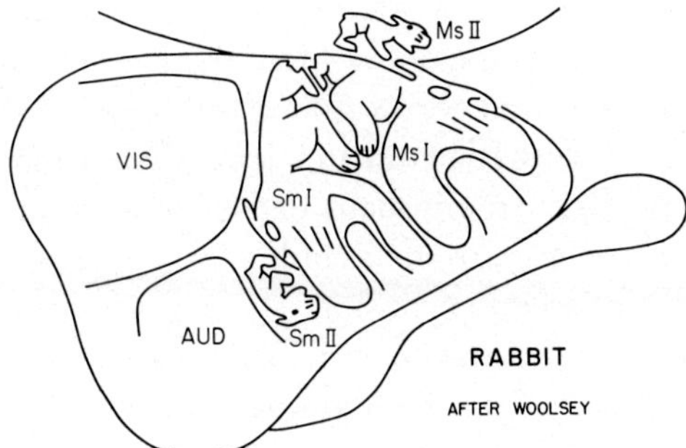

Figure 241 E. Aspect of Lagomorph (Rabbit) telencephalon with neopallial areas and somatotopic animalculus figures (after WOOLSEY, from LENDE, 1969). AUD: auditory area; Ms, I, II: somatic motor-sensory areas I and II; Sm I, II: somatic sensory-motor areas I and II; VIS: visual area.

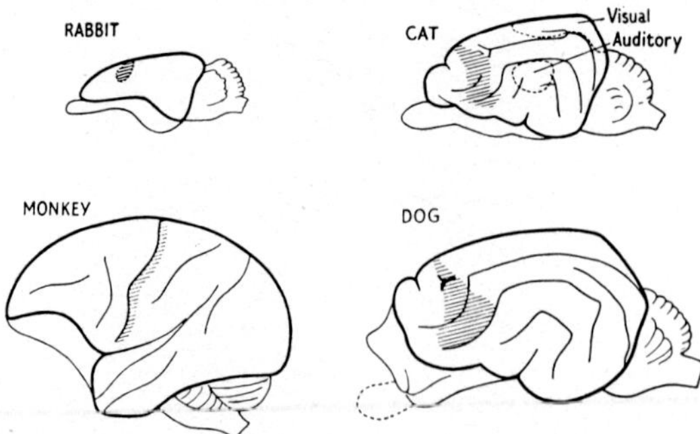

Figure 242 A. Location and relative size of neopallial somatosensory areas in various Mammals, including, in the Cat, the auditory and visual areas (from ADRIAN, 1947).

groups such as e.g. Monotremes, Marsupials, Insectivores, Rodents, Carnivores, Ungulates and Primates, and perhaps even within such groups.[211]

It should also be added that, depending on stimulation parameters, body, head, and eye movements may be produced from cortical areas other than the just mentioned four. Thus, rather nondescript motions have been observed upon stimulations of the parahippocampal cingulate cortex. Various inconclusive interpretations of the diverse 'supplementary' motor areas are summarized and discussed by CROSBY *et al.* (1962).

The *rostral (or frontal) neocortex* anteriorly to the 'somatic motor region' may be related to 'motor integrative functions' of various 'higher' and 'lower' order, besides providing output to pontine nuclei. Considerable differences between Mammalian groups may here obtain. Thus, at least in Primates including Man, there is an agranular premotor cortex and a prefrontal cortex granularis, to be dealt with further below in connection with the Human telencephalon and again in chapter XV of volume 5/II. An unusually extensive 'premotor neopallium' was reported for the Monotreme Echidna (cf. Fig. 241 A), but actually seems to be a rostrolateral expansion, characteristic for Monotremes, of the parahippocampal cortex (cf. vol. 5/II, chapter XV, section 4).

The *occipital neopallial area* of Mammals is generally acknowledged to receive optic input and to represent the visual cortes *('area striata')*.

[211] Thus, LENDE (1969) claims that, in Marsupials, a MsI area does not occur, being substituted by a complete overlap with SmI. This claim, however, appears not altogether convincing.

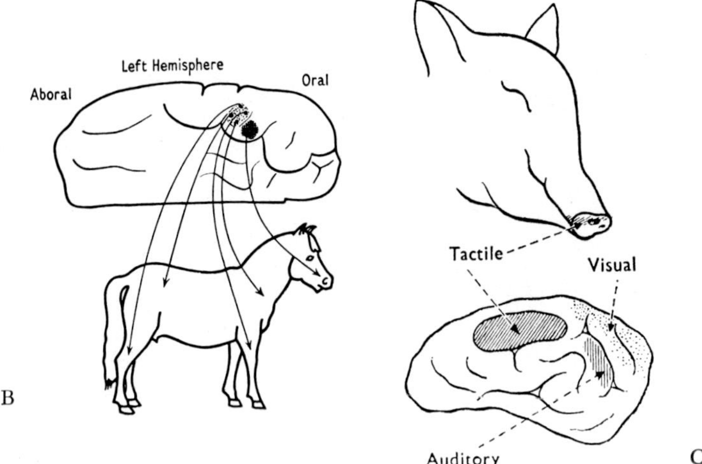

Figure 242 B. Somatic sensory neopallial area in the Horse (Shetland pony). The anterior part is concerned entirely with the representation of the nostril (from ADRIAN, 1947).

Figure 242 C. Sensory neopallial areas in the Ungulate Pig (from ADRIAN, 1947).

The projection of the contralateral and homolateral retinal quadrants upon the lateral geniculate complex and from there upon the visual neopallium of diverse Mammals has been investigated by numerous authors, e.g. HENSCHEN (1903, 1923), MINKOWSKI (1911), PUTNAM and PUTNAM (1926), POLIAK (1932), BODIAN (1937). Figure 242F illustrates the obtained projection mappings, which should be compared with Figure 145 (chapter XII) indicating said projections in Man. Although various uncertainties with respect to details still obtain (cf. e.g. Carnivores in Fig. 242F), a fairly well defined overall similarity of retinal projection areas in Opossum, Rabbit, Dog, and Primates seems to be displayed. The expansion of the visual area upon the medial surface of the hemisphere in various Mammals appears evident. Caudal projection neighborhoods, e.g. in the Rabbit, thereby become rostromedial in Primates, including Man, while rostral projection subareas, including the macular one, then become caudal or rather occipital. Generally speaking, the upper retinal quadrants are represented dorsally, and the lower ones basally. As regards the visual field, it can roughly be said that, in accordance with the variable amount of decussation in the optic chiasma, the right field is projected upon the left visual pallium, and the left field upon the right visual area. As in the retina, the projection

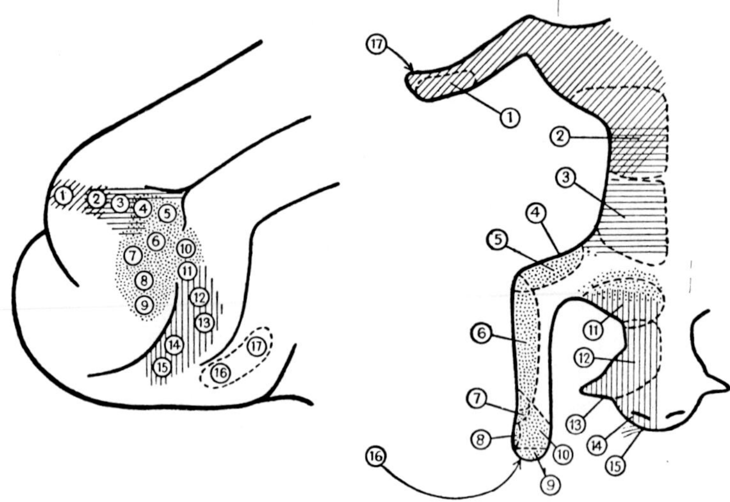

Figure 242 D. Somatosensory neopallial areas of the Cat, including a secondary area as recorded for the digits, marked 16 and 17 (from ADRIAN, 1947).

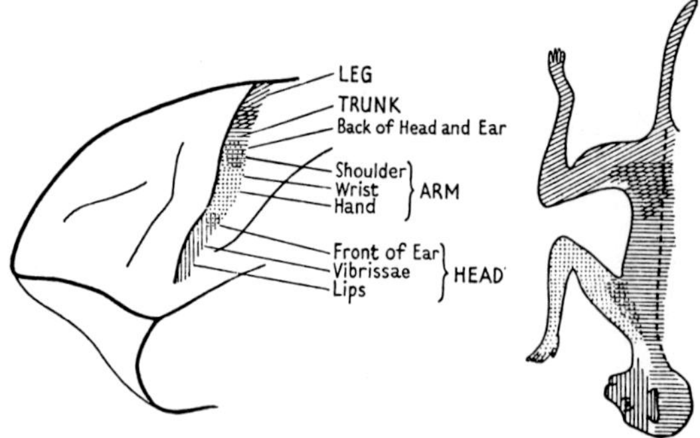

Figure 242E. Neopallial somatosensory area I in the Rhesus Monkey, showing somatotopic representation (from ADRIAN, 1947).

is upside down, as it were, with basal area portion for '*sky vision*', and dorsal (upper) area portion for '*ground vision*'. Each pallial visual projection area receives binocular input, but for a taxonomically variable portion of the peripheral temporal visual field (nasal retinal half) monocular projection obtains (cf. Fig. 251 A). In 'higher Mammals', a more or less developed *secondary visual area* tends to adjoin respectively surround the primary 'projection area'. This region may again, e.g. in

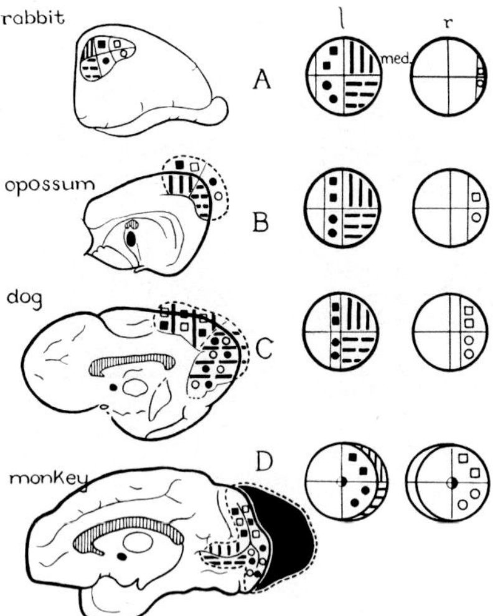

Figure 242 F. Comparison of assumed retino-cortical projection in diverse Mammals (from BODIAN, 1937). A: Rabbit, lateral aspect, modified after PUTNAM and PUTNAM (1926); B: Opossum, medial aspect after BODIAN (1937); C: Dog, medial aspect, modified after MINKOWSKI (1911); D: Monkey, medial aspect, modified after POLIAK (1932). In B, C, and D the lateral expansion of the visual projection area has been added. Left and right retinae with their quadrants (and in D with macula *sive* fovea centralis) are indicated at right and obviously as seen from behind.

Primates, become subdivided into a 'parastriate' and 'peristriate' area. While the primary area, although 'binocular' seems to be contralateral 'unilateral', a directly adjacent 'secondary visual area' seems to be concerned with 'bilateral vision', presumably mediated by commissural fibers of the corpus callosum. Within the visual pallium, respectively its 'association area', an 'occipital eye field' from which eye movements can be elicited by stimulation, has been noted in various Mammals. Such observations agree with the assumed presence of corticopretectal and corticotectal fibers originating in the occipital (visual) region of the neopallium.

Details of cortical structure and of thereto related functional activities, as e.g. recorded in the experimental studies by HUBEL and WIESEL (1962, and their many additional ones), will be dealt with in chapter XV of volume 5/II. Some particulars with respect to Man shall also be considered further below in the present chapter.

The *temporal neopallial area* is likewise well documented as a sensory projection area which receives *auditory input* from homo- and contrala-

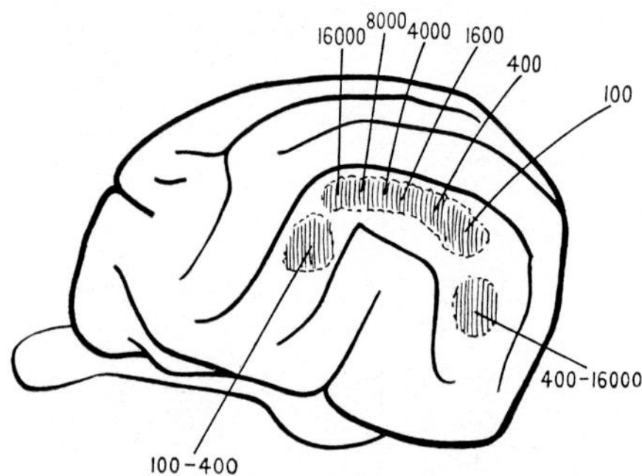

Figure 243 A. Auditory neopallial areas I and II in the Dog. The figures indicate the sound frequencies resulting in maximal recorded potentials (after TUNTURI, 1944, from ADRIAN, 1947).

teral cochleae. Within the auditory cortex, two separate '*primary*' and '*secondary*' neighborhoods have been recognized, the latter one being located basally to the first as noted in Carnivores (Figs. 243 A, B). A discrete 'representation' for different sound pitches has been established, the sequence in the primary area being reversed in the secondary one (WOOLSEY and WALZL, 1942; TUNTURI, 1944; and other authors).

As regards the still poorly understood *insular cortex*, various observations (e.g. by ROBERTS and AKERT, 1963, and by others) suggest that it could represent a '*visceral sensory area*' also including the cortical 'center' for *taste*. The so-called second sensorimotor area may extend toward or upon the insula of some Mammals.

The *piriform lobe* can presumably be evaluated as the main *cortical 'center' for olfaction*, but it is not quite certain to which extent the parahippocampal posterior cortex (so-called area entorhinalis) of the piriform lobe is concerned with olfactory input. Other rhinencephalic grisea being fundus striati, parts of amygdaloid complex and of paraterminal grisea, together perhaps with precommissural neighborhood of parahippocampal and hippocampal formation.[212]

Generally speaking, in 'higher' Mammalian forms, *secondary areas of 'association'* and '*integration*' become added to, or intercalated between, the 'primary' motor and sensory 'projection areas' as discussed on pp. 657–663, section 6, chapter VI of volume 3/II. Figures 241 A to

[212] It should be recalled that these grisea (except for the olfactory bulb) are nevertheless present even in anosmatic Mammals.

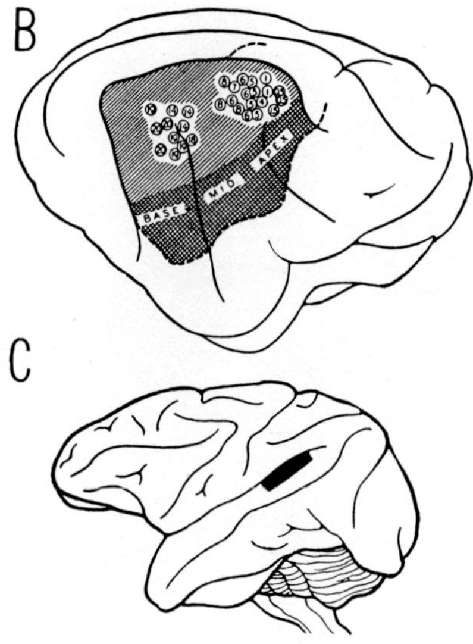

Figure 243 B, C. Auditory neopallial areas in (B) the Cat, and (C) in the Monkey (after Woolsey and Walzl, 1942, and Ades and Felder, 1942, from Fulton, 1949). The figure B indicates stimulated spots of cochlea, the (lower) secondary acoustic area displaying a reverse tonotopic distribution with respect to the (upper) primary area.

243C illustrate diverse aspects of functional localization within the Mammalian neopallium according to various authors. Figure 243D shows Lende's (1969) concept of an hypothetical evolution of these cortical areas. Although differing in some details from the just mentioned concept of 'primary' and 'secondary' areas based on considerations of cytoarchitecture, assumed fiber connections, and myelogenesis, both concepts are not mutually exclusive if the still rather poorly understood relationships between localizatory and equipotential 'principles' manifested by the telencephalic grisea are kept in mind. The diverse motor and sensory projection areas of the Mammalian neopallium were first unequivocally demonstrated in Man by Flechsig (1920, 1927, and many of his previous publications) on the basis of his myelogenetic method.

It should also be mentioned that, some years ago, various neopallial areas of the frontal, parietal, and occipital lobe were interpreted as so-called '*suppressor strips*' or '*bands*' exerting an inhibition of movements. This concept, however, seems to be based on a phenomenon of spreading depression artificially induced by the exposure of cortical

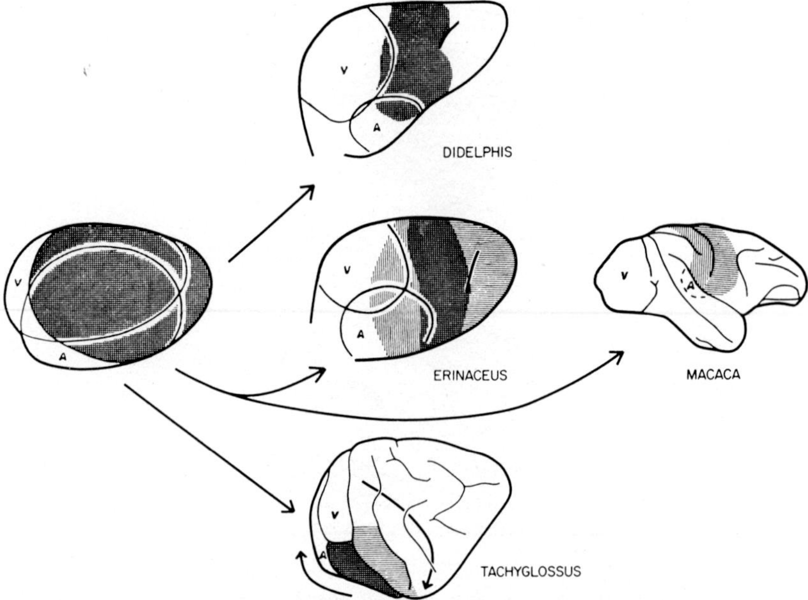

Figure 243 D. Diagrammatic sketch purporting to illustrate a hypothetical concept of the evolution of neopallial sensory and somatic motor neopallial areas in Mammals according to LENDE's interpretation (from LENDE, 1969). Vertical hatching: somatic-sensory; horizontal hatching: somatic-motor; cross-hatching: somatic sensory-motor; A: auditory area; V: visual area. The assumed wide overlaps in the postulated primitive form (left) and in Didelphis and Erinaceus should be noted. Arrows about Tachyglossus (Echidna) areas indicate the direction of assumed rotational displacements.

areas in the course of animal experiments. The actual existence of such suppressor strips can be considered erroneous on the basis of the available evidence.[212a]

The greater extent of Mammalian hippocampal formation and parahippocampal pallium seems to be the essential component of what is now generally designated as the '*limbic system*', concerned with '*emotional behavior*' or '*affective tone*' as well as with '*visceral mechanisms*', and is apparently phylogenetically derived from grisea originally more or less pertaining to the olfactory system. It appears most likely that the still remaining relationships of limbic system to the 'rhinencephalon' may

[212a] This topic, with the relevant bibliographic references, is discussed on pp. 163–164 of the author's monograph on the Human diencephalon (K., 1954).

vary with respect to the diverse Mammalian forms.[213] Doubts about the predominant 'olfactory function' of hippocampal (cingular) cortex were expressed and documented by Brodal (1947b), thus indirectly supporting the concept of Papez (1937) briefly mentioned in Brodal's paper. The importance of the *Papez-circuit*, pointed out in our publications (K. and Haymaker, 1949; K., 1954, 1957) was not yet generally recognized at that time. The subsequent concepts of the limbic system, as e.g. propounded by MacLean (1969) and others, represent further elaborations of the fundamental notions introduced by Papez. Further comments on the limbic system, which is also supposed to play a role in functional activities related to memory, will be given in chapter XV of volume 5/II.

The *paraterminal (septal) grisea* are traversed by the complex fiber bundles of the fornix system as enumerated above and are thus doubtless intimately related to the so-called limbic system as well as to the basal forebrain bundle. Since about 20 years, many experiments with implanted electrodes have been undertaken by Olds (1956, 1958 1960), Olds and Olds (1963), Olds and Milner (1954), Delgado (1955, 1967a, b), Delgado *et al.* (1954, 1956), Bruner (1967), and others, leading to the concepts of '*self-stimulation*', '*approach reactions*' and '*avoidance reactions*'.

A rostral region of 'approach reaction' leading to self-stimulation is located in the paraterminal region ('septum' and fornix).[214] Thus, a little pedal, which the animal could press, thereby closing the electric circuit of the implanted electrode, was e.g. provided in the Rat's cage. The stimulation apparently causes such pleasant sensations that the Rat will continuously push this pedal, without stopping to take available food or water. Roughly comparable effects were recorded in Cats and Monkeys. If a Monkey with such implanted electrode is sharply startled from sleep and becomes angry, closing the circuit will suddenly change his behavior: the animal immediately becomes pleasant and affectionate. Stimulation of 'septum' and 'limbic structures', e.g. hippo-

[213] Evidence that, at least in the Mouse, the precommissural hippocampus receives olfactory input, was pointed out by K. and Wiener-Kirber (1962). Such input presumably also reaches the adjacent rostral (precommissural) parahippocampal cortex.

[214] It is dubious whether this effect results from stimulation of the 'septal' grisea or of fornix system fibers or of both. It is not unlikely that, to some extent, the septal grisea represent interstitial nuclei of the fornix system and that their parcellation into different 'nuclei' is merely a fortuitous result of griseal fragmentation by the fibers of the fornix system (cf. Figs. 236A–D and 239A–E).

campus and parahippocampal cortex produced, among others, 'grooming reactions' and penile erection in Rats, Cats, and Monkeys (cf. MacLean, 1969).[215]

Positive reinforcement localizations, some of which are related to the self-stimulation effect, run caudad from septum and 'anterior commissure region' along the basal forebrain bundle into the lateral hypothalamus as far as the mesencephalic tegmentum. Negative reinforcement localizations are said to run along medial hypothalamus, also along medial lemniscus system and thereto related thalamic grisea. Moreover, stimulation of the intralaminar and midline thalamic nuclei is reported as causing the animals suddenly to 'freeze' in their assumed posture. Negative reinforcing effects are also reported to result from stimulation of neighborhoods in the dorsal mesencephalic reticular formation, and of the adjacent periventricular gray. Nevertheless, ambivalent results are reported, and various discrepancies between the diverse observations obtain, as e.g. discussed by Olds and Olds (1963), and Wilkinson and Peele (1963, in the Cat). Strength and frequency of the stimuli seem to represent parameters upon which differences in behavioral response may depend. Thus, the topic of 'electrically controlled behavior' cannot be said to be reasonably well clarified and still remains poorly understood.[216]

Again, on the basis of experiments on the Rat with conditioned response to a noxious stimulus and subsequent damaging lesions to limbic region, septum, and fornix, Lyon and Harrison (1959), who recorded recovery of the response following the lesion, have expressed doubts that said structures are related to 'affective behavior'. In view

[215] Likewise 'the medial septo-preoptic region and the medial part of the mediodorsal thalamic nucleus appear to be nodal points for erection (MacLean, 1969).

[216] Sensational reports, such e.g. as the editorial in Hospital Tribune (vol. 4, No. 16, p. 18, 1970: 'A radio links Chimp's brain and computer'), claiming that 'a new age in research and therapy on the brain and mind' has been 'introduced', should be evaluated with due scepticism. As reported by New York Times Magazine (Nov. 15, 1970, pp. 46–170) 'Brain Researcher José Delgado' asks: 'What kind of human would we like to construct?' Again (loc.cit.), not to be outdone by von Holst's cackling hens (cf. above section 9, p. 588), Delgado performed in a Spanish bull-ring, waving a red cape at a fierce bull, in an 'inhibitory area' of whose brain he had previously implanted a radio-controlled electrode. 'The bull lowered his read and charged through the dust. But as the animal bore down on him, Delgado pressed a small button on the radio transmitter in his head: the bull braked to a halt.' 'When the professor pressed another button, the bull turned away and trotted docilely toward the high wooden barrier.' Quoting an old German quip, one could here indeed remark: '*Da staunt der Laie, selbst der Fachmann wundert sich, nur der Kenner bleibt gelassen.*'

of the various data clearly suggesting the questioned relationship, the results by the cited authors remain unconvincing, since the extent of their lesions did not eliminate various grisea and channels (e.g. amygdaloid complex, etc.) normally cooperating in the multiplex activities of the 'limbic system', presumably characterized, like many neural networks, by considerable redundancy (cf. also vol. 1, chapter I, sections 4 and 5, pp. 20–24).

Although part of the amygdaloid complex is doubtless still related to olfaction, the substantial relationships of that complex to autonomic and limbic systems is well documented (cf. e.g. GLOOR, 1955, 1960; ZBROZYNA, (1963). Lack of the capacity for anger and for fear are among the diverse manifestations of the *Klüver-Bucy syndrome*[217] resulting from bilateral ablations involving rostrotemporal parts of limbic system (amygdala, piriform lobe, parahippocampal and contiguous portions of hippocampal cortex) in Monkeys. Among other disturbances are hypersexuality of a bizarre nature, loss of discrimination of taste for thirst-quenching fluids, swallowing of any object placed in the mouth, and loss of awareness of what is harmful and painful, such that a Monkey will try to put in his mouth a burning match. The Monkey is thus transformed from a normally aggressive creature to a placid one incapable of anger and fear. Other Mammals, e.g. Cats, seem to exhibit comparable behavior following similar ablations.

The grisea forming the *corpus striatum (nucleus caudatus, putamen)* appear essentially related to the extrapyramidal motor system, although in part, at least, likewise participating in 'vegetative functions'. Thus, fundus striati, including nucleus accumbens, which overlaps with the paraterminal grisea (nucleus basimedialis inferior) are related to the basal forebrain bundle, and to the limbic system.[218] Again, as men-

[217] KLÜVER and BUCY (1937, 1939), AKERT *et al.* (1961), cf. also the comments and summary by HAYMAKER and ANDERSON (1971) and in 'Brain and Consciousness' (K., 1957). The syndrome can be roughly described as characterized by (1) visual, possibly also tactile and auditory agnosia, (2) 'oral tendencies', (3) compulsion of reacting to every visual stimulus, (4) loss of anger and fear reactions, (5) hypersexuality, (6) omnivorous bulimia. The visual agnosia may be related to damage involving the temporo-occipital neocortical 'association region'.

[218] These basal telencephalic grisea are also recently included in a vaguely defined so-called 'telencephalic reticular formation' supposed to include e.g. nucleus accumbens, parts of 'striopallidum' and amygdaloid complex (cf. LOBO, 1972). Be that as it may, one should keep in mind the morphologically relevant distinctions between the mesencephalo-rhombencephalic formatio reticularis tegmenti, the diverse diencephalic grisea of 'reticular' type, and the alleged 'telencephalic reticular formation'.

tioned above, fundus striati and basal cortex (area ventralis anterior respectively tuberculum olfactorium, and area ventrolateralis posterior respectively cortical amygdaloid nucleus) pertain to the olfactory system, discounting totally anosmatic Mammals.

Nothing certain can be stated about the functional significance of the *claustrum*, whose caudobasal portions represent a transition to the amygdaloid complex. It is likely that both connections and functions of the claustral grisea may differ in the various Mammalian groups. The close relation of claustrum to anterior piriform lobe cortex in 'lower' and even some 'intermediate' Mammals, e.g. the Cat, as stressed by BERLUCCHI (1927) suggests, in these forms, the possibility of an 'olfactory' function, from which that griseum might have become 'emancipated' in Primates and microsmatic respectively anosmatic Mammals. The spatial relationships of claustrum to insular cortex as well as its continuity with the amygdaloid cortex are also of interest, but hardly permit specific conclusions. None of the hitherto claimed connections of the claustrum seems unambiguous or specific enough for sufficiently definite clues as to its possible functions (cf. CROSBY *et al.*, 1962).

Turning now to the overall external and internal configuration of the *Human telencephalon* with its frontal, parietal, occipital and temporal lobes, its insula, as well as its limbic lobe and its basal 'rhinencephalic' components, Figures 244 to 247 illustrate the characteristic features, whose general morphologic significance was dealt with on pages 610–668 of chapter VI in volume 3/II.[219]

As regards the pertinent sulci respectively fissures, whose further details are considered in the chapter XIV, volume 5/II, it will here be sufficient to recall the *fossa Sylvii*, related to the hemispheric bend, the *sulcus centralis Rolandi* separating frontal from parietal lobe and approximately 'motor' from 'sensory' cortex, and the *rhinencephalic sulci*, generally not very conspicuous in Man, namely *sulcus rhinalis lateralis* and *sulcus endorhinalis* (cf. Fig. 175D). On the medial aspect the following grooves may also be mentioned: *sulcus cinguli sive calloso-marginalis*, ap-

[219] The configurational relationships at the adult stage cannot be properly understood without an adequate knowledge of the relevant morphogenetic states discussed and illustrated in the cited volume. Of particular importance are, *inter alia*, the transformations concerning hemispheric stalk (Fig. 241, p. 467, vol. 3/II), lateral ventricle (Figs. 78F, G, p. 206/07; 78H, p. 208; 78I, p. 210, vol. 3/II), lamina affixa and sulcus terminalis (Fig. 234A, p. 458, vol. 3/II), as well as corpus callosum (Figs. 337, p. 622; 342, p. 629; 343, p. 630, vol. 3/II).

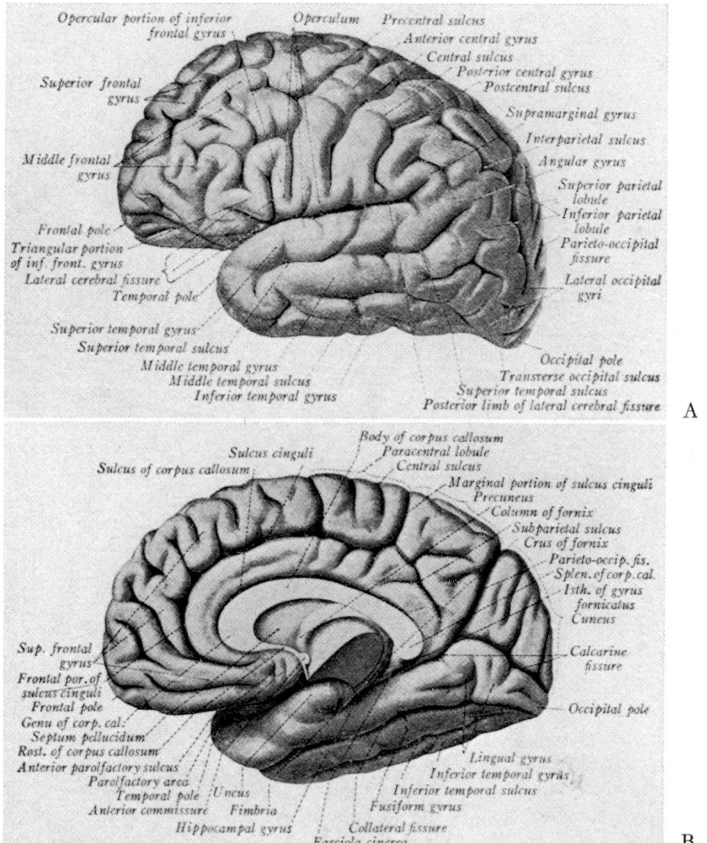

Figure 244 A. Lateral aspect of the Human cerebral hemisphere (after SOBOTTA, 1931, from RANSON, 1943).

Figure 244 B. Medial aspect of the Human cerebral hemisphere. A section through diencephalon has removed a large portion of this latter as well as the entire cerebellum and brain stem (after SOBOTTA, 1931, from RANSON, 1943).

proximately indicating the boundary between the neocortex and the parahippocampal cortex of gyrus cinguli, the *fissura parieto-occipitalis* separating parietal and occipital lobes, moreover the *fissura calcarina* within the visual neocortex (area striata).

As regards *blood supply and drainage* of the Human telencephalon, general morphologic data, particuarly concerning the *circle of Willis* and its branches, and the venous drainage through the sinuses of the

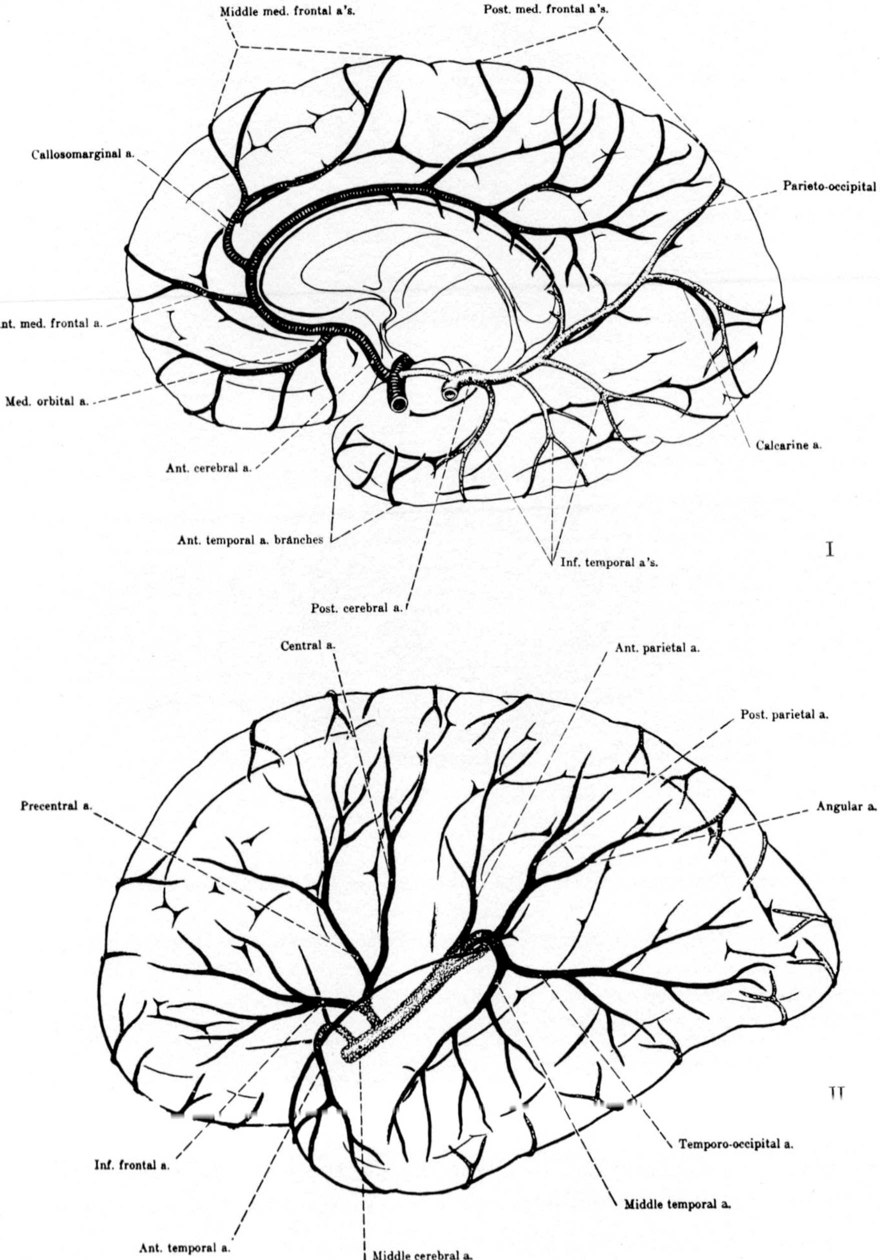

Figure 244 C. Average distribution of the arterial supply upon the cerebral hemisphere as seen in medial (I) and in lateral (II) aspect (from HAYMAKER and BING, 1969).

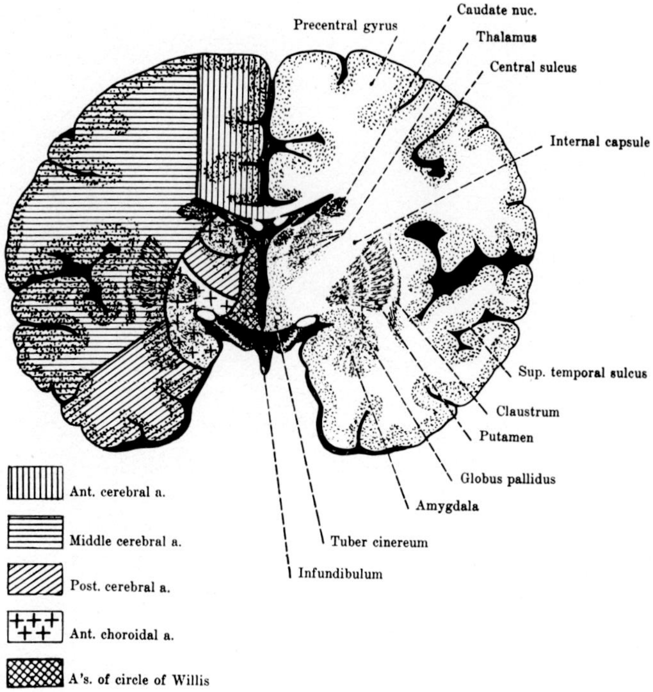

Figure 244 D. Average regions of arterial blood supply traced upon a cross-section of the hemispheres at level of tuber cinereum (from HAYMAKER and BING, 1969).

dura mater, were given in chapter VI, section 7 of volume 3/II. Additional details, insofar as features of forebrain vascularization concern not only diencephalon, but hemispheric stalk, capsula interna and adjacent telencephalic neighborhood, were dealt with in section 9 of chapter XII (cf. Figs. 136B, E–H). Figures 244C–E of the present chapter illustrate the average distribution of the arteries upon surface and some interior regions of the hemisphere. The variability of the vasculature, pointed out in the just mentioned previous passages, should again be recalled.

It will be seen that the *anterior cerebral artery* supplies a large part of the hemisphere's medial aspect as well as the rostrobasal portion and most of the corpus callosum. The *recurrent artery of Heubner* and relevant variations of the anterior communicating artery are depicted in Figures 286 and 287 A, B, p. 714–715 of volume 3/II.

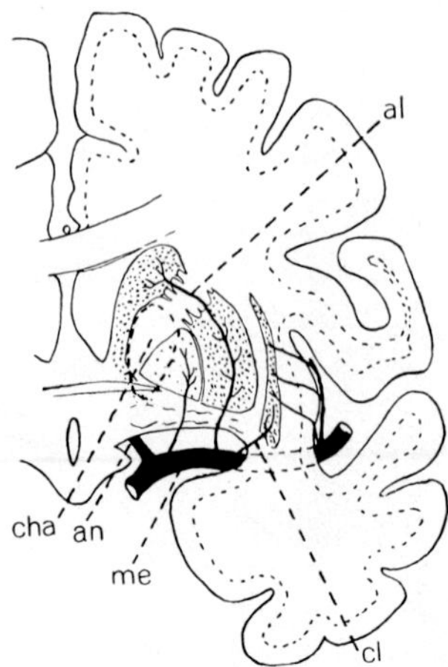

Figure 244 E. Simplified sketch of arterial supply to basal ganglia in a cross-section at level of anterior commissure (from K., 1927). al: one of the lateral lenticulostriate arteries; an: arteria cerebri anterior; cha: region supplied by a chorioidea anterior; cl: basal artery for claustrum (SHELLSHEAR's claustral artery); me: arteria cerebri media.

The *middle cerebral artery (a. cerebri media sive fossae Sylvii)*, besides supplying the insula, extends with its branches over most of the hemisphere's lateral surface.

The *posterior cerebral artery* supplies the medial and inferior aspect of the temporal lobe and the medial parieto-occipital region. A parietal and a calcarine branch (Fig. 244 C) are of particular importance for the vascularization of the visual cortex. Of significance is here the fact that a posterior branch of the middle cerebral artery's ramus temporo-occipitalis commonly reaches the lateral portion of the visual cortex near the occipital pole, thus accounting for the sparing of the macula, projected upon that region (cf. Figs. 145 A–C, chapter XIII), in visual defects caused by occlusion of the posterior cerebral artery's calcarine (or parieto-occipitocalcarine) branch.

Other relevant branches of the posterior cerebral artery are the *posterior choroidal arteries*, dealt with in the discussion of the Human diencephalon (chapter XII) and the *hippocampal artery*, whose significance for the hippocampal formation shall be pointed out in chapter XV of volume 5/II.

An *arteriovenous malformation* involving the posterior cerebral artery is the so-called *aneurysm of the vena magna Galeni*, resulting from a failure of development of intervening capillaries between arterial and venous channels. The blood then shunts directly from arterial to venous system and the vein of Galen becomes aneurysmally dilated. Occlusion of the quadrigeminal cistern and of the aqueduct may result, followed by hydrocephalus. Other major cerebral arteries (anterior and posterior) may likewise be involved in arteriovenous malformations of this type, representing 'vascular tumors' with a variety of clinical symptoms.

With regard to important structural features of cerebral vascularization at the histological level, including the perivascular spaces and the so-called barriers, the discussions and illustrations in section 5, chapter V of volume 3/I should be recalled.

A careful inspection of the above-mentioned Figures 246 and 247 will disclose that all telencephalic grisea enumerated and discussed with regard to other Mammals in the preceding part of the present section 10 can be easily identified. Figures 248 and 249 depict additional details of the subpallial basal grisea,[220] namely, septum, striatum with the adjacent hypothalamic pallidum, and the claustro-amygdaloid complex. The nucleus of the ansa peduncularis is shown in Figure 147D, which also illustrates various grisea of the amygdaloid complex.

As regards the Human *septum pellucidum* with its pars gliosa and its pars nervosa, its variable cavum, including the so-called *ventriculus Vergae* (cavum psalterii), reference to chapter VI, section 6, p. 627 to 635 of volume 3/II will here be sufficient (cf. also Figs. 246A, B, 247A and 248A, B of the present chapter).

The paraventricular portion of the *striatum*, represented by the nucleus caudatus, follows, together with the ventricular lumen, the curvature of the hemispheric bend, displaying a bulky caput in the wall of anterior horn, continuous with the corpus at about the level of interventricular foramen. The corpus, in turn, along the sulcus terminalis, tapers into the cauda nuclei caudati, which curves along the sulcus terminalis around the posterior surface of hemispheric stalk, merging ros trobasally with the amygdaloid complex (Fig. 249B). Variable islands or stripes of gray represent the ill-defined transitional neighborhoods between putamen, caudatum, and amygdaloid complex. Between this latter and caudobasal neighborhoods of the claustrum, similar nondescript griseal clusters form a transition.

---

[220] Cf. also the descriptions and bibliography included in the treatises by CLARA (1959), CROSBY *et al.* (1962), and the papers by ANDY and STEPHAN (1968), BROCKHAUS (1940a, b, 1942), FILIMONOFF (1965/66), HILPERT (1928), KAHLE (1969), RAE (1954a, b), and others quoted above in dealing with Mammals in general.

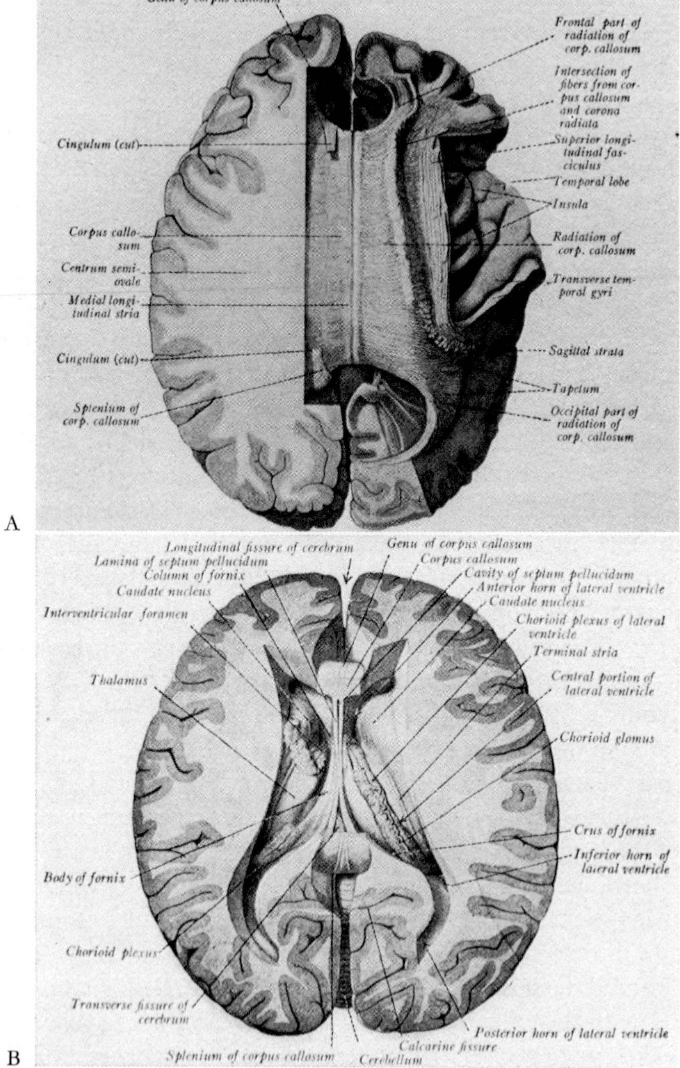

Figure 245 A. Dissection of Human telencephalon in dorsal view, showing radiation of corpus callosum. The stria longitudinalis lateralis has been scraped off (after SOBOTTA, 1931, from RANSON, 1943).

Figure 245 B. Dissection of Human telencephalon in dorsal view. The truncus corporis callosi has been removed, exposing the lateral ventricles (after SOBOTTA, 1931, from RANSON, 1943).

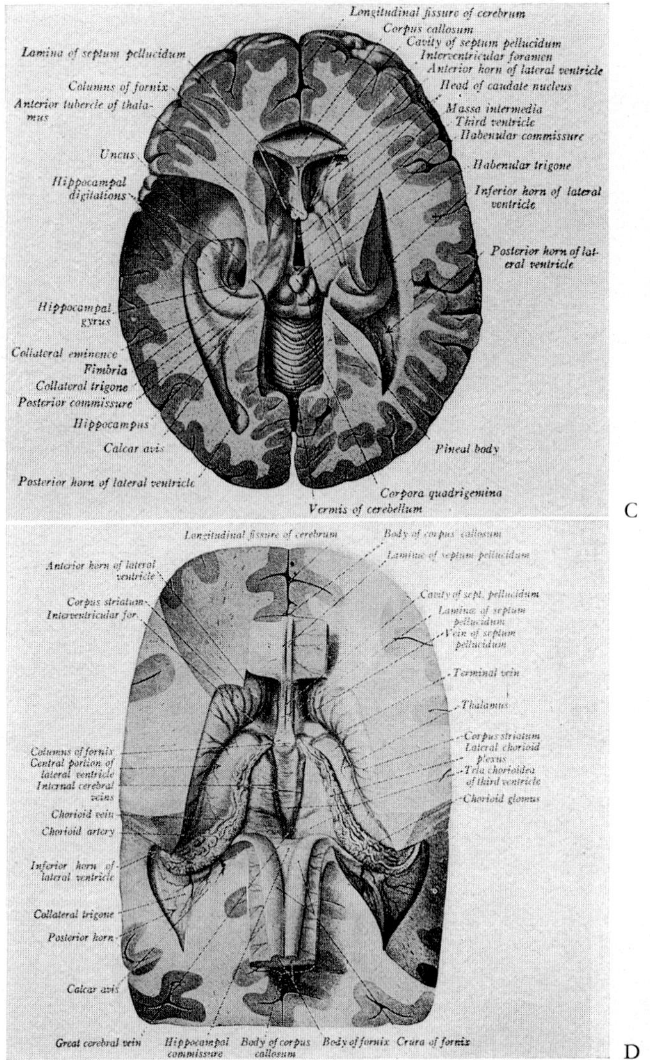

Figure 245 C. Dissection of Human telencephalon in dorsal view. Body of fornix and tela chorioidea ventrical tertii have been removed. Inferior horns of lateral ventricle are exposed, showing rostrobasal end of hippocampus at left (after SOBOTTA, 1931, from RANSON, 1943).

Figure 245 D. Dissection of Human telencephalon in dorsal view. Truncus corporis callosi cut and reflected occipitalward, showing basal surface of truncus with crura fornicis. Tela chorioidea ventriculi tertii and choroid plexuses of lateral ventricles left *in situ* (after SOBOTTA, 1931, from RANSON, 1943).

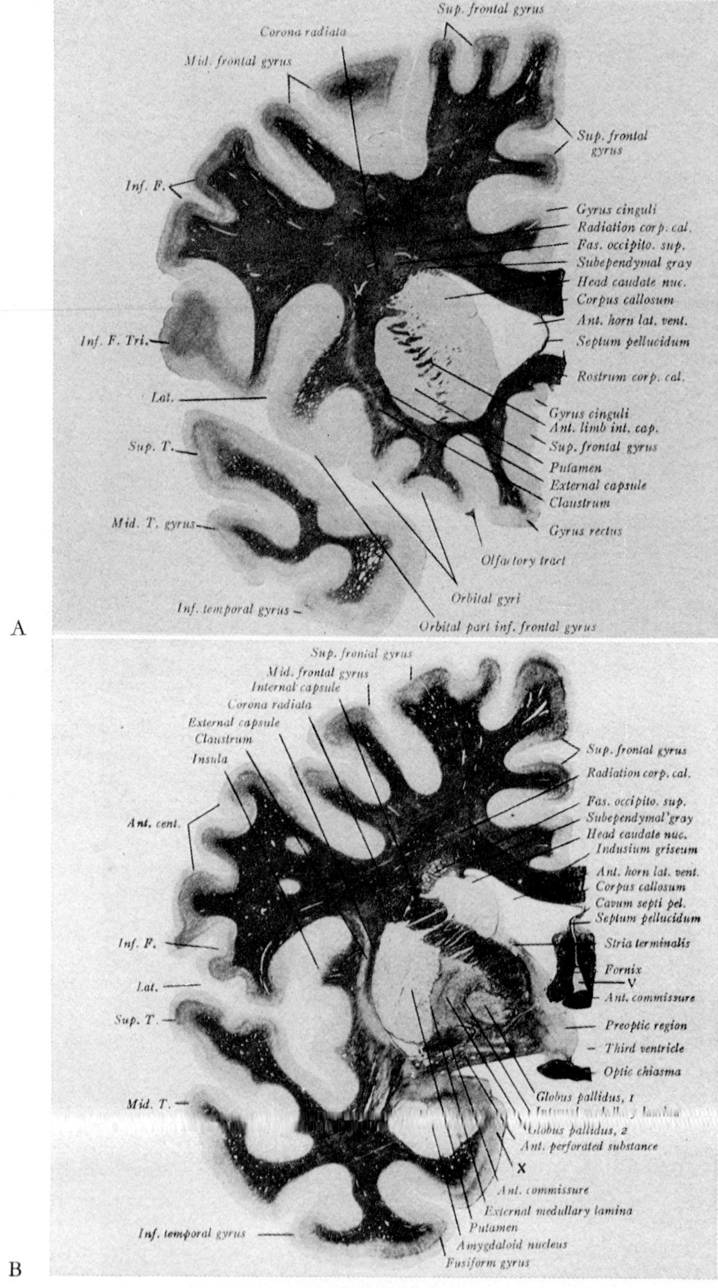

Figure 246 A, B

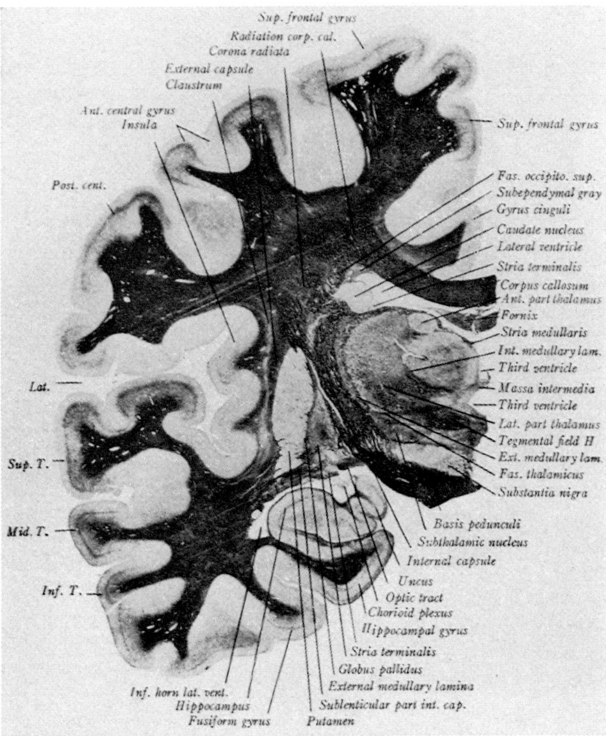

C

Figures 246 A–E. Frontal sections (myelin stain) through a Human cerebral hemisphere in rostro-occipital sequence, showing internal configuration of medullary center, grisea and ventricle (from RANSON, 1943, after JELGERSMA's atlas of the Human brain). Altered designations: u: junction of cella media, cornu inferius and cornu posterius ventriculi lateralis; v: recessus triangularis; x: caudal cortex of piriform lobe (gyrus ambiens); y: transition of hippocampal medulla (alveus) to calcar avis; z: 'tangential section' cutting slice of postcommissural hippocampus; arrow: collateral trigone.

Concerning additional relevant macroscopic features, Figures 245 A–D illustrate relationships of *corpus callosum, fornix and hippocampal formation*. The conventional rough subdivision of corpus callosum into rostrum, genu, truncus, and splenium can easily be recognized. The medially concave rostral and occipital callosal radiations, shown in Figure 245 A are also designated as forceps anterior *sive* minor, and forceps posterior *sive* major,[221] respectively.

[221] The forceps major is included in the bulbus cornu posterioris and in the so-called tapetum, described further below.

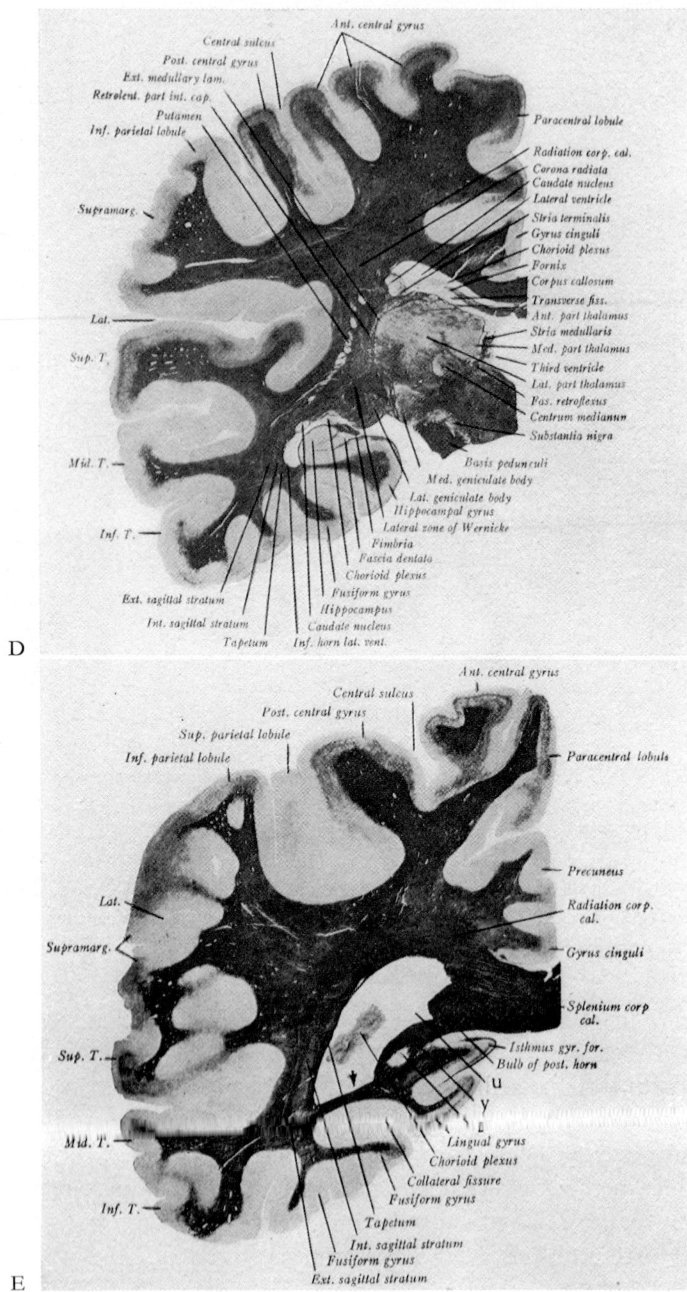

Figure 246 D, E (legend see p. 777)

The various portions of the *fornix*, namely the fimbria, the crus, the corpus, and the column, are likewise shown in these illustrations. The relationship of commissura hippocampi *sive* fornicis to splenium corporis callosi is depicted in Figure 245D, illustrating the so-called *psalterium sive lyra Davidis* formed by the converging antimeric crura fornicis with their interconnecting commissural fibers. Further aspects of fornix system and corpus callosum, as seen in a median sagittal section, are depicted by Figure 249C.

The *recessus triangularis* (BNA, omitted in PNA), between anterior commissure and columns of fornix, is shown in Figure 246B. Although appearing as a part of the third ventricle, said recess pertains morphologically to the rudimentary telencephalon impar.[222] Its roof contains the *subfornical organ*, as depicted in Figure 234, p. 356 of volume 3/I, in which the paraependymal respectively circumventricular 'organs' are dealt with. A description of the Human subfornical organ was also published by Rabl (1966).

Concerning the *hippocampal formation*, details of supracommissural and postcommissural hippocampus[223] are indicated in Figures 223C, III and 245C. The Human supracommissural hippocampus, accompanied by the stria longitudinalis lateralis, extends to a variable degree as indusium spurium upon the surface of corpus callosum. It has been scraped off in Figure 245A, which merely shows the retained *stria longitudinalis medialis sive Lancisii*, related to the indusium verum (supracommissural component of paraterminal grisea). In Figure 245C, the rostrobasalward bend of the hippocampal formation, providing the medial wall of cornu inferius ventriculi lateralis, is depicted (cf. also Fig. 246D). Near the rostral terminal expansion of hippocampal formation, forming the so-called pes hippocampi and the uncus, the variable intraventricular bulges described as digitationes hippocampi can be seen.

The Human *telencephalic ventricle*, whose median unpaired portion (ventriculus impar) is extremely reduced and includes the above-mentioned recessus triangularis, comprises the following paired subdivisions. The *anterior horn*, rostrally to *foramen Monroi*, is devoid of choroid plexus. Its roof is provided by the corpus callosum, its medial wall

---

[222] Cf. Figure 78HII, p 208 of volume 3/II.

[223] Because of its relation to so-called parolfactory area and limbic lobe respectively gyrus fornicatus, the macroscopic aspect of the rudimentary Human precommissural hippocampus shall be pointed out in chapter XIV (vol. 5/II), dealing with the details of Mammalian and Human telencephalic surface morphology.

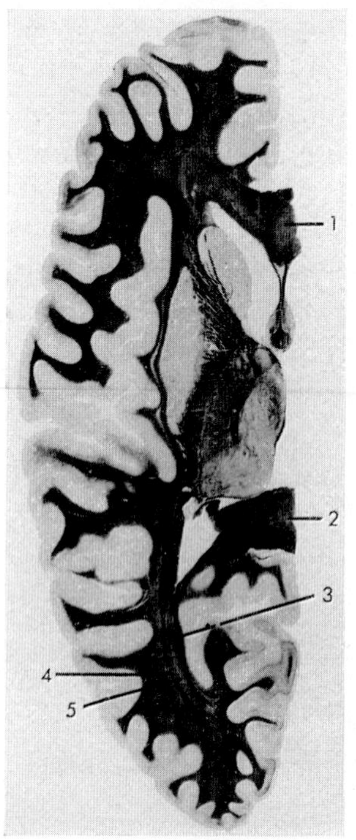

Figure 247A. Horizontal section (myelin stain) through a Human hemisphere at level of floor of interventricular foramen (from JELGERSMA's atlas of the Human brain). 1: rostro-general portion of corpus callosum; 2: splenium corporis callosi with fimbria fornicis and part of commissura hippocampi; 3: tapetum; 4: external sagittal stratum; 5: internal sagittal stratum. The designations have been added. Other structures easily identified by comparison with preceding and following Figures.

likewise by genu and rostrum corporis callosi rostrally, and more caudally by septum and fornix. Its slanting lateral wall is the caudate nucleus, its floor being the groove between caudatum and paraterminal grisea including fornix. The *cella media* extends from *foramen Monroi* along the sulcus terminalis to the caudal bend of hemispheric stalk, where inferior and posterior horn of the ventricle diverge. Roof of the cella media is the corpus callosum, the lateral wall is provided by body respectively tail of caudate nucleus, the floor by the lamina affixa covering the dorsal surface of thalamus, and the medial wall by the body of fornix (with a variable caudal extension of septum).

The size of the *posterior horn*, located within the medulla of occipital lobe, varies greatly, its lumen being not infrequently partly or even sometimes completely obliterated.[224] The radiation of corpus callosum (forceps major) forms a ventricular protrusion, the *bulbus cornus poste-*

[224] This obliteration, as seen in normal ventriculograms, may also affect right and left ventricles to quite different degrees.

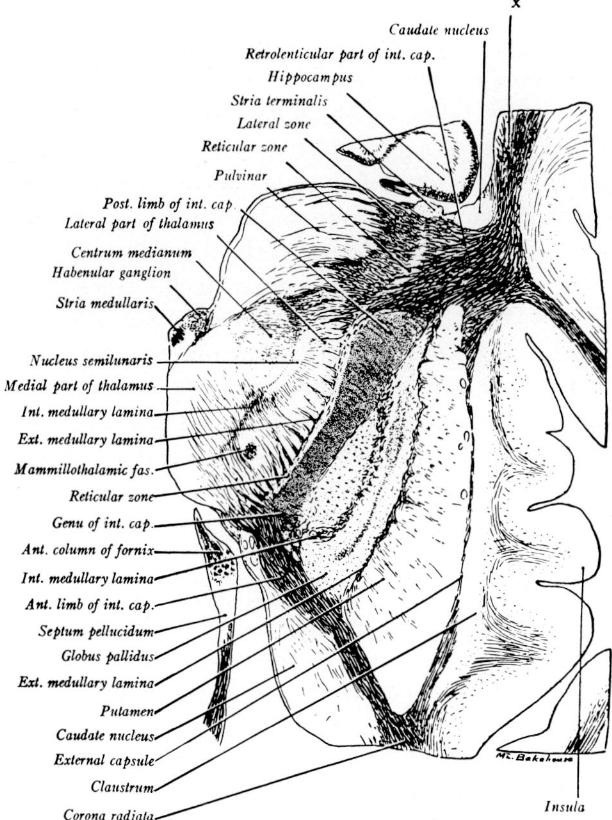

Figure 247B. Approximately horizontal section (myelin stain) through internal capsule and hemispheric stalk of Human forebrain (redrawn after DEJERINE, from RANSON, 1943). x: occipital stalk of thalamus.

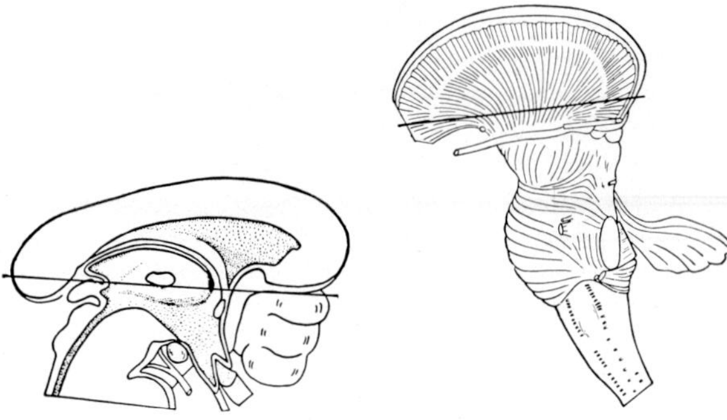

Figure 247C. Sketches indicating approximate plane of section illustrated by Figure 247B (from RANSON, 1943).

rioris (Fig. 246E), basally to which another marked protrusion, related to fissura calcarina respectively its rostral junction with fissure parieto-occipitalis represents the *calcar avis*[225] (cf. Figs. 245C, 246E, 251D).

The *inferior horn*, following the sulcus terminalis in the caudal limb of hemispheric bend along the posterior surface of hemispheric stalk, ends rostrally in the region of amygdaloid complex and uncus hippocampi. The medial wall of cornu inferius is the hippocampus with its alveus. The collateral fissure corresponds to the floor of inferior horn, forming the *trigonum collaterale* respectively *eminentia collateralis* (Figs. 245C, 246E). The roof is here formed by sulcus terminalis with stria terminalis and adjacent tail of caudate nucleus.

The *choroid plexus* of the lateral ventricle begins at the *foramen Monroi* as a continuation of the unpaired plexus of third ventricle and telencephalon impar. It extends occipitalward through the cella media, being attached to the fimbria fornicis dorsally, and to the lamina affixa basally.[226] At the divergence of cornu posterius and inferius, it follows fimbria fornicis and a gradually shortening lamina affixa along the sulcus terminalis, ending at occipital levels of uncus hippocampi. Like cornu anterius, the cornu posterius is devoid of choroid plexus, except for variable minor protrusions at the region of the diverging cornua.[227] Variable clusters of cystic or vesicular enlargements of choroid villi, commonly found near the transition of cella media to cornu inferius are designated as *glomus chorioideum*.[228]

---

[225] The *calcar avis* was also formerly designated as '*hippocampus minor*', the true hippocampal protrusion in cornu inferius being then the '*hippocampus major*'. Cf. the 'Note on the resemblances and differences in the structure of the brain in Man and Apes. By Professor Huxley' included in Darwin's 'Descent of Man' (pp. 557–563 of the Modern Library edition, New York, n.d.). Huxley, in 1874, as here quoted by Darwin, emphasizes here, *inter alia*, the presence of an hippocampus minor in the cornu posterius of Apes. This had been erroneously denied by authors claiming that the hippocampus minor distinguished Man from the Apes.

[226] If torn off the edges of these attachments represent taenia fornicis respectively taenia chorioidea. The term '*fissura chorioidea*' designates the space between the two attachments of lamina epithelialis, through which the vascularized leptomeninx of tela chorioidea ventriculi tertii within fissura transversa cerebri becomes continuous with the mesodermal core of the plexus (cf. Figs. 136B, H, 245D).

[227] At ontogenetic stages, the choroid plexus likewise protrudes into the developing anterior horn, and minor protrusions may also still occur in the adult condition.

[228] Histology and cytology of the choroid plexus were dealt with in chapter V, volume 3/I, section 5 (pp. 336–341) and section 6 (pp. 471–483). The relationships of choroid plexus and cerebrospinal fluid were also discussed in chapter VI, vol. 3/II, section 7

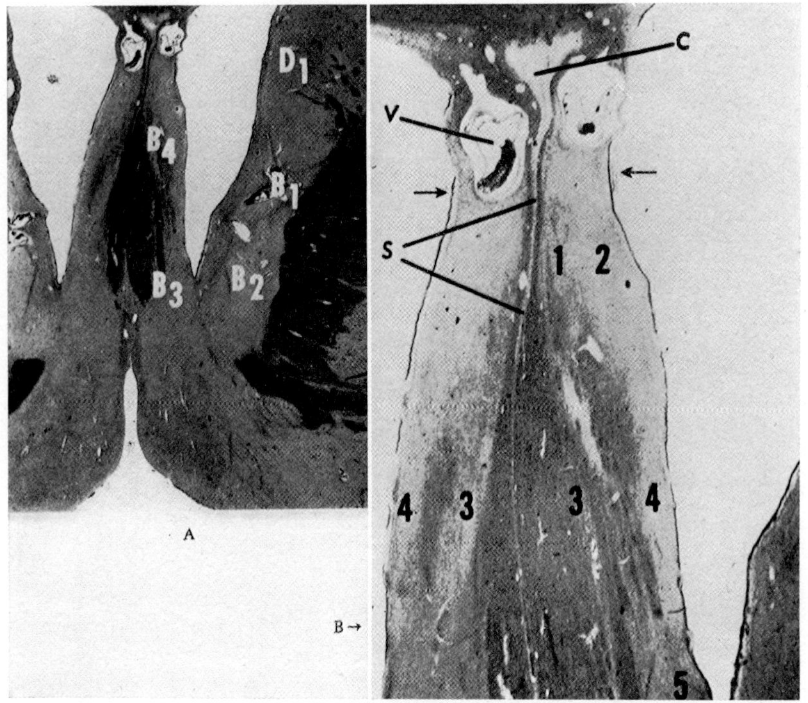

Figure 248 A, B. Cross-section through a fairly representative Human septum telencephali of the type with relatively predominant septum gangliosum (from K., 1969; hematoxylin-eosin stain emphasizing medullated fiber systems; A: ×4; B: ×10, red. $^{2}/_{3}$). The topologic neighborhoods are indicated in A. B shows additional details, the two arrows approximately indicating the boundary zone between septum gliosum (above) and septum gangliosum (below). c: cavum septi pellucidi; s: glial median raphe; v: vena septi pellucidi (BNA, PNA); 1: dorsomedial neighborhood of septal grisea; 2: dorsolateral neighborhood; 3: ventromedial neighborhood; 4: ventrolateral neighborhood; 5: griseum at basis of paraterminal region, representing 'nucleus accumbens septi', respectively a dorsal portion of B_3 (cf. Fig. 248 A).

As regards the *Human telencephalic communication channels*, the lateral, intermediate and medial olfactory tracts, considered in section 1 of the present chapter, moreover basal forebrain bundle, ansa lenticularis, ansa peduncularis, stria terminalis, stria medullaris, and fornix system, dealt with further above and in chapter XII, may here be omitted.

(pp. 694–701). As regards the manifold variations displayed by development and extent of the choroid plexus of telencephalon (and of other brain subdivisions) in the Vertebrate series, cf. also *Graf* HALLER (1922).

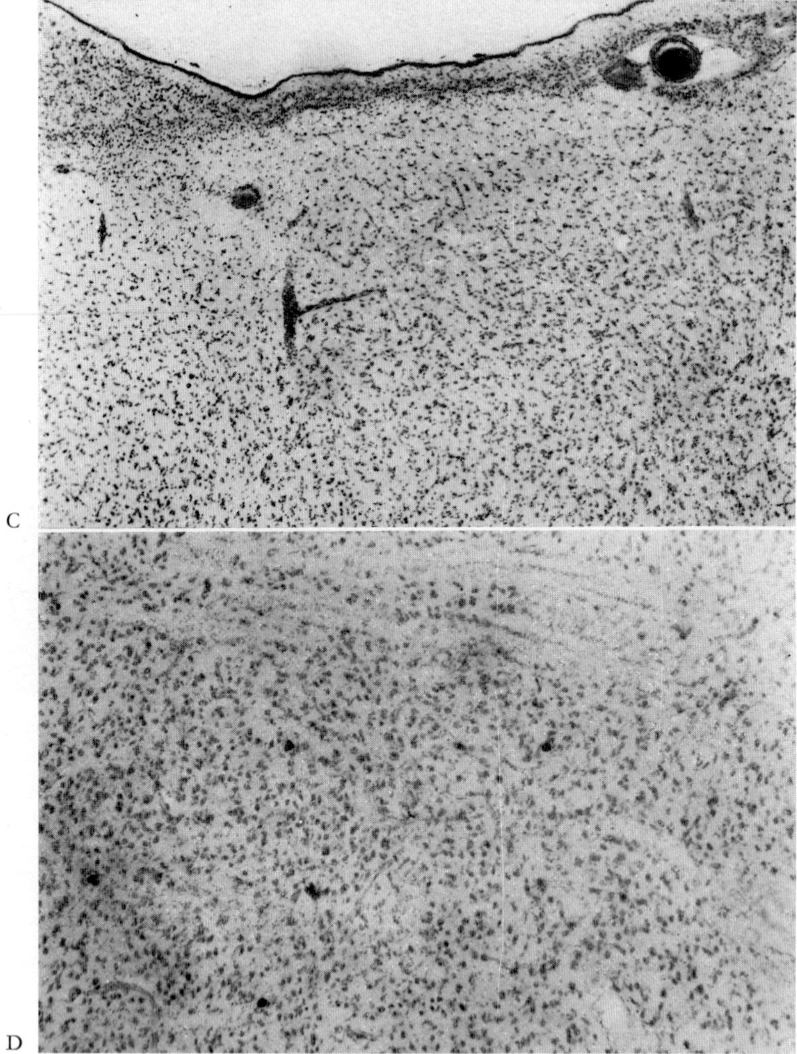

Figure 248 C. Cytoarchitecture of adult Human nucleus caudatus (intermediate level of caput) near lateral ventricular angle *(Nissl stain,* ×60, red. $^2/_3$). Predominantly small nerve cells, with a few scattered larger ones. The ependyma and the conspicuous subependymal cell plate are easily identified. The lateral ventricular angle is indicated by the groove at left.

Figure 248 D. Cytoarchitecture of putamen at same level as Figure 248 C *(Nissl stain,* ×70, red. $^2/_3$). The small nerve cells are slightly larger and somewhat less crowded, and the scattered larger elements are more conspicuous.

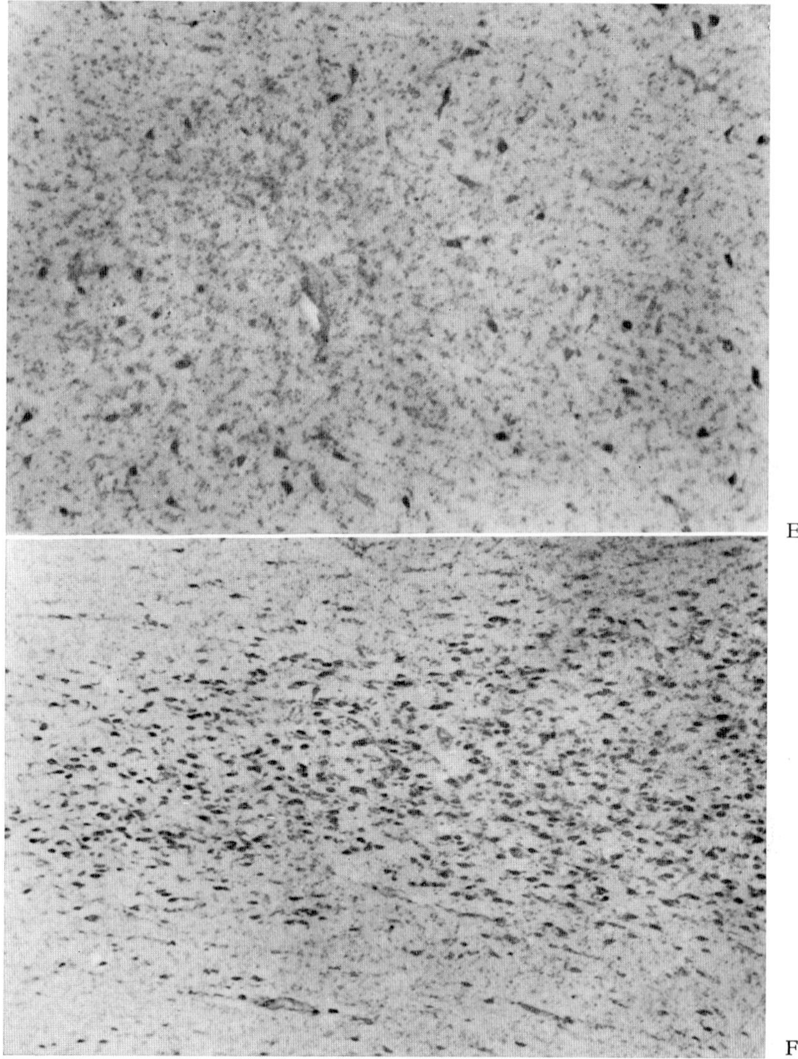

Figure 248 E. Cytoarchitecture of globus pallidus at same level as Figures 248 C and D (*Nissl stain*, ×70, red. $^2/_3$). Only one type of scattered fairly large multipolar nerve cells is present. The other cells are neuroglial elements and mesodermal ones pertaining to the vascular apparatus, whose lumina are collapsed.

Figure 248 F. Cytoarchitecture of claustrum at same level as Figures 248 C–E (*Nissl stain*, ×70, red. $^2/_3$). The capsula externa is at top, capsula extrema at bottom, basal side of claustral cell plate at right. Many of the nerve cells are fusiform, with dorso-basally oriented (i.e. vertical) long axis.

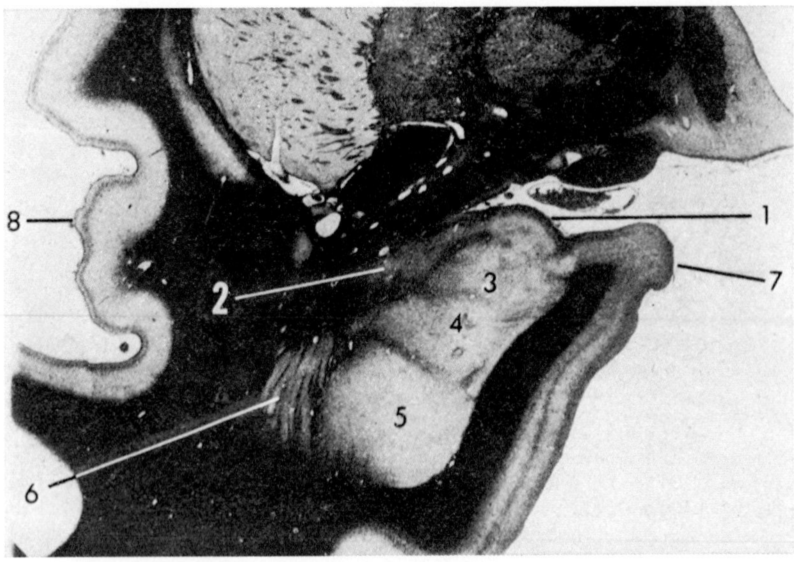

Figure 249 A. Cross-section (myelin stain) through the amygdaloid complex of the Human brain (from LANDAU, 1923). 1: cortical nucleus amygdalae (its medial subdivision delta is barely recognizable); 2: diffuse portion of nucleus amygdalae gamma; 3, 4, 5: subdivisions of main amygdaloid complex (beta grisea); 6: transition of main complex to claustrum (separated by fiber systems of fasciculus uncinatus, occipitofrontalis, and commissura anterior from claustrum proper, which is not labelled but easily identifiable); 7: transition between cortex lobi piriformis and parahippocampal cortex; 8: insula. This Figure should be compared with Figure 147D of chapter XII (cell stain).

Generally speaking, the here relevant pathways located within the extensive medullary core of the telencephalic hemisphere, and forming the so-called centrum semiovale in macroscopic horizontal sections at the level of corpus callosum and above this latter, can roughly be classified as follows.

(1) *Association fibers* interconnecting homolateral cortical regions. (2) *Commissural fibers* of *corpus callosum*, *fornix system*, and *commissura anterior*, interconnecting the antimeric halves of the telencephalon. (3) *Corticopetal* and *corticofugal projection fibers* related to subcortical grisea of telencephalon, diencephalon, brain-stem, and spinal cord. In the narrower sense, the term projection fibers refers only to those connecting the cortex with grisea external to the telencephalon. The projection fibers converge from the diverse cortical regions toward the hemispheric stalk. The expansion of these fibers dorsally to capsula interna

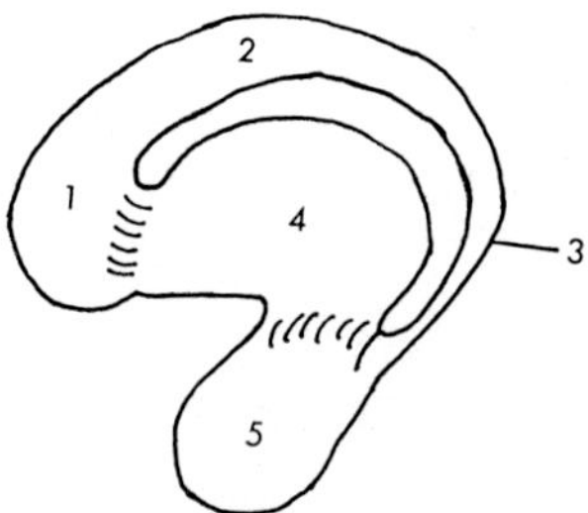

Figure 249 B. Semidiagrammatic sketch roughly illustrating the overall configuration of strio-amygdaloid complex. 1, 2, 3: caput, corpus and cauda nuclei caudati; 4: putamen; 5: amygdaloid complex. In similar figures, RANSON (1943, Fig. 207, p. 260) and LANDAU (1923, Fig. 62, p. 202) considerably underestimated the relative size of the amygdaloid griseum.

forms, together with callosal commissural fibers, the *corona radiata* of macroscopic anatomy.

Discounting details of association and commissural fiber systems to be dealt with in an appraisal of the cerebral cortex (chapter XV, vol. 5/II), Figures 250 A, B and 251 A–E illustrate the here relevant pathways.

The Human *capsula interna* displays in horizontal sections an *anterior limb*, between 'lentiform nucleus' and caput nuclei caudati, a *knee* in the vicinity of interventricular foramen, and a *posterior limb* between 'lentiform nucleus' and thalamus (cf. also Figs. 247 A–C). Occipitalward, the lenticulo-thalamic part of posterior limb is continuous with the *retro-* and *sublenticular* portions of the internal capsule.

On the basis of numerous clinicopathologic observations in combination with data obtained by studies of myelogenesis and of adult myelin-stained sections as well as by the gross dissection of fiber bundles,[229] the approximate location, in capsula interna and hemispheric medulla in general, of the main communication pathway has been reasonably well established (cf. Figs. 250 A, B, 251 A–E).

Thus, through *anterior limb* of capsula interna run frontopontine tract (also known as *Arnold's bundle*) and anterior stalk of thalamus. Through the *genu* pass corticobulbar pathways pertaining to 'upper motor neurons' of efferent cranial nerves, followed, in the *posterior limb*, and in occipitalward sequence, by the cortico-spinal (pyramidal)

---

[229] Cf. in particular the atlas by LUDWIG and KLINGLER (1956).

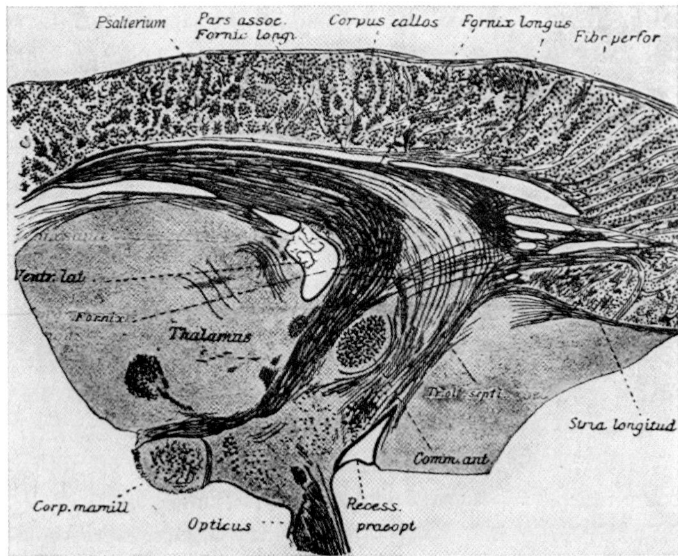

Figure 249 C. Approximately median sagittal section (myelin stain) through fornix and telencephalic commissures in the Human brain (after SHIMAZONO, from KAPPERS, 1921). The designations are self-explanatory.

tract fibers for upper extremity, trunk, and inferior extremity. These systems are apparently intermingled, in the *lenticulo-thalamic* portion, with cortico-rubral connections. Within the occipital part of lenticulo-thalamic portion runs the parietal stalk of the thalamus which includes the sensory input to the cortex mediated by the general lemniscus systems reaching the thalamic grisea. The limbic thalamic radiation, related to the anterior group of thalamic nuclei, presumably joins the posterior part of frontal thalamic radiation near the genu and also penetrates parietal stalk or centroparietal radiation. It then bends mediad and in part caudad through corona radiata towards the gyrus cinguli (limbic lobe).

Through the occipital portion of lenticulo-thalamic part presumably likewise run cortico-pretectal and tectal connections together with the parieto-occipito-temporal corticopontile fibers *(Türck's bundle)*. This latter system also extends into retrolenticular and sublenticular neighborhoods.

The *auditory radiation*, interconnecting the medial geniculate grisea with the transverse temporal *gyri of Heschl*, passes through retrolenticular and sublenticular portions.

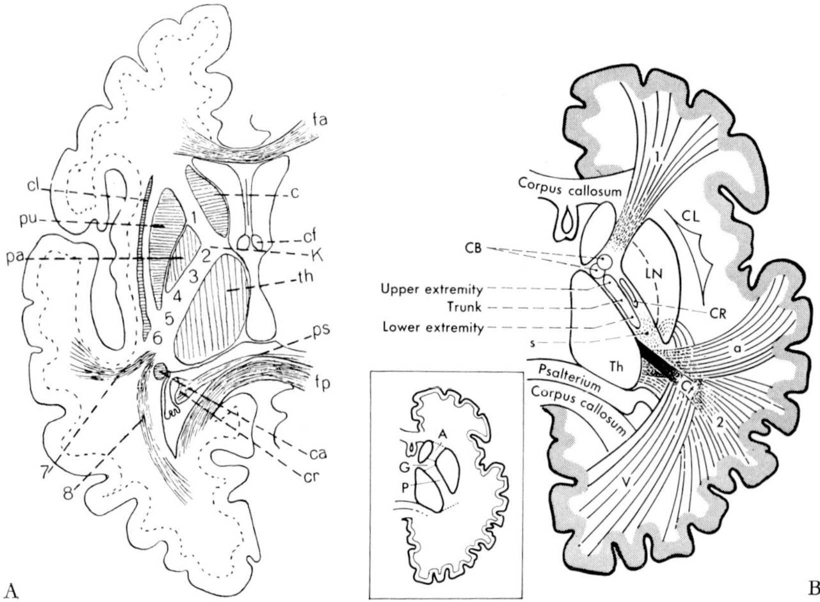

Figure 250 A. Semidiagrammatic sketch of a horizontal section through a Human cerebral hemisphere at level of interventricular foramen, showing subdivisions of internal capsule (from K., 1927). 1: frontopontile tract and frontal thalamic stalk; 2: corticobulbar tracts; 3: corticospinal tract for upper extremity; 4: corticospinal tract for lower extremity; 5: general sensory radiation (parietal thalamic stalk); 6: temporoparieto-pontile tract; 7: central auditory radiation; 8: optic radiation; c: nucleus caudatus (caput); ca: cauda nuclei caudati; cf: columna fornicis; cl: claustrum; cr: crus fornicis; fa: corpus callosum, forceps anterior; fp: corpus callosum, forceps posterior; k: genu capsulae internae; pa: globus pallidus; ps: commissura hippocampi *(psalterium sive lyra Davidis);* pu: putamen; th: thalamus.

Figure 250 B. Semidiagrammatic horizontal section through Human hemisphere showing additional details of fiber systems (from Haymaker and Bing, 1969). CL: claustrum; CB: corticobulbar pathways; CR: assumed location of corticorubral and corticotegmental tracts; LN: lenticular (lentiform) nucleus; Th: thalamus; V: optic radiation; a: auditory radiation; s: sensory radiation to postcentral gyrus; 1: frontopontile tract; 2: temporopontile tract. Assumed corticotectal and corticotegmental channels are indicated by Ct. The inset shows anterior limb (A), genu (G), and posterior limb (P) of internal capsule.

The *optic radiation* or geniculocalcarine tract (Fig. 251 A) also runs through retrolenticular and sublenticular portions, first passing through a neighborhood designated as *Wernicke's field* (Fig. 143 E). Its most basal fibers form a loop, known as the temporal knee, near the tip of cornu inferius ventriculi lateralis and laterally to said ventricle. This

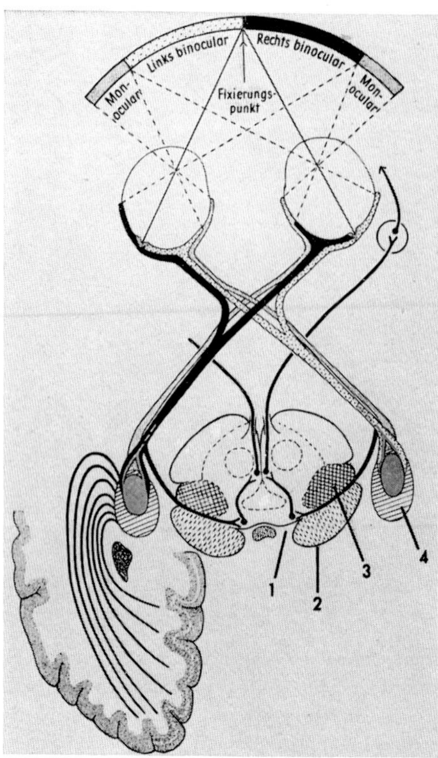

Figure 251 A. Diagram illustrating general features of the Human optic pathway (from GLEES, 1957). Added designations: 1: pretectal region; 2: tectum mesencephali; 3: medial geniculate body; 4: lateral geniculate body.

loop was recognized by FLECHSIG (1920, and his previous publica tions) and well depicted in PFEIFER's (1925, 1930) studies on the optic radiation (cf. Figs. 251 B, C). Said radiation forms the *stratum sagittale externum*[230] (Figs. 246 E, 247 A, 251 D, E) on the lateral side of inferior and posterior horn of the lateral ventricle. It finally bends basad and mediad to reach the visual cortex (area striata) on both lips of the cal-carine fissure. The geniculocalcarine tract ist thus quite distinct from the other fibers in the occipital peduncle of the thalamus which pass through the *stratum sagittale internum*. This component of the occipital

[230] The stratum sagittale externum is identical with the so-called fasciculus longi-tudinalis inferior, which formerly, mainly on the basis of gross dissection was erroneously interpreted as a cortical 'association bundle'.

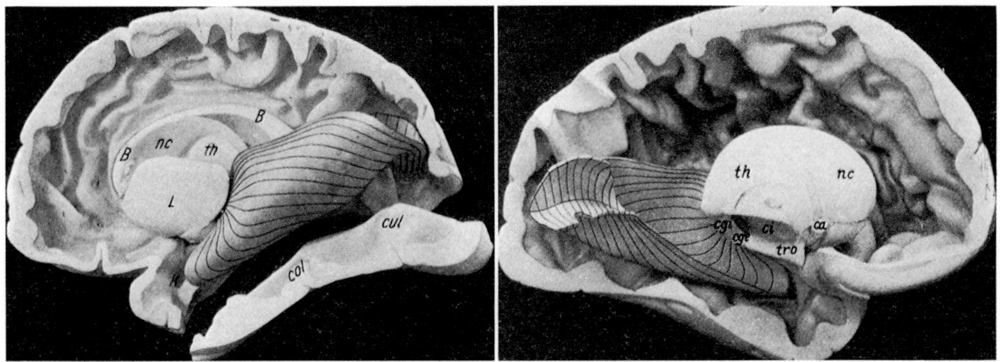

Figure 251 B. Model showing lateral aspect of shape and fiber course of the geniculocalcarine tract *(Sehmarklamelle)* in the Human brain (from PFEIFER, 1925). B: corpus callosum; L: lentiform nucleus; col: eminentia collateralis; cul: culmen of collateral eminence; K: temporal knee (FLECHSIG) of optic radiation; nc: nucleus caudatus; th: thalamus.

Figure 251 C. Model showing medial aspect of geniculocalcarine tract in the Human brain (from PEIFER, 1925). ca: anterior commissure; cge: lateral geniculate body; cgi: medial geniculate body; ci: capsula interna; tro: optic tract; other designations as in Figure 251C.

thalamic stalk consists to a great part of fibers interrelating pulvinar with occipital cortex outside of area striata. Connections between occipital lobe, pretectal grisea and mesencephalic tectum also seem to run through internal sagittal stratum. This latter could be designated as occipital thalamic stalk or peduncle *sensu strictiori*, while that peduncle *sensu latiori* also includes the geniculocalcarine tract (stratum sagittale externum). Internally to stratum internum lies the *tapetum cornus posterioris*, consisting of callosal fibers pertaining to forceps major, presumably also intermingled with association fibers of fasciculus longitudinalis superior.[231]

The rostral part of occipital thalamic radiation is not sharply delimited from the temporal thalamic radiation (peduncle or stalk) which encompasses thalamic connections with parts of temporal and adjacent parietal lobe, including an ill-defined *fasciculus temporothalamicus Arnoldi*. Bundles of these systems appear to traverse *Wernicke's field*. In the wider sense, the temporal thalamic stalk also includes the above-men-

[231] The tapetum *(ventriculi sive corporis callosi)* was also occasionally designated as stratum sagittale mediale. The fasciculus longitudinalis superior and other association systems shall be dealt with in chapter XV of volume 5/II.

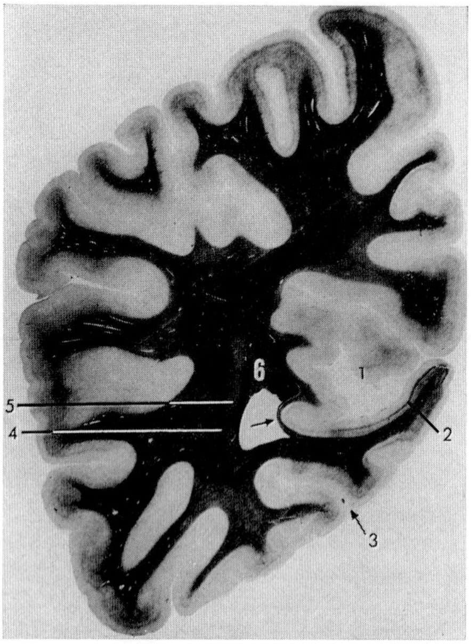

D

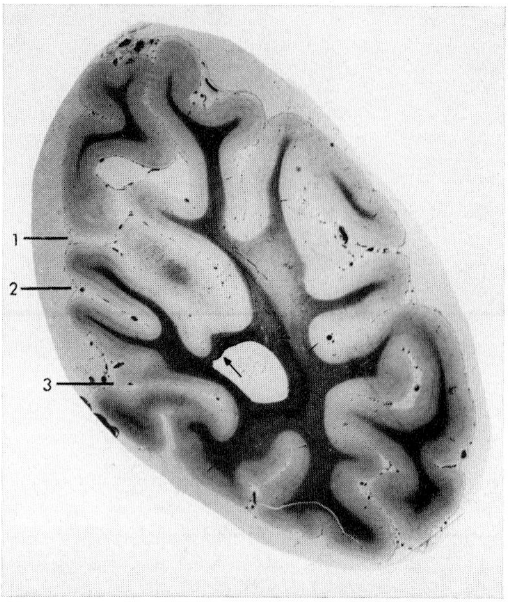

E

Figure 251 D. Cross-section through occipito-parieto-occipital lobe (myelin stain) at rostral level of ventricular posterior horn (from JELGERSMA's atlas of the human brain). 1: Common stem of fissura parieto-occipitalis and fissura calcarina; 2: area striata; 3: fissura collateralis; 4: stratum sagittale externum (optic radiation); 6: tapetum; between 4 and 6 the lighter stratum sagittale internum (5); arrow: rostral begin of calcar avis. The labels have been added.

Figure 251 E. Cross-section (myelin stain) through Human occipital lobe and posterior horn of lateral ventricle. 1: fissura parieto-occipitalis; 2: fissura calcarina; 3: fissura collateralis. The sagittal strata and a transition between bulbus cornus posterioris and calcar avis (arrow) as well as the flat eminentia collateralis are easily identifiable. The pattern differs slightly from that of Figure 251 D.

tioned auditory radiation, some of whose fibers appear to intermingle with the optic radiation.

The region in which somesthetic, auditory, and optic fiber systems commingle is also known as *Charcot's carrefour sensitif*; it includes rostral portions of *Wernicke's field*, which, in turn, has also been designated as *meditullium laterale*.

The inferior thalamic peduncle[232] (Figs. 143 A, B) includes fibers connecting rostral parts of thalamus with temporal lobe and inferior

[232] It will be recalled that the combined loop of ansa lenticularis and of inferior thalamic peduncle is also designated as ansa peduncularis.

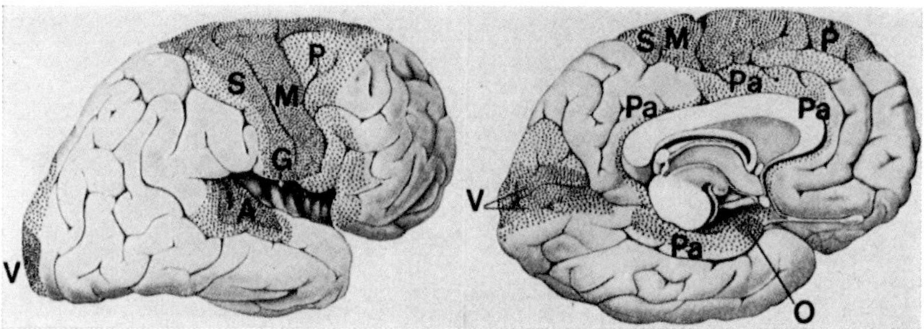

Figure 252 A. Projection areas or 'centers' of the Human telencephalon (after FLECHSIG, 1896, from K., 1957). A: auditory cortex; G: approximate location of gustatory cortex (present interpretation); M: motor cortex; O: olfactory center (present interpretation); P: premotor cortex (present interpretation); Pa: parahippocampal cortex (present interpretation); S: sensory cortex; V: visual cortex. This early mapping, whose labeling has been modified in accordance with the present interpretation, appears superior to FLECHSIG's later highly parcellated charts of 1920. It should be added that P and Pa, despite relatively early myelinization, do not represent primary 'projection areas'.

insular region. On the basal side, this peduncle is covered by the gray of substantia perforata anterior. Within this pathway are fibers of posterior piriform lobe and connections of amygdaloid complex.[233] The inferior thalamic peduncle also intermingles with fibers of the *diagonal band of Broca* (Figs. 148 A, B) which interconnect cortical amygdaloid nucleus and adjacent grisea with gyrus subcallosus. These fibers run in individually variable fasciculi but gather rostrally in a more constant and fairly well circumscribed bundle laterally and rostrally to optic tract, where they can be easily distinguished from the fiber system of pedunculus inferior thalami. These complex pathways, moreover, are traversed by fibers pertaining to the systems of anterior commissure (Fig. 246 B) and of association bundles such as fasciculus uncinatus and occipitofrontalis inferior (cf. Fig. 116 C, chapter XV, vol. 5/II).

In a detailed atlas, in which myelogenetic stages were also included, KRIEG (1973) has recently summarized and depicted his interpretation of Human telencephalic fiber systems.

With regard to cortical regions and their reasonably well document-

[233] A substantial part of amygdaloid connections, however, pass through the stria terminalis system running along the sulcus terminalis. So-called 'olfactohabenular' fibers from the basal rhinencephalon are presumably included in this pathway as well as in the fiber masses of ansa peduncularis.

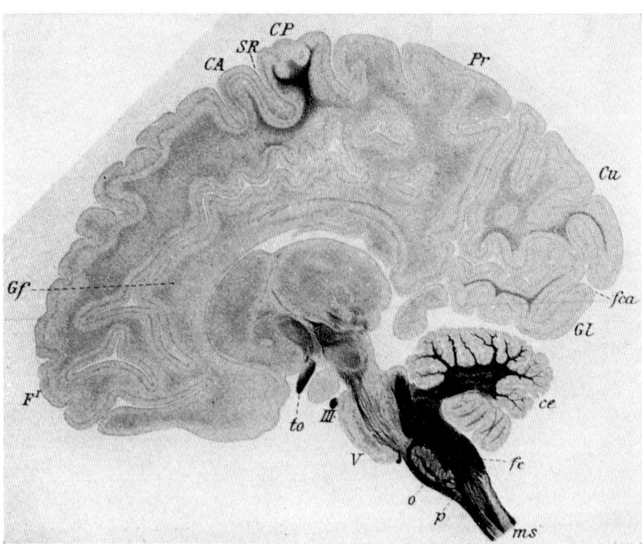

Figure 252B. Sagittal section (myelin stain) through the brain of a prematurely born infant (at 7 months) who lived 48 days (from FLECHSIG, 1920). Designations for Figures 252B–D: CA, CP: gyrus centralis anterior respectively posterior; Cu: cnueus; F^I, F^{II}, F^{III}: gyrus frontalis superior, medius, inferior; Ga: gyrus angularis; Gf: gyrus fornicatus *(sive* cinguli); GH: gyrus hippocampi; Gl: gyrus lingualis; Got: gyrus occipito-temporalis; J: Insula; O^{II}: gyrus occipitalis medius; P^I, P^{II}: gyrus parietalis superior, inferior; Pr: praecuneus; Pu: putamen; Sip: sulcus interparietalis; SR: sulcus centralis *(Rolandi)*; Ti: inner field of temporal lobe; V: pons; ce: cerebellum; fc: fasciculus cuneatus; fca: fissura calcarina; gl: gyrus lingualis; gsa: gyrus subangularis; ms: spinal cord; o: inferior olivary complex; p: pyramidal tract; ra: auditory radiation; roe: optic radiation (external sagittal stratum); roi: internal sagittal stratum and tapetum; saa: gyrus supra-angularis; to: optic tract; III: oculomotor root.

ed subdivision into *projection areas*[234] and *association areas* propounded by FLECHSIG (1896, 1920, and other publications), Figure 252A illustrates the overall features of their location. Figures 252B–E show mye-

[234] The non-committal term area seems preferable to 'center'. Again, the probability that most or even all association areas are provided with subcortical connections, as emphasized by opponents to FLECHSIG's views, do not invalidate the relevant aspects of that author's concepts. The observations by HESS (1954) in the Guinea pig, and similar findings by others seem to indicate that, in a given tract, the fibers that will be the largest are the first to myelinate and to display (central) nodes of Ranvier. This may also obtain in the Human CNS, but can hardly be used as an argument against the basic data recorded by FLECHSIG. Again, the internodal distances in both central and peripheral nervous system are said to be roughly, i.e. more or less, directly proportional to the fiber caliber.

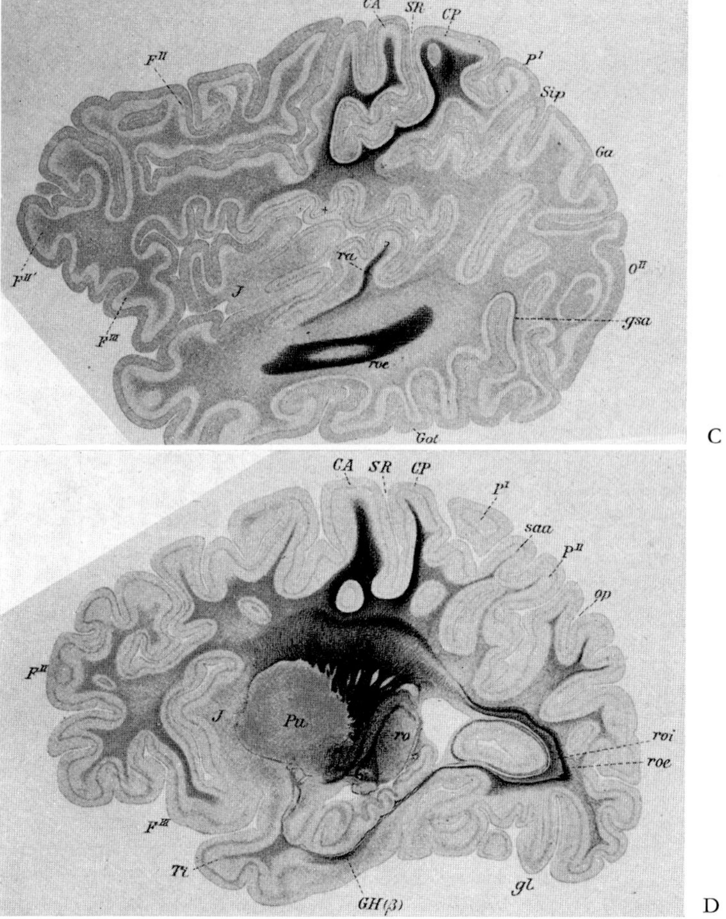

Figure 252 C. Sagittal section (myelin stain) through hemisphere of a newborn at late term, showing myelinization of sensorimotor, acoustic, and optic radiation (from FLECHSIG, 1920).

Figure 252 D. Sagittal section (myelin stain) through hemisphere of a 7 weeks old infant, showing further progress of myelinization (from FLECHSIG, 1920).

logenetic stages of relevant fiber connections which rather convincingly support that concept.

The *main motor projection area*, from which most oft the corticobulbar and corticospinal 'pyramidal tract' can be assumed to originate, is located in the precentral gyrus, characterized by cortex agranularis gigantopyramidalis (cf. chapter XV, vol. 5/II). The effects of electrical

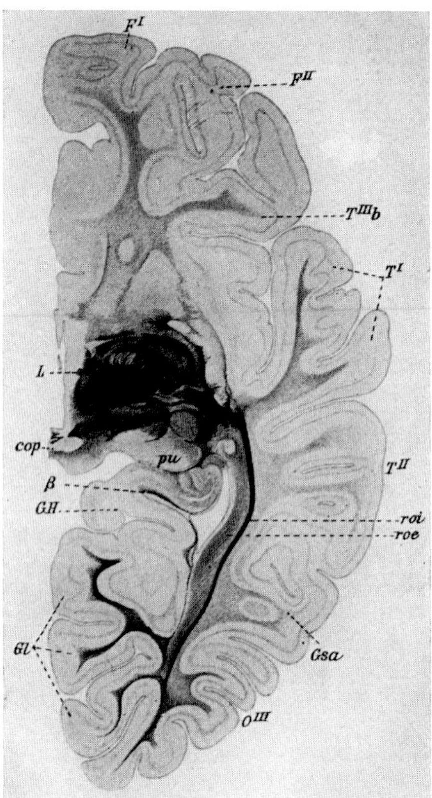

Figure 252E. Horizontal section (myelin stain) through the hemisphere of a 7 weeks old infant (from FLECHSIG, 1920). L: corpus subthalamicum Luysi; O^{III}: inferior occipital gyrus; T^{I}, T^{II}: superior and middle tempora gyrus; $T^{III}b$: inferior frontal gyrus (T misprinted for F); cop: commissura posterior; pu: pulvinar; roe: internal sagittal stratum and tapetum; roi: optic radiation (external sagittal stratum); β: fiber system of parahippocampal cortex and hippocampal formation. Other abbreviations as in Figures 252B–D, except for the here erroneously reversed roe and roi leads.

stimulation performed during brain operations, as recorded in the course of his pioneering surgical procedures by FEDOR KRAUSE (1908, 1911) and subsequently by PENFIELD and many others, have indicated a rather constant somatotopic localization (Figs. 253 A, B), graphically illustrated by PENFIELD's and RASMUSSEN's (1950) *motor homunculus*. As shown in the Figures, the sequence begins basally with swallowing, mastication and tongue movements, ending dorsally on the medial surface of the hemisphere with motions of ankles and toes.

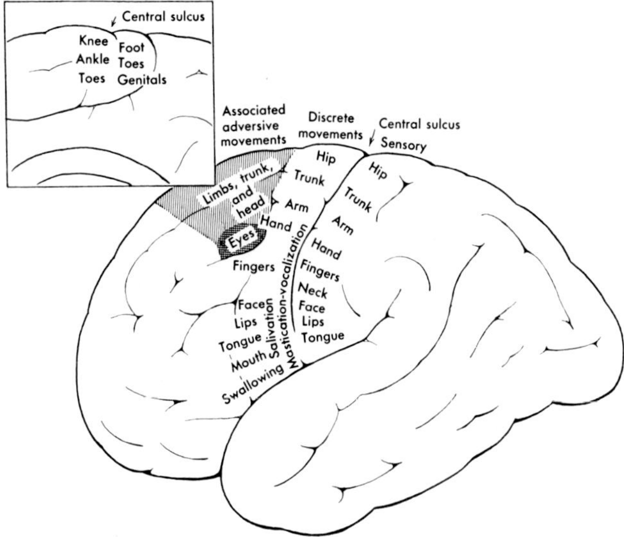

Figure 253 A. Outline of Human hemisphere indicating a now widely accepted interpretation of functional representation. Inset illustrates portion of medial aspect (modified after PENFIELD, 1954, and PENFIELD and RASMUSSEN, 1950, from HAYMAKER and BING, 1969).

In accordance with the predominant crossing of the pyramidal tract, the elicited motions occur, as a rule, on the contralateral side.[235] Whether separate and constant localization for single muscles or rather less rigidly defined representations for functional muscular groups obtain, still remains a moot question, again to be considered in chapter XV of volume 5/II. It is of interest that stimulation of this cortex

[235] WEIR MITCHELL, according to EARNEST (1950), was aware as early as 1860 that one side of the brain innervates the contralateral body side. This was just before the time that MITCHELL took up his work in a hospital established at Philadelphia for casualties of the Civil War (1861–1865). It should, however, be pointed out that this contralateral innervation was already recorded in the *Hippocratic writings*, more than 2200 years ago. Thus, in the treatise 'On wounds of the head' (περὶ τῶν ἐν κεφαλῇ τρωμάτων, XIX, 20) we read:

ἢν μὲν ἐν τῷ ἐπ' ἀρίστερα τῆς κεφαλῆς ἔχῃ τὸ ἕλκος, τὰ ἐπὶ δεξιὰ τοῦ σώματος ὁ σπασμὸς λαμβάνει· ἢν δ' ἐν τῷ ἐπὶ δεξιὰ τῆς κεφαλῆς ἔχῃ τὸ ἕλκος, τὰ ἐπ' ἀριστερὰ τοῦ σώματος ὁ σπασμὸς ἐπιλαμβάνει.

(If the patient has the lesion on the left side of the head, spasm seizes the right side of the body; if he has the lesion on the right side of the head, spasm seizes the left side of the body.)

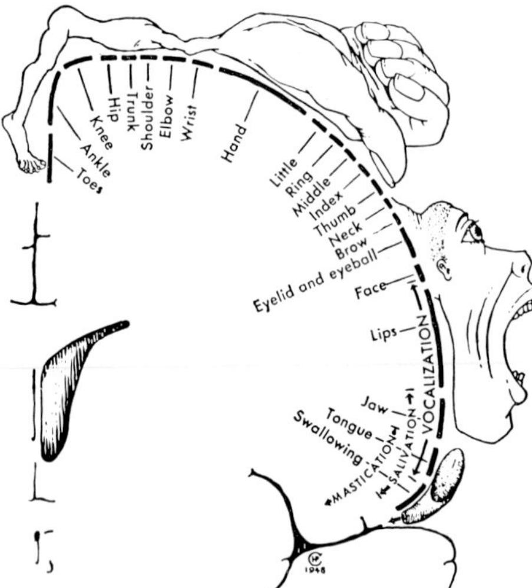

Figure 253 B. Diagrammatic cross-section of hemisphere at level of motor projection area with so-called motor homunculus corresponding to somatotopic localization (from PENFIELD and RASMUSSEN, 1950).

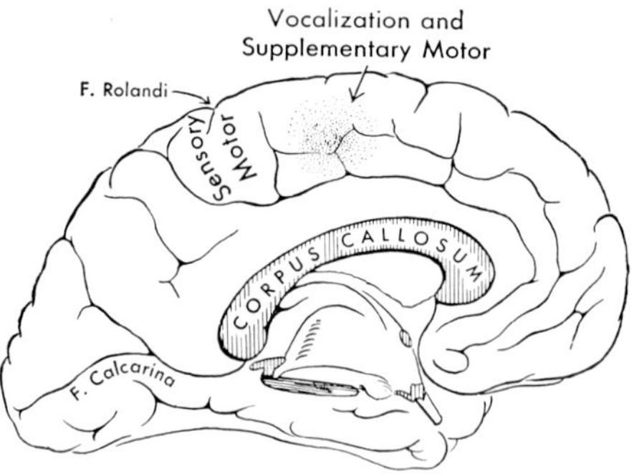

Figure 253 C. Sketch of medial hemisphere surface with so-called supplementary motor area (from PENFIELD and RASMUSSEN, 1950).

elicits movements which the patient (under local anesthesia) cannot prevent. In other words, conscious patients did not feel that they had 'willed' the action (PENFIELD, 1954). Occasionally, arrest of movement or prevention of movement was induced by stimulation. It seems unlikely that the 'intention' to perform a 'voluntary' movement is related to neural processes in the precentral convolution which may be regarded as merely the origin of a 'final' cortical output. PENFIELD (1954)

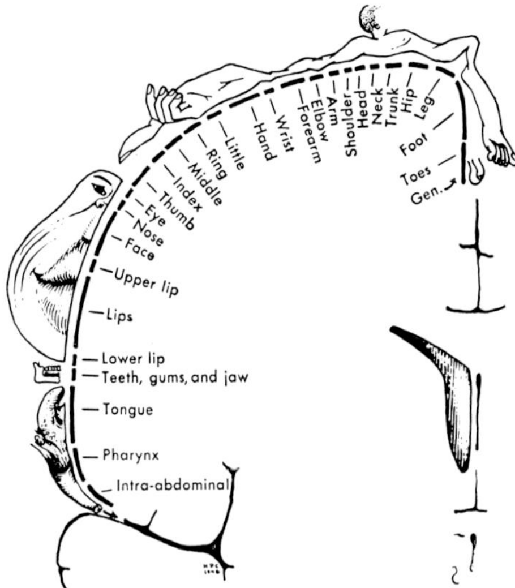

Figure 253 D. Diagrammatic cross-section of hemisphere at level of sensory projection area with so-called sensory homunculus corresponding to assumed somatotopic localization (from PENFIELD and RASMUSSEN, 1950).

justly stresses that the electrode stimulus cannot imitate the use of the precentral gyrus in normal voluntary action but merely demonstrates its peripheral output connections.

Closely contiguous with the 'face-' and 'hand-regions' of the main motor cortex, an 'eye-field' is located in the posterior part of the middle frontal convolution, from which conjugate eye movements toward the opposite side can generally be elicited.

PENFIELD and his associates have furthermore outlined a so-called supplementary motor area,[236] located rostrally to the upper part of the main motor projection area, and extending to the medial hemispheric surface within the fissura longitudinalis cerebri (Figs. 253A, C). Its function seems to differ in some respects from that of the main area, but the findings of the cited authors indicate that it retains some sort of control over the somatic musculature in the absence of an operating area gigantopyramidalis. It is possible that this supplementary area belongs to the category of association centers to be dealt with further below.

---

[236] It may correspond to the so-called MsII area of lower Mammals referred to above and depicted in Figure 241 E.

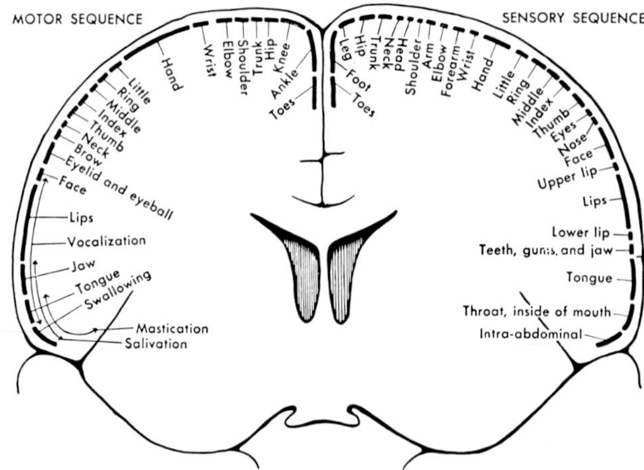

Figure 253 E. Diagrammatic cross-section of hemisphere, for comparison of main motor and sensory sequences (from PENFIELD and RASMUSSEN, 1950).

The *somatic sensory projection area* is located, adjacent to the main motor projection area, and perhaps to some extent functionally overlapping with this latter, in the postcentral convolution, including the posterior wall of sulcus centralis. It is characterized by a granular type of cortex (koniocortex) and receives, through the parietal thalamic stalk, exteroceptive and proprioceptive input, transmitted by the general lemniscus system (medial lemniscus), ventral and lateral spinothalamic tracts, trigeminal lemniscus, and perhaps vestibular channels to the relevant thalamic relay grisea.

The somatotopic representation within the main somatic sensory area, depicted by Figures 253 A and D, corresponds to that in the adjacent motor area (Fig. 253 E), such that sensorimotor representation may roughly be illustrated by a single homunculus (Fig. 253 F). From the leg region on, the sensory somatotopic representation likewise extends upon the medial surface of the hemisphere, facing the fissura longitudinalis cerebri, approximately as far as the sulcus cinguli. The projection, in accordance with the decussation of the lemniscus system, is essentially, but presumably not exclusively, contralateral.

Upon electrical stimulation, contralateral sensations are elicited in conformity with the obtaining sensory sequence along the postcentral convolution. These sensations are described by the patients as numbness, tingling, rarely as 'cold' or as mild 'pain' and occasionally as a

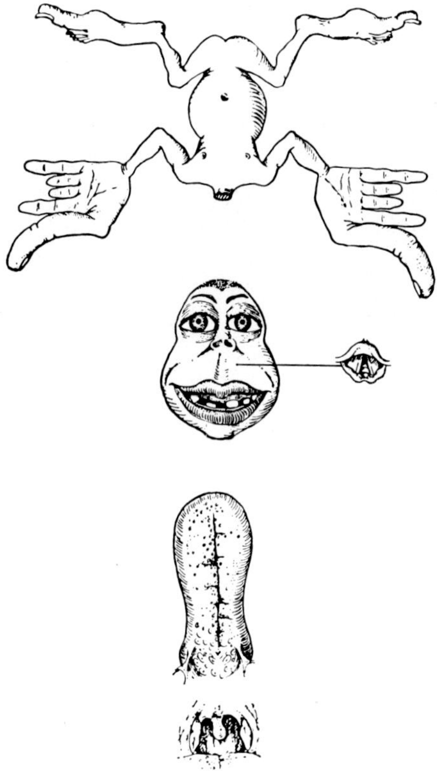

Figure 253F. So-called sensorimotor homunculus roughly indicating assumed distortion of body representation in the projection areas, to be compared with Figures 253B and D (from PENFIELD and RASMUSSEN, 1950).

sensation of movement, although no actual motion could be noticed by the attending observers.

Basally to the main somatic sensorimotor region, PENFIELD and RASMUSSEN (1950) found a second representation of hand and foot, pertaining to a *second sensory and motor*, or perhaps *joint sensorimotor area*,[237] (lower sensorimotor strip). This region seems to be located on the superior bank of the *Sylvian fissure*, within which the exact pattern of localization is not easily ascertained, but the somatotopic sequence is apparently reversed in comparison with the main projection cortex (cf. Fig. 254 B). Again, according to PENFIELD's observations, patients oc-

[237] It may correspond to the so-called SmII area of lower Mammals referred to above and depicted in Figure 241 E.

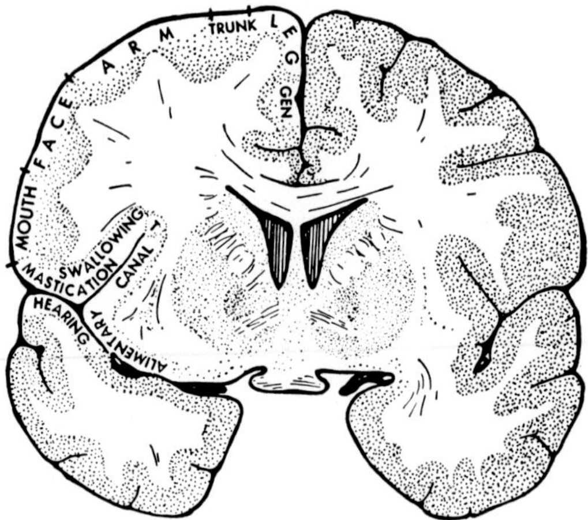

Figure 254 A. Simplified drawing of cross-section through telencephalon at level of assumed sensorimotor somatotopic representation and of auditory projection area (from PENFIELD and RASMUSSEN, 1950). T: tentative localization of taste area.

casionally stated, upon stimulation, that they had a desire to perform the resulting motion.

As regards the *'viscerosensory' projection area* (together perhaps with a rostrally adjacent or overlapping 'visceromotor' one), its location may be assumed within the *insula* (PENFIELD and JASPER, 1954; PENFIELD and FAULK, 1955). The visceral afferent components of the general lemniscus system, containing, *inter alia*, ascending channels related to the gastro-intestinal tract, and reaching their thalamic grisea (cf. chapter XII) are presumably relayed to the insular cortex which also appears to include the taste representation. This would agree with the interpretation of findings in other Mammals.

The Human *auditory projection area*, as first unequivocally demonstrated by FLECHSIG, is located in the region of the *transverse temporal gyri of Heschl*, the greater part of whose surface is hidden within the *fissura Sylvii* (Figs. 252 A, E, 254 A, C). Although its details in Man are not entirely clarified, a discrete tonotopic representation, related to different sound pitches, can be assumed (PFEIFER, 1921). On the basis of comparative physiological evidence, the presence of a smaller adjacent secondary auditory area with reversed distribution of input can be surmised. The auditory projection area receives FLECHSIG's auditory radiation, originating in the medial geniculate complex, which represents the relay griseum for the lateral lemniscus system. Said radiation

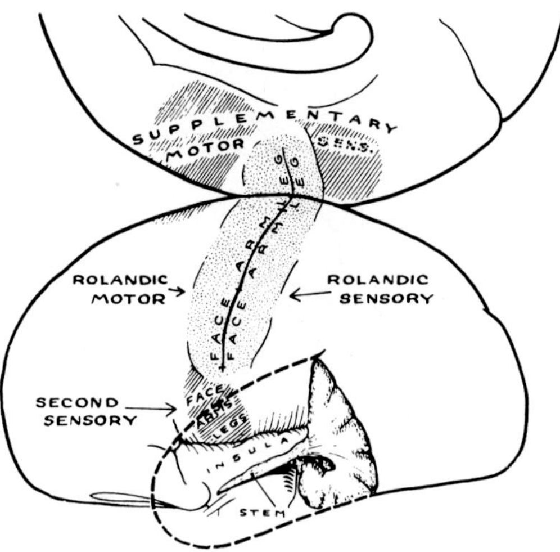

Figure 254B. Sketch indicating primary, second, and supplementary somatosensory areas in Penfield's interpretation (from PENFIELD, 1954, 1958). The cited author adds: 'the extent of the "second" area is not clear and its motor influence requires verification'. It is also not quite clear what the author means by 'stem' in his illustration.

apparently conveys impulses from both (i.e. homolateral and contralateral cochleae). Electrical stimulations in conscious patients either elicit or suppress sound sensations. The thereby artificially evoked sounds are of a nondescript 'elementary nature, described as buzzing, thumping, whistling, chirping, etc.'; 'they may be referred to the homolateral or to the contralateral side' (PENFIELD, 1954).

Vestibular impulses could be included within the ascending cochlear system, and the findings by SPIEGEL (1934) suggest a cortical vestibular 'center' in close vicinity to or relation with, the cochlear projection area. On the other hand, HASSLER (1948) assumed that an ascending secondary vestibular channel, located in the tegmentum (fasciculus tegmenti dorsolateralis, close to or within the central tegmental tract) and reaching a rostral portion of nucleus ventralis posterolateralis thalami, is then projected upon a strip of the sensory cortex of the postcentral convolution. A double or even multiplex vestibular input to neopallium thus does not appear improbable.[238]

---

[238] CLARA (1959) comments: '*Ein Tractus vestibulothalamicus, der von den Endkernen des N. vestibularis zu dem Thalamus zieht, hat sich bis jetzt anatomisch nicht mit völliger Eindeutigkeit nachweisen lassen, was um so merkwürdiger erscheint, als wir da zweifellos Lage- und Bewegungs-*

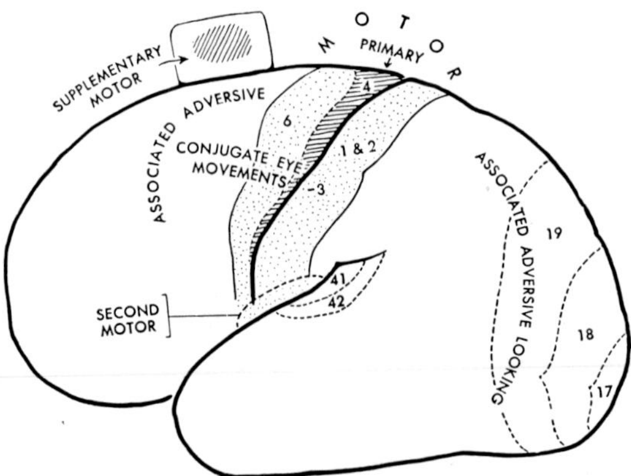

Figure 254C. Sketch indicating overall distribution of some 'functional' areas upon lateral and part of the medial aspect of the hemisphere (from PENFIELD and RASMUSSEN, 1950). The numbers refer to cortical fields in BRODMANN's notation. 1, 2, 3: postcentral sensory projection areas; 4: motor projection area (a. praecentralis gigantopyramidalis; 6: premotor cortex (a. frontalis agranularis); 17: visual projection area (a. striata); 18: area parastriata (a. occipitalis); 19: area peristriata (a. praeoccipitalis); 41, 42: auditory projection area (aa. temporalis transversa interna et externa). Compare with Figure 253A.

The *olfactory projection area* is presumably represented by the piriform lobe cortex in the most rostral part of the gyrus hippocampi, closely adjacent to, but not included in, the true uncus hippocampi (cf. Fig. 175D). Electrical stimulation of the olfactory bulb or of the 'uncinate region' (i.e. of the piriform lobe)[239] elicit olfactory sensations of disagreeable type, likened to manure or burning rubber (PENFIELD and JASPER, 1954).

The *primary optic projection area*, characterized by the peculiar koniocortex of the *area striata*, is located on the medial surface of the occipital lobe, on both lips of the calcarine fissure, extending rostralward into fissura parieto-occipitalis and along lower tip of the common stem of both fissures (Figs. 145A, 250A, B, 252A, E). The area reaches the

empfindungen haben, die in dem Vestibularapparat ausgelöst werden.' Yet, CLARA points out, clinical experiences suggest a cortical vestibular representation in temporal and parietal (presumably postcentral), moreover in the frontal (premotor) eye areas.

[239] The hippocampal formation, including the true uncus hippocampi, moreover the parahippocampal cortex of *Broca's area parolfactoria*, gyrus cinguli, and gyrus hippocampi are most likely not directly related to olfactory functions. At most, some olfactory projection, by way of the medial olfactory stria, to the precommissural hippocampal rudiment could be assumed.

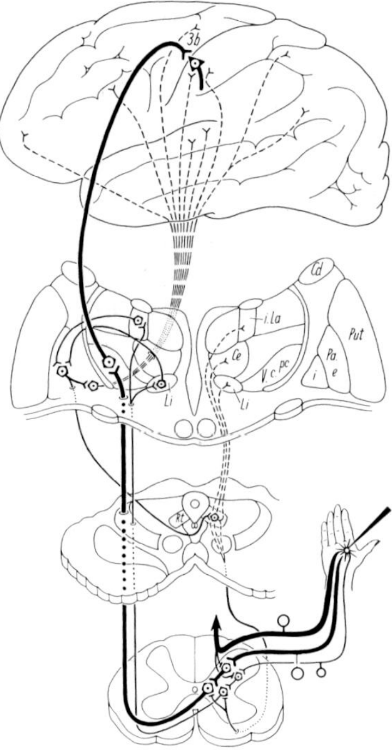

Figure 254 D. Diagram indicating HASSLER's concept of rapid and slow pain conduction (courtesy of Professor HASSLER). Rapid pain conducting system in thick outline, slow conducting systems drawn with thin or interrupted lines. Ce: centrum medianum; Cd: nucleus caudatus; i. La: intralaminar grisea; Li: nucleus limitans; Pa. i, e: globus pallidus with internal and external segments; Put: putamen; Rt: reticular grisea of mesencephalon; V.c.pc: ventrocaudal parvocellular thalamic grisea (nn. vent. posteromed. et posterolat.); 3b: sensory cortex. Cf. p. 412 (chapter XII) and p. 817.

occipital pole and expands to a variable degree toward the lateral surface, where it may or may not rostrally be limited by a sulcus lunatus (cf. Figs. 62, 63, chapter XIV, vol. 5/II). The representation of the retinal quadrants was discussed above with reference to other Mammals (Fig. 242F). Thus, in the calcarine cortex (area striata, Brodmann's field 17) a transposed and distorted, but topologically homeomorphic (isomorphic) projection of superimposed contralateral nasal and homolateral temporal retinal halves, is manifested, with emphasis on the macula. Further details concerning retinal projection and physiology of vision will be found further below in a brief discussion of relevant clinical and pathologic data.

The extensive occipital and occipitolateral representation of the macula is of particular significance and corresponds to the sizeable extension of functionally important representations in the sensorimotor pallium (e.g. thumb, index, fingers, face), depicted by the distortions of the homunculus mappings (Figs. 253B, F). A bilateral cortical representation of each entire macula has been assumed by some authors, but evidence in this respect is not conclusive and extensive unilateral projection of the macular quadrants may be consistent with the observed preservation of central vision in unilateral lesions.[240]

Electrical stimulation of the posterolateral part of the calcarine cortex elicits the perception of colored lights, of stars in motion or flickering, and of comparable nondescript luminous appearances (photic sensations, phosphenes).[241] Similar results were obtained by stimulation of the adjacent, extrastriate occipital cortex (area parastriata and peristriata). It must be kept in mind, however, that such electrical stimulation, compared with normal neural impulse conduction, represents, biologically speaking, a most crude interference. It may be likened to random coarse thumping against a precision mechanism. Again, whether the pallium surrounding the primary projection 'center' (area parastriata and perhaps part of peristriata) should be conceived as a '*second visual area*' for bilateral vision, or rather subsumed under the 'association areas', remains a moot question.

The regions of isocortical pallium not taken up by the projection areas receiving sensory input, or, in the case of the precentral motor projection area not delivering 'upper motor neuronal output', comprise the '*association areas*'[242] presumably concerned with a hierarchy of further processing activities. These latter seem to include, with regard

---

[240] It should also be mentioned that in the optic chiasma of various Mammals, CAJAL (1911) and others have described bifurcating fibers (cf. Fig. 27C) which could be interpreted as macular fibers connecting with the lateral geniculate griseum of both sides. It is a moot question whether these fibers actually represent macular fibers, and whether these findings indeed apply to Primates and Man.

[241] Further aspects of this topic with regard to cortical function and attempts to develop 'visual prostheses' for the blind will be considered in chapter XV of volume 5/II.

[242] The term 'association' for these cortical neighborhoods is a purely anatomical one, and does neither denote nor necessarily connote 'psychologic association' mentioned further below in the text. The anatomical term 'association' merely implies that said areas or 'centers' are substantially characterized by functionally relevant intrahemispheric fiber connections between diverse cortical regions, while the relevant characteristic of projection areas is their particular sensory input, respectively their *upper motor neuron output*. Although, initially, FLECHSIG tended to assume that 'association areas' lacked subcortical

to the *sensory input*, abstraction of invariants by registering complex patterns, regardless of their various transpositions. Such 'discrimination of patterns' is then combined with storage (engraphy) within the 'plastic' neural network. These activities presume 'learning' by the acquisition of repeated information, and complex operations upon the encoded and stored information by repeated circuit activations, related to 'association' and practice.

With regard to the *motor output*, a similar hierarchy of processes seems to obtain, by the organization of complex action patterns which, in turn, are translated into the particular motor patterns of skilled muscular performances triggering the motor projection center. These activities, again, require, in addition to a genetically determined adequate neuronal network organization, repeated practice.

The genetically determined organization or mechanism of association and projection areas is, of course, not given at once, but gradually evolves to 'maturity' thus not precluding significant effects of learning respectively practice upon details of its development.

JOHN LOCKE (1632–1714), with a substantial degree of justification, considered the 'mind', which in the aspect here under consideration actually means its physical mechanism, namely essentially the cortical network, to be a *tabula rasa* at birth. This mechanism, nevertheless, must be conceived as a developing, 'plastic' system. Although LOCKE's conclusion that *nihil est in intellectu, quod non prius fuerit in sensu* can be fully upheld on rather convincing grounds, it must be qualified by the addition: *nisi intellectus ipse*. This, of course, means the presence and availability of the relevant cortical and therewith related registering and processing mechanisms.

Again although all 'knowledge' can be regarded as derived from, and limited by the senses, the logical mechanisms (to be dealt with in chapter XV, vol. 5/II), doubtless may greatly expand said 'knowledge' derived from the sensory input.

Introspectively, i.e. psychologically *sensu strictiori*, there obtains a connection between different conscious events. This psychologic phenomenon, already pointed out by PLATO and ARISTOTLE, was particularly emphasized by LOCKE, who seems to have coined the term '*association of ideas*' and, as it were, became the founder of '*association*

---

connections, i.e. 'projection fibers', the fact that most, if not all 'association areas' are provided with such subcortical connections does not invalidate FLECHSIG's well documented concept of a significant distinction between his 'projection' and 'association' areas.

psychology' further elaborated by BERKELEY (1685–1753), HUME (1711–1776), JAMES MILL (1773–1836), JOHN STUART MILL (1806 to 1873), THEODOR ZIEHEN (1862–1950), and many others.

Roughly speaking, there are natural or original (simultaneous) associations which occur together in compresent perceptions, and are correlated with successive associations characterized by temporal sequences. Abstraction of such event patterns by the 'intellect' leads to a wide variety of acquired associations. The phenomenon of storage respectively engraphy as elaborated by SEMON (1904, 1909, 1920) and dealt with in section 6 of chapter XV of volume 5/II, plays here an important role.

Roughly speaking, one might distinguish a parietal tactile (in the wider sense), an occipital visual, a temporal acoustic, a frontal (premotor) and a prefrontal association area, discounting their doubtless obtaining complex functional interrelationships.[243]

The significance of the *prefrontal association area* characterized by an isocortex frontalis granularis, and located rostrally to the premotor area, has so far defied accurate definition. Yet, following MCLARDY (1950), the prefrontal lobe might be considered to function 'as a whole'. The monograph of FEUCHTWANGER (1923), in which about 200 cases of frontal lobe lesions during World War I were analyzed, still remains one of the best contributions on this subject. In addition, there are numerous subsequent publications based on the study of cases with lobotomy or similar operations,[244] and of frontal lobe lesions (cf. e.g. DALY, 1974; MEYER, 1974; WARREN and AKERT, 1964; HALSTEAD, 1947).

The psychological activities found to be most altered by prefrontal lobe damage appear to be those of emotional drive or active affectivity *(Affektivität und Antrieb)*. PENFIELD and EVANS (1935) reached the conclusion that the most conspicuous detectable change after maximum amputation of right or left frontal lobe is 'an impairment of those mental processes which are prerequisite to planned initiative'. Others

---

[243] Cf. volume 3/II, p. 658–661 and Figures 355 and 356 loc. cit.

[244] The operation of frontal lobotomy, introduced by MONIZ (1936), is based on the assumption that an interruption of prefrontal connections reduces emotional reactivity and especially anxiety states. This result was suggested by the results of experiments in anthropoid Apes. '*Psychosurgery*' thus initiated by MONIZ and further elaborated by others was for some time widely practiced but remains highly controversial. To a large extent, it has been replaced by the medication with 'tranquilizers'. Further comments can be found in the author's monograph 'Brain and Consciousness' (K., 1957, 1973).

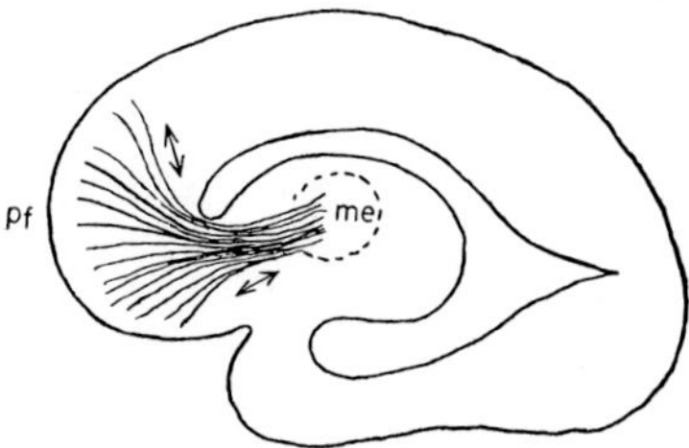

Figure 255 A. Diagram roughly depicting the assumed prefrontal emotional feedback circuit (from K., 1957). me: nucleus medialis (dorsomedialis) thalami; pf: prefrontal lobe.

have used the exceedingly vague term 'biologic intelligence' which is supposed to be represented throughout the cortex, but manifesting a gradient with maximal representation in the prefrontal cortex (HALSTEAD, 1948). Personality defects following lobotomy have been characterized by relatives of the patients with the expressions 'her soul is in some way lost', 'his soul appears to be destroyed', or 'I'm living now with another person' (RYLANDER, 1948).

In damage to prefrontal region, some memory impairments as well as difficulties in concentration have been claimed. A reversal in the patient's character may be manifested, previously gentle individuals can become irritable and prone to fits of rage, while choleric, highly excitable and stubborn persons may become indifferent and docile. Deterioration of judgment is often very pronounced. Moria, an exaggerated tendency toward inane or tactless wisecracks *('Witzelsucht')* is occasionally a symptom of prefrontal lobe damage, especially in cases of neoplasms, but cannot be considered a reliable localisatory symptom. Otherwise fairly normal individuals may manifest this habit in a conspicuous fashion.

Although, roughly speaking, the prefrontal lobe, as e.g. believed by SPATZ (1964), who particularly emphasized the relevance of its basal (orbital) neighborhoods *(basaler Neocortex)*, can be regarded as related to what is vaguely termed 'intelligence', said lobe doubtless also pertains to the grisea concerned with affectivity respectively emotion. As regards such grisea, two main cortical feedback systems modulating affectivity can perhaps be recognized, namely the prefrontal circuit, and the *hippocampal circuit of Papez*, which pertains to the so-called *limbic system*.

The *prefrontal circuit* (Figs. 255 A, B) is mediated by components of the frontal thalamic radiation containing substantial reciprocal connec-

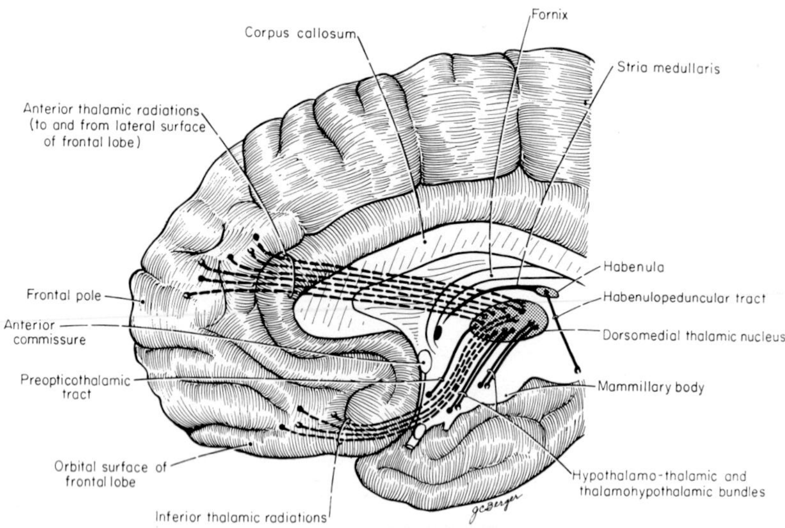

Figure 255 B. Drawing of interconnections pertaining to the circuit depicted in Figure 255 A, as interpreted by Crosby *et al.* (from Crosby *et al.*, 1962). It will be noted that the cited authors depict frontal connections within the inferior thalamic stalk, while I am inclined to include these channels in the frontal thalamic radiation (cf. Fig. A). This radiation (or stalk) is also well illustrated on Tabulae 54–56 of the atlas by Ludwig and Klingler (1956).

tions between prefrontal lobe and nucleus medialis (dorsomedialis) thalami. A fairly precise 'point to point' distribution seems to obtain in this feedback connection (cf. e.g. Meyer *et al.*, 1947; McLardy, 1950; and other authors). The nucleus medialis, in turn, has connections with periventricular system and hypothalamus. The most basal, that is, the orbital portion of prefrontal lobe (regio frontalis granularis) is furthermore assumed to have substantial direct connections with hypothalamic grisea.

The concept of the *hippocampal feedback circuit*, presumably related to the modulation of affective tone, emotion, as well as a diversity of 'programming' activities, is based on the fundamental elaboration by Papez (1937). The validity of this concept can be regarded as reasonably well substantiated by the observations and data of numerous subsequent authors.[245]

[245] It should also be mentioned that Votaw (1959) obtained nondescript bilateral or ipsilateral motions upon electrical stimulation of the hippocampal formation in the

PAPEZ (1937) assumed that the 'central emotive process of cortical origin' may be built up in the hippocampal formation, transferred to the mammillary body and thence through the anterior thalamic nuclei to the cortex of the gyrus cinguli: 'The cortex of the cingular gyrus may be looked on as the receptive region for the experiencing of emotion as the result of impulses coming from the hypothalamic region in the same way as the area striata is considered the receptive cortex for photic excitations coming from the retina. Radiation of the emotive process from the gyrus cinguli to other regions in the cerebral cortex would add emotional coloring to psychic processes occurring elsewhere. This circuit would explain how emotion may arise in two ways: as a result of psychic activity, and as a consequence of hypothalamic activity.' It might, however, be better to say: emotion, a psychic (conscious, parallel) process, could arise in two ways, as a correlate of autochthonous cortical circuit activity, and as a correlate of cortical neural circuit processes triggered by hypothalamic activities. PAPEZ did not specifically introduce the concept of a closed circuit or feedback mechanism in his theory of emotion. Nevertheless, the cortex of the gyrus cinguli appears to be rather intimately connected with the hippocampal formation through fibers of the cingulum, the stria longitudinalis lateralis, and the dense lamina tangentialis. In discussing the hypothalamic functions, we interpreted therefore the relationship between the grisea of PAPEZ' scheme as a closed feedback circuit (K. and HAYMAKER, 1949; K., 1951, 1954). A similar interpretation of the mechanism in question was also independently and almost simultaneously presented by v. BONIN (1950, p. 124), 'introducing the concept of reverberating circuits which had not yet been developed when PAPEZ wrote'. I do not entirely agree with this last remark: it is true that the concept of reverberating circuits and the term 'feedback' were not widely used in neurologic literature when PAPEZ wrote; nevertheless, reverberating circuits and feedback concepts had been introduced into neurologic thinking at least since 1868, as I indicated in my short review of this subject (K., 1954). On the other hand, I completely agree with v. BONIN's (1950) statement that 'the "epistemic correlates" of emotion are neuronal events in circuits' rather than a specific state of any one group of cells. From the viewpoint of parallelism I would phrase this as fol-

---

Macaque, and attempted to elaborate some sort of localization pattern. Since random movements can be elicited by stimulation of the most diverse grisea, the interpretation of the hippocampal formation as another 'supplementary motor area' does not appear very convincing.

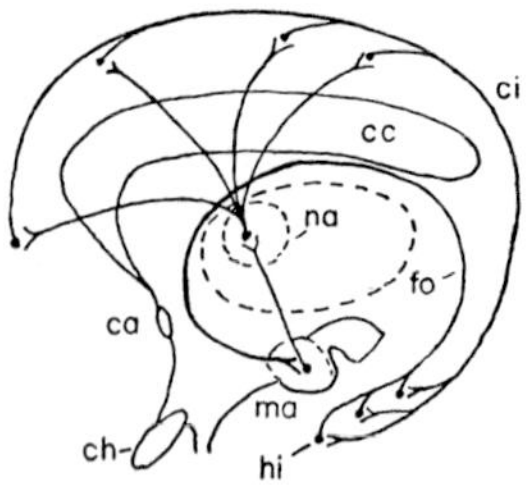

Figure 255 C. Diagram roughly indicating some features of the assumed hippocampal emotional circuit (from K., 1957). cc: corpus callosum; ch: optic chiasma; ci: fiber system of cingulum (originating in, and adjacent to parahippocampal cortex of gyrus cinguli); fo: fornix system; hi: hippocampal formation; ma: mammillary body and its grisea; na: anterior thalamic nuclei; ca: commissura anterior.

lows: emotion is a consciousness-modality in private perceptual space-time, and represents a parallel phenomenon related to certain neural circuit activities in the postulated public physical space-time system.

For purely descriptive purposes, the modulating hippocampal feedback circuit can be regarded as beginning with the cornu Ammonis, discharging through its substantially efferent channel, the fornix. The mammillary body receives the bulk of the fornix bundle, and, in addition, ascending impulses through the mammillary peduncle; moreover, directly or indirectly, ascending and descending impulses reach the mammillary body through medial forebrain bundle, hypothalamic, and other fiber systems. The main discharge path of the mammillary body is provided by mammillo-thalamic and mammillo-tegmental tract, essentially originating as a common, bifurcating bundle, the fasciculus mammillaris princeps. The mammillo-thalamic tract discharges into the anterior nuclei of the thalamus. These nuclei, in turn, project presumably upon the whole length of gyrus cinguli. From this convolution, characterized by parahippocampal cortex, impulses are carried back to the hippocampal formation (fascia dentata and cornu Ammonis), thus closing the feedback loop. Activity within this loop radiates by collateral connections into a large number of additional grisea and is also influenced by numerous neural discharges converging toward that circuit mechanism. It will be recalled that Cajal (1911), who still considered the hippocampus to be an olfactory griseum, already emphasized the parahippocampal-hippocampal connections, and particularly the *voie temporo-ammonique croisée* forming the psalterium dorsale.

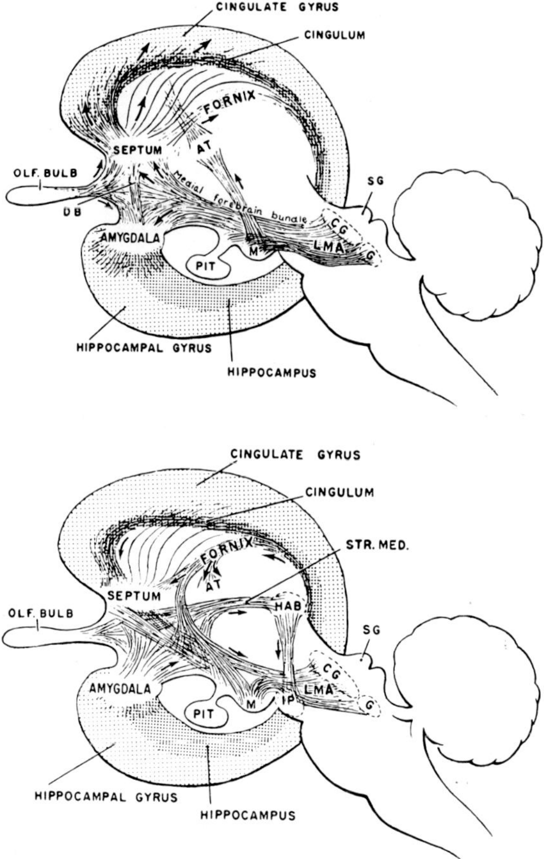

Figure 255 D. The limbic system as interpreted by MacLean (after MacLean, 1958, from Haymaker and Anderson, 1971). AT: anterior thalamic nuclei; CG: central gray of midbrain; DB: *diagonal band of Broca;* G: dorsal and ventral *tegmental nuclei of Gudden;* Hab: habenular grisea; IP: interpeduncular nucleus; LMA: so-called limbic midbrain area; M: mammillary body; PIT: pituitary complex; SG: superior colliculus; STR. MED: stria meddullaris thalami system. The upper drawing depicts the afferent pathways to limbic system, the lower drawing the system's output channels.

Figures 255 C and D illustrate some aspects of the *Papez circuit.* Since, in its mechanism, the parahippocampal cortex of gyrus cinguli and gyrus hippocampi, forming together the gyrus fornicatus, and included in Broca's (1878) *grand lobe limbique*, seems to play an important role, the term limbic system is now generally used for contemporary elaborations of Papez' original concept.

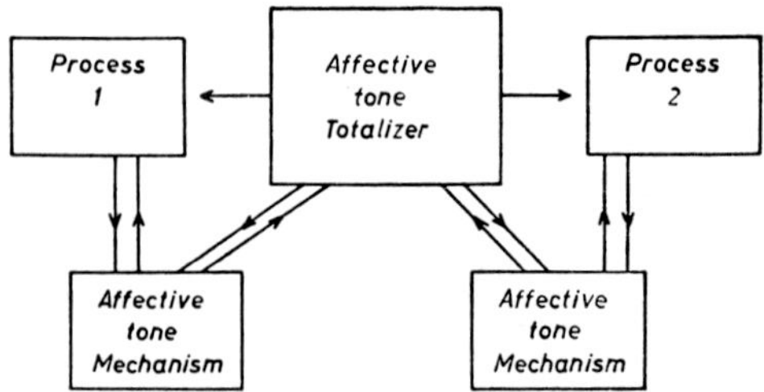

Figure 255E. Diagramm of affective tone feedback mechanisms (modified after WIENER, 1948, from K., 1957).

Among grisea closely related to, or included in the limbic system are, in addition to an undefined number of hypothalamic cell groups, the amygdala and the septal (paraterminal) 'nuclei'. Further comments on the limbic system can be found in the accounts by BRADY (1960), MACLEAN (1969), and in the author's monograph on brain and consciousness (K., 1957, 1973). It is here of particular interest that, from the viewpoint of so-called cybernetics, affective tone can be arranged on some sort of scale with a positive or negative increase from an assumed 'zero point'. Accordingly, increase or decrease in affective tone could influence other activities. Increase may, for instance, favor all processes that are under way at the time, and give them a secondary power to increase affective tone. Conversely, decrease may tend to inhibit such processes and give them a secondary ability to decrease affective tone. WIENER stresses that the mechanism of affective tone is itself a feedback mechanism and a diagram similar to that of Figure 255E is suggested by that author. This view fully agrees with the concept of circuit mechanisms related to emotion as discussed above in connection with problems of localization. It seems evident that 'affective tone' may involve both the *expression* and the *limning processes* of KRETSCHMER.

Summarizing the distribution and functional significance of the Human cortical areas,[246] one may thus roughly distinguish three main categories.

[246] Various relevant older and still valuable summaries of cortical respectively cerebral localization based on reasonably well documented observations combined with a diversity of interpretations are those by FLECHSIG (1896), v. MONAKOW (1914), BERGER (1921, 1927), and FOREL (1922). Some of the contemporary, and in part still controversial views are briefly summarized by DALY (1974), who also refers to cortical organization (dealt with in

(1) *Projection areas* for the five conventional senses of sight, hearing, touch (including body sense in general as well as proprioception), taste, and smell, moreover for upper motor neuron output in the conventional meaning (i.e. bulk of pyramidal tract).

The sensory projection areas, including their '*Randzonen*' pointed out further below, can be presumed to represent the grisea whose function is correlated with the perceptions involving the specific sensory modalities as discussed in section 1 of chapter VII in volume 3/II. In other words, these griseal neighborhoods seem to form the substratum for the phenomenon originally designated as the manifestation of a so-called 'specific energy' *(spezifische Sinnesenergie)* by JOHANNES MÜLLER about 1838.[247]

In this respect, however, the cortical vestibular representation remains somewhat unclarified. Moreover, nothing certain is known concerning cortical representation of sensory qualities such as cold, warmth, pain, and sexual sensations (e.g. *Wollustgefühle)*. Some comments about these two latter categories of sensation will be added further below at the conclusion of this brief 'summary'.

(2) So-called *association areas* related to further processing of input respectively output, and to the relevant operations of storage. Yet, as already recognized by FLECHSIG, there are certain transitional neighborhoods between 'projection' and 'association' areas, provided by what the cited author designated as '*Randzonen der Sinnessphären*' (cf. also vol. 3/II, p. 659). As regards the motor projection area, the so-called premotor cortex seems to represent such a '*Randzone*'.

(3) *Areas concerned with 'affectivity'* and its modulation, respectively with 'emotion'. Parahippocampal cortex and hippocampal formation presumably pertain to that category, together with non-cortical grisea such as paraterminal ones and apparently most of the amygdaloid complex.[248] The prefrontal area may functionally pertain to both category (2) and (3).

The grisea of *corpus striatum*, significantly related to the 'extrapyramidal motor system' display, moreover, particularly but not exclu-

---

chapter XV of the present treatise) More details can be found in a recent symposium on 'cortical localization' edited by ZÜLCH *et al.* (1975), and in a publication edited by DIMOND and BEAUMONT (1974).

[247] Cf. volume 3/II, chapter VII, p. 781 and footnote 4 on that page.

[248] Rather inconclusive data on the various possible functions of the amygdaloid complex, and on nondescript results following its operative ablation in Man can be found in a paper by WILLIAMS (1953).

sively *qua* fundus striati including nucleus accumbens, relationships with the vegetative nervous system.[249]

The involvement of the striatal grisea in parkinsonism was briefly discussed in chapter XI, section 9, p. 993–995 of volume 4. The relevant reduction in dopamine concentration characteristic for said condition was pointed out, and the relief provided by administration of l-dopa was mentioned. This latter substance, a precursor of dopamine, crosses the hemato-encephalic barrier, which dopamine itself cannot pass. With regard to the neuronal channels concerned, it seems that the reciprocal striatonigral and nigrostriatal pathways differ *qua* transmitter substances, the latter pathway being dopaminergic and abnormal in parkinsonism. On the other hand, GABA appears to be the transmitter in the striatonigral pathway, whose function seems to be abnormal in *Huntington's disease*. Comparable neurochemical deficiencies presumably obtain in various other neurologic conditions, e.g. in *Friedreich's ataxia* (cf. chapter VI, section 1 C, p. 282 of volume 3/II, and chapter X, section 9, p. 767 of volume 4). in which pyruvate oxidase might be defective.

As already mentioned above in discussing the telencephalic grisea of Mammals in general, nothing of any significance can be said about the functional significance of the *claustrum*. The inconclusive available

---

[249] Thus, in certain disorders of autonomic function ('primary dysautonomias'), pathologic changes in the striatum (putamen) characterized by astrogliosis, rarefaction of neuropil, and cell loss were recently described in addition to the changes in brain stem and spinal cord 'autonomic' grisea (KLUTZOW *et al.*, 1975). The cited authors distinguish 'primary dysautonomia', which may be familial, from 'secondary dysautonomia' related to other conditions, e.g. among others *Hallervorden-Spatz*, and *Creutzfeldt-Jakob disease* (cf. vol. 3/II, chapter V, section 1 C, pp. 271, 284). Answering a critic (in Ber. Biochem. Biol. *389:* No. 4, 1974) who took exception to my mentioning '*Hallervorden-Spatz Krankheit*, in einem Atemzug mit Agyrie, Mikrogyrie and Porencephalie', I may reply that this was merely done in order to point out diverse pathologic conditions of presumed or suspected hereditary respectively dysgenetic nature. This also applies to my brief mention of the *Creutzfeldt-Jakob syndrome*, which the critic evaluates as a viral spongiform encephalopathy, and of *Pick's disease*, since the hereditary nature of both had been suspected by some geneticists. The critic furthermore states that in my section on disturbances of neurogenesis '*eine Fülle diskrepanter Entwicklungsstörungen, Fehlbildungen und metabolischer Prozesse*' are dealt with '*in etwas eigenwilliger und nicht immer klar verständlicher Weise*'. The impartial reader may find, however, that far from being 'self-willed' or 'obstinate' I have merely reviewed a number of different approaches by various authors to the intrinsically unsatisfactory classification of the relevant numerous disturbances, most of which are indeed not '*klar verständlich*' because no satisfactory knowledge concerning their causal aspects is available.

data have been reviewed by RAE (1954a, b), and, as regards a lower Primate (Macaca), by BERKE (1960).

The paraterminal so-called *septal area* is anatomically ill-defined by most authors and includes rostral parahippocampal, hippocampal, and medial rhinencephalic components, as well as portions of the basal ganglia. It has *inter alia* many hypothalamic connections, and was formerly believed to be a part of the rhinencephalon. Neural activities of that region seem to influence vegetative processes as well as affectivity. The *nucleus amygdalae*, or at least parts of it, may likewise be concerned with visceral and circulatory activities as well as with affectivity. Feelings of fear, anxiety and terror were elicited by electric stimulation in human patients (CHAPMAN *et al.*, 1954). The connections between the Human amygdaloid complex and temporal pole cortex, posterior insula, septal grisea, and hypothalamus were described by KLINGLER and GLOOR (1960).

BALASUBRAMIAM and KANAKA (1975) have recently claimed that unilateral stereotactic amygdalotomy in cases of infantile hemiplegia with 'behavioral disorders' and with or without fits has provided satisfactory relief respectively amelioration. These authors believe that such amygdalotomy can be done as the procedure of first choice with respect to hemispherectomy (discussed further below), which may be indicated in cases of infantile hemiplegia with intractable epilepsy.

The central mechanisms correlated with *pain sensations* are poorly understood and depend, as pointed out in section 9 of chapter XII, at least in part on diencephalic activities. The sensory cortex of postcentral gyrus as well as the adjacent parietal cortex seem to play a role. Figure 254D shows HASSLER's (1972) concept of pain channels, emphasizing the difference between slow and fast pain conduction (cf. p. 801, vol. 3/II). HASSLER uses the term '*Schmerzempfindung*' for the perception of 'fast pain', and designates 'slow pain' sensations as '*Schmerzgefühl*'.

It will be seen that in HASSLER's interpretation of the relevant circuits, slow pain signals are presumed to be channelled by way of nucleus limitans (n. lentiformis mesencephali) and intralaminar thalamic grisea, with additional channels via mesencephalic reticular grisea and centrum medianum, and with further loops extending through globus pallidus. The fast pain signals, *per contra*, are presumed to be channeled by way of the ventral posterolateralis and posteromedialis complex (lemniscal grisea). Be this as it may, the prefrontal lobe may likewise play a role in the experience of pain, since frontal lobotomy was found, in some instances, to bring relief from intractable pain. Thus, the experience of pain could be assessed as a multifactorial phenomenon, in part also related to the mechanisms of affectivity.[250]

Much the same could be said about *sexual sensations*. The genital regions are represented in the main somatosensory area within the fissura longitudinalis cerebri (cf. Fig. 253D); stimulation of that region is reported to elicit sensations in the external genitals (PENFIELD and RASMUSSEN, 1950). These sensations, however, were not of erotic character. It seems likely that lust sensations *(Wollustgefühle)* and associated experiences are related to activities of the limbic circuit and its extensions, as well as to a variety of cortical and diencephalic regions (cf. e.g. MACLEAN, 1969; MACLEAN and PLOOG, 1962). Quite evidently, a number of vegetative and somatic circuits, including various regions and mechanisms of the spinal cord, become involved. Some details of the here relevant so-called '*Sexualreflexe*' are reviewed by CLARA (1959).[251]

Before discussing a few selected topics concerning pathological respectively clinical conditions which are of importance for an understanding of the telencephalon's functional activities, some remarks on *bilateralism* and so-called *dominance* of one cerebral hemisphere seem pertinent. It is generally assumed that in right-handed persons the left hemisphere is dominant with respect to speech and acquired manual skills. In some instances, however, the leading hemisphere has been found to be homolateral with the leading hand. NIELSEN (1955) believes that the incidence of ipsilateral 'brainedness' and 'handedness' may be estimated at about 5 per cent. This author and others prefer the terms major and minor hemisphere: the major hemisphere was believed to do the actual work, but if destroyed early in life, the other was known to assume, in many cases, the entire function. NIELSEN assumed that engrams are formed bilaterally and that the main difference

[250] Further comments on this topic can be found in the author's monograph 'Brain and Consciousness' (K., 1957, 1973) in which the rare condition of congenital insensitivity to pain is also dealt with. Additional reports on this latter topic are those by BAXTER and OLSZEWSKI (1960) and by THRUSH (1973). In contradistinction to the authors assuming an involvement of cortical or corticothalamic mechanisms, THRUSH suggests a 'neural defect' in reticular formation or in dorsal horn or in both.

[251] Since, as pointed out in section 6, chapter VII of volume 3/II, the details of structural connections and functional relationships of the autonomic system to male and female genitals still remain poorly elucidated, specific comments on those combined 'vegetative and 'somatic' reflex activities were omitted from the discussion of reflex activities in volume 4, but nevertheless implied in the generalized account of vegetative functions.

between the two hemispheres is one of training. According to PENFIELD (1954) 'there is evidence that one temporal lobe is so to speak the carbon copy of the other'.

After destruction of an area on the major side, the recovery through training of the corresponding area on the minor side depends on a number of variables. These include age, general health and vigor, natural endowments of the patient, efforts as retraining, and certain local functional characteristics of the particular area involved (NIELSEN, 1955). This latter clinical observation, again, favors the well-established concepts of relative localization.

With regard to bilateralism, the bilateral representation of several types of sensory input, and the bilateral origin of some motor output components might be mentioned. Although normally most impulses via spinothalamic and secondary ascending trigeminal tracts seem to take the circuit to contralateral thalamus and cortex, such ascending impulses may also, according to circumstances, utilize a homolateral pathway. In the lateral lemniscus system, a bilateral transmission of signals from each cochlea appears to be the general role. Despite prevailing contralateral relationships, bilateral efferent pathways are likewise well known; such connections are for instance quite evident in the bilateral cortico-bulbar output reaching those subdivisions of each facial nerve nucleus which innervates the upper facial musculature. These problems of bilateral representation are perhaps less significant for the topics under discussion and need not be elaborated here. It might finally be asked why right-handedness predominates and has an incidence of approximately 90%. No satisfactory answer can be given for this question.

In recent years, experiments in Cats and Monkeys, as well as clinical experiences following complete surgical transection of the Human corpus callosum have added numerous data concerning asymmetric hemispheric functions, 'bilateralism', 'dominance', and 'handedness'. Pertinent detailed but inconclusive reviews of this topic can be found in the *Ciba Foundation Symposium* (1965) on functions of the corpus callosum and in the publications by ZANGWILL (1960), GAZZANIGA (1965, 1970), CUÉNOD (1972), and SCHNABL (1976).

A Cat, a Monkey or a Man with all the commissures between the cerebral hemispheres surgically divided retain all sensory faculties, and are still able to coordinate automatic as well as learned patterns of movement. It is not possible to name one ability which is completely lost as a result of such an operation. The split-brain remains essentially one brain[252] (TREVARTHEN, 1965). Much the same applies to those cases in which apparently quite symptomless congenital absence of the corpus callosum in Man, as discussed in section 1 C, chapter VI of volume 3/II remained undetected during life.[253] We have here a striking instance of redundant and ultrastable systems: it must be assumed that neural 'impulses' essential for a functional correlation of both hemispheres can be reduced to a minimum and shunted over one or more of the remaining commissural systems (commissura anterior, posterior, habenulae, supraoptica, supramammillaris, and, if present, massa intermedia). It seems, however, that in cases with total commissurotomy, involving corpus callosum, hippocampal and anterior commissure, as well as massa intermedia, a significant loss of 'attentional capacity' has been noted by means of 'vigilance' tests (DIMOND, 1976). Such loss was not recorded in partial commissurotomy patients, 'whose performance equalled that of normal man'.

Despite the recent detailed data concerning asymmetric hemispheric functions the thereto related problems remain poorly elucidated. There are, nevertheless, some indications that the concept of dominance should be revised insofar as each hemisphere may be leading as regards some particular functional aspects. Thus, the normally 'minor' right hemisphere seems to lead in spatial pattern recognition and *qua* 'non-verbal memory functions'. Higher abilities accordingly do no longer appear to be the exclusive domain of a single overall dominant hemisphere. PENFIELD's simile of a 'carbon copy', as quoted above therefore does not seem very appropriate, since substantial differences *qua* circuit processes and details of engram formation presumably obtain in the two hemispheres despite apparent general structural similarity at the histologically observable level. Concerning emotional activities, some evidence suggesting generally 'right hemisphere lateraliza-

---

[252] Since, however, certain functions may be confined to one side, tests show that when only one hemisphere of a split-brain subject has been exposed, during training, to the stimuli of a discrimination test, the other hemisphere shows no sign of this learning (cf. the comments on pp. [illegible]–271 in section 1 C, chapter VI, vol. 3/II).

[253] In addition to another group of cases which displayed outright feeble-mindedness and agenesis of corpus callosum detected at autopsy, a third group, afflicted with epilepsy, but otherwise with only minor other symptoms, such as motor clumsiness and mediocre 'intelligence' or slow learning capacity, was diagnosed *intra vitam* by means of ventriculography (cf. e.g. JEEVES, 1965). It is of interest to note that whereas transection of the corpus callosum has been more or less successfully used for the treatment of extreme epileptic conditions, congenital lack of the corpus callosum likewise tends to be associated with a clinical history of epileptic seizures.

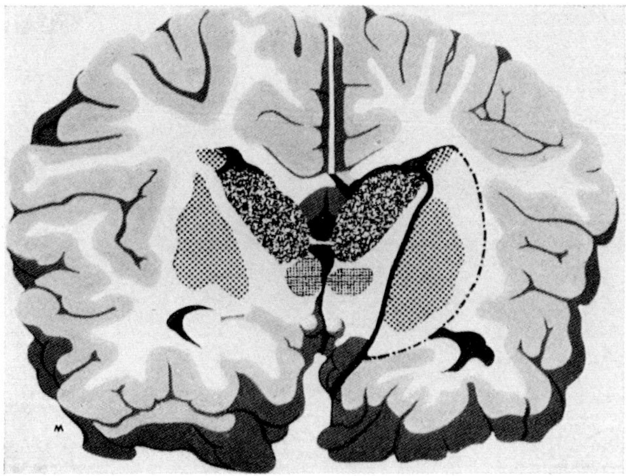

Figure 256 A. Cross-section through cerebrum indicating incisions with removal of basal ganglia (thick line), and with retention (dotted line) of these grisea (from GERLACH *et al.*, 1967).

tion for emotion' has been discussed by SCHWARZ *et al.* (1975). Again, some evidence for specialization of the right hemisphere for spatial processing, with possible sex differences *qua* 'neural plasticity' during development in childhood, was recently claimed by WITELSON (1976).

With regard to bilateralism it is likewise of interest that more or less complete *unilateral hemispherectomy* is compatible not only with survival but may be followed by a remarkable degree of rehabilitation. This operation was first performed by DANDY (1928) for malignant glioma, and subsequently by others, particularly by KRYNAUW (1950) for cerebral palsy (infantile hemiplegia) and intractable epilepsy. It was likewise performed and discussed by GERLACH *et al.* (1967). WHITE (1961) reviewed 269 cases, and WILSON (1970) presented a detailed analysis of 50 cases.

Although the obviously resulting defects include contralateral homonymous hemianopia (dealt with further below) and contralateral hemiplegia, this latter tends to subside, respectively, if already previously obtaining, greatly to improve. Much the same can be said concerning the epileptic manifestations. The best results, as pointed out by GERLACH *et al.* (1967) are recorded following operations on children between 6 and 8 years of age. Speaking ability increases here regardless

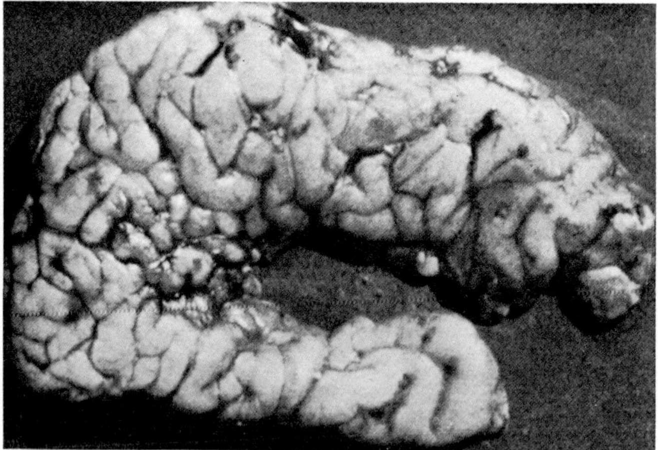

Figure 256 B. Right hemisphere (including basal ganglia) removed by Prof. GERLACH in a case of hemiatrophia cerebri. The thalamus was spared (from GERLACH *et al.*, 1967).

whether the right or the left hemisphere has been removed. Figure 256 A illustrate two different procedures of excision, namely, either with or without removal of striatum and pallidum.[254] Figures 256 B and C show a right hemisphere as removed by Professor GERLACH, and the patient before and after the operation.

In addition to aspects of hemispheric dominance, an ipsilateral sensory 'dominance' has been suggested, such that, when two parts of the body are simultaneously stimulated, perception related to the 'dominant' region induces 'extinction' of that related to the non-dominant ones. A rostrocaudal order of dominance has been suggested, according to which face is dominant to hand, and hand to foot. This 'dominance' is supposed to be particularly noticeable in patients affected by focal or diffuse cerebral lesions, but various uncertainties obtain concerning this 'dominance concept', which was recently reviewed by GAINOTTI *et al.* (1975).

Concerning structure and function of the brain, an 18th century au-

[254] Several problems concerning indication and technique for this indeed very major operation still remain unclarified, in particular as regards sparing or removal of striatum and pallidum. The thalamus is usually not resected, but the question remains whether an atrophic thalamus should or should not be removed. Evoked responses reported to be recorded from the operated side of the head after hemispherectomy (SALETU *et al.*, 1971) are presumably originated by diencephalic grisea.

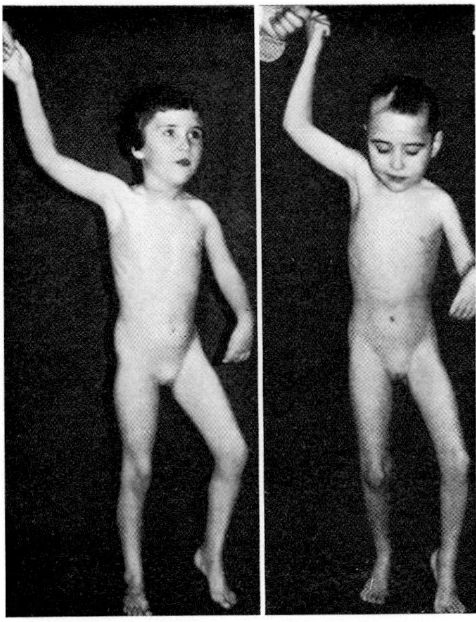

Figure 256 C. Patient from whom hemisphere in B was removed, before (left) and after (right) the operation (from GERLACH *et al.*, 1967). The operation was performed in 1963, and the patient was still living, and doing reasonably well, in 1977, at the time of final revision of the present chapter.

thor stated: *obscura textura, obscuriores morbi, functiones obscurissimae.*[255] Much progress in neurology has been made since that time, but numerous problems remain quite unsettled, particularly with regard to the pathology of the forebrain whose complex functions are performed by intricate network activities based on various not yet significantly understood neural coding systems. It is evident, nevertheless, that a substantially improved first approximation toward the understanding of Human hemispheric activities and their functional dependence on the various griseal structures has been obtained by correlated clinical and pathologic studies. An early proponent of this approach was the Breslau psychiatrist CARL WERNICKE (1848–1904) who published a pioneering *Lehrbuch der Gehirnkrankheiten* in 3 volumes (1881–1883) and emphasized that the separation of neurology and psychiatry was artifi-

[255] Quoted after RAUBER-KOPSCH, *Lehrbuch der Anatomie des Menschen*, vol. V, p. 2 (1914).

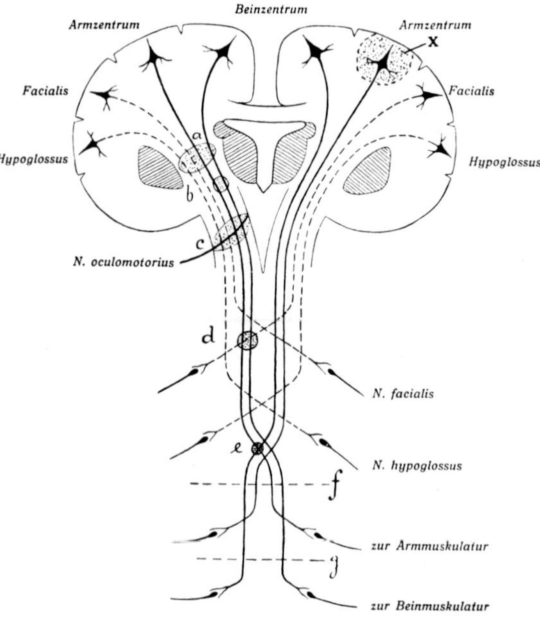

Figure 257 A. Diagram indicating sites of lesions causing upper motor neuron paralyses (from VILLIGER, 1933). a: extensive lesion in capsula interna (contralateral hemiplegia); b: partial lesion in capsula interna (contralateral hemiplegia with sparing of facial and hypoglossal nerve); c: lesion in pes pedunculi with hemiplegia alterna oculomotoria; d: lesion in pons with hemiplegia alterna facialis; e: lesion in caudal oblongata causing hemiplegia cruciata; f, g: levels of spinal cord; x: lesion causing contralateral monoplegia.

cial, since '*Geisteskrankheiten sind Gehirnkrankheiten*'. Among further attempts at establishing overall theories of neurology correlated with the data of brain pathology are those by v. MONAKOW (1905, 1914) and by KLEIST (1934). This latter author, in particular, dealt with the clinicopathologic observations on brain wounds in World War I.[256] A more recent attempt to present an organic background for the higher cerebral functions and their clinical disorders on the basis of the available inconclusive and unsatisfactory data is that by SCHLESINGER (1962). It contains, nevertheless, a large number of thought-provoking discussions in addition to an extensive bibliography, but was dismissed with

[256] Our own much more modest clinicopathologic study on a series of fatal missile-caused craniocerebral injuries in World War II with different survival periods likewise included a few pertinent observations (CAMPBELL *et al.*, 1958).

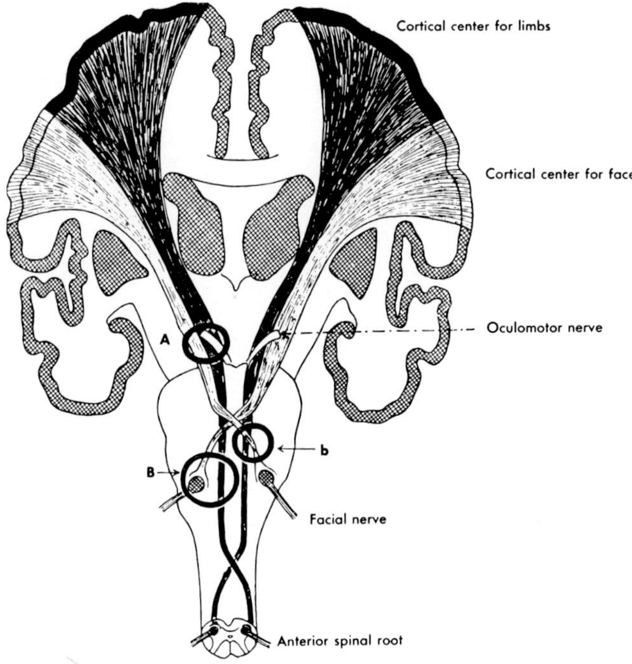

Figure 257B. Sketch indicating sites of lesions causing alternate hemiplegias (from HAYMAKER and BING, 1969). A: oculomotor; B, b: facial.

excessive and undue severity in a very short book review by ROSS ASHBY (J. nerv. ment. Dis. *138:* 195, 1964). Contemporary views on Human hemisphere function, as interpreted by diverse authors, are presented in a publication edited by DIMOND and BEAUMONT (1974).

In the present context, only a few selected topics of *cerebral pathology* illustrating basic functional relationships can briefly be pointed out.

As regards the *motor cortex*, a circumscribed cortical lesion may cause a contralateral upper motor *monoplegia*,[257] e.g. of an arm or leg, or of one of their muscle groups (Fig. 257A). A lesion involving the internal capsule, where the motor channels are crowded together within a narrow space, will cause *contralateral hemiplegia* (Fig. 257A).

Figure 257B illustrates sites of lesions causing alternate hemiplegia, while Figure 257C depicts the *bilateral 'supranuclear innervation'* of

[257] The concepts of upper motor and lower motor neuron paralysis with their respective symptoms, and the diverse crossed paralyses of brain stem nerves indicated in Figure 257 were dealt with in chapters VIII, IX, and XI of volume 4.

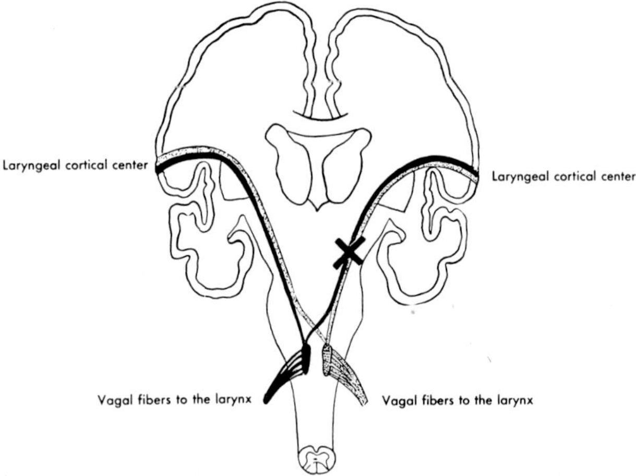

Figure 257C. Sketch indicating presumed supranuclear input to subdivision of nucleus ambiguus innervating laryngeal musculature. A unilateral lesion at x (e.g. in pes pedunculi) does not cause laryngeal paralysis (from HAYMAKER and BING, 1969).

certain cranial nerve nuclei, dealt with in chapter IX of volume 4. It will be seen why a lesion in capsula interna or pes pedunculi does not appreciably affect the bilaterally innervated lower motor griseum.

Figure 257D shows an extensive left cortical lesion caused by embolism occluding branches of the middle cerebral artery supplying the lower portion of precentral motor as well as parts of lower premotor and perhaps prefrontal cortex. The symptoms displayed were spastic paralysis of right arm, upper motor neuron paralysis of the right lower facial musculature, whose motoneurons lack bilateral supranuclear innervation (cf. Fig. 257B), upper motor neuron paralysis of the right half of the tongue, and motor aphasia due to involvement of *Broca's area*. The manifestations of aphasia and the significance of *Broca's area* and other speech 'centers' will be discussed further below.

Figure 257E illustrates an extensive hemorrhagic lesion of the right internal capsule causing left spastic hemiplegia with involvement of the lower facial, and of the tongue musculature (cf. also Fig. 257D). Since the lesion extended well into posterior limb and retrolenticular portions of internal capsule (cf. Figs. 250A, B), there was also sensory impairment over the entire left side of body and head. The sense of posture and passive movement was lost. Tactile sensibility was likewise seriously affected, thermal sensibility to a lesser degree, and pain

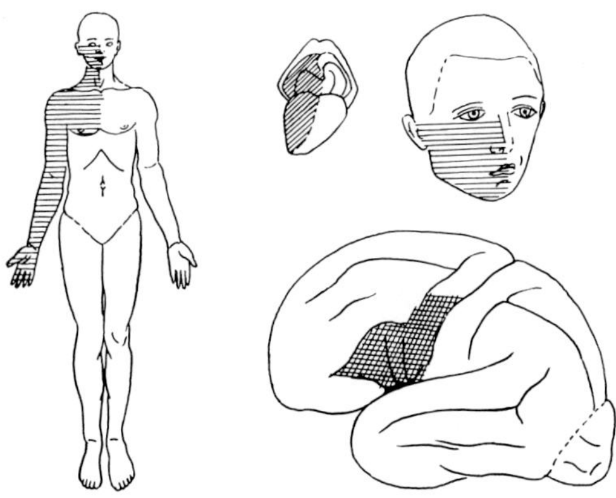

Figure 257 D. Vascular lesion affecting branches of middle cerebral artery supplying *Broca's area* and lower part of precentral gyrus. Right arm, right side of tongue and soft palate, and lower subdivision of right facial nerve (cf. Fig. 257 B) are affected (from RANSON, 1943).

was felt equally well on both sides.[258] In addition, there was a left homonymous hemianopia due to involvement of the right geniculocalcarine tract.

Details of the *cortical optic input channel* and its relay in the lateral geniculate griseum were pointed out in section 9 of chapter XII (cf. Figs. 145 A–C, p. 404). Additional data concerning the geniculocalcarine tract and the optic projection upon the occipital cortex (area 17 sive area striata) were discussed and illustrated further above in the present chapter (Figs. 251 A–E, 252 A, E).

The visual fields of both eyes, as mapped by routine examination with the perimeter, are shown in Figures 258 A and B.[259] Defects, i.e. blind areas of the visual field, are designated as scotomas. In most instances, such defects are not experienced by the patient as dark areas, but, although 'blank' regions, become unnoticeably integrated into the visual space of consciousness (negative scotomas). The blind spot, in

[258] As reported by RANSON (1943), from whose text the fairly typical cases of Figure 257 D and E were taken. In different cases involving the sensory channels of the capsula interna, pain sensation may, however, be affected in various degree and even be abolished at least for some time.

[259] It should also be added that, within the visual field, there is, from the periphery toward the macula, a succession of bands with increasing color sensitivity, the outermost band being almost achromatic, followed by blue, red, and green sensibility, innermost band being thus trichromatic, except for the macula, which is said to be blue-blind.

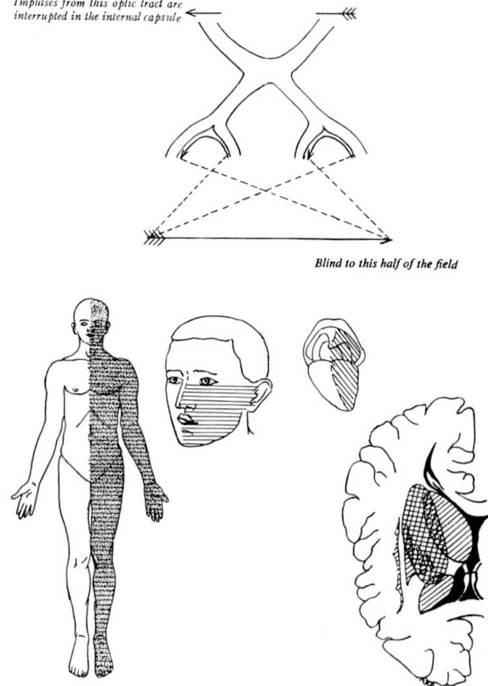

Figure 257E. Extensive hemorrhage from lenticulo-striate artery into right internal capsule causing left hemiplegia with impaired sensibility and left homonymous hemianopia (from RANSON, 1943).

which small visual shapes can be shown to disappear by perimetry or by means of a simple test figure (Fig. 258C) corresponds to the optic disk (cf. p. 59, chapter XII) and represents a physiological *negative scotoma*, located in the temporal visual field. It is compensated in binocular vision, because temporal and nasal fields become here superimposed qua cortical projection. In monocular vision it becomes compensated by the *saccadic movements* (cf. p. 88). A blind area noticed as a dark spot in the visual field is called a positive scotoma.

Binocular vision, characterized by 'fusion of two images', and related to the just mentioned superposition of temporal and nasal visual fields in each area striata, is attributed to bilaterally corresponding retinal points which, when simultaneously stimulated, 'cause' a single visual sensation. The paramount corresponding points are the two central spots of the macula through which the optic axis of each eye pass. The

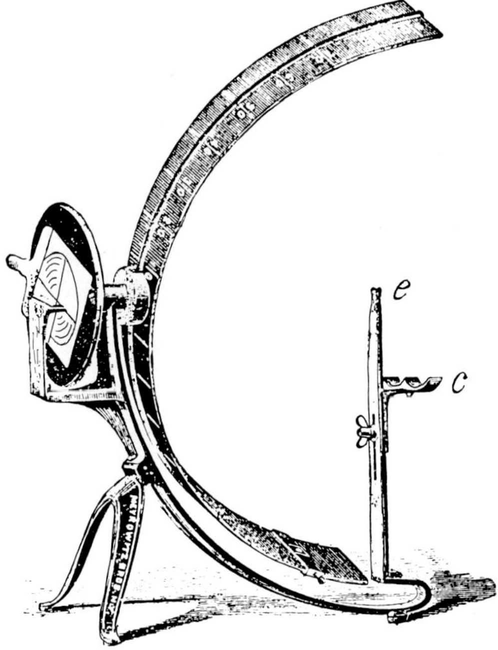

Figure 258 A. A commonly used type of perimeter (from BEST and TAYLOR, 1950). c: adjustable chin-rest; e: position of the eye which is to be examined, and whose fixation point is the midpoint of the arc of circle provided by the graduated metal band rotated around the pivot at the midpoint. A sliding holder on the arc carries the white or colored disk used as test surface along the spherical coordinates.

other corresponding points are those equidistant, in the same direction, from the macular ones. It is evident that, except for corresponding points on the 90–270° meridian, separating right and left visual fields (cf. Fig. 258B), one of the two points of any such corresponding pair must lie in the temporal retinal half, and the other in the nasal one of the antimeric eyes. Images not projected upon corresponding points appear double in perception *(diplopia)*.

A line joining points in space which are projected upon corresponding points of the two retinae at any given position of fixation and convergence is called the *horopter* (JOHANNES MÜLLER, 1834/35). In a first approximation, such line can roughly be conceived as representing a circle passing through the fixation point and the nodal points of the eyes (Fig. 258D). There is a different horopter for each position of convergence. Again, the extraocular portion of the horopter can also

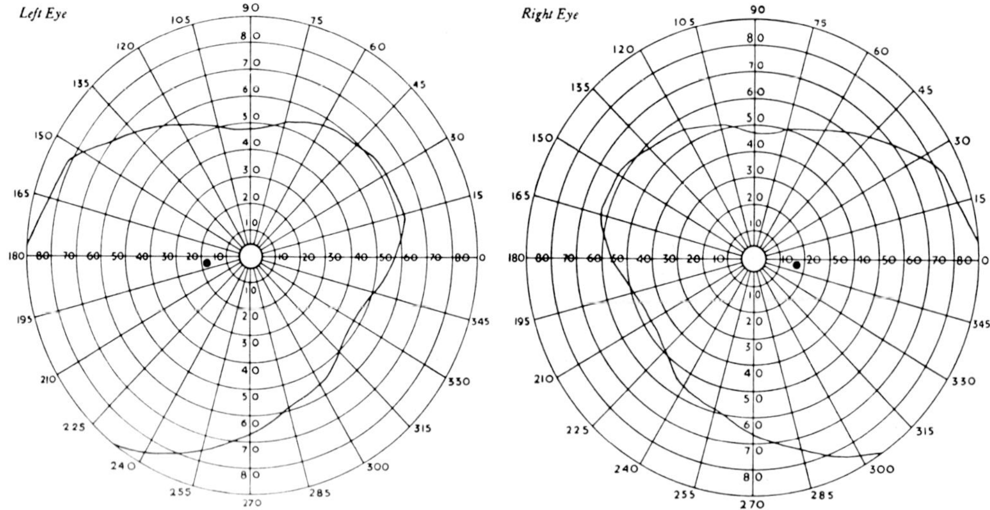

Figure 258 B. Standard perimetry chart for left and right eye as used for routine examinations. The average normal visual fields are outlined. The average location of the blind spot is indicated by a black dot in the temporal field, at a distance of roughly 16° from the macula's center, and is, of course, located in the nasal retinal half. It will be seen that the nasal visual field is less extensive than the temporal one. In this latter, the monocular portion begins peripherally at approximately 60°.

be conceived in a simplified formulation as a line generating a portion of an approximately spherical surface.

If the converged eyes are fixed and focused on a near object, e.g. a pencil held close to the nose, distant objects appear double, the single pencil in turn becoming double upon divergence by focussing on the now again single distant objects. This physiologic diplopia is usually suppressed in the normal visual perception of consciousness, but becomes immediately experienced upon 'giving attention'. Pathologic diplopia occurs if the motions of the eyebulbs are not properly coordinated, e.g. in toxic states or in eye muscle disturbances causing squint *(strabismus)*. The diplopia of strabismus can be imitated by gently pushing, with a finger, one eyebulb out of position.

The diplopia is proportional to the amount of deviation from the norm. The image produced by the eye that fixes the object in the optic axis is called the true image and is more distinct than the false image produced by the deviating eye. In converging strabismus, the false image is on the side of the deviating eye *(homonymous diplopia)*. In diverg-

Figure 258C. The left part illustrates the demonstration of the blind spot. If, with closed left eye, the cross is fixed with the right one, and the figure moved forward and backward, the circle, projected upon the blind spot, disappears (for the emmetropic eye) at a distance of about 25 cm (MARIOTTE's [1620–1684] experiment). The right part illustrates an actual mapping of the right eye's blind spot (from BEST and TAYLOR, 1950).

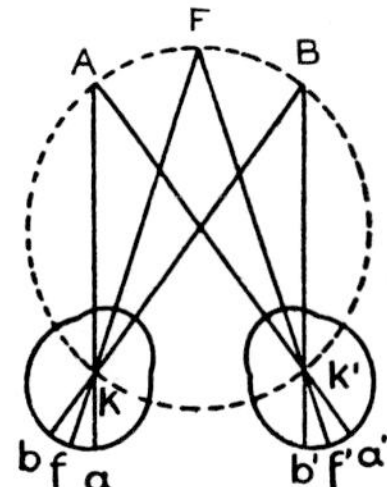

Figure 258 D. Diagram illustrating a simplified tracing of MÜLLER's horopter (after DUKE-ELDER, from BEST and TAYLOR, 1950). F is the fixation point whose image is projected upon the maculae f, f′; the images of A and B likewise fall upon corresponding retinal points a, b, a′, b′: k, k′ are the conjoint nodal points (cf. Fig. 17B, p. 35) of both eyes.

ing strabismus, the opposite effect obtains *(crossed diplopia)*. When the two images are level, the diplopia is horizontal, when the optic axes diverge upward or downward, the diplopia is vertical, the false image being the upper one from the downward deflected eye and the lower image from an eye with upward deflection.

In long standing acquired strabismus, the diplopia gradually disappears, being adjusted by not yet understood compensating retino-cortical mechanisms. In congenital strabismus, no diplopia is experienced. Following corrective operations in such compensated strabismus, an initial transitory period of diplopia results.

Stereoscopic vision, characterized by depth perception in a three-dimensional visual space, is significantly but not exclusively related to the slightly dissimilar images of any given object projected upon the retinae of left and right eye. The 'fusion' of these slightly differing images in the visual cortex, that is, the processing of the slightly differing retinal signals, is assumed to 'produce' the depth perception. This is obviously substantiated by the well-known stereoscope, by which two

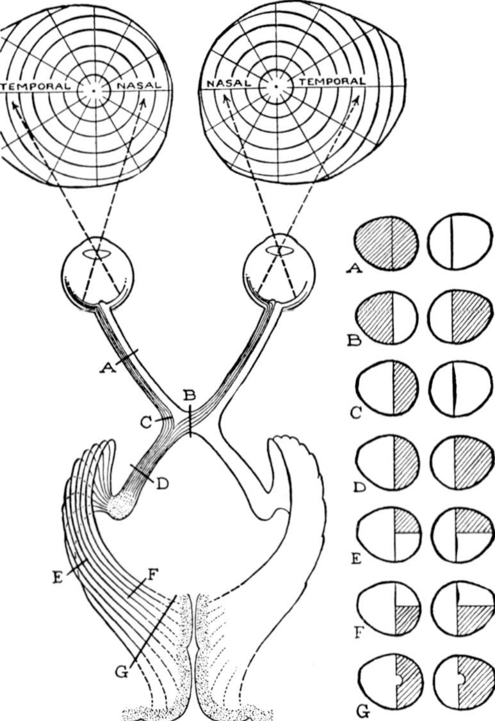

Figure 259 A. Diagram illustrating the visual field defects produced by lesions at different levels of the optic pathway (after HOMANS, from RANSON, 1943). A: complete blindness in left eye; B: heteronymous bitemporal hemianopsia; C: nasal hemianopsia of left eye; D: right homonymous hemianopsia; E, F: right upper respectively lower homonymous quadrant hemianopsia; G: right homonymous hemianopsia with sparing of macula. Cf. Figure 251 A for anatomical details of optic pathway.

slightly differing photographs, corresponding to the images obtained at a slightly different angle by the two eyes, are projected, by prismatic effect, upon corresponding retinal points. Yet, normal stereoscopic vision does not entirely depend upon two dissimilar retinal images projected at different angles. Although the three-dimensional aspect of visual space is much more vivid in binocular vision, it is by no means abolished when one eye is closed. The optic space of monocular vision remains three-dimensional, despite a reduction in the accuracy of depth-perception. Saccadic motion and the effects of 'perspective' may here play a certain role. Normal *visual depth-perception*, correlated with tactile and proprioceptice space perception, results presumably from

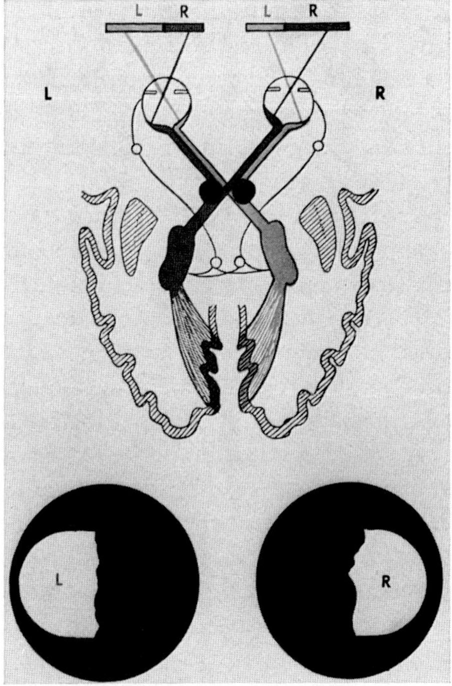

Figure 259B. Diagram illustrating binasal heteronymous hemianopsia, not shown in Figure 259A (from HAYMAKER and BING, 1969).

cerebral processing of early postnatal sensory input, being thus gradually acquired by the infant. The details concerning development of visuotactile, as well as of audiovisuotactile space perception coordination during early infancy remain poorly understood and controversial (cf. e.g. McGURK and LEWIS, 1974).

Figure 259A illustrates the defects in the visual field caused by lesions at various levels along the optic input channel. Defects involving one half or one quadrant of the visual field are *hemianopsias* respectively *quadrant hemianopsias*. It both left or both right visual field halves are affected, the hemianopsia (also spelled hemianopia) is called *homonymous*. If both temporal or both nasal halves are involved, the hemianopsias are *heteronymous (bitemporal* respectively *binasal)*. Binasal hemianopsia (Fig. 259B) is rare, since it must result from roughly bilateral symmetric encroachment upon the optic chiasma by conditions such e.g. as bilateral internal carotid aneurysms or bilateral outgrowths of some pituitary adenomas. If a lesion damaging only one lateral portion of the chiasma is large, ipsilateral blindness combined with contralateral temporal hemianopsia may result. Transection of the chiasma (Fig. 259 A, lesion B) causes bitemporal hemianopsia. Lesions involving optic tract, lateral geniculate complex, and geniculocalcarine tract cause homonymous hemianopsias (Fig. 259 A).

Partial lesions of area striata (area 17 of Brodmann) cause partial defects in the corresponding half of visual field. Complete unilateral destruction of area striata results in crossed homonymous hemianopsia. Complete bilateral destruction of the entire area striata is followed by *cortical blindness*. If bilateral thrombosis of the posterior cerebral artery occurs, almost complete cortical blindness results, but the fixation point of the macula is usually spared. This sparing can be attributed to the fact that the cortical representation of the fixation point is located most laterally, on the lateral aspect of the occipital pole. This region is ordinarily supplied by branches of the middle cerebral artery. Additional problems concerning bilateral representation of the macula (fovea centralis) were mentioned above (cf. also footnote 240).

The cortical blindness caused by complete bilateral destruction of area striata may be accompanied by a peculiar type of anosognosia known as Anton's *syndrome* (1899). Although the patient is totally blind, he is unable to realize that he cannot see. There is no sensation of darkness, and the total situation may be compared to the non-existence of a visual field behind a normal person's head.[259a] The clinical picture is characterized by mental confusion sometimes associated with extreme confabulation.

Destruction of area striata thus seems to eliminate spatially configurated optic shapes and colors, that is visual percepts. It could also be said that this destruction leads to 'an elimination of visual space with its dimensions' (Klüver, 1942). However, if the occipital lobes are not entirely destroyed and adjacent para- or peristriate cortex is preserved, various components of visual consciousness commonly remain, although this consciousness-modality becomes severely affected.

The occasional occurrence of visual hallucinations and visual dreams in cases of cortical blindness raises the question whether spa-

[259a] This has been called '*totales optisches Nichts*' or '*vision nulle*'. There are, however, in this respect differing reports, perhaps due to the diversity in the extent of the obtaining lesions as well as to the complexity of the multifactorial cortical events correlated with the (conscious) perception of 'light' and 'darkness'.

tially configurated optic consciousness-modalities comparable in vividity to those of normal percepts could be regarded as parallel phenomena related to a neural activity which may occur in extra-striate occipital cortex. However, LHERMITTE (1924), who studied an actual case, and discusses the relevant problems, believed that in instances of hallucinations the area striata is not completely destroyed and that the cortical blindness is only apparently complete. According to this author '*les processus destructifs de toute l'aire striée... ne sont jamais accompagnés par l'extériorisation hallucinatoire d'images visuelles*'.

Total cortical blindness can be defined as complete inability to perceive external objects by sight, combined with loss of ability to distinguish between darkness and illumination of the environment, despite normal condition of the eyes and retention of the pupillary reflexes. Although eye motility obtains, there is no convergence as required in fixing the gaze upon a near object. The lid closure reflex to bright illumination and to threatening gestures is abolished.

The damage to the area striata and its surroundings may be caused by trauma, neoplasms, vascular conditions, abscesses, and diverse degenerative or toxic processes. Thus, cortical blindness, also called 'occipital blindness' may result from occlusion of basilar artery or of both posterior cerebral arteries (cf. e.g. SYMONDS and MACKENZIE, 1957; ABRAHAM *et al.*, 1975).

A particularly interesting and carefully analyzed case of progressive cortical blindness was observed by Professor GERLACH and reported by his student KRAUSENECK (1975). In a patient with left homonymous hemianopsia caused by a malignant meningioma of the falx cerebri, GERLACH found it necessary to perform a complete right occipital lobectomy.

Subsequently, disturbances in the remaining right visual field developed, and about 11 months after the first operation, another meningeal neoplasm compressing the left calcarine region had to be removed with some damage to the adjacent cortex. Some moderate degree of vision in the right fields was at first retained, accompanied by highly disagreeable blinding illumination effects caused by any source of light.

Still later, due to a recidivation of the neoplastic growth, the vision in the right fields deteriorated to a mere dim light perception, and a third operation, about 22 months after the second one, became necessary, entailing further damage to the calcarine cortex, and resulting in essentially complete cortical blindness.

The highly intelligent and cooperative patient, a teacher of deaf-mutes, gave very detailed accounts of his experiences and willingly underwent numerous neurological and psychological tests. After the third operation, he displayed a transitory *Anton syndrome* (cf. above, p. 834). Despite other transitory disturbances which might have been initial effects of the operational trauma inflicted upon the brain, no serious defect of intelligence and 'praxies', including stereognosis were manifested. Rather slight disturbances in calculation, in writing and in drawing did occur, presumably related to a deficiency in the capacity of optic representation *(optische Vorstellungsfähigkeit)*. Occasional dreams were reported after the third operation, mostly ideational *(gedankliche Inhalte)* but at times optic and even in color *(farbige Bilder)*.

After the third operation the patient complained again about dazzling light effects which subsequently disappeared, but the occasional impression of a light glow localized on the right side remained when he was near a bright light source. This might have been due to a minor remnant of left incompletely destroyed area striata within the left occipital lobe. There is, of course, the possibility that in accordance with inconclusive hypotheses concerning a so-called accessory or second optic system, optic input (e.g. via pretectal grisea), bypassing the geniculocalcarine pathway, might reach peri- or parastriate cortical areas, mediating some sort of 'light sensation', but such explanation does not seem very convincing.

In a letter about two years after the operation, the patient complained about his darkness-experience: '*Dieses ist bei mir in einer Intensität vorhanden, die durch keinerlei Licht- oder Farbenerinnerung unterbrochen wird, und das Schlimme ist, dass ich das Gefühl habe, diese Dunkelheit beginnt sich mehr und mehr auch auf andere Gebiete meines Geistes auszudehnen.*' A few weeks later, however, he wrote: '*Zum Punkt Dunkelheitserlebnis möchte ich, um weitere Irrtümer auszuschliessen, hinzufügen, dass ich den Ausdruck Dunkelheitserlebnis nicht für ganz zutreffend halte. Ich erlebe nicht das Gegenteil von Helligkeit oder Farbe, dazu müsste ich ja von beiden noch eine Vorstellung haben, was aber nicht der Fall ist, sondern um mich herum ist ein absolutes optisches Nichts.*'

Soon afterwards, the condition of the patient deteriorated, with strong back pain and urinary retention, apparently caused by a tumor metastasis in the lumbar region. Personality changes and strong depression developed. The patient died about 5 years after the first, respectively about 2 years and 5 months after the third operation at the age of 58 years. No regular autopsy was performed, but the brain and a portion of lumbar vertebral column were rather carelessly removed and forwarded in a damaged and improperly fixed condition to Würzburg. Thus, an appropriate study of the brain became impossible, but the gross inspection of the brain confirmed the findings at the operation. The right occipital lobe was completely missing, and the calcarine region of the left one destroyed with a few small 'necrobiotic' remnants. Some portions of extrastriate occipital cortex were preserved. Since this in various respects unusual case has relevant bearings on aspects of cerebral function and pathology dealt with in the present section, as well as on the various difficulties, uncertainties, and ambiguities inherent in neurobiological and clinical interpretations, a discussion of some of its details seemed appropriate in the present context.

As regards lesions of the *auditory cortex*, unilateral destruction of *Heschl's transverse temporal gyri* does not cause cortical deafness but only slight hearing loss, with some difficulty in the localization of sounds. Bilateral destruction, however, results in *cortical deafness*, that is to say loss of auditory percepts.[259b] Since these bilateral lesions must involve the antimeric (i.e. symmetric) structures, such cases are not common, but several have been recorded in the neurological literature, and FLECHSIG (1896) refers to one case which he personally studied. Concerning the *gustatory* and the *olfactory projection centers*, no relevant clinicopathologic observations providing serviceable functional or anatomical data seem to be extant.

Turning now to the '*association areas*', an evaluation of their functions on the basis of available clinicopathologic observations is beset with substantial additional difficulties and uncertainties. While it seems reasonably well established that the activities of said areas are related to the faculties of 'thinking', to recognition respectively interpretation of sensory input, to verbalization respectively speech, and to the performance of skilled, acquired, motor activities,[260] the problem how these behavioral functions are 'localized' has, so far, defied satisfactory solutions.

[259b] It should be recalled that in the terminology which I have adopted, *all percepts* are *conscious*. *Unconscious receptions of stimuli* respectively *signals* by grisea or by hardware instruments etc., are here subsumed under the term '*registration*'. Evidently, in cortical deafness or blindness, the subcortical grisea can register input, and may mediate reflexes.

[260] EDINGER (1912) subsumed these higher cerebral neural activities under the designations *Gnosis* and *Praxien*, the two being interrelated by *Assoziation*. EDINGER's *Gnosis* and *Praxien* roughly correspond to what KRETSCHMER (1926) designated as *Abbildungsvorgänge* (limning processes) and *Ausdrucksvorgänge* (expression processes).

The attempts at constructing models of the relevant cerebral organization has led, roughly speaking, to three different sorts of concepts. One view assumes a network of unapproachable complexity with little discrete localization of function, thereby virtually precluding a significant analysis. The opposite view emphasizes specialized, well localized regions in accordance with oversimplified organizing principles, and assumes a signal flow, as it were, from 'point to point'. A third, compromising view, to which I would subscribe, regards the two apparently contradictory concepts of diffuse network and of localized 'centers' as not mutually exclusive, but as expressing jointly obtaining tendencies, admitting a still undefined degree of 'localization' combined with a likewise still undefined degree of diffuseness and redundancy. Thus, in such network, neither complete equivalence nor minutely detailed rigid localization would obtain.

The cortex, receiving coded input signals (N-events) related to extraneural physical events, models, that is symbolizes, by abstraction of invariants, these physical processes (R-events) by combinations of activated neural networks. The neural processes (N-events) are very likely, as CRAIK (1943) and McCULLOCH and PITTS (1943) suggested, of the same nature as the circuit mechanisms of computers and similar devices, based on *Boolean algebra*.[261] The application of this algebra respectively calculus to relay and switching circuits was established and demonstrated by SHANNON (1938).

Three essential processes have been well summarized by CRAIK. With only minor modifications, they can be enumerated as follows: (1) Translation of sensory input patterns into simple mnestic patterns (primitive symbolization, without words), or into higher mnestic patterns, or into both. The higher patterns are symbols such as words or numbers. (2) Arrival at other symbols by a process corresponding to association, deduction, inference, reasoning, etc. (3) Re-translation of these symbols into motor processes, speaking, writing, performing properly patterned movements in the execution of various other activities. These neural processes are, essentially but not exclusively, funnelled through the output of the cortical efferent projection center of FLECHSIG (precentral gyrus).

The processes under (2) and (3) overlap with limning and expression processes, but cannot be considered as entirely identical with

[261] Further details of this topic will be discussed in chapter XV (vol. 5/II), dealing with structure and function of the cerebral cortex.

these concepts of KRETSCHMER. The correspondence between a pattern of symbols arrived at by the processes under (2) and external events, as in realizing that a prediction concerning such events is fulfilled, involves, I believe, essentially a combination of the processes under (1) and (2); CRAIK, however, classifies that process of recognition under (3).

All three fundamental types of processes are closely interrelated and integrated into a common pattern of activities which cannot be separated by arbitrary boundaries, but which may vary in emphasis. Thus, one may think out loud, or a poorly trained and clumsy reader may silently articulate and move his lips while studying a text. In such instances there is considerable overflow into the motor periphery. In this respect, the significance of sensorimotor or 'operational adjustment' stressed by some authors appears substantiated, although one may not agree with many of the implications postulated by these authors.

Since the neural processes corresponding to the various activities discussed under the preceding classification presumably involve spatially extensive circuits, any attempt at definite localization will result in a somewhat artificial scheme and the concept of centers cannot be upheld in a narrow or rigid sense. As regards these cortical activities, I agree with NIELSEN's (1955) statement that if by 'center' a delimited cortical area with independent functional capacity is meant, such centers cannot be assumed to exist. However, if a center is defined as an anatomically delimited zone essential to a certain function, then most certainly several such centers exist. In a somewhat similar manner, I interpreted in 1927 diverse higher cortical centers as regions indispensable for the ecphory of certain types of engrams, but not as regions in which such engrams should be regarded as localized. It is very likely that any given neuron or group of neurons in any such center may be involved in a variety of structurally different circuits, corresponding to functionally different processes.

Moreover, CRAIK points out that models of the brain on the pattern of circuit exchanges would be much more convincing if they did not postulate any rigidly defined particular connections. Such constancy of connections is very unlikely in view of the numerous individual variations. ASHBY's concept of iterated ultrastable systems, combined with a high degree of structural randomness, appears particularly helpful in an appraisal of the cortical functions under discussion. This concept is compatible with the assumption of a certain degree of equivalence and multiple representation, as well as mass-action.

FLOURENS assumed that the cerebral hemispheres were active as a whole and that after removal of a large amount of their equipotential substance, the remainder had still the capacity to perform most functions. The general reduction of functional capacity was believed to be roughly proportional to the amount of destruction. In recent years, LASHLEY has reached similar although perhaps somewhat less radical conclusions based on his experimental studies concerning the Rat's cerebral cortex.

As regards the cerebral cortex of Man, it can here be repeated that the results of clinico-pathologic observations seem to favor a compromise between the two extreme and opposite views of complete equivalence and of minutely detailed or rigid localization. It has been pointed out by various authors, familiar with the topic, that, if damage to a given area impairs or abolishes a certain 'function', this latter is not necessarily 'located' in such area as its 'center'. One can merely say that said area's normal activity is required for, or is a component of, a particular performance. Thus, 'to locate the damage which destroys speech, and to locate speech are two different things' as already stressed in 1874 by JACKSON (selected writings, 1931, 1932).

The *symbolizing processes*, moreover, are combined with those of affectivity, or, in other words, 'thought' cannot entirely be separated from 'emotion'. Human thought, again, although not identical with language or speech, is, nevertheless, closely related with language. *Speech*, in turn, can be regarded as a communication system. *Language*, whose vocal aspect is speech, and one of whose visual aspects is *writing*, represents a most relevant human attribute. It is generally presumed to be radically different from all forms of animal communication. While some authors therefore emphasize that, in this respect, the gap between animals and man precludes a satisfactory theory of biological language evolution, DARWIN believed that the faculty of articulated speech does not in itself offer any insuperable objection to the plausible hypothesis deriving human language from a more primitive form of animal communication. GESCHWIND (1970), with whom I would agree, assumes that forerunners of language do exist in 'lower forms'.[261a]

The process of abstraction and the formation of concepts are doubtless performed by the neural activities in all animals provided with a

[261a] Some interesting data concerning this topic with reference to 'the voices of the Dolphin' can be found in a slightly sensational-melodramatic publication by LILLY (1963).

sufficiently complex central nervous system. Although such animals seem incapable of developing sharply delineated concepts associated with definite sound symbols, they may, by means of averbal concepts, frame judgments, and reach conclusions, as is evidenced, in various degrees, by their behavior (cf. e.g. RENSCH, 1968, 1971). There are, of course, numerous hierarchical degrees of conceptualization, such, e.g. as indicated by the difference between concrete and abstract concepts or by an arbitrary distinction between 'signs' and 'symbols'. *Animal communication* is, to a large extent, predominantly related to emotion, although, e.g., the 'dance language of Bees' (cf. vol. 2, p. 178) essentially seems to encode 'factual' information.

Again, in *Human communication*, and following HUGHLINGS JACKSON (1931, 1932), a distinction can be made between propositional speech, in which the communication of descriptive or more abstract meanings predominates, and emotional speech, in which the expression of feelings prevails and the propositional value becomes secondary or may even reach the vanishing point. Such emotional speech then becomes gibberish.

The observations and reports of BROCA (1861) initiated, as it were, the elaboration of various concepts concerning the localization of 'speech' and other 'higher' cerebral functions. BROCA showed that damage to the posterior part of the inferior frontal convolution of the leading hemisphere *(Broca's convolution)* caused loss of motor speech, i.e. of language output.[262] In this respect, the inability to convert thought into motor word patterns *(motor aphasia)* might be distinguished from the inability to convert these word patterns into articulate sound by controlling the muscles used in speech *(dysarthria, anarthria)*. In the dysarthric patient, symbolic verbal formulation might be considered normal, only the mechanism of sound production being faulty.

WERNICKE (1874) established the existence of aphasias related to defect of word comprehension, associated with lesions in the posterior part of the superior temporal gyrus. This region, subsequently called *Wernicke's area*, can be evaluated as an auditory 'association area' in FLECHSIG's sense. MARIE (1906) contested BROCA's and WERNICKE's views and claimed that '*la troisième circonvolution frontale gauche ne joue au-*

---

[262] Originally called *aphemia* by BROCA. Subsequently, the term *aphasia* was introduced to designate disturbances of the 'higher' speech functions (e.g. 'internal speech') above the level of coordinate speech musculature control.

cun rôle spécial dans la fonction du langage'. MARIE attributed aphasia to a general defect of intellectual functions involving language in the widest sense and considered lesions of *Broca's area* to cause merely anarthria.

Subsequently, many different theories and numerous highly specialized terms became elaborated in the discussion of aphasia. This is doubtless due to the complex and still poorly understood nature of language, which, from the neurologist's viewpoint, has been reviewed in an interesting monograph by BRAIN (1965).[263]

There is thus no single, generally accepted theory respectively classification of aphasia, about which many authors, including JACKSON (1931, 1932), HEAD (1920, 1926), LIEPMANN (1900, 1905), GOLDSTEIN (1948) and others have expressed their opinions. There is no single, generally accepted nomenclature of the varieties of aphasia, and HEAD (1926) appropriately characterized the status of the aphasia problem as chaotic. He nevertheless added his own totally different terminology to that 'chaos' which, at least for the time being, one might also regard as 'hopeless'.

In addition to BRAIN's monograph (1965) cited above, relevant contemporary elaborations on this topic are those by BENSON and GESCHWIND (1974), GESCHWIND (1970), BROWN (1972), LURIA (1966), SCHILLER (1969), and PRIBRAM (1971).

Reverting to EDINGER's (1912) concepts of *praxis* and *gnosis*, it seems evident that the aphasias can be subsumed under a wider category of disturbances including, besides 'motor' and 'sensory' aphasia, various

---

[263] Additional publications, dealing with the diverse relevant aspects of language are those by GREENBERG (1961, 1963), HÖRMANN (1969), and particularly bei WEIGL (1966, 1968, 1969) and WEIGL and BIERWISCH (1970). The problem of abstract but apparently 'wordless' sphairal thought (*'nicht sprachgebundenes begriffliches Denken'*, e.g. WEIGL and METZE, 1968) is of especial interest. The volume edited by LYONS (1973), 'New horizons in linguistics' brings a survey of present-day concepts and includes a review of the much publicized views expressed by CHOMSKY's theories of so-called *'generative grammar'*. With regard to this latter author's views, it might here be added that, in elaborating on the diverse problems of syntax and meaning, he seems to disregard the fact that languages are merely secondary codes representing transforms of various complex but insufficiently understood neural coding systems. In other words, CHOMSKY's theories, although propounding an unconventional, rather arbitrary semantic approach to grammar and linguistic, have no significant bearing on the intrinsic nature of language *qua* transform of coded neural events. With regard to concepts of neurologic epistemology, some comments on words, language, thought and definition were given by the present author in his contributions to the *Helen Adolf Festschrift* (K., 1968) and in his monograph 'Mind and Matter' (K., 1961).

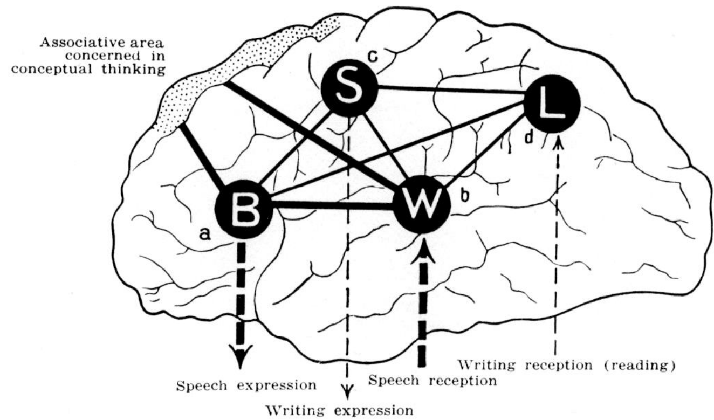

Figure 260 A. The cortical speech areas and their connections, as conceived in classical views on aphasia, apraxia, and related conditions, with emphasis on the location of 'centers' (from HAYMAKER and BING, 1969). B: *Broca's area;* L: 'reading center'; S: 'writing center'; W: *Wernicke's area* ('center for acoustic sound and word understanding'); a: left inferior frontal gyrus; b: superior temporal gyrus; c: premotor region; d: angular gyrus.

apraxias, dysarthria, and diverse agnosias. As regards the 'localization' of symbolizing cerebral activities, the following generalized conclusions can perhaps be drawn.

If the *premotor cortex* (regio frontalis agranularis) is damaged, the performance of skillfull, patterned motor activities becomes impaired. Many variables obtain, including *inter alia* the distance of the lesion from the area gigantopyramidalis of the precentral convolution. Besides spasticity and loss of ability to perform finer movements, various degrees of disturbance in execution and formulation of motor patterns may occur, such as dysarthria and apraxia. *Agraphia* may be caused by a lesion in the posterior part of the middle frontal convolution of the dominant hemisphere. *Apraxia of speech*, also occasionally termed *motor aphasia*, may be produced by damage to the posterior part of the inferior frontal convolution of the leading hemisphere *(Broca's convolution)*. Similar disturbances, more in the nature of dysarthria, may occur in lesions immediately adjacent to or involving the lower part of the dominant precentral convolution. In such speech disturbances of purely apraxic and dysarthric type, internal language is not affected. Observations recorded by PENFIELD and JASPER (1954) seem to indicate that the posterior part of the supplementary motor area is also related to speech functions. This region is designated by the cited authors as 'speech 4', while *Broca's convolution* is referred to as 'speech 1'. Fig-

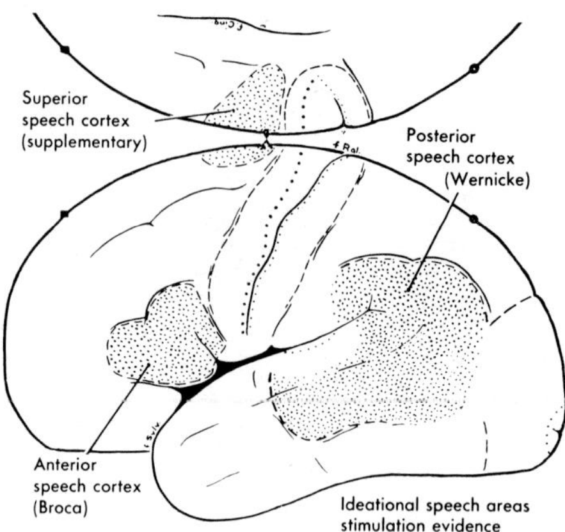

Figure 260 B. Cortical areas of dominant hemisphere involved in 'ideational elaboration of speech' and drawn on the basis of electrical 'speech mapping' (after PENFIELD and ROBERTS, 1959, from HAYMAKER and BING, 1969). It will be seen that the 'posterior speech cortex *(Wernicke)*' does not quite correspond to *Wernicke's area* as originally outlined by that author.

ures 260 A and B illustrate an older and a more recent concept of cortical areas related to speech respectively language functions.

Again, in damage to the premotor region of lobus frontalis, the grasp reflex, respectively forced grasping may be elicited, usually on the contralateral side, and particularly in deep-seated lesions. Since the grasp reflex is normally displayed by infants, its occurrence in adults is generally considered 'a regression to the infantile stage of the function of grasping'.

Although the *corpus callosum* must be considered not essential for skilled performances or symbolic consciousness activities, there are well documented cases of apraxia apparently caused by lesions of that commissural structure.

Damage to the *occipito-parieto-temporal region* of the dominant hemisphere affects to a variable degree the recognition or formulation of sensory patterns. Thus, *visual agnosia* may occur after lesion to extracalcarine cortex and *auditory agnosia* in damage to *Wernicke's area*, located on the superior temporal gyrus in the vicinity of *Heschl's transverse temporal gyri*. If such agnosias are complete, the terms *psychic blindness* and *psychic deafness* are used. For the patient, perceived shapes or sounds have become devoid of meaning. Lesions in the region of the lobulus parietalis superior can produce agnosia of tactile patterns *(astereognosis)*.

However, damage to the occipito-parieto-temporal region may, in certain instances, affect only certain types of symbolic pattern recognition or formulation. Such functional impairment can be manifested as word-deafness or as *word-blindness (alexia)* in which the patient has lost his previous ability to understand the spoken or the written word. Words or letters are still distinguished from other acustic or optic percepts, but have lost their (symbolic) meaning. In higher degrees of disturbances, there is gross impairment of thought and expression. Then again, the recalling of names might be mainly affected, either by lack of ecphory *(amnestic aphasia)*, or by wrong ecphory *(paraphasia);* other disturbances can predominantly involve the formulation of correct word sequencces and sentence schemes. Neural activity in the *gyrus angularis* of the leading hemisphere seems to be an essential functional link in the recognition of higher symbolic visual patterns such as the written word. Thus, *alexia* can be caused by damage to the region of that convolution. Among the different types of alexia are those with or without agraphia (cf. above, p. 843), and with or without aphasia. The term alexia refers to the impairment of previously obtaining reading ability, and is generally distinguished from *dyslexia*, which designates a pronounced difficulty in learning to read during the course of development.

An interesting case of *alexia* with *agraphia* in a Japanese patient was recently presented by YAMADORI (1975). An occlusion of the angular branch of the left middle cerebral artery could be demonstrated by arteriography and presumably damaged the angular gyrus within the so-called posterior or parietal speech area (cf. Figs. 260 A, B). Impoverishment of spontaneous speech, word-finding difficulty, *Gerstmann's syndrome*, autotopagnosia, apraxia, and severe agraphia were manifested. As regards the obtaining alexia, the reading difficulty was extreme for words composed of the syllabic Japanese *phonograms (Kana)* symbolizing sounds, but the reading of the essentially *visual*, i.e. pictorial Sino-Japanese *ideograms (Kanji)* was relatively well preserved. As regards the agraphia, the few words which the patient could still write were only in *Kanji*.

The analysis of this case suggested that the lesion had caused, *inter alia*, some sort of disconnection between visual and 'auditory-oral' mechanisms.

Reverting to *dyslexia*, which has been intensively studied in recent years, the following types, as e.g. enumerated by BENSON and GESCHWIND (1974), are generally distinguished. (1) *Symptomatic dyslexia* associated with obvious neurologic abnormalities causing difficulties in learning to read (mental retardation, hydrocephalus, cerebral palsy, chorea, epilepsies). (2) *Specific or developmental dyslexia*, in which innate inability of various degrees to comprehend or acquire comprehension of written or printed language symbols occurs in the absence of other recognizable neurologic abnormalities. (3) *Secondary dyslexia* or reading disability, in which slow or poor acquisition of reading skill results from poor general health, prolonged absence from school, emotional disturbances or decreased vision.

PENFIELD and JASPER include the angular gyrus in their parietal speech area ('speech 2'), and *Wernicke's area* in their temporal area ('speech 3'). These authors add the appropriate comment that parietal and temporal speech area may be separate or may in reality represent a continuous region. In addition, since complex motor patterns presumably depend upon the formulation of triggering mnestic sensory patterns, apraxia can likewise be caused by lesions in the parietal association area. Thus, a type of agraphia may result from damage to the dominant angular gyrus.

Amusia, as a result of cerebral damage, can be of the apraxic type (expressive) or can be related to the agnosias (receptive amusia). Various associational regions of the dominant hemisphere may be involved in this rather rare symptom.

In *Gerstmann's syndrome* (GERSTMANN, 1927) the patient is unable to recognize and to select individual fingers when looking at both hands. This finger agnosia is combined with confusion of laterality, agraphia, and occasionally *acalculia* (defect in the use of numerical or mathematical symbols). The syndrome occurs after a parieto-occipital lesion, adjacent to, but mostly above the angular gyrus, and usually on the left side. *Autotopagnosia* (failure to recognize a limb as a part of one's own body) may likewise be caused by damage to the parietal region. Like finger agnosia, it seems to result from a disturbance of the mechanisms providing a postural model of the body (*Körperschema* of SCHILDER, 1923).

As already stated above, the various classifications of aphasic and allied disorders are to some extent controversial and not entirely satisfactory. The terminology is often inadequate and even confusing. Various factors contribute to these shortcomings. The symbolizing activities, either as conscious (parallel) phenomena or as neural mechanisms (unconscious cerebration), are exceedingly complex and difficult to analyze in significant terms. Aphasias, apraxias, and agnosias may be of mixed or combined types. Again, differently localized disturbances can manifest an almost identical symptom, such as agraphia in frontal or in parietal lobe lesions. *Paraphasia*, for instance, might be an apraxic disturbance crudely comparable to a wrong gear shift in the premotor association area, or it might, in other instances, be caused by mnestic disturbances mainly related to the occipito-parieto-temporal region. Again, it can be caused by sphairal emotional effects or by thereto related mechanisms of unconscious cerebration (e.g. *Freudian complexes*).

The verbal analysis of the clinical symptoms results in a nomenclature which is perhaps not sufficiently applicable to the actual performance of the neural circuit mechanisms in symbolic cerebration. Again, some authors stress localization while others present an intricate and occasionally exaggeratedly artificial analysis of functional components.

Aphasic patients are often highly confused and may not manifest easily definable patterns of disturbance; different, even very competent neurologists not infrequently disagree with each other in the interpretation of the actual clinical manifestations.

As regards *organized memory patterns*, Penfield (1952) and Penfield and Jasper (1954) have recorded a large number of observations demonstrating that vivid memories, involving often complete situations, can be evoked by electrical stimulation of the cortex on the superior and lateral surfaces of the temporal lobe of either side. As mapped by Penfield and Jasper, the 'memory areas' extend rather far postero-superiorly into the parietal lobe and also comprise the 'inferior peri-insular' and '*inferior Sylvian bank*'. The memories are often in the nature of actual auditory or visual hallucinations, or both, re-enacting certain experiences from the recent or distant past. These recollections also frequently include the emotional component and the thoughts related to the original experience. The vivid memories or hallucinations ceased as soon as the electrode was withdrawn. Renewed stimulation of an area sometimes reproduced a similar memory or dream, and sometimes evoked a different response. Memory defects were occasionally noticed upon unilateral 'posterior hippocampal removal'. The authors suggest that this rare occurrence is related either to a 'neighborhood effect' or to an unsuspected pre-existing lesion on the contralateral side, resulting in an unexpected bilateral deficit following the operation.

Penfield states that he observed these memory phenomena upon electrical stimulation 'only in the case of patients who have had previous epileptic discharges in the temporal region'; he offers the suggestion 'that the epileptic state renders the cortex more susceptible to stimulation'. Although the region indicated as 'memory area' by the cited authors appears fairly large, the observations under discussion do not necessarily imply that the localization or 'recording' of the revived engrams is restricted to this region. It can certainly be assumed that neural processes in the 'memory area' activate the engrams, and it might be furthermore inferred that at least parts of the circuits pertaining to such engrams are localized within this region. PENFIELD qualifies his own concept of localization by remarking that obviously only 'a portion of the neurological mechanism involved in the recording and the recall of a memory is thus localizable'.

Numerous authors have stressed that memory is a complex function. It involves *pattern symbolization* of a situation or event, permanent or at least *durable recording*, *reproduction of the pattern* by reactivation, and *identification*. All these processes can be performed by physical mechanisms in accordance with the principles of communication engineering and circuit algebra or calculus.

There may thus be one aspect of neural circuit mechanisms that has the symbolic value or code value of recognition. In the course of normally functioning processes, recognition will always be related to an adequate set of events, that is to a set of previously experienced events. Nevertheless, just as in paraphasia, circuit processes will not always properly function. Likewise, in memory activities disturbances can occur, such as the *common memory errors* and others. If identification or recognition activities are falsely triggered, the well-known phenomenon designated as '*déjà vu*' or '*déjà entendu*' results.

Appreciation of over-all spatio-temporal relationships, with *Schilder's Körperschema* as a frame of reference, in other words integration of the different modalities into a unified space-time system, is presumably likewise a complex activity of circuits interrelating, perhaps *via* the diencephalon, the various cortical projection areas with cortical association areas. SPIEGEL *et al.* (1955, 1956) have observed transitory disturbances of temporal orientation in the initial stages after dorsomedial thalamotomy. Such disturbances are also noted in lesions causing diffuse impairment of cerebral function, as e.g. in conditions affecting the parieto-occipital region, or even large areas of the brain stem. Since SPIEGEL and his associates observed rather elementary disturbances of temporal orientation, they avoid the term agnosia, preferably reserved for more complex types of dysfunction, and have chosen the designation *chronotaraxis* (confusion in time) to indicate that temporal disorientation is the most prominent symptom.

The patients were confused regarding date, season, and time of the day, furthermore concerning their age or that of their children. They made serious errors regarding the time elapsed since operation, or the duration of their hospitalization. Some patients tended to overestimate, and others to underestimate the duration of their experiences. Dissociation, as observed by some authors in scopolamine poisoning, was likewise noticed: a patient, two months after thalamotomy, understood perfectly well that she had lived in her present home for five years, but felt as if she had resided there only a few weeks. Another patient reported postoperatively the peculiar feeling 'that the day finished within a few hours'. The disturbances involve thus both '*Zeitgedächtnis*' and '*Zeitsinn*'. The authors state that other memory defects, usually concomitant with the temporal disorientation, might be an important component of that condition, but stress that such memory defects alone seem insufficient to explain the syndrome of chronotaraxis.

Interruption of the circuit connecting prefrontal cortex with nucleus medialis thalami seems to play an important role in the genesis of temporal disturbances, but SPIEGEL and his associates assume that other circuits may also participate in the underlying mechanism. Thus, transitory temporal disorientation after lesion of nucleus medialis thalami reappeared several months later, after a new operation in which a more rostral lesion including the anterior nuclei was added.

Certain aspects of *Korsakoff's psychosis*, or better *syndrome*, appear somewhat related to SPIEGEL's *chronotaraxis*. In *Korsakoff's syndrome* there is, however, a more definite disturbance of attention and memory, leading to disorientation in space and time. Memory gaps are furthermore filled by confabulation. This syndrome does not only occur in chronic alcoholism associated with polyneuritis, but in a variety of other pathologic processes affecting the central nervous system, such as arteriosclerosis or cerebral tumors. SPIEGEL and collaborators point out that in some instances a *Korsakoff syndrome* associated with lesions in the mammillary region might have definite features of chronotaraxis. On the other hand, even severe lesions in the mammillary region do not necessarily induce a *Korsakoff syndrome*. SPIEGEL *et al.* (1955, 1956) reach the conclusion that it does not seem possible to relate the mechanisms involved in temporal orientation to single diencephalic nuclei or their cortical connections. Multiple circuits can be assumed to participate in the mechanism of temporal orientation, so that a lesion of a single thalamic nucleus or circuit produces only transitory disturbances. The authors did not observe chronotaraxis following lesions of other subcortical regions such as pallidum, or midbrain in the area of the ascending spinothalamic system.

Emotional behavior, or, introspectively speaking, *emotional consciousness* may be conceived as related to a highly complex pattern of modulation superimposed upon, interwoven with, or even directly triggering other activities of the hemispheric grisea such as limning and expression processes. Verbal analysis and classification of the emotional component (the accompanying 'feeling') as well as of the limning and expression processes and their components are not altogether satisfactory or adequate. From the neurological viewpoint, cortical, reciprocal cortico-thalamic, cortico-hypothalamic, thalamic, hypothalamic, and reciprocal thalamo-hypothalamic circuit mechanisms appear involved. The results of lobotomy respectively of damage to the prefrontal lobe with regard to its assumed emotional circuit were briefly mentioned above on p. 809.

With respect to the *limbic system*, unilateral surgical removal of temporal lobe including hippocampal formation in man, performed in the treatment of temporal lobe epilepsy, does not seem to cause conspicuous changes in emotional behaviour. In cases of bilateral removal, some results similar to those described by KLÜVER and BUCY in Monkeys have apparently been noticed, and one case reproducing almost exactly that syndrome has been published (TERZIAN and DALLE ORE, 1955).

GLEES and GRIFFITH (1952) have examined the brain of a person suffering from loss of memory and subsequent dementia. Initial agitation was followed by 'a vegetative mode of life', characterized by lack of emotional drive. Finally a series of grand mal seizures occurred. It was found that the hippocampal formation had suffered severe damage and that the fornix fibers were depleted to 24 per cent of the normal number. In addition, gyrus hippocampi and gyrus fusiformis were completely destroyed. The authors reach the tentative conclusion that the hippocampal formation of adult man 'seems to be essential for recent memory and for carrying on normal mental activity'.

In evaluating clinical and pathological evidence of this and similar type it must be kept in mind that almost identical symptoms may result from lesions involving different combinations of grisea, and conversely, that different symptoms may result from damage to identical structures, depending on a variety of additional parameters. NATHAN and SMITH (1950) describe a brain of a mentally normal man ('easygoing, never angry', 'he did not have a temper', 'never grumbling') with bilateral gross abnormalities of gyrus fornicatus and hippocampus. This latter was small and no distinct fornix could be identified. The sense of olfaction was normal. Although, looking at their illustrations, I am somewhat dubious concerning additional details of the anatomical interpretation by the authors, this interesting case would agree with the concepts of multiple circuits, stability and ultrastability. There is, however, according to the illustrations, no doubt about the actual presence of (a topologically distorted) hippocampal and parahippocampal cortex.

Operations on the anterior portion of gyrus cinguli, either as stereotaxic *cingulotomy* (FOLTZ and WHITE, 1962) or as open bilateral cingulectomy (WILSON and CHANG, 1974) are said 'to reduce the intensity of feeling' and thereby also to reduce pain, but the results of these procedures cannot be assessed as sufficiently conclusive.

Finally, with regard to cerebral lesions caused by neoplasms, RIGGS and RUPP (1958), in a clinico-pathologic study of 86 patients with supratentorial glioma, found early and conspicuous disturbances of 'behavior and emotion' in 46 instances. Invasion of grisea pertaining to the limbic system could be demonstrated in 89 per cent of the cases with psychiatric disturbances, while some involvement of these structures was present only in 22 per cent of the cases without such disturbances. Discounting some exceptions related to insufficiently understood parameters (e.g. redundancy and bilaterality), the findings of the cited authors can be interpreted to suggest that the presence or absence of emotional or 'psychiatric' symptoms may be dependent on whether or not the 'emotional circuits' are involved by the neoplastic lesions.

Wernicke's dictum (cf. above p. 824) that '*Geisteskrankheiten sind Gehirnkrankheiten*' seems to be a fully justifiable conclusion. Yet, clearly recognizable and relevant pathologic changes have not been detected with regard to most major and minor psychoses,[264] whose causative factors remain unknown and can be assumed to act at the macromolecular level, essentially involving still insufficiently understood biochemical and synaptic processes. All findings and tentative hypotheses, so far, have remained vague, ambiguous and inconclusive.

---

[264] Cf. the comments on pp. 320–324 in 'Brain and Consciousness' (K., 1957). Of the three major psychoses in KRAEPELIN's nosologic schema, namely (1) schizophrenia, (2) the manic-depressive psychoses, and (3) true paranoia, the latter is a rather rare form, although paranoid symptoms of various degree are quite common in many if not most types of psychoses as well as in many instances of fairly 'normal' emotional thinking.

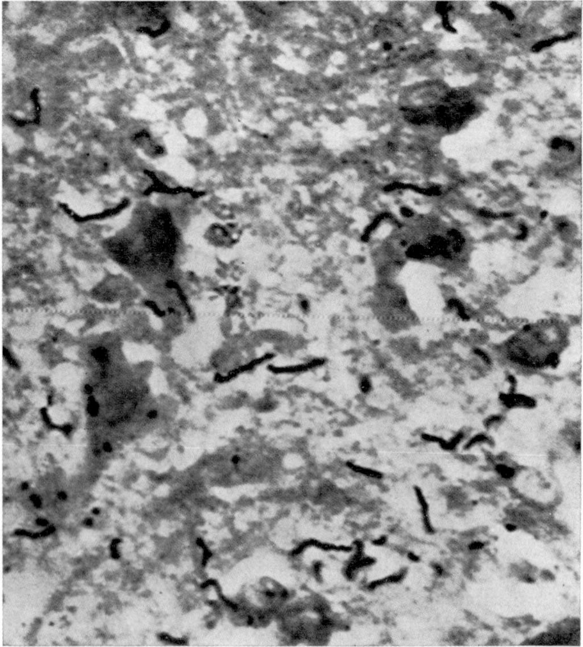

Figure 261. Area of cerebral cortex in general paresis *(Jahnel's silver-pyridine method)*, showing the presence of numerous Treponema organisms (from BIGGART, 1949; $\times$1,000, red. $^1/_1$).

There are, however, miscellaneous psychoses clearly correlated with, and doubtless 'caused' by more or less identifiable disturbances, such as endogenous and exogenous toxins and related substances (alcoholic psychoses, 'psychedelic' and narcotic 'drugs', delirium in infectious diseases, psychoses in endocrine disturbances, etc.). Another subgroup comprises psychoses associated with gross morphologic disturbances or with finer but fairly definable structural histopathologic changes. This entire subgroup comprises psychoses with brain tumor, with cerebral arteriosclerosis or other cerebrovascular conditions, presenile and senile psychoses, psychoses with lues cerebri, dementia paralytica (general paresis), with multiple sclerosis, paralysis agitans, *Huntington's chorea*, *Sydenham's chorea*, with meningitides, various encephalitides, and miscellaneous other conditions of this general type. Depending on details of localization or on unknown variables, some of the enumerated affections (e.g. multiple sclerosis, paralysis agitans, etc.)

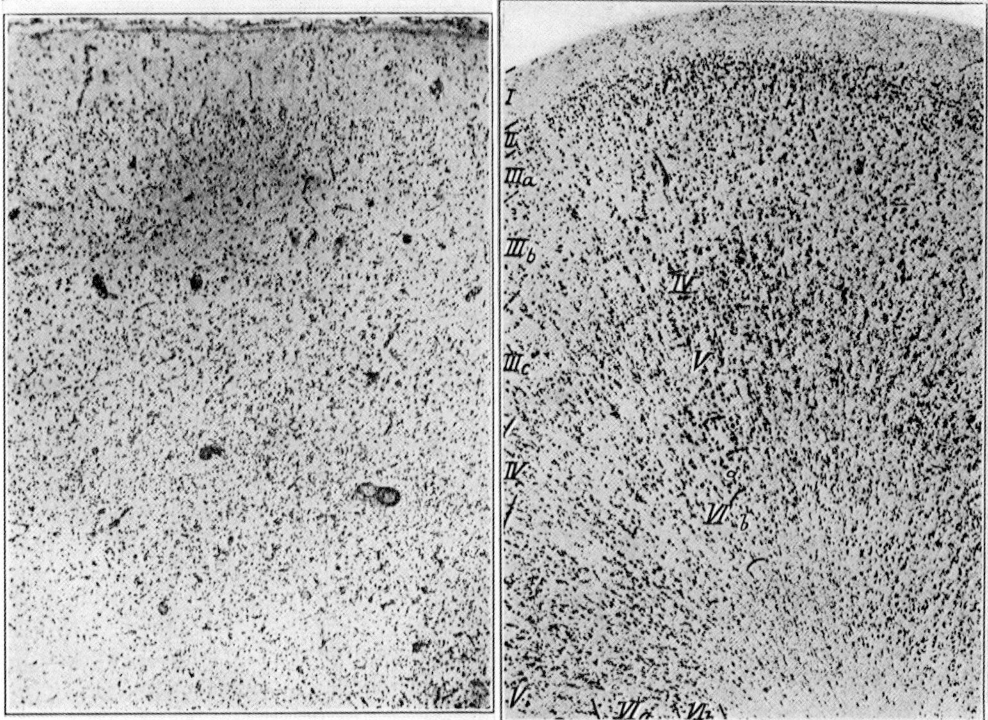

Figure 262. Blurred cytoarchitecture with loss of distinctive lamination in the frontal cortex of a case of general paresis (A) compared with the same area (B) in a normal hemisphere, at transition of premotor to prefrontal cortex. Increase in glia nuclei, and perivascular infiltration is likewise displayed in A (from WEIL, 1945).

may not, despite involvement of the cerebrum, cause 'psychotic' disturbances.

As regards these latter, however, *general paresis* appears of particular theoretical interest, since it represents a major psychosis with well defined substantial histopathologic changes widely diffused throughout the cerebral cortex. General paresis, 'general paralysis of the insane', or dementia paralytica, so named because it was initially regarded as a form of paralysis supervening in persons who had already become insane, is, like tabes dorsalis, a manifestation of metalues, which latter was discussed on pp. 262–264 of chapter VIII in volume 4. It occurs in a small percentage (about 4 to 5 per cent) of untreated luetic infection after a period varying between about 3 to 40 years, and is now a very rare condition,[265] curable, if timely detected, by antibiotic (penicillin) treatment. Males are more liable to general paresis than females approximately in the proportion four to one. Again, in Negroes infected with lues, general paresis occurs much more rarely than in European races.

Although the invasion of the neuroectodermal parenchyma by the Treponema pallidum, first demonstrated about 1913 by NOGUCHI, represents the relevant feature of general paresis (Fig. 261), a conspicuous luetic meningo-encephalitis involving the mesodermal stroma also commonly obtains. Within the brain, iron pigment is found in the vascular wall and the histocytic elements.[266] The microglia cells *(Hortega cells)* become transformed into the typical *rod cells* or *Stäbchenzellen* of NISSL (cf. Figs. 219, 220, p. 330, vol. 3/I). The neuroectodermal ependymal and subependymal ventricular lining usually displays sizeable ependymal granulations characterized by various transformations such as subependymal 'polar spongioblasts' (Figs. 116 A–D, pp. 172–173, vol. 3/I). All types of pathologic nerve cell changes can be seen in the cortex, with randomly retained normal elements in the midst of diseased ones. The laminar cytoarchitectural arrangement of the cortex becomes blurred or even completely unrecognizable (Fig. 262). The basal ganglia and the cerebellum may likewise become affected.

In the aspect here under consideration, an outstanding feature of general paresis is the progressive destruction of the so-called mental faculties, including memory, reasoning, judgment, and manifestations of affectivity. As my teacher HANS BERGER used to say in his lectures and clinicopathologic conferences: '*man kann oft geradezu das allmähliche*

---

[265] It was formerly, when lues was treated with mercury, salvarsan, etc., a statistically important psychosis. Thus, during my activities in Japan, 1924–1927, dementia paralytica ranked, with about 20 per cent, as the third most common type of officially recorded mental diseases (schizophrenia about 36.4 per cent, manic-depressive insanity about 26.4 per cent). It may also be recalled that, before the introduction of antibiotic therapy, a fever therapy by means of inoculation with malaria had been attempted. According to one view, the treponema was supposed to have a low heat resistance, being destroyed at high fever temperatures. According to others, the apparently beneficial effects of fever therapy were due to unknown physiologic and immunologic changes accompanying the fever.

[266] Thus, at the postmortem of dubious cases, an iron reaction displayed along cortical vessels in fresh tissue-spread microscopic preparations allowed an instant diagnosis of general paresis by means of a simple technique *(anatomische Schnelldiagnose der progressiven Paralyse)* introduced about 1921 by SPATZ (1923).

Abbröckeln der Seele beobachten' (one can often observe the gradual crumbling of the soul).

Since, in an apparently random way, manifestations characteristic for manic-depressive, or for schizophrenic psychoses can be displayed, the differential diagnosis, particularly with regard to *schizophrenia*, may initially present substantial difficulties (cf. e.g. also BLEULER, 1930).

In cultured and well-behaved persons, a complete breakdown of the sense of propriety is a frequent symptom and leads to grossly tactless and crudely offensive actions. In addition, the patient becomes slovenly, erratic, his work becomes inefficient and may completely deteriorate. Responsible executives suddenly make utterly absurd major decisions. The affective state is highly labile. Anxiety, irritability and impulsive conduct may lead to acts of violence, and of sexual aberrations. On the other hand, dullness and complete apathy may be manifested. Again, in some forms, hallucinations occur.

The mental instability may also lead to pronounced euphoria combined with moria and to a not uncommon particular so-called *grandiose form* characterized by extreme delusions. The patient claims to be a great inventor or a general with millions of soldiers, ordering to shoot down everybody in sight who refuses to salute him. He claims to be *Napoleon*, king, emperor, pope, God and above God *('Obergott')*,[267] possessing unlimited treasures and power.

The neurologic, in contradistinction to the psychiatric symptoms, can include the different types of apraxia, dysarthria, aphasia as dealt with above,[268] and motor disturbances of convulsive (epileptiform) or of spastic paretic type, these latter particularly in the terminal stage, to

---

[267] In this respect, the following items are perhaps of interest. An uninhibited African sergeant-major, who succeeded in making himself 'general' and dictator of his country, sentenced a British author to be shot because this latter had called him a 'village tyrant' in an unpublished private diary. With much difficulty, said author's release was finally obtained through international public pressure. An American Negro, claiming to be God, called himself *Father Divine* and founded a well-organized sect with a substantial number of black and even white followers. Whether, in either of these cases, merely primitive hyponoic and hypobulic behavior at the lunatic fringe of the 'normal' can be assumed, or whether these cases should be assessed as atypical psychoses (manic ?, schizophrenic ?, paralytic ?) with fairly successful adjustment to a primitive respectively decadent environment, remains a moot question.

[268] Thus, most forms of non-luetic apraxia or dyspraxia, including agraphia, and of expressive and receptive aphasia can be manifested. Echolalia, palilalia, and difficulty in naming objects are common. Since, in general paresis, the involvement of the cerebral

which the term dementia paralytica refers. The bedridden patient, in a state of oligophrenia, unable to move, and incontinent, finally leads a vegetative existence. As in other forms of lues involving the neuraxis, the *Argyll Robertson pupil* (cf. chapter XI, section 9, pp. 990–991, vol. 4) is generally but not always displayed. The patellar reflex may be either absent or increased, but can be normal in some instances. It is evident that the multiform aspects of dementia paralytica, in its untreated forms, run, as it were, a gamut from 'psychotic' and 'neurological' symptoms to complete oligophrenia, that is to say conditions in which all 'higher' mental activities become affected to a degree reaching extinction.

As regards the group of *oligophrenias*, mentally deficient individuals are commonly classified in accordance with the severity of their condition, but regardless of its etiology, as *idiots*, *imbeciles*, and *morons*. The most noticeable defect usually involves the ill-defined group of capacities collectively designated as intelligence. Learning and understanding are particularly impaired, but the emotive reactions performed by the limbic system likewise become affected.

The oligophrenias can be congenital or acquired, e.g. as a result of some of the acute demyelinizing diseases (encephalomyelitides) and of other diseases diffusely involving the cortex and its connections. Such conditions may also be caused by biochemical disturbances, as e.g. in cretinism, in phenylpyruvic idiocy, or in *Tay-Sachs disease*. Prenatal and birth injuries, and developmental anomalies, either caused by hereditary or by secondary disturbing factors, can likewise bring about mental deficiencies. Thus, mental deficiency may be associated with micrencephaly, microgyria, with other gross anatomical defects, with so-called mongolism *(Down's disease)* or with tuberous sclerosis *(Bourneville's disease)*. In other instances, definite anatomical or histologic defects cannot be detected. A general introduction to the topic of morphogenetic and related functional anomalies, including conditions connected with oligophrenia, is given in subsection 1 C, pp. 211–284, chapter VI of volume 3/II.

---

cortex is diffuse, its symptoms are compatible with the diverse localizatory concepts discussed above, but do not support any particular one. It is, moreover, of interest that involvement of the sensory projection areas is manifested by a not uncommon hypalgesia or analgesia, predominantly of the skin. Vision is frequently impaired, but this is in some instances attributable to optic nerve atrophy. Cortical deafness and severe cortical blindness do not seem to be noticeably manifested, except perhaps at the terminal stage of complete oligophrenia, in which their occurrence would remain hidden and rather difficult to detect.

In extreme cases, with reduction or malformation of the cerebral hemispheres and its cortex, such as described by EDINGER and FISCHER (1913), GAMPER (1926) and ourselves (K. *et al.*, 1959, 1964) it is doubtful whether in addition to the remaining behavioral activities there was any occurrence of consciousness at all. These cases, figuratively speaking, can be likened to, and closely correspond with, the results of experiments performed on Mammals, as quoted above (p. 747 ff.), in which telencephalon or cerebral cortex were extirpated. As regards our own, rather exceptional case (vol. 3/II, pp. 259–265) it may be added that the patient died in February 1974, at the age of 19½ years, without any change in condition. The mother, whose peculiar emotional attachment was again pointed out in the just cited case summary, refused autopsy, so that, unfortunately, the brain, whose defect was only roentgenologically definable, did not become available for examination.[269]

In addition to gross laceration of the forebrain in open cranial *trauma*, damage to the cerebrum by *contusion* and *concussion* does occur in 'closed head injuries' with or without fracture of the skull. *Contusion* involves gross bruising on the side of the impact as well as elsewhere by contrecoup, with petechial hemorrhages and edema. There is no general agreement concerning the pathologic changes in 'concussion' and several theories have been suggested (cf. e.g. WARD *et al.*, 1948; HALLERVORDEN and QUADBECK 1957). Minute changes such as thixotropy[270] and microscopic effects of accelerating, decelerating, torsional and shearing motions, inflicting structural damage upon nerve cells and fibers, some of which latter are said to become fragmented, have been assumed by various authors.

Severe head trauma of the aforementioned diverse types is characterized, for a variable duration, by loss of consciousness (cf. e.g. JACOBSON, 1956). This period may be quite short,[271] of a few minutes or

[269] The mother had, nevertheless, reluctantly agreed to a renewed physical and EEG examination of her daughter, meant to supplement our reports of 1959 and 1964, but the patient died before we could complete the arrangements and carry out the intended examination.

[270] *Thixotropy* refers to the property of certain gels of becoming fluid when shaken and subsequently becoming firm again.

[271] Recovery of consciousness can be followed by a period of *post-traumatic amnesia* in which the patient seems to act normally but later does not remember anything of his actions. There may also be *retrograde amnesia* which extends to the accident itself and the period immediately preceding it.

less,[272] or last for an undefined time, up to many months, and even finally ending in death. The protracted deep coma can be followed by spontaneous clearing of consciousness or by semi-conscious stupor and so-called twilight states such as amentia, akinetic mutism, and the 'apathetic syndrome'.

Prolonged unconsciousness and the semi-conscious states might here be, at least in many cases, caused by damage to the cortical fiber systems and the corticothalamic channels. Since relevant cortical activity seems to be eliminated, KRETSCHMER (1940) introduced the term '*apallic syndrome*' for these conditions. Subsequently, ULE *et al.* (1961), GERSTENBRAND (1967), and others have elaborated in detail on KRETSCHMER's apallic syndrome.[273] It seems evident that, in its extreme form, this syndrome is, as far as cortical activities are concerned, functionally comparable if not identical with the condition obtaining in our own, congenital case of 'decorticate' hydranencephaly (K. *et al.*, 1959, 1964), and in other congenital defects such as those studied by EDINGER and FISCHER (1913) and GAMPER (1926). We have here, as it were, congenital 'apallic syndromes'.

As regards the traumatic apallic syndrome, of which I had myself the opportunity to examine some cases as a consultant, the length of survival, respectively the *chance* of improvement or even restitution, substantially depend on the quality and amount of intensive nursing care which is available to the patient.

Irreversible as well as reversible concussion and contusion damage to the brain can also occur in boxing and football activities (transitory '*punch-drunk*', permanent disability, and fatalities.

---

[272] Epidural, subdural, subarachnoid, intracerebral, and intraventricular hemorrhage may occur in closed head injuries with or without cranial fracture. In typical cases of epidural hemorrhage with tearing of arteria meningea media the patient experiences, after a short period of unconsciousness, a 'lucid' or 'latent' interval of one or more hours, followed by progressively increasing symptoms of intracranial pressure, and sinks into coma. Homolateral mydriasis and contralateral convulsions or hemiplegia indicate epidural hemorrhage by involvement of the arteria meningea media. Timely diagnosis and surgical intervention become here essential. Massive subarachnoid as well as intraventricular hemorrhage are usually fatal. Mild forms of subarachnoid hemorrhage merely cause meningism.

[273] It should be added that some authors (e.g. WARD, 1966; BRAIN and WALTON, 1969) stress (or perhaps at times overstress) the significance of the so-called ascending 'alerting' reticular system for the maintenance of consciousness and 'alertness' and emphasize evidence that this system can be reversibly (or irreversibly) blocked by acceleration or effects of concussion respectively contusion. Again, large amounts of acetylcholine have been demonstrated in the cerebrospinal fluid after head injury or concussion.

Severe contusion and concussion, involving not only the cerebrum but also the brain stem with its respiratory and 'activating' grisea may also lead to irreversible coma and abolished cephalic reflexes, followed by so-called *brain death*, characterized by *electrocerebral silence* (ECS), arrest of brain circulation, and respiratory standstill. In such cases, artificial respiration can prolong, for an undefined period, a *vita reducta* which has retained cardiac activity. A detailed study of the so-called 'respirator brain' of patients which died after protracted coma and artificial respiration was recently published by WALKER *et al.* (1975). Such patients die either from spontaneous cardiac arrest or if, on the evidence of 'brain death' the artificial respiration is stopped.

11. References to Chapter XIII

ABBIE, A.A.: The origin of the corpus callosum and the fate of structures related to it. J. comp. Neurol. *70:* 9–44 (1939).

ABRAHAM, F.A.; MELAMED, E., and LAVY, S.: Prognostic value of visual evoked potentials in occipital blindness following basilar artery occlusion. Appl. Neurophysiol. *38:* 126–135 (1975).

ADAMS, A.J. and KING, R.L.: Evoked responses in the chicken telencephalon to auditory, visual, and tactile stimuli. Exp. Neurol. *17:* 498–504 (1967).

ADDENS, J.L. und KUROTSU, T.: Die Pyramidenbahn von Echidna. Proc. kon. Akad. Wetensch. Amsterdam *39:* 1143–1151 (1936).

ADES, H.W. and FELDER, R.: The acoustic area of the monkey. J. Neurophysiol. *5:* 49–54 (1942).

ADRIAN, E.D.: The physical background of perception (Clarendon Press, Oxford 1947).

ADRIAN, E.D.: The sense of smell. Adv. Sci. *4:* 287–292 (1948).

ADRIAN, E.D.: The electrical activity of the mammalian olfactory bulb. Electroenceph. clin. Neurophysiol. *2:* 377–388 (1950).

ADRIAN, E.D.: The mechanism of olfactory stimulation in the mammal. Adv. Sci. *9:* 417–420 (1953).

AGRANOFF, B.W. and APRISON, M.H. (eds.): Advances in neurochemistry (Plenum Press, New York 1975).

AKERT, K.; GRUESEN, R.A.; WOOLSEY, C.N., and MEYER, D.R.: Klüver-Bucy syndrome in monkeys with neocortical ablations of temporal lobe. Brain *84:* 480–498 (1961).

ALLISON, A.C. and WARWICK, R.T.T.: Quantitative observations on the olfactory system of the rabbit. Brain *72:* 186–197 (1949).

ANDY, O.J. and STEPHAN, H.: The septum of the cat (Thomas, Springfield 1964).

ANDY, O.J. and STEPHAN, H.: The septum in the human brain. J. comp. Neurol. *133:* 383–409 (1968).

ANDRES, K.H.: Anatomy and ultrastructure of the olfactory bulb in fish, amphibia, reptiles and mammals; in WOLSTENHOLME and KNIGHT Taste and smell in vertebrates. A Ciba Foundation symposium, p. 177–196 (Churchill, London 1970).

Anton, A.: Über die Selbstwahrnehmung der Herderkrankungen des Gehirns durch die Kranken bei Rindenblindheit und Rindentaubheit. Arch. Psychiat. *32:* 86–127 (1899).

Aronson, L.R.: The central nervous system of sharks and bony fishes with special reference to sensory and integrative mechanisms; in Gilbert Sharks and survival, pp. 165–241 (Heath, Boston 1963).

Aronson, L.R. and Kaplan, H.: Function of the teleostean forebrain; in Ingle The central nervous system and fish behavior, pp. 107–125 (University of Chicago Press, Chicago 1968).

Ashby, W. Ross: Design for a brain; 2nd ed. (Wiley, New York 1960).

Ayers, H.: Vertebrate cephalogenesis. IV. Transformation of the anterior end of the head, resulting in the formation of the 'nose'. J. comp. Neurol. *30:* 323–342 (1919).

Bäckström, K.: Contributions to the forebrain morphology in Selachians. Acta zool. *5:* 123–240 (1924).

Bagley, C. and Richter, C.P.: Electrically excitable region of the forebrain of the alligator. Arch. Neurol. Psychiat. *11:* 257–263 (1924).

Balasubramiam, V. and Kanaka, T.S.: Why hemispherectomy? Appl. Neurophysiol. *38:* 197–205 (1975).

Bargmann, W. and Schadé, J.P. (eds.): The rhinencephalon and related structures. Progress in Brain Research, vol. 3 (Elsevier, Amsterdam 1963).

Baxter, D.W. and Olszewski, J.: Congenital universal insensitivity to pain. Brain *83:* 381–393 (1960).

Beccari, N.: Neurologia comparata anatomo-funzionale dei vertebrati compreso l'uomo (Sansoni, Firenze 1943).

Benson, D.F. and Geschwind, N.: The aphasias and related disturbances; in Baker Clinical neurology; 3rd ed., vol. 1, chapter 8, pp. 1–26 (Hoeber-Harper, New York 1974).

Berger, H.: Psychophysiologie in 12 Vorlesungen (Fischer, Jena 1921).

Berger, H.: Über die Lokalisation im Grosshirn. Jenaer akad. Reden, vol. 4 (Fischer, Jena 1927).

Berke, J.J.: The claustrum, the external capsule and the extreme capsule of Macaca mulatta. J. comp. Neurol. *115:* 297–331 (1960).

Berlucchi, C.: Ricerche di fine anatomia sul claustrum e sull'insula del gatto. Riv. sperim. Freniatria *51:* 3–35 (1927).

Best, C.H. and Taylor, N.B.: The physiological basis of medical practice; 5th ed. (Williams & Wilkins, Baltimore 1950).

Biggart, J.H.: Pathology of the nervous system; 2nd ed. (Livingstone, Edinburgh 1949).

Bindewald, C.A.E.: Das Vorderhirn von Amblystoma mexicanum. Arch. mikr. Anat. *84:* 1–74 (1914).

Black, P. and Meyers, R.E.: Visual function of the forebrain commissures in the chimpanzee. Science *146:* 799–800 (1964).

Bleuler, E.: Lehrbuch der Psychiatrie; 5th Aufl. (Springer, Berlin 1930).

Blinkov, S.M. and Glezer, I.I.: The human brain in figures and tables. A quantitative handbook (Plenum Press, New York 1968).

Bodian, D.: An experimental study of the optic tract and retinal projections in the Virginia opossum. J. comp. Neurol. *66:* 113–144 (1937).

Bone, Q. (1963): cf. References to chapter XII.

Bonin, G. v.: Essay on the cerebral cortex (Thomas, Springfield 1950).

BONIN, G. v.: Some papers on the cerebral cortex. Translated from the French and German (Thomas, Springfield 1960).

BRADY, J.V.: Emotional behavior; in FIELD vol. 3, sect. 1, pp.1529–1552, Handbook of Physiology (Am. Physiol. Soc., Washington 1960).

BRAIN, *Lord:* Speech disorders, aphasia, apraxia, and agnosia; 2nd ed. (Butterworths, Washington 1965).

BRAIN, *Lord* and WALTON, J.N.: *Brain's* diseases of the nervous system; 7th ed. (Oxford University Press, London 1969).

BRAUER, K. and SCHOBER, W.: Katalog der Säugetiergehirne (Fischer VEB. Jena 1970).

BREMER, F.: Cerveau isolé et physiologie du sommeil. C. R. Soc. Biol., Paris *118:* 1235–1242 (1935).

BROADWELL, R.D.: Olfactory relationships of the telencephalon and diencephalon in the rabbit. I. An autoradiographic study of the efferent connections of the main and accessory olfactory bulbs. II. An autoradiographic and horseradish peroxidase study of the efferent connections of the anterior olfactory nucleus. J. comp. Neurol. *163:* 329–345; *164:* 389–409 (1975).

BREATHNACH, A.S.: The olfactory tubercle, prepyriform cortex and precommissural region of the porpoise (Phocaena phocaena). J. Anat. *87:* 96–113 (1953).

BREATHNACH, A.S. and GOLDBY, F.: The amygdaloid nuclei, hippocampus and other parts of the rhinencephalon in the porpoise (Phocaena phocaena). J. Anat. *88:* 267–291 (1954).

BROCA, P.: Remarques sur le siège de la faculté du langage articulé, suivi d'une observation d'aphémie. Bull. Soc. anat., Paris *36:* 330–357 (1861).

BROCA, P.: Anatomie comparée des circonvolutions cérébrales. Le grand lobe limbique et la scissure limbique dans la série des mammifères. Rev. Anthropol. Ser. *2:* 385–498 (1878).

BROCKHAUS, H.: Zur normalen und pathologischen Anatomie des Mandelkerngebietes J. Psychol. Neurol. *49:* 1–136 (1940a).

BROCKHAUS, H.: Die Cyto- und Myeloarchitektonik des Cortex claustralis und des Claustrum beim Menschen. J. Psychol. Neurol. *49:* 249–348 (1940b).

BROCKHAUS, H.: Zur feineren Anatomie des Septum und des Striatum. J. Psychol. Neurol. *51:* 1–56 (1942).

BRODAL, A.: The hippocampus and the sense of smell. A review. Brain *70:* 179–222 (1947a).

BRODAL, A.: The amygdaloid nucleus in the rat. J. comp. Neurol. *87:* 1–16 (1947b).

BRODMANN, K.: Vergleichende Lokalisationslehre der Grosshirnrinde (Barth, Leipzig 1909).

BROMAN, I.: Über die Entwicklung der Geruchsorgane bei den Lungenfischen. Morph. Jb. *83:* 85–106 (1939).

BROOKOVER, C.: The olfactory nerve, the nervus terminalis and the preoptic sympathetic system in Amia calva. J. comp. Neurol. *20:* 49–118 (1910).

BROOKOVER, C.: The development of the olfactory nerve and its associated ganglion in Lepidosteus. J. comp. Neurol. *24:* 113–130 (1914a).

BROOKOVER, C.: The nervus terminalis in adult man. J. comp. Neurol. *24:* 131–135 (1914b).

BROOKOVER, C.: The peripheral distribution of the nervus terminalis in an infant. J. comp. Neurol. *27:* 340–360 (1917).

BROWN, J. W.: Aphasia, apraxia and agnosia (Thomas, Springfield 1972).

BRUNER, A.: Self-stimulation in the rabbit: an anatomical map of stimulation effects. J. comp. Neurol. *131:* 615–629 (1967).

BUDDENBROCK, W. v.: The senses (University of Michigan Press, Ann Arbor 1958).

BUMM, A.: Das Grosshirn der Vögel. Z. wiss. Zool. *38:* 430–467 (1883).

BURNETT, T. C.: Some observations on decerebrate frogs, with especial reference to the formation of associations. Amer. J. Physiol. *30:* 80–87 (1912).

BURNS, B. D.: The uncertain nervous system (Arnold, London 1968).

BURR, H. S.: Regeneration of the brain in Amblystoma. I. The regeneration of the forebrain. J. comp. Neurol. *26:* 203–211 (1916a).

BURR, H. S.: The effects of the removal of the nasal pits in Amblystoma embryos. J. exp. Zool. *20:* 27–57 (1916b).

BURR, H. S.: Some experiments on the transplantation of the olfactory placode in Amblystoma. I. An experimentally produced aberrant cranial nerve. J. comp. Neurol. *37:* 455–479 (1924).

CAIN, W. S. (ed.): Odors: evaluation, utilization, and control. Ann. N. Y. Acad. Sci. *237* (1974).

CAIRNEY, J.: A general survey of the forebrain of Sphenodon punctatum. J. comp. Neurol. *42:* 255–348 (1926).

CAJAL, S. R. Y: Histologie du système nerveux de l'homme et des vertébrés, vol. II (Maloine, Paris 1911).

CALLEJA, C.: La région olfatoria del cerebro (Moya, Madrid 1893).

CAMPBELL, E. H.; KUHLENBECK, H.; CAVENAUGH, R. L., and NIELSEN, A. E.: Clinicopathologic aspects of fatal missile-caused craniocerebral injuries. Chapt. XV, pp. 335–399, in Surgery in World War II, Neurosurgery, vol. I (Dept. of the Army, Washington 1958).

CAREY, J. H.: The nuclear pattern of the telencephalon of the blacksnake, Coluber constrictor constrictor; in HASSLER and STEPHAN Evolution of the forebrain, pp. 73–80 (Thieme, Stuttgart 1966).

CATOIS, E. M.: Recherches sur l'histologie et l'anatomie microscopique de l'encéphale chez les poissons. Bull. scient. France Belg. *36:* 1–166 (1901).

CHAPMAN, W. P.; SCHROEDER, H. R.; GEYER, G.; BRAZIER, M. A. B.; FAGER, C.; POPPEN, J. L.; SOLOMON, H. C., and YAKOVLEV, P. I.: Physiologic evidence concerning importance of the amygdaloid nuclear region in the integration of circulatory function and emotion in man. Science *120:* 949–950 (1954).

Ciba Foundation Study Group No. 20: Functions of the corpus callosum (Little, Brown, Boston 1965).

CLAIRAMBAULT, P.: Le télencéphale de Discoglossus pictus (Oth.). Etude anatomique chez le têtard et chez l'adulte. J. Hirnforsch. *6:* 87:121 (1963/64).

CLAIRAMBAULT, P. et DERER, P.: Contributions à l'étude architectonique du télencéphale des Ranides. J. Hirnforsch. *10:* 123–172 (1968).

CLARA, M.: Das Nervensystem des Menschen; 3. Aufl. (Barth, Leipzig 1959).

CLARK, W. E. LE GROS and WARWICK, R. T. T.: Pattern of olfactory innervation. J. Neurol. Neurosurg. Psychiat. *9:* 101–111 (1946).

CLARK, W. E. LE GROS: Anatomical pattern as the essential basis of sensory discrimination (Blackwell, Oxford 1947).

CLARK, W. E. LE GROS: Projections of the olfactory epithelium on the olfactory bulb: a correction. Nature, Lond. *165:* 452–453 (1950).

CLARK, W.E. LE GROS: The projection of the olfactory epithelium on the olfactory bulb in the rabbit. J. Neurol. Neurosurg. Psychiat. *14:* 1–10 (1951).

COBB, S.: A note on the size of the avian olfactory bulb. Epilepsia *1:* 394–402 (1960a).

COBB, S.: Observations on the comparative anatomy of the avian brain. Perspect. Biol. Med. *3*; 383–408 (1960b).

COHEN, D.H.: Involvement of Avian amygdalar homologue (archistriatum posterior and mediale) in defensively conditioned heart rate change. J. comp. Neurol. *160:* 13–35 (1975.

COHEN, D.H.; DUFF, T.A., and EBBESSON, S.O.E.: Electrophysiological identification of a visual area in shark telencephalon. Science *182:* 492–494 (1973).

COHRS, P.: Das subfornikale Organ des 3. Ventrikels. Nach Untersuchungen bei den Haussäugetieren, einigen Nagern und dem Menschen. Z. Anat. Entwgesch. *105:* 491–518 (1936).

CONEL, J.L.: The development of the brain of Bdellostoma Stouti. II. Internal growth changes. J. comp. Neurol. *52:* 365–499 (1931).

CRAIGIE, E.H.: An introduction to the finer anatomy of the central nervous system based upon that of the albino rat (Blakiston, Philadelphia 1925).

CRAIGIE, E.H.: Observations on the brain of the humming bird (Chrysolampis mosquitus Linn. and Chlorostilbon caribaeus Lawr.) J. comp. Neurol. *45:* 377–481 (1928).

CRAIGIE, E.H.: The vascularity of the cerebral cortex in a specimen of Apteryx. Additional evidence of the presence of a homologue of the mammalian neocortex. Anat. Rec. *43:* 209–214 (1929a).

CRAIGIE, E.H.: The cerebral cortex of Apteryx. Anat. Anz. *68:* 97–105 (1929b).

CRAIGIE, E.H.: Studies on the brain of the kiwi (Apteryx australis). J. comp. Neurol. *49:* 223–357 (1930).

CRAIGIE, E.H.: The cell structure of the cerebral hemisphere of the humming bird. J. comp. Neurol. *56:* 135–168 (1932).

CRAIGIE, E.H.: Multilaminar cortex in the dorsal pallium of the emu, Dromiceius Novae hollandiae. Psych. neurol. Bladen *(Feestbundel Kappers):* 702–711 (1934).

CRAIGIE, E.H.: The cerebral hemispheres of the kiwi and of the emu (Apteryx and Dromiceius). J. Anat. *69:* 380–393 (1935a).

CRAIGIE, E.H.: The hippocampal and parahippocampal cortex of the emu (Dromiceius). J. comp. Neurol. *61:* 563–591 (1935b).

CRAIGIE, E.H.: Some features of the pallium of the cassowary (Casuarius uniappendiculatus). Anat. Anz. *81:* 16–28 (1935c).

CRAIGIE, E.H.: The cerebral cortex of the ostrich (Struthio). J. comp. Neurol. *64:* 389–415 (1936).

CRAIGIE, E.H.: The cerebral cortex of Rhea americana. J. comp. Neurol. *70:* 331–353 (1939).

CRAIGIE, E.H.: The cerebral cortex in some Tinamidae. J. comp. Neurol. *72:* 299–328 (1940a).

CRAIGIE, E.H.: The cerebral cortex in Palaeognathine and Neognathine birds. J. comp. Neurol. *73:* 179–234 (1940b).

CRAIGIE, E.H.: The cerebral cortex of the penguin. J. comp. Neurol. *74:* 353–366 (1941a).

CRAIGIE, E.H.: The capillary bed of the central nervous system of a second genus of Gymnophiona – Siphonops. J. Anat. *76:* 56–64 (1941b).

CRAIK, K.J.W.: The nature of explanation (Cambridge University Press, Cambridge 1943).

CROSBY, E.C.: The forebrain of Alligator mississippiensis. J. comp. Neurol. *27:* 325–402 (1917).

CROSBY, E.C. and HUMPHREY, T.: Studies on the vertebrate telencephalon. I. The nuclear configuration of the olfactory and accessory olfactory formations and of the nucleus olfactorius anterior of certain reptiles, birds, and mammals. J. comp. Neurol. *71:* 121–213 (1939).

CROSBY, E.C. and HUMPHREY, T.: The nuclear pattern of the anterior olfactory nucleus, tuberculum olfactorium and the amygdaloid complex in adult man. J. comp. Neurol. *74:* 309–352 (1941).

CROSBY, E.C.: and HUMPHREY, T.: The amygdaloid complex in the shrew (Blarina brevicauda). J. comp. Neurol. *81:* 285–305 (1944).

CROSBY, E.C.; HUMPHREY, T., and LAUER, E.W.: Correlative anatomy of the nervous system (Macmillan, New York 1962).

CROSBY, E.C.; JONGE, B.R. DE, and SCHNEIDER, R.C.: Evidence for some trends in the phylogenetic development of he vertebrate telencephalon; in HASSLER and STEPHAN. Evolution of the forebrain, pp. 117–135 (Thieme, Stuttgart 1966).

CUÉNOD, M.: Split brain studies. Functional interactions between bilateral central nervous structures; in BOURNE Structure and function of nervous tissue, vol. 5, pp. 455–506 (Academic Press, New York 1972).

CURWEN, A.O.: The telencephalon of Tupinambis migropunctatus. I, II, III. J. comp. Neurol. *66:* 375–404 (1937); *69:* 229–247 (1938); *71:* 613–636 (1939).

DALY, D.D.: Cerebral localization; in BAKER Clinical neurology; 3rd ed., vol. 1, chapter 7, pp. 1–42 (Harper & Row, New York 1974).

DANDY, W.E.: Removal of right cerebral hemisphere for certain tumors with hemiplegia. Preliminary report. J. amer. med. Ass. *90:* 823–825 (1928).

DAVID-WEST, T.S.: and LABZOFFSKY, N.A.: Studies on the site of replication of vesicular stomatitis virus. Arch. ges. Virusforsch. *24:* 30–47 (1968).

DELGADO, J.M.R.: Cerebral structures involved in transmission and elaboration of noxious stimulation. J. Neurophysiol. *18:* 261–275 (1955).

DELGADO, J.M.R.: Social rank and radio-stimulated agressiveness in monkeys. J. nerv. Dis. *144:* 383–390 (1967a).

DELGADO, J.M.R.: Aggression and defense under cerebral radio control; in CLEMENTE and LINDSLEY. Aggression and defense, pp. 171–193 (University of California Press, Berkeley 1967b).

DELGADO, J.M.R.; ROBERTS, W.W., and MILLER, N.E.: Learning motivated by electric stimulation of the brain. Amer. J. Physiol. *179:* 587–593 (1954).

DELGADO, J.M.R.; ROSWOLD, H.E., and LOONEY, E.: Evoking conditioned fear by electrical stimulation of subcortical structures in the monkey brain. J. comp. physiol. Psychol. *48:* 373–380 (1956).

DE LORENZO, A.J.D.: The olfactory neuron and the blood-brain barrier; in WOLSTENHOLME and KNIGHT. Taste and smell in vertebrates, pp. 151–176 (Churchill, London 1970).

DEMSKI, L.S. and KNIGGE, K.M.: The telencephalon of the Bluegill (Lepomis macrochirus): evoked feeding, aggressive and reproductive behavior with representative frontal sections. J. comp. Neurol. *143:* 1–16 (1971).

DENNLER, G.: Zur Morphologie des Vorderhirns der Vögel. Der Sagittalwulst. Folia neurobiol. *12:* 343–362 (1922).

DEXLER, H.: Zur Anatomie des Zentralnervensystems von Elephas indicus. Arb. neurol. Inst. Univers. Wien. *15:* 137–281 (1907).

DIMOND, S.J.: Depletion of attentional capacity after total commissurotomy in man. Brain *99:* 347–356 (1976).

DIMOND, S.J. and BEAUMONT, J.G. (eds.): Hemisphere function in the human brain (Wiley, New York 1974).

DØVING, K.B.: Experiments in olfaction; in WOLSTENHOLME and KNIGHT. Taste and smell in vertebrates, pp. 197–225 (Churchill, London 1970).

DØVING, K.B.: Odorant properties correlated with physiological data. Ann. N. Y. Acad. Sci. *237:* 184–192 (1974).

DRESEL, K.: Die Funktionen eines grosshirn- und striatumlosen Hundes. Klin. Wschr. *3:* 2231–2233 (1924).

DURWARD, A.: The cell masses in the forebrain of Sphenodon punctatum. J. Anat. *65:* 8–44 (1930).

DURWARD, A.: Observations on the cell masses in the cerebral hemisphere of the New Zealand kiwi (Apteryx australis). J. Anat. *66:* 437–477 (1932).

DURWARD, A.: Some observations on the development of the corpus striatum in birds, with special reference to certain stages in the common sparrow (Passer domesticus) J. Anat. *62:* 492–499 (1934).

DUSSER DE BARENNE, J.G.: Recherches expérimentales sur les fonctions du système nerveux central, faites en particulier sur deux chats dont le néopallium avait été enlevé. Arch. Physiol. *4:* 31–123 (1919–1920).

EARNEST, E.: *S. Weir Mitchell*, novelist and physician (University of Pennsylvania Press, Philadelphia 1950).

EBBESSON, S.O.E. and VONEIDA, T.J.: The cytoarchitecture of the pallium in the tegu lizard. Brain Behav. Evol. *2:* 431–466 (1969).

EBBINGHAUS, H.: Abriss der Psychologie (Veit, Leipzig 1919).

EBNER, F.F. and COLONNIER, M.: Synaptic patterns in the visual cortex of turtle: an electron microscopic study. J. comp. Neurol. *160:* 51–79 (1975).

EDINGER, F.: Die Leistungen des Zentralnervensystems beim Frosch, dargestellt mit Rücksicht auf die Lebensweise des Tieres; Inauguraldiss. Heidelberg (1912).

EDINGER, F.: Leistungen des Zentralnervensystems beim Frosch (Sammelreferat). Z. allg. Physiol. *15:* 15–64 (1913).

EDINGER, L.: Untersuchungen über die vergleichende Anatomie des Gehirns. 3. Neue Studien über das Vorderhirn der Reptilien. Abh. Senckenberg. naturf. Ges. *19:* 313–386 (1896a).

EDINGER, L.: Vorlesungen über den Bau der nervösen Zentralorgane; 5. Aufl. (Vogel, Leipzig 1896b).

EDINGER, L.: Über das Gehirn von Myxine glutinosa. Abh. preuss. Akad. Wiss., Anhang *1906:* 1–36 (1906).

EDINGER, L.: Vorlesungen über den Bau der nervösen Zentralorgane des Menschen und der Tiere, vol. I: Das Zentralnervensystem des Menschen und der Säugetiere; 8. Aufl., 1911; vol. II: Vergleichende Anatomie des Gehirns; 7. Aufl., 1908 (Vogel, Leipzig 1908–1911).

EDINGER, L.: Einführung in die Lehre vom Bau und den Verrichtungen des Nervensystems (Vogel, Leipzig 1912).

EDINGER, L. und FISCHER, B.: Ein Mensch ohne Grosshirn. Arch. ges. Physiol. *152:* 1–27 (1913).

EDINGER, L. und WALLENBERG, A.: Untersuchungen über das Gehirn der Tauben. Anat. Anz. *15:* 245–271 (1899).

EDINGER, L.; WALLENBERG, A. und HOLMES, G.: Untersuchungen über die vergleichende Anatomie des Gehirns. 5. Das Vorderhirn der Vögel. Abh. Senckenberg. naturf. Ges. *20:* 343–426 (1903).

ELLENBERGER, W. und BAUM, H.: Handbuch der vergleichenden Anatomie der Haustiere (Springer, Berlin 1926).

ETTLINGER, E.G. (ed.): Functions of the corpus callosum (Little, Brown, Boston 1965).

FAHRENHOLZ, C.: Über die Entwicklung des Gesichts und der Nase bei der Geburtshelferkröte (Alytes obstetricans). Morph. Jb. *54:* 421–503 (1925).

FAUCETTE, J.R.: The olfactory bulb and medial hemisphere wall of the rat-fish, Chimaera. The accessory olfactory bulbs and the lateral telencephalic wall of the rat-fish, Chimaera. J. comp. Neurol. *137:* 377–403, 407–431 (1969).

FAUL, J.: The comparative ontogenetic development of the corpus striatum in Reptiles. Proc. kon. Akad. Wetensch. Amsterdam *29:* 150–162 (1926).

FERRIER, D.: Experimental researches in cerebral physiology and pathology. West Riding lunatic Asylum med. Rep. *3:* 1–50 (1873).

FEUCHTWANGER, E.: Die Funktionen des Stirnhirns (Springer, Berlin 1923).

FIEDLER, K.: Verhaltenswirksame Strukturen im Fischgehirn. Verh. Dt. zool. Ges., Heidelberg 1967. Zool. Anz., suppl. 31, pp. 602–616 (1968).

FILIMONOFF, I.N.: Homologies of the cerebral formations of mammals and reptiles. J. Hirnforsch. *7:* 229–251 (1964/65).

FILIMONOFF, I.N.: The claustrum, its origin and development. J. Hirnforsch. *8:* 503–528 (1965/66).

FINGER, T.E.: The distribution of the olfactory tracts in the bullhead catfish Ictalmus nebulosus. J. comp. Neurol *161:* 125–141 (1975).

FLECHSIG, P.: Gehirn und Seele; 2. Aufl. (Veit, Leipzig 1896).

FLECHSIG, P.: Anatomie des menschlichen Gehirns und Rückenmarks auf myelogenetischer Grundlage (Thieme, Leipzig 1920).

FLECHSIG, P.: Meine myelogenetische Hirnlehre (Springer, Berlin 1927).

FLOURENS, M.J.P.: Recherches expérimentales sur les propriétés et les fonctions du système nerveux dans les animaux vertébrés (Crevot, Paris 1824).

FLOURENS, M.J.P.: Recherches expérimentales sur les propriétés et les fonctions du système nerveux (Baillière, Paris 1852).

FOLTZ, E.L. and WHITE, L.E.: Pain 'relief' by frontal cingulotomy. J. Neurosurg. *19:* 89–100 (1962).

FOREL, A.: Gesammelte hirnanatomische Abhandlungen (Reinhardt, München 1907).

FOREL, A.: Gehirn und Seele; 13. Aufl. (Kroner, Leipzig 1922).

FOREL, A.: Rückblick auf mein Leben (Europa-Verlag, Zürich 1935).

FORTUYN, J.D. and STEFENS, R.: On the anatomical relations of the intralaminar and midline cells of the thalamus. Electroenceph. clin. Neurophysiol. *3:* 393–400 (1951).

FREDERIKSE, A.: The lizard's brain. Acad. Proefschr. Amsterdam (Callenbach, Nijkerk 1931).

FRITSCH, G.: Untersuchungen über den feinen Bau des Fischgehirns. Mit besonderer Berücksichtigung der Homologien bei anderen Wirbeltieren (Gutmann, Berlin 1878).

FRITSCH, G. und HITZIG, E.: Über die elektrische Erregbarkeit der Grosshirnrinde. Arch. Anat., Physiol. wiss. Med. *1870:* 300–332 (1870).

FULTON, J.F.: Physiology of the nervous system; 3rd ed. (Oxford University Press, New York 1949).

GAGE, S.P.: The brain of Diemyctilus viridescens from larval to adult life and comparison with the brain of Amia and of Petromyzon, pp. 259–314 (Wilder Quarter Century Book, Ithaca 1893).

GAINOTTI, G.; CALTAGIRONE, C.; LEMMO, M.A., and MICELI, G.: Pattern of ipsilateral clinical extinction in brain-damaged patients. Appl. Neurophysiol. *38:* 115–125 (1975).

GAMBLE, H.J.: An experimental study of the secondary olfactory connexions in Lacerta viridis. J. Anat. *86:* 180–196 (1952).

GAMBLE, H.J.: An experimental study of the secondary olfactory connexions in Testudo graeca. J. Anat. *90:* 15–29 (1956).

GAMPER, E.: Bau und Leistungen eines menschlichen Mittelhirnwesens (Arhinencephalie mit Encephalocele). Zugleich ein Beitrag zur Teratologie und Fasersystematik. Z. ges. Neurol. Psychiat. *102:* 154–235; *104:* 49–120 (1926).

GANSER, L.: Vergleichend-anatomische Studien über das Gehirn des Maulwurfs. Morph. Jb. *7:* 591–725 (1882).

GARCÍA SANTOS, J: Aportación a la organización telencefálica de la rata blanca. Anales Anat. *24:* 79–107 (1975).

GAWRILENKO, A.: Die Entwicklung des Geruchsorgans bei Salmo salar (Zur Stammesentwicklung des Jacobsonschen Organs. Anat. Anz. *36:* 411–427 (1910).

GAZZANIGA, M.S.: Psychologic properties of the disconnected hemispheres in man. Science *150:* 372 (1965).

GAZZANIGA, M.S.: The bisected brain (Appleton-Centura-Crofts, New York 1970).

GEHUCHTEN, A. VAN: Anatomie du système nerveux de l'homme; 1st ed., 4th ed. (Uystpruyst-Dieudonné, Louvain, 1897, 1906).

GERLACH, J.: Über das Gehirn von Protopterus annectens. Ein Beitrag zur Morphologie des Dipnoerhirnes. Anat. Anz. *75:* 310–406 (1933).

GERLACH, J.: Beiträge zur vergleichenden Morphologie des Selachierhirnes. Anat. Anz. *96:* 79–165 (1947).

GERLACH, J.; JENSEN, H.P.; KOOS, W. und KRAUS, H.: Pädiatrische Neurochirurgie (Thieme, Leipzig 1967).

GERSTENBRAND, E.: Das traumatische apallische Syndrom (Springer, Wien 1967).

GERSTMANN, J.: Fingeragnosie und isolierte Agraphie – ein neues Syndrom. Z. ges. Neurol. Psychiat. *108:* 152–177 (1927).

GESCHWIND, N.: The organization of language and the brain. Science *170:* 940–944 (1970).

GILBERT, P.W. (ed.): Sharks and survival (Heath, Boston 1963).

GLEES, P.: Morphologie und Physiologie des Nervensystems (Thieme, Stuttgart 1957).

GLEES, P. and GRIFFITH, H.B.: Bilateral destruction of hippocampus (cornu ammonis) in a case of dementia. Mthl. Rev. Psychiat. Neurol. *123:* 193–204 (1952).

GLOOR, P.: Electrophysiological studies on the connections of the nucleus amygdalae in the cat. I, II. Electroenceph. clin. Neurophysiol. *7:* 223–242; 243–264 (1955).

GLOOR, P.: Amygdala; in FIELD *et al.* Handbook of physiology, Sect. 1, Neurophysiology, vol. 2, pp. 1395–1420 (Am. Physiol. Soc., Washington 1960).

GOLDBY, F.: The cerebral hemispheres of Lacerta viridis. J. Anat. *68:* 157–215 (1934).

GOLDBY, F.: An experimental investigation of the cerebral hemispheres of Lacerta viridis. J. Anat. *71:* 332–355 (1937).

Goldstein, K.: Untersuchungen über das Vorderhirn und Zwischenhirn einiger Knochenfische. Arch. mikr. Anat. *66:* 135–219 (1905).

Goldstein, K.: Language and language disturbances (Grune & Stratton, New York 1948).

Golgi, C.: Untersuchungen über den feineren Bau der zentralen und peripheren Nervensystems. Transl. by Teuscher, R. (Fischer, Jena 1894).

Goltz, F.: Über die Verrichtungen des Grosshirns. Pflügers Arch. ges. Physiol. *42:* 419–467 (1888).

Goltz, F.: Der Hund ohne Grosshirn. Pflügers Arch. ges. Physiol. *51:* 570–613 (1892).

Goodman, I.J. and Schein, M.W. (eds.): Birds: brain and behavior (Academic Press, New York 1974).

Gray, P.A.: The cortical lamination pattern of the opossum, Didelphys virginiana. J. comp. Neurol. *27:* 221–263 (1924).

Gray, P.A. and Turner, E.L.: The motor cortex of the opossum. J. comp. Neurol. *36:* 375–385 (1924).

Greenberg, J.H.: Essays in linguistics; 3rd ed. (University of Chicago Press, Chicago 1961).

Greenberg, J.H. (ed.): Universals of language (MIT Press, Cambridge 1963).

Greenberg, J.H.: Language universals: a research frontier. Science *166:* 473–478 (1969).

Grey Walter, W.: The living brain (Norton, New York 1953).

Grünbaum, A.S.F. and Sherrington, C.S.: Observations on the physiology of the cerebral cortex in some of the higher apes. Proc. roy. Soc. *69:* 206–209 (1901).

Haefelfinger, H.R.: Beiträge zur vergleichenden Ontogenese des Vorderhirns bei Vögeln (Helbing & Lichtenhahn, Basel 1958).

Hafferl, A.: Ein Beitrag zur Kenntnis der ontogenetischen Entwicklung des Prosencephalon bei Scyllium canicula. Z. Anat. Entwgesch. *73:* 395–412 (1926).

Hahn, O. und Kuhlenbeck, H.: Defektbildungen des Septum pellucidum im Enzephalogramm. Fortschr. Röntgenstr. *41:* 737–742 (1930).

Hainer, R.M.; Emslie, A.G., and Jacobson, A.: An information theory of olfaction. Ann. N. Y. Acad. Sci. *58* (Art. 2)*:* 158–174 (1954).

Hajdu, F.; Hassler, R., and Back, I.J.: Electron microscopic study of the substantia nigra and the strio-nigral projection in the rat. Z. Zellforsch. *146:* 207–221 (1973).

Hall, W.C. and Ebner, F.F.: Thalamotelencephalic projections in the turtle (Pseudemys scripta). J. comp. Neurol. *140:* 101–122 (1970).

Haller, B.: Vom Bau des Wirbeltiergehirns. I. Salmo und Scyllium. Morph. Jb. *26:* 345–641 (1898).

Haller, Graf V.: Die epithelialen Gebilde am Gehirn der Wirbeltiere. Z. Anat. Entwgesch. *64:* 118–202 (1922).

Hallervorden, J. und Quadbeck, G.: Die Hirnerschütterung und ihre Wirkung auf das Gehirn. Dtsch. med. Wschr. *82:* 129–134 (1957).

Halstead, W.C.: Brain and intelligence (University of Chicago Press, Chicago 1947).

Halstead, W.C.: Specialization of behavioral functions and the frontal lobes. Res. Publs Ass. Res. nerv. ment. Dis. *27:* 59–66 (1948).

Hassler, R.: Forel's Haubenfaszikel als verstibuläre Empfindungsbahn mit Bemerkungen über einige andere sekundäre Bahnen des Vestibularis und Trigeminus. Arch. Psychiat. Z. Neurol. *180:* 23–53 (1948).

HASSLER, R.: Afferente Systeme. Über die Zweiteilung der Schmerzleitung in die Systeme der Schmerzempfindung und des Schmerzgefühls; in JANZEN *et al.* Schmerz. Grundlagen, Pharmakologie, Therapie pp. 105–120 (Thieme, Stuttgart 1972).

HAYMAKER, W. and ANDERSON, E.: Disorders of the hypothalamus and pituitary gland; in BAKER Clinical neurology; 3rd ed., vol. 2, chapter 28, pp. 1–78 (Hoeber, Harper & Row, New York 1971).

HAYMAKER, W. and BING, R.: *Bing's* local diagnosis in neurological diseases; 14th ed., 15th ed. (Mosby, St. Louis 1956, 1969).

HEAD, H.: Studies in neurology (Frowde, London 1920).

HEAD, H.: Aphasia and kindred diseases of speech (Cambridge University Press, London 1926).

HEIER, P.: Fundamental principles in the structure of the brain. A study of the brain of Petromyzon fluviatilis (Ohlsson, Lund 1948); also Acta anat., suppl. 6 (1948).

HENNING, H.: Der Geruch. Z. Psychol. *73:* 161–257; *74:* 305–434; *75:* 177–230; *76:* 1–127 (1915–1916).

HENSCHEN, S.E.: La projection de la rétine sur la corticalité calcarine. Semaine méd. *22:* 125–127 (1903).

HENSCHEN, S.E.: 40-jähriger Kampf um das Sehzentrum und seine Bedeutung in der Hirnforschung. Z. ges. Neurol. Psychiat. *87:* 505–535 (1923).

HERMAN, W.: The relations of the corpus striatum and the pallium in Varanus and a discussion of their bearing on birds, mammals and man. Brain *48:* 362–379 (1925).

HERRICK, C.J.: A sketch of the origin of the cerebral hemispheres. J. comp. Neurol. *32:* 429–454 (1921a).

HERRICK, C.J.: The connections of the vomeronasal nerve, accessory bulb and amygdala in amphibia and reptilia. J. compt. Neurol. *20:* 413–547 (1921b).

HERRICK, C.J.: Functional factors in the morphology of the forebrain of fishes. Libro en honor de D. Santiago Ramón y Cajal (Madrid) *I:* 143–202 (1922).

HERRICK, C.J.: The nucleus olfactorius anterior of the opossum. J. comp. Neurol. *37:* 317–359 (1924a).

HERRICK, C.J.: The amphibian forebrain. II. The olfactory bulb of Amblystoma. J. comp. Neurol. *37:* 373–396 (1924b).

HERRICK, C.J.: Brains of rats and men (University of Chicago Press, Chicago 126).

HERRICK, C.J.: An introduction to neurology; 5th ed. (Saunders, Philadelphia 1931).

HERRICK, C.J.: The functions of the olfactory parts of the cerebral cortex. Proc. nat. Acad. Sci., Wash. *19:* 7–14 (1933).

HERRICK, C.J.: The brain of the tiger salamander, Ambystoma tigrinum (University of Chicago Press, Chicago 1948).

HERRICK, C.L.: The commissures and histology of the Teleost brain. Anat. Anz. *6:* 676–681 (1891).

HERTWIG, R.: Lehrbuch der Zoologie; 10. Aufl. (Fischer, Jena 1912).

HESS, A.: Post-natal development and maturation of the nerve fibers of the central nervous system. J. comp. Neurol. *100:* 461–480 (1954).

HILPERT, P.: Der Mandelkern des Menschen. I. Cytoarchitektonik und Faserverbindungen. J. Psychol. Neurol. *36:* 44–74 (1928).

HINDE, R.A.: Animal behaviour; a synthesis of ethology and comparative psychology (McGraw-Hill, New York 1966).

HINDS, J.W.: Early neuron differentiation in the Mouse olfactory bulb. I. Light microscopy. II. Electron microscopy. J. comp. Neurol. *146:* 233–276 (1972).

HINES, M.: The development of the telencephalon in Sphenodon punctatum. J. comp. Neurol. *35:* 483–537 (1923).

HINES, M.: The brain of Ornithorhynchus anatinus. Philos. Trans. roy. Soc., London, Sci. B. *217:* 155–287 (1929).

HOCHSTETTER, F.: Beiträge zur Entwicklungsgeschichte des menschlichen Gehirns. I. Teil (Deuticke, Wien 1919).

HOCHSTETTER, F.: Über das Cavum septi pellucidi. Gegenbaurs morph. Jb. *75:* 269–295 (1935).

HOCHSTETTER, F.: Das Cavum septi pellucidi des Kaninchens. Gegenbaurs morph. Jb. *86:* 498–503 (1941).

HODOS, W.; KARTEN, H.J., and BONBRIGHT, J.C., JR.: Visual intensity and pattern discrimination after lesions of the thalamafugal visual pathway in pigeons. J. comp. Neurol. *148:* 447–467 (1973).

HOFFMAN, H.H.: The hippocampal and septal formations in Anurans; in HASSLER and STEPHAN. Evolution of the forebrain, pp. 61–72 (Thieme, Stuttgart 1966).

HOLMGREN, N.: Zur Anatomie und Histologie des Vorder- und Zwischenhirns der Knochenfische. Acta zool. *1:* 137–315 (1920).

HOLMGREN, N.: Points of view concerning forebrain morphology in lower vertebrates. J. comp. Neurol. *34:* 391–459 (1922).

HOLMGREN, N.: On two embryos of Myxine glutinosa. Acta zool. *27:* 1–90 (1946).

HOLMGREN, N. and HORST, C.J., VAN DER: Contributions to the morphology of the brain in Ceratodus. Acta zool. *6:* 59–165 (1925).

HOLST, E. VON und SAINT PAUL, U. VON: Vom Wirkungsgefüge der Triebe. Naturwissenschaften *47:* 409–422 (1960).

HOLST, E. VON und SAINT PAUL, U. VON: Electrically controlled behavior. Scient. Amer. *206/3:* 50–59 (1962).

HOLST, E. VON and SAINT PAUL, U. VON: On the functional organization of drives. Anim. Behav. *11:* 1–20 (1963).

HOOGENBOOM, K.J. HOCKE: Das Gehirn von Polyodon folium Lacep. Z. mikr. Anat. Forsch. *18:* 311–392 (1929).

HOPKINS, A.E.: The olfactory receptors in vertebrates. J. comp. Neurol. *41:* 253–289 (1926).

HÖRMANN, H.: Psychologie der Sprache (Springer, Berlin 1969).

HORST, C.J., VAN DER: The forebrain of Synbranchidae. Proc. kon. Akad. Wetensch. Amsterdam *20:* 216–228 (1917).

HUBEL, D.H. and WIESEL, T.N.: Receptor fields, binocular interaction and functional architecture in the cat's visual cortex. J. Physiol., Lond. *160:* 106–154 (1962).

HUBER, G.C. and CROSBY, E.C.: The nuclei and fiber paths of the avian diencephalon, with consideration of telencephalic and certain mesencephalic centers and connections. J. comp. Neurol. *48:* 1–225 (1929).

HUBER, G.C. and GUILD, S.R.: Observations on the peripheral distribution of the nervus terminalis in mammals. Anat. Rec. *7:* 253–272 (1913).

HUMPHREY, T.: The telencephalon of the bat. J. comp. Neurol. *65:* 603–711 (1936).

HUMPHREY, T.: The development of the human amygdala during early embryonic life. J. comp. Neurol. *132:* 135–165 (1968).

HUNTER, J.L.: The forebrain of Apteryx australis. Proc. kon. Akad. Wetensch. Amsterdam *26:* 807–824 (1923/24).

INGLE, D. (ed.): The central nervous system and fish behavior (University of Chicago Press, Chicago 1968).

ISSAJEW, P.O.: Ein Fall der Abwesenheit des N. olfactorius. Anat. Anz. *74:* 398–400 (1932).

JACKSON, H.: Selected writings, 2 vols., edited by TAYLOR (Hodder & Stoughton, London 1931, 1932).

JACOBSON, S.A.: Protracted unconsciousness due to closed head injury. Neurology *6:* 281–287 (1956).

JAKOB, Ch.: La filogenia cortical. Sobre la corteza cerebral de gimnofiones y amfisbenas Argentinas. 1° Congreso Argentino de Medicina, Buenos Aires, vol. 4, pp. 81–88 (1918).

JAKOB, Ch. and ONELLI, C.: Vom Tierhirn zum Menschenhirn. Vergleichende morphologische, histologische und biologische Studien zur Entwicklung der Grosshirnhemisphären und ihrer Rinde (Lehmann, München 1911).

JANSEN, J.: The brain of Myxine glutinosa. J. comp. Neurol. *49:* 359–507 (1930).

JANSEN, J. and JANSEN, J.K.S.: The nervous system of Cetacea; in Biology of marine mammals, chapter 7, pp. 176–252 (Academic Press, New York 1968).

JANSSEN, P. et STEPHAN, H.: Recherches sur le cerveau de l'éléphant d'Afrique (Loxodonta africana Blum). Acta neurol. psychiat. belg. *11:* 731–757 (1956).

JEEVES, M.A.: Psychological studies of three cases of congenital agenesis of the corpus callosum; in *Ciba Foundation Study Group* No. 20, Function of the corpus callosum, pp. 73–94 (Little, Brown, Boston 1965).

JELGERSMA, G.: Atlas anatomicum cerebri humani (Scheltema & Holkema, Amsterdam n.d.).

JOHNSTON, J.B.: The olfactory lobes, forebrain, and habenular tracts of Acipenser. Zool. Bull. *1:* 221–241 (1898).

JOHNSTON, J.B.: The brain of Acipenser. Zool. Jb. Abt. Anat. Ontog. *15:* 59–260 (1901).

JOHNSTON, J.B.: The brain of Petromyzon. J. comp. Neurol. *12:* 1–87 (1902).

JOHNSTON, J.B.: The nervous system of vertebrate (Blakiston, Philadelphia 1906).

JOHNSTON, J.B.: The telencephalon of selachians. J. comp. Neurol. *21:* 1–113 (1911a).

JOHNSTON, J.B.: The telencephalon of ganoids and teleosts. J. comp. Neurol. *21:* 489–591 (1911b).

JOHNSTON, J.B.: The telencephalon in cyclostomes. J. comp. Neurol. *22:* 341–404 (1912).

JOHNSTON, J.B.: Nervus terminalis in reptiles and mammals. J. comp. Neurol. *23:* 97–120 (1913a)

JOHNSTON, J.B.: The morphology of the septum, hippocampus and pallial commissures in reptiles and mammals. J. comp. Neurol. *23:* 371–498 (1913b).

JOHNSTON, J.B.: The nervus terminalis in man and mammals. Anat. Rec. *8:* 185–198 (1914).

JOHNSTON, J.B.: The cell masses in the telencephalon of the turtle, Cistudo carolina. J. comp. Neurol. Neurol. *25:* 393–468 (1915).

JOHNSTON, J.B.: Evidence of a motor pallium in the forebrain of reptiles. J. comp. Neurol. *26:* 475–479 (1916a).

JOHNSTON, J.B.: The development of the dorsal ventricular ridge in turtles. J. comp. Neurol. *26:* 481–505 (1916b).

JOHNSTON, J.B.: Further contributions to the study of the evolution of the brain. Parts I–IV. J. comp. Neurol. *35:* 337–481. Part V. J. comp. Neurol. *36:* 143–192 (1923).

JONES, A. W. and LEVI-MONTALCINI, R.: Patterns of differentiation of the nerve centers and fiber tracts in the avian cerebral hemispheres. Arch. ital. Biol. *96:* 231–284 (1958).

JONES, E. G.; BURTON, H.; SAPER, C. B., and SWANSON, L. W.: Midbrain, diencephalic and cortical relationships of the basal nucleus of Meynert and associated structures in primates. J. comp. Neurol. *167:* 385–419 (1976).

KAHLE, W.: Die Entwicklung der menschlichen Grosshirnhemisphären. Schriftenreihe Neurologie, vol. 1 (Springer, Berlin 1969).

Kalischer, O.: Über Grosshirnexstirpation bei Papageien. Sitz. Ber. preuss. Akad. Wiss. phys.-math. Kl. *1900:* 722–726 (1900).

KALISCHER, O.: Weitere Mitteilung zur Grosshirnlokalisation bei den Vögeln. Sitz. Ber. preuss. Akad. Wiss. phys.-math. Kl. *1901:* 428–439 (1901).

KALISCHER, O.: Das Grosshirn des Papageien in anatomischer und physiologischer Beziehung. Abh. preuss. Akad. Wiss. *1905/4:* 1–105 (1905).

KÄLLEN, B.: On the nuclear differentiation during ontogenesis in the avian forebrain. Acta anat. *17:* 72–84 (1953).

KÄLLEN, B.: Embryogenesis of brain nuclei in the chick telencephalon. Erg. Anat. EntwGesch. *36:* 62–82 (1962).

KAPPERS, C. U. A.: The structures of the teleostean and selachian brain. J. comp. Neurol. *16:* 1–112 (1906).

KAPPERS, C. U. A.: Eversion and inversion of the dorso-lateral wall in different parts of the brain. J. comp. Neurol. *18:* 433–436 (1908).

KAPPERS, C. U. A.: Die Furchen am Vorderhirn einiger Teleostier. Nebst Diskussion über den allgemeinen Bauplan des Vertebratenhirns und dessen Kommissursysteme. Anat. Anz. *40:* 1–18 (1911).

KAPPERS, C. U. A.: Die vergleichende Anatomie des Nervensystems der Wirbeltiere und des Menschen. II. Abschnitt, vol. 2 (Bohn, Haarlem 1921).

KAPPERS, C. U. A.: The ontogenetic development of the corpus striatum in birds and a comparison with mammals and man. Proc. kon. Akad. Wetensch. Amsterdam *26:* 135–158 (1922).

KAPPERS, C. U. A.: Three lectures on neurobiotaxis and other subjects (Levin & Munksgaard, Copenhagen 1928).

KAPPERS, C. U. A.: Anatomie comparée du système nerveux, particulièrement de celui des mammifères et de l'homme. Avec la collaboration de E. H. STRASBURGER (Masson, Paris 1947).

KAPPERS, C. U. A. und HAMMER, E.: Das Zentralnervensystem des Ochsenfrosches (Rana catesbyana). Psych. neurol. Bladen (Fesstb. Winkler) *1918:* 368–415 (1918).

KAPPERS, C. U. A.; HUBER, G. C., and CROSBY, E. G.: The comparative anatomy of the nervous system of vertebrates, including man (Macmillan, New York 1936).

KAPPERS, J. A.: The development of the paraphysis cerebri in man with comments on its relationship to the intercolumnar tubercle and its significance for the origin of cystic tumors in the third ventricle. J. comp. Neurol. *102:* 425–509 (1955).

KAPPERS, J. A.: On the development, structure and function of the paraphysis cerebri; in KAPPERS Progress in neurobiology, pp. 130–145 (Elsevier, Amsterdam 1956).

KARPLUS, I. P. und KREIDL, A.: Über Totalexstirpationen einer und beider Grosshirnhemisphären an Affen (Macacus thesus). Arch. Anat. Physio. *1914:* 155–212 (1914).

KARTEN, H. J. and DUBBELDAM, J. L.: The organization and projections of the palaeostriatal complex in the pigeon (Columba livia). J. comp. Neurol. *148:* 61–89 (1973).

KARTEN, J.H. and HODOS, W.: A stereotaxic atlas of the brain of the pigeon (Columba livia). (Hopkins Press, Baltimore 1967).

KARTEN, J.H.; HODOS, W.; NAUTA, W.J.H., and REVZIN, A.M.: Neural connections of the 'visual wulst' of the Avian telencephalon. Experimental studies in the pigeon (Columba livia) and owl (Speotyto cunicularia). J. comp. Neurol. *150:* 253–277 (1973).

KATAOKA, A.; BAK, I.J.; HASSLER, R., and WAGNER, A.: L-Glutamate decarboxylase and choline acetyltransferase activity in the substantia nigra and the striatum after surgical interruption of the strionigral fibres in the baboon. Exp. Brain Res. *19:* 217–227 (1974).

KICLITER, E. and NORTHCUTT, R.G.: Ascending afferents to the telencephalon of Ranid frogs: an anterograde degeneration study. J. comp. Neurol. *161:* 239–253 (1975).

KIESEWALTER, C.: Zur Morphologie der Ganglienkerne im Grosshirn von Lacerta. Jena. Z. Naturwiss. *58:* 488–532 (1922).

KIESEWALTER, C.: Basis und Pallium. Ihre mediale Grenze am Grosshirn der Amphibien und Reptilien. Jena. Z. Naturwiss. *61:* 575–406 (1925).

KIESEWALTER, C.: Zur allgemeinen und speziellen Morphogenie des Hemisphärenhirns der Tetrapoden. Jena. Z. Naturwiss. *64:* 369–454 (1928).

KIM, J.S.; BAK, I.J.; HASSLER, R., and OKADA, Y.: Role of gamma aminobutyric acid (GABA) in the extrapyramidal system. 2. Some evidence for the existence of a type of GABA-rich strio-nigral neurons. Exp. Brain Res. *14:* 95–104 (1971).

KIRSCHE, K. und KIRSCHE, W.: Experimentelle Untersuchung über den Einfluss der Regeneration des Nervus olfactorius auf die Vorderhirnregeneration von Amblystoma mexicanum. J. Hirnforsch. *7:* 315–333 (1964).

KIRSCHE, W.: Die Entwicklung des Telencephalons der Reptilien und deren Beziehung zur Hirn-Bauplanlehre. Nova Acta Leopoldina *37/2:* 1–78 (1972).

KLEIST, K.: Gehirnpathologie (Barth, Leipzig 1934).

KINGLER, J. and GLOOR, P.: Connections of the amygdala and of the anterior temporal cortex in the human brain. J. comp. Neurol. *115:* 333–369 (1960).

KLUTZOW, F.W.; EARLE, K.M., and WEBSTER, D.D.: Disorders of autonomic function (dysautonomias). Milit. Med *140:* 338–344 (1975).

KLÜVER, H.: Functional significance of the geniculo-striate system. Biol. Symp. *7:* 253–299 (1942).

KLÜVER, H.: Brain mechanisms and behavior with special reference to the rhinencephalon. Journal-Lancet, Minneap. *72:* 567–577 (1952).

KLÜVER, H. and BUCY, P.C.: 'Psychic blindness' and other symptoms following bilateral temporal lobectomy in rhesus monkeys. Amer. J. Physiol. *119:* 352–353 (1937).

KLÜVER, H. and BUCY, P.C.: Preliminary analysis of functions of the temporal lobe in monkeys. Arch. Neurol. Psychiat. *42:* 979–1000 (1939).

KNOOK, H.L.: The fibre-connections of the forebrain (Van Gorcum, Assen 1965).

KOLMER, W.: Geruchsorgan; in V. MÖLLENDORFF, Handb. d. mikr. Anat. d. Menschen, vol. III/I, pp. 192–249 (Springer, Berlin 1927).

KOPPANYI, T. and PEARCY, J.F.: Comparative studies on the excitability of the forebrain. Amer. J. Physiol. *71:* 339–343 (1925).

KRABBE, K.H.: Studies on the morphogenesis of the brain in birds (Munksgaard, Copenhagen 1952).

KRAUSE, F.: Chirurgie des Gehirns und Rückenmarks. 2 vols. (Urban & Schwarzenberg, Berlin 1908, 1911).

KRAUSE, R.: Mikroskopische Anatomie der Wirbeltiere in Einzeldarstellungen. I. Säugetiere. II. Vögel und Reptilien. III. Amphibien. IV. Teleostier, Plagiostomen, Zyklostomen und Leptokardier (De Gruyter, Berlin 1921, 1922, 1923, 1923).

KRAUSENECK, P.: Wesentliche Aspekte der Rindenblindheit. Dargestellt an einem Fall dauernder Erblindung nach operativer Entfernung eines doppelseitigen occipitalen Meningeoms, mit Autopsie; Inaugural-Diss. Würzburg (1975).

KRETSCHMER, E.: Medizinische Psychologie; 3. Aufl. (Thieme, Leipzig 1926).

KRETSCHMER, E.: Das apallische Syndrom. Z. ges. Neurol. Psychiat. *169:* 576–579 (1940).

KRIEG, W. S. J.: Connections of the frontal cortex of the monkey (Thomas, Springfield 1954).

KRIEG, W. S. J.: Connections of the cerebral cortex (Brain Books, Evanston 1963).

KRIEG, W. J. S.: Functional neuroanatomy; 3rd ed. (Brain Books, Evanston 1966).

KRIEG, W. J. S.: Architectonics of human cerebral fiber systems (Brain Books, Evanston 1973).

KRUGER, L.: Experimental analyses of the reptilian nervous system. Ann. N. Y. Acad. *167:* 102–117 (1969).

KRUGER, L. and BERKOWITZ, E. C.: The main afferent connections of the reptilian telencephalon as determined by degeneration and electrophysiological methods. J. comp. Neurol. *115:* 125–141 (1960).

KRYNAUW, R. A.: Infantile hemiplegia treated by removing one cerebral hemisphere. J. Neurol. Neurosurg. Psychiat. *13:* 246–267 (1950).

KUDO, K.: Studien zur mikroskopischen Anatomie des Fischgehirns. I. Eine bisher wenig beachtete Zellgruppe im Telencephalon der Knochenfische. Folia anat. jap. *6:* 711–715 (1928).

KUENZI, W.: Versuch einer systematischen Morphologie des Gehirns der Vögel. Rev. Suisse Zool. *26:* 17–112 (1918).

KUHLENBECK, H.: Zur Morphologie des Urodelenvorderhirns. Jena. Z. Naturwiss. *57:* 463–490 (1921a).

KUHLENBECK, H.: Zur Histologie des Anurenpalliums. Anat. Anz. *54:* 280–285 (1921b).

KUHLENBECK, H.: Die Regionen des Anurenvorderhirns. Anat. Anz. *54:* 304–316 (1921c).

KUHLENBECK, H.: Zur Morphologie des Gymnophionengehirns. Jena. Z. Naturwiss. *58:* 453–484 (1922a).

KUHLENBECK, H.: Über den Ursprung der Grosshirnrinde. Eine phylogenetische und neurobiotaktische Studie. Anat. Anz. *55:* 338–365 (1922b).

KUHLENBECK, H.: Über den Ursprung der Basalganglien des Grosshirns. Anat. Anz. *58:* 49–74 (1924a).

KUHLENBECK, H.: Über die Homologien der Zellmassen im Hemisphärenhirn der Wirbeltiere. Folia anat. jap. *2:* 325–364 (1924b).

KUHLENBECK, H.: Vorlesungen über das Zentralnervensystem der Wirbeltiere (Fischer, Jena 1927).

KUHLENBECK, H.: Über die anatomischen Grundlagen nervöser Mechanismen. Psychiat.-neurol. Wschr. *30:* No. 46: 1–3 (1928).

KUHLENBECK, H.: Die Grundbestandteile des Endhirns im Lichte der Bauplanlehre Anat. Anz. *67:* 1–51 (1929).

KUHLENBECK, H.: The ontogenetic development and phylogenteic significance of the cortex telencephali in the chick. J. comp. Neurol. *69:* 273–301 (1938).

KUHLENBECK, H.: The derivatives of thalamus dorsalis and epithalamus in the human brain: their relation to cortical and other centers. Milit. Surg. *108:* 205–256 (1951).

KUHLENBECK, H.: The human diencephalon (Karger, Basel 1954).

KUHLENBECK, H.: Brain and consciousness (Karger, Basel 1957).

KUHLENBECK, H.: Mind and matter. An appraisal of their significance for neurologic theory (Karger, Basel 1961).

KUHLENBECK, H.: Some comments on words, language, thought, and definition; in BUEHNE *et al.* Helen Adolf Festschrift, pp. 9–29 (Unger, New York 1968).

KUHLENBECK, H.: Some comments on the development of the human corpus callosum and septum pellucidum. Acta anat. nippon. *44:* 245–256 (1969).

KUHLENBECK, H.: Gehirn und Bewusstsein. Transl. by Prof. J. GERLACH and Dr. U. PROTZER (Duncker & Humblot, Berlin 1973).

KUHLENBECK, H.; HAFKESBRING, R., and ROSS, M.: Further observations on a living 'decorticate' (hydranencephalic) child. J. amer. med. Women's Ass. *14:* 216–225 (1959).

KUHLENBECK, H. and HAYMAKER, W.: The derivatives of the hypothalamus in the human brain: their relation to the extrapyramidal and autonomic systems. Milit. Surg. *105:* 26–52 (1949).

KUHLENBECK, H. and KIESEWALTER, C.: Zur Phylogenese des Epistriatums. Anat. Anz. *55:* 145–156 (1922).

KUHLENBECK, H. and NIIMI, K.: Further observations on the morphology of the brain in the Holocephalian Elasmobranchs Chimaera and Callorhynchus. J. Hirnforsch. *11:* 267–314 (1969).

KUHLENBECK, H. and SZEKELY, E.G.: Evoked patentials from tectum mesencephali and telencephalon of the chicken after unilateral optic stimulation (Abstract). Anat. Rec. *145:* 332 (1963).

KUHLENBECK, H.; SZEKELY, E.G., and SPULER, H.: Observations on the EEG of a hydranencephalic 'decorticate' child in the resting condition and upon stimulation. Prog. Brain Res. *6:* 198–206 (1964).

KUHLENBECK, H. and WIENER-KIRBER, M.: Some observations on neurotropic effects of vesicular stomatitis virus in the mouse brain. Confin. neurol. *22:* 65–120 (1962).

KUNTZ, A.: The autonomic nervous system; 3rd ed. (Lea & Febiger, Philadelphia 1947).

KUNTZ, A.: Edward Hitzig (1838–1907); in HAYMAKER The founders of neurology; 1st ed., pp. 138–143; 2nd ed., pp. 229–238 (Thomas, Springfield 1953, 1970).

KUPFFER, C. VON: Die Morphogenie des Centralnervensystems; in HERTWIG Handbuch der vergleichenden und experimentellen Entwicklungslehre der Wirbeltiere, vol. 2, 3. Teil, pp. 1–272 (Fischer, Jena 1906).

KUREPINA, M.: Entwicklung der primären Choanen bei Amphibien. I. Teil, Anura. II. Teil, Urodela. Rev. Zool. russe *6:* 72–74 (1926); 28–30 (1927).

LANDAU, E.: The comparative anatomy of the nucleus amygdalae, the claustrum and the insular cortex. J. Anat. *53:* 251–360 (1919).

LANDAU, E.: Anatomie des Grosshirns. Formanalytische Untersuchungen (Bircher, Bern 1923).

LANDIS, D.M.D.; REESE, T.S., and RAVIOLA, E.: Differences in membrane structure between excitatory and inhibitory components of the reciprocal synapse in the olfactory bulb. J. comp. Neurol. *155:* 67–91 (1974).

LANGE, S.J. DE: Das Vorderhirn der Reptilien. Folia neurobiol. *5:* 548–597 (1911).

LANGWORTHY, O.R.: A description of the central nervous system in the porpoise (Tursiops truncatus). J. comp. Neurol. *54:* 437–499 (1932).

LASHLEY, K.S.: Studies of cerebral function in learning. III. The motor areas. Brain *44:* 225–285 (1921).

LASHLEY, K.S.: Brain mechanisms and intelligence (University of Chicago Press, Chicago 1929).

LAUBMANN, W.: Über die Morphogenese vom Hirn und Geruchsorgan der Gymnophionen. Z. Anat. EntwGesch. *84:* 597–637 (1927).

LENDE, R.A.: A comparative approach to the neocortex: localization in monotremes, marsupials, and insectivores. Ann. N. Y. Acad. Sci. *167:* 262–276 (1969).

LEVETEAU, J. et MACLEOD, P.: La discrimination des odeurs par les glomérules olfactifs du lapin: influence de la concentration du stimulus. J. Physiol., Paris *61:* 5–16 (1969).

LEYTON, A.S.F. and SHERRINGTON, C.S.: Observations on the excitable cortex of the chimpanzee, orang-utan, and gorilla. Quart. J. exp. Physiol. *11:* 135–222 (1917).

LHERMITTE, J.: Les fondements biologiques de la psychologie (Gauthier-Villars, Paris n.d., about 1924).

LIEPMANN, H.: Das Krankheitsbild der Apraxie ('motorischen Asymbolie') (Karger, Berlin 1900).

LIEPMANN, H.: Über Störungen des Handelns bei Gehirnkranken (Karger, Berlin 1905).

LILLY, J.C.: Man and dolphin (Doubleday, New York 1963).

LIM, R.K.S.; LIU, C.N., and MOFFITT, R.L.: A stereotaxic atlas of the dog's brain (Thomas, Springfield 1960).

LIVINI, F.: Il proencefalo di un marsupiale (Hypsiprymnus rufescens). Arch. ital. Anat. Embriol. *6:* 549–584 (1907).

LOBO, A.: La formación reticular telencephalica en relación con funciones vegetativas (Aportaciones por via experimental). Anales Anat. *21:* 445–512 (1972).

LOCY, W.A.: On a newly recognized nerve connected with the forebrain of selachians. Anat. Anz. *26:* 33–36, 111–123 (1905).

LOESER, W.: A study of the functions of different parts of the frog's brain. J. comp. Neurol. *15:* 355–373 (1905).

LOHMAN, A.H.M. and LAMMERS, H.J.: On the structure and fibre connections of the olfactory centres in mammals; in ZOTTERMAN Sensory mechanisms. Progress in Brain Research, vol. 23, pp. 65–85 (Elsevier, Amsterdam 1967).

LOO, Y.T.: The forebrain of the opossum, Didelphis virginiana. I. Gross anatomy. II. Histology. J. comp. Neurol. *51:* 13–64 (1930); *52:* 1–148 (1931).

LUDWIG, E. and KLINGLER, J.: Atlas cerebri humani (Karger, Basel 1956).

LURIA, A.R.: Higher cortical functions in man (Basic Books, New York 1966).

LYON, M. and HARRISON, J.M.: The effects of certain neural lesions in the rat on the reaction to a noxious stimulus. I. The limbic region. II. Septal nuclei and fornix components. J. comp. Neurol. *111:* 101–114, 115–131 (1959).

LYONS, J. (ed.): New horizons in linguistics (Penguin Books, Harmondsworth 1973).

MACDONALD, R.L. and COHEN, D.H.: Heart rate and blood pressure response to electrical stimulation of the central nervous system in the pigeon (Columba livia). J. comp. Neurol. *150:* 109–136 (1973).

MACLEAN, P.D.: Contrasting functions of limbic and neocortical systems of the brain and their relevance to psycho-physiological aspects of medicine. Amer J. Med. *25:* 611–626 (1958).

MacLean, P.D.: The hypothalamus and emotional behavior; in Haymaker, Anderson and Nauta, The hypothalamus, pp. 659–678 (Thomas, Springfield 1969).

MacLean, P. and Ploog, D.W.: Cerebral representation of erection. J. Neurophysiol. *25:* 29–55 (1962).

Macrides, F. and Chorover, S.L.: Olfactory bulb units: activity correlated with inhalation cycles and odor quality. Science *175:* 84–87 (1971).

Maier, N.R.F.: The pattern of cortical injury in the rat and its relation to mass action. J. comp. Neurol. *60:* 409–436 (1934).

Margolis, F.L.: Carnosine in the primary olfactory pathway. Science *185:* 909–911 (1974).

Marie, P.: La troisième circonvolution frontale gauche ne joue aucun rôle spécial dans la fonction du langage. Semaine méd., Paris *26:* 241–247 (1906).

McCotter, R.E.: The connections of the vomeronasal nerves with the accessory olfactory bulb in the opossum and other mammals. Anat. Rec. *6:* 299–318 (1912).

McCotter, R.E.: The vomeronasal apparatus in Chrysemys punctata and Rana catesbiana. Anat. Rec. *13:* 57–67 (1917).

McCulloch, W.S. and Pitts, W.: A logical calculus of the ideas immanent in nervous activity. Bull. math. Biophys. *5:* 115–133 (1943).

McGurk, H. and Lewis, M.: Space perception in early infancy: perception within a common auditory-visual space? Science *186:* 649–650 (1974).

McKibben, P.S.: The nervus terminalis in urodele Amphibia. J. comp. Neurol. *21:* 261–309 (1911).

McLardy, T.: Thalamic projection to frontal cortex in man. J. Neurol. Neurosurg. Psychiat. *13:* 198–202 (1950).

Meader, R.E.: The forebrain of bony fishes. Proc. kon. nederl. Akad. Wet. *42:* 657–670 (1939).

Meyer, A.: Über das Vorderhirn einiger Reptilien. Z. wiss. Zool. *55:* 63–133 (1892).

Meyer, A.: The frontal lobe syndrome, the aphasias and related conditions – A contribution to the history of cortical localization. Brain *97:* 565–600 (1974).

Meyer, A.; Beck, E., and McLardy, T.: Prefrontal leucotomy: a neuroanatomical report. Brain *70:* 18–49 (1947).

Meynert, T.: Vom Gehirn der Säugethiere; in Strickers Handbuch der Gewebelehre, Kapitel XXXI, pp. 694–805 (Engelmann, Leipzig 1872).

Miller, R.N.: The telencephalic zonal system of the teleost Corydora paliatus. J. comp. Neurol. *72:* 149–176 (1940).

Millot, J. and Anthony, J.: Anatomie de Latimeria chalumnae. II. Système nerveux et organes des sens (Editions du Centre National de la Recherche Scientifique, Paris 1954).

Millot, J. et Anthony, J.: L'organisation générale du prosencéphale de Latimeria chalumnae Smith (Poisson crossoptérygien coelacanthidé); in Hassler and Stephan, Evolution of the forebrain, pp. 50–60 (Thieme, Leipzig 1966).

Minkowski, M.: Zur Physiologie der Sehsphäre. Arch. ges. Physiol. *141:* 171–327 (1911).

Mittelstrass, H.: Vergleichend-anatomische Untersuchungen über den Mandelkern der Säugetiere. Z. Anat. EntwGesch. *106:* 717–738 (1937).

Miyoshi, K.; Harter, D.H., and Hsu, K.C.: Neuropathological and immunofluorescence studies of experimental vesicular stomatitis virus encephalitis in mice. J. Neuropath. exp. Neurol. *30*: 266–277 (1971).

MONAKOW, C. v.: Gehirnpathologie (Holder, Wien 1905).

MONAKOW, C. v.: Die Lokalisation im Grosshirn und der Abbau der Funktion durch kortikale Herde (Bergmann, Wiesbaden 1914).

MONAKOW, C. v.: Experimentell- und pathologisch-anatomische sowie entwicklungsgeschichtliche Untersuchungen über die Beziehungen des Corpus striatum und des Linsenkerns zu den übrigen Hirnteilen. Schweiz. Arch. Neurol. Psychiat. *16:* 225–234 (1925).

MONIZ, E.: Tentatives opératoires dans le traitement de certaines psychoses (Masson, Paris 1936).

MOSKOWITZ, H.R. and GERBER, C.L.: Dimensional salience of odors. Ann. N. Y. Acad. Sci. *237:* 1–16 (1974).

MOULTON, D.G.: Dynamics of cell populations in olfactory epithelium. Ann. N. Y. Acad. Sci. *237:* 52–61 (1974).

MOULTON, D.G.; CELEBI, G., and FINK, R.P.: Olfaction in mammals; in WOLSTENHOLME and KNIGHT Taste and smell in vertebrates, pp. 227–250 (Churchill, London 1970).

MUNK, H.: Über die Funktionen der Grosshirnrinde (Hirschwald, Berlin 1881).

MÜNZER, E. and WIENER, H.: Beiträge zur Anatomie und Physiologie des Zentralnervensystems der Taube. Mschr. Psychiat. Neurol. *3:* 379–406 (1898).

NATHAN, P.W. and SMITH, M.C.: Normal mentality associated with a maldeveloped 'rhinencephalon'. J. Neurol. Neurosurg. Psychiat. *13:* 191–197 (1950).

NEAL, H.V. and RAND, H.W.: Comparative anatomy (Blakiston, Philadelphia 1936).

NICOLL, R.A.: Recurrent excitation of secondary olfactory neurons. A possible mechanism for signal amplification. Science *171:* 824–826 (1971).

NIELSEN, J.M.: Agnosias, apraxias, speech and aphasias; in Baker Clinical neurology; 1st ed., vol. 1, pp. 352–378 (Hoeber-Harper, New York 1955).

NIEUWENHUYS, R.: Some observations on the structure of the forebrain of bony fishes. Proc. 2nd Int. Meet. Neurobiologists, pp. 144–149 (Elsevier, Amsterdam 1960a).

NIEUWENHUYS, R.: Het telencephalon der Actinopterygii; Academ. Proefschrift Amsterdam (1960b).

NIEUWENHUYS, R.: The comparative anatomy of the actinopterygian forebrain. J. Hirnforsch. *6:* 171–192 (1963).

NIEUWENHUYS, R.: The interpretation of the cell masses in the teleostean forebrain; in HASSLER and STEPHAN Evolution of the forebrain, pp. 32–39 (Thieme, Stuttgart 1966).

NIEUWENHUYS, R.: Comparative anatomy of olfactory centres and tracts; in ZOTTERMAN Sensory mechanisms. Progress in Brain Research, vol. 23, pp. 1–64 (Elsevier, Amsterdam 1967).

NIEUWENHUYS, R.: A survey of the structure of the forebrain in higher bony fishes (Osteichthyes). Ann. N. Y. Acad. Sci. *167:* 31–64 (1969).

NIEUWENHUYS, R. and HICKEY, M.: A survey of the forebrain of the Australian lungfish Neoceratodus forsteri. J. Hirnforsch. *7:* 433–452 (1965).

NOBLE, G.K.: The biology of the Amphibia (McGraw-Hill, New York 1931; Dover, New York 1954).

NOBLE, G.K.: Function of the corpus striatum in the social behavior of fishes (Abstract). Anat. Rec. *64:* 34 (1936).

NOBLE, G.K.: Effects of lesions of the corpus striatum on the brooding behavior of cichlid fishes (Abstract). Anat. Rec. *70:* 58 (1937).

NOBLE, G.K. and BORNE, R.: The effect of forebrain lesions on the sexual and fighting behavior of Betta splendens and other fishes (Abstract). Anat. Rec. *79:* suppl., p. 49 (1941).

NOLTE, A.: Die Abhängigkeit der Proportionierung der Cytoarchitektonik des Gehirns von der Körpergrösse bei Urodelen. Zool. Jb. Abt. allg. Zool. *64:* 538–597 (1953).

NOLTE, W.: Experimentelle Untersuchungen zum Problem der Lokalisation des Assoziations-Vermögen im Fischgehirn. Z. vergl. Physiol. *18:* 255–279 (1932).

NORTHCUTT, R.G.: Architectonic studies of the telencephalon of iguana. J. comp. Neurol. *130:* 109–147 (1967).

NORTHCUTT, R. G.: Some histochemical observations on the telencephalon of the bullfrog, Rana catesbeiana Shaw. J. comp. Neurol. *157:* 379–389 (1974).

NOTTEBOHM, F. and ARNOLD, A.P.: Sexual dimorphism in vocal control areas of the songbird brain. Science *194:* 211–213 (1976).

OBENCHAIN. J.B.: The brains of the South American marsupials Caenolestes and Orolestes. Field Mus. nat. Hist. Publ. zool. Ser. *14:* 175–232 (1925).

OBERSTEINER, H.: Anleitung beim Studium des Baues der nervösen Zentralorgane im gesunden und kranken Zustande; 5. Aufl. (Deuticke, Leipzig 1912).

OLDS, J.: A preliminary mapping of electrical reinforcing effects in the rat brain. J. comp. physiol. Psychol. *49:* 281–285 (1956).

OLDS, J.: Self-stimulation of the brain. Science *127:* 315–324 (1958).

OLDS, J.: Differentiation of reward systems in the brain by self-stimulating technics; in RAMEY and O'DOHERTY Electrical studies on the unanesthetized brain, pp.17–51 (Hoeber, New York 1960).

OLDS, J. and MILNER, P.: Positive reinforcement produced by electrical stimulation of septal area and other regions of the rat brain. J. comp. physiol. Psychol. *47:* 419–427 (1954).

OLDS, M.E. and OLDS, J.: Approach-avoidance analysis of the rat diencephalon. J. comp. Neurol. *120:* 259–295 (1963).

OLIVEROS, N.L.: Observations on the lining of the cavum septi pellucidi in the brain of newborn and adult man. Confin. neurol. *26:* 45–55 (1965).

OTTOSON, D. and SHEPHERD, G.M.: Experiments and concepts in olfactory physiology; in ZOTTERMAN Sensory mechanisms. Progress in Brain, Research, vol. 23, pp. 83–138 (Elsevier, Amsterdam 1967).

PAPEZ, J.W.: Comparative neurology (Crowell, New York 1929).

PAPEZ, J.W.: A proposed mechanism of emotion. Arch. Neurol. Psychiat. *38:* 725–743 (1937).

PARENT, A.: Distribution of monoamine-containing nerve terminals in the brain of the painted turtle, Chrysemys picta. J. comp. Neurol. *148:* 153–165 (1973)

PARIZEK, R.; HASSLER, R., and BAK, I.J.: Light and electron microscopic autoradiography of substantia nigra of rat after intraventricular administration of tritium labelled norepinephrine, dopamine, serotonin and the precursors. Z. Zellforsch. *115:* 139–148 (1971).

PEARLMAN, A.L. and DAW, N.W.: Opponent color cells in the cat lateral geniculate nucleus. Science *167:* 84–86 (1970).

PEARSON, R.: The Avian brain (Academic Press, London 1972).

PENFIELD, W.: Memory mechanisms. Arch. Neurol. Psychiat. *67:* 178–198 (1952).

PENFIELD, W.: Studies on the cerebral cortex of man. A review and interpretation; in DELAFRESNAYE Brain mechanisms and consciousness, pp. 284–309 (Blackwell, Oxford 1954).

PENFIELD, W.: The excitable cortex in conscious man (Thomas, Springfield 1958).

PENFIELD, W. and EVANS, J.: The frontal lobe in man: a clinical study of maximum removal. Brain *58:* 115–133 (1935).

PENFIELD, W. and FAULK, M.E., JR.: The insula. Further observations on its function. Brain *78:* 446–470 (1955).

PENFIELD, W. and JASPER, H.H.: Epilepsy and the functional anatomy of the human brain (Little, Brown, Boston 1954).

PENFIELD, W. and RASMUSSEN, T.: The cerebral cortex of man. A clinical study of localization (Macmillan, New York 1950).

PENFIELD, W. and ROBERTS, L.: Speech and brain-mechanism (Princeton University Press, Princeton 1959).

PETER, R.E. and GILL, V.E.: A stereotaxic atlas and technique for forebrain nuclei of the goldfish, Carassius auratus. J. comp. Neurol. *159:* 69–101 (1975).

PETER, R.E.; MACEY, M.J., and GILL, V.E.: A stereotaxic atlas and technique for forebrain nuclei in killifish, Fundulus heteroclitus. J. comp. Neurol. *159:* 103–127 (1975).

PETTIGREW, J.D. and KONISHI, M.: Neurons selective for orientation and binocular disparity in the visual wulst of the barn owl (Tyto alba). Science *193:* 675–678 (1976).

PFEIFER, R.A.: Myelogenetisch-anatomische Untersuchungen über das kortikale Ende der Hörleitung. Abh. math.-phys. Cl. k. sächs. Ges. d. Wissensch. *37/2:* 1–54 (1920).

PFEIFER, R.A.: Die Lokalisation der Tonskala innerhalb der kortikalen Hörsphäre des Menschen. Mschr. Psychiat. Neurol. *50:* 7–48, 99–108 (1921).

PFEIFER, R.A.: Myelogenetisch-anatomische Untersuchungen über den zentralen Abschnitt der Sehleitung (Springer, Berlin 1925).

PFEIFER, R.A.: Die nervösen Verbindungen des Auges mit dem Zentralorgan; in SCHIEK and BRÜCKNER Kurzes Handbuch der Ophthalmologie, vol. 1, pp. 387–475 (Springer, Berlin 1930).

PIATT, J.: An experimental approach to the problem of pallial differentiation. J. comp. Neurol. *94:* 105–121 (1951).

PINKUS, F.: Über einen noch nicht beschriebenen Hirnnerven des Protopterus annectens. Anat. Anz. *9:* 562–566 (1894).

PILLERI, G.: Die zentralnervöse Rangordnung der Cetacea (Mammalia). Acta anat. *51:* 241–258 (1962).

PILLERI, G.: Morphologie des Gehirnes des 'Southern Right Whale', Eubalaena australis Desmoulins 1822 (Cetacea, Mysticeti, Balaenidae). Acta zool. *45:* 245–272 (1964).

PILLERI, G.: Morphologie des Gehirnes des Seiwals (Cetacea, Mysticeti, Balaenopteridae). Morphologie des Gehirnes des Buckelwals, Megaptera novaeangliae Boronski (Cetacea, Mysticeti, Balaenopteridae). J. Hirnforsch. *8:* 221–267, 437–491 (1966).

PLATE, L.: Allgemeine Zoologie und Abstammungslehre, vol. II: Die Sinnesorgane der Tiere (Fischer, Jena 1924).

PLATEL, R.; BECKERS, H.J.A. et NIEUWENHUYS, R.: Les champs corticaux chez Testudo hermanii (Reptile Chelonien) et chez Caiman crocodilus (Reptile Crocodilien). Acta morphl. neerl. scand. *11:* 121–150 (1973).

POLIAK, S.: The main afferent fiber systems of the cerebral cortex in Primates (University of California Press, Berkeley 1932).

POWELL, T.P.S. and KRUGER, L.: The thalamic projections upon the telencephalon in Lacerta viridis. J. Anat. *94:* 528–542 (1960).

POWERS, J.B. and WINANS, S.S.: Vomeronasal organ: critical role in mediating sexual behavior of the male hamster. Science *187:* 961–963 (1975).

PRIBRAM, K.H.: Languages of the brain (Prentice Hall, Englewood Cliffs 1971).

PRICE, J.L. and SPRICH, W.W.: Observations on the lateral olfactory tract of the rat. J. comp. Neurol. *162:* 321–336 (1975).

PRITZ, M.B.: Anatomical identification of a telencephalic visual area in crocodiles: ascending connections of nucleus rotundus in Caiman crocodilus. J. comp. Neurol. *164:* 323–338 (1975).

PUTNAM, T.J. and PUTNAM, I.K.: Studies on the central visual system. I. The anatomic projection of the retinal quadrants on the striate cortex of the rabbit. Arch. Neurol. Psychiat. *16:* 1–20 (1926).

QUIROGA, J.C.: El sistema nervioso central de Amphisbaena (Reptilia). I. Histoestructura de la corteza dorsal. Arch. Fundación Roux-Ocefa, in press, 1976).

RABL, R.: Das Subfornicalorgan des Menschen. J. Hirnforsch. *8:* 529–545 (1966).

RABL-RÜCKHARD, H.: Zur Deutung und Entwicklung des Gehirns der Knochenfische. Arch. Anat. Physiol., anat. Abt. *1882:* 111–138 (1882).

RABL-RÜCKHARD, H.: Das Grosshirn der Knochenfische und seine Anhangsgebilde. Arch. Anat. Physiol., anat. Abt. *1883:* 279–322 (1883).

RABL-RÜCKHARD, H.: Das Gehirn der Knochenfische. Dtsch. med. Wschr. *1884:* 1–25 (1884).

RABL-RÜCKHARD, H.: Das Vorderhirn der Cranioten. Eine Antwort an F. K. Studnička. Anat. Anz. *9:* 536–547 (1894).

RAE, A.S.L.: The form and structure of the human claustrum. J. comp. Neurol. *100:* 15–39 (1954a).

RAE, A.S.L.: The connections of the claustrum. Confin. neurol. *14:* 211–219 (1954b).

RAMÓN (Y CAJAL), P.: Nuevo estudio del encefalo de los reptiles. Trab. Lab. Invest. biol. Madrid *15:* 83–99 (1917); *16:* 309–333 (1918).

RAMÓN (Y CAJAL), P.: El cerebro de los batracios. Libro en honor de D. Santiago Ramón y Cajal (Madrid) *I:* 13–150 (1922).

RANSON, S.W.: The anatomy of the nervous system from the standpoint of development and function; 7th ed. (Saunders, Philadelphia 1943).

RAUBER KOPSCH: Lehrbuch der Anatomie des Menschen. Abt. 5, Nervensystem, 10. Aufl.; Abt. 6, Sinnesorgane, 10. Aufl.; Abt. 3, Muskeln, Gefässe, 14. Aufl. (Thieme, Leipzig 1914, 1916, 1933).

REESE, T.S.: Olfactory cilia in the frog. J. cell Biol. *25:* 209–230 (1965).

REESE, T.S. and BRIGHTMAN, M.W.: Olfactory surface and central olfactory connexions in some vertebrates; in WOLSTENHOLME and KNIGHT Taste and smell. A Ciba Foundation symposium, pp. 115–149 (Churchill, London 1970).

RENSCH, B.: Increase of learning capability with increase in brain size. Amer. Naturalist *90:* 81–95 (1956).

RENSCH, B.: Biophilosophie auf erkenntnistheoretischer Grundlage (Panpsychistischer Identismus) (Fischer, Stuttgart 1968).

RENSCH, B.: Biophilosophy (Transl. by C.A.M. SYM) (Columbia University Press, New York 1971).

RIESE, W.: Über die Stammganglien der Wale. J. Psychol. Neurol. *23:* 21–28 (1925).

RIGGS, J.E. and RUPP, C.: A clinico-anatomic study of personality and mood disturbances associated with gliomas of the cerebrum. J. Neuropath. exp. Neurol. *17:* 318–345 (1958).

RISS, W.; HALPERN, M., and SCALIA, F.: The quest for clues to forebrain evolution – the study of reptiles. Brain Behav. Evol. *2:* 1–50 (1969).

Riss, W.; Koizumi, K., and McC. Brooks, C. (eds.): Basic thalamic structure and function. Proceedings of a conference. Brain Behav. Evol. *6:* (1972a).

Riss, W.; Pedersen, R. A.; Jakway, J. S., and Ware, C. B.: Levels of function and their representation in the vertebrate thalamus. Brain Behav. Evol. *6:* 26–41 (1972b).

Roberts, T. S. and Akert, K.: Insular and opercular cortex and its projection in Macaca mulatta. Schweiz. Arch. Neurol. Neurochir. Psychiat. *92:* 1–43 (1963).

Rogers, F. T.: Contributions to the physiology of the stomach. XXXIX. The hunger mechanism of the pigeon and its relation to the central nervous system. Amer. J. Physiol. *41:* 555–570 (1916).

Rogers, F. T.: Experimental studies on the optic thalamus and the corpus striatum. J. nerv. ment. Dis. *49:* 1–4 (1919).

Rogers, F. T.: Studies on the brain stem. VI. An experimental study on the corpus striatum of the pigeon as related to various instinctive types of behavior. J. comp. Neurol. *35:* 21–60 (1922a).

Rogers, F. T.: A note on the excitable areas of the cerebral hemispheres of the pigeon. J. comp. Neurol. *35:* 61–66 (1922b).

Rogers, F. T.: An experimental study of the cerebral physiology of the Virginian opossum. J. comp. Neurol. *27:* 265–315 (1924).

Rogers, K. T.: Studies on the chick brain of biochemical differentiation related to morphological development. I. Morphological development. J. exp. Zool. *144:* 77–87 (1960).

Romanoff, A. L.: The avian embryo. Structural and functional development (Macmillan, New York 1960).

Romer, A. S.: The vertebrate body (Saunders, Philadelphia 1950).

Rose, M.: Über die cytoarchitektonische Gliederung des Vorderhirns der Vögel. J. Psychol. Neurol. *21:* 278–352 (1914).

Rose, M.: Histologische Lokalisation des Vorderhirns der Reptilien. J. Psychol. Neurol. *29:* 219–272 (1923).

Rose, M.: Cytoarchitektonischer Atlas der Grosshirnrinde der Maus. J. Psychol. Neurol. *40:* 1–51 (1929).

Rose, M.: Entwicklungsgeschichtliche Einleitung; in Bumke *et al.* Handbuch der Neurologie, vol. 1, pp. 1–34; 588–778 (Springer, Berlin 1935).

Röthig, P.: Beiträge etc. Nr. 5. Die Zellanordnungen im Vorderhirn der Amphibien mit besonderer Berücksichtigung der Septumkerne und ihr Vergleich mit den Verhältnissen bei Testudo und Lacerta. Proc. kon. med. Acad. Wet. tweede Sectie *17:* 1–23 (1912).

Röthig, P.: Einige Erfahrungen mit technischen Methoden zur Untersuchung kleinerer Gehirne. Z. mikr. anat. Forsch. *24:* 399–411 (1931).

[illegible] den Rothmannschen [illegible] Hund nach klinischer und anatomischer Untersuchung. Z. ges. Neurol. Psychiat. *87:* 247–313 (1923).

Rothmann, M.: Demonstration zur Physiologie der Grosshirnrinde (Hund 60 Tage nach totaler Grosshirnexstirpation). Neurol. Centralbl. *28:* 614, 840, 1045 (1909).

Rudebeck, B.: Does an accessory olfactory bulb exist in Dipnoi? Acta zool. *25*; 1–8 (1944).

Rudebeck, B.: Contribution to forebrain morphology in Dipnoi. Acta zool. *26:* 9–156 (1945).

RYLANDER, G.: Personality analysis before and after frontal lobotomy. Res. Publ. Ass. nerv. ment. Dis. *27:* 691–705 (1948).

SABIN, A.B. and OLITSKY, P.: Influence of host factors on neuroinvasiveness of vesicular stomatitis virus. I–IV. J. exp. Med. *66:* 15–57 (1937); *67:* 201–249 (1938).

SAITO, T.: Über das Gehirn des japanischen Flussneunauges (Entosphenus japonicus Martens). Folia anat. jap. *8:* 189–263 (1930).

SALETU, B.; ITIL, T.M., and SALETU, M.: Evoked responses after hemispherectomy. Confin. neurol. *33:* 221–230 (1971).

SANIDES, F.: Untersuchungen über die histologische Struktur des Mandelkerngebietes. J. Hirnforsch. *3:* 56–77 (1957).

SCALIA, F. and WINANS, S.S.: The differential projection of the olfactory bulb and accessory olfactory bulb in Mammals. J. comp. Neurol. *161:* 31–55 (1975).

SCHEPERS, G.W.H.: Evolution of the forebrain. The fundamental anatomy of the telencephalon with special reference to that of Testudo geometrica (Maskew Miller, Cape Town 1948).

SCHIFFMAN, S.S.: Contributions to the physiochemical dimensions of odor: a psychophysical approach. Ann. N. Y. Acad. Sci. *237:* 164–183 (1974).

SCHILDER, P.: Das Körperschema (Springer, Berlin 1923).

SCHILLER, F.: Dysarthria, aphasia and apraxia; in HAYMAKER and BING Local diagnosis in neurological diseases; 15th ed., chapter 23, pp. 397–403 (Mosby, St. Louis 1969).

SCHLESINGER, B.: Higher cerebral functions and their clinical disorders (Grune & Stratton, New York 1962).

SCHNABL, I.: Händigkeit, Hemisphärendominanz-Händigkeit; Inaug. Diss. Würzburg (1976).

SCHNEIDER, R.: Morphologische Untersuchungen am Gehirn der Chiropteren (Mammalia). Senckenberg Abh. *495:* 1–92 (1957).

SCHNITZLEIN, H.N. and CROSBY, E.C.: The telencephalon of the lungfish, Protopterus. J. Hirnforsch. *9:* 105–149 (1967).

SCHOBER, W.: Vergleichende Betrachtungen am Telencephalon niederer Wirbeltiere; in HASSLER and STEPHAN Evolution of the forebrain, pp. 20–31 (Thieme, Stuttgart 1966).

SCHRADER, M.E.G,: Zur Physiologie des Froschgehirns (Vorläufige Mitteilung). Pflügers Arch. ges. Physiol. *41:* 75–90 (1887).

SCHROEDER, K.: Der Faserverlauf im Vorderhirn des Huhns. J. Physiol. Neurol. *18:* 115–173 (1911).

SCHWARZ, G.E.; DAVIDSON, R.J., and MAER, F.: Right hemisphere lateralization in the human brain: interaction with emotion. Science *190:* 286–288 (1975).

SEGAAR, J.: Etho-physiological experiments with male Gasterosteus aculeatus; in TOWER and SCHADÉ Structure and function of the cerebral cortex, pp. 301–305 (Elsevier, Amsterdam 1960).

SEGAAR, J.: Telencephalon and behavior in Gasterosteus aculeatus. Behaviour *28:* 256–287 (1961).

SEGAAR, J.: Behavioral aspects of degeneration and regeneration in fishbrain: a comparison with higher vertebrates. Progress in Brain Research, vol. 14, pp. 143–231 (Elsevier, Amsterdam 1965).

SEGAAR, J. and NIEUWENHUYS, R.: New etho-physiological experiments with male Gasterosteus aculeatus, with anatomical comment. Anim. Behav. *11:* 331–344 (1963).

Seifert, K.: Neue Ergebnisse licht- und elektronenmikroskopischer Untersuchungen am peripheren Geruchsorgan einschliesslich der Bowman-Drüsen und des Organon vomero-nasale. Acta oto-rhino-laryng. belg. *26:* 263–492 (1972).

Sem-Jacobsen, C.; Bickford, R.G.; Dodge, H.W., and Peterson, M.C.: Human olfactory responses recorded by depth electrography. Proc. Mayo Clin. *28:* 166–170 (1953).

Semon, R.: Die Mneme als erhaltendes Princip im Wechsel des organischen Geschehens (Engelmann, Leipzig 1904).

Semon, R.: Die mnemischen Empfindungen in ihren Beziehungen zu den Originalempfindungen (Engelmann, Leipzig 1909).

Semon, R.: Bewusstseinsvorgang und Gehirnprozess (Bergmann, Wiesbaden 1920).

Sewertzoff, A.N.: Morphologische Gesetzmässigkeiten der Evolution (Fischer, Jena 1931).

Shanklin, W.M.: The central nervous system of Chameleon vulgaris. Acta zool. *11:* 425–490 (1930).

Shannon, C.: A symbolic analysis of relay and switching circuits. Trans. amer. Inst. electr. Engineers *57:* 713–723 (1938).

Sheldon, R.E.: The olfactory tracts and centers in teleosts. J. comp. Neurol. *22:* 177–340 (1912).

Shepherd, G.M.: Neuronal systems controlling mitral cell excitability. J. Physiol., Lond. *168:* 101–117 (1963).

Simonetta, B.: Origine e sviluppo del nervo terminale nei Mammiferi e suoi rapporti con l'organo di Jacobson. Z. Anat. EntwGesch. *96:* 425–463 (1932).

Sisson, S. and Grossman, J.D.: The anatomy of the domestic animals (Saunders, Philadelphia 1959).

Slotnick, B.M. and Katz, H.M.: Olfactory learning-set formation in rats. Science *185:* 796–798 (1974).

Smialowski, A.: Amygdaloid complex in the Macaque. Acta Biol. exp., Warszawa *25:* 77–89 (1965).

Smith, G. Elliot: Morphology of the true limbic lobe, corpus callosum, septum pellucidum and fornix. J. Anat. *30:* 185–205 (1895).

Smith, G. Elliot: Structure of the cerebral hemisphere of Ornithorhynchus. J. Anat. Physiol. *30:* 463–487 (1896).

Smith, G. Elliot: The morphology of the indusium and striae Lancisii. Anat. Anz. *13:* 23–27 (1897).

Smith, G. Elliot: The brain in the Edentata. Trans. Linn. Soc., Lond. *7:* part 7, pp. 276–394 (1899).

Smith, G. Elliot: On the morphology of the brain in the mammalia with special reference to that of the lemurs, recent and extinct. Trans. Linn. Soc. Lond. *8:* 319–432 (1903).

Smith, G. Elliot: The cerebral cortex in Lepidosiren, with comparative notes on the interpretation of certain features in the forebrain of other vertebrates. Anat. Anz. *33:* 513–540 (1908).

Smith, G. Elliot: Some problems relating to the evolution of the brain (The Arris and Gale lectures). Lancet *1910:* 1–6, 147–155, 221–227 (1910).

Smith, G. Elliot: A preliminary not on the morphology of the corpus striatum and the origin of the neopallium. J. Anat. *53:* 271–291 (1919).

SNESSAREW, P.: Über die Nervenfasern des Rhinencephalons beim Frosch. J. Psychol. Neurol. *13:* 97–125 (1908).

SNIDER, R.S. and LEE, J.C.: A stereotaxic atlas of the monkey brain (Macaca mulatta) (University of Chicago Press, Chicago 1962).

SOBOTTA, J.: Atlas der deskriptiven Anatomie des Menschen. 3. Teil; 7. Aufl. (Lehmann, München 1931).

SÖDERBERG, G.: Contributions to the forebrain morphology in amphibians. Acta zool. *3:* 65–121 (1922).

SPATZ, H.: Zur anatomischen Schnelldiagnose der progressiven Paralyse. Cbl. allg. Path. path. Anat. *33:* 313–320 (1923).

SPATZ, H.: Der basale Neocortex und seine Bedeutung für den Menschen. Ber. physik.-med. Ges. Würzburg *71:* 7–17 (1962–1964).

SPIEGEL, E.: Die Kerne im Vorderhirn der Säuger. Arb. neurol. Inst. Univers. Wien *22:* 418–497 (1919).

SPIEGEL, E.A.: Labyrinth and cortex. The electroencephalogram of the cortex in the stimulation of the labyrinth. Arch. Neurol. Psychiat. *31:* 469–482 (1934).

SPIEGEL, E.A.; WYCIS, H.T.; ORCHINIK, C.W., and FREED, H.: The thalamus and temporal orientation. Science *121:* 771–772 (1955).

SPIEGEL, E.A.; WYCIS, H.T.; ORCHINIK, C.W., and FREED, H.: Thalamic chronotaraxis. Amer. J. Psychiat. *113:* 97–105 (1956).

STEINER, J.: Die Die Funktionen des Zentralnervensystems und ihre Phylogenese. I. Untersuchungen über die Physiologie des Froschhirns (Vieweg, Braunschweig 1885).

STEINER, J.: Die Funktionen des Zentralnervensystems und ihre Phylogenese. II. Die Fische (Vieweg, Braunschweig 1888).

STEINER, J.: Die Funktionen des Zentralnervensystems und ihre Phylogenese. IV. Reptilien, Rückenmarksreflexe, Vermischtes (Vieweg, Braunschweig 1900).

STEPHAN, H.: Allocortex; in v. MÖLLENDORFF Handb. d. mikr. Anat. d. Menschen, Bd. r, 9. Teil, fortgeführt von W. BARGMANN (Springer, Berlin 1975).

STEPHAN, H. and ANDY, O.J.: The septum. A comparative study of its size in Insectivores and Primates. J. Hirnforsch. *5:* 229–244 (1962).

STINGELIN, W.: Vergleichend morphologische Untersuchungen am Vorderhirn der Vögel auf cytologischer und cytoarchitektonischer Grundlage (Helbing & Lichtenhahn, Basel 1958).

STÖHR/MÖLLENDORFF: Lehrbuch der Histologie; 23. Aufl. (Fischer, Jena 1933).

STOKES, T.M.; LEONARD, C.M., and NOTTEBOHM, F.: The telencephalon, diencephalon, and mesencephalon of the canary (Serinus canaria) in stereotaxic coordinates. J. comp. Neurol. *156:* 337–374 (1974).

STUDNIČKA, F.K.: Zur Lösung einiger Fragen aus der Morphologie des Vorderhirnes der Cranioten. Anat. Anz. *9:* 307–320 (1894).

STUDNIČKA, F.K.: Bemerkungen zu dem Aufsatze: 'Das Vorderhirn der Cranioten' von Rabl-Rückhard. Anat. Anz. *10:* 130–137 (1895).

STUDNIČKA, F.K.: Beiträge zur Anatomie und Entwicklungsgeschichte des Vorderhirns der Cranioten. Sitz. Ber. k. Böhm. Ges. d. Wiss., math. nat. kl. II. Mitt. (1896).

STUMPF, W.E. and GRANT, L.D. (eds.): Anatomical neuroendocrinology (Karger, Basel 1975).

SWANSON, L.W. and COWAN, W.M.: Hippocampo-hypothalamic connections: origin in subicular cortex, not Ammon's horn. Science *189:* 303–304 (1975).

Symonds, C. and MacKenzie, I.: Bilateral loss of vision from cerebral infarction. Brain *80:* 415–455 (1957).

Terzian, H. and Dalle Ore, G.: Syndrome of Klüver and Bucy reproduced in man by bilateral removal of the temporal lobes. Neurology *5:* 373–380 (1955).

Thompson, L. M.: On the cavum septi pellucidi. J. Anat. *67:* 59–77 (1932a).

Thompson, L. M.: On certain abnormal conditions of the septum pellucidum. Univ. California Publ. Anatomy *1:* 21–54 (1932b).

Thrush, D. C.: Congenital insensitivity to pain. Brain *96:* 369–386 (1973).

Tienhoven, A. van and Juhasz, L. P.: The chicken telencephalon, diencephalon, and mesencephalon in stereotaxic coordinates. J. comp. Neurol. *118:* 185–197 (1962).

Trevarthen, C.: Functional interactions between the cerebral hemispheres of the split-brain monkey; in *Ciba Foundation Study Group 20*, Function of the corpus callosum, pp. 24–46 (Little, Brown, Boston 1965).

Tunturi, A. R.: Audio-frequency localization in the acoustic cortex of the dog. Amer. J. Physiol. *141:* 397–403 (1944).

Turner, C. H.: Morphology of the avian brain. J. comp. Neurol. *1:* 39–92, 107–133 (1891).

Ule, G.; Dohner, W. und Bues, E.: Ausgedehnte Hemisphärenmarkschädigung nach gedecktem Hirntrauma mit apallischem Syndrom und partieller Spätrehabilitation. Arch. Psychiat. Nervenkr. *202:* 155–176 (1961).

Ulinski, P. S.: Cytoarchitecture of cerebral cortex in Snakes. J. comp. Neurol. *158:* 243–266 (1974).

Unger, L.: Untersuchungen über die Morphologie und Faserung des Reptiliengehirns. I. Das Vorderhirn des Gecko. Anat. Hefte *31:* 269–438 (1906).

Vanegas, H. and Ebbeson, S. O. E.: Telencephalic projections in two teleost species. J. comp. Neurol. *165:* 181–195 (1976).

Villiger, E.: Gehirn und Rückenmark; 7th ed. (Engelmann, Leipzig 1920).

Villiger, E.: Die periphere Innervation; 6th ed. (Engelmann, Leipzig 1933).

Votaw, C. L.: Certain functional and anatomical relations of the cornu Ammonis of the macaque monkey. I. Functional relations. J. comp. Neurol. *112:* 353–382 (1959).

Völsch, M.: Zur vergleichenden Anatomie des Mandelkerns und seiner Nachbargebilde. I, II. Arch. mikr. Anat. *68:* 573–683 (1906); *76:* 373–523 (1910).

Vries, E. de: Das Corpus striatum der Säugetiere. Anat. Anz. *37:* 385–405 (1910).

Vulpian, E. F. A.: Leçons sur la physiologie générale et comparée du système nerveux faites au Muséum d'histoire naturelle (Germer-Ballière, Paris 1866).

Wächtler, K.: Vergleichend-histochemische Untersuchungen zur Acetylcholesterase-verteilung im Telencephalon der Wirbeltiere; Habilitationsschrift, Tierärztliche Hochschule Hannover (1973).

Walker, A. E.; Diamond, E. L., and Moseley, J.: The neuropathological findings in irreversible coma. A critique of the [illegible] J. [illegible] 295–323 (1975).

Wallenberg, A.: Der Ursprung des Tractus isthmo-striaticus oder bulbo-striaticus der Taube. Neurol. Zentbl. *22:* 98–101 (1903).

Wallenberg, A.: Neue Untersuchungen über den Hirnstamm der Taube. Anat. Anz. *24:* 357–369 (1904).

Wallenberg, A.: Die basalen Äste des Scheidewandbündels der Taube. Anat. Anz. *28:* 394–400 (1906).

Ward, A. A.: The physiology of concussion. Clin. Neurosurg. *12:* 95–111 (1966).